Current Cancer Research

Series Editor

Wafik El-Deiry
University of Pennsylvania School of Medicine, Philadelphia, Pennsylvania, USA

More information about this series at http://www.springer.com/series/7892

Michael Andreeff
Editor

Targeted Therapy of Acute Myeloid Leukemia

Springer

Editor
Michael Andreeff
Dept. Leukemia
University of Texas MD Anderson
Cancer Center
Houston
Texas
USA

ISBN 978-1-4939-1392-3 ISBN 978-1-4939-1393-0 (eBook)
DOI 10.1007/978-1-4939-1393-0
Springer New York Heidelberg Dordrecht London

Library of Congress Control Number: 2014950144

Printed on acid-free paper

Springer is part of Springer Science+Business Media (www.springer.com)

Preface

No other hematopoietic malignancy has attracted the degree of attention and investigative passion as what is still termed "leukemia." The condition was first described by John Hughes Benett in October 1845 and by Rudolf Virchow in November 1845 as the "white blood" disease, following earlier reports by Alfred Velpeau (1827) and others. Virchow coined the term "Leukaemie" in 1847. This was only shortly after Henri Dutrochet had discovered that "the cell is the fundamental element of organization," thus formulating the underlying tenet of cell theory (1824), and Theodor Schwann and Matthias Jakob Schleiden had proposed (in 1839) that cells were the basic units of life. Thus, the discovery of leukemia was closely linked, personally and historically, to the most revolutionary new concepts of modern biology and medicine. Ten years later, Virchow, plagiarizing earlier work of Robert Remak, declared that cells originate from pre-existing cells by cell division and founded the field of cellular pathology. He also distinguished between splenic and lymphatic leukemias, which was of great value for the subsequent subclassification of what was initially thought to be a uniform entity.

Although cytogenetic and molecular analyses have dissected leukemias into an ever expanding universe of distinct maladies, the clinical designation of acute myeloid leukemia (AML) has so far survived further dissection, perhaps because of its life-threatening acuity. The immediate clinical challange to the physician is to establish the correct diagnosis and to cope with complications that can engage all medical subspecialties, ranging from blood-product replacement to the treatment of septic shock and renal failure. In the center of the evolving drama is the individual patient, who, with no antecedent warning has to suddenly face his own mortality.

In the early 1900s, Paul Ehrlich started developing drugs to treat infectious diseases and developed the first "chemotherapy." About 100 years later we are still following the paradigm that he established.

The modern era of chemotherapy evolved with the observation that the accidental exposure of humans to sulfur mustards resulted in the depletion of bone marrows and lymph nodes. Goodman and Gillman at Yale, following experiments in animals, initiated treatment of patients with non-Hodgkin's lymphoma with nitrogen mustard. The first responses were observed in 1943 and stimulated the development of alkylating agents after the war. Farber treated children with leukemia with folate antagonists (methotrexate) and Hitchings and Elion developed inhibitors of adenine metabolism (thiopurines) and observed major, although short-lived responses. Charles Heidelberger introduced 5-fluorouracil, the first "targeted" anticancer therapy. L-asparaginase is another example and once the double-helical structure of DNA was discovered, DNA-targeted agents such as cytosine arabinoside and daunomycin were successfully developed.

These chemotherapies were brought into clinical practice over the past 50 years and were based on evolving insights into the chemistry and biochemistry of the leukemic cell. Then as now, therapeutic strategies are attempting to exploit evolving scientific insights into metabolic and molecular abnormalities of the leukemia cell; therefore most therapies were "targeted" from their inception.

Subsequent steps in the development of targeted therapies were based on the discovery of tumor-associated antigens and monoclonal antibodies, on the plethora of newly identified mutations, evolving insights into the mechanisms of epigenetic gene regulation and the functions of oncogenic kinases. Much effort is now being focused on kinase inhibitors, following the spectacular success of the bcr-abl tyrosine kinase inhibitor Gleevec in Philadelphia-chromosome positive chronic myeloid and acute lymphoblastic leukemias.

What we are facing today is an exponential gain in our knowledge of the cell and molecular biology, immunology and epigenetic gene regulation of leukemias, which is fortunately matched by the zeal of laboratory and clinical investigators to translate scientific insights into clinical success. In addition, numerous treatment options have evolved and are being explored in clinical investigations even without clear identification and understanding of their targets. This approach is fueled by the notion that leukemia cells have an uncanny ability to evade highly specific targeted therapies and that less specific, multipotent "dirty" drugs may be more effective. Both approaches may have merits and are likely to contribute to the development of our therapeutic arsenal for the next decades.

Finally, it is becoming increasingly clear that many of the classical therapies are causing genetic damage in surviving cells that actually contribute to relapse. Hence, the development of nongenotoxic agents is paramount, such as BH3-mimetics and MDM2 inhibitors, that activate apoptosis pathways, presumably without causing DNA damage.

Targeted therapy also needs to define its cellular, not just molecular target. The concept of a highly drug resistant leukemia stem cell that contributes to relapse is now being accepted, but needs more attention by drug developers and clinicians alike It has also become increasingly clear that the leukemia microenvironment is a partner of equal importance to leukemia cell-intrinsic resistance mechanisms for the survival, expansion and relapse of leukemia stem cells, leading to clinical relapse. Early attempts are under way to address both sides of this "Yin-Yang" equation, which is dramatically put in focus by the recent discovery that molecular changes in the microenvironment can cause, and not just support, leukemia development.

In this book it is attempted to summarize and interpret targeted AML therapies by pairing chapters describing the basic molecular biology of major targets with clinical results of targeted therapies, thus rendering the translational and clinical researcher a useful tool to apply recent research data available to him in a more specific way. The exponential growth in knowledge exceeds the ability of any researcher to be up-to-date and we do hope that this collection of 43 chapters from the leading experts in their fields will make a difference in how we approach the ever-growing complexity of AML-directed therapeutics. Topics and targets are evolving with further elucidation of the underlying basic science, as the tools at our disposal.

The introductory chapters present updated contemporary classification systems of myeloid leukemias, their genetic defects and the proteomic alterations characteristic of AML. The basic mechanisms of apoptosis disregulation described in these chapters include the roles of BCl-2, IAP and p53 families of proteins and exemplify the concept of pairing basic science with evolving therapeutics such as BH3 and SMAC mimetics and MDM2 inhibitors that can reactivate p53 signaling.

Tyrosine kinase controlled cell signaling pathways have been elucidated in a large number of investigations following the spectacular success of Gleevec and its successors in the treatment of chronic myeloid leukemias. The book then transitions into a discussion of PIM, FLT3, NPM1, Ras/Raf/MAPK, PI3K/AKT/mTOR and aurora kinase functions and therapeutic targeting—a field which is rapidly developing as exemplified by the recent successes reported for IDH2 inhibitors but could not be included here.

Epigenetic modulators of acetylation and methylation are being covered and discussed along with epigenetic therapies which have already yielded significant therapeutic impact but will only increase with improved specificities and more precise definition of their targets. Although the underlying mechanisms of action are well studied, much work still needs to be done until we will have the ability to selectively affect the epigenetic regulation of specific genes.

PML-RARa and orphan nuclear receptors such as nor1 and nur77 have provided fascinating targets in leukemias, as is best exemplified in the improvement of the survival rates of patients with acute promyelocytic leukemias from 30 to 90 % in just a few years, following the introduction of ATRA and arsenic trioxide. Nur77 may provide another target which is universally dysregulated in AML, not just in APL.

Rapid advances in tumor immunology led to the development of monoclonal antibodies that resulted in impressive response rates, alone or in combination with immunotoxins, in chronic and acute lymphocytic leukemias. These achievements have been more modest in AMLs, but recent advances in leukemic stem cell biology may make antibodies the most effective tools in the elimination of AML stem cells.

The leukemia microenvironment has graduated from supportive bystander to therapeutic target with the discovery that changes in bone marrow stroma cells, such as deletions of DICER which regulates microRNAs, results in the development of AMLs. First attempts to disrupt leukemia/stroma interactions by blockade of CXCR4 and VLA-4 signaling are yielding successes, but this field is clearly in more need of development. The recent identification of pronounced hypoxia in AML bone marrows makes hypoxia and hypoxia-induced genes novel targets in leukemia therapy.

MicroRNAs are also becoming attractive targets, as their complex regulatory control of many pathways of vital importance for leukemia proliferation and survival is being better understood.

All genetic and epigenetic alterations in leukemias finally result in metabolic alterations. Cancer metabolism, a field founded originally by Otto Warburg, has recently evolved in unexpected ways and will become of critical importance for the eradication of leukemias. Glycolysis, oxidative phosphorylation and fatty acid oxidation are at the center of leukemia metabolomics. The above mentioned development of IDH2 mutation inhibitors and their initial activity in patients whose AML cells carry this mutation provide an impressive example on how molecular genetics can be linked to epigenetic and metabolic functions and then be applied to the small subset of patients who may benefit.

The development of specific immunotherapies employing NK^{-}, and T-cells is making its own way in the therapy of AML. CAR T-cell have yielded dramatic and sustained responses in certain leukemias and it is hoped that these concepts can be carried over into the treatment of AMLs. Hematopoietic transplantation has been refined and improved after half a century of unrelenting efforts and may be complemented by strategies utilizing mesenchymal stem cells as well.

All contributing authors hope that this book will serve as a useful guideline and provide inspiration and support to all scientists working in a field with high challenges but also great potential and importance for our collective future.

I wish to thank all authors for their thoughtful and insightful contributions. Special thanks go to Ms. Beate Buske-Kosel for her persistence, diligence and patience, without which this book would not have been possible.

Michael Andreeff

Contents

Part II Signaling

Part III Epigenetics

Contributors

Julien Ablain Université Paris Diderot, Sorbonne Paris Cité, Paris cedex 10, France

INSERM UMR 944, Equipe labellisée par la Ligue Nationale contre le Cancer, Institut Universitaire dHématologie, Paris cedex 10, France

CNRS UMR 7212, Paris cedex 10, France

Sairah Ahmed Department of Stem Cell Transplantation and Cellular Therapy, The University of Texas MD Anderson Cancer Center, Houston, TX, USA

Michael Andreeff Paul and Mary Haas Chair in Genetics, Section of Molecular Hematology and Therapy, The University of Texas MD Anderson Cancer Center, Houston, TX, USA

Department of Leukemia, University of Texas M.D. Anderson Cancer Center, Houston, TX, USA

Frederick R. Appelbaum Seattle Cancer Care Alliance and Clinical Research Division, Fred Hutchinson Cancer Research Center, Seattle, WA, USA

Jörg Bäsecke Department of Medicine, University of Göttingen, Göttingen, Germany

Halvard Bönig University of Washington, Seattle, WA, USA

Institute for Transfusion Medicine and Immunohematology, Department of Cellular Therapeutics/Cell Processing, Goethe University School of Medicine and German Red Cross Blood Service Baden-Württemberg-Hesse, Frankfurt am Main, Germany

Ulrike Bacher MLL Munich Leukemia Laboratory, Munich, Germany

Ramesh Balusu Kansas Masonic Cancer Research Institute, University of Kansas Medical Center, Kansas City, KS, USA

A. John Barrett Stem Cell Allogenic Transplantation Section, Hematology Branch, National Heart, Lung and Blood Institute, National Institutes of Health, Bethesda, MD, USA

Brandon Beagle Department of Molecular Biology & Biochemistry, University of California, Irvine, CA, USA

Pamela S. Becker Division of Hematology and Institute for Stem Cell and Regenerative Medicine, University of Washington School of Medicine, University of Washington, Seattle, WA, USA

Juliana Benito Department of Leukemia, Division of Cancer Medicine, The University of Texas MD Anderson Cancer Center, Houston, TX, USA

Irwin D. Bernstein Clinical Research Division, Fred Hutchinson Cancer Research Center, Department of Pediatrics, University of Washington, Seattle, WA, USA

Kapil N. Bhalla Houston Methodist Research Institute, Houston, TX, USA

Catherine M. Bollard Blood+Marrow Transplantation, Children's National Health System, Washington, DC, USA

Dominique Bonnet Haematopoietic Stem Cell Laboratory Cancer Research UK, London Research Institute, London, UK

Christopher H. Booth Wake Forest Institute of Regenerative Medicine, Wake Forest Comprehensive Cancer Center, Winston-Salem, NC, USA

Prithviraj Bose Massey Cancer Center, Virginia Commonwealth University, Richmond, VA, USA

Edwin Bremer Department of Surgery/Translational Surgical Oncology, University Medical Center Groningen, Groningen, GZ, The Netherlands

Karen Bussard Wake Forest Institute of Regenerative Medicine, Wake Forest Comprehensive Cancer Center, Winston-Salem, NC, USA

George A. Calin Department of Experimental Therapeutics, Houston, USA

Graduate School of Biomedical Sciences, Houston, USA

Center for RNA Interference and Non-Coding RNAs, The University of Texas MD Anderson Cancer Center, Houston, TX, USA

Department of Experimental Therapeutics, Division of Cancer Medicine, The University of Texas MD Anderson Cancer Center, Houston, Texas, USA

Martin Carroll Department of Medicine, University of Pennsylvania, Philadelphia, PA, USA

Bing Z. Carter Department of Leukemia, Section of Molecular Hematology and Therapy, The University of Texas MD Anderson Cancer Center, Houston, TX, USA

Richard Champlin Department of Stem Cell Transplantation and Cellular Therapy, The University of Texas MD Anderson Cancer Center, Houston, TX, USA

Lisa S. Chen Department of Experimental Therapeutics, Division of Cancer Medicine, The University of Texas MD Anderson Cancer Center, Houston, TX, USA

Laurence James Neil Cooper Division of Pediatrics Cell Therapy Section, The University of Texas MD Anderson Cancer Center, Houston, TX, USA

Jorge Eduardo Cortes Department of Leukemia, Division of Cancer Medicine, The University of Texas MD Anderson Cancer Center, Houston, TX, USA

Naval Daver Department of Leukemia, Unit 428, The University of Texas MD Anderson Cancer Center, Houston, TX, USA

Hugues de Thé Université Paris Diderot, Sorbonne Paris Cité, Paris cedex 10, France

INSERM UMR 944, Equipe labellisée par la Ligue Nationale contre le Cancer, Institut Universitaire dHématologie, Paris cedex 10, France

CNRS UMR 7212, Paris cedex 10, France

Pole Sino-Francais des Sciences du Vivant et de Génomique de lHôpital Rui Jin, Rui-Jin Hospital affiliated with Jiao Tong University, Shanghai, China

Department Molecular Pathology and Virology, INSERM U 944 et CNRS UMR 7212, Institut Universitaire dHématologie, Paris Cedex 10, France

Lee Anthony Dean Division of Pediatrics Cell Therapy Section, The University of Texas MD Anderson Cancer Center, Houston, TX, USA

John F. DiPersio Division of Oncology, Washington University School of Medicine, Saint Louis, MO, USA

Elihu Estey University of Washington and Fred Hutchinson Cancer Research Center, Seattle, WA, USA

Warren Fiskus Kansas Masonic Cancer Research Institute, University of Kansas Medical Center, Kansas City, KS, USA

Ciara L. Freeman Department of Haematology, Barts and The London NHS Trust, London, E1 2ES, UK

David A. Fruman Department of Molecular Biology & Biochemistry and Institute for Immunology, University of California, Irvine, CA, USA

Varsha Gandhi Department of Experimental Therapeutics, Division of Cancer Medicine, The University of Texas MD Anderson Cancer Center, Houston, TX, USA

Department of Experimental Leukemia, Division of Cancer Medicine, The University of Texas MD Anderson Cancer Center, Houston, TX, USA

Chezi Ganzel Department of Hematology, Shaare Zedek Medical Center, Jerusalem, Israel

Guillermo Garcia-Manero Department of Leukemia, Division of Cancer Medicine, University of Texas MD Anderson Cancer Center, Houston, TX, USA

Ramiro Garzon College of Medicine, Department of Internal Medicine, Comprehensive Cancer Center, The Ohio State University, Columbus, OH, USA

Margherita Ghisi Cancer Therapeutics Program, Peter MacCallum Cancer Centre, East Melbourne, Victoria, Australia

Francis J. Giles HRB Clinical Research Facility, Galway & Trinity College Dublin, National University of Ireland, Galway, Ireland

Steven Grant Division of Hematology and Oncology, Massey Cancer Center, Virginia Commonwealth University, Richmond, VA, USA

Vera Grossmann MLL Munich Leukemia Laboratory, Munich, Germany

Monica L. Guzman Department of Medicine, Hematology, and Medical Oncology, Pharmacology in Medicine, Weill Medical College of Cornell University, New York, NY, USA

Claudia Haferlach MLL Munich Leukemia Laboratory, Munich, Germany

Torsten Haferlach MLL Munich Leukemia Laboratory, Munich, Germany

Duane C. Hassane Department of Medicine, Hematology, and Medical Oncology, Institute for Computational Biomedicine, Weill Medical College of Cornell University, New York, NY, USA

Wijnand Helfrich Department of Surgery/Translational Surgical Oncology, University Medical Center Groningen, Groningen, GZ, The Netherlands

Ricky W. Johnstone Cancer Therapeutics Program, Peter MacCallum Cancer Centre, East Melbourne, Victoria, Australia

Hiroshi Katayama Department of Molecular Oncology, Okayama University Graduate School of Medicine, Dentistry and Pharmaceutical Sciences, Okayama, Japan

Kevin R. Kelly Institute for Drug Development, Cancer Therapy, and Research Center, University of Texas, Health Science Center, Hematology Clinic—Cancer Therapy & Research Center, San Antonio, TX, USA

Kojima Kensuke Section of Molecular Hematology and Therapy, Department of Leukemia, The University of Texas MD Anderson Cancer Center, Houston, TX, USA

Yong-Mi Kim Department of Pediatrics, Division of Hematology and Oncology, Children's Hospital Los Angeles, University of Southern California, Los Angeles, CA, USA

Marina Konopleva Division of Cancer Medicine—Unit 448, The University of Texas MD Anderson Cancer Center, Houston, TX, USA

Steven M. Kornblau Department of Leukemia and Department of Stem Cell Transplantation, Division of Cancer Medicine, The University of Texas, MD Anderson Cancer Center, Houston, TX, USA

Rob C. Laister Ontario Cancer Institute, Princess Margaret Hospital, Toronto Medical Discovery Tower, Toronto, ON, Canada

Valérie Lallemand-Breittenbach Université Paris Diderot, Sorbonne Paris Cité, Paris cedex 10, France

INSERM UMR 944, Equipe labellisée par la Ligue Nationale contre le Cancer, Institut Universitaire dHématologie, Paris cedex 10, France

CNRS UMR 7212, Paris cedex 10, France

Dean Anthony Lee Division of Pediatrics Cell Therapy Section, The University of Texas MD Anderson Cancer Center, Houston, TX, USA

Syng-Ook Lee Institute of Biosciences and Technology, Texas A&M Health Science Center, Houston, TX, USA

Daniel C. Link Division of Oncology, Department of Medicine, Washington University School of St. Louis, Saint Louis, MO, USA

Tak W. Mak Campbell Family Institute for Breast Cancer Research, Ontario Cancer Institute, Toronto, ON, Canada

Guido Marcucci The Ohio State University, Department of Internal Medicine—Division of Hematology, 898 Biomedical Research Tower, Columbus, OH, USA

Frank C. Marini Wake Forest Institute of Regenerative Medicine, Wake Forest Comprehensive Cancer Center, Winston-Salem, NC, USA

Department of Leukemia, University of Texas M.D. Anderson Cancer Center, Houston, TX, USA

Alberto M. Martelli Dipartimento di Scienze Anatomiche Umane e Fisiopatologia dell'Apparato Locomotore, Università di Bologna, Bologna, Italy

IGM-CNR, Bologna, Italy

James A. McCubrey Department of Microbiology and Immunology, Brody School of Medicine at East Carolina University, Greenville, NC, USA

Cong Meng Department of Veterinary Physiology and Pharmacology, Texas A&M University, College Station, TX, USA

Mark D. Minden Ontario Cancer Institute, Princess Margaret Hospital, Toronto, ON, Canada

Gabriela Motyckova Department of Medical Oncology, Dana-Farber Cancer Institute, Harvard Medical School, Boston, MA, USA

Lysette Mutkus Wake Forest Institute of Regenerative Medicine, Wake Forest Comprehensive Cancer Center, Winston-Salem, NC, USA

Rihab Nasr Department of Anatomy, Cell Biology and Physiological Sciences, American University of Beirut, Beirut, Lebanon

Thalia Papayannopoulou Department of Medicine, Division of Hematology, Fred Hutchinson Cancer Research Center, Seattle, WA, USA

Naveen Pemmaraju Department of Leukemia, Division of Cancer Medicine, The University of Texas MD Anderson Cancer Center, Houston, TX, USA

Department of Leukemia, Division of Cancer Medicine, University of Texas MD Anderson Cancer Center, Houston, TX, USA

Oliver W. Press Fred Hutchinson Cancer Research Center, Department of Medicine/ Division of Medical Oncology, University of Washington, Seattle, WA, USA

Parvathi Ranganathan College of Medicine, Department of Internal Medicine, Comprehensive Cancer Center, The Ohio State University, Columbus, OH, USA

Farhad Ravandi Department of Leukemia, Unit 428, The University of Texas MD Anderson Cancer Center, Houston, TX, USA

3Department of Leukemia, Division of Cancer Medicine, The University of Texas MD Anderson Cancer Center, Houston, TX, USA

John C. Reed Sanford Burnham Institute for Medical Research, La Jolla, CA, USA

Jacob M. Rowe Department of Hematology, Shaare Zedek Medical Center, Jerusalem, Israel

Technion, Israel Institute of Technology, Haifa, Israel

Stephen Safe Department of Veterinary Physiology and Pharmacology, Texas A&M University, College Station, TX, USA

Institute of Biosciences and Technology, Texas A&M Health Science Center, Houston, TX, USA

Ismael Samudio Biochemistry Laboratory # 301, Pontificia Universidad Javeriana, Bogota, Colombia

Susanne Schnittger MLL Munich Leukemia Laboratory, Munich, Germany

Judith S. Sebolt-Leopold Translational Oncology Program, University of Michigan, Comprehensive Cancer Center, Ann Arbor, MI, USA

Subrata Sen Department of Translational Molecular Pathology, Unit 951, The University of Texas M.D. Anderson Cancer Center, Houston, TX, USA

Maitri Y. Shah Department of Experimental Therapeutics, Houston, USA

Graduate School of Biomedical Sciences, Houston, USA

Elizabeth J. Shpall Department of Stem Cell Transplantation and Cellular Therapy, Division of Cancer Medicine, The University of Texas MD Anderson Cancer Center, Houston, TX, USA

Erika Spaeth Department of Leukemia, University of Texas M.D. Anderson Cancer Center, Houston, TX, USA

Linda S. Steelman Department of Microbiology and Immunology, Brody School of Medicine at East Carolina University, Greenville, NC, USA

Eytan M. Stein Medical Oncology and Hematology, Memorial Sloan-Kettering Cancer Center, New York, NY, USA

Richard M. Stone Department of Medical Oncology, Dana-Farber Cancer Institute, Harvard Medical School, Boston, MA, USA

Martin S. Tallman Leukemia Service, Memorial Sloan-Kettering Cancer Center, Weill Cornell Medical Center, New York, NY, USA

Geoffrey L. Uy Washington University School of Medicine, Saint Louis, MO, USA

Juliana Vélez Biochemistry Laboratory # 301, Pontificia Universidad Javeriana, Bogota, Colombia

Lyubomir T. Vassilev Discovery Oncology, Roche Research Center, Hoffmann-La Roche Inc, Nutley, NJ, USA

Alison R. Walker The Ohio State University, Department of Internal Medicine—Division of Hematology, 898 Biomedical Research Tower, Columbus, OH, USA

Roland B. Walter Clinical Research Division, Fred Hutchinson Cancer Research Center; Department of Medicine/ Division of Hematology, University of Washington, Seattle, WA, USA

Gerrit Weber University Children's Hospital, Pediatric Hematology/Oncology, University of Wuerzburg, Houston, TX, USA

William R. Wilson Auckland Cancer Society Research Centre, University of Auckland, Auckland, NZ, USA

Beiyan Zhou Department of Veterinary Physiology and Pharmacology, Texas A&M University, College Station, TX, USA

Jun Zhu Université Paris Diderot, Sorbonne Paris Cité, Paris cedex 10, France

INSERM UMR 944, Equipe labellisée par la Ligue Nationale contre le Cancer, Institut Universitaire dHématologie, Paris cedex 10, France

CNRS UMR 7212, Paris cedex 10, France

Pole Sino-Francais des Sciences du Vivant et de Génomique de lHôpital Rui Jin, Rui-Jin Hospital affiliated with Jiao Tong University, Shanghai, China

Chapter 1
Genetics and Classification of Acute Myeloid Leukemia

Alison R. Walker and Guido Marcucci

Abstract Nonrandom chromosomal abnormalities present at the time of diagnosis of acute myeloid leukemia (AML) are used to direct treatment decisions for patients as they are known predictors of response to therapy and disease-free and overall survival. During recent years, molecular analyses have revealed a number of acquired gene mutations and changes in gene and microRNA expression that allow for classification of patients into cytogenetic and molecular subsets with distinct clinical characteristics. Aside from outcome prediction and stratification to risk-adapted therapies, these discoveries have also provided insight into mechanisms of leukemogenesis and potential targets for therapeutic intervention. While the greatest improvement in risk stratification made by these advances has been in patients with cytogenetically normal (CN) AML, a group that comprises 40–50% of all patients with newly diagnosed disease, recurrent mutations with prognostic significance have also played a key role in other cytogenetic groups (i.e., core binding factor AML and AML with complex karyotype). The importance of these molecular abnormalities has been recognized, and select mutations are now included in classification systems (i.e., World Health Organization and European LeukemiaNet) as a complement to cytogenetics. In this chapter, we review these recent genetic findings and discuss their prognostic significance and clinical implications.

Keywords Cytogenetically normal AML · FLT3 · NPM1 · CEBPA · Mutation analysis

A. R. Walker (✉) · G. Marcucci
The Ohio State University, Department of Internal Medicine—Division of Hematology, 898 Biomedical Research Tower, 460 W. 12th Ave, Columbus, OH 43210, USA
e-mail: Alison.Walker@osumc.edu

G. Marcucci
e-mail: guido.marcucci@osumc.edu

M. Andreeff (ed.), *Targeted Therapy of Acute Myeloid Leukemia*,
Current Cancer Research, DOI 10.1007/978-1-4939-1393-0_1

1.1 Introduction

The diagnosis and classification of a patient with acute myeloid leukemia (AML) has progressed from one determined by conventional morphologic and cytochemical methods (Bennett et al. 1976), to one that incorporates both chromosomal analysis and the presence or absence of somatically acquired gene mutations (Dohner et al. 2010; Swerdlow et al. 2008). Aside from providing consistency in the reporting of descriptions of patients enrolled onto clinical trials, genetic classification allows for prediction of outcome, and in turn, is used to direct treatment decisions. While these mutations may occur in any cytogenetic group, their discovery has had the most prognostic impact in the 40–50% of patients who present with cytogenetically normal (CN) AML, a group whose outcome risk had previously been difficult to define. Once considered a homogeneous group with equal, intermediate genetic risk, patients with CN-AML are now recognized as a diverse group with distinct clinical outcomes. Recurrent mutations with prognostic significance have also been identified in other cytogenetic groups (i.e., core binding factor (CBF) AML and AML with complex karyotype). Although important because of their clinical relevance, the identification of these molecular markers has also led to improved insight into the pathogenesis of disease and the development of novel targeted therapies. In this chapter, we will first review current classification systems and then discuss genetic markers with clinical and therapeutic implications for patients with AML.

1.2 Classification Systems

The recent World Health Organization (WHO) classification continues to define disease entities, including AML, based on a combination of clinical, morphologic, immunophenotypic and genetic features. Added to the category of "AML with recurrent genetic abnormalities" that previously included AML with t(8;21)(q22;q22), AML with inv(16) (p13.1q22) or t(16;16) (p13.1;q22), acute promyelocytic leukemia (APL) with t(15;17) (q22;q12) and AML with t(9;11) (p22;q23), are AML with t(6;9) (p23;q34), AML with inv(3) (q21;q26.2) or t(3;3) (q21;q26.2), and AML with t(1;22) (p13;q13) (Table 1.1; Swerdlow et al. 2008). Each of these new abnormalities is felt to share distinct morphologic and clinical features, meeting the classification's definition of a "clinic–pathologic–genetic entity." Although recognized as important, AML with mutated nucleophosmin 1 (*NPM1*) and CCAAT/enhancer-binding protein α (*CEBPA*) were added as "provisional entities" because of variability in morphology, immunophenotype and clinical features, and because the mutations are not mutually exclusive. AML with mutated FMS-like tyrosine kinase 3 (*FLT3*) was not considered a distinct entity, although the WHO acknowledged its prognostic importance and recommends that *FLT3* mutational testing be determined in all patients.

Table 1.1 Acute myeloid leukemia (Swerdlow et al. 2008)

Categories
Acute myeloid leukemia with recurrent genetic abnormalities
AML with t(8;21)(q22;q22); *RUNX1-RUNX1T1*
AML with inv(16)(p13.1;q22) or t(16;16)(p13.1;q22); *CBFB-MYH11*
APL with t(15;17)(q22;q12); *PML-RARA*
AML with t(9;11)(p22;q23); *MLLT3-MLL*
AML with t(6;9)(p23;q34); *DEK-NUP214*
AML with inv3(q21;q26.2) or t(3;3)(q21;q26.2); *RPN1-EVI1*
AML with t(1;22)(p13;q13); *RBM15;MKL1*
Provisional entity: AML with mutated NPM1
Provisional entity: AML with mutated CEBPA
Acute myeloid leukemia with myelodysplasia-related changes
Therapy-related neoplasms
Acute myeloid leukemia, not otherwise specified (NOS)
Acute myeloid leukemia with minimal differentiation
Acute myeloid leukemia without maturation
Acute myeloid leukemia with maturation
Acute myelomonocytic leukemia
Acute monoblastic/monocytic leukemia
Acute erythroid leukemia
Pure erythroid leukemia
Erythroleukemia, erythroid/myeloid
Acute megakaryoblastic leukemia
Acute basophilic leukemia
Acute panmyelosis with myelofibrosis

In 2010, the European LeukemiaNet (ELN) proposed a standard system for the reporting of genetic abnormalities with clinical outcome that incorporates both cytogenetics and mutation analyses of *NPM1, CEBPA,* and *FLT3* genes, creating four genetic groups: favorable, intermediate-I, intermediate-II, and adverse (Table 1.2; Dohner et al. 2010). Notably, patients with mutated *NPM1* without *FLT3*-ITD CN-AML, or mutated *CEBPA* CN-AML are now classified as having favorable genetic risk, similar to those with CBF AML (i.e., AML harboring t(8;21) (q22;q22) *RUNX1-RUNX1T1* inv(16) (p13.1q22) or t(16;16) (p13.1;q22), or *CBFB-MYH11*). All other *FLT3*-ITD/*NPM1* mutation combinations are considered to have intermediate-1 genetic risk. Although this classification was initially intended as a reporting system for AML trials, the genetic groups have recently proven to have prognostic significance (Mrozek et al. 2011; Rollig et al. 2011). The National Comprehensive Cancer Network (NCCN) has also incorporated molecular markers and chromosomal analysis for risk stratification, similar to the ELN, classifying patients as better risk, intermediate risk or poor risk (2012, www.nccn.org). Patients with mutated *NPM1* without *FLT3*-ITD CN-AML or mutated *CEBPA* CN-AML are classified as better risk, as in the ELN classification, and patients with *FLT3*-ITD mutated CN-AML are classified as poor risk.

Table 1.2 European LeukemiaNet Classification (Dohner et al. 2010)

Genetic group	Subsets
Favorable	t(8;21)(q22;q22); *RUNX1-RUNX1T1*
	inv(16)(p13.1;q22) or t(16;16)(p13.1;q22); *CBFB-MYH11*
	Mutated *NPM1* without *FLT3*-ITD (normal karyotype)
	Mutated *CEBPA* (normal karyotype)
Intermediate-I	Mutated *NPM1* and *FLT3*-ITD (normal karyotype)
	Wild type *NPM1* and *FLT3*-ITD (normal karyotype)
	Wild type *NPM1* without *FLT3*-ITD (normal karyotype)
Intermediate-II	t(9;11)(p22;q23); *MLLT3-MLL*
	Cytogenetic abnormalities not classified as favorable or adverse
Adverse	inv(3)(q21q26) or t(3;3)(q21;q26.2); *RPN1-EVI1*
	t(6;9)(p23;q34); *DEK-NUP214*
	t(v;11)(v;q23); *MLL* rearranged
	-5 or del(5q); −7; abnl(17p); complex karyotype[a]

Three or more chromosome abnormalities in the absence of one of the WHO designated recurring translocations or inversions (i.e., t(15;17), t(8;21), inv(16) or t(16;16), t(9;11), t(v;11)(v;q23), t(6;9), inv(3) or t(3;3))

1.3 Gene Mutations in AML

With implementation of novel discovering techniques, the number of mutations implicated in AML is rapidly growing. In this section, we will review somatically acquired gene mutations that have been identified and have classification and prognostic relevance for patients with AML. We will also focus on mutations discovered in CN-AML and other cytogenetic groups (i.e., CBF AML and AML with complex karyotype). Aside from what has been mentioned previously, our discussion of APL in this chapter will be limited, with a focus on the frequency and impact of *FLT3* mutations in this group.

1.3.1 FLT3 Mutations

Expressed on the surface of hematopoietic progenitor cells, the class III FLT3 receptor is an important regulator of normal cellular growth and differentiation (McKenna et al. 2000). Mutations that occur within the gene have been identified in two functional domains of the FLT3 receptor and confer a proliferative and survival advantage to leukemic blasts (Fenski et al. 2000). An internal duplication (ITD) within exons 14 and 15 that maps to the juxtamembrane domain and leads to ligand-independent receptor dimerization and constitutive kinase activation is the most common form of the *FLT3* mutation (Fenski et al. 2000). The second type of mutation, caused by insertions, deletions, and missense point mutations, occurs within the tyrosine kinase domain (TKD) and primarily involves codons 835 and 836 (Mead et al. 2007; Whitman et al. 2008; Yanada et al. 2005).

Initial studies reported an overall frequency of *FLT3* mutations in 23 % of newly diagnosed patients with non-FAB M3 AML, though it is now known that the mutational frequency varies within genetic subsets and among those with de novo, secondary or therapy-related AML (Frohling et al. 2002; Kiyoi et al. 1999; Nakao et al. 1996; Wiernik 2010). *FLT3* mutations occur in 23–27 % of patients with de novo AML as compared to 11 and 15 % of patients with therapy-related or secondary AML, respectively. *FLT3*-ITD mutations are uncommon in CBF AML or AML with complex karyotype, but do occur in 30–35 % of patients with APL, although do not impact the rate of relapse or their overall survival (OS; Frohling et al. 2002; Kottaridis et al. 2001; Schnittger et al. 2002). Patients with CN-AML and *FLT3*-ITD mutations, 25–35 % of cases, tend to present with higher-total white blood cells (WBC) counts, have a similar rate of achievement of complete remission (CR) as those with wild-type *FLT3,* but have an increased risk for relapse and death (Frohling et al. 2002). Interestingly, the allelic ratio (ratio of mutant to wild-type *FLT3*) of the gene has prognostic impact as patients with a low allelic ratio have an outcome similar to those without an *FLT3*-ITD mutation, while those with an increased allelic ratio have a worse survival (Thiede et al. 2002; Whitman et al. 2001).

The prognostic significance of *FLT3*–TKD mutations, which occur in 5–10 % of all AML cases, is controversial. The Cancer and Leukemia Group B (CALGB) reported a negative impact of *FLT3*–TKD mutations (in the absence of *FLT3*-ITD) on disease-free survival (DFS) in younger patients (age <60 years) with CN-AML as compared to *FLT3* wild-type patients (Whitman et al. 2008). In a report by the United Kingdom Medical Research Council (UK MRC) of patients entered onto either AML10 or AML12 trials, a subset analysis of CN-AML patients did not show a negative impact of *FLT3*–TKD mutations, and in fact a trend for an improved OS was observed in this group (Mead et al. 2007). The difference in outcomes observed in these two trials may have been due to the inclusion of patients with *FLT3*-ITD mutations and patients who had received an allogeneic stem cell transplant (SCT) as post-remission therapy in the MRC trials. Additional studies are needed to determine the prognostic impact of *FLT3*–TKD mutations and presently, *FLT3*–TKD mutational status is not included in genetic risk classification systems.

Constitutive activation of the FLT3 receptor signaling pathway by these mutations has made it an attractive target for the development of small molecules that inhibit the kinase domain. Thus far, clinical trials with tyrosine kinase inhibitors alone or in combination with chemotherapy have not decreased the risk of relapse or improved OS for patients with *FLT3* mutations, although more potent inhibitors of FLT3 are in the early phases of development and may provide greater clinical benefit (Levis et al. 2011). For now, allogeneic SCT has been recommended in patients with *FLT3*-ITD mutations if not entered onto clinical trials (Schlenk et al. 2008).

1.3.2 *NPM1* Mutations

NPM1 is a multifunctional protein that acts as a molecular chaperone, shuttling between the nucleus and cytoplasm, and has both proliferative and growth-suppressive roles in normal cellular function (Grisendi et al. 2006). Mutations within exon 12 of the *NPM1* gene are the most common molecular mutations in patients with CN-AML, occurring in 45–60 % of patients, and result in abnormal localization of the protein within the cytoplasm (Falini et al. 2005). In the initial report that described *NPM1* and its prognostic significance, 35.2 % of all patients with de novo AML had *NPM1* mutations, while 61.7 % of those with CN-AML had the mutation, which was associated with a response to induction chemotherapy and a CR rate of 71 % (Falini et al. 2005). Additional studies with larger cohorts of patients have confirmed the favorable prognostic impact of *NPM1* mutations, as well as an association with elevated WBC counts, myelomonocytic/monocytic morphology, and extramedullary involvement.

An interaction between *NPM1* and *FLT3* mutations was first observed in younger (age <60 years) patients with de novo or secondary CN-AML entered onto multicenter trials of the AML Study Group (AMLSG) where patients with *NPM1* mutated (mut)/*FLT3*-ITD wild-type (wt) AML had the highest remission rates and most favorable OS (Dohner et al. 2005). In each of these trials, post-remission therapy included allogeneic SCT and in a donor versus no donor comparison, patients with *NPM1* mut/*FLT3*-ITD wt CN-AML did not benefit from SCT and had an OS of 60 % at 5 years. Given that treatment outcomes for this group are on par with patients with CBF AML, both the WHO and ELN have reclassified patients with this molecular profile from intermediate to favorable genetic risk categories, reserving allogeneic SCT until the time of relapse (Dohner et al. 2010; Swerdlow et al. 2008).

The impact of *NPM1* mutations as a single marker in older patients (age >60 years) with CN-AML treated with intensive chemotherapy has been reported by the CALGB (Becker et al. 2009). *NPM1* mutations were identified in 83 of the 148 patients (56 %) evaluated, and the clinical features of this older population were similar to what had been observed in younger cohorts, with improved CR rates (83 % vs. 48 %) and a longer survival at 3 years, 35 % versus 8 %, in those with the mutation. Most interesting was the association of *NPM1* mutations and higher CR rates with improved DFS and OS in patients ≥70 years, while in patients aged 60–69 years, *NPM1* mutations did not affect DFS or OS. This finding supports molecular analysis regardless of age in all patients diagnosed with AML and consideration of intensive chemotherapy in these patients.

Recently, all-*trans* retinoic acid (ATRA), a vitamin A derivative with a key role in the management of patients with APL, has been reported to improve the OS of older patients with non-FAB-M3 AML with mutated *NPM1* when given concurrently with induction chemotherapy and following high-dose cytarabine consolidation (Schlenk et al. 2009). The authors suggest that this may be due to an effect of ATRA on the leukemic cells that express mutant *NPM1,* perhaps due to an increase in chemosensitivity, or an effect of the sequence of administration of ATRA in relation to the other chemotherapy agents. A trial in younger patients (age <60 years) found that those with *NPM1*-mutated AML who received ATRA in combination with intensive chemotherapy had both an increase in CR rate and improved event-free survival (EFS), while there was no effect of ATRA in those with wild-type *NPM1* (Schlenk et al. 2011). Additional studies combining ATRA with chemotherapy are needed to further explore the role of this agent in *NPM1*-mutated AML.

Monitoring *NPM1* transcript levels as a marker of minimal residual disease (MRD) and way to follow remission status has been evaluated in patients with *NPM1*-mutated CN-AML (Schnittger et al. 2005). In a study utilizing serial RNA-based real-time quantitative polymerase chain reaction (RQ-PCR) in 245 patients with *NPM1*-mutated AML, Kronke et al. reported that the level of the *NPM1*-mutated transcript at diagnosis did not have an impact on DFS or OS, yet patients with RQ-PCR positivity after two induction cycles or following completion of consolidation chemotherapy had a higher cumulative incidence of relapse (Kronke et al. 2011). In those patients without RQ-PCR evidence of the transcript at the end of consolidation, serial monitoring of transcript levels thereafter did allow for prediction of relapse. Without a cytogenetic abnormality to follow in patients with CN-AML, monitoring of mutated *NPM1* transcripts, and consideration of treatment intensification based on MRD assessment, may offer a way to improve outcomes in this group of patients, although additional studies are needed.

1.3.3 CEBPA Mutations

CEBPA is a basic leucine zipper transcription factor necessary for normal myeloid differentiation (Nerlov 2004). Within the same reading frame of the mRNA, there are two initiation sites which give rise to separate translation products of different length, the shorter p30 isoform and the longer p42 isoform (Nerlov 2004). Each has a domain that allows for interaction with other transcription factors, but only the p42 isoform has a transactivating domain. *CEBPA* mutations occur in 10–15% of patients with CN-AML and may be due to frameshift mutations in the N-terminus that affect the transactivating domain creating a nonfunctioning protein, or insertions or deletions within the C-terminus that affect DNA binding. In two thirds of cases, *CEBPA* mutations are biallelic and result in absent expression of wild-type *CEBPA* due to either homozygous (same mutation on each allele) or heterozygous (different mutations on each allele) mutations. Interestingly, germ-line mutations in *CEBPA* have been identified as causative gene mutations in cases of familial AML (Smith et al. 2004; Sellick et al. 2005).

The presence of *CEBPA* mutations has been associated with a favorable outcome in patients with CN-AML, similar to that of *NPM1* mut/*FLT3*-ITD wt CN-AML (Schlenk et al. 2008). The CALGB evaluated the impact of *CEBPA* mutations in 175 younger patients (age <60 years) with molecular high-risk CN-AML (*NPM1* wt/*FLT3*-ITD mut) and reported results of gene and microRNA (miRNA) expression profiles in those with this mutation (Marcucci et al. 2008). The 5-year EFS of the 32 patients with *CEBPA* mutations was 53 % as compared to 30 % in those with *CEBPA* wild type. Most interesting was the observation that most of the *CEBPA* mutations (29 of the 32 cases) occurred in patients with *NPM1* wt/*FLT3*-ITD mut AML, a group classified as having unfavorable genetic risk (Dohner et al. 2010; Swerdlow et al. 2008), yet the presence of a *CEBPA* mutation predicted for an improved DFS and OS that was independent of *FLT3* or *NPM1* status. Differences in gene expression profiles were also observed in those with *CEBPA* mutations where patients with a single affected allele expressed a gene expression profile that was different from those who had both alleles affected. Patients with one affected allele could not be distinguished from wild-type cases with regard to outcome.

Given its association with a favorable outcome, AML with double *CEBPA* mutations is classified as a low molecular risk subset of patients with CN-AML. Recommendations for post-remission therapy are similar to those with *NPM1* mutations without *FLT3*-ITD and include three to four cycles of high-dose cytarabine, reserving allogeneic SCT until relapse.

1.3.4 DNMT3A Mutations

Aberrant methylation of CpG dinucleotides by DNA methyltransferase enzymes leading to epigenetic silencing of structurally normal genes is believed to play a role in AML pathogenesis (Herman and Baylin 2003). There are three isoforms of the DNA Methyltransferase enzyme. *DNMT1* is responsible for maintenance of methylation patterns, while *DNMT3A* and *DNMT3B* initiate methylation patterns. In AML blasts, all three isoforms are over-expressed when compared to normal bone marrow cells (Mizuno et al. 2001). Yamashiata et al. were the first to identify an amino acid substitution at position 882 of *DNMT3A* in AML blasts, and several groups have gone onto further describe the frequency and prognostic significance of *DNMT3A* mutations in AML (Ley et al. 2010; Marcucci et al. 2012; Metzeler et al. 2011; Ribeiro et al. 2012; Thol et al. 2011; Yamashita et al. 2010).

One of the first descriptions of the prognostic impact of this mutation was in a cytogenetically heterogeneous cohort of 282 patients with de novo AML where mutations in *DNMT3A* were identified in 62 patients (22.1 %). In those classified as having intermediate cytogenetic risk, 56 patients (33.7 %) had a mutation in *DNMT3A,* suggesting that the frequency of this mutation may be higher in this group of patients (Ley et al. 2010). In this study, the presence of a *DNMT3A* mutation was associated with a worse median OS as compared to those without the mutation (12.3 months vs. 41.1 months, $p<0.001$). Patients included in this analysis received a variety of chemotherapy agents including cytarabine- and anthracycline-based regimens, azacitidine or decitabine, and lenalidomide. The type of post-remission therapy also varied and included high-dose cytarabine as well as autologous or allogeneic SCT. To evaluate the impact of *DNMT3A* mutational status in a larger, more homogeneous population of patients treated with intensive chemotherapy, Thol et al. reported an overall mutational frequency of 17.8 %, with 27.2 % of patients with CN-AML harboring *DNMT3A* mutations (Thol et al. 2011). Similar to earlier reports, patients with *DNMT3A* mutations were found to have a shorter OS as compared to patients with wild-type *DNMT3A,* 1.73 versus 3.33 years, although there was no difference in relapse-free survival between the groups. The prognostic impact of *DNMT3A* mutations was also evaluated in patients with high molecular risk, which included those with *NPM1* wt/*FLT3*-ITD AML, and found that those with a *DNMT3A* mutation had a worse OS and RFS.

In patients with CN-AML, the impact of *DNMT3A* mutations appears to be dependent upon age and the type of *DNMT3A* mutation that is present. The CALGB reported the prognostic impact of *DNMT3A* mutations in a cohort of 415 patients with CN-AML who received cytarabine- and daunorubicin-based induction chemotherapy (Marcucci et al. 2012). Missense mutations affecting arginine codon 882 (R882-*DNMT3A*) were the most frequent mutation (62 %) observed, similar to other reports, and all other mutations were classified as non-R882-*DNMT3A* mutations. In younger patients (age <60 years), there was no impact of R882-*DNMT3A* mutations on DFS or OS; however, in older patients, R882-*DNMT3A* mutations were associated with a both a shorter DFS (3 % vs. 21 % $p=0.006$) and OS (4 % vs. 24 % $p=0.01$) at 3 years. The impact of non-R882-*DNMT3A* mutations was also age dependent. In younger patients, non-R882-*DNMT3A* mutations were associated with a shorter DFS (20 % vs. 49 % $p=0.007$) and trend towards a shorter OS (29 % vs. 52 % $p=0.09$) at 3 years, while in older patients, non-R882-*DNMT3A* mutations did not have an effect. The basis for the differences in outcome based on mutation type observed in this study is not clear, although it may be due to the different frequencies of the specific mutations in each age group.

The impact of *DNMT3A* mutations on response to decitabine-based chemotherapy has also been reported (Metzeler et al. 2011). In a group of mostly older patients treated with decitabine, either as a single agent at a dose of 20 mg/m^2 per day for 10 days alone or in combination with bortezomib, six of the eight patients with *DNMT3A* mutations (75 %) achieved a CR as compared to 13 of the 38 *DNMT3A* wild-type patients (34 % $p=0.05$), suggesting an association between *DNMT3A* mutational status and response to decitabine-based therapy. Of note, this was a small, cytogenetically heterogeneous group of patients which limits the interpretation of these results. However, if these findings are confirmed in a larger, homogeneous cohort of patients with AML, *DNMT3A* mutational status could be used to stratify older patients to decitabine-based therapies.

An integrated mutational analysis that evaluated the impact of 18 recurrent somatic mutations in patients enrolled on Eastern Cooperative Oncology Group protocol E1900 reported an impact of *DNMT3A* mutations that was dependent upon the dose of daunorubicin received (Patel et al. 2012). In this study, patients with *DNMT3A* mutations that were randomized to receive the higher dose of daunorubicin (90 mg/m^2) during induction had an improved rate of survival as compared to patients with wild-type *DNMT3A.* If confirmed in additional studies, *DNMT3A* mutational status may be used to direct anthracycline dose intensification.

1.3.5 *IDH1/IDH2* Mutations

A metabolic enzyme, isocitrate dehydrogenase (IDH) catalyzes the oxidative carboxylation of 2,3-isocitrate creating 2-oxoglutarate and carbon dioxide in the Krebs cycle (Ward et al. 2010). When mutations in *IDH1* and *IDH2* arise, this forward reaction cannot occur, and rather a gain of function in the reverse direction that reduces α-ketoglutarate to 2-hydroxyglutarate, an oncogenic molecule, takes place. Mutations in *IDH1* and *IDH2* have only recently been described in AML, having first been discovered in colorectal cancer and then in primary brain tumors including glioblastomas, astrocytomas, and oligodendrogliomas (Boissel et al. 2010; Dang et al. 2010; Marcucci et al. 2010; Mardis et al. 2009; Paschka et al. 2010; Wagner et al. 2010). Interestingly, a significant increase in promoter methylation is observed in AML with *IDH1* or *IDH2* mutations as compared to other AML subtypes, and expression of mutant IDH proteins has been shown to increase DNA methylation (Figueroa et al. 2010a). These findings suggest that *IDH* mutations may play a role in leukemogenesis by dysregulating epigenetic processes within hematopoietic cells, and emerging small molecule inhibitors are being tested to target this process.

In AML, a mutation within *IDH1* was first identified following whole-genome sequencing and validation analysis of a patient with CN-AML, and subsequently, *IDH* mutations were found to be recurrent in a small group of patients, also with CN-AML (Mardis et al. 2009). The CALGB described the frequency and prognostic impact of *IDH* mutations in a cohort of 358 patients with CN-AML (Marcucci et al. 2010). A total of 118 patients (33 %) were found to have mutations in *IDH1* and *IDH2*, and a new mutation in *IDH2* was discovered, R172. While there were no differences in outcome when comparing all patients with *IDH* mutations and *IDH1*/*IDH2* wild type, in specific subsets, each mutation had prognostic relevance. In younger patients (age < 60 years) with *NPM1* mut/*FLT3*-ITD wt low-molecular-risk CN-AML, *IDH1* mutations were associated with a shorter DFS, and in older patients, where R172 *IDH2* mutation were most common, the CR rate was lower in those with the mutation than those without (20 % vs. 67 %). In a larger study of 805 younger patients with AML, a lower aggregate frequency of *IDH* mutations, 16 %, was reported; however, the negative prognostic impact of the mutation in patients with *NPM1* mut/*FLT3* wild-type AML was confirmed (Paschka et al. 2010). Recently, a single nucleotide polymorphism located within the same exon as the R132 *IDH1* mutation site, rs11554137 ($\mathrm{IDH1}^{105GGT}$), has been found in 12 % of patients with CN-AML and shown to have a negative impact on OS in those with *NPM1* wild type or *FLT3*-ITD-mutated AML (Wagner et al. 2010).

IDH mutations appear to have prognostic impact and may help to further molecularly risk stratify patients with CN-AML. Interestingly, when incorporated into an integrated mutational analysis including 17 other recurrent mutations, patients with both *IDH1* or *IDH2* and *NPM1* mutations were found to have an improved 3-year OS (89 % vs. 31 %) as compared to patients with mutant *NPM1* who did not have an *IDH* mutation (Patel et al. 2012). These patients did receive a higher dose of daunorubicin which may have led to the differences in outcome observed in this as compared to other trials. Additional studies are needed to further define the prognostic impact of *IDH* mutations, particularly as they may further define risk in those with favorable molecular risk.

1.3.6 RUNX1 Mutations

Runt-related transcription factor 1 (RUNX1) is a transcription factor that plays a key role in regulating hematopoietic differentiation, and monoallelic germ-line mutations in RUNX1 are known to occur in patients with familial platelet disorders that predispose to the development of AML (Michaud et al. 2002). Chromosomal translocations affecting the normal function of the gene have also been reported in AML (Dohner and Dohner 2008). Acquired *RUNX1* mutations have been identified in 13.2% of patients with de novo AML, and were associated with a lower CR rate and shorter DFS and OS than those without the mutation (Tang et al. 2009). A subsequent study of younger patients (age < 60 years) found a lower frequency of *RUNX1* mutations, 5.6% overall and 6.3% in CN-AML patients, but did confirm an association with a lower CR rate and shorter EFS and OS (Gaidzik et al. 2011). In this study, *RUNX1* mutations were associated with mutations previously reported to have a negative prognostic impact, *MLL*-PTD and *IDH1*/*IDH2* mutations, yet were inversely associated with the more favorable *NPM1* and *CEBPA* mutations.

The CALGB identified *RUNX1* mutations in 12.5% of all patients with CN-AML, and 16% of older CN-AML patients (age > 60 years) treated with cytarabine- and anthracycline-based chemotherapy (Mendler et al. 2012). The negative prognostic impact of *RUNX1* mutations in both younger and older patients was confirmed in this study. Patients with *RUNX1* mutations had lower CR rates and shorter DFS and OS as compared to *RUNX1* wild-type patients, and in younger patients, *RUNX1* mutations were inversely associated with *NPM1* and *CEBPA* mutations. *RUNX1* mutation-associated gene/miRNA expression signatures were analyzed and found to show upregulation of genes normally expressed in hematopoietic progenitor cells and/or B-cell progenitors that have been implicated in chemoresistance.

1.3.7 WT1 Mutations

The Wilms' Tumor 1 (WT1) gene encodes for a tissue-specific transcriptional regulator with an important role in normal cellular growth and development that is able to act as both a tumor suppressor and an oncogene (Call et al. 1990). The C-terminal domain of the protein contains four zinc fingers that allow for binding to DNA sequences, and therefore targets gene expression by repressing or enhancing transcription, while the N- terminal domain interacts with RNA and other proteins. Interactions with other proteins or post-translation modifications of the molecule likely allow for WT1 to function as both a tumor suppressor and an oncogene (Hohenstein and Hastie 2006; Yang et al. 2007). In patients with CN-AML, *WT1* mutations have been reported in 7–12 % with controversial prognostic impact.

In a series of 600 patients with CN-AML, *WT1* mutations were not associated with a different CR or survival rate as compared to patients without *WT1* mutations, although patients with both a *WT1* and *FLT3*-ITD mutation had a lower CR rate and increased rate of refractory disease (Gaidzik et al. 2009). In earlier studies of patients with CN-AML, *WT1* mutations were associated with a shorter OS and increased rate of relapse (Paschka et al. 2008; Virappane et al. 2008). In the study reported by the CALGB, the 3-year DFS of patients with *WT1* mutations was 13 % as compared to 50 % in *WT1* wild-type patients; however, in a subset analysis of older (age ≥60 years) patients, the DFS was similar in those with *WT1* mutations and those without. In the older patients with *WT1* mutations, a trend towards a decrease in CR and OS was observed, but did not reach statistical significance, which may have been because of the overall poor outcome of the group (Becker et al. 2010; Paschka et al. 2008).

A single nucleotide polymorphism in a mutational hotspot of *WT1*, SNP rs17754, was recently identified in the blasts and germline DNA of 25.7 % of patients with CN-AML and 36 % of healthy volunteers (Damm et al. 2009). In this study, the presence of the SNP was associated with a favorable impact on RFS and OS, although given the presence of this polymorphism in healthy volunteers, it may instead reflect chemotherapy sensitivity or metabolism.

1.3.8 MLL Mutations

Chromosomal translocations involving band 11q23 and the mixed-lineage leukemia (MLL) gene occur in both AML and acute lymphoblastic leukemia (ALL). In AML, balanced rearrangements involving 11q23 are heterogeneous; however, the most common translocations are t(9;11) (p22;q23), t(6;11) (q27;q23), and t(11;19) (q23;p13.1) (Krauter et al. 2009). Interestingly, the prognostic impact of 11q23 translocations is dependent on the fusion partner (Krauter et al. 2009; Strout et al. 1999). Indeed, patients with de novo AML and t(9;11) have a superior outcome when compared to patients with other 11q23 translocations.

A duplication of exons 5 through 11 or 5 through 12 of the *MLL* gene that then insert into intron 4 of the full-length gene creates a partial tandem duplication, *MLL*-PTD (Caligiuri et al. 1994; Ernst et al. 2002). The first molecular marker reported in patients with CN-AML, *MLL*-PTD mutations were identified in 11 % of patients with CN-AML (ages 18–84) and associated with a shorter CR duration as compared to those without *MLL*-PTD rearrangements (Caligiuri et al. 1998). Additional studies have confirmed a negative prognostic impact of *MLL*-PTD rearrangements in patients with CN-AML and trisomy 13; however, in a larger study of patients with CN-AML and *MLL*-PTD rearrangements, those who received intensive consolidation with autologous SCT had a similar DFS and OS to those without *MLL*-PTD rearrangements (Dohner and Dohner 2008; Dohner et al. 2002; Whitman et al. 2007). In an integrated mutational analysis that evaluated the impact of *MLL*-PTD arrangements, patients with *NPM1* wt/*FLT3*-ITD wt CN-AML with *MLL*-PTD rearrangements had a worse OS than those without *MLL*-PTD rearrangements, although the OS was improved in patients who received high-dose daunorubicin (90 mg/m^2) as compared to standard-dose daunorubicin (Patel et al. 2012).

The *MLL*-PTD protein retains its normal function and contributes to hypermethylation of tumor suppressor genes (Whitman et al. 2008). Interestingly, in vitro data have shown that this can be reversed following hypomethylating and histone deacetylase inhibitor treatment and a clinical trial with this combination of agents in patients with relapsed or refractory AML is ongoing (Whitman et al. 2005).

1.3.9 TET2 Mutations

Tet oncogene family member 2 (*TET2*) gene mutations have been previously reported in both myelodysplastic syndromes (MDS) and myeloproliferative neoplasms (MPNs) (Delhommeau et al. 2009; Langemeijer et al. 2009). *TET2* mutations impair hydroxylation of 5-methylcytosine which prevents demethylation of DNA and leads to hypermethylation of genes necessary for normal cellular function (Ko et al. 2010). In an animal model, loss of *Tet2* in the hematopoietic compartment was associated with self-renewal and progressive myeloproliferation in vivo (Moran-Crusio et al. 2011).

The CALGB reported an analysis of 427 patients with CN-AML and identified *TET2* mutations in 23 % of patients. In those with the mutation, there was an association with older age and higher pretreatment WBC as compared to those without the mutation, and those with favorable risk CN-AML had a lower CR rate and shorter EFS and OS as compared to those with wild-type *TET2* (Metzeler et al. 2011). There was no impact of *TET2* mutations in patients with ELN intermediate-1 risk (*CEBPA* wt, *FLT3*-ITD and/or *NPM1* wt) CN-AML. A lack of impact of *TET2* mutations in patients with ELN intermediate-1 risk CN-AML was confirmed by the German-Austrian AML Study Group; however, this study did not show an impact of *TET2* mutations in ELN favorable risk patients as had been reported by the CALGB and showed no difference in RFS or OS (Gaidzik et al. 2012). Given that the age of the populations in each of the studies were different, the CALGB study included patients over the age of 60 while the German-Austrian AML Study Group did not, and the intensity of post-remission therapy was different, this may have contributed to the differences in outcome that were observed. When included as part of an integrated mutational analysis (Patel et al. 2012), *TET2* mutations were associated with adverse risk and in fact CN-AML patients with N*PM1* wt/*FLT3* wt/*TET2* mut AML had a worse OS. If confirmed in future studies, *TET2* mutational status may also be used to risk stratify patients with CN-AML.

1.3.10 *ASXL1* Mutations

Mutations in the additional sex combs like −1 (ASXL1) have been identified in MDS, MPNs as well as AML, similar to *TET2* mutations (Carbuccia et al. 2009; Metzeler et al. 2011; Thol et al. 2011). In a large cohort of patients with previously untreated CN-AML, *ASXL1* mutations were fivefold more common in patients 60 years of age or older (16.2 %) than in younger patients (3.2 %), and almost half of the mutations were due to duplication of the guanine nucleotide at position 1934 (c.2934dupgG; Metzeler et al. 2011). *CEBPA* mutations were more common in patients with *ASXL1* mutations (26 % vs. 9 % $p=0.01$) than those with *ASXL1* wild type, and *NPM1* and *FLT3*-ITD mutations rarely occurred concurrently with *ASXL1* mutations. Overall, patients with *ASXL1* mutations had a lower CR rate and shorter DFS and OS than those without the mutation, and in CN-AML patients with ELN favorable genetic risk, *ASXL1* mutations were associated with lower CR rates than wild-type patients, 50 % versus 82 %, $p=0.02$. All patients with *ASXL1* mutations who did achieve a CR relapsed within 13 months and died within 18 months from diagnosis. ELN favorable genetic risk patients without *ASXL1* mutations had a 3-year DFS and OS of 27 and 34 %, respectively, and there was no impact of *ASXL1* mutations on ELN intermediate-1 risk on CR, DFS or OS. Similar to and perhaps with greater impact than *TET2*, *ASXL1* mutations are able to identify high-molecular-risk older patients with CN-AML.

1.3.11 *KIT* Mutations

The KIT receptor is a 145-kDa transmembrane protein that belongs to the type III receptor tyrosine kinase family (Sattler and Salgia 2004). The ligand for KIT, identified as stem cell factor (SCF), induces dimerization of the receptor and activates tyrosine kinase activity with subsequent phosphorylation of proteins involved in normal hematopoiesis. *KIT* mutations lead to constitutive tyrosine kinase activity, are more commonly seen in CBF AML, and have been associated with an increased risk of relapse as compared to wild-type cases in this favorable genetic risk group (Paschka et al. 2006). Indeed, in a series of 110 patients with de novo CBF AML reported by the CALGB, mutations were found in 18 (29.5 %) of 61 patients with inv(16) and in 11 (22 %) of 49 patients with t(8;21) (Paschka et al. 2006). While the CR rates in patients with inv(16) and mutated or wild-type *KIT* were similar, the 5-year CIR was 56 % for those with mutated *KIT* and 29 % for those with wild-type *KIT*. Also, the presence of mutated *KIT* in patients with inv(16) had a negative impact on OS. Similar findings were seen in patients with t(8;21), with no difference in CR and a 5-year CIR of 70 % for patients with t(8;21) and mutated *KIT*, as compared to 36 % for patients with wild-type *KIT*. There was no significant difference in OS for patients with t(8;21) and mutated or wild-type *KIT*.

Although *KIT* mutations are not commonly detected in CN-AML, they remain an important consideration in patients with CBF AML with regard to prognosis. *KIT* mutations also serve as a target for tyrosine kinase inhibitors. Indeed, in both the USA and Europe, multi-institutional clinical trials led by the CALGB and AMLSG, respectively, are combining cytarabine- and anthracycline-based induction chemotherapy and the tyrosine kinase inhibitor dasatinib in all patients with CBF AML. In addition to *KIT* mutations, CBF AML is also characterized by a significantly higher level of *KIT* expression which provides a rationale for the use of a tyrosine kinase inhibitor in all patients with CBF AML (Liu et al. 2010).

1.3.11.1 p53 Mutations

AML with complex karyotype (CK-AML), defined as three or more chromosome abnormalities in the absence of rearrangements listed in the WHO category of "AML with recurrent genetic abnormalities," occurs in 10–12% of patients and has been associated with a poor outcome (Mrozek 2008). Mutations and/or deletions in the tumor suppressor gene *p53* (*p53* alterations) have been associated with CK-AML, and in a cohort of 243 patients with CK-AML, those with *p53* alterations had a lower rate of CR (28% vs. 50%; $p=0.01$) and worse EFS (1% vs. 13%; $p=0.0007$) and OS (3% vs. 28% $p<0.0001$) as compared to those without *p53* alterations (Rucker et al. 2012) The monosomal karyotype, defined as the presence of two or more distinct autosomal monosomies in a karyotype, or a single autosomal monosomy in the presence of one or more structural chromosomal abnormalities, has further stratified patients with CK-AML and is associated with a very poor prognosis and 4-year OS of <5% (Breems et al. 2008). The previous study also evaluated the impact of *p53* mutations on treatment outcome in patients with CK-AML and a monosomal karyotype (Rucker et al. 2012). Interestingly, *p53* alterations were found in 80% of patients with a monsomal karyotype and were associated with a worse OS when compared to those patients with a monosomal karyotype without *p53* alterations. Based on these findings, patients with CK-AML and *p53* alterations should be considered for novel investigational agents for remission induction and allogeneic SCT in CR1.

1.4 Deregulated MiRNA Expression in AML

MiRNAs are noncoding RNAs consisting of 19–25 nucleotides that modulate post-transcriptional activity of mRNA and therefore play a role in protein expression (Bartel 2004). Deregulation of miRNAs prevents normal differentiation and proliferation of hematopoietic cells and therefore contributes to leukemogenesis. Upregulation or downregulation of specific miRNAs has been shown to be able to distinguish myeloid and lymphoid leukemia as well as cytogenetic subtypes of AML (Garzon et al. 2006, 2008; Li et al. 2008; Marcucci et al. 2008; Mi et al. 2007). The CALGB reported a distinct gene expression signature in a cohort of older patients (age ≥60 years) with CN-AML and *NPM1* mutations and found increased expression of *miR-10a, miR-10b, miR-196a, miR-196b* and upregulation of homeobox genes (Becker et al. 2009). Patients with *FLT3*-ITD mutations had an increase in *miR-155* as well as *miR-10a* and *miR-10b* expression. Patients with *CEBPA* mutated CN-AML were noted to have an upregulation of genes involved in erythroid differentiation and an upregulation of miRNA from the *miR-181* family which are known to be involved in erythroid lineage differentiation (Marcucci et al. 2008). Increased expression of *miR-181a* and *miR-181b* in patients with *NPM1* wt/*FLT3*-ITD mut CN-AML was associated with a higher CR rate and longer DFS and OS, providing additional prognostic information for patients with CN-AML (Schwind et al. 2010).

In addition to prognostic information, miRNAs have also been shown to be predictive of response to therapy. Blum et al. reported the predictive value of *miR-29b* in older patients (age ≥60 years) with previously untreated AML treated with a 10-day schedule of decitabine (Blum et al. 2010). In this study, the overall CR rate was 47% and higher pretreatment levels of *miR-29b* were found in responders as opposed to nonresponders ($p=0.02$). Targeting miRNA as a way to improve response to decitabine therapy has been incorporated into future clinical trial design for patients with AML.

1.5 Deregulated Gene Expression in AML

Variable expression of certain genes in subsets of patients with AML has been shown to have prognostic impact, similar to that of chromosomal translocations and molecular mutations. ETS-related gene (*ERG*) is a transcription factor with a role in hematopoietic differentiation (Oikawa 2004). Increased *ERG* expression has been associated with a lower CR rate and shorter EFS in patients with *NPM1* mut/*FLT3* wt AML when compared to low expressers (Marcucci et al. 2005). Chromosomal translocations involving 3q26 result in abnormal *EVI1* expression; however, increased *EVI1* expression can occur without chromosome 3 abnormalities and have a negative prognostic impact on OS (Lugthart et al. 2008). Shorter RFS, EFS, and lower CR rates were most pronounced in patients with intermediate genetic risk and increased *EVI1* expression (Groschel et al. 2010). The brain and acute leukemia, cytoplasmic (*BAALC*) gene has variable expression in CN-AML, and higher expression was associated with lower CR rates and shorter DFS and OS (Baldus et al. 2003; Langer et al. 2008). The meningioma 1 gene (*MN1*) encodes for a nuclear protein that normally functions as part of a transcriptional complex and in patients with CN-AML, increased *MN1* expression was associated with decreased CR and DFS rates (Heuser et al. 2006; Langer et al. 2009). Just as integrated mutational analysis has been performed to determine the overall prognostic impact of certain mutations, the impact of *ERG*, *BAALC* and *MN1* expression in patients with CN-AML has been reported (Metzeler et al. 2009). In this study, high *ERG* expression predicted for a lower likelihood of achieving a CR; however, *BAALC* and *MN1* expression levels did not have independent prognostic impact.

Conclusion

The number of recurrent somatic mutations in AML will likely continue to expand. Analyses that account for the complicated interrelationships of recurrent mutations, gene expression patterns, and miRNAs, as well as the impact of cytogenetics, are necessary to continue to improve risk stratification, especially for patients with CN-AML. Large studies of molecularly well-characterized, similarly treated patients will continue to lead to future discoveries, and the importance of enrollment on clinical trials cannot be overstated.

While the molecular markers discussed here have unmistakable clinical relevance, improving our understanding of pathways involved in disease pathogenesis is also necessary. The importance of epigenetic control of gene expression by DNA methylation and/or histone acetylation is clear, and whole-genome methylation profiling has identified distinct methylation patterns and epigenetic signatures in subsets of AML patients (Figueroa et al. 2010b). Given its role in leukemogenesis, targeting aberrant DNA methylation has been one strategy employed in the treatment of patients with AML (Arai et al. 2000; Blum et al. 2007, 2010, 2012; Cashen et al. 2009). Yan et al. recently reported the first description of genome-wide methylation profiling before and after one cycle of therapy with the hypomethylating agent decitabine (20 mg/m^2/day for 10 days) in untreated older AML patients (Yan et al. 2012). Decitabine induced significant global hypomethylation, predominantly in the CpG islands, which suggests that this is a target of decitabine therapy. A larger randomized phase II clinical trial of decitabine versus decitabine/bortezomib in older AML patients sponsored by the Alliance for Clinical Trials in Oncology is currently ongoing that will provide additional information regarding decitabine-related methylation changes.

Future trials will likely include novel agents that either target unfavorable mutations or increase the sensitivity of myeloid blasts to cytotoxic chemotherapy by converting an unfavorable molecular risk microenvironment to a more favorable one. Only with rational clinical trial design and drug development will we improve the outcomes for patients with AML.

References

Arai Y, Kyo T, Miwa H, Arai K, Kamada N, Kita K, Ohki M (2000) Heterogenous fusion transcripts involving the NUP98 gene and HOXD13 gene activation in a case of acute myeloid leukemia with the t(2;11)(q31;p15) translocation. Leukemia 14(9):1621–1629

Baldus CD, Tanner SM, Ruppert AS, Whitman SP, Archer KJ, Marcucci G, Caligiuri MA, Carroll AJ, Vardiman JW, Powell BL, Allen SL, Moore JO, Larson RA, Kolitz JE, de la Chapelle A, Bloomfield CD (2003) BAALC expression predicts clinical outcome of de novo acute myeloid leukemia patients with normal cytogenetics: a Cancer and Leukemia Group B Study. Blood 102(5):1613–1618

Bartel DP (2004) MicroRNAs: genomics, biogenesis, mechanism, and function. Cell 116(2):281–297.

Becker H, Marcucci G, Maharry K, Radmacher MD, Mrozek K, Margeson D, Whitman SP, Wu YZ, Schwind S, Paschka P, Powell BL, Carter TH, Kolitz JE, Wetzler M, Carroll AJ, Baer MR, Caligiuri MA, Larson RA, Bloomfield CD (2009) Favorable prognostic impact of NPM1 mutations in older patients with cytogenetically normal de novo acute myeloid leukemia and associated gene- and microRNA-expression signatures: a Cancer and Leukemia Group B study. J Clin Oncol 28(4):596–604

Becker H, Marcucci G, Maharry K, Radmacher MD, Mrozek K, Margeson D, Whitman SP, Paschka P, Holland KB, Schwind S, Wu YZ, Powell BL, Carter TH, Kolitz JE, Wetzler M, Carroll AJ, Baer MR, Moore JO, Caligiuri MA, Larson RA, Bloomfield CD (2010) Mutations of the Wilms tumor 1 gene (WT1) in older patients with primary cytogenetically normal acute myeloid leukemia: a Cancer and Leukemia Group B study. Blood 116(5):788–792.

Bennett JM, Catovsky D, Daniel MT, Flandrin G, Galton DA, Gralnick HR, Sultan C (1976) Proposals for the classification of the acute leukaemias. French-American-British (FAB) cooperative group. Br J Haematol 33(4):451–458

Blum W, Klisovic RB, Hackanson B, Liu Z, Liu S, Devine H, Vukosavljevic T, Huynh L, Lozanski G, Kefauver C, Plass C, Devine SM, Heerema NA, Murgo A, Chan KK, Grever MR, Byrd JC, Marcucci G (2007) Phase I study of decitabine alone or in combination with valproic acid in acute myeloid leukemia. J Clin Oncol 25(25):3884–3891

Blum W, Garzon R, Klisovic RB, Schwind S, Walker A, Geyer S, Liu S, Havelange V, Becker H, Schaaf L, Mickle J, Devine H, Kefauver C, Devine SM, Chan KK, Heerema NA, Bloomfield CD, Grever MR, Byrd JC, Villalona-Calero M, Croce CM, Marcucci G (2010) Clinical response and miR –29b predictive significance in older AML patients treated with a 10-day schedule of decitabine. Proc Natl Acad Sci U S A 107(16):7473–7478

Blum W, Schwind S, Tarighat SS, Geyer S, Eisfeld AK, Whitman S, Walker A, Klisovic R, Byrd JC, Santhanam R, Wang H, Curfman JP, Devine SM, Jacob S, Garr C, Kefauver C, Perrotti D, Chan KK, Bloomfield CD, Caligiuri MA, Grever MR, Garzon R, Marcucci G (2012) Clinical and pharmacodynamic activity of bortezomib and decitabine in acute myeloid leukemia. Blood 119(25):6025–6031

Boissel N, Nibourel O, Renneville A, Gardin C, Reman O, Contentin N, Bordessoule D, Pautas C, de Revel T, Quesnel B, Huchette P, Philippe N, Geffroy S, Terre C, Thomas X, Castaigne S, Dombret H, Preudhomme C (2010) Prognostic impact of isocitrate dehydrogenase enzyme isoforms 1 and 2 mutations in acute myeloid leukemia: a study by the Acute Leukemia French Association group. J Clin Oncol 28(23):3717–3723

Breems DA, Van Putten WL, De Greef GE, Van Zelderen-Bhola SL, Gerssen-Schoorl KB, Mellink CH, Nieuwint A, Jotterand M, Hagemeijer A, Beverloo HB, Lowenberg B (2008) Monosomal karyotype in acute myeloid leukemia: a better indicator of poor prognosis than a complex karyotype. J Clin Oncol 26(29):4791–4797

Caligiuri MA, Schichman SA, Strout MP, Mrozek K, Baer MR, Frankel SR, Barcos M, Herzig GP, Croce CM, Bloomfield CD (1994) Molecular rearrangement of the ALL –1 gene in acute myeloid leukemia without cytogenetic evidence of 11q23 chromosomal translocations. Cancer Res 54(2):370–373

Caligiuri MA, Strout MP, Lawrence D, Arthur DC, Baer MR, Yu F, Knuutila S, Mrozek K, Oberkircher AR, Marcucci G, de la Chapelle A, Elonen E, Block AW, Rao PN, Herzig GP, Powell BL, Ruutu T, Schiffer CA, Bloomfield CD (1998) Rearrangement of ALL1 (MLL) in acute myeloid leukemia with normal cytogenetics. Cancer Res 58(1):55–59

Call KM, Glaser T, Ito CY, Buckler AJ, Pelletier J, Haber DA, Rose EA, Kral A, Yeger H, Lewis WH et al (1990) Isolation and characterization of a zinc finger polypeptide gene at the human chromosome 11 Wilms' tumor locus. Cell 60(3):509–520

Carbuccia N, Murati A, Trouplin V, Brecqueville M, Adelaide J, Rey J, Vainchenker W, Bernard OA, Chaffanet M, Vey N, Birnbaum D, Mozziconacci MJ (2009) Mutations of ASXL1 gene in myeloproliferative neoplasms. Leukemia 23(11):2183–2186

Cashen AF, Schiller GJ, O'Donnell MR, DiPersio JF (2009) Multicenter, phase II study of decitabine for the first-line treatment of older patients with acute myeloid leukemia. J Clin Oncol 28(4):556–561

Damm F, Heuser M, Morgan M, Yun H, Grosshennig A, Gohring G, Schlegelberger B, Dohner K, Ottmann O, Lubbert M, Heit W, Kanz L, Schlimok G, Raghavachar A, Fiedler W, Kirchner H, Dohner H, Heil G, Ganser A, Krauter J (2009) Single nucleotide polymorphism in the mutational hotspot of WT1 predicts a favorable outcome in patients with cytogenetically normal acute myeloid leukemia. J Clin Oncol 28(4):578–585

Dang L, Jin S, Su SM (2010) IDH mutations in glioma and acute myeloid leukemia. Trends Mol Med 16(9):387–397

Delhommeau F, Dupont S, Della Valle V, James C, Trannoy S, Masse A, Kosmider O, Le Couedic JP, Robert F, Alberdi A, Lecluse Y, Plo I, Dreyfus FJ, Marzac C, Casadevall N, Lacombe C, Romana SP, Dessen P, Soulier J, Viguie F, Fontenay M, Vainchenker W, Bernard OA (2009) Mutation in TET2 in myeloid cancers. N Engl J Med 360(22):2289–2301

Dohner K, Dohner H (2008) Molecular characterization of acute myeloid leukemia. Haematologica 93(7):976–982

Dohner K, Tobis K, Ulrich R, Frohling S, Benner A, Schlenk RF, Dohner H (2002) Prognostic significance of partial tandem duplications of the MLL gene in adult patients 16–60 years old with acute myeloid leukemia and normal cytogenetics: a study of the Acute Myeloid Leukemia Study Group Ulm. J Clin Oncol 20(15):3254–3261

Dohner K, Schlenk RF, Habdank M, Scholl C, Rucker FG, Corbacioglu A, Bullinger L, Frohling S, Dohner H (2005) Mutant nucleophosmin (NPM1) predicts favorable prognosis in younger adults with acute myeloid leukemia and normal cytogenetics: interaction with other gene mutations. Blood 106(12):3740–3746

Dohner H, Estey EH, Amadori S, Appelbaum FR, Buchner T, Burnett AK, Dombret H, Fenaux P, Grimwade D, Larson RA, Lo-Coco F, Naoe T, Niederwieser D, Ossenkoppele GJ, Sanz MA, Sierra J, Tallman MS, Lowenberg B, Bloomfield CD (2010) Diagnosis and management of acute myeloid leukemia in adults: recommendations from an international expert panel, on behalf of the European LeukemiaNet. Blood 115(3):453–474

Ernst P, Wang J, Korsmeyer SJ (2002) The role of MLL in hematopoiesis and leukemia. Curr Opin Hematol 9(4):282–287

Falini B, Mecucci C, Tiacci E, Alcalay M, Rosati R, Pasqualucci L, La Starza R, Diverio D, Colombo E, Santucci A, Bigerna B, Pacini R, Pucciarini A, Liso A, Vignetti M, Fazi P, Meani N, Pettirossi V, Saglio G, Mandelli F, Lo-Coco F, Pelicci PG, Martelli MF (2005) Cytoplasmic nucleophosmin in acute myelogenous leukemia with a normal karyotype. N Engl J Med 352(3):254–266

Fenski R, Flesch K, Serve S, Mizuki M, Oelmann E, Kratz-Albers K, Kienast J, Leo R, Schwartz S, Berdel WE, Serve H (2000) Constitutive activation of FLT3 in acute myeloid leukaemia and its consequences for growth of 32D cells. Br J Haematol 108(2):322–330

Figueroa ME, Abdel-Wahab O, Lu C, Ward PS, Patel J, Shih A, Li Y, Bhagwat N, Vasanthakumar A, Fernandez HF, Tallman MS, Sun Z, Wolniak K, Peeters JK, Liu W, Choe SE, Fantin VR, Paietta E, Lowenberg B, Licht JD, Godley LA, Delwel R, Valk PJ, Thompson CB, Levine RL, Melnick A (2010a) Leukemic IDH1 and IDH2 mutations result in a hypermethylation phenotype, disrupt TET2 function, and impair hematopoietic differentiation. Cancer Cell 18(6):553–567

Figueroa ME, Lugthart S, Li Y, Erpelinck-Verschueren C, Deng X, Christos PJ, Schifano E, Booth J, van Putten W, Skrabanek L, Campagne F, Mazumdar M, Greally JM, Valk PJ, Lowenberg B, Delwel R, Melnick A (2010b) DNA methylation signatures identify biologically distinct subtypes in acute myeloid leukemia. Cancer Cell 17(1):13–27

Frohling S, Schlenk RF, Breitruck J, Benner A, Kreitmeier S, Tobis K, Dohner H, Dohner K (2002) Prognostic significance of activating FLT3 mutations in younger adults (16–60 years) with acute myeloid leukemia and normal cytogenetics: a study of the AML Study Group Ulm. Blood 100(13):4372–4380

Gaidzik VI, Schlenk RF, Moschny S, Becker A, Bullinger L, Corbacioglu A, Krauter J, Schlegelberger B, Ganser A, Dohner H, Dohner K (2009) Prognostic impact of WT1 mutations in cytogenetically normal acute myeloid leukemia: a study of the German-Austrian AML Study Group. Blood 113(19):4505–4511

Gaidzik VI, Bullinger L, Schlenk RF, Zimmermann AS, Rock J, Paschka P, Corbacioglu A, Krauter J, Schlegelberger B, Ganser A, Spath D, Kundgen A, Schmidt-Wolf IG, Gotze K, Nachbaur D, Pfreundschuh M, Horst HA, Dohner H, Dohner K (2011) RUNX1 mutations in acute myeloid leukemia: results from a comprehensive genetic and clinical analysis from the AML study group. J Clin Oncol 29(10):1364–1372

Gaidzik VI, Paschka P, Spath D, Habdank M, Kohne CH, Germing U, von Lilienfeld-Toal M, Held G, Horst HA, Haase D, Bentz M, Gotze K, Dohner H, Schlenk RF, Bullinger L, Dohner K (2012) TET2 mutations in Acute Myeloid Leukemia (AML): results from a comprehensive genetic and clinical analysis of the AML study group. J Clin Oncol 30(12):1350–1357

Garzon R, Fabbri M, Cimmino A, Calin GA, Croce CM (2006) MicroRNA expression and function in cancer. Trends Mol Med 12(12):580–587

Garzon R, Garofalo M, Martelli MP, Briesewitz R, Wang L, Fernandez-Cymering C, Volinia S, Liu CG, Schnittger S, Haferlach T, Liso A, Diverio D, Mancini M, Meloni G, Foa R, Martelli MF, Mecucci C, Croce CM, Falini B (2008) Distinctive microRNA signature of acute myeloid leukemia bearing cytoplasmic mutated nucleophosmin. Proc Natl Acad Sci U S A 105(10):3945–3950

Grisendi S, Mecucci C, Falini B, Pandolfi PP (2006) Nucleophosmin and cancer. Nat Rev Cancer 6(7):493–505

Groschel S, Lugthart S, Schlenk RF, Valk PJ, Eiwen K, Goudswaard C, van Putten WJ, Kayser S, Verdonck LF, Lubbert M, Ossenkoppele GJ, Germing U, Schmidt-Wolf I, Schlegelberger B, Krauter J, Ganser A, Dohner H, Lowenberg B, Dohner K, Delwel R (2010) High EVI1 expression predicts outcome in younger adult patients with acute myeloid leukemia and is associated with distinct cytogenetic abnormalities. J Clin Oncol 28(12):2101–2107

Herman JG, Baylin SB (2003) Gene silencing in cancer in association with promoter hypermethylation. N Engl J Med 349(21):2042–2054

Heuser M, Beutel G, Krauter J, Dohner K, von Neuhoff N, Schlegelberger B, Ganser A (2006) High meningioma 1 (MN1) expression as a predictor for poor outcome in acute myeloid leukemia with normal cytogenetics. Blood 108(12):3898–3905

Hohenstein P, Hastie ND (2006) The many facets of the Wilms' tumour gene, WT1. Hum Mol Genet 15 Spec No 2:R196–201

Kiyoi H, Naoe T, Nakano Y, Yokota S, Minami S, Miyawaki S, Asou N, Kuriyama K, Jinnai I, Shimazaki C, Akiyama H, Saito K, Oh H, Motoji T, Omoto E, Saito H, Ohno R, Ueda R (1999) Prognostic implication of FLT3 and N-RAS gene mutations in acute myeloid leukemia. Blood 93(9):3074–3080

Ko M, Huang Y, Jankowska AM, Pape UJ, Tahiliani M, Bandukwala HS, An J, Lamperti ED, Koh KP, Ganetzky R, Liu XS, Aravind L, Agarwal S, Maciejewski JP, Rao A (2010) Impaired hydroxylation of 5-methylcytosine in myeloid cancers with mutant TET2. Nature 468 (7325):839–843

Kottaridis PD, Gale RE, Frew ME, Harrison G, Langabeer SE, Belton AA, Walker H, Wheatley K, Bowen DT, Burnett AK, Goldstone AH, Linch DC (2001) The presence of a FLT3 internal tandem duplication in patients with acute myeloid leukemia (AML) adds important prognostic information to cytogenetic risk group and response to the first cycle of chemotherapy: analysis of 854 patients from the United Kingdom Medical Research Council AML 10 and 12 trials. Blood 98(6):1752–1759

Krauter J, Wagner K, Schafer I, Marschalek R, Meyer C, Heil G, Schaich M, Ehninger G, Niederwieser D, Krahl R, Buchner T, Sauerland C, Schlegelberger B, Dohner K, Dohner H, Schlenk RF, Ganser A (2009) Prognostic factors in adult patients up to 60 years old with acute myeloid leukemia and translocations of chromosome band 11q23: individual patient data-based meta-analysis of the German Acute Myeloid Leukemia Intergroup. J Clin Oncol 27(18):3000–3006

Kronke J, Schlenk RF, Jensen KO, Tschurtz F, Corbacioglu A, Gaidzik VI, Paschka P, Onken S, Eiwen K, Habdank M, Spath D, Lubbert M, Wattad M, Kindler T, Salih HR, Held G, Nachbaur D, von Lilienfeld-Toal M, Germing U, Haase D, Mergenthaler HG, Krauter J, Ganser A, Gohring G, Schlegelberger B, Dohner H, Dohner K (2011) Monitoring of minimal residual disease in NPM1-mutated acute myeloid leukemia: a study from the German-Austrian acute myeloid leukemia study group. J Clin Oncol 29(19):2709–2716

Langemeijer SM, Kuiper RP, Berends M, Knops R, Aslanyan MG, Massop M, Stevens-Linders E, van Hoogen P, van Kessel AG, Raymakers RA, Kamping EJ, Verhoef GE, Verburgh E, Hagemeijer A, Vandenberghe P, de Witte T, van der Reijden BA, Jansen JH (2009) Acquired mutations in TET2 are common in myelodysplastic syndromes. Nat Genet 41(7):838–842

Langer C, Radmacher MD, Ruppert AS, Whitman SP, Paschka P, Mrozek K, Baldus CD, Vukosavljevic T, Liu CG, Ross ME, Powell BL, de la Chapelle A, Kolitz JE, Larson RA, Marcucci G, Bloomfield CD (2008) High BAALC expression associates with other molecular prognostic markers, poor outcome, and a distinct gene-expression signature in cytogenetically normal patients younger than 60 years with acute myeloid leukemia: a Cancer and Leukemia Group B (CALGB) study. Blood 111(11):5371–5379

Langer C, Marcucci G, Holland KB, Radmacher MD, Maharry K, Paschka P, Whitman SP, Mrozek K, Baldus CD, Vij R, Powell BL, Carroll AJ, Kolitz JE, Caligiuri MA, Larson RA, Bloomfield CD (2009) Prognostic importance of MN1 transcript levels, and biologic insights from MN1-associated gene and microRNA expression signatures in cytogenetically normal acute myeloid leukemia: a cancer and leukemia group B study. J Clin Oncol 27(19):3198–3204

Levis M, Ravandi F, Wang ES, Baer MR, Perl A, Coutre S, Erba H, Stuart RK, Baccarani M, Cripe LD, Tallman MS, Meloni G, Godley LA, Langston AA, Amadori S, Lewis ID, Nagler A, Stone R, Yee K, Advani A, Douer D, Wiktor-Jedrzejczak W, Juliusson G, Litzow MR, Petersdorf S, Sanz M, Kantarjian HM, Sato T, Tremmel L, Bensen-Kennedy DM, Small D, Smith BD (2011) Results from a randomized trial of salvage chemotherapy followed by lestaurtinib for patients with FLT3 mutant AML in first relapse. Blood 117(12):3294–3301

Ley TJ, Ding L, Walter MJ, McLellan MD, Lamprecht T, Larson DE, Kandoth C, Payton JE, Baty J, Welch J, Harris CC, Lichti CF, Townsend RR, Fulton RS, Dooling DJ, Koboldt DC, Schmidt H, Zhang Q, Osborne JR, Lin L, O'Laughlin M, McMichael JF, Delehaunty KD, McGrath SD, Fulton LA, Magrini VJ, Vickery TL, Hundal J, Cook LL, Conyers JJ, Swift GW, Reed JP, Alldredge PA, Wylie T, Walker J, Kalicki J, Watson MA, Heath S, Shannon WD, Varghese N, Nagarajan R, Westervelt P, Tomasson MH, Link DC, Graubert TA, DiPersio JF, Mardis ER, Wilson RK (2010) *DNMT3*A mutations in acute myeloid leukemia. N Engl J Med 363(25):2424–2433

Li Z, Lu J, Sun M, Mi S, Zhang H, Luo RT, Chen P, Wang Y, Yan M, Qian Z, Neilly MB, Jin J, Zhang Y, Bohlander SK, Zhang DE, Larson RA, Le Beau MM, Thirman MJ, Golub TR, Rowley JD, Chen J (2008) Distinct microRNA expression profiles in acute myeloid leukemia with common translocations. Proc Natl Acad Sci U S A 105(40):15535–15540

Liu S, Wu LC, Pang J, Santhanam R, Schwind S, Wu YZ, Hickey CJ, Yu J, Becker H, Maharry K, Radmacher MD, Li C, Whitman SP, Mishra A, Stauffer N, Eiring AM, Briesewitz R, Baiocchi RA, Chan KK, Paschka P, Caligiuri MA, Byrd JC, Croce CM, Bloomfield CD, Perrotti D, Garzon R, Marcucci G (2010) Sp1/NFkappaB/HDAC/miR-29b regulatory network in KIT-driven myeloid leukemia. Cancer Cell 17(4):333–347

Lugthart S, van Drunen E, van Norden Y, van Hoven A, Erpelinck CA, Valk PJ, Beverloo HB, Lowenberg B, Delwel R (2008) High EVI1 levels predict adverse outcome in acute myeloid leukemia: prevalence of EVI1 overexpression and chromosome 3q26 abnormalities underestimated. Blood 111(8):4329–4337

Marcucci G, Baldus CD, Ruppert AS, Radmacher MD, Mrozek K, Whitman SP, Kolitz JE, Edwards CG, Vardiman JW, Powell BL, Baer MR, Moore JO, Perrotti D, Caligiuri MA, Carroll AJ, Larson RA, de la Chapelle A, Bloomfield CD (2005) Overexpression of the ETS-related gene, ERG, predicts a worse outcome in acute myeloid leukemia with normal karyotype: a Cancer and Leukemia Group B study. J Clin Oncol 23(36):9234–9242

Marcucci G, Maharry K, Radmacher MD, Mrozek K, Vukosavljevic T, Paschka P, Whitman SP, Langer C, Baldus CD, Liu CG, Ruppert AS, Powell BL, Carroll AJ, Caligiuri MA, Kolitz JE, Larson RA, Bloomfield CD (2008) Prognostic significance of, and gene and microRNA expression signatures associated with, CEBPA mutations in cytogenetically normal acute myeloid leukemia with high-risk molecular features: a Cancer and Leukemia Group B Study. J Clin Oncol 26(31):5078–5087

Marcucci G, Maharry K, Wu YZ, Radmacher MD, Mrozek K, Margeson D, Holland KB, Whitman SP, Becker H, Schwind S, Metzeler KH, Powell BL, Carter TH, Kolitz JE, Wetzler M, Carroll

AJ, Baer MR, Caligiuri MA, Larson RA, Bloomfield CD (2010) IDH1 and IDH2 gene mutations identify novel molecular subsets within de novo cytogenetically normal acute myeloid leukemia: a Cancer and Leukemia Group B study. J Clin Oncol 28(14):2348–2355

Marcucci G, Metzeler KH, Schwind S, Becker H, Maharry K, Mrozek K, Radmacher MD, Kohlschmidt J, Nicolet D, Whitman SP, Wu YZ, Powell BL, Carter TH, Kolitz JE, Wetzler M, Carroll AJ, Baer MR, Moore JO, Caligiuri MA, Larson RA, Bloomfield CD (2012) Age-related prognostic impact of different types of *DNMT3*A mutations in adults with primary cytogenetically normal acute myeloid leukemia. J Clin Oncol 30(7):742–750

Mardis ER, Ding L, Dooling DJ, Larson DE, McLellan MD, Chen K, Koboldt DC, Fulton RS, Delehaunty KD, McGrath SD, Fulton LA, Locke DP, Magrini VJ, Abbott RM, Vickery TL, Reed JS, Robinson JS, Wylie T, Smith SM, Carmichael L, Eldred JM, Harris CC, Walker J, Peck JB, Du F, Dukes AF, Sanderson GE, Brummett AM, Clark E, McMichael JF, Meyer RJ, Schindler JK, Pohl CS, Wallis JW, Shi X, Lin L, Schmidt H, Tang Y, Haipek C, Wiechert ME, Ivy JV, Kalicki J, Elliott G, Ries RE, Payton JE, Westervelt P, Tomasson MH, Watson MA, Baty J, Heath S, Shannon WD, Nagarajan R, Link DC, Walter MJ, Graubert TA, DiPersio JF, Wilson RK, Ley TJ (2009) Recurring mutations found by sequencing an acute myeloid leukemia genome. N Engl J Med 361(11):1058–1066

McKenna HJ, Stocking KL, Miller RE, Brasel K, De Smedt T, Maraskovsky E, Maliszewski CR, Lynch DH, Smith J, Pulendran B, Roux ER, Teepe M, Lyman SD, Peschon JJ (2000) Mice lacking flt3 ligand have deficient hematopoiesis affecting hematopoietic progenitor cells, dendritic cells, and natural killer cells. Blood 95(11):3489–3497

Mead AJ, Linch DC, Hills RK, Wheatley K, Burnett AK, Gale RE (2007) FLT3 tyrosine kinase domain mutations are biologically distinct from and have a significantly more favorable prognosis than FLT3 internal tandem duplications in patients with acute myeloid leukemia. Blood 110(4):1262–1270

Members. NAMLP (Published 2012) Clinical practice guidelines in oncology-acute myeloid leukemia. www.nccn.org/professionals/physician_gls/pdf/aml.pdf. Accessed 22 June 2012

Mendler JH, Maharry K, Radmacher MD, Mrózek K, Becker H, Metzeler KH, Schwind S, Whitman SP, Khalife J, Kohlschmidt J, Nicolet D, Powell BL, Carter TH, Wetzler M, Moore JO, Kolitz JE, Baer MR, Carroll AJ, Larson RA, Caligiuri MA, Marcucci G, Bloomfield CD (2012) RUNX1 mutations are associated with poor outcome in younger and older patients with cytogenetically normal acute myeloid leukemia and with distinct gene and MicroRNA expression signatures. J Clin Oncol 30(25):3109–3118

Metzeler KH, Dufour A, Benthaus T, Hummel M, Sauerland MC, Heinecke A, Berdel WE, Buchner T, Wormann B, Mansmann U, Braess J, Spiekermann K, Hiddemann W, Buske C, Bohlander SK (2009) ERG expression is an independent prognostic factor and allows refined risk stratificationin cytogenetically normal acute myeloid leukemia: a comprehensive analysis of ERG, MN1, and BAALC transcript levels using oligonucleotide microarrays. J Clin Oncol 27(30):5031–5038

Metzeler KH, Walker A, Geyer S, Garzon R, Klisovic RB, Bloomfield CD, Blum W, Marcucci G (2011) *DNMT3*A mutations and response to the hypomethylating agent decitabine in acute myeloid leukemia. Leukemia 26(5):1106–1107

Mi S, Lu J, Sun M, Li Z, Zhang H, Neilly MB, Wang Y, Qian Z, Jin J, Zhang Y, Bohlander SK, Le Beau MM, Larson RA, Golub TR, Rowley JD, Chen J (2007) MicroRNA expression signatures accurately discriminate acute lymphoblastic leukemia from acute myeloid leukemia. Proc Natl Acad Sci U S A 104(50):19971–19976

Michaud J, Wu F, Osato M, Cottles GM, Yanagida M, Asou N, Shigesada K, Ito Y, Benson KF, Raskind WH, Rossier C, Antonarakis SE, Israels S, McNicol A, Weiss H, Horwitz M, Scott HS (2002) In vitro analyses of known and novel RUNX1/AML1 mutations in dominant familial platelet disorder with predisposition to acute myelogenous leukemia: implications for mechanisms of pathogenesis. Blood 99(4):1364–1372

Mizuno S, Chijiwa T, Okamura T, Akashi K, Fukumaki Y, Niho Y, Sasaki H (2001) Expression of DNA methyltransferases DNMT1, 3A, and 3B in normal hematopoiesis and in acute and chronic myelogenous leukemia. Blood 97(5):1172–1179

Moran-Crusio K, Reavie L, Shih A, Abdel-Wahab O, Ndiaye-Lobry D, Lobry C, Figueroa ME, Vasanthakumar A, Patel J, Zhao X, Perna F, Pandey S, Madzo J, Song C, Dai Q, He C, Ibrahim S, Beran M, Zavadil J, Nimer SD, Melnick A, Godley LA, Aifantis I, Levine RL (2011) Tet2 loss leads to increased hematopoietic stem cell self-renewal and myeloid transformation. Cancer Cell 20(1):11–24

Mrozek K (2008) Cytogenetic, molecular genetic, and clinical characteristics of acute myeloid leukemia with a complex karyotype. Semin Oncol 35(4):365–377

Mrozek K MG, Maharry K et al (2011) Prognostic utility of the European LeukemiaNet (ELN) genetic-risk classification in adults with de novo Acute Myeloid Leukemia (AML): A study of 1,550 patients (Pts). ASH Annual Meeting Abstracts 18(21):414

Nakao M, Yokota S, Iwai T, Kaneko H, Horiike S, Kashima K, Sonoda Y, Fujimoto T, Misawa S (1996) Internal tandem duplication of the flt3 gene found in acute myeloid leukemia. Leukemia 10(12):1911–1918

Nerlov C (2004) C/EBPalpha mutations in acute myeloid leukaemias. Nat Rev Cancer 4(5):394–400

Oikawa T (2004) ETS transcription factors: possible targets for cancer therapy. Cancer Sci 95(8):626–633

Paschka P, Marcucci G, Ruppert AS, Mrozek K, Chen H, Kittles RA, Vukosavljevic T, Perrotti D, Vardiman JW, Carroll AJ, Kolitz JE, Larson RA, Bloomfield CD (2006) Adverse prognostic significance of KIT mutations in adult acute myeloid leukemia with inv(16) and t(8;21): a Cancer and Leukemia Group B Study. J Clin Oncol 24(24):3904–3911

Paschka P, Marcucci G, Ruppert AS, Whitman SP, Mrozek K, Maharry K, Langer C, Baldus CD, Zhao W, Powell BL, Baer MR, Carroll AJ, Caligiuri MA, Kolitz JE, Larson RA, Bloomfield CD (2008) Wilms' tumor 1 gene mutations independently predict poor outcome in adults with cytogenetically normal acute myeloid leukemia: a cancer and leukemia group B study. J Clin Oncol 26(28):4595–4602

Paschka P, Schlenk RF, Gaidzik VI, Habdank M, Kronke J, Bullinger L, Spath D, Kayser S, Zucknick M, Gotze K, Horst HA, Germing U, Dohner H, Dohner K (2010) IDH1 and IDH2 mutations are frequent genetic alterations in acute myeloid leukemia and confer adverse prognosis in cytogenetically normal acute myeloid leukemia with NPM1 mutation without FLT3 internal tandem duplication. J Clin Oncol 28(22):3636–3643

Patel JP, Gonen M, Figueroa ME, Fernandez H, Sun Z, Racevskis J, Van Vlierberghe P, Dolgalev I, Thomas S, Aminova O, Huberman K, Cheng J, Viale A, Socci ND, Heguy A, Cherry A, Vance G, Higgins RR, Ketterling RP, Gallagher RE, Litzow M, van den Brink MR, Lazarus HM, Rowe JM, Luger S, Ferrando A, Paietta E, Tallman MS, Melnick A, Abdel-Wahab O, Levine RL (2012) Prognostic relevance of integrated genetic profiling in acute myeloid leukemia. N Engl J Med 366(12):1079–1089

Ribeiro AF, Pratcorona M, Erpelinck-Verschueren C, Rockova V, Sanders M, Abbas S, Figueroa ME, Zeilemaker A, Melnick A, Lowenberg B, Valk PJ, Delwel R (2012) Mutant *DNMT3*A: a new marker of poor prognosis in acute myeloid leukemia. Blood 119(24):5824–5831

Rollig C, Bornhauser M, Thiede C, Taube F, Kramer M, Mohr B, Aulitzky W, Bodenstein H, Tischler HJ, Stuhlmann R, Schuler U, Stolzel F, von Bonin M, Wandt H, Schafer-Eckart K, Schaich M, Ehninger G (2011) Long-term prognosis of acute myeloid leukemia according to the new genetic risk classification of the European LeukemiaNet recommendations: evaluation of the proposed reporting system. J Clin Oncol 29(20):2758–2765

Rucker FG, Schlenk RF, Bullinger L, Kayser S, Teleanu V, Kett H, Habdank M, Kugler CM, Holzmann K, Gaidzik VI, Paschka P, Held G, von Lilienfeld-Toal M, Lubbert M, Frohling S, Zenz T, Krauter J, Schlegelberger B, Ganser A, Lichter P, Dohner K, Dohner H (2012) TP53 alterations in acute myeloid leukemia with complex karyotype correlate with specific copy number alterations, monosomal karyotype, and dismal outcome. Blood 119(9):2114–2121

Sattler M, Salgia R (2004) Targeting c-Kit mutations: basic science to novel therapies. Leuk Res 28(Suppl 1):S11–20

Schlenk RF, Dohner K, Krauter J, Frohling S, Corbacioglu A, Bullinger L, Habdank M, Spath D, Morgan M, Benner A, Schlegelberger B, Heil G, Ganser A, Dohner H (2008) Mutations

and treatment outcome in cytogenetically normal acute myeloid leukemia. N Engl J Med 358(18):1909–1918
Schlenk RF, Dohner K, Kneba M, Gotze K, Hartmann F, Del Valle F, Kirchen H, Koller E, Fischer JT, Bullinger L, Habdank M, Spath D, Groner S, Krebs B, Kayser S, Corbacioglu A, Anhalt A, Benner A, Frohling S, Dohner H (2009) Gene mutations and response to treatment with all-trans retinoic acid in elderly patients with acute myeloid leukemia. Results from the AMLSG Trial AML HD98B. Haematologica 94(1):54–60
Schlenk RF, Dohner K, Krauter J, Gaidzik VI, Paschka P, Heuser M, Kindler T, Lubbert M, Martin H, Salih HR, Kuendgen A, Horst HA, von Lilienfeld-Toal M, Gotze K, Nachbaur D, Wattad M, Kohne C-H, Fiedler W, Bentz M, Wulf G, Held G, Hertenstein B, Mergenthaler H-G, Salwender H, Rummel MJ, Raghavachar A, Benner A, Schlegelberger B, Ganser A, Dohner H (2011) All-trans retinoic acid improves outcome in younger adult patients with nucleophosmin –1 Mutated acute myeloid leukemia—results of the AMLSG 07-04 randomized treatment trial. ASH Annual Meeting Abstracts 118(21):80
Schnittger S, Schoch C, Dugas M, Kern W, Staib P, Wuchter C, Loffler H, Sauerland CM, Serve H, Buchner T, Haferlach T, Hiddemann W (2002) Analysis of FLT3 length mutations in 1003 patients with acute myeloid leukemia: correlation to cytogenetics, FAB subtype, and prognosis in the AMLCG study and usefulness as a marker for the detection of minimal residual disease. Blood 100(1):59–66
Schnittger S, Schoch C, Kern W, Mecucci C, Tschulik C, Martelli MF, Haferlach T, Hiddemann W, Falini B (2005) Nucleophosmin gene mutations are predictors of favorable prognosis in acute myelogenous leukemia with a normal karyotype. Blood 106(12):3733–3739
Schwind S, Maharry K, Radmacher MD, Mrozek K, Holland KB, Margeson D, Whitman SP, Hickey C, Becker H, Metzeler KH, Paschka P, Baldus CD, Liu S, Garzon R, Powell BL, Kolitz JE, Carroll AJ, Caligiuri MA, Larson RA, Marcucci G, Bloomfield CD (2010) Prognostic significance of expression of a single microRNA, miR-181a, in cytogenetically normal acute myeloid leukemia: a Cancer and Leukemia Group B study. J Clin Oncol 28(36):5257–5264
Sellick GS, Spendlove HE, Catovsky D, Pritchard-Jones K, Houlston RS (2005) Further evidence that germline CEBPA mutations cause dominant inheritance of acute myeloid leukaemia. Leukemia 19(7):1276–1278
Smith ML, Cavenagh JD, Lister TA, Fitzgibbon J (2004) Mutation of CEBPA in familial acute myeloid leukemia. N Engl J Med 351(23):2403–2407
Strout MP, Marcucci G, Caligiuri MA, Bloomfield CD (1999) Core-binding factor (CBF) and MLL-associated primary acute myeloid leukemia: biology and clinical implications. Ann Hematol 78(6):251–264
Swerdlow SH, Campo E, Harris NL, al. E (2008) WHO classification of tumours of haematopoietic and lymphoid tissues. IARC Press, Lyon
Tang JL, Hou HA, Chen CY, Liu CY, Chou WC, Tseng MH, Huang CF, Lee FY, Liu MC, Yao M, Huang SY, Ko BS, Hsu SC, Wu SJ, Tsay W, Chen YC, Lin LI, Tien HF (2009) AML1/RUNX1 mutations in 470 adult patients with de novo acute myeloid leukemia: prognostic implication and interaction with other gene alterations. Blood 114(26):5352–5361
Thiede C, Steudel C, Mohr B, Schaich M, Schakel U, Platzbecker U, Wermke M, Bornhauser M, Ritter M, Neubauer A, Ehninger G, Illmer T (2002) Analysis of FLT3-activating mutations in 979 patients with acute myelogenous leukemia: association with FAB subtypes and identification of subgroups with poor prognosis. Blood 99(12):4326–4335
Thol F, Damm F, Ludeking A, Winschel C, Wagner K, Morgan M, Yun H, Gohring G, Schlegelberger B, Hoelzer D, Lubbert M, Kanz L, Fiedler W, Kirchner H, Heil G, Krauter J, Ganser A, Heuser M (2011) Incidence and prognostic influence of *DNMT3*A mutations in acute myeloid leukemia. J Clin Oncol 29(21):2889–2896
Virappane P, Gale R, Hills R, Kakkas I, Summers K, Stevens J, Allen C, Green C, Quentmeier H, Drexler H, Burnett A, Linch D, Bonnet D, Lister TA, Fitzgibbon J (2008) Mutation of the Wilms' tumor 1 gene is a poor prognostic factor associated with chemotherapy resistance in normal karyotype acute myeloid leukemia: the United Kingdom Medical Research Council Adult Leukaemia Working Party. J Clin Oncol 26(33):5429–5435

Wagner K, Damm F, Gohring G, Gorlich K, Heuser M, Schafer I, Ottmann O, Lubbert M, Heit W, Kanz L, Schlimok G, Raghavachar AA, Fiedler W, Kirchner HH, Brugger W, Zucknick M, Schlegelberger B, Heil G, Ganser A, Krauter J (2010) Impact of IDH1 R132 mutations and an IDH1 single nucleotide polymorphism in cytogenetically normal acute myeloid leukemia: SNP rs11554137 is an adverse prognostic factor. J Clin Oncol 28(14):2356–2364

Ward PS, Patel J, Wise DR, Abdel-Wahab O, Bennett BD, Coller HA, Cross JR, Fantin VR, Hedvat CV, Perl AE, Rabinowitz JD, Carroll M, Su SM, Sharp KA, Levine RL, Thompson CB (2010) The common feature of leukemia-associated IDH1 and IDH2 mutations is a neomorphic enzyme activity converting alpha-ketoglutarate to 2-hydroxyglutarate. Cancer Cell 17(3):225–234.

Whitman SP, Archer KJ, Feng L, Baldus C, Becknell B, Carlson BD, Carroll AJ, Mrozek K, Vardiman JW, George SL, Kolitz JE, Larson RA, Bloomfield CD, Caligiuri MA (2001) Absence of the wild-type allele predicts poor prognosis in adult de novo acute myeloid leukemia with normal cytogenetics and the internal tandem duplication of FLT3: a cancer and leukemia group B study. Cancer Res 61(19):7233–7239

Whitman SP, Liu S, Vukosavljevic T, Rush LJ, Yu L, Liu C, Klisovic MI, Maharry K, Guimond M, Strout MP, Becknell B, Dorrance A, Klisovic RB, Plass C, Bloomfield CD, Marcucci G, Caligiuri MA (2005) The MLL partial tandem duplication: evidence for recessive gain-of-function in acute myeloid leukemia identifies a novel patient subgroup for molecular-targeted therapy. Blood 106(1):345–352

Whitman SP, Ruppert AS, Marcucci G, Mrozek K, Paschka P, Langer C, Baldus CD, Wen J, Vukosavljevic T, Powell BL, Carroll AJ, Kolitz JE, Larson RA, Caligiuri MA, Bloomfield CD (2007) Long-term disease-free survivors with cytogenetically normal acute myeloid leukemia and MLL partial tandem duplication: a Cancer and Leukemia Group B study. Blood 109(12):5164–5167

Whitman SP, Ruppert AS, Radmacher MD, Mrozek K, Paschka P, Langer C, Baldus CD, Wen J, Racke F, Powell BL, Kolitz JE, Larson RA, Caligiuri MA, Marcucci G, Bloomfield CD (2008) FLT3 D835/I836 mutations are associated with poor disease-free survival and a distinct gene-expression signature among younger adults with de novo cytogenetically normal acute myeloid leukemia lacking FLT3 internal tandem duplications. Blood 111(3):1552–1559

Wiernik PH (2010) FLT3 inhibitors for the treatment of acute myeloid leukemia. Clin Adv Hematol Oncol 8(6):429–436, 444

Yamashita Y, Yuan J, Suetake I, Suzuki H, Ishikawa Y, Choi YL, Ueno T, Soda M, Hamada T, Haruta H, Takada S, Miyazaki Y, Kiyoi H, Ito E, Naoe T, Tomonaga M, Toyota M, Tajima S, Iwama A, Mano H (2010) Array-based genomic resequencing of human leukemia. Oncogene 29(25):3723–3731

Yan P, Frankhouser D, Murphy M, Tam HH, Rodriguez B, Curfman J, Trimarchi M, Geyer S, Wu YZ, Whitman SP, Metzeler K, Walker A, Klisovic R, Jacob S, Grever MR, Byrd JC, Bloomfield CD, Garzon R, Blum W, Caligiuri MA, Bundschuh R, Marcucci G (2012) Genome-wide methylation profiling in decitabine-treated patients with acute myeloid leukemia. Blood 120(12):2466–2474

Yanada M, Matsuo K, Suzuki T, Kiyoi H, Naoe T (2005) Prognostic significance of FLT3 internal tandem duplication and tyrosine kinase domain mutations for acute myeloid leukemia: a meta-analysis. Leukemia 19(8):1345–1349

Yang L, Han Y, Suarez Saiz F, Minden MD (2007) A tumor suppressor and oncogene: the WT1 story. Leukemia 21(5):868–876

Chapter 2
The Use of "Omics" to Guide the Selection of Targeted Therapy

Steven M. Kornblau

Abstract In this book, the role of many different proteins, pathways, or genetic events in affecting the function of leukemic cells, as well as their suitability for therapeutic modulation, is discussed. However, since acute myeloid leukemia (AML) is a very heterogeneous disease, it will be necessary to match the characteristics of individual patients to the selection of the right targeted agent in order to maximize efficacy. In this chapter, we discuss how the application of the numerous molecular profiling arrays that are now available can be utilized to provide the answer to this dilemma. The development of methods to combine data from multiple arrays, termed "integromics" will enable a more sophisticated application of "omic" data that will facilitate the selection of combinations of targeted agents on an individualized basis.

Keywords Omics · Integromics · Gene expression profiling (GEP) · Whole genome sequencing · Single nucleotide polymorphism (SNP) · Reverse phase protein array (RPPA) · Proteomics

2.1 Introduction

A central theme developed in the prior chapter is that acute myeloid leukemia (AML) is a heterogeneous disease with myriad changes occurring at multiple different levels that combine to determine the biology of the leukemic cell. In the chapters that follow, each author will discuss how a single gene (e.g., AKT, P53), or pathway (e.g., Ras/RAF/MEK/ERK), or genetic modification (e.g., methylation) affects AML cell biology and how that individual event might be targeted to improve therapy. The heterogeneous nature of AML suggests that each of these potential targets will be therapeutically efficacious, but in only a fraction of cases of AML. If we as a scientific community are successful at discovering the key points of attack and actually develop functional therapies for each of these, how will we determine when to use which? Furthermore, redundancy in pathways and mechanisms within a cell may require that targeted therapies be used in combination to achieve efficacy.

S. M. Kornblau (✉)
Department of Leukemia and Department of Stem Cell Transplantation,
Division of Cancer Medicine, The University of Texas, MD Anderson Cancer Center,
1515 Holcombe Blvd., Box 448, Houston, TX 77030, USA
e-mail: skornblau@mdanderson.org

M. Andreeff (ed.), *Targeted Therapy of Acute Myeloid Leukemia*,
Current Cancer Research, DOI 10.1007/978-1-4939-1393-0_2

Given the complexity of AML, how will we determine which agents to use in combination? In which cases? By what schedule and in what order? There is a very high risk that promising agents will be discarded as therapeutic failures as a result of faulty study design and the barriers erected by regulatory and proprietary interests. A drug may lack efficacy as single agent, but be of great benefit when used in combination with other chemotherapeutic agents. Current regulation makes it difficult to combine two non-approved agents that are currently in phase I–II evaluation into a single trial, and pharmaceutical companies are highly reluctant to pair a developmental molecule with one from another company. Consequently, combination trials of unapproved agents are nearly impossible to conduct. A second problem is that the typical phase I–II trial typically treats about 15 patients. If the frequency of the population that is susceptible to a given agent is low, then there is a high chance that no patients in a trial of that small size will have the characteristic (for event frequencies of 3, 5, 8, and 10 %, then 63, 46, 29, and 21 % of trials of $n=15$ would have no patients with that characteristic). The activity of a highly efficacious agent can easily be missed if the correct patients do not make it onto the right trial. The successful translation of the discoveries discussed in the rest of this book into effective therapy that improves the outcome of AML is therefore critically dependent on having the ability to match patients on an individualized basis to the correct therapy.

In this chapter, we discuss how the various "omics" (Weinstein 2002) might be used to provide this missing information, to enable us to match patients to therapy, and to rationally combine different targeted therapies (Barretina et al. 2012) from a broad menu into a restricted list for an individual patient. High-throughput array-based technologies have been developed that enable the global characterization of multiple characteristics of a cell, enabling the collection of historically unprecedented amount of information. For virtually every aspect of a cell in which abnormalities could arise, an array now exists for its evaluation. The basic dogma of biology formerly was DNA → RNA → protein, with mutations leading to changes in the sequence or expression level of mRNA, resulting in quantitative or qualitative changes in protein function. But in light of the effects of epigenetics and noncoding RNAs on gene expression, this view is now overly simplistic.

2.2 Types of Alterations That Can Affect Leukemia Biology and Methods Available to Evaluate Them (Table 2.1)

2.2.1 Nucleic Acids

Mutations are theoretically the most stable form of change. A gene is mutated or not, and once a mutation develops, it should be passed on to all progeny. The location in a regulatory region can lead to over- or under-expression of the mRNA. Likewise, copy number abnormalities can lead to overexpression of an unmutated, oncogenic gene. A mutation in the coding region can result in a stop codon, or frameshift

Table 2.1 Currently available arrays that can be utilized in an integromics approach

Analyte	Assay	What is measured
DNA	Whole genome sequencing	Mutations
	Exon sequencing	
	Methylation	DNA methylation
Histone	Methylation	
	Acetylation	
RNA	Gene expression profiling (GEP)	Expression
	Noncoding RNA profiling	Expression
Protein	Reverse phase protein array (RPPA)	Expression, activation status
	Proteomics	Expression, activation status
	Single-cell network profiling	Adaptability
Cytokines and chemokines	Multiplex kits	Expression
Metabolites	Metabalomics	Concentration of chemical substrates
Lipids	Lipidomics	Concentration of lipid moieties in the cell

mutation, resulting in loss or gain of function, or result in an amino acid change that changes the function and regulation of the protein. Translocations can result in novel proteins with oncogenic function. Additional complexity can result from mutations in epigenetic regulators or affecting transcription factors, and by changes in splicing, for which the causative changes may not be known. Each of these can be determined using the currently available technologies, with each base pair analyzed in whole genome sequencing, or just known coding regions assayed in whole exome sequencing, while sequencing of those genes that are actually transcribed can be performed by RNA-seq (also called whole transcriptome shotgun sequencing). In addition to detecting mutations, the relative frequency of the detected fragments provides copy number data as well. With prices rapidly decreasing, it may soon be feasible, from a technical and economic standpoint, to sequence each individual patient's tumor as the standard of care. However, among the important caveats is that the interpretation of such massive sequencing data remains challenging. For every "actionable" or "driver" mutation with a currently known prognostic and potentially therapeutic implication in AML, such as ones affecting FLT3, there may be many more nonsynonymous differences from the reference genome or patient's germ line sequence that could be oncogenic in unknown ways, "passenger" mutations due to genomic instability frequently present in tumors, or single-nucleotide polymorphisms (SNPs). Furthermore, sequencing is subject to technical and sampling variation. Two recent studies demonstrated that there is clonal heterogeneity within both solid tumors (Gerlinger et al. 2012) and myelodysplastic syndrome (MDS) and AML (Walter et al. 2012), suggesting that deep sequencing of multiple sites or samples may be necessary to fully recognize the spectrum of mutations within a patient.

The genome can be altered in ways that affect gene transcription without mutation. SNPs can regulate the level of expression of a gene, or the function of the resulting protein (e.g., SNP309 in MDM2, Pichiorri et al. 2010; Phillips et al. 2010).

Arrays exist that can measure 250,000–600,000 to over a million SNPs. Likewise, DNA can be methylated, and methylation arrays have been developed to assess the status of methylation across the genome. The interaction of DNA with histones can also regulate the availability of the gene to transcription through histone methylation (Butler et al. 2012), acetylation, phosphorylation, and sumoylation. The first two can be assessed by genome-wide location analyses using chromatin immunoprecipitation linked to cDNA arrays or to high-throughput sequencing (Miao and Natarajan 2005) although there are not commercial kits for this readily available. Noncoding RNAs, also known as microRNAs (miRNAs), are small strands of RNAs that can bind to DNA to regulate transcription or to mRNA to regulate translation and mRNA longevity. They are subject to mutation, and variation in their levels has been associated with many malignancies (Garzon et al. 2009; Calin et al. 2004; Calin et al. 2005; Calin et al. 2007; Lee et al. 2012; Taby et al. 2009; Valastyan et al. 2011). They might also function in complex networks with pseudogenes to regulate protein levels (Salmena et al. 2011). Mutations in ncRNA can be detected by whole genome sequencing, and expression levels can be readily ascertained via the many commercially available miRNA arrays. This technology is in rapid flux as the number of ncRNA covered is rapidly increasing. Measurement of global gene expression at the mRNA level has become commonplace with two platforms, Affymetrix and Illumina, currently used widely. Gene expression profiling (GEP) is often thought of as a "detector" of causal changes, in which the changes in gene expression are themselves causative of a phenotype. However, the "unbiased," genome-wide nature of GEP makes it good as a "reflector" of causal changes, in which the changes in gene expression are a downstream readout of changes, such as the activity of a signaling pathway, that are ultimately causative of a phenotype. The changes in gene expression may contribute to the phenotype, but that is not necessarily understood, nor does it need to be, when GEP is used as a detector. This "molecular pattern recognition" has a lot of clinical potential, and is in fact the way in which GEP is finally being Food and Drug Administration (FDA)-approved for clinical use, in breast cancer and myeloma. While many insights into cancer biology have emerged from GEP studies, these results are somewhat hampered by a lack of reproducibility from one dataset to the next (Xu et al. 2008; Michiels et al. 2005) and by the lack of correlation between mRNA expression and protein expression (Gygi et al. 1999; Griffin et al. 2002; Washburn et al. 2003; Greenbaum et al. 2003; Jansen et al. 2002; Kern et al. 2003).

2.2.2 Stroma, Chemokines, and Cytokines

External influences from chemokines and cytokines as well as from interaction with stroma can influence gene expression and protein activation. Multiplex kits are available to measure from 20 to 75 different chemokines and cytokines at once from very small quantities of plasma. In an analysis of the expression of 27 cytokines and chemokines in the serum from AML and MDS patients, we demonstrated that both of these diseases are characterized by recurrent patterns of expression, and surprisingly, these carried prognostic significance (Kornblau et al. 2010a). We have recently generated a reverse phase protein array (RPPA; see below) using mesenchymal stromal cells (MSC) derived from AML patients to define the characteristics of the stroma and determine how this correlates with characteristics of the AML blast.

2.2.3 Protein

Proteins are the predominant effectors of most genes (ribosomes, tRNA, etc., aside), and the net effect of mutations, DNA methylation, histone modifications, ncRNA and external influences, determines the level of expression of individual proteins. But since protein activity is regulated by post-translational modifications, it is not only sufficient to ascertain the protein expression level, but also the activation state must be known as well. The potential for splice variants and posttranslational modifications, including phosphorylation, glycosylation, ubiquitination, cleavage, etc., raise the number of different members of the proteome from the 20,000–40,000 known genes into more than a million potential variants. This complexity has made proteomics the least developed of the array technologies even though it is likely the most critical. Traditional proteomics involves two-dimensional (2-D) gel electrophoresis based on charge and mass, followed by tandem mass spectrometry (so-called 2D-MALDI-TOF). While the entire proteome can be analyzed, this technology is currently expensive, and impractical for individualized patient analysis as the amount of starting material required is high and the throughput low. There are alternatives to mass spectrometry-based proteomics.

2.2.4 Proteomics Methodologies

One approach, in commercial development (SOMA, Colorado), uses DNA-based aptamers, selected for their ability to specifically bind to conformed protein, with a tag that allows for quantitative detection (Gold et al. 2010). While phospho-specific aptamers could theoretically be developed, thus far the company has only developed aptamers to measure total protein levels. A second related approach, known as proximity ligation, uses two antibodies with overlapping tags that bind to each other enabling amplification to permit detection, in an enzyme-linked immunosorbent assay (ELISA)-like approach (Fredriksson et al. 2002). The other methodology uses an antibody-based proteomic approach. Two types of these exist. In forward phase protein arrays (FPPA), "capture" antibodies can be printed on an array in known positions, and these then bind proteins of interest which are then sandwiched by the binding of a second antibody with some form of detection capability. The difficulty of finding two structurally nonoverlapping, highly specific antibodies for each target have prevented this from generalized use. In the second type of antibody array, called RPPA, the protein from hundreds of patients are printed on the array and each slide is probed with a single highly validated antibody. Since only one antibody is utilized at a time, with proper validation, nonspecific binding can be minimized. The drawback of this technology is that samples must be batched to be printed at once, making this methodology impractical for real-time individual patient analysis. A second shortcoming is that it does not give data on all proteins and isoforms (e.g., it is biased as opposed to GEP that are unbiased for all potential open reading frames), and furthermore, only those targets with a validated antibody can be probed. On the plus side, the methodology is sample sparing, requiring only 50,000 cells to test about 150 different antibodies. Our laboratory has optimized the technology to do this in hematological malignancies and has generated arrays for AML, MDS, acute lymphoblastic leukemia (ALL) and with AML and MDS $CD34^+$ and $CD34^+CD38^-$ stem cell-enriched fractions, and recently with AML patient-derived MSC (Kornblau et al. 2009c; Tibes et al. 2006; Neeley et al. 2012; Quintas-Cardama et al. 2012; Kornblau et al. 2011a; Ruvolo et al. 2011a; Ruvolo et al. 2011b; Starkova et al. 2011; Kornblau et al. 2011b; Ravandi et al. 2011; Levin et al. 2010; Kornblau et al. 2010b; Kornblau et al. 2010c; Tsao et al. 2010; Ma et al. 2010a; Carter et al. 2010; Neeley et al. 2009; Poland et al. 2009; Pemmaraju et al. 2009; Lee et al. 2012; Minden et al. 2009; Kadia et al. 2009; Carter et al. 2009; Kornblau et al. 2009a; Kornblau et al. 2009b; Kornblau et al. 2013). Newer technologies like CyTOF (Bendall et al. 2011) offer higher throughput and individual cell analysis with a flow cytometry-like output, but the number of validated antibodies is currently limited. This technology will likely eventually be able to provide real-time throughput that RPPA cannot provide.

2.2.5 Metabolomics

Additional technologies to enable detailed profiling of the relative concentration of metabolic pathway components, including lipids, are also emerging. Ascertaining the levels of these would provide another level of validation that the changes observed at different levels are culminating in the expected changes. The unanticipated discovery that the excess 2-OH-glutarate resulting from isocitrate dehydrogenase (IDH) mutations (Lu et al. 2012) results in alteration of methylation highlights that the cell-wide consequences of a mutation extend beyond the canonical pathway that a gene is "contained" in.

2.2.6 Assessing Dynamic Change

A final consideration is expression in the setting of perturbation. The dynamic response of the cells to stress, such as chemotherapy, or apoptotic inducers, etc., can be a key to whether a cell is responsive or resistant. The dynamic adaptation of leukemic cells has been shown to relate to sensitivity. For example, changes in cholesterol synthesis and uptake correlate with sensitivity to ara-c and anthracyclines, (Banker et al. 2004) and blockade of this adaptation was shown to improve chemosensitivity in a clinical trial (Kornblau et al. 2007). Methods to assess adaptability are complex and still in development. The effect of perturbation with exogenous ligands to cell surface receptors, apoptosis inducers, chemotherapeutic agents on the expression and activation of various pathways, and the effect of this on cell viability has been measured using multiparameter flow cytometry as developed by The Nolan Lab (Krutzik et al. 2004; Irish et al. 2004). Identifying the key parameters to perturb and the critical outcomes to measure remains a daunting task. The development of this technology for common use is currently being pursued by Nodality Inc. In collaboration with Nodality, our group has conducted research aimed at the identification of certain stimulation-readout nodes (Kornblau et al. 2011a; Rosen et al. 2010; Cesano et al. 2009; Kornblau et al. 2009b). We also observed that the expression of cells with responsiveness in the Stat6 pathway in subfractions at diagnosis is predictive of relapse (Kornblau et al. 2011a). With further development, omics approaches for dynamic adaptability hold additional promise for improved identification of key targets on an individualized basis.

2.3 Integromics

The goal, once the information from each of these arrays has been performed, is to integrate information from many levels into a single dataset reflecting what is happening in a particular patient's leukemic cells. This process has been named "integromics" (Fig. 2.1). Once the integrated dataset is generated, then the effects of the different changes on a particular pathway can be determined. Expression of components of a pathway might be affected by different processes that combine to result in a "net" effect on a downstream effector that determines activity. For example, a cell might have an activating rat sarcoma (RAS) mutation, which would lead to expectation that the downstream pathway would be active, but lack mitogen extracellular signal-regulated kinase (MEK) and extracellular signal-regulated kinase (ERK) activity due to the combined action of histone acetylation preventing MEK expression, gene methylation preventing ERK expression, and the overexpression of ncRNA resulting in negligible expression of an ERK target. Conversely, small effects at many levels might combine to lead to highly significant overexpression of ERK protein and cross-activation by a kinase with ERK phosphorylation activity other than MEK. In the first case, detecting a RAS mutation might lead us to consider use of a RAS, MEK, or ERK inhibitor, but such a patient would be unlikely to benefit as all the components downstream of the RAS mutation are already under-expressed. Conversely, in the second example, an ERK inhibitor might be therapeutically highly effective, but simply searching for RAS mutations would not have selected this patient for such a trial. So the multilevel information that can be gained from an integromic approach will be key to applying omics to the direction

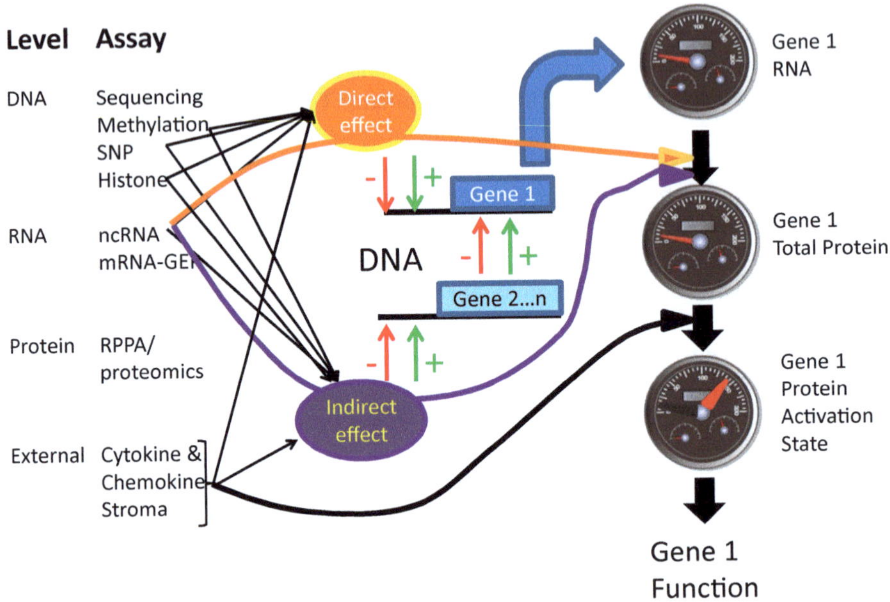

Fig. 2.1 Integromics. The goal is to understand how much function there is of a given protein (*gene1*) and what factors regulate that. The assays that can currently go into an integromics analysis and the level they measure are shown at the *left side*. The combined effects of DNA methylation, Histone modifications, SNPs and non-coding RNA can have direct effects on the transcription of the gene in question. Likewise they can affect the level of expression of other gene(s) that modulate the expression of gene 1 and thereby have an indirect effect. This sets the rheostat for how much *gene 1* mRNA is produced. Non-coding RNAs can also modulate gene 1 mRNA fate, affecting translation along with other cell characteristics that affect the rate of translation to set the rheostat for how much total protein exists. Post translational modification, which is regulated by direct or indirect effects of internal and external factors can affect the activation state of protein 1 thereby setting the rheostat for protein 1 function

of targeted therapy. The integration of datasets is not a trivial computational matter, and there are many technical considerations that must be addressed if this is to be successfully accomplished.

2.3.1 Binary Outcomes Versus Quantitative Measurements

Regardless of the material that is being measured (DNA, RNA, protein miRNA, etc.), there are certain statistical analytic challenges that must be met. The first is a determination of whether detection is adequate or whether relative quantity is needed. In the case of mutations, simply detecting the presence or absence of an event provides one level of answer, but given the heterogeneity of tumor tissues, it may be important to also quantify what percentage of a tumor carries the mutation. For most other analytes, such as mRNA, miRNA, methylation events, protein quantity, etc., relative quantification of the level of expression is required. While the complete absence of expression of a given analyte would provide important information, for many of them it is the relative change in expression level that will determine the functional consequences, and it must be remembered that small coordinated changes in a pathway can be cumulatively important, even if the changes in individual genes do not rise to the level of statistical significance. GEP typically uses a cutoff of threefold expression differences between two samples to define a significant difference. While such cutoffs have the potential to provide higher signal-to-noise ratios, they introduce the potential to miss important changes. Consider the interaction of the pro- and antiapoptotic proteins Bcl2 and Bax, where the relative concentration of each is thought to regulate the apoptosis rheostat. A twofold increase in one accompanied by a twofold decrease in the other would have both being listed as not significantly different, although the relative ratio change would be 4:1 and would likely have a functional consequence. Similarly, if both increased by fourfold, they would be included in the list of significantly different proteins, yet the ratio of each to the other would remain unchanged, and there might be no functional consequences despite a large change in expression.

One approach to generating a more systematic network-based approach to the analysis of GEP data is to evaluate multiple genes simultaneously in sets or signatures linked to some biological feature, such as a signaling pathway. Many gene sets have been prepared in collections that are available online for application to gene expression data, such as Gene Ontology (GO), (Ma et al. 2009) the Molecular Signatures Database (MSigDB), (Subramanian et al. 2005) and the Kyoto Encyclopedia of Genes and Genomes (KEGG) (Ogata et al. 1999). One frequently used approach, so-called gene set enrichment analysis (GSEA), (Gerlinger et al. 2012) has the advantage of not being dependent on the identification of differentially expressed genes (DEGs), according to some arbitrary threshold (fold change, statistical test, etc.). GSEA determines whether the overall expression of a set of genes associated with a particular biological feature differs significantly between two groups of samples, i.e., whether the feature is enriched, even if few or even none of the differences in individual genes meets the threshold of a DEG. Gene set analysis can also be based on DEGs, using the hypergeometric distribution test to determine the significance of enrichment (Subramanian et al. 2005), as in the popular proprietary application Ingenuity Pathway Analysis. While some caveats exist, gene set analysis has been a major advance in the analysis of gene expression data; investigators can now do more than stare at a heat map or list of DEGs and hope for inspiration. However, experience shows that gene set analysis typically only "explains" a small proportion of the differences in gene expression between two groups of samples. Gene set analysis does not eliminate the need for thoughtful interpretation and validation of gene expression data. GSEA serves as a paradigm for how a more integrated analysis can improve the strength of the data emerging from array approaches.

2.3.2 *Defining the "Normal" Comparator*

For each analyte with a quantitative expression measurement, the first question is whether expression is changed from normal. However, defining what is normal is complicated. What is the "normal" comparator for an AML blast, is this a $CD33^+$ cell? A $CD34^+$ cell? A $CD34^+$ $CD38^-$ cell? Normal hematopoietic stem cell (HSC) and leukemia stem cell (LSC) tend to be $CD90^-$/$CD123^+$/$CD117^-$/HLA-DR^- and also express CLL-1, CD44, CD47, and CD96 (Chan and Huntley 2008; Roboz and Guzman 2009; Blair et al. 1998; Blair and Sutherland 2000; Jordan et al. 2000). The restriction of LSC to this phenotype is controversial as some studies have demonstrated LIC with specific molecular abnormalities (del 5q) (Tehranchi et al. 2010) in this pool while other have found persistence of cells harboring specific mutations (NPM1) in the $CD34^-$ fractions (Taussig et al. 2010). Ideally, a study of normal HSC might use $CD34^+$ $CD38^-$ cells, but these cannot be obtained in sufficient numbers even for RPPA. Normal HSC are much rarer than LSC, and it was not possible to purchase sufficient normal $CD34^+$ cells to isolate 2×10^5 normal HSC from available. If normal $CD34^+$ cells are 5 % of normal bone marrow and normal HSC comprise 1:10,000 marrow cells, then there would be 1 HSC per 500 $CD34^+$. Assuming a 50 % recovery, this would require 1000 $CD34^+$ cells to recover a single HSC or that we start with 2×10^8 $CD34^+$ cells (=20 vials × $750/vial = $15,000/sample), enough to engraft a 50 kg patient to recover 2×10^5 cells. If the normal HSC frequency is 1:100,000, then multiply this by 10. For our protein work, we have therefore used a $CD34^+$ cell from a normal bone marrow as our standard comparator (we have successfully used $CD34^+$ cells from AllCells for our RPPA work). We also evaluated granulocyte colony-stimulating factor (G-CSF)-primed peripheral blood derived $CD34^+$ cells by RPPA. As expected, the expression of many proteins was markedly different upon G-CSF stimulation. We have demonstrated that protein expression is markedly different between bulk leukemia cells, $CD34^+$ cells, or cells from the stem cell-enriched $CD34^+$ $CD38^-$ fraction (Kornblau et al. 2013). If the cell that gives rise to resistance comes predominantly from one compartment, then the biology of that compartment needs to be targeted, and it is important to know what is normal for the non-leukemic counterpart from that compartment.

2.3.3 *Effect of Sample Source*

AML samples can come from the blood or the marrow. Does expression change by the origin of the sample? For DNA and epigenetic events, location should not affect the readout, but for quantitative analytes, this could be changed. In our RPPA analyses, we did not observe a statistically significant difference in expression between blood and marrow samples for 85 % of proteins measured, and for most of these, the fold change was less than 20 %, so, therefore, of uncertain functional consequence. However, real differences could have been missed due to dilution of "marrow" samples with blood as our research collection sample is the last aspirate added on at the end of a diagnostic study. If resistance arises from cells in particular environments, such as adjacent to stroma, or in areas of hypoxia, then the key feature may be present in only a small fraction of cells, and it may be important to isolate specific subsets of cells and to maintain those cells in the native condition until the labile fraction can be isolated and stabilized.

2.3.4 Use of Statistical Versus Biologically Functional Thresholds

Another crucial question, that is very difficult to answer, is what is the correlation between the difference in expression and a biologically functional difference? What level correlates to a functional difference? Is this a level that reaches statistical significance by standard methods, or is it a level that is much greater than the threshold for statistical significance? Rigorous testing of each change that appears significantly different by an array to determine the actual functional difference, using other methods for altering level (e.g., siRNA, knock in, knock out), would be costly and time consuming. The level that gives a survival difference on a Kaplan–Meier survival plot may be unrelated to the level that can be pharmacologically modified.

2.4 Issues in Applying Integromics to Patient Care

Despite these obvious potentially confounding issues (and in ignorance of the non-obvious ones), the integromic dataset has the potential to guide the intelligent selection of targeted therapies. Let us consider how having this knowledge might be applied in different circumstances.

2.4.1 Data for Individual Proteins

For an agent targeting one molecule, the first question is whether that molecule is expressed in that patient's leukemic cells. As mutations are recognized and studied, we collectively develop a catalog of the functional consequences of that mutation. Is it activating? Does it eliminate kinase activity, etc.? For non-mutational events, quantity is relevant to therapeutic targeting. Intuitively, a target that is not expressed would not be expected to respond to an agent that inhibits its function. However, if dynamic studies show that the target is significantly upregulated as a defensive adaptation in response to exposure to a therapeutic agent, then even a target with low expression at baseline may be targetable. Both the individual arrays and the integromic approach, especially if dynamic measures are included, can therefore suggest if an individual marker is a candidate.

2.4.2 *Data from Multiple Proteins*

A stronger level of evidence that a pathway is a good target would come from evidence that components up and down a pathway are altered in the same direction. Observing high levels of phospho-RAF, phospho-MEK, phospho-ERK, and phosphorylated ERK targets in a case, with or without RAS mutation, would suggest that this pathway was relevant to the biology of that leukemic cell and a good target. We can envision having heat maps of pathway activation as opposed to heat maps of individual components, showing which pathways are most heavily utilized. Use of available databases such as Ingenuity or Cytoscape can aid in development of lists of genes or proteins that have known canonical associations. However, neither is currently very useful for the incorporation of data on protein activation states. It is important to recognize the novel consequences of oncogenic changes as well, such as the noncanonical activation of one pathway by mutated kinases in another pathway, or the previously mentioned effect of IDH mutations on methylation. The ability to build networks from integromic data will permit the recognition of these novel consequences by differential mapping of actual networks onto canonical pathways. We are attempting to build the technology to do this with RPPA data and have used this to recognize new connections in favorable prognosis core binding factor leukemias (York et al. 2012). Novel connections are likely to be prime targets for therapeutic intervention as the consequences of their blockade are likely to only occur in neoplastic cells and not in the normal cells that lack these connections.

2.4.3 *Data from Multiple Different Analytes*

It will also be important to be able to recognize when more subtle changes at different levels, or changes in multiple components of one level, combine to result in biologically relevant changes in a target. Understanding the consequences of ncRNA expression can be mind-boggling as individual ncRNAs affect many genes and many ncRNA can affect the same gene. Recently, a "ceRNA hypothesis" has been proposed as a major regulator of gene expression and hence protein levels (Salmena et al. 2011). Furthermore, a given target is subject to direct modulation by one ncRNA and indirect modulation by others that act on the other modulators of that gene. Sensing the integrated effect of all the ncRNA on a given target may therefore be more important than measurement of an individual ncRNA. Partial methylation or histone acetylation in combination with the effects of ncRNAs adds another layer of complexity. Detecting when the regulation of a target is arising solely from one layer, or as the integrated effect of multiple levels, can suggest which modulator to target, or whether a combinatorial approach is required. Protein-based analysis can provide the answer on where the rheostat is "set," but each layer must be assessed to determine which hand is controlling the knob.

2.4.4 *How to Test Rationally*

Testing this will require the development of catalogs of ncRNA that can affect a gene and the development of computational methods that can correlate data from the actual levels of all the ncRNA relevant to a given gene with actual mRNA expression and protein level. Likewise, having highly accurate and detailed maps of network interactions of all the proteins that regulate expression of a target must exist before a computational prediction of how epigenetic changes at all the relevant sites combine to affect mRNA and protein levels. Ideally, these predictions would then be validated in controlled systems and the results from those used as feedback on the computational predictions. This can practically be done in cell lines for some interactomes using high-throughput testing systems capable of measuring the consequences of individual and combinatorial multigene modulations. However, in the forthcoming era of individualized medicine, it will not be possible to validate each individual prediction for each patient. It will therefore be necessary to extrapolate from generalized knowledge to individual patients.

2.4.5 *If It Is Low, Can You Elevate?*

The ability to restore expression of a missing gene depends on what mechanism is decreasing expression. Since repression could arise from histone acetylation, histone methylation, gene methylation, or blockade by ncRNA, or by protein-mediated repression, identifying which is the responsible event and the in vitro testing of whether use of an agent to reverse this has the desired result. Posttranslational modulation of wild-type p53 by the MDM2 modulating agent Nutlin and KST185 (discussed later in this book) is a good example of this.

2.4.6 *If It Is High, Can You Successfully Suppress?*

Similarly, omics could identify the mechanism by which a protein is overexpressed or activated and suggest where to target. Additionally, specific ncRNA could be used therapeutically to repress a gene directly, or through the repression on proteins than enhance expression (Ma et al. 2010b). This could be tested in vitro to verify that the modulation produces the desired effect or to identify unexpected consequences in other pathways.

Conclusion

In summary, the ever-growing list of available arrays has the potential to provide an overwhelming amount of data for each patient who can reveal the workings of their leukemia cells at a level and degree of detail that was unimaginable until recently.

Combined with the rapidly expanding list of targeted agents designed to interfere with the expression or function of these genes and proteins, this presents the theoretical opportunity to match targets to patients in real time and in a manner that dramatically improves response and outcome. Achieving this requires the development of integromic computational capabilities to enable us as physicians to see the relevant from among a torrential flood of data. Or, as once stated by Mao Zedong, "There is great chaos under heaven—the situation is excellent."

Acknowledgments Thanks to Dr. Richard Eric Davis for critical reading and suggestions.

References

Banker DE, Mayer SJ, Li HY et al (2004) Cholesterol synthesis and import contribute to protective cholesterol increments in acute myeloid leukemia cells. Blood 104:1816–1824

Barretina J, Caponigro G, Stransky N et al (2012) The Cancer Cell Line Encyclopedia enables predictive modelling of anticancer drug sensitivity. Nature 483:603–607

Bendall SC, Simonds EF, Qiu P et al (2011) Single-cell mass cytometry of differential immune and drug responses across a human hematopoietic continuum. Science 332:687–696

Blair A, Sutherland HJ (2000) Primitive acute myeloid leukemia cells with long-term proliferative ability in vitro and in vivo lack surface expression of c-kit (CD117). Exp Hematol 28:660–671

Blair A, Hogge DE, Sutherland HJ (1998) Most acute myeloid leukemia progenitor cells with long-term proliferative ability in vitro and in vivo have the phenotype $CD34^+/CD71^-/HLA-DR^-$. Blood 92:4325–4335

Butler JS, Koutelou E, Schibler AC, Dent SY (2012) Histone-modifying enzymes: regulators of developmental decisions and drivers of human disease. Epigenomics 4:163–177

Calin GA, Liu CG, Sevignani C et al (2004) MicroRNA profiling reveals distinct signatures in B cell chronic lymphocytic leukemias. Proc Natl Acad Sci U S A 101:11755–11760

Calin GA, Ferracin M, Cimmino A et al (2005) A MicroRNA signature associated with prognosis and progression in chronic lymphocytic leukemia. N Engl J Med 353:1793–1801

Calin GA, Liu CG, Ferracin M et al (2007) Ultraconserved regions encoding ncRNAs are altered in human leukemias and carcinomas. Cancer Cell 12:215–229

Carter BZ, Zhang N, Coombes KR et al (2009) Expression of apoptosis repressor with caspase recruitment domain (ARC), an antiapoptotic protein, is strongly prognostic in acute myeloid leukemia (AML) [abstract]. Blood 114:166

Carter BZ, Qiu YH, Zhang N et al (2010) Expression of ARC (apoptosis repressor with caspase recruitment domain), an antiapoptotic protein, is strongly prognostic in AML. Blood 117(3):780–787

Cesano A, Woronicz J, Cohen AC et al (2009) Single cell network profiling as a tool to identify AML chemotherapy resistant cell phenotypes under in vivo therapeutic pressure [abstract]. Blood 114:165

Chan WI, Huntly BJ (2008) Leukemia stem cells in acute myeloid leukemia. Semin Oncol 35:326–335

Fredriksson S, Gullberg M, Jarvius J et al (2002) Protein detection using proximity-dependent DNA ligation assays. Nat Biotechnol 20:473–477

Garzon R, Heaphy CE, Havelange V et al (2009) MicroRNA 29b functions in acute myeloid leukemia. Blood 114:5331–5341

Gerlinger M, Rowan AJ, Horswell S et al (2012) Intratumor heterogeneity and branched evolution revealed by multiregion sequencing. N Engl J Med 366:883–892

Gold L, Ayers D, Bertino J et al (2010) Aptamer-based multiplexed proteomic technology for biomarker discovery. PLoS ONE 5:e15004

Greenbaum D, Colangelo C, Williams K, Gerstein M (2003) Comparing protein abundance and mRNA expression levels on a genomic scale. Genome Biol 4:117

Griffin TJ, Gygi SP, Ideker T et al (2002) Complementary profiling of gene expression at the transcriptome and proteome levels in *Saccharomyces cerevisiae*. Mol Cell Proteomics 1:323–333

Gygi SP, Rochon Y, Franza BR, Aebersold R (1999) Correlation between protein and mRNA abundance in yeast. Mol Cell Biol 19:1720–1730

Irish JM, Hovland R, Krutzik PO et al (2004) Single cell profiling of potentiated phospho-protein networks in cancer cells. Cell 118:217–228

Jansen R, Greenbaum D, Gerstein M (2002) Relating whole-genome expression data with protein-protein interactions. Genome Res 12:37–46

Jordan CT, Upchurch D, Szilvassy SJ et al (2000) The interleukin-3 receptor alpha chain is a unique marker for human acute myelogenous leukemia stem cells. Leukemia 14:1777–1784

Kadia T, Kornblau SM, Kantarjian H et al (2009) Clinical characterization and proteomic consequences of mutated Ras in acute myeloid leukemia [abstract]. Blood 114:139–140

Kern W, Kohlmann A, Wuchter C et al (2003) Correlation of protein expression and gene expression in acute leukemia. Cytometry B Clin Cytom 55:29–36

Kornblau SM, Banker DE, Stirewalt D et al (2007) Blockade of adaptive defensive changes in cholesterol uptake and synthesis in AML by the addition of pravastatin to idarubicin + high dose Ara-C: a phase I study. Blood 109(7):2999–3006

Kornblau SM, Singh N, Qiu YH et al (2009a) Highly phosphorylated FOXO3 A is an adverse prognostic factor in acute myeloid leukemia [abstract]. Blood 114:166–167

Kornblau SM, Minden MD, Rosen DB et al (2009b) Single cell network profiles in non-M3 AML associated with patient response to standard induction therapy [abstract]. Blood 114:633

Kornblau SM, Tibes R, Qiu YH et al (2009c) Functional proteomic profiling of AML predicts response and survival. Blood 113:154–164

Kornblau SM, McCue D, Singh N et al (2010a) Recurrent expression signatures of cytokines and chemokines are present and are independently prognostic in acute myelogenous leukemia and myelodysplasia. Blood 116(20):4251–4261

Kornblau SM, Minden A, Hogge D, Chohen A, Cesano A (2010b) Insights into acute myeloid leukemia via single cell network profiling. Clin Lab Int 1:12–15

Kornblau SM, Minden M, Rosen D et al (2010c) Dynamic single cell network profiles in AML are associated with patient response to standard induction therapy. Clin Can Res 16(14):3721–3733

Kornblau SM, Covey T, Putta S et al (2011a) Signaling changes in the stem cell factor-AKT-S6 pathway in diagnostic AML samples are associated with disease relapse. Blood Cell J 1(2):e3

Kornblau SM, Qiu YH, Zhang N et al (2011b) Abnormal expression of FLI1 protein is an adverse prognostic factor in acute myeloid leukemia. Blood 118:5604–5612

Kornblau SM, Qutub A, Yao H et al (2013) Proteomic profiling identifies distinct protein patterns in acute myelogenous leukemia CD34+CD38- stem-like cells. PLoS One 8(10):e78453

Krutzik PO, Irish JM, Nolan GP, Perez OD (2004) Analysis of protein phosphorylation and cellular signaling events by flow cytometry: techniques and clinical applications. Clin Immunol 110:206–221

Lee DW, Futami M, Carroll M et al (2012) Loss of SHIP-1 protein expression in high-risk myelodysplastic syndromes is associated with miR-210 and miR-155. Oncogene 31:4085–4094 (Epub)

Levin VA, Panchabhai SC, Shen L et al (2010) Different changes in protein and phosphoprotein levels result from serum starvation of high-grade glioma and adenocarcinoma cell lines. J Proteome Res 9:179–191

Lu C, Ward PS, Kapoor GS et al (2012) IDH mutation impairs histone demethylation and results in a block to cell differentiation. Nature 483:474–478

Ma W, Yang D, Gu Y et al (2009) Finding disease-specific coordinated functions by multi-function genes: insight into the coordination mechanisms in diseases. Genomics 94:94–100

Ma W, Wang M, Wang ZQ et al (2010a) Effect of long-term storage in TRIzol on microarray-based gene expression profiling. Cancer Epidemiol Biomarkers Prev 19:2445–2452

Ma L, Reinhardt F, Pan E et al (2010b) Therapeutic silencing of miR-10b inhibits metastasis in a mouse mammary tumor model. Nat Biotechnol 28:341–347

Miao F, Natarajan R (2005) Mapping global histone methylation patterns in the coding regions of human genes. Mol Cell Biol 25:4650–4661

Michiels S, Koscielny S, Hill C (2005) Prediction of cancer outcome with microarrays: a multiple random validation strategy. Lancet 365:488–492

Minden MD, Kornblau SM, Rosen DB et al (2009) FLT3 ITD signaling profiles in AML samples harboring mutations [abstract]. Blood 114:635

Neeley ES, Kornblau SM, Coombes KR, Baggerly KA (2009) Variable slope normalization of reverse phase protein arrays. Bioinformatics 25:1384–1389

Neeley ES, Baggerly KA., Kornblau SM (2012) Surface adjustment of reverse phase protein arrays using positive controls spots. Cancer Inform 2012:77–86

Ogata H, Goto S, Sato K et al (1999) KEGG: Kyoto Encyclopedia of Genes and Genomes. Nucleic Acids Res 27:29–34

Pemmaraju N, Kantarjian HM, Ravandi F et al (2009) FLT3 inhibitor therapy for patients with myelodysplastic syndromes (MDS) and acute myeloid leukemia (AML): impact on survival according to FLT3 status [abstract]. Blood 114:424–425

Phillips CL, Gerbing R, Alonzo T et al (2010) MDM2 polymorphism increases susceptibility to childhood acute myeloid leukemia: a report from the Children's Oncology Group. Pediatr Blood Cancer 55:248–253

Pichiorri F, Suh SS, Rocci A et al (2010) Downregulation of p53-inducible microRNAs 192, 194, and 215 impairs the p53/MDM2 autoregulatory loop in multiple myeloma development. Cancer Cell 18:367–381

Poland KS, Shardy DL, Azim M et al (2009) Overexpression of ZNF342 by juxtaposition with MPO promoter/enhancer in the novel translocation t(17;19)(q23;q13.32) in pediatric acute myeloid leukemia and analysis of ZNF342 expression in leukemia. Genes Chromosomes Cancer 48:480–489

Quintas-Cardama A, Qiu YH, Post SM et al (2012) Reverse phase protein array profiling reveals distinct proteomic signatures associated with chronic myeloid leukemia progression and with chronic phase in the CD34-positive compartment. Cancer 118:5283–5292 (Epub)

Ravandi F, Patel K, Luthra R et al (2011) Prognostic significance of alterations in IDH enzyme isoforms in patients with AML treated with high-dose cytarabine and idarubicin. Cancer 118:(10):2665–2673

Roboz GJ, Guzman M (2009) Acute myeloid leukemia stem cells: seek and destroy. Expert Rev Hematol 2:663–672

Rosen DB, Minden MD, Kornblau SM et al (2010) Functional characterization of FLT3 receptor signaling deregulation in acute myeloid leukemia by single cell network profiling (SCNP). PLoS ONE 5:e13543

Ruvolo PP, Qui YH, Coombes KR et al (2011a) Low expression of PP2 A regulatory subunit B55alpha is associated with T308 phosphorylation of AKT and shorter complete remission duration in acute myeloid leukemia patients. Leukemia 25:1711–1717

Ruvolo PP, Zhou L, Watt JC et al (2011b) Targeting PKC-mediated signal transduction pathways using enzastaurin to promote apoptosis in acute myeloid leukemia-derived cell lines and blast cells. J Cell Biochem 112:1696–1707

Salmena L, Poliseno L, Tay Y, Kats L, Pandolfi PP (2011) A ceRNA hypothesis: the Rosetta Stone of a hidden RNA language? Cell 146:353–358

Starkova J, Gadgil S, Qiu YH et al (2011) Up-regulation of homeodomain genes, DLX1 and DLX2, by FLT3 signaling. Haematologica 96:820–828

Subramanian A, Tamayo P, Mootha VK et al (2005) Gene set enrichment analysis: a knowledge-based approach for interpreting genome-wide expression profiles [abstract]. Proc Natl Acad Sci U S A 102:15545–15550

Taby RF, Yellapragada SV, Kroeger H et al (2009) Validation of a DNA methylation signature of favorable prognosis in newly diagnosed acute myeloid leukemia [abstract]. Blood 114:645

Taussig DC, Vargaftig J, Miraki-Moud F et al (2010) Leukemia-initiating cells from some acute myeloid leukemia patients with mutated nucleophosmin reside in the CD34⁻ fraction. Blood 115:1976–1984

Tehranchi R, Woll PS, Anderson K et al (2010) Persistent malignant stem cells in del(5q) myelodysplasia in remission. N Engl J Med 363:1025–1037

Tibes R, Qiu YH, Lu Y et al (2006) Reverse phase protein array (RPPA): validation of a novel proteomic technology and utility for analysis of primary leukemia specimens and hematopoietic stem cells (HSC). Mol Cancer Ther 5:2512–2521

Tsao T, Kornblau S, Safe S et al. (2010) Role of peroxisome proliferator-activated receptor-gamma and its coactivator DRIP205 in cellular responses to CDDO (RTA-401) in acute myelogenous leukemia. Cancer Res 70:4949–4960

Valastyan S, Chang A, Benaich N, Reinhardt F, Weinberg RA (2011) Activation of miR-31 function in already-established metastases elicits metastatic regression. Genes Dev 25:646–659

Walter MJ, Shen D, Ding L et al (2012) Clonal architecture of secondary acute myeloid leukemia. N Engl J Med 366:1090–1098

Washburn MP, Koller A, Oshiro G et al (2003) Protein pathway and complex clustering of correlated mRNA and protein expression analyses in *Saccharomyces cerevisiae*. Proc Natl Acad Sci U S A 100:3107–3112

Weinstein JN (2002) 'Omic' and hypothesis-driven research in the molecular pharmacology of cancer. Curr Opin Pharmacol 2:361–365

Xu L, Tan AC, Winslow RL, Geman D (2008) Merging microarray data from separate breast cancer studies provides a robust prognostic test. BMC Bioinformatics 9:125

York H, Kornblau SM, Qutub AA (2012) Network analysis of reverse phase protein expression data: Characterizing protein signatures in acute myeloid leukemia cytogenetic categories t(8;21) and inv(16). Proteomics 12:2084–2093

Part I
Apoptosis

Chapter 3
Roles of Apoptosis-Regulating Bcl-2 Family Genes in AML

John C. Reed

Abstract Among the cardinal features of malignancy is abrogation of cell death mechanisms, thus endowing cancer and leukemia cells with a selective survival advantage relative to normal cells. Genetic and epigenetic lesions that result in defects in cell death regulation represent an essential characteristic of acute myeloid leukemia (AML), promoting accumulation of leukemia cells by conferring tolerance to oncogene activation, cell cycle checkpoint defects, and genetic instability. Defects in cell death mechanisms also greatly contribute to resistance to cytotoxic anticancer drugs. Bcl-2 family proteins are central regulators of cell life and death, impacting both apoptotic and non-apoptotic cell death. The Bcl-2 family includes both cell survival- and death-promoting members, with the relative levels and activities of these proteins becoming imbalanced in favor of cell survival in AML and most other malignancies. The fundamental mechanisms of Bcl-2 family proteins and some of their roles in AML are reviewed in this chapter.

Keywords Bcl-2 · Cell death · Apoptosis · Cell stress · Autophagy · Chemoresistance

3.1 Introduction

Defects in the normal apoptosis mechanisms that keep blood cell numbers in check commonly occur during the pathogenesis and progression of hematopoietic malignancies. Most physiological cell death occurs via apoptosis, a type of programmed cell death essential for normal development and adult tissue homeostasis. In addition to promoting clonal cell expansion by extending cell life span, apoptosis defects also contribute to myriad aspects of tumor biology, including (1) factor-independent growth, allowing hematopoietic cells to survive in absence of trophic support from lymphokines or colony-stimulating factors (CSFs); (2) oncogene activation, negating the pro-apoptotic activity of oncoproteins such as c-Myc and Cyclin-D1 that drive cell division and cell death unless complemented by anti-apoptotic mechanisms; (3) tissue infiltration, where invasive properties of leukemias and lymphomas are supported by promoting cell survival in extra-vascular and extra-nodal

J. C. Reed (✉)
Sanford Burnham Institute for Medical Research, 10901 North Torrey Pines Road,
La Jolla, CA 92037, USA
e-mail: jreed@sanfordburnham.org

M. Andreeff (ed.), *Targeted Therapy of Acute Myeloid Leukemia*,
Current Cancer Research, DOI 10.1007/978-1-4939-1393-0_3

locations; (4) resistance of immune surveillance mechanisms, especially blunting the pathways that cytolytic T-cells (CTL) and natural killer (NK) cells rely upon for killing tumor target cells; and (5) chemoresistance, where defects in cell death mechanisms raise the threshold for cytotoxicity resulting from macromolecule-damaging drugs that attack DNA, microtubules, and other cellular structures. Thus, restoring sensitivity to apoptosis is an attractive strategy for eliminating malignant cells, which takes advantage in part of the intrinsic abnormalities in neoplastic cells to selectively trigger their death.

From an understanding of the core components of the apoptosis machinery at the molecular and structural levels, and from an emerging knowledge of the signal transduction pathways that link to this core machinery, many potential new therapies for leukemia are beginning to emerge, and new insights into the mechanisms by which currently available therapies work (and why they fail) are accumulating. This knowledge base provides the foundation upon which more effective therapies for hematopoietic malignancies may become a near-term reality.

3.1.1 Apoptosis Inducers and Effectors

Apoptosis is caused by the activation of intracellular proteases, known as caspases. These cysteine proteases cleave their cellular targets at aspartic acid residues residing in the context of tetrapeptide motifs within polypeptide substrates (Boatright and Salvesen 2003). The human genome encodes ten caspases (Reed et al. 2004; Saleh et al. 2004). Numerous cellular substrates of caspases have been identified, which in aggregate produce the characteristic morphology we call "apoptosis" when cleaved. Several pathways for triggering caspase activation exist, though two have been elucidated in great detail and have been the center of much attention in recent years. These two pathways for apoptosis are commonly referred to as the *intrinsic* and the *extrinsic* pathways (Salvesen 2002).

The intrinsic pathway centers on mitochondria as initiators of cell death. Multiple signals converge on mitochondria, including DNA damage, hypoxia, and oxidative stress, causing these organelles to release cytochrome *c* (cyt *c*) and other apoptogenic proteins into cytosol. In the cytosol, cyt *c* binds caspase-activating protein Apaf1, triggering its oligomerization into a heptameric complex that binds pro-caspase-9, forming a multi-protein structure known as the "apoptosome" (Salvesen and Renatus 2002). Physical binding of Apaf1 to pro-caspase-9 is mediated by their caspase recruitment domains (CARDs), through homotypic CARD–CARD binding. Activation of apoptosome-associated cell death protease caspase-9 then initiates a proteolytic cascade, where activated caspase-9 cleaves and activates downstream effector proteases, such as pro-caspase-3.

In contrast, the extrinsic apoptotic pathway relies on tumor necrosis factor (TNF) family death receptors for triggering apoptosis. A subgroup of the TNF family receptors contains a cytosolic death domain (DD) that enables their intracellular interaction with downstream adapter proteins, which link these receptors to specific caspases. Upon ligand binding, TNF family receptors containing cytosolic DDs (e.g., Fas, TNFR, TRAIL-R1, and TRAIL-R2) cluster in membranes, recruiting caspase-binding adaptor proteins, including the bipartite adapter Fas-associated protein with death domain (FADD) that contains both a DD and a death effector domain (DED) (Wallach et al. 1999). The DED of FADD binds DED-containing procaspases (e.g., caspases-8 and -10), forming a "death-inducing signaling complex" (DICS) and resulting in caspase activation by an "induced proximity" mechanism (Boatright and Salvesen 2003).

Aside from the intrinsic and extrinsic pathways, multiple additional routes to caspase activation are possible, though for some the pathophysiological relevance is less well established (reviewed in (Reed 2005; Xu et al. 2005)). Also it is important to note that while these and the aforementioned can all lead to caspase activation and apoptosis, some of them also trigger parallel caspase-independent cell death mechanisms, which nevertheless kill cells via non-apoptotic mechanisms. For example, mitochondria not only release caspase-activating proteins such as cyt *c*, but also release endonuclease G (Endo G) and a chromatin-modifying protein apoptosis-inducing factor (AIF) that promote genome digestion and cell death independent of caspases (Penninger and Kroemer 2003). Thus, some types of cell death stimuli can induce parallel paths to apoptotic (caspase-dependent) and non-apoptotic (caspase-independent) cell demise.

3.1.2 Apoptosis Blockers

Given the critical importance of making the correct choices about cell life–death decisions in complex multicellular organisms, it is not surprising that the pathways governing caspase activation are under exquisite control by networks of proteins that directly or indirectly communicate with these proteases. A delicate balance between pro-apoptotic and anti-apoptotic regulators of apoptosis pathways is at play on a continual basis, ensuring the survival of long-lived cells and the proper turnover of short-lived cells in a variety of tissues, including the bone marrow, thymus, and peripheral lymphoid tissues. However, imbalances in this delicate dance of pro- and anti-apoptotic proteins occur in disease scenarios, including cancer where the scales tip in favor of anti-apoptotic proteins and endow cells with a selective survival advantage that promotes neoplasia and malignancy.

The anti-apoptotic proteins responsible for creating roadblocks to apoptosis have been mapped to specific pathways, providing insights into the defective cell death mechanisms that contribute to malignancy. Among these apoptosis blockers are members of the Bcl-2 family, a large group of proteins ($n \geq 25$ in humans) that control mitochondria-dependent steps in cell death pathways (Reed et al. 2004), including dictating whether cyt *c* is or is not released from these organelles. These proteins control the intrinsic pathway (Kroemer and Reed 2000). Bcl-2 family proteins however are also capable of modulating other cell death and cell survival pathways and mechanisms, including the endoplasmic reticulum (ER) pathway for cell death (Demaurex and Distelhorst 2003), the extrinsic (death receptor) pathway, and autophagy (see below).

The Bcl-2 family proteins can be contrasted with other types of anti-apoptotic proteins known to create apoptosis roadblocks in cancer cells, which are discussed in other chapters in this book. Examples of alternative blockers of apoptosis include (a) c-FLIP, a protein that competes with pro-caspase-8 and -10 for interactions with TNF/Fas family death receptors complexes (Tschopp et al. 1998), and (b) IAP family proteins that thwart cell death by directly binding to and suppressing the activity of certain caspases (Deveraux and Reed 1999; Deveraux et al. 1999; Salvesen and Duckett 2002), in addition to other mechanisms.

3.2 The Bcl-2 Family—Diversity of Players

The human genome encodes at least 25 Bcl-2 family proteins. Only six of these are anti-apoptotic and thus represent logical targets for cancer therapy. The six anti-apoptotic members of the family are Bcl-2, Bcl-XL, Mcl-1, Bcl-W, Bfl-1, and Bcl-B. Overexpression of several of these anti-apoptotic Bcl-2-family proteins has been documented in various hematopoietic malignancies (reviewed in Kitada et al. 2002). The anti-apoptotic members of the Bcl-2 are comprised of alpha-helical bundles with structural similarity to the pore-forming proteins of bacteria (e.g., colicins, diphtheria toxin, etc.). Sequence homology motifs called Bcl-2 homology (BH) domains are recognizable in these proteins, with anti-apoptotic members possessing BH1, BH2, BH3, and BH4 domains. These family members typically also have a C-terminal hydrophobic transmembrane domain that anchors them into intracellular membranes—primarily mitochondria and ER membranes.

Pro-apoptotic Bcl-2 family members can be subgrouped into different categories based on sequence, structure, and function properties, making this type of Bcl-2 family member much more diverse than the anti-apoptotic branch of the family. By definition, all pro-apoptotic family members possess the BH3 domain, an amphipathic alpha-helix that mediates their interactions with anti-apoptotic family members. One subset, characterized by Bax, Bak, and probably Bok, appears to have evolved directly from the same ancestor gene as the anti-apoptotic branch of the family, giving these proteins a similar 3D structure (namely alpha-helical bundled with similar to pore-forming bacterial proteins) and possessing several BH domains (typically BH1, BH2, and BH3, but not BH4). This pro-apoptotic group of the family is sometimes called the "multidomain" branch (reviewed in Reed 2006). The other subset of pro-apoptotic Bcl-2 family proteins typically has little overall structural similarity, with the BH3 domain representing the only commonality. This branch of the family has thus been dubbed the "BH3-only" proteins (reviewed in Bouillet and Strasser 2002). Among the BH3-only proteins are those that (a) bind to the anti-apoptotic family members to neutralize them and (b) those that bind to both anti-apoptotic and the multidomain pro-apoptotic family members, antagonizing the anti-apoptotic and activating the pro-apoptotic proteins.

3.3 Dimerization of Bcl-2 Family Proteins

Bcl-2 family proteins participate in hand-to-hand combat to control cell life-and-death decisions. The surface of the anti-apoptotic members of the family forms a crevice that serves as a receptor-like structure for binding the ligand-like BH3 domains of the pro-apoptotic proteins. The BH3 domain consists of an amphipathic alpha-helix that binds this hydrophobic crevice on anti-apoptotic Bcl-2-family members, negating their cytoprotective activity (Fesik 2000). Proof-of-concept experiments using BH3 peptides have suggested that compounds docking at this regulatory site on Bcl-2 and its related anti-apoptotic proteins could provide a route to effective suppression of these proteins, thereby promoting apoptosis of malignant cells (Holinger et al. 1999). Indeed, several chemicals have been identified that bind this pocket on anti-apoptotic Bcl-2-family proteins and that promote apoptosis (reviewed in Reed 2005; Pellecchia and Reed 2004; Reed 2008). Small-molecule inhibitors that directly interact with Bcl-2 or related anti-apoptotic proteins via the BH3-binding pocket have entered clinical trials for cancer and leukemia, a topic reviewed elsewhere in this book.

Additionally, some pro-apoptotic BH3-only members of the family use their BH3 domains to activate multidomain pro-apoptotic proteins Bax and Bak (Reed 2006). The structural basis for this protein interaction has been elucidated (Gavathiotis et al. 2010), thus suggesting a possible path to production of agonists of Bax and Bak for oncology therapeutics development.

Various BH3-mimicking compounds have been reported as a strategy for therapeutics development (Vogler et al. 2009; Kang and Reynolds 2009). A critical but unanswered question about the various chemical antagonists of Bcl-2 is to what extent they inhibit various anti-apoptotic members of the Bcl-2-family ($n = 6$), and whether broad-spectrum versus selective inhibitors would provide the optimal path forward for clinical applications, where efficacy is balanced against toxicity. Moreover, the structure activity relation (SAR) characteristics of the optimal compound may vary depending on the type of cancer or leukemia one wishes to attack, given that the repertoire of Bcl-2 family proteins differs among different types of malignancies. For instance, for optimal treatment of acute myeloid leukemia (AML), one might anticipate that the best compounds will have SAR characteristics that consistently promote apoptosis of AML cells (especially chemorefractory AML cells) but that do not kill normal hematopoietic stem cells.

3.4 Phenotype Conversion by Bcl-2 Family Proteins

Conditions have been identified where anti-apoptotic Bcl-2 family members can seemingly switch their phenotype, converting into pro-apoptotic proteins. The mechanism underlying this phenomenon involves exposure of the BH3 domain, which in proteins such as Bcl-2 and Bcl-XL is normally buried in the protein, with the interaction surface of the BH3 domain oriented towards the hydrophobic core of the protein. Mechanisms for phenotypic conversion include proteolytic cleavage and interactions with other proteins that promote conformational changes in anti-apoptotic Bcl-2 family proteins. Cleavage of Bcl-2 and Bcl-XL, for example, by caspases removes the N-terminal region containing the BH4 domain, presumably causing an opening of the protein structures that exposes the BH3 domain. Interaction of Bcl-2 with the orphan nuclear receptor, Nur77 (also known as TR3 and NR4A1), has also been identified as a mechanism for phenotype conversion. Stimuli that cause Nur77 to leave the nucleus and traffic to mitochondria cause apoptosis in a Bcl-2-dependent manner (Li et al. 2000). In the cytosol, Nur77 binds the Bcl-2 protein, thus accounting for accumulation of this nuclear receptor on the surface of mitochondria. Nur77 induces a profound conformational change of Bcl-2, causing exposure of this BH3 domain, and converting Bcl-2 from a protector to a killer (Lin et al. 2004). The Nur77-mediated mechanism of phenotypic conversion of Bcl-2 may be particularly relevant to AML, given that Nur77 is frequently epigenetically silenced in AML and considering that double knockout of Nur77 and its close relative Nor1 in mice causes AML (Mullican et al. 2007).

The revelation that the Bcl-2 protein can have two opposing phenotypes, depending upon its interactions with other proteins such as Nur77, suggests a possible explanation for the paradoxical association of higher Bcl-2 levels with favorable (not worse) clinical outcome in some types of cancer (Reed 1996), including some subtypes of AML (Kornblau et al. 1999). Examination of Bcl-2 and Nur77 expression in clinical specimens may help to select patients most likely to benefit from therapeutic strategies designed to stimulate the Nur77 pathway for Bcl-2 conversion, since cells with higher Bcl-2 are more sensitive to Nur77 activators (Lin et al. 2004).

3.5 Mitochondrial Mechanisms at the Core of Bcl-2 Family Function

Though Bcl-2 family proteins possess diverse mechanisms by which they can impact cell life-and-death decisions (see below), their central mechanism is believed to relate to regulation of permeability of the outer mitochondrial membrane, thus controlling the release (or sequestration) of apoptogenic proteins stored in the inner membrane space between the inner and outer mitochondria membranes (reviewed in (Chipuk and Green 2008)). When activated, multidomain pro-apoptotic Bcl-2 family members Bax and Bak insert in the outer membrane, oligomerize, evidently forming lipid pores directly or indirectly, cause release of proteins from mitochondria, thus constituting the phenomenon of mitochondrial outer membrane permeabilization (MOMP) (reviewed in Reed 1997; Jurgensmeier et al. 1998; Qian et al. 2008).

Anti-apoptotic Bcl-2 family proteins bind Bax and Bak, preventing their oligomerization in mitochondrial membranes (Fig. 3.1). The BH3-only proteins bind anti-apoptotic proteins, thus neutralizing them and preventing them from binding to and negating the pore-forming proteins Bax and Bak. In addition, a few of the BH3-only proteins (e.g., Bid, Bim, and Puma) not only antagonize the anti-apoptotic

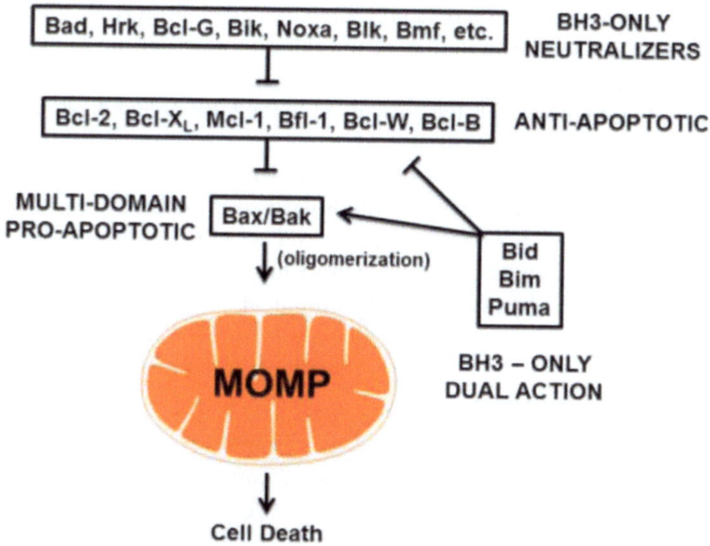

Fig. 3.1 Hierarchy of Bcl-2 family interactions in the regulation of mitochondrial outer membrane permeabilization (*MOMP*)

members of the family (e.g., Bcl-2 and Bcl-XL), but also bind to and activate the multidomain pro-apoptotic Bax and Bak proteins, stimulating their oligomerization in mitochondrial membranes.

Some studies have also suggested a role for Bcl-2 family proteins in regulating aspects of inner mitochondrial membrane function, particularly as pertains to perturbations in the electrochemical gradient (Chen et al. 2011). This aspect of Bcl-2 protein function is still poorly understood. Roles for interactions of Bcl-2 family proteins with resident mitochondrial proteins such as the voltage-dependent anion channels (VDACs) have also been implicated in mitochondrial control of apoptosis, but their overall importance is unclear (Cheng et al. 2003; Shimizu et al. 1999). Bcl-2 family proteins such as pro-apoptotic Bax have also been reported to impact mitochondrial ultrastructure by regulating membrane fission/fusion proteins such as Opa1 and Dynamin, a feature that has been correlated with remodeling of cristae of mitochondria to affect accessibility of cyt *c* for release (Yamaguchi et al. 2008; Cassidy-Stone et al. 2008).

3.6 Bcl-2 Family Proteins at the ER

The ER also plays an important role in regulating cell life-and-death decisions in the context of various cell stress scenarios. At least two aspects of ER biology are relevant. First, the ER is the major storage site for intracellular Ca^{2+}. Control of release of Ca^{2+} from the ER has a number of important implications for cell death regulation (reviewed in Demaurex and Distelhorst 2003; Kim et al. 2008a). Second, conditions that cause accumulation of unfolded proteins in the lumen of the ER invoke an evolutionarily conserved signaling program, called the unfolded protein response (UPR). Signaling proteins involved in the UPR are connected to the Bcl-2 family. These mechanisms are described below.

Conversely, the ER appears to be capable of regulating the functions of Bcl-2 family proteins. For example, sphingolipids derived from the ER and delivered into mitochondrial membranes (at sites of direct contact of ER with mitochondria) have been implicated in creating conditions in mitochondrial membranes that are permissive for pore formation by Bax and Bak (Chipuk et al. 2012; Lee et al. 2011).

3.6.1 Ca^{2+} *Regulation*

Disturbances in intracellular Ca^{2+} regulation contribute to cell life and death. For example, acute release of Ca^{2+} from the ER can trigger a variety of signaling mechanisms that promote cell death (Kim et al. 2008b). Conversely, pulses of Ca^{2+} delivered via inositol triphosphate receptors (IP3Rs) at contact sites of ER with mitochondria promote mitochondrial bioenergetics, sustaining adenosine triphosphate (ATP) and cell survival (Cardenas et al. 2010). Several Bcl-2 family members, particularly the anti-apoptotic members of the family, reside within ER membranes. At least some Bcl-2 family proteins regulate ER Ca^{2+} homeostasis, with anti-apoptotic proteins Bcl-2 and Bcl-XL reducing the steady-state levels of luminal [Ca^{2+}] (Chae et al. 2004; Xu et al. 2008; Kim et al. 2008a; Hunsberger et al. 2011). Various studies have shown that Bcl-2 and Bcl-XL increase passive leak of Ca^{2+} ions from the ER via a mechanisms that is dependent upon IP3Rs. A picture has emerged that envisions anti-apoptotic Bcl-2 proteins associating with IP3Rs, causing passive leak of Ca^{2+} to achieve a lower resting (free) Ca^{2+} concentration in the ER (Rong et al. 2008). In contrast, Bax and Bak increase ER Ca^{2+} levels (Oakes et al. 2005).

Bcl-2 and Bcl-XL interact in ER membranes with BI-1 (Bax inhibitor-1), a multi-membrane ER protein that also regulates ER Ca^{2+} homeostasis, in a manner that largely phenocopies Bcl-2 (Chae et al. 2004; Xu et al. 2008). Interestingly, gene ablation studies suggest that BI-1 is required for ER Ca^{2+} regulation by Bcl-XL (Xu et al. 2008). BI-1 also reportedly binds IP3Rs (Kiviluoto et al. 2012) and controls Ca^{2+} transport from ER to mitochondria via IP3Rs (Sano et al. 2012). Thus, BI-1 appears to collaborate with Bcl-2/Bcl-XL to regulate ER Ca^{2+}. The impact of ER Ca^{2+} on tumor cell survival during ER and metabolic stress requires further investigation.

3.6.2 UPR Regulation

Accumulation of unfolded proteins in the ER stimulates the UPR, also called the ER stress response. ER stress can be induced by myriad stimuli that perturb protein folding, including hypoxia and oxidative stress (via effects on protein disulfide bonding), nutrient perturbations (hypoglycemia/hyperglycemia), and disturbances to cellular protein homeostasis that overwhelm proteasome function and molecular chaperones (Kim et al. 2008b; Ma and Hendershot 2004). (Dong et al. 2005; Shuda et al. 2003; Fernandez et al. 2000). While UPR signaling events clearly help cells (malignant and normal) to adapt to inhospitable microenvironments (Jamora et al. 1996; Li and Lee 2006; Lee 2007; Romero-Ramirez et al. 2004), when sustained or excessive, ER stress triggers cell death, usually by apoptosis but also by non-apoptotic mechanisms (Xu et al. 2005; Kim et al. 2008). Therefore, malignant cells often develop barriers to ER stress-induced apoptosis, providing novel targets for therapeutic intervention. For example, ER stress is causally involved in the cytotoxic activity of proteasome inhibitors used for cancer treatment (Lee et al. 2003). Also, some experimental agents now in clinical testing induce ER stress as their primary mechanism of action (Zou et al. 2008).

While three major signal transduction nodes have been identified as components of the UPR (initiated by IRE1, PERK, ATF6), it is the IRE1 that has been most clearly linked to the mechanism of Bcl-2 family proteins IRE1α is a transmembrane protein that contains both a Ser/Thr-kinase domain and an endoribonuclease domain, the latter of which processes an intron from X box-binding protein-1 (XBP-1) mRNA to produce the 41 kDa XBP-1 protein (a bZIP family transcription factor). XBP-1 binds to promoters of several genes involved in UPR and ERAD (ER-assisted degradation) (Rao and Bredesen 2004), and thus seems to be generally protective. In contrast, the protein kinase activity of IRE1α has been linked to cell death induction, initiated by apoptotic signaling kinase-1 (Ask1), which causes Jun-N-terminal kinase (JNK) activation (Nishitoh et al. 2002).

Bax and Bak reportedly bind IRE1 and activate it, thus inducing a cascade of stress kinase activation (Hetz et al. 2006) (Fig. 3.2). Within these kinase cascades, among the apoptosis-relevant substrates of JNK are Bcl-2 and Bim, which are inhibited and activated, respectively, by JNK phosphorylation (Lei and Davis 2003; Putcha et al. 2003; Srivastava et al. 1998; Wei et al. 2008). Substrates of p38 mitogen-activated protein kinase (MAPK) include transcription factor C/EBP homologous protein (CHOP), which represses expression of the gene encoding Bcl-2 and induces expression of genes encoding Bim and DR5 (TRAIL-R2) (Zou et al. 2008; Lei and Davis 2003; Putcha et al. 2003; McCullough et al. 2001; Puthalakath et al. 2007; Wang and Ron 1996). Prolonged CHOP activity can also promote non-apoptotic cell death via induction of ER oxidase-1a (ERO1a) (Li et al. 2009; Ozcan and Tabas 2012; Marciniak et al. 2004). Thus, IRE1 is a focal point for the ER-relevant activities of Bcl-2-family proteins, with pro-apoptotic Bax and Bak proteins interacting with and activating IRE1 (Hetz et al. 2006). Conversely, cytoprotective Bcl-2-interacting ER membrane protein, BI-1 (Tmbim6) (Xu and Reed 1998; Reimers et al. 2008) binds to and suppresses IRE1 signaling in cultured cells and in mice (Bailly-Maitre et al. 2006; Lisbona et al. 2009; Bailly-Maitre et al. 2010).

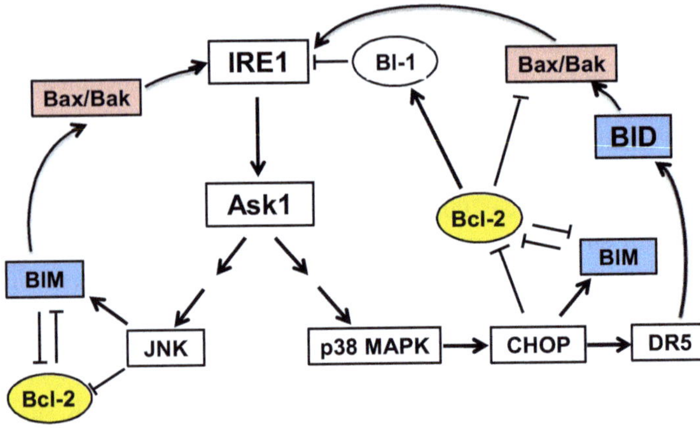

Fig. 3.2 Networks connecting Bcl-2 family proteins to the UPR (ER stress) machinery. MAPK mitogen-activated protein kinase, *CHOP* C/EBP homologous protein, *JNK* Jun-N-terminal kinase, *Ask1* apoptotic signaling kinase-1

3.7 Autophagy

Autophagy is a catabolic cellular process for lysosome-mediated degradation of senescent proteins and organelles (Bernales et al. 2006; Kruse et al. 2006). Autophagy is implicated in cancer primarily as a survival mechanism, where it provides substrates for maintaining ATP levels during times of nutrient deprivation and hypoxia. However, autophagy that is excessive in nature may also contribute or induce to cell death in some contexts (Levine and Yuan 2005; Kroemer et al. 2010). Among the Bcl-2-interacting proteins is beclin (ATG6), an essential component of the autophagy machinery. The beclin protein has a BH3-like domain that mediates interactions with Bcl-2. Bcl-2 inhibits beclin's participation in protein complexes that generate autophagic vesicles. Phosphorylation of Bcl-2 by JNK provides a means of freeing beclin to allow autophagy to proceed (Wei et al. 2008), in addition to displacement of beclin from Bcl-2 by various pro-apoptotic BH3-containing proteins.

Interestingly, ER stress also induces autophagy (Bernales et al. 2006; Momoi 2006; Yorimitsu et al. 2006; Ogata et al. 2006), potentially involving a diversity of mechanisms, including (a) changes in ER Ca^{2+}, perhaps mediated by IP3Rs (Hoyer-Hansen and Jaattela 2007; Criollo et al. 2007; Lam et al. 2008; Vicencio et al. 2009); (b) activation of Ca^{2+} -dependent kinases, including CamKK and DAPK1 (Gozuacik et al. 2008; Zalckvar et al. 2009; Sakaki et al. 2008); and (c) JNK-mediated phosphorylation of Bcl-2 (which causes its release of Beclin-1) occurring downstream of IRE1 signaling (Pattingre et al. 2005; He et al. 2012). Thus, the ability of Bcl-2 family proteins to regulate ER Ca^{2+} via interactions with IP3Rs and to modulate UPR signaling via interactions with IRE1 appears to establish a complex network of regulatory mechanisms that link the Bcl-2 family to autophagy and cellular resilience.

3.8 Interacting Proteins and Posttranslational Modifications

A plethora of proteins have been reported to interact with Bcl-2 family members, suggesting that the Bcl-2 family connects to a variety of protein networks and cellular processes as a mechanism for sensing the cellular status. Bcl-2-interacting proteins range from nuclear receptors that exit the nucleus to bind Bcl-2 and promote apoptosis (Nur77, SHP), to molecular chaperones and co-chaperones (FK506-binding proteins, FKBPs; BAG family Hsp70/Hsc70 co-chaperone) that collaborate with Bcl-2 to promote cell survival, to NLR family innate immunity proteins involved in activation of pro-inflammatory caspases, to kinases and phosphatases (reviewed in Chipuk et al. 2010). Pro-apoptotic family members such as Bax have been reported to interact with Hsp27 family chaperone clusterin, innate immunity adapter protein ASC, the cytoprotective peptide humanin, cathepsin family proteases, and more (reviewed in Reed 2006). Much remains unknown about the cellular contexts in both health and disease where these various protein interactions play fundamentally important roles in cell life-and-death decisions.

Post-translational modifications of Bcl-2 family proteins have not been systemically studied, but multiple examples are found in the literature. For example, phosphorylation of Bcl-2 has been shown to modulate its activity (Konopleva et al. 2002), with MEK1 inhibitors suppressing Bcl-2 phosphorylation and displaying robust synergy with chemical inhibitors of Bcl-2 in AML cells (Milella et al. 2002). Akt (PKB) and several other kinases phosphorylate pro-apoptotic protein BAD (a BH3-only protein), causing its sequestration in a complex with 14-3-3 proteins (Khwaja 1999). Dephosphorylation of pro-apoptotic Bak protein on specific tyrosines has been reported to be required for its activation (Fox et al. 2010).

Proteolytic cleavage of Bcl-2 family proteins is another functionally important post-translational modification, with cleavage of pro-apoptotic BH3-only protein Bid serving as a prime example. Bid becomes activated by caspase-8-mediated cleavage in the context of signaling by TNF family death receptors (extrinsic pathway), resulting in its translocation from cytosol to mitochondrial membranes and exposure of its BH3 domain to enable interaction with other Bcl-2 family members (reviewed in Korsmeyer et al. 2000). Bid is also cleaved and activated by other classes of intracellular cysteine proteases in various pathological contexts (Droga-Mazovec et al. 2008; Upton et al. 2008; Stoka et al. 2001; Chen et al. 2001).

Deamidation of Bcl-XL, converting asparagine to isoaspartic acid residues, has been associated with DNA damage responses (Deverman et al. 2002). Deamidation of Bcl-XL impairs its ability to bind various BH3-containing proteins and to reduce its anti-apoptotic activity. Interestingly, defects in Bcl-XL deamidation have been associated with myeloproliferative disorders (Zhao et al. 2008).

Ubiquitination is an important post-translational modification for regulating the stability of Bcl-2 family proteins. For example, anti-apoptotic protein Mcl-1 is the substrate of an E3 ligase (MULE/ARF-BP1) that contains a BH3-like domain, mediating Mcl-1 degradation in the context of DNA damage (Zhong et al. 2005).

Phosphorylation impacts Mcl-1 protein stability, with phosphorylation by Erk1 slowing and GSK3β accelerating degradation (Domina et al. 2004; Maurer et al. 2006). In contrast, ubiquitination of Bfl-1 occurs constitutively, keeping levels of Bfl-1 protein low—a mechanism that becomes defective in lymphomas (Fan et al. 2010).

3.9 Deregulated Expression of Bcl-2 Family Genes in AML

Many examples exist of alterations in the expression of either apoptosis-suppressing or apoptosis-inducing members of the Bcl-2 family in human cancers (reviewed in Reed et al. 2004; Levine and Yuan 2005; Kroemer et al. 2010; Momoi 2006). The explanations for overexpression of anti-apoptotic proteins of the Bcl-2 family in cancer and leukemia cells vary, but documented mechanisms include chromosomal translocations, gene amplification, loss of microRNAs that target Bcl-2 family genes, gene hypomethylation, transcriptional upregulation, and perhaps altered protein stability (reviewed in (Kitada et al. 2002)).

Bcl-2, Bcl-XL, Mcl-1, and, occasionally, other anti-apoptotic members of the Bcl-2 family are commonly overexpressed in AML (Kornblau et al. 1999; Andreeff et al. 1999; Campos et al. 1993;Konopleva and Andreeff 2002; Kasimirbauer et al. 1998; Deng et al. 1998; Lauria et al. 1997; Schaich et al. 2001). In AML, higher levels of Bcl-2 protein or mRNA have been associated with poor responses to chemotherapy and/or shorter overall survival (Campos et al. 1993; Kasimirbauer et al. 1998; Deng et al. 1998; Lauria et al. 1997; Bincoletto et al. 1999; Campos et al. 1997; Rochitz et al. 1999; Maung et al. 1994; Karakas et al. 1998).

Conversely, pro-apoptotic genes that oppose these cytoprotective proteins can become inactivated in malignant cells, through gene deletion, somatic mutation, gene hypermethylation, and transcriptional downregulation and probably other mechanisms. For example, inactivating mutations in the pro-apoptotic Bcl-2 family gene *Bax* occur in some hematological malignancies including AML (Meiijerink et al. 1995; Brimmell et al. 1998; Meijerink et al. 1998).

The tumor microenvironment undoubtedly contributes to the regulation of Bcl-2 family proteins in AML. Several cytokines, including granulocyte colony-stimulating factor (G-CSF) and Flt3, have been reported to upregulate expression of Bcl-2 and promote apoptosis resistance in freshly isolated AML cells in vitro, implying that the Bcl-2 gene is subject to regulation by inputs from cytokine/lymphokine receptor pathways of the tumor microenvironment (Lisovsky et al. 1996; Bradbury et al. 1994). Lymphokines that activate STAT3 stimulate expression of Bcl-XL. Retinoids also can modulate the expression of Bcl-2 family proteins, including downregulating Bcl-2 and Bcl-XL in AML (Andreeff et al. 1999; Ahmed et al. 1999; Elstner et al. 1996; Dipietrantonio et al. 1996; Delia et al. 1995). Thus, numerous signal transduction and transcriptional pathways may be capable of modulating the expression of Bcl-2 family genes in AML.

3.10 Cytotoxic Chemotherapy and Bcl-2 Family Proteins

In many instances, cell death induced by anticancer drugs appears to occur via the mitochondrial pathway that involves release of cytochrome *c*. It remains unclear how damage induced to DNA or other macromolecules by anticancer drug triggers mitochondrial release of cyt *c*. However, some specific mechanisms have been revealed in recent years. For example, in some types of cells, DNA-damaging agents can induce activation of p53, which in turn binds to cis-acting elements located within the Bax gene (Miyashita and Reed 1995) and the Bax–activity gene Puma and Bim. The p53 protein has also been reported to modulate the activity of Bcl-2 family proteins in the cytosol and on the surface of mitochondria. Another example of how anticancer drugs activate the mitochondrial (intrinsic) pathway is found in the pro-apoptotic Bcl-2 family protein Bim, which can be activated by anti-microtubule drugs. The Bim protein is normally sequestered at microtubules but can be released into the cytosol when it translocates to the surface of mitochondria, heterodimerizing with Bcl-2 or Bcl-XL and triggering cytochrome *c* release.

Conclusions

Bcl-2 family proteins are intricately involved in the biology of AML. From pathogenesis to progression and resistance to therapy, Bcl-2 family proteins are inextricably linked to AML. The advent of experimental therapeutics targeting anti-apoptotic Bcl-2 family proteins (small molecules) and mRNA (antisense) creates hope that are more effective strategies for treating AML may be near.

References

Ahmed N, Sammons J, Hassan H (1999) Bcl-2 protein in human myeloid leukemia cells and its down-regulation during chemotherapy-induced apoptosis. Oncol Rep 6(2):403–407

Andreeff M, Jiang SW, Zhang X, Konopleva M, Estrov Z, Snell VE et al (1999) Expression of Bcl-2-related genes in normal and AML progenitors: changes induced by chemotherapy and retinoic acid. Leukemia 13:1881–1892

Bailly-Maitre B, Fondevila C, Kaldas F, Droin N, Luciano F, Ricci JE et al (2006) Cytoprotective gene bi-1 is required for intrinsic protection from endoplasmic reticulum stress and ischemia-reperfusion injury. Proc Natl Acad Sci U S A 103:2809–2814

Bailly-Maitre B, Belgardt BF, Jordan SD, Coornaert B, von Freyend MJ, Kleinridders A et al (2010) Hepatic Bax inhibitor-1 inhibits IRE1alpha and protects from obesity-associated insulin resistance and glucose intolerance. J Biol Chem 285(9):6198–6207

Bernales S, McDonald KL, Walter P (2006) Autophagy counterbalances endoplasmic reticulum expansion during the unfolded protein response. PLoS Biol 4(12):e423

Bincoletto C, Saad STO, Soares de Silva E, Queiroz MLS (1999) Haematopoietic response and bcl-2 expression in patients with acute myeloid leukaemia. Eur J Haematol 62:38–42

Boatright KM, Salvesen GS (2003) Mechanisms of caspase activation. Curr Opin Cell Biol 15:725–731

Bouillet P, Strasser A (2002) BH3-only proteins—evolutionarily conserved proapoptotic Bcl-2 family members essential for initiating programmed cell death. J Cell Sci 115(Pt 8):1567–1574
Bradbury D, Zhu Y-M, Russell N (1994) Regulation of Bcl-2 expression and apoptosis in acute myeloblastic leukemia cells by granulocyte-macrophage colony-stimulating factor. Leukemia 8:786–791
Brimmell M, Mendiola R, Mangion J, Packham G (1998) Bax frameshift mutations in cell lines derived from human haemopoietic malignacies are associated with resistance to apoptosis and microsatellite instability. Oncogene 16:1803–1812
Campos L, Roualult J-P, Sabido O, Roubi N, Vasselon C, Archimbaud E et al (1993) High expression of Bcl-2 protein in acute myeloid leukemia cells is associated with poor response to chemotherapy. Blood 81:3091–3096
Campos L, Oriol P, Sabido O, Guyotat D (1997) Simultaneous expression of P-glycoprotein and bcl-2 in acute. Leuk Lymphoma 27(1–2):119–125
Cardenas C, Miller RA, Smith I, Bui T, Molgo J, Muller M et al (2010) Essential regulation of cell bioenergetics by constitutive InsP3 receptor Ca^{2+} transfer to mitochondria. Cell 142(2):270–283
Cassidy-Stone A, Chipuk JE, Ingerman E, Song C, Yoo C, Kuwana T et al (2008) Chemical inhibition of the mitochondrial division dynamin reveals its role in Bax/Bak-dependent mitochondrial outer membrane permeabilization. Dev Cell 14(2):193–204
Chae HJ, Kim HR, Xu C, Bailly-Maitre B, Krajewska M, Krajewski S et al (2004) BI-1 regulates an apoptosis pathway linked to endoplasmic reticulum stress. Mol Cell 15(3):355–356
Chen M, He H, Zhan S, Krajewski S, Reed JC, Gottlieb RA (2001) Bid is cleaved by calpain to an active fragment in vitro and during myocardial ischemia/ reperfusion. J Biol Chem 276:30724–30728
Chen YB, Aon MA, Hsu YT, Soane L, Teng X, McCaffery JM et al (2011) Bcl-xL regulates mitochondrial energetics by stabilizing the inner membrane potential. J Cell Biol 195(2):263–276
Cheng EH, Sheiko TV, Fisher JK, Craigen WJ, Korsmeyer SJ (2003) VDAC2 inhibits BAK activation and mitochondrial apoptosis. Science 301:513–517
Chipuk JE, Green DR (2008) How do BCL-2 proteins induce mitochondrial outer membrane permeabilization? Trends Cell Biol 18(4):157–164
Chipuk JE, Moldoveanu T, Llambi F, Parsons MJ, Green DR (2010) The BCL-2 family reunion. Mol Cell 37(3):299–310
Chipuk J, McStay G, Bharti A, Kuwana T, Clarke C, Siskind L et al (2012) Sphingolipid metabolism cooperates with BAK and BAX to promote the mitochondrial pathway of apoptosis. Cell 148(5):845–848
Criollo A, Maiuri MC, Tasdemir E, Vitale I, Fiebig AA, Andrews D et al (2007) Regulation of autophagy by the inositol trisphosphate receptor. Cell Death Differ 14(5):1029–1039
Delia D, Aiello A, Formelli F, Fontanella E, Costa A, Miyashita T et al (1995) Regulation of apoptosis induced by N-(4-hydroxyphenyl) retinamide and effect of Bcl-2. Blood 85:359–367
Demaurex N, Distelhorst C (2003) Apoptosis—the calcium connection. Science 300(5616):65–67
Deng GL ane C, Kornblau S, Goodacre A, Snell V, Andreeff M et al (1998) Ratio of bcl-xshort to bcl-xlong is different in good- and poor-prognosis subsets of acute myeloid leukemia. Mol Med 4(3):158–164
Deveraux QL, Reed JC (1999) IAP family proteins: suppressors of apoptosis. Genes Dev 13:239–252
Deveraux QL, Leo E, Stennicke HR, Welsh K, Salvesen GS, Reed JC (1999) Cleavage of human inhibitor of apoptosis protein XIAP results in fragments with distinct specificities for caspases. EMBO J 18:5242–5251
Deverman BE, Cook BL, Manson SR, Niederhoff RA, Langer EM, Rosova I et al (2002) Bcl-xL deamidation is a critical switch in the regulation of the response to DNA damage. Cell 111:51–62
Dipietrantonio A, Hsieh T-C, Wu JM (1996) Differential effects of retinoic acid (RA) and *N* –(4-hydroxyphenyl)retinamide (4-HPR) on cell growth, induction of differentiation, and changes in $p34^{cdc2}$, bcl-2, and actin expression in the human promyelocytic hl-60 leukemic cells. Biochem Biophys Res Commun 224:837–842

Domina AM, Vrana JA, Gregory MA, Hann SR, Craig RW (2004) MCL1 is phosphorylated in the PEST region and stabilized upon ERK activation in viable cells, and at additional sites with cytotoxic okadaic acid or taxol. Oncogene 23(31):5301–5315

Dong D, Ko B, Baumeister P, Swenson S, Costa F, Markland F et al (2005) Vascular targeting and antiangiogenesis agents induce drug resistance effector GRP78 within the tumor microenvironment. Cancer Res 65(13):5785–5791

Droga-Mazovec G, Bojic L, Petelin A, Ivanova S, Romih R, Repnik U et al (2008) Cysteine cathepsins trigger caspase-dependent cell death through cleavage of bid and antiapoptotic Bcl-2 homologues. J Biol Chem 283(27):19140–19150

Elstner E, Linker-Israeli M, Umiel T, Le J, Grillier I, Said J et al (1996) Combination of a potent 20-epi-vitamin D3 analogue (KH 1060) with 9-cis-retinoic acid irreversibly inhibits clonal growth, decreases bcl-2 expression, and induces apoptosis in HL-60 leukemic cells. Cancer Res 56(15):3570–3576

Fan G, Simmons MJ, Ge S, Dutta-Simmons J, Kucharczak J, Ron Y et al (2010) Defective ubiquitin-mediated degradation of antiapoptotic Bfl-1 predisposes to lymphoma. Blood 115(17):3559–3569

Fernandez PM, Tabbara SO, Jacobs LK, Manning FC, Tsangaris TN, Schwartz AM et al (2000) Overexpression of the glucose-regulated stress gene GRP78 in malignant but not benign human breast lesions. Breast Cancer Res Treat 59(1):15–26

Fesik SW (2000) Insights into programmed cell death through structural biology. Cell 103:273–282

Fox JL, Ismail F, Azad A, Ternette N, Leverrier S, Edelmann MJ et al (2010) Tyrosine dephosphorylation is required for Bak activation in apoptosis. EMBO J 29(22):3853–3868

Gavathiotis E, Reyna DE, Davis ML, Bird GH, Walensky LD (2010) BH3-Triggered Structural Reorganization Drives the Activation of Proapoptotic BAX. Mol Cell 40(3):481–492

Gozuacik D, Bialik S, Raveh T, Mitou G, Shohat G, Sabanay H et al (2008) DAP-kinase is a mediator of endoplasmic reticulum stress-induced caspase activation and autophagic cell death. Cell Death Differ 15(12):1875–1886

He C, Bassik MC, Moresi V, Sun K, Wei Y, Zou Z et al (2012) Exercise-induced BCL2-regulated autophagy is required for muscle glucose homeostasis. Nature 481(7382):511–515

Hetz C, Bernasconi P, Fisher J, Lee AH, Bassik MC, Antonsson B et al (2006) Proapoptotic BAX and BAK modulate the unfolded protein response by a direct interaction with IRE1alpha. Science 312(5773):572–576

Holinger E, Chittenden T, Lutz R (1999) Bak BH3 Peptides antagonize Bcl-x_L Function and induce apoptosis through cytochrome c-independent Activation of Caspases. J Biol Chem 274:13298–13304

Hoyer-Hansen M, Jaattela M (2007) Connecting endoplasmic reticulum stress to autophagy by unfolded protein response and calcium. Cell Death Differ 14(9):1576–1582

Hunsberger JG, Machado-Vieira R, Austin DR, Zarate C, Chuang DM, Chen G et al (2011) Bax inhibitor 1, a modulator of calcium homeostasis, confers affective resilience. Brain Res 1403:19–27

Jamora C, Dennert G, Lee AS (1996) Inhibition of tumor progression by suppression of stress protein GRP78/BiP induction in fibrosarcoma B/C10ME. Proc Natl Acad Sci U S A 93(15):7690–7694

Jurgensmeier JM, Xie Z, Deveraux Q, Ellerby L, Bredesen D, Reed JC (1998) Bax directly induces release of cytochrome c from isolated mitochondria. Proc Natl Acad Sci U S A 95:4997–5002

Kang MH, Reynolds CP (2009) Bcl-2 inhibitors: targeting mitochondrial apoptotic pathways in cancer therapy. Clin Cancer Res 15(4):1126–1132

Karakas T, Maurer U, Weidmann E, Miething C, Hoelzer D, Bergmann L (1998) High expression of bcl-2 mRNA as a determinant of poor prognosis in acute myeloid leukemia. Ann Oncol 9(2):159–165

Kasimirbauer S, Ottinger H, Meusers P, Beelen D, Brittinger G, Seeber S et al (1998) In acute myeloid leukemia, coexpression of at least two proteins, including P-glycoprotein, the multidrug resistance-related protien, bcl-2, mutant p53, and heat-shock protein 27, is predictive of the response to induction chemotherapy. Exp Hematol 26(12):1111–1117

Khwaja A (1999) Akt is more than just a bad kinase. Nature 401:33–34
Kim I, Xu W, Reed JC (2008) Cell death and endoplasmic reticulum stress: disease relevance and therapeutic opportunities. Nat Rev Drug Discov 7(12):1013–1030
Kim HR, Lee GH, Ha KC, Ahn T, Moon JY, Lee BJ et al (2008) Bax Inhibitor-1 Is a pH-dependent regulator of Ca^{2+} channel activity in the endoplasmic reticulum. J Biol Chem 283(23):15946–15955
Kitada S, Pedersen IM, Schimmer A, Reed JC (2002) Dysregulation of apoptosis genes in hematopeietic malignancies. Oncogene 21:3459–3474
Kiviluoto S, Schneider L, Luyten T, Vervliet T, Missiaen L, De Smedt H et al (2012) Bax Inhibitor-1 is a novel IP(3) receptor-interacting and -sensitizing protein. Cell Death Dis 3:e367
Konopleva M, Andreeff M (2002) Mechanisms in drug resistance in AML. In: Murray D, Andersson BS (eds) Clinically relevant resisitance in cancer chemotherapy, Kluwer Academic Publishers, Norwell, pp 237–262
Konopleva M, Tsao T, Ruvolo P, Stiouf I, Estrov Z, Leysath CE et al (2002) Novel triterpenoid CDDO-Me is a potent inducer of apoptosis and differentiation in acute myelogenous leukemia. Blood 99:326–335
Kornblau SM, Thall PF, Estrov Z, Walterscheid M, Patel S, Theriault A et al (1999) The prognostic impact of BCL2 protein expression in acute myelogenous leukemia varies with cytogenetics. Clin Cancer Res 5:1758–1766
Korsmeyer SJ, Wei MC, Saito M, Weiler S, Oh KJ, Schlesinger PH (2000) Pro-apoptotic cascade activates BID, which oligomerizes BAK or BAX into pores that result in the release of cytochrome c. Cell Death Differ 7:1166–1173
Kroemer G, Reed JC (2000) Mitochondrial control of cell death. Nat Med 6:513–519
Kroemer G, Marino G, Levine B (2010) Autophagy and the integrated stress response. Mol Cell 40(2):280–293
Kruse KB, Dear A, Kaltenbrun ER, Crum BE, George PM, Brennan SO et al (2006) Mutant fibrinogen cleared from the endoplasmic reticulum via endoplasmic reticulum-associated protein degradation and autophagy: an explanation for liver disease. Am J Pathol 168(4):1299–1308; quiz 1404–1295
Lam D, Kosta A, Luciani MF, Golstein P (2008) The Inositol 1,4,5-Trisphosphate receptor is required to signal autophagic cell death. Mol Biol Cell 19(2):691–700
Lauria F, Raspadori D, Rondelli D, Ventura M, Fiacchini M, Visani G et al (1997) High bcl-2 expression in acute myeloid leukemia cells correlates with CD34 positivity and complete remission rate. Leukemia 11(12):2075–2078
Lee AS (2007) GRP78 induction in cancer: therapeutic and prognostic implications. Cancer Res 67(8):3496–3499
Lee AH, Iwakoshi NN, Anderson KC, Glimcher LH (2003) Proteasome inhibitors disrupt the unfolded protein response in myeloma cells. Proc Natl Acad Sci U S A 100(17):9946–9951
Lee H, Rotolo JA, Mesicek J, Penate-Medina T, Rimner A, Liao WC et al (2011) Mitochondrial ceramide-rich macrodomains functionalize Bax upon irradiation. PLoS One 6(6):e19783
Lei K, Davis RJ (2003) JNK phosphorylation of Bim-related members of the Bcl2 family induces Bax-dependent apoptosis. Proc Natl Acad Sci U S A 100(5):2432–2437
Levine B, Yuan J (2005) Autophagy in cell death: an innocent convict? J Clin Invest 115(10):2679–2688
Li J, Lee AS (2006) Stress induction of GRP78/BiP and its role in cancer. Curr Mol Med 6(1):45–54
Li H, Kolluri SK, Gu J, Dawson MI, Cao X, Hobbs PD et al (2000) Cytochrome c release and apoptosis induced by mitochondrial targeting of nuclear orphan receptor TR3. Science 289:1159–1164
Li G, Mongillo M, Chin KT, Harding H, Ron D, Marks AR et al (2009) Role of ERO1-alpha-mediated stimulation of inositol 1,4,5-triphosphate receptor activity in endoplasmic reticulum stress-induced apoptosis. J Cell Biol 186(6):783–792
Lin B, Kolluri SK, Lin F, Liu W, Han Y-H, Cao X et al (2004) Conversion of Bcl-2 from protector to killer by interaction with nuclear orphan receptor TR3/NGFI-B/Nur Cell 116:527–540
Lisbona F, Rojas-Rivera D, Thielen P, Zamorano S, Todd D, Martinon F et al (2009) BAX inhibitor-1 is a negative regulator of the ER stress sensor IRE1alpha. Mol Cell 33(6):679–691

Lisovsky M, Extrov Z, Zhang X, Consoli U, Sanchez-Williams G, Snell V et al (1996) Flt3 ligand stimulates proliferation and inhibits apoptosis of acute myeloid leukemia cells: regulation of Bcl-2 and Bax. Blood 88(10):3987–3997

Ma Y, Hendershot LM (2004) The role of the unfolded protein response in tumour development: friend or foe? Nat Rev Cancer 4(12):966–977

Marciniak SJ, Yun CY, Oyadomari S, Novoa I, Zhang Y, Jungreis R et al (2004) CHOP induces death by promoting protein synthesis and oxidation in the stressed endoplasmic reticulum. Genes Dev 18(24):3066–3077

Maung ZT, MacLean FR, Reid MM, Pearson ADJ, Proctor SJ, Hamilton PJ et al (1994) The relationship between Bcl-2 expression and response to chemotherapy in acute leukemia. Br J Haematol 88:105–109

Maurer U, Charvet C, Wagman AS, Dejardin E, Green DR (2006) Glycogen synthase kinase-3 regulates mitochondrial outer membrane permeabilization and apoptosis by destabilization of MCL-1. Mol Cell 21:749–760

McCullough KD, Martindale JL, Klotz LO, Aw TY, Holbrook NJ (2001) Gadd153 sensitizes cells to endoplasmic reticulum stress by down-regulating Bcl2 and perturbing the cellular redox state. Mol Cell Biol 21(4):1249–1259

Meiijerink JPP, Smetsers TFCM, Slöetjes AW, Linders EHP, Mensink EJBM (1995) Bax mutations in cell lines derived from hematological malignancies. Leukemia 9:1828–1832

Meijerink JP, Mensink EJ, Wang K, Sedlak TW, Sloetjes AW, de Witte T et al (1998) Hematopoietic malignancies demonstrate loss-of-function mutations of BAX. Blood 91:2991–2997

Milella M, Estrov Z, Kornblau SM, Carter BZ, Konopleva M, Tari A et al (2002) Synergistic induction of apoptosis by simultaneous disruption of the Bcl-2 and MEK/MAPK pathways in acute myelogenous leukemia. Blood 99:3461–3464

Miyashita T, Reed JC (1995) Tumor suppressor p53 is a direct transcriptional activator of human Bax gene. Cell 80:293–299

Momoi T (2006) Conformational diseases and ER stress-mediated cell death: apoptotic cell death and autophagic cell death. Curr Mol Med 6(1):111–118

Mullican S, Zhang S, Konopleva M, Ruvolo V, Andreeff M, Milbrandt J et al (2007) Abrogation of nuclear receptors Nr4a3 and Nr4a1 leads to development of acute myeloid leukemia. Nat Med 13(6):730–735

Nishitoh H, Matsuzawa A, Tobiume K, Saegusa K, Takeda K, Inoue K et al (2002) ASK1 is essential for endoplasmic reticulum stress-induced neuronal cell death triggered by expanded polyglutamine repeats. Genes Dev 16(11):1345–1355

Oakes SA, Scorrano L, Opferman JT, Bassik MC, Nishino M, Pozzan T et al (2005) Proapoptotic BAX and BAK regulate the type 1 inositol trisphosphate receptor and calcium leak from the endoplasmic reticulum. Proc Natl Acad Sci U S A 102:105–110

Ogata M, Hino S, Saito A, Morikawa K, Kondo S, Kanemoto S et al (2006) Autophagy is activated for cell survival after endoplasmic reticulum stress. Mol Cell Biol 26(24):9220–9231

Ozcan L, Tabas I (2012) Role of endoplasmic reticulum stress in metabolic disease and other disorders. Annu Rev Med 63:317–328

Pattingre S, Tassa A, Qu X, Garuti R, Liang XH, Mizushima N et al (2005) Bcl-2 antiapoptotic proteins inhibit beclin 1-dependent autophagy. Cell 122:927–939

Pellecchia M, Reed JC (2004) Inhibition of anti-apoptotic Bcl-2 family proteins by natural polyphenols: new avenues for cancer chemoprevention and chemotherapy. Curr Pharm Des 10:1387–1398

Penninger JM, Kroemer G (2003) Mitochondria, AIF and caspases–rivaling for cell death execution. Nat Cell Biol 5:97–99

Putcha GV, Le S, Frank S, Besirli CG, Clark K, Chu B et al (2003) JNK-mediated BIM phosphorylation potentiates BAX-dependent apoptosis. Neuron 38(6):899–914

Puthalakath H, O'Reilly L, Gunn P, Lee L, Kelly P, Huntington N et al (2007) ER stress triggers apoptosis by activating BH3-only protein bim. Cell 129:1337–1349

Qian S, Wang W, Yang L, Huang HW (2008) Structure of transmembrane pore induced by Bax-derived peptide: evidence for lipidic pores. Proc Natl Acad Sci U S A 105(45):17379–17383

Rao RV, Bredesen DE (2004) Misfolded proteins, endoplasmic reticulum stress and neurodegeneration. Curr Opin Cell Biol 16:653–662
Reed JC (1996) Balancing cell life and death: bax, apoptosis, and breast cancer. J Clin Invest 97: 2403–2404
Reed JC (1997) Double identity for proteins of the Bcl-2 family. Nature 387:773–776
Reed JC (2005) Pellecchia M. apoptosis-based therapies for hematological malignancies. Blood 106:408–418
Reed JC (2006) Proapoptotic multidomain Bcl-2/Bax-family proteins: mechanisms, physiological roles, and therapeutic opportunities. Cell Death Differ 13(8):1378–1386
Reed JC (2008) Bcl-2-family proteins and hematologic malignancies: history and future prospects. Blood 111(7):3322–3330
Reed JC, Doctor KS, Godzik A (2004) The domains of apoptosis: a genomics perspective. Science STKE 239:RE9
Reimers K, Choi CY, Bucan V, Vogt PM (2008) The Bax Inhibitor-1 (BI-1) Family in Apoptosis and Tumorigenesis. Curr Mol Med 8(2):148–156
Rochitz C, Lohri A, Bacchi M, Schmidt M, Nagel S, Fopp M et al (1999) Axl expression is associated with adverse prognosis and with expression of Bcl-2 and CD34 in de novo acute myeloid leukemia (AML): results from a multicenter trial of the swiss group for clinical cancer research. Leukemia 13(9):1352–1358
Romero-Ramirez L, Cao H, Nelson D, Hammond E, Lee AH, Yoshida H et al (2004) XBP1 is essential for survival under hypoxic conditions and is required for tumor growth. Cancer Res 64(17):5943–5947
Rong YP, Aromolaran AS, Bultynck G, Zhong F, Li X, McColl K et al (2008) Targeting Bcl-2-IP3 Receptor Interaction to Reverse Bcl-2's Inhibition of Apoptotic Calcium Signals. Mol Cell 31(2):255–265
Sakaki K, Wu J, Kaufman RJ (2008) Protein Kinase C{theta} is required for autophagy in response to stress in the endoplasmic reticulum. J Biol Chem 283(22):15370–15380
Saleh M, Vaillancourt JP, Graham RK, Huyck M, Srinivasula SM, Alnemri ES et al (2004) Differential modulation of endotoxin responsiveness by human caspase-12 polymorphisms. Nature 429:75–79
Salvesen GS (2002) Caspases: opening the boxes and interpreting the arrows. Cell Death Differ 9:3–5
Salvesen GS, Duckett CS (2002) IAP proteins: blocking the road to death's door. Nat Rev Mol Cell Biol 3:401–410
Salvesen GS, Renatus M (2002) Apoptosome: the seven-spoked death machine. Develop Cell 2:256–257
Sano R, Hou Y-CC, Hedvat M, Correa RG, Shu CW, Krajewska M et al (2012) Endoplasmic Reticulum Protein BI-1 regulates Ca^{2+} -mediated bioenergetics to promote autophagy Genes Dev 26(10):1041–1054
Schaich M, Illmer T, Seitz G, Mohr B, Schakel U, Beck JF et al (2001) The prognostic value of Bcl-XL gene expression for remission induction is influenced by cytogenetics in adult acute myeloid leukemia. Haematologica 86:470–477
Shimizu S, Narita M, Tsujimoto Y (1999) Bcl-2 family proteins regulate the release of apoptogenic cytochrome *c* by the mitochondrial channel VDAC. Nature 399:483–487
Shuda M, Kondoh N, Imazeki N, Tanaka K, Okada T, Mori K et al (2003) Activation of the ATF6, XBP1 and grp78 genes in human hepatocellular carcinoma: a possible involvement of the ER stress pathway in hepatocarcinogenesis. J Hepatol 38(5):605–614
Srivastava RK, Srivastava AR, Korsmeyer SJ, Nesterova M, Cho-Chung YS, Longo D (1998) Involvement of microtubules in the regulation of Bcl-2 phosphorylation and apoptosis through cyclic AMP-dependent protein kinase. Mol Cell Biol 18:3509–3517
Stoka V, Turk B, Schendel SL, Kim TH, Cirman T, Snipas SJ et al (2001) Lysosomal protease pathways to apoptosis. Cleavage of bid, not pro-caspases, is the most likely route. J Biol Chem 276(5):3149–3157
Tschopp J, Irmler M, Thome M (1998) Inhibition of Fas death signals by FLIPs. Curr Opin Immunol 10:552–558

Upton JP, Austgen K, Nishino M, Coakley KM, Hagen A, Han D et al (2008) Caspase-2 cleavage of BID is a critical apoptotic signal downstream of endoplasmic reticulum stress. Mol Cell Biol 28(12):3943–3951

Vicencio JM, Ortiz C, Criollo A, Jones AW, Kepp O, Galluzzi L et al (2009) The inositol 1,4,5-trisphosphate receptor regulates autophagy through its interaction with Beclin 1. Cell Death Differ 16(7):1006–1017

Vogler M, Dinsdale D, Dyer MJ, Cohen GM (2009) Bcl-2 inhibitors: small molecules with a big impact on cancer therapy. Cell Death Differ 16(3):360–367

Wallach D, Varfolomeev EE, Malinin NL, Goltsev YV, Kovalenko AV, Boldin MP (1999) Tumor necrosis factor receptor and Fas signaling mechanisms. Ann Rev Immunol 17:331–367

Wang XZ, Ron D (1996) Stress-induced phosphorylation and activation of the transcription factor CHOP (GADD153) by p38 MAP Kinase. Science 272:1347–1349

Wei Y, Sinha S, Levine B (2008) Dual role of JNK1-mediated phosphorylation of Bcl-2 in autophagy and apoptosis regulation. Autophagy 4(7):949–951

Xu Q, Reed JC (1998) BAX inhibitor-1, a mammalian apoptosis suppressor identified by functional screening in yeast. Mol Cell 1:337–346

Xu C, Bailly-Maitre B, Reed JC (2005) Endoplamic reticulum stress: cell life and death decisions. J Clin Invest 115:2656–2664

Xu C, Xu W, Palmer AE, Reed JC (2008) BI-1 regulates endoplasmic reticulum Ca^{2+} homeostasis downstream of Bcl-2 family proteins. J Biol Chem 283(17):11477–11484

Yamaguchi R, Lartigue L, Perkins G, Scott RT, Dixit A, Kushnareva Y et al (2008) Opa1-mediated cristae opening is Bax/Bak and BH3 dependent, required for apoptosis, and independent of Bak oligomerization. Mol Cell 31(4):557–569

Yorimitsu T, Nair U, Yang Z, Klionsky DJ (2006) Endoplasmic reticulum stress triggers autophagy. J Biol Chem 281(40):30299–30304

Zalckvar E, Berissi H, Eisenstein M, Kimchi A (2009) Phosphorylation of Beclin 1 by DAP-kinase promotes autophagy by weakening its interactions with Bcl-2 and Bcl-XL. Autophagy 5(5):720–722

Zhao R, Follows GA, Beer PA, Scott LM, Huntly BJ, Green AR et al (2008) Inhibition of the Bcl-xL deamidation pathway in myeloproliferative disorders. N Engl J Med 359(26):2778–2789

Zhong Q, Gao W, Du F, Wang X (2005) Mule/ARF-BP1, a BH3-only E3 ubiquitin ligase, catalyzes the polyubiquitination of Mcl-1 and regulates apoptosis. Cell 121:1085–1095

Zou W, Yue P, Khuri FR, Sun SY (2008) Coupling of endoplasmic reticulum stress to CDDO-Me-induced up-regulation of death receptor 5 via a CHOP-dependent mechanism involving JNK activation. Cancer Res 68(18):7484–7492

Chapter 4
Bcl-2 Family: Translational Aspects

Prithviraj Bose and Steven Grant

Abstract The Bcl-2 family of antiapoptotic and proapoptotic proteins serve as key regulators of the mitochondrial pathway of apoptosis. Multiple signals from a variety of cell death stimuli converge upon mitochondria to trigger the intrinsic apoptotic cascade. Bcl-2 family proteins are intimately related to prognosis and therapeutic resistance in acute myeloid leukemia (AML), making them rational targets for drug development. Also, directly targeting Bcl-2 family proteins circumvents many of the problems associated with targeting upstream molecules. The Bcl-2 antisense oligonucleotide oblimersen failed to live up to its initial promise in large phase III trials. The discovery of ABT-737, a novel, small-molecule inhibitor of specific protein–protein interactions, gave a much-needed impetus to the field of "BH3 mimetic" research. The demonstration that Mcl-1, an antiapoptotic Bcl-2 family protein that is not inhibited by ABT-737 or its analogs, is of crucial importance in AML, underscores the need for rational drug combinations that simultaneously target multiple arms of the apoptotic regulatory machinery.

Keywords Apoptosis · Bcl-2 · Mcl-1 · AML · Oblimersen · ABT-737 · ABT-199 · Navitoclax · Obatoclax · GX15-070

4.1 Introduction

The term "apoptosis" was first coined by Andrew Wyllie in 1972 to describe a morphologically distinct form of cell death in which the cell actively participates in its own demise (Irvine et al. 2002). Bcl-2 (B-cell lymphoma-2) was the first antiapoptotic gene discovered, a milestone that effectively launched a new era in cell death research (Reed 2008). In this chapter, we begin with an overview of the apoptotic

P. Bose (✉)
Massey Cancer Center, Virginia Commonwealth University, 1201 E Marshall St, MMEC −11th floor, Room 213, P.O. Box 980070, Richmond, VA 23298, USA
e-mail: pbose@mcvh-vcu.edu

S. Grant
Division of Hematology and Oncology, Massey Cancer Center, Virginia Commonwealth University, 401 College Street, Box 980035, Richmond, VA 23298, USA
e-mail: stgrant@vcu.edu

M. Andreeff (ed.), *Targeted Therapy of Acute Myeloid Leukemia*, Current Cancer Research, DOI 10.1007/978-1-4939-1393-0_4

pathways. The functions of the different members of the Bcl-2 family of apoptosis regulators are then described. Next, we discuss the profound importance of the Bcl-2 family of proteins in acute myeloid leukemia (AML) pathogenesis, prognosis, and response to therapy. Different strategies that have been employed to target therapeutically the Bcl-2 family in the context of AML represent the focus of the remainder of the chapter. In these sections, we discuss antisense oligonucleotides, peptide and nonpeptide BH3 mimetics, and other strategies, as well as rational combinations of targeted drugs that are synergistically lethal to AML cells. Finally, a separate section is devoted to Mcl-1 (myeloid cell leukemia-1), given the growing importance and unique features of this antiapoptotic protein.

4.2 Overview of the Apoptotic Pathways

Apoptosis, or programmed cell death, is central to the development and homeostasis of the hematopoietic system. AML cells invariably have abnormalities in one or more apoptotic pathways, determining a survival advantage for these cells over their normal counterparts. Furthermore, abnormalities in the apoptotic response also play a role in the development of drug resistance by leukemic cells (Testa and Riccioni 2007). Apoptosis is morphologically characterized by cell shrinkage, membrane blebbing, and nuclear condensation (Schimmer 2007) Cells which die by apoptosis are dismantled in an orderly fashion. The cell contents are packaged up in membrane-bound vesicles called apoptotic bodies which can then be removed by phagocytosis. The proteolytic enzymes within the apoptotic cell are not released and cannot provoke inflammation, in contrast to the situation with necrotic cell death (Irvine et al. 2002). Most, if not all, anticancer drugs kill malignant cells through induction of apoptosis, and most tumor cells do retain their apoptotic machinery (Davids and Letai 2012). Defects in apoptotic pathways thus represent an ideal target for therapeutic intervention (Davids and Letai 2012).

At a molecular level, apoptosis is caused by the activation of caspases, a family of intracellular cysteine proteases that cleave critical intracellular proteins upon activation and effectively demolish the cell (Irvine et al. 2002; Testa and Riccioni 2007; Schimmer 2007). There are two pathways that induce *initiator* caspase (caspases 2, 8, 9, and 10) activation, which ultimately leads to the activation of *effector*-caspases (caspases 3, 6, and 7). The *extrinsic,* or death receptor pathway is initiated by engagement of death receptors of the tumor necrosis factor (TNF) superfamily on the cell membrane by their respective ligands, such as Fas ligand (FasL) and TNF or TNF-related apoptosis-inducing ligand (TRAIL). Upon ligand binding, the receptors oligomerize and recruit adaptor proteins such as Fas-associated protein with death domain (Fadd/Mort-1) and caspase-8, forming a death-inducing signaling complex (DISC) that leads to the cascade of caspase activation. Cellular FLICE-like inhibitory protein (c-FLIP) is an endogenous inhibitor of this pathway. Strategies to promote apoptosis in AML cells via the death receptor pathway have included the use of TRAIL, TRAIL receptor agonistic antibodies, and the synthetic triterpenoid 2-cyano-3, 12-dioxooleana-1, 9-dien-28-oic acid (CDDO), that downregulates c-FLIP (Testa and Riccioni 2007). Such strategies are discussed in detail elsewhere in this volume.

The *intrinsic,* or mitochondrial pathway of apoptosis is triggered by various cellular stresses, such as growth factor deprivation, DNA or microtubule damage, hypoxia, etc., and is the mechanism by which apoptosis is induced by cytotoxic chemotherapy (Davids and Letai 2012). The seminal event in this pathway is mitochondrial outer membrane permeabilization (MOMP), which is controlled by the Bcl-2 family of proteins and results from the formation of pores in the mitochondrial outer membrane (Davids and Letai 2012). MOMP constitutes a "point of no return" in the mitochondrial apoptotic process and leads to the release of cytochrome *c* and other proapoptotic molecules such as second mitochondria-derived activator of caspases/direct inhibitor of apoptosis (IAP)-binding protein with low pI (SMAC/DIABLO) and HtrA2/Omi. An "apoptosome" is then formed in the cytosol by cytochrome *c,* apoptotic protease-activating factor-1 (APAF-1) and procaspase-9. The latter is activated in the presence of ATP, and triggers massive activation of downstream caspases.

A link between the extrinsic and intrinsic pathways is provided by the cleavage of the proapoptotic Bcl-2 family "BH3-only" (see following section) protein Bid by caspase-8 to produce a truncated molecule, tBid, that is able to directly activate the "multi-domain" (see following section) proapoptotic Bcl-2 family molecules Bcl-2 associated protein X (Bax) and Bcl-2 antagonist/killer (Bak). In addition, the multi-domain proapoptotic protein Bak is generally held in check (tethered) by both Mcl-1 and Bcl-X_L. Consequently, release of Bak from both of these antiapoptotic proteins is required for maximal induction of apoptosis (Willis et al. 2005). Mitochondrial translocation of Bax (normally a cytosolic protein) and oligomerization of Bax and Bak lead to MOMP(Pope 2007).

The IAPs are cytoplasmic proteins that inhibit the action of caspases. The most important member of this family, X-linked inhibitor of apoptosis (XIAP), potently inhibits active caspases-9, -3 and -7, thus blocking apoptosis triggered by multiple caspase activation pathways. Inhibition of this downstream blockade of apoptosis by SMAC-mimetics, small-molecule inhibitors of XIAP and specific antisense oligonucleotides represents another promising strategy in AML (Testa and Riccioni 2007) that is addressed in other chapters of this volume.

4.3 The Bcl-2 Family

The *BCL-2* gene was discovered in 1985 in the context of the t (14;18) translocation commonly observed in follicular lymphoma. Its oncogenic potential was first demonstrated in 1988, and shortly thereafter, its antiapoptotic activity was discovered. Bcl-2 is the founding member of a family of proteins that controls an evolutionarily conserved pathway for regulating programmed cell death (Reed 2008). The Bcl-2 family contains at least 25 members localized to mitochondria, smooth endoplasmic reticulum, and perinuclear membranes in hematopoietic cells (Kang and Reynolds 2009). The antiapoptotic members are Bcl-2, Bcl-X_L, Bcl-w, Mcl-1, Bfl-1/A1, and Bcl-B; these share sequence homology in four α-helical Bcl-2 homology (BH) domains, BH1-4 (with the exception of Mcl-1, which only contains three BH domains). The "multi-domain" proapoptotic members are Bax, Bak and Bcl-2-related ovarian killer (Bok), which share homology in domains BH1-3 (Labi et al. 2008). The other proapoptotic members share only the 9–16 amino acid BH3 domain and are termed "BH3-only" proteins; these include Bim (Bcl-2 interacting mediator of cell death), Bid (BH3-interacting domain death agonist), Bad (Bcl-2 antagonist of cell death), Bik (Bcl-2 interacting killer), Hrk (activator of apoptosis harakiri), Bmf (Bcl-2 modifying factor), Puma (p53-upregulated modulator of apoptosis), and Noxa. The BH3 domain forms an amphipathic α-helix of about 24 residues that enables binding to a hydrophobic groove on the surface of the antiapoptotic Bcl-2 molecules. The multi-domain proteins Bax and Bak are essential for the proapoptotic function of the BH3-only proteins (Labi et al. 2008). How Bcl-2 family proteins interact to induce MOMP remains controversial (Kang and Reynolds 2009). Two models have been proposed to explain the activation of Bax and Bak to form oligomers in the mitochondrial membranes upon receipt of a death signal. In the *direct activation model,* apart from binding to antiapoptotic Bcl-2 proteins, the "activator" BH3-only proteins (Bim, tBid, and possibly Puma) are also able to directly activate Bax and Bak, while the remaining "sensitizer" BH3-only proteins cannot do so (Davids and Letai 2012). In this model, the latter inhibit the antiapoptotic Bcl-2 proteins from engaging with activator proteins or with Bax/Bak, thereby sequestering activators and Bax/Bak (Davids and Letai 2012). The *indirect activation model* challenges the notion of direct binding to and activation of Bax/Bak by any BH3-only proteins (Kang and Reynolds 2009; Labi et al. 2008). Indeed, evidence for direct binding of Bax and Bak by BH3-only proteins remains scant (Adams and Cory 2007).

Individual BH3-only proteins differ markedly in their ability to bind to different antiapoptotic Bcl-2 relatives. The so-called "promiscuous" BH3-only proteins Bim and Puma bind to all antiapoptotic Bcl-2 family members with high affinity. Others are more selective. tBid binds with high affinity to Bcl-X_L, Bcl-w, Mcl-1, and Bfl-1/A1, but weakly to Bcl-2. Bik and Hrk bind to Bcl-X_L, Bcl-w, and Mcl-1, but poorly to Bcl-2 or Bfl-1/A1. Bad and Bmf bind strongly with Bcl-2, Bcl-X_L, and Bcl-w, and Noxa binds only with high affinity to Mcl-1 and Bfl-1/A1. Thus, BH3-only proteins with different selective binding patterns (e.g., Bad plus Noxa) are synergized in apoptosis induction, an important finding from co-transfection assays of great translational relevance (Kuroda and Taniwaki 2009).

The regulation of the Bcl-2 family of proteins by posttranslational modification adds an additional layer of complexity (Reed 2008). Such modifications include phosphorylation, proteolysis, ubiquitination, myristoylation, and deamidation. Phosphorylation of proapoptotic Bcl-2 family proteins can either neutralize (e.g., Bad) or potentiate (e.g., Bim) their functions. In addition, phosphorylation of proapoptotic proteins can play a major role in their disposition and interactions with other (antiapoptotic) proteins. For example, phosphorylation of Bim by the mitogen-activated protein kinase kinase (MEK) 1/2-extracellular signal-regulated protein (ERK) 1/2 pathway promotes Bim degradation as well as alterations in its association with Mcl-1, accompanied by enhanced cell survival (Ewings et al. 2007). On the other hand, the effects of phosphorylation on Bcl-2 function are variable and controversial. Different studies have found conflicting results even when examining phosphorylation at the same amino acid residue.

Cancer cells are subject to diverse stresses and are thus "primed" for death (Chonghaile and Letai 2008), depending upon antiapoptotic protein(s) to stay alive. Three classes of "road blocks to apoptosis" have been proposed by the proponents of the *direct activation model* discussed above: loss of "activator" BH3-only proteins (class A block), loss or inactivation of the effectors of mitochondrial apoptosis, Bax/Bak (class B block), and increased expression of an antiapoptotic protein such as Bcl-2 or Mcl-1 (class C block). BH3 profiling is a technique that can characterize the class of apoptotic block a cancer cell has evolved and can be useful in identifying and understanding cellular dependence on individual antiapoptotic proteins (Del and Letai 2012). While thus far validated only in lymphoma cell lines, this technique could be helpful in personalizing therapy for AML as multiple BH3 mimetics enter the clinical arena.

The functions of Bcl-2 family proteins extend beyond the regulation of mitochondrial apoptosis. Thus, Bax and Bak are required for the function of IRE1α, an effector of the unfolded protein response, which tumor cells often rely on for survival in the face of endoplasmic reticulum (ER) stress. IRE1α binds to the same pocket in Bax and Bak that mediates binding to BH3-only proteins. Also, ER stress activates Bim by inducing its transcription and preventing its degradation. Finally, Bcl-2 downregulation causes autophagy in a caspase-independent manner in human HL-60 leukemia cells (Saeki et al. 2000). The evolutionarily conserved autophagy inducer Beclin-1 is a BH3-only protein that binds to Bcl-2 and Bcl-X_L, similar to proapoptotic proteins. BH3-only proteins can displace Beclin-1 from Bcl-2 and Bcl-X_L, and BH3 mimetics activate several proautophagic pathways (Malik et al. 2011). However, while therapy can manipulate autophagy, whether its promotion or inhibition will be more desirable remains to be seen. These concepts have been reviewed (Adams and Cory 2007).

Finally, it should be recognized that in addition to classic apoptosis, transformed cells, including leukemia cells, may experience multiple forms of nonapoptotic cell death, including programmed necrosis (necroptosis), autophagy, mitotic catastrophe, senescence, and paraptosis, among others (Galluzzi et al. 2012). Furthermore, these processes may also be governed by complex regulatory mechanisms as observed in the case of apoptosis (Weinlich et al. 2011). Given the cross talk that exists between the various forms of cell death, it is highly likely that agents that promote apoptosis will modulate these processes as well. However, for simplicity, the present chapter will focus primarily on more classic forms of apoptotic cell death.

4.4 Importance of the Bcl-2 Family of Proteins in AML

In AML, multiple studies have shown that high expression of Bcl-2 is associated with CD34-positivity (Stijn and Kok 2003; Venditti et al. 2004; Lauria et al. 1997), a low complete remission (CR) rate after intensive chemotherapy (Lauria et al. 1997); Campos et al. 1993) and poor survival (Campos et al. 1993), while expression of the proapoptotic protein Bcl-X_S was predictive of a high remission rate (Campos et al. 1999) and higher relapse-free (RFS) and overall survival (OS) (Yamaguchi et al. 2002). In AML patients with poor-risk cytogenetics (e.g., monosomy 5 or 7), Bcl-2 is the predominant member expressed, and the ratio of antiapoptotic Bcl-X_L to proapoptotic Bcl-X_S is higher than in patients with favorable cytogenetics (e.g., inv 16, t(8;21) abnormality) (Deng et al. 1998). The Bcl-X_L/Bcl-X_S ratio is also higher in quiescent CD34$^+$ AML cells than in quiescent CD34$^+$ normal cells, and Bcl-X_L and Bcl-2 are overexpressed in quiescent compared to proliferating CD34$^+$ AML cells (Konopleva et al. 2002b), possibly explaining the greater drug resistance of this population. While loss or mutation of p53 is not common in AML, high levels of Bcl-2 suppress wild-type p53 through phosphorylation (Irish et al. 2007).

High Bax expression has been shown to be a good prognostic indicator in AML in some studies (Ong et al. 2000), while others have shown no correlation between Bax expression and response to induction chemotherapy or survival (Kornblau et al. 2000). Fms-like tyrosine kinase 3 (FLT3) ligand exerts its antiapoptotic effect on AML cells by phosphorylating Bad (Minami et al. 2003) and decreasing the expression of Bax (Lisovsky et al.1996). Overexpression of Bcl-X_L is a critical mediator of drug resistance in *FLT3*-mutated AML (Minami et al. 2003; Bagrintseva et al. 2005; Bagrintseva et al. 2005). Through multiple signal transduction pathways, *FLT3*-internal tandem duplication (ITD) mutations prevent apoptosis of AML cells via suppression of Bim (Scheijen et al. 2004) and phosphorylation of Bad (Kim et al. 2006; Yang et al. 2005). Conversely, inhibition of phosphatidylinositol 3-kinase (PI3K)/Akt dephosphorylates Bad and promotes apoptosis in AML (Zhao et al. 2004), and FLT3 tyrosine kinase inhibitors (TKIs) exert their growth-inhibitory and proapototic effects through upregulation of Bim (Nordigarden et al. 2009).

In other studies, the amount of spontaneous apoptosis detected by the Bax/Bcl-2 ratio has been found to be predictive of outcome in AML in terms of CR rates, overall and disease-free survival (Del Poeta et al. 2003). Low Bax/Bcl-2 ratios correlated with higher CD34-positivity and poor-risk cytogenetics. The favorable molecular profile *FLT3-ITD* negative/*NPM1* (nucleophosmin)-mutated is also significantly associated with a high (e.g.,>0.35) Bax/Bcl-2 ratio (Del Poeta and Lavorgna 2010). In a study of elderly patients with AML, the Bcl-2/Bax ratio was significantly higher in CD34$^+$ blasts compared to CD34$^-$ blasts, providing a potential explanation for the greater resistance to chemotherapy observed in elderly AML patients, who often have a CD34$^+$ homogeneous blast cell population (Suarez et al. 2001, 2005). Last but not least, minimal residual disease in AML is predicted by an apoptosis-resistant profile (Bcl-2, Bcl-2/Bax and a combination thereof, designated antiapoptotic index) at diagnosis (van Stijn et al. 2005).

An interesting observation is that higher Bcl-2 expression may portend a poor prognosis only in patients with favorable or intermediate-risk cytogenetics, and actually be a good prognostic factor in patients with poor-risk cytogenetics (Kornblau et al. 1999). Similarly, while in AML patients ≤ 60 years of age with intermediate-risk cytogenetics, Bcl-X_L expression was an independent negative prognostic factor for response to induction therapy, no prognostic impact of Bcl-X_L expression on treatment response was seen among patients with high-risk cytogenetic features (Schaich et al. 2001). However, the prognostic impact of Bcl-2 expression in AML is modulated by its dimerization with Bax and phosphorylation by protein kinase C-alpha (PKC-α), and the ratios of Bcl-2 to Bax and PKC-α phosphorylated Bcl-2 to Bax were highly prognostic in one study (Kornblau et al. 2000). Clearly, the interplay of different Bcl-2 family members and the various posttranslational modifications they can undergo in predicting outcome in AML is complex, and gene-expression profiling studies have shown that high expression of *both* pro and antiapoptotic genes predict poor overall survival in AML, which postulates a mechanism of activation of the apoptosis pathway as a whole (Hess et al. 2007).

Inhibition of the intrinsic but not the extrinsic apoptosis pathway accelerates and drives *myc*-driven tumorigenesis towards AML (Hogstrand et al. 2012). In AML associated with the t(8;21) abnormality that gives rise to the AML1–ETO fusion product, Bcl-2 may play an etiological role in leukemogenesis (Chou et al. 2011). Indeed, deregulated apoptotic signaling may explain the heterogeneity within the group of patients with the "core-binding factor" AML subtypes that have traditionally been felt to carry a favorable prognosis (Luck et al. 2011). Bcl-2 is particularly abundant in leukemias with the t (4;11) abnormality, that leads to the mixed lineage leukemia (*MLL*) gene rearrangement and a poor prognosis (Robinson et al. 2008). Among patients with myelodysplastic syndrome (MDS), Bcl-2 expression by early myeloid precursors correlated with initial MDS stage, progressed over time independent of granulocyte colony-stimulating factor (G-CSF) therapy, and was associated with evolution to AML (Davis and Greenberg 1998). While excessive apoptosis is the hallmark of early MDS, leukemic transformation arises through inhibition of apoptosis rather than excessive cell growth (Parker et al. 2000). $CD34^+$ myeloid cells from patients with AML and high-risk MDS display significantly higher expression of Bcl-2 compared to those with low-risk MDS and normal controls (Suarez et al. 2004).

Inhibited apoptosis is associated with pan-resistance to antileukemic chemotherapy both in vitro and in vivo (Smith et al. 1998). Marked potentiation of chemotherapy-induced AML cell death by downregulation of Bcl-2 using antisense oligonucleotides(Campos et al. 1994; Keith et al. 1995; Konopleva et al. 2000) or all-*trans* retinoic acid (ATRA)(Konopleva et al. 2002a; Andreeff et al. 1999) has been demonstrated, both in vitro and in vivo. Additionally, it has been shown that Bcl-X_L expression can confer a multidrug resistant phenotype (Minn et al. 1995). Perhaps most significantly, a recent study has elegantly demonstrated that Mcl-1, the expression of which is known to be elevated at the time of AML relapse (Kaufmann et al. 1998), may be more critical than Bcl-2, Bcl-X_L, or Bcl-w for the development and maintenance of AML (Glaser et al. 2012). Mcl-1 haploinsufficiency protects mice from *myc*-induced AML (Xiang et al. 2010), and *FLT3-ITD* upregulates Mcl-1 to promote survival of AML stem cells (Yoshimoto et al. 2009).

4.5 Targeting Bcl-2 Family Members for Therapy of AML

Strategies to overcome the cytoprotective effects of Bcl-2 and related antiapoptotic proteins in AML include: (1) interfering with gene transcription, (2) inducing mRNA degradation with antisense oligonucleotides, (3) directly inhibiting the proteins with small-molecule drugs, (4) activating endogenous antagonists of Bcl-2

Table 4.1 Clinically relevant Bcl-2 antagonists in development for the treatment of AML

Agent	Manufacturer	Drug class (for mechanism of action, see text)	Bcl-2 family member(s) targeted	Phase of clinical testing	Phase of testing in AML	References (AML)
Oblimersen (G3139, Genasense®)	Genta (Berkeley Heights, NJ)	Antisense oligo-nucle-otide against Bcl-2	Bcl-2	III	III	Marcucci et al. 2007; Konopleva et al. 2000
AT-101 (levo-gossy-pol)	Ascenta therapeutics (Malvern, PA)	Small-molecule BH3-mimetic (aromatic phenol), may also directly upregulate Noxa and Puma)	Bcl-2, Bcl-X_L, Mcl-1, Bcl-w	II	Preclinical	Jarvis et al. 1994
Obatoclax (GX15-070)	Teva Phar-maceutical industries limited (Israel)	Small-molecule BH3-mimetic (synthetic prodiginine)	Bcl-2, Bcl-X_L, Mcl-1, Bcl-w, Bfl-1/A1, Bcl-B	II	II	Konopleva et al. 2008
Navitoclax (ABT-263)	Abbvie, Inc. (North Chi-cago, IL)	Small-molecule BH3-mimetic	Bcl-2, Bcl-X_L, Bcl-w	II	Preclinical (ABT-737)	Konopleva et al. 2006
ABT-199	AbbVie Inc. (North Chi-cago, IL)	Small-molecule BH3-mimetic	Bcl-2	I/II	Preclinical	Pan et al. 2014

(Reed and Pellecchia 2005), and (5) immunologic approaches. These are covered in the following subsections. Additional strategies specifically focusing on Mcl-1 are discussed separately. Table 4.1. summarizes from an AML perspective the Bcl-2 family antagonists that have entered the clinic.

4.5.1 Drugs Regulating Bcl-2 Family Gene Expression

Retinoids such as ATRA (Andreeff et al.1999; Bradbury et al. 1996; Pisani et al. 1997; Zheng et al. 1999) and 9-*cis* retinoic acid (Elstner et al. 1996) downregulate Bcl-2 mRNA and protein in AML cells, and synergism, sometimes sequence-dependent, with Ara-C (Konopleva et al. 2002b; Andreeff et al.1999), vitamin D analogs (Elstner et al. 1996) and inhibitors of MEK/ERK (Milella et al. 2007) has been shown in vitro. In a phase II trial of ATRA plus Ara-C-based timed sequential therapy in AML, the CR rate was 60% and the median disease-free survival (DFS) for patients achieving a CR was 11.2 months (32% at 3-plus years). Outcomes were considerably better for younger patients (<60 years), those with de novo AML and nonadverse cytogenetics (Lisovsky et al. 1996). Unfortunately, a randomized phase II trial evaluating the addition of ATRA, G-CSF, or both to a backbone of fludarabine, Ara-C, and idarubicin (FAI) in patients with newly diagnosed AML or high-risk MDS and one of several poor prognostic features failed to show any benefit to this approach (Luck et al. 2011). Inorganic (arsenic trioxide) and organic (melarsoprol) arsenic compounds downregulate Bcl-2 and induce apoptosis in AML cell lines in a *PML* and *PML-RAR*-independent manner (Malik et al. 2011). Histone deacetylase inhibitors (HDACIs) also reduce the expression of *BCL-2, BCL-X* and *MCL-1* at a transcriptional level in some leukemia cell lines and primary cultured leukemia cells (Kojima et al. 2009). The small molecule harmine-derivative compound 6 may promote cell death by inhibiting the transcription of Bcl-2 and other genes (Bolanos-Meade et al. 2003). The nucleoside analog ARC, a transcriptional cyclin-dependent kinase inhibitor (CDKI), triggers apoptosis in AML cells via downregulation of Mcl-1 (Mandelin and Pope 2007).

4.5.2 Antisense Approaches Against Bcl-2 mRNA

In 1990, the first demonstrations that knocking down Bcl-2 expression induces leukemia cell death and inhibits leukemia growth in vivo were made, using antisense methods (Marcucci et al. 2003). Subsequent studies (Kaufmann et al. 1998; Keith et al. 1995; Kessel et al. 2007) showed that Bcl-2 antisense oligodeoxynucleotides decreased Bcl-2 protein expression and induced apoptosis in multiple AML cell lines, even in the presence of other antiapoptotic proteins (Kessel et al. 2007). Additionally, enhancement of daunorubicin and Ara-C-induced cytotoxicity was seen. Antisense Bcl-2 (G3139, oblimersen sodium, Genasense, Genta, Berkeley Heights, NJ) is an 18-mer phosphorothioate oligonucleotide designed to bind to the first six codons of human Bcl-2 mRNA (Marcucci et al. 2005), leading to its degradation. Several antisense and nonantisense mechanisms of action have been proposed for this agent. These include the induction of both apoptotic and autophagic cell death via Bcl-2 downregulation, and an immunostimulatory effect of its two CpG motifs. In a phase I study in 20 patients with relapsed or refractory acute leukemia (Kojima et al. 2010), oblimersen was administered in combination with fludarabine, Ara-C, and G-CSF (FLAG). The combination was found to be safe, and 9 (45 %) patients achieved either a CR or CR with incomplete count recovery (CRi). Bcl-2 mRNA levels were downregulated in 9 of 12 (75 %) evaluable patients. The safety of this drug in combination with daunorubicin and Ara-C was subsequently established in a phase I trial in 29 untreated elderly ($\geq$ 60 years) patients with AML (Konopleva et al. 2000). In all, 14 patients achieved CR. Bcl-2 mRNA levels significantly decreased in responders. Considering that Bcl-2 downregulation is likely to depend on achievable intracellular concentrations of oblimersen rather than on plasma concentrations, it has been shown that measurable intracellular levels of the drug are achievable in vivo in AML patients (Marcucci et al. 2007). A phase II trial of oblimersen in combination with gemtuzumab ozogamycin in elderly patients ($\geq$60 years) with AML then followed (Konopleva et al. 2002a). Of 48, 12 (25 %) patients achieved a CR or CR with incomplete platelet recovery (CRp), and no additional serious toxicity was observed. These promising results led to a randomized phase III trial of intensive induction and consolidation chemotherapy with or without oblimersen being conducted by the Cancer and Leukemia Group B (CALGB) in 503 elderly (>60 years) patients with AML (Marcucci et al. 2007). Although no differences in toxicity were noted between the two arms, an interim futility analysis showed no differences in CR rates, OS, DFS, or event-free survival (EFS). The lack of benefit in this trial, along with the failure of phase III trials of oblimersen in multiple myeloma and melanoma to meet their primary end points, has led to waning of the initial enthusiasm for this drug. Of note, however, a randomized phase III trial in chronic lymphocytic leukemia (CLL) revealed a significant survival benefit at 5 years for oblimersen in combination with fludarabine and cyclophosphamide (FC) in responders and those with fludarabine-sensitive disease (O'Brien et al. 2009). Several other antisense molecules targeting Bcl-2, Bcl-X_L, or both are currently being studied preclinically in different tumor types.

4.5.3 Drugs Inhibiting Bcl-2 and Related Antiapoptotic Proteins (BH3 mimetics)

Most research in recent years has concentrated on the identification/discovery of drugs that can mimic the action of the BH3-only proteins of the Bcl-2 family. These drugs bind to Bcl-2 and related antiapoptotic proteins at the same binding site as the BH3-only proapoptotic proteins and Bax/Bak, thus freeing the apoptosis activators and effectors to trigger apoptosis.

4.5.3.1 Peptide-Based Approaches

BH3 peptides longer than 14 amino acids can retain an α-helical structure, and in cell-free assays, bind to the hydrophobic grooves of antiapoptotic proteins, disrupting complexes formed between pro and antiapoptotic Bcl-2 family proteins. Similar to BH3-only proteins, BH3 peptides can activate Bax and Bak and induce MOMP. However, poor cell permeability, bioavailability, solubility, and metabolic stability limit their clinical application (Zhang et al. 2007b). Several modifications can circumvent some of these problems. For example, a Bid-derived BH3 peptide can be chemically "stapled" to maintain its α-helical conformation. Such "stapled peptides" are helical, protease-resistant, and cell-permeable molecules that bind with increased affinity to multi-domain Bcl-2 member pockets and induce apoptosis in leukemia cells in vitro and in vivo (Walensky et al. 2004). cpm-1285 is a cell-permeable peptide designed by chemically attaching a fatty acid to a peptide derived from the proapoptotic protein Bad that is able to enter human HL-60 leukemia cells, bind Bcl-2, and induce apoptosis in vitro (Wang et al. 2000b). Furthermore, cpm-1285 had in vivo activity in slowing human myeloid leukemia growth in severe combined immunodeficient (SCID) mice. Similarly, the α-helical conformation of a Bak BH3 peptide can be stabilized using unnatural amino acid residues (Zhang et al. 2007c). A novel Bim-derived peptide Bcl-X_L inhibitor, 072RB, induced apoptosis and dose-dependent growth inhibition of AML cell lines and patient-derived AML cells (Ponassi et al. 2008). Though promising, none of these approaches has proven effective clinically, and the focus has shifted towards the small-molecule BH3 mimetics.

4.5.3.2 HA14-1 and Other Small-Molecule Inhibitors

The first small-molecule, nonpeptide ligand of a Bcl-2 surface pocket, HA14-1, was discovered using an in silico screening strategy based on the predicted structure of the Bcl-2 protein (Wang et al. 2000a). HA14-1 effectively induced apoptosis of human myeloid leukemia cell lines, especially those overexpressing Bcl-2 protein, and their multidrug resistant P-glycoprotein-expressing counterparts (Wang et al. 2000a; Oliver et al. 2007). The drug efficiently induced cell death in a broad spectrum of patient-derived AML blasts, including chemotherapy-resistant ones (Oliver et al. 2007) Not surprisingly, inhibition of Bcl-2 and Bcl-X_L function by HA14-1 also induces autophagy in leukemia cells (Kessel and Reiners 2007), presumably by disrupting the inhibitory effects of Bcl-2 on Beclin-1 (Pattingre et al. 2005). However, the lethal effects of HA14-1 on leukemia cells appear not to be restricted to its inhibition of Bcl-2 family antiapoptotic proteins. For example, HA14-1 can uncouple mitochondrial respiration, thereby sensitizing leukemic cells to TRAIL-induced apoptosis (Hao et al. 2004). Similarly sHA14-1, a stable analog of HA14-1, kills human leukemia cells by targeting both Bcl-2 antiapoptotic and smooth endoplasmic reticulum Ca^{2+}-ATPase (SERCA) proteins (Hermanson et al. 2009). While sHA14-1, unlike HA14-1, does not lead to the production of reactive oxygen species (ROS) by spontaneous decomposition, its induction of apoptosis in human leukemia cell lines might involve both the intrinsic and extrinsic pathways (Tian et al. 2008). 4-(3-methoxy-phenylsulfannyl)-7-nitro-benzofurazan-3-oxide (MNB) is another novel small-molecule Bcl-2 inhibitor that induces apoptosis in human HL-60 leukemia cell lines and xenograft models, as well as in primary AML cells (Zhang et al. 2007c). However, none of these agents have been tested clinically. Other inhibitors of Bcl-2 antiapoptotic proteins include the cell-permeable BH3 domain analogs BH3I-1 and BH3I-2, terphenyl derivatives, NSC365400 (compound 6), and the natural compounds antimycin A, chelerythrine (better known as a PKC inhibitor), and the theaflavins and epigallechatechins in black and green tea (Kang and Reynolds 2009; Kuroda and Taniwaki 2009; Zhang et al. 2007b). Low binding affinities for Bcl-2 and Bcl-X_L, and unfavorable pharmacological properties make these drugs poor clinical candidates (Zhang et al. 2007b).

4.5.3.3 Gossypol and Analogs

The aromatic phenol gossypol, extracted from the cotton plant (*Gossypium* species) and the tropical tree *Thespesia populnea,* has been extensively evaluated as a reversible male contraceptive agent. However, the compound exhibits a wide range of antitumor activity and is a powerful antioxidant (Dodou et al. 2005). The antitumor activity of gossypol can be attributed to a plethora of biological actions, including inhibition of a large number of enzymes, cell-cycle regulatory and Bcl-2 antiapoptotic proteins, and interaction with cellular and mitochondrial membranes (Dodou et al. 2005). Natural gossypol is a racemic mixture, and the (−)-enantiomer has been found to be more potent in most biological systems (Kang and Reynolds 2009; Dodou et al. 2005). Gossypol is a BH3-mimetic agent that inhibits Bcl-2, Bcl-X_L, and Mcl-1 at submicromolar concentrations, while TW-37, a rationally designed gossypol derivative based on computer modeling of the BH3 domain of Bim, binds to these antiapoptotic proteins with even higher affinity (Kang and Reynolds 2009; Zhang et al. 2007b). Several phase I and II clinical trials of oral gossypol and levo gossypol, AT-101 (Ascenta Therapeutics, Malvern, PA) have been conducted in patients with a variety of solid tumors (Kang and Reynolds 2009; Dodou et al. 2005), both as single agents and in combination with cytotoxic drugs. In general, gossypol and AT-101 were well tolerated, the reported toxicities being primarily gastrointestinal, but disappointing in terms of efficacy (Dodou et al. 2005; Khaw et al. 2011). Similar findings were reported in early phase trials in CLL and follicular lymphoma (in combination with rituxmab) (Khaw et al. 2011). Apogossypol, a less reactive derivative of gossypol in which two aldehydes have been eliminated, may have superior efficacy and reduced toxicity (Kitada et al. 2008). In primary AML cells, levo-gossypol induces apoptosis via inhibition of NF-κB and downregulation of Mcl-1(Pepper et al. 2009). No clinical trials of these drugs in patients with AML have been conducted. Similarly, the preclinical experience with these molecules is largely restricted to other tumor types. The pan-Bcl-2 inhibitor sabutoclax is an apogossypol derivative that has primarily been studied in solid tumors to date but may sensitize bone marrow-resident chronic myeloid leukemia (CML) stem cells to TKIs at doses that spare normal progenitors (Goff et al. 2013) A pan-Bcl-2 inhibitor renders bone-marrow-resident human leukemia stem cells sensitive to tyrosine kinase inhibition. *Cell Stem Cell* 12:316–328). There appear to be no immediate plans for clinical development of sabutoclax, however.

4.5.3.4 Obatoclax (GX15-070)

Obatoclax mesylate (Teva Pharmaceutical Industries Limited, Israel, originally Gemin X Pharmaceuticals, Montreal, QC, Canada) is a novel BH3 mimetic pan-Bcl-2 inhibitor effective against Bcl-2, Bcl-X_L, Mcl-1, Bcl-w, Bfl-1/A1, and Bcl-B (Kuroda and Taniwaki 2009). Significantly, obatoclax potently interferes with the sequestration of the proapoptotic protein Bak by Mcl-1, untethering Bak and overcoming resistance to apoptosis conferred by Mcl-1(Nguyen et al. 2007). Obatoclax induced apoptosis and inhibited the growth of AML cell lines and primary AML samples, and synergized with ABT-737 and cytarabine (Konopleva et al. 2008). Apoptosis induced by obatoclax in these cells was preceded by the release of Bak from Mcl-1, liberation of Bim from both Bcl-2 and Mcl-1, and the formation of an active Bak/Bax complex. At growth-inhibitory doses that did not induce apoptosis or decrease viability, obatoclax induced an S-G_2 cell-cycle block (Konopleva et al. 2008). Obatoclax primarily induces autophagic cell death in acute lymphoblastic leukemia (ALL) cells, and whether the drug predominantly induces apoptotic or nonapoptotic cell death may thus depend on cell type (Khaw et al. 2011).

Forty-four patients with refractory leukemias or MDS received obatoclax by continuous intravenous (IV) infusions in a phase I study (Schimmer et al. 2008). The recommended phase II dose (RPTD) was 28 mg/m^2 over 24 h for up to 4 consecutive days every 2 weeks. The drug was well tolerated, with grade 1–2 neurologic symptoms being the most common toxicities. One patient with AML and an *MLL* gene rearrangement experienced a CR which lasted 8 months. Three of 14 patients with MDS had hematologic improvement. A trial in previously untreated older patients with AML found the maximum tolerated doses (MTDs) to be 20 mg over 3 h or 60 mg over 24 h, both administered daily for 3 consecutive days every 2 weeks (Raza et al. 2009). The most common toxicities were reversible, infusion-related somnolence, euphoria, and ataxia. In phase I studies conducted in patients with CLL and solid tumors or lymphoma, the MTDs were 28 mg/m^2 over 3 h every 3 weeks, and 20 mg/m^2 over 3 h every week, respectively, and the toxicity profiles were very similar (Joudeh and Claxton 2012). Modest single-agent activity was noted in heavily pretreated patients with advanced CLL, and obatoclax has been granted orphan drug status by the US Food and Drug Administration (FDA) for this indication (Kang and Reynolds 2009). Phase II studies in myelofibrosis (Joudeh and Claxton 2012) and Hodgkin's lymphoma (Oki et al. 2012) employing a 24-h infusion of 60 mg every 2 weeks, however, yielded no responses. Based on preclinical evidence of synergism with bortezomib in mantle cell lymphoma, a phase I study was conducted, with encouraging results (Joudeh and Claxton 2012). Phase I and II trials in small-cell lung cancer (SCLC) in combination with chemotherapy have been completed (Joudeh and Claxton 2012). However, the continued development of this agent is presently uncertain.

4.5.3.5 ABT-737 and Analogs

ABT-737 (Abbott Laboratories, Abbott Park, IL, now AbbVie, Inc., North Chicago, IL) is a specific small-molecule inhibitor of Bcl-2, Bcl-X_L, and Bcl-w discovered through nuclear magnetic resonance (NMR)-based screening, parallel synthesis and structure-based design (Oltersdorf et al. 2005). ABT-737 binds to these antiapoptotic proteins with high affinity and much more weakly to Mcl-1, Bcl-B, and Bfl-1/A1, thus mimicking the actions of Bad (Labi et al. 2008; Zhang et al. 2007b). Of its targets, Bcl-2 might be the most critical in vivo (Merino et al. 2012). Notably, ABT-737 selectively targeted transformed cells, arguing that such cells may be more addicted to antiapoptotic proteins compared to their normal counterparts. ABT-737 and its analogs are probably the only bona fide BH3 mimetics, triggering apoptosis only in the presence of Bax and Bak (Delft, Mason 2006) and releasing Bad from Bcl-2, an effect not shared by other "BH3 mimetics" such as obatoclax, gossypol, apogossypol, HA14-1, antimycin A3 and S1 which, instead, may induce apoptosis via off-target effects such as upregulation of Noxa and induction of ER stress (Albershardt et al. 2011). ABT-737 acts as a "sensitizer" BH3 mimetic, releasing activators (Bim) and effectors (Bax) from Bcl-2 and triggering conformational alteration of Bax, but not directly activating it (Labi et al. 2008; Zhang et al. 2007c). Besides disrupting the inhibitory interaction between the BH3 domain of Beclin-1 and Bcl-2/Bcl-X_L, ABT-737 stimulates multiple proautophagic signal transduction pathways (Malik et al. 2011). ABT-737 effectively kills AML blast, progenitor and stem cells without affecting normal hematopoietic cells by inducing disruption of the Bcl-2/Bax complex and Bak-dependent but Bim-independent activation of the intrinsic apoptotic pathway (Konopleva et al. 2006). In infant leukemia cells bearing the *MLL* gene rearrangement, ABT-737 induced cell death and synergized with sunitinib (Jayanthan et al. 2011). An analog of ABT-737, A-385358 is a small molecule with relative selectivity for binding to Bcl-X_L versus Bcl-2 that, like ABT-737, promotes apoptosis and potentiates the activity of cytotoxic drugs in several tumor cell lines and xenograft models (Shoemaker et al. 2006).

ABT-263 (navitoclax) is a potent, orally bioavailable analog of ABT-737 that is currently in clinical trials (Tse et al. 2008). In a phase I trial in patients with relapsed or refractory CLL evaluating both intermittent (14/21) and continuous (21/21) once daily dosing schedules of navitoclax, 9 of 29 patients achieved a partial response, and 7 others had stable disease for > 6 months (Roberts et al. 2012). Median PFS was 25 months. Activity was seen in poor-prognosis subsets, and the RPTD was 250 mg daily. Low Mcl-1 expression and high Bim: Mcl-1 or Bim: Bcl-2 ratios in leukemic cells correlated with response. Phase I trials in patients with solid tumors, especially SCLC, and lymphoma have also been completed, and encouraging responses have been seen (Gandhi et al. 2011; Wilson et al. 2010), although single-agent activity in relapsed SCLC was limited in the phase II setting (Rudin et al. 2012). Dose-dependent thrombocytopenia, due to platelet apoptosis caused by high-affinity inhibition of Bcl-X_L (Mason et al. 2007; White et al. 2012; Zhang et al. 2007a), has been a consistent finding in all these trials, and a 150-mg daily lead-in phase has been used to mitigate this. Some studies have found that Bcl-X_L inhibition by BH3 mimetics may undermine platelet hemostatic function (Schoenwaelder et al. 2011; Vogler et al. 2011) and that the intrinsic apoptosis pathway of megakaryocytes must be restrained in order for them to survive and produce platelets (Josefsson et al. 2011); however, thrombocytopenia or bleeding refractory to platelet transfusion has not been observed in clinical trials (personal communication, Jennifer R Brown, Dana Farber Cancer Institute, Boston, MA). Nevertheless, concern regarding the thrombocytopenia induced by on-target inhibition of Bcl-X_L by navitoclax has prompted the company to develop ABT-199, a highly potent and selective Bcl-2 inhibitor, by reverse engineering of navitoclax (Souers et al. 2013). In in vitro studies, ABT-199 exhibited potent mechanism-dependent induction of apoptosis in multiple NHL cell lines, particularly those with gains/amplifications of Bcl-2, or t (14;18), as well as in AML and ALL cell lines. ABT-199 induced regression of hematologic tumors in a variety of xenograft models of human lymphoid malignancies, which are frequently addicted to Bcl-2 for survival, either alone or in combination with bendamustine and rituximab. ABT-199 displayed reduced effects on platelets compared to navitoclax, consistent with the former's lack of Bcl-X_L inhibition. The drug induced concentration-dependent apoptosis in primary CLL cells, and caused tumor lysis in three patients with refractory CLL (Souers et al. 2013). Significant single-agent activity of ABT-199 has been observed in patients with CLL and small lymphocytic lymphoma (Anonymous 2014). In preclinical studies, ABT-199 causes on-target cell death in AML and synergizes with azacytidine (Pan et al. 2014; Bogenberger et al. 2014).

4.5.4 Activating Endogenous Antagonists of Bcl-2

An orphan member of the retinoid/steroid family of nuclear receptors, TR3 (Nur77), translocates from the nucleus to the cytosol in response to certain death stimuli, binds to a regulatory domain within Bcl-2, and induces a profound conformational change of Bcl-2, exposing its BH3 domain and converting Bcl-2 from a protector to a killer. 3C1-AHPC is a compound that can induce this process by phosphorylating TR3, possibly via Jun N-terminal kinase (JNK). It has demonstrated activity against cultured leukemia cells and in preclinical animal models of AML, even against retinoid refractory leukemia cell lines (Reed and Pellecchia 2005). Akt/protein kinase B opposes the proapoptotic effects of TR3, thus invoking a role for Akt inhibition. Indeed, Akt inhibitors sensitize established myeloid leukemia cell lines to apoptosis induced by a wide variety of agents. However, TR3 is only one of many apoptosis-relevant substrates of Akt, such as Bid, caspase-9, the p53 antagonist murine double minute homologue (Mdm2), etc. 3C1-AHPC may also reduce Akt activity in tumor cells (Reed and Pellecchia 2005). Finally, heat-shock protein 90 (hsp90) inhibitors such as 17-allylamino-17-demethoxygeldanamycin (17-AAG) cause reductions in Akt levels and trigger apoptosis in HL-60 cells that is markedly enhanced by Bcl-2 antagonists such as HA14-1 (Nimmanapalli et al. 2003).

4.5.5 Immunologic Approaches

Spontaneous T-cell reactivity against Bcl-2 in peripheral blood from AML patients has been demonstrated (Andersen et al. 2005). Mcl-1(139) is an HLA-A2-restricted epitope from Mcl-1 recognized spontaneously in cancer patients. Cytotoxic T-lymphocytes recognizing this epitope are able to kill *ex vivo* enriched leukemia cells (Andersen et al. 2008). Retroviral immunotoxin gene therapy of leukemia in mice using leukemia-specific T-cells transduced with an interleukin-3/Bax fusion protein gene has been successful (Vallera et al. 2003).

4.6 Mcl-1 as a Therapeutic Target in AML

The antiapoptotic Bcl-2 family protein Mcl-1 is a key therapeutic target in AML (Glaser et al. 2012). Mcl-1 is an indispensable regulator of self-renewal in human hematopoietic and leukemic stem cells (Campbell et al. 2010). Mcl-1 sequesters the apoptotic effector Bak on the mitochondrial surface, protecting against Bak oligomerization and the induction of MOMP. Mcl-1 also interacts with and sequesters the proapoptotic proteins Bim and tBid (Mandelin and Pope 2007). Noxa, which is induced by the pan-Bcl-2 family inhibitors, obatoclax and gossypol, antagonizes Mcl-1, freeing Bak and Bim and promoting the proteosomal degradation of Mcl-1 (Mazumder et al. 2012). The very short half-life of Mcl-1 enhances our ability to downregulate it using inhibitors of translation such as sorafenib(Rahmani et al. 2005; Yu et al. 2005) or homoharringtonine (Tang et al. 2006), transcriptional repressors such as cyclin-dependent kinase inhibitors and inhibitors of the deubiquitinase USP9X, that promote its degradation (Gores and Kaufmann 2012). In addition, HDACIs diminish Mcl-1 expression (Rosato et al. 2003). Upregulation of Mcl-1 and its phosphorylation (leading to stabilization of Mcl-1 and its interaction with Bim) represent the primary mechanisms of leukemia cell resistance to ABT-737(Konopleva et al. 2006; Delft and Mason 2006; Mazumder et al. 2012), which does not bind to Mcl-1 because of structural differences that distinguish Mcl-1 from Bcl-2, Bcl-X_L, and Bcl-w (Lee et al. 2007). This forms the basis for many synergistic strategies (see Sect. 5.7).

4.7 Other Apoptosis-Inducing Agents Impacting Bcl-2 Family Proteins in AML Cells

The Bcl-2 family is intimately linked to multiple signal transduction pathways, and a complete discussion of all agents known to trigger apoptosis in AML cells that affect the Bcl-2 family of proteins in some way is outside the scope of this chapter. For example, PKC inhibitors induce apoptosis in HL-60 cells by inhibiting the phosphorylation of Bcl-2, without affecting its expression (Jarvis et al. 1994). Among other proapoptotic actions, MEK inhibitors downregulate Mcl-1 and Bcl-X_L in AML cell lines (Milella et al. 2001). The novel synthetic triterpenoid methyl ester CDDO-Me induces apoptosis in AML cell lines and primary AML cells via induction of Bax and inhibition of both Bcl-2 phosphorylation and ERK activation, and synergizes with retinoids (Konopleva et al. 2002a). Among natural products, the protein synthesis inhibitor homoharringtonine induces apoptosis in AML cells by reducing the levels of the antiapoptotic proteins Mcl-1, XIAP, and survivin (Yin et al. 2011), and upregulating Bax (Yinjun et al. 2004). Bcl-X_L plays an important role in protecting against homoharringtonine-induced apoptosis, suggesting that the combination of a Bcl-X_L antagonist with homoharringtonine might be a promising therapeutic strategy in AML (Yin et al. 2011). The diterpenoid triptolide induces apoptosis in AML cell lines and primary AML blasts at nanomolar concentrations through downregulation of Mcl-1 and XIAP (Carter et al. 2006). A derivative, MRx102, induces apoptosis through similar mechanisms in $CD34^{+}CD38^{-}$ stem/progenitor cells derived from AML patients at low nanomolar concentrations despite coculture with bone marrow stromal cells, as well as in a nonobese diabetic (NOD)/SCID mouse model (Carter et al. 2012).

4.8 Manipulating Bcl-2 Family Proteins with Rational Drug Combinations in AML

CDKIs downregulate Mcl-1 via inhibition of transcription (Gores and Kaufmann 2012). Multiple studies from our group have shown CDKI-induced downregulation of Mcl-1 and XIAP to be markedly potentiated by HDACIs (Rosato et al. 2002a, 2005, 2007), proteasome inhibitors (Dai et al. 2003), and PI3 kinase inhibitors (Yu et al. 2003), leading to pronounced enhancement of apoptosis in AML cell lines and primary AML blasts. Based on several of these observations, a phase I clinical trial of flavopiridol and vorinostat in patients with relapsed, refractory or poor-prognosis AML or MDS was conducted (Holkova et al. 2013). Although no objective responses were seen, sustained plasma concentrations of flavopiridol high enough to mimic the preclinical in vitro findings were achieved (Holkova et al. 2013). While the CDKI RO-3306 cooperatively induced Bax activation and mitochondrial apoptosis in AML cells along with Nutlin 3, an inhibitor of the Mdm2-p53 protein–protein interaction by, among other mechanisms, downregulating Bcl-2 (Kojima et al. 2009), all patients in a clinical trial of flavopiridol, Ara-C and mitoxantrone (FLAM) in adults with refractory AML unexpectedly showed upregulation of Bcl-2 by flavopiridol, leading the authors to propose that a combination of flavopiridol with a Bcl-2 antagonist might be synergistic (Nelson et al. 2011). The CDKI roscovitine and ABT-737 synergistically untether Bak from Mcl-1 and Bcl-X_L, respectively, triggering Bak activation and Bax translocation, leading to apoptosis in human AML cells (Chen et al. 2007). HDACIs upregulate Bim in human AML cells, leading to its sequestration by Bcl-2 and Bcl-X_L, from which it is released by ABT-737, thus providing a mechanistic framework by which Bcl-2 antagonists potentiate the lethality of agents, such as HDACIs, which upregulate Bim (Chen et al. 2009). The combination of HDACIs and obatoclax has been shown to synergistically induce both autophagy and apoptosis in AML cell lines and primary AML cells (Wei et al. 2010).

Downregulation of Mcl-1 by inhibition of translation of this short-lived anti-apoptotic protein by sorafenib (Zhang et al. 2008) or silvestrol (Cencic et al. 2010) has been exploited for synergism with ABT-737 in AML cell lines. Sorafenib upregulates Bim, increases Bad, Bax, and Bak protein levels, and decreases Mcl-1, XIAP, and survivin levels (Zhang et al. 2008). Sorafenib synergizes with TRAIL to cause apoptosis of AML cell lines and primary AML blasts but not of normal $CD34^{+}$ bone marrow cells via down regulation of Mcl-1, Bcl-$X_{L,}$ and c-FLIP (Rosato et al. 2007b). Mcl-1 serves as a buffer for Bak, Bim, and Puma, and its down regulation by sorafenib thus provides a mechanistic basis for TRAIL sensitization (Meng et al.

2007). Most recently, marked synergism between sorafenib and obatoclax has been observed in multiple AML cell lines and primary AML cells, both *FLT3*-wild type and mutated, but not normal $CD34^+$ cells, in vitro and in vivo (Rahmani et al. 2012). Sorafenib triggered rapid and pronounced Mcl-1 down regulation accompanied by enhanced binding of Bim to Bcl-2 and Bcl-X_L, effects that were abolished by obatoclax coadministration. Marked induction of autophagy was noted; however, this was cytoprotective in nature, as inhibition of this process (e.g., with chloroquine) significantly increased lethality of the sorafenib/obatoclax regimen. In line with the known FLT3-inhibitory action of sorafenib, synergism occurred at much lower sorafenib concentrations in *FLT3*-mutated cells. Overexpression of Bcl-2 causes resistance to FLT3 inhibitors in *FLT3*-mutated AML, and ABT-737 synergizes efficiently with FLT3 inhibitors in AML samples with activating *FLT3* mutations (Kohl et al. 2007). Furthermore, in *FLT3*-ITD AML, upregulation of Mcl-1 is dependent on FLT3 signaling and an essential effector of *FLT3*-ITD-mediated drug resistance, and down regulation of Mcl-1 sensitizes *FLT3*-ITD hematopoietic cells to cytotoxic and targeted therapies (Kasper et al. 2012). Although there is substantial enthusiasm surrounding the platelet-sparing Bcl-2-selective BH3 mimetic ABT-199, especially in lymphoid malignancies (Davids and Letai 2013; Vandenberg and Cory 2013), untethering of Bak from both Bcl-X_L and Mcl-1 (Willis et al. 2005), could be critical in AML. Thus, dual inhibition of Bcl-2 and Bcl-X_L by ABT-737 was recently demonstrated to strikingly enhance apoptosis caused by the dual PI3K/mTOR (mammalian target of rapamycin) inhibitors BEZ235 and PI-103 in AML cell lines and patient-derived samples, and to significantly diminish tumor growth in xenograft models of AML (Rahmani et al. 2013). PI3K/mTOR inhibitors markedly downregulated Mcl-1 but increased Bim binding to Bcl-2/Bcl-X_L; the latter effect was abrogated by ABT-737. Combined treatment also markedly diminished Bax/Bak binding to Mcl-1, Bcl-2, or Bcl-X_L. This strategy might be particularly effective in the setting of basal Akt activation and/or addiction (Rahmani et al. 2013).

MEK inhibitors strikingly potentiate the apoptosis induced by HA14-1 or by Bcl-2 antisense oligonucleotides in AML cell lines with constitutively active MAPK and in primary AML samples (Milella et al. 2002). It was recently found that ABT-737 induces Mcl-1 and activates ERK in AML cells (Konopleva et al. 2012). The combination of a MEK inhibitor with ABT-737 resulted in potent synergistic killing of AML-derived cell lines, primary AML blasts and $CD34^+CD38^-CD123^+$ progenitor/stem cells and exerted major antileukemia activity in a murine xenograft model (Konopleva et al. 2012). More recent studies have shown enhancement of apoptosis induction by ABT-737 in AML cells by the PI3K inhibitor GDC-0941 (Jin et al. 2013) and by the combination of the MEK inhibitor selumetinib and the mTOR inhibitor AZD8055 (Zhang et al. 2014). Combined targeting of the Mdm2–p53 interaction by Nutlin-3 and of Bcl-2 by ABT-737 induced Bax conformational change and mitochondrial apoptosis in AML cells in a strikingly synergistic fashion (Kojima et al. 2006). The two drugs also complemented each other, predominantly inducing apoptosis in different phases of the cell cycle. Nutlin-3 also synergized with the FLT3 inhibitor FI-700, which induced rapid proteasome-mediated degradation of Mcl-1, resulting in reduced ability of Mcl-1 to sequester proapoptotic Bim (Kojima et al. 2010). Nutlin-3 induced Noxa, which displaced Bim from Mcl-1. The

FI-700/Nutlin-3 combination profoundly activated Bax and induced apoptosis in AML cells. Maritoclax (marinopyrolle A) is a selective antagonist of Mcl-1 that disrupts its binding to Bim and induces Mcl-1 degradation via the proteasome system (Doi et al. 2012). Maritoclax selectively kills Mcl-1-dependent, but not Bcl-2- or Bcl-X_L-dependent, leukemia cells and markedly enhances the efficacy of ABT-737. ARRY-520, a selective inhibitor of kinesin-spindle protein (KSP), a microtubule-associated motor protein essential for cell cycle progression that is overexpressed in many cancers, potently induces cell cycle block and subsequent death in HL-60 cells via the mitochondrial pathway and synergizes with ABT-737 (Carter et al. 2009). ARRY-520 significantly inhibited tumor growth of xenografts in SCID mice and may have the potential to eradicate AML progenitor cells. The Rac pathway, which positively regulates expression of Bcl-2 family proteins toward a prosurvival balance, is critical to *MLL-AF9* leukemogenesis. *MLL-AF9* cells are exquisitely sensitive to ABT-737, and concurrent exposure to NSC23766, a small-molecule inhibitor of Rac activation, increases the apoptotic effect of ABT-737 (Mizukawa et al. 2011). Inhibition of XIAP sensitizes human leukemia cell lines to apoptosis induced by HA14-1 (Skommer et al. 2011). Pharmacologic inhibitors of fatty acid oxidation sensitized human leukemia cells—cultured alone or on bone marrow stromal cells—to apoptosis induction by ABT-737 and the combination provided substantial therapeutic benefit in a murine model of leukemia (Samudio et al. 2010). Furthermore, these drugs decreased the number of quiescent leukemia progenitor cells in approximately 50% of primary human AML samples. Finally, 2-deoxyglucose, which can dissociate the Bak-Mcl-1 complex in highly glycolytic cells (such as most cancer cells), may represent an ideal partner for combination with ABT-737 (Yamaguchi and Perkins 2012).

Conclusion

Multiple signal transduction pathways in normal and malignant cells converge on the Bcl-2 family proteins, which stand at the crossroads of cellular survival and programmed cell death. An in-depth understanding of Bcl-2 family biology has allowed us to intervene pharmacologically in this complex system in various ways, giving us the ability to flip the life-death switch in cancer cells. Little progress has been made in AML therapy since anthracycline–cytarabine induction regimens were first described almost 40 years ago. AML cells, like most malignant cells, are particularly vulnerable to the lethal consequences of simultaneous disruption of multiple cooperative prosurvival pathways. Bcl-2 antagonists, especially BH3 mimetics, thus have the potential to unleash the power of inhibitors of oncogenic kinases and other targeted therapies, as discussed in this chapter. The focus in the treatment of AML is clearly moving towards a combination of targeted drugs. There are virtually limitless possibilities for rational combinations involving agents that modulate the expression and/or function of Bcl-2 family proteins. Choosing the best combinations to take forward into clinical trials is a major challenge for investigators. In the coming years, Bcl-2 antagonists can be expected to play an increasingly prominent role in the targeted therapy of AML.

Acknowledgments This work was supported in part by the following awards to Dr. Grant: RO1 CA093738, RO1 CA100866, P50 CA130805, P50 CA142509, and RC2 CA148431 from the National Institutes of Health, and award #6181–10 from the Leukemia and Lymphoma Society.

References

Anonymous (2014) BCL-2 inhibitor yields high response in CLL and SLL. Cancer Discov 4(2):OF5

Adams JM, Cory S (2007) Bcl-2-regulated apoptosis: mechanism and therapeutic potential. Curr Opin Immunol 19:488–496

Albershardt TC, Salerni BL, Soderquist RS et al (2011) Multiple BH3 mimetics antagonize anti-apoptotic MCL1 protein by inducing the endoplasmic reticulum stress response and up-regulating BH3-only protein NOXA. J Biol Chem 286:24882–24895

Andersen MH, Svane IM, Kvistborg P et al (2005) Immunogenicity of Bcl-2 in patients with cancer. Blood 105:728–734

Andersen RS, Wenandy L, Sorensen RB, thor Straten P, Andersen MH (2008) Mcl-1 and anticancer vaccination: identification of an HLA-A2-restricted epitope. Leukemia 22:668–9

Andreeff M, Jiang S, Zhang X et al (1999) Expression of Bcl-2-related genes in normal and AML progenitors: changes induced by chemotherapy and retinoic acid. Leukemia 13:1881–1892

Bagrintseva K, Geisenhof S, Kern R et al (2005) FLT3-ITD-TKD dual mutants associated with AML confer resistance to FLT3 PTK inhibitors and cytotoxic agents by overexpression of Bcl-x(L). Blood 105:3679–3685

Bhat UG, Gartel AL (2010) Nucleoside analog ARC targets Mcl-1 to induce apoptosis in leukemia cells. Leukemia 24:851–855

Bogenberger JM, Kornblau SM, Pierceall WE, Lena R, Chow D, Shi CX, Mantei J, Ahmann G, Gonzales IM, Choudhary A, Valdez R, Camoriano J, Fauble V, Tiedemann RE, Qiu YH, Coombes KR, Cardone M, Braggio E, Yin H, Azorsa DO, Mesa RA, Stewart AK, Tibes R (2014) BCL-2 family proteins as 5-Azacytidine-sensitizing targets and determinants of response in myeloid malignancies. Leukemia 28(8):1657–1665

Bolanos-Meade J, Karp JE, Guo C et al (2003) Timed sequential therapy of acute myelogenous leukemia in adults: a phase II study of retinoids in combination with the sequential administration of cytosine arabinoside, idarubicin and etoposide. Leuk Res 27:313–321

Bradbury DA, Aldington S, Zhu YM, Russell NH (1996) Down-regulation of Bcl-2 in AML blasts by all-trans retinoic acid and its relationship to CD34 antigen expression. Br J Haematol 94:671–675

Campbell CJ, Lee JB, Levadoux-Martin M et al (2010) The human stem cell hierarchy is defined by a functional dependence on Mcl-1 for self-renewal capacity. Blood 116:1433–1442

Campos L, Rouault JP, Sabido O et al (1993) High expression of Bcl-2 protein in acute myeloid leukemia cells is associated with poor response to chemotherapy. Blood 81:3091–3096

Campos L, Sabido O, Rouault JP, Guyotat D (1994) Effects of BCL-2 antisense oligodeoxynucleotides on in vitro proliferation and survival of normal marrow progenitors and leukemic cells. Blood 84:595–600

Campos L, Sabido O, Viallet A, Vasselon C, Guyotat D (1999) Expression of apoptosis-controlling proteins in acute leukemia cells. Leuk Lymphoma 33:499–509

Carter BZ, Mak DH, Schober WD et al (2006) Triptolide induces caspase-dependent cell death mediated via the mitochondrial pathway in leukemic cells. Blood 108:630–637

Carter BZ, Mak DH, Woessner R et al (2009) Inhibition of KSP by ARRY-520 induces cell cycle block and cell death via the mitochondrial pathway in AML cells. Leukemia 23:1755–1762

Carter BZ, Mak DH, Shi Y et al (2012) MRx102, a triptolide derivative, has potent antileukemic activity in vitro and in a murine model of AML. Leukemia 26:443–450

Cencic R, Carrier M, Trnkus A, Porco JA, Jr, Minden M, Pelletier J (2010) Synergistic effect of inhibiting translation initiation in combination with cytotoxic agents in acute myelogenous leukemia cells. Leuk Res 34:535–541

Chen S, Dai Y, Harada H, Dent P, Grant S (2007) Mcl-1 down-regulation potentiates ABT-737 lethality by cooperatively inducing Bak activation and Bax translocation. Cancer Res 67:782–791

Chen S, Dai Y, Pei XY, Grant S (2009) Bim upregulation by histone deacetylase inhibitors mediates interactions with the Bcl-2 antagonist ABT-737: evidence for distinct roles for Bcl-2, Bcl-X_L, and Mcl-1. Mol Cell Biol 29:6149–6169

Chonghaile TN, Letai A (2008) Mimicking the BH3 domain to kill cancer cells. Oncogene 27(1):149–157

Chou FS, Wunderlich M, Griesinger A, Mulloy JC (2011) N-Ras(G12D) induces features of stepwise transformation in preleukemic human umbilical cord blood cultures expressing the AML1-ETO fusion gene. Blood 117:2237–2240

Dai Y, Rahmani M, Grant S (2003) Proteasome inhibitors potentiate leukemic cell apoptosis induced by the cyclin-dependent kinase inhibitor flavopiridol through a SAPK/JNK- and NF-kappaB-dependent process. Oncogene 22:7108–7122

Dai G, Chan KK, Liu S et al (2005) Cellular uptake and intracellular levels of the bcl-2 antisense g3139 in cultured cells and treated patients with acute myeloid leukemia. Clin Cancer Res 11:2998–3008

Davids MS, Letai A (2012) Targeting the B-cell lymphoma/leukemia 2 family in cancer. J Clin Oncol 30(25):3127–3135

Davids MS, Letai A (2013) ABT-199: taking dead aim at Bcl-2. Cancer Cell 23:139–141

Davis RE, Greenberg PL (1998) Bcl-2 expression by myeloid precursors in myelodysplastic syndromes: relation to disease progression. Leuk Res 22:767–777

Del Gaizo Moore V, Letai A (2012) BH3 profiling—Measuring integrated function of the mitochondrial apoptotic pathway to predict cell fate decisions. Cancer Lett [Epub ahead of print]

Del Poeta G, Venditti A, Del Principe MI et al (2003) Amount of spontaneous apoptosis detected by Bax/Bcl-2 ratio predicts outcome in acute myeloid leukemia (AML). Blood 101:2125–2131

Del Poeta G, Ammatuna E, Lavorgna S et al (2010) The genotype nucleophosmin mutated and FLT3-ITD negative is characterized by high bax/bcl-2 ratio and favourable outcome in acute myeloid leukaemia. Br J Haematol 149:383–387

van Delft MF, Wei, Mason KD et al (2006) The BH3 mimetic ABT-737 targets selective Bcl-2 proteins and efficiently induces apoptosis via Bak/Bax if Mcl-1 is neutralized. Cancer Cell 10:389–399

Deng G, Lane C, Kornblau S et al (1998) Ratio of bcl-xshort to bcl-xlong is different in good- and poor-prognosis subsets of acute myeloid leukemia. Mol Med 4:158–164

Dodou K, Anderson RJ, Small DA, Groundwater PW (2005) Investigations on gossypol: past and present developments. Expert Opin Investig Drugs 14:1419–1434.

Doi K, Li R, Sung SS et al (2012) Discovery of Marinopyrrole A (Maritoclax) as a Selective Mcl-1 Antagonist that Overcomes ABT-737 Resistance by Binding to and Targeting Mcl-1 for Proteasomal Degradation. J Biol Chem 287:10224–10235

Elstner E, Linker-Israeli M, Umiel T et al (1996) Combination of a potent 20-epi-vitamin D3 analogue (KH 1060) with 9-cis-retinoic acid irreversibly inhibits clonal growth, decreases bcl-2 expression, and induces apoptosis in HL-60 leukemic cells. Cancer Res 56:3570–3576

Estey EH, Thall PF, Pierce S et al (1999) Randomized phase II study of fludarabine + cytosine arabinoside + idarubicin + /- all-trans retinoic acid + /- granulocyte colony-stimulating factor in poor prognosis newly diagnosed acute myeloid leukemia and myelodysplastic syndrome. Blood 93:2478–2484

Ewings KE, Hadfield-Moorhouse K, Wiggins CM et al (2007) ERK1/2-dependent phosphorylation of BimEL promotes its rapid dissociation from Mcl-1 and Bcl-X_L. EMBO J 26:2856–2867

Galluzzi L, Vitale I, Abrams JM et al (2012) Molecular definitions of cell death subroutines: recommendations of the Nomenclature Committee on Cell Death 2012. Cell Death Differ 19:107–120

Gandhi L, Camidge DR, Ribeiro de Oliveira M et al (2011) Phase I study of Navitoclax (ABT-263), a novel Bcl-2 family inhibitor, in patients with small-cell lung cancer and other solid tumors. J Clin Oncol 29:909–916
Glaser SP, Lee EF, Trounson E et al (2012) Antiapoptotic Mcl-1 is essential for the development and sustained growth of acute myeloid leukemia. Genes Dev 26:120–5
Goff et al (2013) Phase I Trial of Vorinostat (SAHA) in Combination with Alvocidib (Flavopiridol) in Patients with Relapsed, Refractory or (Selected) Poor Prognosis Acute Leukemia or Refractory Anemia with Excess Blasts-2 (RAEB-2). ASH Annual Meeting Abstracts 112:2986 ASH Dec 6–9, 2008
Gores GJ, Kaufmann SH (2012) Selectively targeting Mcl-1 for the treatment of acute myelogenous leukemia and solid tumors. Genes Dev 26:305–311
Hao JH, Yu M, Liu FT, Newland AC, Jia L (2004) Bcl-2 inhibitors sensitize tumor necrosis factor-related apoptosis-inducing ligand-induced apoptosis by uncoupling of mitochondrial respiration in human leukemic CEM cells. Cancer Res 64:3607–3616
Hermanson D, Addo SN, Bajer AA et al (2009) Dual mechanisms of sHA 14–1 in inducing cell death through endoplasmic reticulum and mitochondria. Mol Pharmacol 76:667–678
Hess CJ, Berkhof J, Denkers F et al (2007) Activated intrinsic apoptosis pathway is a key related prognostic parameter in acute myeloid leukemia. J Clin Oncol 25:1209–1215
Hogstrand K, Hejll E, Sander B, Rozell B, Larsson LG, Grandien A (2012) Inhibition of the intrinsic but not the extrinsic apoptosis pathway accelerates and drives myc-driven tumorigenesis towards acute myeloid leukemia. PLoS ONE 7:e31366
Holkova B, Supko JG, Ames MM, Reid JM, Shapiro GI, Perkins EB, Ramakrishnan V, Tombes MB, Honeycutt C, McGovern RM, Kmieciak M, Shrader E, Wellons MD, Sankala H, Doyle A, Wright J, Roberts JD, Grant S (2013) A phase I trial of vorinostat and alvocidib in patients with relapsed, refractory, or poor prognosis acute leukemia, or refractory anemia with excess blasts-2. Clin Cancer Res 19(7):1873–1883
Irish JM, Anensen N, Hovland R et al (2007) Flt3 Y591 duplication and Bcl-2 overexpression are detected in acute myeloid leukemia cells with high levels of phosphorylated wild-type p53. Blood 109:2589–2596
Irvine AE, McMullin MF, Ong YL (2002) Bcl-2 family members as prognostic indicators in AML. Hematology 7:21–31
Jarvis WD, Turner AJ, Povirk LF, Traylor RS, Grant S (1994) Induction of apoptotic DNA fragmentation and cell death in HL-60 human promyelocytic leukemia cells by pharmacological inhibitors of protein kinase C. Cancer Res 54:1707–1714
Jayanthan A, Incoronato A, Singh A et al (2011) Cytotoxicity, drug combinability, and biological correlates of ABT-737 against acute lymphoblastic leukemia cells with MLL rearrangement. Pediatr Blood Cancer 56:353–360
Jin L, Tabe Y, Kojima K, Shikami M, Benito J, Ruvolo V, Wang RY, McQueen T, Ciurea SO, Miida T, Andreeff M, Konopleva M (2013) PI3K inhibitor GDC-0941 enhances apoptotic effects of BH-3 mimetic ABT-737 in AML cells in the hypoxic bone marrow microenvironment. J Mol Med (Berl) 91(12):1383–1397
Josefsson EC, James C, Henley KJ et al (2011) Megakaryocytes possess a functional intrinsic apoptosis pathway that must be restrained to survive and produce platelets. J Exp Med 208:2017–2031
Joudeh J, Claxton D (2012) Obatoclax mesylate: pharmacology and potential for therapy of hematological neoplasms. Expert Opin Investig Drugs 21:363–373
Kang MH, Reynolds CP (2009) Bcl-2 inhibitors: targeting mitochondrial apoptotic pathways in cancer therapy. Clin Cancer Res 15:1126–1132
Kasper S, Breitenbuecher F, Heidel F et al (2012) Targeting MCL-1 sensitizes FLT3-ITD-positive leukemias to cytotoxic therapies. Blood Cancer Journal 2:e60
Kaufmann SH, Karp JE, Svingen PA et al (1998) Elevated expression of the apoptotic regulator Mcl-1 at the time of leukemic relapse. Blood 91:991–1000
Keith FJ, Bradbury DA, Zhu YM, Russell NH (1995) Inhibition of bcl-2 with antisense oligonucleotides induces apoptosis and increases the sensitivity of AML blasts to Ara-C. Leukemia 9:131–138

Kessel D, Reiners J Jr (2007) Initiation of apoptosis and autophagy by the Bcl-2 antagonist HA14–1. Cancer Lett 249:294–299

Khaw SL, Huang DC, Roberts AW (2011) Overcoming blocks in apoptosis with BH3-mimetic therapy in haematological malignancies. Pathology 43:525–535

Kim KT, Levis M, Small D (2006) Constitutively activated FLT3 phosphorylates BAD partially through pim-1. Br J Haematol 134:500–509

Kim R, Emi M, Matsuura K, Tanabe K (2007) Antisense and nonantisense effects of antisense Bcl-2 on multiple roles of Bcl-2 as a chemosensitizer in cancer therapy. Cancer Gene Ther 14:1–11

Kitada S, Kress CL, Krajewska M, Jia L, Pellecchia M, Reed JC (2008) Bcl-2 antagonist apogossypol (NSC736630) displays single-agent activity in Bcl-2-transgenic mice and has superior efficacy with less toxicity compared with gossypol (NSC19048). Blood 111:3211–3219

Kohl TM, Hellinger C, Ahmed F et al (2007) BH3 mimetic ABT-737 neutralizes resistance to FLT3 inhibitor treatment mediated by FLT3-independent expression of BCL2 in primary AML blasts. Leukemia 21:1763–1772

Kojima K, Konopleva M, Samudio IJ, Schober WD, Bornmann WG, Andreeff M (2006) Concomitant inhibition of MDM2 and Bcl-2 protein function synergistically induce mitochondrial apoptosis in AML. Cell Cycle 5:2778–2786

Kojima K, Shimanuki M, Shikami M, Andreeff M, Nakakuma H (2009) Cyclin-dependent kinase 1 inhibitor RO-3306 enhances p53-mediated Bax activation and mitochondrial apoptosis in AML. Cancer Sci 100:1128–1136

Kojima K, Konopleva M, Tsao T et al (2010) Selective FLT3 inhibitor FI-700 neutralizes Mcl-1 and enhances p53-mediated apoptosis in AML cells with activating mutations of FLT3 through Mcl-1/Noxa axis. Leukemia 24:33–43

Konopleva M, Tari AM, Estrov Z et al (2000) Liposomal Bcl-2 antisense oligonucleotides enhance proliferation, sensitize acute myeloid leukemia to cytosine-arabinoside, and induce apoptosis independent of other antiapoptotic proteins. Blood 95:3929–3038

Konopleva M, Tsao T, Ruvolo P et al (2002a) Novel triterpenoid CDDO-Me is a potent inducer of apoptosis and differentiation in acute myelogenous leukemia. Blood 99:326–335

Konopleva M, Zhao S, Hu W et al (2002b) The antiapoptotic genes Bcl-X(L) and Bcl-2 are overexpressed and contribute to chemoresistance of non-proliferating leukaemic CD34+ cells. Br J Haematol 118:521–534

Konopleva M, Contractor R, Tsao T et al (2006) Mechanisms of apoptosis sensitivity and resistance to the BH3 mimetic ABT-737 in acute myeloid leukemia. Cancer Cell 10:375–388

Konopleva M, Watt J, Contractor R et al (2008) Mechanisms of antileukemic activity of the novel Bcl-2 homology domain-3 mimetic GX15–070 (obatoclax). Cancer Res 68:3413–3420

Konopleva M, Milella M, Ruvolo P et al (2012) MEK inhibition enhances ABT-737-induced leukemia cell apoptosis via prevention of ERK-activated MCL-1 induction and modulation of MCL-1/BIM complex. Leukemia 26:778–787

Kornblau SM, Thall PF, Estrov Z et al (1999) The prognostic impact of BCL2 protein expression in acute myelogenous leukemia varies with cytogenetics. Clin Cancer Res 5:1758–1766

Kornblau SM, Vu HT, Ruvolo P et al (2000) BAX and PKCalpha modulate the prognostic impact of BCL2 expression in acute myelogenous leukemia. Clin Cancer Res 6:1401–1409

Kuroda J, Taniwaki M (2009) Involvement of BH3-only proteins in hematologic malignancies. Crit Rev Oncol Hematol 71:89–101

Labi V, Grespi F, Baumgartner F, Villunger A (2008) Targeting the Bcl-2-regulated apoptosis pathway by BH3 mimetics: a breakthrough in anticancer therapy? Cell Death Differ 15:977–987

Lauria F, Raspadori D, Rondelli D et al (1997) High bcl-2 expression in acute myeloid leukemia cells correlates with CD34 positivity and complete remission rate. Leukemia 11:2075–2078

Lee EF, Czabotar PE, Smith BJ et al (2007) Crystal structure of ABT-737 complexed with Bcl-X_L: implications for selectivity of antagonists of the Bcl-2 family. Cell Death Differ 14:1711–1713

Lisovsky M, Estrov Z, Zhang X et al (1996) Flt3 ligand stimulates proliferation and inhibits apoptosis of acute myeloid leukemia cells: regulation of Bcl-2 and Bax. Blood 88:3987–3997

Luck SC, Russ AC, Botzenhardt U et al (2011) Deregulated apoptosis signaling in core-binding factor leukemia differentiates clinically relevant, molecular marker-independent subgroups. Leukemia 25:1728–1738

Malik SA, Orhon I, Morselli E et al (2011) BH3 mimetics activate multiple pro-autophagic pathways. Oncogene 30:3918–3029

Mandelin AM 2nd, Pope RM (2007) Myeloid cell leukemia-1 as a therapeutic target. Expert Opin Ther Targets 11:363–373

Marcucci G, Byrd JC, Dai G et al (2003) Phase 1 and pharmacodynamic studies of G3139, a Bcl-2 antisense oligonucleotide, in combination with chemotherapy in refractory or relapsed acute leukemia. Blood 101:425–432

Marcucci G, Stock W, Dai G et al (2005) Phase I study of oblimersen sodium, an antisense to Bcl-2, in untreated older patients with acute myeloid leukemia: pharmacokinetics, pharmacodynamics, and clinical activity. J Clin Oncol 23:3404–3411

Marcucci G, Moser B, Blum W et al (2007) A phase III randomized trial of intensive induction and consolidation chemotherapy {+/-} oblimersen, a pro-apoptotic Bcl-2 antisense oligonucleotide in untreated acute myeloid leukemia patients > 60 years old. ASCO Meeting Abstracts 25:7012. ASCO 1–5, 2007

Mason KD, Carpinelli MR, Fletcher JI et al (2007) Programmed anuclear cell death delimits platelet life span. Cell 128:1173–1186

Mazumder S, Choudhary GS, Al-Harbi S, Almasan A (2012) Mcl-1 phosphorylation defines ABT-737 resistance that can be overcome by increased NOXA expression in leukemic B-cells. Cancer Res 72(12):3069–3079

Meng XW, Lee SH, Dai H et al (2007) Mcl-1 as a buffer for proapoptotic Bcl-2 family members during TRAIL-induced apoptosis: a mechanistic basis for sorafenib (Bay 43–9006)-induced TRAIL sensitization. J Biol Chem 282:29831–29846

Merino D, Khaw SL, Glaser SP et al (2012) Bcl-2, Bcl-X_L and Bcl-w are not equivalent targets of ABT-737 and Navitoclax (ABT-263) in lymphoid and leukemic cells. Blood 119(24):5807–5816

Milella M, Kornblau SM, Estrov Z et al (2001) Therapeutic targeting of the MEK/MAPK signal transduction module in acute myeloid leukemia. J Clin Invest 108:851–859

Milella M, Estrov Z, Kornblau SM et al (2002) Synergistic induction of apoptosis by simultaneous disruption of the Bcl-2 and MEK/MAPK pathways in acute myelogenous leukemia. Blood 99:3461–3464

Milella M, Konopleva M, Precupanu CM et al (2007) MEK blockade converts AML differentiating response to retinoids into extensive apoptosis. Blood 109:2121–2129

Minami Y, Yamamoto K, Kiyoi H, Ueda R, Saito H, Naoe T (2003) Different antiapoptotic pathways between wild-type and mutated FLT3: insights into therapeutic targets in leukemia. Blood 102:2969–2975

Minn AJ, Rudin CM, Boise LH, Thompson CB (1995) Expression of Bcl-X_L can confer a multidrug resistance phenotype. Blood 86:1903–1910

Mizukawa B, Wei J, Shrestha M et al (2011) Inhibition of Rac GTPase signaling and downstream prosurvival Bcl-2 proteins as combination targeted therapy in MLL-AF9 leukemia. Blood 118:5235–5245

Moore J, Seiter K, Kolitz J et al (2006) A Phase II study of Bcl-2 antisense (oblimersen sodium) combined with gemtuzumab ozogamicin in older patients with acute myeloid leukemia in first relapse. Leuk Res 30:777–783

Nelson DM, Joseph B, Hillion J, Segal J, Karp JE, Resar LM (2011) Flavopiridol induces BCL-2 expression and represses oncogenic transcription factors in leukemic blasts from adults with refractory acute myeloid leukemia. Leuk Lymphoma 52:1999–2006

Nguyen M, Marcellus RC, Roulston A et al (2007) Small molecule obatoclax (GX15–070) antagonizes MCL-1 and overcomes MCL-1-mediated resistance to apoptosis. Proc Natl Acad Sci USA 104:19512–19517

Nimmanapalli R, O'Bryan E, Kuhn D, Yamaguchi H, Wang HG, Bhalla KN (2003) Regulation of 17-AAG-induced apoptosis: role of Bcl-2, Bcl-X_L, and Bax downstream of 17-AAG-mediated down-regulation of Akt, Raf-1, and Src kinases. Blood 102:269–275

Nordigarden A, Kraft M, Eliasson P et al (2009) BH3-only protein Bim more critical than Puma in tyrosine kinase inhibitor-induced apoptosis of human leukemic cells and transduced hematopoietic progenitors carrying oncogenic FLT3. Blood 113:2302–2311

O'Brien S, Moore JO, Boyd TE et al (2009) 5-year survival in patients with relapsed or refractory chronic lymphocytic leukemia in a randomized, phase III trial of fludarabine plus cyclophosphamide with or without oblimersen. J Clin Oncol 27:5208–5212

Oki Y, Copeland A, Hagemeister F et al (2012) Experience with obatoclax mesylate (GX15–070), a small molecule pan-Bcl-2 family antagonist in patients with relapsed or refractory classical Hodgkin lymphoma. Blood 119:2171–2172

Oliver L, Mahe B, Gree R, Vallette FM, Juin P (2007) HA14–1, a small molecule inhibitor of Bcl-2, bypasses chemoresistance in leukaemia cells. Leuk Res 31:859–863

Oltersdorf T, Elmore SW, Shoemaker AR et al (2005) An inhibitor of Bcl-2 family proteins induces regression of solid tumours. Nature 435:677–681

Ong YL, McMullin MF, Bailie KE, Lappin TR, Jones FG, Irvine AE (2000) High bax expression is a good prognostic indicator in acute myeloid leukaemia. Br J Haematol 111:182–189

Pan R, Hogdal LJ, Benito JM, Bucci D, Han L, Borthakur G, Cortes J, DeAngelo DJ, Debose L, Mu H, Döhner H, Gaidzik VI, Galinsky I, Golfman LS, Haferlach T, Harutyunyan KG, Hu J, Leverson JD, Marcucci G, Müschen M, Newman R, Park E, Ruvolo PP, Ruvolo V, Ryan J, Schindela S, Zweidler-McKay P, Stone RM, Kantarjian H, Andreeff M, Konopleva M, Letai AG (2014) Selective BCL-2 inhibition by ABT-199 causes on-target cell death in acute myeloid leukemia. Cancer Discov 4(3):362–375

Parker JE, Mufti GJ, Rasool F, Mijovic A, Devereux S, Pagliuca A (2000) The role of apoptosis, proliferation, and the Bcl-2-related proteins in the myelodysplastic syndromes and acute myeloid leukemia secondary to MDS. Blood 96:3932–3938

Pattingre S, Tassa A, Qu X et al (2005) Bcl-2 antiapoptotic proteins inhibit Beclin 1-dependent autophagy. Cell 122:927–939

Pepper CJ, Osman HY, Hewamana S, Walsby EJ, Burnett AK, Knapper SL (2009) Gossypol inhibits NF- {kappa}B, down regulates Mcl-1 and induces apoptosis of primary acute myeloid leukaemia cells. ASH Annual Meeting Abstracts 114:4813; ASH December 7–8, 2009

Pisani F, Del Poeta G, Aronica G, Venditti A, Caravita T, Amadori S (1997) In vitro down-regulation of bcl-2 expression by all-trans retinoic acid in AML blasts. Ann Hematol 75:145–7

Ponassi R, Biasotti B, Tomati V et al (2008) A novel Bim-BH3-derived Bcl-X_L inhibitor: biochemical characterization, in vitro, in vivo and ex-vivo antileukemic activity. Cell Cycle 7:3211–3224

Rahmani M, Davis EM, Bauer C, Dent P, Grant S (2005) Apoptosis induced by the kinase inhibitor BAY 43–9006 in human leukemia cells involves down-regulation of Mcl-1 through inhibition of translation. J Biol Chem 280:35217–35227

Rahmani M, Aust MM, Attkisson E, Williams DC, Jr, Ferreira-Gonzalez A, Grant S (2012) Inhibition of Bcl-2 anti-apoptotic members by obatoclax potently enhances sorafenib-induced apoptosis in human myeloid leukemia cells through a Bim-dependent process. Blood 119:6089–6098

Rahmani M, Aust MM, Attkisson E, Williams DC Jr, Ferreira-Gonzalez A, Grant S (2013) Dual inhibition of Bcl-2 and Bcl-X_L strikingly enhances PI3K-inhibition-induced apoptosis in human myeloid leukemia cells through a GSK3- and Bim-dependent mechanism. Cancer Res 73:1340–1351

Raza A, Galili N, Borthakur G et al (2009) A safety and schedule seeking trial of Bcl-2 inhibitor obatoclax in previously untreated older patients with acute myeloid leukemia (AML). ASCO Meeting Abstracts 27:3579 ASCO May 29–June 2, 2009

Reed JC (2008) Bcl-2-family proteins and hematologic malignancies: history and future prospects. Blood 111:3322–3330

Reed JC, Pellecchia M (2005) Apoptosis-based therapies for hematologic malignancies. Blood 106:408–418

Reed JC, Stein C, Subasinghe C et al (1990) Antisense-mediated inhibition of BCL2 protooncogene expression and leukemic cell growth and survival: comparisons of phosphodiester and phosphorothioate oligodeoxynucleotides. Cancer Res 50:6565–6570

Roberts AW, Seymour JF, Brown JR et al (2012) Substantial susceptibility of chronic lymphocytic leukemia to BCL2 inhibition: results of a phase I study of navitoclax in patients with relapsed or refractory disease. J Clin Oncol 30:488–496

Robinson BW, Behling KC, Gupta M et al (2008) Abundant antiapoptotic BCL-2 is a molecular target in leukaemias with t(4;11) translocation. Br J Haematol 141:827–839
Rosato RR, Almenara JA, Cartee L, Betts V, Chellappan SP, Grant S (2002) The cyclin-dependent kinase inhibitor flavopiridol disrupts sodium butyrate-induced p21WAF1/CIP1 expression and maturation while reciprocally potentiating apoptosis in human leukemia cells. Mol Cancer Ther 1:253–266
Rosato RR, Almenara JA, Grant S (2003) The histone deacetylase inhibitor MS-275 promotes differentiation or apoptosis in human leukemia cells through a process regulated by generation of reactive oxygen species and induction of p21CIP1/WAF1 1. Cancer Res 63:3637–3645
Rosato RR, Almenara JA, Maggio SC et al (2005) Potentiation of the lethality of the histone deacetylase inhibitor LAQ824 by the cyclin-dependent kinase inhibitor roscovitine in human leukemia cells. Mol Cancer Ther 4:1772–1785
Rosato RR, Almenara JA, Kolla SS et al (2007a) Mechanism and functional role of XIAP and Mcl-1 down-regulation in flavopiridol/vorinostat antileukemic interactions. Mol Cancer Ther 6:692–702
Rosato RR, Almenara JA, Coe S, Grant S (2007b) The multikinase inhibitor sorafenib potentiates TRAIL lethality in human leukemia cells in association with Mcl-1 and cFLIPL down-regulation. Cancer Res 67:9490–9500
Rudin CM, Hann CL, Garon EB et al (2012) Phase 2 Study of Single Agent Navitoclax (ABT-263) and Biomarker Correlates in Patients with Relapsed Small Cell Lung Cancer. Clin Cancer Res 18(11):3163–3169
Saeki K, Yuo A, Okuma E et al (2000) Bcl-2 down-regulation causes autophagy in a caspase-independent manner in human leukemic HL60 cells. Cell Death Differ 7:1263–1269
Samudio I, Harmancey R, Fiegl M et al (2010) Pharmacologic inhibition of fatty acid oxidation sensitizes human leukemia cells to apoptosis induction. J Clin Invest 120:142–156
Schaich M, Illmer T, Seitz G et al (2001) The prognostic value of Bcl-X_L gene expression for remission induction is influenced by cytogenetics in adult acute myeloid leukemia. Haematologica 86:470–477
Scheijen B, Ngo HT, Kang H, Griffin JD (2004) FLT3 receptors with internal tandem duplications promote cell viability and proliferation by signaling through Foxo proteins. Oncogene 23:3338–3349
Schimmer AD (2007) Novel therapies targeting the apoptosis pathway for the treatment of acute myeloid leukemia. Curr Treat Options Oncol 8:277–286
Schimmer AD, O'Brien S, Kantarjian H et al (2008) A phase I study of the pan bcl-2 family inhibitor obatoclax mesylate in patients with advanced hematologic malignancies. Clin Cancer Res 14:8295–8301
Schoenwaelder SM, Jarman KE, Gardiner EE et al (2011) Bcl-X_L-inhibitory BH3 mimetics can induce a transient thrombocytopathy that undermines the hemostatic function of platelets. Blood 118:1663–1674
Shoemaker AR, Oleksijew A, Bauch J et al (2006) A small-molecule inhibitor of Bcl-X_L potentiates the activity of cytotoxic drugs in vitro and in vivo. Cancer Res 66:8731–9
Skommer J, Das SC, Nair A, Brittain T, Raychaudhuri S (2011) Nonlinear regulation of commitment to apoptosis by simultaneous inhibition of Bcl-2 and XIAP in leukemia and lymphoma cells. Apoptosis 16:619–626
Smith BD, Bambach BJ, Vala MS et al (1998) Inhibited apoptosis and drug resistance in acute myeloid leukaemia. Br J Haematol 102:1042–1049
Souers AJ, Leverson JD, Boghaert ER et al (2013) ABT-1999, a potent and selective Bcl-2 inhibitor, achieves antitumor activity while sparing platelets. Nat Med 19:202–208
van Stijn A, van der Pol MA, Kok A et al (2003) Differences between the CD34+ and CD34− blast compartments in apoptosis resistance in acute myeloid leukemia. Haematologica 88:497–508
van Stijn A, Feller N, Kok A, van der Pol MA, Ossenkoppele GJ, Schuurhuis GJ (2005) Minimal residual disease in acute myeloid leukemia is predicted by an apoptosis-resistant protein profile at diagnosis. Clin Cancer Res 11:2540–6

Suarez L, Vidriales B, Garcia-Larana J et al (2001) Multiparametric analysis of apoptotic and multidrug resistance phenotypes according to the blast cell maturation stage in elderly patients with acute myeloid leukemia. Haematologica 86:1287–1295

Suarez L, Vidriales MB, Garcia-Larana J et al (2004) CD34+ cells from acute myeloid leukemia, myelodysplastic syndromes, and normal bone marrow display different apoptosis and drug resistance-associated phenotypes. Clin Cancer Res 10:7599–7606

Suarez L, Vidriales MB, Moreno MJ et al (2005) Differences in antiapoptotic and multidrug resistance phenotypes in elderly and young acute myeloid leukemia patients are related to the maturation of blast cells. Haematologica 90:54–59

Tang R, Faussat AM, Majdak P et al (2006) Semisynthetic homoharringtonine induces apoptosis via inhibition of protein synthesis and triggers rapid myeloid cell leukemia-1 down-regulation in myeloid leukemia cells. Mol Cancer Ther 5:723–731

Testa U, Riccioni R (2007) Deregulation of apoptosis in acute myeloid leukemia. Haematologica 92:81–94

Tian D, Das SG, Doshi JM, Peng J, Lin J, Xing C (2008) sHA 14–1, a stable and ROS-free antagonist against antiapoptotic Bcl-2 proteins, bypasses drug resistances and synergizes cancer therapies in human leukemia cell. Cancer Lett 259:198–208

Tse C, Shoemaker AR, Adickes J et al (2008) ABT-263: a potent and orally bioavailable Bcl-2 family inhibitor. Cancer Res 68:3421–8

Vallera DA, Jin N, Shu Y, Panoskaltsis-Mortari A, Kelekar A, Chen W (2003) Retroviral immunotoxin gene therapy of leukemia in mice using leukemia-specific T-cell transduced with an interleukin-3/Bax fusion protein gene. Hum Gene Ther 14:1787–1798

Vandenberg CJ, Cory S (2013) ABT-199, a Bcl-2-specific BH3 mimetic, has in vivo efficacy against aggressive Myc-driven mouse lymphomas without provoking thrombocytopenia. Blood 121:2285–2288

Venditti A, Del Poeta G, Maurillo L et al (2004) Combined analysis of bcl-2 and MDR1 proteins in 256 cases of acute myeloid leukemia. Haematologica 89:934–9

Vogler M, Hamali HA, Sun XM et al (2011) BCL2/BCL-X(L) inhibition induces apoptosis, disrupts cellular calcium homeostasis, and prevents platelet activation. Blood 117:7145–7154

Walensky LD, Kung AL, Escher I et al (2004) Activation of apoptosis in vivo by a hydrocarbon-stapled BH3 helix. Science 305:1466–1470

Wang ZG, Rivi R, Delva L et al (1998) Arsenic trioxide and melarsoprol induce programmed cell death in myeloid leukemia cell lines and function in a PML and PML-RARalpha independent manner. Blood 92:1497–1504

Wang JL, Liu D, Zhang ZJ et al (2000a) Structure-based discovery of an organic compound that binds Bcl-2 protein and induces apoptosis of tumor cells. Proc Natl Acad Sci U S A 97:7124–7129

Wang JL, Zhang ZJ, Choksi S et al (2000b) Cell permeable Bcl-2 binding peptides: a chemical approach to apoptosis induction in tumor cells. Cancer Res 60:1498–1502

Wei Y, Kadia T, Tong W et al (2010) The combination of a histone deacetylase inhibitor with the Bcl-2 homology domain-3 mimetic GX15–070 has synergistic antileukemia activity by activating both apoptosis and autophagy. Clin Cancer Res 16:3923–3932

Weinlich R, Dillon CP, Green DR (2011) Ripped to death. Trends Cell Biol 21:630–637

White MJ, Schoenwaelder SM, Josefsson EC et al (2012) Caspase-9 mediates the apoptotic death of megakaryocytes and platelets, but is dispensable for their generation and function. Blood 119:4283–4290

Willis SN, Chen L, Dewson G et al (2005) Proapoptotic Bak is sequestered by Mcl-1 and Bcl-X_L, but not Bcl-2, until displaced by BH3-only proteins. Genes Dev 19:1294–1305

Wilson WH, O'Connor OA, Czuczman MS et al (2010) Navitoclax, a targeted high-affinity inhibitor of BCL-2, in lymphoid malignancies: a phase 1 dose-escalation study of safety, pharmacokinetics, pharmacodynamics, and antitumour activity. Lancet Oncol 11:1149–1159

Xiang Z, Luo H, Payton JE et al (2010) Mcl1 haploinsufficiency protects mice from Myc-induced acute myeloid leukemia. J Clin Invest 120:2109–2118

Yamaguchi H, Inokuchi K, Dan K (2002) The study for loss of bcl-xs expression as a prognostic factor in acute myeloid leukemia. Leuk Res 26:1119–1123

Yamaguchi R, Perkins G (2012) Finding a Panacea among combination cancer therapies. Cancer Res 72:18–23

Yang X, Liu L, Sternberg D et al (2005) The FLT3 internal tandem duplication mutation prevents apoptosis in interleukin-3-deprived BaF3 cells due to protein kinase A and ribosomal S6 kinase 1-mediated BAD phosphorylation at serine 112. Cancer Res 65:7338–7347

Yin S, Wang R, Zhou F, Zhang H, Jing Y (2011) Bcl-X_L is a dominant antiapoptotic protein that inhibits homoharringtonine-induced apoptosis in leukemia cells. Mol Pharmacol 79:1072–1083

Yinjun L, Jie J, Weilai X, Xiangming T (2004) Homoharringtonine mediates myeloid cell apoptosis via upregulation of pro-apoptotic bax and inducing caspase-3-mediated cleavage of poly(ADP-ribose) polymerase (PARP). Am J Hematol 76:199–204

Yoshimoto G, Miyamoto T, Jabbarzadeh-Tabrizi S et al (2009) FLT3-ITD up-regulates MCL-1 to promote survival of stem cells in acute myeloid leukemia via FLT3-ITD-specific STAT5 activation. Blood 114:5034–5043

Yu C, Rahmani M, Dai Y et al (2003) The lethal effects of pharmacological cyclin-dependent kinase inhibitors in human leukemia cells proceed through a phosphatidylinositol 3-kinase/Akt-dependent process. Cancer Res 63:1822–1833

Yu C, Bruzek LM, Meng XW et al (2005) The role of Mcl-1 downregulation in the proapoptotic activity of the multikinase inhibitor BAY 43–9 Oncogene 24:6861–6869

Zhang H, Nimmer PM, Tahir SK et al (2007a) Bcl-2 family proteins are essential for platelet survival. Cell Death Differ 14:943–951

Zhang L, Ming L, Yu J (2007b) BH3 mimetics to improve cancer therapy; mechanisms and examples. Drug Resist Updat 10:207–217

Zhang M, Ling Y, Yang CY et al (2007c) A novel Bcl-2 small molecule inhibitor 4-(3-methoxyphenylsulfannyl)-7-nitro-benzofurazan-3-oxide (MNB)-induced apoptosis in leukemia cells. Ann Hematol 86:471–481

Zhang W, Konopleva M, Ruvolo VR et al (2008) Sorafenib induces apoptosis of AML cells via Bim-mediated activation of the intrinsic apoptotic pathway. Leukemia 22:808–818

Zhang W, Ruvolo VR, Gao C, Zhou L, Bornmann W, Tsao T, Schober WD, Smith P, Guichard S, Konopleva M, Andreeff M (2014) Evaluation of apoptosis induction by concomitant inhibition of MEK, mTOR, and Bcl-2 in human acute myelogenous leukemia cells. Mol Cancer Ther 13(7):1848–1859

Zhao S, Konopleva M, Cabreira-Hansen M et al (2004) Inhibition of phosphatidylinositol 3-kinase dephosphorylates BAD and promotes apoptosis in myeloid leukemias. Leukemia 18:267–275

Zheng A, Mantymaa P, Saily M, Siitonen T, Savolainen ER, Koistinen P (1999) An association between mitochondrial function and all-trans retinoic acid-induced apoptosis in acute myeloblastic leukaemia cells. Br J Haematol 105:215–224

Chapter 5
IAP Family of Proteins as Therapeutic Targets for Acute Myeloid Leukemia

Bing Z. Carter and Michael Andreeff

Abstract The inhibitor of apoptosis protein (IAP) family is a group of proteins that are critical regulators of cell survival. They also play roles in cell cycle, cell signaling, and cell migration. The expression of many IAPs is deregulated in acute myeloid leukemia (AML) and other malignant cells and contributes to tumor formation, progression, and maintenance; drug resistance and treatment failure; and poor prognosis. In the last decade, abundant knowledge has been accumulated regarding the structure, expression, regulation, function, and mechanism of action of IAPs, which has led to significant progress in developing strategies to antagonize IAPs. Many candidate compounds have been investigated and tested in the preclinical setting, and some of them have moved to clinical development. This chapter highlights current knowledge of the expression and roles of IAPs in AML and discusses current strategies exploiting IAPs as therapeutic targets for AML.

Keywords AML · Survivin · XIAP · cIAPs · Apoptosis · Antisense oligonucleotide · SMAC mimetics

5.1 Introduction

Chemotherapy is the primary treatment for patients with acute myeloid leukemia (AML). Chemoresistance leading to disease relapse is a major obstacle to curing these patients and is often attributed to the deregulation of apoptosis in AML cells, particularly AML stem cells. Consequently, anti-apoptotic proteins can be excellent targets for cancer therapy. The inhibitor of apoptosis protein (IAP) family is an evolutionarily conserved group of proteins that have one to three copies of a well-conserved domain of about 70 amino acid (i.e., the baculovirus IAP repeat

---

B. Z. Carter (✉) · M. Andreeff
Department of Leukemia, Section of Molecular Hematology and Therapy,
The University of Texas MD Anderson Cancer Center, 1515 Holcombe Blvd # 448,
77030-4009 Houston, TX, USA
e-mail: bicarter@mdanderson.org

M. Andreeff
e-mail: mandreef@mdanderson.org

M. Andreeff (ed.), *Targeted Therapy of Acute Myeloid Leukemia*,
Current Cancer Research, DOI 10.1007/978-1-4939-1393-0_5

(BIR) domain) that are an important part of the apoptosis machinery. They directly and indirectly regulate caspase activity and cell survival pathways through their BIR domains and/or ubiquitin-ligase RING zinc finger activity. The expression of IAPs is frequently deregulated in many human cancers because of genetic aberrations; inappropriate induction by growth factors, cytokines, oncogenes, or multiple signaling pathways that are often constitutively active in malignant cells and their microenvironments; or the loss of tumor suppressors or endogenous IAP inhibitors. The pro-survival function of IAPs, their overexpression in leukemic cells, and their roles in tumorigenesis, progression, and treatment resistance make them attractive therapeutic targets for AML and cancer in general. This chapter will discuss the expression patterns, roles, and therapeutic targeting of survivin, X-linked inhibitor of apoptosis (XIAP), cellular inhibitors of apoptosis proteins (cIAPs), the most thoroughly characterized IAPs in AML.

5.2 Expression and Roles of IAPs in AML

5.2.1 Survivin

Survivin is a unique IAP in terms of structure, function, and regulation. Survivin is the smallest IAP, with only one BIR domain (Ambrosini et al. 1997) and is the only IAP that plays roles in both cell cycle regulation and inhibition of cell death, both of which favor cancer survival. It is part of the Aurora B-survivin-INCENP-Borealin/Dasra B complex, the chromosomal passenger complex essential for cell cycle progression and cytokinesis (Gassmann et al. 2004; Honda et al. 2003). The mechanism by which survivin suppresses cell death is not fully elucidated, but probably relies on the interaction between survivin and its partner proteins (Carter and Andreeff 2008; Dohi et al. 2004; Song et al. 2003).

The survivin gene promoter contains GC-rich sequences for its basal Sp1-induced transcription and multiple cell cycle response elements, and its expression is developmentally regulated and cell cycle dependent (Adida et al. 1998; Li et al. 1998; Li and Altieri 1999; Xu et al. 2007). Survivin expression becomes undetectable in most terminally differentiated normal tissues but is reactivated in most malignant cells (Ambrosini et al. 1997). In fact, survivin is one of the most frequently upregulated transcripts in cancer (Velculescu et al. 1999). Survivin expression is also regulated in a cell cycle-independent manner: survivin expression is repressed by wild-type p53 and upregulated by oncogenes and multiple signaling pathways (PI3K, MAPK, Bcr-Abl, STAT3, Wnt/β-catenin, and NF-κB) and microenvironmental factors (hypoxia and angiogenesis; Carter et al. 2001, 2006a; Kim et al. 2003; Lu et al. 2011; Mahboubi et al. 2001; Mitsiades et al. 2002; Papapetropoulos et al. 2000; Pennati et al. 2007; Yang et al. 2004; You et al. 2004). Lack of functional p53 occurs frequently in malignant cells either by TP53 gene mutations or by aberrant p53 signaling. Constitutive activation of either single or multiple signaling pathways is a common phenomenon in cancer cells and malignant hematopoietic cells, and the tumor microenvironment secretes a variety of growth factors and cytokines and is highly hypoxic. Therefore, cell cycle-independent regulation of survivin constitutes one major mechanism of survivin expression in neoplastic cells.

Overexpression of survivin has been recognized in a number of hematological malignancies including AML (Fulda 2009). Survivin is highly expressed in AML blasts, and its expression is regulated by hematopoietic cytokines through MAPK and PI3K signaling (Carter et al. 2001). Recently, survivin expression was found to be stimulated by the AML1/ETO fusion protein in AMLs carrying the t(8;21)(q22;q22) chromosomal translocation (Balkhi et al. 2008). Survivin was reported to also be upregulated by FLT3-ITD and to regulate the expansion of FLT3-ITD-transformed hematopoietic progenitor cells with self-renewal capacity and the development of FLT3-ITD acute leukemia in mice (Fukuda et al. 2009). The role of survivin in promoting leukemogenesis was further supported by a study showing that overexpression of survivin initiates hematologic malignancies in transgenic mice (Small et al. 2010). Interestingly, the expression of survivin in the bone marrow (BM) of patients with acute mixed-lineage leukemia, as determined by quantitative reverse transcription polymerase chain reaction (RT-PCR), was even higher than that in patients with either AML or acute lymphocytic leukemia (ALL) and the expression levels correlated with the chemoresistance of the disease (Nakagawa et al. 2005). Survivin was also found to be associated with disease progression in myelodysplastic syndrome (MDS), and interestingly, survivin levels were higher in patients with secondary AML following MDS than those with de novo AML (Yoshida et al. 2012).

Although survivin is highly expressed in AML, its prognostic impact is not fully established. Some found that survivin expression predicts poor clinical outcome, whereas others found that it did not (Adida et al. 2000; Carter et al. 2003a; Oto et al. 2007; Sugahara et al. 2004). This lack of a definitive result is most likely related to small sample sizes. Taking advantage of robustness and small sample requirement of the reverse-phase protein array (RPPA) method, we recently profiled survivin expression in samples from 511 newly diagnosed AML patients and found that survivin levels predict poor clinical outcomes in AML (i.e., higher survivin levels predicted shorter overall and event-free survival as determined by multivariate Cox model analysis) (Carter et al. 2011c). We also discovered that more primitive AML cells had correspondingly higher survivin levels. $CD34^{+}38^{-}$ AML stem/progenitor cells have significantly higher survivin expression than $CD34^{+}$ and bulk AML cells (Carter et al. 2011c), which suggests that survivin may have roles in AML stem cell survival, self-renewal, and disease relapse. Survivin's high expression and pro-survival role in AML cells, overexpression in AML stem cells, and association with poor clinical outcome in AML patients make it an attractive target for AML therapy. Indeed, survivin inhibition by antisense oligonucleotides (ASOs) induces cell proliferation defects and subsequent cell death in AML cells (Carter et al. 2003b).

5.2.2 XIAP

XIAP contains three BIR domains and is the best-characterized IAP in terms of its structure and mechanisms of action. XIAP is a widely expressed IAP (Deveraux and Reed 1999) and the only IAP that directly binds and inhibits caspases (Eckelman et al. 2006). XIAP acts by inhibiting caspase-3 and -7 through its BIR1-BIR2 linker region and BIR2 domain and caspase-9 through its BIR3 domain (Deveraux et al. 1999; Riedl et al. 2001; Scott et al. 2005; Shiozaki et al. 2003). It also degrades caspases through its E3 ligase activity (Morizane et al. 2005; Suzuki et al. 2001b).

XIAP has received great attention as a therapeutic target for numerous reasons. For example, it is the most potent IAP that blocks both the intrinsic and extrinsic apoptosis pathways. Furthermore, XIAP expression is elevated in various cancers and cancer cell lines. For example, XIAP is highly expressed in leukemia and lymphomas (Carter et al. 2003, 2003a; Tamm et al. 2000). Its overexpression is also associated with chemoresistance (Notarbartolo et al. 2004; Wagenknecht et al. 1999), and inhibiting XIAP sensitizes nonsmall cell lung carcinoma to low-dose γ-irradiation and increases apoptosis in chemoresistant human ovarian cancer (Holcik et al. 2000; Sasaki et al. 2000). XIAP inhibition also induces apoptosis and enhances the chemosensitivity of human prostate cancer cells (Amantana et al. 2004). Targeting XIAP with ASOs delayed tumor growth in a lung cancer xenograft animal model (Hu et al. 2003). Interestingly, although XIAP is ubiquitously expressed in normal tissues (Duckett et al. 1996; Liston et al. 1996), XIAP knockout mice were normal at birth and displayed no significant pathological abnormalities or apoptotic defects (Harlin et al. 2001). Despite reports that XIAP deficiency increases sympathetic neuron susceptibility to cytochrome *c*-induced cell death (Potts et al. 2003) and delays lobuloalveolar development in the mouse mammary gland (Olayioye et al. 2005), XIAP inhibition seems to be tolerable to normal adult cells, which further increases its appeal as a potential cancer treatment.

XIAP is highly expressed in AML and upregulated by cytokines and the MAPK/PI3K signaling pathways supporting its pro-survival roles in this leukemia (Carter et al. 2003). The prognostic impact of XIAP expression in AML remains debatable. Two studies reported that low levels of XIAP expression correlated with better overall survival and favorable cytogenetics in AML (Tamm et al. 2000, 2004), whereas another study found no prognostic impact of XIAP expression in AML (Carter et al. 2003a). Nevertheless, inhibition of XIAP by ASOs induces apoptosis in and enhances the chemosensitivity of leukemic cells (Carter et al. 2003), and the inhibition of XIAP activity by small-molecule XIAP inhibitors promotes apoptotic cell death in both leukemic cell lines and primary AML blasts (Carter et al. 2005). Furthermore, XIAP is a downstream effector of multiple signal transduction cascades. It also activates NF-κB (Hofer-Warbinek et al. 2000; Levkau et al. 2001) and PI3K/AKT (Asselin et al. 2001) signals, which suggests that inhibition of XIAP has broad potential therapeutic implications. Clearly, XIAP is a promising therapeutic target for various cancers and leukemia.

5.2.3 cIAPs

Although most early studies focused on survivin and XIAP, more recent work has supported critical roles of cIAPs in cell survival and NF-κB signal transduction pathways. Unlike XIAP, cIAP1 and cIAP2 do not directly inhibit caspases (Eckelman et al. 2006). Rather, they prevent the assembly of the apoptotic death receptor complex and primarily regulate the caspase-8-mediated extrinsic apoptosis pathway. They also enhance XIAP-mediated caspase inhibition by binding to endogenous IAP antagonist Second Mitochondria-derived Activator of Caspases (SMAC) with high affinity. Like in XIAP, the E3 ubiquitin ligase activity in the RING domain of the cIAPs allows cIAPs to regulate the stability of many cell death regulators that directly affect cell survival, such as SMAC (Hu and Yang 2003; Vucic et al. 2011). cIAP1 and cIAP2 regulate TNFα-mediated canonical NF-κB activation by acting as ubiquitin E3 ligases for RIP1 and noncanonical NF-κB signaling by promoting ubiquitination and proteasomal degradation of the NF-κB-inducing kinase (NIK), thereby activating the pro-survival canonical NF-κB signaling pathway and inhibiting the noncanonical NF-κB signaling pathway (Varfolomeev and Vucic 2008).

Amplifications of the 11q21–11q23 region, which harbors both cIAP1 and cIAP2, have been identified in various solid tumors (Bashyam et al. 2005; Dai et al. 2003; Imoto et al. 2001, 2002; Zender et al. 2006). The t(11;18) (q21;q21) translocation results in fusion of the BIR domain of cIAP2 with the paracaspase mucosa-associated lymphoid tissue lymphoma translocation protein 1 (MALT1), and occurs frequently in mucosa-associated lymphoid tissue lymphoma (Akagi et al. 1999; Dierlamm et al. 1999), in which the fusion protein constitutively activates the NF-κB signaling pathway (Morgan et al. 1999; Varfolomeev et al. 2006; Zhou et al. 2005). Upregulation of cIAP2 was found in human T-cell leukemia virus type 1-transformed adult T-cell leukemia (Waldele et al. 2006). Inconsistent results in chronic lymphocytic leukemia (CLL) showing that expression of cIAPs predicted clinical outcomes and resistance to apoptosis induction (Grzybowska-Izydorczyk et al. 2010; Vallat et al. 2003) and that cIAPs did not correlate with prognostic factors and treatment response (Silva et al. 2006) were reported.

The expression and function of cIAP1 and cIAP2 in AML have not been extensively investigated. Because NF-κB and cIAPs positively regulate each other and NF-κB is constitutively active in primitive human AML cells (Guzman et al. 2001), it is likely that cIAPs are overexpressed in primitive human AML cells as well. In a study using multiple apoptosis-related transcripts to identify the combination of genes that best predicted overall survival, a three-gene signature including high cIAP2 mRNA expression correlated with poor overall survival in AML (Hess et al. 2007). A recent study illustrated that deregulated IAP expression and apoptotic signaling differentiate clinically relevant core-binding factor (CBF) AML. A subgroup of CBF AML was resistant to IAP antagonist SMAC mimetic BV-6, and the gene expression profile of this subgroup resembles that of a previously identified unfavorable CBF-subgroup (Luck et al. 2011), which suggests that expression of IAPs has an impact on the clinical outcome of CBF AML. Using RPPA, we recently discovered that cIAP1 was variably expressed in samples from 511 newly diagnosed AML patients. Importantly, we discovered that $CD34^{+}38^{-}$ AML stem/progenitor cells expressed significantly higher cIAP1 than bulk and $CD34^{+}$ AML cells (Carter et al. 2011a), which suggests that cIAP1 plays a role in AML stem/progenitor cells and that cIAP1 is a potential target for AML cells including AML stem/progenitor cells.

The prognostic values of IAPs in AML and any other cancer need to be interpreted with caution. Many studies did not have large enough sample sizes to give statistically meaningful results, which is exacerbated by the fact that AML is a heterogeneous disease. Whether samples were taken from newly diagnosed, treated, or resistant/relapsed patients was not mentioned in most cases. Antibodies were also from different sources, and different techniques were used by different investigators (such as western blot versus immunohistochemistry and measurement of protein levels versus RNA levels). Current and future technical breakthroughs will help us gain a deeper and richer understanding of the roles of IAPs in leukemogenesis, disease progression, treatment responses, and clinical outcome and their regulations in genetic, epigenetic, noncoding RNA, mRNA, and protein levels with large sample sizes and minimal amounts of material required for each sample. Regardless of their prognostic impact, high expression, pro-survival activity, contributions to drug resistance, and upregulation by oncogenes and multiple signaling pathways that are constitutively activated in AML and cancer in general make IAPs ideal therapeutic targets.

5.2.4 Negative Regulators of IAPs

IAPs are negatively regulated by a number of cellular proteins including SMAC (also known as DIABLO; Du et al. 2000; Srinivasula et al. 2000), XIAP-associated factor 1 (XAF1; Liston et al. 2001), apoptosis-related protein in the TGFβ signaling pathway (ARTS, also known as SEPT4; Gottfried et al. 2004), and high-temperature requirement protein A2 (HtrA2, also known as Omi; Hegde et al. 2002; Martins et al. 2002; Suzuki et al. 2001a; Verhagen et al. 2002). This group of proteins has direct impact on the activities of IAPs and is the basis for the development of some IAP inhibitors, thereby warranting discussion here.

SMAC, a mitochondrial protein, is the most-studied and best-understood endogenous inhibitor of IAPs. Once matured in the cytoplasm, SMAC binds to IAP proteins predominantly via its N-terminal tetrapeptide AVPI sequence, suppresses XIAP-mediated caspase inhibition, and causes cIAP auto-ubiquitination and degradation leading to apoptosis (Chai et al. 2000). Several SMAC mimetics, a new class of anticancer agents, designed based on the N-terminal tetrapeptide AVPI sequence, have entered clinical testing. Low or no expression of SMAC has been reported in various solid tumors and correlated with poor clinical outcomes (Mizutani et al. 2005; Pluta et al. 2011; Sekimura et al. 2004). Expression of SMAC was evaluated in one study by flow cytometry in newly diagnosed AML patients treated with conventional chemotherapy and was found to be an independent prognostic factor in AML: high levels of SMAC protein expression were correlated with longer overall survival and higher complete remission rate and were associated with favorable karyotype (Pluta et al. 2010). Using RPPA, we discovered that SMAC levels were significantly lower and cIAP1 levels significantly higher in $CD34^{+}38^{-}$ AML stem/progenitor cells than in bulk and $CD34^{+}$ AML cells (Carter et al. 2011a), which suggest that SMAC mimetics can target AML cells, particularly AML stem/progenitor cells. These findings agree with reports demonstrating that IAPs are ubiquitin ligase for cytosolic SMAC (Hu and Yang 2003; MacFarlane et al. 2002; Morizane et al. 2005).

XAF1 binds to XIAP and blocks its activity. Low expression, in part due to epigenetic modification, of XAF-1 has been found in many cancer cells (Byun et al. 2003; Fong et al. 2000) and often correlated with disease stage, tumor grade, and poor clinical outcome (Chung et al. 2007; Huang et al. 2010; Kempkensteffen et al. 2009; Lee et al. 2006). We found that the XAF-1 protein level increases in AML cells treated with demethylating agents, and combinations of SMAC mimetics and demethylating agents synergistically promote apoptosis in AML cell lines, primary AML cells, and AML stem/progenitor cells (Carter et al. 2011a). In addition to antagonizing XIAP, XAF1 activates p53 via posttranslational modification, resulting in nuclear accumulation and increased transcriptional activity of p53 and enhanced p53-dependent apoptosis (Zou et al. 2011). Interestingly, the expression levels of XAF1 were shown to modulate the cytotoxic response of B-CLL cells to nutlin3a (Secchiero et al. 2010), consistent with its proposed pro-apoptotic role. If XAF-1 indeed antagonizes XIAP, this result agrees with our finding that inhibition of XIAP and activation of p53 by nutlin3a enhanced apoptosis induction in AML cells (Carter et al. 2010).

ARTS is a mitochondrial released protein that promotes apoptosis through direct binding and degradation of XIAP in a mechanism distinct from other IAP-antagonists (Bornstein et al. 2011; Gottfried et al. 2004). Evidence from human and mouse studies indicates that ARTS functions as an XIAP antagonist and a tumor suppressor protein and plays an important role in leukemogenesis. ARTS-deficient mice have higher XIAP levels, have more stem/progenitor cells, are more resistance to cell death, and develop spontaneous hematopoietic tumors; all of these phenotypes can be suppressed by XIAP inactivation (Garcia-Fernandez et al. 2010). XIAP in turn promotes degradation of the ARTS protein (Bornstein et al. 2012). ARTS is known to be deleted or methylated in many malignant cells, and expression of ARTS is absent in lymphoblasts from most ALL patients (Elhasid et al. 2004). ARTS expression and function in AML cells have not been explored. Inhibition of XIAP by ASOs in a clinical trial preferentially promoted death of $CD34^{+}38^{-}$ AML stem/progenitor cells (Carter et al. 2011b), which suggests that AML stem/progenitor cells may rely more on XIAP than bulk AML cells for survival and that these cells may have decreased levels of ARTS. It will be interesting to determine ARTS levels in AML blasts and stem/progenitor cells. Because of the potential role of ARTS in cancer stem cells, ARTS-based compounds could be a good choice for targeting AML stem cells and worth pursuing. To prove the concept, peptides mimicking the unique ARTS-XIAP binding sites have been made, and they were able to decrease XIAP levels and promote apoptosis in several cancer cell lines (Edison et al. 2012).

HtrA2 is a member of the HtrA family of serine proteases. The mature form of HtrA2 contains an N-terminal tetrapeptide motif (AVPS) that mediates its interaction with IAPs through their BIR domains, which leads to caspase activation and apoptosis induction (Hegde et al. 2002; Martins et al. 2002; Verhagen et al. 2002). However, whether HtrA2 has a role in cancer is inconclusive. Increased and decreased expressions of HtrA2 have both been reported in cancer cells (Zurawa-Janicka et al. 2010). A recent study on CLL examining the expression and prognostic significance of IAPs and their antagonists showed no difference in HtrA2 levels between samples from CLL and normal controls; conversely, levels of SMAC were decreased in CLL patients, and the levels were even lower in patients with progressive disease than in those with stable CLL (Grzybowska-Izydorczyk et al. 2010). The expression and role of HtrA2 in AML have not been reported.

5.3 IAP Family of Proteins as Therapeutic Targets for AML

5.3.1 Survivin

Because of its selective expression in cancer cells and role in tumor cell biology, survivin has emerged as an attractive target for cancer therapy. Survivin expression is downregulated by p53 and upregulated by multiple pro-survival signaling pathways that are constitutively active in cancer, including AML. Therefore, activation of p53 and inhibition of pro-survival signaling pathways can broadly suppress survivin levels. In this section, we will focus on compounds selectively targeting survivin that have been tested clinically, including ASOs, survivin peptide-derived vaccines, and the small-molecule survivin suppressors YM155 and terameprocol.

5.3.1.1 ASO

LY2181308 is a survivin ASO originally developed by ISIS Pharmaceuticals (ISIS23722) that binds to survivin mRNA and inhibits its expression. Preclinical investigation of ISIS23722 demonstrated that it induces apoptosis, inhibits cell proliferation and tumor growth, and sensitizes chemo- and radiotherapies in a variety of malignant cells and in human xenografts in animal models. We showed that in AML cells, survivin is essential for cell cycle progression, and targeting survivin expression with ISIS23722 induces cell proliferation defects and subsequent cell death (Carter et al. 2003b).

A dose escalation study of LY2181308 in 40 patients with various solid tumors showed accumulation of LY2181308 within tumor tissues, reduction of both survivin RNA and protein levels, and restoration of apoptosis signaling in tumor cells in vivo at the recommended phase II dose of 750 mg (Talbot et al. 2010). Phase I study of LY2181308 in 14 patients with advanced solid tumors showed that LY2181308 was well tolerated at doses up to 750 mg with favorable safety, tolerability, pharmacokinetics, and phamarcodynamic profiles (Tanioka et al. 2011). Phase II studies combining LY2181308 and conventional chemotherapeutic agents are ongoing in patients with nonsmall cell lung cancer and hormone-refractory prostate cancer. Recently, a phase II trial of LY2181308 in combination with idarubicin and cytarabine in patients with refractory or relapsed AML was completed and results have not been reported.

Recent studies demonstrated that survivin expression is also regulated by microRNAs (miRNAs). miRNA-203, for example, was found to directly target survivin (Bian et al. 2012), which suggests the applicability of anti-survivin miRNAs in cancer therapies. Although ASOs have advantages of target specificity over small-molecule inhibitors, miRNAs have great potential for therapy because they target a number of protein-coding genes that often belong to the same pathway at the same time, thus making gene targeting more efficient. However, as for ASOs, delivery of miRNAs is a major challenge. Clinical successes of oligonucleotide-based therapies rely on overcoming this technical hurdle. With rapid advances in technology, we anticipate that the great therapeutic potentials of miRNA will become reality in the near future.

5.3.1.2 Vaccine

Vaccines derived from survivin peptides have been developed and several of them have entered trials in patients with advanced cancers. No patients with AML, however, have been included in these studies. Reports from phase I trials in advanced solid tumors showed the vaccines were safe and warrant further development (Miyazaki et al. 2011). However, no further studies following the trials have been announced at this time.

5.3.1.3 Small-Molecule Inhibitors

Despite great efforts, small molecular survivin inhibitors for clinical development are limited. YM155 is a novel, small imidazolium-based compound that has been characterized as a selective survivin suppressant using a survivin gene promoter activity assay and has demonstrated antitumor activity in cell lines and in xenografts of human tumors (Nakahara et al. 2007). YM155 as a single agent in phase I clinical trials did not reveal significant toxicity. It has entered phase II clinical trials in various solid tumors and lymphoma and showed limited single-agent activity. Combination trials are ongoing (Cheson et al. 2011; Giaccone et al. 2009; Lewis et al. 2011; Tolcher et al. 2012). Although YM155 has shown activity in a broad spectrum of cancer cell lines and xenografts of human cancer, its effect in AML, in either cell lines, primary patient samples, or animal models of human AML xenografts, has not been evaluated.

Terameprocol (EM-1421) is a derivative of nordihydroguaiaretic acid originally isolated from *Larrea tridentata*. Studies demonstrated that EM-1421 exerts its activity by selective inhibition of SP-1 regulated proteins, including survivin and Cdc2, and therefore inhibits cell cycle progression and induces apoptosis in various solid tumor cells (Heller et al. 2001; Park et al. 2005). The compound is currently in clinical trials for the treatment of solid tumors. Evaluation of EM-1421 in AML cells showed that it inhibited cell growth and induced cell death in both leukemic cell lines and AML patient samples but failed to inhibit Cdc2 and survivin (Mak et al. 2007). A phase I/II clinical trial of terameprocol as a single agent was carried out in patients with various hematologic malignancies, including AML. The drug was safe, the maximum tolerated dose was established, and potential clinical activity in AML was observed (Tibes et al. 2009). The update of the trial has not been reported.

Other anticancer agents acting at least in part by inhibiting survivin have been described. For example, survivin is a client protein of Hsp90. Inhibition of Hsp90 by 17-AAG decreases survivin level. In fact, shepherdin, an engineered cell-permeable peptidomimetic modeled on the binding interface between Hsp90 and survivin, inhibited human tumor growth in mice without toxicity (Plescia et al. 2005). Shepherdin potently induced apoptosis in AML blasts in vitro and abolished the growth of AML xenografts in mice (Gyurkocza et al. 2006). Whether this approach will enter clinical development remains to be seen. The Wnt/β-catenin pathway is as recognized essential for the self-renewal of leukemia stem cells (Wang et al. 2010). Survivin is a known downstream target of the Wnt/β-catenin pathway (Lu et al. 2011; You et al. 2004). The finding that survivin levels are particularly high in AML stem/progenitor cells is intriguing (Carter et al. 2011c). Inhibitors of the Wnt/β-catenin pathway are under clinical development for the treatment of both solid tumors and leukemia. Targeting this pathway may hold promise for the inhibition of survivin and for the elimination of AML stem cells. It is important to determine whether Wnt/β-catenin inhibitors decrease survivin expression in AML stem/progenitor cells and whether this reduction translates into clinical benefits for patients.

5.3.2 XIAP

Potent caspase inhibition activity, suppression of both intrinsic and extrinsic apoptosis cascades, and high levels of expression in malignant cells make XIAP an ideal target for cancer therapy. Developing ASOs and identifying small-molecule inhibitors are the major strategies.

5.3.2.1 ASO

Currently, XIAP ASOs are the most advanced agents under clinical developments that selectively target XIAP. Preclinical studies demonstrated that AEG35156, a second-generation XIAP ASO, reduced XIAP protein levels and sensitized cancer cells to diverse chemotherapies in vitro and in murine models of human cancer xenografts (LaCasse et al. 2006; McManus et al. 2004). A phase I trial of AEG35156 revealed favorable profiles of tolerability, toxicity, and pharmacokinetics as well as clinical evidence of antitumor activity in patients with refractory lymphoma, melanoma, and breast cancers (Dean et al. 2009). A phase I//II trial of AEG35156 in combination with idarubicin and cytarabine was conducted in patients with relapsed or refractory AML. AEG35156 treatment led to dose-dependent reduction of XIAP mRNA. At a dose of 350 mg/m^2, AEG35156 was effective in knocking down XIAP in both RNA and protein levels and inducing apoptosis in circulating blasts. Interestingly, the apoptosis induction was more prevalent in $CD34^+38^-$ AML stem/progenitor cell compartments (Carter et al. 2011b). AEG35156 was also well tolerated. At a dose of 350 mg/m^2 AEG35156 and chemotherapy, 15 of 32 (47%) patients achieved complete response (CR) or CR with incomplete platelet count recovery. Among them, 10 of 11 (91%) were refractory to a single induction chemotherapy regimen, which suggests that AEG35156 when combined with chemotherapy is particularly effective in patients with AML refractory to a single induction regimen (Schimmer et al. 2009). On the basis of this outcome, an open-label randomized phase II trial was conducted to compare the efficacy of reinduction chemotherapy in patients with AML who did not achieve remission with initial induction chemotherapy with or without AEG35156. Although the addition of AEG35156 to reinduction chemotherapy was well tolerated, it, unfortunately, did not improve remission rates (Schimmer et al. 2011). In this phase II trial, however, no correlative study was conducted. Whether target knockdown and apoptosis induction, particularly in $CD34^+38^-$ AML stem/progenitor cells, were achieved (as in the original trial) is unknown; therefore, no definite conclusion can be drawn.

5.3.2.2 Small-Molecule Inhibitors

Over the past decade, a number of small XIAP inhibitors designed to selectively target XIAP have been described (Orzaez et al. 2011; Schimmer et al. 2004; Wu et al. 2003) but none have been successful to date. Among them, polyphenylureas are a class of small-molecule nonpeptide XIAP antagonists that were selected on the basis of their ability to overcome XIAP-mediated suppression of caspase-3 and -7, induce apoptosis of many types of tumor cell lines, and sensitize cancer cells to chemotherapeutic drugs in vitro and to suppress growth of tumors in xenograft models in mice (Schimmer et al. 2004). They also effectively induced apoptosis of leukemia cells and blasts from AML patients via activation of caspase-3 and -7 (Carter et al. 2005). However, recently recognized off-target activities prevented their further developments.

A number of natural products demonstrating anticancer activities in part through inhibition of XIAP have been described. Embelin, from the Japanese *Ardisia* herb, binds to the BIR3 domain of XIAP, activates caspase-9, induces apoptosis, inhibits cell growth, and overcomes the anti-apoptotic effect of XIAP (Nikolovska-Coleska et al. 2004). It induced apoptosis in HL-60 human AML cells through downregulation of XIAP (Hu et al. 2011). Triptolide, a diterpenoid isolated from a Chinese herb *Tripterygium wilfordii* Hook.f, was shown to have antitumor properties by suppressing the growth of and inducing apoptosis in a broad spectrum of human tumor cells (Chen 2001; Kiviharju et al. 2002; Shamon et al. 1997; Yang et al.1998, 2003). It decreased XIAP RNA and protein levels and potently induced apoptosis in AML cells (Carter et al. 2006b). A water-soluble triptolide derivative, PG490–88Na (F60008) (Fidler et al. 2003; Kiviharju et al. 2002; Kitzen et al. 2009), has entered clinical development and has shown strong antitumor activity. Importantly, complete remissions in patients with refractory or relapsing acute leukemias were reported in a phase I study (Harousseau et al. 2008). MRx102 is another derivative of triptolide currently under preclinical evaluation. Studies in AML demonstrated that MRx102 decreases XIAP protein levels, has potent antileukemic activity both in vitro and in a murine xenograft model of human AML, and has the potential to eliminate AML stem/progenitor cells and overcome microenvironmental protection of leukemic cells (Carter et al. 2012) and therefore warrants clinical investigation. Spongisatin 1, a therapeutic agent isolated from marine sponges, induces apoptosis by degradation of XIAP with high efficiency in cells from patients with primary acute leukemia (Schyschka et al. 2008).

5.3.3 cIAPs

SMAC antagonizes multiple IAPs. It binds with the SMAC-binding groove of XIAP-BIR3 (Shiozaki et al. 2003) and also with the binding groove on the surface of XIAP-BIR2 (Scott et al. 2005) and therefore directly blocks the interaction of XIAP with caspase-9, -3, and -7. cIAP1 and cIAP2 also have high affinities for SMAC, presumably because they promote XIAP-mediated caspase inhibition by sequestering SMAC (Duckett 2005; Shin et al. 2005). SMAC-derived peptides sensitize various tumor cell lines to apoptosis induced by a variety of pro-apoptotic molecules both in vitro and in vivo (Arnt et al. 2002; Fulda et al. 2002; Guo et al. 2002).

Increasing understanding of the roles of SMAC, cIAP1, and cIAP2 and the interactions between SMAC and IAPs has prompted the development of SMAC mimetics as a strategy to target IAPs in cancer therapy. Over the past decade, more than 50 patents and patent applications pertaining to IAP antagonists have been filed and some of these SMAC mimetics have entered human clinical trials.

SMAC mimetics are a class of agents that mimic the amino terminus of the endogenous SMAC protein and thus block critical IAP protein interactions. In fact, the biological function of cIAPs was characterized because of investigations into the mechanism of action of SMAC mimetics and the preclinical testing of these compounds. SMAC mimetics were initially designed to imitate the binding of endogenous SMAC with XIAP and thereby to prevent the XIAP-mediated inhibition of caspases. Further investigation revealed that they have more profound effects on the levels of cIAPs, especially cIAP1. Their binding to cIAPs changed the conformation of cIAPs and led to induction of cIAP1/cIAP2 E3 ligase activity and rapid auto-ubiquitination and proteasomal degradation (Varfolomeev et al. 2007; Vince et al. 2007).

In the absence of cIAPs, TNFα induces activation of caspase-8 and the extrinsic apoptosis. Clearly, the single-agent activity of SMAC mimetics depends on the autocrine production of TNFα, which explains the lack of activity of these compounds in many in vitro cell line-based studies, especially in hematopoietic malignancies. In vivo, however, TNFα production from the microenvironment may be sufficient to assist SMAC mimetics for activation of the death receptor-mediated apoptosis.

SMAC mimetics can be divided into three groups: monovalent SMAC-derived peptides, monovalent SMAC-derived peptidomimetics, and bivalent SMAC-derived peptidomimetics. The peptide-based compounds lack the physicochemical properties that favor clinical development because they are unstable and low in cell permeability. Consequently, recent studies have been focused on identifying peptidomimetics with improved pharmacological properties and IAP binding affinities.

5.3.3.1 Monovalent SMAC-Derived Peptidomimetic Compounds

Monovalent SMAC-derived peptidomimetic compounds were designed to target the BIR3 domain of XIAP. A number of them have been described by both academic groups and pharmaceutical companies. The compounds that showed activities and had potential as AML therapies will be discussed.

LBW242, developed by Novartis, is one of the most well-known SMAC mimetics. It has been tested in various cancer cells and showed antitumor activity in cells derived from a number of hematological malignancies. LBW242 sensitized human precursor B-cell leukemia derived Nalm-6 cells to prednisone-induced apoptosis (Hundsdoerfer et al. 2010). It also promoted apoptosis in multiple myeloma cell lines and samples from patients that were resistant to bortezomib and conventional therapies and overcame the protective effect of BM stromal cells without significant toxicity to normal lymphocytes in vitro and was well tolerated, inhibited tumor growth, and prolonged in vivo survival in a human multiple myeloma xenograft murine model (Chauhan et al. 2007). Preclinical study in AML showed that the combination of LBW242 with either tyrosine kinase inhibitor PKC412 or chemotherapeutic agents such as doxorubicin and Ara-C enhanced the in vitro killing of both PKC412-sensitive and PKC412-resistant cell lines expressing mutant FLT3. Furthermore, the combination of LBW242 and PKC412 was able to overcome stroma-induced resistance in Molm13 cells to PKC412 and lowered the tumor burden in mice harboring Ba/F3FLT3-ITD cells significantly more than either agent alone (Weisberg et al. 2007).

LCL161 is an orally bioavailable structural analog of LBW242 with improved potency and pharmacokinetics and is a candidate for clinical development. Phase I trials to evaluate safety, establish the maximum tolerated dose of LCL161 in solid tumors, and appraise the effect of its combination with weekly paclitaxel in adult patients with advanced solid tumors have been conducted. Early reports demonstrated reduction of cIAP1 protein levels and showed that LCL161 was well tolerated in patients (Infante et al. 2010). Preclinical studies in cell lines, primary AML patient samples, and a xenograft murine model of human AML demonstrated that LCL161 has antileukemic effect regardless of FLT3 mutation status and potentiates the effects of tyrosine kinase inhibitors in vitro (even in the presence of stromal protection) and in vivo (Weisberg et al. 2010). The drug has yet to be tested clinically in AML.

AT-406 is an orally active SMAC mimetic developed by Ascenta Therapeutics. AT-406 has demonstrated strong single-agent antitumor activity in multiple xenograft solid tumor models of human cancer, including breast cancer, pancreatic cancer, prostate cancer, and lung cancer. It has also been shown to work synergistically with conventional chemotherapeutic and targeted agents (such as TRAIL and tyrosine kinase inhibitors) in preclinical tumor models. Clinical trials of AT-406 in patients with a variety of solid tumors and lymphomas are being conducted in the USA by Ascenta. A clinical trial of AT-406 for patients with hematologic malignancies including AML is currently under planning.

5.3.3.2 Bivalent SMAC-Derived Peptidomimetic Compounds

Bivalent SMAC-derived peptidomimetics consist of monovalent counterparts connected via appropriate linkers. They were generated to increase affinity to IAPs due to their ability to interact simultaneously with the BIR2 and BIR3 domains of XIAP. Several of them, such as HGS1029, BV-6, SM-164, and TL37211, have been described as showing activity in vitro and in vivo in mouse models. Because of their increased molecular mass, bivalent SMAC-derived peptidomimetic compounds require intravenous administration that may diminish their effectiveness, whereas monovalent SMAC-derived peptidomimetics are orally bioavailable and allow for easier dosing. However, bivalent compounds are, in general, more potent than monovalent compounds for apoptosis induction. Clinical trials will decide which group of compounds has better pharmacodynamic and pharmacokinetic profiles and therapeutic efficacy in patients.

TL32711 (birinapant) is a SMAC peptidomimetic selected from a library of over 2500 rationally designed synthetic IAP antagonists developed by TetroLogic Pharmaceuticals. TL32711 caused tumor regression as a single agent and had potent antitumor activity when combined with chemotherapies and death receptor ligands in preclinical studies. TL32711 also exhibited superior cytotoxicity against AML cells. It induced apoptosis in AML cells, including $CD34^{+}38^{-}$ AML stem/progenitor cells, alone and in combination with chemotherapy (Carter et al. 2011a).

TL32711 is thought to have improved tolerability and therapeutic index compared to other SMAC mimetics and is the most advanced SMAC mimetic in terms of clinical development. A single-agent phase I study in solid tumors and lymphomas and a phase Ib five-arm combination clinical study in solid tumors are nearing completion. In phase I clinical studies as a single agent and in combination with standard-of-care chemotherapies, TL32711 was able to cause potent and sustained suppression of cIAP1 protein in patient's peripheral blood mononuclear cells and tumor biopsies and has demonstrated strong correlations between drug exposure, target reduction, and apoptosis induction in tumors at well-tolerated doses and promising antitumor activity in patients (Amaravadi et al. 2011). TL32711 is entering phase II trials as a single agent and part of a combination therapy for solid tumors and a phase I/II clinical study for AML. A phase II open label non-randomized study of TL32711 in combination with 5-azacytidine in elderly patients with AML has also been planned.

BV-6, another bivalent SMAC-derived peptidomimetic developed by Genentech, had no single-agent activity in AML cells but sensitized AML cells to therapeutic agent-induced apoptosis (Mak et al. 2010). Genentech does not plan to further develop the compound.

SMAC mimetics are superior to compounds selectively targeting a single IAP not only because they antagonize multiple IAPs and have broader activity but also because IAP proteins may functionally compensate each other (e.g., the levels of cIAP1 and cIAP2 were found to increase in XIAP-deficient mice; Harlin et al. 2001). In addition to SMAC, IAPs are negatively regulated by other cellular molecules such as ARTS and XAF-1. Their potential in therapeutic interventions is largely unexplored. Elucidating the precise mechanism of their interaction will provide options to design additional and possibly better molecules to target IAPs.

5.3.4 Combination of IAP Targeting with Various Therapies

IAPs play critical roles in cancer cell survival and drug resistance. Inhibiting IAPs not only directly activates apoptosis machinery but also lowers the threshold of cell death induction. Therefore, it is believed that inhibiting IAP expression or function sensitizes cells to other therapeutic drugs, which is supported by a vast amount of preclinical data. Many clinical trials discussed above combine IAP targeting agents with standard-of-care chemotherapeutic drugs or plan to do so in forthcoming trials. Future studies should emphasize combinations of IAP antagonists with other molecular targeted therapies to further improve therapeutic efficacy with less toxicity. Designing rational combinations requires deeper understanding of the mechanism of action of IAP inhibitors and the regulation of apoptosis and related signaling pathways. cIAPs primarily suppress the activation of the extrinsic apoptosis pathway, and SMAC mimetics should logically enhance apoptosis induced by death receptor ligands or any agents modulating key protein components in the extrinsic apoptosis pathway. Indeed, IAP inhibition sensitizes cells to TRAIL-induced cell death (Fulda and Vucic 2012). Mechanistic studies have recognized that cell death induction by SMAC mimetics depends on TNFα released by autocrine production or paracrine supply (Li et al. 2004; Petersen et al. 2007; Varfolomeev et al. 2007; Vince et al. 2007; Wang et al. 2008). Furthermore, simultaneous inhibition of XIAP and activation of p53 enhances the activation of apoptosis signaling pathways and synergistically promotes apoptosis of AML cells in part through the following mechanisms: XIAP inhibition potentiates p53-induced apoptosis by decreasing p53-induced p21 level; and p53 activation enhances XIAP inhibition-induced cell death by promoting mitochondrial release of SMAC and by inducing the expression of caspase-6 (Carter et al. 2010). p53 is known to upregulate death receptor DR5. Combination of SMAC mimetics with p53-activating agent nutlin3a could have a synergistic effect on the killing of leukemia cells. We recently found that combinations of SMAC mimetics and demethylating agents greatly sensitized AML cells to apoptosis in part via demethylating agent-mediated inhibition of IAPs and NF-κB signaling (Carter et al. 2011a). Importantly, p53 activation by nutlin3a as therapeutic strategy in AML is under clinical development, and demethylating agents are being used for elderly patients with AML. A phase II open-label non-randomized study of TL32711 in combination with 5-azacytidine in elderly patients with AML has been planned. In addition, a combination of SMAC mimetics with tyrosine kinase inhibitors overcame resistance and enhanced the effect of tyrosine kinase inhibitors (Bai et al. 2012; Weisberg et al. 2007, 2010). Some of these inhibitors are in clinical use or under clinical development.

It is important to point out that the BM microenvironment plays critical roles in leukemogenesis, disease progression, and drug resistance (Konopleva and Jordan 2011). Combinations of SMAC mimetics with other therapeutic agents are likely more effective in overcoming microenvironmental protection of leukemia cells. In particular, the effect of combining SMAC mimetics and agents that interrupt leukemia and stroma interaction, such as CXCR4 antagonists, should be explored. Targeting the leukemia microenvironment by CXCR4 inhibition was able to overcome resistance to various therapeutic agents in AML cells in vitro, in in vivo mouse models, and in patients with AML (Konopleva et al. 2009; Uy et al. 2012; Zeng et al. 2009). Combinations of SMAC with the aforementioned agents should be feasible in the near future and are expected to achieve higher efficacy.

It is also essential to develop biomarkers that can be used in preclinical study to assist in designing optimal therapeutic combinations and in clinical settings to help identify patients that are most likely to benefit from the trial drug and combinations. Furthermore, for combination treatments, carefully designed scheduling and drug sequencing are needed to achieve desired clinical benefit.

Conclusion

In recent years, IAPs have emerged as promising targets for cancer therapy, including AML treatment, because of their elevated expression, their roles in cancer cell survival and drug resistance, and their association with poor clinical outcomes. ASOs against these proteins and several small-molecule IAP antagonists, in particular SMAC mimetics, have been developed and are currently being evaluated in clinical trials in AML and other malignancies. Future studies should emphasize better understanding of other natural IAP inhibitors and their interaction with IAPs to develop improved IAP antagonists and rationally combine SMAC mimetics with other therapeutic agents to maximize apoptotic induction and minimize drug resistance and microenvironmental protection to leukemic cells. We hope to see all these research efforts translate into better treatment strategies and improved outcomes for patients with AML in the near future. IAP-based targeted therapy holds promise for enhancing and improving current chemotherapies for AML.

References

Adida C, Crotty PL, McGrath J, Berrebi D, Diebold J Altieri DC (1998) Developmentally regulated expression of the novel cancer anti-apoptosis gene survivin in human and mouse differentiation. Am J Pathol 152(1):43–49

Adida C, Recher C, Raffoux E, Daniel MT, Taksin AL, Rousselot P, Sigaux F, Degos L, Altieri DC, Dombret H (2000) Expression and prognostic significance of survivin in de novo acute myeloid leukaemia. Br J Haematol 111:196–203. (available from: PM:11091201)

Akagi T, Motegi M, Tamura A, Suzuki R, Hosokawa Y, Suzuki H, Ota H, Nakamura S, Morishima Y, Taniwaki M, Seto M (1999) A novel gene, MALT1 at 18q21, is involved in t(11;18)(q21;q21) found in low-grade B-cell lymphoma of mucosa-associated lymphoid tissue. Oncogene 18(42):5785–5794. (available from: PM:10523859)

Amantana A, London CA, Iversen PL, Devi GR (2004) X-linked inhibitor of apoptosis protein inhibition induces apoptosis and enhances chemotherapy sensitivity in human prostate cancer cells. Mol Cancer Ther 3(6):699–707

Amaravadi RK, Schilder RJ, Dy GK, Ma WW, Fetterly GJ, Weng DE, Graham MA, Burns JM, Chunduru SK, Condon SM, McKinlay MA, Adjel AA (2011) Phase 1 study of the Smac TL32711 in adult subjects with advanced solid tumors and lymphoma to mimetic evaluate safety, pharmacokinetics, pharmacodynamics, and antitumor activity. AACR March 2011

Ambrosini G, Adida C, Altieri DC (1997) A novel anti-apoptosis gene, survivin, expressed in cancer and lymphoma. Nat Med 3(8):917–921. (available from: PM:9256286)

Arnt CR, Chiorean MV, Heldebrant MP, Gores GJ, Kaufmann SH (2002) Synthetic Smac/DIABLO peptides enhance the effects of chemotherapeutic agents by binding XIAP and cIAP1 in situ. J Biol Chem 277:44236–44243

Asselin E, Mills GB, Tsang BK (2001) XIAP regulates Akt activity and caspase -3-dependent cleavage during cisplatin-induced apoptosis in human ovarian epithelial cancer cells. Cancer Res 61(5):1862–1868. (available from: PM:11280739)

Bai L, McEachern D, Yang CY, Lu J, Sun H, Wang S (2012) LRIG1 modulates cancer cell sensitivity to Smac Mimetics by regulating TNFalpha expression and receptor tyrosine kinase signaling. Cancer Res 72(5):1229–1238. (available from: PM:22241084)

Balkhi MY, Christopeit M, Chen Y, Geletu M, Behre G (2008) AML1/ETO-induced survivin expression inhibits transcriptional regulation of myeloid differentiation. Exp Hematol 36(11):1449–1460

Bashyam MD, Bair R, Kim YH, Wang P, Hernandez-Boussard T, Karikari CA, Tibshirani R, Maitra A, Pollack JR (2005) Array-based comparative genomic hybridization identifies localized DNA amplifications and homozygous deletions in pancreatic cancer. Neoplasia 7(6):556–562. (available from: PM:16036106)

Bian K, Fan J, Zhang X, Yang XW, Zhu HY, Wang L, Sun JY, Meng YL, Cui PC, Cheng SY, Zhang J, Zhao J, Yang AG, Zhang R (2012) MicroRNA-203 leads to G1 phase cell cycle arrest in laryngeal carcinoma cells by directly targeting survivin. FEBS Lett 586(6):804–809. (available from: PM:22306317)

Bornstein B, Gottfried Y, Edison N, Shekhtman A, Lev T, Glaser F, Larisch S (2011) ARTS binds to a distinct domain in XIAP-BIR3 and promotes apoptosis by a mechanism that is different from other IAP-antagonists. Apoptosis 16(9):869–881. (available from: PM:21695558)

Bornstein B, Edison N, Gottfried Y, Lev T, Shekhtman A, Gonen H, Rajalingam K, Larisch S (2012) X-linked inhibitor of apoptosis protein promotes the degradation of its antagonist, the pro-apoptotic ARTS protein. Int J Biochem Cell Biol 44(3):489–495. (available from: PM:22185822)

Byun DS, Cho K, Ryu BK, Lee MG, Kang MJ, Kim HR, Chi SG (2003) Hypermethylation of XIAP-associated factor 1, a putative tumor suppressor gene from the 17p13.2 locus, in human gastric adenocarcinomas. Cancer Res 63(21):7068–7075. (available from: PM:14612497)

Carter BZ, Andreeff M (2008) Targeting survivin in leukemia. Oncol Rev 2:19–28

Carter BZ, Milella M, Altieri DC, Andreeff M (2001) Cytokine-regulated expression of *survivin* in myeloid leukemia. Blood 97:2784–2790

Carter BZ, Milella M, Tsao T, McQueen T, Schober WD, Hu W, Dean NM, Steelman L, McCubrey JA, Andreeff M (2003) Regulation and targeting of antiapoptotic XIAP in acute myeloid leukemia. Leukemia 17:2081–2089

Carter BZ, Kornblau SM, Tsao T, Wang RY, Schober WD, Milella M, Sung HG, Reed JC, Andreeff M (2003a) Caspase-independent cell death in AML: caspase inhibition in vitro with pan-caspase inhibitors or in vivo by XIAP or Survivin does not affect cell survival or prognosis. Blood 102:4179–4186

Carter BZ, Wang R-Y, Schober WD, Milella M, Chism D, Andreeff M (2003b) Targeting survivin expression induces cell proliferation defect and subsequent cell death involving the mitochondrial pathway in myeloid leukemic cell. Cell Cycle 2:488–493

Carter BZ, Gronda M, Wang Z, Welsh K, Pinilla C, Andreeff M, Schober WD, Nefzi A, Houghten RA, Brandwein J, Minden MD, Schuh A, Wells RA, Chun K, Reed JC, Schimmer AD (2005) Small-molecule XIAP inhibitors derepress downstream effector caspases and induce apoptosis of acute myeloid leukemia cells. Blood 105:4043–4050

Carter BZ, Mak D, Schober WD, Cabreira-Hansen M, Beran M, McQueen T, Chen W, Andreeff M (2006a) Regulation of survivin expression through bcr-abl/MAPK cascade: targeting survivin overcomes Imatinib resistance and increases Imatinib sensitivity in Imatinib responsive CML cells. Blood 107:1555–1563

Carter BZ, Mak DH, Schober WD, McQueen T, Harris D, Estrov Z, Evans RL, Andreeff M (2006b) Triptolide induces caspase-dependent cell death mediated via the mitochondrial pathway in leukemic cells. Blood 108:630–637

Carter BZ, Mak DH, Schober WD, Estrove Z, Koller E, Pinilla C, Vassilev LT, Reed JC, Andreeff M (2010) Simultaneous activation of p53 and inhibition of XIAP enhance the activation of apoptosis signaliing pathways in AML. Blood 115(2):306–314

Carter BZ, Mak DH, Qiu Y, Kornblau SM, Mak PY, Weng D, McKinlay MA, Andreeff M (2011a) Antagonizing IAPs by SMAC Mimetic TL32711 induces Apoptosis in AML cells Including AML Stem/Progenitor Cells Alone and in Combination with Chemotherapy. Blood 118(21):32

Carter BZ, Mak DH, Morris SJ, Borthakur G, Estey E, Byrd AL, Konopleva M, Kantarjian H, Andreeff M (2011b) XIAP antisense oligonucleotide (AEG35156) achieves target knockdown and induces apoptosis preferentially in CD34^{+}38^{-} cells in a phase 1/2 study of patients with relapsed/refractory AML. Apoptosis 16(1):67–74. (available from: PM:20938744)

Carter BZ, Qiu Y, Huang X, Dia L, Zhang N, Coombes KR, Mak DH, Konopleva M, Cortes JE, Kantarjian HM, Andreeff M, Kornblau SM (2011c) Survivin is highly expressed in AML stem cells and predicts poor clinical outcome. Blood 118(21):108–109

Carter B, Mak DH, Shi Y, Fidler JM, Chen R, Ling X, Plunkett W, Andreeff M (2012) MRx102, a triptolide derivative, has potent antileukemic activity in vitro and in a murine model of AML. Leukemia 26(3):443–450. (available from: PM:21904380)

Chai J, Du C, Wu JW, Kyin S, Wang X, Shi Y (2000) Structural and biochemical basis of apoptotic activation by Smac/DIABLO. Nature 406:855–862

Chauhan D, Neri P, Velankar M, Podar K, Hideshima T, Fulciniti M, Tassone P, Raje N, Mitsiades C, Mitsiades N, Richardson P, Zawel L, Tran M, Munshi N, Anderson KC (2007) Targeting mitochondrial factor Smac/DIABLO as therapy for multiple myeloma (MM). Blood 109(3):1220–1227

Chen BJ (2001) Triptolide, a novel immunosuppressive and anti-inflammatory agent purified from a Chinese herb Tripterygium wilfordii Hook F. Leuk Lymphoma 42(3):253–265

Cheson BD, Bartlett NL, Vose JM, Lopez-Hernandez A, Seiz AL, Keating AT, Shamsili S (2011) A phase II study of the survivin suppressant YM155 in patients with refractory diffuse large B-cell lymphoma. Cancer 118(12):128–134. (available from: PM:22006123)

Chung SK, Lee MG, Ryu BK, Lee JH, Han J, Byun DS, Chae KS, Lee KY, Jang JY, Kim HJ, Chi SG (2007) Frequent alteration of XAF1 in human colorectal cancers: implication for tumor cell resistance to apoptotic stresses. Gastroenterology 132(7):2459–2477. (available from: PM:17570219)

Dai Z, Zhu WG, Morrison CD, Brena RM, Smiraglia DJ, Raval A, Wu YZ, Rush LJ, Ross P, Molina JR, Otterson GA, Plass C (2003) A comprehensive search for DNA amplification in lung cancer identifies inhibitors of apoptosis cIAP1 and cIAP2 as candidate oncogenes. Hum Mol Genet 12(7):791–801. (available from: PM:12651874)

Dean E, Jodrell D, Connolly K, Danson S, Jolivet J, Durkin J, Morris S, Jowle D, Ward T, Cummings J, Dickinson G, Aarons L, LaCasse E, Robson L, Dive C, Ranson M (2009) Phase I trial of AEG35156 administered as a 7-day and 3-day continuous intravenous infusion in patients with advanced refractory cancer. J Clin Oncol 27(10):1660–1666

Deveraux QL, Reed JC (1999) IAP family proteins—suppressors of apoptosis. Genes Dev 13: 239–252. (available from: PM:9990849)

Deveraux QL, Leo E, Stennicke HR, Welsh K, Salvesen GS, Reed JC (1999) Cleavage of human inhibitor of apoptosis protein XIAP results in fragments with distinct specificities for caspases. EMBO J 18:5242–5251

Dierlamm J, Baens M, Wlodarska I, Stefanova-Ouzounova M, Hernandez JM, Hossfeld DK, De Wolf-Peeters C, Hagemeijer A, Van den Berghe H, Marynen P (1999) The apoptosis inhibitor gene API2 and a novel 18q gene, MLT, are recurrently rearranged in the t(11;18)(q21;q21) associated with mucosa-associated lymphoid tissue lymphomas. Blood 93(11):3601–3609

Dohi T, Okada K, Xia F, Wilford CE, Samuel T, Welsh K, Marusawa H, Zou H, Armstrong R, Matsuzawa S, Salvesen GS, Reed JC, Altieri DC (2004) An IAP-IAP complex inhibits apoptosis. J Biol Chem 279(33):34087–34090. (available from: PM:15218035)

Du C, Fang M, Li Y, Li L, Wang X (2000) Smac, a mitochondrial protein that promotes cytochrome *c*-dependent caspase activation by eliminating IAP inhibition. Cell 102(1):33–42. (available from: PM:10929711)

Duckett CS, Nava VE, Gedrich RW, Clem RJ, Van Dongen JL, Gilfillan MC, Shiels H, Hardwick JM, Thompson CB (1996) A conserved family of cellular genes related to the baculovirus IAP gene and encoding apoptosis inhibitors. EMBO J 15(11):2685–2694

Duckett CS (2005) IAP proteins: sticking it to Smac. Biochem J 385 (Pt 1) e1–2. (available from: PM:15588250)

Eckelman BP, Salvesen GS, Scott FL (2006) Human inhibitor of apoptosis proteins: why XIAP is the black sheep of the family. EMBO Rep 7(10):988–994. (available from: PM:17016456)

Edison N, Reingewertz TH, Gottfried Y, Lev T, Zuri D, Maniv I, Carp MJ, Shalev G, Friedler A, Larisch S (2012) Peptides mimicking the unique ARTS-XIAP binding site promote apoptotic cell death in cultured cancer cells. Clin Cancer Res 18(9):2569–2578. (available from: PM:22392914)

Elhasid R, Sahar D, Merling A, Zivony Y, Rotem A, Ben-Arush M, Izraeli S, Bercovich D, Larisch S (2004) Mitochondrial pro-apoptotic ARTS protein is lost in the majority of acute lymphoblastic leukemia patients. Oncogene 23(32):5468–5475. (available from: PM:15122323)

Fidler JM, Li K, Chung C, Wei K, Ross JA, Gao M, Rosen GD (2003) PG490-88, a derivative of triptolide, causes tumor regression and sensitizes tumors to chemotherapy. Mol Cancer Ther 2(9):855–862

Fong WG, Liston P, Rajcan-Separovic E, St Jean M, Craig C, Korneluk RG (2000) Expression and genetic analysis of XIAP-associated factor 1 (XAF1) in cancer cell lines. Genomics 70(1):113–122. (available from: PM:11087668)

Fukuda S, Singh P, Moh A, Abe M, Conway EM, Boswell HS, Yamaguchi S, Fu XY, Pelus LM (2009) Survivin mediates aberrant hematopoietic progenitor cell proliferation and acute leukemia in mice induced by internal tandem duplication of Flt3. Blood 114(2):394–403. (available from: PM:19411632)

Fulda S (2009) Inhibitor of apoptosis proteins in hematological malignancies. Leukemia 23(3):467–476

Fulda S, Vucic D (2012) Targeting IAP proteins for therapeutic intervention in cancer. Nat Rev Drug Discov 11(2):109–124. (available from: PM:22293567)

Fulda S, Wick W, Weller M, Debatin KM (2002) Smac agonists sensitize for Apo2L/TRAIL- or anticancer drug-induced apoptosis and induce regression of malignant glioma in vivo. Nat Med 8:808–815

Garcia-Fernandez M, Kissel H, Brown S, Gorenc T, Schile AJ, Rafii S, Larisch S, Steller H (2010) Sept4/ARTS is required for stem cell apoptosis and tumor suppression. Genes Dev 24(20):2282–2293. (available from: PM:20952537)

Gassmann R, Carvalho A, Henzing AJ, Ruchaud S, Hudson DF, Honda R, Nigg EA, Gerloff DL, Earnshaw WC (2004) Borealin: a novel chromosomal passenger required for stability of the bipolar mitotic spindle. J Cell Biol 166(2):179–191. (available from: PM:15249581)

Giaccone G, Zatloukal P, Roubec J, Floor K, Musil J, Kuta M, van Klaveren RJ, Chaudhary S, Gunther A, Shamsili S (2009) Multicenter phase II trial of YM155, a small-molecule suppressor of survivin, in patients with advanced, refractory, non-small-cell lung cancer. J Clin Oncol 27(27):4481–4486. (available from: PM:19687333)

Gottfried Y, Rotem A, Lotan R, Steller H, Larisch S (2004) The mitochondrial ARTS protein promotes apoptosis through targeting XIAP. EMBO J 23(7):1627–1635. (available from: PM:15029247)

Grzybowska-Izydorczyk O, Cebula B, Robak T, Smolewski P (2010) Expression and prognostic significance of the inhibitor of apoptosis protein (IAP) family and its antagonists in chronic lymphocytic leukaemia. Eur J Cancer 46(4):800–810. (available from: PM:20045309)

Guo F, Nimmanapalli R, Paranawithana S, Wittman S, Griffin D, Bali P, O'Bryan E, Fumero C, Wang HG, Bhalla K (2002) Ectopic overexpression of second mitochondria-derived activator of caspases (Smac/DIABLO) or cotreatment with N-terminus of Smac/DIABLO peptide potentiates epothilone B derivative-(BMS 247550) and Apo-2L/TRAIL-induced apoptosis. Blood 99(9):3419–3426. (available from: PM:11964312)

Guzman ML, Neering SJ, Upchurch D, Grimes B, Howard DS, Rizzieri DA, Luger SM, Jordan CT (2001) Nuclear factor-kappaB is constitutively activated in primitive human acute myelogenous leukemia cells. Blood 98(8):2301–2307

Gyurkocza B, Plescia J, Raskett CM, Garlick DS, Lowry PA, Carter BZ, Andreeff M, Meli M, Colombo G, Altieri DC (2006) Antileukemic activity of shepherdin and molecular diversity of hsp90 inhibitors. J Natl Cancer Inst 98(15):1068–1077. (available from: PM:16882944)

Harlin H, Reffey SB, Duckett CS, Lindsten T, Thompson CB (2001) Characterization of XIAP-deficient mice. Mol Cell Biol 21:3604–3608

Harousseau JLH, Dombret HD, Pigneux AP, Michallet MM, Brandely MB (2008) Phase I study of F60008, a triptolide derivative, in patients with refractory or relapsing acute leukemias. Hematol J 93:14–15.

Hegde R, Srinivasula SM, Zhang Z, Wassell R, Mukattash R, Cilenti L, DuBois G, Lazebnik Y, Zervos AS, Fernandes-Alnemri T, Alnemri ES (2002) Identification of Omi/HtrA2 as a mitochondrial apoptotic serine protease that disrupts inhibitor of apoptosis protein-caspase interaction. J Biol Chem 277(1):432–438

Heller JD, Kuo J, Wu TC, Kast WM, Huang RC (2001) Tetra-O-methyl nordihydroguaiaretic acid induces G2 arrest in mammalian cells and exhibits tumoricidal activity in vivo. Cancer Res 61(14):5499–5504

Hess CJ, Berkhof J, Denkers F, Ossenkoppele GJ, Schouten JP, Oudejans JJ, Waisfisz Q, Schuurhuis GJ (2007) Activated intrinsic apoptosis pathway is a key related prognostic parameter in acute myeloid leukemia. J Clin Oncol 25(10):1209–1215. (available from: PM:17401010)

Hofer-Warbinek R, Schmid JA, Stehlik C, Binder BR, Lipp J, de Martin R (2000) Activation of NF-kappa B by XIAP, the X chromosome-linked inhibitor of apoptosis, in endothelial cells involves TAK1. J Biol Chem 275:22064–22068. (available from: PM:10807933)

Holcik M, Yeh C, Korneluk RG, Chow T (2000) Translational upregulation of X-linked inhibitor of apoptosis (XIAP) increases resistance to radiation induced cell death. Oncogene 19:4174–4177. (available from: PM:10962579)

Honda R, Korner R, Nigg EA (2003) Exploring the functional interactions between Aurora B, INCENP, and survivin in mitosis. Mol Biol Cell 14:3325–3341

Hu S, Yang X (2003) Cellular inhibitor of apoptosis 1 and 2 are ubiquitin ligases for the apoptosis inducer Smac/DIABLO. J Biol Chem 278(12):10055–10060. (available from: PM:12525502)

Hu Y, Cherton-Horvat G, Dragowska V, Baird S, Korneluk RG, Durkin JP, Mayer LD, LaCasse EC (2003) Antisense oligonucleotides targeting XIAP induce apoptosis and enhance chemotherapeutic activity against human lung cancer cells in vitro and in vivo. Clin Cancer Res 9(7):2826–2836

Hu R, Zhu K, Li Y, Yao K, Zhang R, Wang H, Yang W, Liu Z (2011) Embelin induces apoptosis through down-regulation of XIAP in human leukemia cells. Med Oncol 28(4)1584–1588. (available from: PM:20625944)

Huang J, Yao WY, Zhu Q, Tu SP, Yuan F, Wang HF, Zhang YP, Yuan YZ (2010) XAF1 as a prognostic biomarker and therapeutic target in pancreatic cancer. Cancer Sci 101(2):559–567. (available from: PM:19922503)

Hundsdoerfer P, Dietrich I, Schmelz K, Eckert C, Henze G (2010) XIAP expression is post-transcriptionally upregulated in childhood ALL and is associated with glucocorticoid response in T-cell ALL. Pediatr Blood Cancer 55(2):260–266. (available from: PM:20582956)

Imoto I, Yang ZQ, Pimkhaokham A, Tsuda H, Shimada Y, Imamura M, Ohki M, Inazawa J (2001) Identification of cIAP1 as a candidate target gene within an amplicon at 11q22 in esophageal squamous cell carcinomas. Cancer Res 61(18):6629–6634. (available from: PM:11559525)

Imoto I, Tsuda H, Hirasawa A, Miura M, Sakamoto M, Hirohashi S, Inazawa J (2002) Expression of cIAP1, a target for 11q22 amplification, correlates with resistance of cervical cancers to radiotherapy. Cancer Res 62:4860–4866

Infante JR, Dees EC, Burris HA, Zawel L, Sager JA, Stevenson C, Clarke K, Dhuria S, Porter D, Sen SK, Zannou E, Sharma S, Cohen RB, Cannon S (2010) A phase I study of LCL161, an oral IAP inhibitor, in patients with advanced cancer. AACR March

Kempkensteffen C, Fritzsche FR, Johannsen M, Weikert S, Hinz S, Dietel M, Riener MO, Moch H, Jung K, Krause H, Miller K, Kristiansen G (2009) Down-regulation of the pro-apoptotic XIAP associated factor-1 (XAF1) during progression of clear-cell renal cancer. BMC Cancer 9:276. (available from: PM:19664236)

Kim PJ, Plescia J, Clevers H, Fearon ER, Altieri DC (2003) Survivin and molecular pathogenesis of colorectal cancer. Lancet 362:205–209

Kitzen JJ, de Jonge MJ, Lamers CH, Eskens FA, van der Biessen D, van Doorn L, Ter SJ, Brandely M, Puozzo C, Verweij J (2009) Phase I dose-escalation study of F60008, a novel apoptosis inducer, in patients with advanced solid tumours. Eur J Cancer 45(10):1764–1772

Kiviharju TM, Lecane PS, Sellers RG, Peehl DM (2002) Antiproliferative and proapoptotic activities of triptolide (PG490), a natural product entering clinical trials, on primary cultures of human prostatic epithelial cells. Clin Cancer Res 8(8):2666–2674

Konopleva M, Tabe Y, Zeng Z, Andreeff M (2009) Therapeutic targeting of microenvironmental interactions in leukemia: mechanisms and approaches. Drug Resist Updat 12(4–5):103–113. (available from: PM:19632887)

Konopleva MY, Jordan CT (2011) Leukemia stem cells and microenvironment: biology and therapeutic targeting. J Clin Oncol 29(5):591–599. (available from: PM:21220598)

LaCasse EC, Cherton-Horvat GG, Hewitt KE, Jerome LJ, Morris SJ, Kandimalla ER, Yu D, Wang H, Wang W, Zhang R, Agrawal S, Gillard JW, Durkin JP (2006) Preclinical characterization of AEG35156/GEM 640, a second-generation antisense oligonucleotide targeting X-linked inhibitor of apoptosis. Clin Cancer Res 12(17):5231–5241

Lee MG, Huh JS, Chung SK, Lee JH, Byun DS, Ryu BK, Kang MJ, Chae KS, Lee SJ, Lee CH, Kim JI, Chang SG, Chi SG (2006) Promoter CpG hypermethylation and downregulation of XAF1 expression in human urogenital malignancies: implication for attenuated p53 response to apoptotic stresses. Oncogene 25(42):5807–5822. (available from: PM:16909101)

Levkau B, Garton KJ, Ferri N, Kloke K, Nofer JR, Baba HA, Raines EW, Breithardt G (2001) xIAP induces cell-cycle arrest and activates nuclear factor-kappaB: new survival pathways disabled by caspase-mediated cleavage during apoptosis of human endothelial cells. Circ Res 88:282–290. (available from: PM:11179195)

Lewis KD, Samlowski W, Ward J, Catlett J, Cranmer L, Kirkwood J, Lawson D, Whitman E, Gonzalez R (2011) A multi-center phase II evaluation of the small molecule survivin suppressor YM155 in patients with unresectable stage III or IV melanoma. Invest New Drugs 29(1):161–166. (available from: PM:19830389)

Li F, Altieri DC (1999) Transcriptional analysis of human survivin gene expression. Biochem J 344(Pt 2):305–311. (available from: PM:10567210)

Li F, Ambrosini G, Chu EY, Plescia J, Tognin S, Marchisio PC, Altieri DC (1998) Control of apoptosis and mitotic spindle checkpoint by survivin. Nature 396(6711):580–584

Li L, Thomas RM, Suzuki H, De Brabander JK, Wang X, Harran PG (2004) A small molecule Smac mimic potentiates TRAIL- and TNFalpha-mediated cell death. Science 305(5689):1471–1474

Liston P, Roy N, Tamai K, Lefebvre C, Baird S, Cherton-Horvat G, Farahani R, McLean M, Ikeda JE, MacKenzie A, Korneluk RG (1996) Suppression of apoptosis in mammalian cells by NAIP and a related family of IAP genes. Nature 379:349–353. (available from: PM:0008552191)

Liston P, Fong WG, Kelly NL, Toji S, Miyazaki T, Conte D, Tamai K, Craig CG, McBurney MW, Korneluk RG (2001) Identification of XAF1 as an antagonist of XIAP anti-Caspase activity. Nat Cell Biol 3(2):128–133. (available from: PM:11175744)

Lu R, Bian F, Zhang X, Qi H, Chuang EY, Pflugfelder SC, Li DQ (2011) The beta-catenin/Tcf4/survivin signaling maintains a less differentiated phenotype and high proliferative capacity of human corneal epithelial progenitor cells. Int J Biochem Cell Biol 43(5):751–759. (available from: PM:21292023)

Luck SC, Russ AC, Botzenhardt U, Paschka P, Schlenk RF, Dohner H, Fulda S, Dohner K, Bullinger L (2011) Deregulated apoptosis signaling in core-binding factor leukemia differentiates clinically relevant, molecular marker-independent subgroups. Leukemia 25(11):1728–1738. (available from: PM:21701487)

MacFarlane M, Merrison W, Bratton SB, Cohen GM (2002) Proteasome-mediated degradation of Smac during apoptosis: XIAP promotes Smac ubiquitination in vitro. J Biol Chem 277(39):36611–36616. (available from: PM:12121969)

Mahboubi K, Li F, Plescia J, Kirkiles-Smith NC, Mesri M, Du Y, Carroll JM, Elias JA, Altieri DC, Pober JS (2001) Interleukin-11 up-regulates survivin expression in endothelial cells through a signal transducer and activator of transcription-3 pathway. Lab Invest 81(3):327–334

Mak DH, Schober WD, Chen W, Heller J, Andreeff M, Carter BZ (2007) Tetra-O-methyl nordihydroguaiaretic acid inhibits growth and induces death of leukemia cells independent of Cdc2 and survivin. Leuk Lymphoma 48(4):774–785

Mak DH, Manton C, Andreeff, M, Carter BZ (2010) SMAC-Mimetic BV-6 Sensitizes Therapeutic Agents-Induced Apoptosis in AML Cells. Blood 116(21):899

Martins LM, Iaccarino I, Tenev T, Gschmeissner S, Totty NF, Lemoine NR, Savopoulos J, Gray CW, Creasy CL, Dingwall C, Downward J (2002) The serine protease Omi/HtrA2 regulates apoptosis by binding XIAP through a reaper-like motif. J Biol Chem 277:439–444

McManus DC, Lefebvre CA, Cherton-Horvat G, St Jean M, Kandimalla ER, Agrawal S, Morris SJ, Durkin JP, LaCasse EC (2004) Loss of XIAP protein expression by RNAi and antisense approaches sensitizes cancer cells to functionally diverse chemotherapeutics. Oncogene 23:8105–8117

Mitsiades CS, Mitsiades N, Poulaki V, Schlossman R, Akiyama M, Chauhan D, Hideshima T, Treon SP, Munshi NC, Richardson PG, Anderson KC (2002) Activation of NF-kappaB and upregulation of intracellular anti-apoptotic proteins via the IGF –1/Akt signaling in human multiple myeloma cells: therapeutic implications. Oncogene 21(37):5673–5683

Miyazaki A, Kobayashi J, Torigoe T, Hirohashi Y, Yamamoto T, Yamaguchi A, Asanuma H, Takahashi A, Michifuri Y, Nakamori K, Nagai I, Sato N, Hiratsuka H (2011) Phase I clinical trial of survivin-derived peptide vaccine therapy for patients with advanced or recurrent oral cancer. Cancer Sci 102(2) :324–329. (available from: PM:21143701)

Mizutani Y, Nakanishi H, Yamamoto, K, Li YN, Matsubara H, Mikami K, Okihara K, Kawauchi A, Bonavida B, Miki T (2005) Downregulation of Smac/DIABLO expression in renal cell carcinoma and its prognostic significance. J Clin Oncol 23(3):448–454. (available from: PM:15572731)

Morgan JA, Yin Y, Borowsky AD, Kuo F, Nourmand N, Koontz JI, Reynolds C, Soreng L, Griffin CA, Graeme-Cook F, Harris NL, Weisenburger D, Pinkus GS, Fletcher JA, Sklar J (1999) Breakpoints of the t(11;18)(q21;q21) in mucosa-associated lymphoid tissue (MALT) lymphoma lie within or near the previously undescribed gene MALT1 in chromosome 18. Cancer Res 59(24):6205–6213. (available from: PM:10626814)

Morizane Y, Honda R, Fukami K, Yasuda H (2005) X-linked inhibitor of apoptosis functions as ubiquitin ligase toward mature caspase-9 and cytosolic Smac/DIABLO. J Biochem 137(2):125–132. (available from: PM:15749826)

Nakagawa Y, Hasegawa M, Kurata M, Yamamoto K, Abe S, Inoue M, Takemura T, Hirokawa K, Suzuki K, Kitagawa M (2005) Expression of IAP-family proteins in adult acute mixed lineage leukemia (AMLL). Am J Hematol 78(3):173–180. (available from: PM:15726601)

Nakahara T, Takeuchi M, Kinoyama I, Minematsu T, Shirasuna K, Matsuhisa A, Kita A, Tominaga F, Yamanaka K, Kudoh M, Sasamata M (2007) YM155, a novel small-molecule survivin suppressant, induces regression of established human hormone-refractory prostate tumor xenografts. Cancer Res 67(17):8014–8021

Nikolovska-Coleska Z, Xu L, Hu Z, Tomita Y, Li P, Roller PP, Wang R, Fang X, Guo R, Zhang M, Lippman ME, Yang D, Wang S (2004) Discovery of embelin as a cell-permeable, small-molecular weight inhibitor of XIAP through structure-based computational screening of a traditional herbal medicine three-dimensional structure database. J Med Chem 47(10):2430–2440. (available from: PM:15115387)

Notarbartolo M, Cervello M, Poma P, Dusonchet L, Meli M, D'Alessandro N (2004) Expression of the IAPs in multidrug resistant tumor cells. Oncol Rep 11:133–136

Olayioye MA, Kaufmann H, Pakusch M, Vaux DL, Lindeman GJ, Visvader JE (2005) XIAP-deficiency leads to delayed lobuloalveolar development in the mammary gland. Cell Death Differ 12:87–90

Orzaez M, Gortat A, Sancho M, Carbajo RJ, Pineda-Lucena A, Palacios-Rodriguez Y, Perez-Paya E (2011) Characterization of dequalinium as a XIAP antagonist that targets the BIR2 domain. Apoptosis 16(5):460–467. (available from: PM:21340509)

Oto OA, Paydas S, Tanriverdi K, Seydaoglu G, Yavuz S, Disel U (2007) Survivin and EPR-1 expression in acute leukemias: prognostic significance and review of the literature. Leuk Res 31;(11):1495–1501. (available from: PM:17328950)

Papapetropoulos A, Fulton D, Mahboubi K, Kalb RG, O'Connor DS, Li F, Altieri DC, Sessa WC (2000) Angiopoietin-1 inhibits endothelial cell apoptosis via the Akt/survivin pathway. J Biol Chem 275(13):9102–9105.

Park R, Chang CC, Liang YC, Chung Y, Henry RA, Lin E, Mold DE, Huang RC (2005) Systemic treatment with tetra-O-methyl nordihydroguaiaretic acid suppresses the growth of human xenograft tumors. Clin Cancer Res 11(12):4601–4609

Pennati M, Folini M, Zaffaroni N (2007) Targeting survivin in cancer therapy: fulfilled promises and open questions. Carcinogenesis 28(6):1133–1139. (available from: PM:17341657)

Petersen SL, Wang L, Yalcin-Chin A, Li L, Peyton M, Minna J, Harran P, Wang X (2007) Autocrine TNFalpha signaling renders human cancer cells susceptible to Smac-mimetic-induced apoptosis. Cancer Cell 12(5):445–456. (available from: PM:17996648)

Plescia J, Salz W, Xia F, Pennati M, Zaffaroni N, Daidone MG, Meli M, Dohi T, Fortugno P, Nefedova Y, Gabrilovich DI, Colombo G, Altieri DC (2005) Rational design of shepherdin, a novel anticancer agent. Cancer Cell 7(5):457–468

Pluta A, Wrzesien-Kus A, Cebula-Obrzut B, Wolska A, Szmigielska-Kaplon A, Czemerska M, Pluta P, Robak T, Smolewski P, Wierzbowska A (2010) Influence of high expression of Smac/DIABLO protein on the clinical outcome in acute myeloid leukemia patients. Leuk Res 34(10):1308–1313. (available from: PM:20061022)

Pluta P, Cebula-Obrzut B, Ehemann V, Pluta A, Wierzbowska A, Piekarski J, Bilski A, Nejc D, Kordek R, Robak T, Smolewski P, Jeziorski A (2011) Correlation of Smac/DIABLO protein expression with the clinico-pathological features of breast cancer patients. Neoplasma 58(5):430–435. (available from: PM:21744997)

Potts PR, Singh S, Knezek M, Thompson CB, Deshmukh M (2003) Critical function of endogenous XIAP in regulating caspase activation during sympathetic neuronal apoptosis. J Cell Biol 163(4):789–799

Riedl SJ, Renatus M, Schwarzenbacher R, Zhou Q, Sun C, Fesik SW, Liddington RC, Salvesen GS (2001) Structural basis for the inhibition of caspase-3 by XIAP. Cell 104(5):791–800. (available from: PM:11257232)

Sasaki H, Sheng Y, Kotsuji F, Tsang BK (2000) Down-regulation of X-linked inhibitor of apoptosis protein induces apoptosis in chemoresistant human ovarian cancer cells. Cancer Res 60:5659–5666. (available from: PM:11059757)

Schimmer AD, Welsh K, Pinilla C, Wang Z, Krajewska M, Bonneau MJ, Pedersen IM, Kitada S, Scott FL, Bailly-Maitre B, Glinsky G, Scudiero D, Sausville E, Salvesen G, Nefzi A, Ostresh JM, Houghten RA, Reed JC (2004) Small-molecule antagonists of apoptosis suppressor XIAP exhibit broad antitumor activity. Cancer Cell 5(1):25–35

Schimmer AD, Estey EH, Borthakur G, Carter BZ, Schiller GJ, Tallman MS, Altman JK, Karp JE, Kassis J, Hedley DW, Brandwein J, Xu W, Mak DH, LaCasse E, Jacob C, Morris SJ, Jolivet J, Andreeff M (2009) Phase I/II trial of AEG35156 X-linked inhibitor of apoptosis protein antisense oligonucleotide combined with idarubicin and cytarabine in patients with relapsed or primary refractory acute myeloid leukemia. J Clin Oncol 27(28):4741–4746

Schimmer AD, Herr W, Hanel M, Borthakur G, Frankel A, Horst HA, Martin S, Kassis J, Desjardins P, Seiter K, Fiedler W, Noppeney R, Giagounidis A, Jacob C, Jolivet J, Tallman MS, Koschmieder S (2011) Addition of AEG35156 XIAP antisense oligonucleotide in reinduction chemotherapy does not improve remission rates in patients with primary refractory acute myeloid leukemia in a randomized phase II study. Clin Lymphoma Myeloma Leuk 11(5):433–438. (available from: PM:21729686)

Schyschka L, Rudy A, Jeremias I, Barth N, Pettit GR, Vollmar AM (2008) Spongistatin 1: a new chemosensitizing marine compound that degrades XIAP. Leukemia 22(9):1737–1745. (available from: PM:18548102)

Scott FL, Denault JB, Riedl SJ, Shin H, Renatus M, Salvesen GS (2005) XIAP inhibits caspase-3 and-7 using two binding sites: evolutionarily conserved mechanism of IAPs. EMBO J 24(3):645–655. (available from: PM:15650747)

Secchiero P, di Iasio MG, Melloni E, Voltan R, Celeghini C, Tiribelli M, Dal BM, Gattei V, Zauli G (2010) The expression levels of the pro-apoptotic XAF-1 gene modulate the cytotoxic response to Nutlin-3 in B chronic lymphocytic leukemia. Leukemia 24(2):480–483. (available from: PM:19847196)

Sekimura A, Konishi A, Mizuno K, Kobayashi Y, Sasaki H, Yano M, Fukai I, Fujii Y (2004) Expression of Smac/DIABLO is a novel prognostic marker in lung cancer. Oncol Rep 11(4):797–802. (available from: PM:15010875)

Shamon LA, Pezzuto JM, Graves JM, Mehta RR, Wangcharoentrakul S, Sangsuwan R, Chaichana S, Tuchinda P, Cleason P, Reutrakul V (1997) Evaluation of the mutagenic, cytotoxic, and antitumor potential of triptolide, a highly oxygenated diterpene isolated from Tripterygium wilfordii. Cancer Lett 112(1):113–117

Shin H, Renatus M, Eckelman BP, Nunes VA, Sampaio CA, Salvesen GS (2005) The BIR domain of IAP-like protein 2 is conformationally unstable: implications for caspase inhibition. Biochem J 385(Pt 1):1–10

Shiozaki EN, Chai J, Rigotti DJ, Riedl SJ, Li P, Srinivasula SM, Alnemri ES, Fairman R, Shi Y (2003) Mechanism of XIAP-mediated inhibition of caspase-9. Mol Cell 11(2):519–527. (available from: PM:12620238)

Silva KL, Vasconcellos DV, Castro ED, Coelho AM, Linden R, Maia RC (2006) Apoptotic effect of fludarabine is independent of expression of IAPs in B-cell chronic lymphocytic leukemia. Apoptosis 11(2):277–285. (available from: PM:16502265)

Small S, Keerthivasan G, Huang Z, Gurbuxani S, Crispino, JD (2010) Overexpression of survivin initiates hematologic malignancies in vivo. Leukemia 24(11):1920–1926

Song Z, Yao X, Wu M (2003) Direct interaction between survivin and Smac/DIABLO is essential for the anti-apoptotic activity of survivin during taxol-induced apoptosis. J Biol Chem 278(25):23130–23140. (available from: PM:12660240)

Srinivasula SM, Datta P, Fan XJ, Fernandes-Alnemri T, Huang Z, Alnemri ES (2000) Molecular determinants of the caspase-promoting activity of Smac/DIABLO and its role in the death receptor pathway. J Biol Chem 275(46):36152–36157

Sugahara K, Uemura A, Harasawa H, Nagai H, Hirakata Y, Tomonaga M, Murata K, Sohda H, Nakagoe T, Shibasaki S, Yamada Y, Kamihira S (2004) Clinical relevance of survivin as a biomarker in neoplasms, especially in adult T-cell leukemias and acute leukemias. Int J Hematol 80(1):52–58. (available from: PM:15293568)

Suzuki Y, Imai Y, Nakayama H, Takahashi K, Takio K, Takahashi R (2001a) A serine protease, HtrA2, is released from the mitochondria and interacts with XIAP, inducing cell death. Mol Cell 8:613–621

Suzuki Y, Nakabayashi Y, Takahashi R (2001b) Ubiquitin-protein ligase activity of X-linked inhibitor of apoptosis protein promotes proteasomal degradation of caspase-3 and enhances its anti-apoptotic effect in Fas-induced cell death. Proc Natl Acad Sci USA 98(15):8662–8667. (available from: PM:11447297)

Talbot DC, Ranson M, Davies J, Lahn M, Callies S, Andre V, Kadam S, Burgess M, Slapak C, Olsen AL, McHugh PJ, de Bono JS, Matthews J, Saleem A, Price P (2010) Tumor survivin is downregulated by the antisense oligonucleotide LY2181308: a proof-of-concept, first-in-human dose study. Clin Cancer Res 16(24):6150–6158

Tamm I, Kornblau SM, Segall H, Krajewski S, Welsh K, Kitada S, Scudiero DA, Tudor G, Qiu YH, Monks A, Andreeff M, Reed JC (2000) Expression and prognostic significance of IAP-family genes in human cancers and myeloid leukemias. Clin Cancer Res 6:1796–1803

Tamm I, Richter S, Scholz F, Schmelz K, Oltersdorf D, Karawajew L, Schoch C, Haferlach T, Ludwig WD, Wuchter C (2004) XIAP expression correlates with monocytic differentiation in adult de novo AML: impact on prognosis. Hematol J 5(6):489–495. (available from: PM:15570290)

Tanioka M, Nokihara H, Yamamoto N, Yamada Y, Yamada K, Goto Y, Fujimoto T, Sekiguchi R, Uenaka K, Callies S, Tamura T (2011) Phase I study of LY2181308, an antisense oligonucleotide against survivin, in patients with advanced solid tumors. Cancer Chemother Pharmacol 68(2):505–511. (available from: PM:21079959)

Tibes R, McDonagh KT, Lekakis L, Frazer N, Mohrland S, Dawn B, Garcia R, Schroeder K, Shanmugam V, Carpten J, von Hoff D, Shea TC (2009) Phase I Study of the Novel Surivin and cdc2/CDK1 Inhibitor Terameprocol in Patients with Advanced Leukemias. Blood 114(22):430–430

Tolcher AW, Quinn DI, Ferrari A, Ahmann F, Giaccone G, Drake T, Keating, A, de Bono JS (2012) A phase II study of YM155, a novel small-molecule suppressor of survivin, in castration-resistant taxane-pretreated prostate cancer. Ann Oncol 23(4):968–973. (available from: PM:21859898)

Uy GL, Rettig MP, Motabi IH, McFarland K, Trinkaus KM, Hladnik LM, Kulkarni S, Abboud CN, Cashen AF, Stockerl-Goldstein KE, Vij R, Westervelt P, DiPersio JF (2012) A phase I/II study of chemosensitization with the CXCR4 antagonist plerixafor in relapsed or refractory acute myeloid leukemia. Blood 119(17):3917–3924. (available from: PM:22308295)

Vallat L, Magdelenat H, Merle-Beral H, Masdehors P, Potocki de MG, Davi F, Kruhoffer M, Sabatier L, Orntoft TF, Delic J (2003) The resistance of B-CLL cells to DNA damage-induced apoptosis defined by DNA microarrays. Blood 101(11):4598–4606. (available from: PM:12586635)

Varfolomeev E, Wayson SM, Dixit VM, Fairbrother WJ, Vucic D (2006) The inhibitor of apoptosis protein fusion c-IAP2.MALT1 stimulates NF-kappaB activation independently of TRAF1 AND TRAF2. J Biol Chem 281(39):29022–29029. (available from: PM:16891304)

Varfolomeev E, Blankenship JW, Wayson SM, Fedorova AV, Kayagaki N, Garg P, Zobel K, Dynek JN, Elliott LO, Wallweber HJ, Flygare JA, Fairbrother WJ, Deshayes K, Dixit VM, Vucic D (2007) IAP antagonists induce autoubiquitination of c-IAPs, NF-kappaB activation, and TNFalpha-dependent apoptosis. Cell 131(4):669–681.

Varfolomeev E, Vucic D (2008) (Un)expected roles of c-IAPs in apoptotic and NFkappaB signaling pathways. Cell Cycle 7(11):1511–1521. (available from: PM:18469528)

Velculescu VE, Madden SL, Zhang L, Lash AE, Yu J, Rago C, Lal A, Wang CJ, Beaudry GA, Ciriello KM, Cook BP, Dufault MR, Ferguson AT, Gao Y, He TC, Hermeking H, Hiraldo SK, Hwang PM, Lopez MA, Luderer HF, Mathews B, Petroziello JM, Polyak K, Zawel L, Kinzler KW (1999) Analysis of human transcriptomes. Nat Genet 23(4):387–388

Verhagen AM, Silke J, Ekert PG, Pakusch M, Kaufmann H, Connolly LM, Day CL, Tikoo A, Burke R, Wrobel C, Moritz RL, Simpson RJ, Vaux DL (2002) HtrA2 promotes cell death through its serine protease activity and its ability to antagonize inhibitor of apoptosis proteins. J Biol Chem 277(1):445–454. (available from: PM:11604410)

Vince JE, Wong WW, Khan N, Feltham R, Chau D, Ahmed AU, Benetatos CA, Chunduru SK, Condon SM, McKinlay M, Brink R, Leverkus M, Tergaonkar V, Schneider P, Callus BA, Koentgen F, Vaux DL, Silke J (2007) IAP antagonists target cIAP1 to induce TNFalpha-dependent apoptosis. Cell 131(4):682–693

Vucic D, Dixit VM, Wertz, I.E (2011) Ubiquitylation in apoptosis: a post-translational modification at the edge of life and death. Nat Rev Mol Cell Biol 12(7):439–452. (available from: PM:21697901)

Wagenknecht B, Glaser T, Naumann U, Kugler S, Isenmann S, Bahr M, Korneluk R, Liston P, Weller M (1999) Expression and biological activity of X-linked inhibitor of apoptosis (XIAP) in human malignant glioma. Cell Death Differ 6(4):370–376

Waldele K, Silbermann K, Schneider G, Ruckes T, Cullen BR, Grassmann R (2006) Requirement of the human T-cell leukemia virus (HTLV-1) tax-stimulated HIAP-1 gene for the survival of transformed lymphocytes. Blood 107(11):4491–4499. (available from: PM:16467195)

Wang L, Du F, Wang X (2008) TNF-alpha induces two distinct caspase-8 activation pathways. Cell 133(4):693–703

Wang Y, Krivtsov AV, Sinha AU, North TE, Goessling W, Feng Z, Zon LI, Armstrong SA (2010) The Wnt/beta-catenin pathway is required for the development of leukemia stem cells in AML. Science 327(5973):1650–1653. (available from: PM:20339075)

Weisberg E, Kung AL, Wright RD, Moreno D, Catley L, Ray A, Zawel L, Tran M, Cools J, Gilliland G, Mitsiades C, McMillin DW, Jiang J, Hall-Meyers E, Griffin JD (2007) Potentiation of antileukemic therapies by Smac mimetic, LBW242: effects on mutant FLT3-expressing cells. Mol Cancer Ther 6(7):1951–1961. (available from: PM:17620426)

Weisberg E, Ray A, Barrett R, Nelson E, Christie AL, Porter D, Straub C, Zawel L, Daley JF, Lazo-Kallanian S, Stone R, Galinsky I, Frank D, Kung, AL, Griffin JD (2010) Smac mimetics: implications for enhancement of targeted therapies in leukemia. Leukemia 24(12):2100–2109

Wu TY, Wagner KW, Bursulaya B, Schultz PG, Deveraux QL (2003) Development and characterization of nonpeptidic small molecule inhibitors of the XIAP/caspase –3 interaction. Chem Biol 10(8):759–767. (available from: PM:12954335)

Xu R, Zhang P, Huang J, Ge S, Lu J, Qian G (2007) Sp1 and Sp3 regulate basal transcription of the survivin gene. Biochem Biophys Res Commun 356(1):286–292. (available from: PM:17350596)

Yang Y, Liu Z, Tolosa E, Yang J, Li L (1998) Triptolide induces apoptotic death of T lymphocyte. Immunopharmacology 40(2):139–149

Yang S, Chen J, Guo Z, Xu XM, Wang L, Pei XF, Yang J, Underhill CB, Zhang L (2003) Triptolide inhibits the growth and metastasis of solid tumors. Mol Cancer Ther 2(1):65–72

Yang L, Cao Z, Li F, Post DE, Van Meir EG, Zhong H, Wood WC (2004) Tumor-specific gene expression using the survivin promoter is further increased by hypoxia. Gene Ther 11(15):1215–1223. (available from: PM:15141159)

Yoshida A, Zokumasu K, Wano Y, Yamauchi T, Imamura S, Takagi K, Kishi S, Urasaki Y, Tohyama K, Ueda T (2012) Marked upregulation of Survivin and Aurora-B kinase are associated with disease progression in the myelodysplastic syndromes. Haematologica March 14. (available from: PM:22419576)

You L, He B, Xu Z, Uematsu K, Mazieres J, Fujii N, Mikami I, Reguart N, McIntosh JK, Kashani-Sabet M, McCormick F, Jablons DM (2004) An anti-Wnt-2 monoclonal antibody induces apoptosis in malignant melanoma cells and inhibits tumor growth. Cancer Res 64(15):5385–5389. (available from: PM:15289346)

Zender L, Spector MS, Xue W, Flemming P, Cordon-Cardo C, Silke J, Fan ST, Luk JM, Wigler M, Hannon GJ, Mu D, Lucito R, Powers S, Lowe SW (2006) Identification and validation of oncogenes in liver cancer using an integrative oncogenomic approach. Cell 125(7):1253–1267. (available from: PM:16814713)

Zeng Z, Shi YX, Samudio IJ, Wang RY, Ling X, Frolova O, Levis M, Rubin JB, Negrin RR, Estey EH, Konoplev S, Andreeff M, Konopleva M (2009) Targeting the leukemia microenvironment by CXCR4 inhibition overcomes resistance to kinase inhibitors and chemotherapy in AML. Blood 113(24):6215–6224. (available from: PM:18955566)

Zhou H, Du MQ, Dixit VM (2005) Constitutive NF-kappaB activation by the t(11;18)(q21;q21) product in MALT lymphoma is linked to deregulated ubiquitin ligase activity. Cancer Cell 7(5):425–431. (available from: PM:15894263)

Zou B, Chim CS, Pang R, Zeng H, Dai Y, Zhang R, Lam CS, Tan VP, Hung IF, Lan HY, Wong BC (2011) XIAP-associated factor 1 (XAF1), a novel target of p53, enhances p53-mediated apoptosis via post-translational modification. Mol Carcinog 51(5):5422–5432. (available from: PM:21678496)

Zurawa-Janicka D, Skorko-Glonek J, Lipinska B (2010) HtrA proteins as targets in therapy of cancer and other diseases. Expert Opin Ther Targets 14(7):665–679. (available from: PM:20469960)

Chapter 6
TP53 Mutations in Acute Myeloid Leukemia

Ulrike Bacher, Claudia Haferlach, Vera Grossmann, Susanne Schnittger and Torsten Haferlach

Abstract Mutations of the *TP53* gene show a low frequency in overall acute myeloid leukemia (AML). However, they were found at frequencies of 60–80 % in complex karyotype AML, and are strongly associated with therapy-related AML. *TP53* mutations are considered to represent a separate functional category independent from the typical class I and class II mutations. The mutations are heterogeneous and are distributed across the *TP53* gene with clustering of the mutations in exons 5–8. *TP53* mutations confer an adverse prognostic impact in patients with AML. High-throughput sequencing facilities are now available for rapid screening for *TP53* mutations at diagnosis of AML aiming to identify patients who may have a benefit from early allogeneic hematopoietic stem cell transplantation or other alternative therapeutic approaches.

Keywords *TP53* mutation · *TP53* copy number alteration · Acute myeloid leukemia (AML) · Complex karyotype · Risk stratification

6.1 Introduction

Wild-type p53 is a major tumor suppressor which is affected by mutations in most human cancers (Van Dyke 2007). A huge amount of information has been collected with respect to its functions in cell cycle, DNA repair, apoptosis, or in signaling. In addition, the p53 protein was shown to transcriptionally activate a multitude of genes whose products counteract apoptosis (Jänicke et al. 2008). Assuming that *TP53* is a classical tumor suppressor gene, mutations in the *TP53* gene are expected to lead to loss of wild-type p53 function. Thus, a combination of a *TP53* mutation and a deletion of the remaining allele or a *TP53* mutation in both alleles or a homozygous *TP53* mutation due to copy neutral loss of heterozygosity can result in total loss of p53 function. Another mechanism is the dominant negative effect of certain

T. Haferlach (✉) U. Bacher · C. Haferlach · V. Grossmann · S. Schnittger
MLL Munich Leukemia Laboratory, Max-Lebsche-Platz 31, 81377 Munich, Germany
e-mail: torsten.haferlach@mll.com

M. Andreeff (ed.), *Targeted Therapy of Acute Myeloid Leukemia*,
Current Cancer Research, DOI 10.1007/978-1-4939-1393-0_6

p53 mutant forms which form mixed tetramers with the wild-type p53 protein that are incapable of DNA binding and transactivation. In addition, it was suggested that cancer-associated *TP53* mutations also endow the mutant protein with new activities that contribute actively to various stages of tumor progression and to increased resistance to anticancer treatments (Oren and Rotter 2010).

TP53 mutations can be found in different myeloid and lymphatic acute and chronic leukemias, albeit at lower frequencies as compared to solid cancers (Xu-Monette et al. 2012): The respective mutations were described, e.g., in relapsed pediatric B- and T-lineage acute lymphoblastic leukemia (ALL; Hof et al. 2011), chronic lymphocytic leukemia (B-CLL; Zenz et al. 2010), or in myelodysplastic syndromes (MDS; Bejar et al. 2011;Jaedersten et al. 2011). Independent from the leukemia subtype, presence of *TP53* mutations were reported to confer an adverse prognostic impact and to be associated with disease progression.

In acute myeloid leukemia (AML), *TP53* mutations were described to show a close association to specific cytogenetic subgroups (e.g., complex karyotypes with 70–80 % of cases bearing *TP53* mutations), whereas a low frequency of only 2 % was found in noncomplex AML (Haferlach et al. 2008; Table 6.1). Further on, a close association of *TP53* mutations to loss of the short arm of one chromosome 17 was described (Fenaux et al. 1991). *TP53* mutations (detected by single-stranded conformation polymorphism analysis, SSCP) were identified as adverse prognostic parameters which were associated with lower complete remission rates and worse survival outcome in patients with AML (Wattel et al. 1994). The wild-type p53 protein has a short half-life and therefore is not detectable by immunohistochemistry on tissue sections. Mutated p53, on the other hand, has a prolonged half-life and becomes detectable by this method, so that its detection by immunohistochemistry is almost synonymous with mutation (Lepelley et al. 1994).

This chapter provides an overview on the frequency, the clinical impact, and screening strategies for *TP53* mutations and *TP53* gene copy number alterations in AML.

6.2 Structure of *TP53* Mutations

TP53 mutations are heterogeneous in structure and localization. They are most frequently represented by missense mutations. Besides, nonsense mutations, deletions, insertions, or splice site mutations can occur. They are distributed across exons 3–10 of the *TP53* gene. Clustering can be observed at exons 5–8 (Haferlach et al. 2008; Rücker et al. 2012; Fig. 6.1, data MLL). In the study from Stirewalt et al. who investigated 140 AML patients aged over 55 years, the frequency of *TP53* mutations was 9 %, and most (11 of 12 mutations) were missense mutations in exons 5–8 (Stirewalt et al. 2001).

Table 6.1 Results of previous studies on the frequency and the prognostic impact of *TP53* gene alterations in AML

Reference	Patients	Diagnostic techniques	Results
Haferlach et al. (Haferlach et al. 2008)	*Cohort 1*: 107 CK-AML *Cohort 2*: 235 AML (different cytogenetic subgroups)	Chromosome banding analysis Interphase FISH (*TP53*) *TP53*mut screening (GeneChip assays or denaturating high-performance liquid chromatography; direct sequencing of aberrant cases)	*Cohort 1*: Loss of one *TP53* allele: $n=57$ *TP53*mut on the remaining allele: $n=50/57$ with loss of 1 *TP53* allele; $n=33/50$ with 2 *TP53* copies Overall *TP53*mut rate in CK-AML: 78 % *Cohort 2*: Overall *TP*53mut rate: $n=33/235$ (14 %) AML without complex karyotypes: $n=4/193$ (2.1 %)
Seifert et al. (Seifert et al. 2009)	2272 AML patients participating in 2 multicenter trials	Interphase FISH (*TP53*)	*TP53* allele loss: 105/2272 (5 %) AML *TP53* deletion without multiple aberrations was an independent negative prognostic factor for OS ($p<0.001$), DFS ($p<0.001$), and relapse risk ($p=0.028$) in multivariable analysis
Parkin et al. (Parkin et al. 2010)	114 AML	SNP 6.0 microarray studies Cytogenetics/FISH Sequencing studies	Presence of *TP53*mut and a combined variable consisting of *TP53*mut or 17p-LOH were independent negative prognostic indicators for OS ($p<0.001$) in multivariable analysis
Rücker et al. (Rücker et al. 2012)	234 CK-AML	*TP53*mut screening Array-based genomic profiling	*TP53* mut: 141/234 (60 %) CK-AML *TP53* allele loss: 94/234 (40 %) CK-AML *TP53* alterations: 164/234 (70 %) CK-AML Presence of *TP53* alteration was an independent adverse prognostic parameter for therapy response ($p=0.05$) and OS ($p=0.0001$) in multivariable analysis

CK-AML complex karyotype AML, *TP53mut TP53* mutations, *LOH* loss of heterozygosity, *OS* overall survival, *DFS* disease-free survival, *DHPLC* denaturating high-performance liquid chromatography

6.3 Detection of Alterations of the *TP53* Gene

TP53 mutation detection can be facilitated by previous screening for mutated cases, e.g., by use of denaturating high-performance liquid chromatography (DHPLC) assays. Amplification by polymerase chain reaction (PCR) may be performed for exons 3–9 and introns 2–9 as to target those regions of the *TP53* gene where the mutations are more frequently localized. Aberrant amplicons are subsequently analyzed by direct Sanger sequencing in both orientations. This results in a sensitivity level of around 20 %. Also, sequencing array technology can be applied to investigate *TP53* mutations (Haferlach et al. 2008).

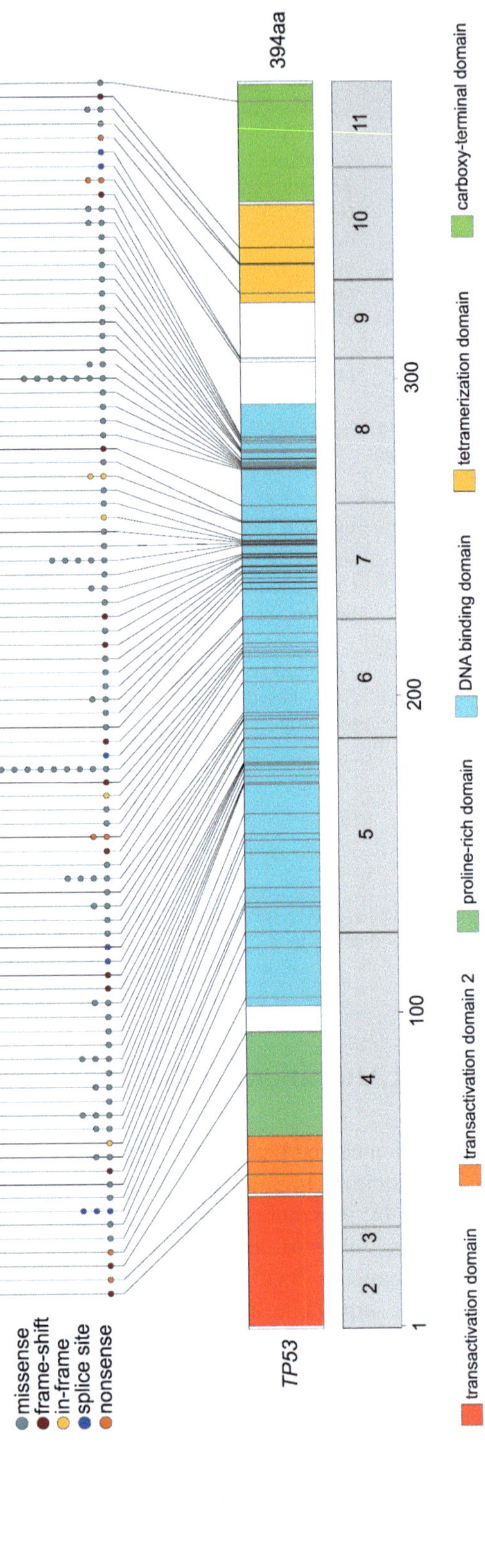

Fig. 6.1 Distribution of 165 mutations (17 frame-shift mutations, 7 in-frame mutations, 122 missense mutations, 9 nonsense mutations, 10 splice site mutations) across the different domains of the *TP53* gene in 139 affected AML patients according to the Munich Leukemia Laboratory's (MLL) database

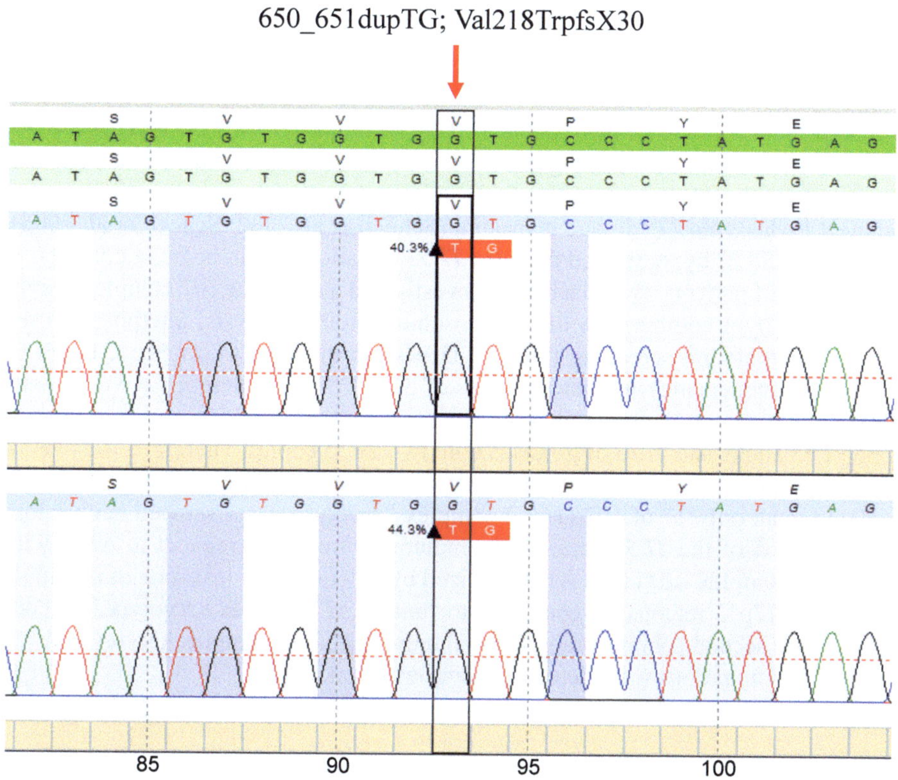

Fig. 6.2 Example of a *TP53* frameshift mutation as detected by high-throughput sequencing in a patient with AML

Today, more sensitive mutation detection can be obtained using high-throughput sequencing (Bejar et al. 2011; Jaedersten et al. 2011). This allows reliable detection of mutant clones down to 1–2% relative mutation abundance (RMA), defined as the proportion of sequence reads containing the mutation (Jaedersten et al. 2011; Fig. 6.2). However, the clinical impact of small subclones carrying a *TP53* mutation remains to be clarified.

The *TP53* allele status can be analyzed by different methods: One option is interphase fluorescence in situ hybridization (FISH) with probes for the *TP53* gene. Single nucleotide polymorphism (SNP) microarray analysis or array comparative genomic hybridization (array CGH) are alternative options for the detection of *TP53* copy number alterations. In addition, SNP microarrays can detect copy neutral loss of heterozygosity which is found in patients with *TP53* mutations and is one mechanism that leads to loss of function of both *TP53* alleles.

6.4 Frequency and Prognostic Impact of *TP53* Mutations in AML

Overall, only a low overall frequency of *TP53* mutations was observed in AML. However, chromosome banding analyses in AML patients revealed that predominantly patients with complex karyotypes show deletion of the short arm of chromosome 17 where the *TP53* gene is located (Schoch et al. 2002; Schoch et al. 2005). Thus, it could be hypothesized that also *TP53* mutations might be more common in this subset of patients. Haferlach et al. investigated a cohort of 107 complex karyotype AML by a combination of chromosome banding analysis, interphase FISH, and *TP53* mutation screening (using a GeneChip method; Affymetrix), DHPLC, and direct sequencing) and detected loss of one *TP53* copy number in 57 patients. *TP53* mutations on the remaining allele were found in 50 of 57 cases with loss of one *TP53* allele, and in 33 of 50 cases with two *TP53* copies. Therefore, the overall frequency of *TP53* mutations in AML with complex karyotype was 78% (Figs. 6.3, 6.4). In total, only 17 of 107 (16%) complex karyotype cases showed neither loss nor mutation of the *TP53* gene. *TP53* mutations are more frequent in AML with "typical" complex karyotype (characterized by deletion of at least one of the 5q31, 7q31, and 17p13 regions) as compared to "untypical" complex karyotypes. In contrast, in the second cohort of this study, in only 2% of noncomplex karyotype AML cases *TP53* mutations were detected (Haferlach et al. 2008).

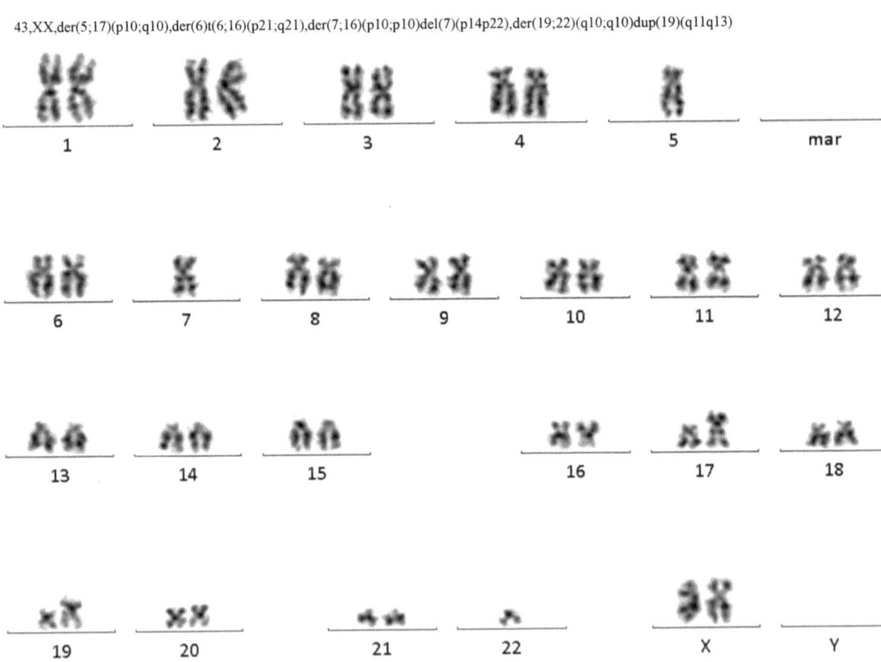

Fig. 6.3 Results of chromosome-banding analysis

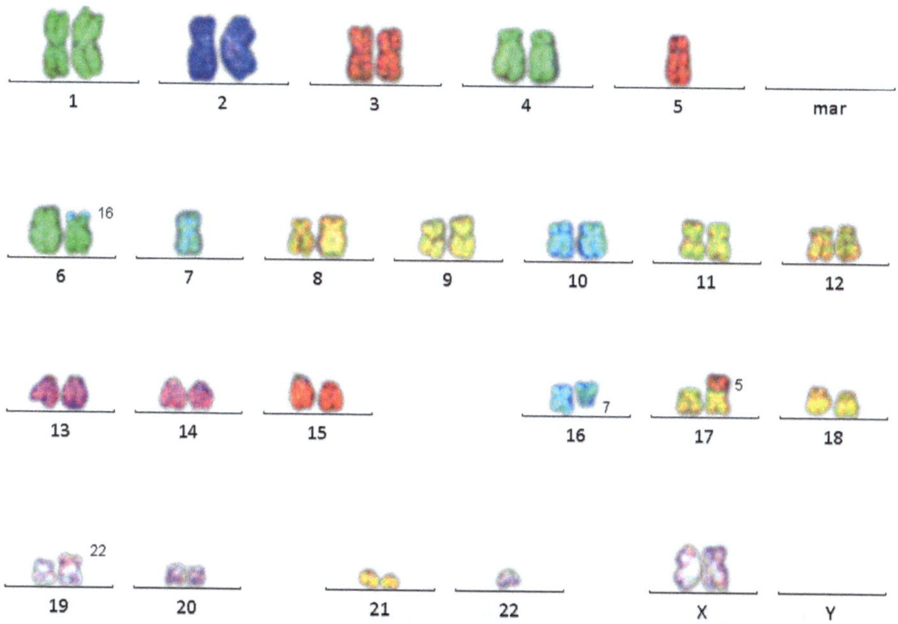

Fig. 6.4 24-color FISH in a case with complex karyotype involving the 17p-region. The karyotype formula was: 43, XX, der(5;17) (p10;q10), der(6)t(6;16) (p21;q21), der(7;16) (p10;p10) del(7) (p14p22), der(19;22) (q10;q10) dup(19) (q11q13)

Recently, Rücker et al. combined *TP53* mutational screening and array-based genomic profiling in 234 complex karyotype AML patients. The *TP53* mutation rate was 141/234 (60 %) in complex karyotype AML, and *TP53* allele loss was detected in 94 of 234 (40 %) cases. Thus, the overall frequency of *TP53* alterations was 164/234 (70 %) in complex karyotype AML. Moreover, *TP53*-altered complex karyotype cases were characterized by a higher degree of genomic complexity as defined by the number of clonal alterations per case and the frequency of specific copy number alterations. Presence of *TP53* alterations was an adverse prognostic factor, which held true in multivariable analysis when parameters such as white blood cell counts or age were considered (Rücker et al. 2012).

Additionally, *TP53* mutations were described at higher frequencies in therapy-related myeloid malignancies. In the Copenhagen series, Pedersen-Bjergaard et al. found *TP53* mutations in 9/51 (18 %) t-AML and 25/89 (28 %) t-MDS. Therefore, *TP53* mutations appeared as the most common genetic abnormality in therapy-related myeloid malignancies (Pedersen-Bjergaard et al. 2008). Side et al. investigated 26 patients with t-AML/t-MDS with abnormalities of chromosomes 5 and/or 7 for molecular mutations. The TP53 mutation rate was 5 in 23 patients (22 %) investigated for this molecular marker. Besides, albeit at lower frequencies, *NRAS*/*KRAS* (n = 3/23 or 13 %) and *FLT3*-internal tandem duplication (ITD; n = 1/23 or 4 %) mutations were detected. One case had a TP53 and KRAS2 mutation in parallel. The authors concluded that RAS and FLT3 mutations were less common in t-AML/t-MDS as compared to de novo disease, whereas *TP53* mutations were more frequent in therapy-related myeloid malignancies (Side et al. 2004). Leonard et al. investigated *TP53* mutations (exons 5–8) in patients with t-AML/t-MDS and a history of ovarian cancer, which is frequently treated by platinum analogues. Adequate sample material was available in 17 of 82 cases of t-AML/t-MDS which were identified in a multicenter, population-based cohort of 23,170 female patients with ovarian cancer. The majority (11 of 17 patients) had a history of platinum compounds and other alkylating agents and/or radiation. Six received other alkylating agents, combined with radiation in one case. *TP53* mutations were detected in 9 of 17 patients investigated (in two cases, two different *TP53* mutations were observed). Therefore, *TP53* mutations were common in therapy-related myeloid malignancies after multi-agent therapy for ovarian cancer. Finally, patients who had a history of platinum treatment showed a propensity for G to A transitions and the authors speculated that this might reflect a specific DNA damage in myeloid malignancies after treatment with platinum analogues (Leonard et al. 2002).

6.5 The Role of the *TP53* Allele Status

Parkin et al. performed investigation of *TP53* gene alterations in patients with AML by combining SNP microarray studies, cytogenetics/FISH, and sequencing. In 114 previously untreated patients with AML, the results from flow cytometry sorted AML cells and buccal cells were compared and were subsequently evaluated for their prognostic impact. In total, 23 cases (20.2%) had either *TP53* mutations or 17p-loss of heterozygosity (LOH) or both: In detail, five cases had isolated *TP53* mutations without 17p-LOH, five cases had *TP53* mutations and 17p-deletion, eight cases had *TP53* mutations and aUPD (acquired uniparental disomy) at 17p, and five cases had wild-type *TP53* exon 2–9 and 17p-LOH. Presence of *TP53* exon 2 through 9 mutations remained an independent adverse prognostic parameter in multivariable analysis, as was a combined variable consisting of *TP53* mutations and loss of 17p-LOH. The genomic lesion load and specific chromosomal abnormalities were evaluated by using SNP 6.0 array profiling (Affymetrix, Santa Clara, CA). Presence of ≥ 2 genomic lesions was an independent adverse risk parameter even when the age group (by a threshold of 60 years) or the cytogenetic risk groups were included in the multivariable analysis (Parkin et al. 2010).

Seifert et al. evaluated a large cohort of 2272 AML patients participating in two multicenter trials for *TP53* deletions by interphase FISH. Loss of *TP53* was detected in 105 out of 2272 (5%) AML patients. In 85 of these cases, the *TP53* deletions were accompanied by multiple aberrations, but 20 patients presented the *TP53* deletion as single aberration or with only one additional chromosomal abnormality. *TP53* deletions without multiple aberrations were confirmed as independent negative prognostic factor in multivariable analysis. Therefore, it was recommended to consider single *TP53* deletions as a high-risk factor for treatment strategies in AML (Seifert et al. 2009).

6.6 Interaction of *TP53* Mutations with Other Genetic Alterations

In the study by Haferlach et al., *TP53* mutations occurred only rarely in combination with *FLT3*-ITD and *MLL*-partial tandem duplication (PTD), and showed no coincidence with the *NPM1* mutations (Haferlach et al. 2008). Ishikawa et al. investigated 144 patients with newly diagnosed de novo AML for a broad panel of molecular markers. Ten of 11 *TP53* mutated cases showed additional cytogenetic abnormalities, and six showed additional molecular mutations which involved both class I and class II mutations (Ishikawa et al. 2009). Therefore, *TP53* mutations seem to represent an extra class of mutations which have no equivalent to the typical class I or II mutations in AML (Pedersen-Bjergaard et al. 2008). This separate class refers to mutations affecting genes implicated in cell cycle regulation or apoptosis. The *NPM1* mutations are also included in this (third) category (Renneville et al. 2008).

6.7 Identification of a Novel *TP53* Cancer Susceptibility Mutation Through Whole-Genome Sequencing

Link et al. performed whole-genome sequencing in a 42-year-old female patient with t-AML and complex karyotype. Increased cancer susceptibility was suspected because of a history of early-onset breast and ovarian cancer. The karyotype of the AML was complex involving 5q-deletion, monosomy 7, and two marker chromosomes. Whole-genome sequencing revealed a novel, heterozygous 3-kilobase (germ line) deletion removing exons 7–9 of *TP53* in the patient's normal skin DNA. Sequence analysis of DNA from the leukemic cells revealed a 17.6-megabase region of uniparental disomy (UPD) on chromosome 17 that resulted in homozygous deletion of exons 7–9 of TP53. Analysis of the sequence of the skin genome of this individual identified no single nucleotide variants (SNVs) or small insertion/deletions (indels) in common cancer susceptibility genes, including BRCA1 and BRCA2 (Link et al. 2011).

6.8 Clinical Impact of p53 Protein Biosignatures

Different p53 protein isoforms exist as a consequence of different promoters and alternative splicing of transcripts. Anensen et al. performed bioinformatic analysis of p53 isoform expression in 68 patients with AML. Distinct p53 protein biosignatures were correlated with clinical outcome. Furthermore, *FLT3* mutations with their adverse prognostic impact were shown to be associated with expression of full-length p53. In contrast, the favorable *NPM1* mutations correlated with expression of the β/γ isoforms. High expression of the p53β/γ isoform indicated favorable response to chemotherapy and long survival whereas presence of the p53 full-length isoform was an adverse prognostic indicator (Anensen et al. 2012).

Conclusions

TP53 mutations seem to represent a separate functional category in AML which have no equivalent to the typical class I or II mutations (Pedersen-Bjergaard et al. 2008; Renneville et al. 2008). The high frequency of *TP53* mutations in complex karyotype AML (Haferlach et al. 2008; Rücker et al. 2012) contributes to explain the frequent phenomenon of chemoresistance in this cytogenetic subgroup. Whereas the traditional molecular work-up for detection of the heterogeneous *TP53* mutations was highly laborious, investigation of *TP53* mutations is considerably facilitated by high-throughput sequencing (Bejar et al. 2011; Jaedersten et al. 2011). A rapid screening for *TP53* mutations eventually combined with investigation of the *TP53* allele status by FISH and/or array-CGH at diagnosis of the AML can identify high-risk patients who may have a benefit from alternative treatment strategies (e.g., an early allogeneic HSCT or novel strategies targeting the *TP53* pathway in clinical studies).

References

Anensen N, Hjelle SM, Van Belle W, Haaland I, Silden E, Bourdon JC, Hovland R, Tasken K, Knappskog S, Lonning PE, Bruserud O, Gjertsen BT (2012) Correlation analysis of p53 protein isoforms with NPM1/FLT3 mutations and therapy response in acute myeloid leukemia. Oncogene 3:1533–1545

Bejar R, Stevenson K, Abdel-Wahab O, Galili N, Nilsson B, Garcia-Manero G, Kantarjian H, Raza A, Levine RL, Neuberg D, Ebert BL (2011) Clinical effect of point mutations in myelodysplastic syndromes. N Engl J Med 364:2496–2506

Fenaux P, Jonveaux P, Quiquandon I, Lai JL, Pignon JM, Loucheux-Lefebvre MH, Bauters F, Berger R, Kerckaert JP (1991) P53 gene mutations in acute myeloid leukemia with 17p monosomy. Blood 78:1652–1657

Haferlach C, Dicker F, Herholz H, Schnittger S, Kern W, Haferlach T (2008) Mutations of the TP53 gene in acute myeloid leukemia are strongly associated with a complex aberrant karyotype. Leukemia 22:1539–1541

Hof J, Krentz S, van Schewick C, Körner G, Shalapour S, Rhein P, Karawajew L, Ludwig WD, Seeger K, Henze G, von Stackelberg A, Hagemeier C, Eckert C, Kirschner-Schwabe R (2011) Mutations and deletions of the TP53 gene predict nonresponse to treatment and poor outcome in first relapse of childhood acute lymphoblastic leukemia. J Clin Oncol 29:3185–3193

Ishikawa Y, Kiyoi H, Tsujimura A, Miyawaki S, Miyazaki Y, Kuriyama K, Tomonaga M, Naoe T (2009) Comprehensive analysis of cooperative gene mutations between class I and class II in de novo acute myeloid leukemia. Eur J Haematol 83:90–98

Jaedersten M, Saft L, Smith A, Kulasekararaj A, Pomplun S, Gohring G, Hedlund A, Hast R, Schlegelberger B, Porwit A, Hellstrom-Lindberg E, Mufti GJ (2011) TP53 mutations in low-risk myelodysplastic syndromes with del(5q) predict disease progression. J Clin Oncol 29:1971–1979

Jänicke RU, Sohn D, Schulze-Osthoff K (2008) The dark side of a tumor suppressor: anti-apoptotic p53. Cell Death Differ 15:959–976

Leonard DG, Travis LB, Addya K, Dores GM, Holowaty EJ, Bergfeldt K, Malkin D, Kohler BA, Lynch CF, Wiklund T, Stovall M, Hall P, Pukkala E, Slater DJ, Felix CA (2002) p53 mutations in leukemia and myelodysplastic syndrome after ovarian cancer. Clin Cancer Res 8:973–985

Lepelley P, Preudhomme C, Vanrumbeke M, Quesnel B, Cosson A, Fenaux P (1994) Detection of p53 mutations in hematological malignancies: comparison between immunocytochemistry and DNA analysis. Leukemia 8:1342–1349

Link DC, Schuettpelz LG, Shen D, Wang J, Walter MJ, Kulkarni S, Payton JE, Ivanovich J, Goodfellow PJ, Le Beau M, Koboldt DC, Dooling DJ, Fulton RS, Bender RH, Fulton LL, Delehaunty KD, Fronick CC, Appelbaum EL, Schmidt H, Abbott R, O'Laughlin M, Chen K, McLellan MD, Varghese N, Nagarajan R, Heath S, Graubert TA, Ding L, Ley TJ, Zambetti GP, Wilson RK, Mardis E R (2011) Identification of a novel TP53 cancer susceptibility mutation through whole-genome sequencing of a patient with therapy-related AML. JAMA 305:1568–1576

Oren M, Rotter V (2010) Mutant p53 gain-of-function in cancer. Cold Spring Harb Perspect Biol 2:a001107

Parkin B, Erba H, Ouillette P, Roulston D, Purkayastha A, Karp J, Talpaz M, Kujawski L, Shakhan S, Li C, Shedden K, Malek SN (2010) Acquired genomic copy number aberrations and survival in adult acute myelogenous leukemia. Blood 116:4958–4967

Pedersen-Bjergaard J, Andersen MK, Andersen MT, Christiansen DH (2008) Genetics of therapy-related myelodysplasia and acute myeloid leukemia. Leukemia 22:240–248

Renneville A, Roumier C, Biggio V, Nibourel O, Boissel N, Fenaux P, Preudhomme C (2008) Cooperating gene mutations in acute myeloid leukemia: a review of the literature. Leukemia 22:915–931

Rücker FG, Schlenk RF, Bullinger L, Kayser S, Teleanu V, Kett H, Habdank M, Kugler CM, Holzmann K, Gaidzik VI, Paschka P, Held G, von Lilienfeld-Toal M, Luebbert M, Fröhling S, Zenz T, Krauter J, Schlegelberger B, Ganser A, Lichter P, Döhner K, Döhner H (2012) TP53 alterations in acute myeloid leukemia with complex karyotype correlate with specific copy number alterations, monosomal karyotype, and dismal outcome. Blood 119:2114–2121

Schoch C, Haferlach T, Bursch S, Gerstner D, Schnittger S, Dugas M, Kern W, Loffler H, Hiddemann W (2002) Loss of genetic material is more common than gain in acute myeloid leukemia with complex aberrant karyotype: a detailed analysis of 125 cases using conventional chromosome analysis and fluorescence in situ hybridization including 24-color FISH. Genes Chromosomes Cancer 35:20–29

Schoch C, Kern W, Kohlmann A, Hiddemann W, Schnittger S, Haferlach T (2005) Acute myeloid leukemia with a complex aberrant karyotype is a distinct biological entity characterized by genomic imbalances and a specific gene expression profile. Genes Chromosomes Cancer 43: 227–238

Seifert H, Mohr B, Thiede C, Oelschlagel U, Schäkel U, Illmer T, Soucek S, Ehninger G, Schaich M (2009) The prognostic impact of 17p (p53) deletion in 2272 adults with acute myeloid leukemia. Leukemia 23:656–663

Side LE, Curtiss NP, Teel K, Kratz C, Wang PW, Larson RA, Le Beau MM, Shannon KM (2004) RAS, FLT3, and TP53 mutations in therapy-related myeloid malignancies with abnormalities of chromosomes 5 and 7. Genes Chromosomes Cancer 39:217–223

Stirewalt DL, Kopecky KJ, Meshinchi S, Appelbaum FR, Slovak ML, Willman CL, Radich JP (2001) FLT3, RAS, and TP53 mutations in elderly patients with acute myeloid leukemia. Blood 97:3589–3595

Van Dyke T (2007) p53 and tumor suppression. N Engl J Med 356:79–81

Wattel E, Preudhomme C, Hecquet B, Vanrumbeke M, Quesnel B, Dervite I, Morel P, Fenaux P (1994) p53 mutations are associated with resistance to chemotherapy and short survival in hematologic malignancies. Blood 84:3148–3157

Xu-Monette ZY, Medeiros LJ, Li Y, Orlowski RZ, Andreeff M, Bueso-Ramos CE, GreinerTC, McDonnell TJ, Young KH (2012) Dysfunction of the TP53 tumor suppressor gene in lymphoid malignancies. Blood 119:3668–3683

Zenz T, Eichhorst B, Busch R, Denzel T, Häbe S, Winkler D, Bühler A, Edelmann J, Bergmann M, Hopfinger G, Hensel M, Hallek M, Döhner H, Stilgenbauer S (2010) TP53 mutation and survival in chronic lymphocytic leukemia. J Clin Oncol 28:4473–4479

Chapter 7
Targeting p53 Tumor Suppressor for AML Therapy

Kojima Kensuke and Lyubomir T. Vassilev

Abstract The p53 tumor suppressor gene is mutated in over half of human solid cancers. However, in hematological malignancies the frequency of p53 mutation is relatively low, and the majority of tumors express a wild-type, transcriptionally competent protein. Nevertheless, leukemia cells have abnormalities in p53 regulatory proteins (e.g., MDM2 or MDMX overexpression, CDKN2A/ARF deletion, and ATM inactivation), frequently resulting in inactivation of the p53 pathway. Unleashing the powerful antitumor activity of the tumor suppressor by nongenotoxic pharmacological intervention in p53 regulation has been considered an attractive strategy. Recently, p53-targeted therapies have been developed with the potential to significantly impact cancer therapy. RG7112 and APR-246 have been tested in phase I trials in patients with hematologic malignancies. RG7112, a new member of the Nutlin family of MDM2 antagonists, prevents wild-type p53 from binding to its negative regulator MDM2, leading to p53 stabilization and activation. PRIMA-1 derivative APR-246 restores wild-type function in cells with mutant p53. The accumulated knowledge about p53 function and its regulation, combined with the advances in drug discovery technologies should translate into novel p53-based therapeutics in the clinical management of leukemia.

Keywords P53 · Mutation · MDM2

7.1 Introduction

The p53 protein was discovered in 1979 and initially called an oncogene, but was later established as a *bona fide* tumor suppressor (Brown et al. 2009; Vousden and Prives 2009; Cheok et al. 2011). Now, p53 has been found to serve as a central

K. Kensuke (✉)
Section of Molecular Hematology and Therapy, Department of Leukemia,
The University of Texas MD Anderson Cancer Center, 1515 Holcombe Blvd.,
Unit 081, FC3.3054, Houston, TX 77030, USA
e-mail: kkojima@mdanderson.org

L. T. Vassilev
Discovery Oncology, Roche Research Center, Hoffmann-La Roche Inc.,
340 Kingsland Street, Nutley, NJ 07110, USA
e-mail: lyubomir.vassilev@roche.com

M. Andreeff (ed.), *Targeted Therapy of Acute Myeloid Leukemia*,
Current Cancer Research, DOI 10.1007/978-1-4939-1393-0_7

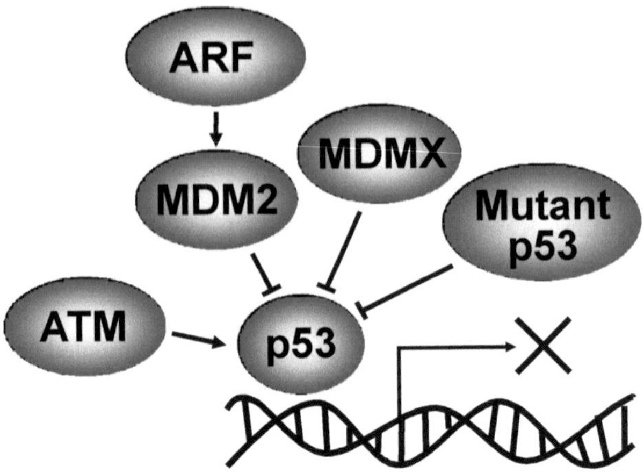

Fig. 7.1 Impaired p53 response in cancer. p53 transcriptional activity is suppressed by MDM2/MDMX overexpression, ARF/ATM inactivation, or p53 mutants- Downstream p53 signaling could also be altered in cancer cells

node in cell signaling involved in diverse biological processes, including metabolism, longevity, fertility, germ-line fidelity, stem-cell reprogramming, autophagy, necrosis, angiogenesis, and tumor cell invasion. However, the overwhelming data accumulated to date emphasize the critical role of p53 in human cancer.

The p53 tumor suppressor has evolved to protect multicellular organisms from cancer development that can be initiated by a variety of oncogenic events (Brown et al. 2009; Vousden and Prives 2009; Cheok et al. 2011). In response to DNA damage or other forms of cellular stress, p53 blocks cell-cycle progression and promotes apoptotic cell death, thus preventing the propagation of damaged cells. *TP53* is the most frequently altered gene in human cancer found mutated in approximately 50% of all human solid tumors. Mutations in p53 usually abrogate its DNA binding and transcriptional activity. The other half of human cancers retains wild-type p53 but acquire defects in the p53 pathway rendering p53 incapable of exerting its tumor-protective role (Fig. 7.1). Therefore, it is thought that the p53 pathway is inactivated in almost all human cancers.

In contrast to solid tumors, hematological malignancies have shown a rather low incidence of *TP53* mutations (Krug et al. 2002) with an approximately 5% mutational rate in both acute myeloid leukemia (AML) and acute lymphoblastic leukemia (ALL) (Krug et al. 2002; Rücker et al. 2012; Seifert et al. 2009). *TP53* mutations are more frequently detected in therapy-related AML/myelodysplastic syndromes (~30%) (Christiansen et al. 2001; Horiike et al. 1999). The incidence of *TP53* mutations in the chronic phase of chronic myelogenous leukemia (CML-CP) is very low (i.e., 0–3%) (Di Bacco et al. 2005). However, CML frequently shows a loss of the short arm of chromosome 17, often through formation of an isochromosome 17q and acquires a p53 mutation during progression to blast crisis (~20%). *TP53* is located on the short arm of chromosome 17 (17p13.1). Similarly, *TP53* mutations are detected in less than 2% of Philadelphia (Ph) chromosome-negative chronic myeloproliferative disorders (Ph^- MPD) in chronic phase but in ~40% of Ph^- MPD in blast crisis (Gaidano et al. 1997).*TP53* mutations occur in 15% of B-cell chronic lymphocytic leukemia (CLL), become more frequent as disease progresses, and predict aggressive disease that does not respond to *alkylating* agents or purine analogue-based therapy (Zenz et al. 2010). About 40 % of fludarabine-refractory patients have been found to carry *TP53* mutations or 17p deletion.

In *de novo* AML, *TP53* mutations are cumulatively detected in patients with complex chromosome abnormalities (60 %) (Rücker et al. 2012; Seifert et al. 2009). *TP53* mutations and complex chromosome abnormalities are independent adverse factors for response to chemotherapy and prognosis in AML. Although *TP53* mutations are rare in AML without complex chromosome abnormalities, inactivation of wild-type p53 protein frequently occurs as a result of overexpression of its negative regulator, MDM2 (Bueso-Ramos et al. 1993; Faderl et al. 2000). MDM2 is a p53-specific E3 ubiquitin ligase, which mediates the ubiquitin-dependent degradation of p53. MDM2 has been found frequently overexpressed in AML, a process that can enhance tumorigenic potential and resistance to chemotherapy-induced apoptosis. Therefore, targeting MDM2 in p53 wild-type cells could offer an attractive approach to the treatment of AML. Here, we focus on p53-activating cancer therapies that either havereached the clinic, demonstrated preclinical efficacy or represent novel approaches, paying special attention to therapies for AML.

7.2 p53 Pathway Abnormalities in Hematological Malignancies

Impairment of p53 function in cancer cells results from either *TP53* mutation/deletion or defective p53 regulation. *TP53* mutations are relatively rare in leukemias but leukemia cells have been reported to have abnormalities in p53 regulatory proteins, including: (1) MDM2 overexpression, (2) MDMX (known also as MDM4) overexpression, (3) CDKN2A/ARF deletion, and (4) ATM inactivation (deletion or mutations). Importantly, these abnormalities lead to upregulation of MDM2 (Fig. 7.2).

7.2.1 MDM2 Overexpression

MDM2 and p53 mutually regulate their protein levels through a negative feedback loop in which MDM2 is transcriptionally activated by p53 while MDM2, an E3 ubiquitin ligase for p53, targets p53 for ubiquitin-dependent degradation. Additionally, MDM2 protein binds the N-terminal transactivation domain of p53 and inhibits its transcriptional activity. MDM2 overexpression has been reported in AML and CLL, but the underlying mechanisms remain unknown (Bueso-Ramos et al. 1993; Faderl et al. 2000; Haidar et al.1997; Merup et al. 1997; Winkler et al. 2005). Although MDM2 gene is located on chromosome 12q15, and trisomy 12 is the most common cytogenetic change in CLL (i.e., 20–25 %), CLL cells with trisomy 12 have not been reported to express high levels of MDM2 (Merup et al. 1997; Winkler et al. 2005).

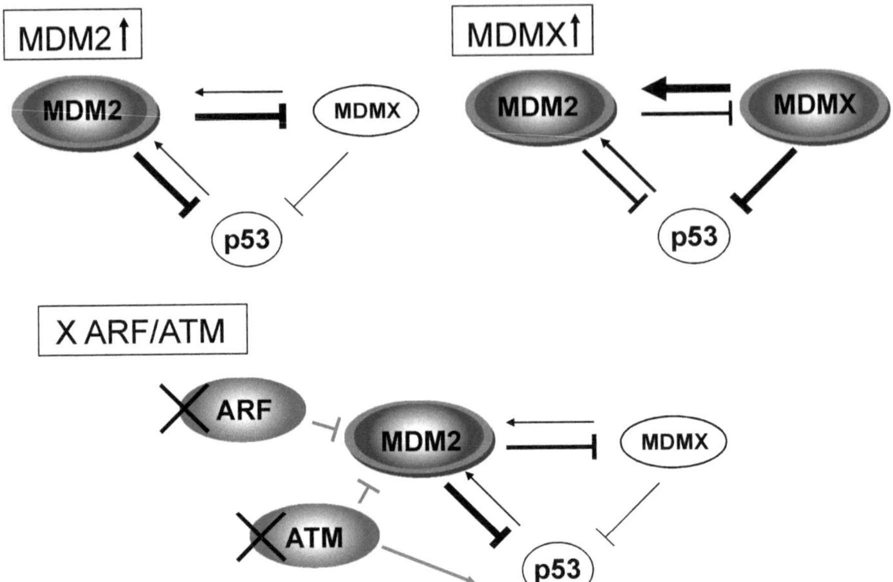

Fig. 7.2 MDM2 is frequently overexpressed in leukemias. Wild-type p53 activity is suppressed by MDM2 or MDMX overexpression or ARF/ATM inactivation

7.2.2 MDMX Overexpression

MDMX is another negative p53 regulator close in structure and activity to MDM2 (Brown et al. 2009; Vousden and Prives 2009; Cheok et al. 2011). Like MDM2, it also binds p53 and inhibits its transcriptional activity but does not possess intrinsic ubiquitin ligase activity. MDMX hetero-oligomerizes with MDM2 and stimulates its E3 ligase activity. MDMX can also be targeted for MDM2-dependent degradation. MDMX overexpression has been reported in ALL and CLL (Bo et al. 2010; Han et al. 2007) but the underlying mechanisms remain unknown. MDMX gene is located on 1q32.

7.2.3 CDKN2A/ARF Deletion

The *CDKN2A* tumor suppressor locus on 9p21 encodes two tumor suppressor genes: $p16^{INK4A}$ and $p14^{ARF}$ ($p19^{ARF}$ in the mouse). $p16^{INK4A}$ is an inhibitor of the cyclin D-dependent kinases. $p14^{ARF}$ binds MDM2, sequesters it in the nucleolus and abrogates its p53 regulating function. *p14/ARF* deletion leads to MDM2 up-regulation and p53 inactivation. Deletion of *CDKN2A/ARF* has been shown to occur in approximately 50 % of ALL, with homozygous deletions as the most frequent mechanism of inactivation (Mullighan et al. 2008; Iacobucci et al. 2011). *CDKN2A/ARF* deletion has also been associated with lymphoid blast crisis in CML. *CDKN2A/ARF* deletion has not been reported in CML-CP or myeloid blast crisis, underscoring its lineage-specific occurrence.

7.2.4 ATM Inactivation (Deletion or Mutations)

ATM is a central component of the signal transduction pathway that is activated in response to DNA double-strad breaks and synchronizes DNA repair with the induction of p53-dependent apoptosis. ATM inactivation leads to increased radiosensitivity, radioresistant DNA synthesis, loss of cell-cycle checkpoints, and p53 dysfunction. A deletion of the long arm of chromosome 11 is found in 10–20% of CLL patients and identifies a group with a poor outcome (Austen et al. 2007). The deletion includes *ATM* gene located at 11q22.3 to q23.1. ATM is mutated in a minority of CLL patients (i.e., 30–40% of CLL cases with 11q deletion). Complete loss of ATM function (biallelic *ATM* defects either by deletions or by mutations) has been associated with poor clinical outcome. *ATM* mutant subclones may develop during disease progression and give rise to additional expansion of the 11q-deleted subclone.

7.3 Outcome of Wild-Type p53 Activation in Cells

It has been thought that transient activation of wild-type p53 preferentially induces apoptosis in tumor cells while causing a reversible cell-cycle arrest in their normal counterparts (Brown et al. 2009; Vousden and Prives 2009; Cheok et al. 2011). Studies have sought to uncover the molecular mechanisms underlying the control of cell response to p53 activation. Available evidence suggests that the decision between life and death can be determined by cell type, the extent of damage or stress duration. Low levels of transient stress associated with repairable DNA damage elicit the survival response, whereas high levels of sustained stress accompanied by irreparable damage lead to cell death or senescence (Bensaad and Vousden 2007). One model pays attention to differences in p53 expression between cancer and normal cells (Cheok et al. 2011). Cancer cells usually have substantial levels of intrinsic DNA damage, which may be associated with elevated levels of post-transcriptionally modified p53 (e.g., phosphorylation at Ser^{15}, Ser^{46}, and Ser^{392}), but the levels of the "proapoptotic" p53 are not sufficient to induce cell death. Normal cells do not have intrinsic DNA damage and p53 is unmodified. Stabilization and activation of p53 after exposure to radiation or exogenous cytotoxic drugs may preferentially lead to cell death or senescence in cancer cells but cell-cycle arrest and stimulation of DNA repair in normal cells. Although current models do not explain satisfactorily the complex and heterogeneous responses to p53 activation, the possibility for modulating p53 response to promote death in tumor cells while sparing normal cells is very attractive.

Most chemotherapeutic agents used in leukemia therapy (e.g., anthracyclines, 1-β-D-arabinofuranosylcytosine (Ara-C), clofarabine, fludarabine, cyclophosphamide) damage DNA, and DNA-damage-induced p53 is believed to be a major contributor for tumor response. However, p53 is also largely responsible for radiation or chemotherapy-induced adverse effects. Although this has often been viewed as an unfortunate but necessary side effect of activating cell-protective functions of p53, it is disadvantageous to cancer patients. In addition to acute adverse effects, chemotherapy-induced DNA damage contributes to resistance or relapse of primary cancer and the development of secondary malignancies (Sill et al. 2011; Ding et al. 2012). It has been recently reported that AML relapse is associated with the addition of new mutations and clonal evolution, which is frequently caused by chemotherapy-induced DNA damage (Ding et al. 2012). Nongenotoxic p53 activation therapies may be able to achieve anti-leukemia effect without the collateral damage and consequent adverse events.

7.4 Restoring or Activating p53 Functions

Several strategies have been developed to restore/activate p53 functions in cancer cells (Guo and Xin 2006; Crompton and Kirn 2007; Tyler et al. 2009; Park et al. 2011; Met et al. 2011; Leffers et al. 2009; Chiappori et al. 2010; Lambert et al. 2009; Ali et al. 2011; Nahi et al. 2004; Bykov et al. 2005; Zache et al. 2008; Xu et al. 2010; Sutherland et al. 2011; Vassilev et al. 2004; Shangary et al. 2008; Shen and Maki 2011; Kojima et al. 2005, 2006, 2010; Coll-Mulet et al. 2006; Secchiero et al. 2006; Gu et al. 2008; Tabe et al. 2009; Saddler et al. 2008; Long et al. 2010; Peterson et al. 2011; Samudio et al. 2010; Issaeva et al. 2004; de Lange et al. 2011; Saha et al. 2012). They include: (1) *TP53* gene therapy, (2) p53 vaccines, (3) rescue of mutant p53 function by small molecules, and (4) p53 stabilization and activation by blocking p53-MDM2 interaction (Fig. 7.3). *TP53* gene therapy and p53-based vaccination have not been studied much in hematological malignancies.

7.4.1 TP53 Gene Therapy

Replication-defective, p53-expressing adenovirus (Ad5CMV-p53, Gendicine) has been approved for clinical use in China in 2003 (Guo and Xin 2006). Recently, conditionally replicating adenoviruses (CRAs: ONYX-015) that can replicate only in p53 mutant cells have been developed and tested in the clinic (Crompton and Kirn 2007). However, due to inefficient systemic delivery and limited intratumoral dissemination, and moderate potency, multiple phase II/III clinical trials have failed to demonstrate single-agent activity of ONYX-015 or synergistic effects with chemotherapeutic agents. *TP53* gene therapy has not been granted approval in the USA, and recent preclinical studies have used mesenchymal and neural stem cells as delivery vehicles for oncolytic viruses (Tyler et al. 2009).

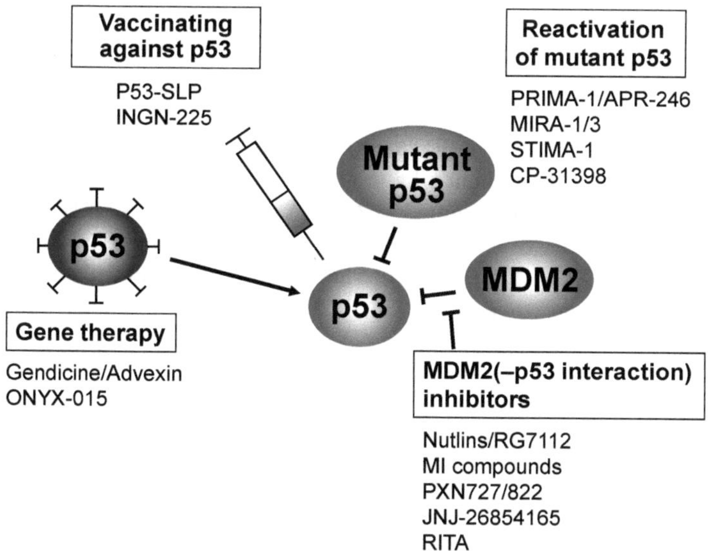

Fig. 7.3 Therapeutics targeting the p53 pathway

7.4.2 *Vaccines Against Mutant p53*

Several studies have suggested p53 as a candidate tumor antigen. The majority of p53 mutants lose their DNA binding and ability to transactivate the *MDM2* gene, leading to a dysfunctional feedback loop and accumulation of high levels of mutant p53 protein in cancer cells. Mutant p53 can be perceived by the immune system as target antigen and p53 auto-antibodies and p53-reactive $CD4^+$ and $CD8^+$ T-cells have been found in cancer patients (Park et al. 2011; Met et al. 2011). Clinical validation of this approach is still pending but early results are encouraging (Leffers et al. 2009; Chiappori et al. 2010). Vaccination of cancer patients with a mixture of ten long synthetic p53-dervied peptides (p53-SLP) and a p53-modified adenovirus-transduced dendritic cell vaccine (INGN-225) has shown p53-immune responses and some antitumor activity (Leffers et al. 2009; Chiappori et al. 2010).

7.4.3 *Reactivation of Mutant p53*

Only a minority of hematological malignancies harbor mutant p53. Although most frequent p53 mutants have abnormalities in the DNA-binding domain that are difficult to restore, some p53 mutants have acquired conformational changes that may be reversible. Small molecules have been reported to restore wild-type function to mutant p53. p53 reactivation and induction of massive apoptosis-1 (PRIMA-1) and its derivative PRIMA-1^{MET}/APR-246 have been shown to react covalently with thiols in mutant p53 and restore, at least partially, wild-type function to p53 (Lambert et al. 2009). PRIMA-1 has shown cytotoxic effects on AML and CLL cells (Ali et al. 2011; Nahi et al. 2004) and has been evaluated in clinical phase I studies. In December 2011, Aprea AB announced continuation of phase I/II-study of APR-246 in patients with refractory AML and CLL (www.aprea.com). Other mutant p53-targeting compounds have also been published like mutant p53-dependent induction of rapid apoptosis-1 (MIRA-1)/MIRA-3, STIMA-1 (SH group targeting and induction of massive apoptosis), and CP-31398 (Bykov et al. 2005; Zache et al. 2008; Xu et al. 2010; Sutherland et al. 2011), but there mechanism of action is not well understood (Bykov et al. 2005; Zache et al. 2008; Xu et al. 2010; Sutherland et al. 2011).

7.4.4 Activation of Wild-Type p53 by MDM2 Inhibitors

Disruption of MDM2–p53 binding has been shown to affectively stabilize p53, activate p53 functions, and sensitize tumors to chemotherapy or radiotherapy (Vassilev et al. 2004; Shangary et al. 2008; Shen and Maki 2011). This strategy requires the expression of wild-type p53 for activity. In hematological malignancies, induction of p53 by MDM2 inhibitors can activate the apoptotic response (Kojima et al. 2005, 2006, 2010; Coll-Mulet et al. 2006; Secchiero et al. 2006; Gu et al. 2008; Tabe et al. 2009; Saddler et al. 2008; Long et al. 2010) while normal cells have higher tolerance to MDM2 inhibition. Several inhibitors of p53-MDM2 interaction have been developed, and some of them are currently in clinical investigation. JNJ-26854165, reported as MDM2 antagonist, has been assessed in phase I studies as an oral agent for advanced or refractory solid tumors; however, its inhibitory activity against the MDM2–p53 interactions has not been demonstrated, and the mechanism of action is poorly understood (Kojima et al. 2010). RG7112, a potent and selective new member of the Nutlin family of MDM2 antagonists (Kojima et al. 2005, 2006; Coll-Mulet et al. 2006; Secchiero et al. 2006; Gu et al. 2008; Tabe et al. 2009) is being tested in phase I trials in patients with hematologic malignancies (NCT00623870) and advanced solid tumors (NCT00559533). MI compounds (MI-43, MI-63, MI-219, and MI-319) and PXN727/822 represent different series of MDM2–p53 binding inhibitors, currently at preclinical stage of investigation (Saddler et al. 2008; Long et al. 2010; Peterson et al. 2011; Samudio et al. 2010). Another p53 activator that reportedly blocks MDM2–p53 interaction is the small-molecule reactivation of p53 and induction of tumor cell apoptosis (RITA) (Issaeva et al. 2004). RITA has been shown to induce p53-mediated apoptosis in AML and CLL cells (Nahi et al. 2008). The proposed mechanism of RITA activity is that it disrupts the MDM2–p53 interaction by binding to p53 (not MDM2); however, such mechanism of action has not been well documented and is still controversial (Issaeva et al. 2004; de Lange et al. 2011; Saha et al. 2012).

7.5 Pharmacological Activation of p53 by MDM2 Antagonists in Leukemias

7.5.1 *Nutlins*

The first class of potent and selective inhibitors of MDM2–p53 binding, termed Nutlins, were reported in 2004 (Vassilev et al. 2004). Nutlin-3a (the active enantiomer of Nutlin-3) is a *cis*-imidazoline compound that binds specifically and with high affinity to the p53 domain of MDM2 by mimicking critical residues of p53 essential for MDM2 binding (i.e., Phe19, Trp23, Leu26). It is the most well-studied MDM2 inhibitor and an excellent molecular tool for studying p53 functions and therapeutic intervention in p53 regulation (Vassilev et al. 2004). Nutlin-3 induces apoptosis in primary leukemia cells but is less toxic to normal hematopoietic cells (Kojima et al. 2005, 2006; Coll-Mulet et al. 2006; Secchiero et al. 2006). This favorable activity profile suggests that induction of the p53 response in transformed cells may offer a novel approach to treat tumors retaining functional p53 pathway. Nutlin activity is dependent on the presence of wild-type p53, as p53 mutant AML, ALL, and CLL cells are resistant to Nutlin-induced cytotoxicity (Kojima et al. 2005, 2006; Coll-Mulet et al. 2006; Gu et al. 2008). Other reported determinants of response to Nutlin are MDM2 and MDMX expression levels. High levels of MDM2 are associated with higher susceptibility to Nutlin-induced apoptosis in AML and ALL (Kojima et al. 2005; Gu et al. 2008). On the other hand, high MDMX levels are associated with low sensitivity of CLL cells (Bo et al. 2010). Inactivation of ARF or ATM has not affected Nutlin activity in leukemia cells (Kojima et al. 2006; Sugihara et al. 2011). Nutlin-3a has shown synergism with conventional chemotherapeutic agents, anthracyclines, Ara-C, fludarabine, chlorambucil and imatinib.(Kojima et al. 2005, 2006; Coll-Mulet et al. 2006; Secchiero et al. 2006; Kurosu et al. 2010).

In leukemia cells, p53 induces apoptosis by transcriptional activation of apoptosis regulators in the extrinsic (death-receptor-dependent) and the intrinsic, (i.e., mitochondria-mediated) apoptosis signaling pathways. The conventional view of p53-mediated apoptosis emphasized its role as a transcription factor. However, recent studies have reported a transcription-independent mechanism that initiates apoptosis by direct p53 binding to anti-apoptotic BCL-2 family proteins at the mitochondrial membrane, resulting in Bax activation and apoptosis (Chipuk et al. 2004; Wolff et al. 2008). Although recent studies have indicated that nontranscriptional effects of p53 play an important role in Nutlin-induced apoptosis in leukemias (Kojima et al. 2005, 2006; Vaseva et al. 2009; Steele et al. 2008), the relative contribution of transcription-independent mitochondrial p53 program to overall pro-apoptotic effects of p53 has not been fully understood. The most advanced clinically relevant Nutlin analog, RG7112, is being tested in phase I trials in patients with hematologic malignancies (NCT00623870) (Andreeff et al. 2010, 2011).

7.5.2 *MI Compounds*

The spiro-oxindole MI compounds (MI-43, MI-63, MI-219, and MI-319) represent a class of MDM2 inhibitors, which similarly to Nutlins, block the MDM2-p53 interaction by mimicking critical residues of MDM2–p53 complex interface. Oral administration of MI-219, a more potent and pharmacologically optimized derivative of the originally described MI-63, led to tumor regression in mouse leukemia models at nontoxic doses, which are particularly sensitive to p53-mediated radiation-induced apoptosis (Shangary et al. 2008). The dependence of MI compound activity on the p53 status has been reported in AML and CLL cells (Saddler et al. 2008; Long et al. 2010). MI-63 showed synergistic apoptotic effect with doxorubicin and Ara-C in AML (Peterson et al. 2011). The mechanism of MI compound-induced leukemia cell apoptosis seems to involve the transcription-independent mitochondrial functions of p53 (Peterson et al. 2011; Mohammad et al. 2009).

7.6 Future Directions

Mutations in p53 in leukemia are rare, and p53 activation plays an important role in the response to currently used genotoxic therapies (Krug et al. 2002; Rücker et al. 2012; Seifert et al. 2009; Christiansen et al. 2001; Horiike et al. 1999; Di Bacco et al. 2005; Gaidano et al. 1997; Zenz et al. 2010). However, the severe adverse effects and genotoxic burden experienced by patients necessitate development of more efficacious and less toxic treatments. Direct and nongenotoxic activation approaches may offer a viable alternative to the current standard of care in leukemia. MDM2 antagonism is the most advanced approach that has entered clinical development. MDM2 inhibitors have induced effective growth arrest and apoptosis in leukemia with fewer toxic effects observed in mouse xenograft models (Vassilev et al. 2004; Shangary et al. 2008). These experiments suggest that pharmacological inhibition of MDM2–p53 interaction may differ from deletion of the *MDM2* gene, shown to be highly toxic in genetic mouse models of p53 activation (Ringshausen et al. 2006; Montes de Oca Luna et al. 1995). One plausible explanation is that *MDM2* gene deletion leads to a complete loss of MDM2 protein and its functions compared to temporary shutoff during pharmacological inhibition. Blocking of MDM2–p53 binding disrupts the negative feedback circuit and leads to generation of elevated levels of MDM2 protein that may retain key p53-independent functions (Tovar et al. 2006; Xia et al. 2008).

It has been suggested that targeting MDM2 alone may not be sufficient and simultaneous inhibition of MDMX is needed for maximal efficacy (Hu et al. 2006; Wade et al. 2006; Patton et al. 2006). Nutlins and MI compounds inhibit MDM2–p53 binding, but are ineffective of disrupting MDMX–p53 binding. Therefore, MDMX could still be bound to p53 and inhibit p53 activity in the presence of MDM2-specific antagonists. Recently developed small-molecule inhibitors of both

MDM2–p53 and MDMX–p53 binding (Graves et al. 2012) have been used to address this hypothesis. Simultaneous inhibition of MDM2 and MDMX have substantially increased p53-dependent cytotoxicity to solid tumor cells overexpressing MDMX but had relatively small enhancing effect on cells with "normal" MDMX levels. Further experiments are needed to assess the validity of these observations in leukemia. Studies with Nutlin-3 in solid tumor epithelial cells have indicated that all tested lines have retained the cell-cycle arrest function of p53 but a substantial fraction have no apoptotic response likely due to defective p53-dependent apoptotic signaling (Tovar et al. 2006). These defects may arise from activation of oncogenic pathways that counteract the p53 pathway in tumor development. Long et al. have recently found that the presence of FLT3-ITD correlates with response of AML cells to the MDM2 inhibitor MI-219 (Long et al. 2010). Characterization of large sets of patient-derived leukemia samples may help to identify markers of response and suggest new combination strategies. Inevitably, combination approaches are essential to maximize response rate and minimize the possibility of acquired resistance to p53 activating agents (Aziz et al. 2011; Michaelis et al. 2011).

Conclusions

The rate of p53 mutation in leukemia is low and the majority of tumors express wild-type, transcriptionally competent protein. Therefore, unleashing the powerful antitumor molecule by nongenotoxic pharmacological intervention in p53 regulation has been considered an attractive strategy. Recently, p53-targeted therapies have been developed and are being investigated in preclinical and clinical studies, which have the potential to significantly impact cancer therapy. The most advanced approach targets p53-negative regulators MDM2 and MDMX to stabilize and activate wild-type p53. MDM2 antagonists have already entered clinical development and in the near future should allow addressing therapeutic utility of these agents in leukemia. The accumulated knowledge about p53 function and its regulation, combined with the great advances in drug discovery technologies should translate into novel p53-based therapeutics in the clinical practice of leukemia.

References

Ali D, Jönsson-Videsäter K, Deneberg S, Bengtzén S, Nahi H, Paul C, Lehmann S (2011) APR-246 exhibits anti-leukemic activity and synergism with conventional chemotherapeutic drugs in acute myeloid leukemia cells. Eur J Haematol 86(3):206–215. doi:10.1111/j.1600-0609.2010.01557.x (Epub 2011 Jan 11. PubMed PMID: 21114538)

Andreeff M, Kojima K, Padmanabhan S, Strair R, Kirschbaum M, Maslak P, Hillmen P, O'Brien S, Samaniego F, Borthakur G, Konopleva M, Vassilev L, Nichols G (2010) A multi-center, open-label, phase I study of single agent RG7112: a first in class p53-MDM2 antagonist, in patients with relapsed/refractory acute myeloid and lymphoid leukemias (AML/ALL) and

refractory chronic lymphocytic leukemia/small cell lymphocytic lymphomas (CLL/SCLL). Annual Meeting, American Society of Hematology, Orlando, FL, 12/6/2010 (Abstract # 657)
Andreeff M, Kojima K, Ruvolo V, Younes A, Wei W, Duvvuri S, Konopleva M, Huang X, Wu L, Jukofsky L, Zhi J, Geho D, Vassilev L, Nichols G (2011) Pharmacodynamic biomarker analysis from the phase 1 Trial of RG7112, a small-molecule MDM2 antagonist, in leukemia. Annual Meeting, American Society of Hematology, San Diego, CA, 12/10/2011 (Abstract # 1545)
Austen B, Skowronska A, Baker C, Powell JE, Gardiner A, Oscier D, Majid A, Dyer M, Siebert R, Taylor AM, Moss PA, Stankovic T (2007) Mutation status of the residual ATM allele is an important determinant of the cellular response to chemotherapy and survival in patients with chronic lymphocytic leukemia containing an 11q deletion. J Clin Oncol 25(34):5448–5457 (Epub 2007 Oct 29. PubMed PMID: 17968022)
Aziz MH, Shen H, Maki CG (2011) Acquisition of p53 mutations in response to the non-genotoxic p53 activator Nutlin-3. Oncogene 30(46):4678–4686. doi:10.1038/onc.2011.185 (Epub 2011 Jun 6. PubMed PMID: 21643018; PubMed Central PMCID: PMC3347888)
Bensaad K, Vousden KH (2007) p53: new roles in metabolism. Trends Cell Biol 17(6):286–291 (Epub 2007 May 3. Review PubMed PMID: 17481900)
Bo MD, Secchiero P, Degan M, Marconi D, Bomben R, Pozzato G, Gaidano G, Del Poeta G, Forconi F, Zauli G, Gattei V (2010) MDM4 (MDMX) is overexpressed in chronic lymphocytic leukaemia (CLL) and marks a subset of p53wild-type CLL with a poor cytotoxic response to Nutlin-3. Br J Haematol 150(2):237–239 (Epub 2010 Apr 28. PubMed PMID: 20507307)
Brown CJ, Lain S, Verma CS, Fersht AR, Lane DP (2009) Awakening guardian angels: drugging the p53 pathway. Nat Rev Cancer 9(12):862–873 (Review PubMed PMID: 19935675)
Bueso-Ramos CE, Yang Y, deLeon E, McCown P, Stass SA, Albitar M (1993) The human MDM-2 oncogene is overexpressed in leukemias. Blood 82(9):2617–2623 (PubMed PMID: 8219216)
Bykov VJ, Issaeva N, Zache N, Shilov A, Hultcrantz M, Bergman J, Selivanova G, Wiman KG (2005) Reactivation of mutant p53 and induction of apoptosis in human tumor cells by maleimide analogs. J Biol Chem 280(34):30384–30391 (Epub 2005 Jul 1. PubMed PMID: 15998635)
Cheok CF, Verma CS, Baselga J, Lane DP (2011) Translating p53 into the clinic. Nat Rev Clin Oncol 8(1):25–37 (Epub 2010 Oct 26. Review PubMed PMID: 20975744)
Chiappori AA, Soliman H, Janssen WE, Antonia SJ, Gabrilovich DI (2010) INGN-225: a dendritic cell-based p53 vaccine (Ad.p53-DC) in small cell lung cancer: observed association between immune response and enhanced chemotherapy effect. Expert Opin Biol Ther 10(6):983–991 (Review PubMed PMID: 20420527; PubMed Central PMCID: PMC3146364)
Chipuk JE, Kuwana T, Bouchier-Hayes L, Droin NM, Newmeyer DD, Schuler M, Green DR (2004) Direct activation of Bax by p53 mediates mitochondrial membrane permeabilization and apoptosis. Science 303(5660):1010–1014 (PubMed PMID: 14963330)
Christiansen DH, Andersen MK, Pedersen-Bjergaard J (2001) Mutations with loss of heterozygosity of p53 are common in therapy-related myelodysplasia and acute myeloid leukemia after exposure to alkylating agents and significantly associated with deletion or loss of 5q, a complex karyotype, and a poor prognosis. J Clin Oncol 19(5):1405–1413 (PubMed PMID: 11230485)
Coll-Mulet L, Iglesias-Serret D, Santidrián AF, Cosialls AM, de Frias M, Castaño E, Campàs C, Barragán M, de Sevilla AF, Domingo A, Vassilev LT, Pons G, Gil J (2006) MDM2 antagonists activate p53 and synergize with genotoxic drugs in B-cell chronic lymphocytic leukemia cells. Blood 107(10):4109–4114 (Epub 2006 Jan 26. PubMed PMID: 16439685)
Crompton AM, Kirn DH (2007) From ONYX-015 to armed vaccinia viruses: the education and evolution of oncolytic virus development. Curr Cancer Drug Targets 7(2):133–139 (Review PubMed PMID: 17346104)
Di Bacco A, Keeshan K, McKenna SL, Cotter TG (2005) Molecular abnormalities in chronic myeloid leukemia: deregulation of cell growth and apoptosis. Oncologist 5(5):405–415 (Review. PubMed PMID: 11040277)
Ding L, Ley TJ, Larson DE, Miller CA, Koboldt DC, Welch JS, Ritchey JK, Young MA, Lamprecht T, McLellan MD, McMichael JF, Wallis JW, Lu C, Shen D, Harris CC, Dooling DJ, Fulton RS, Fulton LL, Chen K, Schmidt H, Kalicki-Veizer J, Magrini VJ, Cook L, McGrath SD, Vickery TL, Wendl MC, Heath S, Watson MA, Link DC, Tomasson MH, Shannon WD, Payton JE,

Kulkarni S, Westervelt P, Walter MJ, Graubert TA, Mardis ER, Wilson RK, DiPersio JF (2012) Clonal evolution in relapsed acute myeloid leukaemia revealed by whole-genome sequencing. Nature 481(7382):506–510. doi:10.1038/nature10738 (PubMed PMID: 22237025; PubMed Central PMCID: PMC3267864)

Faderl S, Kantarjian HM, Estey E, Manshouri T, Chan CY, Rahman Elsaied A, Kornblau SM, Cortes J, Thomas DA, Pierce S, Keating MJ, Estrov Z, Albitar M (2000) The prognostic significance of p16(INK4a)/p14(ARF) locus deletion and MDM-2 protein expression in adult acute myelogenous leukemia. Cancer 89(9):1976–1982 (PubMed PMID: 11064355)

Gaidano G, Pastore C, Santini V, Nomdedeu J, Gamberi B, Capello D, Vischia F, Resegotti L, Mazza U, Ferrini PR, Lo Coco F, Saglio G (1997) Genetic lesions associated with blastic transformation of polycythemia vera and essential thrombocythemia. Genes Chromosomes Cancer 19(4):250–255 (PubMed PMID: 9258660)

Graves B, Thompson T, Xia M, Janson C, Lukacs C, Deo D, Di Lello P, Fry D, Garvie C, Huang KS, Gao L, Tovar C, Lovey A, Wanner J, Vassilev LT (2012) Activation of the p53 pathway by small-molecule induced MDM2 and MDMX dimerization. Proc Natl Acad Sci U S A 109:11788–11793

Gu L, Zhu N, Findley HW, Zhou M (2008) MDM2 antagonist nutlin-3 is a potent inducer of apoptosis in pediatric acute lymphoblastic leukemia cells with wild-type p53 and overexpression of MDM2. Leukemia 22(4):730–739 (Epub 2008 Feb 14. PubMed PMID: 18273046)

Guo J, Xin H (2006) Chinese gene therapy. Splicing out the West? Science 314(5803):1232–1235 (PubMed PMID: 17124300)

Haidar MA, El-Hajj H, Bueso-Ramos CE, Manshouri T, Glassman A, Keating MJ, Maher A (1997) Expression profile of MDM-2 proteins in chronic lymphocytic leukemia and their clinical relevance. Am J Hematol 54(3):189–195 (PubMed PMID: 9067496)

Han X, Garcia-Manero G, McDonnell TJ, Lozano G, Medeiros LJ, Xiao L, Rosner G, Nguyen M, Fernandez M, Valentin-Vega YA, Barboza J, Jones DM, Rassidakis GZ, Kantarjian HM, Bueso-Ramos CE (2007) HDM4 (HDMX) is widely expressed in adult pre-B acute lymphoblastic leukemia and is a potential therapeutic target. Mod Pathol 20(1):54–62 (Epub 2006 Nov 24. PubMed PMID: 17143258)

Horiike S, Misawa S, Kaneko H, Sasai Y, Kobayashi M, Fujii H, Tanaka S, Yagita M, Abe T, Kashima K, Taniwaki M (1999) Distinct genetic involvement of the TP53 gene in therapy-related leukemia and myelodysplasia with chromosomal losses of Nos 5 and/or 7 and its possible relationship to replication error phenotype. Leukemia 13(8):1235–1242 (PubMed PMID: 10450752)

Hu B, Gilkes DM, Farooqi B, Sebti SM, Chen J (2006) MDMX overexpression prevents p53 activation by the MDM2 inhibitor Nutlin. J Biol Chem 281(44):33030–33035 (Epub 2006 Aug 11. PubMed PMID: 16905541)

Iacobucci I, Ferrari A, Lonetti A, Papayannidis C, Paoloni F, Trino S, Storlazzi CT, Ottaviani E, Cattina F, Impera L, Abbenante MC, Vignetti M, Vitale A, Potenza L, Paolini S, Soverini S, Pane F, Luppi M, FoÃ R, Baccarani M, Martinelli G (2011) CDKN2A/B alterations impair prognosis in adult BCR-ABL1-positive acute lymphoblastic leukemia patients. Clin Cancer Res 17(23):7413–7423 (PubMed PMID: 22134481)

Issaeva N, Bozko P, Enge M, Protopopova M, Verhoef LG, Masucci M, Pramanik A, Selivanova G (2004) Small molecule RITA binds to p53, blocks p53-HDM-2 interaction and activates p53 function in tumors. Nat Med 10(12):1321–1328 (Epub 2004 Nov 21. PubMed PMID: 15558054)

Kojima K, Konopleva M, Samudio IJ, Shikami M, Cabreira-Hansen M, McQueen T, Ruvolo V, Tsao T, Zeng Z, Vassilev LT, Andreeff M (2005) MDM2 antagonists induce p53-dependent apoptosis in AML: implications for leukemia therapy. Blood 106(9):3150–3159 (Epub 2005 Jul 12. PubMed PMID: 16014563; PubMed Central PMCID: PMC1895324)

Kojima K, Konopleva M, McQueen T, O'Brien S, Plunkett W, Andreeff M (2006) Mdm2 inhibitor Nutlin-3a induces p53-mediated apoptosis by transcription-dependent and transcription-independent mechanisms and may overcome Atm-mediated resistance to fludarabine in chronic lymphocytic leukemia. Blood 108(3):993–1000 (Epub 2006 Mar 16. PubMed PMID: 16543464; PubMed Central PMCID: PMC1895860)

Kojima K, Burks JK, Arts J, Andreeff M (2010) The novel tryptamine derivative JNJ-26854165 induces wild-type p53- and E2F1-mediated apoptosis in acute myeloid and lymphoid leukemias. Mol Cancer Ther 9(9):2545–57 (Epub 2010 Aug 24. PubMed PMID: 20736344; PubMed Central PMCID: PMC2949269)

Krug U, Ganser A, Koeffler HP (2002) Tumor suppressor genes in normal and malignant hematopoiesis. Oncogene 21(21):3475–3495 (Review PubMed PMID: 12032783)

Kurosu T, Wu N, Oshikawa G, Kagechika H, Miura O (2010) Enhancement of imatinib-induced apoptosis of BCR/ABL-expressing cells by nutlin-3 through synergistic activation of the mitochondrial apoptotic pathway. Apoptosis 15(5):608–620 (PubMed PMID: 20094798)

Lambert JM, Gorzov P, Veprintsev DB, Söderqvist M, Segerbäck D, Bergman J, Fersht AR, Hainaut P, Wiman KG, Bykov VJ (2009) PRIMA-1 reactivates mutant p53 by covalent binding to the core domain. Cancer Cell 15(5):376–388 (PubMed PMID: 19411067)

de Lange J, Verlaan-de Vries M, Teunisse AF, Jochemsen AG (2011) Chk2 mediates RITA-induced apoptosis. Cell Death Differ doi:10.1038/cdd.2011.182 (Epub ahead of print. PubMed PMID: 22158418)

Leffers N, Lambeck AJ, Gooden MJ, Hoogeboom BN, Wolf R, Hamming IE, Hepkema BG, Willemse PH, Molmans BH, Hollema H, Drijfhout JW, Sluiter WJ, Valentijn AR, Fathers LM, Oostendorp J, van der Zee AG, Melief CJ, van der Burg SH, Daemen T, Nijman HW (2009) Immunization with a P53 synthetic long peptide vaccine induces P53-specific immune responses in ovarian cancer patients, a phase II trial. Int J Cancer 125(9):2104–13 (PubMed PMID: 19621448)

Long J, Parkin B, Ouillette P, Bixby D, Shedden K, Erba H, Wang S, Malek SN (2010) Multiple distinct molecular mechanisms influence sensitivity and resistance to MDM2 inhibitors in adult acute myelogenous leukemia. Blood 116(1):71–80 (Epub 2010 Apr 19. PubMed PMID: 20404136; PubMed Central PMCID: PMC2904583)

Merup M, Juliusson G, Wu X, Jansson M, Stellan B, Rasool O, Röijer E, Stenman G, Gahrton G, Einhorn S (1997) Amplification of multiple regions of chromosome 12, including 12q13–15, in chronic lymphocytic leukaemia. Eur J Haematol 58(3):174–180 (PubMed PMID: 9150711)

Met O, Balslev E, Flyger H, Svane IM (2011) High immunogenic potential of p53 mRNA-transfected dendritic cells in patients with primary breast cancer. Breast Cancer Res Treat 125(2):395–406 (Epub 2010 Mar 25. PubMed PMID: 20336365)

Michaelis M, Rothweiler F, Barth S, Cinatl J, van Rikxoort M, Löschmann N, Voges Y, Breitling R, von Deimling A, Rödel F, Weber K, Fehse B, Mack E, Stiewe T, Doerr HW, Speidel D, Cinatl J Jr (2011) Adaptation of cancer cells from different entities to the MDM2 inhibitor nutlin-3 results in the emergence of p53-mutated multi-drug-resistant cancer cells. Cell Death Dis 2:e243. doi:10.1038/cddis.2011.129 (PubMed PMID: 22170099; PubMed Central PMCID: PMC3252738)

Mohammad RM, Wu J, Azmi AS, Aboukameel A, Sosin A, Wu S, Yang D, Wang S, Al-Katib AM (2009) An MDM2 antagonist (MI-319) restores p53 functions and increases the life span of orally treated follicular lymphoma bearing animals. Mol Cancer 8:115 (PubMed PMID: 19958544; PubMed Central PMCID: PMC2794250)

Montes de Oca Luna R, Wagner DS, Lozano G (1995) Rescue of early embryonic lethality in mdm2-deficient mice by deletion of p53. Nature 378(6553):203–206 (PubMed PMID: 7477326)

Mullighan CG, Williams RT, Downing JR, Sherr CJ (2008) Failure of CDKN2A/B (INK4A/B-ARF)-mediated tumor suppression and resistance to targeted therapy in acute lymphoblastic leukemia induced by BCR-ABL. Genes Dev 22(11):1411–1415 (PubMed PMID: 18519632; PubMed Central PMCID: PMC2732413)

Nahi H, Lehmann S, Mollgard L, Bengtzen S, Selivanova G, Wiman KG, Paul C, Merup M (2004) Effects of PRIMA-1 on chronic lymphocytic leukaemia cells with and without hemizygous p53 deletion. Br J Haematol 127(3):285–291 (PubMed PMID: 15491287)

Nahi H, Selivanova G, Lehmann S, Möllgård L, Bengtzen S, Concha H, Svensson A, Wiman KG, Merup M, Paul C (2008) Mutated and non-mutated TP53 as targets in the treatment of leukaemia. Br J Haematol 141(4):445–453 (Epub 2008 Mar 12. PubMed PMID: 18341636)

Park Y, Kim Y, Lee JH, Lee EY, Kim HS (2011) Usefulness of serum anti-p53 antibody assay for lung cancer diagnosis. Arch Pathol Lab Med 135(12):1570–1575 (PubMed PMID: 22129186)

Patton JT, Mayo LD, Singhi AD, Gudkov AV, Stark GR, Jackson MW (2006) Levels of HdmX expression dictate the sensitivity of normal and transformed cells to Nutlin-3. Cancer Res 66(6):3169–3176 (PubMed PMID: 16540668)

Peterson LF, Mitrikeska E, Giannola D, Lui Y, Sun H, Bixby D, Malek SN, Donato NJ, Wang S, Talpaz M (2011) p53 stabilization induces apoptosis in chronic myeloid leukemia blast crisis cells. Leukemia 25(5):761–769 (Epub 2011 Feb 25. PubMed PMID: 21350558)

Ringshausen I, O'Shea CC, Finch AJ, Swigart LB, Evan GI (2006) Mdm2 is critically and continuously required to suppress lethal p53 activity in vivo. Cancer Cell 10(6):501–514 (PubMed PMID: 17157790)

Rücker FG, Schlenk RF, Bullinger L, Kayser S, Teleanu V, Kett H, Habdank M, Kugler CM, Holzmann K, Gaidzik VI, Paschka P, Held G, von Lilienfeld-Toal M, Lübbert M, Fröhling S, Zenz T, Krauter J, Schlegelberger B, Ganser A, Lichter P, Döhner K, Döhner H (2012) TP53 alterations in acute myeloid leukemia with complex karyotype correlate with specific copy number alterations, monosomal karyotype, and dismal outcome. Blood 119(9):2114–21121 (Epub 2011 Dec 20. PubMed PMID: 22186996)

Saddler C, Ouillette P, Kujawski L, Shangary S, Talpaz M, Kaminski M, Erba H, Shedden K, Wang S, Malek SN (2008) Comprehensive biomarker and genomic analysis identifies p53 status as the major determinant of response to MDM2 inhibitors in chronic lymphocytic leukemia. Blood 111(3):1584–1593 (Epub 2007 Oct 30. PubMed PMID: 17971485)

Saha MN, Jiang H, Yang Y, Zhu X, Wang X, Schimmer AD, Qiu L, Chang H (2012) Targeting p53 via JNK pathway: a novel role of RITA for apoptotic signaling in multiple myeloma. PLoS One 7(1):e30215 (Epub 2012 Jan 20. PubMed PMID: 22276160; PubMed Central PMCID: PMC3262803)

Samudio IJ, Duvvuri S, Clise-Dwyer K, Watt JC, Mak D, Kantarjian H, Yang D, Ruvolo V, Borthakur G (2010) Activation of p53 signaling by MI-63 induces apoptosis in acute myeloid leukemia cells. Leuk Lymphoma 51(5):911–919 (PubMed PMID: 20423286)

Secchiero P, Barbarotto E, Tiribelli M, Zerbinati C, di Iasio MG, Gonelli A, Cavazzini F, Campioni D, Fanin R, Cuneo A, Zauli G (2006) Functional integrity of the p53-mediated apoptotic pathway induced by the nongenotoxic agent nutlin-3 in B-cell chronic lymphocytic leukemia (B-CLL). Blood 107(10):4122–4129 (Epub 2006 Jan 26. PubMed PMID: 16439677)

Seifert H, Mohr B, Thiede C, Oelschlägel U, Schäkel U, Illmer T, Soucek S, Ehninger G, Schaich M (2009) Study Alliance Leukemia (SAL). The prognostic impact of 17p (p53) deletion in 2272 adults with acute myeloid leukemia. Leukemia 23(4):656–663 (Epub 2009 Jan 8. PubMed PMID: 19151774)

Shangary S, Qin D, McEachern D, Liu M, Miller RS, Qiu S, Nikolovska-Coleska Z, Ding K, Wang G, Chen J, Bernard D, Zhang J, Lu Y, Gu Q, Shah RB, Pienta KJ, Ling X, Kang S, Guo M, Sun Y, Yang D, Wang S (2008) Temporal activation of p53 by a specific MDM2 inhibitor is selectively toxic to tumors and leads to complete tumor growth inhibition. Proc Natl Acad Sci U S A 105(10):3933–3938 (Epub 2008 Mar 3. PubMed PMID: 18316739; PubMed Central PMCID: PMC2268798)

Shen H, Maki CG (2011) Pharmacologic activation of p53 by small-molecule MDM2 antagonists. Curr Pharm Des 17(6):560–568 (Review. PubMed PMID: 21391906)

Sill H, Olipitz W, Zebisch A, Schulz E, Wölfler A (2011) Therapy-related myeloid neoplasms: pathobiology and clinical characteristics. Br J Pharmacol 162(4):792–805. doi:10.1111/j.1476-5381.2010.01100.x (Review PubMed PMID: 21039422; PubMed Central PMCID: PMC3042191)

Steele AJ, Prentice AG, Hoffbrand AV, Yogashangary BC, Hart SM, Nacheva EP, Howard-Reeves JD, Duke VM, Kottaridis PD, Cwynarski K, Vassilev LT, Wickremasinghe RG (2008) p53-mediated apoptosis of CLL cells: evidence for a transcription-independent mechanism. Blood 112(9):3827–3834 (Epub 2008 Aug 5. PubMed PMID: 18682598)

Sugihara E, Shimizu T, Kojima K, Onishi N, Kai K, Ishizawa J, Nagata K, Hashimoto N, Honda H, Kanno M, Miwa M, Okada S, Andreeff M, Saya H (2011) Ink4a and Arf are crucial factors in the determination of the cell of origin and the therapeutic sensitivity of Myc-induced mouse lymphoid tumor. Oncogene. doi:10.1038/onc.2011.462 (Epub ahead of print. PubMed PMID: 21986948; PubMed Central PMCID: PMC3271180)

Sutherland HS, Hwang IY, Marshall ES, Lindsay BS, Denny WA, Gilchrist C, Joseph WR, Greenhalgh D, Richardson E, Kestell P, Ding A, Baguley BC (2011) Therapeutic reactivation of mutant p53 protein by quinazoline derivatives. Invest New Drugs (in press) (Epub 2011 Sep 13. PubMed PMID: 21912889)

Tabe Y, Sebasigari D, Jin L, Rudelius M, Davies-Hill T, Miyake K, Miida T, Pittaluga S, Raffeld M (2009) MDM2 antagonist nutlin-3 displays antiproliferative and proapoptotic activity in mantle cell lymphoma. Clin Cancer Res 15(3):933–942 (PubMed PMID: 19188164)

Tovar C, Rosinski J, Filipovic Z, Higgins B, Kolinsky K, Hilton H, Zhao X, Vu BT, Qing W, Packman K, Myklebost O, Heimbrook DC, Vassilev LT (2006) Small-molecule MDM2 antagonists reveal aberrant p53 signaling in cancer: implications for therapy. Proc Natl Acad Sci U S A 103(6):1888–1893 (Epub 2006 Jan 27. PubMed PMID: 16443686; PubMed Central PMCID: PMC1413632)

Tyler MA, Ulasov IV, Sonabend AM, Nandi S, Han Y, Marler S, Roth J, Lesniak MS (2009) Neural stem cells target intracranial glioma to deliver an oncolytic adenovirus in vivo. Gene Ther 16(2):262–278 (Epub 2008 Dec 11. PubMed PMID: 19078993; PubMed Central PMCID: PMC2642530)

Vaseva AV, Marchenko ND, Moll UM (2009) The transcription-independent mitochondrial p53 program is a major contributor to nutlin-induced apoptosis in tumor cells. Cell Cycle 8(11):1711–1719 (Epub 2009 Jun 30. PubMed PMID: 19411846; PubMed Central PMCID: PMC2823264)

Vassilev LT, Vu BT, Graves B, Carvajal D, Podlaski F, Filipovic Z, Kong N, Kammlott U, Lukacs C, Klein C, Fotouhi N, Liu EA (2004) In vivo activation of the p53 pathway by small-molecule antagonists of MDM2. Science 303(5659):844–848 (Epub 2004 Jan 2. PubMed PMID: 14704432)

Vousden KH, Prives C (2009) Blinded by the Light: The Growing Complexity of p53. Cell 137(3):413–431 (Review PubMed PMID: 19410540)

Wade M, Wong ET, Tang M, Stommel JM, Wahl GM (2006) Hdmx modulates the outcome of p53 activation in human tumor cells. J Biol Chem 281(44):33036–33044 (Epub 2006 Aug 13. PubMed PMID: 16905769)

Winkler D, Schneider C, Kröber A, Pasqualucci L, Lichter P, Döhner H, Stilgenbauer S (2005) Protein expression analysis of chromosome 12 candidate genes in chronic lymphocytic leukemia (CLL). Leukemia 19(7):1211–1215 (PubMed PMID: 15902296)

Wolff S, Erster S, Palacios G, Moll UM (2008) p53's mitochondrial translocation and MOMP action is independent of Puma and Bax and severely disrupts mitochondrial membrane integrity. Cell Res 18(7):733–744 (PubMed PMID: 18504456; PubMed Central PMCID: PMC2830652)

Xia M, Knezevic D, Tovar C, Huang B, Heimbrook DC, Vassilev LT (2008) Elevated MDM2 boosts the apoptotic activity of p53-MDM2 binding inhibitors by facilitating MDMX degradation. Cell Cycle 7(11):1604–1612 (Epub 2008 Mar 17. PubMed PMID: 18520179)

Xu J, Timares L, Heilpern C, Weng Z, Li C, Xu H, Pressey JG, Elmets CA, Kopelovich L, Athar M (2010) Targeting wild-type and mutant p53 with small molecule CP-31398 blocks the growth of rhabdomyosarcoma by inducing reactive oxygen species-dependent apoptosis. Cancer Res 70(16):6566–6576 (Epub 2010 Aug 3. PubMed PMID: 20682800; PubMed Central PMCID: PMC2922473)

Zache N, Lambert JM, Rökaeus N, Shen J, Hainaut P, Bergman J, Wiman KG, Bykov VJ (2008) Mutant p53 targeting by the low molecular weight compound STIMA-1. Mol Oncol 2(1):70–80 (Epub 2008 Mar 7. PubMed PMID: 19383329)

Zenz T, Mertens D, Küppers R, Döhner H, Stilgenbauer S (2010) From pathogenesis to treatment of chronic lymphocytic leukaemia. Nat Rev Cancer 10(1):37–50 (Epub 2009 Dec 3. Review PubMed PMID: 19956173)

Chapter 8
AML-Selective Apoptosis Induction by Rationally Designed Death Ligand Fusion Proteins

Edwin Bremer and Wijnand Helfrich

Abstract In the past decade, antibody-based therapeutics have started to make good on their promise with potent clinical activity in several human malignancies, including acute myeloid leukemia (AML). In this chapter, various classes of antibody-based agents designed to selectively induce apoptotic cell death in AML will be discussed, including naked antibodies, antibody–drug conjugates, and immunotoxins. Moreover, the rationale for exploiting several of the body's own immune effector molecules for the targeted elimination of AML cells is highlighted. In particular, this involves the use of the death ligand TNF-related apoptosis-inducing ligand (TRAIL). Recombinant fusion proteins in which an antitumor antibody fragment (scFv) is fused to soluble TRAIL proved to be essentially inactive while en route, but gain potent pro-apoptotic antitumor activity after selective binding to a predefined tumor-associated cell surface antigen. The use of death ligand fusion proteins may be of clinical significance in targeted approaches in AML.

Keywords Targeted therapy · Immunotoxin · TRAIL · Fusion proteins · Apoptosis

Abbreviations

TRAIL	TNF-related apoptosis-inducing ligand
sTRAIL	soluble TRAIL
scFv	single chain of the variable fragments
MAb	monoclonal antibodies
CDC	complement-dependent cytotoxicity
ADCC	antibody-dependent cellular cytotoxicity
ADCP	antibody-dependent cellular phagocytosis
NK	natural killer
FcRn	neonatal Fc receptor

W. Helfrich (✉) · E. Bremer
Department of Surgery/Translational Surgical Oncology, University Medical Center Groningen, Hanzeplein 1/BA44, Groningen, GZ 9713, The Netherlands
e-mail: w.helfrich@umcg.nl

E. Bremer
e-mail: e.bremer@umcg.nl

M. Andreeff (ed.), *Targeted Therapy of Acute Myeloid Leukemia*,
Current Cancer Research, DOI 10.1007/978-1-4939-1393-0_8

FLT3	FMS-related tyrosine kinase 3
bsAb	bispecific antibodies
MHC	major histocompatibility complex
BiTEs	bispecific T-cell engager molecules
MNC	mononuclear cell
GO	Gemtuzumab Ozogamicin
VPA	valproic acid
DT	Diphtheria toxin
CTCL	Cutaneous T-cell lymphoma
SCID	Severe Combined Immune Deficiency
PE	Pseudomonas exotoxin
ETA	Pseudomonas exotoxin A
eEF2	eukaryotic Elongation Factor 2
IL-3Rα	alpha subunit of the interleukin-3 receptor
CLL-1	C-type lectin-like molecule-1
RI	RNAse inhibitor
CsA	cyclosporine
FasL	Fibroblast associated cell surface ligand
DD	death domains
OPG	osteoprotegerin
RANKL	Receptor Activator of NFκB ligand
DICS	death inducing signaling complex
LSCs	Leukemic stem cells
DARPins	Designed Ankyrin Repeat Proteins

8.1 Introduction

Already in the early nineteen hundreds, Paul Ehrlich postulated that the immune system may play a key role in the suppression of cancer. In the 1950s, this notion led Burnet and Thomas to formulate the so-called cancer immune surveillance theory. This theory states that immune cells act as sentinels in recognizing and eliminating continuously arising pre-cancerous cells. Since then, numerous scientific reports have provided evidence that immune surveillance is a key defense mechanism that protects the host against the formation of tumors by safeguarding normal cellular homeostasis (Dunn et al. 2004). Understandably, considerable effort has been made to exploit cells and molecules of the immune system to treat or retard the progression of cancer. Several so-called immunotherapies have already made their way into clinical practice. Arguably, the most successful of these immunotherapeutic approaches thus far is the use of monoclonal antibodies (MAbs). Of particular interest is the use of the cancer selectivity of MAbs for the targeted delivery of toxic compounds to malignant cells, for instance, in the format of antibody–drug conjugates and recombinant immunotoxins.

Unfortunately, authentic tumor-specific target antigens for MAb-mediated therapy are very rare. In fact, most of the clinical anticancer MAb are merely directed at over-expressed tumor-associated antigens that to some extent are also expressed on normal cells. Therefore, conventional MAb-based approaches are often not specific enough to allow for the delivery of extremely toxic agents at clinically relevant concentrations. This concern is evidenced by the sometimes-serious adverse effects observed for the antibody-targeted delivery of intrinsically toxic compounds, such as microbial and plant-derived toxins, oncolytic drugs, and radionuclides.

8.2 Antibody-Based Approaches in Acute Myeloid Leukemia

8.2.1 Naked Antibodies

The therapeutic effects of unmodified or "naked" MAb are largely attributable to the natural effector functions of the human immune response, such as complement-dependent cytotoxicity (CDC), antibody-dependent cellular cytotoxicity (ADCC), and antibody-dependent cellular phagocytosis (ADCP; Fig. 8.1). Moreover, binding and cross-linking of certain target antigens by naked antitumor MAbs may also result in a therapeutic (e.g., pro-apoptotic) antitumor effect in its own right. ADCC and ADCP are mediated by the interaction of the Fc-domain of a given therapeutic MAb with Fc receptors (FcRs) on the cell surface of myeloid effector cells and natural killer (NK) cells, while CDC is triggered by interactions between these Fc-domains and catalytic components of the complement cascade.

Leukemia is particularly suited for MAb-based approaches due to the accessibility of malignant cells in the circulation, bone marrow, and lymphoid organs. Moreover, the well-defined immunophenotypes of the various lineages and differentiation stages of malignant hematopoietic transformations may be exploited for targeted therapy. Various MAb-based targeted approaches have been developed that are directed to cell surface differentiation markers selectively overexpressed on acute myeloid leukemia (AML) cells, including CD66 (Koenecke et al. 2008), CD64 (Thepen et al. 2009), CD47 (Majeti et al. 2009), CD45 (Pagel et al. 2011), CD44 (Jin et al. 2006), CD32 (Saito et al. 2010), CLL-1 (Bakker et al. 2004), and in particular CD33 (Linenberger 2005).

The expression of CD33 on leukemic blasts is found in approximately 80–90 % of AML patients, whereas its expression on normal cells is limited to the myeloid cell lineage. The restricted expression profile of CD33 has fueled the clinical development of lintuzumab, a naked anti-CD33 MAb, and the MAb–drug conjugate gemtuzumab ozogamicin (GO; see Sect. 11.4.1).

Lintuzumab was constructed by grafting the complementarity determining regions of the murine anti-CD33 MAb M195 onto a human IgG1 antibody (Caron et al. 1992). The mode of action of lintuzumab predominantly involves ADCC and ADCP. Furthermore, in vitro studies indicate that ligation of CD33 by lintuzumab

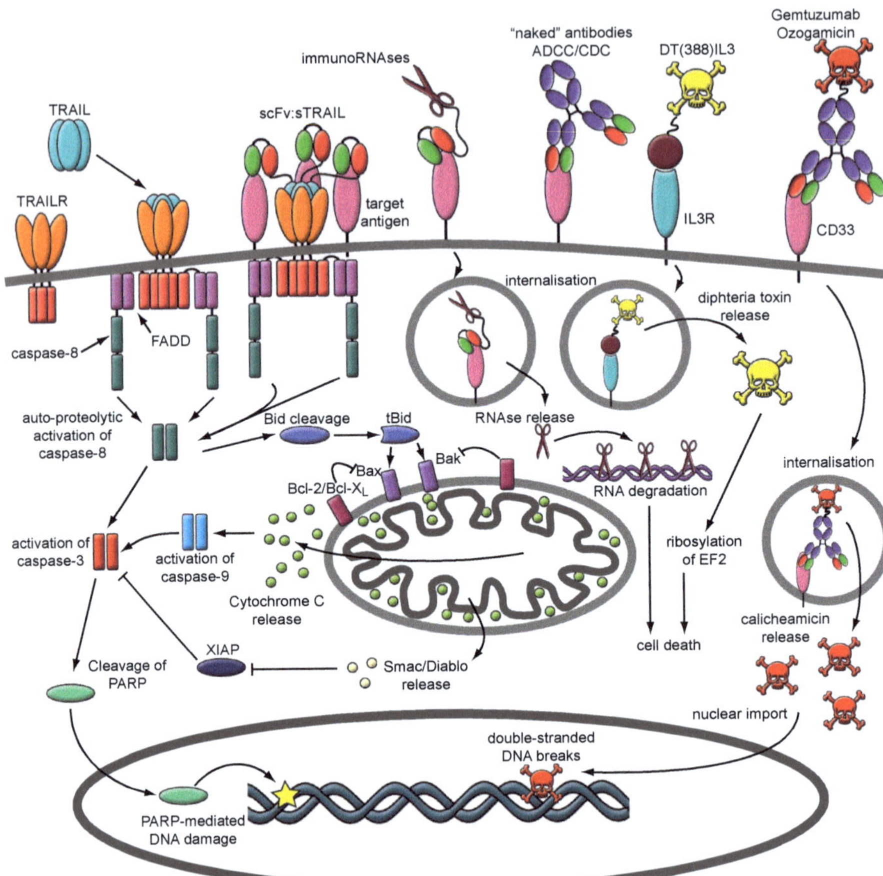

Fig. 8.1 Targeted approaches for the elimination of AML cells. Therapeutic strategies targeting the death receptors, TNF-related apoptosis-inducing ligand (TRAIL)-R1/R2 in AML, include the use of recombinant soluble TRAIL (sTRAIL) and single chain of the variable fragments (scFv):sTRAIL fusion proteins. Activation of TRAIL-R1/-R2 receptors results in the recruitment of the adaptor molecule Fas-associated death domain (FADD) and caspase-8 to the death domains of these receptors. Assembly of this so-called death-inducing signaling complex (DISC) leads to sequential activation of caspase-8 and caspase-3, ultimately resulting in DNA damage and apoptosis, i.e., poly ADP ribose polymerase (PARP) cleavage. In addition, caspase-8 can cleave Bid to tBid and thereby activate a mitochondrial amplification loop. This loop is characterized by cytochrome C release from the mitochondria and the subsequent activation of caspase-9. In turn, caspase-9 can activate caspase-3. Treatment of AML cells with naked antibodies can result in antibody-dependent cell-mediated cytotoxicity (ADCC) and complement-dependent cytotoxicity (CDC). The antibody–drug conjugate gemtuzumab ozogamicin (GO) selectively binds to CD33-positive AML cells, internalizes and is routed to the lysosomes. In the acidic milieu of the lysosomes, the calicheamicin moiety is hydrolytically released. Next, calicheamicin translocates to the nucleus and intercalates with the DNA, resulting in double-stranded breaks followed by apoptotic cell death. In immuno ribonucleases (RNAses), an antibody fragment allows for targeted delivery and internalization, after which the RNase moiety exerts its RNA degrading activity, readily leading to the apoptotic cell death of the targeted cell. DT(388)IL3 is a targeted toxin in which the catalytic and translocation domains DT are fused to interleukin-3 (IL3). The antitumor effects of DT-fusion proteins require internalization by receptor-mediated endocytosis. Subsequently, DT needs to be processed into its active form after which it kills the cell by catalyzing irreversible ADP ribosylation and subsequent inactivation of elongation factor 2. Therefore, the number of DT molecules needed to induce cell death is remarkably low. XIAP X-linked inhibitor of apoptosis protein

can also block proliferation and induce apoptosis in AML cells. This effect of lintuzumab appears to proceed through activation of the intracellular SHP-1 and SHP-2 domains of CD33 and the subsequent recruitment of Syk protein tyrosine kinase. Moreover, preclinical models of AML demonstrated that lintuzumab modulated the cytokine milieu in the tumor microenvironment (Sutherland et al. 2009).

Typically, targeted therapies for AML have low single agent activity and thus must be combined with one or a combination of chemotherapeutic agents to mount clinical efficacy. Of note, in clinical evaluation, lintuzumab failed to improve patient outcomes when combined with conventional chemotherapy (Jurcic 2012).

8.2.2 Engineering of Naked Anti-AML MAbs

Recent data demonstrated that the efficacy of naked human and humanized MAbs can be increased by engineering their Fc-domain for improved neonatal Fc receptor (FcRn)-mediated storage (Kuo and Aveson 2011) and enhanced capability to recruit Fc receptor-bearing immune effector cells. This can be achieved by defined genetic modifications of the glycosylation pattern or the amino-acid sequence of the Fc domain. In a recent report, a first anti-AML Fc-engineered MAb, designated 4G8SDIEM, was preclinical characterized. MAb 4G8SDIEM carries the amino-acid exchanges S239D and I332E (hence SDIE modification), which markedly enhances ADCC. MAb 4G8SDIEM is directed against FMS-related tyrosine kinase 3 (FLT3), a cell surface antigen that is expressed on the leukemic blasts in a vast majority of AML patients. The expression levels of FLT3 on AML cells are substantially higher than those detected on normal hematopoietic precursor cells and dendritic cells. The Fc-engineered MAb 4G8SDIEM showed markedly enhanced ADCC, both towards FLT3-expressing AML cell lines and malignant blasts derived from AML patients. Importantly, MAb 4G8SDIEM-mediated ADCC towards normal FLT3-positive cells was not observed (Hofmann et al. 2012).

8.3 Antibody-Formats to Redirect Immune Effector Cells

Using bispecific antibodies (bsAbs) it is possible to retarget large numbers of predefined immune effector cells to a predefined target cells, such as AML cells. BsAbs are designed to have two specificities, with one being directed at an activating cell surface molecule that is expressed on a cytotoxic immune effector cell population and another directed at a tumor cell surface-expressed target antigen. In this approach, an entire cytotoxic effector cell population may be recruited to kill predefined target cells. Most researchs has focused on the retargeting of cytotoxic T-cell as they are highly motile and are endowed with powerful cytotoxic effector mechanisms. T-cell retargeting bsAb is designed to engage and activate pre-existing polyclonal T-cell populations via the TCR-CD3 complex, preferably only in the presence of the pre-selected tumor antigen-positive tumor cells. In other words, T cell retargeting bsAb is capable of connecting any kind of cytotoxic T-cell to a cancer cell, independently of T-cell receptor specificity, co-stimulation, or major histocompatibility complex (MHC) restriction. Of particular clinical interest are the so-called bispecific T-cell engager molecules (BiTEs). BiTEs are recombinant fusion proteins consisting of two scFvs, one of which binds to T-cells via the TCR-CD3 complex whereas the other binds to a pre-selected tumor antigen. A BiTE binds monovalently to T-cells with low affinity. Therefore, BiTE binding does not trigger T-cell signaling via CD3, unless the BiTE is presented to the T-cell in a multivalent fashion by a target cell (Brischwein et al. 2007). Unlike other bsAb formats, BiTEs do not require co-stimulation to induce T-cell cytotoxicity. Moreover, BiTEs are active at low concentrations and at lower effector to target cell ratios compared with other bsAb formats. Their efficacy has been attributed to the formation of a close to authentic immunological synapse between effector and target cells, in which granzymes and perforins are released resulting in the rapid elimination of the targeted cell by osmotic lysis and apoptosis induction. In early clinical trials, a CD19-targeted BiTE, called blinatumomab, showed high response rates and durable remissions in non-Hodgkin's lymphoma, highlighting the promise of this approach.

Alternatively, bsAbs with combined specificity for one of the human Fc-receptors (FcRs; e.g., CD16 (Silla et al. 1995)), or CD64 (Balaian and Ball 2004), and CD33 may be of therapeutic interest in AML. As described above, FcRs are expressed on various immune effector cells, such as NK cells, monocytes, macrophages, neutrophils, and dendritic cells. Triggering of FcRs typically results in ADCC, phagocytosis, and cytokine release. For efficient retargeting of FcR-positive effector cells, bsAbs have been developed that bind outside of the Ig-ligand binding site of the respective FcR, enabling the recruitment of these effector cells even in the presence of high concentrations of serum Ig.

A more intricate class of agents that is of interest for AML are the so-called triplebodies. A triplebody comprises three scFv antibody fragments genetically linked in a single polypeptide. Recently, an anti-AML tri-specific triplebody was described that consisted of three scFv's specific for CD123, CD33, and the Fcγ-receptor III (CD16), respectively. This triplebody induced potent ADCC of two different AML-derived CD33+ and CD123+ cell lines in the low picomolar range using isolated mononuclear cells (MNCs) as effector cells. In addition, this triplebody showed a high potency in mediating ADCC of primary leukemia cells isolated from peripheral blood or bone marrow of seven AML patients and as such may represent an attractive candidate for further development.

8.4 Antibody-Mediated Delivery of Toxic Compounds

8.4.1 Antibody–Drug Conjugates

In general, antibody–drug conjugates are produced by the chemical coupling (e.g., by disulfide bonds or thioether bonds) of a cytotoxic drug to a naked MAb. The main goal of antibody–drug conjugates is not so much the recruitment of immune effector cells via its Fc-domain, but rather the induction of cancer cell death after its internalization and the intracellular release of the coupled cytotoxic drug. Typically, the cytotoxic drug induces apoptosis in the targeted cell. Apoptosis hinges on the activation of cellular proteases called caspases. Once activated, these caspases proteolytically cleave a host of cellular proteins leading to demise of the cell. A variety of toxic drugs have been used for to produce anti-AML antibody–drug conjugates, including calicheamicins, maytansinoids (Lapusan et al. 2012), and auristatins (Singh and Erickson 2009).

The first antibody–drug conjugate to gain approval by the US Food and Drug Administration (FDA) was GO, also known under its trade name "mylotarg." GO comprises a humanized anti-CD33 MAb (IgG4-kappa) that is chemically coupled to a highly cytotoxic calicheamicin derivative through a hydrolysable linker. Calicheamicins belong to the class of enediyne antibiotics that are derived from the bacterium Micromonospora echinospora. Calicheamicins are among the most potent and toxic antitumor agents known. Their toxicity is directly related to their potent ability to induce double-strand DNA breaks.

Upon antibody-mediated selective binding to cell surface-expressed CD33, GO internalizes, after which the calicheamicin moiety is hydrolytically released in the acidic milieu of the lysosome. Next, calicheamicin translocates to the nucleus in which it induces sequence-specific double strand DNA breaks resulting in apoptotic cell death (Fig. 8.1). Treatment with GO improved the survival in a subset of AML patients when combined with standard chemotherapy. Unfortunately, treatment with GO may result in serious adverse effects, including severe and sometimes life-threatening myelosuppression. This toxicity is thought to be related to poor serum stability due to the pH-sensitive linker technology that is used to link calicheamicin to the antibody. This instability may lead to inadvertent release of calicheamicin into the circulation, resulting in toxicity in various normal cells and organs due to CD33-independent uptake of calicheamicin. GO was withdrawn from the US market in June 2010 after clinical trials demonstrated an increase in patient death and lack of benefit over conventional AML therapy. Nevertheless, a recent study indicated that fractionated lower doses of GO, leading to delivery of a higher cumulative dose, improved outcome in AML. Thus, reassessment of GO as front-line therapy for AML is to be anticipated (Castaigne et al. 2012).

From the above, it is clear that single agent treatment with GO is unlikely to be curative in AML. Therefore, evaluation of combination treatment of GO with other therapeutics is warranted in order to reduce toxicity while enhancing its therapeutic efficacy. Previously, we hypothesized that the histone deacetylase inhibitor valproic acid (VPA) may be used to sensitize AML cells to GO. We reasoned that VPA-induced histone hyperacetylation, known to open the chromatin structure (Marchion et al. 2005), should augment DNA intercalation by calicheamicin, resulting in enhanced DNA degradation. We found that at clinically relevant nontoxic concentrations, VPA potently augmented the tumoricidal activity of GO towards AML cell lines and primary AML blasts. Indeed, we could demonstrate that VPA treatment augmented DNA intercalation by calicheamicin and enhanced DNA degradation. Importantly, the observed synergy of co-treatment was restricted to CD33+ AML cells. This indicates that co-treatment of AML with VPA and GO may be of clinical importance (ten Cate et al. 2007). Importantly, improvements in linker design and coupling strategies have resulted in various aurastatin-based antibody–drug conjugates with promising activity. For instance, treatment with the antibody–drug conjugates brentuximab vedotin, which delivers a modified aurastatin to CD30 expressed on Hodgkin's lymphoma, induced objective responses in 75 % of patients with relapsed or refractory disease with manageable toxicity. Remarkably, durable complete responses were observed that, upon publication of these results, were approaching 2 years (Chen et al. 2012). Thus, novel linker design, coupling strategy, as well as effector moieties may help to translate these promising results towards AML.

8.4.2 Immunotoxins

Typically, immunotoxins are chimeric recombinant fusion proteins consisting of a tumor-selective binding domain (growth factor, cytokine, peptide, or antibody fragment) genetically linked to a complete or modified form of a catalytic toxin protein of microbial or plant origin. Usually, the toxin domain is engineered to be devoid of its natural cell binding activity, which allows it to be redirected after genetic fusion to a tumor-selective binding domain. Most immunotoxins are designed to bind to preselected cell surface receptors that are overexpressed on the cell surface of malignant cells. Subsequently, immunotoxins must be internalized by endocytosis followed by intracellular processing and routing in order to exert their cytotoxic activity. The toxin moiety of immunotoxins usually acts as intracellular enzymes that block cytosolic protein synthesis either by modifying eukaryotic elongation factor-2 eEF2 (e.g., DT and PE) or ribosomal RNA (e.g., ricin, gelonin, saporin), or by degrading total ribonuclease (RNA).

Usually, the catalytic activity of immunotoxins results in high substrate turnover. In theory, this may allow for the efficient killing of a tumor cell by only one single internalized toxin molecule. Importantly, protein toxins must stay intact and avoid lysosomal degradation. Various recombinant immunotoxins have been developed for targeted approaches of hematologic malignancies (Kreitman 2009), of which some prime examples will be discussed below.

8.4.3 Targeted Delivery of Diphtheria Toxin

Truncated forms of diphtheria toxin (DT) have been fused to a number of cytokines in order to target malignant cells (Williams et al. 1987; Chan et al. 1995; Lakkis et al. 1991; Chadwick et al. 1993; Jean and Murphy 1991). The antitumor effects of DT-fusion proteins require receptor/target binding and subsequent internalization by receptor-mediated endocytosis. Subsequently, DT needs to be processed into its active form. It kills the cell by catalyzing the irreversible ADP ribosylation and subsequent inactivation of elongation factor 2. Therefore, the number of DT molecules needed to induce cell death is remarkably low.

Currently, Denileukin diftitox (trade name Ontak), is the only clinically approved immunotoxin. Ontak is a recombinant fusion protein in which DT lacking its endogenous cell binding domain is genetically fused to human interleukin-2 (IL2). Via its IL2 domain, Ontak binds to the high-affinity IL2 receptor (CD25) that is expressed on an array of abnormal cells, including malignant cells, resulting in internalization and intracellular delivery of DT. Ontak was approved by the FDA in 1999 for treatment of cutaneous T-cell lymphoma (CTCL). In general, advanced-stage primary CTCL has an unfavorable prognosis and low survival rates. Aggressive treatment of CTCL with chemotherapy is not curative and causes considerable side effects. Treatment of these patients with Ontak frequently results in disease regression and may extend the life expectancy of patients from 8 months to more than a year.

For AML, a similar type of DT immunotoxin has been developed in which the catalytic and translocation domains DT are fused to interleukin-3 (IL3; Fig. 8.1). This immunotoxin, designated DT(388)IL3, showed potent anti-AML activity in in vitro and in vivo studies. A subsequent phase I clinical trial with 40 evaluable AML patients treated with DT(388)IL3 resulted in one complete responder and one partial responder. Thus, DT(388)IL3 produced remissions in patients with relapsed or refractory AML, although in only 2 out of 40 patients, with limited toxicities (Frankel et al. 2008).

Another example of an AML-targeted DT-immunotoxin is DT-GMCSF. Fusion protein DT-GMCSF consists of DT genetically fused to human granulocyte-macrophage colony-stimulating factor (GM-CSF). In in vitro experiments, this immunotoxin was found to also eliminate a subset of AML cells, the so-called AML stem cells, considered to be the driving force of AML. Furthermore, DT-GMCSF eliminated AML cells with leukemia-initiating potential in severe combined immune deficient (SCID) mice. Repopulation by this subset of AML cells is critical for the relapse of AML. Interestingly, in addition to the anticipated activation of caspase-dependent apoptosis, DT-GMCSF also kills AML cells by simultaneously activating caspase-independent necroptosis. These mechanisms depend on the ability of the targeted toxin to inhibit protein synthesis, and are not affected by the receptor that is targeted or the mechanism through which protein synthesis is blocked (Horita et al. 2008).

8.4.4 Targeted Delivery of Pseudomonas Exotoxin

Pseudomonas exotoxin A (PE, also known as ETA) is a highly potent bacterial toxin that is secreted by Pseudomonas aeruginosa. PE requires a KDEL-like sequence in order to traffic to the endoplasmic reticulum from which it can then enter the cytosol. Like DT, the potent cytotoxic activity of PE comes from its ability to bind and subsequently irreversibly inactivate eEF2.

Already in the late 1990s, studies were undertaken to evaluate the applicability of PE for the targeted elimination of leukemia. A recombinant IL6-PE fusion protein (IL6-PE4E) was produced in which the toxicity and specificity of the chimeric toxin was enhanced by substitution of four glutamine residues (hence 4E) in the PE domain I. Pediatric AML cells expressing IL-6 receptors proved to be sensitive to IL6-PE4E. Furthermore, treatment with IL6-PE4E abrogated tumor growth in a mouse tumor model.

Recently, a novel recombinant immunotoxin, designated H22(scFv)-ETA', was described in which a humanized scFv antibody against CD64 was genetically fused to a truncated form of ETA. CD64 is a promising target antigen in AML, because it is highly expressed on monocytoid differentiated AML subtypes. H22(scFv)-ETA' potently induced cell death in CD64+ AML-derived tumor cell lines and primary patient-derived AML cells, both in vitro and ex vivo. Moreover, H22(scFv)-ETA' showed promising activity in an AML xenograft tumor model, providing the basis for a novel therapeutic strategy for the treatment of AML patients.

Similarly, a set of novel scFv-ETA fusion protein have been reported that are directed against the human leukocyte surface antigen CD123 (Stein et al. 2010) and CD33 (Schwemmlein et al. 2006), respectively. In these fusion proteins, the ETA domain was modified with the eukaryotic endoplasmatic reticulum retention motif KDEL mediating improved retrograde transport to the endoplasmic reticulum. CD123, the α subunit of the interleukin-3 receptor (IL-3Rα), is expressed on a variety of hematopoietic cells, including myeloid cells. Intriguingly, CD123 as well as C-type lectin-like molecule-1 (CLL-1), CD44, CD33, and CD96 are also expressed on AML-leukemia stem cells. The CD123-directed PE fusion protein induced potent apoptosis in AML-derived MOLM-13 and SKNO-1 cells at nanomolar concentrations.

An analogous CD33-directed scFv-ETA fusion protein has been reported by the same research group. The scFvCD33:ETA fusion protein showed potent apoptotic activity in both AML cell lines as well as primary patient-derived AML cells as present in blood or bone marrow at low nanomolar concentrations. The antigen-restricted cytotoxicity of both novel fusion proteins makes them promising candidates for further evaluation of their therapeutic potential.

8.4.5 ImmunoRNAses

In recent years, also mammalian enzymes or active fragments thereof have been used as effector domains in immunotoxins. Prominent examples are the so-called immunoRNAses in which the effector moiety is a nontoxic RNAse (Schirrmann et al. 2009). Normally, RNases catalyze the rapid degradation of RNA and are present in a broad range of species, from simple prokaryotes to mammals. Controlled degradation of RNA is a key factor in gene expression, cell maturation, and cellular turnover, both in normal and malignant cells. Certain RNases proved to have remarkable antitumor activities in their own right. In this respect, one of the most promising RNases, Ranpirnase (trademark Onconase), isolated from northern leopard frog (Rana pipiens) is being developed for various malignancies (Ardelt et al. 2008). In immunoRNAses, an antibody fragment allows for targeted delivery and internalization, after which the RNase moiety exerts its RNA-degrading activity, readily leading to the apoptotic cell death of the targeted cell. Importantly, immunoRNAses gain substantial activity when engineered to evade the cytosolic RNAse inhibitor (RI). RI is a 50-kDa protein found exclusively in the cytosol of mammalian cells. Current RI-evasion strategies involve the modification of the RNAse domain to perturb its interaction with RI, while fully retaining its RNA-degrading activity (Lomax et al. 2012).

Thus far, immunogenicity of amphibian Onconase has not been reported as a serious problem. However, by selecting a human RNase, e.g., eosinophil-derived RNAse or pancreatic RNAse, and a human or humanized antibody fragment as building blocks, immunoRNAses would be even less immunogenic. In this respect, an entirely human immunoRNAse was constructed that was directed against CD30+ lymphomas. Proliferation of a CD30+ lymphoma cell line was strongly inhibited by this CD30-specific immunoRNAse at an IC50 of 3.3 nM (Menzel et al. 2008). ImmunoRNAses appear to be promising candidates for the immunotherapy of AML because of the expected low immunogenicity, good production yields, and potent effector function upon target cell binding and internalization.

8.4.6 Enhancing the Efficacy of Immunotoxins

A prominent efficacy-limiting step for immunotoxins is lysosomal degradation due to an inefficient endosomal escape into the cytosol. This necessitates infusion of relatively high doses of immunotoxins, frequently triggering serious adverse side effects such as vascular leak syndrome (Kreitman and Pastan 2006). In this respect, certain plant secondary metabolites known as saponins were recently shown to synergistically increase the efficacy of selected immunotoxins by facilitating their endosomal escape. Purified saponin SA1641 isolated from the ornamental plant Gypsophila paniculata L proved to augment the antitumor activity of PE- and DT-based immunotoxins by at least tenfold. More strikingly, saporin-based and dianthin-based immunotoxins were augmented more than a million-fold by SA1641 (Weng et al. 2012). This indicates that different toxins are affected differently by SA1641, which likely reflects differences in their intracellular routing. The dramatic enhancement of the efficacy of saporin-based and dianthin-based immunotoxins by saponins may be paralleled by a significant decrease of side effects.

8.4.7 Enhancing the Efficacy of Immunotoxins by Small Molecular Drugs

Combinations of immunotoxins with small inhibitory drugs with anticancer activity may result in synergistic antitumor effects. The inhibition of protein synthesis by immunotoxins typically results in apoptosis induction in targeted cancer cells via activation of the so-called mitochondrial or intrinsic pathway. Unfortunately, many cancer cells are, to a varying degree, resistant to this form of apoptosis induction. Mitochondrial sensitivity to apoptotic insults is regulated by the Bcl-2 family of pro- and antiapoptotic proteins. Overexpression of the antiapoptotic members of this family is frequently observed and contributes to apoptosis resistance. With that in mind, it is of interest to inhibit or block these proteins. A prominent example hereof is the BH3-only mimetic ABT-737. In mammalian cells, ABT-737 neutralizes the pro-survival proteins Bcl-2, Bcl-xl, and Bcl-w, which proves to be lethal for tumor cells that depend on one of these proteins for their survival. Indeed, in vitro resistance to a CD22-directed PE-based immunotoxin treatment can be overcome when combined with ABT-737 (Fitzgerald et al. 2011). Moreover, in selected cases, co-treatment resulted in additive or even synergistic antitumor effects. These data warrants further development of this approach for clinical trials.

8.4.8 Counteracting the Immunogenicity of Immunotoxins

The clinical development of immunotoxins has been impeded by the induction of strong immune response directed primarily against the microbial or plant toxin moiety. The immunogenicity of immunotoxins prevents their repeated long-term application. Consequently, it is necessary to apply immunotoxins in a limited number of rather high doses to achieve clinical efficacy. However, this, in turn, can lead to considerable toxic side effects. Indeed, clinical trials showed that a high percentage of patients develop neutralizing antibodies already after one injection cycle of an immunotoxin. Usually, these neutralizing antibodies are directed against the toxin portion of immunotoxin. Protein immunogenicity can be ameliorated by mutating sequences likely to be recognized by the naive antibody repertoire (B-cell epitopes) or sequences that are bound by MHC-II molecules and thus can elicit T-cell-dependent immune responses (Cantor et al. 2011).

To prevent immune responses, potential B- and T-cell epitopes on the toxin domains have to be identified and mutated accordingly in order to create an equally potent yet nonimmunogenic variant, allowing for repeated administrations (Pastan et al. 2011).

Alternatively, anti-immunotoxin antibody responses may be prevented when applied in combination with immunosuppressive agents such as cyclophosphamide and fludarabine (Kushner et al. 2007) or cyclosporine (CsA; Andersson et al. 2009).

8.5 Exploiting Effector Molecules of the Human Immune System

From the above, it is evident that current MAb-based targeted approaches in AML are still far from perfect. There is a need for improved antitumor activity combined with a more favorable toxicity profile. In this respect, much can be learned from the human immune system, which tightly controls potentially dangerous immune effector mechanisms, both spatially and temporally. Indeed, immune cells are perfectly equipped with powerful effector molecules for cancer cell elimination. Strategies that make use of these immune effector molecules, therefore, hold considerable appeal. Integral parts of the immune cell armament are the so-called death inducing ligands of the tumor necrosis factor (TNF) ligand superfamily.

8.5.1 Immune Effector Molecules of the TNF Family

Members of the TNF super family are involved in diverse biologic activities that range from the selective induction of apoptotic cell death in aged, potentially dangerous or superfluous cells to providing pivotal co-stimulatory signals that help mount an effective immune response (Bodmer et al. 2002). The majority of TNF ligands are type II transmembrane proteins, but proteolytic processing or alternative splicing typically yields soluble forms in vivo. The most prominently featured ligands are: TNF, fibroblast associated cell surface ligand (FasL), and TNF-related apoptosis inducing ligand (TRAIL; reviewed in Almasan and Ashkenazi 2003; Wajant et al. 2005). All three ligands have well-established tumoricidal pro-apoptotic activity. In particular, TRAIL appears to be suitable for exploitation as effector moiety in the magic bullet concept as proposed by Paul Ehrlich approximately 100 years ago.

8.5.2 TRAIL-Based Therapeutics

TRAIL interacts with a complex receptor set of five distinct receptors. TRAIL receptors TRAIL-R1/DR4 and TRAIL-R2/DR5 are agonistic receptors which can initiate apoptotic cell death via their intracellular death domains (DDs). TRAIL also binds, albeit with lower affinity, to TRAIL-R3/DcR1 and TRAIL-R4/DcR2, both of which lack a functional death domain. TRAIL can also bind to a soluble receptor named osteoprotegerin (OPG; Ashkenazi 2002). The main function of OPG appears to be in bone homeostasis, where it functions as a soluble receptor for the TNF family member receptor activator of NFκB ligand (RANKL; Boyce and Xing 2008). However, reports suggest that interaction of OPG with TRAIL may affect vascular biology (Corallini et al. 2008) as well as bone turnover (Robinson et al. 2007).

Apoptotic TRAIL-signaling via TRAILR1 and TRAILR2 is characterized by the formation of the so-called death inducing signaling complex (DISC) at their intracellular DDs [30]. The DISC comprises TRAILR1 and/or TRAILR2, the adaptor protein Fas-associated death domain (FADD) and the pro-form of initiator caspase-8. Upon assembly of the DISC, proximity-induced activation of pro-caspase-8 into mature caspase-8 occurs (Boatright et al. 2003). Caspase-8 subsequently proteolytically cleaves effector caspases, such as caspase-3. These effector caspases proteolytically degrade a multitude of cellular proteins leading to apoptotic cell death (Fig. 8.1). Of note, cell death induction via TRAIL receptors is subject to negative regulation at the level of the DISC, e.g., recruitment of cFLIP or PED (reviewed in Kataoka 2005). These proteins are structurally related to caspase-8, lack enzymatic activity and inhibit different steps in DISC-associated activation and processing of caspase-8.

The fact that TRAILR3 and TRAILR4 lack a functional death domain has led researchers to suggest that both may act as so-called decoy receptors that negatively regulate apoptosis induction by TRAIL. However, in many cancer types, expression of both agonistic and antagonistic TRAIL receptors often lacks predictive value in terms of sensitivity to TRAIL-mediated apoptosis induction. Nevertheless, recent studies indicated that AML patients with high expression of TRAILR1 on blasts and T-cell were characterized by a prolonged survival (Pordzik et al. 2011), whereas high expression of TRAILR3 on leukemic blasts is associated with poor outcome and apoptosis resistance (Chamuleau et al. 2011).

Currently, the complexity of TRAIL/TRAILR biology is still incompletely elucidated. Recent reports indicate that mixed trimerization of TRAILR2 with TRAILR4 can also occur (Clancy et al. 2005; Merino et al. 2006). Heterotrimeric TRAIL-receptors may well represent yet another level of regulation of the cellular response to TRAIL and warrants further in depth research.

8.5.3 Recombinant Non-Targeted sTRAIL in Cancer Therapy

TRAIL is an immune effector molecule that is expressed as a type II transmembrane protein, e.g., NK-cells. On these NK-cells, TRAIL is involved in immune surveillance of cancerous cells (Takeda et al. 2001; Cretney et al. 2002). Thus, TRAIL is an intrinsic part of the immune defense against cancer. Importantly, TRAIL can be proteolytically cleaved from the membrane, resulting in a soluble form of the protein (sTRAIL). sTRAIL retains tumoricidal activity and can be recombinantly produced. Recombinant human sTRAIL has generated tremendous enthusiasm as a potential anticancer drug based on the fact that many different types of cancer are sensitive to apoptotic elimination by sTRAIL. In contrast, most if not all, normal human cell types appear to be resistant to sTRAIL. A large body of evidence has been generated in preclinical studies that demonstrate pro-apoptotic antitumor activity of sTRAIL in vivo with essentially no toxicity towards normal cells and tissues.

Several phase 1 and 2 studies have been undertaken using recombinant human sTRAIL, which has revealed a very favorable toxicity profile. This is evidenced by the fact that no maximal tolerated dose was reached (Herbst et al. 2010). However, clinical responses upon single agent therapy with sTRAIL are disappointing, whereas combinatorial strategies have yielded some promise.

The effectiveness of TRAIL as anticancer drug may be hampered by the occurrence of intrinsic or acquired resistance to TRAIL. Importantly, TRAIL resistance can often be overcome by combination with chemotherapeutics, radiation, or other novel therapeutic drugs such as proteasome inhibitors. For instances, low doses of

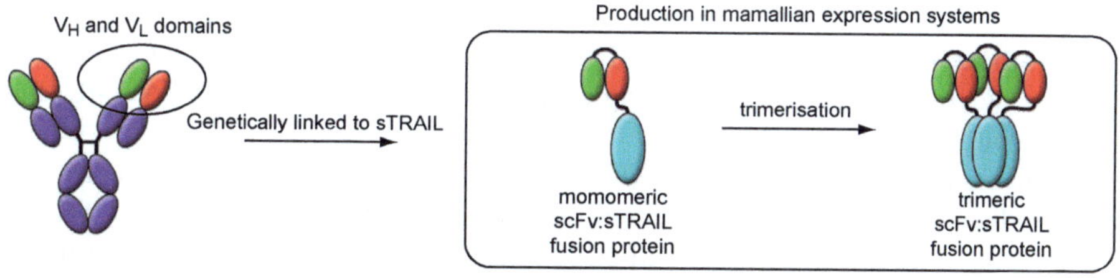

Fig. 8.2 Design and eukaryotic production of scFv:sTRAIL fusion proteins. Variable heavy (VH) and variable light (VL) chains that comprise the antigen-recognition domains of an anti-AML antibody are genetically engineered into a single chain of the variable fragments (scFv) antibody fragment and subsequently genetically fused to soluble TRAIL (sTRAIL). The resulting scFv:sTRAIL fusion protein self-trimerizes during production in mammalian expression systems and is secreted as a homotrimeric protein of ~180 kDa

the proteasome inhibitor bortezomib was recently shown to prime refractory AML cells for enhanced TRAIL-mediated killing. Co-treatment with TRAIL and bortezomib resulted in upregulation of both TRAILR1 and TRAILR2 and a downregulation of Bax. These results suggest that TRAIL-based approaches could be effective for treating AML patients, even for patients who are refractory to conventional chemotherapy (Conticello et al. 2008).

However, such combinatorial strategies may also inadvertently sensitize normal cells to TRAIL-mediated apoptosis. In this respect, combined treatment with sTRAIL and proteasome inhibition yielded enhanced toxicity towards hepatocytes, although there still was a therapeutic window for hepatoma-specific killing by TRAIL and proteasome inhibition. With regard to potential side-effects of TRAIL, the most likely side-effect is indeed hepatotoxicity, as hepatocytes have been shown to be sensitive to aggregated TRAIL preparations (Gores and Kaufmann 2001).

8.5.4 Targeted Delivery of sTRAIL

To fully exploit its therapeutic potential, the characteristics of both the TRAIL receptor system and soluble TRAIL (sTRAIL) should be taken into account: first, the widespread expression of the various TRAIL receptors throughout the human body; second, the differential binding affinities and cross-linking requirements of the agonistic receptors TRAIL-R1 and TRAIL-R2; and third, the solution behavior of particular sTRAIL preparations. Therefore, we constructed a series of novel TRAIL fusion protein, designated scFv:sTRAIL, comprising anticancer scFv antibody fragment genetically linked to the N-terminus of human sTRAIL (Fig. 8.2). The scFv:sTRAIL fusion proteins are designed to induce apoptosis by cross-linking of agonistic TRAIL receptors only after specific binding of scFv:sTRAIL, a preselected tumor-associated cell surface antigen. Target antigen-restricted apoptosis induction by scFv:sTRAIL fusion proteins was demonstrated for various malignancies including AML (ten Cate et al. 2007).

Target antigen binding converts a soluble scFv:sTRAIL fusion protein into a membrane-bound form of TRAIL. Consequently, tumor cell surface bound scFv:sTRAIL is capable of signaling apoptosis not only through TRAIL-R1 but also through TRAIL-R2.

Size-exclusion chromatography demonstrated that scFv:sTRAIL fusion proteins are produced as stable and homogeneous trimers in the absence of detectable TRAIL aggregates. Studies in primates and early clinical trials revealed that conventional sTRAIL has only a half-life of ~30 min, which is most likely due its relatively small size leading to a rapid first-pass clearance by kidney filtration (Kelley et al. 2001). In contrast, homotrimeric scFv:sTRAIL fusion proteins have a molecular weight of ~180 kDa. Therefore, the half-life of scFv:sTRAIL is likely to be significantly improved.

Stable trimeric scFv:sTRAIL fusion proteins contains three identical tumor targeting scFv domains, which enhances its accretion to target antigen-positive tumor cells by the associated avidity effect. Enhanced avidity has been shown to be beneficial for in vivo tumor targeting in many antibody-based therapeutic strategies (Power and Hudson 2000).

Since the number of target molecules usually greatly exceeded that of the TRAIL receptors on the same cell, a surplus of sTRAIL domains is available for subsequent cross-linking of agonistic TRAIL receptors on neighboring tumor cells in a bi- or even multicellular fashion. In this way, neighboring tumor cells that lack the relevant target antigen can also be eliminated by the so-called bystander effect (Fig. 8.3). The feasibility of this novel approach has been shown in vitro and in vivo for both solid tumors (Bremer et al. 2004; Bremer et al. 2005b; de Bruyn et al. 2010) and various leukemias (Stieglmaier et al. 2008).

With respect to AML, we recently reported on fusion protein scFvCD33:sTRAIL in which an anti-CD33 scFv was genetically fused to sTRAIL. scFvCD33:sTRAIL triggered apoptosis in AML cell lines and primary patient-derived AML blasts, whereas CD33-positive monocytes proved to be fully resistant. Furthermore, in a side-by-side comparison with GO, scFvCD33:sTRAIL displayed up to 30-fold higher apoptotic activity. Importantly, the CD33-restricted antileukemia activity of scFvCD33:sTRAIL remained stable during prolonged storage at 37°C, whereas GO showed a rapid increase in CD33-independent cytotoxicity. Taken together, fusion protein scFvCD33:sTRAIL proved to be superior to GO in terms of tumor selectivity, activity, and stability, warranting its further development for the treatment of CD33+ leukemias (ten Cate et al. 2009).

Similarly, we reported on an analogous AML-targeting TRAIL fusion protein, designated scFvCD7:sTRAIL, that contains a scFv antibody fragment specific for human CD7. CD7 is a cell surface glycoprotein that is abundantly expressed on most T-cell malignancies and ~10% of AMLs (Del Poeta et al. 1995). Expression of CD7 on normal blood cells is limited to T- and myeloid cells in early hematopoietic cell ontogeny, thymocytes, NK cells, and a large distinct subset of peripheral blood T-cell. Fusion protein scFvCD7:sTRAIL shows enhanced and target antigen-restricted apoptotic activity towards CD7-positive leukemic cells with no toxicity to normal cells. The tumor selectivity of scFvCD7:sTRAIL

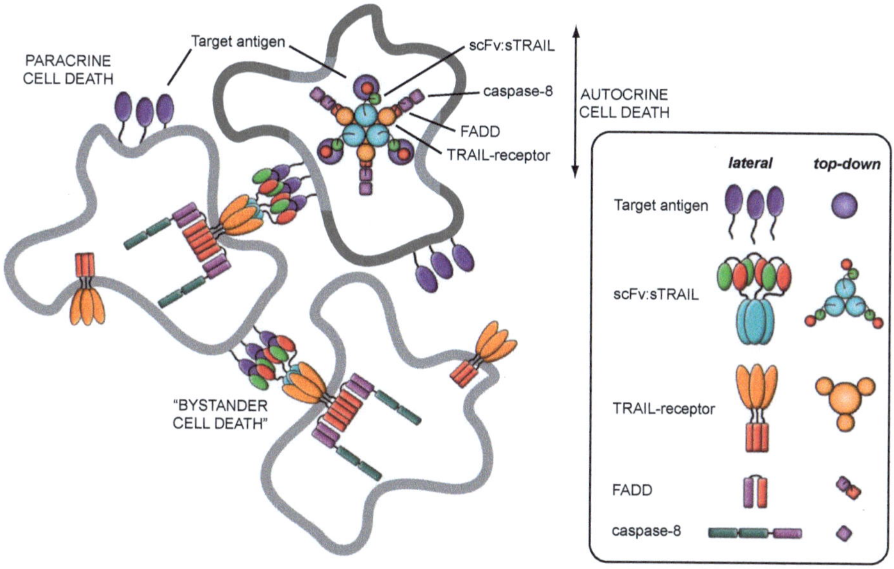

Fig. 8.3 Tumor cell-restricted induction of apoptosis by single chain of the variable fragments : soluble TNF-related apoptosis-inducing ligand (scFv:sTRAIL) fusion proteins. Specific binding of scFv:sTRAIL to an AML-associated target antigen results in its accretion at the tumor cell surface. Subsequently, apoptosis can be induced in an autocrine fashion by cross-linking TRAIL receptors on the same tumor cell. Moreover, tumor cell surface-bound scFv:sTRAIL fusion proteins can induce apoptosis by reciprocal cross-linking of TRAIL-R1/R2 between two (or more) neighboring AML cells (paracrine apoptosis induction). Importantly, apoptosis is also induced in neighboring AML cells that have lost expression of the target antigen, resulting in bystander apoptosis induction. Of note, cancer cells with partial loss of target antigen expression can escape from conventional antibody-based therapies. FADD Fas-associated death domain

was directly compared with the ETA-based immunotoxin scFvCD7:ETA. At equimolar concentrations, scFvCD7:sTRAIL was significantly more potent than scFvCD7:ETA. Furthermore, resting peripheral blood lymphocytes were resistant to prolonged (72 h) treatment with scFvCD7:sTRAIL, whereas identical treatment with scFvCD7:ETA resulted in a marked increase in apoptosis. Activated T-cell showed an even more pronounced apoptotic response to scFvCD7:ETA treatment (Bremer et al. 2005a).

In short, scFv:sTRAIL fusion proteins appear to be a favorable platform for the targeted and safe elimination of AML cells. ScFv:sTRAIL fusion proteins are conditionally inactive while on route and its pro-apoptotic activity is activated only when delivered on the outside of the targeted cell. Off-target binding (e.g., on normal cells) does not result in any appreciable cytotoxic effects, as most normal cells are resistant to TRAIL. Internalization or intracellular processing of scFv:sTRAIL is not necessary to activate its therapeutic activity, nor does it appear to inhibit its activity. Importantly, the antitumor activity of scFv:sTRAIL fusion proteins can be synergized with numerous anticancer agents that are already part of clinical practice and those that are still under investigation.

8.6 Targeted Elimination of AML Stem Cells

In AML, it is well established that a small population of therapy-resistant malignant cells exist that has self-renewal capacity and the ability to generate large numbers of more differentiated leukemic blasts. This was evidenced by the fact that only a small sub-fraction of leukemic cells obtained from AML patients were able to propagate the disease after xenografting in immunocompromised mice. These leukemia-initiating cells are commonly referred to as leukemia stem cells (LSCs). The targeted elimination of LSCs may have profound clinical implications. Not surprisingly, contemporary leukemia research has focused on ways to specifically eliminate LSCs by antibody- or ligand-based cell surface delivery of therapeutics such as naked antibodies, immunotoxins, and immunocytokines. Recent studies have identified that selected cell surface antigens including CD33, CD44, CD47 CD96, CD123, and CLL-1 are expressed on LSCs. The restricted expression patterns of these antigens may be exploited for LSC-selective therapy, while largely sparing normal hematopoietic stem cells. In this respect, we have constructed a novel fusion protein, designated scFvCLL-1:sTRAIL, in which a human anti-CLL-1 scFv antibody fragment is genetically fused to human sTRAIL. Its preclinical evaluation is currently ongoing (unpublished results).

8.7 Perspectives

Initially, the use of immunotoxins and antibody–drug conjugates as a strategy to deliver an intrinsically cytotoxic payload to AML cells was not very successful due to lack of tumor specificity of the antibody, reduced activity of the cytotoxic conjugate, adverse side effects of the toxic moiety and high immunogenicity.

After more than three decades of preclinical optimization, the number of clinical trials evaluating new generations of immunotoxins and antibody–drug conjugates has recently increased. Several of these trials have shown impressively improved clinical responses and a gradual reduction in adverse effects observed.

Tumor targeting antibodies and antibody fragments such as scFv's have practical limitations due to relative poor productions yield, stability, and aggregation tendency. A solution may come from the use of non-IgG scaffolds like designed ankyrin repeat proteins (DARPins; Stumpp and Amstutz 2007). DARPins are small, single domain proteins (14 kDa) which can be selected by ribosome display and phage display to bind any given target protein with high affinity and specificity. DARPins consist of repeat sequence motifs forming a contiguous protein domain for target binding flanked by N- and C-terminal capping repeats. DARPins have been developed into a robust and versatile scaffold for binding proteins. Due to their modular design, high stability, and favorable biophysical properties, DARPins tolerate engineering procedures not applicable to antibodies or antibody fragments. Currently, no AML-selective DARPins have been reported in literature.

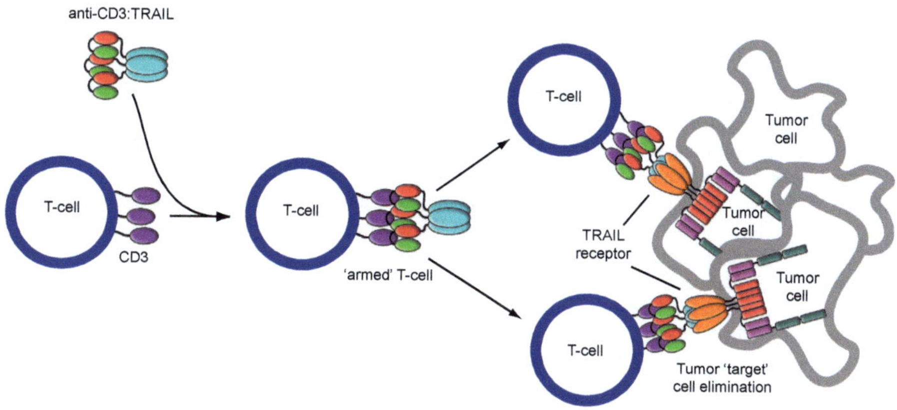

Fig. 8.4 Cell surface delivery of TNF-related apoptosis-inducing ligand (TRAIL) augments the tumoricidal activity of T-cells. Fusion protein anti-CD3:TRAIL was designed to selectively bind to the T-cell activation marker CD3. This approach results in enhanced TRAIL levels coated at the cell surface of strongly activated T-cells. The tumoricidal activity of anti-CD3:TRAIL-armed T-cells is dramatically increased

As detailed in this chapter, the antigen-restricted delivery and activation of apoptosis-inducing ligands such as sTRAIL to AML cells may hold particular promise. Recently, we developed another promising approach in which two novel TRAIL fusion proteins, designated anti-CD3:TRAIL and K12:TRAIL, were designed to harness human T-cells with additional cell surface-bound TRAIL (Fig. 8.4). Normally, T-cells show very low endogenous levels of TRAIL at their cell surface. Arming human T-cells with anti-CD3:TRAIL or K12:TRAIL resulted in a >500-fold increase in tumoricidal activity towards both cancer cell lines and primary patient-derived malignant cells. Moreover, treatment of xenografted mice with K12:TRAIL armed T-cells significantly prolonged their survival time. This strategy may represent an easy-to-incorporate approach to enhance the efficacy of various forms of adoptive T-cell therapy in cancer (de Bruyn et al. 2011). Although not evaluated for AML, such a T-cell arming approach may well contribute to therapeutic outcome of various T-cell-based approaches in AML.

Taken together, we are currently at the beginning of a new and exciting era of evaluating novel targeted approaches in AML in clinical practice. Additional studies are currently being designed to define case to case the optimal dose, schedule, and combinations with established or newly developed anti-AML drugs. There is ample reason to believe that considerably more potent and safer targeted approaches in the combat against AML are to be expected in the next 10 years (Figs. 8.2, 8.3 and 8.4).

References

Almasan A, Ashkenazi A (2003) Apo2L/TRAIL: apoptosis signaling, biology, and potential for cancer therapy. Cytokine Growth Factor Rev 14:337–348
Andersson Y, Engebraaten O, Fodstad O (2009) Synergistic anticancer effects of immunotoxin and cyclosporin in vitro and in vivo. Br J Cancer 101:1307–1315
Ardelt W, Shogen K, Darzynkiewicz Z (2008) Onconase and amphinase, the antitumor ribonucleases from Rana pipiens oocytes. Curr Pharm Biotechnol 9:215–225
Ashkenazi A (2002) Targeting death and decoy receptors of the tumour-necrosis factor superfamily. Nat Rev Cancer 2:420–430
Bakker AB, van den Oudenrijn S, Bakker AQ, Feller N, van Meijer M, Bia JA, Jongeneelen MA, Visser TJ, Bijl N, Geuijen CA, Marissen WE, Radosevic K, Throsby M, Schuurhuis GJ, Ossenkoppele GJ, de Kruif J, Goudsmit J, Kruisbeek AM (2004) C-type lectin-like molecule-1: a novel myeloid cell surface marker associated with acute myeloid leukemia. Cancer Res 64:8443–8450
Balaian L, Ball ED (2004) Inhibition of acute myeloid leukemia cell growth by mono-specific and bi-specific anti-CD33 x anti-CD64 antibodies. Leuk Res 28:821–829
Boatright KM, Renatus M, Scott FL, Sperandio S, Shin H, Pedersen IM, Ricci JE, Edris WA, Sutherlin DP, Green DR, Salvesen GS (2003) A unified model for apical caspase activation. Mol Cell 11:529–541
Bodmer JL, Schneider P, Tschopp J (2002) The molecular architecture of the TNF superfamily. Trends Biochem Sci 27:19–26
Boyce BF, Xing L (2008) Functions of RANKL/RANK/OPG in bone modeling and remodeling. Arch Biochem Biophys 473:139–146
Bremer E, Kuijlen J, Samplonius D, Walczak H, de Leij L, Helfrich W (2004) Target cell-restricted and -enhanced apoptosis induction by a scFv:sTRAIL fusion protein with specificity for the pancarcinoma-associated antigen EGP2. Int J Cancer 109:281–290
Bremer E, Samplonius DF, Peipp M, van Genne L, Kroesen BJ, Fey GH, Gramatzki M, de Leij LF, Helfrich W (2005a) Target cell-restricted apoptosis induction of acute leukemic T cells by a recombinant tumor necrosis factor-related apoptosis-inducing ligand fusion protein with specificity for human CD7. Cancer Res 65:3380–3388
Bremer E, Samplonius DF, van Genne L, Dijkstra MH, Kroesen BJ, de Leij LF, Helfrich W (2005b) Simultaneous inhibition of epidermal growth factor receptor (EGFR) signaling and enhanced activation of tumor necrosis factor-related apoptosis-inducing ligand (TRAIL) receptor-mediated apoptosis induction by an scFv:sTRAIL fusion protein with specificity for human EGFR. J Biol Chem 280:10025–10033
Brischwein K, Parr L, Pflanz S, Volkland J, Lumsden J, Klinger M, Locher M, Hammond SA, Kiener P, Kufer P, Schlereth B, Baeuerle PA (2007) Strictly target cell-dependent activation of T cells by bispecific single-chain antibody constructs of the BiTE class. J Immunother 30:798–807
Cantor JR, Yoo TH, Dixit A, Iverson BL, Forsthuber TG, Georgiou G (2011) Therapeutic enzyme deimmunization by combinatorial T-cell epitope removal using neutral drift. Proc Natl Acad Sci USA 108:1272–1277
Caron PC, Co MS, Bull MK, Avdalovic NM, Queen C, Scheinberg DA (1992) Biological and immunological features of humanized M195 (anti-CD33) monoclonal antibodies. Cancer Res 52:6761–6767
Castaigne S, Pautas C, Terre C, Raffoux E, Bordessoule D, Bastie JN, Legrand O, Thomas X, Turlure P, Reman O, de Revel T, Gastaud L, de Gunzburg N, Contentin N, Henry E, Marolleau JP, Aljijakli A, Rousselot P, Fenaux P, Preudhomme C, Chevret S, Dombret H, Acute Leukemia French Association (2012) Effect of gemtuzumab ozogamicin on survival of adult patients with de-novo acute myeloid leukaemia (ALFA-0701): a randomised, open-label, phase 3 study. Lancet 379:1508–1516

Chadwick DE, Jean LF, Jamal N, Messner HA, Murphy JR, Minden MD (1993) Differential sensitivity of human myeloma cell lines and normal bone marrow colony forming cells to a recombinant diphtheria toxin-interleukin 6 fusion protein. Br J Haematol 85:25–36

Chamuleau ME, Ossenkoppele GJ, van Rhenen A, van Dreunen L, Jirka SM, Zevenbergen A, Schuurhuis GJ, van de Loosdrecht AA (2011) High TRAIL-R3 expression on leukemic blasts is associated with poor outcome and induces apoptosis-resistance which can be overcome by targeting TRAIL-R2. Leuk Res 35:741–749

Chan CH, Blazar BR, Eide CR, Kreitman RJ, Vallera DA (1995) A murine cytokine fusion toxin specifically targeting the murine granulocyte–macrophage colony-stimulating factor (GM-CSF) receptor on normal committed bone marrow progenitor cells and GM-CSF-dependent tumor cells. Blood 86:2732–2740

Chen R, Palmer JM, Thomas SH, Tsai NC, Farol L, Nademanee A, Forman SJ, Gopal AK (2012) Brentuximab vedotin enables successful reduced-intensity allogeneic hematopoietic cell transplantation in patients with relapsed or refractory Hodgkin lymphoma. Blood 119:6379–6381

Clancy L, Mruk K, Archer K, Woelfel M, Mongkolsapaya J, Screaton G, Lenardo MJ, Chan FK (2005) Preligand assembly domain-mediated ligand-independent association between TRAIL receptor 4 (TR4) and TR2 regulates TRAIL-induced apoptosis. Proc Natl Acad Sci USA 102:18099–18104

Conticello C, Adamo L, Vicari L, Giuffrida R, Iannolo G, Anastasi G, Caruso L, Moschetti G, Cupri A, Palumbo GA, Gulisano M, De Maria R, Giustolisi R, Di Raimondo F (2008) Antitumor activity of bortezomib alone and in combination with TRAIL in human acute myeloid leukemia. Acta Haematol 120:19–30

Corallini F, Rimondi E, Secchiero P (2008) TRAIL and osteoprotegerin: a role in endothelial physiopathology?. Front Biosci 13:135–147

Cretney E, Takeda K, Yagita H, Glaccum M, Peschon JJ, Smyth MJ (2002) Increased susceptibility to tumor initiation and metastasis in TNF-related apoptosis-inducing ligand-deficient mice. J Immunol 168:1356–1361

de Bruyn M, Rybczynska AA, Wei Y, Schwenkert M, Fey GH, Dierckx RA, van Waarde A, Helfrich W, Bremer E (2010) Melanoma-associated Chondroitin Sulfate Proteoglycan (MCSP)-targeted delivery of soluble TRAIL potently inhibits melanoma outgrowth in vitro and in vivo. Mol Cancer 9:301

de Bruyn M, Wei Y, Wiersma VR, Samplonius DF, Klip HG, van der Zee AG, Yang B, Helfrich W, Bremer E (2011) Cell surface delivery of TRAIL strongly augments the tumoricidal activity of T cells. Clin Cancer Res 17:5626–5637

Del Poeta G, Stasi R, Venditti A, Cox C, Aronica G, Masi M, Bruno A, Simone MD, Buccisano F, Papa G (1995) CD7 expression in acute myeloid leukemia. Leuk Lymphoma 17:111–119

Dunn GP, Old LJ, Schreiber RD (2004) The immunobiology of cancer immunosurveillance and immunoediting. Immunity 21:137–148

Fitzgerald DJ, Moskatel E, Ben-Josef G, Traini R, Tendler T, Sharma A, Antignani A, Mussai F, Wayne A, Kreitman RJ, Pastan I (2011) Enhancing immunotoxin cell-killing activity via combination therapy with ABT-737. Leuk Lymphoma 52 Suppl 2:79–81

Frankel A, Liu JS, Rizzieri D, Hogge D (2008) Phase I clinical study of diphtheria toxin-interleukin 3 fusion protein in patients with acute myeloid leukemia and myelodysplasia. Leuk Lymphoma 49:543–553

Gores GJ, Kaufmann SH (2001) Is TRAIL hepatotoxic? Hepatology 34:3–6

Herbst RS, Eckhardt SG, Kurzrock R, Ebbinghaus S, O'Dwyer PJ, Gordon MS, Novotny W, Goldwasser MA, Tohnya TM, Lum BL, Ashkenazi A, Jubb AM, Mendelson DS (2010) Phase I dose-escalation study of recombinant human Apo2L/TRAIL, a dual proapoptotic receptor agonist, in patients with advanced cancer. J Clin Oncol 28:2839–2846

Hofmann M, Grosse-Hovest L, Nubling T, Pyz E, Bamberg ML, Aulwurm S, Buhring HJ, Schwartz K, Haen SP, Schilbach K, Rammensee HG, Salih HR, Jung G (2012) Generation, selection and preclinical characterization of an Fc-optimized FLT3 antibody for the treatment of myeloid leukemia. Leukemia 26:1228–1237

Horita H, Frankel AE, Thorburn A (2008) Acute myeloid leukemia-targeted toxin activates both apoptotic and necroptotic death mechanisms. PLoS One 3:e3909.

Jean LF, Murphy JR (1991) Diphtheria toxin receptor-binding domain substitution with interleukin 6: genetic construction and interleukin 6 receptor-specific action of a diphtheria toxin-related interleukin 6 fusion protein. Protein Eng 4:989–994

Jin L, Hope KJ, Zhai Q, Smadja-Joffe F, Dick JE (2006) Targeting of CD44 eradicates human acute myeloid leukemic stem cells. Nat Med 12:1167–1174

Jurcic JG (2012) What happened to anti-CD33 therapy for acute myeloid leukemia?. Curr Hematol.Malig Rep 7:65–73.

Kataoka T (2005) The caspase-8 modulator c-FLIP. Crit Rev Immunol 25:31–58

Kelley SK, Harris LA, Xie D, Deforge L, Totpal K, Bussiere J, Fox JA (2001) Preclinical studies to predict the disposition of Apo2L/tumor necrosis factor-related apoptosis-inducing ligand in humans: characterization of in vivo efficacy, pharmacokinetics, and safety. J Pharmacol Exp Ther 299:31–38

Koenecke C, Hofmann M, Bolte O, Gielow P, Dammann E, Stadler M, Franzke A, Boerner AR, Eder M, Ganser A, Knapp W, Hertenstein B (2008) Radioimmunotherapy with [188Re]-labelled anti-CD66 antibody in the conditioning for allogeneic stem cell transplantation for high-risk acute myeloid leukemia. Int J Hematol 87:414–421

Kreitman RJ (2009) Recombinant immunotoxins containing truncated bacterial toxins for the treatment of hematologic malignancies. BioDrugs 23:1–13

Kreitman RJ, Pastan I (2006) Immunotoxins in the treatment of hematologic malignancies. Curr Drug Targets 7:1301–1311

Kuo TT, Aveson VG (2011) Neonatal Fc receptor and IgG-based therapeutics. MAbs 3:422–430

Kushner BH, Cheung IY, Kramer K, Modak S, Cheung NK (2007) High-dose cyclophosphamide inhibition of humoral immune response to murine monoclonal antibody 3F8 in neuroblastoma patients: broad implications for immunotherapy. Pediatr Blood Cancer 48:430–434

Lakkis F, Steele A, Pacheco-Silva A, Rubin-Kelley V, Strom TB, Murphy JR (1991) Interleukin 4 receptor targeted cytotoxicity: genetic construction and in vivo immunosuppressive activity of a diphtheria toxin-related murine interleukin 4 fusion protein. Eur J Immunol 21:2253–2258

Lapusan S, Vidriales MB, Thomas X, de Botton S, Vekhoff A, Tang R, Dumontet C, Morariu-Zamfir R, Lambert JM, Ozoux ML, Poncelet P, San Miguel JF, Legrand O, Deangelo DJ, Giles FJ, Marie JP (2012) Phase I studies of AVE9633, an anti-CD33 antibody-maytansinoid conjugate, in adult patients with relapsed/refractory acute myeloid leukemia. Invest New Drugs 30:1121–1131

Linenberger ML (2005) CD33-directed therapy with gemtuzumab ozogamicin in acute myeloid leukemia: progress in understanding cytotoxicity and potential mechanisms of drug resistance. Leukemia 19:176–182

Lomax JE, Eller CH, Raines RT (2012) Rational design and evaluation of mammalian ribonuclease cytotoxins. Methods Enzymol 502:273–290

Majeti R, Chao MP, Alizadeh AA, Pang WW, Jaiswal S, Gibbs KD, Jr, van Rooijen N, Weissman IL (2009) CD47 is an adverse prognostic factor and therapeutic antibody target on human acute myeloid leukemia stem cells. Cell 138:286–299

Marchion DC, Bicaku E, Daud AI, Sullivan DM, Munster PN (2005) Valproic acid alters chromatin structure by regulation of chromatin modulation proteins. Cancer Res 65:3815–3822

Menzel C, Schirrmann T, Konthur Z, Jostock T, Dubel S (2008) Human antibody RNase fusion protein targeting CD30+ lymphomas. Blood 111:3830–3837

Merino D, Lalaoui N, Morizot A, Schneider P, Solary E, Micheau O (2006) Differential inhibition of TRAIL-mediated DR5-DISC formation by decoy receptors 1 and 2. Mol Cell Biol 26:7046–7055

Pagel JM, Kenoyer AL, Back T, Hamlin DK, Wilbur DS, Fisher DR, Park SI, Frayo S, Axtman A, Orgun N, Orozco J, Shenoi J, Lin Y, Gopal AK, Green DJ, Appelbaum FR, Press OW (2011) Anti-CD45 pretargeted radioimmunotherapy using bismuth-213: high rates of complete remission and long-term survival in a mouse myeloid leukemia xenograft model. Blood 118:703–711

Pastan I, Onda M, Weldon J, Fitzgerald D, Kreitman R (2011) Immunotoxins with decreased immunogenicity and improved activity. Leuk Lymphoma 52 Suppl 2:87–90

Pordzik S, Petrovici K, Schmid C, Kroell T, Schweiger C, Kohne CH, Schmetzer H (2011) Expression and prognostic value of FAS receptor/FAS ligand and TrailR1/TrailR2 in acute myeloid leukemia. Hematology 16:341–350

Power BE, Hudson PJ (2000) Synthesis of high avidity antibody fragments (scFv multimers) for cancer imaging. J Immunol Methods 242:193–204

Robinson LJ, Borysenko CW, Blair HC (2007) Tumor necrosis factor family receptors regulating bone turnover: new observations in osteoblastic and osteoclastic cell lines. Ann NYAcad Sci 1116:432–443

Saito Y, Kitamura H, Hijikata A, Tomizawa-Murasawa M, Tanaka S, Takagi S, Uchida N, Suzuki N, Sone A, Najima Y, Ozawa H, Wake A, Taniguchi S, Shultz LD, Ohara O, Ishikawa F (2010) Identification of therapeutic targets for quiescent, chemotherapy-resistant human leukemia stem cells. Sci Transl Med 2:17ra9

Schirrmann T, Krauss J, Arndt MA, Rybak SM, Dubel S (2009) Targeted therapeutic RNases (ImmunoRNases). Expert Opin Biol Ther 9:79–95

Schwemmlein M, Peipp M, Barbin K, Saul D, Stockmeyer B, Repp R, Birkmann J, Oduncu F, Emmerich B, Fey GH (2006) A CD33-specific single-chain immunotoxin mediates potent apoptosis of cultured human myeloid leukaemia cells. Br J Haematol 133:141–151

Silla LM, Chen J, Zhong RK, Whiteside TL, Ball ED (1995) Potentiation of lysis of leukaemia cells by a bispecific antibody to CD33 and CD16 (Fc gamma RIII) expressed by human natural killer (NK) cells. Br J Haematol 89:712–718

Singh R, Erickson HK (2009) Antibody-cytotoxic agent conjugates: preparation and characterization. Methods Mol Biol 525:445–467, xiv

Stein C, Kellner C, Kugler M, Reiff N, Mentz K, Schwenkert M, Stockmeyer B, Mackensen A, Fey GH (2010) Novel conjugates of single-chain Fv antibody fragments specific for stem cell antigen CD123 mediate potent death of acute myeloid leukaemia cells. Br J Haematol 148:879–889

Stieglmaier J, Bremer E, Kellner C, Liebig TM, ten Cate B, Peipp M, Schulze-Koops H, Pfeiffer M, Buhring HJ, Greil J, Oduncu F, Emmerich B, Fey GH, Helfrich W (2008) Selective induction of apoptosis in leukemic B-lymphoid cells by a CD19-specific TRAIL fusion protein. Cancer Immunol Immunother 57:233–246

Stumpp MT, Amstutz P (2007) DARPins: a true alternative to antibodies. Curr Opin Drug Discov Devel 10:153–159

Sutherland MK, Yu C, Lewis TS, Miyamoto JB, Morris-Tilden CA, Jonas M, Sutherland J, Nesterova A, Gerber HP, Sievers EL, Grewal IS, Law CL (2009) Anti-leukemic activity of lintuzumab (SGN-33) in preclinical models of acute myeloid leukemia. MAbs 1:481–490

Takeda K, Hayakawa Y, Smyth MJ, Kayagaki N, Yamaguchi N, Kakuta S, Iwakura Y, Yagita H, Okumura K (2001) Involvement of tumor necrosis factor-related apoptosis-inducing ligand in surveillance of tumor metastasis by liver natural killer cells. Nat Med 7:94–100

ten Cate B, Samplonius DF, Bijma T, de Leij LF, Helfrich W, Bremer E (2007) The histone deacetylase inhibitor valproic acid potently augments gemtuzumab ozogamicin-induced apoptosis in acute myeloid leukemic cells. Leukemia 21:248–252

ten Cate B, Bremer E, de Bruyn M, Bijma T, Samplonius D, Schwemmlein M, Huls G, Fey G, Helfrich W (2009) A novel AML-selective TRAIL fusion protein that is superior to Gemtuzumab Ozogamicin in terms of in vitro selectivity, activity and stability. Leukemia 23:1389–1397

Thepen T, Huhn M, Melmer G, Tur MK, Barth S (2009) Fcgamma receptor 1 (CD64), a target beyond cancer. Curr Pharm Des 15:2712–2718

Wajant H, Gerspach J, Pfizenmaier K (2005) Tumor therapeutics by design: targeting and activation of death receptors. Cytokine Growth Factor Rev 16:55–76

Weng A, Thakur M, Beceren-Braun F, Bachran D, Bachran C, Riese SB, Jenett-Siems K, Gilabert-Oriol R, Melzig MF, Fuchs H (2012) The toxin component of targeted antitumor toxins determines their efficacy increase by saponins. Mol Oncol 6:323–332

Williams DP, Parker K, Bacha P, Bishai W, Borowski M, Genbauffe F, Strom TB, Murphy JR (1987) Diphtheria toxin receptor binding domain substitution with interleukin-2: genetic construction and properties of a diphtheria toxin-related interleukin-2 fusion protein. Protein Eng 1:493–498

Part II
Signaling

Chapter 9
PIM Kinases in AML

Lisa S. Chen and Varsha Gandhi

Abstract PIM proto-oncogenes encode for a family of serine/threonine/tyrosine kinases (Pim-1, Pim-2 and Pim-3) that are overexpressed in a number of solid tumors and hematological malignancies including acute myeloid leukemia (AML). Pim kinases are constitutively expressed and are involved in critical AML signaling pathways including leukemogenesis, cell proliferation, apoptosis inhibition, microenvironment signaling, and drug resistance. Pim kinases play a key role in FLT3-mediated leukemogenesis, and, in particular, are upregulated by FLT3-ITD activating mutation and contribute to cell survival and proliferation. The fact that there was minimal phenotype in mice lacking all three Pim kinase functionality strongly suggested these kinases as therapeutic targets. Recent efforts in both industry and academic institutions to identify and develop Pim kinase inhibitors highlight the utility of targeting, converging, and overlapping pathways through Pim kinases. In this chapter, we give an overview of Pim kinase signaling pathways relevant to AML and discuss the therapeutic strategies that are under development against these pathways.

Keywords Pim kinases · Acute myeloid leukemia · FLT3 · Drug resistance · Pim kinase inhibitors · Myc · 4EBP1

V. Gandhi (✉) L. S. Chen
Department of Experimental Therapeutics, Division of Cancer Medicine,
The University of Texas MD Anderson Cancer Center, P.O. Box 301429
Unit Number: 1950 Houston, TX 77230-1429, USA
e-mail: vgandhi@mdanderson.org

V. Gandhi
Department of Experimental Leukemia, Division of Cancer Medicine, The University of Texas MD Anderson Cancer Center, 1515 Holcombe Blvd., Unit, Houston, TX 77030, USA

L. S. Chen
e-mail: lschen@mdanderson.org

M. Andreeff (ed.), *Targeted Therapy of Acute Myeloid Leukemia*,
Current Cancer Research, DOI 10.1007/978-1-4939-1393-0_9

Abbreviations

AML	Acute myeloid leukemia
CML	Chronic myeloid leukemia
CLL	Chronic lymphocytic leukemia
ER	Endoplasmic reticulum
miRNA	MicroRNA
FLT3	FMS-like tyrosine kinase 3

9.1 Introduction

Provirus integration site for Moloney murine leukemia (*PIM*) virus proto-oncogenes encode for a family of serine/threonine/tyrosine kinases that are constitutively active when expressed (Qian et al. 2005). *PIM1* was first identified in 1984 (Cuypers et al. 1984) in murine leukemia virus-induced lymphomas as a small chromosome region that contains virus insertion site resulting in activation of Pim. In humans, three PIM kinases have been identified to date; genes *PIM1*, *PIM2*, and *PIM3* are located on chromosome 6p21, Xp11.23, and 22q13, respectively (Nagarajan et al. 1986; The International Radiation Hybrid Mapping Consortium). At the amino acid level, there is substantial homology between Pim-1 and Pim-2 (53 %) (van der Lugt et al. 1995) and Pim-3 (69 %) (Konietzko et al. 1999), and both Pim-1 and Pim-2 have multiple isoforms that arise from alternative start sites (Mikkers et al. 2004). The various Pim kinases have overlapping functions and compensate for one another, and Pim-1 knockout experiments identified high frequency of activation of Pim-2 (Allen and Berns 1996). Pim kinases have been identified to be overexpressed in several solid tumors (reviewed in Nawijn 2011), lymphomas, and leukemias (reviewed by Shah et al. 2008). In addition to overexpression, the oncogenic functions of Pim kinases are accelerated as they cooperate with proto-oncogene Myc and are generally co-overexpressed (Verbeek et al. 1991). Overexpression of Pim kinases results in increased function activity as these genes are constitutively active (Fox et al. 2005). Pim kinases can act as survival factors (Fox et al. 2005), and act as primary response genes since their expression can be stimulated by chemokines, growth factors (Wang et al. 2001), and microenvironment. Furthermore, Pim kinases appear to play critical roles in proliferation, tumorigenesis/leukemogenesis, and resistance to therapy.

9.2 Pim Kinase Substrates

The protein substrates of Pim kinases are numerous (reviewed in Nawijn et al. 2004; Alvarado et al. 2012). Pim kinase targets are involved with a number of cellular processes including transcription, protein translation, energy regulation (Beharry

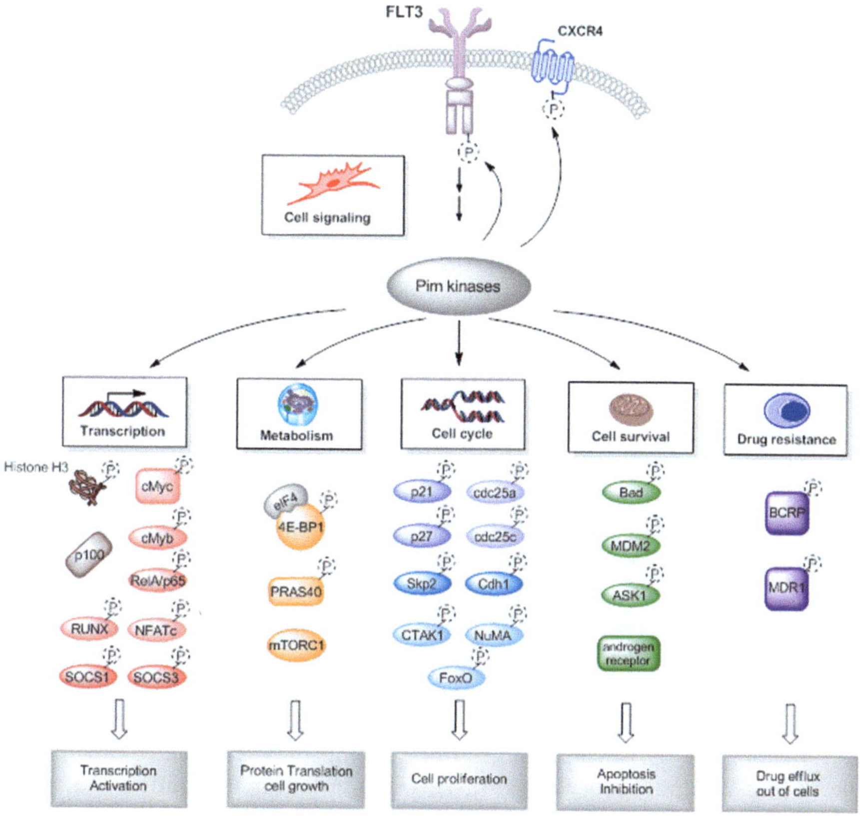

Fig. 9.1 Pim kinase cell signaling and downstream effects. Pim kinase targets include regulators of transcription, metabolism, cell cycle, cell survival, and drug resistance

et al. 2011), cell cycle, and cell survival, which, along with selected substrates, is summarized in Fig. 9.1

9.2.1 *Role in Transcription*

Literature is replete with reports of cooperativity between Myc and Pim kinases, and their overexpression has been identified in several cancers including in hematological malignancies (Alizadeh et al. 2000; Dave et al. 2006; Gomez-Abad et al. 2011; Poulsen et al. 2005; Tamburini et al. 2009; Wright et al. 2003; Zhu et al. 2002). Pim and Myc gene synergistic cooperation was observed in several mouse model systems (Breuer et al. 1991; Schmidt et al. 1998; van Lohuizen et al. 1989; Verbeek et al. 1991). In the Eu-myc transgenic mice, lack of Pim-1 leads to compensatory activation of Pim-2, further demonstrating collaboration between these two oncogenes (van der Lugt et al. 1995). Functionally, Pim kinases phosphorylate

Myc at two serine sites, which protects Myc from protein degradation (Zhang et al. 2008). Furthermore, Pim acts as a coactivator of Myc-driven transcription by phosphorylating Ser10 on histone H3, which is essential for Myc/Max heterodimer binding to the E-box and polymerase II-mediated transcription (Naud and Eilers 2007; Zippo et al. 2007; Zippo et al. 2009). Similar to Myc, Myb also has shown collaborative interactions with Pim directly or through transcriptional coactivator p100 (Leverson et al. 1998; Weston 1999; Winn et al. 2003). In addition to Myc and Myb, Pim kinases activate several other transcription factors such as RelA (Nihira et al. 2009), RUNX (Aho et al. 2006), NFATc (Rainio et al. 2002), SOCS1, and SOCS3 (Chen et al. 2002; Peltola et al. 2004).

9.2.2 *Role in Translation and Metabolism*

Most of the mRNAs that are synthesized in cells are temporally and spatially regulated for translation (Richter and Sonenberg 2005). Among the well-studied protein synthesis process is cap (modified guanosine at the mRNA transcript 5′-end)-dependent translation. This protein synthesis process is tightly regulated by eukaryotic translation initiation factor (eIF4E) inhibitory proteins (Richter and Sonenberg 2005) that is inhibited by 4E-binding protein1 (4E-BP1). In cancers, eIF4E act as oncogenic proteins while 4E-BP1 functions as a tumor suppressor protein. The mTORC1 complex, which regulates cellular metabolism, promotes protein synthesis by phosphorylating 4E-BP1 and repressing its inhibitory effect on eIF4E. Several studies from Thompson's group have established a role of Pim-2 in phosphorylating 4E-BP1 and initiating translation (Amaravadi and Thompson 2005; Fox et al. 2005; Hammerman et al. 2005; Woodland et al. 2008). This Pim activity was independent of Akt regulation of protein translation in hematological malignancies (Hammerman et al. 2005). Specifically in AML blasts, rapamycin-insensitive (inhibitor of mTORC1) but Pim kinase inhibition-sensitive protein synthesis was identified suggesting critical role of Pim in protein translation in this disease (Tamburini et al. 2009). Consistent with this observation, pharmacological inhibition of Pim by small molecule inhibitor SGI-1776 abrogated protein synthesis in AML cell lines and primary blasts (Chen et al. 2011). More recently, Pim-1 has been reported to regulate expression of receptor tyrosine kinase MET through phosphorylation of eukaryotic initiation factor 4B (eIF4B) on Ser406 (Cen et al. 2014). Loss of all three Pim protein kinases in mouse embryo fibroblasts resulted in decreased glycolysis and mitochondrial biogenesis and a decline in the cellular ATP pool (Beharry et al. 2011). Consistent with these data, AMP-activated protein kinase (AMPK) was activated; activation of this protein is due to change in the ratio of AMP to ATP (Beharry et al. 2011). In line with this, recent studies have shown role of Pim 1 in preservation of mitochondrial morphology (Din et al. 2013). Collectively, these actions of Pim demonstrate the role of these kinases in regulation of protein translation and cellular bioenergy.

9.2.3 Role in Cell-Cycle Progression and Proliferation

Pim kinases phosphorylate and inactivate cyclin-dependent kinase inhibitor proteins p21 and p27. Pim also regulates p27 both at transcriptional level and post-translational modifications leading to a decline in both mRNA and protein. These proteins are checkpoints for cell-cycle progression and their inactivation leads to dysregulated cell cycle (reviewed in (Besson et al. 2008). In addition, cycle-regulatory phosphatases (CDC25A and C) are phosphorylated and inactivated by Pim-1 (Bachmann et al. 2006; Mochizuki et al. 1999). NuMA protein interacts with other proteins during mitosis and is involved in the formation of spindle apparatus. Pim-mediated phosphorylation of this protein activates this process and hence mitosis (Bhattacharya et al. 2002). More recently, Pim kinases have been reported to regulate cell-cycle checkpoint and DNA damage response protein Chk1 in AML cells (Yuan et al. 2014). Pim1/2 phosphorylation of Chk1(Ser280) did not alter Chk1 subcellular localization, but enabled its cell cycle and damage response functions. Collectively these reports establish a role of Pim kinases on cell cycle regulation resulting in increased proliferation.

9.2.4 Role in Cell Survival

With respect to its prosurvival actions, Pim resembles the Akt pathway (White 2003). Pro-apoptotic protein Bad, a member of the Bcl-2 family of proteins, is a Pim kinase substrate (Aho et al. 2004; Fox et al. 2005; Macdonald et al. 2006). Phosphorylation of this protein and its dissociation from 14-3-3 proteins initiate a cascade of phosphorylation leading to separation of Bcl-2 and Bcl-X_L antiapoptotic proteins that mediate survival of cells. Binding to 14-3-3 protein can also be affected by Mdm-2 protein (Hogan et al. 2008; Wood et al. 2009). However, decline in Bad phosphorylation in leukemia cells treated with SGI-1776 was not recapitulated consistently (Chen et al. 2009; Chen et al. 2011).

9.2.5 Role in Drug Resistance

The above-mentioned roles of Pim kinases that include induction of gene transcription, protein translation, induction or increased functionality of survival proteins, augmentation of cell proliferation through cell cycle effects strongly suggest role of Pim kinases in maintenance and proliferation of cancer cells. These processes also render cells resistant to chemotherapeutic agents. In addition, Pim kinases directly affect drug resistance efflux transporters that remove chemotherapeutic drugs from the intracellular component. Multidrug resistant protein-1 (MDR-1, also known as ABCB1, p-glycoprotein) and breast cancer resistant protein (BCRP, also known as ABCG2, MXR, ABCP) are protected by

Pim-1 resulting in an increased drug resistance (Xie et al. 2010; Xie et al. 2008; Natarajan et al. 2013a) for agents that use these pumps.

9.2.6 *Others*

There are additional substrates of Pim kinases (For detailed information on substrate phosphorylation sites and involvement of each Pim kinase, refer to the reviews by Alvarado et al. 2012; Merkel et al. 2012; Nawijn et al. 2011).

9.3 Pim Kinase Regulation

Several upstream pathways regulate Pim expression. Since Pim transcripts short-lived, their production depends on stimuli generated through signaling pathways. With this regard, Pim are early-response genes and their transcription is triggered by several factors.

9.3.1 *By Growth Factors*

Pim kinase expression can be stimulated by growth factors, (Wang et al. 2001), and they can act as survival factors (Fox et al. 2005). B-lymphocyte-stimulating factor (BLyS) that acts as a ligand for several B-cell activating factor (BAFF) family receptors in B-cell ontogeny, have been shown to induce both Pim-1 and Pim-2. Similar to B-cell response, in myeloid lineage, IL-3 and GM-CSF were able to induce Pim 1 expression (Lilly et al. 1992). IL-2 and IL-3-mediated augmentation of Pim was also observed in murine hematopoietic cell lines (Dautry et al. 1988). A multitude of growth factors have been reported to increase Pim expression (Wang et al. 2001). Similarly, Pim-1 mRNA and protein expression was induced after stimulation of MO7e cell line with interferon gamma with and without steel factor and Pim-1 transcript stability was augmented (Yip-Schneider et al. 1995). In myeloid leukemia cells, knockdown of Pim kinases impaired the survival of resistant forms of FLT3- and BCR-ABL-transformed cells (Adam et al. 2006).

9.3.2 *By Transcription Factors*

The stimulation of Pim expression at protein and message levels by growth factors, chemokines, and cytokines was generally due to downstream signaling through transcription factors. Upon ligand binding to its cognate receptors, several signaling processes initiate resulting in transcription of genes. Pim kinases are early response genes and are regulated by several transcription factors such as HOXA9, STATs, and RUNX. Many of these have been evaluated in AML setting. This section is detailed under Sect. 4.3 of this chapter.

9.3.3 Negative Regulation of Pim Kinases: Heat-Shock Proteins and MicroRNAs

Pim-1 is degraded through the ubiquitin-proteasome pathway, and when Pim-1 is ubiquitylated it is associated with heat-shock protein (HSP) 70. In contrast, Pim-1 protein is stabilized by HSP90 that protects it from proteasomal degradation. (Mizuno et al. 2001; Shay et al. 2005). Several miRNAs have been proposed to downregulate Pim-1 including miR-1 in lung cancer cells, (Nasser et al. 2008) miR-33a in various leukemic and solid tumor cell lines (Thomas et al. 2012), miR-328 in chronic myeloid leukemia (CML; Eiring et al. 2010), and hypoxia-induced miR-210 in solid tumor cell lines (Huang et al. 2009). With regards to AML, in murine myeloid FDC-P1 cells expressing mutated FLT3-ITD mir-16 was found to be downregulated in FLT3-ITD expressing cells, and upregulated following FLT3 inhibition with lestaurtinib and sunitinib (Kim et al. 2012). MiR-16 suppressed Pim-1 reporter gene expression, and overexpression of mir-16 in FLT3-ITD expressing FDC-P1 cells suppressed Pim-1 expression, suggesting that mir-16 repression may be involved with Pim-1 regulation in FLT3-ITD AML.

9.4 Pim Kinase Expression

9.4.1 Pim Kinase Expression in Cancers

Since Pim kinase proteins do not have a regulatory site, expression of mRNA usually suggests constitutively active proteins (Fox et al. 2005). Among hematological malignancies, Pim-1 is upregulated in B-cell malignancies (Gomez-Abad et al. 2011) including diffuse large B-cell lymphomas (Alizadeh et al. 2000; Pasqualucci et al. 2001; Poulsen et al. 2005; Rossi et al. 2006), mantle cell lymphoma (de Vos et al. 2003; Schatz et al. 2011; Zhu et al. 2002), Burkitt's lymphoma (Dave et al. 2006), and mediastinal large B-cell lymphoma (Martelli et al. 2008). Similar to B-cell lymphomas, other B-lymphoid leukemias such as CLL (Chen et al. 2009; Cohen and Kieff. 1991; Huttmann et al. 2006) show increased expression of Pim kinases. Studies are limited regarding protein expression of Pim kinases in primary cells from hematological malignancies, but in CLL (Chen et al. 2009). Pim kinases are overexpressed compared to normal lymphocytes. For Pim expression among epithelial cancers, prostate cancer has been widely-studied, however gastric, colorectal, pancreatic tumor cells show induction of these kinases (for review see Alvarado et al. 2012; Nawijn et al. 2004).

9.4.2 Pim Kinase Expression in AML

Similar to other cancers, majority of data reflect expression of Pim kinase at mRNA level. Specifically both Pim 1 and Pim 2 are overexpressed at transcript levels in AML blasts (Amson et al. 1989; Tamburini et al. 2009) . While each Pim kinase has one transcript, because of alternative initiation site for translation, there are different size proteins for Pim-1 and Pim-2. All these Pim-1 and Pim-2 isoforms contain a kinase domain and are functionally active (Saris et al. 1991) and AML cell lines express Pim kinases (Kelly et al. 2011; Kim et al. 2005).

9.4.3 Mechanism of Pim Upregulation in AML: RUNX, HOXA, STAT and ERG Transcription Factors

In addition to their overexpression, Pim kinases are involved with a number of pathways that are critical for AML biology, and several lines of evidence suggest that Pim is a target. Studies have elucidated that for M0-AML, MDS-AML, and therapy related-AML, sporadic mutations in RUNX family transcription factors play a major role; however, mutation in RUNX alone does not result in AML and other cooperating proteins are needed (Osato 2004). Pim-1 phosphorylates RUNX and enhances its transcription activity (Aho et al. 2006). Similarly, STAT family transcription factors have been shown to induce Pim kinases (Bachmann et al. 2006; Nosaka et al. 1999; Shirogane et al. 1999). Co-operativity with these two proteins was also reported (Mizuki et al. 2000). Another important transcription factor in AML is HOXA9 and Pim-1 is a transcriptional target of HOXA9 homeoprotein (Hu et al. 2007), which is involved in hematopoiesis and is frequently upregulated in AML (Golub et al. 1999; Lawrence et al. 1999; Verhaak et al. 2005). Pim kinases also are transcriptional targets of STAT3 and STAT5. STAT5 is downstream of a number of tyrosine kinase receptors including FMS-like tyrosine kinase 3 (FLT3) and constitutive STAT activation has been observed in a number of hematologic malignancies including AML (Benekli et al. 2009). Of significance in AML, STAT5 signaling is constitutively active in FLT3-ITD, however FLT3-ligand stimulation is needed for STAT5 activation in wild-type FLT3 (Yao et al. 2005). Accordingly, Pim is highly expressed in FLT3-ITD AML (Kim et al. 2005), which may be due to STAT activation that promotes Pim-1 and Pim-2 activity (Choudhary et al. 2009). ETS transcription factor ERG, which is overexpressed in AML and correlated with poor prognosis, was demonstrated to induce Pim-1 expression in transgenic mice (Goldberg et al. 2013). The combined high expression of both ERG and Pim together was also correlated with decreased survival rate, which further emphasizes Pim as a potential target for AML.

9.5 AML Biology: Impact on Pim Kinases

9.5.1 *FMS-like Tyrosine Kinase 3*

One of the most clinically significant genetic mutations in AML is in the *FLT3* gene, and is found in ~25 % of AML cases (reviewed by Stirewalt and Radich, 2003). The most frequent FLT3-activating mutation is internal tandem duplication (ITD), which accounts for ~15–20 % of mutations and tyrosine kinase domain (TKD) mutations, with the most prevalent being the D835 point mutation accounting for ~7 % of mutations. FLT3-ITD mutation is associated with inferior disease-free survival (DFS) and increased relapse rate (Thiede et al. 2002), and biallelic FLT-ITD is a strong adverse prognostic factor in de novo AML in patients with normal cytogenetics. (Whitman et al. 2001) While both ITD and TKD can confer factor-independent growth in IL3-dependent cell lines in liquid culture, only ITD strongly activate STAT5 and has transforming capacity for cells to form colonies in soft agar (Choudhary et al. 2005).

9.5.1.1 FLT3 and Pim Expression

As described in Sect. 4.3, Pim kinases are downstream of FLT3, thus activating mutations in principle would result in higher levels of *PIM* expression. Indeed, Pim-1 is expressed more highly in BaF3/FLT3-ITD transformed cells compared to FLT3 WT transformed cells stimulated with FLT-ligand (FL). These findings were also observed in AML primary cells, and cells from patients with FLT3-ITD mutations had higher levels of Pim-1 mRNA compared with FLT3 WT, which were both higher than normal bone marrow samples (Kim et al. 2005). Consistent with these observations, FLT3 inhibition using CEP-701 treatment of AML cell lines and primary cells resulted in reduction of both Pim-1 mRNA and protein levels (Kim et al. 2005). Similarly, in FLT3 AML, Pim-2 complements FLT3-WT to transform 32D cells (Agrawal et al. 2008). These reports establish a link between FLT3 and Pim kinases.

9.5.1.2 FLT3 Protein Maturation

Pim-1 has also been reported to be involved with FLT3 protein maturation and post-translational modifications. Figure 9.2a summarizes the general structure and cellular localization of FLT3 proteins that have been identified. Two major forms of human FLT3 have been reported; an unglycosylated form (~130 kDa) and an N-linked glycosylated form that is on the cell membrane (~150 kDa) (Carow et al. 1996; Lyman et al. 1993). FLT3-ITD exists mainly as the immature 130 kDa isoform that is retained in the endoplasmic reticulum (ER) through complex formation with ER-resident chaperone calnexin, whereas FLT3 WT exists

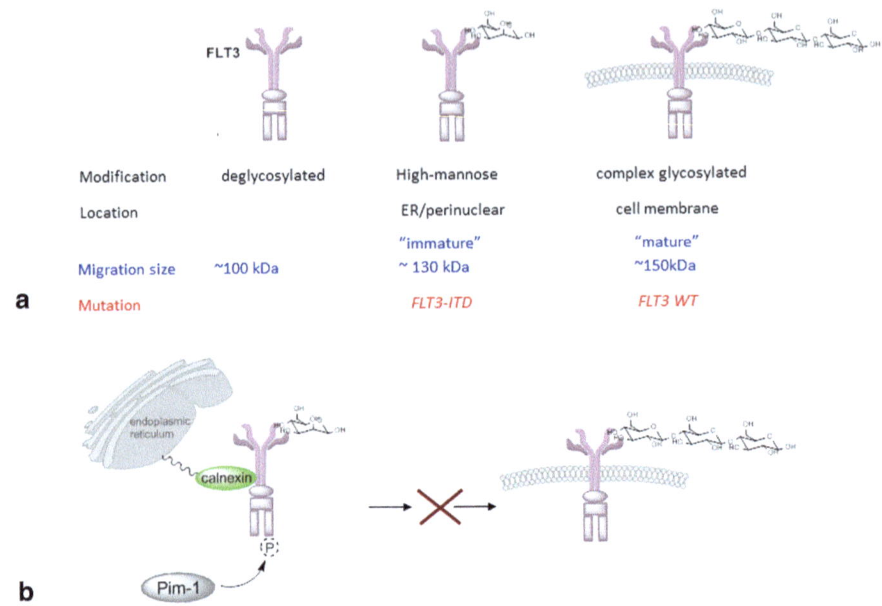

Fig. 9.2 Modulation of FLT3 by Pim-1 kinase. **a** Post-translational FLT3 protein maturation states. Following protein translation, FLT3 undergoes glycosylation to formhigh-mannose FLT3 and then further modified to form the mature complex N-glycosylated form With FLT3-WT, the equilibrium is towards the mature complex glycosylated protein and is largely located at the cell membrane, whereas in FLT3-ITD, the lower molecular weight high-mannose form is favored and located mainly in the endoplasmic reticulum (*ER*) sor perinuclear region (Koch et al., 2008; Schmidt-Arras et al., 2009). **b** Pim-1 inhibition shifts the equilibrium from the immature FLT3-ITD form to the larger mature isoform, which may be through disruption of FLT3-ITD binding with calnexin in the endoplasmic reticulum (ER) (Natarajan et al. 2013a). Pim may stabilize FLT3-ITD in the ER and prevent maturation and subsequent localization to the cell membrane

predominantly as the 150 kDa isoform (Schmidt-Arras et al. 2005). Pim-1 inhibition with quercetagetin disrupted FLT3-ITD binding with calnexin and increased the formation of the 150-kDa isoform, suggesting that Pim-1 can stabilize the 130 kDa isoform and prevent its glycosylation in to the 150 kDa form (Fig. 9.2b) (Natarajan et al. 2013b). Work done in our group was in agreement with these observations and treatment of FLT3-ITD cell line MV-4-11 with a small molecule Pim kinase-FLT3 inhibitor, SGI-1776, resulted in an increase in the 150 kDa isoform (Chen et al. 2011).

9.5.1.3 FLT3 Cellular Localization and Signaling

In addition to affecting FLT3 protein maturation, Pim-1 may also play a role in FLT3-ITD signaling through STATs. In elegant work done by Schmidt-Arras et al. engineered FLT3 mutants were anchored to the ER and their investigations demonstrated that while FLT3-WT signaling is largely at the cell surface, FLT3-ITD is capable of activating STAT signaling from the cell surface as well as intracellularly (Schmidt-Arras et al. 2009). Moreover, localization of FLT3-ITD in the cell affected which signaling pathways are activated. When localized in the ER, FLT3-ITD activated STAT5 and upregulated Pim-1/2, however membrane targeting of FLT3-ITD resulted in MAPK/PI3K pathway activation with diminished STAT5 phosphorylation (Choudhary et al. 2009). Inhibition of Pim-1 with quercetagetin resulted in decreased phosphorylation of FLT3 at Tyr591 (Natarajan et al. 2013a), which, interestingly, is one of the docking sites for STAT5 with FLT3-ITD (Tyr589/591) that is required for FLT3-ITD transforming potential in vivo (Rocnik et al. 2006). Of further interest, a recent report comparing FLT3 mutations identified that FLT3-ITD, but not FLT3-TKD, uses SRC to activate STAT5 that is contingent on phosphorylation of the same Tyr591 FLT3-ITD site (Leischner et al. 2012). Since Pim-1 is upregulated by constitutively active FLT3-ITD (Kim et al. 2005), these two proteins may be in a feedback loop and Pim-1 may contribute to the strong factor-independent activation of STAT5 in FLT3-ITD (Mizuki et al. 2000).

9.5.1.4 Clinical Significance

One of the challenges of FLT3 inhibition therapy is the development of resistance, and Pim-1 has been shown to contribute to increased resistance to FLT3 inhibition-mediated apoptosis (Kim et al. 2005). In FLT3-ITD-dependent cell line, BaF3/ITD, enforced expression of Pim-1 rendered the cells less sensitive to kinase inhibitor CEP-701-induced cytotoxicity and induction of apoptosis (Kim et al. 2005).

9.5.2 *p27, Skp2, and FoxO*

As introduced in Sect. 2.3, cell-cycle inhibitor p27 regulates cell-cycle control, apoptosis, cell proliferation and migration (reviewed in Besson et al. 2008). The expression and regulation of p27 is in part through FoxO transcription factors and Skp2 E3-ligase, which are also Pim kinase targets and are interconnected.

9.5.2.1 Transcriptional and Post-Translational Regulation of p27 by Pim Kinases

Pim kinases have been shown to downregulate p27 at both the transcriptional and post-translational levels (Morishita et al. 2008). At the gene expression level, Pim kinases suppress p27 by phosphorylating and inactivating forkhead transcription factors FoxO1a and FoxO3a, that both drive p27 expression (Medema et al. 2000). Post-translationally, Pim kinases can affect p27 degradation both directly and through other signaling proteins. All three Pim kinases can directly bind and phosphorylate p27, which leads to binding with 14-3-3 protein and subsequent nuclear export and proteasome-dependent degradation (Morishita et al. 2008). In addition, p27 is degraded by Skp2 E3 ligase (Carrano et al. 1999), which itself is stabilized by Pim-mediated phosphorylation that controls its functional ability to degrade p27 (Cen et al. 2010). Pim-1 was also demonstrated to be required for Skp2 cell-cycle signaling into S-phase entry (Cen et al. 2010). Thus, the consequence of Pim activity is sustained Skp2 protein stability and functional activity, in addition to Pim-mediated p27 protein degradation and gene expression or suppression.

9.5.2.2 Clinical Significance

High p27 expression had been reported as a significant prognostic factor for disease-free survival in a study examining 72 cases of AML patients (Yokozawa et al. 2000), although there has been conflicting data on its correlation with overall survival (Haferlach et al. 2011). In the context of AML disease progression, AML patients harboring FLT3 mutations have been found to have increased levels of FoxO3a phosphorylation, which was correlated with higher rates of primary resistance and shorter remission durations (Kornblau et al. 2010). In primary blast cells, from patients with AML, the ratio of cytoplasmic to nuclear p27 was strongly correlated with Skp2 protein levels and Skp2 expression was suggested as an independent marker for a poor prognosis AML (Min et al. 2004). Thus, Pim kinase modulation of FoxO3a and Skp2 may play a significant role in AML biology.

Given the findings discussed in this section, Pim kinase inhibition may be a valuable strategy to target even AML with FLT3-activating mutations and underscores the need for the identification and development of Pim kinase inhibitors.

9.6 Pim Kinase Inhibitors: Pre-clinical Investigations in AML

As described previously, Pim kinases play several and critical role in survival of cancer cells, and these pathways could be blocked by targeting Pim kinases using small-molecule inhibitors. A number of Pim kinase inhibitors of varying potency and specificity have been identified and developed in both pharmaceutical companies and academic settings and have been investigated in various cell types (reviewed in Isaac et al. 2011; Merkel et al. 2012). Inhibitors include small-molecule natural products (Holder et al. 2007) and synthetic compounds, and also a Pim-1-specific monoclonal antibody (Hu et al. 2009). The optimum strategy for Pim kinase inhibitors may be in combination with other cytotoxic agents, and there have been several preclinical combination that have been investigated. There is substantial overlap and redundancy in cancer signaling pathways, possibly as a survival mechanism. Treatment with agents against multiple signaling pathways may be a strategy to improve response to therapy or circumvent resistance. The Pim kinase inhibitors that have been evaluated preclinically as single agents or in combination with other therapeutics and those that have been evaluated in AML will be discussed in this section.

9.6.1 SGI-1776

The first Pim kinase inhibitor to enter clinical trial was SGI-1776 (SuperGen, now Astex Pharmaceuticals, see Sect. 7), which was effective in inducing apoptosis in primary cells from patients with CLL (Chen et al. 2009), AML primary cells, and cell lines (Chen et al. 2011). In both cases, there was a dose-dependent increase in cell death with increasing drug concentration, with a parallel decrease in RNA synthesis and Mcl-1 protein levels. SGI-1776 also inhibits FLT3, and there has been some debate regarding the extent to which Pim versus FLT3 inhibition contributes to SGI-1776 cytotoxicity in AML (Hospital et al. 2012). In MV4-11 cells with stable FLT3 knockdown (FLT3 shRNA), SGI-1776 decreased the viable cells as measured by MTT assay, suggesting that Pim inhibition by SGI-1776 may be a factor (Kelly et al. 2011). There is evidence that Pim inhibition alone in the absence of FLT3 inhibition is cytostatic rather than cytotoxic (Natarajan et al 2012), however, apoptosis induction by SGI-1776 was detected in both FLT3-WT and FLT3-ITD cell lines and primary cells (Chen et al. 2011). Consistent with the results obtained in AML cells ex vivo, AML xenograft model studies with mice bearing MV-4-11 tumors treated with SGI-1776 demonstrated tumor volume reduction. SGI-1776 has also been evaluated in combination with 1-β-D-arabinofuranosylcytosine (Ara-C), which is standard of care in AML induction therapy (Kelly et al. 2011) in paired Ara-C-resistant HL-60 AML cell line. Ara-C-resistant HL60 expressed higher levels of Pim-1 and Pim-3 protein and mRNA, however, SGI-1776 reduced viability in both cell lines and augmented Ara-C cytoxicity in the resistant cell line. The combination of SGI-1776 with Ara-C was also investigated in both FLT3-WT and FLT3-ITD cell lines, and in all cases the combination of both drugs together was greater than additive in reducing cell viability and clonogenic survival in HL60 (FLT-WT), MOLM-13 (heterozygous FLT3-ITD), and MV-4-11 (homozygous FLT3-ITD). These results suggest that Pim inhibition in combination may be a strategy to treat chemoresistant AML and/or may augment standard induction therapy.

SGI-1776 has also been investigated in combination with a number of therapeutic agents in other cancer cell types. SGI-1776 was evaluated in multiple myeloma cells and resulted in the induction of apoptosis as well as autophagy (Cervantes-Gomez F et al. 2012). SGI-1776 also sensitizes prostate cancer cell lines to taxanes (Mumenthaler et al. 2009) and Pim inhibition, with either M-110 or SGI-1776, in combination with EGFR inhibitor, Gefitinib, in prostate cancer cell lines had synergistic effects on inhibition of cell proliferation (Siu et al. 2011). SGI-1776 was also evaluated in combination with bendamustine in mantle cell lymphoma cells and the two agents were additive in cell killing (Yang et al. 2012a, b).

9.6.2 AZD1208

An inhibitor developed by AstraZeneca, AZD1208, inhibits all three Pim kinases at nanomolar levels and does not inhibit FLT3 (Keeton et al. 2014). Preliminary investigations indicate that it inhibits AML cell-line growth as well as MOLM-16 and KG-1a xenograft tumors in vivo, and the mechanism of action may involve 4EBP-1 or p70S6K signaling (Keeton et al. 2014) or eIF4B phosphorylation (Cen et al. 2014). Investigations in our laboratory suggest PRAS40, S6 and mTOR pathway proteins may be involved with AZD1208 mechanism (Chen LS et al. 2013). A clinical trial in AML has been initiated and is ongoing (see Sect. 7). Another Pim kinase inhibitor under development, AZD1897, was investigated in combination with and Akt inhibitor and was reported to be synergistic (Meja K et al. 2014).

9.6.3 SMI-4a

Pim inhibitor, SMI-4a, was developed at the Medical University of South Carolina (Xia et al. 2009) and has been demonstrated to be cytotoxic in a number of hematological and solid tumor cell lines (Beharry et al. 2011; Beharry et al. 2009; Lin et al. 2010; Song and Kraft, 2012; Xia et al. 2009). SMI-4a treatment induced apoptosis in various myeloid cell lines, and increased the proportion of cells in G1-phase cell cycle in AML cell line MV-4-11. SMI-4a has also been evaluated in combination treatment, and addition with rapamycin in MV-4-11 resulted in greater than additive growth inhibition compared with each single agent alone (Beharry et al. 2009). SMI-4a was also shown to be synergistic with MEK1/2 inhibitor, U0126, in precursor T-cell lymphoblastic leukemia/lymphoma cell lines (Lin et al. 2010). More recently, SMI-4a was also demonstrated to sensitize prostate cancer cell lines to apoptosis triggered by Bcl-2 family inhibitor ABT-737 (Song and Kraft, 2012).

9.6.4 ETP-39010/ETP-45299

Pan-Pim kinase inhibitors, ETP-39010 and ETP-45299, were developed by the Experimental Therapeutics Program at the Spanish National Cancer Research Centre (Wozniak et al. 2010).

ETP-39010 as a single agent inhibited the cell growth of AML cell lines MV-4-11 and HL-60, as well as other hematological malignancy cell lines, however its selectivity profile is similar to SGI-1776 and K00135 (see Sect. 6.6) and inhibits receptor tyrosine kinases FLT3, Kit and PDGFR1 (Blanco-Aparicio et al. 2011). Combination of ETP-39010 with histone deacetylase inhibitor vorinostat (suberoylanilidehydroxamic acid, SAHA) has been investigated in T-cell lymphoma (Wozniak et al. 2010) with varying results, but has not been evaluated in AML.

A second-generation derivative of ETP-39010 was developed, ETP-45299 (Gomez-Abad et al. 2011) that unlike ETP-39010, has little activity against FLT3, PDGFR1, or Kit (Blanco-Aparicio et al. 2011). ETP-45299 has demonstrated anti-proliferative effects as a single agent in MV-4-11 and HL-60 as well as other hematological and solid tumor cell lines. ETP-45299 has also been investigated in combination with PI3K kinase inhibitor, GDC-094, and was demonstrated to be synergistic in MV-4-11 cells (Blanco-Aparicio et al. 2011).

9.6.5 CX-6258

Cylene pharmaceuticals also has developed CX-6258, a nanomolar pan-Pim kinase inhibitor that does not inhibit FLT3 and inhibits cell proliferation of a number of cell lines including AML. CX-6258 was effective in reducing tumor volume in MV-4-11 xenograft models and preclinical investigations are ongoing to determine its clinical viability (Haddach et al. 2012).

9.6.6 Other Pim Inhibitors in the Pipeline for AML

K00135 treatment resulted in inhibition of clonogenic growth in a number of AML cell lines (Pogacic et al. 2007). K00135 was also shown to sensitize ALL primary tumors to ionizing radiation (unpublished data). Abbott Laboratories has developed a series of benzothienopyrimidinone Pim-inhibitors that reduce cell viability in MV-4-11 (Tao et al. 2009). Another Pim inhibitor, DHPCC-9, was academically synthesized at CNRS (Aubière, France) (Akue-Gedu et al. 2009) and was evaluated in murine IL-dependent myeloid FDCP1 cell line that stably express Pim-1 (FD/Pim44) or neomycin (FD/neo; Santio et al. 2010). DHPCC-9 treatment reduced cell viability in FD/Pim44 as determined by MTT assay but not FD/neo, indicating that DHPCC-9 can enter cells and impair Pim-1 anti-apoptotic activity. Recent Pim kinase inhibitors developed by Sanofi were evaluated in AML cells and resulted in a decline in Myc protein and identified CD25 as a biomarker for sensitivity to Pim inhibitors (Guo et al. 2014).

9.6.7 Pim Inhibitors with FLT3 Inhibitors

Given the significance of FLT3 in AML and that Pim-1 and FLT3 may be in a signaling loop through each other, a combination of FLT3 and Pim inhibitors may be a rational strategy. As discussed previously, SGI-1776 inhibits all three Pim kinases and FLT3 and was cytotoxic in both FLT3-WT and FLT3-ITD AML primary cells and cell lines, and partially restored sensitivity in Ara-C-resistant HL-60 cells. In agreement, quercetagetin, a flavonol Pim-1 kinase inhibitor, in combination with

FLT3 inhibitor PKC412 had a synergistic effect in inducing apoptosis of Ba/F3-ITD cells (Xie et al. 2009). In AML cell lines/cells, combination of Array Biopharma's Pim kinase inhibitor AR00459339 with FLT3 inhibitor (AR00454200) resulted in mildly synergistic cytotoxicity (Fathi et al. 2012). These data support the design of Pim and FLT3 inhibitor combinations for AML therapy.

9.6.8 Pim-1 Monoclonal Antibody

Evidence suggests that Pim-1 is expressed on the cell surface (Hu et al. 2009), and that Pim-kinases activity is present in the cell membrane fraction or inner leaflet of the membrane (Saris et al. 1991; Xie et al. 2006). A Pim-1-specific monoclonal antibody (P9) has been developed; although it has not been reported to be evaluated in AML, P9 inhibited proliferation activation of PHA-stimulated human PBMC and induced apoptosis with little effect on freshly isolated human blood T-lymphocytes that express relatively lower levels Pim-1. In DU145 prostate cancer cell line, P9 mAb-induced cancer cell apoptosis and synergistically inhibited cell proliferation when used in combination with cisplatin and epirubicin (Hu et al. 2009; Li et al., 2011).

9.7 Pim Kinase Inhibitors: Clinical Investigations in AML

The first in class Pim kinase inhibitor to enter clinical trial was SGI-1776 for refractory prostate cancer and relapsed/refractory non-Hodgkin's lymphoma, which was developed by SuperGen (now Astex Pharmaceuticals), however the trial was terminated due to dose-limiting toxicity of cardiac QTc prolongation (http://clinicaltrials.gov/ct2/show/NCT00848601). Second-generation Pim inhibitors have been developed by Astex with reduced hERG activity and are under investigation (Foulks et al. 2012). Currently, AstraZeneca's Pim kinase inhibitor AZD1208 is in Phase I clinical trial and is the first evaluation of a Pim kinase inhibitor in AML (http://clinicaltrials.gov/ct2/show/NCT01489722) at University of Texas MD Anderson Cancer Center (Houston, TX), Dana Farber Cancer Center (Boston, MA), and Ontario Cancer Institute (Toronto, Canada). There is also currently an observational clinical trial investigating Pim kinases as a treatment target in multiple myeloma (http://clinicaltrials.gov/ct2/show/NCT01410981), and Novartis initiated a clinical trial with their clinical candidate LGH447 in patients with relapsed and/or refractory multiple myeloma(http://clinicaltrials.gov/show/NCT01456689). Recently, a clinical trial investigating LGH447 in patients with AML or high risk myelodysplastic syndrome (MDS) (http://clinicaltrials.gov/ct2/show/NCT02078609). Novartis' related Pim kinase inhibitor LGB321 was synergistic with cytarabine in the KG-1 AML xenograft model (Garcia et al. 2014), and the results of these trials will provide insight into the clinical utility of Pim kinase inhibitors in cancer therapy.

9.8 Future Perspectives

There is a large body of information is available for Pim kinases, however, much still needs to be explored. For example, among the Pim kinase substrates, it is not clear which ones are most important. There may be diverse critical targets of Pim depending on the disease and at present we do not know which Pim kinase is important in which disease. This is further complicated by the similar substrate specificities of the Pim kinases. In AML specifically, it is further complicated given the interconnection between Pim kinases and FLT3. Despite these uncertainties, there is also interest in developing and testing Pim kinase inhibitors since mouse models lacking all three Pim kinases demonstrated minimal phenotype changes, suggesting that Pim kinases are valid targets. Second generation highly selective, and potent inhibitors have been generated and are under investigation. The outcome of the Pim inhibitor clinical trials will be a key in determining the "bench to bedside" significance and whether the preclinical observations translate into effects in patients during therapy. In parallel to these clinical investigations, biomarker studies will be key and further investigations comparing the different signaling pathways that are affected by the different Pim kinase inhibitors may provide insight into cancer cell survival mechanisms and AML biology. Finally, as with other signal transduction inhibitors or survival pathway blockers, Pim kinase inhibitors will be used most likely in combination. Several combination strategies are under development and will eventually move to the clinic.

References

Adam M, Pogacic V et al (2006) Targeting PIM kinases impairs survival of hematopoietic cells transformed by kinase inhibitor-sensitive and kinase inhibitor-resistant forms of Fms-like tyrosine kinase 3 and BCR/ABL. Cancer Res 66:3828–3835

Agrawal S, Koschmieder S et al (2008) Pim2 complements Flt3 wild-type receptor in hematopoietic progenitor cell transformation. Leukemia 22:78–86

Aho TL, Sandholm J et al (2004) Pim-1 kinase promotes inactivation of the pro-apoptotic Bad protein by phosphorylating it on the Ser112 gatekeeper site. FEBS Lett 571:43–9

Aho TL, Sandholm J et al (2006) Pim-1 kinase phosphorylates RUNX family transcription factors and enhances their activity. BMC Cell Biol 7:21

Akue-Gedu R, Rossignol E et al (2009) Synthesis, kinase inhibitory potencies, and in vitro antiproliferative evaluation of new Pim kinase inhibitors. J Med Chem 52:6369–6381

Alizadeh AA, Eisen MB et al (2000) Distinct types of diffuse large B-cell lymphoma identified by gene expression profiling. Nature 403:503–511

Allen JD, Berns A (1996) Complementation tagging of cooperating oncogenes in knockout mice. Semin Cancer Biol 7:299–306

Alvarado Y, Giles FJ et al (2012) The PIM kinases in hematological cancers. Expert Rev Hematol 5:81–96

Amaravadi R, Thompson CB (2005) The survival kinases Akt and Pim as potential pharmacological targets. J Clin Invest 115:2618–2624

Amson R, Sigaux F et al (1989) The human protooncogene product p33pim is expressed during fetal hematopoiesis and in diverse leukemias. Proc Natl Acad Sci U S A 86:8857–8861

Bachmann M, Kosan C et al (2006) The oncogenic serine/threonine kinase Pim-1 directly phosphorylates and activates the G2/M specific phosphatase Cdc25C. Int J. Biochem Cell Biol 38:430–443

Beharry Z, Zemskova M et al (2009) Novel benzylidene-thiazolidine-2,4-diones inhibit Pim protein kinase activity and induce cell cycle arrest in leukemia and prostate cancer cells. Mol Cancer Ther 8:1473–1483

Beharry Z, Mahajan S et al (2011) The Pim protein kinases regulate energy metabolism and cell growth. Proc Natl Acad Sci U S A 108:528–533

Benekli M, Baumann H et al (2009) Targeting signal transducer and activator of transcription signaling pathway in leukemias. J Clin Oncol 27:4422–4432

Banerjee S, Lu J et al (2014) EBNA3C Augments Pim-1 Mediated Phosphorylation and Degradation of p21 to Promote B-Cell Proliferation. PLoS Pathog. 10:e1004304

Besson A, Dowdy SF et al (2008) CDK inhibitors: cell cycle regulators and beyond. Dev Cell 14:159–169

Bhattacharya N, Wang Z et al (2002) Pim-1 associates with protein complexes necessary for mitosis. Chromosoma 111:80–95

Blanco-Aparicio C, Collazo AM et al (2011) Pim 1 kinase inhibitor ETP-45299 suppresses cellular proliferation and synergizes with PI3 K inhibition. Cancer Lett 300:145–153

Breuer M, Wientjens E et al (1991) Carcinogen-induced lymphomagenesis in pim-1 transgenic mice: dose dependence and involvement of myc and ras. Cancer Res 51:958–963

Carow CE, Levenstein M et al (1996) Expression of the hematopoietic growth factor receptor FLT3 (STK-1/Flk2) in human leukemias. Blood 87:1089–1096

Carrano AC, Eytan E et al (1999) SKP2 is required for ubiquitin-mediated degradation of the CDK inhibitor p27. Nature Cell Biol 1:193–199

Cervantes-Gomez F, Chen LS et al (2012) Biological effects of the Pim kinase inhibitor, SGI-1776, in multiple myeloma. Clin Lymphoma Myeloma Leuk Suppl 2:S317–29

Cen B, Mahajan S et al (2010) Regulation of Skp2 levels by the Pim-1 protein kinase. J Biol Chem 285:29128–29137

Cen B, Xiong Y et al (2014) The Pim-1 protein kinase is an important regulator of MET receptor tyrosine kinase levels and signaling. Mol Cell Biol 34:2517–32

Chen LS, Redkar S et al (2009) Pim kinase inhibitor, SGI-1776, induces apoptosis in chronic lymphocytic leukemia cells. Blood 114:4150–4157

Chen LS, Redkar S et al (2011) Mechanisms of cytotoxicity to Pim kinase inhibitor, SGI-1776, in acute myeloid leukemia. Blood 118:693–702

Chen LS, Cortes JE et al (2013) Mechanisms of action of Pim kinase inhibitor, AZD1208, in acute myeloid leukemia cells. In Proceedings: AACR 104th Annual Meeting Washington, DC. Abstract 1050:6–10

Chen XP, Losman JA et al (2002) Pim serine/threonine kinases regulate the stability of Socs-1 protein. Proc Natl Acad Sci U S A 99:2175–2180

Choudhary C, Schwable J et al (2005) AML-associated Flt3 kinase domain mutations show signal transduction differences compared with Flt3 ITD mutations. Blood 106:265–273

Choudhary C, Olsen JV et al (2009) Mislocalized activation of oncogenic RTKs switches downstream signaling outcomes. Mol Cell 36:326–339

Cohen JI, Kieff E (1991) An Epstein-Barr virus nuclear protein 2 domain essential for transformation is a direct transcriptional activator. J Virol 65:5880–5885

Cuypers HT, Selten G et al (1984) Murine leukemia virus-induced T-cell lymphomagenesis: integration of proviruses in a distinct chromosomal region. Cell 37:141–150

Dautry F, Weil D et al (1988) Regulation of pim and myb mRNA accumulation by interleukin 2 and interleukin 3 in murine hematopoietic cell lines. J Biol Chem 263:17615–17620

Dave SS, Fu K et al (2006) Molecular diagnosis of Burkitt's lymphoma. N Eng J Med 354:2431–2442

de Vos S, Krug U et al (2003) Cell cycle alterations in the blastoid variant of mantle cell lymphoma (MCL-BV) as detected by gene expression profiling of mantle cell lymphoma (MCL) and MCL-BV. Diagn Mol Pathol 12:35–43
Din S, Mason M, Völkers M et al (2013) Pim-1 preserves mitochondrial morphology by inhibiting dynamin-related protein 1 translocation. Proc Natl Acad Sci USA 110:5969–74
Eiring AM, Harb JG et al (2010) miR-328 functions as an RNA decoy to modulate hnRNP E2 regulation of mRNA translation in leukemic blasts. Cell 140:652–665
Fathi AT, Arowojolu O et al (2012) A potential therapeutic target for FLT3-ITD AML: PIM1 kinase. Leuk Res 36:224–231
Foulks JM, Xu Y et al (2012) Second generation PIM inhibitors exhibit improved activity in solid tumor models. In Proceedings: AACR 103rd Annual Meeting 2012– Mar 31-Apr 4, 2012; Chicago, IL, pp. Abstract 3762
Fox CJ, Hammerman PS et al (2005) The Pim kinases control rapamycin-resistant T cell survival and activation. J Exp Med 201:259–266
Garcia PD, Langowski JL et al (2014) Pan-PIM kinase inhibition provides a novel therapy for treating hematologic cancers. Clin Cancer Res 20:1834–45
Goldberg L, Tijssen MR et al (2013) Genome-scale expression and transcription factor binding profiles reveal therapeutic targets in transgenic ERG myeloid leukemia. Blood 122:2694–703
Golub TR, Slonim DK et al (1999) Molecular classification of cancer: class discovery and class prediction by gene expression monitoring. Science 286:531–537
Gomez-Abad C, Pisonero H et al (2011) PIM2 inhibition as a rational therapeutic approach in B-cell lymphoma. Blood 118:5517–5527
Guo Z, Wang A et al (2014) PIM inhibitors target CD25-positive AML cells through concomitant suppression of STAT5 activation and degradation of MYC oncogene. Blood, in press
Haddach M, Michaux J et al (2012) Discovery of CX-6258. A Potent, Selective, and Orally Efficacious pan-Pim Kinases Inhibitor. ACS Med Chem Lett 3:135–139
Haferlach C, Bacher U et al (2011) CDKN1B, encoding the cyclin-dependent kinase inhibitor 1B (p27), is located in the minimally deleted region of 12p abnormalities in myeloid malignancies and its low expression is a favorable prognostic marker in acute myeloid leukemia. Haematologica 96:829–836
Hammerman PS, Fox CJ et al (2005) Pim and Akt oncogenes are independent regulators of hematopoietic cell growth and survival. Blood 105:4477–4483
Hogan C, Hutchison C et al (2008) Elevated levels of oncogenic protein kinase Pim-1 induce the p53 pathway in cultured cells and correlate with increased Mdm2 in mantle cell lymphoma. J Biol Chem 283:18012–18023
Holder S, Zemskova M et al (2007) Characterization of a potent and selective small-molecule inhibitor of the PIM1 kinase. Mol Cancer Ther 6:163–172
Hospital MA, Green AS et al (2012) The FLT3 and Pim kinases inhibitor SGI-1776 preferentially target FLT3-ITD AML cells. Blood 119:1791–1792
Hu XF, Li J et al (2009) PIM-1-specific mAb suppresses human and mouse tumor growth by decreasing PIM-1 levels, reducing Akt phosphorylation, and activating apoptosis. J Clin Invest 119:362–375
Hu YL, Passegue E et al (2007) Evidence that the Pim1 kinase gene is a direct target of HOXA9. Blood 109:4732–4738
Huang X, Ding L et al (2009) Hypoxia-inducible mir-210 regulates normoxic gene expression involved in tumor initiation. Mol Cell 35:856–867
Huttmann A, Klein-Hitpass L et al (2006) Gene expression signatures separate B-cell chronic lymphocytic leukaemia prognostic subgroups defined by ZAP-70 and CD38 expression status. Leukemia 20:1774–1782
Isaac M, Siu A et al (2011) The oncogenic PIM kinase family regulates drug resistance through multiple mechanisms. Drug Resist Updat 14:203–211
Keeton EK, McEachern K, et al (2014) AZD1208, a potent and selective pan-Pim kinase inhibitor, demonstrates efficacy in preclinical models of acute myeloid leukemia. Blood 123:905–713

Kelly KR, Espitia CM et al (2011) Targeting PIM kinase activity significantly augments the efficacy of cytarabine. Br J Haematol 156:129–132

Kim KT, Baird K et al (2005) Pim-1 is upregulated by constitutively activated FLT3 and plays a role in FLT3-mediated cell survival. Blood 105:1759–1767

Kim KT, Carroll AP et al (2012) MicroRNA-16 is down-regulated in mutated FLT3 expressing murine myeloid FDC-P1 cells and interacts with Pim-1. PLoS One 7:e44546

Koch S, Jacobi A et al (2008) Abnormal localization and accumulation of FLT3-ITD, a mutant receptor tyrosine kinase involved in leukemogenesis. Cells, tissues, organs 188:225–235

Konietzko U, Kauselmann G et al (1999) Pim kinase expression is induced by LTP stimulation and required for the consolidation of enduring LTP. EMBO J 18:3359–3369

Kornblau SM, Singh N et al (2010) Highly phosphorylated FOXO3A is an adverse prognostic factor in acute myeloid leukemia. Clin Cancer Res 16:1865–1874

Lawrence HJ, Rozenfeld S et al (1999) Frequent co-expression of the HOXA9 and MEIS1 homeobox genes in human myeloid leukemias. Leukemia 13:1993–1999

Leischner H, Albers C et al (2012) SRC is a signaling mediator in FLT3-ITD- but not in FLT3-TKD-positive AML. Blood 119:4026–4033

Leverson JD, Koskinen PJ et al (1998) Pim-1 kinase and p100 cooperate to enhance c-Myb activity. Mol Cell 2:417–425

Li J, Hu XF et al (2009) Pim-1 expression and monoclonal antibody targeting in human leukemia cell lines. Exp Hematol 37:1284–1294

Li J, Loveland BE et al (2011) Anti-Pim-1 mAb inhibits activation and proliferation of T lymphocytes and prolongs mouse skin allograft survival. Cell Immunol 272:87–93

Lilly M, Le T et al (1992) Sustained expression of the pim-1 kinase is specifically induced in myeloid cells by cytokines whose receptors are structurally related. Oncogene 7:727–732

Lin YW, Beharry ZM et al (2010) A small molecule inhibitor of Pim protein kinases blocks the growth of precursor T-cell lymphoblastic leukemia/lymphoma. Blood 115:824–833

Lyman SD, James L et al (1993) Characterization of the protein encoded by the flt3 (flk2) receptor-like tyrosine kinase gene. Oncogene 8:815–822

Macdonald A, Campbell DG et al (2006) Pim kinases phosphorylate multiple sites on Bad and promote 14-3-3 binding and dissociation from Bcl-XL. BMC Cell Biol 7:1

Martelli M, Ferreri AJ et al (2008) Primary mediastinal large B-cell lymphoma. Crit Rev Oncol Hematol 68:256–263

Medema RH, Kops GJ et al (2000) AFX-like Forkhead transcription factors mediate cell-cycle regulation by Ras and PKB through p27kip1. Nature 404:782–787

Meja K, Stengel C et al (2014) PIM and AKT kinase inhibitors show synergistic cytotoxicity in acute myeloid leukaemia that is associated with convergence on mTOR and MCL1 pathways. Br J Haematol, in press

Merkel AL, Meggers E et al (2012) PIM1 kinase as a target for cancer therapy. Expert Opin Investig Drugs 21:425–436

Mikkers H, Nawijn M et al (2004) Mice deficient for all PIM kinases display reduced body size and impaired responses to hematopoietic growth factors. Mol Cell Biol 24:6104–6115

Min YH, Cheong JW et al (2004) Elevated S-phase kinase-associated protein 2 protein expression in acute myelogenous leukemia: its association with constitutive phosphorylation of phosphatase and tensin homologue protein and poor prognosis. Clin Cancer Res 10:5123–5130

Mizuki M, Fenski R et al (2000) Flt3 mutations from patients with acute myeloid leukemia induce transformation of 32D cells mediated by the Ras and STAT5 pathways. Blood 96:3907–3914

Mizuno K, Shirogane T et al (2001) Regulation of Pim-1 by Hsp90. Biochem Biophys Res Commun 281:663–369

Mochizuki T, Kitanaka C et al (1999) Physical and functional interactions between Pim-1 kinase and Cdc25A phosphatase. Implications for the Pim-1-mediated activation of the c-Myc signaling pathway. J Biol Chem 274:18659–18666

Morishita D, Katayama R et al (2008) Pim kinases promote cell cycle progression by phosphorylating and downregulating p27Kip1 at the transcriptional and posttranscriptional levels. Cancer Res 68:5076–5085

Mumenthaler SM, Ng PY et al (2009) Pharmacologic inhibition of Pim kinases alters prostate cancer cell growth and resensitizeschemoresistant cells to taxanes. Mol Cancer Ther 8:2882–2893

Nagarajan L, Louie E et al (1986) Localization of the human pim oncogene (PIM) to a region of chromosome 6 involved in translocations in acute leukemias. Proc Natl Acad Sci U S A 83:2556–2560

Nasser MW, Datta J et al (2008) Downregulation of micro-RNA-1 (miR-1) in lung cancer. Suppression of tumorigenic property of lung cancer cells and their sensitization to doxorubicin-induced apoptosis by miR-1. J Biol Chem 283:33394–33405

Natarajan K, Burcu M et al (2012) Pim-1 kinase inhibition in the absence of FLT3 inhibition results in cytostatic effects in acute myeloid leukemia cells Proceedings: AACR 103rd Annual Meeting 2012– Mar 31-Apr 4, 2012; Chicago, IL72:Abstract 4581

Natarajan K, Bhullar J et al (2013a) The Pim kinase inhibitor SGI-1776 decreases cell surface expression of P-glycoprotein (ABCB1) and breast cancer resistance protein (ABCG2) and drug transport by Pim-1-dependent and -independent mechanisms. Biochem Pharmacol 85:514–24

Natarajan K, Xie Y et al (2013b) Pim-1 kinase phosphorylates and stabilizes 130 kDa FLT3 and promotes aberrant STAT5 signaling in acute myeloid leukemia with FLT3 internal tandem duplication. PLoS One 8:e74653

Naud JF, Eilers M (2007) PIM1 and MYC: a changing relationship? Nat Cell Biol 9:873–875

Nawijn MC, Alendar A et al (2011) For better or for worse: the role of Pim oncogenes in tumorigenesis. Nat Rev Cancer 11:23–34

Nihira K, Ando Y et al (2009) Pim-1 controls NF-kappaB signaling by stabilizing RelA/p65. Cell Death Differ

Nosaka T, Kawashima T et al (1999) STAT5 as a molecular regulator of proliferation, differentiation and apoptosis in hematopoietic cells. EMBO J 18:4754–4765

Osato M (2004) Point mutations in the RUNX1/AML1 gene: another actor in RUNX leukemia. Oncogene 23:4284–4296

Pasqualucci L, Neumeister P et al (2001) Hypermutation of multiple proto-oncogenes in B-cell diffuse large-cell lymphomas. Nature 412:341–346

Peltola KJ, Paukku K et al (2004) Pim-1 kinase inhibits STAT5-dependent transcription via its interactions with SOCS1 and SOCS3. Blood 103:3744–3750

Pogacic V, Bullock AN et al (2007) Structural analysis identifies imidazo[1,2-b]pyridazines as PIM kinase inhibitors with in vitro antileukemic activity. Cancer Res 67:6916–6924

Poulsen CB, Borup R et al (2005) Microarray-based classification of diffuse large B-cell lymphoma. Eur J Haematol 74:453–465

Qian KC, Wang L et al (2005) Structural basis of constitutive activity and a unique nucleotide binding mode of human Pim-1 kinase. J Biol Chem 280:6130–6137

Rainio EM, Sandholm J et al (2002) Cutting edge: Transcriptional activity of NFATc1 is enhanced by the Pim-1 kinase. J Immunol 168:1524–1527

Richter JD, Sonenberg N (2005) Regulation of cap-dependent translation by eIF4E inhibitory proteins. Nature 433:477–480

Rocnik JL, Okabe R et al (2006) Roles of tyrosine 589 and 591 in STAT5 activation and transformation mediated by FLT3-ITD. Blood 108:1339–1345

Rossi D, Berra E et al (2006) Aberrant somatic hypermutation in transformation of follicular lymphoma and chronic lymphocytic leukemia to diffuse large B-cell lymphoma. Haematologica 91:1405–1409

Santio NM, Vahakoski RL et al (2010) Pim-selective inhibitor DHPCC-9 reveals Pim kinases as potent stimulators of cancer cell migration and invasion. Mol Cancer 9:279

Saris CJ, Domen J et al (1991) The pim-1 oncogene encodes two related protein-serine/threonine kinases by alternative initiation at AUG and CUG. EMBO J 10:655–664

Schatz JH, Oricchio E et al (2011) Targeting cap-dependent translation blocks converging survival signals by AKT and PIM kinases in lymphoma. J Exp Med 208:1799–1807

Schmidt-Arras DE, Bohmer A et al (2005) Tyrosine phosphorylation regulates maturation of receptor tyrosine kinases. Mol Cell Biol 25:3690–3703

Schmidt-Arras D, Bohmer SA et al (2009) Anchoring of FLT3 in the endoplasmic reticulum alters signaling quality. Blood 113:3568–3576

Schmidt T, Karsunky H et al (1998) Zinc finger protein GFI-1 has low oncogenic potential but cooperates strongly with pim and myc genes in T-cell lymphomagenesis. Oncogene 17:2661–2667

Shah N, Pang B et al (2008) Potential roles for the PIM1 kinase in human cancer - a molecular and therapeutic appraisal. Eur J Cancer 44:2144–2151

Shay KP, Wang Z et al (2005) Pim-1 kinase stability is regulated by heat shock proteins and the ubiquitin-proteasome pathway. Mol Cancer Res 3:170–181

Shirogane T, Fukada T et al (1999) Synergistic roles for Pim-1 and c-Myc in STAT3-mediated cell cycle progression and antiapoptosis. Immunity 11:709–719

Siu A, Virtanen C et al (2011) PIM kinase isoform specific regulation of MIG6 expression and EGFR signaling in prostate cancer cells. Oncotarget 2:1134–1144

Song JH, Kraft AS (2012) Pim kinase inhibitors sensitize prostate cancer cells to apoptosis triggered by Bcl-2 family inhibitor ABT-737. Cancer Res 72:294–303

Stirewalt DL, Radich JP (2003) The role of FLT3 in haematopoietic malignancies. Nat Rev Cancer 3:650–665

Tamburini J, Green AS et al (2009) Protein synthesis is resistant to rapamycin and constitutes a promising therapeutic target in acute myeloid leukemia. Blood 114:1618–1627

Tao ZF, Hasvold LA et al (2009) Discovery of 3H-benzo[4,5]thieno[3,2-d]pyrimidin-4-ones as potent, highly selective, and orally bioavailable inhibitors of the human protooncogeneproviral insertion site in moloney murine leukemia virus (PIM) kinases. J Med Chem 52:6621–6636

Thiede C, Steudel C et al (2002) Analysis of FLT3-activating mutations in 979 patients with acute myelogenous leukemia: association with FAB subtypes and identification of subgroups with poor prognosis. Blood 99:4326–4335

Thomas M, Lange-Grunweller K et al (2012) The proto-oncogene Pim-1 is a target of miR-33a. Oncogene 31:918–928

van der Lugt NM, Domen J et al (1995) Proviral tagging in E mu-myc transgenic mice lacking the Pim-1 proto-oncogene leads to compensatory activation of Pim-2. EMBO J 14:2536–2544

vanLohuizen M, Verbeek S et al (1989) Predisposition to lymphomagenesis in pim-1 transgenic mice: cooperation with c-myc and N-myc in murine leukemia virus-induced tumors. Cell 56:673–682

Verbeek S, van Lohuizen M et al (1991) Mice bearing the E mu-myc and E mu-pim-1 transgenes develop pre-B-cell leukemia prenatally. Mol Cell Biol 11:1176–1179

Verhaak RG, Goudswaard CS et al (2005) Mutations in nucleophosmin (NPM1) in acute myeloid leukemia (AML): association with other gene abnormalities and previously established gene expression signatures and their favorable prognostic significance. Blood 106:3747–3754

Wang Z, Bhattacharya N et al (2001) Pim-1: a serine/threonine kinase with a role in cell survival, proliferation, differentiation and tumorigenesis. J Vet Sci 2:167–179

Weston K (1999) Reassessing the role of C-MYB in tumorigenesis. Oncogene 18:3034–8

White E (2003) The pims and outs of survival signaling: role for the Pim-2 protein kinase in the suppression of apoptosis by cytokines. Genes Dev 17:1813–6

Whitman SP, Archer KJ et al (2001) Absence of the wild-type allele predicts poor prognosis in adult de novo acute myeloid leukemia with normal cytogenetics and the internal tandem duplication of FLT3: a cancer and leukemia group B study. Cancer Res 61:7233–9

Winn LM, Lei W et al (2003) Pim-1 phosphorylates the DNA binding domain of c-Myb. Cell Cycle 2:258–262

Wood NT, Meek DW et al (2009) 14-3-3 Binding to Pim-phosphorylated Ser166 and Ser186 of human Mdm2–Potential interplay with the PKB/Akt pathway and p14(ARF). FEBS Lett 583:615–620

Woodland RT, Fox CJ et al (2008) Multiple signaling pathways promote B lymphocyte stimulator dependent B-cell growth and survival. Blood 111:750–760

Wozniak MB, Villuendas R et al (2010) Vorinostat interferes with the signaling transduction pathway of T-cell receptor and synergizes with phosphoinositide-3 kinase inhibitors in cutaneous T-cell lymphoma. Haematologica 95:613–621

Wright G, Tan B et al (2003) A gene expression-based method to diagnose clinically distinct subgroups of diffuse large B cell lymphoma. Proc Natl Acad Sci U S A 100:9991–6

Xia Z, Knaak C et al (2009) Synthesis and evaluation of novel inhibitors of Pim-1 and Pim-2 protein kinases. J Med Chem 52:74–86

Xie Y, Xu K et al (2006) The 44 kDa Pim-1 kinase directly interacts with tyrosine kinase Etk/BMX and protects human prostate cancer cells from apoptosis induced by chemotherapeutic drugs. Oncogene 25:70–78

Xie Y, Xu K et al (2008) The 44-kDa Pim-1 kinase phosphorylates BCRP/ABCG2 and thereby promotes its multimerization and drug-resistant activity in human prostate cancer cells. J Biol Chem 283:3349–3356

Xie Y, Burcu M et al (2010) Pim-1 kinase protects P-glycoprotein from degradation and enables its glycosylation and cell surface expression. Mol Pharmacol 78:310–8

Yang Q, Chen LS et al (2012a) Transcription and translation are primary targets of Pim kinase inhibitor SGI-1776 in mantle cell lymphoma. Blood 120:3491–500

Yang Q, Chen LS et al (2012b) Combination of Pim kinase inhibitor SGI-1776 and bendamustine in B-cell lymphoma. Clin Lymphoma Myeloma Leuk Suppl 2:S355–62

Yao Q, Nishiuchi R et al (2005) Human leukemias with mutated FLT3 kinase are synergistically sensitive to FLT3 and Hsp90 inhibitors: the key role of the STAT5 signal transduction pathway. Leukemia 19:1605–1612

Yip-Schneider MT, Horie M et al (1995) Transcriptional induction of pim-1 protein kinase gene expression by interferon gamma and posttranscriptional effects on costimulation with steel factor. Blood 85:3494–3502

Yokozawa T, Towatari M et al (2000) Prognostic significance of the cell cycle inhibitor p27Kip1 in acute myeloid leukemia. Leukemia 14:28–33

Yuan LL, Green AS et al (2014) Pim kinases phosphorylate Chk1 and regulate its functions in acute myeloid leukemia. Leukemia 28:293–301

Zhang Y, Wang Z et al (2008) Pim kinase-dependent inhibition of c-Myc degradation. Oncogene 27:4809–4819

Zhu Y, Hollmen J et al (2002) Investigatory and analytical approaches to differential gene expression profiling in mantle cell lymphoma. Br J Haematol 119:905–915

Zippo A, De Robertis A et al (2007) PIM1-dependent phosphorylation of histone H3 at serine 10 is required for MYC-dependent transcriptional activation and oncogenic transformation. Nat Cell Biol 9:932–944

Zippo A, Serafini R et al (2009) Histone crosstalk between H3S10ph and H4K16ac generates a histone code that mediates transcription elongation. Cell 138:1122–1136

Chapter 10
Development of Midostaurin as a Tyrosine Kinase Inhibitor

Gabriela Motyckova and Richard M. Stone

Abstract Midostaurin (PKC-412) is a multitargeted kinase inhibitor that is now in development primarily as a FLT3 inhibitor for patients with acute myeloid leukemia (AML). Blasts from 5 to 10% of patients with AML carry a point mutation in the FLT3 activation loop; another 25% display a repeat of 3 to over 300 base pairs in the region of the gene encoding the juxtamembrane region (internal tandem duplication, ITD), which markedly enhances the risk for relapse and death. AML patents with *FLT3*-ITD mutations, for whom allogeneic transplantation must be strongly considered, are in need of better medical therapy. Putative FLT3 inhibitors such as midostaurin, tandutinib, lestaurtinib, quizartinib, KW02449, and sorafenib have been studied in patients with AML either alone or in combination with chemotherapy. While the FLT3 inhibitors vary widely with regard to potency, specificity, and protein binding, one of the most extensively studied to date is midostaurin (developmental name PKC412). Midostaurin may also be useful as a single agent in systemic mastocytosis due to its ability to inhibit the D816V c-KIT mutation. A key study to determine if midostaurin can improve survival when added to chemotherapy in adults under age 60 with mutant FLT3 AML has completed its accrual.

Keywords Midostaurin · Acute myeloid leukemia · FLT3 · Tyrosine kinase inhibitor

---

R. M. Stone (✉) · G. Motyckova
Department of Medical Oncology, Dana-Farber Cancer Institute, Harvard Medical School, 450 Brookline Ave, Boston, MA 02115, USA
e-mail: rstone@partners.org

G. Motyckova
e-mail: gmotyckova@partners.org

M. Andreeff (ed.), *Targeted Therapy of Acute Myeloid Leukemia*,
Current Cancer Research, DOI 10.1007/978-1-4939-1393-0_10

10.1 Introduction

The fms-like tyrosine kinase (FLT3, also known as fetal liver kinase 2 or stem cell kinase 1) is important for the development of hematopoietic stem cells and the immune system. The *FLT3* gene on chromosome 13q12 encodes a member of the class III receptor tyrosine kinase family that also includes *c-KIT, c-fms,* and the platelet derived growth receptors. FLT3 is normally expressed by immature hematopoietic cells (Knapper 2007; Wiernik 2010). The ligand for the FLT3 receptor (referred to as FL or FLT3 ligand) is expressed on many cell types and likely stimulates hematopoietic cells synergistically with other cytokines (Knapper 2007). The FLT3 receptor consists of the following domains: extracellular ligand-binding domain, transmembrane domain, juxtamembrane domain (JMD), and tyrosine kinase domain (TKD). In an unmutated state, the FLT3 receptor dimerizes upon binding to the FLT3 ligand, undergoing autophosphorylation and activation of the tyrosine kinase activity thereby turning on downstream signaling pathways. The pathways important in FLT3 signaling that mediate cell proliferation include RAS/MAP kinase, PI3 kinase/AKT, and phospholipase C gamma (Knapper 2007; Pratz and Levis 2010).

10.2 FLT3 Mutations

The JMD normally negatively regulates FLT3 activity; mutations in this region result in FLT3 activation. The most common FLT3 mutations result in repeats of between one and greater than 40 amino acids in the JMD which disrupt the domain's negative regulatory activity (Pratz and Levis 2010). This mutant *FLT3*-internal tandem duplication (ITD) produces a constitutively activated receptor tyrosine kinase, due to disruption of the auto regulatory mechanism of the receptor by the ITD. The constitutively active mutant *FLT3*-ITD receptor can undergo dimerization and autophosphorylation in the absence of ligand, which results in activation of downstream signaling pathways important in survival, differentiation, and proliferation (Hayakawa 2000). Patients with *FLT3*-ITD mutations are more frequently diagnosed with de novo acute myeloid leukemia (AML), have higher bone marrow blast percentage and higher white blood cell count, increased relapse rate, and decreased survival compared to wild-type *FLT3* AML (Bienz 2005; Mrozek 2007).

Another class of mutations are point mutations in the kinase activation loop (TKD), generally at residue 835 (Whitman et al. 2008). Point mutations in the JMD (*FLT3*-JMD) and extracellular domain have also been described (Reindl et al. 2006; Frohling 2007). Among the *FLT3* mutations, the *FLT3*-ITD mutations are most common, appearing in about 25 % of AML cases (Schlenk et al. 2008). Both *FLT3*-ITD- and *FLT3*-TKD-mutated proteins can activate several signaling pathways through ERK1/2, AKT, or Shc (Pratz and Levis 2010). *FLT3*-ITD mutations result in upregulation of PIM-1, activation of STAT5 and repression of myeloid transcription factors Pu.1 and CEBPα (Pratz and Levis 2010). Pim-1 is a serine/threonine kinase important in proliferation and survival (Kim et al. 2005). Based on vitro studies, the *FLT3*-ITD and *FLT3*-TKD D835 activating mutations were able to cause growth-factor-independent proliferation of cell lines (Zhao et al. 2000; Yamamoto et al. 2001). In a mouse bone marrow transplantation model, FLT3 mutations induced an oligoclonal myeloproliferative disorder with splenomegaly and leukocytosis, but were not sufficient to yield an AML phenotype (Kelly et al. 2002b). Therefore, additional mutations may be required to recapitulate the AML phenotype in this mouse model.

The leukemogenic properties of *FLT3* mutations are therefore multifactorial, but certainly involve activation of signaling pathways and factors important in proliferation and growth-factor-independent survival in leukemic cells. In addition to direct inhibition of the mutant FLT3 protein, interruption of downstream pathways alone or in combination represents another potential therapeutic avenue. Although the wild-type *FLT3* is over expressed in 80% of AML (Knapper 2007; Stirewalt et al. 2006), its role in AML pathogenesis is unclear.

FLT3-TKD mutation occurs in 7% of patients with AML (Yamamoto et al. 2001). The effect of the *FLT3*-TKD mutation on prognosis of AML patient is not clear, but it may (Gale et al. 2008) or may not (Whitman et al. 2008) have better prognosis than *FLT3*-ITD mutation. The *FLT3*-JMD mutant may have a lower level of autophosphorylation and possibly weaker transforming ability (Reindl et al. 2006).

The *FLT3* mutant allele burden varies in patients. Higher burden of mutant allele is generally recognized as prognostically worse than a lower allele level of mutant allele (Gale et al. 2008), but there are some data to the contrary (How et al. 2012). Moreover, blasts from patients with relapsed AML may also have higher *FLT3* allele burden as well as higher level of FL expression as a counter-regulatory response to chemotherapy or FLT3 inhibitors (Sato et al. 2011). In vitro studies suggest that samples from relapsed AML *FLT3*-ITD blasts and those with a high *FLT3*-ITD allele burden may respond relatively well to FLT3 inhibition (Pratz et al. 2010).

FLT3-ITD mutations can activate CXCR4 signaling (Rombouts et al. 2004) which in turn may diminish the clinical activity of FLT3 inhibitors by enhancing the ability of the so-called protective stem cell to maintain leukemic cell survival (Zeng et al. 2009). CXCR4 inhibition increases the sensitivity of FLT3-mutated cells to sorafenib in a mouse model, perhaps by inhibiting key AML survival signals mediated by the bone marrow microenvironment (Zeng et al. 2009). In vitro studies demonstrated increased FLT3 surface expression in cell lines treated with the FLT3 inhibitor midostaurin (Weisberg et al. 2011) as well as upregulation of anti-apoptotic genes and down regulation of proapoptotic genes, possibly contributing to midostaurin resistance (Stolzel et al. 2010).

AML, including *FLT3* mutant disease, is genetically complex. Moreover, *FLT3* mutations may represent a "late hit" that promotes proliferation, but not critical for self-renewal or survival (Mardis et al. 2009). Thus, inhibition of FLT3 alone without adjunctive therapy may not result in long-term control of AML. On the other hand, a potent and prolonged inhibition of the mutant enzyme could be useful in certain clinical settings and/or if the leukemic cell is "addicted" to survival based on an activated FLT3 protein. FLT3 inhibitors reduce anti apoptotic Mcl-1 levels and increased p53-dependent apoptosis in *FLT3*-ITD positive AML blasts

(Zhang et al. 2008). Inhibitors of other pathways important in leukemogenesis such as the MEK, PI3K, JAK/STAT, and other signaling pathways in combination with FLT3 inhibitor therapy may be needed to achieve better treatment outcomes and also help prevent and overcome AML resistance to existing FLT3 inhibitors.

FLT3 may still be a viable target despite concerns about its centrality in leukemia pathogenesis. High level of wild-type *FLT3* expression occurs in up to 80% of AML patients (Knapper 2007; Stirewalt et al. 2006) and may contribute to worse outcomes (Ozeki et al. 2004; Bullinger et al. 2004) and some wild-type *FLT3* patients can respond to FLT3 inhibitors (Ozeki et al. 2004). Therefore, *FLT3* over expression but mainly activation mutations in *FLT3* are reasonable candidates for targeted therapy. Many potential tyrosine kinase inhibitors in addition to midostaurin have been studied preclinically and clinically in mutant FLT3 AML including sorafenib (Auclair et al. 2007; Rahmani et al. 2005; Rosato et al. 2007; von Bubnoff et al. 2009; Lierman et al. 2007; Metzelder et al. 2009), sunitinib (Yee et al. 2004; O'Farrell et al. 2003; Fiedler et al. 2005; Kancha et al. 2007), lestaurtinib (Smith et al. 2004; Knapper et al. 2006a,b; Levis et al. 2011), tandutinib (Kelly et al. 2002a; Pandey et al. 2002), and quizartinib (Zarrinkar et al. 2009) (Table 10.1).

10.3 Midostaurin (PKC412)

Midostaurin (PKC412) was synthesized as a derivative of staurosporine (CGP41251) and showed selectivity for protein kinase C inhibition (Meyer et al. 1989). PKC412 had an in vitro antiproliferative effect and an in vivo activity against T-24 human bladder carcinoma in a mouse xenograft model (Meyer et al. 1989). The drug affected T-cell function (Alkan et al. 1993) and had anti-solid tumor activity in animal and tissue culture models (Ikegami et al. 1995; Ikegami et al. 1996). Additional laboratory studies demonstrated that PKC412 was able to sensitize cells with the multidrug resistance (MDR) phenotype due to P-glycoprotein to cytotoxic agents (Utz et al. 1994; Budworth et al. 1996). PKC412 also has potent inhibiting activity against class III tyrosine kinases such as VEGF-1, c-Kit, PDGFRα, and β (Fabbro et al. 2000).

The drug was first studied (as a protein kinase C inhibitor) in the context of a phase-1 trial in 32 patients with solid tumors who received doses of 12.5–300 mg daily in 28-day cycles (Propper et al. 2001). The drug could be safely administered; doses of 150 mg/day or lower were well tolerated which corresponded to plasma concentrations 5–10 times higher than the cellular IC_{50}, even accounting for heavy protein binding. Patients received a median of two cycles with a range of 1–12 cycles. The most common treatment-related toxicities were nausea, vomiting, diarrhea, and fatigue. There were significant reductions in circulating lymphocyte and monocyte counts in patients receiving doses ≥ 100 mg/day. Two patients had myelosuppression and two patients with cholangiocarcinoma had stable disease for 4.5 months and partial response, respectively (Propper et al. 2001).

Table 10.1 Selected FLT3 inhibitors

FLT3 inhibitor	Chemical class	Targets	IC_{50}[a] for wt vs. FLT3-ITD (Zarrinkar et al. 2009)	References
Sorafenib BAY 43-9006	Bi-aryl urea	FLT3 B-RAF PDGF VEGFR	3.2 vs. 2.0	(Knapper 2007; Wilhelm et al. 2004)
Sunitinib SU11248	3-substituted indolinone	FLT3 KIT PDGFR VEGFR	9.9 vs. 34	(Knapper 2007)
Midostaurin PKC412	Indolocarbazole	FLT3 PDGFRα and β KIT	15 vs. 13	(Knapper 2007)
Lestaurtinib CEP-701	Indolocarbazole	FLT3 TrkA	2.6 vs. 1.5	(Knapper 2007; Knapper et al. 2006a)
Tandutinib MLN518	Piperazinyl quinazoline	FLT3 KIT PDGFR β	170 vs. 33	(Knapper 2007; Kelly et al. 2002a; Pandey et al. 2002)
Quizartinib AC220	Novel structure	FLT3 KIT CSR1R RET PDGFR	4.2 vs. 1.1	(Zarrinkar et al. 2009)

[a] IC_{50} refers to the concentration required for 50 % inhibition of wild-type (wt) or FLT3-ITD phosphorylation (Zarrinkar et al. 2009)

Preclinical studies reporting the ability of midostaurin to directly inhibit the FLT3 tyrosine kinase, to kill Ba/F3 cell lines expressing mutant FLT3 (Weisberg et al. 2002), and to prevent progression of leukemia in Balb/c mice which were transplanted with bone marrow expressing *FLT3*-ITD via a retrovirus (Brown et al. 2004) ushered in the development of this drug as a FLT3 inhibitor in AML. Subsequent work demonstrating the ability of the FLT3 inhibitor to prolong the survival of mice with model mutant FLT3 myeloid neoplasm (Meyer et al. 1989; Zheng et al. 2004) spurred its development further.

Would midostaurin be able to deal with potential resistance mechanisms in mutant FLT3 AML? Proposed mechanisms of resistance to TKI in AML include FLT3 upregulation, upregulation of FLT3 ligand, and secondary FLT3 mutations. Midostaurin is effective against some of the FLT3 point mutations that are not inhibited by most of the other agents in development. For example, quizartinib treatment of patients with mutant FLT3 selects for the development of TKD mutations biochemically and clinically resistant to the trial drug (Smith et al. 2012). These FLT3 mutations will be sensitive to midostaurin (Barry et al. 2007). However, there are secondary FLT3 mutations that can emerge and account for clinical resistance (Heidel et al. 2006) to midostaurin. The development of such secondary mutations in the presence of an inhibitor provides further evidence that mutant FLT3 is critical in maintaining the survival of AML cells.

10.3.1 Midostaurin: Clinical Trials

Because of the preclinical studies suggesting potent anti leukemic effect in mutant *FLT3* AML, a phase-2 study with PKC412 at 75 mg orally three times a day administered to 20 patients with *FLT3*-mutation (relapsed or refractory AML and one chronic myelomonocytic leukemia) was performed (Stone et al. 2005) (Table 10.2). PKC212 decreased the peripheral blast count in 14 patients. In six patients, PKC412 also reduced the bone marrow blast count by 50 % or more (including three patients with bone marrow blast percentage decrease to <5 %). The median duration of response was 13 weeks with a range of 9–47 weeks (Stone et al. 2005). One patient (with a FLT3-TKD mutation) sustained a near complete response (CR; all CR criteria met except for marrow hypocellularity). The drug was usually well tolerated with most common side effects including grade 1 or 2 nausea or vomiting. Pharmacokinetic studies revealed that the levels of PKC412 decreased by 70 % at steady state after an initial peak plasma concentration. The levels of the active midostaurin metabolite, CGP52541, gradually increased and plateaued at 28 days, whereas the levels of the parent drug declined over this time (Stone et al. 2005). The CGP52541 metabolite is less bound to alpha-1 acid glycoprotein (AAG) and, appears to be less selective and is present at higher plasma concentrations. A second active midostaurin metabolite CGP62221 had similar pharmacokinetics to PKC412 (Stone et al. 2005). Additionally, although some tyrosine kinase inhibitors effect cardiac repolarization and prolong the QTc, it was demonstrated that midostaurin did not exhibit this property in healthy individuals (del Corral et al. 2012).

In a phase-2B study of 95 patients with relapsed, refractory AML or high-risk MDS, 57 wild-type FLT3 and 35 FLT3 mutant patients were evaluated after treatment with midostaurin twice daily with oral doses of 50 or 100 mg (Fischer et al. 2010). Out of 92 evaluated patients, blast reduction (peripheral blood and/or bone marrow) was detected in 71 % of FLT3 mutant patients and in 42 % of FLT3 wild-type patients. The drug was generally well tolerated at these doses, there was no clear dose response, but only one patient (FLT3 mutant AML at 100 mg twice daily dosing) had a partial response (Fischer et al. 2010). The frequent biological activity, but rare CR/PR rate, prompted the combination of midostaurin with other agents active in AML, most notably chemotherapy.

10.3.2 Midostaurin Combination Trials

Preclinical studies demonstrated that midostaurin and other FLT3 inhibitors synergistically killed AML cells in combination with typical chemotherapeutic agents (Tam and Gilliland 2008). Dosing schedule is very important in that pretreatment

Table 10.2 Midostaurin: single agent clinical trials

PKC 412 clinical trial	Patient population	N total, N FLT3 mut	Dosing	Activity	Common toxicity	Reference
Phase 1	Advanced, refractory solid cancers	32	Seven dose levels 12.5–300 mg daily, 28-day cycles	150 mg/day tolerated, SD for 4.5 months and PR on 2 patients with cholangio-carcinoma	Nausea, vomiting, fatigue, diarrhea, anorexia, headache	Propper et al. (2001)
Phase 2	Relapsed, refractory AML, 1 CMML	20, FLT3 mut	75 mg PO TID	Decrease PB blast count in 14, decreased BM blasts in 6 patients by 50 % (including 2 patients with BM blasts <5 %)	Nausea, vomiting	Stone et al. (2005)
Phase 2B	Relapsed, refractory AML, not candidate for IC, high-risk MDS	95 but 92 evaluated; 57 WT; 35 FLT3 mut	50 or 100 mg PO BID	71 % BR FLT3-mutant AML and 42 % BR in FLT3 wt AML. One PR in FLT3-mutant AML at 100 mg.	Mild to moderate nausea and vomiting	Fischer et al. (2010)

SD stable disease, *PR* partial response, *AML* acute myelogenous leukemia, *CMML* chronic myelomonocytic leukemia, *BR* reduction in peripheral blood or bone marrow blasts by ≥ 50 %, *IV* intravenous, *N* number of patients, *IC* induction chemotherapy, *BM* bone marrow, *PB* peripheral blood, *FTL3mut* FLT3 mutant, *WT* wild type

of AML cells with an FLT3 inhibitor could abrogate the effect of cytotoxic chemotherapy (Levis et al. 2004) presumably due to the inability of such drugs to kill non-cycling cells. Thus, several clinical trials have employed midostaurin in combination with chemotherapy in AML. One such trial, the results of which were recently published, explored the feasibility of combining cytarabine and daunorubicin and midostaurin in newly diagnosed adult patients with AML below age 60 (Stone et al. 2012). The phase-1B trial investigated several schedules and doses of midostaurin in combination with induction chemotherapy with cytarabine and daunorubicin and post-remission therapy with high-dose cytarabine. Dose schedules of 100 mg twice daily starting either on day one of chemotherapy (concomitant) or at day, eight relative to the 7-day induction (sequential) given in a prolonged or 14-day schedule were poorly tolerated due to nausea and vomiting. However, the 50 mg twice daily dosing given on cycle days 8–21 or 1–7, and 15–22 was well tolerated; 40 patients received this dose. The CR rate for midostaurin 50 mg PO BID was 80 %, specifically, for FLT3 wild-type patients 20/27 (74 %), and 12/13 (92 %) for those with mutant *FLT3* AML. Overall survival probabilities for patients with *FLT3* mutant AML was 0.85 and 0.62 at 1 and 2 years, respectively, and 0.78 and 0.52 at 1 and 2 years for FLT3 wild-type AML patients (Stone et al. 2012). These results were considered encouraging since the outcome for *FLT3* mutant patients was not inferior to that for wild-type *FLT3* AML, in contrast to most prior studies which have shown an inferior outcome for mutant *FLT3* AML (Kottaridis et al. 2001; Yanada et al. 2005).

Based on the ability to combine midostaurin (50 mg PO BID) with standard chemotherapy, an international prospective randomized phase-3 trial comparing midostaurin versus placebo plus induction, post-remission chemotherapy, and maintenance in adults age 18–60 with mutant *FLT3* AML was undertaken (CALGB 10603, NCT00651261). The study accrued 714 patients with overall survival (no censoring for transplant) as the primary endpoint. The study completed accrual in October 2011 and final analysis is expected in early 2015. Since only patients with mutant *FLT3* AML were eligible and registration had to take place before chemotherapy started, rapid ascertainment of *FLT3* mutational status was accomplished via assays in several reference laboratories worldwide.

In addition to single agent trials in pediatric leukemia, other AML trials have combined midostaurin with novel regimens, for example, with cladribine, ATRA, and GCSF, a combination with acceptable toxicity yielding a maximal tolerated dose of midostaurin of 50 mg PO BID (Ramsingh et al. 2011). In this study, 11 patients were enrolled and 9 were evaluable, of which 3/9 (33.3 %) achieved complete remission/complete remission with incomplete count recovery (CR/CRi). Five patients died (four from progressive disease and one from pulmonary hemorrhage) (Ramsingh et al. 2011). Additional ongoing combination studies include midostaurin with bortezomib (with or without chemotherapy) for relapsed refractory AML and many others (Table 10.3).

Preliminary results of a combination of midostaurin with decitabine in 16 patients > 60 years old with relapsed, refractory AML were recently presented (Williams et al. 2011). Only 2/16 (13 %) patients had a FLT3-ITD mutation. Out of 12 evaluable patients, 10 (83 %) had stable disease or better, 3/12 (25 %) had compete response (CR or CRi) after one cycle. One patient with FLT3-ITD was evaluable for response and achieved CR. The maximum tolerated dose was determined to be decitabine 20 mg/m2 IV daily days 1–5, with midostaurin 50 mg PO BID days 8–21, in 28-day cycles (Williams et al. 2011). Another trial explored inhibition of a pathway downstream to activated FLT3, mTOR/AKT (Mohi et al. 2004), and used a combination of RAD001 plus PKC412 in relapsed refractory or poor prognosis AML or high-risk MDS (Table 10.3).

Table 10.3 Midostaurin: Ongoing clinical trials

Disease	Phase	Agents	Patient age	Trial
AML, FLT3 mut	2	Midostaurin and induction (cytarabine, daunorubicin), consolidation, allo SCT	18–70	NCT01477606
AML	1	Midostaurin and induction (cytarabine, daunorubicin), consolidation	18–60	NCT00093600
AML, relapsed, refractory	1	Midostaurin, cladribine, cytarabine, ATRA, G-CSF	18 or older	NCT01161550
AML, high-risk MDS, wt and mut FLT3	1/2	Midostaurin 50 or 100 mg PO BID	14 and older	NCT00977782
AML, MDS, relapsed, refractory	2	Midostaurin 50 or 100 mg PO BID	18 and older	NCT00045942
AML, high-risk MDS	1/2	Midostaurin and itraconazole	18 and older	NCT00045578
AML	1/2	5-azacitidine and midostaurin	60 and older	NCT01093573
AML and relapsed, refractory AML	1	Midostaurin and decitabine	60 and older	NCT01130662
AML, MDS, relapsed, refractory	1/2	Midostaurin and 5-azacitidine	18 and older	NCT01202877
AML, MDS, relapsed, refractory	1	RAD001 and PKC412	18 and older	NCT01093573
AML, ALL, relapsed, refractory	1	Midostaurin	2 months to 18 years	NCT00866281
AML relapsed, refractory	1	Midostaurin and bortezomib; midostaurin, bortezomib and MEC	18 and older	NCT01174888
AML, MDS, relapsed, refractory	2	Midostaurin	18 and older	NCT00045942
AML	3	Midostaurin vs. placebo with and daunorubicin and cytarabine	18–59	NCT00651261
AML, MDS, relapsed refractory	1/2	Midostaurin	18 and older	NCT00045942
Locally advanced rectal cancer	1	Midostaurin with XRT and 5-FU starting week 2	18 year and older	NCT01282502
Systemic mastocytosis	2	Midostaurin	18 and older	NCT00233454
Systemic mastocytosis and mast cell leukemia	2	Midostaurin	18 and older	NCT00782067
Normal and impaired hepatic function	1	Midostaurin	18–70	NCT01429337

Allo SCT allogeneic stem cell transplant, XRT radiation, 5-FU 5-fluorouracil, wt wild type, MEC mitoxantrone, etoposide, cytarabine

10.4 Mastocytosis

Mast cells from patients with systemic mastocytosis usually carry the KIT D816V mutation which is resistant to imatinib and most other available kinases (Growney et al. 2005) but is responsive to midostaurin. Aggressive systemic mastocytosis is a challenging disease with no standard disease-remitting therapy. In a case report by Gotlib and colleagues, a patient with mast cell leukemia (with an associated myelodysplastic syndrome/myeloproliferative disorder) had a partial response with significant resolution of liver function abnormalities (Gotlib et al. 2005). There was a decline in peripheral blood mast cells and serum histamine level, decrease in KIT phosphorylation, as well as KIT D816 V mutation prevalence. The patient died after 3 months of therapy due to progression to AML (Gotlib et al. 2005).

Preliminary results from a study of 26 patients treated with midostaurin for systemic mastocytosis and mast cell leukemia were promising (Gotlib et al. 2010). Midostaurin at 100 mg PO BID was administered in a continuous 28-day cycle until progression or toxicity. Responses were seen in 69 % (18 of 26 patients), including 10 patients (38 %) with a major response (six incomplete remissions and four clinical responses), 5 patients (19 %) with a good partial response, and three patients with a minor partial response. Moreover, four patients had stable disease and four patients had progressive disease. Midostaurin was given for a median of 15 cycles (1.2 years) for responding patients and was well tolerated. In a preliminary report of a separate compassionate use study including 11 patients with mastocytosis, midostaurin was well tolerated; most patients showed clinical responses with improvement in systemic mediator symptoms and quality of life (Knapper et al. 2011).

Conclusion

Patients with FLT3-ITD-mutated AML have an inferior outcome compared to patients without the mutation. FLT3 inhibitors including midostaurin have shown activity not only in FLT3 mutated but to a lesser extent in wild-type AML. However, midostaurin as a single agent has had limited activity in AML and MDS with resistance mechanisms suspected to include both pharmacokinetic and pharmacodynamic issues. A phase-1B clinical trial showed tolerability and feasibility of induction chemotherapy with cytarabine and daunorubicin in addition to midostaurin. A phase-3 international trial is underway to determine a potential survival benefit for the addition of midostaurin to induction and post-remission chemotherapy in FLT3 mutant AML. Midostaurin's efficacy in mastocytosis may indeed lead to its widespread use in this condition before AML. Approaches to capitalize on FLT3 signaling inhibition include more potent FLT3 inhibitors, combinations with other signal pathway inhibitors, anti FLT3 antibodies, and anti FLT3 ligand agents.

References

Alkan SS et al (1993) Effects of a new protein kinase C inhibitor CGP 41251 on T cell functions: inhibition of activation, growth, and target cell killing. Cell Immunol 150(1):137–148

Auclair D et al (2007) Antitumor activity of sorafenib in FLT3-driven leukemic cells. Leukemia 21(3):439–445

Barry EV et al (2007) Uniform sensitivity of FLT3 activation loop mutants to the tyrosine kinase inhibitor midostaurin. Blood 110(13):4476–4479

Bienz M et al (2005) Risk assessment in patients with acute myeloid leukemia and a normal karyotype.[erratum appears in Clin Cancer Res 2005 Aug 1;11(15):5659]. Clin Cancer Res 11(4):1416–1424

Brown P et al (2004) Pediatric AML primary samples with FLT3/ITD mutations are preferentially killed by FLT3 inhibition. Blood 104(6):1841–1849

Budworth J et al (1996) Comparison of staurosporine and four analogues: their effects on growth, rhodamine 123 retention and binding to P-glycoprotein in multidrug-resistant MCF-7/Adr cells. Br J Cancer 73(9):1063–1068

Bullinger L et al (2004) Use of gene-expression profiling to identify prognostic subclasses in adult acute myeloid leukemia.[see comment]. New Engl J Med 350(16):1605–1616

del Corral A et al (2012) Midostaurin does not prolong cardiac repolarization defined in a thorough electrocardiogram trial in healthy volunteers. Cancer Chemother Pharmacol 69(5):1255–1263

Fabbro D et al (2000) PKC412–a protein kinase inhibitor with a broad therapeutic potential. Anticancer Drug Des 15(1):17–28

Fiedler W et al (2005) A phase 1 study of SU11248 in the treatment of patients with refractory or resistant acute myeloid leukemia (AML) or not amenable to conventional therapy for the disease. Blood 105(3):986–993

Fischer T et al (2010) Phase IIB trial of oral Midostaurin (PKC412), the FMS-like tyrosine kinase 3 receptor (FLT3) and multi-targeted kinase inhibitor, in patients with acute myeloid leukemia and high-risk myelodysplastic syndrome with either wild-type or mutated FLT3. J Clin Oncol 28(28):4339–4345

Frohling S et al (2007) Identification of driver and passenger mutations of FLT3 by high-throughput DNA sequence analysis and functional assessment of candidate alleles. Cancer Cell 12(6):501–513

Gale RE et al (2008) The impact of FLT3 internal tandem duplication mutant level, number, size, and interaction with NPM1 mutations in a large cohort of young adult patients with acute myeloid leukemia. Blood 111(5):2776–2784

Gotlib J et al (2005) Activity of the tyrosine kinase inhibitor PKC412 in a patient with mast cell leukemia with the D816V KIT mutation. Blood 106(8):2865–2870

Gotlib J, De Angelo D, George TI (2010) Kit inhibitor midostaurin exhibits a high rate of clinically meaningful and durable responses in advanced systemic mastocytosis: report of a fully accrued phase II trial. ASH 2011 Annual Meeting Abstracts. Blood 116(21):316a

Growney JD et al (2005) Activation mutations of human c-KIT resistant to imatinib mesylate are sensitive to the tyrosine kinase inhibitor PKC4 Blood 106(2):721–724

Hayakawa F et al (2000) Tandem-duplicated Flt3 constitutively activates STAT5 and MAP kinase and introduces autonomous cell growth in IL-3-dependent cell lines. Oncogene 19(5):624–631

Heidel F et al (2006) Clinical resistance to the kinase inhibitor PKC412 in acute myeloid leukemia by mutation of Asn-676 in the FLT3 tyrosine kinase domain. Blood 107(1):293–300

How J et al (2012) Influence of FLT3-internal tandem duplication allele burden and white blood cell count on the outcome in patients with intermediate-risk karyotype acute myeloid leukemia. Cancer 118:6110–6117

Ikegami Y et al (1995) Antitumor activity of the new selective protein kinase C inhibitor 4'-N-benzoyl staurosporine on murine and human tumor models. Arzneimittelforschung 45(11):1225–1230

Ikegami Y, Yano S, Nakao K (1996) Antitumor effect of CGP41251, a new selective protein kinase C inhibitor, on human non-small cell lung cancer cells. Jpn J Pharmacol 70(1):65–72

Kancha RK et al (2007) Sensitivity toward sorafenib and sunitinib varies between different activating and drug-resistant FLT3-ITD mutations. Exp Hematol 35(10):1522–1526

Kelly LM et al (2002a) CT53518, a novel selective FLT3 antagonist for the treatment of acute myelogenous leukemia (AML). Cancer Cell 1(5):421–432

Kelly LM et al (2002b) FLT3 internal tandem duplication mutations associated with human acute myeloid leukemias induce myeloproliferative disease in a murine bone marrow transplant model. Blood 99(1):310–318

Kim KT et al (2005) Pim-1 is up-regulated by constitutively activated FLT3 and plays a role in FLT3-mediated cell survival. Blood 105(4):1759–1767

Knapper S (2007) FLT3 inhibition in acute myeloid leukaemia. Br J Haematol 138(6):687–699

Knapper S et al (2006a) A phase 2 trial of the FLT3 inhibitor lestaurtinib (CEP701) as first-line treatment for older patients with acute myeloid leukemia not considered fit for intensive chemotherapy. Blood 108(10):3262–3270

Knapper S et al (2006b) The effects of lestaurtinib (CEP701) and PKC412 on primary AML blasts: the induction of cytotoxicity varies with dependence on FLT3 signaling in both FLT3-mutated and wild-type cases. Blood 108(10):3494–3503

Knapper S et al (2011) Midostaurin a multi-targeted oral kinase inhibitor in systemic mastocytosis: report of an Open-Label Compassionate Use Program in the United Kingdom. ASH 2011Annual Meeting Abstracts 118(21):5145a

Kottaridis PD et al (2001) The presence of a FLT3 internal tandem duplication in patients with acute myeloid leukemia (AML) adds important prognostic information to cytogenetic risk group and response to the first cycle of chemotherapy: analysis of 854 patients from the United Kingdom Medical Research Council AML 10 and 12 trials. Blood 98(6):1752–1759

Levis M et al (2004) In vitro studies of a FLT3 inhibitor combined with chemotherapy: sequence of administration is important to achieve synergistic cytotoxic effects. Blood 104(4):1145–1150

Levis M et al (2011) Results from a randomized trial of salvage chemotherapy followed by lestaurtinib for FLT3 mutant AML patients in first relapse. Blood 117(11):3294–3301

Lierman E et al (2007) The ability of sorafenib to inhibit oncogenic PDGFRbeta and FLT3 mutants and overcome resistance to other small molecule inhibitors. Haematologica 92(1):27–34

Mardis ER et al (2009) Recurring mutations found by sequencing an acute myeloid leukemia genome. N Engl J Med 361(11)1058–1066

Metzelder S et al (2009) Compassionate use of sorafenib in FLT3-ITD-positive acute myeloid leukemia: sustained regression before and after allogeneic stem cell transplantation. Blood 113(26):6567–6571

Meyer T et al (1989) A derivative of staurosporine (CGP 41 251) shows selectivity for protein kinase C inhibition and in vitro anti-proliferative as well as in vivo anti-tumor activity. Int J Cancer 43(5):851–856

Mohi MG et al (2004) Combination of rapamycin and protein tyrosine kinase (PTK) inhibitors for the treatment of leukemias caused by oncogenic PTKs. Proc Natl Acad Sci U S A 101(9):3130–3135

Mrozek K et al (2007) Clinical relevance of mutations and gene-expression changes in adult acute myeloid leukemia with normal cytogenetics: are we ready for a prognostically prioritized molecular classification? Blood 109(2):431–448

O'Farrell AM et al (2003) An innovative phase I clinical study demonstrates inhibition of FLT3 phosphorylation by SU11248 in acute myeloid leukemia patients. Clin Cancer Res 9(15):5465–5476

Ozeki K et al (2004) Biologic and clinical significance of the FLT3 transcript level in acute myeloid leukemia. Blood 103(5):1901–1908

Pandey A et al (2002) Identification of orally active, potent, and selective 4-piperazinylquinazolines as antagonists of the platelet-derived growth factor receptor tyrosine kinase family. J Med Chem 45(17):3772–3793

Pratz KW, Levis MJ (2010) Bench to bedside targeting of FLT3 in acute leukemia. Curr Drug Targets 11(7):781–789

Pratz KW et al (2010) FLT3-mutant allelic burden and clinical status are predictive of response to FLT3 inhibitors in AML. Blood 115(7):1425–1432

Propper DJ et al (2001) Phase I and pharmacokinetic study of PKC412, an inhibitor of protein kinase C. J Clin Oncol 19(5):1485–1492

Rahmani M et al (2005) Apoptosis induced by the kinase inhibitor BAY 43-9006 in human leukemia cells involves down-regulation of Mcl-1 through inhibition of translation. J Biol Chem 280(42):35217–35227

Ramsingh G et al (2011) Phase I Study of Cladribine (2-chlorodeoxyadenosie), Cytarabine and G-CSF Based Induction Therapy (CLAG) with ATRA (All-trans retinoic acid) and Midostaurin for Relapsed/Refractory AML. ASH 2011Annual Meeting Abstracts 118(21):3609a

Reindl C et al (2006) Point mutations in the juxtamembrane domain of FLT3 define a new class of activating mutations in AML. Blood 107(9):3700–3707

Rombouts EJ et al (2004) Relation between CXCR-4 expression, Flt3 mutations, and unfavorable prognosis of adult acute myeloid leukemia. Blood 104(2):550–557

Rosato RR et al (2007) The multikinase inhibitor sorafenib potentiates TRAIL lethality in human leukemia cells in association with Mcl-1 and cFLIPL down-regulation. Cancer Res 67(19):9490–9500

Sato T et al (2011) FLT3 ligand impedes the efficacy of FLT3 inhibitors in vitro and in vivo. Blood 117(12):3286–3293

Schlenk RF et al (2008) Mutations and treatment outcome in cytogenetically normal acute myeloid leukemia.[see comment]. New Engl J Med 358(18):1909–1918

Smith BD et al (2004) Single-agent CEP-701, a novel FLT3 inhibitor, shows biologic and clinical activity in patients with relapsed or refractory acute myeloid leukemia. Blood 103(10):3669–3676

Smith CC et al (2012) Validation of ITD mutations in FLT3 as a therapeutic target in human acute myeloid leukaemia. Nature 485(7397):260–263

Stirewalt DL et al (2006) Size of FLT3 internal tandem duplication has prognostic significance in patients with acute myeloid leukemia.[see comment]. Blood 107(9): pp. 3724–3726

Stolzel F et al (2010) Mechanisms of resistance against PKC412 in resistant FLT3-ITD positive human acute myeloid leukemia cells. Ann Hematol 89(7): 653–662

Stone RM et al (2005) Patients with acute myeloid leukemia and an activating mutation in FLT3 respond to a small-molecule FLT3 tyrosine kinase inhibitor, PKC4 Blood 105(1):54–60

Stone RM et al (2012) Phase IB study of the FLT3 kinase inhibitor midostaurin with chemotherapy in younger newly diagnosed adult patients with acute myeloid leukemia. Leukemia 26(9):2061–2068

Tam WF, Gilliland DG (2008) Can FLT3 inhibitors overcome resistance in AML? Best Pract Res Clin Haematol 21(1):13–20

Utz I et al (1994) The protein kinase C inhibitor CGP 41251, a staurosporine derivative with antitumor activity, reverses multidrug resistance. Int J Cancer 57(1):104–110

von Bubnoff N et al (2009) FMS-like tyrosine kinase 3-internal tandem duplication tyrosine kinase inhibitors display a nonoverlapping profile of resistance mutations in vitro. Cancer Res 69(7):3032–3041

Weisberg E et al (2002) Inhibition of mutant FLT3 receptors in leukemia cells by the small molecule tyrosine kinase inhibitor PKC4 Cancer Cell 1(5):433–443

Weisberg E et al (2011) Reversible resistance induced by FLT3 inhibition: a novel resistance mechanism in mutant FLT3-expressing cells. PLoS One 6(9): e25351

Whitman SP et al (2008) FLT3 D835/I836 mutations are associated with poor disease-free survival and a distinct gene-expression signature among younger adults with de novo cytogenetically normal acute myeloid leukemia lacking FLT3 internal tandem duplications.[see comment]. Blood 111(3):1552–1559

Wiernik PH (2010) FLT3 inhibitors for the treatment of acute myeloid leukemia. Clin Adv Hematol Oncol 8(6):429-436

Wilhelm SM et al (2004) BAY 43-9006 exhibits broad spectrum oral antitumor activity and targets the RAF/MEK/ERK pathway and receptor tyrosine kinases involved in tumor progression and angiogenesis. Cancer Res 64(19):7099–7109

Williams CB et al (2011) Results of a phase I open-label study of decitabine in combination with midostaurin (PKC412) for elderly (Age >= 60) newly diagnosed or relapsed/refractory adult patients with acute myeloid leukemia. ASH 2011 Annual Meeting Abstracts 118(21):3610a

Yamamoto Y et al (2001) Activating mutation of D835 within the activation loop of FLT3 in human hematologic malignancies.[see comment]. Blood 97(8):2434–2439

Yanada M et al (2005) Prognostic significance of FLT3 internal tandem duplication and tyrosine kinase domain mutations for acute myeloid leukemia: a meta-analysis. Leukemia 19(8):1345–1349

Yee KW et al (2004) Synergistic effect of SU11248 with cytarabine or daunorubicin on FLT3 ITD-positive leukemic cells. Blood 104(13):4202–4209

Zarrinkar PP et al (2009) AC220 is a uniquely potent and selective inhibitor of FLT3 for the treatment of acute myeloid leukemia (AML). Blood 114(14):2984–2992

Zeng Z et al (2009) Targeting the leukemia microenvironment by CXCR4 inhibition overcomes resistance to kinase inhibitors and chemotherapy in AML.[see comment]. Blood 113(24):6215–6224

Zhang W et al (2008) Mutant FLT3: a direct target of sorafenib in acute myelogenous leukemia. J Natl Cancer Inst 100(3):184–198

Zhao M et al (2000) In vivo treatment of mutant FLT3-transformed murine leukemia with a tyrosine kinase inhibitor. Leukemia 14(3):374–378

Zheng R et al (2004) Internal tandem duplication mutation of FLT3 blocks myeloid differentiation through suppression of C/EBPalpha expression. Blood 103(5):1883–1890

Chapter 11
FLT3 in AML

Naval Daver and Farhad Ravandi

Abstract FMS-like tyrosine kinase 3 (FLT3) receptor and its ligand play a significant role in human hematopoiesis and the proliferation and malignant transformation of primitive hematopoietic cells. *FLT3* mutations are observed in approximately 30 % of adult AML, 1–3 % of adult B-ALL and 2–5 % of MDS patients. Binding of *FLT3* ligand to *FLT3* receptor results in autophosphorylation and activation of downstream pathways that promote cell proliferation including the MAPK/ERK, Pi3K/AKT, signal transducer activator of transcription (STAT5), and BCL2-associated X protein (BAX) pathways. There are two major types of *FLT3* mutations: The *FLT3* internal tandem duplication (ITD) results from duplication and insertion of a fragment of the juxtamembrane domain coding sequence, whereas the less common tyrosine kinase domain (TKD) results from a missense point mutation within the activation loop of the second TKD. Studies suggest that patients with *FLT3*-ITD have significantly elevated peripheral blood white cell counts and increased bone marrow blasts at diagnosis. Furthermore, they have a significantly higher induction death rate, increased relapse risk, inferior event-free survival, and decreased overall survival. Recent studies have further indicated that *FLT3*-ITD mutations may be a significant prognostic indicator in patients with AML and diploid karyotype, but not in those with core-binding factor AML or AML with unfavorable cytogenetics. Also, presence of *FLT3*-ITD mutation may be a negative prognostic marker, not only at diagnosis but also at first relapse. Several small-molecule FLT3 inhibitors are currently in clinical trials and may improve responses for patients with *FLT3*-mutated AML.

Keywords FLT3 mutation · AML · FLT3 inhibitors · FLT3 ligand · FLT3 receptor

---

F. Ravandi (✉) · N. Daver
Department of Leukemia, Unit 428, The University of Texas MD Anderson Cancer Center, 1515 Holcombe Blvd, Houston, TX 77030, USA
email: fravandi@mdanderson.org

M. Andreeff (ed.), *Targeted Therapy of Acute Myeloid Leukemia*,
Current Cancer Research, DOI 10.1007/978-1-4939-1393-0_11

11.1 Introduction

Leukemogenesis results from dysregulation of pathways that regulate proliferation, differentiation, and cell death of hematopoietic cells (Stirewalt and Radich 2003). The *FMS-like Tyrsoine Kinase 3* (FLT3) receptor and its ligand have been shown to play a significant role in human hematopoiesis and the survival, proliferation, and malignant transformation of primitive hematopoietic cells (Lyman et al. 1993a; Zeigler et al. 1994; Hirayama et al. 1995; Hudak et al. 1995; Piacibello et al. 1995). In normal human cells, *FLT3* is predominantly expressed on early myeloid and B-lymphoid progenitors. *FLT3* expression on leukemic cells reflects this expression. Subsequently, high levels of FLT3 receptor expression are seen in 70–100% of cases of AML (Nakao et al. 1996), a vast majority of the cases of progenitor B cell acute lymphoblastic leukemia (B-ALL), some cases of T-ALL, and chronic myeloid leukemia (CML) in lymphoid blast phase (Carow et al. 1996; Rosnet et al. 1996; Drexler 1996). Additionally, activating mutations in *FLT3* are observed in approximately 30% of adult AML patients (Odenike et al. 2011), 1–3% of adult B-ALL patients, (Xu et al. 1999) and 2–5% of myelodysplastic syndrome (MDS) patients (Horiike et al. 1997; Yokota et al. 1997).

11.2 Class II Receptor Tyrosine Kinases

The *W* (dominant White spotting) locus resulting in white spots on the bellies of afflicted mice was first described in the early 1900s (Russell 1979). Subsequently, the W locus was isolated to chromosome 5, and numerous allelic variants involving multiple organ systems were identified. The W locus was found to encode a tyrosine kinase receptor known as c-kit (Geissler et al. 1988; Chabot et al. 1988). This discovery led to the identification of a new group of receptors for signaling molecules called the receptor tyrosine kinase (RTK) family of proteins. RTKs and their signaling ligands play a central role in hematopoiesis (Ullrich and Schlessinger 1990). FLT3 is a member of the class III RTK family (Rosnet et al. 1993a; Small et al. 1994). Other members of this family include receptors for colony-stimulating factor 1 (CSF-1) (Coussens et al. 1986; Rothwell and Rohrschneider 1987; Woolford et al. 1988; Sherr 1990) and steel factor (Williams et al. 1992), respectively, encoded by the *FMS* and *KIT* (Wagner and Alexander 1991) proto-oncogenes and the two receptors for platelet-derived growth factor receptors (PDGFR α and β) (Yarden et al. 1986; Matsui et al. 1989). The RTKs have significant sequence homology. Each of these receptors is approximately 1000 amino acids in length, has five immunoglobulin-like extracellular domains, a transmembrane (TM) domain, a juxtamembrane (JM) domain, and a split intracellular catalytic domain that phosphorylates tyrosine residues in specific target proteins after activation of the receptor by the ligand (Lyman and Jacobsen 1998).

11.2.1 FLT3 Gene

FLT3, also known as FLK-2 (fetal liver kinase-2) and STK-1 (human stem cell kinase-1) was first cloned by Mathews et al. in enriched mouse fetal liver cells in 1991 (Matthews et al. 1991). They described FLT3 as a novel RTK related to the W locus gene product *C-KIT*. They found the FLT3 RTK to be specifically expressed in stem cells and early uncommitted hematopoietic progenitor cells, with no expression in more mature cells. Almost simultaneously, Rosnet et al. also isolated and characterized murine *FLT3*. They found the gene to be expressed in placenta, in various adult tissues including gonads and brain, and in hematopoietic cells in mice (Rosnet et al. 1991). This was followed closely by the cloning and description of human *FLT3* gene in 1993 (Rosnet et al. 1993a). The human *FLT3* gene located on chromosome 13q12 has 85 % amino acid sequence homology with mouse *FLT3* (Rosnet et al. 1993b). The *FLT3* gene is expressed in precursors of both myeloid and B-lymphoid lineage(Rosnet et al. 1996;Brasel et al. 1995;Turner et al. 1996). Expression is usually restricted to early precursors, including CD34$^+$ cells with high levels of CD117 (C-KIT) expression (Rosnet et al. 1996; Rasko et al. 1995). Cells of erythroid and megakaryocytic series do not express *FLT3* gene (Gabbianelli et al. 1995; Ratajczak et al. 1996). The *FLT3* gene is also expressed by placenta, liver, spleen, thymus, and gonads (deLapeyriere et al. 1995; Maroc et al. 1993).

11.2.2 FLT3 Ligand

The *FLT3* ligand (*FLT3L*) was cloned by Lyman et al. in 1993. (Lyman et al. 1993a) The FLT3L is a type 1 TM protein, which contains an amino-terminal signaling peptide, four extracellular helical domains, spacer and tether regions, a TM domain, and a small cytoplasmic domain. (Lyman and Jacobsen 1998; Lyman et al. 1994b) Ligand activity resides in the extracellular domain with the intracytoplasmic domain playing minimal to no role in FLT3 activation and function. (Lyman et al. 1993b) The ligand can also be released as a soluble homodimeric protein. Both the soluble form and the TM form are capable of activating the FLT3 receptor (Hannum et al. 1994; Lyman et al. 1995). The FLT3L is expressed by hematopoietic cells (spleen, thymus, and bone marrow), placenta, lung, colon, kidney, prostate, ovary, testis, and heart. Of note, the brain does not seem to express FLT3L. In the bone marrow microenvironment, FLT3L is expressed predominantly by myeloid cells, B and T lymphoid cells, (Brasel et al. 1995) and bone marrow fibroblasts (Lisovsky et al. 1996a). FLT3L on its own is not efficient at inducing hematopoietic progenitor cell proliferation (Lyman 1995; Lyman et al. 1994a; Ebihara et al. 1997; Ray et al. 1996). However, in synergy with other growth factors, it can act as a potent stimulus for CD34$^+$ progenitor proliferation in vitro and in vivo(Rasko et al. 1995; Rusten et al. 1996; McKenna et al. 1995; McKenna 2001; Maraskovsky et al. 2000; Broxmeyer et al. 1995; Hunte et al. 1996).

11.2.3 FLT3 Receptor

Unlike the c-kit receptor which is ubiquitously distributed and found in high concentration on cells of both normal and malignant tissue, the FLT3 receptor has a more restricted distribution and expression pattern. (Matthews et al. 1991; Rosnet et al. 1991; Hannum et al. 1994; Birg et al. 1992; DaSilva et al. 1994) *FLT3* messenger RNA (mRNA) is found in early hematopoietic cells and in human myeloid leukemia blasts. Turner et al. first characterized FLT3 receptors on normal and leukemic human marrow cells in terms of receptor density, binding affinity, dimerization, and ligand internalization (Turner et al. 1996). In comparison to c-kit, the cells that expressed FLT3 receptors expressed relatively lower number of receptors on the cell surface. The highest expression was found in human myeloid cell lines. This is consistent with prior studies that demonstrated expression of FLT3 receptor mRNA predominantly on cells of myeloid/monocytic and pre-B cell lineage (Brasel et al. 1995; Meierhoff et al. 1995). Human erythroleukemia cells, mast cells, and megakaryocytes which strongly express c-kit receptors (Turner et al. 1995) have little to no FLT3 receptor surface expression (Ratajczak et al. 1996; Lyman et al. 1994a).

FLT3 receptors are usually found as monomers on the plasma membrane in an inactive state. Like other members of the class of proteins (Li and Stanley 1991; Blume-Jensen et al. 1991), ligand binding results in dimerization of the membrane-bound FLT3 and exposure of the TKD phosphoryl acceptor sites (Turner et al. 1996). Dimerization further stabilizes the conformational change of membrane-bound FLT3, resulting in enhanced activation of the receptor (Weiss and Schlessinger 1998). After dimerization, FLT3 dimer-phosphate complex is internalized and either degraded or recycled, whereas the FLT3L is usually degraded. Internalization usually begins at 5 min and reaches a peak at 15 min. Degradation products can be detected as early as 20 min post stimulation (Turner et al. 1996). Thus, similar to other RTKs, the entire process of activation, internalization, and degradation of FLT3 occurs within a matter of minutes.

11.3 FLT3 Signal Transduction

Binding of FLT3L to FLT3 receptor results in autophosphorylation of the receptor and sets off a chain of events resulting in downstream signaling of pathways that promote cell proliferation and inhibit apoptosis. These include the mitogen-activated protein kinase/extracellular signal-related kinase (MAPK/ERK), Pi3K/AKT, signal transducer activator of transcription (STAT5), and BCL2-associated X protein (BAX) (Srinivasa and Doshi 2002; Zhang and Broxmeyer 1999,2000; Zhang et al. 1999, 2000; Lisovsky et al. 1996b). Initial studies on FLT3 signal transduction were done prior to cloning of the FLT3L. These studies used a chimeric receptor containing the extracellular binding domains of human FLT3 with the intracellular catalytic domains of murine FLT3 (Dosil et al. 1993; Rottapel et al.

1994). The chimeric receptor was able to undergo independent activation in the absence of other growth factors. This activation resulted in direct association of the chimeric receptor with Src-homology 2 (SH2) domain of the p85 subunit of phosphatidylinositol 3-kinase (Pi3K), growth factor receptor-bound protein 2 (Grb2), and SH2-domain-containing inositol phosphatase (SHIP) (Marchetto et al. 1999). These associations led to phosphorylation of downstream effectors involved in the Ras-Raf-Mek-Erk pathway (Zhang et al. 1999) and other modulators of hematopoiesis such as Vav (Bustelo et al. 1992).

Cloning of the murine and human FLT3L has greatly enhanced our understanding of the signal transduction process (Lyman et al. 1993a, 1994b). Unlike chimeric FLT3 receptors, the wild-type (WT) human FLT3 did not directly associate with the p85 subunit of PI3K (Zhang and Broxmeyer 1999). Instead, it associates with numerous proteins including GRB2, GAB2, SHIP, SHP2, CBL, and CBLB (CBL-related protein) that subsequently bind to and activate p85 (Zhang and Broxmeyer 1999; Zhang et al. 1999). These proteins in association with FLT3 can stimulate the Ras-Raf-Mek-Erk pathway. Further research supports the hypothesis that the association between FLT3 and RAS pathways may play a crucial role in leukemogenesis (Lisovsky et al. 1996b; Nakagawa et al. 1992). The activated FLT3L–receptor complex can also activate STAT 5a, but not STAT 1 through 4 or STAT 5B. FLT3-mediated STAT activation occurs independent of Jannus Kinase (JAK) (Zhang et al. 2000). Thus, FLT3 plays a crucial role in differentiation, proliferation, and apoptosis of hematopoietic cells.

11.3.1 FLT3 in Immune Function

FLT3 seems to play an important role in immune regulation. Disruption of FLT3 gene in mice results in significant reduction in the number of dendritic cells, natural killer (NK) cells, and myeloid and B-lymphoid progenitors with resultant impairment of the immune system (McKenna et al. 2000). Conversely, FLT3L stimulation promotes dendritic-cell development resulting in the clonal expansion of dendritic cells in the bone marrow, lymph nodes, and peripheral blood (Morse et al. 2000; Manfra et al. 2003). FLT3 in synergy with granulocyte-macrophage colony-stimulating factor (GM-CSF) and interleukin-4 (IL-4) can induce dendritic cell differentiation. Similarly, FLT3 can stimulate IL-2 resulting in proliferation of an early $CD34^+$ NK-cell progenitor (Maraskovsky et al. 1996; Yu et al. 1998). Attempts to harness the dendritic and NK cell proliferative activity of FLT3L for antineoplastic approach have met with mixed success. In murine models, FLT3L has been used successfully to induce tumor shrinkage in murine fibrosarcoma, lymphoma, melanoma, and breast cancer models (Lynch et al. 1997; Esche et al. 1998; Chen et al. 1997). Human studies have yielded mixed results. In a phase 1 human study, 11 patients with cancer (10 breast and 1 ovarian) were vaccinated with HER2/NEU peptide followed by infusion of FLT3L alone or FLT3L in combination with GM-CSF. No significant antitumor effect was noted; however, some patients developed autoimmune manifestations possibly due to FLT3L-mediated expansion of interferon-γ T-cells (Disis et al. 2002).

11.3.2 FLT3 in Normal Hematopoiesis

Activated FLT3, both independently and in concert with other growth factors, plays a significant role in the proliferation and differentiation of primitive hematopoietic cells (Ray et al. 1996; Rusten et al. 1996; Veiby et al. 1996). Surprisingly, targeted disruption of FLT3 in mice resulted in healthy adult mice with relatively normal appearing hematopoietic populations (Mackarehtschian et al. 1995). However, further investigations confirmed deficiencies in pro-B cell and pre-B cell compartments. Also, transplantation experiments have shown that FLT3 nonfunctioning stem cells are unable to reconstitute myeloid and T-lymphoid cell lineages (Mackarehtschian et al. 1995). Similarly, mice with both FLT3 and C-KIT knocked out develop severe hematopoietic deficiencies (Mackarehtschian et al. 1995).

In normal human bone marrow, FLT3 is expressed predominantly on early myeloid and B-lymphoid precursors and fibroblasts (Brasel et al. 1995; Turner et al. 1996). Expression is restricted to early precursors, especially those with high CD34 expression (Rosnet et al. 1996; Rasko et al. 1995). FLT3 can promote differentiation of early hematopoietic precursors in the absence of other growth factors as shown by Gabbiani et al. (Gabbianelli et al. 1995; Rusten et al. 1996). In their experiments, FLT3 stimulation led to monocytic differentiation of hematopoietic receptors with muted proliferative response. However, FLT3-induced differentiation and proliferation of granulocyte-monocyte lineage is greatly enhanced in the presence of co-stimulatory growth factors such as IL-3, KIT-ligand, GM-CSF, G-CSF, and erythropoietin (EPO) (Gabbianelli et al. 1995; Rusten et al. 1996; Shah et al. 1996). Similarly, activated FLT3 in combination with IL-7 stimulates stromal-cell-independent expansion of human fetal pro-B cells and promotes differentiation of pro-B cells to pre-B cells (Namikawa et al. 1996). Furthermore, activated FLT3 in the presence of IL-3, IL-6, and IL-7 stimulates the growth and differentiation of early thymic progenitors of the T-cell lineage (Moore and Zlotnik 1997). As mentioned previously, FLT3 receptors are poorly expressed on erythroid and megakaryocytic precursors and FLT3 stimulation does not play a significant role in the proliferation or differentiation of these lineages (Gabbianelli et al. 1995).

11.3.3 FLT3 in Human Leukemia

Murine *FLT3* was cloned by Matthew et al. in 1991 (Matthews et al. 1991). This was followed closely by analysis and description of FLT3 expression in human leukemias by Berg et al. in 1992 (Birg et al. 1992). They performed northern blot analysis on RNA obtained from patients with AML, ALL, biphenotypic leukemia, and CML. Forty-four AML samples were analyzed to determine *FLT3* expression.

FLT3 gene was expressed in 41 of 44 samples (93 %). This included eight of eight French–American–British (FAB) M1 cases, five of seven M2 cases, four of four M3 cases, nine of nine M4 cases, nine of ten M5 cases, four, two of two of the M7 subtype, and four of four of M0 (undifferentiated forms of AML of M1 subtype). Eight of eight B-ALL and four of six T-ALL samples also expressed FLT3. FLT3 was also expressed in all of eight biphenotypic (hybrid) leukemia samples. Interestingly, *FLT3* transcript was detected in 0 of 11 chronic phase CML samples. In contrast, two of seven accelerated CML and twelve of sixteen blast-phase CML expressed FLT3 transcript (Birg et al. 1992).

Further studies have confirmed that *FLT3* is expressed in 70–100 % of cases of AML and B-ALL (Rosnet et al. 1996; Carow et al. 1996; Meierhoff et al. 1995; Birg et al. 1994; Stacchini et al. 1996). CLL cells and a majority of chronic phase CML phase cells do not express *FLT3*. Similarly, *FLT3* expression is uncommon in T-ALL (Uckun et al. 1997).

11.3.4 FLT3 Mutations in Acute Myeloid Leukemia

As previously mentioned, high levels of FLT3 receptor expression are seen in 70–100 % of cases of AML (Nakao et al. 1996). Nakao et al. first described *FLT3*-ITD in AML in 1996. (Nakao et al. 1996) They observed *FLT3* expression in 22 of 30 (73 %) AML and 39 of 50 (78 %) ALL samples. Five patients with AML showed unexpectedly longer transcripts on primer amplification of the TM through JM domain. On genomic amplification, these patients with abnormal FLT3 transcripts also expressed abnormal longer polymerase chain reaction (PCR) products. Sequence analysis of these abnormal reverse transcriptase-polymerase chain reaction (RT-PCR) products revealed tandem duplication of partial sequences. They concluded that ITD of the *FLT3* gene is a somatic change detected preferentially in AML and may play a significant role in the pathology of AML. Subsequently, other groups of investigators have confirmed these findings (Xu et al. 1999; Horiike et al. 1997; Yokota et al. 1997; Kiyoi et al. 1997; Stirewalt et al. 2001; Meshinchi et al. 2001; Schnittger et al. 2002; Thiede et al. 2002; Iwai et al. 1997, 1999).

FLT3-ITD mutations are found in 15–35 % of patients with AML (Xu et al. 1999; Horiike et al. 1997; Yokota et al. 1997; Kiyoi et al. 1997; Stirewalt et al. 2001; Meshinchi et al. 2001; Schnittger et al. 2002; Thiede et al. 2002; Iwai et al. 1997, 1999; Kottaridis et al. 2001). Pooled data from >5000 newly diagnosed AML patients show an overall *FLT3*-ITD incidence of 23 % (Levis and Small 2003). *FLT3*-ITD mutations have been identified in MDS and ALL, albeit at a much lower frequency (Xu et al. 1999; Horiike et al. 1997; Yokota et al. 1997; Fenaux 2001). *FLT3*-ITD mutations most frequently occur in exons 14 and 15 (previously known as exons 11 and 12). The *FLT3*-ITD results from duplication of a fragment of the JM domain coding sequence and its insertion in a direct head-to-tail orientation. The length of the ITD can vary from 3 to >400 base pairs, but the reading frame is always preserved (Schnittger et al. 2002).

The second most common type of *FLT3* mutation is the TKD mutation, seen in 7–10% of patients with AML (Thiede et al. 2002). This results from a missense point mutation in exon 20 (previously known as exon 17) of the TKD (Thiede et al. 2002; Spiekermann et al. 2002). These include single base substitutions, small deletions, or insertions within the activation loop of the second TKD (Thiede et al. 2002; Yamamoto et al. 2001; Abu-Duhier et al. 2001). These mutations are also called "activation loop mutations" as they cause the tyrosine kinase loop to adopt an activated configuration allowing access to the kinase. Activating mutations have been identified at amino acid position Asp816 of KIT and were subsequently identified at corresponding locations in other RTKs, including RET and MET. In *FLT3*, the corresponding aspartate residue is located at position 835. The most common nucleotide substitution involves substitution of aspartic acid 835 with tyrosine (D835Y). Other less frequent substitutions included Asp835Val, Asp835His, Asp835Glu, and Asp835Asn (Yamamoto et al. 2001). *FLT3*-TKD mutations have been identified in MDS (3%) and ALL (3%) (Yamamoto et al. 2001; Griffin 2001). Usually, the FLT3-ITD and FLT3-TKD mutations occur independently, although combined expression has been identified on rare occasions (Thiede et al. 2002; Meshinchi et al. 2003).

More recently, an additional class of FLT3 mutations has been identified in AML patients in which isoleucine 836 is either deleted (FLT3-Ile836del) or substituted with methionine and arginine (FLT3-Ile836Met + Arg) (Thiede et al. 2002).

Thus, approximately 25–40% of the patients with AML will harbor FLT3 mutations (ITD or TKD), making FLT3 mutation one of the most frequent recurring molecular aberrations in this disease.

11.4 Incidence and Demographics of FLT3-ITD Mutations

FLT3-ITD mutations are clinically relevant in AML. The frequency of FLT3 mutations in pediatric AML is 10–15% (Xu et al. 1999; Meshinchi et al. 2001; Iwai et al. 1999; Kondo et al. 1999), suggesting that these mutations occur more frequently in adults. The frequency seems to increase with age with a higher incidence of *FLT3*-ITD mutations noted in elderly populations (Stirewalt et al. 2001). *FLT3* mutations are found in all FAB subtypes; however, they are found with increased frequency in FAB M5a and M5b and are relatively uncommon in M2 and M6 subtypes (Thiede et al. 2002). *FLT3*-ITD mutations are more frequent in patients with normal cytogenetics, and those with t(15:17) and t(6:9) (Thiede et al. 2002; Slovak et al. 2000). Thiede et al. reported that FLT3 mutations were three times more frequent in patients with normal karyotype. They noted that *FLT3*-ITD mutations were significantly more frequent in patients with t(15:17) and occurred in nearly 30% of the patients harboring this translocation (Thiede et al. 2002). Similarly, other investigators have reported *FLT3*-ITD incidences of 20.3, 28.6, and 36% in patients with t(15:17) (Kiyoi et al. 1997; Kottaridis et al. 2001; Schnittger et al. 2000). However, patients with core-binding factor leukemias (i.e., t(8:21) and inv(16) mutations) have a low incidence of *FLT3*-ITD mutations (Kottaridis et al. 2001). As mentioned previously, FLT3 mutations are occasionally identified in MDS and ALL (Xu et al. 1999; Horiike et al. 1997; Yokota et al. 1997; Fenaux 2001). As regards other hematopoietic malignancies, *FLT3* mutations have been detected in a very small number of cases of CML, adult T-cell ALL, CLL, lymphoma, and multiple myeloma (Xu et al. 1999; Horiike et al. 1997; Iwai et al. 1997). *FLT3* mutations have not been identified in solid tumors (Baldwin et al. 2001) or in normal hematopoietic cells (Ishii et al. 1999).

11.5 Biological Activity of FLT3-ITD Mutations in AML

FLT3-ITD mutations result in ligand-independent dimerization and autophosphorylation resulting in constitutive tyrosine kinase activation and activation of signaling pathways downstream of FLT3 (Hayakawa et al. 2000; Mizuki et al. 2000; Kiyoi et al. 2002). Although the length of the duplicated sequence may vary in humans, it is always in frame and results in elongation of the JM domain. Investigators have shown that *FLT3*-ITD-transfected murine IL-3-dependent cell lines, such as Ba/F3 and 32D, are able to proliferate independent of IL-3 when inoculated in syngeneic mice (Hayakawa et al. 2000; Zhao et al. 2002).

The exact mechanisms by which *FLT3*-ITD mutations promote constitutive activation remain unknown. It is known that binding of FLT3-ligand to extracellular domain of FLT3 results in receptor dimerization and juxtaposition of intracellular cytoplasmic domains. This results in phosphorylation of tyrosine kinase residues on the JM domain and activation of downstream signal-transduction molecules such as SHC, MAP kinase, and STAT5a. In the absence of FLT3-ligand, the kinase activity is suppressed by tyrosine phosphatases that maintain the tyrosine residues in a dephosphorylated state. Investigators have recently elucidated the crystal structure of the auto-inhibited, dephosphorylated form of *FLT3* (Griffith et al. 2004) and shown that, in normal resting state, the JM domain inhibits the autophosphorylation of WT *FLT3* in absence of FLT3-ligand. However, in *FLT3*-ITD-mutated samples, dimerization can occur without the addition of ligand and can activate WT *FLT3* (Kiyoi et al. 2002). Additional experiments support this hypothesis by showing that tyrosine residues within the JM domain were not essential for the signal transduction in the length-mutated FLT3 (Kiyoi et al. 2002).

FLT3-TKD mutations also result in constitutive activation of the FLT3 receptor through stabilization of the activation loop in its open binding configuration. Interestingly, in mouse transplant models, *FLT3*-ITD or *FLT3*-TKD mutations alone are not sufficient to cause overt AML. In fact, *FLT3*-ITD mutations alone induce a myeloproliferative (Kelly et al. 2002) state and *FLT3*-TKD mutations alone induce an oligoclonal lymphoproliferative state (Grundler et al. 2005). Thus, the exact mechanism by which loss of repression and constitutive activation of the FLT3 tyrosine kinase domains ultimately result in leukemogenesis remains poorly understood.

11.6 Clinical Relevance and Prognostic and Predictive Implications of FLT3-ITD Mutations in AML

Initial studies reported conflicting data regarding the prognostic and predictive values of *FLT3*-ITD mutations in AML. An initial study from Japan suggested that *FLT3* mutations did not influence the complete remission (CR) rate, but adversely predicted for overall survival (OS) (Kiyoi et al. 1999). A Dutch study found that *FLT3* mutations were associated with both lower CR rate and increased relapse rate (Rombouts et al. 2000). Subsequently, two large studies examined the clinical impact of *FLT3*-ITD mutations in a well-defined population. The first of these, by Kottaridis et al., included 854 patients, mostly 60 years of age or younger, treated in the United Kingdom Medical Research Council (UKMRC) in 10 and 12 trials (Kottaridis et al. 2001). Patients with *FLT3*-ITD had significantly elevated peripheral blood white cell counts and increased bone marrow blasts at diagnosis. Furthermore, they had a significantly higher induction death rate, increased relapse risk, inferior event-free survival (EFS), and decreased OS. There was a borderline association with a lower complete remission rate ($P=0.05$). Thiede et al. analyzed the clinical and prognostic significance of *FLT3*-ITD and *FLT3*-TKD mutations in 979 patients treated according to the AML-96 multicenter protocol of the German Suddeutsche Hamoblastose group (SHG) (Thiede et al. 2002). Again, both ITD and TKD mutations were associated with elevated leukocyte count and bone marrow blasts. Presence of FLT3 mutations had no association with remission rates. Both ITD and TKD mutations were associated with significantly increased risk for relapse. *FLT3*-ITD mutations had significant association for disease-free survival (DFS) but not for OS, whereas the converse was true for TKD mutations. They also analyzed the effect of FLT3 mutant to WT ratio on outcome. Patients below the age of 60 years with diploid cytogenetics had significantly shorter OS and DFS if their FLT3 ratios were above the preselected threshold median value of 0.78. They had a relative risk of relapse of 1.6. On further sub-classifying patients on the basis of mutant to WT ratio, it was noted that patients with the highest ratio had the worst outcome with a highly significant shorter DFS and OS. Thus, in addition to the analysis for presence or absence of mutation, quantification of the mutant to WT FLT3 alleles may provide additional prognostic information. Whitman et al. noted that patients with *FLT3*-ITD mutation with concomitant *FLT3* allelic loss had a worse prognosis than patients with WT *FLT3* or *FLT3* mutation without allelic loss (Whitman et al. 2001). In addition to quantification of mutant to WT burden, the presence of single versus multiple *FLT3*-ITD mutation variants may be of prognostic importance. Borthakur et al. noted that among patients with AML with diploid karyotype, CR duration was significantly longer in patients who had multiple *FLT3*-ITD mutations ($P=0.03$; Borthakur et al. 2012). However, OS and EFS did not seem to be affected by the number of *FLT3*-ITD mutation variants.

Although *FLT3*-ITD mutations are associated with worse outcomes in AML in general, the precise prognostic impact of these mutations in different cytogenetic subclasses of AML remains poorly understood. In a study by Santos et al., 481 AML patients were divided into three cytogenetic subgroups: good risk/core-binding factor AML, normal karyotype AML and poor risk/unfavorable cytogenetics (Santos et al. 2011). The presence of *FLT3*-ITD mutation did not affect the EFS in patients with core-binding factor AML or poor risk cytogenetics. However, EFS was significantly worse for patients with normal karyotype AML who had the *FLT3*-ITD mutation ($P<0.001$). Thus, the presence of *FLT3*-ITD mutations is a significant prognostic indicator in patients with AML and diploid karyotype, but not in those with core-binding factor AML or AML with unfavorable cytogenetics.

Recent data suggest that the presence of FLT3-ITD mutation may be associated with negative outcomes, not only at diagnosis but also at first relapse. Patients with FLT3-ITD mutations at relapse have a reduced likelihood of achieving second CR and a shortened survival (Ravandi et al. 2010). The feasibility of FLT3-ITD mutations at CR as a marker for minimal residual disease (MRD) has also been investigated. Nazha et al. noted that FLT3-ITD mutations were unstable and lacked consistency at follow-up (Nazha et al. 2012). Furthermore, FLT3 mutations could occur for the first time at relapse. They concluded that FLT3-ITD is not a reliable marker for MRD monitoring in AML. Prognostic impact of FLT3 mutations in the presence of other concomitant mutations also remains poorly understood. Among these, the best studied is the concurrent occurrence of *FLT3*-ITD and *NPM1* mutations. Presence of *FLT3*-ITD in patients with *NPM1* mutations abrogates the positive effects on CR rate and OS associated with mutated NPM1 (Schnittger et al. 2005; Gale et al. 2008).

FLT3-ITD occurs more frequently in older patients with AML but does not adversely affect outcomes. In a large study by Stirewelt et al. *FLT3* mutations were identified in 34% of evaluable elderly patients with AML (Stirewalt et al. 2001). The *FLT3*-ITD mutation was associated with higher absolute white cell counts, higher peripheral blast percentages, normal cytogenetics, and less disease resistance. *FLT3* mutations were not associated with inferior clinical outcomes. It is possible that FLT3 mutations may not be as significant a prognostic marker in the elderly AML population as in the younger patients due to the already poor outcome of the former.

The high frequency and negative impact of FLT3 mutations, particularly the FLT3-ITD mutation, make it an ideal candidate for targeted therapy in AML patients. Several small molecule inhibitors have been developed and are currently in clinical trials. A full discussion of the mechanism of actions, efficacy, and side effects of a number of these drugs are included elsewhere in this publication. There is hope that such targeted therapies will improve responses without increasing toxicity in the treatment of patients with FLT3-mutated AML.

References

Abu-Duhier FM, Goodeve AC, Wilson GA et al (2001) Identification of novel FLT-3 Asp835 mutations in adult acute myeloid leukaemia. Br J Haematol 113:983–988

Baldwin BR, Zheng R, Small D (2001) FLT3 is not frequently mutated in solid tumors. Blood 98:156b–157b

Birg F, Courcoul M, Rosnet O et al (1992) Expression of the FMS/KIT-like gene FLT3 in human acute leukemias of the myeloid and lymphoid lineages. Blood 80:2584–2593
Birg F, Rosnet O, Carbuccia N et al (1994) The expression of FMS, KIT and FLT3 in hematopoietic malignancies. Leuk Lymphoma 13:223–227
Blume-Jensen P, Claesson-Welsh L, Siegbahn A et al (1991) Activation of the human c-kit product by ligand-induced dimerization mediates circular actin reorganization and chemotaxis. EMBO J 10:4121–4128
Borthakur G, Kantarjian H, Patel KP et al (2012) Impact of numerical variation in FMS-like tyrosine kinase receptor 3 internal tandem duplications on clinical outcome in normal karyotype acute myelogenous leukemia. Cancer 118:5819–5822. (Epub ahead of print, May 17, 2012)
Brasel K, Escobar S, Anderberg R et al (1995) Expression of the flt3 receptor and its ligand on hematopoietic cells. Leukemia 9:1212–1218
Broxmeyer HE, Lu L, Cooper S et al (1995) Flt3 ligand stimulates/costimulates the growth of myeloid stem/progenitor cells. Exp Hematol 23:1121–1129
Bustelo XR, Ledbetter JA, Barbacid M (1992) Product of vav proto-oncogene defines a new class of tyrosine protein kinase substrates. Nature 356:68–71
Carow CE, Levenstein M, Kaufmann SH et al (1996) Expression of the hematopoietic growth factor receptor FLT3 (STK-1/Flk2) in human leukemias. Blood 87:1089–1096
Chabot B, Stephenson DA, Chapman VM et al (1988) The proto-oncogene c-kit encoding a transmembrane tyrosine kinase receptor maps to the mouse W locus. Nature 335:88–89
Chen K, Braun S, Lyman S et al (1997) Antitumor activity and immunotherapeutic properties of Flt3-ligand in a murine breast cancer model. Cancer Res 57:3511–3516
Coussens L, Van Beveren C, Smith D et al (1986) Structural alteration of viral homologue of receptor proto-oncogene fms at carboxyl terminus. Nature 320:277–280
DaSilva N, Hu ZB, Ma W et al (1994) Expression of the FLT3 gene in human leukemia-lymphoma cell lines. Leukemia 8:885–888
Disis ML, Rinn K, Knutson KL et al (2002) Flt3 ligand as a vaccine adjuvant in association with HER-2/neu peptide-based vaccines in patients with HER-2/neu-overexpressing cancers. Blood 99:2845–2850
Dosil M, Wang S, Lemischka IR (1993) Mitogenic signalling and substrate specificity of the Flk2/Flt3 receptor tyrosine kinase in fibroblasts and interleukin 3-dependent hematopoietic cells. Mol Cell Biol 13:6572–6585
Drexler HG (1996) Expression of FLT3 receptor and response to FLT3 ligand by leukemic cells. Leukemia 10:588–599
Ebihara Y, Tsuji K, Lyman SD et al (1997) Synergistic action of Flt3 and gp130 signalings in human hematopoiesis. Blood 90:4363–4368
Esche C, Subbotin VM, Maliszewski C et al (1998) FLT3 ligand administration inhibits tumor growth in murine melanoma and lymphoma. Cancer Res 58:380–383
Fenaux P (2001) Chromosome and molecular abnormalities in myelodysplastic syndromes. Int J Hematol 73:429–437
Gabbianelli M, Pelosi E, Montesoro E et al (1995) Multi-level effects of flt3 ligand on human hematopoiesis: expansion of putative stem cells and proliferation of granulomonocytic progenitors/monocytic precursors. Blood 86:1661–1670
Gale RE, Green C, Allen C et al (2008) The impact of FLT3 internal tandem duplication mutant level, number, size, and interaction with NPM1 mutations in a large cohort of young adult patients with acute myeloid leukemia. Blood 111:2776–2784
Geissler EN, Ryan MA, Housman DE (1988) The dominant-white spotting (W) locus of the mouse encodes the c-kit proto-oncogene. Cell 55:185–192
Griffin JD (2001) Point mutations in the FLT3 gene in AML. Blood 97:2193A–2193
Griffith J, Black J, Faerman C et al (2004) The structural basis for autoinhibition of FLT3 by the juxtamembrane domain. Mol Cell 13:169–178
Grundler R, Miething C, Thiede C et al (2005) FLT3-ITD and tyrosine kinase domain mutants induce 2 distinct phenotypes in a murine bone marrow transplantation model. Blood 105:4792–4799

Hannum C, Culpepper J, Campbell D et al (1994) Ligand for FLT3/FLK2 receptor tyrosine kinase regulates growth of haematopoietic stem cells and is encoded by variant RNAs. Nature 368:643–648

Hayakawa F, Towatari M, Kiyoi H et al (2000) Tandem-duplicated Flt3 constitutively activates STAT5 and MAP kinase and introduces autonomous cell growth in IL-3-dependent cell lines. Oncogene 19:624–631

Hirayama F, Lyman SD, Clark SC et al (1995) The flt3 ligand supports proliferation of lymphohematopoietic progenitors and early B-lymphoid progenitors. Blood 85:1762–1768

Horiike S, Yokota S, Nakao M et al (1997) Tandem duplications of the FLT3 receptor gene are associated with leukemic transformation of myelodysplasia. Leukemia 11:1442–1446

Hudak S, Hunte B, Culpepper J et al (1995) FLT3/FLK2 ligand promotes the growth of murine stem cells and the expansion of colony-forming cells and spleen colony-forming units. Blood 85:2747–2755

Hunte BE, Hudak S, Campbell D et al (1996) flk2/flt3 ligand is a potent cofactor for the growth of primitive B cell progenitors. J Immunol 156:489–496

Ishii E, Zaitsu M, Ihara K et al (1999) High expression but no internal tandem duplication of FLT3 in normal hematopoietic cells. Pediatr Hematol Oncol 16:437–441

Iwai T, Yokota S, Nakao M et al (1997) Internal tandem duplication in the juxtatransmembrane domain of the flt3 is not involved in blastic crisis of chronic myeloid leukemia. Leukemia 11:1992–1993

Iwai T, Yokota S, Nakao M et al (1999) Internal tandem duplication of the FLT3 gene and clinical evaluation in childhood acute myeloid leukemia. The Children's Cancer and Leukemia Study Group, Japan. Leukemia 13:38–43

Kelly LM, Liu Q, Kutok JL et al (2002) FLT3 internal tandem duplication mutations associated with human acute myeloid leukemias induce myeloproliferative disease in a murine bone marrow transplant model. Blood 99:310–318

Kiyoi H, Naoe T, Yokota S et al (1997) Internal tandem duplication of FLT3 associated with leukocytosis in acute promyelocytic leukemia. Leukemia Study Group of the Ministry of Health and Welfare (Kohseisho). Leukemia 11:1447–1452

Kiyoi H, Naoe T, Nakano Y et al (1999) Prognostic implication of FLT3 and N-RAS gene mutations in acute myeloid leukemia. Blood 93:3074–3080

Kiyoi H, Ohno R, Ueda R et al (2002) Mechanism of constitutive activation of FLT3 with internal tandem duplication in the juxtamembrane domain. Oncogene 21:2555–2563

Kondo M, Horibe K, Takahashi Y et al (1999) Prognostic value of internal tandem duplication of the FLT3 gene in childhood acute myelogenous leukemia. Med Pediatr Oncol 33:525–529

Kottaridis PD, Gale RE, Frew ME et al (2001) The presence of a FLT3 internal tandem duplication in patients with acute myeloid leukemia (AML) adds important prognostic information to cytogenetic risk group and response to the first cycle of chemotherapy: analysis of 854 patients from the United Kingdom Medical Research Council AML 10 and 12 trials. Blood 98:1752–1759

deLapeyriere O, Naquet P, Planche J et al (1995) Expression of Flt3 tyrosine kinase receptor gene in mouse hematopoietic and nervous tissues. Differentiation 58:351–359

Levis M, Small D (2003) FLT3: ITDoes matter in leukemia. Leukemia 17:1738–1752

Li W, Stanley ER: (1991) Role of dimerization and modification of the CSF-1 receptor in its activation and internalization during the CSF-1 response. EMBO J 10:277–288

Lisovsky M, Braun SE, Ge Y et al (1996a) Flt3-ligand production by human bone marrow stromal cells. Leukemia 10:1012–1018

Lisovsky M, Estrov Z, Zhang X et al (1996b) Flt3 ligand stimulates proliferation and inhibits apoptosis of acute myeloid leukemia cells: regulation of Bcl-2 and Bax. Blood 88:3987–3997

Lyman SD (1995) Biology of flt3 ligand and receptor. Int J Hematol 62:63–73

Lyman SD, Jacobsen SE (1998) c-kit ligand and Flt3 ligand: stem/progenitor cell factors with overlapping yet distinct activities. Blood 91:1101–1134

Lyman SD, James L, Vanden Bos T et al (1993a) Molecular cloning of a ligand for the flt3/flk-2 tyrosine kinase receptor: a proliferative factor for primitive hematopoietic cells. Cell 75:1157–1167
Lyman SD, James L, Zappone J et al (1993b) Characterization of the protein encoded by the flt3 (flk2) receptor-like tyrosine kinase gene. Oncogene 8:815–822
Lyman SD, Brasel K, Rousseau AM et al (1994a) The flt3 ligand: a hematopoietic stem cell factor whose activities are distinct from steel factor. Stem Cells 12 Suppl 1:99–107 (discussion 108–10, 1994)
Lyman SD, James L, Johnson L et al (1994b) Cloning of the human homologue of the murine flt3 ligand: a growth factor for early hematopoietic progenitor cells. Blood 83:2795–2801
Lyman SD, Stocking K, Davison B et al (1995) Structural analysis of human and murine flt3 ligand genomic loci. Oncogene 11:1165–1172
Lynch DH, Andreasen A, Maraskovsky E et al (1997) Flt3 ligand induces tumor regression and antitumor immune responses in vivo. Nat Med 3:625–31
Mackarehtschian K, Hardin JD, Moore KA et al (1995) Targeted disruption of the flk2/flt3 gene leads to deficiencies in primitive hematopoietic progenitors. Immunity 3:147–161
Manfra DJ, Chen SC, Jensen KK et al (2003) Conditional expression of murine Flt3 ligand leads to expansion of multiple dendritic cell subsets in peripheral blood and tissues of transgenic mice. J Immunol 170:2843–2852
Maraskovsky E, Brasel K, Teepe M et al (1996) Dramatic increase in the numbers of functionally mature dendritic cells in Flt3 ligand-treated mice: multiple dendritic cell subpopulations identified. J Exp Med 184:1953–1962
Maraskovsky E, Daro E, Roux E et al (2000) In vivo generation of human dendritic cell subsets by Flt3 ligand. Blood 96:878–884
Marchetto S, Fournier E, Beslu N et al (1999) SHC and SHIP phosphorylation and interaction in response to activation of the FLT3 receptor. Leukemia 13:1374–1382
Maroc N, Rottapel R, Rosnet O et al (1993) Biochemical characterization and analysis of the transforming potential of the FLT3/FLK2 receptor tyrosine kinase. Oncogene 8:909–918
Matsui T, Heidaran M, Miki T et al (1989) Isolation of a novel receptor cDNA establishes the existence of two PDGF receptor genes. Science 243:800–804
Matthews W, Jordan CT, Wiegand GW et al (1991) A receptor tyrosine kinase specific to hematopoietic stem and progenitor cell-enriched populations. Cell 65:1143–1152
McKenna HJ (2001) Role of hematopoietic growth factors/flt3 ligand in expansion and regulation of dendritic cells. Curr Opin Hematol 8:149–154
McKenna HJ, de Vries P, Brasel K et al (1995) Effect of flt3 ligand on the ex vivo expansion of human CD34+ hematopoietic progenitor cells. Blood 86:3413–3420
McKenna HJ, Stocking KL, Miller RE et al (2000) Mice lacking flt3 ligand have deficient hematopoiesis affecting hematopoietic progenitor cells, dendritic cells, and natural killer cells. Blood 95:3489–3497
Meierhoff G, Dehmel U, Gruss HJ et al (1995) Expression of FLT3 receptor and FLT3-ligand in human leukemia-lymphoma cell lines. Leukemia 9:1368–1372
Meshinchi S, Woods WG, Stirewalt DL et al (2001) Prevalence and prognostic significance of Flt3 internal tandem duplication in pediatric acute myeloid leukemia. Blood 97:89–94
Meshinchi S, Stirewalt DL, Alonzo TA et al (2003) Activating mutations of RTK/ras signal transduction pathway in pediatric acute myeloid leukemia. Blood 102:1474–1479
Mizuki M, Fenski R, Halfter H et al (2000) Flt3 mutations from patients with acute myeloid leukemia induce transformation of 32D cells mediated by the Ras and STAT5 pathways. Blood 96:3907–3914
Moore TA, Zlotnik A (1997) Differential effects of Flk-2/Flt-3 ligand and stem cell factor on murine thymic progenitor cells. J Immunol 158:4187–4192
Morse MA, Nair S, Fernandez-Casal M et al (2000) Preoperative mobilization of circulating dendritic cells by Flt3 ligand administration to patients with metastatic colon cancer. J Clin Oncol 18:3883–3893

Nakagawa T, Saitoh S, Imoto S et al (1992) Multiple point mutation of N-ras and K-ras oncogenes in myelodysplastic syndrome and acute myelogenous leukemia. Oncology 49:114–122
Nakao M, Yokota S, Iwai T et al (1996) Internal tandem duplication of the flt3 gene found in acute myeloid leukemia. Leukemia 10:1911–1918
Namikawa R, Muench MO, de Vries JE et al (1996) The FLK2/FLT3 ligand synergizes with interleukin-7 in promoting stromal-cell-independent expansion and differentiation of human fetal pro-B cells in vitro. Blood 87:1881–1890
Nazha A, Cortes J, Faderl S et al (2012) Activating mutations of the FMS-like tyrosine kinase-3 internaltandem duplication (FLT3-ITD) at complete response and relapsein patients with acute myeloid leukemia. Haematologica 97:1242–1245
Odenike O, Thirman MJ, Artz AS et al (2011) Gene mutations, epigenetic dysregulation, and personalized therapy in myeloid neoplasia: are we there yet? Semin Oncol 38:196–214
Piacibello W, Fubini L, Sanavio F et al (1995) Effects of human FLT3 ligand on myeloid leukemia cell growth: heterogeneity in response and synergy with other hematopoietic growth factors. Blood 86:4105–4114
Rasko JE, Metcalf D, Rossner MT et al (1995) The flt3/flk-2 ligand: receptor distribution and action on murine haemopoietic cell survival and proliferation. Leukemia 9:2058–2066
Ratajczak MZ, Ratajczak J, Ford J et al (1996) FLT3/FLK-2 (STK-1) Ligand does not stimulate human megakaryopoiesis in vitro. Stem Cells 14:146–150
Ravandi F, Kantarjian H, Faderl S et al (2010) Outcome of patients with FLT3-mutated acute myeloid leukemia in first relapse. Leuk Res 34:752–756
Ray RJ, Paige CJ, Furlonger C et al (1996) Flt3 ligand supports the differentiation of early B cell progenitors in the presence of interleukin-11 and interleukin-7. Eur J Immunol 26:1504–1510
Rombouts WJ, Blokland I, Lowenberg B et al (2000) Biological characteristics and prognosis of adult acute myeloid leukemia with internal tandem duplications in the Flt3 gene. Leukemia 14:675–683
Rosnet O, Marchetto S, deLapeyriere O et al (1991) Murine Flt3, a gene encoding a novel tyrosine kinase receptor of the PDGFR/CSF1R family. Oncogene 6:1641–1650
Rosnet O, Schiff C, Pebusque MJ et al (1993a) Human FLT3/FLK2 gene: cDNA cloning and expression in hematopoietic cells. Blood 82:1110–1119
Rosnet O, Stephenson D, Mattei MG et al (1993b) Close physical linkage of the FLT1 and FLT3 genes on chromosome 13 in man and chromosome 5 in mouse. Oncogene 8:173–179
Rosnet O, Buhring HJ, Marchetto S et al (1996) Human FLT3/FLK2 receptor tyrosine kinase is expressed at the surface of normal and malignant hematopoietic cells. Leukemia 10:238–248
Rothwell VM, Rohrschneider LR (1987) Murine c-fms cDNA: cloning, sequence analysis and retroviral expression. Oncogene Res 1:311–324
Rottapel R, Turck CW, Casteran N et al (1994) Substrate specificities and identification of a putative binding site for PI3K in the carboxy tail of the murine Flt3 receptor tyrosine kinase. Oncogene 9:1755–1765
Russell ES (1979) Hereditary anemias of the mouse: a review for geneticists. Adv Genet 20:357–459
Rusten LS, Lyman SD, Veiby OP et al (1996) The FLT3 ligand is a direct and potent stimulator of the growth of primitive and committed human CD34+ bone marrow progenitor cells in vitro. Blood 87:1317–1325
Santos FP, Jones D, Qiao W et al (2011) Prognostic value of FLT3 mutations among different cytogenetic subgroups in acute myeloid leukemia. Cancer 117:2145–2155
Schnittger S, Schoch C, Kern W et al (2000) FLT3 length mutations in AML: correlation to cytogenetics, FAB-subtype, and prognosis in 652 patients. Blood 96:826a–826a
Schnittger S, Schoch C, Dugas M et al (2002) Analysis of FLT3 length mutations in 1003 patients with acute myeloid leukemia: correlation to cytogenetics, FAB subtype, and prognosis in the AMLCG study and usefulness as a marker for the detection of minimal residual disease. Blood 100:59–66

Schnittger S, Schoch C, Kern W et al (2005) Nucleophosmin gene mutations are predictors of favorable prognosis in acute myelogenous leukemia with a normal karyotype. Blood 106:3733–3739
Shah AJ, Smogorzewska EM, Hannum C et al (1996) Flt3 ligand induces proliferation of quiescent human bone marrow CD34+CD38- cells and maintains progenitor cells in vitro. Blood 87:3563–3570
Sherr CJ (1990) Colony-stimulating factor-1 receptor. Blood 75:1–12
Slovak ML, Kopecky KJ, Cassileth PA et al (2000) Karyotypic analysis predicts outcome of preremission and postremission therapy in adult acute myeloid leukemia: a Southwest Oncology Group/Eastern Cooperative Oncology Group Study. Blood 96:4075–4083
Small D, Levenstein M, Kim E et al (1994) STK-1, the human homolog of Flk-2/Flt-3, is selectively expressed in CD34+ human bone marrow cells and is involved in the proliferation of early progenitor/stem cells. Proc Natl Acad Sci U S A 91:459–463
Spiekermann K, Bagrintseva K, Schoch C et al (2002) A new and recurrent activating length mutation in exon 20 of the FLT3 gene in acute myeloid leukemia. Blood 100:3423–3425
Srinivasa SP, Doshi PD (2002) Extracellular signal-regulated kinase and p38 mitogen-activated protein kinase pathways cooperate in mediating cytokine-induced proliferation of a leukemic cell line. Leukemia 16:244–253
Stacchini A, Fubini L, Severino A et al (1996) Expression of type III receptor tyrosine kinases FLT3 and KIT and responses to their ligands by acute myeloid leukemia blasts. Leukemia 10:1584–1591
Stirewalt DL, Radich JP (2003) The role of FLT3 in haematopoietic malignancies. Nat Rev Cancer 3:650–665
Stirewalt DL, Kopecky KJ, Meshinchi S et al (2001) FLT3, RAS, and TP53 mutations in elderly patients with acute myeloid leukemia. Blood 97:3589–3595
Thiede C, Steudel C, Mohr B et al (2002) Analysis of FLT3-activating mutations in 979 patients with acute myelogenous leukemia: association with FAB subtypes and identification of subgroups with poor prognosis. Blood 99:4326–4335
Turner AM, Bennett LG, Lin NL et al (1995) Identification and characterization of a soluble c-kit receptor produced by human hematopoietic cell lines. Blood 85:2052–2058
Turner AM, Lin NL, Issarachai S et al (1996) FLT3 receptor expression on the surface of normal and malignant human hematopoietic cells. Blood 88:3383–3390
Uckun FM, Gaynon PS, Sensel MG et al (1997) Clinical features and treatment outcome of childhood T-lineage acute lymphoblastic leukemia according to the apparent maturational stage of T-lineage leukemic blasts: a Children's Cancer Group study. J Clin Oncol 15:2214–2221
Ullrich A, Schlessinger J (1990) Signal transduction by receptors with tyrosine kinase activity. Cell 61:203–212
Veiby OP, Lyman SD, Jacobsen SE (1996) Combined signaling through interleukin-7 receptors and flt3 but not c-kit potently and selectively promotes B cell commitment and differentiation from uncommitted murine bone marrow progenitor cells. Blood 88:1256–1265
Wagner EF, Alexander WS (1991) Of kit and mouse and man. Curr Biol 1:356–358
Weiss A, Schlessinger J (1998) Switching signals on or off by receptor dimerization. Cell 94:277–280
Whitman SP, Archer KJ, Feng L et al (2001) Absence of the wild-type allele predicts poor prognosis in adult de novo acute myeloid leukemia with normal cytogenetics and the internal tandem duplication of FLT3: a cancer and leukemia group B study. Cancer Res 61:7233–7239
Williams DE, de Vries P, Namen AE et al (1992) The Steel factor. Dev Biol 151:368–376
Woolford J, McAuliffe A, Rohrschneider LR (1988) Activation of the feline c-fms proto-oncogene: multiple alterations are required to generate a fully transformed phenotype. Cell 55:965–977
Xu F, Taki T, Yang HW et al (1999) Tandem duplication of the FLT3 gene is found in acute lymphoblastic leukaemia as well as acute myeloid leukaemia but not in myelodysplastic syndrome or juvenile chronic myelogenous leukaemia in children. Br J Haematol 105:155–162
Yamamoto Y, Kiyoi H, Nakano Y et al (2001) Activating mutation of D835 within the activation loop of FLT3 in human hematologic malignancies. Blood 97:2434–2439

Yarden Y, Escobedo JA, Kuang WJ et al (1986) Structure of the receptor for platelet-derived growth factor helps define a family of closely related growth factor receptors. Nature 323:226–232

Yokota S, Kiyoi H, Nakao M et al (1997) Internal tandem duplication of the FLT3 gene is preferentially seen in acute myeloid leukemia and myelodysplastic syndrome among various hematological malignancies. A study on a large series of patients and cell lines. Leukemia 11:1605–1609

Yu H, Fehniger TA, Fuchshuber P et al (1998) Flt3 ligand promotes the generation of a distinct CD34(+) human natural killer cell progenitor that responds to interleukin-15. Blood 92:3647–3657

Zeigler FC, Bennett BD, Jordan CT et al (1994) Cellular and molecular characterization of the role of the flk-2/flt-3 receptor tyrosine kinase in hematopoietic stem cells. Blood 84:2422–2430

Zhang S, Broxmeyer HE (1999) p85 subunit of PI3 kinase does not bind to human Flt3 receptor, but associates with SHP2, SHIP, and a tyrosine-phosphorylated 100-kDa protein in Flt3 ligand-stimulated hematopoietic cells. Biochem Biophys Res Commun 254:440–445

Zhang S, Broxmeyer HE (2000) Flt3 ligand induces tyrosine phosphorylation of gab1 and gab2 and their association with shp-2, grb2, and PI3 kinase. Biochem Biophys Res Commun 277:195–199

Zhang S, Mantel C, Broxmeyer HE (1999) Flt3 signaling involves tyrosyl-phosphorylation of SHP-2 and SHIP and their association with Grb2 and Shc in Baf3/Flt3 cells. J Leukoc Biol 65:372–380

Zhang S, Fukuda S, Lee Y et al (2000) Essential role of signal transducer and activator of transcription (Stat)5a but not Stat5b for Flt3-dependent signaling. J Exp Med 192:719–728

Zhao M, Kiyoi H, Yamamoto Y et al (2002) In vivo treatment of mutant FLT3-transformed murine leukemia with a tyrosine kinase inhibitor. Leukemia 14:374–378

Chapter 12
FLT3-ITD. Clinical (Sorafenib/AC220)

Naveen Pemmaraju and Jorge Eduardo Cortes

Abstract A plethora of new molecular abnormalities has been identified to be associated with acute myelogenous leukemia (AML). These markers have contributed to a more accurate prognostic stratification of these patients, and have nurtured hopes for effective targeted therapies. One such abnormality is the FMS-like tyrosine kinase 3, or FLT3 gene mutation, present in approximately 30 % of AML patients. The presence of the FLT3 internal tandem duplication (FLT3-ITD) confers a poorer prognosis with higher risk of relapse for this subset of AML patients. Rapid development of a large number of FLT3 tyrosine kinase inhibitors (TKIs) has enabled an exciting era of research and treatment for patients with these abnormalities. The expectation is that FLT3 inhibition may offer improvements in outcomes in this poor prognosis group of patients. Early data suggest that this promise may be materializing for patients. This chapter will focus on two of these FLT3 inhibitors, sorafenib (Nexavar®) and AC220 (quizartinib).

Keywords FLT3 · Tyrosine kinase inhibitor · Sorafenib · Quizartinib · AML

12.1 Introduction

Over the past decade, a growing number of molecular abnormalities have been elucidated in the diagnosis and treatment of acute myelogenous leukemia (AML). One of these discoveries, in particular the FMS-like tyrosine kinase 3 (FLT3) mutation, has led to increased research avenues into the biology, pathogenesis, prognostication, and treatment options for patients with AML. The aim of this chapter is to describe the clinical relevance of the FLT3 mutation and the background and development of FLT3 inhibitors from bench to bedside in treatment of AML, and

J. E. Cortes (✉)
Department of Leukemia, Division of Cancer Medicine, The University of Texas MD Anderson Cancer Center, 1515 Holcomb Boulevard Suite R81000, Houston, TX 77030, USA
e-mail: jcortes@mdanderson.org

N. Pemmaraju
Department of Leukemia, Division of Cancer Medicine, The University of Texas MD Anderson Cancer Center, 1515 Holcombe Blvd. Unit 0428, Houston, TX 77030, USA
e-mail: Npemmaraju@mdanderson.org

M. Andreeff (ed.), *Targeted Therapy of Acute Myeloid Leukemia*,
Current Cancer Research, DOI 10.1007/978-1-4939-1393-0_12

to discuss future directions, clinical trials, and translational studies that may help lead to better understanding of the biology and treatment of AML. The focus of this chapter will be to discuss two FLT3 inhibitors now available in clinical trials, sorafenib and AC220 (quizartinib).

12.2 Sorafenib

12.2.1 Drug Development

The development of Sorafenib (Nexavar®, formerly BAY 43-9006, Bayer Pharmaceuticals Co, West Haven, CT) was originally based on the premise of inhibition of the RAS/RAF pathway governing cellular growth pathways. Since the RAS/RAF pathway had been identified as a potentially important pathway for malignant cell growth, a chemical optimization series exploration was initiated and sorafenib was ultimately identified on the basis of its potent kinase inhibition properties against Raf-1. Sorafenib, a member of the bis-aryl urea class of drugs, was designed as an oral medication, and was the first drug of its class to be tested in human clinical trials. Early trials featured pharmacokinetic data that revealed a large amount of variability between patients. Early studies also suggested that dividing the dose of the drug into a twice a day dosing led to an increased AUC and Cmax (Hotte and Hirte 2002; Wilhelm and Chien 2002). The metabolism of the drug is hepatic, primarily mediated via CYP3A4 and UGT1A9 enzymatic pathways. The half-life of sorafenib is approximately 24–48 h, and its bioavailability is approximately 38–49 %. The time to peak plasma concentration in early trials was found to be 2–7 h (Awada et al. 2005; Strumberg et al. 2005). Sorafenib is now recognized as a multikinase inhibitor/anti-angiogenic agent against FLT3, RET, Raf, c-KIT, VEGFR, PDGFR-beta among other targets (Wright and Kapoor 2011; Wilhelm et al. 2004; Carlomagno et al. 2006).

After the completion of two separate multicenter phase III trials in hepatocellular cancer (HCC; SHARP trial) (Llovet et al. 2008) and in renal cell carcinoma (RCC; TARGET trial) (Escudier et al. 2007, 2009), sorafenib gained FDA approval for hepatocellular cancer and renal cell carcinoma, respectively. In the SHARP trial, Llovet and colleagues evaluated 602 untreated HCC patients who had advanced stage disease. Patients were randomized to either sorafenib 400 mg orally twice daily or placebo and primary outcomes were measured as overall survival (OS) and the time to progression of symptomatic disease. The investigators demonstrated a median OS of 10.7 months in the sorafenib-treated cohort versus 7.9 months in the placebo-treated cohort ($p < 0.001$) and no significant difference between the groups in terms of median time to symptomatic progression of disease. Partial response was found in seven (2 %) of sorafenib-treated patients versus two (1 %) patients in the placebo-treated group. No patients on the trial experienced a complete response (Llovet et al. 2008). The TARGET trial was a multicenter phase IIII trial in which Escudier and colleagues evaluated 903 patients with diagnosis of RCC. All patients were previously treated with one systemic therapy prior to enrollment. Patients were randomized to a sorafenib-treated arm (dose 400 mg orally twice daily) versus placebo-treated arm. Crossover was allowed to patients on the placebo arm to the sorafenib arm if progression-free survival (PFS) benefit was demonstrated for sorafenib treatment. The primary end point of this trial was OS assessment with secondary end points including PFS and response rate. The TARGET trial investigators concluded that although the OS analysis was not significant, perhaps due in part to the fact that crossover was allowed, PFS was doubled in the sorafenib-treated group compared to the placebo-treated patients (Escudier et al. 2007, 2009).

Because the Raf/MEK/Erk signaling pathway had been demonstrated to be important and potentially activated in AML cells, the pro-apoptotic qualities of sorafenib were considered for investigation in AML. Sorafenib was shown to result in not only the inhibition of Raf/MEK/Erk signaling in AML cell lines but also the induction of apoptosis in a Bim-dependent fashion in AML patient samples from a phase I clinical trial with sorafenib-treated AML patients (Zhang et al. 2008a).

The prospects for potential use of sorafenib in AML changed with the discovery of the anti-FLT3 activity of sorafenib. In experiments by Auclair et al., investigators demonstrated that sorafenib inhibited FLT3 cellular activity and caused arrest of tumor development in athymic mice harboring FLT3-ITD. In this study, sorafenib resulted in decreased FLT3 signaling and decreased tumor cell growth in both EOL-1 and MV4-11 leukemia cell lines (Auclair et al. 2007). Another group investigating sorafenib's activity against mutated FLT3 found the drug to cause apoptosis in two FLT3-ITD-containing cell lines, MOLM-13 and MV4-11; importantly they showed no such inhibition by sorafenib in four other tested non-FLT3-ITD containing AML cell lines (Lierman et al. 2007). Another study also demonstrated that sorafenib was up to 3000 times more potent in inducing apoptosis and growth inhibition in FLT3-ITD- or D835G-containting Ba/F3 cells compared to FLT3 wild-type cells. Sorafenib was found to have decreased the blast percentage in both bone marrow and peripheral blood of FLT3-ITD AML patient samples when compared to cells from patients with FLT3 wild-type AML (median percent before and after sorafenib for peripheral blood: 81 vs. 7.5 %, $p = 0.16$, and for bone marrow: 75.5 vs. 34 %, $p = 0.05$, respectively; Zhang et al. 2008b) Importantly, this study showed that sorafenib had minimal or no inhibitory activity against normal hematopoietic progenitors. These and other studies led researchers to hypothesize that sorafenib may have clinical benefit specifically for those AML patients harboring the FLT3 mutation.

12.2.2 Sorafenib in AML and Development as FLT3 Inhibitor

12.2.2.1 Phase I Trials: Sorafenib Single-Agent Clinical Trials in AML

In a randomized phase I trial by the National Cancer Institute of Canada (NCIC), 42 patients with relapsed/refractory AML were evaluated at multiple centers. All patients had one prior induction regimen or were 65 years or older with untreated

MDS or secondary AML. In this trial, sorafenib was administered as a single agent, orally in one of the following schedules: (a) 14 days on drug followed by 14 days off drug or (b) 28 days continuously. Four separate doses were tested: 100, 200, 300, or 400 mg orally twice daily. Among 27 patients who were found to have enough messenger RNA (mRNA) available for FLT3 mutation testing, nine were found to have FLT3-ITD and 18 had wild-type FLT3 status. The only patient to achieve a complete remission (CR) in this study was a patient with FLT3-ITD mutation and normal cytogenetics. Overall, three of the nine patients with FLT3-ITD experienced a reduction in bone marrow and peripheral blood blasts while none of the 18 patients in the FLT3-wild type group experienced any significant reduction in blasts (Crump et al. 2010).

A second phase I trial, conducted by Borthakur and colleagues, evaluated 50 patients with relapsed and refractory MDS and AML treated with single-agent sorafenib. Two different dosing schemas were investigated. The first dosing schedule followed a 21-day cycle and included sorafenib administration either once or twice daily, for 5 days per week; the second dosing schedule contained once or twice daily dosing for 2 weeks every 21 days. The authors demonstrated CR or CRi in five (10 %) patients. Responses were usually transient, although one patient had a response lasting 18 months. Importantly, all responding patients demonstrated the presence of FLT3-ITD. In addition, 17 patients (34 %) experienced a marked lowering of bone marrow or peripheral blood blasts, all of them carrying an FLT3-ITD. The authors concluded that the recommended dosing schedule to test in phase II testing for hematologic cancers should be 400 mg orally twice daily and that, at this dose, sorafenib could be given safely and had significant clinical efficacy among patients with relapsed and refractory AML, especially patients harboring the FLT3-ITD mutation (Borthakur et al. 2011).

12.2.2.2 Combination Phase I Clinical Trials

In a phase I study, the combination of sorafenib and vorinostat was tested in a cohort of relapsed/refractory AML patients 70 years of age and older (Sayar et al. 2010). The investigators used a 21-day schedule which featured 14 days of active drug treatment followed by a 7-day rest period. Dosing of sorafenib was 400 mg orally twice daily and vorinostat was dosed in escalating fashion at doses 200, 300, and 400 mg daily in cohorts of three patients in a 3×3 design fashion. Thirteen of 15 patients enrolled were evaluable for response. After the first cycle of therapy, six (46 %) patients experienced a partial response (PR) and one achieved a CR. Two (33 %) of the six patients who achieved PR had FLT3-ITD, with both patients having experienced disappearance of blasts in the peripheral blood in the first week of therapy. One of these patients also demonstrated a lowering in the bone marrow blasts from 90 to 10 % at the end of two cycles. The study established sorafenib 400 mg twice daily and vorinostat 200 mg orally daily for 14 days on and 7 days off on a 21-day schedule to be the optimal dosing schedule, and found the combination to be safe with evidence for clinical activity (Sayar et al. 2010).

The combination of sorafenib, with clofarabine and cytarabine, was recently reported in a pediatric cohort of AML patients with relapsed/refractory disease (Inaba et al. 2011). Twelve patients (ages 6–17 years) were given single agent sorafenib on days 1–7 and continued in 28-day cycles in combination with cytarabine and clofarabine, these given on days 8–12. Sorafenib was administered at two different planned dose schedules: 150 mg/m^2 twice daily and 200 mg/m^2 twice daily. At the higher dose level, two of four patients (50 %) receiving clofarabine 40 mg/m^2 and one of two patients receiving clofarabine at 20 mg/m^2 experienced a grade 3 hand-foot skin rash as dose-limiting toxicity (DLT). The investigators found that by end of first week on therapy (i.e., prior to the start of clofarabine and cytarabine), the blast percent had decreased in 10 of 12 (83 %) patients. With the introduction of the combination therapy, the results ultimately demonstrated six patients with CR, with three of these patients exhibiting FLT3-ITD. In addition, two patients (both with FLT3-ITD) achieved CRi, and one achieved PR (FLT3-WT). The authors concluded that the combination of sorafenib with cytarabine and clofarabine in this pediatric cohort was safe and showed promising clinical activity in patients with relapsed/refractory AML.

12.2.2.3 Phase II Clinical Trials: Sorafenib Combination Therapy

In a single-institution phase I/II clinical trial, Ravandi et al. demonstrated the safety and clinical activity of the combination of sorafenib with standard cytotoxic chemotherapy in the treatment of adult AML patients (Ravandi et al. 2010). In this study, the investigators used sorafenib in combination with idarubicin and cytarabine in patients with AML aged 65 and younger. The initial phase I portion, which included ten patients with relapsed AML, were given sorafenib in escalating dose fashion combined with the standard chemotherapy combination in order to establish the safety of the combination. Once and acceptable dose was identified, the trial proceeded to the phase II portion in which 51 patients were treated. With a median age of 53 years of age, patients with untreated AML received frontline treatment with sorafenib, 400 mg orally twice daily for 7 days, in addition to idarubicin 12 mg/m^2 i.v. daily for 3 days, cytarabine 1.5 g/m^2 continuous infusion i.v. daily for 4 days (or 3 days treatment if aged 60 years or greater). Fifteen patients had mutations in FLT3 (13 were FLT3-ITD, and 2 point mutations in the kinase domain—FLT3-TKD). Four (40 %) of the ten patients included in the phase I portion of the trial achieved CR (three of these patients had FLT3-ITD), with all four of them successfully transitioned to allogeneic stem cell transplantation (SCT) and were alive at time of publication. In the phase II portion of the trial, 38 (75 %) of the 51 patients treated achieved a CR. Importantly, all patients with FLT3-ITD included in the phase II portion of the study achieved CR: 12 full CR and one complete remission with incomplete platelet recovery (CRp). In addition, 2/2 (100 %) patients with FLT3-TKD and 24/36 (66 %) of patients with FLT3-WT also achieved a CR. There was a significant difference in CR rates favoring FLT3-mutated patients ($p=0.33$) compared to those with FLT3-WT. At the time of the publication, with a median follow-up of 62 weeks (range 10–76 weeks), 10/15 FLT3-mutated patients (67%) had experienced relapse and 5/15 (33%) were still in CR(Ravandi et al. 2010).

In a multi-institution phase II trial, investigators studied sorafenib in combination with standard cytotoxic chemotherapy as initial therapy in an unselected population of AML patients age 60 and greater (Serve et al. 2010). The clinical trial was a placebo-controlled, randomized, double-blind study in which 197 patients with AML were randomized to chemotherapy combined with either sorafenib or placebo. Both groups received anthracycline plus cytarabine-based chemo (3 plus 7) at standard doses and up to two cycles of consolidation chemotherapy with intermediate dose cytarabine (six doses at 1 g/m^2). Sorafenib dosing was given as 400 mg orally twice daily during induction and consolidation and at completion of chemotherapy as maintenance for up to 1 year. The comparison of event-free survival (EFS) between the two groups was established as the primary endpoint. Overall, out of the 197 patients, 102 (52%) received chemotherapy plus sorafenib. The results showed no statistically significant difference in EFS and OS between the two groups. In extended exploratory analysis of subgroups, the researchers found FLT3-ITD mutation in 28/197 patients (14%). The subset analysis found no significant difference in EFS, OS, or CR duration in this cohort. The investigators concluded that the combination of sorafenib with standard chemotherapy appeared to be safe and well tolerated in AML patients age 60 and older but they were unable to show any improvement in EFS, OS, or CR duration. As mentioned, however, this was an unselected group of patients with only a small number of patients carrying the FLT3-ITD. These patients are the ones that would be expected to benefit from the addition of sorafenib given the lack of preclinical activity of sorafenib against FLT3 WT AML. Thus, future studies should focus on patients with FLT3-ITD who are most likely to benefit from the addition of sorafenib.

Another approach that has been pursued to enhance the clinical benefit with sorafenib is to explore combinations with non-cytotoxic agents such as hypomethylating drugs. This strategy is of particular interest based on the novel findings of Sato (Sato et al. 2011) that FLT3 ligand levels increased during repeated rounds of cytotoxic chemotherapy and may lead to resistance to FLT3 inhibition. In an attempt to avoid this upregulation of FLT3 ligand, a phase II study with a combination of sorafenib and 5-azacitadine was designed for patients with relapsed or refractory AML (Alattar ML ASCO abstract 2012 abstract 6558). The investigators enrolled patients age 18 years or older with relapsed/refractory AML and administered sorafenib in dose escalation fashion, with the starting dose level at 200 mg orally twice daily continuously with 5-azacitidine at the standard dose of 75 mg/m^2 daily for 7 days, days 1–7 of each cycle. Thirty-two patients have been analyzed thus far. Median age for the group was 61 years (range 24–87 years). Thirty (94%) of the 32 patients treated had FLT3-ITD mutation at baseline. Ten of 32 patients (31%) were not evaluable in this preliminary analysis because they had not reached the 1-month evaluation timepoint. Out of the 22 evaluable patients, the researchers reported that 11 (50%) achieved CR or CRi (two CR, nine CRi) with median CR/CRi duration of 3 months (range < 1–5.5 months). The median time to CR/CRi was 3 months (range 1–4 months). Three patients thus far had gone on to allogeneic SCT at the time the preliminary results were presented. These results offer the possibility of clinical benefit of combination of sorafenib with hypomethylating agents and may provide an alternative to cytotoxic chemotherapy combinations, especially for older FLT3-ITD mutation-positive patients in the future.

12.2.2.4 Sorafenib and SCT in FLT3-ITD-Mutated Patients

Because of the poor prognosis associated with the FLT3-ITD mutation, several groups have investigated the utilization of allogeneic SCT in an attempt to overcome this poor prognosis. In a retrospective review, investigators confirmed that although (SCT) offers the best option for long-term durable remissions for these patients, FLT3-ITD continues to carry an adverse outcome profile despite undergoing allogeneic SCT. These results have triggered an interest in administering FLT3 inhibitors in either pretransplant or posttransplant settings in an attempt to improve the cure potential for these patients after SCT (Brunet et al. 2012).

Safaian et al. reported one instance of a patient with FLT3-ITD-mutated AML with extramedullary disease after allogeneic SCT who echoed a molecular response after treatment with sorafenib (Safaian et al. 2009). Subsequently, sorafenib has been explored in clinical trials both in the pretransplant and posttransplant setting for the treatment of patients with AML FLT3-ITD.

Metzelder and colleagues reported results of the treatment of sorafenib among patients with FLT3-ITD AML on a compassionate use basis (Metzelder et al. 2009). They administered sorafenib to six patients undergoing SCT and showed rapid responses in these six patients in different phases of SCT (two patients prior to SCT, three patients post SCT, and one patient received sorafenib both prior to and post SCT). The investigators found that two of the four patients who received sorafenib after SCT remained in complete molecular remission at the time of publication and that two of the three patients with relapsed disease who were treated with sorafenib were able to proceed to SCT. The authors concluded that single-agent sorafenib therapy demonstrated clinical benefit in the pre- and post-SCT settings and that prospective studies, for both single agent sorafenib and in combination with other agents, were warranted both to facilitate SCT and as maintenance thereafter in FLT3-ITD patients.

In the phase I/II combination study of idarubicin, cytarabine, and sorafenib discussed previously (Ravandi et al. 2010), four of ten patients (40 %) in the phase I portion of the trial achieved a CR, with three of these patients carrying an FLT3-ITD. All four patients who achieved a CR went on to receive SCT and were alive at the time of study publication. In the phase II experience, seven of the 51 patients included received an SCT in first CR with three of these patients exhibiting FLT3-ITD and one with FLT3-TKD. These results suggest feasibility in a subset of AML patients receiving intensive chemotherapy in combination with sorafenib to be able to proceed to SCT with utilization of sorafenib-based therapy.

In a retrospective single-institution review of 16 AML patients with FLT3-ITD mutation positivity who relapsed after SCT, investigators were unable to demonstrate a definable outcomes benefit with use of sorafenib (Sharma et al. 2011). Half of the patients received sorafenib alone and the other half in combination with cytotoxic agents. Although the majority of patients had a decrease in blasts (80 % of patients), none of the patients treated achieved CR. There were three patients with PR and two patients who were able to be bridged to a second SCT; however, both patients experienced relapses quickly after SCT (3 months or less). Limitations to this study included its retrospective nature, small sample of patients, and that only 3/16 of the patients (19 %) were in CR at the time of transplantation.

In contrast, another group examined sorafenib single-agent therapy in the post-SCT setting with a larger number of patients and found clinical activity with sorafenib (Metzelder et al. 2012). Sixty-five AML patients with FLT3-ITD were treated with sorafenib in a multicenter institutional study. Of these patients, 63 (97 %) were chemotherapy refractory/relapsed and had previously received a median of three treatments for AML. Twenty-nine of the 65 patients (45 %) had received prior SCT. For the total population, (SCT and non-SCT groups) results included CR in 5 (8 %) patients, CRi in 15 (23 %), and molecular remission of FLT3-ITD in 10 (15 %) patients. The investigators reported that sorafenib resistance developed less frequently and occurred later in the SCT than in non-SCT group ($p=0.03$). Also, when sorafenib was used after SCT, remissions were more sustained. This analysis suggests the possibility of a synergistic alloimmune effect of sorafenib and allo-SCT that will result in more durable remissions than in sorafenib treated patients who have not undergone SCT and will need to be further tested in future studies.

12.2.3 Future Directions with Sorafenib

12.2.3.1 CXCR4 Inhibition and Sorafenib Combination Therapy

Many groups have elucidated the importance of the stroma-secreted chemokine stromal-derived factor 1 alpha (SDF-1alpha) and the receptor CXCR4 in leukemic cell pathways (Konopleva et al. 2002; Burger et al. 2000; Panayiotidis et al. 1996; Mohle et al. 2000). FLT3 mutations have been demonstrated to activate CXCR4 signaling, and CXCR4 has been shown to be overexpressed in AML FLT3-ITD mutation positive samples (Fukuda et al. 2005). Zeng et al. demonstrated in in vivo studies that CXCR4 inhibition (with AMD3465) either by itself or combined with granulocyte colony-stimulating factor (G-CSF) increased the antileukemia effect of concomitantly administered chemotherapy and sorafenib in animal studies. These experiments demonstrated increased OS in animal studies and suggested the feasibility of pursuing stem cell mobilization/microenvironment disruption with CXCR4 inhibitors and sorafenib in AML patients, especially those harboring FLT3-ITD mutations (Zeng et al. 2009). This work has led to an ongoing phase I/II clinical trial at MD Anderson (PI: Dr. Michael Andreeff) investigating the combination of G-CSF, pleraxifor (a CXCR4 inhibitor), and sorafenib, in FLT3-ITD AML.

12.2.3.2 Mechanisms of FLT3 Resistance

Although sorafenib has indeed resulted in the achievement of reduction of blasts and ultimately CR/CRi for a subset of AML patients, as discussed in this chapter, most frequently, these responses have been transient. Therefore, an ongoing area of investigation in FLT3 inhibitor research is the mechanisms of resistance to FLT3 inhibitors. Some studies have shown that inhibition of FLT3 by blocking FLT3 autophorphylation with an FLT3 inhibitor does not reliably lead to leukemic cell death in FLT3-ITD samples, and that relapsed samples often carry higher FLT3 mutant allele burdens at relapse (Pratz et al. 2010). A recent study highlighted the biological–clinical problem of sorafenib resistance (Man et al. 2012). In this study, although 12 of 13 patients (92 %) with relapsed/refractory AML achieved CR/CRi with sorafenib, FLT3-ITD remained detectable in the majority of patients. This is in contrast to cytotoxic chemotherapy-treated patients who often will have FLT3-ITD disappearance. Patients treated with sorafenib ultimately lose their response and the mechanisms driving this secondary resistance are only now starting to be elucidated. Recent studies suggest that FLT3 signaling remains intact in most of these patients, indicating a possibility of an alternate or escape pathway for resistance. Gene expression profiling was performed in one study, and researchers found several mutations that had developed in these sorafenib-resistant samples, including potentially newly implicated genes JAK3, MMP15, ALDH1A1, and the most common point mutation in FLT3 mutated patients, the emergence of the D835 point mutation. Further studies will need to be performed on sorafenib-treated FLT3-ITD AML patients to continue to investigate the clonal evolution and molecular mechanisms of resistance to sorafenib and other FLT3 inhibitors.

12.3 AC220

12.3.1 Drug Development of AC220 as FLT3 Inhibitor

AC220, a tyrosine kinase inhibitor (TKI), now known as quizartinib, is a highly potent inhibitor of FLT3 and other kinases including CSF1R, RET, PDGFR, and KIT (Pemmaraju et al. 2011; Chao et al. 2009). AC220 was initially identified and designed by Chao et al. (Chao et al. 2009) as "Compound 7," named *N*-(5-tert-butyl-isoxazol-3-yl)-*N*-{4-[7-(2-morpholin-4-yl-ethoxy)imidazo[2,1-b][1,3]benzothiazol-2-yl]phenyl}urea dihydrochloride. This group was able to synthesize a series of very active FLT3 inhibitor agents that featured excellent oral bioavailability. The first compound in the series, "Compound 1" (AB-530) was found to be effective, as it induced tumor regression in a mouse xenograft model with human leukemia cell line MV4-11 cells (Patel et al. 2009). However, the oral PK qualities of this drug at the higher doses tested did not lend itself to proceed with clinical development as this formulation demonstrated suboptimal solubility, deemed inadequate for a clinical agent. By removing the carboxamide group in "compound 1" and with the addition of a solubilizing water moiety, the investigators were able to overcome the PK shortcomings. These changes led to the successful development of several FLT3 inhibitors, ultimately leading to the design of AC220.

12.3.1.1 Binding Affinity, Inhibition of Proliferation, and Compound Series Optimization

Chao and colleagues analyzed this series of compounds for their binding affinity to the FLT3 catalytic domain. They utilized an assay that featured competition binding with an ATP-competitive FLT3 inhibitor that was bound to a solid surface. The compounds were indeed confirmed to have a high degree of binding to the FLT3 kinase.

In order to measure the inhibition of leukemic cell proliferation, the series of compounds was studied against the FLT3-ITD mutation positive MV4-11, an FLT3-dependent human leukemia cell line. Most of the series of compounds demonstrated more potency in the cell proliferation assays than in the in vitro binding assays; notably, sunitinib has an approximately tenfold less potent activity in cellular proliferation assays than binding assays. For reference, sorafenib appears to reveal in vitro properties closer to this series of compounds with $K_d = 13$ nM and cell $IC_{50} = 0.87$ nM (Karaman et al. 2008; Zarrinkar et al. 2009).

12.3.1.2 Development of Kinase Selectivity

With the aid of KinomeScan sequencing (Fabian et al. 2005), this series of compounds was evaluated for the selectivity of kinase inhibition against a series of 227 kinases. Each compound screened was assigned a coding for the level of selectivity that it conferred, which was derived by the kinase targets of each compound divided by the overall number of kinases that were contained in the panel. This yielded a selectivity score, with the lowest scores indicating a more selective compound for FLT3 inhibition (Karaman et al. 2008). AC220 was found to have a very selective rating, a score of 0.079 in the 227 kinase panel testing which was consistent with a fairly specific target inhibition of FLT3. This scoring ranked AC220 to be more selective than sunitinib (selectivity score 0.47) and also more selective than other FLT3 inhibitors in development, notably midostaurin and lestarutainib (Karaman et al. 2008; Zarrinkar et al. 2009).

12.3.2 Activity of AC220 as FLT3 Inhibitor

Based on preclinical work demonstrating the activity of AC220 in AML cell lines, peripheral blasts of an FLT3-ITD AML patient were tested. These blasts were treated with AC220 for 1 h. FLT3 autophosphorylation was potently inhibited after exposure to AC220. Four additional patient samples were also tested, with all five samples revealing comparable FLT3 inhibition and AML primary cell inhibition results in line with several other FLT3 inhibitors (Zarrinkar et al. 2009). This study suggested that AC220 might represent a more specific, more potent inhibitor of FLT3 compared to several other FLT3 inhibitors already in development. Thus, these results indicated that AC220 would be a desirable agent to develop as an AML therapeutic agent.

12.3.2.1 Phase I Clinical Trial Results

In a multi-institutional phase I study, Cortes et al. presented the results of a first-in-human experience with treatment of AC220 in relapsed/refractory AML patients (Cortes et al. 2009). The trial was conducted in a 3 plus 3 standard dose escalating design. Seventy-six patients, 46 of them male, were treated with AC220. Dose escalations in increments of 50% were planned. AC220 was administered orally, once daily, as an oral solution. Dosing schedule was intermittent with drug given on days 1–14 on, 14 days off, with one cycle constituting 28 days on therapy. In this intermittent dosing schema, with a starting dose of 12 mg/day, the dose was incrementally escalated up to 450 mg/day. In addition, there was evaluation of a continuous dosing schedule at doses of 200 and 300 mg daily, administered continuously.

The median age of the study population was 60 years (range 23–86 years). Twelve patients had received a previous SCT. FLT3-ITD mutation was present in 18 of 76 (24%) patients treated. The adverse event profile was favorable, with the most common events being gastrointestinal (GI) disturbances, edema, and changes in taste, all of them grade 2 or below, transient and manageable. DLT was prolongation of QTc and established at 300 mg daily continuous dosing schedule, with the MTD established at 200 mg po daily. Clinical activity was observed at several dose levels. Of note, most of the responses appeared to occur quickly, within first cycle of therapy. Nine (12%) patients achieved CR (5 CRi, 2CRp, 2 CR) and another 14 (18%) experienced PR. Among the FLT3-ITD patients, 10 of 18 (56%) achieved a response (1 CR, 3 CRi, 6PR). In the 200 mg continuous dosing group, four of six patients with FLT3-ITD that were included responded (1 CR, 1 CRi, 1 CRp, 1 PR). Interestingly, two of the four responders had already failed prior treatment with an FLT3 inhibitor (sorafenib in both cases). This trial demonstrated safety and efficacy for AC220 as single agent therapy in relapsed/refractory AML, particularly among patients with FLT3-ITD. This led to the development of a phase II trial of AC220 in patients with refractory-relapsed AML.

12.3.2.2 Phase II Clinical Trial Results

Based on the promising results of the phase I trial discussed above, Cortes et al. initiated an open-label phase II clinical trial to further investigate the efficacy of AC220 as single agent therapy in AML (Cortes et al. 2011). Given the drug's FLT3 inhibitory activity in preclinical models, and its clinical benefit in FLT3-ITD-mutated AML patients from the phase I experience, the investigators focused the phase

II trial on patients with relapsed/refractory AML carrying an FLT3-ITD mutation. Two cohorts were defined: one for age 60 or greater and who had experienced failure after first-line chemotherapy, and a second cohort for patients age 18 years and older, with relapsed status after two lines of chemotherapy or after SCT. The study included two planned groups: (1) an exploratory group of 62 patients and (2) a confirmatory group of 240 patients. To date, only the results of the 62-patient exploratory cohort have been reported. In this group, criteria for being evaluable for analysis of efficacy included FLT3-ITD status confirmed by central laboratory analysis, patient being administered at least one dose of AC220, a posttreatment assessment of response, and no major protocol deviations. A total of 53 patients were evaluable based on these criteria. The results demonstrated 24 of these 53 patients (45 %) achieving a CR/CRp/CRi (CRi 22, CRp 2). In addition, PR was achieved in 13 (24 %) patients. In addition, 2 of 25 (8 %) patients in Cohort 1 and 11 of 37 (30 %) in Cohort 2 were able to proceed to SCT; these patients were censored for assessment of CR duration. The adverse event profile revealed most prominently QTc prolongation in 21 patients (34 %) and it was grade 3 in 11 (18 %) patients. Importantly, this was transient and reversed with temporary drug interruption and dose reductions. Other adverse events included nausea/vomiting, fatigue, taste changes, febrile neutropenia, anorexia, and GI disturbances; these were usually mild and manageable. In view of the presence of QTc prolongation and the observed correlation of frequency of this adverse event with gender, the study was amended to change the starting dose from 200 mg daily to 135 mg in males and 90 mg in females. The authors concluded that these results confirm the significant clinical activity of AC220 in patients with FLT3 ITD AML with a favorable toxicity profile warranting further development of this agent.

12.3.3 AC220: Future Directions

12.3.3.1 Combination Studies of AC220 with Cytotoxic Chemotherapy

Preclinical data examining the role of AC220 in combination with standard AML cytotoxic chemotherapy has been reported. In a recent study, the investigators demonstrated a potential role for daunorubicin and cytarabine to be given in combination with AC220 (Belli et al. 2011) in which the authors treated animals with MV4-11 homozygous FLT3-ITD AML cells with this combination. The authors found that there is enhancement of clinical activity of the cytotoxic chemotherapy combination with addition of AC220 in this FLT3-ITD-containing model system, in a dosing pattern of intermittent administration after cytotoxic chemotherapy or continuous dosing of AC220 with cytotoxic chemotherapy. Based on these data, investigators are now exploring an ongoing phase I clinical trial of daunorubicin, cytarabine, and AC220 as initial therapy for AML among patients aged 60 and younger. This study will help us to better understand the role of AC220 given in combination with standard cytotoxic AML therapy.

12.3.3.2 Emergence of AC220-Resistant FLT3 Kinase Domain Mutations

The development of AC220-resistant FLT3 clones is beginning to be observed and studied. In one study, an isogenic BaF3 cell line model system was created. This cell line was transfected with various isoforms of KIT, FLT3, or BCR/ABL and was developed to better understand the differences in mutant isoforms with regard to activity of AC220 (Kampa-Schittenhelm et al. 2010). While potent inhibition of FLT3 was observed as expected, as in other FLT3 inhibitors, data reveal the development of unique resistance pattern towards some domain 2 mutations, and the authors suggest possibly checking mutational screening analysis to be performed prior to FLT3 inhibition therapy. These concepts will need to be further explored prospectively and in clinical trials. Further, confirming the emergence of AC220-specific resistant FLT3 mutations, it has been demonstrated that there is indeed a pattern of resistance-conferring kinase domain mutations demonstrated in vitro (Fiskus et al. 2011; Smith et al. 2011a). Mutations at positions F691 and D835 have been found in 9/9 FLT3-ITD positive patients who experienced relapse after bone marrow blast clearance (Smith et al. 2011a; Perl et al. 2011). Importantly, these AC220-resistant mutations conferred cross-resistance to sorafenib. Studies have shown that the gatekeeper and activation loop substitutions in FLT3-ITD are leading to the cross-resistance patterns emerging in FLT3 inhibitor-treated patients to date. Ponatinib, an ABL and FLT3 inhibitor, appears to have in vitro efficacy against FLT3 F691, the "gatekeeper" mutation; however, other mutational configurations appear to confer resistance in vitro to ponatinib as well, suggesting that AC220 resistance will present a multifactorial mutational panel that will warrant better characterization and more specific profiling (Smith et al. 2011b).

In an effort to identify future targets for drug development and to better characterize the biology of AC220 resistant mutations, investigators aimed to delineate specific point mutations within the kinase domain (Smith et al. 2012, 2011c). Using patient samples from the ongoing phase II AC220 single-agent clinical trial (discussed above), researchers identified, via in vitro saturation mutagenesis assay, AC220 resistance mutations were found at four residues in the FLT3 kinase domain: the "gatekeeper" residue (F691), two activation loop residues (D835 and Y842), and E608K (which was found to not confer a high level of resistance and was not investigated further in this analysis). Notably, one novel mutation was elucidated—D835F. This was found in one patient sample and featured both AC220 resistance and sorafenib cross-resistance. The researchers concluded that the possibility of emergence of mutations at F691 and D835 will create higher risk situations for FLT3-ITD patients and ultimately elaborate resistance to both AC220 and sorafenib. Current research is focusing on developing new agents that may overcome these mutations and discovering novel mechanisms of resistance to FKT3 inhibitors.

Conclusion

The standard backbone of treatment for AML (i.e., anthracycline- and cytarabine-based therapies) has remained largely unchanged over the past 40 years, and outcomes remain overall poor in this hematologic malignancy. Therapeutic strategies inhibiting the kinase activity of FLT3 in patients with AML with FLT3 mutations have yielded promising results thus far, as this approach represents one of the first attempts at rational, biologically based personalized treatment in patients with AML. However, significant challenges still remain as most of the responses are transient and mechanisms of resistance need to be overcome and discovered. The optimal ways to combine FLT3 inhibitors with other agents need to be investigated and is likely to vary from patient to patient. Further translational and clinical research is warranted to continue to investigate FLT3 inhibitors in the clinic with particular attention to the biology of FLT3 and better characterization of the various mechanisms of FLT3 resistance. Still, FLT3 inhibitors are likely to become an important tool in our quest to conquer AML.

References

Auclair D, Miller D, Yatsula V, Pickett W, Carter C, Chang Y et al (2007) Antitumor activity of sorafenib in FLT3-driven leukemic cells. Leukemia 21(3):439–445

Awada A, Hendlisz A, Gil T, Bartholomeus S, Mano M, de Valeriola D et al (2005) Phase I safety and pharmacokinetics of BAY 43-9006 administered for 21 days on/7 days off in patients with advanced, refractory solid tumours. Br J Cancer 92(10):1855–1861

Belli BA, Dao A, Wierenga W, Armstrong RC (2011) AC220, a potent and specific FLT3 inhibitor, enhances the activity of combined cytarabine and daunorubicin chemotherapy in a FLT3-ITD Model of AML. ASH Annual Meeting Abstracts 118(21):1538. 10–13 Dec 2011

Borthakur G, Kantarjian H, Ravandi F, Zhang W, Konopleva M, Wright JJ et al (2011) Phase I study of sorafenib in patients with refractory or relapsed acute leukemias. Haematologica 96(1):62–68

Brunet S, Labopin M, Esteve J, Cornelissen J, Socie G, Iori AP et al (2012) Impact of FLT3 internal tandem duplication on the outcome of related and unrelated hematopoietic transplantation for adult acute myeloid leukemia in first remission: a retrospective analysis. J Clin Oncol 30(7):735–741

Burger JA, Tsukada N, Burger M, Zvaifler NJ, Dell'Aquila M, Kipps TJ (2000) Blood-derived nurse-like cells protect chronic lymphocytic leukemia B cells from spontaneous apoptosis through stromal cell-derived factor-1. Blood 96(8):2655–2663

Carlomagno F, Anaganti S, Guida T, Salvatore G, Troncone G, Wilhelm SM et al (2006) BAY 43-9006 inhibition of oncogenic RET mutants. J Natl Cancer Inst 98(5):326–334

Chao Q, Sprankle KG, Grotzfeld RM, Lai AG, Carter TA, Velasco AM et al (2009) Identification of N-(5-tert-butyl-isoxazol-3-yl)-N′-{4-[7-(2-morpholin-4-yl-ethoxy)imidazo[2,1-b][1,3]benzothiazol-2-yl]phenyl}urea dihydrochloride (AC220), a uniquely potent, selective, and efficacious FMS-like tyrosine kinase-3 (FLT3) inhibitor. J Med Chem 52(23):7808–7816

Cortes J, Foran J, Ghirdaladze D, DeVetten MP, Zodelava M, Holman P et al (2009) AC220, a potent, selective, second generation FLT3 receptor tyrosine kinase (RTK) inhibitor, in a first-in-human (FIH) phase 1 AML study. ASH Annual Meeting Abstracts 114(22). 5–8 Dec 2009

Cortes JE, Perl AE, Smith CC, Kovacsovics T, Dombret H, Dohner H et al (2011) A phase II open-label, AC220 monotherapy efficacy study in patients with refractory/relapsed FLT3-ITD positive acute myeloid leukemia: updated interim results. ASH Annual Meeting Abstracts 118(21). 10–13 Dec 2011
Crump M, Hedley D, Kamel-Reid S, Leber B, Wells R, Brandwein J et al (2010) A randomized phase I clinical and biologic study of two schedules of sorafenib in patients with myelodysplastic syndrome or acute myeloid leukemia: a NCIC (National Cancer Institute of Canada) clinical trials group study. Leuk Lymphoma 51(2):252–260
Escudier B, Eisen T, Stadler WM, Szczylik C, Oudard S, Siebels M et al (2007) Sorafenib in advanced clear-cell renal-cell carcinoma. N Engl J Med 356(2):125–134
Escudier B, Eisen T, Stadler WM, Szczylik C, Oudard S, Staehler M et al (2009) Sorafenib for treatment of renal cell carcinoma: final efficacy and safety results of the phase III treatment approaches in renal cancer global evaluation trial. J Clin Oncol 27(20):3312–3318
Fabian MA, Biggs WH, 3rd, Treiber DK, Atteridge CE, Azimioara MD, Benedetti MG et al (2005) A small molecule-kinase interaction map for clinical kinase inhibitors. Nat Biotechnol 23(3):329–336
Fiskus W, Smith CC, Smith J, Wise SC, Lasater E, Damon LE et al (2011) Activity of allosteric, switch-pocket, ABL/FLT3 kinase inhibitor DCC2036 against cultured and primary AML progenitors with FLT-ITD or FLT3 kinase domain mutations. ASH Annual Meeting Abstracts 118(21):2611. 10–13 Dec 2011
Fukuda S, Broxmeyer HE, Pelus LM (2005) FLT3 ligand and the FLT3 receptor regulate hematopoietic cell migration by modulating the SDF-1alpha(CXCL12)/CXCR4 axis. Blood 105(8):3117–3126
Hotte SJ, Hirte HW (2002) BAY 43–9006: early clinical data in patients with advanced solid malignancies. Curr Pharm Des 8(25):2249–2253
Inaba H, Rubnitz JE, Coustan-Smith E, Li L, Furmanski BD, Mascara GP et al (2011) Phase I pharmacokinetic and pharmacodynamic study of the multikinase inhibitor sorafenib in combination with clofarabine and cytarabine in pediatric relapsed/refractory leukemia. J Clin Oncol 29(24):3293–3300
Kampa-Schittenhelm KM, Trikha M, Schlenk RF, Doehner H, Doehner K, Heinrich MC et al (2010) AC220, a potent second generation class I/III tyrosine kinase inhibitor, displays a distinct inhibition profile on mutant-FLT3 as well as -KIT isoforms. ASH Annual Meeting Abstracts 116(21):291. 4–7 Dec 2010
Karaman MW, Herrgard S, Treiber DK, Gallant P, Atteridge CE, Campbell BT et al (2008) A quantitative analysis of kinase inhibitor selectivity. Nat Biotechnol 26(1):127–132
Konopleva M, Konoplev S, Hu W, Zaritskey AY, Afanasiev BV, Andreeff M (2002) Stromal cells prevent apoptosis of AML cells by up-regulation of anti-apoptotic proteins. Leukemia 16(9):1713–1724
Lierman E, Lahortiga I, Van Miegroet H, Mentens N, Marynen P, Cools J (2007) The ability of sorafenib to inhibit oncogenic PDGFRbeta and FLT3 mutants and overcome resistance to other small molecule inhibitors. Haematologica 92(1):27–34
Llovet JM, Ricci S, Mazzaferro V, Hilgard P, Gane E, Blanc JF et al (2008) Sorafenib in advanced hepatocellular carcinoma. N Engl J Med 359(4):378–390
Man CH, Fung TK, Ho C, Han HH, Chow HC, Ma AC et al (2012) Sorafenib treatment of FLT3-ITD+ acute myeloid leukemia: favorable initial outcome and mechanisms of subsequent nonresponsiveness associated with a D835 mutation. Blood 119(22):5133–5143
Metzelder S, Wang Y, Wollmer E, Wanzel M, Teichler S, Chaturvedi A et al (2009) Compassionate use of sorafenib in FLT3-ITD-positive acute myeloid leukemia: sustained regression before and after allogeneic stem cell transplantation. Blood 113(26):6567–6571
Metzelder S, Schroeder T, Finck A, Scholl S, Fey M, Gotze K et al (2012) High activity of sorafenib in FLT3-ITD-positive acute myeloid leukemia (AML) synergizes with alloimmune effects to induce sustained responses. Leukemia 2012 Apr 16. (epub ahead of print)
Mohle R, Schittenhelm M, Failenschmid C, Bautz F, Kratz-Albers K, Serve H et al (2000) Functional response of leukaemic blasts to stromal cell-derived factor-1 correlates with preferential

expression of the chemokine receptor CXCR4 in acute myelomonocytic and lymphoblastic leukaemia. Br J Haematol 110(3):563–572

Panayiotidis P, Jones D, Ganeshaguru K, Foroni L, Hoffbrand AV (1996) Human bone marrow stromal cells prevent apoptosis and support the survival of chronic lymphocytic leukaemia cells in vitro. Br J Haematol 92(1):97–103

Patel HK, Grotzfeld RM, Lai AG, Mehta SA, Milanov ZV, Chao Q et al (2009) Arylcarboxyamino-substituted diaryl ureas as potent and selective FLT3 inhibitors. Bioorg Med Chem Lett 19(17):5182–5185

Pemmaraju N, Kantarjian H, Ravandi F, Cortes J (2011) FLT3 inhibitors in the treatment of acute myeloid leukemia: the start of an era? Cancer 117(15):3293–3304

Perl AE, Jeschke GR, Smith CC, Mangan JK, Luger SM, Carroll M (2011) Phospho-specific flow cytometry of fixed whole blood demonstrates in vivo FLT3 inhibition in circulating leukemic blasts during AC220 therapy and accurately detects the development of therapeutic resistance. ASH Annual Meeting Abstracts 118(21):3502. 10–13 Dec 2011

Pratz KW, Sato T, Murphy KM, Stine A, Rajkhowa T, Levis M (2010) FLT3-mutant allelic burden and clinical status are predictive of response to FLT3 inhibitors in AML. Blood 115(7):1425–1432

Ravandi F, Cortes JE, Jones D, Faderl S, Garcia-Manero G, Konopleva MY et al (2010) Phase I/II study of combination therapy with sorafenib, idarubicin, and cytarabine in younger patients with acute myeloid leukemia. J Clin Oncol 28(11):1856–1862

Safaian NN, Czibere A, Bruns I, Fenk R, Reinecke P, Dienst A et al (2009) Sorafenib (Nexavar) induces molecular remission and regression of extramedullary disease in a patient with FLT3-ITD+ acute myeloid leukemia. Leuk Res 33(2):348–350

Sato T, Yang X, Knapper S, White P, Smith BD, Galkin S et al (2011) FLT3 ligand impedes the efficacy of FLT3 inhibitors in vitro and in vivo. Blood 117(12):3286–3293

Sayar H, Cripe L, Cangany M, Weisenbach J, Sargent KJ, Goswami C et al (2010) Cyclic administration of combination of sorafenib and vorinostat in poor-risk AML: a pharmacodynamically-oriented extended phase I trial. ASH Annual Meeting Abstracts 116(21):3272. 4–7 Dec 2010

Serve H, Wagner R, Sauerland C, Brunnberg U, Krug U, Schaich M et al (2010) Sorafenib in combination with standard induction and consolidation therapy in elderly AML patients: results from a randomized, placebo-controlled phase II trial. ASH Annual Meeting Abstracts 116(21):333. 4–7 Dec 2010

Sharma M, Ravandi F, Bayraktar UD, Chiattone A, Bashir Q, Giralt S et al (2011) Treatment of FLT3-ITD-positive acute myeloid leukemia relapsing after allogeneic stem cell transplantation with sorafenib. Biol Blood Marrow Transplant 17(12):1874–1877

Smith CC, Perl AE, Lasater E, Zhang C, Jeschke GR, Damon LE et al (2011a) PLX3397 is an investigational selective FLT3 inhibitor that retains activity against the clinically-relevant FLT3-ITD/F691 L "gatekeeper" mutation in vitro. ASH Annual Meeting Abstracts 118(21):764. 10–13 Dec 2011

Smith CC, Damon LE, Zhu X, Salerno S, Shah N (2011b) Analysis of in vitro activity of the clinically-active ABL/FLT3 inhibitor ponatinib (AP24534) against AC220-resistant FLT3-ITD mutants. ASH Annual Meeting Abstracts 118(21):930. 10–13 Dec 2011

Smith CC, Chin J, Wang Q, Salerno S, Damon LE, Hunt JP et al (2011c) Validation of FLT3-ITD as a therapeutic target in human acute myeloid leukemia. ASH Annual Meeting Abstracts 118(21):937. 10–13 Dec 2011

Smith CC, Wang Q, Chin CS, Salerno S, Damon LE, Levis MJ et al (2012) Validation of ITD mutations in FLT3 as a therapeutic target in human acute myeloid leukaemia. Nature 485(7397):260–263

Strumberg D, Richly H, Hilger RA, Schleucher N, Korfee S, Tewes M et al (2005) Phase I clinical and pharmacokinetic study of the Novel Raf kinase and vascular endothelial growth factor receptor inhibitor BAY 43-9006 in patients with advanced refractory solid tumors. J Clin Oncol 23(5):965–972

Wilhelm S, Chien DS (2002) BAY 43-9006: preclinical data. Curr Pharm Des 8(25):2255–2257

Wilhelm SM, Carter C, Tang L, Wilkie D, McNabola A, Rong H et al (2004) BAY 43-9006 exhibits broad spectrum oral antitumor activity and targets the RAF/MEK/ERK pathway and receptor tyrosine kinases involved in tumor progression and angiogenesis. Cancer Res 64(19):7099–7109

Wright I, Kapoor A (2011) Current systemic management of metastatic renal cell carcinoma—first line and second line therapy. Curr Opin Support Palliat Care 5(3):211–221

Zarrinkar PP, Gunawardane RN, Cramer MD, Gardner MF, Brigham D, Belli B et al (2009) AC220 is a uniquely potent and selective inhibitor of FLT3 for the treatment of acute myeloid leukemia (AML). Blood 114(14):2984–2992

Zeng Z, Shi YX, Samudio IJ, Wang RY, Ling X, Frolova O et al (2009) Targeting the leukemia microenvironment by CXCR4 inhibition overcomes resistance to kinase inhibitors and chemotherapy in AML. Blood 113(24):6215–6224

Zhang W, Konopleva M, Ruvolo VR, McQueen T, Evans RL, Bornmann WG et al (2008a) Sorafenib induces apoptosis of AML cells via Bim-mediated activation of the intrinsic apoptotic pathway. Leukemia 22(4):808–818

Zhang W, Konopleva M, Shi YX, McQueen T, Harris D, Ling X et al (2008) Mutant FLT3: a direct target of sorafenib in acute myelogenous leukemia. J Natl Cancer Inst 100(3):184–198

Chapter 13
Nucleophosmin (NPM1)

Ramesh Balusu, Warren Fiskus and Kapil N. Bhalla

Abstract Nucleophosmin (NPM) 1 is a multifunctional nucleolar protein, which as an oligomer and a molecular chaperone serves vital roles in cell proliferation, including ribosome biogenesis, maintenance of genomic stability, and DNA transcription and repair. Heterozygous mutations in exon 12 of NPM1 are common and pathogenetic in acute myeloid leukemia (AML). Mutant (mt) NPM1 may also be co-expressed with the internal tandem duplications in FLT3 (FLT3-ITD), which confers poor prognosis in AML. Additional genetic mutations that coexist with mt-NPM1 are described. This chapter also covers the domain structure, genetic alterations, as well as the pathogenetic role of mt-NPM1 in AML. Recently, knockdown of NPM1 has been shown to promote differentiation and sensitize AML blasts to differentiation-inducing agents. These reports highlight the anti-AML activity of targeted depletion of levels or function of NPM1 in AML. This chapter also describes the potential for developing novel anti-NPM1-based combinations with differentiation-inducing agents or other targeted agents in the therapy of AML.

Keywords Nucleophosmin · FLT3-ITD · Acute myeloid leukemia · Oligomerization

13.1 Introduction

Nucleophosmin 1 (NPM1, also referred to as B23 or numatrin in mammals, and NO38 in amphibians) is a ubiquitously expressed nucleolar phosphoprotein and is a member of the nucleoplasmin family of nuclear chaperons (Mamrack et al. 1979; Chang et al. 1988). NPM1 was first identified in rat liver and Novikoff hepatoma

K. N. Bhalla (✉)
Houston Methodist Research Institute,
6670 Bertner Ave R9-113, Houston, TX 77030, USA
e-mail: knbhalla@houstonmethodist.org

R. Balusu · W. Fiskus
Kansas Masonic Cancer Research Institute, University of Kansas Medical Center,
3901 Rainbow Blvd, Kansas City, KS 66160, USA
e-mail: rbalusu@kumc.edu

W. Fiskus
e-mail: wcfiskus@gmail.com

M. Andreeff (ed.), *Targeted Therapy of Acute Myeloid Leukemia*,
Current Cancer Research, DOI 10.1007/978-1-4939-1393-0_13

ascites cells using two-dimensional gel electrophoresis analyses (Orrick et al. 1973; Kang et al. 1975). The protein is named as B23 because it was the 23rd protein in region B of the gel slab when the protein spots were numbered in order of decreasing mobility in both electrophoretic dimensions (Orrick et al. 1973). Chang and collaborators for the first time characterized the nucleotide sequence of the cDNA encoding human nucleophosmin (NPM) (Chang et al. 1988). NPM1 is found to be more abundantly expressed in proliferating cells than resting cells (Okuwaki 2008).

NPM1 plays multiple roles in cell growth and proliferation by participating in diverse biologic processes, including ribosome biogenesis, maintenance of genomic stability, centrosome duplication, regulation of cell cycle, DNA repair, inhibition of apoptosis, chaperoning of histones, regulation of transcriptional, and regulation of the ARF-p53 tumor suppressor pathway (Okuda 2002; Bertwistle et al. 2004; Kurki et al. 2004; Swaminathan et al. 2005; Yu et al. 2006; Murano et al. 2008; Falini et al. 2011). Nuclear and cytoplasmic shuttling of NPM1 is critical for most of its biological functions (Grisendi et al. 2006; Naoe et al. 2006; Meani and Alcalay 2009). Under native conditions, NPM1 exists as a dimer or larger oligomers in cells (Herrera et al. 1996; Colombo et al. 2002; Qi et al. 2008; Balusu et al. 2011). The multifunctional characteristic of NPM1 appears to be dictated not only by its subcellular localization and its binding partners but also by post-translational modifications including phosphorylation, acetylation, poly(ADP-ribosyl)ation, ubiquitination, and sumoylation (Ramsamooj et al. 1995; Tokuyama et al. 2001; Sato et al. 2004; Liu et al. 2007; Shandilya et al. 2009). NPM1 contains distinct structural domains that account for its ability to act as a multifunctional protein (Fig. 13.1) (Hingorani et al. 2000; Okuwaki 2008; Qi et al. 2008; Yip et al. 2011).

Acute myeloid leukemia (AML) is a cytogenetically and molecularly heterogeneous disease characterized by the accumulation of acquired, somatic genetic alterations in hematopoietic progenitor cells that alter normal mechanisms of self-renewal, proliferation, and differentiation (Fröhling et al. 2005; Döhner and Döhner 2008). More than 50 % of adult AML exhibit normal karyotype, and their biological features are poorly understood (Falini et al. 2006). Several NPM1 gene alterations have been reported in a significant proportion of adults with different types of hematologic malignancies (Meani and Alcalay 2009). AML with mutated NPM1 (NPM1c^{+} AML-cytoplasmic NPM1 positive) subtype shows unique biologic and clinical features. These include unique gene expression profiles, high association with FLT3-ITD, low CD34 expression, and multilineage involvement (Grisendi et al. 2006; Falini et al. 2009a). NPM1 gene mutations or chromosomal translocations result in the dislocation of NPM1 into the cytoplasm (Verhaak et al. 2005). The cytoplasmic dislocation of NPM1 reduces the nucleolar component of NPM1, which affects its normal functions as the binding partner and chaperone protein—an early event in leukemogenesis (Vassiliou et al. 2011). Generally, NPMc^{+} AML exhibits good response to induction therapy (Falini et al. 2007). However, the overall outcome is poorer when FLT3-ITD is co-expressed with NPMc^{+} (Falini et al. 2011).

In this chapter, we will review NPM1 structure/function, gene mutations, coexistence of mt-NPM1 with mutations in other genes, mt-NPM1 gene expression signature, in vivo disease models and possible targeted therapies.

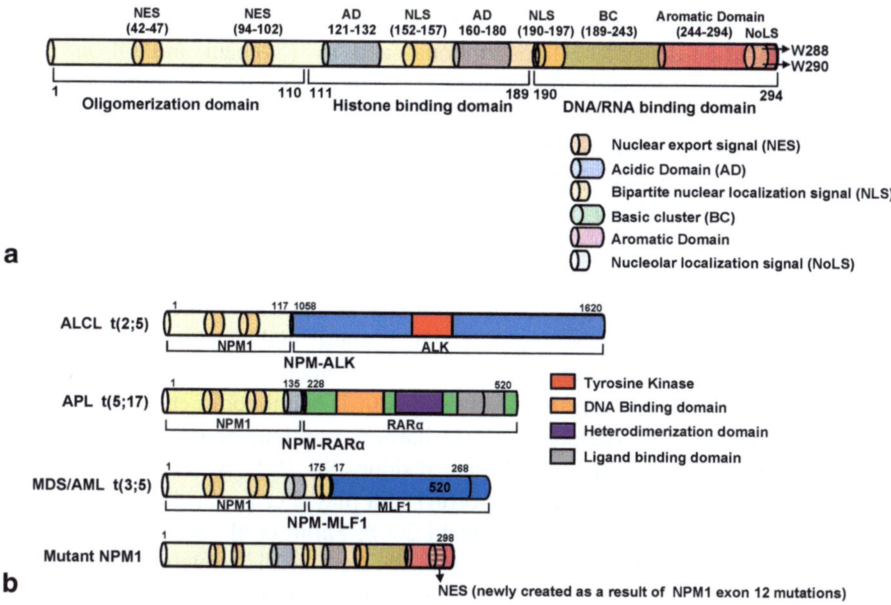

Fig. 13.1 Structure and functional domains of nucleophosmin (NPM1) and NPM1 fusion proteins. **a** NPM1. **b** NPM1 fusion proteins

13.2 Structural and Functional Domains of NPM1

The human NPM1 gene spans a genomic region of about 23 kb on chromosome 5q35 and has 12 exons. Two splice variants of NPM, namely NPM1 (B23.1) and NPM1.2 (B23.2) have been reported (Wang et al. 1993). In the long splice-variant NPM1, exon 9 is spliced to exon 11 and the coding sequence stops at exon 12, resulting in a protein of 294 amino acids (Chang and Olson 1989). Except for the C-terminal 35 amino acids which are only present in NPM1, and are essential for nucleolar localization, NPM1 and NPM1.2 share common amino acid sequence. NPM1 is the most prevalent form of NPM and is expressed in all the tissues and localized mainly in the nucleolus. The biological significance and physiological function of NPM1.2 remain mostly unclear (Colombo et al. 2011).

The NPM1 protein sequence contains several domains (Fig. 13.1) with distinct biochemical, structural, and functional properties (Hingorani et al. 2000; Okuwaki 2008; Qi et al. 2008; Yip et al. 2011). In addition to structural domains, NPM1 also contain several sequence motifs, including the nuclear export signal (NES), nuclear localization signal (NLS), and nucleolar localization signal (NoLS) (Grisendi et al. 2006; Okuwaki 2008; Falini 2010).

NPM1 shares a conserved N-terminal core structure (oligomerization domain) with other NPM family proteins (Herrera et al. 1996). This core domain is essential for oligomerization and the molecular chaperone activity of NPM1. The core domain suppresses the misfolding and aggregation of the target proteins in the crowded environment of the nucleolus to favor histone and nucleosome assembly and to increase acetylation-dependent transcriptional activity (Szebeni and Olson 1999; Okuwaki et al. 2001; Swaminathan et al. 2005). The functional role of this domain in ribosomal biogenesis and p53-mediated tumor suppression has also been described (Colombo et al. 2011; Murano et al. 2008). Lee et al. (2007) reported on the X-ray crystal structure of the human NPM1 core region (residues 9–122) which reveals a compact domain consisting of two four-stranded β-sheets packed in a β-sandwich topology. The core region forms a tight pentameric structure through hydrophobic interactions between the monomeric subunits. In the crystallization state, two pentameric rings align along their fivefold symmetry axis and associate in head-to-head fashion to form a decameric structure. The monomeric and pentameric structures of the human NPM1 core domain are very similar to other NPM orthologs, including those of *Xenopus* and *Drosophila* (Dutta et al. 2001; Namboodiri et al. 2004).

The N-terminal core region of NPM1 harbors two leucine-rich nuclear export signal (NES) motifs (residues 42–47 and residues 94–102), which are targeted by the nuclear export receptor Crm1/exportin1 and play a critical role in nuclear and cytoplasmic shuttling of NPM1 (Okuwaki 2008; Falini et al. 2010). The core domain of NPM1 protein is followed by the two long acidic stretches A1 (residues 121–132) and A2 (residues 160–180). These acidic regions contain clusters of aspartic acid and glutamic acid residues. The distinct electrostatic property of this region serves as a binding site for the positively charged basic proteins, such as histones and ribosomal proteins in the nucleolus to facilitate nucleosome assembly and chromatin remodeling (Dutta et al. 2001; Namboodiri et al. 2004). The region between the two acidic stretches has the ribonuclease activity (critical for ribosome biogenesis) and one bipartite nuclear localization signal (NLS residues 190–197), which is responsible for the nuclear import of NPM1 (Dingwall et al. 1987; Hingorani et al. 2000).

Another important domain (residues 189–243) in NPM1, the basic domain, is enriched with basic residues lysine and arginine. This basic domain is important for nucleic acid binding because its favorable electrostatic properties promote the association of NPM1 with DNA/RNA molecules (Hingorani et al. 2000). This nucleic acid binding may explain the role of NPM1 in nucleosome assembly by coordinating the packaging of DNA molecules onto core histones. NPM1 is also involved in ribosomal biogenesis in cells. Peri-ribosomal particles bind to NPM1, and this association facilitates ribosome transport from the nucleus into the cytoplasm. NPM1 interacts directly with a subset of ribosomal proteins including RPL5, RPS9, and RPL23 (Yu et al. 2006; Lindstrom et al. 2008; Wanzel et al. 2008). NPM1 and p53 interaction at the basic region leads to enhanced stability of p53 and upregulation of its tumor suppressor function (Colombo et al. 2002). NPM1 interacts with and stabilizes p14ARF, thereby inactivating HDM2 and preventing the proteasomal degradation of p53 (Colombo et al. 2002; Yang et al. 2002; Maiguel et al. 2004).

The C-terminal end of NPM1 (residues 244–294) is unique to the isoform 1 of NPM family members. This region has the nucleolar localization signal (NoLS) and the tryptophan residues at positions 288 and 290, which are critical for retaining NPM1 in the nucleolus. The two N-terminal NES motifs and C-terminal NoLS motif of NPM1 play a key role in regulating the shuttling of NPM1 across the nucleolus/nucleoplasm and the cytosol (Falini et al. 2010).

13.3 NPM1 Gene Alterations Cause Cytoplasmic Dislocation of Nucleophosmin

13.3.1 FrameShift Mutations of the NPM1 Gene

About 35 % of adult AML exhibit aberrant cytoplasmic localization of NPM1 (NPM1c^{+}) due to exon-12 mutations of the NPM1 gene. NPM1 mutation is associated with a wide spectrum of morphological subtypes of AML from M0 to M7 except M3 (Naoe et al. 2006). The mutations of exon 12 have been found in a significant proportion of de novo AML, especially in those with a normal karyotype. The most common exon-12 mutation in AML is the type-A, which consists of TCTG, a tetranucleotide tandem duplication at position 956–959 of the NPM1 coding sequence and accounts for 70–80 % of cases. Type-B and type D together account for 15–20 % of cases, which causes a frame shift of the downstream coding sequence (Naoe et al. 2006; Balusu et al. 2011). This mutation causes the loss of tryptophans 288 and 290 (or 290 alone) from the C-terminus and the creation of an additional leucine-rich nuclear export motif in the NPM1 protein, resulting in the aberrant cytoplasmic dislocation of NPM1c^{+} (Falini et al. 2007a, 2009; Ahmad et al. 2009). To date, about 39 different exon-12 mutations have been reported in the NPM1 gene (Table 13.1). Evidence suggests that the presence of NPM1c^{+} is critical for leukemogenesis. NPM1 mutations in AML patients are always heterozygous and never coexist with uniparental disomy (Falini et al. 2007a, 2007b) or deletion of the wild-type allele. A fraction of wild-type NPM1 seems to be necessary for the survival of leukemic cells (Bolli et al. 2009).

13.3.2 Chromosomal Translocations Involving the NPM1 Gene

Several NPM1 gene alterations have been reported in different types of hematologic malignancies (Meani and Alcalay 2009). The NPM1 gene located on chromosome 5q35 is frequently involved in recurrent chromosomal translocations in lymphomas, but rarely in leukemia. NPM1 chromosomal translocations or gene mutations result in the dislocation of NPM1 into the cytoplasm (Verhaak et al. 2005). The cytoplasmic dislocation of NPM1 may affect its normal functions as a binding partner and transporter protein, which is an early event in leukemogenesis. In all of these translocations, the NPM1 N-terminal region is conserved (Verhaak et al. 2005). The three reported NPM1-fusion proteins are characterized by chromosomal translocations (e.g., NPM-ALK t(2;5) (p23;q35), NPM-RARα t(5;17) (q35;q21), and NPM-MLF1 t(3;5) (q25;q35) mutations) (Falini et al. 2007b).

Table 13.1 Nucleophosmin exon 12 mutations reported in AML patients and the associated amino acid sequence alterations in tryptophan 288 and/or 290 create a nuclear export signal (NES) motif. Color code: Red letters: nucleotides inserted by each type of mutation; Blue boxes: leucine/valine-rich NES motif; Yellow: tryptophan (W) residues (Falini et al. 2007a; Ahmed et al. 2009)

Wild Type	D gat	L ctc	W tgg	Q cag	W tgg	R agg	K aag	S tct	L ctt	Stop taa	gaa	aat	ag	
Type-A	D gat	L ctc	C tgT	L CTG	A gca	V gtg	E gag	E gaa	V gtc	S tct	L tta	R aga	K aaa	Stop tag
Type-B	D gat	L ctc	C tgC	M ATG	A gca	V gtg	E gag	E gaa	V gtc	S tct	L tta	R aga	K aaa	Stop tag
Type-C	D gat	L ctc	C tgC	V GTG	A gca	V gtg	E gag	E gaa	V gtc	S tct	L tta	R aga	K aaa	Stop tag
Type-D	D gat	L ctc	C tgC	L CTG	A gca	V gtg	E gag	E gaa	V gtc	S tct	L tta	R aga	K aaa	Stop tag
Type-E	D gat	L ctc	W tgg	Q cag	S tCT	L CTT	A GCC	Q Caa	V gtc	S tct	L tta	R aga	K aaa	Stop tag
Type-E*	D gat	L ctc	W tgg	Q cag	S tCC	L CTC	A GCC	Q Caa	V gtc	S tct	L tta	R aga	K aaa	Stop tag
Type-F	D gat	L ctc	W tgg	Q cag	S tCC	L CTG	E GAG	K Aaa	V gtc	S tct	L tta	R aga	K aaa	Stop tag
Type-F1	D gat	L ctc	C tgC	W TGG	R AGa	V gtg	E gag	E gaa	V gtc	S tct	L tta	R aga	K aaa	Stop tag
Type-F2	D gat	L ctc	W tgg	E cag	C tGC	F TTT	A GCC	K Aaa	V gtc	S tct	L tta	R aga	K aaa	Stop tag
Type-G*	D gat	L ctc	W tgg	Q cag	C tGC	F TTC	A GCC	Q Caa	V gtc	S tct	L tta	R aga	K aaa	Stop tag
Type-G†	D gat	L ctc	C tgT	F TTC	A gca	V gtg	E gag	E gaa	V gtc	S tct	L tta	R aga	K aaa	Stop tag
Type-Gm	D gat	L ctc	C tgC	R AGG	A gca	V gtg	E gag	E gaa	V gtc	S tct	L tta	R aga	K aaa	Stop tag
Type-H†	D gat	L ctc	C tgC	L TTG	A gca	V gtg	E gag	E gaa	V gtc	S tct	L tta	R aga	K aaa	Stop tag
Type-H*	D gat	L ctc	W tgg	Q cag	C tGT	F TTT	S TCA	K Aaa	V gtc	S tct	L tta	R aga	K aaa	Stop tag
Type-I⊥	D gat	L ctc	C tgT	K AAG	A gca	V gtg	E gag	E gaa	V gtc	S tct	L tta	R aga	K aaa	Stop tag
Type-IΔ	D gat	L ctc	C tgC	R AGA	A gca	V gtg	E gag	E gaa	V gtc	S tct	L tta	R aga	K aaa	Stop tag
Type-J	D gat	L ctc	W tgg	Q cag	S tCT	L CTT	S TCT	K Aaa	V gtc	S tct	L tta	R aga	K aaa	Stop tag
Type-J¥	D gat	L ctc	C tgT	M ATG	A gca	V gtg	E gag	E gaa	V gtc	S tct	L tta	R aga	K aaa	Stop tag
Type-K	D gat	L ctc	W tgg	Q cag	S tCC	L CTT	S TCC	K Aaa	V gtc	S tct	L tta	R aga	K aaa	Stop tag

Table 13.1 (continued)

Wild Type	D	L	W	Q	W	R	K	S	L	Stop						
	gat	ctc	tgg	cag	tgg	agg	aag	tct	ctt	taa	gaa	aat	ag			
Type-Km	D	L	C	R	A	V	E	E	V	S	L	R	K	Stop		
	gat	ctc	tgC	CGG	gca	gtg	gag	gaa	gtc	tct	tta	aga	aaa	tag		
Type-L	D	L	S	R	A	V	E	E	V	S	L	R	K	Stop		
	gat	ctc	tCC	CGg	gca	gtg	gag	gaa	gtc	tct	tta	aga	aaa	tag		
Type-Lm	D	L	C	Q	A	V	E	E	V	S	L	R	K	Stop		
	gat	ctc	tgC	CGC	GGa	gtg	gag	gaa	gtc	tct	tta	aga	aaa	tag		
Type-M	D	L	C	T	A	V	E	E	V	S	L	R	K	Stop		
	gat	ctc	tgT	AGC	gca	gtg	gag	gaa	gtc	tct	tta	aga	aaa	tag		
Type-N	D	L	C	H	A	V	E	E	V	S	L	R	K	Stop		
	gat	ctc	tgC	CAC	gca	gtg	gag	gaa	gtc	tct	tta	aga	aaa	tag		
Type-Nm	D	L	C	Q	A	V	E	E	V	S	L	R	K	Stop		
	gat	ctc	tgC	CAG	gca	gtg	gag	gaa	gtc	tct	tta	aga	aaa	tag		
Type-Om	D	L	C	L	A	V	E	E	V	S	L	R	K	Stop		
	gat	ctc	tgT	TTG	gca	gtg	gag	gaa	gtc	tct	tta	aga	aaa	tag		
Type-P	D	L	C	T	F	L	E	E	V	S	L	R	K	Stop		
	gat	ctc	tgT	ACC	TTC	Ctg	gag	gaa	gtc	tct	tta	aga	aaa	tag		
Type-Q	D	L	W	Q	R	M	E	E	V	S	L	R	K	Stop		
	gat	ctc	tgg	cag	AGG	Atg	gag	gaa	gtc	tct	tta	aga	aaa	tag		
Type-Qm	D	L	C	R	A	V	E	E	V	S	L	R	K	Stop		
	gat	ctc	tgT	CGG	gca	gtg	gag	gaa	gtc	tct	tta	aga	aaa	tag		
Type-R	D	L	C	S	A	V	E	E	V	S	L	R	K	Stop		
	gat	ctc	tgC	TCG	gca	gtg	gag	gaa	gtc	tct	tta	aga	aaa	tag		
Type-1	D	L	W	Q	S	M	E	E	V	S	L	R	K	Stop		
	gat	ctc	tgg	cag	TCC	Atg	gag	gaa	gtc	tct	tta	aga	aaa	tag		
Type-3	D	L	C	H	A	V	E	E	V	S	L	R	K	Stop		
	gat	ctc	tgT	CAT	gca	gtg	gag	gaa	gtc	tct	tta	aga	aaa	tag		
Type-4	D	L	C	L	A	V	E	E	V	S	L	R	K	Stop		
	gat	ctc	tgC	TTG	gca	gtg	gag	gaa	gtc	tct	tta	aga	aaa	tag		
Type-6	D	L	W	Q	D	F	L	N	R	L	F	K	K	I	V	Stop
	gat	ctc	tgg	caA	GAT	TTC	TTA	AAT	Cgt	ctc	ttt	aag	aaa	ata	gtt	taa
Type-7	D	L	C	L	A	V	E	E	V	S	L	R	K	Stop		
	gat	ctA	TGC	ctg	gca	gtg	gag	gaa	gtc	tct	tta	aga	aaa	tag		
Type-10	D	L	W	Q	C	C	S	Q	V	S	L	R	K	Stop		
	gat	ctc	tgg	cag	tgC	TGC	TCC	Caa	gtc	tct	tta	aga	aaa	tag		
Type-12	D	L	C	A	A	V	E	E	V	S	L	R	K	Stop		
	gat	ctc	tgG	CCC	gca	gtg	gag	gaa	gtc	tct	tta	aga	aaa	tag		
Type-13	D	L	C	K	A	V	E	E	V	S	L	R	K	Stop		
	gat	ctc	tgT	AAG	gca	gtg	gag	gaa	gtc	tct	tta	aga	aaa	tag		
Type-14	D	L	W	C	C	C	S	Q	V	S	L	R	K	Stop		
	gat	ctc	tgg	cag	tTA	TTT	TCC	Caa	gtc	tct	tta	aga	aaa	tag		

NPM-ALK The t(2;5) chromosomal translocation is associated with approximately 60% of the cases of anaplastic large-cell lymphoma (ALCL). In this translocation, the oligomerization domain of NPM1 fuses with part of the anaplastic lymphoma kinase (ALK) gene that encodes for the entire cytoplasmic region of the ALK protein (Morris et al. 1994; Drexler et al. 2000; Dupuis-Coronas et al. 2011; Bonvini et al. 2012). This brings ALK gene under the control of NPM1 promoter, which in turn constitutively drives the NPM–ALK gene. The NPM1 moiety of the NPM-ALK chimeric protein helps to form homodimers or heterodimers (Meani and Alcalay 2009). The formation of homodimers leads to constitutive activation of the catalytic domain of ALK(Bischof et al. 1997; Dupuis-Coronas et al. 2011), a major trigger of lymphomagenesis, by mimicking receptor dimerization and activation (Duyster et al. 2001).

NPM-RARα This extremely rare translocation forms a chimeric protein NPM-RARα, composed of the oligomerization domain of NPM1 and a major portion of the retinoic acid receptor (RARα) protein (Redner et al. 1996; Grisendi et al. 2006; Chen et al. 2010). The phenotype of NPM-RARα leukemia is that of a typical acute promyelocytic leukemia (APL, FAB M3) bearing RARα rearrangements. The NPM-RARα fusion protein forms homodimers and heterodimers with wild-type NPM1 and RXR. The transforming potential of NPM-RARα may derive from its interaction with wild-type NPM1, either by delocalizing or by altering its function in the nucleolus (Falini et al. 2007b). APL with NPM-RARα translocation shows good response to differentiation therapy.

NPM1-MLF1 This is a rare chromosomal translocation (less than 1%) occurring in AML and generates the NPM-MLF1 (myelodysplasia/myeloid leukemia factor 1) fusion protein (Yoneda-Kato et al. 1996). In this chimeric protein, the N-terminal portion of NPM1 is fused with almost the full-length MLF1 protein, minus the first 16 amino acids at the N-terminus. The function of MLF1 is not fully characterized, however, by inhibiting its E3 ubiquitin ligase activity, a recent report demonstrated its role in stabilizing p53 (Yoneda-Kato et al. 1996, 2008). According to the French–American–British classification, NPM-MLF1 AML is classified as acute erythroid leukemia (M6; Raimondi et al. 1989). The exact mechanism(s) by which the NPM-MLF1 protein induces leukemic transformation is not yet clear. Immunostaining studies demonstrate that MFL1 protein is mainly localized in the cytoplasm, whereas the NPM-MLF1 fusion protein is localized in the nucleus. The NPM1 moiety of the NPM-MLF1 chimeric protein still contains the nuclear localization signal, and the wild-type NPM1 protein forms heterodimers and translocates the fusion protein into the nucleus, which may alter its function. Decreased levels of expression of the wild-type NPM1 (due to the loss of one *NPM1* functional allele) and its cytoplasmic dislocation due to heterodimerization may also contribute to leukemogenesis (Falini et al. 2006; Lee et al. 2012).

In mutation screening studies, Jeong et al. (2007) observed that NPM1 exon-12 mutations are absent in common solid tumors. NPM1 is overexpressed in many types of human solid tumors and elevated levels of NPM1 are considered to be a tumor marker (Grisendi et al. 2006; Nozawa et al. 1996; Subong et al. 1999). The direct interaction of NPM1 with c-Myc has been shown to regulate the expression of c-Myc target genes at the promoter (Zeller et al. 2001; Li et al. 2008). The ability of NPM1 to suppress apoptosis and to promote DNA repair plays a significant pro-survival role during tumor development (Lindstrom 2011). However, there remains a need to explore the role of the wild-type NPM1 as a proto-oncogene in solid tumors.

13.4 Associated Mutations in NPMc⁺ AML

13.4.1 Co-existence of NPM1 Mutations with other Mutations in AML

Leukemogenesis is a complex and multistep process. Gilliland and Griffin (2002) proposed the "two-hit" model in which the class-I mutations activate signaling pathways that are involved in the proliferation of hematopoietic stem or progenitor cells, while the class II mutations affect transcription factors and primarily impair hematopoietic differentiation (Fröhling et al. 2005; Rocquain et al. 2010). Identification of mutations in genes such as FMS-related tyrosine kinase 3 (FLT3), CAAT enhancer-binding protein alpha (C/EBPα), runt-related transcription factor 1 (Runx1), myeloid-lymphoid or mixed lineage leukemia (MLL), Wilms' tumor (WT1), NPM1, DNA (cytosine-5)-methyltransferase 3 A (DNMT3A), additional sex comb-like 1 (ASXL1), isocitrate dehydrogenase, and IDH, 1 and 2 genes, in AML has led to a new understanding of the molecular mechanisms that lead to leukemogenesis (Patel et al. 2012; Takahashi 2011).

FLT3 (FMS-like tyrosine kinase 3) is a receptor tyrosine kinase (RTK) involved in the proliferation, differentiation, and apoptosis of hematopoietic cells. It is mainly expressed by early myeloid and lymphoid progenitor cells (Stirewalt and Radich 2003; Kindler et al. 2010). *FLT3* mutations are the most frequent genetic lesion seen in acute myeloid leukemia (AML) (Smith et al. 2012). Two types of mutations are commonly found in the FLT3 gene, an internal tandem duplication (ITD; 15–35 %) of the juxtamembrane domain-coding sequence, and a missense mutation of D835 (5–10 %) within the kinase domain. Both types of mutations in the *FLT3* gene result in ligand-independent, constitutive activation of the receptor and activation of its downstream signaling pathways.

Currently, expression of NPM1 and FLT3 mutations is commonly considered while deciding therapeutic strategies for cytogenetically normal CN-AML patients. Out of 600 mutant NPM1 AML cases analyzed, FLT3-ITD was also detected in 36.5 % of the patients (Falini et al. 2005). Boissel et al. determined the prevalence of NPM1 mutations in a cohort of 106 patients with normal-karyotype AML (Boissel et al. 2005). NPM1 mutations were identified in 47 % of the patients. Among the NPM1 mutant group, 38 % of cases coexisted with FLT3-ITD. In another study, Boonthimat et al. screened 400 leukemia patients and NPM1 mutations were identified in 26.25 % of cases of AML, but in no cases of acute lymphoid leukemia or chronic myeloid leukemia (CML; Boonthimat et al. 2008). NPM1 mutations mainly coexisted with FLT3-ITD mutation, which were detected in 56.8 % of cases of mutated NPM1, as compared to 25.6 % of the cases with the wild-type NPM1. Only one case in the mutated NPM1 group and 25 cases in the wild-type NPM1 group had RAS mutation. No RUNX1 mutation was found among the mutated NPM1 group, whereas 11 cases of RUNX1 mutation were found in the wild-type NPM1 group. The higher frequency of FLT3 mutations in cases with mutated NPM1 suggests a possible pathogenic link between these two gene mutations. The coexistence of *NPM1* and *FLT3* mutations could suggest that the mutant NPM1 protein may serve to impair differentiation of hematopoietic cells in the same way as RUNX1 and C/EBPα.

IDH1 and IDH2 are the family of isocitrate dehydrogenase (IDH) enzymes that catalyze oxidative decarboxylation of isocitrate into α-ketoglutarate by using nicotinamide adenine dinucleotide (NAD) or NAD phosphate (NADP) as a cofactor to yield NADH or NADPH, respectively (Paschka et al. 2010). IDH1 and IDH2 enzymes are important in the cellular defense of oxidative damage and lipid metabolism (Koh et al. 2004). A study by Paschka et al. (2010) in AML ($n = 805$) revealed that IDH mutations occur in ~16 % of AML patients with IDH1 mutations occurring in 7.6 %, and IDH2 in 8.7 %. A very small proportion of AML patients carry both IDH1 and IDH2 mutations (2 out of 805). The most common *IDH1* mutation results in R132 H substitution, while IDH2 mutations result in R140Q or R172 K substitution. *IDH* mutations were associated with older age, CN-AML and *NPM1* mutations, in particular with the genotype of mutated *NPM1* without *FLT3* internal tandem duplication.

DNA methyltransferases (DNMTs) are enzymes that catalyze the addition of a methyl group to the cytosine residue of CpG dinucleotides that regulates gene expression (Fiskus et al. 2009; Ley et al. 2010). Alterations of DNA methylation patterns could lead to cancer by disrupting normal physiological gene expression programs (Chik and Szyf 2011). The *DNMT3A* gene is a member of the DNMT family, and mutations in DNMT3A have been reported in about 20 % of de novo adult AML cases, mostly associated with normal karyotype (Ley et al. 2010; Yan et al. 2011). DNMT3A mutations are usually heterozygous, and the presence of *DNMT3A* mutations is associated with poor outcome (Tiacci et al. 2012; Yan et al. 2011). To date, 33 separate DNMT3A mutations have been identified, and all of these mutations reside in the methyltransferase domain of the protein (Ley et al. 2010). Recent studies by Tiacci and colleagues (2012) reported that OCI-AML3 cells which harbor NPM1c$^+$ protein also possess the DNMT3A point mutation R882 C. This finding is similar to the observation that AML patients with *NPM1* mutations frequently harbor *DNMT3A* mutations (Ley et al. 2010; Yan et al. 2011).

C/EBPα is a transcription factor that coordinates cellular differentiation with growth arrest (Nerlov 2004). The *C/EBPα* gene is mutated in approximately 9 % of AML cases. C/EBPα mutations were significantly less frequent (10 %) in NPM1 mutant group compared to NPM1 wild-type group (27 %).

Table 13.2 Occurrence and co-existence of other mutations in NPMc+myeloid malignancies

Type of myeloid malignancy	% of NPM1 mutations	Co-existence of other mutations	Reference
Acute myeloid leukemia (AML)	40–50%	FLT3-ITD, IDH1, IDH2, DNMT3A, CEBP-α	Boissel et al. 2005 Boonthimat et al. 2008 Paschka et al. 2010 Yan et al. 2011 Tiacci et al. 2012
Secondary AML from Myelodysplastic syndrome (MDS)	13.1%	FLT3-ITD, NRAS, RUNX1	Oki et al. 2006 Schnittger et al. 2011a
Secondary AML from Myeloproliferative neoplasm (MPN)	9%	KIT D816V, ETV6-PDGFRB; JAK2 V617F FLT3-ITD, FLT3-TKD NRAS, RUNX1	Pasqualucci et al. 2008 Schnittger et al. 2011a
Myelodysplastic syndrome (MDS)	~5%	N.D.	Zhang et al. 2007

13.4.2 Presence of Mutant NPM1 in other Myeloproliferative Neoplasms

The NPM1 gene is located on chromosome 5q, a region which is altered in many myeloid malignancies (Grisendi et al. 2006). The 5q-syndrome is a subtype of myelodysplastic syndrome (MDS) with distinct features (Ebert 2011). NPM1 mutations are rarely found in MDS, CML, or lymphoid malignancies (Shiseki et al. 2007; Falini et al. 2005; Table 13.2). About 25 % of patients with MDS show acute transformation with blasts >20 % during the course of disease (Schnittger et al. 2011a). This transformation from MDS to AML is assumed to be a multistep process, mediated by the acquisition of different mutations of the *NRAS, FLT3,* and *CEBPA* genes, as well as the *MLL*-PTD. Schnittger and colleagues (2011) detected *NPM1* mutations in 13.1 % of patients with secondary AML following MDS or chronic myelomonocytic leukemia (CMML). In the 67 patients with acutely transformed MPNs, 9.0 % were positive for NPM1 mutation. Seventy percent of these cases were found to be NPM1 type-A mutation, 30 % type-B, and 11 % were of type-D mutation. Oki et al. (2006) also screened various myeloid malignancies including MDS, CML, CMML, and MPN for NPM1 mutations. Out of 50 CMML samples, two (4 %) were positive for NPM1 exon-12 mutations: one with type-A and the other with type-D mutations. Zhang and colleagues (2007) were the first to report the presence of NPM1 mutations in MDS. Out of 38 primary MDS samples screened, NPM1 type-A mutations were detected in two patients (5.2 %). Pasqualucci et al. (2008) reported data on one patient who developed AML with NPM1 mutation (type A) 6 years after JAK2-V617 F positive primary myelofibrosis (PMF). AMLs developing from chronic MPNs are likely to be genetically and prognostically heterogeneous based on the coexistence of the various genetic alterations (Walter et al. 2012). Direct gene sequencing analysis of 48 MRC-AML (AML with myelodysplasia-related changes) patients, revealed mutations in ASXL1 (35 %), RUNX1 (17 %) TET2 (15 %), IDH1 (4 %), IDH2 (21 %), DNMT3A (8 %), NPM1 (8 %), and FLT3 (2 %) (Devillier et al. 2012). Further studies involving whole-genome/exome sequencing are required to fully characterize the spectrum of genetic alterations associated with NPM1 mutations in MPN and its transformation to AML.

13.5 In Vitro and In Vivo Models to Study Leukemogenic Potential of Mutant NPM1

13.5.1 In Vitro Model Human Cell Lines to Study the Leukemogenicity of Mutant NPM1-expressing AML

Human cell lines bearing NPM1 mutation have served as excellent models to elucidate the pathogenetic role of the mutated NPM1 gene in leukemogenesis. OCI-AML3 is the first reported human cell line bearing an NPM1 exon-12 mutation (Quentmeier et al. 2005). OCI-AML3 cells harbor a heterozygous TCTG tandem duplication at exon 12 of the NPM1 gene, which corresponds to the type-A mutation. The OCI-AML3 cell line exhibits all of the typical molecular and biological features of primary NPM1c^{+} AML blast cells including the cytoplasmic localization of mt-NPM1 and substantial engraftment in the NOD/SCID mice. Recently, another cell line IMS-M2, has been reported that bears the type-A NPM1 mutation along with the ETV6-NTRC fusion gene (Chi et al. 2010). These cell lines are currently being used as the in vitro models to determine the interactions of NPM1c^{+} with other proteins, as well as for high-throughput screening of small-molecule libraries for NPM1 antagonists.

13.5.2 In Vivo Models to Study Leukemogenicity of Mutated NPM1 AML

Several in vivo models have been developed to study leukemogenecity of mt-NPM1. In the first experimental in vivo model, Cheng et al. (2010) generated and characterized transgenic mice expressing the NPM1 type-A mutation driven by the myeloid-specific human *MRP8* promoter. Interestingly, *hMRP8-NPMc*$^{+}$ transgenic mice developed myeloproliferation in bone marrow and spleen, whereas the non-transgenic littermates and *hMRP8-NPM* transgenic mice remained disease free. These findings provided the first in vivo evidence indicating that NPMc^{+} confers a proliferative advantage in the myeloid lineage.

In another in vivo model, the zebrafish (*Danio rerio*) was used to elucidate the underlying mechanisms leading to leukemogenesis in mt-NPM1-expressing AML. Zebrafish express two genes orthologous to human NPM1, referred to as npm1a and npm1b (Bolli et al. 2010). Both genes are ubiquitously expressed, and their knockdown produces a reduction in myeloid cell numbers that is specifically rescued by human NPM1 overexpression. In zebrafish, wild-type human NPM1 is nucleolar while mutant NPMc$^+$ is cytoplasmic, as in human AML, and both interact with endogenous zebrafish NPM1a and NPM1b. Forced NPMc$^+$ expression in zebrafish causes an increase in pu.1$^+$ primitive early myeloid cells. A more marked perturbation of myelopoiesis occurs in the p53 null embryos expressing NPMc$^+$, where mpx$^+$ and csf1r$^+$ cell numbers are also expanded. Importantly, NPMc$^+$ expression results in increased numbers of definitive hematopoietic cells, including erythromyeloid progenitors in the posterior blood island and c-myb/cd41$^+$ cells in the ventral wall of the aorta. These results are likely to be relevant to human NPMc$^+$ AML, where the observed NPMc$^+$ multilineage expression pattern implies transformation of a multipotent stem or progenitor cell.

A third in vivo model was generated to further explain the molecular, cellular, and leukemogenic effects of NPM1c$^+$. Vassiliou et al. (2011) generated a conditional knock-in mouse model that expressed the common, type-A NPM1 mutation. The activation of the humanized NPM1c$^+$ knock-in allele in mouse hemopoietic stem cells caused *Hox* gene overexpression, enhanced self-renewal, and expanded myelopoiesis. One third of mice developed delayed-onset AML, suggesting a requirement for cooperating mutations such as *CSF2, FLT3,* or *RASGRP1*. This study also demonstrated recurrent integrations in known and newly discovered leukemia genes including *NF1, BACH2, DLEU2,* and *NUP98* (Vassiliou et al. 2011).

Development of these in vivo models is crucial to developing pathogenetic insights and aid in the development of novel therapeutics that selectively target NPMc$^+$-expressing AML.

13.6 Distinct MicroRNA-expression Signature of AML with Cytoplasmic NPM1

AML carrying cytoplasmic NPM1 mutations exhibit a unique gene expression profile. The microRNA (miRNA) expression profile of AML cells expressing NPMc$^+$ shows a distinct signature compared to AML cells expressing wild-type NPM1. The expression signature is characterized by the upregulation of miR10a, miR10b, and several family members of let−7 and miR−29, as well as downregulation of miR−204 and miR−128a (Garzon et al. 2008). The latter miRNAs are known to target several HOX (homeobox) genes and MEIS1 (Meis homeobox). Therefore, in AML expressing mt-NPM1, downregulation of miR204 could deregulate and/or increase the expression of HOX genes and MEIS1. This could lead to the development of AML by arresting cellular differentiation of hematopoietic progenitors. Additionally, upregulation of miR−10a expression in NPMc$^+$ AML can impact cellular defense mechanism by downregulating MDM4, a known mediator of cellular stress that interferes with p53 activation (Colombo et al. 2002; Ovcharenko et al. 2011). Gene-expression profiling studies (Mulligan et al. 2007) using a microarray platform also confirmed that AML cells expressing mt-NPM1 exhibit increased expression of multiple homeobox genes including HOXA9, A10, B2, B6, and MEIS1.

13.7 Targeted Therapy in AML Harboring Mutated NPM1

13.7.1 Targeting NPM1 Levels in Leukemic Cells

One of the major therapeutic hallmarks of cancer cells is uncontrolled cell proliferation (Hanahan and Weinberg 2011). The current anti-cancer strategies are focused on inhibiting cell proliferation, promoting differentiation, or the induction of apoptosis in cancer cells. NPM1 plays a role in regulation of cell growth and proliferation of cells. NPM1 is involved in regulating the sensitivity of leukemia cells to agents that induce apoptosis. Overexpression of NPM1 has been shown to decrease all-*trans* retinoic acid (ATRA)-mediated sensitivity of HL-60 leukemic cells (Hsu and Yung 2000) and blocks induction of differentiation (Hsu and Yung 2003). Downregulation of NPM1 increased the sensitivity of AML cells to sodium butyrate-induced apoptosis (Liu et al. 1999). In AML cells, mt-NPM1 dislocates into the cytoplasm, also promoting the translocation of the wild-type NPM1, with which it forms heterodimers, into the cytosol. AML cells expressing NPMc$^+$ also overexpress HOXA9 and Meis1, which are important regulators of hematopoiesis, and are expressed early in embryogenesis. Previous studies have shown that expression of HOXA9 and Meis1 is depleted during differentiation (Mulligan et al. 2007; Hu et al. 2009). Upregulation of HOXA9 and Meis1 has been implicated in leukemogenesis (Hu et al. 2009; Wang et al. 2010). Our recent studies demonstrated that knockdown of total NPM1 in NPMc$^+$ leukemia cells induced p53, p21, and C/EBP-α and depleted HOXA9, Meis1 mRNA, and protein. These alterations induced morphologic differentiation and diminished clonogenic survival of AML cells expressing mt-NPM1 (Fig. 13.2; Balusu et al. 2011). Additionally, knockdown of NPM1 significantly sensitized cultured NPMc$^+$ AML cells to ATRA, and the anti-leukemia agent Ara-C (Balusu et al. 2011; Qin and Shao 2011). Our in vivo studies demonstrated that NOD/SCID mice implanted with OCI-AML3 transduced with NPM1 shRNA, and demonstrating NPM1 knockdown, exhibited dramatically delayed leukemia initiation. Importantly, mice bearing OCI-AML3 cells transduced with NPM1 shRNA, showing knockdown of NPM1, also exhibited improved survival, as compared to the control mice ($p<0.01$).

13.7.2 Targeting Disruption of NPM1 Oligomerization and Nucleolar Assembly

NPM1 protein contains distinct functional domains through which it performs various biological functions. The crystal structure of the full-length NPM1 has not been elucidated, but the N-terminal oligomerization domain and the C-terminal region of the mutant and wild-type NPM1 have been well studied (Scaloni et al. 2010; Grummitt et al. 2008; Lee et al. 2007). The NPM1 oligomerization domain plays

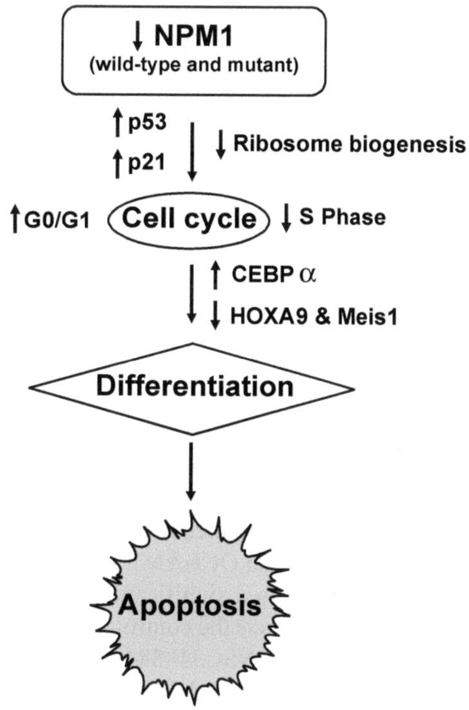

Fig. 13.2 Knockdown of total NPM1 (wild-type and mutant) causes cell cycle growth arrest, promotes differentiation, and induces apoptosis of NPMc⁺ AML cells

a critical role in chaperone activity and provides a dimeric interface for its binding partners. NPM1 is also a key element in the regulation of nucleolar functions. Targeted disruption of the oligomerization of NPM1 or abrogating its role in the regulation of nucleolar assembly may be a rational therapeutic strategy to treat AML cells harboring NPM1 mutations (Perea et al. 2009; Balusu et al. 2011)

13.7.2.1 Disruption of NPM1 Oligomerization

In the cell, NPM1 exists as dimeric and oligomeric forms through its oligomerization domain (Grisendi et al. 2006). The N-terminal oligomerization domain is critical for NPM1 chaperone activities (Fig. 13.1; Meani and Alcalay 2009). Since NPM1 gene mutations are heterozygous, the mutated NPM1 likely acts as a dominant negative mutant, causing the retention of wild-type NPM1 in the cytoplasm by heterodimerization through its oligomerization domain (Chen et al. 2006). The crystal structures of the carboxy-terminus NPM1c$^+$ (Scaloni et al. 2010; Grummitt et al. 2008) and its oligomerization domain (Lee et al. 2007) would aid in the design of small-molecule inhibitors of NPM1 oligomerization (Meani and Alcalay 2009). Screening of small-molecule inhibitors that can target the interactions between NPM1c$^+$ and its binding partners at the dimer interface would cause inhibition of NPM1 function. NSC348884 is the first small molecule inhibitor, screened in silico, that has been shown to disrupt oligomerization of NPM1. NSC348884 disrupts NPM1 oligomers and dimers and induces p53-mediated apoptosis in several solid tumor cell lines (Qi et al. 2008). We determined the effects of treatment with NSC348884 against AML cells with or without the expression of mt-NPM1 (Balusu et al. 2011). Treatment with NSC348884 disrupted oligomers and dimers of NPM1 in AML cells. Additionally, treatment with NSC348884, at lower concentrations, induced significantly more apoptosis in cultured and primary AML cells expressing mutant NPM1, as compared to AML or normal cells with homozygous expression of the wild-type NPM1 gene. Thus, AML cells expressing NPMc$^+$ are more sensitive to NSC348884-mediated disruption of NPM1 oligomers in the nucleus and cytoplasm and induction of apoptosis than AML cells with homozygous wild-type *NPM1* expression. The possible explanation for this differential sensitivity of AML cells expressing mt-NPM1 could be that these cells have less wild-type NPM1 protein in their nucleoli. Our studies also demonstrated that knockdown of NPM1 sensitized AML cells expressing mt-NPM1 to ATRA-induced differentiation, as indicated by an increase in the % CD11b positive cells, or by morphologic evaluation. Additionally, siRNA-mediated NPM1 knockdown enhanced ATRA-induced p53, p21, and C/EBPα in OCI-AML3 cells. Treatment with ATRA dose-dependently induced apoptosis to a significantly greater extent in OCI-AML3 cells treated with NPM1 siRNA versus the control siRNA-treated cells ($p < 0.05$) (Balusu et al. 2011). Co-treatment with NSC348884 and ATRA also induced more apoptosis in primary AML cells with mutant NPM1 than each agent alone (Balusu and Bhalla, unpublished observations).

Presence of NPM1 mutations have been shown to predict for a better response to AML induction therapy and for better overall survival. However, the overall outcome remains poor especially in the elderly AML patients and in AML patients with co-expression of mt-NPM1 and mutant FLT3 (Falini et al. 2009b, 2011). In NPM1c$^+$ AML, presence of FLT3-ITD is twofold more likely compared to AML negative for mutant NPM1 (Gale et al. 2008). However, co-expression of mt-NPM1 with FLT3-ITD confers a poor prognosis for remission duration and survival in AML, especially when the FLT3-ITD to wt-FLT3 allelic ratio is ≥ 0.5 (Becker et al. 2010; Schnittger et al. 2011b). AML blasts with high FLT3-ITD to wt-FLT3 allelic ratio are even more addicted to FLT3 ligand, the levels of which have been demonstrated to increase during treatment of AML with chemotherapy (Zheng et al. 2011; Sato et al. 2011). Additionally, FLT3 ligand has been demonstrated to impede the efficacy of FLT3 tyrosine kinase inhibitors in vitro and in vivo (Zheng et al. 2011; Sato et al. 2011). Balusu et al. (2011) demonstrated that sensitivity to NSC348884-induced apoptosis is diminished in the cultured and primary AML cells which co-express mt-NPM1 and FLT3-ITD. These observations have led to our ongoing studies, which are focused on determining the in vitro and in vivo activity of co-treatment with NSC348884 and FLT3 antagonist AC220 or ponatinib against cultured and primary AML cells co-expressing mt-NPM11 and FLT3-ITD (Balusu and Bhalla, unpublished observations).

13.7.2.2 Targeting Nucleolar Assembly

Casein kinase-2 (CK2) is a highly ubiquitous, pleiotropic, and constitutively active serine/threonine kinase that phosphorylates approximately 300 substrates (Meggio and Pinna 2003) and accounts for substantial proportion of the eukaryotic phosphoproteome (Pagano et al. 2006). The role of CK2 in cancer biology includes cell survival, chromatin remodeling, angiogenesis, and suppression of apoptosis (Perea et al. 2011). NPM1 is a known CK2 substrate (Falini et al. 2007). CIGB-300 is a peptide-based drug that targets the phosphoacceptor site and binds to CK2 substrates. Interestingly, NPM1 was also identified as a major target for CIGB−300 (Perera et al. 2009). The binding of CIGB-300 to NPM1 in cancer cells induces apoptosis. Impairment of CK2-mediated phosphorylation of NPM1 by CIGB-300 caused nucleolar disassembly and resulted in altered ribosomal biogenesis, which led to rapid and massive apoptosis (Perera et al. 2009).

13.7.3 Blocking Translocation of Mutant NPM1

Previous studies have demonstrated that CRM1-dependent translocation of mt-NPM1 from the nuclear compartment to the cytosol is a very crucial step in mt-NPM1-induced leukemogenesis (Falini et al. 2009b). Therefore, blocking mt-NPM1 dislocation to the cytoplasm may be an attractive therapeutic approach against AML. Leptomycin B is a potent inhibitor of CRM1 function that irreversibly binds with the sulfhydryl group of a Cys residue near or within the cargo-binding domain of CRM1 (alkylating Cys 528) (Ossareh-Nazari et al. 1997). In vitro studies have shown that treatment of mt-NPM1-expressing AML cells with leptomycin B redirected mt-NPM1 into the nucleus (Falini et al. 2006; Balusu et al. 2009). In vivo use of CRM1 inhibitors in NPM1c$^+$ AML is associated with certain limitations. Leptomycin B redirects NPM1 mutant proteins into the nucleoplasm, but not into the nucleolus, due to loss of the nucleolar localization domain (loss of tryptophan 288 and/or tryptophan 290). The other functional limitation of leptomycin B is its lack of specificity, since CRM1 is critical for the export of RNA and proteins containing a nuclear export sequence (NES) in normal and neoplastic cells (Falini et al. 2011). Developing CRM1 inhibitors with more specificity may help improve the therapeutic window and index (Mutka et al. 2009).

Conclusions

Over the past few decades, research has identified the structure and functional domains of NPM1 and its biologic role in cells. Since the discovery of NPM1 mutations in AML, remarkable progress has been made in biologic, diagnostic, and clinical characterization of mt-NPM1 in AML and other myeloid malignancies (Falini et al. 2005, Abdel-Wahab et al. 2011). The export of mt-NPM1 from the nucleus into the cytoplasm has been shown to affect multiple cellular pathways and is implicated in the pathogenesis of AML (Leong et al. 2010). Hence, targeting NPM1 appears to be a legitimate therapeutic approach in AML expressing mt-NPM1. As discussed in this chapter, targeting NPM1 levels, disrupting NPM1 oligomerization, or blocking translocation of mt-NPM1 could potentially be utilized to achieve this goal. This also underscores the need to develop more effective "drug-like" NPM1 antagonists and utilize these in combination with other differentiation therapies such as ATRA as a therapy against mt-NPM1-expressing AML, especially in elderly patients. Co-expression of mt-NPM1 with FLT3-ITD confers a poor prognosis for remission duration and survival in AML (Falini et al. 2009b). Treatment with a FLT3-TK inhibitor (TKI) alone, or in combination with chemotherapy, is currently under investigation in patients with AML. Based on the evidence presented above, in AML co-expressing mt-NPM1 and FLT3-ITD, a strong rationale has emerged for developing and testing the combination of novel NPM1 antagonist with FLT3-TKI. In AML where mt-NPM1 is co-expressed with other "epimutations" such as mutant DNMT3A, ASXL1 and IDH1 & 2, there also appears to be a rationale for testing the activity of NPM1 antagonist with agents that target epigenetic mechanisms (e.g., DNMT inhibitors and/or histone methyltransferases inhibitors).

References

Abdel-Wahab O, Patel J et al (2011) Clinical implications of novel mutations in epigenetic modifiers in AML. Hematol Oncol Clin North Am 25:1119–1133

Ahmad F, Mandava S et al (2009) Mutations of NPM1 gene in de novo acute myeloid leukaemia: determination of incidence, distribution pattern and identification of two novel mutations in Indian population. Hematol Oncol 27:90–97

Balusu R, Fiskus WC et al (2009) Targeting levels, aberrant localization or oligomerization of mutant nucleophosmin induces differentiation and loss of survival of human AML cells with mutant NPM1. Blood (ASH Annual Meeting Abstracts) December 5–8, 2009 114:2656

Balusu R, Fiskus W et al (2011) Targeting levels or oligomerization of nucleophosmin 1 induces differentiation and loss of survival of human AML cells with mutant NPM1. Blood 118:3096–3106

Becker H, Marcucci G et al (2010) Favorable prognostic impact of NPM1 mutations in older patients with cytogenetically normal de novo acute myeloid leukemia and associated gene- and microRNA-expression signatures: a cancer and leukemia Group B study. J Clin Oncol. 28:596–604

Bertwistle D, Sugimoto M et al (2004) Physical and functional interactions of the Arf tumor suppressor protein with nucleophosmin/B23. Mol Cell Biol 24:985–996

Bischof D, Pulford K et al (1997) Role of the nucleophosmin (NPM) portion of the non-Hodgkin's lymphoma-associated NPM-anaplastic lymphoma kinase fusion protein in oncogenesis. Mol Cell Biol 17:2312–2325

Boissel N, Renneville A et al (2005) Prevalence, clinical profile, and prognosis of NPM mutations in AML with normal karyotype. Blood 106:3618–3620

Bolli N, De Marco MF et al (2009) A dose-dependent tug of war involving the NPM1 leukaemic mutant, nucleophosmin, and ARF. Leukemia 23:501–509

Bolli N, Payne EM et al (2010) Expression of the cytoplasmic NPM1 mutant (NPMc+) causes the expansion of hematopoietic cells in zebrafish. Blood 115:3329–3340

Bonvini P, Zorzi E et al (2012) Consequences of heat shock protein 72 (Hsp72) expression and activity on stress-induced apoptosis in CD30+ NPM-ALK+ anaplastic large-cell lymphomas. Leukemia 26:1375–1382

Boonthimat C, Thongnoppakhun W et al (2008) Nucleophosmin mutation in Southeast Asian acute myeloid leukemia: eight novel variants, FLT3 coexistence and prognostic impact of NPM1/FLT3 mutations. Haematologica 93:1565–1569

Chang JH, Olson MO (1989) A single gene codes for two forms of rat nucleolar protein B23 mRNA. J Biol Chem 264:11732–11737

Chang JH, Dumbar TS et al (1988) cDNA and deduced primary structure of rat protein B23, a nucleolar protein containing highly conserved sequences. J Biol Chem 263:12824–12827

Chen W, Rassidakis GZ et al (2006) High frequency of NPM1 gene mutations in acute myeloid leukemia with prominent nuclear invaginations ('cuplike' nuclei). Blood 108:1783–1784

Chen Y, Gu L et al (2010) Relapsed APL patient with variant NPM-RARalpha fusion responded to arsenic trioxide-based therapy and achieved long-term survival. Int J Hematol 91:708–710

Cheng K, Sportoletti P et al (2010) The cytoplasmic NPM mutant induces myeloproliferation in a transgenic mouse model. Blood 115: 3341–3345

Chi HT, Vu HA et al (2010) Detection of exon 12 type A mutation of NPM1 gene in IMS-M2 cell line. Leuk Res 34:261–262

Chik F, Szyf M (2011) Effects of specific DNMT gene depletion on cancer cell transformation and breast cancer cell invasion; toward selective DNMT inhibitors. Carcinogenesis 32:224–232

Colombo E, Marine JC et al (2002) Nucleophosmin regulates the stability and transcriptional activity of p53. Nat Cell Biol 4:529–533

Colombo E, Alcalay M et al (2011) Nucleophosmin and its complex network: a possible therapeutic target in hematological diseases. Oncogene 30:2595–2609

Devillier R, Gelsi-Boyer V et al (2012) Acute myeloid leukemia with myelodysplasia-related changes are characterized by a specific molecular pattern with high frequency of ASXL1 mutations. Am J Hematol 87:659–662

Dingwall C, Dilworth SM et al (1987) Nucleoplasmin cDNA sequence reveals polyglutamic acid tracts and a cluster of sequences homologous to putative nuclear localization signals. EMBO J 6:69–74

Döhner K, Döhner H (2008) Molecular characterization of acute myeloid leukemia. Haematologica 93:976–982

Drexler HG, Gignac SM et al (2000) Pathobiology of NPM-ALK and variant fusion genes in anaplastic large cell lymphoma and other lymphomas. Leukemia 14:1533–1559

Dupuis-Coronas S, Lagarrigue F et al (2011) The Nucleophosmin-Anaplastic Lymphoma Kinase Oncogene Interacts, Activates, and Uses the Kinase PIKfyve to Increase Invasiveness. J Biol Chem 286:32105–32114

Dutta S, Akey IV et al (2001) The crystal structure of nucleoplasmin-core: implications for histone binding and nucleosome assembly. Mol Cell 8:841–853

Duyster J, Bai RY et al (2001) Translocations involving anaplastic lymphoma kinase (ALK). Oncogene 20:5623–5637

Ebert BL (2011) Molecular dissection of the 5q deletion in myelodysplastic syndrome. Semin Oncol 38:621–626

Falini B, Mecucci C et al (2005) Cytoplasmic nucleophosmin in acute myelogenous leukemia with a normal karyotype. N Engl J Med 352:254–266

Falini B, Martelli MP et al (2006) Immunohistochemistry predicts nucleophosmin (NPM) mutations in acute myeloid leukemia. Blood 108:1999–2005

Falini B, Nicoletti I et al (2007a) Acute myeloid leukemia carrying cytoplasmic/mutated nucleophosmin (NPMc + AML): biologic and clinical features. Blood 109:874–885

Falini B, Nicoletti I et al (2007b) Translocations and mutations involving the nucleophosmin (NPM1) gene in lymphomas and leukemias. Haematologica 92:519–532

Falini B, Bolli N et al (2009a) Altered nucleophosmin transport in acute myeloid leukaemia with mutated NPM1: molecular basis and clinical implications. Leukemia 23:1731–1743

Falini B, Sportoletti P et al (2009b) Acute myeloid leukemia with mutated NPM1: diagnosis, prognosis and therapeutic perspectives. Curr Opin Oncol 21:573–581

Falini B, Macijewski K et al (2010) Multilineage dysplasia has no impact on biologic, clinicopathologic, and prognostic features of AML with mutated nucleophosmin (NPM1). Blood 115:3776–3786

Falini B, Gionfriddo I et al (2011) Acute myeloid leukemia with mutated nucleophosmin (NPM1): any hope for a targeted therapy? Blood Rev 25:247–254

Falini B, Martelli MP et al (2011) Acute myeloid leukemia with mutated nucleophosmin (NPM1): is it a distinct entity? Blood 117:1109–1120

Fiskus W, Buckley K et al (2009) Panobinostat treatment depletes EZH2 and DNMT1 levels and enhances decitabine mediated de-repression of JunB and loss of survival of human acute leukemia cells. Cancer Biol Ther 8:939–950

Fröhling S, Scholl C et al (2005) Genetics of myeloid malignancies: pathogenetic and clinical implications. J Clin Oncol 23:6285–6295

Gale RE, Green C et al (2008) Medical Research Council Adult Leukaemia Working Party. The impact of FLT3 internal tandem duplication mutant level, number, size, and interaction with NPM1 mutations in a large cohort of young adult patients with acute myeloid leukemia. Blood 111:2776–2784

Garzon R, Garofalo M et al (2008) Distinctive microRNA signature of acute myeloid leukemia bearing cytoplasmic mutated nucleophosmin. Proc Natl Acad Sci U S A 105:3945–3950

Gilliland DG, Griffin JD (2002) The roles of FLT3 in hematopoiesis and leukemia. Blood 100:1532–1542

Grisendi S, Mecucci C et al (2006) Nucleophosmin and cancer. Nat Rev Cancer 6:493–505

Grummitt CG, Townsley FM et al (2008) Structural consequences of nucleophosmin mutations in acute myeloid leukemia. J Biol Chem 283:23326–23332

Hanahan D, Weinberg RA (2011) Hallmarks of cancer: the next generation. Cell 144:646–674

Herrera JE, Correia JJ et al (1996) Sedimentation analyses of the salt- and divalent metal ion- induced oligomerization of nucleolar protein B23. Biochemistry 35:2668–2673

Hingorani K, Szebeni A et al (2000) Mapping the functional domains of nucleolar protein B23. J Biol Chem 275:24451–24457

Hsu CY, Yung BY (2000) Over-expression of nucleophosmin/B23 decreases the susceptibility of human leukemia HL –60 cells to retinoic acid-induced differentiation and apoptosis. Int J Biometeorol (Journal international du cancer) 88:392–400

Hsu CY, Yung BY (2003) Involvement of nucleophosmin/B23 in TPA-induced megakaryocytic differentiation of K562 cells. Br J Cancer 89:1320–1326

Hu Y-L, Fong S et al (2009) HOXA9 Modulates Its Oncogenic Partner Meis1 To Influence Normal Hematopoiesis. Mol Cell Biol 29:5181–5192

Jeong EG, Lee SH et al (2007) Absence of nucleophosmin 1 (NPM1) gene mutations in common solid cancers. APMIS 115:341–346

Kang YJ, Olson MO et al (1975) Nucleolar phosphoproteins of normal rat liver and Novikoff hepatoma ascites cells. Cancer Res 35:1470–1475

Kindler T, Lipka DB et al (2010) FLT3 as a therapeutic target in AML: still challenging after all these years. Blood 116:5089–5102

Koh HJ, Lee SM et al (2004) Cytosolic NADP+ -dependent isocitrate dehydrogenase plays a key role in lipid metabolism. J Biol Chem 279:39968–39974

Kurki S, Peltonen K et al (2004) Nucleophosmin, HDM2 and p53—Players in UV damage incited nucleolar stress response. Cell Cycle 3:976–979

Lee HH, Kim HS et al (2007) Crystal structure of human nucleophosmin-core reveals plasticity of the pentamer-pentamer interface. Proteins 69:672–678

Lee WH, Salek-Ardakani S et al (2012) NPM-MLF1 synergizes with Npm haploinsufficiency to enhance myeloid progenitor activity. Leukemia 26:1110–1112

Leong SM, Tan BX et al (2010) Mutant nucleophosmin deregulates cell death and myeloid differentiation through excessive caspase –6 and –8 inhibition. Blood 116:3286–3296
Ley TJ, Ding L et al (2010) DNMT3A mutations in acute myeloid leukemia. New Engl J Med 363:2424–2433
Li Z, Boone D et al (2008) Nucleophosmin interacts directly with c-Myc and controls c-Myc-induced hyperproliferation and transformation. Proc Natl Acad Sci U S A 105:18794–18799
Lindstrom MS (2011) NPM1/B23: A multifunctional chaperone in ribosome biogenesis and chromatin remodeling. Biochem Res Int 2011:195–209
Lindstrom MS, Zhang Y (2008) Ribosomal protein S9 is a novel B23/NPM-binding protein required for normal cell proliferation. J Biol Chem 283:15568–15576
Liu WH, Hsu CY et al (1999) Nucleophosmin/B23 regulates the susceptibility of human leukemia HL –60 cells to sodium butyrate-induced apoptosis and inhibition of telomerase activity. Int J Cancer 83:765–771
Liu X, Liu Z et al (2007) Sumoylation of nucleophosmin/B23 regulates its subcellular localization, mediating cell proliferation and survival. Proc Natl Acad Sci U S A 104:9679–9684
Maiguel DA, Jones L et al (2004) Nucleophosmin sets a threshold for p53 response to UV radiation. Mol Cell Biol 24:3703–3711
Mamrack MD, Olson MO et al (1979) Amino acid sequence and sites of phosphorylation in a highly acidic region of nucleolar nonhistone protein C23. Biochemistry 18:3381–3386
Meani N, Alcalay M (2009) Role of nucleophosmin in acute myeloid leukemia. Expert Rev Anticancer Ther 9:1283–1294
Meggio F, Pinna LA (2003) One-thousand-and-one substrates of protein kinase CK2? FASEB J 17: 349–368
Morris SW, Kirstein MN et al (1994) Fusion of a kinase gene, ALK, to a nucleolar protein gene, NPM, in non-Hodgkin's lymphoma. Science 263:1281–1284
Mullighan CG, Kennedy A et al (2007) Pediatric acute myeloid leukemia with NPM1 mutations is characterized by a gene expression profile with dysregulated HOX gene expression distinct from MLL-rearranged leukemias. Leukemia 21:2000–2009
Murano K, Okuwaki M et al (2008) Transcription regulation of the rRNA gene by a multifunctional nucleolar protein, B23/nucleophosmin, through its histone chaperone activity. Mol Cell Biol 28:3114–3126
Mutka SC, Yang WQ et al (2009) Identification of nuclear export inhibitors with potent anticancer activity in vivo. Cancer Res 69:510–517
Namboodiri VM, Akey IV et al (2004) The structure and function of Xenopus NO38-core, a histone chaperone in the nucleolus. Structure 12:2149–2160
Naoe T, Suzuki T et al (2006) Nucleophosmin: a versatile molecule associated with hematological malignancies. Cancer Sci 97:963–969
Nerlov C (2004) C/EBP[alpha] mutations in acute myeloid leukaemias. Nat Rev Cancer 4:394–400
Nozawa Y, Van Belzen N et al (1996) Expression of nucleophosmin/B23 in normal and neoplastic colorectal mucosa. J Pathol 178:48–52
Oki Y, Jelinek J et al (2006) Mutations and promoter methylation status of NPM1 in myeloproliferative disorders. Haematologica 91:1147–1148
Okuda M (2002) The role of nucleophosmin in centrosome duplication. Oncogene 21:6170–6174
Okuwaki M (2008) The structure and functions of NPM1/Nucleophsmin/B23, a multifunctional nucleolar acidic protein. J Biochem 143:441–448
Okuwaki M, Matsumoto K et al (2001) Function of nucleophosmin/B23, a nucleolar acidic protein, as a histone chaperone. Febs Lett 506:272–276
Orrick LR, Olson MO et al (1973) Comparison of nucleolar proteins of normal rat liver and Novikoff hepatoma ascites cells by two-dimensional polyacrylamide gel electrophoresis. Proc Natl Acad Sci U S A 70:1316–1320
Ossareh-Nazari B, Bachelerie F et al (1997) Evidence for a role of CRM1 in signal-mediated nuclear protein export. Science 278:141–144

Ovcharenko D, Stolzel F et al (2011) miR-10a overexpression is associated with NPM1 mutations and MDM4 downregulation in intermediate-risk acute myeloid leukemia. Exp Hematol 39:1030–1042.e7

Pagano MA, Cesaro L et al (2006) Protein kinase CK2: a newcomer in the 'druggable kinome'. Biochem Soc Trans 34:1303–1306

Paschka P, Schlenk RF et al (2010) IDH1 and IDH2 mutations are frequent genetic alterations in acute myeloid leukemia and confer adverse prognosis in cytogenetically normal acute myeloid leukemia with NPM1 mutation without FLT3 internal tandem duplication. J Clin Oncol 28:3636–3643

Pasqualucci L, Li S et al (2008) NPM1-mutated acute myeloid leukaemia occurring in JAK2-V617 F + primary myelofibrosis: de-novo origin? Leukemia 22:1459–1463

Patel JP, Gonen M et al (2012) Prognostic relevance of integrated genetic profiling in acute myeloid leukemia. N Engl J Med 366:1079–1089

Perea SE, Baladron I et al (2011) CIGB-300, a synthetic peptide-based drug that targets the CK phosphoaceptor domain. Translational and clinical research. Mol Cell Biochem 356:45–50

Perera Y, Farina HG et al (2009) Anticancer peptide CIGB-300 binds to nucleophosmin/B23, impairs its CK2-mediated phosphorylation, and leads to apoptosis through its nucleolar disassembly activity. Mol Cancer Ther 8:1189–1196

Qi W, Shakalya K et al (2008) NSC348884, a nucleophosmin inhibitor disrupts oligomer formation and induces apoptosis in human cancer cells. Oncogene 27:4210–4220

Qin FX, Shao HY et al (2011) Knockdown of NPM1 by RNA interference inhibits cells proliferation and induces apoptosis in leukemic cell line. Int J Med Sci 8:287–294

Quentmeier H, Martelli MP et al (2005) Cell line OCI/AML3 bears exon-12 NPM gene mutation-A and cytoplasmic expression of nucleophosmin. Leukemia 19: 1760–1767

Raimondi SC, Dube ID et al (1989) Clinicopathologic manifestations and breakpoints of the t(3;5) in patients with acute nonlymphocytic leukemia. Leukemia 3:42–47

Ramsamooj P, Notario V et al (1995) Modification of nucleolar protein B23 after exposure to ionizing radiation. Radiat Res 143:158–164

Redner RL, Rush EA et al (1996) The t(5;17) variant of acute promyelocytic leukemia expresses a nucleophosmin retinoic acid receptor fusion. Blood 87:882–886

Rocquain J, Carbuccia N et al (2010) Combined mutations of ASXL1, CBL, FLT3, IDH1, IDH2, JAK2, KRAS, NPM1, NRAS, RUNX1, TET2 and WT1 genes in myelodysplastic syndromes and acute myeloid leukemias. BMC Cancer 10:401

Sato K, Hayami R et al (2004) Nucleophosmin/B23 is a candidate substrate for the BRCA1-BARD1 ubiquitin ligase. J Biol Chem 279:30919–30922

Sato T, Yang X et al (2011) FLT3 ligand impedes the efficacy of FLT3 inhibitors in vitro and in vivo. Blood 117:3286–3293

Scaloni F, Federici L et al (2010) Deciphering the folding transition state structure and denatured state properties of nucleophosmin C-terminal domain. Proc Natl Acad Sci U S A 107:5447–5452

Schnittger S, Bacher U et al (2011a) Characterization of NPM1-mutated AML with a history of myelodysplastic syndromes or myeloproliferative neoplasms. Leukemia 25: 615–621

Schnittger S, Bacher U et al (2011b) Prognostic impact of FLT3-ITD load in NPM1 mutated acute myeloid leukemia. Leukemia 25:1297–1304

Shandilya J, Swaminathan V et al (2009) Acetylated NPM1 localizes in the nucleoplasm and regulates transcriptional activation of genes implicated in oral cancer manifestation. Mol Cell Biol 29:5115–5127

Shiseki M, Kitagawa Y et al (2007) Lack of nucleophosmin mutation in patients with myelodysplastic syndrome and acute myeloid leukemia with chromosome 5 abnormalities. Leuk Lymphoma 48: 2141–2144

Smith CC, Wang Q et al (2012) Validation of ITD mutations in FLT3 as a therapeutic target in human acute myeloid leukaemia. Nature 485:260–263

Stirewalt DL, Radich JP (2003) The role of FLT3 in haematopoietic malignancies. Nat Rev Cancer 3:650–665

Subong EN, Shue MJ et al (1999) Monoclonal antibody to prostate cancer nuclear matrix protein (PRO:4–216) recognizes nucleophosmin/B23. The Prostate 39:298–304

Swaminathan V, Kishore AH et al (2005) Human histone chaperone nucleophosmin enhances acetylation-dependent chromatin transcription. Mol Cell Biol 25:7534–7545

Szebeni A, Olson MO (1999) Nucleolar protein B23 has molecular chaperone activities. Protein Sci 8:905–912

Takahashi S (2011) Current findings for recurring mutations in acute myeloid leukemia. J Hematol Oncol 4:36

Tiacci E, Spanhol-Rosseto A et al (2012) The NPM1 wild-type OCI-AML2 and the NPM1-mutated OCI-AML3 cell lines carry DNMT3A mutations. Leukemia 26: 554–557

Tokuyama Y, Horn HF et al (2001) Specific phosphorylation of nucleophosmin on Thr(199) by cyclin-dependent kinase 2-cyclin E and its role in centrosome duplication. J Biol Chem 276: 21529–21537

Vassiliou GS, Cooper JL et al (2011) Mutant nucleophosmin and cooperating pathways drive leukemia initiation and progression in mice. Nat Genet 43:470–475

Verhaak RGW, Goudswaard CS et al (2005) Mutations in nucleophosmin (NPM1) in acute myeloid leukemia (AML): association with other gene abnormalities and previously established gene expression signatures and their favorable prognostic significance. Blood 106:3747–3754

Walter MJ, Shen D et al (2012) Clonal architecture of secondary acute myeloid leukemia. N Engl J Med 366:1090–1098

Wang D, Umekawa H et al (1993) Expression and subcellular locations of two forms of nucleolar protein B23 in rat tissues and cells. Cell Mol Biol Res 39:33–42

Wang Z, Iwasaki M et al (2010) GSK-3 promotes conditional association of CREB and its coactivators with MEIS1 to facilitate HOX-mediated transcription and oncogenesis. Cancer Cell 17:597–608

Wanzel M, Russ AC et al (2008) A ribosomal protein L23-nucleophosmin circuit coordinates Miz1 function with cell growth. Nat Cell Biol 10:1051–1061

Yan XJ, Xu J et al (2011) Exome sequencing identifies somatic mutations of DNA methyltransferase gene DNMT3A in acute monocytic leukemia. Nat Genet 43:309–315

Yang C, Maiguel DA et al (2002) Identification of nucleolin and nucleophosmin as genotoxic stress-responsive RNA-binding proteins. Nucleic Acids Res 30:2251–2260

Yip S, Siu P et al (2011) The Multifunctional Nucleolar Protein Nucleophosmin/NPM/B23 and the Nucleoplasmin Family of Proteins. In: Olson MOJ (ed) The nucleolus. Springer, New York, p 213–252

Yoneda-Kato N, Kato JY (2008) Shuttling imbalance of MLF1 results in p53 instability and increases susceptibility to oncogenic transformation. Mol Cell Biol 28:422–434

Yoneda-Kato N, Look AT et al (1996) The t(3;5)(q25.1;q34) of myelodysplastic syndrome and acute myeloid leukemia produces a novel fusion gene, NPM-MLF1. Oncogene 12:265–275

Yu Y, Maggi LB et al (2006) Nucleophosmin is essential for ribosomal protein L5 nuclear export. Mol Cell Biol 26:3798–3809

Zeller KI, Haggerty TJ et al (2001) Characterization of nucleophosmin (B23) as a Myc target by scanning chromatin immunoprecipitation. J Biol Chem 276:48285–48291

Zhang Y, Zhang M et al (2007) NPM1 mutations in myelodysplastic syndromes and acute myeloid leukemia with normal karyotype. Leuk Res 31:109–111

Zheng R, Bailey E et al (2011) Further activation of FLT3 mutants by FLT3 ligand. Oncogene 30:4004–4014

Chapter 14
Raf/MEK/ERK Signaling

James A. McCubrey, Linda S. Steelman, Jörg Bäsecke and Alberto M. Martelli

Abstract The Ras/Raf/MEK/ERK cascade is often activated by genetic alterations in upstream signaling molecules such as receptor tyrosine kinases (RTK). Certain components of these pathways, *RAS, NF1, BRAF, and MEK1* are also deregulated by mutations. Upstream mutations in one signaling pathway or even in downstream components of the same pathway can alter the sensitivity of the cells to certain small molecule inhibitors. This pathway has profound effects on proliferative, apoptotic and differentiation pathways. Dysregulation of components of this pathway can contribute to: resistance to other pathway inhibitors and chemotherapeutic drugs. Intense scientific, commercial, and clinical interest has developed over the past decade in elucidating the functions of this pathway in leukemia and other blood cancers. Targeting this pathways is often complex and can result in pathway activation depending on the presence of upstream mutations (e.g., Raf inhibitors induce Raf activation in cells with wild type *RAF* in the presence of mutant, activated *RAS*). Furthermore, targeting with inhibitors directed at two constituents of the same pathway may be a more effective approach. This chapter will first describe these pathways and then evaluate potential uses of Raf and MEK inhibitors that have been investigated in pre-clinical and clinical investigations.

Keywords Targeted therapy · Therapy resistance · Mutation · Raf · B-Raf · MEK

J. A. McCubrey (✉) · L. S. Steelman
Department of Microbiology and Immunology, Brody School of Medicine at East Carolina University, Brody Building 5N98C, Greenville, NC 27858, USA
e-mail: mccubreyj@ecu.edu

J. Bäsecke
Department of Medicine, University of Göttingen, Göttingen, Germany

A. M. Martelli
Dipartimento di Scienze Anatomiche Umane e Fisiopatologia dell'Apparato Locomotore, Università di Bologna, Bologna, Italy

IGM-CNR, Sezione di Bologna, C/o IOR, Bologna, Italy

M. Andreeff (ed.), *Targeted Therapy of Acute Myeloid Leukemia*,
Current Cancer Research, DOI 10.1007/978-1-4939-1393-0_14

14.1 Introduction

Since the discovery of the *RAS, RAF, MEK* oncogenes and neurofibromin 1 (*NF1*) tumor suppressor genes, the Ras/Raf/MEK/ERK signaling cascade has been extensively investigated with the ultimate goal of determining how these genes become activated/inactivated and whether it is possible to suppress their activity in leukemia and other cancers (Chappell et al. 2011). Furthermore this pathway is also implicated in the resistance and sometimes sensitivity to therapy (McCubrey et al. 2011). There have been breakthroughs in the discovery of complex interacting pathway components, their genetic and epigenetic regulation, the complicated mechanisms by which they relay their signals, the key post-translational modifications of these proteins and how mutations of these components can lead to aberrant signaling, uncontrolled proliferation and in some cases confer sensitivity to targeted therapy. Research has also resulted in the development of inhibitors that target critical components of these pathways with the ultimate goal to increase patient survival (Solit et al. 2006). This chapter will review some of the current inhibitors, their targets and how they are being used to treat blood cancers and overcome therapeutic resistance.

Before we discuss the Ras/Raf/MEK/ERK signaling cascade, it is important to discuss some genetic terms as they are critical to understanding the importance of these pathways and the classes of mutations that occur (Greenman et al. 2007; Weinstein et al. 2008; Torkamani et al. 2009; Negrini et al. 2010; Whittaker et al. 2010a; Sharma et al. 2010; Caponigro et al. 2011; Chan et al. 2011). First is the concept of a driver mutation. This is a mutation that is statistically enriched in a particular cancer and usually thought to be one of the first events in the transformation of those particular cells to cancer cells (Greenman et al. 2007; Torkamani et al. 2009). Examples of driver mutations in these two pathways include *RAS*, *NF1*, *BRAF* and *MEK1*. Passenger mutations have no known effects on the fitness of transformed cells (so far) but may be linked with the progression of the cell containing the driver mutation from a pre-malignant cell into a full blown malignant tumor (Greenman et al. 2007; Chan et al. 2011). Caretaker genes are involved in genomic stability and function to suppress the mutation rate (Greenman et al. 2007; Negrini et al. 2010). Caretaker mutations occur mainly in tumor suppressor genes, such as *TP53* (p53). *TP53* is both a driver and a caretaker genes. Caretaker genes help maintain the integrity of the genome (Greenman et al. 2007).

Gatekeeper genes directly regulate cell growth and their loss can lead to tumorigenesis. They encode critical proteins which can regulate growth or the induction of apoptosis. Many genes fall into this class including: *MAPK3/MAPK1* (ERK1/ERK2), *TP53*, and *NF1*. Obviously some genes can fall into multiple classifications (Whittaker et al. 2010a).

Gatekeeper mutations often occur in genes (often protein kinases) in the ATP binding domain. They are detected in *BRAF*, *BCR-ABL*, and *EGFR* and can mediate resistance to small molecule inhibitors as that is often where they bind and inhibit activity.

When there is an activated oncogene or inactivated tumor suppressor gene in a cell which results in the proliferation of the cells, a synthetic lethal mutation may occurs at a second gene which results in the death of the transformed cell (Chan et al. 2011). In essence, there is the loss of a biochemical interaction between the mutant oncogene or tumor suppressor gene and the second gene and the cell dies. Hence the second mutation is referred to as a synthetic lethal mutation. In terms of the Ras/Raf/MEK/ERK pathway, which proliferate in response to mutant *KRAS*, silencing of genes such as *v*oltage-*d*ependent *a*nion *c*hannel (*VDAC1*), *s*erine/*t*hreonine *k*inase 33 (*STK33*), *T*ANK- *b*inding *k*inase 1 (TBK) or *p*olo-*l*ike *k*inase-1 (*PLK1*) results in synthetic lethal interactions (Chan et al. 2011).

Oncogene addiction is a widely-used term indicating that the cells with a mutation at a particular oncogene, or correspondingly, inactivation of a tumor suppressor gene become addicted to the consequences of that mutation and grow under conditions where a normal cell would not persist (Weinstein et al. 2008). Chronic myeloid leukemia cells become addicted to mutant *BCR-ABL* for proliferation (Luo et al. 2009). Non-oncogene addiction is a more recently devised term to describe the addiction of a cell on another protein is which not an oncogene per se (Luo et al. 2009). Interesting it has been observed that upon inactivation of the oncogene responsible for survival, the pro-survival and pro-apoptotic signals decay at different rates. In absence of the oncogene responsible for the oncogene addiction phenotype, the pro-survival signals decayed more rapidly than the pro-apoptotic signal. This has lead to the concept of oncogenic shock and provides the basics for the success of certain inhibitors in suppressing the growth of oncogene-transformed cells (Sharma et al. 2010). Now that we have discussed some general genetic terms, we can discuss in more detail the Ras/Raf/MEK/ERK pathway.

14.2 The Ras/Raf/MEK/ERK Pathway

Usually signaling commences upon ligation of a growth factor/cytokine/interleukin/mitogen (ligand) to its cognate receptor at the cell surface. This event can result in the activation of many downstream signaling cascades including the Ras/Raf/MEK/ERK pathway which can further transmit their signals to the nucleus to control gene expression, to the translational apparatus to enhance the translation of "weak" mRNAs, to the apoptotic machinery to regulate apoptosis or to other events involved in the regulation of cellular proliferation (for example, interactions with the p53 pathway to regulate cell cycle progression). Regulation of the Ras/Raf/MEK/ERK pathway is mediated by a series of kinases, phosphatases, GTP:GDP exchange and scaffolding proteins. There are also many tumor suppressor proteins which interact with this cascade which frequently serve to fine tune or limit activity (e.g., RKIP, PP2A). Mutations occur in many of the genes in these pathway leading to uncontrolled regulation and aberrant signaling. An overview of the effects of mutations and the activation of the Ras/Raf/MEK/ERK signaling pathway is presented in Fig. 14.1. Also presented in this figure is the concept that ERK is pivotal in the regulation of gene transcription, mRNA translation and apoptosis.

Following stimulation of a receptor with a ligand, a *S*rc *h*omology 2 domain *c*ontaining protein (Shc) adaptor protein becomes associated with the C-terminus of the activated *g*rowth *f*actor *r*eceptor (GFR). Shc recruits the growth *f*actor *r*eceptor-*b*ound protein *2* (Grb2) protein and the *s*on *o*f *s*evenless (SOS) homolog protein [a *g*uanine nucleotide *e*xchange *f*actor (GEF)], resulting in the loading of the membrane-bound GDP:GTP exchange protein (GTPase) Ras with GTP (McCubrey et al. 2011). *RAS* is considered a driver gene and is frequently mutated blood cancers. GEFs promote Ras activation by displacing GDP from Ras which leads to GTP binding. Ras activation is suppressed by the *G*TPase activating *p*roteins (GAPs) that stimulate the GTPase activity of Ras. There are two prominent GAP proteins, p120GAP and *n*euro*f*ibromin *1* (NF1). *NF1* is a tumor suppressor and has both driver and gatekeeper gene functions. Germline mutations at *NF1* lead to neurofibromatosis (Simmons et al. 2011). Ras can also be activated by *g*rowth *f*actor *r*eceptor *t*yrosine *k*inases (GFRTK), such as *i*nsulin *r*eceptor (IR), via intermediates like *i*nsulin *r*eceptor *s*ubstrate (IRS) proteins that bind Grb2 (Hayashi et al. 2004). Ras:GTP then recruits the serine/threonine (S/T) kinase Raf to the membrane where it becomes activated, likely via a Src-family tyrosine (Y) kinase (McCubrey et al. 2011).

Both *RAS* and *RAF* are members of multi-gene families and there are three Ras members (*KRAS*, *NRAS* and *HRAS*) and three *RAF* members [*BRAF*, *RAF1* (a.k.a c-Raf) and *ARAF*) (McCubrey et al. 2011). Activation of B-Raf and Raf-1 differ in that Raf-1 requires activation by a Src family kinase while B-Raf does not (McCubrey et al. 2011). Raf-1 can be regulated by dephosphorylation by the protein serine/threonine phosphatase 2A (PP2A) and others (Mischak et al. 1996; Abraham et al. 2000). PP2A has been reported to positively and negatively regulate Raf-1. PP2A is also considered a tumor suppressor gene and has gatekeeper gene functions. PP2A is considered a drugable target in CML (Perrotti et al. 2008).

Raf is responsible for S/T phosphorylation of mitogen-activated protein kinase kinase-1 (MEK1) (a dual specificity kinase (T/Y) (McCubrey et al. 2011). Other proteins such as kinase suppressor of Ras (KSR) have been shown recently to phosphorylate MEK1 (Brennan et al. 2011; McKay et al. 2011). KSR has scaffolding properties and interacts with Raf, MEK and ERK. These interactions regulate ERK activation. KSR can form dimers with various Raf proteins which alter the effects of Raf inhibitors. KSR competes with Raf-1 for Raf inhibitor-induced binding to B-Raf which decreases the normal ERK activation observed after Raf-inhibitor treatment (McKay et al. 2011).

MEK1 phosphorylates *e*xtracellular signal *r*egulated *k*inases 1/2 (ERK1 and 2) at specific T and Y residues (McCubrey et al. 2011). *MEK1* was originally not thought to be mutated frequently in human cancer. However, recent large scale mutation screening studies and studies aimed at determining mechanisms of resistance to small molecule inhibitors have observed that *MEK1* is mutated in certain human cancers and also is mutated in certain inhibitor resistant cells. *MEK1* is also considered to be a driver oncogene in certain cancers (Pao et al. 2011). Rac (Ras related gene) and *p*21-*a*ctivating *k*inases (PAK) can also regulate MEK/ERK activation (Eblen et al. 2002).

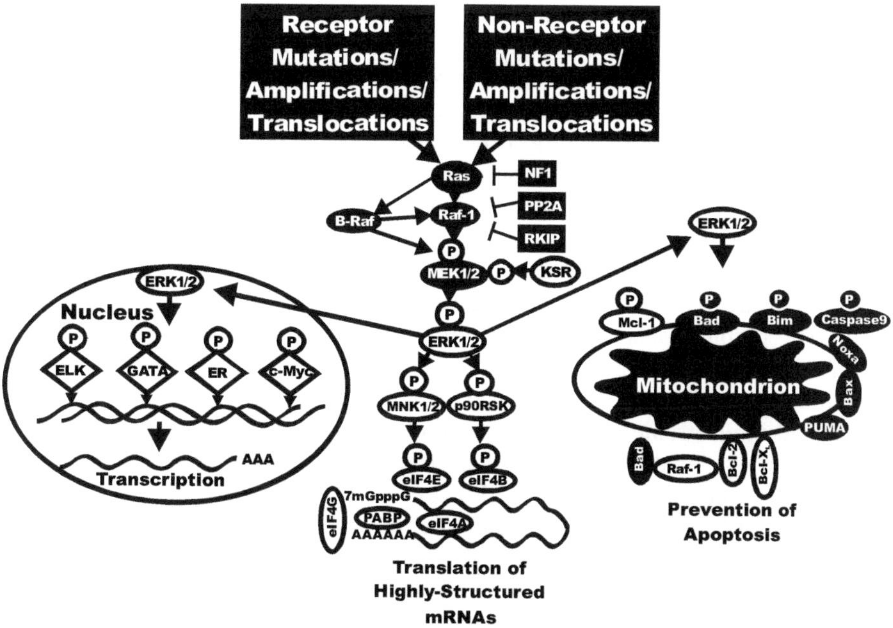

Fig. 14.1 Dysregulated expression of upstream receptors and kinases can result in activation of the Ras/Raf/MEK/ERK pathway. Sometimes dysregulated expression of growth factor receptors occurs by increased expression, genetic translocations or genomic amplifications which can lead to activation of the Ras/Raf/MEK/ERK pathways. Alternatively chromosomal translocations can occur in non-receptor kinases and other genes which result in activation of these pathways. Genes in the Ras/Raf/MEK/ERK pathway that have activating mutations detected in human cancer and proliferative diseases are indicated in *blue* ovals. Neurofibromin (NF1) is a GTPase activating protein which is a tumor suppress. Loss of NF-1 activity results in elevated Ras activation. Germline mutations of NF-1 result in neurofibromatosis. Genes mutated are indicated in *ovals*. Transcription factors are indicated in *diamonds*. Tumor suppressor genes mutated in human cancer are indicated in *rectangles*. Other key genes are indicated in *ovals*. *Arrows* indicate activating events in pathways. *Blocked arrows* indicate inactivating events in pathway. *PABP* poly A binding protein

Activated ERK1 and ERK2 S/T kinases phosphorylate and activate a variety of substrates, including p90 Ribosomal six kinase-1 ($p90^{Rsk1}$) (McCubrey et al. 2011). ERK1, 2 are considered by some as gatekeeper genes. Figure 14.1 illustrates how ERK can regulate many important events involved in cell proliferation. ERK also phosphorylates MAPK signal integrating kinases (Mnk1/2) which can in turn phosphorylate (eukarytotic translation initiation factor 4E) eIF4E, a key protein involved in the translation of difficult mRNAs (Topisirovic et al. 2011). *EIF4E* is considered to be a gatekeeper gene.

$p90^{Rsk1}$ can activate the *c*AMP *r*esponse *e*lement *b*inding protein (CREB) transcription factor as well as proteins involved in regulation of protein translation (e.g., Mnk-1, *p70* ribosomal *S6 k*inase (p70S6K), *e*ukaryotic translation *i*nitiation *f*actor *4B*, (eIF4B), and *r*ibosomal *p*rotein *S6* (rpS6) (Xing et al. 1996; McCubrey et al. 2011; Topisirovic et al. 2011).

The number of ERK1/2 substrate/targets is easily in the hundreds. These substrate/targets include different types of molecules including: other kinases, transcription factors, or proteins involved in protein translation or apoptosis. Suppression of MEK and ERK can have profound effects on cell growth, inflammation and aging. Activated ERK can also phosphorylate "upstream" Raf-1 and MEK1 which alter their activity. Depending upon the site phosphorylated on Raf-1, ERK phosphorylation can either enhance (Balan et al. 2006) or inhibit (Dougherty et al. 2005) Raf-1 activity. In contrast, some studies have indicated that when MEK1 is phosphorylated by ERK, its activity decreases (Catalanotti et al. 2009). Recent studies indicate that ERK does not negatively feedback inhibit B-Raf (Sturm et al. 2010). By computational and basic scientific experiments, it has been demonstrated that the mammalian Raf/MEK/ERK pathway exhibits properties of a *n*egative *f*eedback *a*mplifier (NFA). In essence, NFA signaling is similar in biological design to those used in electronic circuits. NFAs in electronic circuits optimize robustness, stabilization of signal and linearization of non linear signal amplification. These properties of the Raf/MEK/ERK NFA are important in determining activation kinetics, response to drugs and various other downstream effects of activated ERK (Sturm et al. 2010).

These phosphorylation events induced by ERK serve to alter the stability and/or activities of the proteins. These examples of feed-back loops become important in consideration of whether to just target MEK or to target both Raf and MEK in blood cancers. It is important that the reader realize that certain phosphorylation events can either inhibit or repress the activity of the affected protein. This often depends on the particular residue on the protein phosphorylated which can confer a different configuration to the protein or target the protein to a different subcellular localization that may result in proteasomal degradation or association with certain scaffolding proteins (e.g., 14-3-3) (Sorokina et al. 2011), *M*EK *p*artner-*1* (MP-1) (Schaeffer et al. 1998), *h*eat *s*hock *p*rotein-*90* (HSP-90) (Grammatikakis et al. 1999), KSR (Goettel et al. 2011) *R*af *k*inase *i*nhibitory *p*rotein (RKIP) (Yeung et al. 2000). RKIP is also considered a metastasis suppressor gene in certain cancers and has gatekeeper and caretaker effects. Heat shock proteins such as HSP-90 are considered caretakers as they normally serve to protect the activity of client proteins. Mutations at *KRAS* will confer sensitivity to HSP90 inhibitors such as geldanamycin, documenting the importance of HSP90 in regulation of this pathway (Sos et al. 2009; Trepel et al. 2010). Raf-1 activation by Ras has been shown to be dependent on the prohibitin protein, an ubiquitously expressed protein which may also serve as a chaperonin protein (Rajalingam et al. 2005).

The regulation of ERK1,2 activity in the nucleus and cytoplasm is complex as the p38MAPKα splice isoform *M*ax-*i*nteracting protein (Mxi-2) can bind ERK1,2, and ensure its translocation into the nucleus and prevent both *M*AP*K p*hosphatase-*1* (*MKP1*) and *d*ual *s*pecificity *p*hosphatase *5* (*DUSP5*) from dephosphorylating ERK1,2 in the nucleus and not the cytoplasm (Casar et al. 2012). Most phosphatases will probably eventually be shown to be tumor suppressor genes. Upon Mxi-2 binding ERK1,2, enhanced ERK1,2 activity is detected in the nucleus. Mxi-2 prevents the dephosphorylation of ERK1,2 by MKP1 and DUSP5. This allows activated ERK1,2 to phosphorylate c-Myc and other critical substrates (Casar et al. 2012).

In the nucleus ERK can phosphorylate transcription factors, such as: *E* twenty-six-*like* transcription factor 1 (Elk-1), *e*strogen *r*eceptor (ER), Fos, *g*lobin *t*ranscription f*a*ctor 1 (Gata-1), c-Myc, *S*ignal *t*ransducer *a*ctivation of *t*ranscription 1 & 3 (STAT1 & 3) and others (Chang et al. 2003; Cheng et al. 2007). These transcription factors bind promoters of many genes, including growth factor and cytokine genes that are important in promoting growth and preventing apoptosis of multiple cell types (Fig. 14.1).

ERK can also phosphorylate and modulate the activity of the Twist, Snail, Slug, and Zeb1 transcription factors either directly or indirectly which can regulate cellular proliferation, survival and some can modulate *e*pithelial *m*esenchymal *t*ransition (EMT) (Graham et al. 2008; Chen et al. 2009; Hong et al. 2011; Nagarajan et al. 2012). Phosphorylation of the transcription factors by ERK1,2, or in some cases the related MAPK, p38MAPK, prevents their ubiquitination and results in their stabilization and increased activity in the nucleus and ability to promote EMT (Graham et al. 2008; Chen et al. 2009; Shin et al. 2010; Hong et al. 2011; Nagarajan et al. 2012).

In the nucleus, ERK can also phosphorylate *m*itogen- and *s*tress-activated protein *k*inases (MSKs) (Kyriakis et al. 2001; Roux et al. 2004) which in turn can phosphorylate transcription factors including *a*ctivator *t*ranscription *f*actor-1 (ATF-1) that are important in the regulation of many immediate early genes controlled by *a*ctivating *p*rotein-*1* (AP-1) (Gupta et al. 2002). The *t*ernary *c*omplex *f*actors (TCF) such as Elk-1, Sap-1 and Net are also phosphorylated by ERK which results in their activation (Costello et al. 2004; Costello et al. 2010). The TCFs form complexes with *s*erum *r*esponsive *f*actor (SRF) and activate many genes through their *s*erum *r*esponsive *e*lements (SRE) in their promoter regions (Mylona et al. 2010).

MSKs also phosphorylate many proteins involved in modulating chromatin structure including: Histone H3, and (*h*igh-*m*obility-group protein-*14*) HMG14 which can result in the transcription of immediate early genes after mitogens/growth factor stimulation (Arthur 2008). ERK1/2 can phosphorylate many proteins critical for cytoskeletal structure/reorganization including: *calpain* (Capn) (Cuevas et al. 2003), *F*ocal *A*dhesion *K*inase (FAK) (Zheng et al. 2009), *M*yosin *L*ight *P*olypeptide *K*inase (MLCK) and *Pa*xillin-*6* (Pax6) (Webb et al. 2004; Zheng et al. 2011). Sometimes phosphorylation by ERK of FAK can result in FAK dephosphylation (Webb et al. 2004; Zheng et al. 2011). The Ras/Raf/MEK/ERK pathway is important in determining cellular shape and mobility/invasion. Under certain circumstances, aberrant regulation of this pathway can contribute to abnormal cellular growth, mobility and invasion which may lead to many abnormalities including various leukemias and other blood cancers.

Thus the reader begins to understand how the Ras/Raf/MEK/ERK pathway can regulate the expression of many genes involved in the response to growth factors and mitogens. Furthermore many of the genes in this pathway, as well as other genes that regulate the activity of this pathway, have various properties to influence cancer development. Components of this pathway can sometimes be driver, gatekeeper or caretaker genes.

14.3 Mutations or Altered Expression of the Ras/Raf/MEK/ERK Pathway Can Lead to Sensitivity to Therapy

An overview of some of the sites where mutations influence the activity of the Ras/Raf/MEK/ERK pathway is depicted in Fig. 14.2. Sites of intervention with signal transduction inhibitors are also presented. Some of the inhibitors are currently being used to treat patients with specific cancers and others have been or are being evaluated in numerous clinical trials with many different types of cancer patients. First we will discuss how mutations will make cells sensitive to targeted therapy, later we will discuss how certain mutations make cells resistant to targeted therapy. We will also discuss studies performed on solid cancers as they can serve as models for leukemia in terms of how mutations contribute to sensitivity to therapy. Mutations at the *RAS* family of genes are among the most frequently detected in human cancer (Downward 2003; Steelman et al. 2004; McCubrey et al. 2007; Steelman et al. 2008; Buday et al. 2008). The frequency of *RAS* mutations varies according to the tissue type with some types of cancer having very frequent mutations (e.g., pancreatic, colon, hematopoietic cancers) with other types of human cancer having lower frequencies (Downward 2003; Steelman et al. 2004; McCubrey et al. 2007; Steelman et al. 2008; Buday et al. 2008). Mutations in *KRAS* make cells sensitive to HSP90 inhibitors (37,38) *BRAF* is mutated frequently in melanomas (50–70%) (Davies et al. 2002), Langerhans'-cell histiocytosis (57%) (Badalian-Very et al. 2010), papillary thyroid cancers (40%) (Puxeddu et al. 2004) and colon cancer (8%) (Tol et al. 2009). BRAF is mutated to lesser extent (2–3%) in *n*on *s*mall *c*ell *l*ung *c*ancers (NSCLC) (Brose et al. 2002). Recently *BRAF* has been observed to be frequently mutated (100%) in hairy cell leukemia (Tiacci et al. 2011). *BRAF* has been observed to be mutated in 8 of 199 patients (4%) with *m*ultiple *m*yeloma (MM) and 4 of those were mutant at *BRAF* V600E (Chapman et al. 2011). A clinical trial with B-Raf inhibitors and MM is in progress. Other B-ALL and peripheral B cell lymphomas have been observed to have low (>3%) frequencies of *BRAF* mutations, but none of these mutants produced the B-Raf V600E protein (Case et al. 2008). Similar mutations were not detected in the Tiacci *et al.* study with similar leukemias and lymphomas. A recent study detected *BRAF* mutations in 2/55 (3.6%) of large B-cell lymphoma (DLBCL) patient samples and the authors postulated that they may be considered as driver mutations for those DLBCL (Lohr et al. 2012). Cancer patients with the *BRAF* driver mutations are postulated to be sensitive to B-Raf inhibitors such as vemurafenib, dabrafenib, and GDC-0879. Some cancer cells carrying *BRAF* mutations were highly sensitive to MEK inhibitors, while cells lacking these *BRAF* mutations or containing *RAS* or *EGFR* mutations were resistant (Solit et al. 2006; Pratilas et al. 2008). Although it was initially thought that *BRAF* was not frequently mutated in blood cancers, more recent studies have shown that it is mutated in some blood cancers and hence it may be an appropriate target in those blood cancers where it is mutated and *RAS* is not mutated (see below).

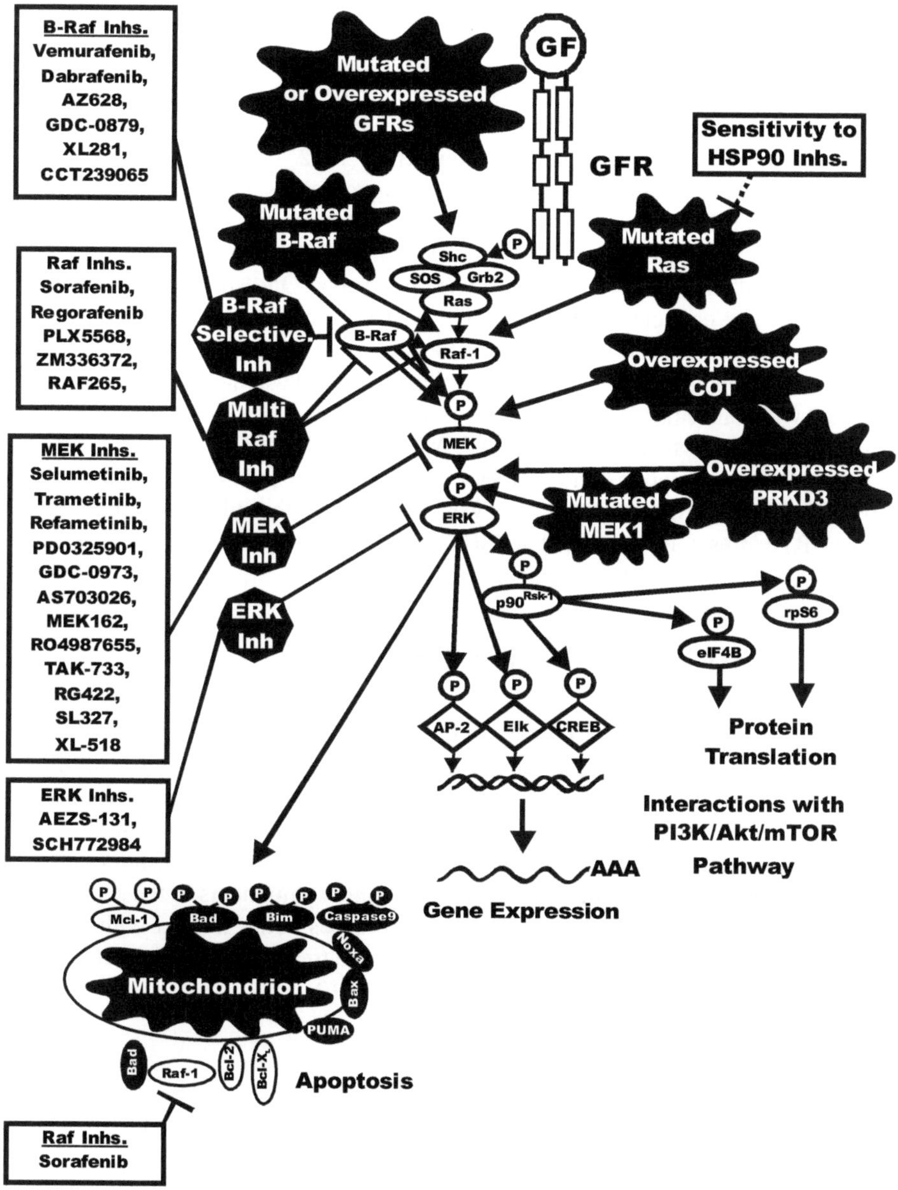

Fig. 14.2 Conceptual overview of targeting the Raf/MEK/ERK pathway to suppress malignant growth. Sites of actions of various types inhibitors are indicated in *black octagons*. Specific inhibitors are listed in *clear boxes* to the *left*. Types of mutations which either activate this pathway or confer resistance (in some cases they can do both) are indicated in *black irregular* shaped objects

14.4 Predicting Sensitivity to Small Molecule Inhibitors

Recent studies have examined panels of cell line for mutations of genes implicated in cancer for their sensitivity to various inhibitors and chemotherapeutic drugs commonly used to treat cancers (Barretina et al. 2012; Garnett et al. 2012). The cell lines were examined by expression profiling, chromosome copy number, deep sequencing, biostatistical and system analyses. Both studies indicated that sensitivity to inhibitors was often associated with genetic mutations at key elements in the Ras/Raf/MEK/ERK and some other pathways. One study has generated a "cancer cell line encyclopedia" which will be useful for predictive modeling of inhibitor sensitivity (Barretina et al. 2012). Sensitivity to Raf and MEK inhibitors was investigated. Sensitivity to the B-Raf inhibitor PLX4720 was shown to be highly associated with particular mutations at *BRAF* (*V600E*). Sensitivity to MEK inhibitors was shown to be associated with *BRAF*, *NRAS* as well as *PTEN*, *PTPN5*, *SPRY2*, *DUSP4*, *DUSP6* mutations and to a lesser extent mutations at *KRAS*. Sensitivity to MEK inhibitors in *NRAS* mutant lines was associated with *a*ryl *h*ydrocarbon *r*eceptor (AHR) expression (Garnett et al. 2012).

14.5 Raf/MEK/ERK Pathway Inhibitors

Effective inhibitors specific for many of the key components of the Raf/MEK/ERK pathway have been developed (Chappell et al. 2011). In many cases, these inhibitors have been examined in clinical trials. Furthermore, inhibitors that target the mutant protein more than the wild type (WT) protein of various genes (e.g., *BRAF*) either have been or are being characterized. Thus specific inhibitors have been made and some are currently used in the clinic. Targeting some components of these pathways has proven clinically effective. Some of the diseases being treated with targeted therapy have a very large market with few effective treatments (e.g., sorafenib and *h*epato*c*ellular *c*arcinoma (HCC) (Cervello et al. 2012).

14.5.1 Raf Inhibitors

Raf inhibitors have been developed and some are being used for therapy while others are being evaluated in clinical trials. Raf inhibitors have in general exhibited greater response rates in clinical trails than MEK inhibitors which may be related to the broader therapeutic index of Raf inhibitors that suppress ERK activity in a mutant-allele specific fashion as opposed to MEK inhibitors which suppress MEK activity in tumor and normal cells (Poulikakos and Solit 2011).

14.5.1.1 Sorafenib

Some inhibitors (i.e., sorafenib, Bayer) were initially thought to inhibit specifically Raf but have been subsequently shown to have multiple targets (e.g., VEGF-R, Flt-3, PDGF-R) (Wilhelm et al. 2004). However, that does not preclude their usefulness in cancer therapy. Sorafenib is approved for the treatment of certain cancers (e.g., renal cell carcinoma (RCC) and patients with unresectable HCC. Sorafenib was evaluated in the sorafenib hepatocellular carcinoma assessment randomized protocol (SHARP) trial, which demonstrated that the drug was effective in prolonging median survival and time-to-progression in patients with advanced HCC (Ratain et al. 2006; Llovet et al. 2008; Rimassa et al. 2009). This led to its approval for treatment of HCC patients by the FDA. Sorafenib is generally well tolerated in HCC patients with a manageable adverse events profile (Rimassa et al. 2009). While sorafenib is not considered effective for the treatment of most melanomas with *BRAF V600E* mutations, it may be effective in the treatment of a minority of melanomas with *G469E* and *D594G* mutations which express constitutive ERK1/2 but low levels of MEK. These melanomas are sensitive to sorafenib, potentially because they signal through Raf-1 (Smalley et al. 2009). MEK inhibitors have also been examined for treating HCC in mouse models (Huynh et al. 2007) but they do not appear to be as effective as sorafenib, most likely due to the broad specificity of sorafenib, which inhibits other targets besides Raf.

The effects of sorafenib on AML were investigated as the Raf/MEK/ERK signaling pathway is activated in many AMLs (Zhang et al. 2008a). Sorafenib inhibited the phosphorylation of Raf and downstream MEK and ERK and induced apoptosis in the cells. Sorafenib also inhibited colony formation of AML cell lines and primary AML patient samples. Sorafenib induced pro-apoptotic Bim as well as Bad, Bax and Bak proteins. In contrast, sorafenib decreased anti-apoptotic MCL-1, XIAP and survivin levels. The effects of sorafenib were shown to be linked to Bim as knockdown of Bim abrogated the effects of sorafenib to induce apoptosis. A phase clinical trial evaluating the ability of sorafenib to treat AML patients was performed.

14.5.1.2 PLX4032/Vemurafenib

PLX4032 (Zelborab, vemurafenib, Plexxikon/Roche) is a B-Raf inhibitor that has and is being evaluated in many clinical trials (Flaherty et al. 2010; Bollag et al. 2010; Chapman et al. 2011). Vemurafenib has been approved by the US Food and Drug Administration (FDA) for the treatment of patients with unresectable or metastatic melanoma carrying the *BRAF V600E* mutation. For vemurafenib to be clinically effective, it needs to suppress downstream ERK activation essentially completely (Bollag et al. 2010). NCT01524978 is a Phase II clinical trial to evaluate the effects of vemurafenib on patients with MM and other cancers containing the *BRAF V600E* mutation. Given the high level of *BRAF* mutations in hairy cell leukemia (Tiacci et al. 2011), it is likely that there will be a clinical trial to examine the effects of vermurafenib on patients with this leukemia.

14.5.1.3 GSK2118436/Dabrafenib

GSK2118436 (Dabrafenib) is an ATP-competitive inhibitor of mutant B-Raf, WT B-Raf and WT Raf-1 developed by GlaxoSmithKlein (GSK) which is in clinical trials (Greger et al. 2012). CCT239065 is a mutant B-Raf-selective inhibitor developed at the Institute of Cancer Research in London, UK (Whittaker et al. 2010b). It inhibits *BRAF V600E* signaling and proliferation more so than cells containing *WT BRAF*. Its effects are more selective for cells containing mutant *BRAF* than *WT BRAF*. CCT239065 is well tolerated in mice and had good oral bioavailability. It suppressed tumors containing *BRAF* mutant gene but not *WT BRAF* tumors in mice tumor xenograft studies (Whittaker et al. 2010b). GDC-0879 is a *BRAF* mutant allele selective inhibitor developed by Genentech (Hoeflich et al. 2009b).

14.5.1.4 PLX4458

PLX5568 is a selective Raf kinase inhibitor developed by Plexicon. It is being examined for the treatment of *p*olycystic *k*idney *d*isease (PKD). In the kidney Raf-1 is localized to the tubular cells where it is linked to many physiologically important functions. PLX5568 suppressed cyst enlargement in a rat model of PKD but did not improve kidney function as fibrosis was not suppressed (Buchholz et al. 2011).

14.5.1.5 Raf-265

Raf-265 is an ATP-competitive pan-Raf inhibitor developed by Novartis. Treatment of bronchus carcinoid NCI-H727 and CM insulinoma cells with Raf-265 enhanced sensitivity to TRAIL-induced apoptosis. These cells were normally resistant to PI3K/mTOR inhibitors when combined with TRAIL. Raf-265 was shown to decrease Bcl-2 levels which correlated with their sensitivity to TRAIL-mediated apoptosis. This approach may be effective in the therapy of neuroendocrine tumors (Zitzmann et al. 2011).

14.5.2 Clinical Trials with Raf Inhibitors

Numerous clinical trials have been performed with Raf inhibitors and sorafenib and vemurafenib are currently used to treat patients with certain cancers including: *r*enal *c*ell *c*arcinoma (RCC), HCC and malignant melanoma.

14.5.2.1 Results of Clinical Trials with Sorafenib

One of first clinical trials with Raf inhibitors were with Sorafenib (Nexavar) in metastatic *R*enal *C*ell *C*arcinoma (Ratain et al. 2006). Clinical trials with melanoma were also done around the same time period (Eisen et al. 2006). The clinical trials with melanoma patients and Sorafinib as a single agent did not yield encouraging results (Ott et al. 2010). Due to the broad specificity of Sorafenib (Nexavar™), this drug has been evaluated for the therapy of diverse cancers, including RCC, some melanomas and HCC [due to the involvement of the Raf/MEK/ERK cascade, as well as altered *v*ascular *e*ndothelial *g*rowth *f*actor (VEGF) pathway in these cancers] and *g*astro-*i*ntestinal *s*tromal *t*umors (GIST) (due to the involvement of *KIT* mutations in this cancer) (Takimoto et al. 2008). NCT01371981 was a Phase II/III with Sorafenib combining the proteosomal inhibitor bortezomib as well as various drugs including: asparaginease, cytarabine, daunorubicin and mitoxantrone in patients with *a*cute *m*yeloid *l*eukemia (AML) and yielded variable results with low response rates (Takimoto et al. 2008). Sorafenib has been approved for the treatment of renal cancer, including RCC in 2005 and for HCC in 2007 (Ratain et al. 2006; Escudier et al. 2007; Rimassa et al. 2009). There are approximately 19 clinical trials with various combinations of sorafenib and chemotherapeutic drugs that either have been performed or are in progress in the treatment of AML patients. The effectiveness of sorafenib on AMLs may be due to the ability of sorafenib to suppress mutant *F*ms-*l*ike *t*yrosine kinase 3 signaling which is upregulated in AMLs due to mutation at FLT3-*i*nternal *t*andem *d*uplication (*FLT3-ITD*), a driver mutation for some AMLs (Zhang et al. 2008b). A clinical trial with sorafenib advanced myelodysplastic syndrome and relapsed or refractory acute leukemias was performed (Borthakur et al. 2011). Complete remissions or complete remissions with incomplete recovery of platelets were observed in 5 (10%) patients and they all had the *FLT3-ITD*. The authors concluded that sorafenib was active and well tolerated in AMLs with *FLT3-ITD*.

14.5.2.2 Results of Clinical Trials with Vemurafenib

Phase I, II and III clinical trials with vemurafenib have been performed. A greaten than 90% reduction in active ERK was necessary for clinical response (Bollag et al. 2010). In the Phase III clinical trial comparing vemurafenib with the standard of care chemotherapeutic drug decarbazine, the trial was prematurely terminated as it became apparent that vemurafenib was more effective than decarbazine (Chapman et al. 2011). Vemurafenib was approved for the treatment of unresectable metastatic *BRAF* mutant melanoma in 2011. Vemurafenib induces clinical responses in greater than 50% of previously treated mutant *BRAF* (*V600E* or *V600K*) melanoma patients (Chapman et al. 2011). The median overall survival was extended to approximately 16 months (Sosman et al. 2012). There is one clinical trial listed to examine the effects of vemurafenib on MM (NCT01524978).

14.5.2.3 Clinical Trials with Dabrafenib (GSK2118436)

Dabrafenib has also displayed positive results in Phase 1/II trials (Kefford et al. 2010). Dabrafenib is in ongoing Phase II clinical trials as a single agent in patients with *BRAF* mutant melanoma.

14.5.3 Need for Genetic Screening Before Treatment with Raf Kinase Inhibitors

It is critical to determine the genetic status at both *BRAF* and *RAS* before treatment with Raf inhibitors (Joseph et al. 2010; Poulikakos et al. 2010; Poulikakos and Rosen 2011). Class I B-Raf inhibitors (active conformation inhibitors) such as (PLX4720 and GSK-2118436) will inhibit *BRAF* mutants, however these ATP-competitive B-Raf inhibitors will not inhibit WT *BRAF* or mutant *RAS*. In fact, these B-Raf inhibitors can activate Raf-1 in these cells in the presence of active Ras. The Raf inhibitors can induce B-Raf binding to Raf-1. PLX4720 can, to a lesser extent, induce B-Raf binding to Raf-1 when the ERK-mediated negative feedback loop on B-Raf was suppressed with a MEK inhibitor. These binding events were determined to require the present of activated Ras (WT or mutant), which may be necessary for the translocation from the cytoplasm to the membrane and assembly into the signaling complex. This has therapeutic implications, as in patients with mutant *RAS*, if they are treated with certain B-Raf inhibitors, B-Raf can bind and activate Raf-1 and promote the oncogenic pathway. In fact, even kinase-dead *BRAF* mutations, which have been observed in human cancer (Davies et al. 2002; Heidorn et al. 2010), the mutant B-Raf proteins can dimerize with Raf-1, when stimulated by the mutant Ras protein and activate the Raf/MEK/ERK cascade. For Raf-selective inhibitors to be therapeutically useful, prior screening of patients for *RAS* mutations will be necessary, as well as perhaps additional screening during treatment. Otherwise resistance may develop and lead to further stimulation of the Raf/MEK/ERK cascade.

The Raf activation observed after treatment of certain cancer patients with Raf inhibitors also can lead to skin diseases such as *k*erato*a*canthomas (KAs) and *c*utaneous *s*quamous *c*ell *c*arcinomas (cSCCs) in patients containing mutations which activate Ras. These results indicate that co-targeting with Raf and MEK inhibitors may be appropriate in certain situations where patients have active Raf and B-Raf together (Su et al. 2012; Oberholzer et al. 2012).

14.5.4 Resistance to Raf Inhibitors

We will discuss the mechanism of resistance to Raf inhibitors with melanoma, CRC and lung cancers as they serve as models to what may also occur in blood cancers upon treatment with Raf inhibitors. A problem with treatment of melanoma patients with mutant BRAF is the emergence of inhibitor-resistance which occurs frequently relatively rapidly after treatment with the Raf inhibitors (2–18 months) (Wagle et al. 2011). This may be due to the persistence of melanoma *c*ancer *i*nitiating *c*ells (CICs) some of which have other mutations besides B-Raf.

There are many different genetic mechanisms by which melanoma cells can become resistant to Raf inhibitors. Unlike resistance mechanisms observed in some other cancers such as Imatinib-resistant chronic myeloid leukemia (CML) where the resistant cells often have mutations at the gatekeeper residues in *BCR-ABL*, other mechanism for Raf inhibitor-resistance are more frequently observed. Gatekeeper mutations in *BRAF* can be created experimentally, and the cells are resistant to the B-Raf specific inhibitors, but these mutations do not appear to occur frequently in B-Raf resistant clinical specimens. (Whittaker et al. 2010a; Nazarian et al. 2010; Wagle et al. 2011).

14.5.4.1 Raf Inhibitor Resistance Due to Defective Splicing

Poulikakos and colleagues demonstrated a novel resistance mechanism which involves a splice variant in the mutated *BRAF* allele that leads to a loss of the Ras binding domain in B-Raf that prevents dimerization. This mutant form of *BRAF V600E* elicits enhanced dimerization in cells which contain low levels of active Ras, in comparison to cells containing the full-length *BRAF V600E* mutation. Thus the truncated BRAF V600E kinase can dimerize with Raf-1 and induce downstream MEK/ERK in the absence of activating *RAS* mutations and the cells are resistant to the Raf inhibitors (Poulikakos and Persaud et al. 2011). This splicing mutation was determined to be present in *BRAF V600E* in six of nineteen patients treated with PLX4032 which had undergone relapse to therapy.

14.5.4.2 Raf Inhibitor Resistance Due to Alterations at CDK4 or Cyclin-D1

Many different types of gene deregulation events have been observed in B-Raf inhibitor resistant cells (Kudchadkar et al. 2012; Rebecca et al. 2012). Mutations at cyclin-dependent kinase 4 (CDK4) and amplification of cyclin-D1 have been observed in clinical specimens from B-Raf inhibitor treated patients which underwent remission (Smalley et al. 2008).

14.5.4.3 Raf Inhibitor Resistance Due to *BRAF* Gene Amplification

Amplification of the *BRAF* gene has been reported in some MEK inhibitor-resistant cells (Corcoran et al. 2010). The *BRAF* gene was determined to be amplified in some treatment-naïve cells. The authors of this study determined that treatment with B-Raf and MEK inhibitors eliminated resistance of the cells. A more recent study also demonstrated that the mutant *BRAF V600E* gene was amplified in 4 out of 20 melanoma patients which were resistant to B-Raf inhibitors (Shi et al. 2012). This mechanism of B-Raf inhibitor-resistance is distinct from resistance generated by *NRAS* mutations or overexpression as this mechanism was independent of Raf-1 expression while N-Ras mediated inhibitor-resistance is dependent on Raf-1 expression.

14.5.4.4 Raf Inhibitor Resistance Due to Increased MAP3K8/COT/Tp12 Expression

In an attempt to identify genes which encode resistance to B-Raf inhibitors, one group expressed a panel of approximately 600 kinase related open reading frames in cells in normally B-Raf inhibitor-sensitive A375 melanoma cells which contain the *BRAF V600E* mutation. They identified mitogen-activated protein kinase kinase kinase 8 (*MAP3K8*) which encodes the serine-threonine protein kinase COT/Tp12 (cancer Osaka thyroid oncogene/ tumor progression locus-2) as a MAPK pathway agonist which drives resistance to Raf inhibition in *BRAF* mutant cell lines. COT was demonstrated to induce ERK via MEK but independent of Raf (Johannessen et al. 2010). COT expression was observed to inversely correlate with *BRAF V600E* expression which may suggest that B-Raf may downregulate COT protein levels by destabilizing the protein. When *BRAF V600E* expression decreases due to B-Raf inhibitor treatment, the levels of COT are predicted to rise. Combining B-Raf and MEK inhibitors could overcome the resistance to the B-Raf inhibitors in the cells which overexpressed COT. The genomic region surrounding *MAP3K8* (COT) was amplified in 2 out of 38 *BRAF* mutant cell lines and these lines had not previously been treated with B-Raf inhibitors. These lines with amplified *MAP3K8* (COT) were in fact resistant to B-Raf inhibitors. COT expression was determined to be increased in expression in some relapse patients. COT inhibitors are being developed and may be effective in overcoming the resistance in some B-Raf inhibitor-resistant tumors (Cusack et al. 2009).

14.5.4.5 Raf Inhibitor Resistance Due to *MEK1* or *NRAS* Mutations

In another study, the investigators examined DNA from tumor cells isolated from a patient that initially was sensitive to the B-Raf inhibitor PLX4032 which became resistant after treatment. The DNA sequences of 138 cancer genes were examined (Wagle et al. 2011). This study observed that there was a mutation in *MEK1* (*C121S*) in the PLX4032-resistant tumor which was not present in the original tumor. The *MEK1 C121S* mutation conferred resistance to both Raf and MEK inhibitors.

Dabrafenib-resistant A375 melanoma cells were isolated by culturing the cells in dabrafenib. The resistant cells were also resistant to PLX4032 and the MEK inhibitor trametinib (GSK112012, an allosteric MEK inhibitor), in frame deletions of *MEK1* and mutations at *NRAS* mutations were detected. The in frame deletions of *MEK* occurred at *MEK1 K59del*, the *NRAS* mutations occurred at *NRAS Q61K* and *NRAS A146T* with and without the *MEK1 P387S* mutation in A375 *BRAF V600E* line and *NRAS Q61K* in the YUSIT1 *BRAF V600K* cell background (Greger et al. 2012). The combination of dabrafenib and trametinib suppressed cell growth in the resistant lines. These results are somewhat surprising as some of the resistant lines had *NRAS* mutations. N-Ras could potentially activate PI3K/PTEN/Akt/mTOR pathway which could promote resistance to these inhibitors. The combination of the PI3K inhibitor GSK2126458 and either B-Raf or MEK inhibitors enhanced growth suppression and decreased ribosomal S6 protein phosphorylation (Greger et al. 2012). Combination clinical trials are planned based on these results.

14.5.4.6 Raf Inhibitor Resistance Due to *NRAS* Mutations Leading to Increased PDGF-Rβ

Expression In another study with B-Raf inhibitor-resistant patient samples, the resistant cells were observed to have mutations at *NRAS* or overexpress PDGF-Rβ (Nazarian et al. 2010). These authors indicated that resistance to B-Raf inhibitors was not due to secondary mutations at *BRAF*, but activation of addition signaling pathways by PDGFRβ or by N-Ras activation of the Raf/MEK/ERK pathway. PDGFRβ was observed to be hyperphosphorylated in the cells from one B-Raf inhibitor-resistant line, but surprising the cells were not sensitive to imatinib which can target PDGFRβ.

14.5.4.7 Raf Inhibitor Resistance Due to Switching of Raf Isoforms and Increased IGF-1R

Expression Other studies have indicated that the switching of Raf isoforms may confer resistance of B-Raf inhibitors. These results were observed after incubation of melanoma cells containing the *BRAF V600E* mutation in the presence of the B-Raf inhibitor GSK2118436 for prolonged periods of time and inhibitor-resistant cells were recovered. In these inhibitor-resistant cells, they were observed to express other isoforms of Raf (e.g., Raf-1 or A-Raf) (Villanueva et al. 2010). In this study some inhibitor-resistant cells were also observed to overexpress IGF-1R which can also induce the expression of the PI3K/PTEN/Akt/mTOR pathway. Combined treatment with IGF-1R/PI3K and MEK inhibitors eliminated the resistance of the cells. Increased expression of IGF-1R and activation of Akt was also demonstrated in one of five pared specimens obtained from post-relapse PLX4032-treated patients as compared to the patient sample prior to treatment.

14.5.4.8 Raf Inhibitor Resistance Due to Suppression of Bim via Mutant *PTEN*

Suppression of pro-apoptotic Bim expression is a mechanism of resistance to B-Raf inhibitors (Paraiso et al. 2010). Suppression of Bim expression occurs in PTEN mutant cells. PLX4720 increases Bim expression in PTEN WT cells. The involvement of Akt-3 and FOXO3a was reported in these studies. Combining B-Raf and PI3K inhibitors enhanced Bim expression via FOXO3a in the PTEN mutant cells.

14.5.4.9 Raf Inhibitor Resistance Due to Protein Kinase D3 Expression

In a study of Raf265-resistant melanomas with the *BRAF V600E* mutation, it was observed that *pr*otein *k*inase *D3* (PRKD3) mediated resistance to both Raf and MEK inhibitors and siRNA knockdown of PRKD3 cooperated with Raf265 in suppressing the growth of the resistant melanoma cells (Chen et al. 2011). CID755673 is a PRKD3 inhibitor (Sharlow et al. 2008). Potentially CID755673 could be combined with B-Raf inhibitors to suppress the growth of B-Raf inhibitor-resistant melanomas.

14.5.5 MEK Inhibitors

Specific inhibitors of MEK have been developed:, AS703026 (EMD Serono) GDC-0973 (Genentech), MEK162/ARRY-162 (Novartis), (Pfizer), PD98059, PD184352 (CI-1040), PD0325901 (Pfizer), RDEA119/Refametinib (Ardea Biosciences/Bayer), RG422, RO4987655 (Roche), Selumetinib (a.k.a., ARRY-142886, AZD6244) (Astra-Zeneca), Trametinib (GSK112012, GlaxoSmithKlein), SL327, and TAK-733 (Takeda San Diego, Millennium Pharmaceuticals, Inc), and U0126 (DuPont), (Chappell et al. 2011).

MEK inhibitors differ from most other kinase inhibitors as they do not compete with ATP binding (non-ATP competitive), which confers a high specificity (Davies et al. 2000; Ohren et al. 2004). Most MEK inhibitors are specific and do not inhibit many different protein kinases (McCubrey et al. 2010) although as will be discussed below, certain MEK inhibitors are more specific than others. The crystal structures of MEK1 and MEK2 have been solved as ternary complexes with ATP and PD184352, and have revealed that both MEK1 and MEK2 have unique inhibitor binding sites located on a hydrophobic pocket adjacent to, but not overlapping with, the ATP-binding site (Ohren et al. 2004). Furthermore, effective targeting of MEK1/MEK2 is highly specific, as ERK1/ERK2 are the only well-described downstream targets. A distinct advantage of inhibiting MEK is that it can be targeted without knowledge of the precise genetic mutation that results in its aberrant activation. This is not true with targeting Raf as certain Raf inhibitors will activate Raf and also certain B-Raf specific inhibitors will not be effective in the presence of Ras mutations as discussed above. An advantage of targeting MEK is that the Ras/Raf/MEK/ERK pathway is a convergence point where a number of upstream signaling pathways can be blocked with the inhibition of MEK.

14.5.5.1 Selumetinib

Selumetinib (AZD6244) has been investigated for the treatment of pancreatic cancers, breast cancers, and other cancers such as hematopoietic malignancies, including MM (Tai et al. 2007). ClinicalTrials.gov lists 49 clinical trials for selumetinib, either as a single agene or combined with another inhibitor or combined with chemotherapy or radiotherapy.

Selumetinib inhibits MEK1 *in vitro* with an IC_{50} value of 14.1 ± 0.79 nM (Davies et al. 2007; Huynh et al. 2007; Yeh et al. 2007); it is specific for MEK1 as it did not appear to inhibit any of the approximately 40 other kinases in the panel tested. Selumetinib is not competitive with ATP. Molecular modeling studies indicate that selumetinib binds to an allosteric binding site on MEK1/MEK2. The binding sites on MEK1/MEK2 are relatively unique to these kinases and may explain the high specificity of MEK inhibitors. This binding may lock MEK1/2 in an inactivate conformation that enables binding of ATP and substrate, but prevents the molecular interactions required for catalysis and access to the ERK activation loop. Selumetinib did not prevent the activation of the related ERK5 that occurs with some older MEK1 inhibitors, which are not being pursued in clinical trials. Inhibition of ERK1/2 suppresses their ability to phosphorylate and modulate the activity of Raf-1, and MEK1 but not MEK2 as MEK2 lacks the ERK1/ERK2 phosphorylation site. In essence, by inhibiting ERK1/2 the negative loop of Raf-1 and MEK phosphorylation is suppressed and hence there will be an accumulation of activated Raf-1 and MEK (Davies et al. 139). This biochemical feedback loop may provide a rationale for combining Raf and MEK inhibitors in certain therapeutic situations.

14.5.5.2 PD-0325901

PD-0325901 (Pfizer) was developed after the earlier MEK inhibitors PD-98059 and PD-184352, both of which have been extensively examined in preclinical investigations to determine the role of MEK in various biochemical processes (Sebolt-Leopold 2008). PD-184352 was the first MEK inhibitor to enter clinical trials and it demonstrated inhibition of activated ERK and anti-tumor activity in patients (Sebolt-Leopold 2008); however, subsequent multicenter, phase II studies with patients with diverse solid tumors did not demonstrate encouraging results. This was probably due to low oral bioavailability and high metabolism, which led to plasma drug levels that were inadequate to suppress tumor growth (Sebolt-Leopold 2008).

The PD-0325901 MEK inhibitor is an orally-active, potent, specific, non-ATP competitive inhibitor of MEK. PD-0325901 demonstrated improved pharmacological and pharmaceutical properties compared with PD-184352, including a greater potency for inhibition of MEK, and higher bioavailability and increased metabolic stability. PD-0325901 has a K_i value of 1 nM against MEK1 and MEK2 in *in vitro* kinase assays. PD-0325901 inhibits the growth of cell lines that proliferate in response to elevated signaling of the Raf/MEK/ERK pathways (Solit et al. 2006). Clinical trials with PD-0325901 have documented some successes and some adverse side effects (Sebolt-Leopold 2008, Haura et al. 2010, LoRusso et al. 2010). MEK inhibitors may be appropriate to treat only those cancers that proliferate in response to activation of the Raf/MEK/ERK pathway (Sebolt-Leopold 2008; McCubrey et al. 2010). Furthermore, it may also be important to include an additional pathway inhibitor, chemotherapeutic drug or radiation treatment to induce death of the cancer cell. There is a phase I clinical trial (NCT01347866) examining the effects of combining PD-0329501 with the PI3K/mTOR inhibitor PF-04691502. Initially this phase I trial will examine toxicity in patients with advanced cancers. If tolerable toxicity levels are observed, then additional studies will be performed with CRC patients containing mutant *KRAS* genes that have had previous therapy but relapsed.

14.5.5.3 RDEA119/Refametinib

Refametinib (RDEA119, BAY 86-9766) is a recently described MEK inhibitor developed by Ardea Biosciences and Bayer (Iverson et al. 2009; Chang et al. 2010; Liu et al. 2010). It is a highly selective MEK inhibitor that displays a > 100-fold selectivity in kinase inhibition in a panel of 205 kinases. In contrast, in the same kinase specificity analysis, other recently developed MEK inhibitors (e.g., PD0325901) also inhibited the Src and RON kinases.

14.5.5.4 AS703026

AS703026 (EMD Serono) is a recently described MEK inhibitor and has been shown to have cytotoxic effects on MM (Kim et al. 2010). Importantly AS703026 sensitized MM cells to a variety of conventional (dexamethasone, melphalan), and novel (lenalidomide, perifosine, bortezomib, rapamycin) drugs used to treat MM.

14.5.6 MEK Inhibitors in Clinical Trials

There are approximately 84 clinical trials with MEK inhibitors listed on the ClinicalTrials.gov website. Clinical trials have been and are being performed with various cancer patients and selumetinib (AZD6244), PD0325901, CI-1040, GSK1120212, TAK-733, RO4987655, MEK162, AS703026 and RHEA119. The MEK inhibitors may be appropriate for the treatment of certain melanomas which have mutant *BRAF* or other cancers which have mutant *RAS* (Tanios Bekaii-Saab et al. 2011; Kudchadkar et al. 2012; Patel et al. 2012). NCT00957580 is a clinical trial with AS703026. The phase I clinical trial will evaluate the effects of AS703026 on patients advanced hematopoietic malignancies. The phase II trial is a continuation of the trial with AS703026 for elderly AML patients who are not good candidates for chemotherapy.

Overall, there are approximately 84 clinical trials with MEK inhibitors listed on the Clinical.Trials.gov website. There are 15 trials with MEK inhibitors and lung cancer, 14 trials with MEK inhibitors and pancreatic cancer, 10 trials with MEK inhibitors and breast cancer, 10 trials with MEK inhibitors and colon cancers, 4 trials with MEK inhibitors and leukemias, 4 trials with MEK inhibitors and HCC, 4 trials with MEK inhibitors and brain cancers, 2 trials with MEK inhibitors and breast cancer. Initial results from clinical trials have not yielded overwhelming support for the use of MEK inhibitors as a single therapeutic agent in cancer patients who are not pre-screened for pre-existing activation of the Ras/Raf/MEK/ERK pathway. Indeed, there are 21 clinical trials listed on the Clinical.Trials.gov website with MEK inhibitors and melanoma patients which often have mutation of *BRAF* and hence activation of downstream MEK. The proper pre-identification of cancer patients who display activation of the Raf/MEK/ERK pathway may be necessary for prescribing MEK inhibitors as part of their therapy, as well as in combination with a cytotoxic drug, as MEK inhibitors are cytostatic and not cytotoxic.

14.5.7 MEK Inhibitor Resistance

In the following section, we discuss mechanisms or resistance to MEK inhibitors. Most of the large studies have not been performed on blood cancers. We have included these studies with solid tumors as when MEK inhibitors are used to treat certain blood cancers, similar mechanisms of resistance may arrive. Some tumors are resistant to MEK inhibitors because they contain *EGFR, KRAS, PI3KCA* or *PTEN* mutations (Pratilas et al. 2008; Hoeflich et al. 2009b; Wee et al. 2009). Some cells with *EGFR* or *KRAS* mutations are resistant to MEK inhibitors since they can also activate the Ras/PI3K/Akt/mTOR pathway (Hoeflich et al. 2009b). These studies, which were performed *in vitro* with cells lines and *in vivo* using xenografts, also demonstrated that PI3K activation and PTEN inactivation were not always equivalent in terms of inhibitor sensitivity. The authors suggested that a possible reason for this phenomenon could be that PTEN has other functions besides the regulation of Akt (e.g., protein phosphatase activity). Furthermore these studies demonstrated that the combination of MEK and PI3K pathway inhibitors could be an effective approach to treat certain cancers that had activation of both pathways.

Only certain types of breast cancer are sensitive to MEK inhibitors (Hoeflich et al. 2009b). Breast cancers can be classified into three types: luminal breast cancers which are usually estrogen receptor positive and have a relatively good prognosis and response rate to hormonal based therapies, HER2-positive breast cancers which have a poor prognosis if untreated but are initially responsive to the HER2-targeting monoclonal antibody Herceptin, and basal-like breast cancers which have a poor prognosis and lack expression of HER2, estrogen and progesterone receptors (referred to as "triple-negative"). Many basal breast cancers express high levels of EGFR which results in activation of the Ras/Raf/MEK/ERK cascade. Hoeflich and colleagues found that basal cell breast cancers expressed a Ras-like expression profile and tested their hypothesis that these breast cancers could be sensitive to MEK inhibitors, providing that they do not have *PI3KCA* mutations or *PTEN* deletions. In contrast many luminal and HER2-amplified tumors are resistant to MEK inhibitors. They also determined that *PTEN* loss was a negative predictor factor for response to MEK inhibitors. Furthermore, treatment with MEK inhibitors often led to an increase in activated Akt expression, providing the rationale to examine the consequences of co-addition of MEK and PI3K inhibitors. The authors also determined that co-administration of MEK and PI3K inhibitors enhanced killing of the certain breast cancers. Thus the studies by Wee *et al,* and Hoeflich *et al.*, have shown the concept that elevated PI3K/Akt/mTOR expression will confer resistance to MEK inhibitors. These studies illuminate the critical role of genetics in determining the sensitivity to targeted therapy.

Other studies have also indicated that some tumors with *EGFR* mutations are resistant to MEK inhibitors. Mutations at the *BRAF*, *KRAS*, *EGFR* genes or the chromosomal fusion between *a*naplastic *l*ymphoma *k*inase (*ALK*) and *ROS* tyrosine kinases are detected in approximately 50 % of NSCLC. NSCLC cells with *BRAF* mutations where shown to be more sensitive to MEK inhibitors than NSCLC with mutations in *EGFR*, *KRAS*, or the chimeric fusion between *ALK* and *ROS* (Pratilas et al. 2008). This was determined by screening a large panel of cell lines ($n = 87$)

and tumors (n=916). In this study, cells with mutations at *EGFR* were resistant to MEK inhibitors. This may have resulted from the ability of EGFR to activate the PI3K/PTEN/Akt/mTOR pathway which has some crucial overlapping targets with the Raf/MEK/ERK pathway. NSCLC patients with *EGFR* mutations should not be treated with MEK (or B-Raf) inhibitors as the respective therapies would be ineffectual.

In some MEK inhibitor-resistant melanoma cells which contained either the *G469E* or *D594G* mutant *BRAF* alleles, activation of Raf-1 by the mutant B-Raf proteins was observed to confer resistance to MEK inhibitors (Smalley et al. 2009). The *G469E* and *D594G BRAF* mutants are considered weak B-Raf mutations and signal through Raf-1. In these cells, survival is mediated by the *G469E*- and *D594G*-mutant B-Raf alleles stimulating Raf-1 which becomes mitochondrial localized and regulates apoptosis though phosphorylation of Bad and enhancement of the anti-apoptotic properties of Bcl-2. Sorafenib induced a reduction of Bad phosphorylation and Bcl-2 expression in the *D594G/G469E* melanoma cells. The effects of Raf-1 on the prevention of apoptosis were demonstrated in the *D594G/G469E* but not *V600E* melanoma cells by shRNA knock down of Raf-1. These studies indicate that sorafenib may be appropriate in the treatment of a minority of melanoma that survive in response to Raf-1 activation and are essentially MEK inhibitor-resistant. It is relevant to discuss these mutations and their effects on sensitivity to sorafenib as certain leukemias and MM are now known to contain mutations at *BRAF*.

Amplification of mutant *BRAF* in selumetinib-resistant CRCs was observed in studies which selected for selumetinib resistance in vitro (Corcoran et al. 2010). The sensitivity of the cells to the MEK inhibitor could be restored by treatment with low doses of the B-Raf inhibitor. In this study, the authors demonstrated that the amplified mutant *BRAF* gene was present in a small minority of treatment-naïve cells. In another study by a different group of investigators, resistance to selumetinib was observed in CRC lines harboring mutations in *BRAF* (COLO205 and HT29) or *KRAS* (HCT116, LoVo). The selumetinib-resistant lines did not appear to have mutations in either *MEK1* or *MEK2* but had upregulation of B-Raf or K-Ras respectively due to intrachromosomal amplification of their respective driving oncogenes, *BRAF V600E* or *KRAS G13D* which the authors demonstrated was responsible for their resistance to selumetinib (Little et al. 2011).

Mutations in the allosteric binding pocket of the *MEK1* gene were observed in a different study which isolated MEK-inhibitor resistant cells from MDA-MB-231 basal breast cancer cells (Hatzivassiliou et al. 2012). Basal breast cancer cells are often sensitivity to MEK inhibitors and the MDA-MB-231 cell line has mutations at *BRAF G464V* and *KRAS G13D*. The MEK inhibitor resistance could be overcome by treatment with ERK inhibitors, even in the resistant cell line with *KRAS* amplification.

Additional MEK inhibitor-resistant lines were derived from HCT-116 and LoVo colorectal cells (Hatzivassiliou et al. 2012). The MEK inhibitor-resistant HCT-116 lines also had mutations in the allosteric binding pocket mutations in *MEK1* while the MEK inhibitor-resistant LoVo cells had mutations in the allosteric binding pocket mutations in *MEK2*. One MEK inhibitor-resistant HCT-116 cell line also had the allosteric binding pocket mutation as well as amplification of *KRAS* but remained sensitive to growth inhibition upon treatment with the ATP-competitive ERK inhibitor, ERKi. These studies also demonstrated the effectiveness of inhibiting ERK in overcoming resistance to MEK inhibitors even if *BRAF* or *KRAS* are amplified or mutated. Furthermore the combination of MEK and ERK inhibitors can be beneficial in treating certain resistant cells.

14.5.8 Combining Raf and MEK Inhibitors

The possibility of treating certain patients with Raf and MEK inhibitors is a concept which is gaining more acceptance as it may be a therapeutic possibility to overcome resistance (Poulikakos and Solit 2011). Raf inhibitors induce Raf activity in cells with WT *RAF* if Ras is active, however, the addition of a MEK inhibitor would suppress the activation of MEK and ERK in the normal cells of the cancer patient. Thus B-Raf would be suppressed by the B-Raf-selective inhibitor in the cancer patient while the consequences of Raf activation in the normal cells would be suppressed by the MEK inhibitor. These concepts are being examined in clinical trials (NCT01072175 and NCT01231594). NCT01072175 is a clinical trial with the Raf inhibitor GSK2118436 in combination with the MEK Inhibitor GSK1120212 in metastatic melanoma patients containing a mutant *BRAF* gene. NCT01352273 is a clinical trial with combinations of MEK162 and RAF265 examining the effects on adult patients with solid tumors with either *RAS* or *BRAF V600E* mutations. The MEK inhibitor RDEA119/refametinib and sorafenib have been combined in Phase I/II clinical trials (NCT00785226) with patients having various types of advanced cancer.

14.5.9 Combining Raf with Bcl-2 Inhibitors to treat AML

As stated previously the Raf/MEK/ERK pathway is activated in many AMLs. Activation of this pathway often results in the activation of anti-apoptotic molecules. Combination of sorafenib with the novel small molecule Bcl-2 inhibitor ABT-737 or cytarabine synergistically induced cell death in AML cell lines (Zhang et al. 2008a).

14.5.10 Combining MEK with Bcl-2 Inhibitors to treat AML

The effects of combining MEK inhibitors with Bcl-2/Bcl-XL have been examined in pre-clinical studies with AML cell lines and patient samples (Konopleva et al. 2012). The Bcl-2 inhibitor ABT-737 was observed to induce ERK activation and Mcl-1 expression. However, when the ABT-737 inhibitor was combined with the MEK inhibitor PD0325901, a synergistic response was observed in terms of the induction of cell death both on AML cell lines and primary tumor cells with the properties of leukemia stem cells. Furthermore these studies were also extended into tumor transplant models with the MOLT-13 cell line and synergy between ABT-737 and PD0325901 was also observed *in vivo*.

14.5.11 ERK Inhibitors

There are at least two ERK molecules regulated by the Raf/MEK/ERK cascade, ERK1 and ERK2. Little is known about the differential *in vivo* targets of ERK1 and ERK2. The development of specific ERK1 and ERK2 inhibitors is ongoing and may be useful in the treatment of certain diseases such as those leukemias where elevated ERK activation is associated with a poor prognosis (e.g., AML, ALL) (Ricciardi et al. 2012). ERK inhibitors have been described (Aronov et al. 2009). AEZS-131 has been reported on the internet to be a highly selective ERK1/2 inhibitor developed by Aeterna Zentaris. Other ERK inhibitors (ERKi) have also been developed and evaluated for their use in overcoming MEK inhibitor-resistance (Hatzivassiliou et al. 2012).

14.5.12 PP2A Activators

Successful targeting of the protein phosphatases has in general not proceeded as rapidly as tarteting of protein kinases. FTY720 (fingolimod) is a PP2A activator which has been approved as an immunomodulator for oral use in patients with multiple sclerosis (Killestein et al. 2011)). Reactivation of PP2A activity by FTY720 suppressed cell growth, enhanced apoptosis, impaired clonogenicity, and decreased *in vivo* leukemogenesis of imatinib- and dasatinib-sensitive and -resistant Ph^+ B-ALL cells, as well as Ph^+ B-ALL progenitors ($CD34^+/CD19^+$). Importantly, healthy $CD34^+$ and $CD34^+/CD19^+$ bone marrow cells were unaffected by FTY720. Furthermore, pharmacologic doses of FTY720 suppressed *in vivo BCR-ABL*-driven leukemogenesis (including leukemogenesis promoted by the *BCR-ABL T315I* mutant which is resistant to imatinib and second generation TKIs without exerting any toxicity in mice (Neviani et al. 2007).

Conclusions

Pharmaceutical companies have developed inhibitors to the Ras/Raf/MEK/ERK pathway. Initially MEK inhibitors were demonstrated to have the most specificity. However, these inhibitors may have limited effectiveness in treating human cancers, unless the particular cancer proliferates directly in response to the Raf/MEK/ERK pathway. Moreover, MEK inhibitors are often cytostatic as opposed to cytotoxic, thus their ability to function as effective anti-cancer agents in a monotherapeutic setting may be limited, and they may be more effective when combined with chemo or radiotherapy. Raf inhibitors have also been developed and some are being used to treat various cancer patients (e.g., sorafenib). This particular Raf inhibitor also inhibits other receptors and kinases which may be required for the growth of the particular cancer. This promiscuous nature of Sorafenib has contributed to the effectiveness of this particular Raf inhibitor for certain cancers. Raf inhibitors such as vemurafenib, dabrafenib, and GDC-0879 are promising for the treatment of melanoma, CRC, thyroid and other solid cancers and leukemias/lymphomas/myelomas which have mutations at *BRAF V600E*. However, problems have been identified with certain *BRAF* mutant allele inhibitors as they will also result in Raf-1 activation if *RAS* is mutated/amplified of if an exon of *BRAF* is deleted, or if *BRAF* is amplified or if there are mutations at *MEK1* and other genetic mechanisms. Combination therapy with either a traditional drug/physical treatment or another inhibitor that targets a specific molecule in either the same or different signal transduction pathway are also key approaches for improving the effectiveness and usefulness of MEK and Raf inhibitors.

Some scientists and clinicians have considered that the simultaneous targeting of Raf and MEK by individual inhibitors may be more effective in cancer therapy than just targeting Raf or MEK by themselves (Ricciardi et al. 2012). This is based in part on the fact that there are intricate feed-back loops from ERK which can inhibit Raf and MEK. For example when MEK1 is targeted, ERK1,2 is inhibited and the negative feed-back loop on MEK is broken and activated MEK accumulates. However, if Raf is also inhibited, it may be possible to completely shut down the pathway. This is a rationale for treatment with both MEK and Raf inhibitors. Finally, an emerging concept is the dual targeting of two different signal transduction pathways, Raf/MEK/ERK and PI3K/PTEN/Akt/mTOR for example. This has been explored in some preclinical models and is being investigated in clinical trials. The rationale for the targeting of both pathways may be dependent on the presence of driver mutations in either/or both pathways or in upstream Ras in the particular cancer which can activate both pathways.

Acknowledgments AMM was supported in part by grants from: Fondazione del Monte di Bologna e Ravenna, MinSan 2008 “Molecular therapy in pediatric sarcomas and leukemias against IGF-IR system: new drugs, best drug-drug interactions, mechanisms of resistance and indicators of efficacy”, MIUR PRIN 2008 (2008THTNLC), and MIUR FIRB 2010 (RBAP10447J-003) and 2011 (RBAP11ZJFA_001).

References

Abraham D, Podar K et al (2000) Raf-1-associated protein phosphatase 2A as a positive regulator of kinase activation. J Biol Chem 275:22300–22304

Aronov AM, Tang Q et al (2009) Structure-guided design of potent and selective pyrimidylpyrrole inhibitors of extracellular signal-regulated kinase (ERK) using conformational control. J Med Chem 52:6362–6368

Arthur JS (2008) MSK activation and physiological roles. Front Biosci 13:5866–5879

Badalian-Very G, Vergilio JA et al (2010) Recurrent BRAF mutations in Langerhans cell histiocytosis. Blood 116:1919–1923

Balan V, Leicht DT et al (2006) Identification of novel in vivo Raf-1 phosphorylation sites mediating positive feedback Raf-1 regulation by extracellular signal-regulated kinase. Mol Biol Cell 17:1141–1153

Barretina J, Caponigro G et al (2012) The cancer cell line encyclopedia enables predictive modelling of anticancer drug sensitivity. Nature 483:603–607

Bollag G, Hirth P et al (2010) Clinical efficacy of a RAF inhibitor needs broad target blockade in BRAF-mutant melanoma. Nature 467:596–599

Borthakur G, Kantarjian H et al (2011) Phase I study of sorafenib in patients with refractory or relapsed acute leukemias. Haematologica 96:62–68

Brennan DF, Dar AC et al (2011) A Raf-induced allosteric transition of KSR stimulates phosphorylation of MEK. Nature 472:366–369

Brose MS, Volpe P et al (2002) BRAF and RAS mutations in human lung cancer and melanoma. Cancer Res 62:6997–7000

Buchholz B, Klanke B et al (2011) The Raf kinase inhibitor PLX5568 slows cyst proliferation in rat polycystic kidney disease but promotes renal and hepatic fibrosis. Nephrol Dial Transplant 26:3458–3465

Buday L, Downward J (2008) Many faces of Ras activation. Biochimica et Biophysica Acta 1786:178–187

Caponigro G, Sellers WR (2011) Advances in the preclinical testing of cancer therapeutic hypotheses. Nat Rev Drug Discov 10:179–187

Casar B, Rodríguez J et al (2012) Mxi2 sustains ERK1/2 phosphorylation in the nucleus by preventing ERK1/2 binding to phosphatases. Biochem J 441:571–578

Case M, Matheson E et al (2008) Mutation of genes affecting the RAS pathway is common in childhood acute lymphoblastic leukemia. Cancer Res 68:6803–6809

Catalanotti F, Reyes G et al (2009) A Mek1-Mek2 heterodimer determines the strength and duration of the Erk signal. Nat Struct Mol Biol 16:294–303

Cervello M, McCubrey JA et al (2012) Targeted therapy for hepatocellular carcinoma: novel agents on the horizon. Oncotarget 3:236–260

Chan DA, Giaccia AJ (2011) Harnessing synthetic lethal interactions in anticancer drug discovery. Nat Rev Drug Discov 10:351–364

Chang F, Steelman LS et al (2003) Signal transduction mediated by the Ras/Raf/MEK/ERK pathway from cytokine receptors to transcription factors: potential targeting for therapeutic intervention. Leukemia 17:1263–1293

Chang Q, Chapman MS et al (2010) Antitumour activity of a potent MEK inhibitor RDEA119/BAY 869766 combined with rapamycin in human orthotopic primary pancreatic cancer xenografts. BMC Cancer 10:515

Chapman MA, Lawrence MS et al (2011) Initial genome sequencing and analysis of multiple myeloma. Nature 471:467–472

Chapman PB, Hauschild A et al (2011) BRIM-3 study group improved survival with vemurafenib in melanoma with BRAF V600E mutation. N Engl J Med 364:2507–2516

Chappell WH, Steelman LS et al (2011) Ras/Raf/MEK/ERK and PI3K/PTEN/Akt/mTOR Inhibitors: rationale and importance to inhibiting these pathways in human health. Oncotarget 2:135–164

Chen H, Zhu G et al (2009) Extracellular signal-regulated kinase signaling pathway regulates breast cancer cell migration by maintaining slug expression. Cancer Res 69:9228–9235

Chen J, Shen Q et al (2011) Protein kinase D3 sensitizes RAF inhibitor RAF265 in melanoma cells by preventing reactivation of MAPK signaling. Cancer Res 71:4280–4291

Cheng J, Zhang C et al (2007) A functional serine 118 phosphorylation site in estrogen receptor-et al alpha is required for down-regulation of gene expression by 17beta-estradiol and 4-hydroxytamoxifen. Endocrinology 148:4634–4641

Corcoran RB, Dias-Santagata D et al (2010) BRAF gene amplification can promote acquired resistance to MEK inhibitors in cancer cells harboring the BRAF V600E mutation. Sci Signal 3:ra84

Costello PS, Nicolas RH et al (2004) Ternary complex factor SAP-1 is required for Erk-mediated thymocyte positive selection. Nat Immunol 5:289–298

Costello P, Nicolas R et al (2010) Ternary complex factors SAP-1 and Elk-1, but not net, are functionally equivalent in thymocyte development. J Immunol 185:1082–1092

Cuevas BD, Abell AN et al (2003) MEKK1 regulates calpain-dependent proteolysis of focal adhesion proteins for rear-end detachment of migrating fibroblasts. EMBO J 22:3346–3355

Cusack K, Allen H et al (2009) Identification of a selective thieno[2,3-c]pyridine inhibitor of COT kinase and TNF-alpha production. Bioorg Med Chem Lett 19:1722–1725

Davies BD, Logie A et al (2007) AZD6244 (ARRY 142886) a potent inhibitor of mitogen-activated protein kinase/extracellular signal-related kinase kinase 1/2 kinases: mechanism of action in vivo, pharmacokinetic/pharmacodynamic relationship and potential for combination in preclinical models. Mol Cancer Ther 6:2209–2219

Davies H, Bignell GR et al (2002) Mutations of the BRAF gene in human cancer. Nature 417:949–954

Davies SP, Reddy H et al (2000) Specificity and mechanism of action of some commonly used protein kinase inhibitors. Biochem J 351:95–105

Dougherty MK, Muller J et al (2005) Regulation of Raf-1 by direct feedback phosphorylation. Mol Cell 17:215–224

Downward J (2003) Targeting Ras signaling pathways in cancer therapy. Nature Reviews Cancer 3:11–22

Eblen ST, Slack JK et al (2002) Rac-PAK signaling stimulates extracellular signal-regulated kinase (ERK) activation by regulating formation of MEK1-ERK complexes. Mol Cell Biol 22:6023–6033

Eisen T, Ahmad T et al (2006) Sorafenib in advanced melanoma: a Phase II randomised discontinuation trial analysis. Br J Cancer 95:581–586

Escudier B, Eisen T et al (2007) TARGET Study Group Sorafenib in advanced clear-cell renal-cell carcinoma. N Engl J Med 356:125–134

Flaherty KT, Puzanov I et al (2010) Inhibition of mutated, activated BRAF in metastatic melanoma. N Engl J Med 363:809–819

Garnett MJ, Edelman EJ et al (2012) Systematic identification of genomic markers of drug sensitivity in cancer cells. Nature 483:570–575

Goettel JA, Liang D et al (2011) KSR1 is a functional protein kinase capable of serine autophosphorylation and direct phosphorylation of MEK1 find rest of reference. Exp Cell Res 317:452–463

Graham TR, Zhau HE et al (2008) Insulin-like growth factor-I-dependent up-regulation of ZEB1 drives epithelial-to-mesenchymal transition in human prostate cancer cells. Cancer Res 68:2479–2488

Grammatikakis N, Lin JH et al (1999) p50(cdc37) acting in concert with Hsp90 is required for Raf-1 function. Mol Cell Biol 19:1661–1672

Greenman C, Stephens P et al (2007) Patterns of somatic mutation in human cancer genomes. Nature 446:153–158

Greger J, Eastman S et al (2012) Combinations of BRAF, MEK, and PI3K/mTOR inhibitors overcome acquired resistance to the BRAF inhibitor GSK2118436 dabrafenib, mediated by NRAS or MEK mutations. Mol Cancer Ther 11:909–920

Gupta P, Prywes R (2002) ATF1 phosphorylation by the ERK MAPK pathway is required for epidermal growth factor-induced c-jun expression. J Biol Chem 277:50550–50556

Hatzivassiliou G, Liu B et al (2012) ERK inhibition overcomes acquired resistance to MEK inhibitors. Mol Cancer Ther 11:1143–1154

Haura EB, Ricart AD et al (2010) A phase II study of PD-0325901, an oral MEK inhibitor, in previously treated patients with advanced non-small cell lung cancer. Clin Cancer Res 16:2450–2457

Hayashi K, Shibata K et al (2004) Insulin receptor substrate-1/SHP-2 interaction, a phenotype-dependent switching machinery of insulin-like growth factor-I signaling in vascular smooth muscle cells. J Biol Chem 279:40807–40818

Heidorn SJ, Milagre C et al (2010) Kinase-dead BRAF and oncogenic RAS cooperate to drive tumor progression through CRAF. Cell 140:209–221

Hoeflich KP, Herter S et al (2009a) Antitumor efficacy of the novel RAF inhibitor GDC-0879 is predicted by BRAFV600E mutational status and sustained extracellular signal-regulated kinase/mitogen-activated protein kinase pathway suppression. Cancer Research 69: 3042–3051
Hoeflich KP, O'Brien C et al (2009b) In vivo antitumor activity of MEK and phosphatidylinositol 3-kinase in basal-like breast cancer models. Clin Cancer Res 15:4649–4664
Hong J, Zhou J et al (2011) Phosphorylation of serine 68 of Twist1 by MAPKs stabilizes Twist1 protein and promotes breast cancer cell invasiveness. Cancer Res 71:3980–3990.
Huynh H, Soo KC et al (2007) Targeted inhibition of the extracellular signal-regulated kinases kinase pathway with AZD-6244 (ARRY-142886) in the treatment of hepatocellular carcinoma. Mol Cancer Ther 6:138–146
Iverson C, Larson G et al (2009) RDEA119/BAY 869766: a potent, selective, allosteric inhibitor of MEK1/2 for the treatment of cancer. Cancer Res 69:6839–6847
Johannessen CM, Boehm JS et al (2010) COT drives resistance to RAF inhibition through MAP kinase pathway reactivation. Nature 468:968–972
Joseph EW, Pratilas CA et al (2010) The RAF inhibitor PLX4032 inhibits ERK signaling and tumor cell proliferation in a V600E BRAF-selective manner. Proc Natl Acad Sci USA 107:14903–14908
Kefford R, Arkenau H et al (2010) Phase I/II study of GSK2118436, a selective inhibitor of on-cogenic mutant BRAF kinase, in patients with metastatic melanoma and other solid tumors. J Clin Oncol 28:15s
Killestein J, Rudick RA et al (2011) Oral treatment for multiple sclerosis. Lancet Neurol 10:1026–1034
Kim K, Kong SY et al (2010) Blockade of the MEK/ERK signalling cascade by AS703026, a novel selective MEK1/2 inhibitor, induces pleiotropic anti-myeloma activity in vitro and in vivo. Br J Haematol. 149:537–549.
Konopleva M, Milella M et al (2012) MEK inhibition enhances ABT-737-induced leukemia cell apoptosis via prevention of ERK-activated MCL-1 induction and modulation of MCL-1/BIM complex. Leukemia 26:778–787
Kudchadkar R, Paraiso KH et al (2012) Targeting mutant BRAF in melanoma: current status and future development of combination therapy strategies. Cancer J 18:124–131
Kyriakis JM, Avruch J (2001) Mammalian mitogen-activated protein kinase signal transduction pathways activated by stress and inflammation. Physiol Rev 81:807–869
Little S, Balmanno K et al (2011) Amplification of the driving oncogene, KRAS or BRAF, underpins acquired resistance to MEK1/2 inhibitors in colorectal cancer cells. Sci Signal 4:ra17
Liu D, Xing J, Trink B et al (2010) BRAF mutation-selective inhibition of thyroid cancer cells by the novel MEK inhibitor RDEA119 and genetic-potentiated synergism with the mTOR inhibitor temsirolimus. Int J Cancer 127:2965–2973
Llovet JM, Ricci S et al (2008) SHARP investigators study group. Sorafenib in advanced hepato-cellular carcinoma. N Engl J Med 359:378–390
Lohr JG, Stojanov P et al (2012) Discovery and prioritization of somatic mutations in diffuse large B-cell lymphoma (DLBCL) by whole-exome sequencing. Proc Natl Acad Sci USA 109:3879–3884
LoRusso PM, Krishnamurthi SS et al (2010) Phase I pharmacokinetic and pharmacodynamic study of the oral MAPK/ERK kinase inhibitor PD-0325901 in patients with advanced cancers. Clin Cancer Res 16:1924–1937
Luo J, Solimini NL et al (2009) Principles of cancer therapy: oncogene and non-oncogene addic-tion. Cell 136:823–837
McCubrey JA, Steelman LS et al (2007) Roles of the Raf/MEK/ERK pathway in cell growth, malignant transformation and drug resistance. Biochem Biophys Acta 773:1263–1284
McCubrey JA, Steelman LS et al (2010) Emerging MEK inhibitors. Exp Opin Emerging Drugs 15:203–223
McCubrey JA, Steelman LS et al (2011) Therapeutic resistance resulting from mutations in Raf/MEK/ERK and PI3K/PTEN/Akt/mTOR signaling pathways. J Cell Physiol 226:2762–2781
McKay MM, Ritt DA et al (2011) RAF inhibitor-induced KSR1/B-RAF binding and its effects on ERK cascade signaling. Curr Biol 21:563–568

Mischak H, Seitz T et al (1996) Negative regulation of Raf-1 by phosphorylation of serine 621. Mol Cell Biol 10:5409–5418
Mylona A, Nicolas R et al (2010) The essential function for SRF in T cell development reflects its specific coupling to ERK signalling. Mol Cell Biol 31:267–276
Nagarajan D, Melo T et al (2012) ERK/GSK3β/Snail signaling mediates radiation-induced alveolar epithelial-to-mesenchymal transition. Free Radic Biol Med 52:983–992
Nazarian R, Shi H et al (2010) Melanomas acquire resistance to B-RAF(V600E) inhibition by RTK or N-RAS upregulation. Nature 468:973–977
Negrini S, Gorgoulis VG et al (2010) Genomic instability–an evolving hallmark of cancer. Nat Rev Mol Cell Biol 11:220–228
Neviani P, Santhanam R et al (2007) FTY720, a new alternative for treating blast crisis chronic myelogenous leukemia and philadelphia chromosome-positive acute lymphocytic leukemia. J Clin Invest 117:2408–2421
Oberholzer PA, Kee D et al (2012) RAS mutations are associated with the development of cutaneous squamous cell tumors in patients treated with RAF inhibitors. J Clin Oncol 30:316–321
Ohren JF, Chen H et al (2004) Structures of human MAP kinase kinase 1 (MEK1) and MEK2 describe novel noncompetitive kinase inhibition. Nat Struct Mol Biol 11:1192–1197
Ott PA, Hamilton A et al (2010) A phase II trial of sorafenib in metastatic melanoma with tissue correlates. PLoS One 5:e15588
Pao W, Girard N (2011) New driver mutations in non-small-cell lung cancer. Lancet Oncol 12:175–180
Paraiso KH, Fedorenko IV et al (2010) Recovery of phospho-ERK activity allows melanoma cells to escape from BRAF inhibitor therapy. Br J Cancer 102:1724–1730
Paraiso KH, Xiang Y et al (2011) PTEN loss confers BRAF inhibitor resistance to melanoma cells through the suppression of BIM expression. Cancer Res 71:2750–2760
Patel SP, Kim KB (2012) Selumetinib (AZD6244; ARRY-142886) in the treatment of metastatic melanoma. Expert Opin Investig Drugs 21:531–539
Perrotti D, Neviani P (2008) Protein phosphatase 2A (PP2A), a drugable tumor suppressor in Ph1(+) leukemias. Cancer Metastasis Rev 27:159–168
Poulikakos PI, Persaud Y et al (2011) RAF inhibitor resistance is mediated by dimerization of aberrantly spliced BRAF(V600E). Nature 480:387–390
Poulikakos PI, Rosen N (2011) Mutant BRAF melanomas–dependence and resistance. Cancer Cell 19:11–15
Poulikakos PI, Solit DB (2011) Resistance to MEK inhibitors: should we co-target upstream? Sci Signal 4:pe16
Poulikakos PI, Zhang C et al (2010) RAF inhibitors transactivate RAF dimers and ERK signalling in cells with wild-type BRAF. Nature 464:427–430
Pratilas CA, Hanrahan AJ et al (2008) Genetic predictors of MEK dependence in non-small cell lung cancer. Cancer Res 68:9375–9383
Puxeddu E, Moretti S et al (2004) BRAF(V599E) mutation is the leading genetic event in adult sporadic papillary thyroid carcinomas. J Clin Endocrinol Metab 89:2414–2420
Rajalingam K, Wunder C et al (2005) Prohibitin is required for Ras-induced Raf-MEK-ERK activation and epithelial cell migration. Nat Cell Biol 7:837–843
Ratain MJ, Eisen T et al (2006) Phase II placebo-controlled randomized discontinuation trial of sorafenib in patients with metastatic renal cell carcinoma. J Clin Oncol 24:2505–2512
Rebecca VW, Sondak VK et al (2012) A brief history of melanoma: from mummies to mutations Melanoma Res 22:114–122
Ricciardi MR, Scerpa MC et al (2012) Therapeutic potential of MEK inhibition in acute myelogenous leukemia: rationale for "vertical" and "lateral" combination strategies. J Mol Med (Berl) [Epub ahead of print]
Rimassa L, Santoro A (2009) Sorafenib therapy in advanced hepatocellular carcinoma: the SHARP trial. Expert Rev Anticancer Ther 9:739–745
Roux PP, Blenis J (2004) ERK and p38 MAPK-activated protein kinases: a family of protein kinases with diverse biological functions. Microbiol Mol Biol Rev 68:320–344

Schaeffer HJ, Catling AD et al (1998) MP1: a MEK binding partner that enhances enzymatic activation of the MAP kinase cascade. Science 281:1668–1671

Sebolt-Leopold JS (2008) Advances in the development of cancer therapeutics directed against the Ras-mitogen-activated protein kinase pathway. Clin Cancer Res 14:3651–3656

Sharlow ER, Giridhar KV et al (2008) Potent and selective disruption of protein kinase D functionality by a benzoxoloazepinolone. J Biol Chem 283:33516–33526

Sharma A, Tran MA et al (2006) Targeting mitogen-activated protein kinase/extracellular signal-regulated kinase kinase in the mutant (V600E) B-Raf signaling cascade effectively inhibits melanoma lung metastases. Cancer Res 66:8200–8209

Sharma SV, Settleman J (2010) Exploiting the balance between life and death: targeted cancer therapy and "oncogenic shock". Biochem Pharmacol 80:666–673

Shi H, Moriceau G et al (2012) Melanoma whole-exome sequencing identifies (V600E)B-RAF amplification-mediated acquired B-RAF inhibitor resistance Nat Commun 3:724. doi:10.1038/ncomms1727

Shin S, Dimitri CA et al (2010) ERK2 but not ERK1 induces epithelial-to-mesenchymal transformation via DEF motif-dependent signaling events. Mol Cell 38:114–1127

Simmons GW, Pong WW et al (2011) Neurofibromatosis-1 heterozygosity increases microglia in a spatially and temporally restricted pattern relevant to mouse optic glioma formation and growth. J Neuropathol Exp Neurol 70:51–62

Smalley KS, Lioni M et al (2008) Increased cyclin D1 expression can mediate BRAF inhibitor resistance in BRAF V600E-mutated melanomas. Mol Cancer Ther 7:2876–2883

Smalley KS, Xiao M et al (2009) CRAF inhibition induces apoptosis in melanoma cells with non-V600E BRAF mutations. Oncogene 28:85–94

Solit DB, Garraway LA et al (2006) BRAF mutation predicts sensitivity to MEK inhibition. Nature 439:358–362

Sorokina EM, Feinstein SI et al (2011) Intracellular targeting of peroxiredoxin 6 to lysosomal organelles requires MAPK activity and binding to 14-3-3ε. Am J Physiol Cell Physiol 300:C1430–1441

Sos ML, Michel K et al (2009) Predicting drug susceptibility of non-small cell lung cancers based on genetic lesions. J Clin Invest 119:1727–1740

Sosman JA, Kim KB et al (2012) Survival in BRAF V600-mutant advanced melanoma treated with vemurafenib. N Engl J Med 366:707–714

Steelman LS, Pohnert SC et al (2004) JAK/STAT, RAF/MEK/ERK, PI3K/AKT and BCR-ABL in cell cycle progression and leukemogenesis. Leukemia 18:189–218

Steelman LS, Abrams SL et al (2008) Contributions of the Raf/MEK/ERK, PI3K/PTEN/Akt/mTOR and JAK/STAT pathways to leukemia. Leukemia 22:686–707

Sturm OE, Orton R et al (2010) The mammalian MAPK/ERK pathway exhibits properties of a negative feedback amplifier. Sci Signal 3:ra90

Su F, Viros A, Milagre C et al (2012) RAS mutations in cutaneous squamous-cell carcinomas in patients treated with BRAF inhibitors. N Engl J Med 366:207–215

Tai YT, Fulciniti M et al (2007) Targeting MEK induces myeloma-cell cytotoxicity and inhibits osteoclastogenesis. Blood 110:1656–1663

Takimoto CH, Awada A (2008) Safety and anti-tumor activity of sorafenib (Nexavar) in combination with other anti-cancer agents: a review of clinical trials. Cancer Chemother Pharmacol 61:535–548

Tanios Bekaii-Saab T, Phelps MA et al (2011) Multi-institutional phase II study of selumetinib in patients with metastatic biliary cancers. JCO 29:2357–2363

Tiacci E, Trifonov V et al (2011) BRAF mutations in hairy-cell leukemia. N Engl J Med 364:2305–2315

Tol J, Nagtegaal ID et al (2009) BRAF mutation in metastatic colorectal cancer. N Engl J Med 361:98–99

Topisirovic I, Sonenberg N (2011) mRNA translation and energy metabolism in cancer: the role of the MAPK and mtorc1 pathways. Cold Spring Harb Symp Quant Biol [Epub ahead of print]

Torkamani A, Schork NJ (2009) Identification of rare cancer driver mutations by network reconstruction Genome Res 19:1570–1578

Trepel J, Mollapour M et al (2010) Targeting the dynamic HSP90 complex in cancer. Nature Rev Cancer 10:537–549

Villanueva J, Vultur A et al (2010) Acquired resistance to BRAF inhibitors mediated by a RAF kinase switch in melanoma can be overcome by cotargeting MEK and IGF-1R/PI3K. Cancer Cell 18:683–695

Wagle N, Emery C et al (2011) Dissecting therapeutic resistance to RAF inhibition in melanoma by tumor genomic profiling. J Clin Oncol 29:3085–3096

Webb DJ, Donais K et al (2004) FAK-Src signalling through paxillin, ERK and MLCK regulates adhesion disassembly. Nat Cell Biol 6:154–156

Wee S, Jagani Z et al (2009) PI3K pathway activation mediates resistance to MEK inhibitors in KRAS mutant cancers. Cancer Res 69:4286–4293

Weinstein IB, Joe A (2008) Oncogene addiction. Cancer Res 68:3077–3080

Whittaker S, Kirk R et al (2010a) Gatekeeper mutations mediate resistance to BRAF-targeted therapies. Sci Transl Med 2:35ra41

Whittaker S, Ménard D et al (2010b) A novel, selective, and efficacious nanomolar pyridopyrazinone inhibitor of V600EBRAF. Cancer Res 70:8036–8044

Wilhelm SM, Carter C et al (2004) BAY 43–9006 Exhibits broad spectrum oral antitumor activity and targets the RAF/MEK/ERK pathway and receptor tyrosine kinases involved in tumor progression and angiogenesis. Cancer Research 64:7099–7109

Xing J, Ginty DD et al (1996) Coupling of the Ras-MAPK pathway to gene activation by Rsk2, a growth factor regulated CREB kinase. Science 273:959–963

Yeh TC, Marsh V et al (2007) Biological characterization of ARRY-142886 (AZD6244), a potent, highly selective mitogen-activated protein kinase kinase 1/2 inhibitor. Clin Cancer Res 13:1576–1583

Yeung K, Janosch P et al (2000) Mechanism of suppression of the Raf/MEK/extracellular signal-regulated kinase pathway by the raf kinase inhibitor protein. Mol Cell Biol 20:3079–3085

Zhang W, Konopleva M, Ruvolo V et al (2008a) Sorafenib induces apoptosis of AML cells via Bim-mediated activation of the intrinsic apoptotic pathway. Leukemia 22:808–818.

Zhang W, Konopleva M, Shi YX et al (2008b). Mutant FLT3: a direct target of sorafenib in acute myelogenous leukemia. J Natl Cancer Inst 100:184–198.

Zheng Y, Xia Y et al (2009) FAK phosphorylation by ERK primes Ras-induced tyrosine dephosphorylation of FAK mediated by PIN1 and PTP-PEST. Mol Cell 35:11–25

Zheng Y, Yang W et al (2011) Ras-induced and extracellular signal-regulated kinase 1 and 2 phosphorylation-dependent isomerization of protein tyrosine phosphatase (PTP)-PEST by PIN1 promotes FAK dephosphorylation by PTP-PEST. Mol Cell Biol 31:4258–4269

Zitzmann K, de Toni E et al (2011) The novel Raf inhibitor Raf265 decreases Bcl-2 levels and confers TRAIL-sensitivity to neuroendocrine tumour cells. Endocr Relat Cancer 18:277–285

Chapter 15
The Role of Translational Medicine in Optimization of Therapies Targeting the RAS-MAP Kinase Pathway

Judith S. Sebolt-Leopold

Abstract Activation of the RAS-MAP kinase (MAPK) pathway is a common feature in hematological malignancies, including acute myelogenous leukemia (AML). Based on its central role in regulating growth and survival of a broad spectrum of human cancers, much attention has focused in recent years on pharmacologic targeting of the RAS–RAF–MEK–MAPK intracellular signaling cascade. Multiple agents now exist in various stages of clinical development that are capable of effectively inhibiting the downstream kinases in this pathway. The emergence of these agents into the clinical trial arena now allows us to turn our attention to how best to deploy them for maximal therapeutic impact in AML patients. Optimization of these therapies will ultimately depend not only on the availability of promising clinical agents but also on the identification of patients most likely to respond. Toward this goal, technological advances have led to comprehensive molecular profiling of AML samples to facilitate the identification of prognostic biomarkers and insights into the rational design of combination strategies.

Keywords RAS · ERK · RAF · MEK · Biomarker · Translational

15.1 Introduction

The application of translational medicine to the implementation of individualized treatment strategies is highly dependent upon the availability of validated biomarkers. Critical biomarkers include markers to assess expression and modulation of the molecular target in addition to outcome-based prognostic markers for predicting the responder population. The RAS-MAP kinase pathway, frequently activated in acute myelogenous leukemia (AML), is particularly amenable to design of targeted therapies that can prospectively be tailored to those patients most likely to respond. This review will cover various therapeutic approaches for targeting this critical signaling

---

J. S. Sebolt-Leopold (✉)
Translational Oncology Program, University of Michigan, Comprehensive Cancer Center, NCRC, 1600 Huron Parkway, Ann Arbor, MI 48109, USA
e-mail: jssl@med.umich.edu

M. Andreeff (ed.), *Targeted Therapy of Acute Myeloid Leukemia*,
Current Cancer Research, DOI 10.1007/978-1-4939-1393-0_15

pathway. It will also cover a multitude of combination strategies to further increase therapeutic outcome.

15.2 Therapeutic Approaches to Targeting the RAS-MAP Kinase Pathway

The impetus for therapeutic targeting of the RAS-MAP kinase signaling cascade to treat human cancers has been reviewed elsewhere (Sebolt-Leopold and Herrera 2004; Sebolt-Leopold 2008; Steelman et al. 2011). The emergence of small-molecule inhibitors of various steps along the pathway has led to decidedly different pharmacological profiles among the target classes, not predicted before these biological tools became available. Inhibitors now exist that specifically target each of the downstream kinases. However, efforts to target the GTP-binding protein RAS have largely been unsuccessful. A number of research programs identified farnesyltransferase inhibitors, e.g., tipifarnib, in an attempt to prevent the membrane localization of RAS, an event necessary for activation of the MAPK (ERK) pathway. However, clinical activity of farnesyltransferase inhibitors in AML patients has been modest at best (Karp et al. 2001; Lancet et al. 2007; Harousseau et al. 2007), perhaps owing to an alternative prenylation route, specifically geranylgeranylation, utilized by RAS (Whyte et al. 1997). Of those patients who did show a response to treatment with tipifarnib, no correlation was observed with baseline levels of pERK (Lancet et al. 2007). More recently, the use of NMR in fragment-based discovery programs, has tackled direct targeting of RAS to provide a new direction for potential therapeutic leads. Consequently, small-molecule ligands have been identified that are capable of binding to a distinct RAS-binding pocket, subsequently interfering with RAS–SOS interactions (Maurer et al. 2012). These discovery programs remain at an early development stage as significant medicinal chemistry effort will be required to address selectivity and potency issues. In the meantime, kinase-targeted small molecules directed against the downstream players in the RAS-MAPK pathway are rapidly advancing into clinical evaluation.

Efforts to develop RAF-targeted agents were spurred by the discovery that BRAF is an oncogene and mutated in a large percentage of human cancers. Sorafenib received early attention as an ATP-competitive inhibitor of CRAF and subsequently gained regulatory approval on the basis of its clinical activity against renal cancer. However, the multitargeted selectivity profile of this agent complicates assessment of the contribution of RAF inhibition to clinical outcome, since it also exhibits inhibitory activity against vascular endothelial growth factor receptor-2 (VEGFR2), platelet-derived growth factor-β (PDGFR), FMS-like tyrosine kinase receptor-3 (FLT3), and c-Kit (Wilhelm et al. 2006). Nonetheless, inhibition of the RAF/MEK/ERK pathway by sorafenib has been demonstrated in phase I AML clinical samples resulting in apoptosis induction (Zhang et al. 2008). The ability of sorafenib to target FLT3 is particularly relevant for the AML patient population, where it is constitutively activated in roughly 30 % of cases owing to internal tandem duplications (ITD). It is therefore not surprising that the complete response reported in a phase I

AML clinical study was observed in a patient with FLT3-ITD (Crump et al. 2010). Further study of sorafenib in FLT3-ITD AML patients has shown a remarkable initial response in 12 of 13 patients. However, with time, the leukemia recurred, likely due to sorafenib-resistant double-mutant subclones harboring both FLT3-ITD and tyrosine kinase domain D835 mutations (Man et al. 2012).

Vemurafanib is the first selective RAF inhibitor to gain regulatory approval based on its striking clinical activity in the BRAF mutant subpopulation of melanoma patients. However, as will be discussed later in this review, the absence of BRAF mutations in AML does not support use of vemurafanib or other BRAF-targeted agents in a monotherapy setting for this particular malignancy.

15.2.1 Overview of MEK Inhibitors

Unlike RAF inhibitors, MEK inhibitors are generally non-ATP-competitive and therefore possess a high degree of specificity. The presence of a unique interior hydrophobic binding pocket adjacent to but distinct from the ATP binding site (Ohren et al. 2004) is the basis for the observed high degree of selectivity of most reported MEK inhibitors. PD98059 was the first published small molecule MEK inhibitor and subsequently served as a valuable biological tool for demonstrating that blockade of MEK function resulted in inhibition of AML cell lines (Milella et al. 2001; Morgan et al. 2001). CI-1040 later emerged as the first orally active MEK inhibitor and was shown to effectively impair tumor growth in vivo in preclinical model systems (Sebolt-Leopold et al. 1999). This agent advanced to become the first MEK inhibitor to be tested in patients. As the first agent of this target class to enter the clinic, CI-1040 became a clinical tool to address how strongly ERK activation could be suppressed without incurring unacceptable toxicities. Translational analysis of phase I biopsy samples revealed that steady-state levels of CI-1040 elicited a 70 % reduction in pERK expression and an acceptable safety profile (Lorusso et al. 2005). Efforts quickly turned to improving the pharmaceutical profile of CI-1040 by incorporating chemical changes in the molecule to enhance its potency, bioavailability, and metabolic stability. AZD6244 and PD0325901 followed on the heels of CI-1040 clinical trials, representing second-generation structural analogs of CI-1040, with significantly improved biological profiles. Today, there are nearly a dozen structurally similar MEK inhibitors at various stages of clinical evaluation against a broad range of human cancers (Table 15.1).

Trametinib (GSK1120212), the first clinically approved MEK inhibitor, exhibits subnanomolar enzyme potency and favorable pharmacokinetics as evidenced by a long half-life and shallow C_{max} to C_{trough} profile (Gilmartin et al. 2011). The degree of target inhibition measured in clinical samples in response to this new generation of MEK inhibitors is >90 %, representing a significant improvement from the early clinical days with CI-1040 (Infante et al. 2012). Among the ongoing clinical trials with GSK1120212 are phase I/II trials in relapsed or refractory leukemias including AML. Separate cohorts of patients are being studied in these trials for the effect of RAS status (www.clinicaltrials.gov).

Table 15.1 MEK inhibitors entered into clinical trials

Agent	Sponsor	Status
CI-1040	Pfizer	Dropped
PD0325901	Pfizer	Phase I
Selumetinib (AZD6244)	Astra-Zeneca	Phase III
AZD8330	Astra-Zeneca	Phase I
Benimetinib (MEK162)	Novartis/Array	Phase III
RDEA119	Bayer/Ardea	Phase II
E6201	Eisai	Phase I
Trametinib (GSK1120212)	GlaxoSmithKline	Approved
WX554	Wilex AG	Phase I/II
Cobimetinib (RG7420 (XL518;GDC-0973))	Genentech/Exelixus	Phase III
RG7421 (GDC-0623)	Genentech	Phase I
RG7304 (RO5126766)	Roche/Chugai	Phase I
RG7167 (RO4987655)	Roche/Chugai	Phase I
Pimasertib (AS703026)	EMD Serono	Phase II
TAK-733	Takeda/Millenium	Phase I

15.3 Role of Molecular Profiling to Guide Treatment Design

Despite extensive efforts to develop new, molecular-targeted agents for the treatment of AML, cytarabine, and anthracyclines remain the mainstay of current therapeutic regimens. The goal of designing individualized treatment strategies begins with defining the underlying genetic aberrations leading to AML pathogenesis. Application of whole-genome sequencing of AML samples has led to important discoveries of common mutations in AML, including therapeutically actionable events. As reviewed elsewhere, whole-genome sequencing has also led to the identification of new mutations in AML, such as IDH1, now the focus of new therapeutic discovery programs (Welch and Link 2011).

Genomic information, while highly informative of dysregulated signaling pathways, nonetheless does not provide a full picture of the dynamics of pathway expression. Kinase-mediated phosphorylation events are a critical component of the molecular signature of an individual cancer. Therefore, assessment of pathway flux is especially important when gauging potential susceptibility to kinase-directed ERK pathway inhibitors. In the early days of MEK inhibitor trials, translational analysis was restricted to pERK assessment in peripheral blood mononuclear cells as well as a limited number of tumor biopsies. The data gleaned from these analyses proved informative and provided encouraging evidence that sufficient target exposure was being achieved to elicit significant reductions in pERK expression (Lorusso et al. 2005). However, compensatory signaling events occurring in response to MEK inhibition were not assessed.

Until recently, technological hurdles precluded the application of a proteomics approach to assess the relative activity of multiple pathways in small sample sizes.

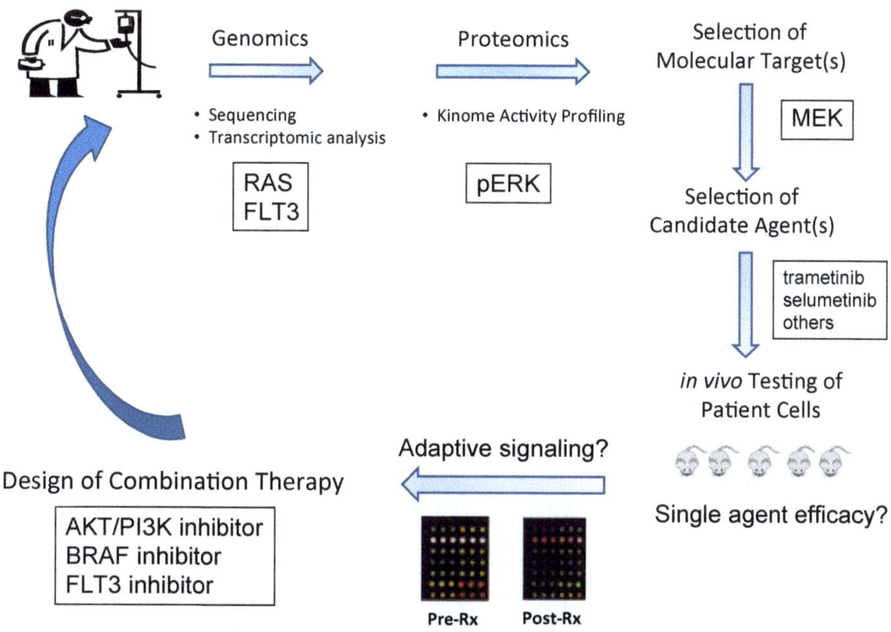

Fig. 15.1 Bedside to bench and back: key stages of therapy optimization. Representative examples of markers, targets, and agents pertinent to the development of MEK inhibitors and kinase-based combination treatment strategies are shown

As reviewed elsewhere, significant improvements in proteomic technologies are enabling the mapping of kinase signaling pathways in cancer (Kolch and Pitt 2010). Quantitative proteomics has been successfully employed to demonstrate that parallel kinase pathways do in fact contribute to resistance of AML cells to inhibitors of various kinases, including MEK (Alcolea et al. 2012). Reverse-phase protein microarray (RPMA) is a highly sensitive and high throughput multiplexed platform that provides quantitative analysis of the phosphorylation/activation state of a given kinase or pathway in patient samples. RPMA analysis of AML specimens has shown that differences in protein expression could be detected in as few as three-cell equivalents (Tibes et al. 2006). The protein expression signatures generated in this fashion provide potentially valuable prognostic information as well as data to guide the selection of targeted therapies (Kornblau et al. 2009). Quantitative phosphoproteomic approaches are providing increasingly more insight into how the kinome is rapidly re-programmed in response to MEK inhibition. In one study focusing on triple-negative breast cancer, multiple receptor tyrosine kinases were found to be activated in response to MEK inhibitor treatment, pointing to the inherent limitations of a monotherapy approach with these agents (Duncan et al. 2012). Figure 15.1 provides a schematic overview of the critical testing steps and decision points in optimization of molecular-targeted therapy as exemplified by MEK inhibitor-based treatment regimens.

15.4 Prognostic Markers: Predicting the Responder Population

15.4.1 RAS

The incidence of oncogenic RAS mutations in AML is generally in the range of 10–25 % and are mainly found in NRAS (codons 12, 13, or 61) or KRAS (codon 12; Neubauer et al. 1994). Point mutations in NRAS are among the most frequent genetic aberrations found in AML along with FLT3 mutations (Kiyoi et al. 1999). However, elevation in RAS activity does not always accompany these mutations, perhaps owing to negative regulators, such as members of the Sprouty gene family and the RAF kinase inhibitor (RKIP). Measurements of RAS activity in a large AML study population revealed that only 22 % of mutated RAS samples were accompanied by activation (Illmer et al. 2005). Conversely, these investigators further showed that strong RAS activity is not restricted to patients harboring RAS mutations. Recent profiling of GSK1120212, the first clinically approved MEK inhibitor, revealed that 11 of 12 AML cell lines showed sensitivity despite only six of lines possessing a mutation in RAS/RAF (Jing et al. 2012). This result is perhaps not surprising in light of multiple signaling pathways intersecting through RAS. For example, internal tandem duplications of FLT3 frequently occur in AML and have been shown to signal through the MAPK pathway (Takahashi 2006). Collectively, these data suggest that RAS mutational status alone is insufficient to judge whether an AML patient is a candidate for therapeutic intervention directed against the RAS-MAPK pathway.

15.4.2 RAF Kinase Inhibitor

Among its multiple functions, RKIP acts as a negative regulator of the RAS–MAPK signaling cascade. By inhibiting the interaction between CRAF and MEK, RKIP suppresses phosphorylation of MEK (Yeung et al. 1999). In two independent cohorts of AML patients ($n=103$ and 285), loss of RKIP was shown to be associated with a monocytic phenotype (Zebisch et al. 2012), a finding that is consistent with association of this phenotype with dysregulation of the RAS-MAPK pathway (Tyner et al. 2009; Bowen et al. 2005). Importantly, AML patients exhibiting loss of RKIP proved to have a significantly longer relapse-free survival leading to enhanced overall survival (Zebisch et al. 2012). Therefore, loss of this regulatory protein may serve as a useful biomarker for predicting treatment response.

15.4.3 Phosphorylated ERK

The MAP kinase ERK has been reported to be strongly elevated in greater than 50% of primary AMLs representing relapsed and refractory as well as newly diagnosed disease (Ricciardi et al. 2005). In this study, normal CD34$^+$ progenitor cells exhibited low-level expression of pERK, importantly providing the basis for a therapeutic index in response to pathway intervention strategies. A critical question is whether low-level expression of pERK in an AML patient would preclude therapeutic benefit from ERK pathway inhibition. Experimental evidence supporting the assumption that high pERK expression signifies pathway dependency comes from two independent studies. Blast cell proliferation in response to MEK inhibition was measured in 14 cases of AML with high ERK activity and four cases with undetectable or low pERK, leading to the observation that only the high expressers were adversely affected (Lunghi et al. 2003). Correlation between sensitivity to MAP kinase pathway inhibition and endogenous levels of phosphorylated ERK has been shown by others (Milella et al. 2001).

15.4.4 RAF

Activating mutations of BRAF occur at an especially high incidence in melanomas (roughly 60%) and at a lower frequency in a number of other solid tumors (Garnett and Marais 2004). However, a screen of 82 AML samples did not produce any evidence for the presence of these mutations in this patient population (Zebisch et al. 2006). Consistent with previous reports, greater than half of the AML samples in this study exhibited constitutive activation of ERK. Sequence analysis of the entire coding region of the RAF genes led to the discovery of heterozygous single-base germ-line mutations in CRAF (S427G or I448 V) in two of the patient samples. Both mutations were detected in patients with therapy-related AML (t-AML), occurring after chemotherapy, and/or radiotherapy for primary disease (Pedersen-Bjergaard 1992) and possibly constitute a tumor-predisposing factor (Zebisch et al. 2006).

The implication of these findings is that AML patients who exhibit constitutive activation of RAS-MAPK signaling are likely to receive greater benefit from single-agent treatment with an MEK inhibitor as opposed to a BRAF inhibitor. The basis for this supposition is that there is a growing body of evidence suggesting that treatment with BRAF inhibitors be restricted to patients harboring mutations of BRAF. Non-BRAF-mutated cancers exhibit induction rather than inhibition of ERK signaling in response to BRAF inhibition in a manner that is independent of their RAS mutational status (Poulikakos et al. 2010). The mechanistic basis for this observation likely hinges on differential activation of CRAF in RAF heterodimers that is dependent upon the state of the BRAF protein (Poulikakos et al. 2010; Garnett et al. 2005; Hatzivassiliou et al. 2010).

15.5 Combination Strategies

In most cases, single-agent therapy with MAP kinase signaling antagonists, e.g., MEK inhibitors, leads to cytostatic rather than cytotoxic effects. Efforts to improve the clinical activity of MEK inhibitors have focused on the design of combination regimens that exploit the pleiotropic nature of the RAS-MAPK pathway. This pathway imparts functionality to a broad array of cellular processes, including differentiation, immune response, proliferation, and survival. Consequently, a number of distinctly different approaches have generated promising signs of synergistic activity in MEK inhibitor-based combinations.

15.5.1 Retinoids: Targeting the Role of MAP Kinase in Differentiation

Several lines of evidence indicate that MAP kinase activity is necessary to trigger myeloid differentiation in response to retinoid treatment (Miranda et al. 2002; Yen et al. 1998; Glasow et al. 2005). Phosphorylation of retinoid X receptor alpha occurs at serine 260, a consensus MAPK site, thereby rendering this receptor a downstream target of MAPK (Solomon et al. 1999). The interconnection between these two pathways is further supported by data showing that MEK inhibition sensitized AML cells in vitro to retinoid-induced apoptosis (Milella et al. 2001; Milella et al. 2007). The proapoptotic interaction between MEK inhibitors and retinoids was shown to be strongly synergistic and required functional RAR and RXR receptors (Milella et al. 2007). Disruption of Bcl-2-mediated maintenance of mitochondrial homeostasis was evoked as a potential mechanism to explain this strong synergy. Identification of the subpopulation of AML patients who would benefit from this combination strategy could be guided by *ex vivo* assessment of the expression levels of RARβ in response to all-*trans* retinoic acid (ATRA) treatment, which have been found to correlate with retinoid sensitivity (Lehmann et al. 2001), as well as ERK phosphorylation levels (Ricciardi et al. 2005).

15.5.2 Immunotherapy: Targeting the Role of MAP Kinase in Toll-Like Receptor Signaling

AML blast cells can escape autologous or allogenic immune responses by various mechanisms, including expression of the B7–H1 protein. In a cohort of 79 AML patients, B7–H1 expression was high in 18 % of patients (Berthon et al. 2010). Expression of toll-like receptor TLR2 and TLR4 ligands or interferon induced by B7–H1 serves to protect AML cells from lysis mediated by cytotoxic T-lymphocytes (CTLs). MEK inhibition was found to reduce B7–H1 expression and effectively restored CTL-mediated lysis of blast cells. These findings support further exploration into whether a MEK-inhibitor-based combination strategy with immunotherapy improves therapeutic outcome in a subpopulation of AML patients.

15.5.3 Targeting Multiple Steps Along the Same Pathway: Vertical Strategies to Increase Growth Inhibition in Response to MEK Inhibition

As previously discussed, single-agent therapy with BRAF inhibitors are potentially contraindicated for the treatment of non-BRAF-mutated malignancies, since they induce a paradoxical increase in the expression of pERK in these cancers. Dual targeting of both RAF and MEK may be a rational approach to attenuate the adverse effects seen with RAF inhibitors alone while also increasing the magnitude of therapeutic response, in part by delaying the onset of resistance. Studies are now ongoing with multiple RAF and MEK inhibitors and are being carried out in multiple cancers to explore the feasibility of this combination approach. Preliminary data have been encouraging for this combination strategy. Preclinical evaluation of the BRAF inhibitor dabrafenib (GSK2118436) and the MEK inhibitor trametinib (GSK1120212) has been reported to result in strongly synergistic growth inhibition, especially in RAS-mutated tumors (Incani et al. 2012). Clinical activity of this combination against V600 BRAF mutant melanomas is encouraging and has led to a phase III trial (Weber et al. 2012). Synergistic growth inhibition and enhanced induction of apoptosis has also been reported for AML cells treated with the MEK inhibitor PD0325901 in combination with sorafenib (Ricciardi et al. 2012). With the caveat that sorafenib is a multitargeted RAF inhibitor, these findings support further exploration of this combination approach in AML, using more selective and potent RAF inhibitors now available.

Another vertical strategy has been directed toward the combination of inhibitors targeting both farnesyltransferase and 3-hydroxy-3-methylglutaryl CoA reductase (HMG-CoA reductase). Elevation of low-density lipoprotein receptor (LDL) activity leads to hypocholesterolemia and a higher rate of cholesterol synthesis in AML patients (Ho et al. 1978; Vitols et al. 1984). Statins, which inhibit HMG-CoA reductase, and therefore lower cholesterol levels, also inhibit downstream production of byproducts of the mevalonate pathway, including the farnesyl and geranylgeranyl isoprenoids. Since geranylgeranylation of RAS may play a role in compromising single-agent effectiveness of farnesyltransferase inhibitors in AML patients, the use of a statin to block production of both isoprenoids provides a rationale for this combination approach. Combination of the statin simvastatin with a farnesyltransferase inhibitor, specifically tipifarnib, has been shown to result in enhanced cytotoxicity in the $CD34^+$ but not the $CD34^-$ subfraction of AML cells (van der Weide et al. 2009). This finding supports a CD34-based biomarker enrichment strategy for future clinical trials with this drug combination.

15.5.4 PI3K/AKT Pathway Inhibitors: Targeting MAP Kinase Pathway Cross Talk

A number of recent studies suggest that RAS-mutant cancers require dual inhibition of both MEK and PI3K pathways to achieve maximal inhibition of tumor growth (Engelman et al. 2008; Yu et al. 2008; Wee et al. 2009). Negative feedback loops have been shown to lead to activation of the PI3K or ERK pathway when the other pathway is inhibited (Mirzoeva et al. 2009; Carracedo et al. 2008). Demonstration of cross talk between these two pathways in AML cells comes from transcriptomic and proteomic analysis of MEK inhibitor-resistant lines, revealing marked upregulation of PI3K, AKT, and S6K expression as well as increased phosphorylation of AKT and S6 (Ricciardi et al. 2012). Consequently, these cells proved especially sensitive to the combination of the MEK inhibitor PD0325901 and the mTOR inhibitor temsirolimus. It should be pointed out that PTEN status will likely be an important prognostic determinant of response for this combination strategy. Novel cross-talk mechanisms have been reported in AML whereby constitutive ERK expression leads to suppression of PTEN expression and conversely, inhibition of ERK expression results in restoration of its activity followed by PI3K pathway inhibition (Ciuffreda et al. 2012).

15.5.5 Role of MAP Kinase Pathway in Modulation of Chemotherapy Response

As reviewed elsewhere, MEK inhibition has been shown to potentiate the effects of a broad spectrum of chemotherapeutic drugs, including 1-β-D-Arabinofuranosylcytosine (Ara-C), taxanes, vinblastine, cisplatin, and topoisomerase I and II inhibitors (Dent and Grant 2001; Milella et al. 2005). The rationale for combination of RAS-MAPK pathway inhibitors with chemotherapeutic agents is largely derived from preclinical models treated with MEK inhibitors. However, there are clinical data that support the role of ERK activation in resistance of AML to ARA-C treatment. This comes from the observation that AML patients who exhibit constitutive ERK phosphorylation at diagnosis are less likely to respond to high-dose ARA-C-based induction regimens (Milella et al. 2005). Furthermore, encouraging data have been generated from analysis of tumor cells acquired from newly diagnosed AML patients subsequently treated ex vivo with ARA-C. Laser-scanning cytometry measurements of pERK and apoptosis regulatory proteins such as Bcl-2 and Bax revealed that surviving cells exhibiting pro-apoptotic net changes, including a decrease in pERK expression, were more likely to originate from patients who survived longer (Milella et al. 2005).

15.5.6 Apoptosis Modulators

15.5.6.1 Bcl-2 Antagonists

Bcl-2 has been established to be a prognostic factor dictating therapeutic outcome in AML (Campos et al. 1993; Kornblau et al. 2000), consistent with observations that inhibition of Bcl-2 leads to induction of apoptosis and chemosensitization of AML cells (Andreeff et al. 1999; Konopleva et al. 2000). Cross talk is believed to occur between the Bcl-2 and ERK pathways, which jointly contribute to leukemogenesis (Blalock et al. 2000). The use of small molecules to simultaneously inhibit both Bcl-2 and MEK leads to the synergistic induction of apoptosis in AML models exhibiting constitutive activation of ERK (Milella et al. 2002). Importantly, this finding was confirmed in clonogenic assays carried out in primary AML samples, providing further support for this combination strategy for future clinical trials.

15.5.6.2 Targeting MDM2 to Activate Proapoptotic Proteins

TP53 mutations occur in greater than 50% of patients diagnosed with AML with complex karyotype (CK-AML) (Rucker et al. 2012). CK-AML account for roughly 10–15% of AML cases and TP53 mutations rarely occur in the broader AML population. However, inactivation of wild-type p53 by overexpression of its negative regulator Mdm2 is frequently observed in this cancer (Seliger et al. 1996). It was earlier reported that incubation of AML cells with Nutlin-3a, a known antagonist of Mdm-2, leads to growth inhibition and induction of p53-mediated apoptosis (Kojima et al. 2005). More recently, the interaction of the RAS/MAPK and p53 pathways in AML was explored leading to the observation that simultaneous blockade of both pathways results in enhanced induction of apoptosis (Kojima et al. 2007). Mechanistic data support a role for the ERK pathway in regulating the subcellular localization of p53, thereby impacting p53-mediated apoptosis. Co-targeting MEK and Mdm2 signaling was further shown to result in the up-regulation of the pro-apoptotic BH3-only proteins Puma and Bim (Zhang et al. 2010). Consistent with these findings, it was recently reported that inhibition of ERK signaling in conjunction with BH3-mimetic therapy proved synergistic in preclinical models, providing further support for clinical testing of this combination strategy (Konopleva et al. 2012).

15.5.7 Overcoming Resistance to FLT3 Inhibitors

As previously discussed, FLT3 is frequently mutated in AML. Clinical evaluation of FLT3 inhibitors is currently ongoing. However, the emergence of drug-resistant leukemic blast cells in AML patients undergoing therapy with these agents is problematic, and responses typically last for only a few months. Among the candidate pathways playing a contributory role to the development of resistance is the RAS–MAPK cascade (Piloto et al. 2007). Consistent with this resistance mechanism, MEK inhibition was reported to strongly impair the growth of FLT3-ITD cells (Takahashi 2006). Therefore, dual targeting of FLT3 and MEK may prove advantageous in FLT3-mutated patients. Preclinical data supporting this combination approach has been reported for the MEK inhibitor AZD6244, which was found to inhibit the proliferation of mutant FLT3-expressing cells, and was synergistic with sunitinib, a multitargeted FLT3 inhibitor (Nishioka et al. 2008).

15.6 Innate Resistance to MEK Inhibitors

Insertional mutagenesis in genetically defined mouse cancer models has proven to be a powerful tool to aid in the identification of candidate genes that contribute to the resistance of RAS–MAPK pathway intervention (Lauchle et al. 2009). In this study, Dr. Shannon's team first initiated myeloproliferative disorder (MPD) by inactivating NF1 in mouse bone marrow. NF1 is a GTPase-activating protein that normally acts as a tumor suppressor, triggering deregulated RAS signaling when inactivated. Retrovirus was then injected in these mice to identify mutations that would cooperate with inactivation of NF1 to induce progression of MPD to AML. MEK inhibition was found to induce regressions in many of the NF1-deficient AMLs. The underlying basis for eventual relapse of these leukemias was explored by identifying candidate resistance genes. One gene identified was Rasgrp1, which exhibited a 1000-fold increase in DNA copy number and tenfold increase in mRNA levels in four independent AMLs. It is noteworthy that Rasgrp1 was subsequently discovered to be a component of a 2-gene transcript that positively predicted clinical response of AML patients to the farnesyltransferase inhibitor tipifarnib (Raponi et al. 2008; Karp et al. 2001). The other component was APTX, which encodes the DNA excision repair protein aprataxin. The contrasting findings that Rasgrp 1 expression correlates with MEK inhibitor resistance as well as FTI sensitivity may reflect differences in the status of NF1. In the former study, NF1 was inactivated in all cases. In a study of 95 adult AML patients, the incidence of copy number alterations in NF1 was 12% (Parkin et al. 2010).

Analysis of transcriptional pathway activation signatures has further led to the identification of a panel of 18 genes thought to dictate MEK inhibitor responsiveness (Dry et al. 2010). A common component of these signatures is the DUSP6 gene, which encodes a dual specificity phosphatase known to dephosphorylate and inactivate ERK (Pratilas et al. 2009; Jing et al. 2012). High expression of DUSP6 may prove to be a useful biomarker to predict response to MEK inhibition. Furthermore, based on the sharp reduction of DUSP6 expression observed after MEK inhibitor treatment, this marker also has the potential to serve as a pharmacodynamics readout (Jing et al. 2012).

Conclusion

The RAS-MAPK pathway has been the subject of intense investigation for the development of therapeutic opportunities to impair the progression of a broad range of malignancies including AML. The ultimate clinical utility of agents targeting this pathway will depend on our ability to prospectively select those patients most likely to benefit from this strategy. The relative ease of obtaining pure samples of tumor cells from AML patients will facilitate the design of individualized therapeutic strategies based on the compilation of comprehensive molecular profiling data. We now have the tools to complement genomic analysis with prospective multiplexed evaluation of kinase activation patterns to provide critical information on dominant signaling pathways in individual cancers. These tools will consequently facilitate selection of AML patients most likely to benefit from RAS-MAPK pathway intervention and will further guide the design of combination strategies to potentiate single-agent effectiveness. As we work toward improvement of AML survival rates, we have made great strides in developing the requisite agents and technologies. We are now faced with the challenge of incorporating them into the practice of molecular medicine for greatest therapeutic impact.

References

Alcolea MP, Casado P, Rodriguez-Prados JC, Vanhaesebroeck B, Cutillas PR (2012) Phosphoproteomic analysis of leukemia cells under basal and drug-treated conditions identifies markers of kinase pathway activation and mechanisms of resistance. Mol Cell Proteomics 11(8):453–466

Andreeff M, Jiang S, Zhang X, Konopleva M, Estrov Z, Snell VE, Xie Z, Okcu MF, Sanchez-Williams G, Dong J, Estey EH, Champlin RC, Kornblau SM, Reed JC, Zhao S (1999) Expression of Bcl-2-related genes in normal and AML progenitors: changes induced by chemotherapy and retinoic acid. Leukemia 13(11):1881–1892.

Berthon C, Driss V, Liu J, Kuranda K, Leleu X, Jouy N, Hetuin D, Quesnel B (2010) In acute myeloid leukemia, B7-H1 (PD-L1) protection of blasts from cytotoxic T cells is induced by TLR ligands and interferon-gamma and can be reversed using MEK inhibitors. Cancer Immunol Immunother 59(12):1839–1849.

Blalock WL, Moye PW, Chang F, Pearce M, Steelman LS, McMahon M, McCubrey JA (2000) Combined effects of aberrant MEK1 activity and BCL2 overexpression on relieving the cytokine dependency of human and murine hematopoietic cells. Leukemia 14(6):1080–1096.

Bowen DT, Frew ME, Hills R, Gale RE, Wheatley K, Groves MJ, Langabeer SE, Kottaridis PD, Moorman AV, Burnett AK, Linch DC (2005) RAS mutation in acute myeloid leukemia is associated with distinct cytogenetic subgroups but does not influence outcome in patients younger than 60 years. Blood 106(6):2113–9.

Campos L, Rouault JP, Sabido O, Oriol P, Roubi N, Vasselon C, Archimbaud E, Magaud JP, Guyotat D (1993) High expression of bcl-2 protein in acute myeloid leukemia cells is associated with poor response to chemotherapy. Blood 81(11):3091–6

Carracedo A, Ma L, Teruya-Feldstein J, Rojo F, Salmena L, Alimonti A, Egia A, Sasaki AT, Thomas G, Kozma SC, Papa A, Nardella C, Cantley LC, Baselga J, Pandolfi PP (2008) Inhibition of mTORC1 leads to MAPK pathway activation through a PI3K-dependent feedback loop in human cancer. J Clin Invest 118(9):3065–3074

Ciuffreda L, Di Sanza C, Cesta Incani U, Eramo A, Desideri M, Biagioni F, Passeri D, Falcone I, Sette G, Bergamo P, Anichini A, Sabapathy K, McCubrey JA, Ricciardi MR, Tafuri A, Blandino G, Orlandi A, De Maria R, Cognetti F, Del Bufalo D, Milella M (2012) The mitogen-activated protein kinase (MAPK) cascade controls phosphatase and tensin homolog (PTEN) expression through multiple mechanisms. J Mol Med (Berl) 90(6):667–679

Crump M, Hedley D, Kamel-Reid S, Leber B, Wells R, Brandwein J, Buckstein R, Kassis J, Minden M, Matthews J, Robinson S, Turner R, McIntosh L, Eisenhauer E, Seymour L (2010) A randomized phase I clinical and biologic study of two schedules of sorafenib in patients with myelodysplastic syndrome or acute myeloid leukemia: a NCIC (National Cancer Institute of Canada) Clinical Trials Group Study. Leuk Lymphoma 51(2):252–260

Dent P, Grant S (2001) Pharmacologic interruption of the mitogen-activated extracellular-regulated kinase/mitogen-activated protein kinase signal transduction pathway: potential role in promoting cytotoxic drug action. Clin Cancer Res 7(4):775–783

Dry JR, Pavey S, Pratilas CA, Harbron C, Runswick S, Hodgson D, Chresta C, McCormack R, Byrne N, Cockerill M, Graham A, Beran G, Cassidy A, Haggerty C, Brown H, Ellison G, Dering J, Taylor BS, Stark M, Bonazzi V, Ravishankar S, Packer L, Xing F, Solit DB, Finn RS, Rosen N, Hayward NK, French T, Smith PD (2010) Transcriptional pathway signatures predict MEK addiction and response to selumetinib (AZD6244). Cancer Res 70(6):2264–2273

Duncan JS, Whittle MC, Nakamura K, Abell AN, Midland AA, Zawistowski JS, Johnson NL, Granger DA, Jordan NV, Darr DB, Usary J, Kuan PF, Smalley DM, Major B, He X, Hoadley KA, Zhou B, Sharpless NE, Perou CM, Kim WY, Gomez SM, Chen X, Jin J, Frye SV, Earp HS, Graves LM, Johnson GL (2012) Dynamic reprogramming of the kinome in response to targeted MEK inhibition in triple-negative breast cancer. Cell 149(2):307–321

Engelman JA, Chen L, Tan X, Crosby K, Guimaraes AR, Upadhyay R, Maira M, McNamara K, Perera SA, Song Y, Chirieac LR, Kaur R, Lightbown A, Simendinger J, Li T, Padera RF, Garcia-Echeverria C, Weissleder R, Mahmood U, Cantley LC, Wong KK (2008) Effective use of PI3K and MEK inhibitors to treat mutant Kras G12D and PIK3CA H1047R murine lung cancers. Nat Med 14(12):1351–6

Garnett MJ, Marais R (2004) Guilty as charged: B-RAF is a human oncogene. Cancer Cell 6(4):313–9

Garnett MJ, Rana S, Paterson H, Barford D, Marais R (2005) Wild-type and mutant B-RAF activate C-RAF through distinct mechanisms involving heterodimerization. Mol Cell 20(6):963–9

Gilmartin AG, Bleam MR, Groy A, Moss KG, Minthorn EA, Kulkarni SG, Rominger CM, Erskine S, Fisher KE, Yang J, Zappacosta F, Annan R, Sutton D, Laquerre SG (2011) GSK1120212 (JTP-74057) is an inhibitor of MEK activity and activation with favorable pharmacokinetic properties for sustained in vivo pathway inhibition. Clin Cancer Res 17(5):989–1000

Glasow A, Prodromou N, Xu K, von Lindern M, Zelent A (2005) Retinoids and myelomonocytic growth factors cooperatively activate RARA and induce human myeloid leukemia cell differentiation via MAP kinase pathways. Blood 105(1):341–9

Harousseau JL, Lancet JE, Reiffers J, Lowenberg B, Thomas X, Huguet F, Fenaux P, Zhang S, Rackoff W, De Porre P, Stone R (2007) A phase 2 study of the oral farnesyltransferase inhibitor tipifarnib in patients with refractory or relapsed acute myeloid leukemia. Blood 109(12):5151–6

Hatzivassiliou G, Song K, Yen I, Brandhuber BJ, Anderson DJ, Alvarado R, Ludlam MJ, Stokoe D, Gloor SL, Vigers G, Morales T, Aliagas I, Liu B, Sideris S, Hoeflich KP, Jaiswal BS, Seshagiri S, Koeppen H, Belvin M, Friedman LS, Malek S (2010) RAF inhibitors prime wild-type RAF to activate the MAPK pathway and enhance growth. Nature 464(7287):431–5

Ho YK, Smith RG, Brown MS, Goldstein JL (1978) Low-density lipoprotein (LDL) receptor activity in human acute myelogenous leukemia cells. Blood 52(6):1099–1114

Illmer T, Thiede C, Fredersdorf A, Stadler S, Neubauer A, Ehninger G, Schaich M (2005) Activation of the RAS pathway is predictive for a chemosensitive phenotype of acute myelogenous leukemia blasts. Clin Cancer Res 11(9):3217–3224

Incani UC, Del Curatolo A, Di Sanza C, Falcone I, Cognetti F, Ciuffreda L, Milella M, Elena R (2012) Synergistic activity of vertical combinations of agents targeting multiple steps along the RAF/MEK/ERK cascade as a therapeutic strategy in human tumors. J Clin Oncol 30 (suppl):e13572. (Abstract)

Infante JR, Fecher LA, Falchook GS, Nallapareddy S, Gordon MS, Becerra C, DeMarini DJ, Cox DS, Xu Y, Morris SR, Peddareddigari VG, Le NT, Hart L, Bendell JC, Eckhardt G, Kurzrock R, Flaherty K, Burris HA 3rd, Messersmith WA (2012) Safety, pharmacokinetic, pharmacodynamic, and efficacy data for the oral MEK inhibitor trametinib: a phase 1 dose-escalation trial. Lancet Oncol 13(8):773–781

Jing J, Greshock J, Holbrook JD, Gilmartin A, Zhang X, McNeil E, Conway T, Moy C, Laquerre S, Bachman K, Wooster R, Degenhardt Y (2012) Comprehensive predictive biomarker analysis for MEK inhibitor GSK1120212. Mol Cancer Ther 11(3):720–9

Karp JE, Lancet JE, Kaufmann SH, End DW, Wright JJ, Bol K, Horak I, Tidwell ML, Liesveld J, Kottke TJ, Ange D, Buddharaju L, Gojo I, Highsmith WE, Belly RT, Hohl RJ, Rybak ME, Thibault A, Rosenblatt J (2001) Clinical and biologic activity of the farnesyltransferase inhibitor R115777 in adults with refractory and relapsed acute leukemias: a phase 1 clinical-laboratory correlative trial. Blood 97(11):3361–9

Kiyoi H, Naoe T, Nakano Y, Yokota S, Minami S, Miyawaki S, Asou N, Kuriyama K, Jinnai I, Shimazaki C, Akiyama H, Saito K, Oh H, Motoji T, Omoto E, Saito H, Ohno R, Ueda R (1999) Prognostic implication of FLT3 and N-RAS gene mutations in acute myeloid leukemia. Blood 93(9):3074–3080

Kojima K, Konopleva M, Samudio IJ, Shikami M, Cabreira-Hansen M, McQueen T, Ruvolo V, Tsao T, Zeng Z, Vassilev LT, Andreeff M (2005) MDM2 antagonists induce p53-dependent apoptosis in AML: implications for leukemia therapy. Blood 106(9):3150–9

Kojima K, Konopleva M, Samudio IJ, Ruvolo V, Andreeff M (2007) Mitogen-activated protein kinase kinase inhibition enhances nuclear proapoptotic function of p53 in acute myelogenous leukemia cells. Cancer Res 67(7):3210–9

Kolch W, Pitt A (2010) Functional proteomics to dissect tyrosine kinase signalling pathways in cancer. Nat Rev Cancer 10(9):618–629

Konopleva M, Tari AM, Estrov Z, Harris D, Xie Z, Zhao S, Lopez-Berestein G, Andreeff M (2000) Liposomal Bcl-2 antisense oligonucleotides enhance proliferation, sensitize acute myeloid leukemia to cytosine-arabinoside, and induce apoptosis independent of other antiapoptotic proteins. Blood 95(12):3929–3938

Konopleva M, Milella M, Ruvolo P, Watts JC, Ricciardi MR, Korchin B, McQueen T, Bornmann W, Tsao T, Bergamo P, Mak DH, Chen W, McCubrey J, Tafuri A, Andreeff M (2012) MEK inhibition enhances ABT-737-induced leukemia cell apoptosis via prevention of ERK-activated MCL-1 induction and modulation of MCL-1/BIM complex. Leukemia 26(4):778–787

Kornblau SM, Vu HT, Ruvolo P, Estrov Z, O'Brien S, Cortes J, Kantarjian H, Andreeff M, May WS (2000) BAX and PKCalpha modulate the prognostic impact of BCL2 expression in acute myelogenous leukemia. Clin Cancer Res 6(4):1401–9

Kornblau SM, Tibes R, Qiu YH, Chen W, Kantarjian HM, Andreeff M, Coombes KR, Mills GB (2009) Functional proteomic profiling of AML predicts response and survival. Blood 113(1):154–164

Lancet JE, Gojo I, Gotlib J, Feldman EJ, Greer J, Liesveld JL, Bruzek LM, Morris L, Park Y, Adjei AA, Kaufmann SH, Garrett-Mayer E, Greenberg PL, Wright JJ, Karp JE (2007) A phase 2 study of the farnesyltransferase inhibitor tipifarnib in poor-risk and elderly patients with previously untreated acute myelogenous leukemia. Blood 109(4):1387–1394

Lauchle JO, Kim D, Le DT, Akagi K, Crone M, Krisman K, Warner K, Bonifas JM, Li Q, Coakley KM, Diaz-Flores E, Gorman M, Przybranowski S, Tran M, Kogan SC, Roose JP, Copeland NG, Jenkins NA, Parada L, Wolff L, Sebolt-Leopold J, Shannon K (2009) Response and resistance to MEK inhibition in leukaemias initiated by hyperactive Ras. Nature 461(7262):411–4

Lehmann S, Paul C, Torma H (2001) Retinoid receptor expression and its correlation to retinoid sensitivity in non-M3 acute myeloid leukemia blast cells. Clin Cancer Res 7(2):367–373

Lorusso PM, Adjei AA, Varterasian M, Gadgeel S, Reid J, Mitchell DY, Hanson L, DeLuca P, Bruzek L, Piens J, Asbury P, Van Becelaere K, Herrera R, Sebolt-Leopold J, Meyer MB (2005) Phase I and pharmacodynamic study of the oral MEK inhibitor CI-1040 in patients with advanced malignancies. J Clin Oncol 23(23):5281–5293

Lunghi P, Tabilio A, Dall'Aglio PP, Ridolo E, Carlo-Stella C, Pelicci PG, Bonati A (2003) Down-modulation of ERK activity inhibits the proliferation and induces the apoptosis of primary acute myelogenous leukemia blasts. Leukemia 17(9):1783–1793

Man CH, Fung TK, Ho C, Han HH, Chow HC, Ma AC, Choi WW, Lok S, Cheung AM, Eaves C, Kwong YL, Leung AY (2012) Sorafenib treatment of FLT3-ITD+ acute myeloid leukemia: favorable initial outcome and mechanisms of subsequent non-responsiveness associated with a D835 mutation. Blood 119(22):5133–5143

Maurer T, Garrenton LS, Oh A, Pitts K, Anderson DJ, Skelton NJ, Fauber BP, Pan B, Malek S, Stokoe D, Ludlam MJ, Bowman KK, Wu J, Giannetti AM, Starovasnik MA, Mellman I, Jackson PK, Rudolph J, Wang W, Fang G (2012) Small-molecule ligands bind to a distinct pocket in Ras and inhibit SOS-mediated nucleotide exchange activity. Proc Natl Acad Sci U S A 109(14):5299–5304

Milella M, Kornblau SM, Estrov Z, Carter BZ, Lapillonne H, Harris D, Konopleva M, Zhao S, Estey E, Andreeff M (2001) Therapeutic targeting of the MEK/MAPK signal transduction module in acute myeloid leukemia. J Clin Invest 108(6):851–9

Milella M, Estrov Z, Kornblau SM, Carter BZ, Konopleva M, Tari A, Schober WD, Harris D, Leysath CE, Lopez-Berestein G, Huang Z, Andreeff M (2002) Synergistic induction of apoptosis by simultaneous disruption of the Bcl-2 and MEK/MAPK pathways in acute myelogenous leukemia. Blood 99(9):3461–4

Milella M, Precupanu CM, Gregorj C, Ricciardi MR, Petrucci MT, Kornblau SM, Tafuri A, Andreeff M (2005) Beyond single pathway inhibition: MEK inhibitors as a platform for the development of pharmacological combinations with synergistic anti-leukemic effects. Curr Pharm Des 11(21):2779–2795

Milella M, Konopleva M, Precupanu CM, Tabe Y, Ricciardi MR, Gregorj C, Collins SJ, Carter BZ, D'Angelo C, Petrucci MT, Foa R, Cognetti F, Tafuri A, Andreeff M (2007) MEK blockade converts AML differentiating response to retinoids into extensive apoptosis. Blood 109(5):2121–9

Miranda MB, McGuire TF, Johnson DE (2002) Importance of MEK-1/-2 signaling in monocytic and granulocytic differentiation of myeloid cell lines. Leukemia 16(4):683–692

Mirzoeva OK, Das D, Heiser LM, Bhattacharya S, Siwak D, Gendelman R, Bayani N, Wang NJ, Neve RM, Guan Y, Hu Z, Knight Z, Feiler HS, Gascard P, Parvin B, Spellman PT, Shokat KM, Wyrobek AJ, Bissell MJ, McCormick F, Kuo WL, Mills GB, Gray JW, Korn WM (2009) Basal subtype and MAPK/ERK kinase (MEK)-phosphoinositide 3-kinase feedback signaling determine susceptibility of breast cancer cells to MEK inhibition. Cancer Res 69(2):565–572

Morgan MA, Dolp O, Reuter CW (2001) Cell-cycle-dependent activation of mitogen-activated protein kinase kinase (MEK –1/2) in myeloid leukemia cell lines and induction of growth inhibition and apoptosis by inhibitors of RAS signaling. Blood 97(6):1823–1834

Neubauer A, Dodge RK, George SL, Davey FR, Silver RT, Schiffer CA, Mayer RJ, Ball ED, Wurster-Hill D, Bloomfield CD, Liu ET (1994) Prognostic importance of mutations in the ras proto-oncogenes in de novo acute myeloid leukemia. Blood 83(6):1603–1611

Nishioka C, Ikezoe T, Yang J, Takeshita A, Taniguchi A, Komatsu N, Togitani K, Koeffler HP, Yokoyama A (2008) Blockade of MEK/ERK signaling enhances sunitinib-induced growth inhibition and apoptosis of leukemia cells possessing activating mutations of the FLT3 gene. Leuk Res 32(6):865–872

Ohren JF, Chen H, Pavlovsky A, Whitehead C, Zhang E, Kuffa P, Yan C, McConnell P, Spessard C, Banotai C, Mueller WT, Delaney A, Omer C, Sebolt-Leopold J, Dudley DT, Leung IK, Flamme C, Warmus J, Kaufman M, Barrett S, Tecle H, Hasemann CA (2004) Structures of human MAP kinase kinase 1 (MEK1) and MEK2 describe novel noncompetitive kinase inhibition. Nat Struct Mol Biol 11(12):1192–1197

Parkin B, Ouillette P, Wang Y, Liu Y, Wright W, Roulston D, Purkayastha A, Dressel A, Karp J, Bockenstedt P, Al-Zoubi A, Talpaz M, Kujawski L, Shedden K, Shakhan S, Li C, Erba H, Malek SN (2010) NF1 inactivation in adult acute myelogenous leukemia. Clin Cancer Res 16(16):4135–4147

Pedersen-Bjergaard J (1992) Radiotherapy- and chemotherapy-induced myelodysplasia and acute myeloid leukemia. A review. Leuk Res 16(1):61–65

Piloto O, Wright M, Brown P, Kim KT, Levis M, Small D (2007) Prolonged exposure to FLT3 inhibitors leads to resistance via activation of parallel signaling pathways. Blood 109(4):1643–1652
Poulikakos PI, Zhang C, Bollag G, Shokat KM, Rosen N (2010) RAF inhibitors transactivate RAF dimers and ERK signalling in cells with wild-type BRAF. Nature 464(7287):427–430
Pratilas CA, Taylor BS, Ye Q, Viale A, Sander C, Solit DB, Rosen N (2009) (V600E)BRAF is associated with disabled feedback inhibition of RAF-MEK signaling and elevated transcriptional output of the pathway. Proc Natl Acad Sci U S A 106(11):4519–4524
Raponi M, Lancet JE, Fan H, Dossey L, Lee G, Gojo I, Feldman EJ, Gotlib J, Morris LE, Greenberg PL, Wright JJ, Harousseau JL, Lowenberg B, Stone RM, De Porre P, Wang Y, Karp JE (2008) A 2-gene classifier for predicting response to the farnesyltransferase inhibitor tipifarnib in acute myeloid leukemia. Blood 111(5):2589–2596.
Ricciardi MR, McQueen T, Chism D, Milella M, Estey E, Kaldjian E, Sebolt-Leopold J, Konopleva M, Andreeff M (2005) Quantitative single cell determination of ERK phosphorylation and regulation in relapsed and refractory primary acute myeloid leukemia. Leukemia 19(9):1543–9.
Ricciardi MR, Scerpa MC, Bergamo P, Ciuffreda L, Petrucci MT, Chiaretti S, Tavolaro S, Mascolo MG, Abrams SL, Steelman LS, Tsao T, Marchetti A, Konopleva M, Del Bufalo D, Cognetti F, Foa R, Andreeff M, McCubrey JA, Tafuri A, Milella M (2012) Therapeutic potential of MEK inhibition in acute myelogenous leukemia: rationale for "vertical" and "lateral" combination strategies. J Mol Med (Berl) 90:1133-1144.
Rucker FG, Schlenk RF, Bullinger L, Kayser S, Teleanu V, Kett H, Habdank M, Kugler CM, Holzmann K, Gaidzik VI, Paschka P, Held G, von Lilienfeld-Toal M, Lubbert M, Frohling S, Zenz T, Krauter J, Schlegelberger B, Ganser A, Lichter P, Dohner K, Dohner H (2012) TP53 alterations in acute myeloid leukemia with complex karyotype correlate with specific copy number alterations, monosomal karyotype, and dismal outcome. Blood 119(9):2114–2121
Sebolt-Leopold JS (2008) Advances in the development of cancer therapeutics directed against the RAS-mitogen-activated protein kinase pathway. Clin Cancer Res 14(12):3651–3656
Sebolt-Leopold JS, Herrera R (2004) Targeting the mitogen-activated protein kinase cascade to treat cancer. Nat Rev Cancer 4(12):937–947
Sebolt-Leopold JS, Dudley DT, Herrera R, Van Becelaere K, Wiland A, Gowan RC, Tecle H, Barrett SD, Bridges A, Przybranowski S, Leopold WR, Saltiel AR (1999) Blockade of the MAP kinase pathway suppresses growth of colon tumors in vivo. Nat Med 5(7):810–6
Seliger B, Papadileris S, Vogel D, Hess G, Brendel C, Storkel S, Ortel J, Kolbe K, Huber C, Huhn D, Neubauer A (1996) Analysis of the p53 and MDM-2 gene in acute myeloid leukemia. Eur J Haematol 57(3):230–240
Solomon C, White JH, Kremer R (1999) Mitogen-activated protein kinase inhibits 1,25-dihydroxyvitamin D3-dependent signal transduction by phosphorylating human retinoid X receptor alpha. J Clin Invest 103(12):1729–1735
Steelman LS, Franklin RA, Abrams SL, Chappell W, Kempf CR, Basecke J, Stivala F, Donia M, Fagone P, Nicoletti F, Libra M, Ruvolo P, Ruvolo V, Evangelisti C, Martelli AM, McCubrey JA (2011) Roles of the Ras/Raf/MEK/ERK pathway in leukemia therapy. Leukemia 25(7):1080–1094
Takahashi S (2006) Inhibition of the MEK/MAPK signal transduction pathway strongly impairs the growth of Flt3-ITD cells. Am J Hematol 81(2):154–5
Tibes R, Qiu Y, Lu Y, Hennessy B, Andreeff M, Mills GB, Kornblau SM (2006) Reverse phase protein array: validation of a novel proteomic technology and utility for analysis of primary leukemia specimens and hematopoietic stem cells. Mol Cancer Ther 5(10):2512–2521
Tyner JW, Erickson H, Deininger MW, Willis SG, Eide CA, Levine RL, Heinrich MC, Gattermann N, Gilliland DG, Druker BJ, Loriaux MM (2009) High-throughput sequencing screen reveals novel, transforming RAS mutations in myeloid leukemia patients. Blood 113(8):1749–1755.
van der Weide K, de Jonge-Peeters SD, Kuipers F, de Vries EG, Vellenga E (2009) Combining simvastatin with the farnesyltransferase inhibitor tipifarnib results in an enhanced cytotoxic effect in a subset of primary CD34 + acute myeloid leukemia samples. Clin Cancer Res 15(9):3076–3083.

Vitols S, Gahrton G, Ost A, Peterson C (1984) Elevated low density lipoprotein receptor activity in leukemic cells with monocytic differentiation. Blood 63(5):1186–1193.

Weber JS, Flaherty KT, Infante Jr, Falchook GS, Kefford R, Daud A, Hamid O, Gonzalez R, Kudchadkar RR, Lawrence DP, Burris HA, Long GV, Algazi AP, Lewis KD, Kim KB, Puzanov I, Sun P, Little SM, Patel K, Sosman JA (2012) Updated safety and efficacy results from a phase I/II study of the oral BRAF inhibitor dabrafenib (GSK2118436) combined with the oral MEK 1/2 inhibitor trametinib (GSK1120212) in patients with BRAFi-naïve metastatic melanoma. J Clin Oncol 30(suppl):8510. (Abstract)

Wee S, Jagani Z, Xiang KX, Loo A, Dorsch M, Yao YM, Sellers WR, Lengauer C, Stegmeier F (2009) PI3K pathway activation mediates resistance to MEK inhibitors in KRAS mutant cancers. Cancer Res 69(10):4286–4293

Welch JS, Link DC (2011) Genomics of AML: clinical applications of next-generation sequencing. Hematology Am Soc Hematol Educ Program 2011:30–35

Whyte DB, Kirschmeier P, Hockenberry TN, Nunez-Oliva I, James L, Catino JJ, Bishop WR, Pai JK (1997) K- and N-Ras are geranylgeranylated in cells treated with farnesyl protein transferase inhibitors. J Biol Chem 272(22):14459–14464

Wilhelm S, Carter C, Lynch M, Lowinger T, Dumas J, Smith RA, Schwartz B, Simantov R, Kelley S (2006) Discovery and development of sorafenib: a multikinase inhibitor for treating cancer. Nat Rev Drug Discov 5(10):835–844

Yen A, Roberson MS, Varvayanis S, Lee AT (1998) Retinoic acid induced mitogen-activated protein (MAP)/extracellular signal-regulated kinase (ERK) kinase-dependent MAP kinase activation needed to elicit HL-60 cell differentiation and growth arrest. Cancer Res 58(14):3163–3172

Yeung K, Seitz T, Li S, Janosch P, McFerran B, Kaiser C, Fee F, Katsanakis KD, Rose DW, Mischak H, Sedivy JM, Kolch W (1999) Suppression of Raf-1 kinase activity and MAP kinase signalling by RKIP. Nature 401(6749):173–7

Yu K, Toral-Barza L, Shi C, Zhang WG, Zask A (2008) Response and determinants of cancer cell susceptibility to PI3K inhibitors: combined targeting of PI3K and Mek1 as an effective anticancer strategy. Cancer Biol Ther 7(2):307–315

Zebisch A, Staber PB, Delavar A, Bodner C, Hiden K, Fischereder K, Janakiraman M, Linkesch W, Auner HW, Emberger W, Windpassinger C, Schimek MG, Hoefler G, Troppmair J, Sill H (2006) Two transforming C-RAF germ-line mutations identified in patients with therapy-related acute myeloid leukemia. Cancer Res 66(7):3401–8

Zebisch A, Wolfler A, Fried I, Wolf O, Lind K, Bodner C, Haller M, Drasche A, Pirkebner D, Matallanas D, Rath O, Blyth K, Delwel R, Taskesen E, Quehenberger F, Kolch W, Troppmair J, Sill H (2012) Frequent loss of RAF kinase inhibitor protein expression in acute myeloid leukemia. Leukemia 26(8):1842–1849.

Zhang W, Konopleva M, Ruvolo VR, McQueen T, Evans RL, Bornmann WG, McCubrey J, Cortes J, Andreeff M (2008) Sorafenib induces apoptosis of AML cells via Bim-mediated activation of the intrinsic apoptotic pathway. Leukemia 22(4):808–818

Zhang W, Konopleva M, Burks JK, Dywer KC, Schober WD, Yang JY, McQueen TJ, Hung MC, Andreeff M (2010) Blockade of mitogen-activated protein kinase/extracellular signal-regulated kinase kinase and murine double minute synergistically induces Apoptosis in acute myeloid leukemia via BH3-only proteins Puma and Bim. Cancer Res 70(6):2424–2434

Chapter 16
Clinical Use of Farnesyltransferase Inhibitors

Chezi Ganzel and Jacob M. Rowe

Abstract The Ras-RAF-MEK pathway activates Erk kinase and is an important route for delivering a signal from outside the cell into the nucleus. RAS mutations, which are not rare in AML, can result in constitutive activation of RAS proteins independent of ligand activation. Farnesyltransferase inhibitors have the ability to prevent the activity of RAS by blocking the rate limiting step of its post-translation process. There are several types of Farnesyltransferase inhibitors. Tipifarnib (Zarnestra), an oral methylquinolone analog, has been the most extensively investigated. Despite promising phase I and II results using tipifarnib as a single agent in relapse/refractory AML, a phase III study could not show an advantage for tipifarnib versus best supportive care, in newly diagnosed AML patients, greater than 70 years. The role of tipifarnib as maintenance therapy in AML needs to be determined.

Keywords Farnesyltransferase inhibitors · Tipifarnib · Acute myeloid leukemia · Refractory · Maintenance · De novo

16.1 Ras-RAF-MAPK Pathway

16.1.1 Farnesylation

Ras and other proteins are terminated in a sequence which comprises cysteine, an aliphatic amino acid and another amino acid (CAAX). These proteins pass a unique posttranslational processing which consists of three steps: isoprenylation, proteolysis, and methylation (Lobell et al. 2001). In the first step, farnesyltransferase, an

J. M. Rowe (✉) · C. Ganzel
Department of Hematology, Shaare Zedek Medical Center,
P.O. Box 3235, 91031 Jerusalem, Israel
e-mail: rowe@jimmy.harvard.edu

J. M. Rowe
Technion, Israel Institute of Technology, Haifa, Israel

C. Ganzel
e-mail: ganzelct@yahoo.com

M. Andreeff (ed.), *Targeted Therapy of Acute Myeloid Leukemia*,
Current Cancer Research, DOI 10.1007/978-1-4939-1393-0_16

intracellular enzyme, catalyzes the transfer of 15-carbon farnesyl isoprenoid from farnesyl pyrophosphate (FPP) to a cysteine of the CAAX sequence (End 1999; Basso et al. 2006). This is the rate-limiting step of the entire process. The farnesylated Ras is inserted into the endoplasmic reticulum and the AAX group is removed through a proteolytic process (second step). The last modification before Ras is transported to the plasma membrane is carboxymethylation (third step).

16.1.2 Ras Mutations in Acute Myeloid Leukemia

Ras mutations can result in constitutive activation of Ras proteins, independent of ligand activation. Ras mutations, predominantly N-Ras as well as K-Ras, can be seen in 12–27 % of acute myeloid leukemia (AML) patients (Neubauer et al. 2008). The prognostic implication of the presence of these mutations is as yet uncertain (Kiyoi et al. 1999; Rockova et al. 2011; Bacher et al. 2006; Bowen et al. 2005; Neubauer et al. 2008).

16.1.3 Farnesyltransferase Inhibitors

The rationale of using farnesyltransferase inhibitors (FTIs) as antitumor therapy is to prevent the activity of Ras by blocking the rate-limiting step in its posttranslational process. The limitation of this model is the presence of an alternative prenylation pathway which bypasses the farnesyltransferase. By geranylgeranyl transferase I, Ras can preserve its ability to be attached to the plasma membrane and to influence the downstream signal transduction (Lobell et al. 2001; Whyte et al. 1997).

16.1.4 FTIs Classification

FTIs are classified according to the mechanism of inhibition. The three first groups are compounds that mimic elements of the farnesyltransferase substrate. The first group, tetrapeptidic pseudosubstrates, is a group of peptidomimetics. The compounds are peptide analogues which mimic the typical CAAX motif of the farnesyltransferase substrate and compete with it. The second group, FPP analogues, includes molecules that can compete with the FPP. The bisubstrate analogues, the third group, have properties of both the CAAX motif and FPP. In contrast, the fourth group includes non-peptidomimetics. It is a large heterogenic group of molecules which were discovered during the course of screening of large drug libraries. All the drugs which have been investigated clinically in AML patients, belong to this group.

16.2 Clinical Data

The most widely investigated FTI is tipifarnib, an oral methylquinolone analog, from the non-peptidomimetic group (Table 16.1).

FTIs have been studied in three different scenarios. Initial studies, as is common, were conducted in relapsed or refractory patients. Once clinical activity was demonstrated, several groups studied the application of this clinical activity to newly diagnosed high-risk or older patients. Finally, the role of tipifarnib as maintenance therapy in AML has also been studied.

Phase-1 studies began in the early years of the last decade. The first study, reported in 2001, described 35 adult patients with different types of poor-risk acute leukemia including relapsed/refractory or poor-prognosis AML, acute lymphoblastic leukemia (ALL), and blast crisis CML. Tipifarnib was well tolerated with no dose-limiting toxicity through doses up to 900 mg twice daily. The drug was shown to accumulate preferentially in the bone marrow, in a dose-dependent manner. The overall response (OR) rate was 29%, including two cases of complete remission (CR). ERK, one of the Ras effectors, became undetectable in four out of eight patients where it was found to be in its activated form, before starting treatment (Karp et al. 2001; Lancet et al. 2002). In 2011, another phase-1 study of tipifarnib in AML patients, was reported. It included 44 patients with relapsed/refractory AML or those over 70 years of age. The tipifarnib was administered, in this study, on a week-on week-off schedule. The maximal tolerated dose was determined to be 1200 mg twice daily (Kirschbaum et al. 2011).

The results of these two phase-2 studies, one in refractory/relapsed AML and the other in newly diagnosed elderly AML patients, raised considerable expectations in the hematologic community. The French Farnesulation Inhibitor Global Human Trials (FIGHT) Acute Myeloid Leukemia Study (Harousseau et al. 2007) evaluated tipifarnib on 252 adults with relapsed or refractory AML. Tipifarinib was given as 600 mg twice daily for 21 out of 28 days. The median age was 62 years. Four percent of these advanced patients achieved CR or CR with incomplete platelet recovery (CRp). Eight percent, including those who survived CR/CRp, had a reduction in bone marrow blast count to less than 5%. The median survival for patients who achieved CR/CRp was 1 year. The most common adverse effect was myelosuppression. Nonhematologic toxicities were mostly fever, nausea, and hypokalemia. A multi-institutional trial (Lancet et al. 2007) conducted in the USA, studied tipifarnib in 158 older adults with newly diagnosed poor-risk AML. The medication was given in a dose of 600 mg twice a day for 21 days followed by a rest period of up to 42 days. Only patients with observed stable disease or clinical response were eligible to receive a second cycle of therapy. The median age was 74. The OR was 23% with CR of 14%. Among patients who achieved CR, the median duration was 7.3 months and overall survival (OS) was 18 months.

Table 16.1 Clinical trials for FTIs in AML. (Adapted and updated from Braun and Fenaux 2008)

Study	Phase	Medication	Combination	Dose	Disease	No.	Response		
							OR	CR (CRp)	PR/HI
Lancet et al. 2007	2	Tipifarnib	–	600 mg X2	AML (new)	158	23 %	14 %	9 %
Harousseau et al. 2009	3	Tipifarnib	–	600 mg X2, 21/28d	AML (new)	457	17 %	8 %	9 %
Karp et al. 2009	1	Tipifarnib	VP − 16	300–600 mgX2, 14–21d	AML (new)	84	40 %	25 %	15 %
Jabbour et al. 2011	1–2	Tipifarnib	Ida + ARAC	200–300 mgX2, 21/28d	AML (new), MDS	95	74 %	64 % (9 %)	ND
Karp et al. 2012	2	Tipifarnib	VP − 16	400–600 mgX2, 14d	AML (new)	84	36 %	25 %	11 %
Harousseau et al. 2007	2	Tipifarnib	–	600 mg X2	AML (relapse/ refractory)	252	10 %	3 % (1 %)	6 %
Karp et al. 2001	1	Tipifarnib	–	100–200 mg X2/d	AML (new/relapse/ refractory)	25	32 %	8 %	24 %
Cortes et al. 2005	1	BMS − 214662	–	42–300 mg/m^2	AML (new/relapse/ refractory), ALL, MDS	30	17 %	0 (7 %)	10 %
Kirschbaum et al. 2011	1	Tipifarnib	–	400–1600 mgX2, week-on, week-off	AML (new/relapse/ refractory)	44	ND	7 %	9 %
Burnett et al. 2012	ND	Tipifarnib	Low dose ARAC	ND	AML (ND), MDS	33	18 %	15 % (3 %)	ND
Ravoet et al. 2008	2	Lonafarnib	–	200 mgX2, 28d	AML—secondary	4	25 %	0	25 %
Karp et al. 2008	2	Tipifarnib	–	400 mgX2, 14/21d	Maintenance for AML in CR	48	NR	NR	NR
Luger et al. 2010	3	Tipifarnib	–	300–400 mgX2	Maintenance for AML in CR	144	NR	NR	NR

OR overall response, *CR* complete response, *CRp* complete response with incomplete platelet recovery, *PR* partial response, *HI* hematologic improvement, *ND* not determined, *NR* not relevant, *AML* acute myeloid leukemia, *MDS* myelodysplastic syndrome, *Ida* Idarubicin, *new* newly diagnosed, *ALL* acute lymphoblastic leukemia

Both of these studies demonstrated that tipifarnib has unequivocal activity in AML; the large French study showing activity in patients with very advanced disease and suggesting that in such a population tipifarnib might best be used as an adjunct to other effective chemotherapy regimens. The US study that demonstrated unequivocal activity in newly diagnosed older patients also suggested that this be a less toxic modality to be used in an appropriate population.

In contrast, the large international FIGHT-AML−301 study was a randomized phase 3 trial of tipifarnib compared with the best supportive care, including hydroxyurea, in the treatment of newly diagnosed AML patients 70 years or older. Tipifarnib, 600 mg orally twice a day was administered for the first 21 consecutive days, in 28-day cycles. The median survival was 107 days for the tipifarnib and 109 days for the best supportive care arm. The disappointing results from this randomized study, showing that tipifarnib treatment did not result in an increased survival compared with best supportive care, put a damper on the high expectations for the ability to deliver less toxic therapy to an older patient population that desperately needs such an approach (Harousseau et al. 2009).

The limited anti-proliferative effect of tipifarnib may be attributed partially to the enzymatic action of geranylgeranyl transferase I, which preserves the activity of Ras despite the inhibition of the farnesyltransferase (FT). In other cases, the malignant cell may develop resistance through selective adaptation or selected FT mutations (Karp et al. 2008).

16.2.1 Directions for Improving Use of Tipifarnib

There are three main directions that may enhance the benefit of tipifarnib, despite the above disappointing results.

16.2.1.1 Combinations of FTI with Chemotherapy

A phase-1 trial based on in vitro studies on cell lines and AML isolates (Karp et al. 2009), concluded that tipifarnib synergistically enhances the anti-proliferative effect of etoposide. In this study, 84 older AML patients (>70 years) who were not candidates for conventional therapy, received different combinations of tipifarnib and etoposide. The overall CR rate with these combinations of tipifarnib and etoposide was 25 %. In contrast, in a previous single-agent study with a similar population, the CR rate was of 14 % ($P=0.036$; Karp et al. 2009). A multi-institutional phase-2 trial (Karp et al. 2012) of the same combination, in two dose schedules, once again reported CR rate of 25 %.

In another study (Jabbour et al. 2011), tipifarnib was combined with a classic protocol of idarubicin and cytarabine (IA). Tipifarnib 300 mg, twice daily, was added to IA in induction and consolidation and also was given as maintenance for 21 days, every 4–6 weeks, for 6 months. Using historical control patients that were treated with IA alone, the authors reported superior results for the combination. On the other hand, the British AML 16 trial (Burnett et al. 2012) failed to demonstrate an advantage for combining tipifarnib with low dose cytarabine (LDA) over LDA alone. This randomized study of elderly AML patients was prematurely discontinued by the independent data monitoring and ethics committee (DMC) and the conclusions were that the addition of tipifarnib had no effect on response, toxicity, and survival.

16.2.1.2 Defining Markers that Predict for a Good Response to FTIs

In an attempt to predict response to FTIs, the gene-expression profiles of elderly AML patients who participated in a phase-2 study were examined (Lancet et al. 2007; Raponi et al. 2008). This study looked for a correlation between gene expression and the clinical behavior of patients, and demonstrated that the relationship between the expression of two genes, RASGRP1 and APTX, can predict the survival in response to therapy with FTIs. RASGRP1 is a guanine nucleotide exchange factor that specifically activates Ras, whereas APTX is involved in DNA excision repair. Similar results were found in a phase-2 study which investigated the combination of etoposide and tipifarnib. AML patients with a RASGRP1/APTX ratio of more than 5.2 had a 78% CR rate with a negative predictive value of 87%. In patients treated with conventional chemotherapy, this ratio did not predict the prognosis (Karp et al. 2012).

16.2.1.3 Using Tipifarnib in Different Stages of the Treatment

Although maintenance therapy has not been proven to be effective in AML and is generally out of favor, it may well be that select targeted agents, such as tipifarnib, can change this state. This was initially investigated in a phase-2 study where (Karp et al. 2008) 48 adults with poor-risk AML, who achieved first CR, received 400 mg of tipifarnib twice daily, for 14 of 21 days, for a maximum of 16 cycles. The median disease-free survival (DFS) was 13.5 months with 30% having DFS > 2 years. Comparison of CR durations among 25 patients who received tipifarnib maintenance therapy and 23 historical controls who received the same induction and consolidation therapies but without maintenance, suggested a better DFS for the patients who received maintenance ($P=0.02$). The Eastern Cooperative Oncology Groups (ECOG) conducted a prospective controlled randomized study of tipifarnib as maintenance therapy in AML. A total of 144 patients were randomized to receive 300–400 mg of tipifarnib twice a day for 21 days and this was compared to a control group. The control patients, representing standard of care, did not receive any maintenance therapy. While the tolerability and feasibility of this approach has been reported (Luger et al. 2010), the definitive efficacy results are eagerly awaited. In the era of increasing importance of minimal residual disease (MRD) in acute leukemia treatment, it may well be that such a maintenance regimen will prove in the future to be efficacious for a patient population with a positive MRD.

16.2.2 Other FTIs in Clinical Studies of AML Patients

While the preponderance of clinical trials of FTIs used tipifarnib, there are two other FTIs which were investigated in clinical trials of AML, albeit in a very small number of patients. Lonafarnib, a tricyclic inhibitor, was examined in a phase-2 study, but only 2 out of 16 patients had AML (secondary) and all the others suffered from myelodysplastic syndrome (MDS). One of the two AML patients achieved partial response. Most of the adverse effects were in the GI tract, including: diarrhea, nausea, and anorexia (Ravoet et al. 2008). BMS–214662, another FTI from a fourth group, was investigated in a phase-1 study of relapsed/refractory/poor chemotherapy candidate AML or MDS patients. It was administered once weekly with different doses and different infusion durations. The investigators reported on five patients with evidence of antileukemia effect and concluded that prolonged exposure may improve the efficacy and the toxicity (Cortes et al. 2005).

16.2.3 Multi-Tyrosine Kinase Inhibitors

Sorafenib is a multi-tyrosine kinase inhibitor, known for its anti-FLT3 effect. It should be noted that among its demonstrated clinical activities, this drug can also inhibit the Raf/Merk/Erk pathway. Further details of these tyrosine kinase inhibitors, are discussed in the relevant chapters of this book.

Conclusion

Clinical studies have shown that FTIs are clearly active in AML. Unfortunately, like many other targeted agents recently employed for AML, despite a potent rationale, alternative resistant pathways exist that limit their efficacy. Based on the data presented in this chapter, it would seem that there may be a clinical role for tipifarnib for patients with advanced disease who have failed to respond to other therapies. Tipifarnib, used as a single agent for newly diagnosed AML, is not as effective as conventional therapy and studies of combinations using tipifarnib as an adjunct to standard therapy, are underway. As such this agent may add to the global armamentarium of targeted agents in AML, many with potent in vitro activity, but with a more limited clinical role. The role of tipifarnib as maintenance therapy is yet to be reported.

References

Bacher U, Haferlach T et al (2006) Implications of NRAS mutations in AML: a study of 2502 patients. Blood 107:3847–3853
Basso AD, Kirschmeier P et al (2006) Lipid posttranslational modifications. Farnesyltransferase inhibitors. J Lipid Res 47:15–31
Bowen DT, Frew ME et al (2005) RAS mutation in acute myeloid leukemia is associated with distinct cytogenetic subgroups but does not influence outcome in patients younger than 60 years. Blood 106:2113–9
Braun T, Fenaux P (2008) Farnesyltransferase inhibitors and their potential role in therapy for myelodysplastic syndromes and acute myeloid leukaemia. Br J Haematol 141:576–586
Burnett AK, Russell NH et al (2012) The addition of the farnesyl transferase inhibitor, tipifarnib, to low dose cytarabine does not improve outcome for older patients with AML. Br J Haematol 158:519–522
Cortes J, Faderl S et al (2005) Phase I study of BMS-214662, a farnesyl transferase inhibitor in patients with acute leukemias and high-risk myelodysplastic syndromes. J Clin Oncol 23:2805–2812
End DW (1999) Farnesyl protein transferase inhibitors and other therapies targeting the Ras signal transduction pathway. Invest New Drugs 17:241–258
Harousseau JL, Lancet JE et al (2007) A phase 2 study of the oral farnesyltransferase inhibitor tipifarnib in patients with refractory or relapsed acute myeloid leukemia. Blood 109:5151–6
Harousseau JL, Martinelli G et al (2009) A randomized phase 3 study of tipifarnib compared with best supportive care, including hydroxyurea, in the treatment of newly diagnosed acute myeloid leukemia in patients 70 years or older. Blood 114:1166–1173
Jabbour E, Kantarjian H et al (2011) A phase 1–2 study of a farnesyltransferase inhibitor, tipifarnib, combined with idarubicin and cytarabine for patients with newly diagnosed acute myeloid leukemia and high-risk myelodysplastic syndrome. Cancer 117:1236–1244
Karp, JE, Flatten K et al (2009) Active oral regimen for elderly adults with newly diagnosed acute myelogenous leukemia: a preclinical and phase 1 trial of the farnesyltransferase inhibitor tipifarnib (R115777, Zarnestra) combined with etoposide. Blood 113:4841–4852
Karp JE, Lancet JE et al (2001) Clinical and biologic activity of the farnesyltransferase inhibitor R115777 in adults with refractory and relapsed acute leukemias: a phase 1 clinical-laboratory correlative trial. Blood 97:3361–9
Karp JE, Smith BD et al (2008) Phase II trial of tipifarnib as maintenance therapy in first complete remission in adults with acute myelogenous leukemia and poor-risk features. Clin Cancer Res 14:3077–3082
Karp JE, Vener TI et al (2012) Multi-institutional phase 2 clinical and pharmacogenomic trial of tipifarnib plus etoposide for elderly adults with newly diagnosed acute myelogenous leukemia. Blood 119:55–63
Kirschbaum MH, Synold T et al (2011) A phase 1 trial dose-escalation study of tipifarnib on a week-on, week-off schedule in relapsed, refractory or high-risk myeloid leukemia. Leukemia 25:1543–7
Kiyoi H, Naoe T et al (1999) Prognostic implication of FLT3 and N-RAS gene mutations in acute myeloid leukemia. Blood 93:3074–3080
Lancet JE, Gojo I et al (2007) A phase 2 study of the farnesyltransferase inhibitor tipifarnib in poor-risk and elderly patients with previously untreated acute myelogenous leukemia. Blood 109:1387–1394
Lancet JE, Rosenblatt JD et al (2002) Farnesyltransferase inhibitors and myeloid malignancies: phase I evidence of Zarnestra activity in high-risk leukemias. Semin Hematol 39:31–35
Lobell RB, Omer CA et al (2001) Evaluation of farnesyl: protein transferase and geranylgeranyl:protein transferase inhibitor combinations in preclinical models. Cancer Res 61:8758–8768

Luger S, Yao X et al (2010) Tipifarnib is well tolerated as maintance therapy in acute myeloid leukemia (AML). Significant, but non-fatal, hematologic toxicity not ameliorated by dose reduction. Preliminary results of the Phase III Intergroup Trial E2902. Blood 116a:3315

Neubauer A, Maharry K et al (2008) Patients with acute myeloid leukemia and RAS mutations benefit most from postremission high-dose cytarabine: a Cancer and Leukemia Group B study. J Clin Oncol 26:4603

Parikh C, Subrahmanyam R et al (2007) Oncogenic NRAS, KRAS, and HRAS exhibit different leukemogenic potentials in mice. Cancer Res 67:7139–7146

Raponi M, Lancet JE et al (2008) A 2-gene classifier for predicting response to the farnesyltransferase inhibitor tipifarnib in acute myeloid leukemia. Blood 111:2589–2596

Ravoet C, Mineur P et al (2008) Farnesyl transferase inhibitor (lonafarnib) in patients with myelodysplastic syndrome or secondary acute myeloid leukaemia: a phase II study. Ann Hematol 87:881–5

Rockova V, Abbas S et al (2011) Risk stratification of intermediate-risk acute myeloid leukemia: integrative analysis of a multitude of gene mutation and gene expression markers. Blood 118:1069–1076

Whyte DB, Kirschmeier P et al (1997) K- and N-Ras are geranylgeranylated in cells treated with farnesyl protein transferase inhibitors. J Biol Chem 272:14459–14464

Zebisch A, Czernilofsky AP et al (2007) Signaling through RAS-RAF-MEK-ERK: from basics to bedside. Curr Med Chem 14:601–623

Chapter 17
The PI3K-AKT-mTOR Signaling Network in AML

Brandon Beagle and David A. Fruman

Abstract Acute myelogenous leukemia (AML) is the most common leukemia in adults. Currently, AML patients are treated mainly with combinations of cytotoxic chemotherapies. There is an urgent need for targeted therapies in AML that improve patient survival while limiting side effects. Small molecules that inhibit crucial signaling components have potential to block proliferation and survival of leukemia cells in a selective manner. A cellular signaling network involving phosphoinositide 3-kinase (PI3K), protein kinase B (AKT), and mechanistic target of rapamycin (mTOR), is highly active in most AML cells and contains several components that are druggable. This chapter will focus on the biology of PI3K/AKT/mTOR signaling in AML cells and the potential of novel inhibitors to disrupt oncogenic pathways in leukemia cells.

Keywords Leukemia · Acute myelogenous leukemia · Kinase · PI3K · AKT · mTOR · Kinase inhibitor

17.1 Introduction

The phosphoinositide 3-kinase (PI3K)/protein kinase B (AKT)/ mechanistic target of rapamycin (mTOR) signaling network promotes oncogenic processes in both solid and hematological malignancies and is also activated by elements of the tumor microenvironment (Engelman 2009; Laplante and Sabatini 2012; Martelli et al. 2010). Acute myelogenous leukemia (AML) cells consistently exhibit elevated activity of multiple components in the PI3K/AKT/mTOR network (Altman et al. 2011b; Martelli et al. 2010; Park et al. 2010; Vu and Fruman 2010). Furthermore, drugs targeting various aspects of the PI3K/AKT/mTOR network have shown antileukemic

D. A. Fruman (✉)
Department of Molecular Biology & Biochemistry and Institute for Immunology,
University of California, Irvine, McGaugh Hall, Room 3242, Irvine, CA 92697-3900, USA
e-mail: dfruman@uci.edu

B. Beagle
Department of Molecular Biology & Biochemistry, University of California, Irvine,
Laboratory David Fruman, McGaugh Hall, Room 3407, Irvine, CA 92697-3900, USA
e-mail: brandonbeagle@yahoo.com

M. Andreeff (ed.), *Targeted Therapy of Acute Myeloid Leukemia*,
Current Cancer Research, DOI 10.1007/978-1-4939-1393-0_17

efficacy in vitro and in vivo (Altman et al. 2011b; Chapuis et al. 2010c; Martelli et al. 2010; Park et al. 2010; Vu and Fruman 2010).

To foster a better understanding of the PI3K/AKT/mTOR network in AML, this chapter will discuss the most relevant data concerning pathway activity as well as its therapeutic targeting in preclinical models of AML. To understand these issues, one must also know the major components of the network and their potential oncogenic function, as is discussed in the next section.

17.2 PI3K/AKT/mTOR Signaling Network

PI3K, AKT, and mTOR are key components in a complex network of signaling proteins that promotes cell growth, proliferation, and survival (Engelman et al. 2006; Laplante and Sabatini 2012; Manning and Cantley 2007). In normal cells, low-level activation of the pathway through trophic factors maintains cell survival but does not drive growth or cell division. Sensing of extracellular growth factors triggers elevated PI3K/AKT/mTOR activity that is essential for cell cycle progression and for the biosynthetic processes that support cell growth and division (Fig. 17.1). Tumor suppressor genes (e.g., phosphatase and tensin homolog deleted from chromosome 10, PTEN; tuberous sclerosis complex 2, TSC2) encode proteins that limit PI3K/AKT/mTOR network activity when growth factors are limiting or cells sense stress. Activation of the network opposes the function of additional tumor suppressor proteins (e.g., forkhead box O (FOXO) transcription factors). Most cancer cells maintain elevated activity of the PI3K/AKT/mTOR network, which can be driven by activated oncogenes or by loss of tumor suppressor genes. The sections below outline key aspects of the three central signaling nodes: PI3K, AKT, and mTOR.

17.2.1 PI3K

PI3Ks are lipid kinases that function downstream of receptor tyrosine kinases (RTKs) and/or G-protein-coupled receptors (GPCRs) (Fruman 2012; Vanhaesebroeck et al. 2010). Class I PI3K enzymes are the most commonly studied of the three different PI3K classes in AML, as well as most cancers, and will be the only class discussed here. Class I PI3Ks (PI3K) are further divided into class IA and IB. Both PI3K subclasses transduce signals from various growth factors and cytokines into intracellular messages by converting phosphatidylinositol-4,5-bisphosphate (PI-4,5-P_2) into phosphatidylinositol-3,4,5-trisphosphate (PIP_3), which, in turn, activates AKT and secondary downstream effectors such as mTORC1 (Fig. 17.1).

Class IA PI3Ks are heterodimers consisting of a p110 catalytic subunit (p110α, p110β, or p110δ) and a regulatory subunit (p85α, p55α, p50α, p85β or p55γ) (Fruman 2012; Vanhaesebroeck et al. 2010). In response to extracellular growth factor/cytokine stimulation and subsequent activation of RTKs, class I PI3K is recruited to the membrane through interaction of its regulatory subunit with phosphorylated

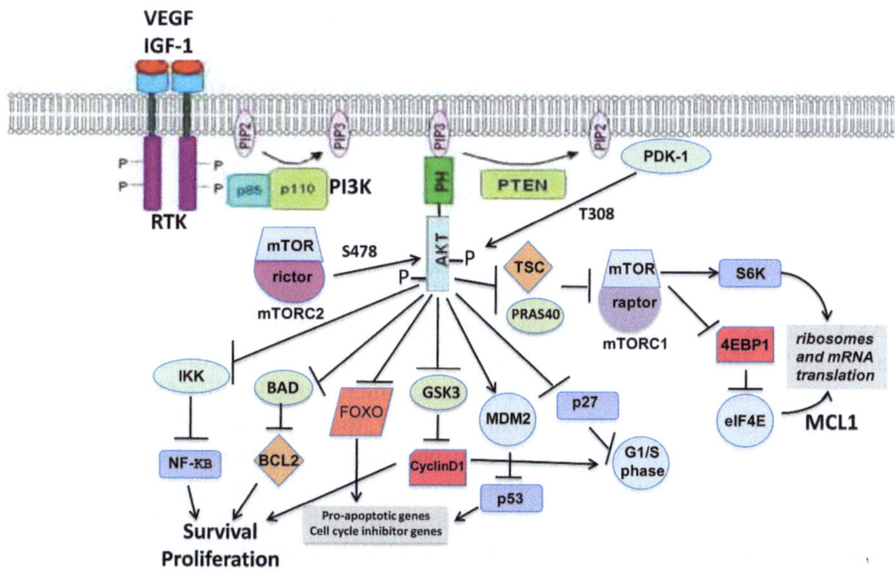

Fig. 17.1 A simplified diagram of the phosphoinositide 3-kinase/AKT/mechanistic target of rapamycin (PI3K/AKT/mTOR) network and associated mechanism(s) that drive oncogenesis in AML cells. Activation of PI3K by a ligand-activated receptor tyrosine kinase (RTK), as well as GPCR (not shown), results in formation of phosphatidylinositol-3,4,5-trisphosphate (PIP_3). Inactive AKT binds PIP_3 which enables 3-phosphoinositide dependent kinase-1 (PDK1) and mTOR in the mTORC2 complex to phosphorylate and activate AKT at Thr308 and Ser473. Activated AKT phosphorylates/inhibits TSC1/2 and/or PRAS40 resulting in downstream activation of mTOR kinase in the mTORC1 complex. Activated mTORC1 phosphorylates substrates such as S6 kinase (S6K) and 4EBP1 that promote ribosome biogenesis and cap-dependent translation of proteins such as anti-apoptotic MCL1. Activated AKT also phosphorylates and inhibits the tumor suppressing transcription factor, FOXO, as well as substrates that are fundamental for cell proliferation and survival. *Arrows* indicate activating phosphorylation events, whereas *perpendicular lines* indicate inhibitory events

tyrosine motifs on activated RTKs (e.g., platelet-derived growth factor receptor, PDGFR) or receptor-associated adaptor protein(s) (e.g., insulin receptor substrate 1, IRS1). The catalytic p110 subunit then catalyzes the phosphorylation of PI-4,5-P_2 into PIP_3.

Class IB PI3Ks are heterodimers composed of a p110γ catalytic subunit and a regulatory subunit (p101 or p87) (Fruman 2012; Vanhaesebroeck et al. 2010). In contrast to class IA PI3Ks, class IB PI3Ks are mainly activated by GPCRs through interaction of the regulatory subunit with the Gβγ subunit of trimeric G proteins. Both class IA and IB PI3K catalytic subunits bind to the small GTPase RAS in its active, guanosine triphosphate (GTP)-bound form. This further enhances class I PI3K activity downstream of both RTKs and GPCRs.

Because PIP_3 is the pivotal second messenger that drives multiple downstream signaling cascades, its levels are tightly regulated in nonmalignant cells. The putative tumor suppressor PTEN (phosphatase and tensin homolog deleted from chromosome 10), is the primary lipid phosphatase that catalyzes PIP_3 dephosphorylation into PIP_2 (Hollander et al. 2011; Song et al. 2012). Loss of PTEN expression and/or function is a common genetic (or epigenetic) aberration seen in a broad range of cancers, and enables hyperactivation of the PI3K/AKT/mTOR network.

17.2.2 AKT

Also known as protein kinase B, AKT is a family of serine/threonine kinases (AKT1, AKT2, and AKT3) encoded by three different genes (Fayard et al. 2010; Manning and Cantley 2007). AKT activation is initiated by translocation to the plasma membrane where the PH domain of AKT binds to a membrane localized PIP_3. The interaction induces a conformational change in AKT, thereby allowing sequential phosphorylation of AKT at Thr308 by 3-phosphoinositide dependent kinase-1 (PDK1) and at Ser473 by mTOR in the mTORC2 complex (Fig. 17.1). Maximal AKT activation requires dual phosphorylation. Depending upon its phosphorylation status, protein interaction, substrate availability, and localization, AKT can phosphorylate numerous substrates that promote protein synthesis, survival, proliferation, and metabolism, to name a few (Fayard et al. 2010; Manning and Cantley 2007). In many cases, AKT-mediated phosphorylation inactivates a protein with tumor suppressor function. For example, AKT phosphorylation inhibits the proapoptotic protein B-cell lymphoma 2 (Bcl-2)-associated death promoter (BAD), various cell cycle inhibitors (e.g., $p27^{Kip1}$), and the FOXO family (FOXO1, FOXO3a, FOXO4) of tumor suppressing transcription factors (Fayard et al. 2010; Manning and Cantley 2007).

17.2.3 mTOR

mTOR plays a pivotal role in cell growth and proliferation by stimulating translation and lipid synthesis as well as monitoring nutrient availability, cellular energy levels, oxygen levels, and mitogenic signals (Dazert and Hall 2011; Laplante and Sabatini 2012; Yecies and Manning 2011). mTOR is a Ser/Thr kinase that exists in two distinct complexes, mTORC1 and mTORC2. The mTORC1 complex is composed of the mTOR catalytic subunit, regulatory associated protein of mTOR (Raptor), the protein mLST8/GβL, and two negative regulators proline-rich AKT substrate 40 kDa (PRAS40) and DEP domain-containing mTOR-interacting protein (DEPTOR). mTORC2 is composed of mTOR, rapamycin insensitive companion of mTOR (Rictor), mammalian stress-activated protein kinase interacting protein 1 (mSIN1) mLST8/GβL, protein observed with Rictor (PROTOR), and DEPTOR. AKT can activate mTORC1 by phosphorylating both PRAS40 and TSC2 to attenuate their inhibitory effects on mTORC1. Thus, mTORC1 activity is partly controlled by AKT whereas AKT itself is partly controlled by mTORC2 (Fig. 17.1). mTORC2 can also phosphorylate/activate other members of the AGC superfamily of kinases including serum and glucocorticoid-inducible kinase (SGK) and protein kinase C

(PKC) (Cybulski and Hall 2009; Dazert and Hall 2011; Laplante and Sabatini 2012; Sparks and Guertin 2010). While the regulation of mTORC2 is less clear than for mTORC1, data suggest that interfering with mTORC2 is selectively toxic to cells "addicted" to elevated PI3K/AKT/mTOR activity, such as cancer cells, relative to normal cells responding to physiological stimuli (Guertin et al. 2009; Nardella et al. 2009).

The best-characterized downstream targets of mTORC1 are ribosomal S6 kinases (S6K1 and S6K2) and eukaryotic initiation factor 4E (eIF4E) binding proteins (4E-BP1, 2, 3) (Dazert and Hall 2011; Laplante and Sabatini 2012; Yecies and Manning 2011). mTORC1-mediated phosphorylation activates S6Ks that promote mRNA translation, ribosome biogenesis, and lipid synthesis. mTORC1-mediated phosphorylation inhibits 4E-BP1, thereby allowing cap-dependent translation to proceed. Aberrant mRNA translation is commonly seen in blood cancers, further supporting mTOR as an attractive target for therapy (Carroll 2009; Hagner et al. 2010; Silvera et al. 2010; Tamburini et al. 2009). Another way that mTORC1 supports cell growth is by suppressing autophagy, in part through phosphorylation of ULK1 (Chan 2009). Notably, mTORC1 and S6K1 also participate in negative feedback loops that dampen upstream PI3K signaling (Dazert and Hall 2011; Laplante and Sabatini 2012; Yea and Fruman 2011). Consequently, selective mTORC1 inhibition tends to enhance PI3K/AKT activity.

17.3 PI3K/AKT/mTOR Signaling in AML

Constitutive activation of the PI3K/AKT/mTOR network has consistently been identified in malignant myeloid blast cells from AML patients (AML blasts) (Brandts et al. 2005; Cheong et al. 2003a; Cheong et al. 2003b; Gallay et al. 2009; Grandage et al. 2005; Kornblau et al. 2006, 2010; Min et al. 2003; Muranyi et al. 2009; Tazzari et al. 2004; Xu et al. 2003). Similar to most cancers, AML cells arise by a combination of aberrations in proto-oncogenes and tumor suppressors, which have the potential to affect the PI3K/AKT/mTOR network. Signals from the microenvironment and from autocrine growth factors also support PI3K/AKT/mTOR activity. This section will review the various mechanisms of network activation in AML cells.

17.3.1 Activation of PI3K/AKT/mTOR Signaling in AML

In AML, the most frequently identified transforming events are activating mutations in the receptor tyrosine kinase, Fms-like tyrosine kinase 3 (FLT3) (Choudhary et al. 2005; Chua and Small 2009; Kindler et al. 2010; Minami et al. 2003; Weisberg et al. 2010). FLT3 activation promotes several downstream signaling pathways including PI3K/AKT/mTOR. The ability of the FLT3 internal tandem duplication

(FLT3-ITD) mutation to increase PI3K/AKT/mTOR activity was shown in a mouse myeloid progenitor cell line (32D cells), in which ectopic expression of FLT3-ITD resulted in constitutive AKT activation (Brandts et al. 2005). This correlated with phosphorylation and inhibition of FOXO3a. Restoring FOXO3a activity reversed FLT3-ITD-mediated growth and expression of a dominant-negative AKT prevented FLT3-ITD-mediated cytokine-independence of 32D cells. FLT3-ITD was also shown to promote mTORC1 activation and survival of multiple human AML cell lines and mouse cells (Chen et al. 2010). Indeed, activation of the PI3K/AKT/mTOR network does promote resistance to FLT3 inhibitors (Chu and Small 2009; Piloto et al. 2007; Weisberg et al. 2010). Interestingly, proteomic analysis or single cell network profiling showed a correlation between chemotherapeutic efficacy in certain AML patients and their level of FLT3 ligand-dependent AKT phosphorylation (Kornblau et al. 2009, 2010; Rosen et al. 2010).

The receptor tyrosine kinase c-KIT drives PI3K activation in response to its ligand, stem cell factor. The gene encoding c-KIT is frequently mutated in AML patients that possess the t(8;21) [RUNX1-CBFA2T1] or inv(16)/t(16;16) [CBFB-MYH11] fusion genes (Frohling et al. 2005). These gain-of-function (GOF) mutations in c-KIT promote its activation and have been associated with the activation of multiple oncogenic pathways including the PI3K/AKT/mTOR network (Kohl et al. 2005; Lennartsson and Ronnstrand 2006; Ma et al. 2012; Munugalavadla et al. 2007; Ning et al. 2001). For example, the Kasumi-1 human AML cell line that contains a gain of function (GOF) point mutation in c-KIT (Asn822Lys) displays PI3K/AKT activation (Beghini et al.2005; Larizza et al.2005). In vitro and in vivo mouse models of myeloproliferative disease (MPD) showed that several GOF mutations previously identified in human AML blasts promote MPD through the PI3K/AKT/mTOR network (Ma et al. 2012).

Vascular endothelial growth factor (VEGF) is a secreted angiogenic factor that is a validated therapeutic target in solid tumor therapy (bevacizumab), and is also associated with hematological malignancies (Cao et al. 2011; Ribatti et al. 2004; Scavelli et al. 2004; Waldner and Neurath 2012). AML blasts secrete VEGF and highly express receptors VEGFR-1 and VEGFR-2 (Gutierrez et al. 2005). In human AML cell lines, VEGF promotes clonal outgrowth through a PI3K/AKT-dependent mechanism (List et al. 2004). Similarly, in AML cell lines and blasts, angiopoietins activate PI3K signaling through an autocrine mechanism (Wakabayashi et al. 2004). Therefore, at least in some AML cases, upregulation of the PI3K/AKT/mTOR network might be due to the production of angiogenic factors, such as VEGF and angiopoietins, in an autocrine and/or paracrine manner.

IGF-1 is a potent growth factor that can activate the PI3K/AKT/mTOR network in a broad range of cell types (Pollak 2012). Multiple large-scale case-control studies have also reported positive associations between high circulating levels of IGF-1 and risk for different types of cancer (Grimberg 2003). The role of IGF-1 in promoting proliferation, survival, and drug resistance of multiple myeloma cells through PI3K/AKT/mTOR signaling has also been shown (McMillin et al. 2009; Mitsiades et al. 2004). Importantly, several studies have shown that IGF-1 signaling promotes growth and survival of human AML cell lines and/or blasts in part through activation of the PI3K/AKT/mTOR network (Chapuis et al. 2010a; Doepfner et al. 2007; Neri et al. 2003; Tamburini et al. 2008; Tazzari et al. 2007b; Wahner Hendrickson et al. 2009).

An elevation in PIP_3 levels driven by PTEN deletion is relatively common in solid tumors and T-cell acute lymphoblastic leukemia (Gutierrez et al. 2009; Hollander et al. 2011; Song et al. 2012), but is a rare phenomenon in AML (Gallay et al. 2009; Kornblau et al. 2005; Martelli et al. 2010; Xu et al. 2003). However, other mechanisms may reduce PTEN activity. PTEN can undergo phosphorylation, which attenuates phosphatase activity towards PIP_3 (Vazquez et al. 2000). Indeed, such phosphorylation events have been identified in AML blasts but there are conflicting data concerning the pathological significance of PTEN phosphorylation in AML (Cheong et al. 2003b; Gallay et al. 2009; Kornblau et al. 2009). A role for PTEN inactivation in AML development might also be inferred from mouse models of PTEN-null hematopoietic stem cells, which rapidly proliferate, have decreased self-renewal potential, and begin migrating out of the bone marrow into distant organs resulting in a leukemic-like disease (Yilmaz et al.2006; Zhang et al. 2006). Importantly, these effects were mediated in part by mTOR, as rapamycin not only depleted leukemic stem cells (LSCs) but also restored normal hematopoietic stem cell function (Yilmaz et al. 2006). A dominant-negative mutation in the lipid phosphatase Src homology 2 domain-containing inositol-5-phosphate-1 (SHIP-1), which dephosphorylates the 5′-phosphate of PIP_3 and plays a key role in normal hematopoiesis (Hazen et al. 2009; Ong et al. 2007), has been associated with AKT activation in AML blasts (Luo et al. 2003).

In solid tumors, PI3K signaling output is often elevated by a direct mechanism through activating mutations in PI3K genes (*PIK3CA*, *PIK3R1*, *PIK3R2*) (Cheung et al. 2011; Jaiswal et al. 2009; Samuels et al. 2004). Activating mutations in AKT genes have also been identified in some solid tumors (Steelman et al. 2008). Activating mutations in PI3K or AKT genes have not been identified in AML (Bousquet et al. 2005; Tibes et al. 2008). However, the PI3K catalytic subunit p110δ is highly expressed in most human AML cell lines and blasts and is implicated in AKT activation as well as proliferation (Billottet et al. 2006; Sujobert et al. 2005). Although expression of the p110β isoform is variable in AML blasts, RNAi-mediated knockdown of both p110δ and p110β cooperated to reduce AKT activation as well as proliferation and survival of AML cell lines and blasts (Doepfner et al. 2007).

Additional factors in the bone marrow microenvironment can enhance PI3K activity in AML cells (Ayala et al. 2009). Interaction between leukemic cells and bone marrow stromal cells through the GPCR, chemokine receptor type 4 (CXCR4), and its endogenous ligand, chemokine ligand 12 (CXCL12), is associated with PI3K/AKT/mTOR activation in AML blasts and FLT3-ITD overexpressing mouse cells (Fiegl et al. 2009; Fierro et al. 2009; Zeng et al. 2009). Furthermore, interactions between β1 integrins on AML cells and stromal fibronectin could lead to pathway activation (Matsunaga et al. 2003, 2008), possibly through upregulation of integrin-linked kinase 1 (ILK1) which is involved in AKT phosphorylation (Ser473) in a PI3K-dependent manner in AML cells (Tabe et al. 2007). The ability of ILK1 to facilitate AKT-Ser473 phosphorylation could be related to the fact that ILK1 interacted with Rictor and was required for AKT phosphorylation by mTORC2 on Ser473 (McDonald et al. 2008; Lennartsson and Ronnstrand 2006).

Oncogenic activation of the RAS protein family is frequently observed in cancer (most frequently N-RAS in AML) and has been shown to promote activation of AKT/mTOR (Mendoza et al. 2011; Santarpia et al. 2012; Young et al. 2009). However, there are insufficient data to show such crosstalk occurs in AML.

In summary, the PI3K/AKT/mTOR network is activated in AML cells through a variety of mechanisms including upstream oncogenes (FLT3-ITD, KIT), autocrine/paracrine growth factors (VEGF, IGF-1), altered expression (p110δ) or phosphorylation (PTEN) of pathway components, and microenvironmental signals including chemokines and adhesion molecules. This knowledge emphasizes that blocking multiple inputs to the network is likely required to produce adequate therapeutic suppression of PI3K-dependent pro-survival signals. The next section will consider the multiple mechanisms by which the PI3K/AKT/mTOR network promotes survival and proliferation of AML cells.

17.3.2 Oncogenic Survival Through PI3K/AKT/mTOR

AKT is a central effector of PI3K-dependent survival signals in many cancers including AML (Martelli et al. 2010; Park et al. 2010; Steelman et al. 2008). One of the first mechanisms established was the AKT-mediated phosphorylation of BAD, a proapoptotic member of the Bcl-2 family (Youle and Strasser 2008). AKT phosphorylation prevents BAD from inhibiting anti-apoptotic BCL-2 family members (e.g., BCL-2, BCL-X_L), thereby suppressing apoptosis. Inhibiting AKT-mediated BAD phosphorylation in AML blasts resulted in apoptosis (Zhao et al. 2004). In a separate mechanism, phosphorylation of the Yes-associated protein by AKT leads to repression of the p73 transcription factor and reduced expression of the proapoptotic protein BAX (Basu et al. 2003). However, there are conflicting data concerning the expression of both BAX and p73 in AML blasts (Olesen et al. 2005; Rizzo et al. 2004; Tamm et al. 2005; van Stijn et al. 2003). In non-Hodgkin's lymphoma and in AML cell lines, PI3K/AKT/mTOR signaling is required for the expression of a potent anti-apoptotic effector, myeloid cell leukemia 1 (MCL-1) (Akgul 2009; Coloff et al. 2011; Yecies et al. 2010). A major mechanism is mTORC1-dependent translation of MCL-1 mRNA through a cap-dependent process. MCL-1 expression is observed in human AML cells (Glaser et al. 2012; Yu et al. 2003) and inhibition of the PI3K/AKT/mTOR network through PI3K reduces MCL-1 expression in APL cell lines (Billottet et al. 2009). A recent study confirmed that MCL-1 is critical for survival of human AML cells and its deletion alleviates disease in a mouse model of AML (Glaser et al. 2012).

AKT can also promote phosphorylation and nuclear translocation of Mdm2, an E3 ubiquitin ligase that mediates proteasome-dependent degradation of the tumor suppressor protein, p53, thus attenuating and blocking p53-mediated tumor suppression (Mayo and Donner 2002). Inhibition of the PI3K/AKT/mTOR network significantly increased p53 activity in AML blasts with a subsequent reduction in survival (Grandage et al. 2005). This indicates that in some AML cases, p53 is suppressed through PI3K/AKT/mTOR-dependent signaling and this pathway could be a mechanism to promote resistance to cytotoxic agents.

The nuclear factor (NF)-κB transcription factor promotes survival through expression of anti-apoptotic target proteins (e.g., X-linked inhibitor of apoptosis (XIAP), tumor necrosis factor receptor-associated factors (TRAFs), etc.) (Aggarwal 2004; Shishodia and Aggarwal 2002). In an AML cell line and blasts, activated AKT indirectly promotes downstream NF-κB transactivation resulting in NF-κB-dependent gene expression and survival (Birkenkamp et al. 2004; Bortul et al. 2003; Frelin et al. 2005; Grandage et al. 2005; Tabellini et al. 2005).

Perhaps one of the most potent proleukemic mechanisms mediated by AKT is suppression of the FOXO transcription factors (Fu and Tindall 2008; Tothova and Gilliland 2007; Zhang et al. 2011). FOXO family members (primarily FOXO-1 and FOXO-3a in AML) augment transcription of target genes encoding cell cycle inhibitory (e.g., p27, p130, etc.) and proapoptotic (e.g., Fas ligand, FASL; BIM; TNF-related apoptosis-inducing ligand, TRAIL; etc.) proteins in hematopoietic cell lineages. FOXO factors can also act to suppress transcription at certain loci; for example, reduced cyclin D1 and D2 expression contributes to the FOXO cell cycle arrest program (Ramaswamy et al. 2002; Schmidt et al. 2002). Phosphorylation of FOXO factors by AKT promotes nuclear exclusion and cytosolic sequestration (Calnan and Brunet 2008; Zhang et al. 2011). FOXO factor inhibition requires nuclear localization of active AKT, which has been documented in mouse (32D cells) and human (HL60) AML cell lines (Brandts et al. 2005; Cappellini et al. 2003). AKT activation was also associated with significantly elevated levels of phospho-FOXO1 and phospho-FOXO3a in AML blast cells (Cheong et al. 2003a; Grandage et al. 2005).

17.3.3 Growth and Proliferation Through PI3K/AKT/mTOR

Activation of the PI3K/AKT/mTOR network promotes AML cell cycle progression, as well as biosynthetic processes that drive cell growth. Activated AKT inhibits $p27^{Kip1}$ (p27) through cytosolic sequestration (Cappellini et al. 2003; Medema et al. 2000; Min et al. 2004; Stahl et al. 2002), which blocks the ability of p27 to promote G1 cell cycle arrest. A direct relationship between AKT activity and cytoplasmic localization of p27 has been demonstrated in a human AML cell line as well as primary AML blasts (Cappellini et al. 2003; Medema et al. 2000; Min et al. 2004). Levels of cyclin D1 were also upregulated in an AML cell line due to suppression of FOXO factors by a constitutively active PI3K/AKT/mTOR network (Cappellini et al. 2003).

mTORC1 promotes cell growth (increased biomass and size) through ribosome biogenesis, protein synthesis, lipid synthesis, and suppression of autophagy (Dazert and Hall 2011; Laplante and Sabatini 2012). Elevated basal phosphorylation of mTORC1 substrates has been detected in the vast majority (~90 %) of AML patient blast samples (Tamburini et al. 2009), and is likely an important mechanism to support the hyperproliferative state. While AKT provides an important input to drive mTORC1 activity, other signaling components can contribute to overall mTORC1 output. For example, Erk and the Src family tyrosine kinase, Lyn, are consistently active in some AML patient blasts (Chow et al. 2006; Dos Santos et al. 2008) and can promote mTORC1 activity.

The central roles of AKT and mTOR in AML cell growth, proliferation, and survival have validated this signaling network as a prime target for therapeutic targeting.

17.4 Targeting PI3K/AKT/mTOR in AML Therapy

Small-molecule inhibitors targeting the PI3K/AKT/mTOR network have demonstrated significant efficacy for reducing cell proliferation, inducing apoptosis and/or enhancing the efficacy of other therapies in various preclinical models of AML (i.e., cell lines, blasts, and/or animal models). Clinical trials involving such inhibitors are discussed in chapter 20. Therefore, this chapter will focus on small-molecule inhibitors in preclinical models of AML as outlined in Table 17.1.

17.4.1 PI3K Inhibitors

Wortmannin and LY294002 were the first PI3K inhibitors to be discovered, and have been commonly used to test the role of PI3K in various tumor cells (Fruman and Rommel 2011; Workman et al. 2010; Yap et al. 2008). Wortmannin irreversibly inhibits PI3K (with an *in vitro* (IC_{50}) of 2–5 nM) by covalent modification of the catalytic subunit (Wymann et al. 1996) whereas LY294002 is a reversible ATP-competitive inhibitor that inhibits class I PI3Ks with an IC_{50} of 1–20 μM (Gharbi et al. 2007). Both inhibitors have numerous off-target effects on lipid and protein kinases and other cellular enzymes. Despite this lack of specificity, both wortmannin and LY294002 have been tested extensively in human AML cell lines and blasts as well as AML xenograft models. Both compounds consistently induce significant cytostatic effects but only moderate cytotoxicity. Some reports indicate that wortmannin or LY294002 sensitize AML cells to proapoptotic effects of other drugs (Gao et al. 2011, 2012a, 2012b). However, both compounds have severe limitations that have largely precluded their clinical application including insolubility, poor pharmacokinetics for in vivo studies, and toxicity due to off-target effects (Fruman and Rommel 2011; Garcia-Echeverria and Sellers 2008; Gharbi et al. 2007; Workman et al. 2010; Yap et al. 2008). Consequently, development of PI3K inhibitors more suitable for clinical use has been the focus of drug development efforts for many years.

Table 17.1 Inhibition of the PI3K/AKT/mTOR network in AML

Inhibitor	Target(s)	Model(s)	References
LY294002	p110/PI3K	PB, CL	(Birkenkamp et al. 2004; Tabellini et al. 2005; Billottet et al. 2009; Takahashi 2010; Cheong et al. 2010; Gao et al. 2011, 2012a, 2012b; Bortul et al. 2003; Neri et al. 2003; Xu et al. 2003; Zhao et al. 2004)
Wortmannin	p110/PI3K	PB, CL	(List et al. 2004; Tabellini et al. 2005; Skladanowski et al. 2007; Tazzari et al. 2007a)
IC87114	p110δ	PP, CL	(Billottet et al. 2006, 2009; Sujobert et al. 2005
TGX-115	p110β	PP	(Billottet et al. 2009)
Perifosine	AKT	PB, CL	(Castillo et al. 2004; Rahmani et al. 2005, 2006; Papa et al. 2008; Martelli et al. 2003; Tabellini et al. 2004; Tazzari et al. 2008)
Deguelin	AKT	PB, CL	(Bortul et al. 2005)
Akt-I-1/2	AKT −1, −2	PB, CL	(DeFeo-Jones et al. 2005; Gallay et al. 2009)
Rapalogs	mTORC1	PB, CL, X	(Altman et al. 2011a; Weisberg et al. 2008; Feldman et al. 2009; Lopez-Fauqued et al. 2010; Maira et al. 2008; Recher et al. 2005; Tamburini et al. 2008; Thoreen et al. 2009; Thoreen and Sabatini 2009; Willems et al. 2011; Xu et al. 2003, 2005a, 2005b; Zeng et al. 2007)
AZD8055	mTOR -(mTORC1/2)	PB, X	(Willems et al. 2011)
OSI-127	mTOR -(mTORC1/2)	PB, CL	(Altman et al. 2011a)
NVP-BEZ235	PI3K/p110 mTOR-(mTORC1/2)	PB, CL	(Chapuis et al. 2010b; Maira et al. 2008)
PI-103	PI3K/p110 mTOR-(mTORC1/2)	PB, CL	(Kojima et al. 2008; Park et al. 2008)
UCN-01/KRX0601	PDK1	PB, CL	(Sampath et al. 2006; Dai et al. 2005; Pei et al. 2005)
KP372–1	FLT3, AKT, PDK1	PB, CL	(Zeng et al. 2006)
BAG956	PI3K/PDK1	PB, mCL	(Weisberg et al. 2008)

PB primary patient AML blast, *CL* AML cell line, *PP* promyelocytic myeloma, *X* AML xenograft, *mCL* mouse AML cell model

Several small-molecule inhibitors that selectively target different isoforms of the class I PI3K catalytic subunit have been developed (Fruman and Rommel 2011; Garcia-Echeverria and Sellers 2008; Ihle and Powis 2009; Workman et al. 2010; Yap et al. 2008). One of the first was the p110δ specific inhibitor, IC87114 (IC_{50} = 0.5 μM and >50-fold selectivity over the other class I PI3K isoforms) (Bilancio et al. 2006; Sadhu et al. 2003). IC87114 reduced phosphorylation of AKT and FOXO3a, diminished AML blast proliferation and synergized with etoposide to reduce the viability of AML blasts showing high expression of the p110δ isoform (Billottet et al. 2006). Moreover, IC87114 suppressed both constitutive and FLT-3–stimulated AKT activation in AML blasts to the same extent as LY294002 and also attenuated blast proliferation without affecting the proliferation of normal hematopoietic progenitor cells (Sujobert et al. 2005). Emerging clinical data indicate that p110δ-selective inhibitors are highly effective in chronic lymphocytic leukemia (Fruman and Rommel 2011), yet their efficacy in AML has not been established. Preclinical studies of AML show that p110δ-specific inhibitors promote a cytostatic rather than a cytotoxic response (Billottet et al. 2006; Sujobert et al. 2005). In primary APL cells, combined treatment with IC87114 and TGX-115 (a p110β-selective inhibitor) triggered apoptosis in the presence or in the absence of the differentiating agent, all-*trans*-retinoic acid (ATRA) (Billottet et al. 2009). These findings suggest that targeting multiple PI3K isoforms might be more effective than inhibiting p110δ alone, even though this isoform appears to be the most abundant in many AML cases.

17.4.2 AKT Inhibitors

Perifosine is a water soluble, orally bioavailable synthetic alkylphosphocholine that interacts with the AKT PH domain, preventing the membrane recruitment and activation of AKT (Kondapaka et al. 2003; van Blitterswijk and Verheij 2008). Perifosine reduced cell proliferation and induced apoptosis in a FAS/FASL-dependent manner in human AML cell lines (Papa et al. 2008). In addition, perifosine synergized with etoposide in AML blasts, and reduced the clonogenic activity of $CD34^+$ cells from AML patients but not from healthy donors (Papa et al. 2008). Perifosine also synergized with histone deacetylase inhibitors (HDACi) (Rahmani et al. 2005) or TRAIL (Martelli et al. 2003; Tazzari et al. 2008) in AML cell lines and blasts displaying constitutive AKT activation. Notably, perifosine targets both mTORC1 and mTORC2 by reducing protein levels of mTOR, raptor, rictor, S6K1, and 4E-BP1, owing to their enhanced degradation (Fu et al. 2009). Moreover, perifosine targeted the mitogen-activated protein kinase (MEK)/extracellular-signal-regulated kinase (ERK) pathway (Papa et al. 2008; Rahmani et al. 2005) and activated proapoptotic JNK (Chiarini et al. 2008; Papa et al. 2008; Tazzari et al. 2008). Therefore, its specificity for AKT is uncertain.

Another nonspecific AKT inhibitor called deguelin (mechanism of inhibition unclear) induced cell cycle arrest or apoptosis, and enhanced the sensitivity of AML cell lines or blasts to etoposide or cytarabine while having relatively minimal effects on normal human CD34+ cells (Bortul et al. 2005).

Akt-I-1/2 is a synthetic, reversible allosteric inhibitor that forms a PH domain-dependent inactive conformation with AKT1 and AKT2 (Barnett et al. 2005). Akt-I-1/2 inhibited cell proliferation and clonogenic properties, and induced apoptosis in AML blasts (Gallay et al. 2009). There are AKT1 or AKT2 isoform-specific inhibitors but it is not clear whether one isoform is predominant in AML cells, and maximal apoptotic effects were seen in various cancer cell lines when the dual AKT1/2 inhibitor was used (DeFeo-Jones et al. 2005).

17.4.3 mTOR Inhibitors

Aberrant mRNA translation is commonly observed in tumor cells and plays a key role in AML pathogenesis (Carroll 2009; Hagner et al. 2010; Silvera et al. 2010; Tamburini et al. 2009). mTORC1 has a dominant role in facilitating protein synthesis and specifically promoting cap-dependent translation. Combined with the ubiquitous activation of mTORC1 in cancers such as AML, significant resources have been devoted towards developing mTOR inhibitors (Altman et al. 2011b; Bhagwat and Crew 2010; Chapuis et al. 2010c; Vu and Fruman 2010). The classical mTOR inhibitors are rapamycin (sirolimus) and its derivatives (known as rapalogs) CCI-779 (temsirolimus), RAD001 (everolimus), and AP23573 (deforolimus) (Guertin and Sabatini 2009). Rapalogs inhibit mTORC1 through an allosteric mechanism by forming a complex with FK506 binding protein 12 (FKBP12). The FKBP12/rapalog complex partially blocks mTORC1 function and does not acutely inhibit mTORC2. Despite their partial effects on mTOR activity, rapalogs have made an important contribution to elucidating mTOR biology. In addition, rapalogs have been FDA-approved for treatment of advanced renal cell carcinoma and mantle cell lymphoma (Abraham 2009). In preclinical studies of human AML and APL cell lines or blasts, rapalogs enhance the efficacy of anticancer therapies including etoposide, HDACi, or ATRA (Hahn et al. 2005; Nishioka et al. 2008, 2009; Weisberg et al. 2008; Xu et al. 2005b).

A drawback of rapalogs in the context of cancer therapy is that these agents are potent immunosuppressants that inhibit lymphocyte proliferation and promote tolerance (Janes and Fruman 2009; Thomson et al. 2009). It has also been shown for a variety of cancers that rapalogs (1) block mTORC1-mediated phosphorylation of S6Ks but not 4EBP1 (2) do not acutely inhibit mTORC2 in most cells (3) and promote PI3K/AKT activation via enhanced growth factor signaling (e.g., IGF1/IGF-1R) due to loss of negative feedback (Altman et al. 2011a; Laplante and Sabatini 2012; Tamburini et al. 2008; Thoreen and Sabatini 2009). Consequently, rapalogs primarily induce a cytostatic response in human AML cell lines and blasts (Recher et al. 2005; Weisberg et al. 2008; Willems et al. 2011).

Due to these drawbacks, many groups have developed ATP-competitive mTOR inhibitors that effectively block mTOR in both the mTORC1 and mTORC2 complexes. This new class of compounds, here termed mTOR kinase inhibitors (also known as TOR-KIs, TORKinibs, asTORi, and TORC1/2 inhibitors), inhibit rapalog-resistant outputs of mTOR including AKT (via mTORC2) and 4EBP1 (Bhagwat and Crew 2010; Chapuis et al. 2010c; Feldman and Shokat 2010; Janes and Fruman 2010; Vu and Fruman 2010; Wander et al. 2011). mTOR kinase inhibitors display enhanced anticancer efficacy relative to rapamycin/rapalogs in cell lines, primary cells, and xenograft models of solid and hematological malignancies (Altman et al. 2011a; Chresta et al. 2010; Janes et al. 2010; Yu et al. 2009, 2010). Surprisingly, mTOR kinase inhibitors seem to be less myelosuppressive and immunosuppressive than rapalogs at doses with anti-leukemic activity (Janes et al. 2010).

The first reports of mTOR kinase inhibitor efficacy in leukemia focused on Philadelphia Chromosome-positive leukemias (Carayol et al. 2010; Janes et al. 2010). Initial studies indicate that mTOR kinase inhibitors are also effective in preclinical AML models (Altman et al. 2011a; Willems et al. 2011). Treatment of AML cell lines and blast samples with the mTOR kinase inhibitor AZD8055 inhibited both mTORC1 and mTORC2, accompanied by a significant decrease in AML blast proliferation and cell cycle progression (Willems et al. 2011). AZD8055 reduced clonogenic capacity of primary leukemic progenitors and induced caspase-dependent apoptosis in AML cells but not in normal $CD34^{+}$cells. Further, AZD8055 markedly increased the survival of AML transplanted mice through a significant reduction of tumor growth, without apparent toxicity. A second mTOR kinase inhibitor, OSI-127, was also effective at inhibiting downstream signaling events for both mTOR complexes in AML cell lines and blasts (Altman et al. 2011a). OSI-027 diminished the proliferation of several AML cell lines as well as leukemic progenitors from AML patients and enhanced the antiproliferative effects of Ara-C *in vitro*. Death analysis was not reported in the latter study.

Several mTOR kinase inhibitors have entered clinical trials in oncology. Further preclinical testing of their efficacy and toxicity in single or combined therapy will help inform the design of clinical trials for AML patients.

17.4.4 Dual PI3K/mTOR Inhibitors

The new mTOR kinase inhibitors exhibit superior antileukemic properties relative to rapalogs. However, the negative feedback loops between mTORC1/S6K and growth factor receptors might limit the efficacy of this class of compounds (Chandarlapaty et al. 2011; Rodrik-Outmezguine et al. 2011). The "rebound" increase in growth factor signaling can stimulate PI3K and maintain AKT activity through elevated phosphorylation of Thr308. To mitigate the rebound effect, less selective adenosine triphosphate (ATP)-competitive inhibitors that target both PI3K and mTOR have been designed. A prototype panPI3K/mTOR inhibitor, PI-103, is a pyridonylfuranopyrimidine that represses the activity of class IA and IB PI3Ks, as well as both mTOR complexes (Raynaud et al. 2007). PI-103 alone displayed limited proapoptotic activity but acted synergistically with Nutlin-3, an MDM2 inhibitor, to induce apoptosis in a p53-dependent manner in AML cell lines and blasts (Kojima et al. 2008). Another group demonstrated that PI-103 was mainly cytostatic for AML cell lines (Park et al. 2008). However, in AML blast cells, PI-103 inhibited leukemic proliferation and clonogenicity, induced mitochondrial apoptosis, and synergized with etoposide.

A number of pharmacologically optimized panPI3K/mTOR inhibitors have entered clinical trials; these include NVP-BEZ235, GDC-0980, and XL765 (Zaytseva et al. 2012). NVP-BEZ235 reduced the proliferation rate in leukemic progenitors and AML blast cells and promoted apoptosis without affecting normal CD34$^+$ survival (Chapuis et al. 2010b). Whether panPI3K/mTOR inhibitors provide improved efficacy in AML patients compared to selective mTOR kinase inhibitors, without increased toxicity, will be interesting to investigate.

17.4.5 PDK1 Inhibitors

PDK1 phosphorylates AKT on Thr308 and is responsible for the non-PI3K-dependent phosphorylation of other AGC family kinases. Some of those kinases, including PKC, SGK and S6K1, promote cancer progression. Thus, PDK1 inhibitors might have added value as cancer therapies compared to AKT inhibitors. The staurosporine derivative UCN-01/KRX0601 (7-hydroxy-staurosporine) inhibits PDK1 activity (Komander et al. 2003; Sampath et al. 2006; Sato et al. 2002), but UCN-01/KRX0601 also nonspecifically targets protein kinase C and CHK1 (Sampath et al. 2006). *In vitro*, UCN-01 cooperates with cytarabine or a farnesyltransferase inhibitor, L744832, to decrease AKT phosphorylation and induce apoptosis in primary multiple myeloma and/or AML cell lines as well as blasts (Dai et al. 2005; Pei et al. 2005; Sampath et al. 2006).

17.4.6 Multi-Targeted Inhibitors

Drug discovery programs often focus on developing highly selective inhibitors of specific validated targets. This can limit toxicity but can also reduce efficacy. "Polypharmacology" is another attractive strategy that aims to target multiple oncogenic proteins in cancer cells (Apsel et al. 2008). One such compound, KP372-1, directly inhibits FLT3, AKT, and PDK-1 in AML cell lines and blast cells (Zeng et al. 2006). Inhibition led to mitochondrial dysfunction and apoptosis of AML cell lines and blast samples (irrespective of FLT3 status) but not normal immature CD34$^+$ cells.

A dual PI3K/PDK-1 inhibitor, BAG956, was shown to reduce AKT activation in FLT3 and non-FLT3 overexpressing mouse cells (Weisberg et al. 2008). BAG956 also potentiated the antiproliferative effects of tyrosine kinase inhibitors against FLT3-ITD-overexpressing mouse cells. Importantly, BAG956 attenuated the clonogenic capacity of human AML bone marrow cells but had only a modest effect on normal human bone marrow cells.

Conclusion

Activation of the PI3K/AKT/mTOR network is now considered a hallmark of cancer cells that promotes survival and proliferation through multiple mechanisms (Hanahan and Weinberg 2011). Consistent evidence from AML cell lines and patient AML blasts clearly shows that the PI3K/AKT/mTOR network is an oncogenic driver in AML and that pharmacological targeting of the network has demonstrable antileukemic effects. Nevertheless, AML is a heterogeneous disease with multiple oncogenic mechanisms that vary among patients. As in other hematological malignancies, inhibition of a single network is not likely to provide a uniformly strong cytotoxic effect. Further studies are required to determine the most effective combinations to eradicate AML blasts and leukemia initiating cells while minimizing toxicity to normal tissue. Biomarker analysis can aid in determining which patients will benefit from a particular inhibitor combination. Genomic and proteomic strategies will help to identify patients most likely to benefit from PI3K/AKT/mTOR network suppression and which companion therapies will unleash the greatest antileukemic response.

References

Abraham RT (2009) Regulation of the mTOR signaling pathway: from laboratory bench to bedside and back again. F1000 Biol Rep 1:8. doi:10.3410/B1-8

Aggarwal BB (2004) Nuclear factor-kappaB: the enemy within. Cancer Cell 3:203–208. doi:10.1016/j.ccr.2004.09.003S1535610804002442 [pii]

Akgul C (2009) Mcl-1 is a potential therapeutic target in multiple types of cancer. Cell Mol Life Sci 8:1326–1336. doi:10.1007/s00018-008-8637-6

Altman JK, Sassano A, Kaur S, Glaser H, Kroczynska B, Redig AJ, Russo S, Barr S, Platanias LC (2011a) Dual mTORC2/mTORC1 targeting results in potent suppressive effects on acute myeloid leukemia (AML) progenitors. Clin Cancer Res 13:4378–4388. doi:10.1158/1078-0432.CCR-10-2285

Altman JK, Sassano A, Platanias LC (2011b) Targeting mTOR for the treatment of AML. New agents and new directions. Oncotarget 6:510–517

Apsel B, Blair JA, Gonzalez B, Nazif TM, Feldman ME, Aizenstein B, Hoffman R, Williams RL, Shokat KM, Knight ZA (2008) Targeted polypharmacology: discovery of dual inhibitors of tyrosine and phosphoinositide kinases. Nat Chem Biol 11:691–9. doi:nchembio.117 [pii]10.1038/nchembio.117

Ayala F, Dewar R, Kieran M, Kalluri R (2009) Contribution of bone microenvironment to leukemogenesis and leukemia progression. Leukemia 12:2233–2241. doi:10.1038/leu.2009.175

Barnett SF, Defeo-Jones D, Fu S, Hancock PJ, Haskell KM, Jones RE, Kahana JA, Kral AM, Leander K, Lee LL, Malinowski J, McAvoy EM, Nahas DD, Robinson RG, Huber HE (2005) Identification and characterization of pleckstrin-homology-domain-dependent and isoenzyme-specific Akt inhibitors. Biochem J Pt 2:399–408. doi:10.1042/BJ20041140

Basu S, Totty NF, Irwin MS, Sudol M, Downward J (2003) Akt phosphorylates the Yes-associated protein, YAP, to induce interaction with 14-3-3 and attenuation of p73-mediated apoptosis. Mol Cell 1:11–23. doi:S1097276502007761 [pii]

Beghini A, Bellini M, Magnani I, Colapietro P, Cairoli R, Morra E, Larizza L (2005) STI 571 inhibition effect on KITAsn822Lys-mediated signal transduction cascade. Exp Hematol 6:682–688. doi:S0301-472X(05)00131-1 [pii] 10.1016/j.exphem.2005.03.007

Bhagwat SV, Crew AP (2010) Novel inhibitors of mTORC1 and mTORC2. Curr Opin Investig Drugs 6:638–645

Bilancio A, Okkenhaug K, Camps M, Emery JL, Ruckle T, Rommel C, Vanhaesebroeck B (2006) Key role of the p110delta isoform of PI3K in B-cell antigen and IL-4 receptor signaling: comparative analysis of genetic and pharmacologic interference with p110delta function in B cells. Blood 2:642–650

Billottet C, Grandage VL, Gale RE, Quattropani A, Rommel C, Vanhaesebroeck B, Khwaja A (2006) A selective inhibitor of the p110delta isoform of PI 3-kinase inhibits AML cell proliferation and survival and increases the cytotoxic effects of VP16. Oncogene 50:6648–6659. doi:10.1038/sj.onc.1209670

Billottet C, Banerjee L, Vanhaesebroeck B, Khwaja A (2009) Inhibition of class I phosphoinositide 3-kinase activity impairs proliferation and triggers apoptosis in acute promyelocytic leukemia without affecting atra-induced differentiation. Cancer Res 3:1027–1036. doi:0008-5472.CAN-08-2608 [pii] 10.1158/0008-5472.CAN-08-2608

Birkenkamp KU, Geugien M, Schepers H, Westra J, Lemmink HH, Vellenga E (2004) Constitutive NF-kappaB DNA-binding activity in AML is frequently mediated by a Ras/PI3-K/PKB-dependent pathway. Leukemia 1:103–112. doi:10.1038/sj.leu.2403145

van Blitterswijk WJ, Verheij M (2008) Anticancer alkylphospholipids: mechanisms of action, cellular sensitivity and resistance, and clinical prospects. Curr Pharm Des 21:2061–2074

Bortul R, Tazzari PL, Cappellini A, Tabellini G, Billi AM, Bareggi R, Manzoli L, Cocco L, Martelli AM (2003) Constitutively active Akt1 protects HL60 leukemia cells from TRAIL-induced apoptosis through a mechanism involving NF-kappaB activation and cFLIP(L) up-regulation. Leukemia 2:379–389. doi:10.1038/sj.leu.24027932402793 [pii]

Bortul R, Tazzari PL, Billi AM, Tabellini G, Mantovani I, Cappellini A, Grafone T, Martinelli G, Conte R, Martelli AM (2005) Deguelin, A PI3K/AKT inhibitor, enhances chemosensitivity of leukaemia cells with an active PI3K/AKT pathway. Br J Haematol 5:677–686. doi:BJH5504 [pii] 10.1111/j.1365-2141.2005.05504.x

Bousquet M, Recher C, Queleen C, Demur C, Payrastre B, Brousset P (2005) Assessment of somatic mutations in phosphatidylinositol 3-kinase gene in human lymphoma and acute leukaemia. Br J Haematol 3:411–413. doi:BJH5784 [pii] 10.1111/j.1365-2141.2005.05784.x

Brandts CH, Sargin B, Rode M, Biermann C, Lindtner B, Schwable J, Buerger H, Muller-Tidow C, Choudhary C, McMahon M, Berdel WE, Serve H (2005) Constitutive activation of Akt by Flt3 internal tandem duplications is necessary for increased survival, proliferation, and myeloid transformation. Cancer Res 21:9643–9650. doi:10.1158/0008-5472.CAN-05-0422

Calnan DR, Brunet A (2008) The FoxO code. Oncogene 16:2276–2288. doi:10.1038/onc.2008.21

Cao Y, Arbiser J, D'Amato RJ, D'Amore PA, Ingber DE, Kerbel R, Klagsbrun M, Lim S, Moses MA, Zetter B, Dvorak H, Langer R (2011) Forty-year journey of angiogenesis translational research. Sci Transl Med 114:114rv113. doi:10.1126/scitranslmed.3003149

Cappellini A, Tabellini G, Zweyer M, Bortul R, Tazzari PL, Billi AM, Fala F, Cocco L, Martelli AM (2003) The phosphoinositide 3-kinase/Akt pathway regulates cell cycle progression of HL60 human leukemia cells through cytoplasmic relocalization of the cyclin-dependent kinase inhibitor p27(Kip1) and control of cyclin D1 expression. Leukemia 11:2157–2167. doi:10.1038/sj.leu.24031112403111 [pii]

Carayol N, Vakana E, Sassano A, Kaur S, Goussetis DJ, Glaser H, Druker BJ, Donato NJ, Altman JK, Barr S, Platanias LC (2010) Critical roles for mTORC2- and rapamycin-insensitive mTORC1-complexes in growth and survival of BCR-ABL-expressing leukemic cells. Proc Natl Acad Sci U S A 28:12469–12474. doi:1005114107 [pii] 10.1073/pnas.1005114107

Carroll M (2009) Taking aim at protein translation in AML. Blood 8:1458–1459. doi:10.1182/blood-2009-06-224220

Castillo SS, Brognard J, Petukhov PA, Zhang C, Tsurutani J, Granville CA, Li M, Jung M, West KA, Gills JG, Kozikowski AP, Dennis PA (2004) Preferential inhibition of Akt and killing of Akt-dependent cancer cells by rationally designed phosphatidylinositol ether lipid analogues. Cancer Res 8:2782–2792

Chan EY (2009) mTORC1 phosphorylates the ULK1-mAtg13-FIP200 autophagy regulatory complex. Sci Signal 2(84):pe51. doi:10.1126/scisignal.284pe51

Chandarlapaty S, Sawai A, Scaltriti M, Rodrik-Outmezguine V, Grbovic-Huezo O, Serra V, Majumder PK, Baselga J, Rosen N (2011) AKT inhibition relieves feedback suppression of receptor tyrosine kinase expression and activity. Cancer Cell 1:58–71. doi:10.1016/j.ccr.2010.10.031

Chapuis N, Tamburini J, Cornillet-Lefebvre P, Gillot L, Bardet V, Willems L, Park S, Green AS, Ifrah N, Dreyfus F, Mayeux P, Lacombe C, Bouscary D (2010a) Autocrine IGF-1/IGF-1R signaling is responsible for constitutive PI3K/Akt activation in acute myeloid leukemia: therapeutic value of neutralizing anti-IGF-1R antibody. Haematologica 3:415–423. doi:10.3324/haematol.2009.010785

Chapuis N, Tamburini J, Green AS, Vignon C, Bardet V, Neyret A, Pannetier M, Willems L, Park S, Macone A, Maira SM, Ifrah N, Dreyfus F, Herault O, Lacombe C, Mayeux P, Bouscary D (2010b) Dual inhibition of PI3K and mTORC1/2 signaling by NVP-BEZ235 as a new therapeutic strategy for acute myeloid leukemia. Clin Cancer Res 22:5424–5435. doi:10.1158/1078-0432.CCR-10-1102

Chapuis N, Tamburini J, Green AS, Willems L, Bardet V, Park S, Lacombe C, Mayeux P, Bouscary D (2010c) Perspectives on inhibiting mTOR as a future treatment strategy for hematological malignancies. Leukemia 10:1686–1699. doi:10.1038/leu.2010.170

Chen W, Drakos E, Grammatikakis I, Schlette EJ, Li J, Leventaki V, Staikou-Drakopoulou E, Patsouris E, Panayiotidis P, Medeiros LJ, Rassidakis GZ (2010) mTOR signaling is activated by FLT3 kinase and promotes survival of FLT3-mutated acute myeloid leukemia cells. Mol Cancer 9:292. doi:1476-4598-9-292 [pii] 10.1186/1476-4598-9-292

Cheong JW, Eom JI, Maeng HY, Lee ST, Hahn JS, Ko YW, Min YH (2003a) Constitutive phosphorylation of FKHR transcription factor as a prognostic variable in acute myeloid leukemia. Leuk Res 2712:1159–1162

Cheong JW, Eom JI, Maeng HY, Lee ST, Hahn JS, Ko YW, Min YH (2003b) Phosphatase and tensin homologue phosphorylation in the C-terminal regulatory domain is frequently observed in acute myeloid leukaemia and associated with poor clinical outcome. Br J Haematol 1223:454–456

Cheong JW, Min YH, Eom JI, Kim SJ, Jeung HK, Kim JS (2010) Inhibition of CK2{alpha} and PI3K/Akt synergistically induces apoptosis of CD34^{+} CD38^{-} leukaemia cells while sparing haematopoietic stem cells. Anticancer Res 11:4625–4634

Cheung LW, Hennessy BT, Li J, Yu S, Myers AP, Djordjevic B, Lu Y, Stemke-Hale K, Zhang F, Ju Z, Cantley LC, Scherer SE, Liang H, Lu KH, Broaddus RR, Mills GB (2011) High frequency of PIK3R1 and PIK3R2 mutations in endometrial cancer elucidates a novel mechanism for regulation of PTEN protein stability. Cancer Discov 2:170–185. doi:10.1158/2159-8290.CD-11-0039

Chiarini F, Del Sole M, Mongiorgi S, Gaboardi GC, Cappellini A, Mantovani I, Follo MY, McCubrey JA, Martelli AM (2008) The novel Akt inhibitor, perifosine, induces caspase-dependent apoptosis and downregulates P-glycoprotein expression in multidrug-resistant human T-acute leukemia cells by a JNK-dependent mechanism. Leukemia 6:1106–1116. doi:10.1038/leu.2008.79

Choudhary C, Schwable J, Brandts C, Tickenbrock L, Sargin B, Kindler T, Fischer T, Berdel WE, Muller-Tidow C, Serve H (2005) AML-associated Flt3 kinase domain mutations show signal transduction differences compared with Flt3 ITD mutations. Blood 1:265–273. doi:2004-07-2942 [pii] 10.1182/blood-2004-07-2942

Chow S, Minden MD, Hedley DW (2006) Constitutive phosphorylation of the S6 ribosomal protein via mTOR and ERK signaling in the peripheral blasts of acute leukemia patients. Exp Hematol 9:1183–1191. doi:10.1016/j.exphem.2006.05.002

Chresta CM, Davies BR, Hickson I, Harding T, Cosulich S, Critchlow SE, Vincent JP, Ellston R, Jones D, Sini P, James D, Howard Z, Dudley P, Hughes G, Smith L, Maguire S, Hummersone M, Malagu K, Menear K, Jenkins R, Jacobsen M, Smith GC, Guichard S, Pass M (2010) AZD8055 is a potent, selective, and orally bioavailable ATP-competitive mammalian target of rapamycin kinase inhibitor with in vitro and in vivo antitumor activity. Cancer Res 1:288–298. doi:10.1158/0008-5472.CAN-09-1751

Chu SH, Small D (2009) Mechanisms of resistance to FLT3 inhibitors. Drug Resist Updat 1–2:8–16. doi:S1368-7646(08)00080-0 [pii] 10.1016/j.drup.2008.12.001

Coloff JL, Macintyre AN, Nichols AG, Liu T, Gallo CA, Plas DR, Rathmell JC (2011) Akt-dependent glucose metabolism promotes mcl-1 synthesis to maintain cell survival and resistance to Bcl-2 inhibition. Cancer Res 15:5204–5213. doi:10.1158/0008-5472.CAN-10-4531

Cybulski N, Hall MN (2009) TOR complex 2: a signaling pathway of its own. Trends Biochem Sci 12:620–627. doi:10.1016/j.tibs.2009.09.004

Dai Y, Rahmani M, Pei XY, Khanna P, Han SI, Mitchell C, Dent P, Grant S (2005) Farnesyltransferase inhibitors interact synergistically with the Chk1 inhibitor UCN-01 to induce apoptosis in human leukemia cells through interruption of both Akt and MEK/ERK pathways and activation of SEK1/JNK. Blood 4:1706–1716. doi:10.1182/blood-2004-07-2767

Dazert E, Hall MN (2011) mTOR signaling in disease. Curr Opin Cell Biol 23(6):744–755. doi:10.1016/j.ceb.2011.09.003

DeFeo-Jones D, Barnett SF, Fu S, Hancock PJ, Haskell KM, Leander KR, McAvoy E, Robinson RG, Duggan ME, Lindsley CW, Zhao Z, Huber HE, Jones RE (2005) Tumor cell sensitization to apoptotic stimuli by selective inhibition of specific Akt/PKB family members. Mol Cancer Ther 2:271–9

Doepfner KT, Spertini O, Arcaro A (2007) Autocrine insulin-like growth factor-I signaling promotes growth and survival of human acute myeloid leukemia cells via the phosphoinositide 3-kinase/Akt pathway. Leukemia 9:1921–1930. doi:10.1038/sj.leu.2404813

Dos Santos C, Demur C, Bardet V, Prade-Houdellier N, Payrastre B, Recher C (2008) A critical role for Lyn in acute myeloid leukemia. Blood 4:2269–2279. doi:10.1182/blood-2007-04-082099

Engelman JA (2009) Targeting PI3K signalling in cancer: opportunities, challenges and limitations. Nat Rev Cancer 8:550–562. doi:10.1038/nrc2664

Engelman JA, Luo J, Cantley LC (2006) The evolution of phosphatidylinositol 3-kinases as regulators of growth and metabolism. Nat Rev Genet 8:606–619. doi:10.1038/nrg1879

Fayard E, Xue G, Parcellier A, Bozulic L, Hemmings BA (2010) Protein kinase B (PKB/Akt), a key mediator of the PI3K signaling pathway. Curr Top Microbiol Immunol 346:31–56. doi:10.1007/82_2010_58

Feldman ME, Shokat KM (2010) New inhibitors of the PI3K-Akt-mTOR pathway: insights into mTOR signaling from a new generation of Tor Kinase Domain Inhibitors (TORKinibs). Curr Top Microbiol Immunol 347:241–262. doi:10.1007/82_2010_64

Feldman ME, Apsel B, Uotila A, Loewith R, Knight ZA, Ruggero D, Shokat KM (2009) Active-site inhibitors of mTOR target rapamycin-resistant outputs of mTORC1 and mTORC2. PLoS Biol 2:e38. doi:10.1371/journal.pbio.1000038

Fiegl M, Samudio I, Clise-Dwyer K, Burks JK, Mnjoyan Z, Andreeff M (2009) CXCR4 expression and biologic activity in acute myeloid leukemia are dependent on oxygen partial pressure. Blood 7:1504–1512. doi:10.1182/blood-2008-06-161539

Fierro FA, Brenner S, Oelschlaegel U, Jacobi A, Knoth H, Ehninger G, Illmer T, Bornhauser M (2009) Combining SDF-1/CXCR4 antagonism and chemotherapy in relapsed acute myeloid leukemia. Leukemia 2:393–6. doi:10.1038/leu.2008.182

Frelin C, Imbert V, Griessinger E, Peyron AC, Rochet N, Philip P, Dageville C, Sirvent A, Hummelsberger M, Berard E, Dreano M, Sirvent N, Peyron JF (2005) Targeting NF-kappaB activation via pharmacologic inhibition of IKK2-induced apoptosis of human acute myeloid leukemia cells. Blood 2:804–811. doi:10.1182/blood-2004-04-14632004-04-1463 [pii]

Frohling S, Scholl C, Gilliland DG, Levine RL (2005) Genetics of myeloid malignancies: pathogenetic and clinical implications. J Clin Oncol 26:6285–6295. doi:23/26/6285 [pii] 10.1200/JCO.2005.05.010

Fruman DA, Rommel C (2011) PI3Kδ inhibitors in cancer: rationale and serendipity merge in the clinic. Cancer Discov 1: 562–572

Fu Z, Tindall DJ (2008) FOXOs, cancer and regulation of apoptosis. Oncogene 16:2312–9. doi:10.1038/onc.2008.24

Fu L, Kim YA, Wang X, Wu X, Yue P, Lonial S, Khuri FR, Sun SY (2009) Perifosine inhibits mammalian target of rapamycin signaling through facilitating degradation of major components in the mTOR axis and induces autophagy. Cancer Res 23:8967–8976. doi:10.1158/0008-5472.CAN-09-2190

Gallay N, Dos Santos C, Cuzin L, Bousquet M, Simmonet Gouy V, Chaussade C, Attal M, Payrastre B, Demur C, Recher C (2009) The level of AKT phosphorylation on threonine 308 but not on serine 473 is associated with high-risk cytogenetics and predicts poor overall survival in acute myeloid leukaemia. Leukemia 6:1029–1038. doi:10.1038/leu.2008.395

Gao N, Budhraja A, Cheng S, Liu EH, Chen J, Yang Z, Chen D, Zhang Z, Shi X (2011) Phenethyl isothiocyanate exhibits antileukemic activity in vitro and in vivo by inactivation of Akt and activation of JNK pathways. Cell Death Dis 2:e140. doi:10.1038/cddis.2011.22

Gao N, Cheng S, Budhraja A, Gao Z, Chen J, Liu EH, Huang C, Chen D, Yang Z, Liu Q, Li P, Shi X, Zhang Z (2012a) Ursolic acid induces apoptosis in human leukaemia cells and exhibits anti-leukaemic activity in nude mice through the PKB pathway. Br J Pharmacol 6:1813–1826. doi:10.1111/j.1476-5381.2011.01684.x

Gao N, Cheng S, Budhraja A, Liu EH, Chen J, Chen D, Yang Z, Luo J, Shi X, Zhang Z (2012b) 3,3′-Diindolylmethane exhibits antileukemic activity in vitro and in vivo through a Akt-dependent process. PloS One 2:e31783. doi:10.1371/journal.pone.0031783

Garcia-Echeverria C, Sellers WR (2008) Drug discovery approaches targeting the PI3K/Akt pathway in cancer. Oncogene 27 (41):5511–5526. doi:10.1038/onc.2008.246

Gharbi SI, Zvelebil MJ, Shuttleworth SJ, Hancox T, Saghir N, Timms JF, Waterfield MD (2007) Exploring the specificity of the PI3K family inhibitor LY294002. Biochem J 1:15–21. doi:10.1042/BJ20061489

Glaser SP, Lee EF, Trounson E, Bouillet P, Wei A, Fairlie WD, Izon DJ, Zuber J, Rappaport AR, Herold MJ, Alexander WS, Lowe SW, Robb L, Strasser A (2012) Anti-apoptotic Mcl-1 is essential for the development and sustained growth of acute myeloid leukemia. Genes Dev 2:120–5. doi:26/2/120 [pii] 10.1101/gad.182980.111

Grandage VL, Gale RE, Linch DC, Khwaja A (2005) PI3-kinase/Akt is constitutively active in primary acute myeloid leukaemia cells and regulates survival and chemoresistance via NF-kappaB, Mapkinase and p53 pathways. Leukemia 4:586–594. doi:10.1038/sj.leu.2403653

Grimberg A (2003) Mechanisms by which IGF-I may promote cancer. Cancer Biol Ther 6:630–5. doi:678 [pii]

Guertin DA, Sabatini DM (2009) The pharmacology of mTOR inhibition. Sci Signal 67:pe24. doi:10.1126/scisignal.267pe24

Guertin DA, Stevens DM, Saitoh M, Kinkel S, Crosby K, Sheen JH, Mullholland DJ, Magnuson MA, Wu H, Sabatini DM (2009) mTOR complex 2 is required for the development of prostate cancer induced by Pten loss in mice. Cancer Cell 2:148–159

Gutierrez NC, Lopez-Perez R, Hernandez JM, Isidro I, Gonzalez B, Delgado M, Ferminan E, Garcia JL, Vazquez L, Gonzalez M, San Miguel JF (2005) Gene expression profile reveals deregulation of genes with relevant functions in the different subclasses of acute myeloid leukemia. Leukemia 3:402–9. doi:2403625 [pii] 10.1038/sj.leu.2403625

Gutierrez A, Sanda T, Grebliunaite R, Carracedo A, Salmena L, Ahn Y, Dahlberg S, Neuberg D, Moreau LA, Winter SS, Larson R, Zhang J, Protopopov A, Chin L, Pandolfi PP, Silverman LB, Hunger SP, Sallan SE, Look AT (2009) High frequency of PTEN, PI3K, and AKT abnormalities in T-cell acute lymphoblastic leukemia. Blood 3:647–650. doi:10.1182/blood-2009-02-206722

Hagner PR, Schneider A, Gartenhaus RB (2010) Targeting the translational machinery as a novel treatment strategy for hematologic malignancies. Blood 115(11):2127–2135. doi:10.1182/blood-2009-09-220020

Hahn M, Li W, Yu C, Rahmani M, Dent P, Grant S (2005) Rapamycin and UCN-01 synergistically induce apoptosis in human leukemia cells through a process that is regulated by the Raf-1/MEK/ERK, Akt, and JNK signal transduction pathways. Mol Cancer Ther 3:457–470. doi:10.1158/1535-7163.MCT-04-0137

Hanahan D, Weinberg RA (2011) Hallmarks of cancer: the next generation. Cell 5:646–674. doi:10.1016/j.cell.2011.02.013

Hazen AL, Smith MJ, Desponts C, Winter O, Moser K, Kerr WG (2009) SHIP is required for a functional hematopoietic stem cell niche. Blood 13:2924–2933. doi:blood-2008-02-138008 [pii] 10.1182/blood-2008-02-138008

Hollander MC, Blumenthal GM, Dennis PA (2011) PTEN loss in the continuum of common cancers, rare syndromes and mouse models. Nat Rev Cancer 4:289–301. doi:10.1038/nrc3037

Ihle NT, Powis G (2009) Take your PIK: phosphatidylinositol 3-kinase inhibitors race through the clinic and toward cancer therapy. Mol Cancer Ther 1:1–9. doi:10.1158/1535-7163.MCT-08-0801

Jaiswal BS, Janakiraman V, Kljavin NM, Chaudhuri S, Stern HM, Wang W, Kan Z, Dbouk HA, Peters BA, Waring P, Dela Vega T, Kenski DM, Bowman KK, Lorenzo M, Li H, Wu J, Modrusan Z, Stinson J, Eby M, Yue P, Kaminker JS, de Sauvage FJ, Backer JM, Seshagiri S (2009) Somatic mutations in p85alpha promote tumorigenesis through class IA PI3K activation. Cancer Cell 6:463–474. doi:S1535-6108(09)00385-7 [pii] 10.1016/j.ccr.2009.10.016

Janes MR, Fruman DA (2009) Immune regulation by rapamycin: moving beyond T cells. Sci Signal 67:pe25. doi:10.1126/scisignal.267pe25

Janes MR, Fruman DA (2010) Targeting TOR dependence in cancer. Oncotarget 1:69–76

Janes MR, Limon JJ, So L, Chen J, Lim RJ, Chavez MA, Vu C, Lilly MB, Mallya S, Ong ST, Konopleva M, Martin MB, Ren P, Liu Y, Rommel C, Fruman DA (2010) Effective and selective targeting of leukemia cells using a TORC1/2 kinase inhibitor. Nat Med 2:205–213. doi:10.1038/nm.2091

Kindler T, Lipka DB, Fischer T (2010) FLT3 as a therapeutic target in AML: still challenging after all these years. Blood 24:5089–5102. doi:10.1182/blood-2010-04-261867

Kohl TM, Schnittger S, Ellwart JW, Hiddemann W, Spiekermann K (2005) KIT exon 8 mutations associated with core-binding factor (CBF)-acute myeloid leukemia (AML) cause hyperactivation of the receptor in response to stem cell factor. Blood 8:3319–3321. doi:2004-06-2068 [pii] 10.1182/blood-2004-06-2068

Kojima K, Shimanuki M, Shikami M, Samudio IJ, Ruvolo V, Corn P, Hanaoka N, Konopleva M, Andreeff M, Nakakuma H (2008) The dual PI3 kinase/mTOR inhibitor PI-103 prevents p53 induction by Mdm2 inhibition but enhances p53-mediated mitochondrial apoptosis in p53 wild-type AML. Leukemia 9:1728–1736. doi:10.1038/leu.2008.158

Komander D, Kular GS, Bain J, Elliott M, Alessi DR, Van Aalten DM (2003) Structural basis for UCN-01 (7-hydroxystaurosporine) specificity and PDK1 (3-phosphoinositide-dependent protein kinase-1) inhibition. Biochem J Pt 2:255–262. doi:10.1042/BJ20031119

Kondapaka SB, Singh SS, Dasmahapatra GP, Sausville EA, Roy KK (2003) Perifosine, a novel alkylphospholipid, inhibits protein kinase B activation. Mol Cancer Ther 11:1093–1103

Kornblau SM, Womble M, Cade JS, Lemker E, Qiu YH (2005) Comparative analysis of the effects of sample source and test methodology on the assessment of protein expression in acute myelogenous leukemia. Leukemia 9:1550–1557. doi:10.1038/sj.leu.2403845

Kornblau SM, Womble M, Qiu YH, Jackson CE, Chen W, Konopleva M, Estey EH, Andreeff M (2006) Simultaneous activation of multiple signal transduction pathways confers poor prognosis in acute myelogenous leukemia. Blood 7:2358–2365. doi:10.1182/blood-2006-02-003475

Kornblau SM, Tibes R, Qiu YH, Chen W, Kantarjian HM, Andreeff M, Coombes KR, Mills GB (2009) Functional proteomic profiling of AML predicts response and survival. Blood 1:154–164. doi: blood-2007-10-119438 [pii] 10.1182/blood-2007-10-119438

Kornblau SM, Minden MD, Rosen DB, Putta S, Cohen A, Covey T, Spellmeyer DC, Fantl WJ, Gayko U, Cesano A (2010) Dynamic single-cell network profiles in acute myelogenous leukemia are associated with patient response to standard induction therapy. Clin Cancer Res 14:3721–3733. doi:1078-0432.CCR-10-0093 [pii] 10.1158/1078-0432.CCR-10-0093

Laplante M, Sabatini DM (2012) mTOR signaling in growth control and disease. Cell 1492:274–293. doi:10.1016/j.cell.2012.03.017

Larizza L, Magnani I, Beghini A (2005) The Kasumi -1 cell line: a t(8;21)-kit mutant model for acute myeloid leukemia. Leuk Lymphoma 2:247–255. doi:UNFPV047Y73GH23Q[pii]10.1080/10428190400007565

Lennartsson J, Ronnstrand L (2006) The stem cell factor receptor/c-Kit as a drug target in cancer. Curr Cancer Drug Targets 1:65–75

List AF, Glinsmann-Gibson B, Stadheim C, Meuillet EJ, Bellamy W, Powis G (2004) Vascular endothelial growth factor receptor-1 and receptor-2 initiate a phosphatidylinositide 3-kinase-dependent clonogenic response in acute myeloid leukemia cells. Exp Hematol 6:526–535. doi:10.1016/j.exphem.2004.03.005

Lopez-Fauqued M, Gil R, Grueso J, Hernandez-Losa J, Pujol A, Moline T, Recio JA (2010) The dual PI3K/mTOR inhibitor PI-103 promotes immunosuppression, in vivo tumor growth and increases survival of sorafenib-treated melanoma cells. Int J Cancer 7:1549–1561. doi:10.1002/ijc.24926

Luo JM, Yoshida H, Komura S, Ohishi N, Pan L, Shigeno K, Hanamura I, Miura K, Iida S, Ueda R, Naoe T, Akao Y, Ohno R, Ohnishi K (2003) Possible dominant-negative mutation of the SHIP gene in acute myeloid leukemia. Leukemia 1:1–8. doi:10.1038/sj.leu.2402725

Ma P, Mali RS, Martin H, Ramdas B, Sims E, Kapur R (2012) Role of intracellular tyrosines in activating KIT-induced myeloproliferative disease. Leukemia Epub 26(7):1499–1506. Feb 2. doi:10.1038/leu.2012.22

Maira SM, Stauffer F, Brueggen J, Furet P, Schnell C, Fritsch C, Brachmann S, Chene P, De Pover A, Schoemaker K, Fabbro D, Gabriel D, Simonen M, Murphy L, Finan P, Sellers W, Garcia-Echeverria C (2008) Identification and characterization of NVP-BEZ235, a new orally available dual phosphatidylinositol 3-kinase/mammalian target of rapamycin inhibitor with potent in vivo antitumor activity. Mol Cancer Ther 7:1851–1863. doi:10.1158/1535-7163.MCT-08-0017

Manning BD, Cantley LC (2007) AKT/PKB signaling: navigating downstream. Cell 1297:1261–1274. doi:10.1016/j.cell.2007.06.009

Martelli AM, Tazzari PL, Tabellini G, Bortul R, Billi AM, Manzoli L, Ruggeri A, Conte R, Cocco L (2003) A new selective AKT pharmacological inhibitor reduces resistance to chemotherapeutic drugs, TRAIL, all-*trans*-retinoic acid, and ionizing radiation of human leukemia cells. Leukemia 9:1794–1805. doi:10.1038/sj.leu.2403044

Martelli AM, Evangelisti C, Chiarini F, McCubrey JA (2010) The phosphatidylinositol 3-kinase/Akt/mTOR signaling network as a therapeutic target in AML patients. Oncotarget 2:89–103

Matsunaga T, Takemoto N, Sato T, Takimoto R, Tanaka I, Fujimi A, Akiyama T, Kuroda H, Kawano Y, Kobune M, Kato J, Hirayama Y, Sakamaki S, Kohda K, Miyake K, Niitsu Y (2003) Interaction between leukemic-cell VLA-4 and stromal fibronectin is a decisive factor for minimal residual disease of acute myelogenous leukemia. Nat Med 9:1158–1165. doi:10.1038/nm909

Matsunaga T, Fukai F, Miura S, Nakane Y, Owaki T, Kodama H, Tanaka M, Nagaya T, Takimoto R, Takayama T, Niitsu Y (2008) Combination therapy of an anticancer drug with the FNIII14 peptide of fibronectin effectively overcomes cell adhesion-mediated drug resistance of acute myelogenous leukemia. Leukemia 2:353–360. doi:10.1038/sj.leu.2405017

Mayo LD, Donner DB (2002) The PTEN, Mdm2, p53 tumor suppressor-oncoprotein network. Trends Biochem Sci 27 (9):462–467. doi:S0968000402021667 [pii]

McDonald PC, Oloumi A, Mills J, Dobreva I, Maidan M, Gray V, Wederell ED, Bally MB, Foster LJ, Dedhar S (2008) Rictor and integrin-linked kinase interact and regulate Akt phosphorylation and cancer cell survival. Cancer Res 6:1618–1624. doi:10.1158/0008-5472.CAN-07-5869

McMillin DW, Ooi M, Delmore J, Negri J, Hayden P, Mitsiades N, Jakubikova J, Maira SM, Garcia-Echeverria C, Schlossman R, Munshi NC, Richardson PG, Anderson KC, Mitsiades CS (2009) Antimyeloma activity of the orally bioavailable dual phosphatidylinositol 3-kinase/mammalian target of rapamycin inhibitor NVP-BEZ235. Cancer Res 14:5835–5842. doi:10.1158/0008-5472.CAN-08-4285

Medema RH, Kops GJ, Bos JL, Burgering BM (2000) AFX-like Forkhead transcription factors mediate cell-cycle regulation by Ras and PKB through p27kip1. Nature 6779:782–787. doi:10.1038/35008115

Mendoza MC, Er EE, Blenis J (2011) The Ras-ERK and PI3K-mTOR pathways: cross-talk and compensation. Trends Biochem Sci 6:320–328. doi:10.1016/j.tibs.2011.03.006

Min YH, Eom JI, Cheong JW, Maeng HO, Kim JY, Jeung HK, Lee ST, Lee MH, Hahn JS, Ko YW (2003) Constitutive phosphorylation of Akt/PKB protein in acute myeloid leukemia: its significance as a prognostic variable. Leukemia 5:995–997. doi:10.1038/sj.leu.2402874

Min YH, Cheong JW, Kim JY, Eom JI, Lee ST, Hahn JS, Ko YW, Lee MH (2004) Cytoplasmic mislocalization of p27Kip1 protein is associated with constitutive phosphorylation of Akt or protein kinase B and poor prognosis in acute myelogenous leukemia. Cancer Res 15:5225–5231. doi:10.1158/0008-5472.CAN-04-0174

Minami Y, Yamamoto K, Kiyoi H, Ueda R, Saito H, Naoe T (2003) Different antiapoptotic pathways between wild-type and mutated FLT3: insights into therapeutic targets in leukemia. Blood 8:2969–2975. doi:10.1182/blood-2002-12-38132002-12-3813 [pii]

Mitsiades CS, Mitsiades NS, McMullan CJ, Poulaki V, Shringarpure R, Akiyama M, Hideshima T, Chauhan D, Joseph M, Libermann TA, Garcia-Echeverria C, Pearson MA, Hofmann F, Anderson KC, Kung AL (2004) Inhibition of the insulin-like growth factor receptor-1 tyrosine kinase activity as a therapeutic strategy for multiple myeloma, other hematologic malignancies, and solid tumors. Cancer Cell 3:221–230

Munugalavadla V, Sims EC, Borneo J, Chan RJ, Kapur R (2007) Genetic and pharmacologic evidence implicating the p85 alpha, but not p85 beta, regulatory subunit of PI3K and Rac2 GTPase in regulating oncogenic KIT-induced transformation in acute myeloid leukemia and systemic mastocytosis. Blood 5:1612–1620. doi: blood-2006-10-053058[pii]10.1182/blood-2006-10-053058

Muranyi AL, Dedhar S, Hogge DE (2009) Combined inhibition of integrin linked kinase and FMS-like tyrosine kinase 3 is cytotoxic to acute myeloid leukemia progenitor cells. Exp Hematol 4:450–460. doi: 10.1016/j.exphem.2009.01.002

Nardella C, Carracedo A, Alimonti A, Hobbs RM, Clohessy JG, Chen Z, Egia A, Fornari A, Fiorentino M, Loda M, Kozma SC, Thomas G, Cordon-Cardo C, Pandolfi PP (2009) Differential requirement of mTOR in postmitotic tissues and tumorigenesis. Sci Signal 55:ra2. doi: 10.1126/scisignal.2000189

Neri LM, Borgatti P, Tazzari PL, Bortul R, Cappellini A, Tabellini G, Bellacosa A, Capitani S, Martelli AM (2003) The phosphoinositide 3-kinase/AKT1 pathway involvement in drug and all-*trans*-retinoic acid resistance of leukemia cells. Mol Cancer Res 3:234–246

Ning ZQ, Li J, McGuinness M, Arceci RJ (2001) STAT3 activation is required for Asp(816) mutant c-Kit induced tumorigenicity. Oncogene 33:4528–4536. doi: 10.1038/sj.onc.1204590

Nishioka C, Ikezoe T, Yang J, Koeffler HP, Yokoyama A (2008) Blockade of mTOR signaling potentiates the ability of histone deacetylase inhibitor to induce growth arrest and differentiation of acute myelogenous leukemia cells. Leukemia 12:2159–2168. doi: 10.1038/leu.2008.243

Nishioka C, Ikezoe T, Yang J, Gery S, Koeffler HP, Yokoyama A (2009) Inhibition of mammalian target of rapamycin signaling potentiates the effects of all-*trans* retinoic acid to induce growth arrest and differentiation of human acute myelogenous leukemia cells. Int J Cancer 7:1710–1720. doi: 10.1002/ijc.24472

Olesen LH, Aggerholm A, Andersen BL, Nyvold CG, Guldberg P, Norgaard JM, Hokland P (2005) Molecular typing of adult acute myeloid leukaemia: significance of translocations, tandem duplications, methylation, and selective gene expression profiling. Br J Haematol 4:457–467. doi: BJH5791 [pii] 10.1111/j.1365-2141.2005.05791.x

Ong CJ, Ming-Lum A, Nodwell M, Ghanipour A, Yang L, Williams DE, Kim J, Demirjian L, Qasimi P, Ruschmann J, Cao LP, Ma K, Chung SW, Duronio V, Andersen RJ, Krystal G, Mui AL (2007) Small-molecule agonists of SHIP1 inhibit the phosphoinositide 3-kinase pathway in hematopoietic cells. Blood 6:1942–9. doi:blood-2007-03-079699 [pii] 10.1182/blood-2007-03-079699

Papa V, Tazzari PL, Chiarini F, Cappellini A, Ricci F, Billi AM, Evangelisti C, Ottaviani E, Martinelli G, Testoni N, McCubrey JA, Martelli AM (2008) Proapoptotic activity and chemosensitizing effect of the novel Akt inhibitor perifosine in acute myelogenous leukemia cells. Leukemia 1:147–160. doi:10.1038/sj.leu.2404980

Park S, Chapuis N, Bardet V, Tamburini J, Gallay N, Willems L, Knight ZA, Shokat KM, Azar N, Viguie F, Ifrah N, Dreyfus F, Mayeux P, Lacombe C, Bouscary D (2008) PI-103, a dual inhibitor of Class IA phosphatidylinositide 3-kinase and mTOR, has antileukemic activity in AML. Leukemia 9:1698–1706. doi:10.1038/leu.2008.144

Park S, Chapuis N, Tamburini J, Bardet V, Cornillet-Lefebvre P, Willems L, Green A, Mayeux P, Lacombe C, Bouscary D (2010) Role of the PI3K/AKT and mTOR signaling pathways in acute myeloid leukemia. Haematologica 5:819–828. doi:10.3324/haematol.2009.013797

Pei XY, Dai Y, Rahmani M, Li W, Dent P, Grant S (2005) The farnesyltransferase inhibitor L744832 potentiates UCN-01-induced apoptosis in human multiple myeloma cells. Clin Cancer Res 12:4589–4600. doi: 10.1158/1078-0432.CCR-04-2346

Piloto O, Wright M, Brown P, Kim KT, Levis M, Small D (2007) Prolonged exposure to FLT3 inhibitors leads to resistance via activation of parallel signaling pathways. Blood 4:1643–1652. doi: blood-2006-05-023804 [pii]10.1182/blood-2006-05-023804

Pollak M (2012) The insulin and insulin-like growth factor receptor family in neoplasia: an update. Nat Rev Cancer 12(3):159–169. doi: 10.1038/nrc3215

Rahmani M, Reese E, Dai Y, Bauer C, Payne SG, Dent P, Spiegel S, Grant S (2005) Coadministration of histone deacetylase inhibitors and perifosine synergistically induces apoptosis in human leukemia cells through Akt and ERK1/2 inactivation and the generation of ceramide and reactive oxygen species. Cancer Res 6:2422–2432. doi: 10.1158/0008-5472.CAN-04-2440

Rahmani M, Anderson A, Habibi JR, Crabtree TR, Mayo M, Harada H, Ferreira-Gonzalez A, Dent P, Grant S (2009) The BH3-only protein Bim plays a critical role in leukemia cell death triggered by concomitant inhibition of the PI3K/Akt and MEK/ERK1/2 pathways. Blood 20:4507–4516. doi: 10.1182/blood-2008-09-177881

Ramaswamy S, Nakamura N, Sansal I, Bergeron L, Sellers WR (2002) A novel mechanism of gene regulation and tumor suppression by the transcription factor FKHR. Cancer Cell 1:81–91. doi: S1535610802000867 [pii]

Raynaud FI, Eccles S, Clarke PA, Hayes A, Nutley B, Alix S, Henley A, Di-Stefano F, Ahmad Z, Guillard S, Bjerke LM, Kelland L, Valenti M, Patterson L, Gowan S, de Haven Brandon A, Hayakawa M, Kaizawa H, Koizumi T, Ohishi T, Patel S, Saghir N, Parker P, Waterfield M, Workman P (2007) Pharmacologic characterization of a potent inhibitor of class I phosphatidylinositide 3-kinases. Cancer Res 12:5840–5850. doi: 10.1158/0008-5472.CAN-06-4615

Recher C, Beyne-Rauzy O, Demur C, Chicanne G, Dos Santos C, Mas VM, Benzaquen D, Laurent G, Huguet F, Payrastre B (2005) Antileukemic activity of rapamycin in acute myeloid leukemia. Blood 6:2527–2534. doi: 10.1182/blood-2004-06-2494

Ribatti D, Scavelli C, Roccaro AM, Crivellato E, Nico B, Vacca A (2004) Hematopoietic cancer and angiogenesis. Stem Cells Dev 5:484–495. doi: 10.1089/1547328042417354

Rizzo MG, Giombini E, Diverio D, Vignetti M, Sacchi A, Testa U, Lo-Coco F, Blandino G (2004) Analysis of p73 expression pattern in acute myeloid leukemias: lack of DeltaN-p73 expression is a frequent feature of acute promyelocytic leukemia. Leukemia 11:1804–9. doi: 10.1038/sj.leu.24034832403483 [pii]

Rodrik-Outmezguine VS, Cao N, Chandarlapaty S, She QB, Ye Q, Smith P, Guichard S, Rosen N (2011) AZD8055 is an effective inhibitor of mTOR kinase signaling and breast cancer growth while relieving feedback inhibition of HER-kinase signaling. Cancer Disc 1:248–259

Rosen DB, Minden MD, Kornblau SM, Cohen A, Gayko U, Putta S, Woronicz J, Evensen E, Fantl WJ, Cesano A (2010) Functional characterization of FLT3 receptor signaling deregulation in acute myeloid leukemia by single cell network profiling (SCNP). PloS One 5(10):e13543. doi:10.1371/journal.pone.0013543

Sadhu C, Masinovsky B, Dick K, Sowell CG, Staunton DE (2003) Essential role of phosphoinositide 3-kinase delta in neutrophil directional movement. J Immunol 5:2647–2654

Sampath D, Cortes J, Estrov Z, Du M, Shi Z, Andreeff M, Gandhi V, Plunkett W (2006) Pharmacodynamics of cytarabine alone and in combination with 7-hydroxystaurosporine (UCN-01) in AML blasts in vitro and during a clinical trial. Blood 6:2517–2524. doi: 10.1182/blood-2005-08-3351

Samuels Y, Wang Z, Bardelli A, Silliman N, Ptak J, Szabo S, Yan H, Gazdar A, Powell SM, Riggins GJ, Willson JK, Markowitz S, Kinzler KW, Vogelstein B, Velculescu VE (2004) High frequency of mutations of the PIK3CA gene in human cancers. Science 5670:554. doi:10.1126/science.10965021096502 [pii]

Santarpia L, Lippman SM, El-Naggar AK (2012) Targeting the MAPK-RAS-RAF signaling pathway in cancer therapy. Expert Opin Ther Targets 1:103–119. doi: 10.1517/14728222.2011.645805

Sato S, Fujita N, Tsuruo T (2002) Interference with PDK1-Akt survival signaling pathway by UCN-01 (7-hydroxystaurosporine). Oncogene 11:1727–1738. doi: 10.1038/sj.onc.1205225

Scavelli C, Vacca A, Di Pietro G, Dammacco F, Ribatti D (2004) Crosstalk between angiogenesis and lymphangiogenesis in tumor progression. Leukemia 6:1054–8. doi: 10.1038/sj.leu.24033552403355 [pii]

Schmidt M, Fernandez de Mattos S, van der Horst A, Klompmaker R, Kops GJ, Lam EW, Burgering BM, Medema RH (2002) Cell cycle inhibition by FoxO forkhead transcription factors involves downregulation of cyclin D. Mol Cell Biol 22:7842–7852

Shishodia S, Aggarwal BB (2002) Nuclear factor-kappaB activation: a question of life or death. J Biochem Mol Biol 1:28–40

Silvera D, Formenti SC, Schneider RJ (2010) Translational control in cancer. Nat Rev Cancer 4:254–266. doi: 10.1038/nrc2824

Skladanowski A, Bozko P, Sabisz M, Larsen AK (2007) Dual inhibition of PI3K/Akt signaling and the DNA damage checkpoint in p53-deficient cells with strong survival signaling: implications for cancer therapy. Cell Cycle 18:2268–2275

So L, Fruman DA (2012) PI3K signalling in B- and T-lymphocytes: new developments and therapeutic advances. Biochem J 4423:465–481. doi: 10.1042/BJ20112092

Song MS, Salmena L, Pandolfi PP (2012) The functions and regulation of the PTEN tumour suppressor. Nat Rev Mol Cell Biol 5:283–296. doi: 10.1038/nrm3330

Sparks CA, Guertin DA (2010) Targeting mTOR: prospects for mTOR complex 2 inhibitors in cancer therapy. Oncogene 26:3733–3744. doi:10.1038/onc.2010.139

Stahl M, Dijkers PF, Kops GJ, Lens SM, Coffer PJ, Burgering BM, Medema RH (2002) The forkhead transcription factor FoxO regulates transcription of p27Kip1 and Bim in response to IL-2. J Immunol 10:5024–5031

Steelman LS, Stadelman KM, Chappell WH, Horn S, Basecke J, Cervello M, Nicoletti F, Libra M, Stivala F, Martelli AM, McCubrey JA (2008) Akt as a therapeutic target in cancer. Expert Opin Ther Targets 9:1139–1165. doi:10.1517/14728222.12.9.1139

van Stijn A, van der Pol MA, Kok A, Bontje PM, Roemen GM, Beelen RH, Ossenkoppele GJ, Schuurhuis GJ (2003) Differences between the $CD34^+$ and $CD34^-$ blast compartments in apoptosis resistance in acute myeloid leukemia. Haematologica 88(5):497–508

Sujobert P, Bardet V, Cornillet-Lefebvre P, Hayflick JS, Prie N, Verdier F, Vanhaesebroeck B, Muller O, Pesce F, Ifrah N, Hunault-Berger M, Berthou C, Villemagne B, Jourdan E, Audhuy B, Solary E, Witz B, Harousseau JL, Himberlin C, Lamy T, Lioure B, Cahn JY, Dreyfus F, Mayeux P, Lacombe C, Bouscary D (2005) Essential role for the p110delta isoform in phosphoinositide 3-kinase activation and cell proliferation in acute myeloid leukemia. Blood 3:1063–6. doi:10.1182/blood-2004-08-3225

Tabe Y, Jin L, Tsutsumi-Ishii Y, Xu Y, McQueen T, Priebe W, Mills GB, Ohsaka A, Nagaoka I, Andreeff M, Konopleva M (2007) Activation of integrin-linked kinase is a critical prosurvival pathway induced in leukemic cells by bone marrow-derived stromal cells. Cancer Res 2:684–694. doi: 10.1158/0008-5472.CAN-06-3166

Tabellini G, Tazzari PL, Bortul R, Billi AM, Conte R, Manzoli L, Cocco L, Martelli AM (2004) Novel 2′-substituted, 3′-deoxy-phosphatidyl-myo-inositol analogues reduce drug resistance in human leukaemia cell lines with an activated phosphoinositide 3-kinase/Akt pathway. Br J Haematol 4:574–582. doi: 10.1111/j.1365-2141.2004.05073.x

Tabellini G, Cappellini A, Tazzari PL, Fala F, Billi AM, Manzoli L, Cocco L, Martelli AM (2005) Phosphoinositide 3-kinase/Akt involvement in arsenic trioxide resistance of human leukemia cells. J Cell Physiol 2:623–634. doi:10.1002/jcp.20153

Takahashi S (2010) Combination therapy with arsenic trioxide for hematological malignancies. Anticancer Agents Med Chem 6:504–510

Tamburini J, Chapuis N, Bardet V, Park S, Sujobert P, Willems L, Ifrah N, Dreyfus F, Mayeux P, Lacombe C, Bouscary D (2008) Mammalian target of rapamycin (mTOR) inhibition activates phosphatidylinositol 3-kinase/Akt by up-regulating insulin-like growth factor-1 receptor signaling in acute myeloid leukemia: rationale for therapeutic inhibition of both pathways. Blood 1:379–382. doi:10.1182/blood-2007-03-080796

Tamburini J, Green AS, Chapuis N, Bardet V, Lacombe C, Mayeux P, Bouscary D (2009) Targeting translation in acute myeloid leukemia: a new paradigm for therapy? Cell Cycle 23:3893–3899

Tamm I, Wagner M, Schmelz K (2005) Decitabine activates specific caspases downstream of p73 in myeloid leukemia. Ann Hematol Suppl 1:47–53. doi: 10.1007/s00277-005-0013-0

Tazzari PL, Cappellini A, Grafone T, Mantovani I, Ricci F, Billi AM, Ottaviani E, Conte R, Martinelli G, Martelli AM (2004) Detection of serine 473 phosphorylated Akt in acute myeloid leukaemia blasts by flow cytometry. Br J Haematol 5:675–681. doi: 10.1111/j.1365-2141.2004.05121.x

Tazzari PL, Cappellini A, Ricci F, Evangelisti C, Papa V, Grafone T, Martinelli G, Conte R, Cocco L, McCubrey JA, Martelli AM (2007a) Multidrug resistance-associated protein 1 expression is under the control of the phosphoinositide 3 kinase/Akt signal transduction network in human acute myelogenous leukemia blasts. Leukemia 3:427–438. doi: 10.1038/sj.leu.2404523

Tazzari PL, Tabellini G, Bortul R, Papa V, Evangelisti C, Grafone T, Martinelli G, McCubrey JA, Martelli AM (2007b) The insulin-like growth factor-I receptor kinase inhibitor NVP-AEW541 induces apoptosis in acute myeloid leukemia cells exhibiting autocrine insulin-like growth factor-I secretion. Leukemia 5:886–896. doi: 10.1038/sj.leu.2404643

Tazzari PL, Tabellini G, Ricci F, Papa V, Bortul R, Chiarini F, Evangelisti C, Martinelli G, Bontadini A, Cocco L, McCubrey JA, Martelli AM (2008) Synergistic proapoptotic activity of recombinant TRAIL plus the Akt inhibitor Perifosine in acute myelogenous leukemia cells. Cancer Res 22:9394–9403. doi:10.1158/0008-5472.CAN-08-2815

Thomson AW, Turnquist HR, Raimondi G (2009) Immunoregulatory functions of mTOR inhibition. Nat Rev Immunol 5:324–337. doi: 10.1038/nri2546

Thoreen CC, Sabatini DM (2009) Rapamycin inhibits mTORC1, but not completely. Autophagy 5:725–6

Thoreen CC, Kang SA, Chang JW, Liu Q, Zhang J, Gao Y, Reichling LJ, Sim T, Sabatini DM, Gray NS (2009) An ATP-competitive mammalian target of rapamycin inhibitor reveals rapamycin-resistant functions of mTORC1. J Biol Chem 12:8023–8032. doi:10.1074/jbc.M900301200

Tibes R, Kornblau SM, Qiu Y, Mousses SM, Robbins C, Moses T, Carpten JD (2008) PI3K/AKT pathway activation in acute myeloid leukaemias is not associated with AKT1 pleckstrin homology domain mutation. Br J Haematol 3:344–347. doi: BJH6920[pii]10.1111/j.1365-2141.2007.06920.x

Tothova Z, Gilliland DG (2007) FoxO transcription factors and stem cell homeostasis: insights from the hematopoietic system. Cell Stem Cell 2:140–152. doi:10.1016/j.stem.2007.07.017

Vanhaesebroeck B, Guillermet-Guibert J, Graupera M, Bilanges B (2010) The emerging mechanisms of isoform-specific PI3K signalling. Nat Rev Mol Cell Biol 5:329–341. doi:nrm2882[pii]10.1038/nrm2882

Vazquez F, Ramaswamy S, Nakamura N, Sellers WR (2000) Phosphorylation of the PTEN tail regulates protein stability and function. Mol Cell Biol 20(14):5010–5018

Vu C, Fruman DA (2010) Target of rapamycin signaling in leukemia and lymphoma. Clin Cancer Res 22:5374–5380. doi:10.1158/1078-0432.CCR-10-0480

Wahner Hendrickson AE, Haluska P, Schneider PA, Loegering DA, Peterson KL, Attar R, Smith BD, Erlichman C, Gottardis M, Karp JE, Carboni JM, Kaufmann SH (2009) Expression of insulin receptor isoform A and insulin-like growth factor-1 receptor in human acute myelogenous leukemia: effect of the dual-receptor inhibitor BMS-536924 in vitro. Cancer Res 19:7635–7643. doi:10.1158/0008-5472.CAN-09-0511

Wakabayashi M, Miwa H, Shikami M, Hiramatsu A, Ikai T, Tajima E, Yamamoto H, Miura K, Satoh A, Itoh M, Imamura A, Mihara H, Katoh Y, Nitta M (2004) Autocrine pathway of angiopoietins-Tie2 system in AML cells: association with phosphatidyl-inositol 3 kinase. Hematol J 4:353–360. doi:10.1038/sj.thj.62004106200410[pii]

Waldner MJ, Neurath MF (2012) Targeting the VEGF signaling pathway in cancer therapy. Expert Opin Ther Targets 1:5–13. doi:10.1517/14728222.2011.641951

Wander SA, Hennessy BT, Slingerland JM (2011) Next-generation mTOR inhibitors in clinical oncology: how pathway complexity informs therapeutic strategy. J Clin Invest 4:1231–1241. doi:10.1172/JCI44145

Weisberg E, Banerji L, Wright RD, Barrett R, Ray A, Moreno D, Catley L, Jiang J, Hall-Meyers E, Sauveur-Michel M, Stone R, Galinsky I, Fox E, Kung AL, Griffin JD (2008) Potentiation of antileukemic therapies by the dual PI3K/PDK-1 inhibitor, BAG956: effects on BCR-ABL- and mutant FLT3-expressing cells. Blood 7:3723–3734. doi:10.1182/blood-2007-09-114454

Weisberg E, Sattler M, Ray A, Griffin JD (2010) Drug resistance in mutant FLT3-positive AML. Oncogene 37:5120–5134. doi:10.1038/onc.2010.273

Willems L, Chapuis N, Puissant A, Maciel TT, Green AS, Jacque N, Vignon C, Park S, Guichard S, Herault O, Fricot A, Hermine O, Moura IC, Auberger P, Ifrah N, Dreyfus F, Bonnet D, Lacombe C, Mayeux P, Bouscary D, Tamburini J (2011) The dual mTORC1 and mTORC2 inhibitor AZD8055 has anti-tumor activity in acute myeloid leukemia. Leukemia 26(6):1195–1202 doi:10.1038/leu.2011.339

Workman P, Clarke PA, Raynaud FI, van Montfort RL (2010) Drugging the PI3 kinome: from chemical tools to drugs in the clinic. Cancer Res 6:2146–2157. doi:10.1158/0008-5472.CAN-09-4355

Wymann MP, Bulgarelli-Leva G, Zvelebil MJ, Pirola L, Vanhaesebroeck B, Waterfield MD, Panayotou G (1996) Wortmannin inactivates phosphoinositide 3-kinase by covalent modification of Lys-802, a residue involved in the phosphate transfer reaction. Mol Cell Biol 4:1722–1733

Xu Q, Simpson SE, Scialla TJ, Bagg A, Carroll M (2003) Survival of acute myeloid leukemia cells requires PI3 kinase activation. Blood 3:972–980. doi:10.1182/blood-2002-11-3429

Xu Q, Thompson JE, Carroll M (2005a) mTOR regulates cell survival after etoposide treatment in primary AML cells. Blood 13:4261–4268. doi:10.1182/blood-2004-11-4468

Xu RH, Pelicano H, Zhang H, Giles FJ, Keating MJ, Huang P (2005b) Synergistic effect of targeting mTOR by rapamycin and depleting ATP by inhibition of glycolysis in lymphoma and leukemia cells. Leukemia 12:2153–8. doi:10.1038/sj.leu.2403968

Yap TA, Garrett MD, Walton MI, Raynaud F, de Bono JS, Workman P (2008) Targeting the PI3K-AKT-mTOR pathway: progress, pitfalls, and promises. Curr Opin Pharmacol 4:393–412. doi:10.1016/j.coph.2008.08.004

Yea SS, Fruman DA (2011) Cell signaling. New mTOR targets Grb attention. Science 6035:1270–1271. doi:10.1126/science.1208071

Yecies JL, Manning BD (2011) Transcriptional control of cellular metabolism by mTOR signaling. Cancer Res 71 (8):2815–2820. doi:10.1158/0008-5472.CAN-10-4158

Yecies D, Carlson NE, Deng J, Letai A (2010) Acquired resistance to ABT-737 in lymphoma cells that up-regulate MCL-1 and BFL-1. Blood 16:3304–3313. doi:blood-2009-07-233304[pii]10.1182/blood-2009-07-233304

Yilmaz OH, Valdez R, Theisen BK, Guo W, Ferguson DO, Wu H, Morrison SJ (2006) Pten dependence distinguishes haematopoietic stem cells from leukaemia-initiating cells. Nature 7092:475–482. doi:nature04703 [pii] 10.1038/nature04703

Youle RJ, Strasser A (2008) The BCL-2 protein family: opposing activities that mediate cell death. Nat Rev Mol Cell Biol 1:47–59. doi:10.1038/nrm2308

Young A, Lyons J, Miller AL, Phan VT, Alarcon IR, McCormick F (2009) Ras signaling and therapies. Adv Cancer Res 102:1–17. doi:10.1016/S0065-230X(09)02001-6

Yu C, Rahmani M, Dai Y, Conrad D, Krystal G, Dent P, Grant S (2003) The lethal effects of pharmacological cyclin-dependent kinase inhibitors in human leukemia cells proceed through a phosphatidylinositol 3-kinase/Akt-dependent process. Cancer Res 8:1822–1833

Yu K, Toral-Barza L, Shi C, Zhang WG, Lucas J, Shor B, Kim J, Verheijen J, Curran K, Malwitz DJ, Cole DC, Ellingboe J, Ayral-Kaloustian S, Mansour TS, Gibbons JJ, Abraham RT, Nowak P, Zask A (2009) Biochemical, cellular, and in vivo activity of novel ATP-competitive and selective inhibitors of the mammalian target of rapamycin. Cancer Res 15:6232–6240. doi:10.1158/0008-5472.CAN-09-0299

Yu K, Shi C, Toral-Barza L, Lucas J, Shor B, Kim JE, Zhang WG, Mahoney R, Gaydos C, Tardio L, Kim SK, Conant R, Curran K, Kaplan J, Verheijen J, Ayral-Kaloustian S, Mansour TS, Abraham RT, Zask A, Gibbons JJ (2010) Beyond rapalog therapy: preclinical pharmacology and antitumor activity of WYE – 125132, an ATP-competitive and specific inhibitor of mTORC1 and mTORC2. Cancer Res 2:621–631. doi:10.1158/0008-5472.CAN-09-2340

Zaytseva YY, Valentino JD, Gulhati P, Mark Evers B (2012) mTOR inhibitors in cancer therapy. Cancer Lett 319:1–7.doi:10.1016/j.canlet.2012.01.005

Zeng Z, Samudio IJ, Zhang W, Estrov Z, Pelicano H, Harris D, Frolova O, Hail N, Jr., Chen W, Kornblau SM, Huang P, Lu Y, Mills GB, Andreeff M, Konopleva M (2006) Simultaneous inhibition of PDK1/AKT and Fms-like tyrosine kinase 3 signaling by a small-molecule KP372-1 induces mitochondrial dysfunction and apoptosis in acute myelogenous leukemia. Cancer Res 7:3737–3746. doi:10.1158/0008-5472.CAN-05-1278

Zeng Z, Sarbassov dos D, Samudio IJ, Yee KW, Munsell MF, Ellen Jackson C, Giles FJ, Sabatini DM, Andreeff M, Konopleva M (2007) Rapamycin derivatives reduce mTORC2 signaling and inhibit AKT activation in AML. Blood 8:3509–3512. doi:10.1182/blood-2006-06-030833

Zeng Z, Shi YX, Samudio IJ, Wang RY, Ling X, Frolova O, Levis M, Rubin JB, Negrin RR, Estey EH, Konoplev S, Andreeff M, Konopleva M (2009) Targeting the leukemia microenvironment by CXCR4 inhibition overcomes resistance to kinase inhibitors and chemotherapy in AML. Blood 24:6215–6224. doi:blood-2008-05-158311 [pii]10.1182/blood-2008-05-158311

Zhang J, Grindley JC, Yin T, Jayasinghe S, He XC, Ross JT, Haug JS, Rupp D, Porter-Westpfahl KS, Wiedemann LM, Wu H, Li L (2006) PTEN maintains haematopoietic stem cells and acts in lineage choice and leukaemia prevention. Nature 441(7092):518–522. doi:nature04747 [pii] 10.1038/nature04747

Zhang X, Tang N, Hadden TJ, Rishi AK (2011) Akt, FoxO and regulation of apoptosis. Biochim Biophys Acta 11:1978–1986. doi:10.1016/j.bbamcr.2011.03.010

Zhao S, Konopleva M, Cabreira-Hansen M, Xie Z, Hu W, Milella M, Estrov Z, Mills GB, Andreeff M (2004) Inhibition of phosphatidylinositol 3-kinase dephosphorylates BAD and promotes apoptosis in myeloid leukemias. Leukemia 2:267–275. doi:10.1038/sj.leu.24032202403220 [pii]

Chapter 18
Targeting the PI3 Kinase-mTOR Signaling Pathway in AML

Martin Carroll

Abstract Multiple lines of evidence have demonstrated over the last 10 years that alterations in signal transduction pathways add to the pathogenesis of acute myeloid leukemia (AML). In particular, the PI3K-Akt-mTOR pathway is constitutively activated in AML cell lines and primary cells. However, in contrast to the dramatic efficacy of ABL kinase inhibitors in the therapy of chronic myeloid leukemia (CML), single agent signal transduction inhibitors (STIs) have had modest success in the therapy of AML in most patients. The pattern that emerges is of intermittent success that is not sustained. These observations suggest that, in contrast to the relative homogeneity of signaling in CML cells, the control of growth deregulation in AML is complex with both patient to patient variation and cell to cell variation in a single patient (tumor diversity). In this chapter, we will summarize the results of clinical trials of mTOR inhibitors and other inhibitors of protein translation. We will, in particular, summarize the recent methodological advances that will allow for better definition of variations in signaling and may provide an opportunity for development of biomarkers for selection of appropriate signaling inhibitors for specific patients. Finally, we propose that the way forward in STI-based therapy of AML will be through combinations of inhibitors that attack the signaling heterogeneity in AML allowing for more effective and sustained responses.

Keywords Acute myeloid leukemia · Signal transduction inhibitors · Mammalian target of rapamycin · mTOR · Phosphotidylinositol 3 phosphate (PI3)

M. Carroll (✉)
Department of Medicine, University of Pennsylvania, Room 708, BRB II/III, 421 Curie Blvd., Philadelphia 19104, PA, USA
email: carroll2@mail.med.upenn.edu

M. Andreeff (ed.), *Targeted Therapy of Acute Myeloid Leukemia*,
Current Cancer Research, DOI 10.1007/978-1-4939-1393-0_18

18.1 Introduction

Signal transduction pathways are constitutively activated in acute myeloid leukemia (AML) cell lines and primary AML cells. Activated pathways may include the phosphotidyl-inositol-3 kinase (PI3K)-Akt-mTOR signaling pathway, the mitogen activated kinase (MAPK) pathway, nuclear factor kB (NFkB) and others. However, in contrast to chronic myeloid leukemia (CML) and the dramatic successes with ABL kinase inhibitors, signal transduction therapy has not become established in AML. The most well-described signaling pathway activated in AML cells is the PI3K pathway (reviewed in Chap. 17). Multiple lines of evidence document activation of this pathway in AML cells. We and others have used inhibitors of the mammalian target of rapamycin (mTOR) in clinical trials. As described below, these trials are beginning to show some success. More importantly, these trials have begun to describe the diversity of signaling networks both between different patients and within individual patient cells. The ribosomal S6 protein has proven to be a robust biomarker for pathway activation that can be used for patient selection for future trials. Overall, these trials have begun to teach us about the complexity of targeting signaling pathways in a biologically complex disease such as acute myeloid leukemia

18.2 What Do We Think We Know?

It is difficult to under-estimate the impact of the robust success of imatinib in CML on the studies of other hematologic malignancies including AML (Druker et al. 2001a, b). Importantly, studies of BCR/ABL function, many of which were done in cell lines, suggested that BCR/ABL signaling reflected cytokine receptor signaling at least in its broad outlines. For example, numerous studies have documented that interleukin 3 (IL3), the cytokine necessary for growth of the popular Ba/F3 cell line, activates signal transducer and activator of transcription 5 (STAT5), PI3K, and MAPK (Carlesso et al. 1996; Tauchi et al. 1994; Skorski et al. 1995). When these cells are transfected with BCR/ABL and selected for growth without IL3, the same pathways are activated and depend on BCR/ABL kinase activity for growth. Thus, a paradigm was established that oncogenic signaling in hematopoietic malignancies would reflect cytokine receptor signaling. In 2002, Kelley and Gilliland proposed that activated tyrosine kinase signaling was an essential component of oncogenic transformation in AML (Kelly and Gilliland 2002). Taken together, much of the field, including my laboratory, has pursued the hypothesis that activated cytokine receptor like signaling would drive cell proliferation and regulate cell survival in AML. It was in fact this hypothesis that lead us and others to originally study the PI3K signaling pathway in AML (Xu et al. 2003). Importantly, the combination of this model with the powerful influence of imatinib studies suggested to many of us that AML signaling, like CML signaling, would be similar from patient to patient and would model cytokine receptor signaling. The latter idea is not yet disproven but the lack of evidence for this model raises the question as to its validity. More immediately, increasing data demonstrates that there is diversity in AML signaling both from patient to patient and within individual's populations of malignant cells.

18.2.1 How Do We Know What We Know (or Think We Know)?

A fundamental question for studies of signal transduction in AML is whether cell lines or murine model systems accurately reflect the human disease. The majority of work describing the biologic function of signal transduction pathways in leukemia's have been done in such model systems. This is because of several challenges in working with primary cells. Such cells are not routinely available to all investigators. Many tissue banks do not preserve cells in manners that allow for thawing and studies of signaling pathways. Importantly, not all methodologies allow for distinction of truly purified leukemic blasts. Most laboratories use mononuclear cell fractions, which include monocytes and lymphocytes that may have quite diverse signaling activation compared to the malignant cells of interest. Primary cells do not survive well in culture and are difficult to transfect or infect with viral vectors. Thus, much of what we know in this disease is in fact not known at the highest level of evidence but implied by studies of surrogate models.

18.2.2 Go with the Flow: A Higher-Resolution Picture of Signal Transduction in Primary Leukemic Cells

The Nolan laboratory originally demonstrated the feasibility of studying heterogeneous populations of primary malignant cells from patients with leukemia using intracellular phosphor-flow methodologies (Irish et al. 2004). Similar work was also described by investigators at Ontario Cancer Institute using modified protocols (Chow et al. 2005). Chow and colleagues first described a robust approach to studying signaling in AML and demonstrated that activation of ribosomal S6 phosphorylation can be defined in AML cells using phospho-flow (Chow et al. 2006). Biologically, this was consistent with the concept that the PI3K pathway is activated in AML. The group used simultaneous cell surface and intracellular phospho-flow to "gate" on only the leukemic blasts, thus avoiding the complicating variables of Ficoll fractionation. Importantly, the Hedley group demonstrated that at any given point in time, not all the cells within a highly defined malignant cell population have detectable phosphorylation of ribosomal S6 (Chow et al. 2006). In fact, in most samples, our laboratory has gone on to demonstrate that the most common scenario is that less than 50 % of cells have activation of ribosomal S6 phosphorylation at any given time (Perl et al. 2012). The reason for this observation is not clear but stands in stark contrast to CML and other models in which there is homogeneous activation of signaling. This and other studies suggest that AML cells may cycle their signaling pathways on and off over time, a hypothesis that suggests that sustained inhibition of signaling pathways is a likely prerequisite for success with signaling inhibitors in AML.

18.3 PI3K-Akt-mTOR Activation in Primary AML Cells

Full review of this topic is provided in Chap. 17 and we will only briefly review references of direct relevance to clinical development of drugs in this pathway. Original studies in the Carroll laboratory and others used indirect measures to imply activation of PI3 kinase itself in primary AML cells. We initially used LY294002 as an inhibitor and demonstrated that LY294002 effectively induces cell death of primary AML cells in liquid culture (Xu et al. 2003). In addition, we performed Akt kinase assays and demonstrated that the majority of AML samples have activation of Akt kinase using a pan-Akt antibody. Subsequent work demonstrated that LY294002 has targets other than PI3 kinase itself raising questions as to the interpretation of our original experiments. However, subsequent studies with the more specific inhibitor, PI-103, are also consistent with the hypothesis that PI3 kinase is activated in primary AML cells and necessary for the survival of most patient samples (Park et al. 2008; Kojima et al. 2008). Akt activation has been confirmed by a number of groups using phosphorylation of downstream markers although preliminary results in our group do not show robust activity of Akt inhibitors in inhibiting cell survival in AML cells (Perry and Carroll, unpublished).

Subsequent reports have focused on the downstream target of PI3K-Akt, mTOR. Initial studies were really studies of transducers of regulated CREB 1 (TORC1) as they relied heavily on the TORC1 inhibitor, rapamycin. Recher and colleagues initially reported activation of mTOR in leukemic cells and suggested that rapamycin would be cytotoxic to AML cells (Recher et al. 2005). In fact, this initial report described clinical responses to single agent, rapamycin in patients with AML (Recher et al. 2005) although later studies with other rapalogs did not confirm a single agent response activity (Yee et al. 2006). We suggested instead that rapamycin may act as a cytostatic agent in AML and that therapeutic benefit may be best achieved by using rapamycin as a chemosensitizing agent (Xu et al. 2005). As described in more detail below, this is an approach that we have and others have studied in the clinic. More recent work suggests that optimal effects on AML cell survival may in fact be achieved with combination inhibition of PI3K and TORC1/2 with the compound, BEZ-235 (Chapuis et al. 2010). This is an important translational study as previous work from Tamburini and colleagues has demonstrated an important contribution of protein translation, likely through TORC2, to AML cell survival (Chapuis et al. 2010). Overall, these data have supported clinical trials of rapalogs in AML as described further below.

18.4 Other Ways to Target Protein Translation in AML

An important conclusion of the above studies is that protein translation is one of the therapeutic targets in primary AML cells. The Borden group has also focused on protein translation, focusing more specifically on the translational complex itself. The group first demonstrated a contribution of the protein translation initiation factor eukaryotic translation initiation factor 4E (eIF4E) to AML cell survival (Topisirovic et al. 2003) and went on to demonstrate that the anti-viral drug, ribavirin, is a competitive inhibitor of eIF4E (Kentsis et al. 2004). The group has gone on to perform clinical trials with single agent ribavirin in AML patients and, consistent with other results with signal transduction inhibitors (STIs), have shown intermittent, robust responses (Assouline et al. 2009). Combinations of ribavirin with chemotherapy or other signaling inhibitors have not yet been undertaken although the rationale for such an approach is compelling.

18.4.1 Perifosine

To our knowledge, only one clinical trial has been undertaken with a stated goal of targeting Akt in AML. Perifosine is an alkylphospholipid that can inhibit Akt as well as other kinases (Alam et al. 2012). Preclinical studies suggested that combination of perifosine with MEK inhibition would be effective in killing AML cells (Rahmani et al. 2009) and an intriguing combination signaling inhibitor phase I trial with perifosine and the staurosporine derivative, UCN-01, has been undertaken at the University of Maryland and the University of Pennsylvania. This trial is one of the first trials to test combinations of signaling inhibitors in a simultaneous dose escalation approach (see Clinical Trials.gov (NCT00301938) for details). Results of this trial are anticipated later this year.

18.4.2 European Temsirolimus Study

Amadori and colleagues have recently reported on the European experience using the rapalog, temsirolimus, in combination with induction clofarabine in elderly patients with high-risk AML (Amadori et al. 2012). Important aspects of the trial are that the targeted population was older than age 60 with relapsed or primary refractory AML. Patients received Temsirolimus 25 mg on days 1, 8, and 15 with clofarabine 20 mg/m^2 on days 1–5. Toxicities were mostly consistent with what is to be expected on chemotherapy induction trials in this age group. Importantly, an increase in fungal infections compared to historical controls was not seen demonstrating that the immunosuppressive properties of temsirolimus did not exacerbate the infection complications of therapy. The overall response rate was a modest 21 % consistent with previous results with single agent clofarabine. However, an important observation in the trial was that not all patients had inhibition of ribosomal S6 phosphorylation after temsirolimus. Importantly, in the small group of patients who had greater than 50 % inhibition of S6 phosphorylation ($n=12$), the response rate is 75 %. This important trial demonstrates that, as stated above, signal transduction is heterogeneous in AML patients and that the use of biomarkers for patient selection is likely a critical aspect for design of future trials.

18.4.3 UPENN/Jeff Siro/MEC Studies

Over the last eight years, we have explored the addition of rapamycin to MEC chemotherapy in high-risk patients seen at the University of Pennsylvania and Jefferson Medical College. We initially reported that the drug combination was safe in a phase I trial (Perl et al. 2009). As described in that report, simultaneous initiation of sirolimus (the commercial name for rapamycin in the USA) with chemotherapy generated challenges for pharmacodynamics monitoring. Blasts quickly disappeared from peripheral blood limiting available sample for study; and chemotherapy induced cell death, which appeared to lead to proteolysis even when adequate samples were harvested. For these reasons, in subsequent trials, we initiated sirolimus therapy three days before chemotherapy and performed pharmacokinetic and pharmacodynamics monitoring before chemotherapy was administered. Interestingly, preliminary results reported at the American Society of Hematology meeting in 2011 (Kasner et al. 2012) suggest an increase in response rate when sirolimus is initiated prior to chemotherapy. Certainly, pharmacodynamic monitoring was enhanced. We have found that there are three groups of patients as defined by ribosomal S6 biomarker analysis. In our experience, only two thirds of patients demonstrate detectable S6 phosphorylation by flow cytometry evaluation. Of those who do have S6 phosphorylation, we also have found that some patients have activity that is suppressable by rapamycin whereas others do not respond pharmacodynamically to the drug. Preliminary results, consistent with the study by Amadori, suggest a substantially improved response rate in patients who have ribosomal S6 phosphorylation that is suppressed by Sirolimus. Thus, this study appears consistent with the Amadori study in demonstrating that patient selection for STI therapy is critical for future studies of these drugs in patients with high-risk AML. Of note, also, is that both our studies and the Amadori study have focused on patients with relapsed and refractory disease—disease that has never been demonstrated to be curable by conventional chemotherapy. It would also be of interest to see the effects of such monitoring and rapalog base therapy in newly diagnosed patients for whom conventional chemotherapy is more likely to be effective.

Conclusion

Studies of STIs in AML have yet to achieve the dramatic success of ABL kinase inhibitors in AML. Nonetheless, the biologic rationale for such therapy is compelling. Innumerable research studies demonstrate that all blood cells need activated signaling pathways for growth and survival. The PI3K-Akt-mTOR pathway clearly contributes to survival in many AML patients. However, we have little understanding of how signaling pathways are activated in AML. Until basic studies develop and enhanced picture of signal transduction regulation in AML, it remains likely that there are levels of redundancy that these early studies have not addressed. Thus, with studies using rapamycin or ribavirin, we are likely pinching off pieces of the signal transduction network in AML cells but not crippling the survival pathways in these cells. Clinical trials of rapalogs with chemotherapy have proceeded appropriately cautiously given the potential toxicities of the drug combinations. However, now that rapalog combination therapy with chemotherapy has documented safety in studies from two different groups, we would propose that these studies form a platform on which to build more sophisticated, multi-drug combination studies. The development of increasingly robust phosphor-flow cytometry methodologies should provide for appropriate biomarker monitoring in these studies and allow the use of STIs in AML to gather momentum and improve patient outcomes in the coming years.

Acknowledgments Dr. Carroll is supported by R01 CA149566 and by VA Merit Award 1I01BX000918.

Suggested Reading

Gojo I, Perl A, Luger S, Baer MR, Norsworthy KJ, Bauer KS, Tidwell M, Fleckinger S, Carroll M, Sausville EA (2013) Phase I study of UCN-01 and perifosine in patients with relapsed and refractory acute leukemias and high-risk myelodysplastic syndrome. Invest New Drugs 31(5):1217–27. PMID: 23443507 PMCID: PMC3723766

Referencess

Alam MM et al (2012) Synthesis and biological evaluation of cyclopentane-linked alkyl phosphocholines as potential anticancer agents that act by inhibiting Akt phosphorylation. Eur J Med Chem 47(1):485–492

Amadori S et al (2012) Temsirolimus, an mTOR inhibitor, in combination with lower-dose clofarabine as salvage therapy for older patients with acute myeloid leukaemia: results of a phase II GIMEMA study (AML-1107). Br J Haematol 156(2):205–212

Assouline S et al (2009) Molecular targeting of the oncogene eIF4E in acute myeloid leukemia (AML): a proof-of-principle clinical trial with ribavirin. Blood 114(2):257–260

Carlesso N, Frank DA, Griffin JD (1996) Tyrosyl phosphorylation and DNA binding activity of signal transducers and activators of transcription (STAT) proteins in hematopoietic cell lines transformed by Bcr/Abl. J Exp Med 183(3):811–820

Chapuis N et al (2010) Dual inhibition of PI3K and mTORC1/2 signaling by NVP-BEZ235 as a new therapeutic strategy for acute myeloid leukemia. Clin Cancer Res 16(22):5424–5435

Chow S et al (2005) Whole blood fixation and permeabilization protocol with red blood cell lysis for flow cytometry of intracellular phosphorylated epitopes in leukocyte subpopulations. Cytometry A 67(1):4–17

Chow S, Minden MD, Hedley DW (2006) Constitutive phosphorylation of the S6 ribosomal protein via mTOR and ERK signaling in the peripheral blasts of acute leukemia patients. Exp Hematol 34(9):1183–1191

Druker BJ et al (2001a) Activity of a specific inhibitor of the BCR-ABL tyrosine kinase in the blast crisis of chronic myeloid leukemia and acute lymphoblastic leukemia with the Philadelphia chromosome. New Engl J Med 344(14):1038–1042

Druker BJ et al (2001b) Efficacy and safety of a specific inhibitor of the BCR-ABL tyrosine kinase in chronic myeloid leukemia. New Engl J Med 344(14):1031–1037

Irish JM et al (2004) Single cell profiling of potentiated phospho-protein networks in cancer cells. Cell 118(2):217–228

Kasner M, Weiss M, Flomenberg N, Luger SM, Loren AW, Carabasi M, Perl AE (2012) Sirolimus plus MEC chemotherapy has significant activity in high risk AML patients especially those who exhibit in vivo inhibition of the mTOR pathway. ASH Annual Meeting Abstracts 120(21):143

Kelly LM, Gilliland DG (2002) Genetics of myeloid leukemias. Annu Rev Genomics Hum Genet 3:179–198

Kentsis A et al (2004) Ribavirin suppresses eIF4E-mediated oncogenic transformation by physical mimicry of the 7-methyl guanosine mRNA cap. Proc Natl Acad Sci U S A 101(52):18105–18110

Kojima K et al (2008) The dual PI3 kinase/mTOR inhibitor PI-103 prevents p53 induction by Mdm2 inhibition but enhances p53-mediated mitochondrial apoptosis in p53 wild-type AML. Leukemia 22(9):1728–1736

Park S et al (2008) PI-103, a dual inhibitor of Class IA phosphatidylinositide 3-kinase and mTOR, has antileukemic activity in AML. Leukemia 22(9):1698–1706

Perl AE et al (2009) A phase I study of the mammalian target of rapamycin inhibitor sirolimus and MEC chemotherapy in relapsed and refractory acute myelogenous leukemia. Clin Cancer Res 15(21):6732–6739

Perl AE et al (2012) Single-cell pharmacodynamic monitoring of S6 ribosomal protein phosphorylation in AML blasts during a clinical trial combining the mTOR inhibitor sirolimus and intensive chemotherapy. Clin Cancer Res 18(6):1716–1725

Rahmani M et al (2009) The BH3-only protein Bim plays a critical role in leukemia cell death triggered by concomitant inhibition of the PI3K/Akt and MEK/ERK1/2 pathways. Blood 114(20):4507–4516

Recher C et al (2005) Antileukemic activity of rapamycin in acute myeloid leukemia. Blood 105(6):2527–2534

Skorski T et al (1995) Phosphatidylinositol-3 kinase activity is regulated by BCR/ABL and is required for the growth of Philadelphia chromosome-positive cells. Blood 86(2):726–736

Tauchi T et al (1994) Coupling between p210bcr-abl and Shc and Grb2 adaptor proteins in hematopoietic cells permits growth factor receptor-independent link to ras activation pathway. J Exp Med 179(1):167–175

Topisirovic I et al (2003) Aberrant eukaryotic translation initiation factor 4E-dependent mRNA transport impedes hematopoietic differentiation and contributes to leukemogenesis. Mol Cell Biol 23(24):8992–9002

Xu Q et al (2003) Survival of acute myeloid leukemia cells requires PI3 kinase activation. Blood 102(3):972–980

Xu Q, Thompson JE, Carroll M (2005) mTOR regulates cell survival after etoposide treatment in primary AML cells. Blood 106(13):4261–4268

Yee KW et al (2006) Phase I/II study of the mammalian target of rapamycin inhibitor everolimus (RAD001) in patients with relapsed or refractory hematologic malignancies. Clin Cancer Res 12(17):5165–5173

Chapter 19
Aurora Kinases

Subrata Sen and Hiroshi Katayama

Abstract Aurora kinase family of serine/threonine kinases, frequently overexpressed in human cancers and implicated in tumorigenesis, induce chromosomal instability and oncogenic transformation when expressed at elevated levels in mammalian cells *in vitro* and *in vivo*. Recent findings on the interactions of Aurora kinases with tumor suppressor gene and oncogene-regulated networks as well as involvement in other nonmitotic processes such as ciliary disassembly affecting important signaling pathways and developmental disorders termed ciliopathies, have led to a greater recognition of the functional significance of these kinases in development and disease. Among the three members of the kinase family, Aurora-A, -B, and -C identified in humans, Aurora-A and -B have been reported to express at detectable levels in most proliferating somatic cells and characterized in detail for their involvement in cellular pathways relevant to cell proliferation and development of cancer-associated phenotypes. As a result, Aurora-A and -B are being investigated as potential targets for cancer therapy and multiple agents targeting the kinases are in early phase clinical trials with some having yielded encouraging results. This chapter discusses functional involvement of Aurora kinase-A and -B in the regulation of cell proliferation and cancer-relevant cellular pathways validating their significance as cancer therapeutic targets.

Keywords Aurora kinases · Centrosomal anomalies · Chromosomal instability · Tumor suppressor proteins · Oncoproteins · Cancer cell signaling · Ciliopathies

S. Sen (✉)
Department of Translational Molecular Pathology, Unit 951, The University of Texas M.D. Anderson Cancer Center, 1515 Holcombe Boulevard, Houston, TX 77030, USA
e-mail: ssen@mdanderson.org

H. Katayama
Department of Molecular Oncology, Okayama University Graduate School of Medicine, Dentistry and Pharmaceutical Sciences, 2-5-1 Shikata-cho, Kita-ku, Okayama, 700-8558, Japan

M. Andreeff (ed.), *Targeted Therapy of Acute Myeloid Leukemia*,
Current Cancer Research, DOI 10.1007/978-1-4939-1393-0_19

19.1 Introduction

19.1.1 Mitosis Regulatory Aurora Kinases as Cancer Therapeutic Targets

Cancer cells undergoing malignant transformation acquire abnormal growth properties and progressively increasing genomic instability. Current therapeutic approaches are designed to disrupt the oncogenic signaling pathways underlying the development of malignant phenotypes. In most instances, however, cancer cells exposed to a drug targeting an aberrantly expressing signaling molecule rewire the signaling events to continue proliferation in a deregulated manner. This becomes possible since abnormal growth properties of malignant cells result from complex functional networks comprising cross-talking signaling pathways, which can be redirected to overcome the drug effects and continue activating the growth regulatory processes of DNA replication and mitosis. It is, therefore, not unexpected that the mechanisms of action of a number of effective chemotherapeutic agents involves disrupting the replication/repair of DNA and/or mitotic cell division process. For example, cancer chemotherapeutic drugs, fluoropyramidines, gemcitabine, and topoisomerase inhibitors interfere with the DNA replication, while platinum analogs and cyclophosphamide generate un-resolvable lesions in replicating DNA leading to cell death. Another class of drugs, such as, taxanes, vinca alkaloids, and epothilones cause tumor growth inhibition by disrupting the microtubule cytoskeleton essential for mitotic proliferation of cells (Gautschi et al. 2008). These drugs, however, cause undesirable damage to healthy tissues and accompanying toxicity due to their indiscriminate effects on proliferating normal cells. It is, therefore, anticipated that pharmacological targeting of cell cycle regulatory proteins differentially expressed in tumor cells should help alleviate or minimize the toxic side effects associated with nonspecific exposure of normal cells to antimitotic agents. A number of small molecule drugs inhibiting mitotic proteins, preferentially expressed in tumor cells, have been moving into clinical trials (Garber 2005). Among the mitotic regulatory kinases, members of the evolutionarily conserved Aurora family of serine/threonine kinases are being investigated as potentially viable targets for anticancer drug discovery. These studies have been triggered as a consequence of the findings that Aurora kinases are overexpressed in a number of human tumors including those of the hematopoietic origin and functionally interact with multiple tumor suppressor and oncoprotein-regulated pathways. We discuss in this chapter functional interactions of Aurora kinases, both in the regulation of mitosis and in other signaling pathways deregulated in malignant cells.

19.1.2 Functional and Structural Conservation in Aurora Kinases

Somatic cell proliferation involves equal segregation of chromosome and division of the nucleus (mitosis) followed by partitioning of the cytoplasm (cytokinesis) into two daughter cells during mitosis. Aurora kinase family are evolutionarily conserved important regulators of mitosis with essential roles in the progressive stages from mitotic entry to cytokinesis (Carmena et al. 2009), ensuring accurate coordination of chromosomal and cytoskeletal events including centrosome maturation and separation, bipolar spindle assembly, and cytokinesis leading to faithful equal partitioning of the chromosomes into daughter cells. Yeast has one prototypic member of the Aurora kinase family, Ipl1/Ark1, while in majority of higher eukaryotes the family has two related members, Aurora-A and -B, represented by conserved orthologs in different species. Aurora-A and -B kinases display different subcellular localizations and functions. A third member of the kinase family, Aurora-C, is present only in mammals and predominantly expressed in testes has also been reported to display subcellular localization similar to Aurora-B and rescue chromosome misalignment and segregation defects developed in absence of Aurora-B in mammalian cells *in vitro* and *in vivo*. The latter findings indicate functional overlap between the two kinases and an essential role for Aurora-C during mammalian development (Sasai et al. 2004; Fenandez-Miranda et al. 2011). All the three Aurora kinases have been reported to overexpress in human cancers but the significance of Aurora-C, expressed at minimally detectable level in adult somatic cells, in the regulation of mitotic cell division cycle and in cancer cells remain somewhat unresolved. This chapter, therefore, discusses the functional characteristics of only Aurora-A and -B in the regulation of mitosis and in pathways relevant to the development of malignant phenotypes.

Structurally, members of the Aurora kinase family share similar carboxyl terminus catalytic domains but divergent amino terminal ends of variable lengths. Aurora-A and -B share about 70% identity in the carboxyl terminus catalytic domain and three conserved Aurora box motifs (A-box I, A-box II, and A-box III) at their varying amino terminal ends. There is suggestive evidence that the A-box motifs are involved in substrate recognition and subcellular localization of the two kinases. The C-terminus has a conserved functional destruction box (D-box) motif that plays a critical role in regulating the stability and degradation of the kinases. Despite sharing a few substrates and co-activators, Aurora-A and -B manifest predominantly different subcellular localizations and interactions with distinct set of proteins through the mitotic cell division cycle. Aurora-A is localized primarily on spindle poles and transiently along the spindle microtubules as cells progress through mitosis, while Aurora-B as a component of the chromosomal passenger complex (CPC) localizes to the inner centromere during prophase through metaphase and then transfers to the spindle midzone and the midbody during late mitosis and cytokinesis. The divergent localization and functions of the two related kinases are regulated by multiple protein-binding cofactors. Activation of Aurora-A is facilitated by binding to several proteins including TPX2, Ajuba/JUB, NEDD9 (also known as HEF1), and BORA. Activation involves autophosphorylation of Thr288 in the T-loop. The spindle assembly factor TPX2 is released from an importin α/β inhibitory complex following activation of GTPase Ran at nuclear envelope breakdown, and TPX2 binding to Aurora-A has been shown to move the activation loop to a position protected from dephosphorylation by PP1 on entry into mitosis (Bayliss et al. 2003; Eyers et al. 2003). Aurora-B activation involves autophosphorylation of Thr232 in the T-loop and requires interaction with the CPC consisting of the inner centromere

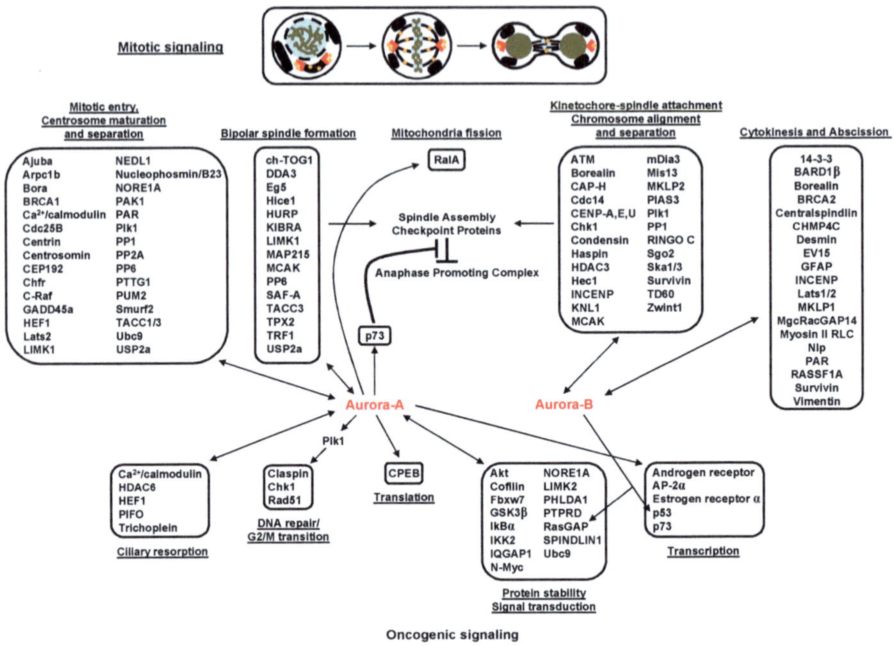

Fig. 19.1 A list of proteins interacting with Aurora-A and -B in the regulation of different mitotic and oncogenic signaling events

protein (INCENP), Survivin, and the Borealin/DasraB proteins (Adams et al. 2000; Jeyaprakash et al. 2007). Intriguingly, a single amino acid difference in the catalytic domain of the two kinases (G198 in human Aurora-A and N142 in human Aurora-B) was shown to be critical in controlling the selective activation of Aurora-A by its binding partner and activator TPX2. Site-directed mutant of Aurora-A residue (G198N) revealed classical Aurora-B localization and association with the CPC components INCENP and Survivin partially rescuing Aurora-B loss of function (Fu et al. 2009; Hans et al. 2009). Such finely tuned functional interactions of the two structurally related kinases are expected to have critical consequences in terms of proliferation and chromosomal ploidy when aberrantly expressed in tumor cells.

Frequent overexpression of Aurora-A and -B in many human cancers indicate that high expression of these kinases play roles in the development of cancer-associated phenotypes. While oncogenic function of Aurora-A has been demonstrated following induced overexpression of the kinase in mammalian cells *in vitro* and in transgenic rodent models (Bischoff and Anderson 1998; Zhou et al. 1998), there is also indirect evidence of high Aurora-B expression being oncogenic *in vivo* (Ota et al. 2002). A large array of proteins interacting with Aurora-A and -B are involved in the regulation of different mitotic and oncogenic signaling events as summarized in Fig. 19.1. For the purpose of this chapter, we focus our discussion on those interactions, which have direct implications on the origin of chromosomal instability and cancer-associated phenotypes.

19.2 Aurora-A: Mitotic, Nonmitotic Signaling and Oncogenic Pathways

19.2.1 Aurora-A in Mitosis

Detailed characterization of Aurora-A functional interactions in proliferating cells have revealed that the kinase plays important roles in the initiation and progression of mitosis. While the events in mitosis regulated by Aurora-A include centrosome maturation–separation, bipolar spindle assembly, chromosome alignment–segregation, and cytokinesis–abscission, proteolysis of Aurora-A accompanies exit from mitosis and entry into G1. Aurora-A gain of function due to overexpression deregulates mitotic entry, centrosome maturation and spindle assembly allowing aberrant progression through mitosis due to hyperactive centrosomes and multipolar spindle assembly in addition to promoting recovery from spindle damage in cells treated with spindle-poisoning drugs.

19.2.2 Aurora-A in Centrosome Maturation

Centrosomes duplicated in the S phase undergo a maturation process at the transition from late G2 to early M phase before the start of microtubule nucleation to form the mitotic spindle. During the maturation process centrosomes acquire a pericentriolar mass (PCM) comprising γ tubulin, the γ tubulin ring complex (γ-TURC), and a number of additional proteins, several of which are recruited with the help of Aurora-A. Recruitment and activation of Aurora-A at the centrosome, in turn, is facilitated by the centrosome protein, Cep192/spindle defective 2 (Spd-2), which also regulates the localization of other PCM proteins associated with the centrosome maturation process (Joukov et al. 2010). Aurora-A is involved in the recruitment of LATS2, NDEL1, and TACC to the centrosome. Phosphorylation of LATS2 and NDEL1 facilitate their localization to the centrosome and the latter helps targeting of TACC3 to the centrosome (Mori et al. 2007).

19.2.3 Aurora-A in Mitotic Entry

Aurora-A has been recognized as an important regulator of mitosis initiation and progression by interacting with a number of substrates on and away from the centrosome. At the initiation of mitosis entry, Aurora-A is involved in the activation of Cdk1/cyclin B complex at the centrosome involving Plk1 and Bora in a negative feedback regulatory manner. For the purpose, Aurora-A in association with Bora activates Plk1 that promotes recruitment of Aurora-A to the centrosome in late G2 phase. Bora binding to Plk1 controls Aurora-A access to Plk1T-loop, where Aurora-A phosphorylation leads to full Plk1 activation. Interestingly, Plk1 phosphorylation of Bora regulates Aurora-A function by creating a recognition site for SCF-betaTrCP E3 ubiquitin ligase on Bora and its degradation. Activated Aurora-A phosphorylates the Cdk activating phosphatase Cdc25B while Plk1 induces degradation of Cdk inhibitory Wee1 kinase and activation of Cdc25C phosphatase leading to the activation of Cdk1/cyclin B that promotes mitotic entry (Seki and Coppinger 2008; Macůrek et al. 2008). Aurora-A has also been shown to interact with and phosphorylate the BRCA1 protein for the latter's centrosomal localization and M phase entry of the cells (Ouchi et al. 2004; Brodie and Henderson 2012). A recent study has demonstrated that Aurora-A phosphorylation of the Ras family GTPase RalA causes its translocation to the mitochondria where it complexes with its effector RalBP1 and dynamin-related GTPase DRP1, whose mitochondrial translocation is facilitated by activated Cdk1/cyclin B. This complex regulates the mitochondrial fission process for equal segregation of mitochondria to the daughter cells post mitosis (Kashatus et al. 2011).

19.2.4 Aurora-A in Bipolar Spindle Assembly, Chromosome Segregation, and Cytokinesis

Nucleation of the spindle microtubules arise predominantly from the centrosomes and chromatin. Aurora-A plays a major role in the spindle microtubule nucleation not only from the centrosomes but also from the chromatin as well as in the absence of nucleating function of the centrosome or chromatin (Tsai et al. 2005). This activity involves Aurora-A interactions with its activating cofactor TPX2 and the proteins, HURP, XMAP215, and Eg5, stimulated by Ran-GTP enriched in the vicinity of kinetochore and chromatin (Gruss and Vernos 2004). Aurora-A interactions with the TACC family proteins TACC1 and TACC3 have been implicated in the stabilization of microtubules at the centrosomes that requires binding of the TACC proteins with the conserved ch-TOG/XMAP215 family proteins, which in turn inactivates the microtubule destabilizing kinesin MCAK (Barros et al. 2005). Interestingly, direct phosphorylation of MCAK by Aurora-A has also been shown to control spindle bipolarity (Zhang et al. 2008). More recent studies have revealed the role of Aurora-A phosphorylation of kinesin, Cenp-E, and the centromere protein A (Cenp-A) in the stable biorientation of chromosomes, chromosome alignment, and segregation (Kim et al. 2010; Hégarat et al. 2011). RASSF1A, another substrate of Aurora-A, following phosphorylation is released from its growth inhibitory binding to the microtubules and helps activate APC/Cdc20 allowing progression through the M phase (Rong et al. 2007). Aurora-A sequentially translocates from the centrosome to the spindle and then to the midzone before being degraded at mitotic exit. Aurora-A-TPX2 complex is moved to the spindle microtubules in association with scaffold attachment factor A (SAF-A) and this complex influences chromosome congression, stability of kinetochore microtubules, and the kinetochore–microtubule attachments (Ma et al. 2011). Finally, Aurora-A phosphorylation negatively regulates the activity of the RhoA activator GEF-H1, required for cleavage furrow formation during abscission, that is achieved when Aurora-A declines at the time of mitotic exit (Birkenfeld et al. 2007).

Aurora-A undergoes proteasomal degradation at the end of mitosis involving interaction with the anaphase-promoting complex/cyclosome (APC/C) in association with the APC/C coactivator subunits Cdc20 and Cdh1 (Barford 2011). Two different sequences in the A-box and D-box are required for Cdh1-dependent degradation. Dephosphorylation of a serine residue in the A-box II is reportedly required for degradation of Aurora-A (Littlepage et al. 2002). The A-box containing amino terminal domain of Aurora-A has also been shown to interact with the E3 ubiquitin ligase Chfr (checkpoint protein with forkhead and ring domain) for ubiquitination and degradation (Yu et al. 2005).

19.2.5 Aurora-A in Nonmitotic Signaling

Recent characterization of Aurora-A functional interactions has revealed involvement of the kinase in important nonmitotic signaling pathways regulating disassembly of cilia (Pugacheva et al. 2007), neurite extension (Mori et al. 2009), and intracellular calcium signaling (Plotnikova et al. 2010). These findings have indicated that Aurora-A activation is achieved by mechanisms other than those well characterized in mitotic cells. There is now evidence that Aurora-A may function in a kinase-independent manner in some instances (Toya et al. 2011).

The role of Aurora-A in the disassembly of cilia is significant since in most cells, a single non-motile cilia acts as the receiver of extracellular mechanical and chemical signals. Defects in cilia are associated with many clinically important diseases, referred to as "ciliopathies," including polycystic kidney disease, nephropathies, Bardet-Biedl syndromes and others (Nikonova et al. 2012). A number of signaling pathways, such as, Hedgehog, Notch, Wnt, Par3-aPKC, and PDGFa, often found deregulated in cancer, also act through cilia. Therefore, it is becoming increasingly clear that regulation of cilia function may play critical roles in both developmental disorders and cancer. Disassembly of cilia by Aurora-A mediated phosphorylation and activation of the tubulin deacetylase HDAC6 (Pugacheva et al. 2007) indicates possible involvement of the kinase in nonmitotic cells with a different mechanistic perspective relevant to human disease. The proteins involved in activating Aurora-A mediated cilia disassembly is shown in Fig. 19.2. Another novel Aurora-A activation mechanism became evident from the finding that interaction of Ca^{2+}/CaM, induced by multiple different stimuli, can elevate Aurora-A kinase activity (Plotnikova et al. 2010). A more recent study revealed that Ca^{2+}/CaM controls activation of Aurora-A in cilia disassembly in mitosis and potentiates Aurora-A/NEDD9 interaction (Plotnikova et al. 2012). A nonmitotic role for highly expressed Aurora-A in postmitotic neurons was also recently reported (Mori et al. 2009). The study revealed co-localization with and phosphorylation by a PKC of Aurora-A on the residue T287 of its activation loop, that was found essential for microtubule-dependent neurite extension. These findings, taken together, offer interesting compelling

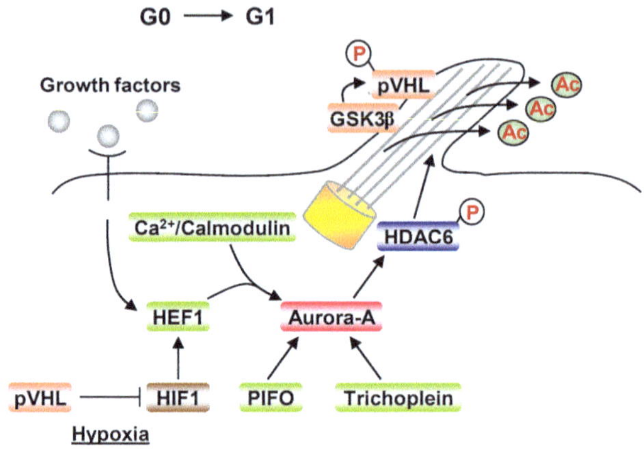

Fig. 19.2 Aurora-A functional interactions in cilia disassembly

evidence in support of previously unknown Aurora-A functional interactions, independent of its role in mitosis, which are likely to have significant clinical implications in multiple developmental disorders and chronic diseases including cancer.

19.2.6 Aurora-A in Oncogenic Signaling

Aurora-A overexpression frequently observed in tumor cells is expected to interfere with centrosome maturation, chromosome biorientation, bipolar spindle assembly, and aberrant cytokinesis resulting in centrosomal anomalies and chromosome segregation errors, commonly detected in human cancers. It is noteworthy, in this context, that a number of transcription factors regulating Aurora-A expression like Erk responsive Ets, STAT5, estrogen responsive GATA3, HIF1 are frequently overexpressed in cancer (Karthigeyan et al. 2011). Furthermore, Aurora-A interacting/activating proteins, such as, NEDD9/HEF1 and TPX2 are elevated, while proteins involved in degradation of Aurora-A, such as Chfr, are downregulated in cancer cells with consequential elevated expression of Aurora-A in all these instances even in the absence of amplification of the gene encoding the kinase. The extent to which centrosomal aberrations and chromosomal ploidy alterations in Aurora-A overexpressing cells play roles in the malignant transformation process remains to be resolved but their strong correlation with progression of disease and development of resistance to DNA and spindle damaging therapeutic agents is now fairly well accepted. The latter possibility is also implicit from the findings that Aurora-A facilitates checkpoint recovery and mitotic reentry after spindle damage by activating Plk-1 (Macůrek et al. 2008; Seki and Coppinger 2008) as well as by mediating the formation of kinetochore/chromatin-associated microtubule assembly (Katayama et al. 2008).

In addition to mitosis specific substrates, Aurora-A also interacts with proteins involved in pathways frequently deregulated in cancer. Aurora-A phosphorylation of the tumor suppressor proteins BRCA1, p53 and p73 have been reported to cause their loss of function making cells resistant to DNA damage and override checkpoint response (Katayama et al. 2004; Liu et al. 2004; Ouchi et al. 2004; Katayama et al. 2012). Thus, tumors with elevated expression of the kinase is expected to acquire loss of function phenotypes for these two critical tumor suppressor pathways even in the presence of wild type alleles of the respective tumor suppressor proteins.

Aurora-A is directly involved in multiple oncogenic pathways likely influencing proliferation and transformation processes. Aurora-A was reported to upregulate telomerase reverse transcriptase mRNA through c-Myc (Yang et al. 2004). There appears to be a positive feedback loop between Aurora-A and c-Myc in certain tumors. A recent study of a mouse model of metastatic skin carcinoma with induced expression of a gain of function $p53^{R172H}$ mutation revealed that the squamous cell tumors were characterized by genomic instability with accompanying c-Myc amplification and Aurora-A overexpression (Torchia et al. 2012). Aurora-A has also been implicated in activating the Akt pathway and directly phosphorylate GSK3β leading to activation of the β-catenin-TCF transcription complex inducing c-Myc and other cancer-relevant genes including cyclin D1, VEGF, and FGF18 (Dar et al. 2009) underscoring a role for Aurora-A in this oncogenic pathway. More recently, loss of the tumor suppressor PTEN was shown to cause stabilization of Aurora-A by attenuating FBXW7 ubiquitin ligase-dependent degradation of Aurora-A through the GSK3β/Akt pathway (Kwon et al. 2012). In addition, positive regulation of NF-κB signaling has been demonstrated following Aurora-A mediated IκBα phosphorylation leading to activation of the NF-κB complexes (Briassouli et al. 2007). IκB Kinase2, a component of the IκK complex responsible for physiologic phosphorylation and degradation of IκB negatively regulate stability of Aurora-A protein and influence bipolar spindle assembly (Irelan et al. 2007). These findings suggest that under normal physiological conditions IκB Kinase2 positively regulates NF-κB but antagonizes Aurora-A signaling to ensure proper mitotic assembly and chromosome segregation while overexpression of Aurora-A leads to deregulation of this network. Furthermore, oncogenic effects of Aurora-A also involves phosphorylation of RalA in the Ras signaling pathway (Wu et al. 2005; Lim et al. 2010) and the tumor suppressor NORE1A that functions as a regulatory node between Ras signaling and microtubule nucleation (Bee et al. 2010). Gain of function of Aurora-A, therefore, appears to be a seminal defect in cancer cells that can cause aberrant activation of the Myc, Ras, and the NF-κB signaling pathways besides induction of mitotic chromosome segregation errors. Finally, a kinase-activity-independent role of Aurora-A in N-Myc amplified neuroblastoma tumors, showing frequent amplification of Aurora-A was revealed that involved stabilization of the oncogenic N-Myc protein (Otto et al. 2009). Intriguingly, though, an Aurora-A inhibitor demonstrated antitumor activity in all neuroblastomas in N-Myc-independent manner (Maris et al. 2010). These findings suggest Aurora-A's possible involvement in multiple cross-talking genetic networks and plausible kinase-independent pleiotropic role in oncogenic signaling pathways, several of which may be remaining to be elucidated. While more in-depth analyses of Aurora-A-regulated pathways in cells from different tissue types may still be needed to develop a comprehensive idea about the role of this kinase in various malignancies, existing data provide more than adequate justification for the current interest in developing effective Aurora kinase inhibitors for anticancer therapy.

19.3 Aurora-B: Mitotic Signaling and Oncogenic Signaling

19.3.1 Aurora-B in Mitosis

Aurora-B is a catalytic subunit of the CPC comprising INCENP, Survivin, and Borealin, which are required for activation of Aurora-B through autophosphorylation of Thr232 in the T-loop and targeting of the complex to the inner centromere by prometaphase–metaphase, to the central spindle microtubules at metaphase and cleavage furrow-midbody in cytokinesis. It is currently known that Aurora B localization to centromeres in early mitosis relies on Haspin-dependent histone H3 phosphorylation and on Bub1-dependent histone H2A phosphorylation which interacts with Borealin through a Shugoshin intermediate (Wang et al. 2011). Consistent with its dynamic localization during mitosis and cytokinesis, Aurora-B plays multiple essential roles in mitotic spindle assembly checkpoint (SAC) by regulating kinetochore–spindle microtubule attachment for segregation of chromatids and central spindle assembly as well as in abscission checkpoint during cytokinesis by regulating cleavage furrow ingression to maintain chromosome stability (Carmena et al. 2009; Ruchaud et al. 2007; van der Waal et al. 2012a).

19.3.1.1 Establishment of Correct Kinetochore–Microtubule Attachment

Aurora-B–CPC functions in recruitment of both centromeric and kinetochore proteins and correction of improper erroneous attachment of spindle microtubules to kinetochore proteins to ensure proper attachment of kinetochore–spindle microtubule and chromosome alignment during prometaphase and metaphase. Microtubule attachment error gives rise to defects in equal chromatid segregation in anaphase resulting in activation of abscission checkpoint (Thompson and Compton 2011). Aurora-B–CPC specifically corrects syntelic and merotelic microtubule attachments which do not generate tension across centromeres and therefore Aurora-B–CPC located at inner centromere can phosphorylate KMN network (composed of KNL1, MIS12, and NDC80) serving as the main microtubule-binding unit at the kinetochore (Welburn et al. 2010) which reduces its microtubule-binding affinity and inhibits the binding of PP1 phosphatase to the kinetochore thus facilitating the release of erroneously attached microtubules (Cheeseman et al. 2006; DeLuca et al. 2006; Liu et al. 2010; Welburn et al. 2010). When correct bipolar attachments of microtubule to kinetochore are reassembled and the tension is generated, kinetochore is pulled away from Aurora-B and PP1 is no longer inhibited which allows dephosphorylation of the KMN network and recruitment of Astrin, SKAP, and Ska complex to stabilize microtubule–kinetochore attachment (Daum et al. 2009; Welburn et al. 2009; Schmidt et al. 2010; Chan et al. 2012).

19.3.1.2 Aurora-B in Spindle Checkpoint Control

Amid corrections of erroneous attachment of kinetochores to microtubules, SAC complex at the kinetochores produce a "wait anaphase" signal to prevent activation of APC-Cdc20 E3 ligase that targets cyclin B and securin for proteasome degradation to delay anaphase onset. The studies in the past few years have established direct role of Aurora-B–CPC in SAC signaling in addition to error correction. Aurora B and Ndc80 are required for efficient recruitment of Mps1 to unattached kinetochores, allowing rapid Mps1 activation at the onset of mitosis (Saurin et al. 2011). More recent study revealed that Mps1 phosphorylates Borealin and enhances Bub1 mediated histone H2A phosphorylation critical for Sgo1 recruitment to centromeres, thereby promoting recruitment and activation of Aurora B at the centromeres, indicating the presence of a feedback regulation of Aurora B–Mps1 recruitment at the centromeres (Jelluma et al. 2008, van der Waal et al. 2012b). In addition, Aurora-B phosphorylates and activates ATM in mitosis in absence of DNA damage, and activated ATM subsequently phosphorylates and activates Bub1 (Yang et al. 2011). Aurora-B is proposed to act downstream of Mad1–Mad2 recruitment in the mitotic checkpoint signaling cascade to maintain the APC/C inhibitory signal (Maldonado and Kapoor 2011). Taken together, these studies clearly indicate dual functions of Aurora-B in destabilization of improper microtubule attachments as well as in the activation and maintenance of SAC.

19.3.1.3 Aurora-B in Cytokinesis and Abscission

When SAC is inactivated following proper chromosome alignment leading to sister chromatids separation, Aurora-B–CPC translocates to the midzone microtubules from centromeres by interacting with MKLP2 to initiate an actomyosin contractile ring formation and cleavage furrow ingression along the cell membrane. In this process, while Aurora-B phosphorylation of centralspindlin complex of MKLP1 enhances microtubule bundling activity, phosphorylation of MgcRacGAP activates RhoA GTPase resulting in the targeting of RhoGEF ECT2 to the contractile ring. Finally, Aurora-B phosphorylation of myosin regulatory light chain-2 enhances ingression of actomyosin contractile ring (Glotzer 2009; Douglas et al. 2010). Recent studies reported involvement of Aurora-B–CPC in abscission checkpoint which delays abscission to avoid tetraploidization caused by progression of cytokinesis until the chromosome bridge connecting the two daughter cells has been resolved (Norden et al. 2006; Steigemann et al. 2009). In this checkpoint, Aurora-B phosphorylates charged multivesicular body (MVB) protein 4C (CHMP4C), a component of the endosomal sorting complex required for transport subunit (ESCRT-III) mediating inhibition of abscission via interaction with Borealin. Phosphorylated

CHMP4C is targeted to the midbody where it inhibits abscission through activation of NoCut (Carlton et al. 2012).

19.3.2 Role of Aurora-B in Oncogenesis

19.3.2.1 Aurora-B in Mutant Ras Pathway

Aurora-B is overexpressed in a various types of tumors including acute myeloid leukemia (AML) similar to Aurora-A. However, in contrast to the clear emerging evidence of direct involvement of Aurora-A in multiple oncogenic signaling pathways described above, it is currently unclear whether functional aberration of Aurora-B is directly involved in oncogenesis while its physical interactions with some tumor suppressor proteins such as p53 and BRCA2 have been identified. Thus, we discuss potential implication of Aurora-B in oncogenic signaling pathways. First, such a role for Aurora-B was implicated since overexpression of Aurora-B and its activator INCENP was reported to augment transformation of normal murine cells in presence of oncogenic Ras (Kanda et al. 2005). Tumors harboring oncogenic Ras mutations almost invariably retain expression of a wild type pRB protein and cell proliferation depends on functional pRB (Williams et al. 2006). pRB is a critical protein involved in postmitotic checkpoint, which prevents endoreduplication and polyploidy after aberrant mitosis due to cytokinesis failure in absence of E2F activation. Intriguingly, Aurora-B has been shown to phosphorylate pRB which leads to stable pRB–E2F1 complex formation, resulting in inhibition of endoreduplication, indicating a role for Aurora B in negatively regulating DNA synthesis (Nair et al. 2009). Although the mechanism by which Aurora-B phosphorylated pRB may augment oncogenic Ras signaling is not yet elucidated, one hypothesis is that Aurora-B phosphorylated pRB inhibits E2F activity mediated gene expressions required for cell cycle arrest and apoptosis, thereby allowing aneuploid cells to proliferate and survive. Further investigation is required to elucidate the molecular mechanisms underlying synergistic effects of Aurora-B deregulation and Ras activation in cancer cells, which could help develop effective therapeutic strategies for cancers with oncogenic Ras and hyperactive Aurora-B.

19.3.2.2 Inhibition of p53 by Aurora-B

Similar to functional repression of p53 by Aurora-A, it has recently been revealed that Aurora-B interacts with and phosphorylates p53 at multiple sites in the DNA-binding domain with serine-215 residue being a conserved phosphorylation site for both Aurora-A and -B. p53 phosphorylation by Aurora-B results in loss of DNA-binding activity as well as acceleration of the degradation of p53 through the MDM2-mediated polyubiquitination–proteasome pathway, thus functionally suppressing the expression of p53 target genes involved in cell cycle arrest and apoptosis (Wu and Ma 2011; Gully et al. 2012). p53 phosphorylation of Mps1 and BubR1 mitotic checkpoint kinases in response to spindle damage has been shown to stabilize and activate p53 and in turn to be essential for postmitotic checkpoint to eliminate the tetraploid cells induced by prolonged mitotic arrest with subsequent adaptation and mitotic exit without cytokinesis (Ha et al. 2007; Huang et al. 2009). Although the direct effect of Aurora-kinases- mediated p53 phosphorylation in mitotic progression remains to be investigated, the opposing control of p53 protein stability by Aurora kinases and mitotic spindle checkpoint kinases which interplay in mitosis suggest that the fate of postmitotic cells following prolonged mitotic spindle checkpoint activation depends on activities of these protein kinases.

19.3.2.3 Aurora-B–BRCA2–BARD1β in Cytokinesis

BRCA1 interacts with BARD1 to achieve its many biological functions such as DNA repair and DNA and spindle damage induced cell cycle checkpoints. While both BRCA1 and BARD1 are co-localized with Aurora-A at centrosome in interphase and at mitotic spindle pole in mitosis, intriguingly BARD1 splicing isoform β, which lack the BRCA1-interacting RING finger and is overexpressed in breast, ovarian, and endometrial cancer cells, co-localizes with Aurora-B at midbody where BARD1β interacts with and stabilizes Aurora-B facilitating a complex formation with BRCA2 (Ryser et al. 2009). BRCA2-deficient cells reportedly display defect in midbody formation and abscission due to abnormalities in myosin II organization (Daniels et al. 2004). Recent work from Modal et al., revealed that BRCA2 is involved in the recruitment of ESCRT-associated proteins, Alix and Tsg101, and formation of CEP55-Alix and CEP55-Tsg101 complexes during abscission whose complex formations are also dependent on Aurora-B mediated phosphorylation of MKLP1 (Mondal and Rowley 2012). Thus, although BARD1β appears to be involved in coordinating the functions of Aurora-B and BRCA2 in cleavage furrow formation and abscission of midbody, consequence of Aurora-B overexpression on BRCA2 and BARD1β function need to be investigated. Since Aurora-B phosphorylates CHMP4C in response to the formation of anaphase chromosome bridge, it will be interesting to investigate whether BRCA2 and BARD1β cooperatively function with Aurora-B in abscission checkpoint control.

Taken together, these findings collectively suggest that the potential signaling pathways compromised by deregulated Aurora-B expression which lead to chromosome instability and tumorigenesis involve postmitotic G1 phase checkpoint to prevent endoreduplication after an aberrant mitosis which depends on p53 and pRb (Borel et al. 2002). It is widely known that pRB expression is low while p53 mutations are rare in AML whereas inactivation of wild-type p53 protein frequently occurs through overexpression of its negative regulator MDM2. Thus, one proposed effective therapeutic strategy against AML may be a combinational usage of Aurora-B inhibitor and MDM2 inhibitor, leading to activation of p53 signaling pathways and subsequent induction of cell death.

19.4 Significance and Future of Aurora Kinases as Therapeutic Targets

Involvement of Aurora kinases in deregulated proliferation and multiple tumor-relevant pathways elicited interest in their evaluation as potential therapeutic targets for human malignancies. Preclinical success with Aurora kinase inhibitors in attenuating growth of tumor cells led to the initiation of early phase clinical trials and development of next generation small molecule inhibitors of Aurora kinases designed to achieve greater specificity and therapeutic efficacy. A number of such inhibitors are at various stages of preclinical investigations or early phase clinical trials. Whether these drugs will be effective on their own or in combination with additional drugs targeting other oncogenic pathways would have to be decided based on the outcomes of the trials currently under way. Since the small molecules inhibit Aurora kinases by targeting the ATP-binding pocket, many of these inhibitors are not specific for one member of the Aurora kinase family and also have "off target" effects as the ATP-binding pockets are highly homologous across the protein kinase family (Scapin et al. 2006). Since therapeutic targeting of Aurora kinases in clinical setting is reviewed in depth separately, we are only briefly discussing the implications of a few Aurora-kinase-targeted therapeutic approaches relevant to this chapter. Multiple Aurora kinase inhibitors are currently undergoing clinical evaluation in patients with AML, Ph^{+}ALL, or CML (Farag 2011) and the results seem to indicate that single agent activity is modest compared with the response rates achieved with combination therapy as reported for the Aurora kinase inhibitor AZD1152 with low dose cytabine (Kantarjian et al., Blood Abstract, 2010). Early phase clinical trials suggest a role for Aurora kinase inhibitors in Ph^{+}ALL and CML, even in patients with compound Abl kinase domain mutations that render patients resistant to multiple Bcr–Abl kinase inhibitors (Shah et al. 2007). Although the mechanism of this efficacy is not conclusively documented, available data appear to reflect that "off target" effect of Aurora inhibitors along with Aurora inhibition is important in treating Ph^{+}ALL and CML. A similar "off target" effect seems to underlie the efficacy of Aurora inhibitors in the treatment of multiple myeloma (MM), where Aurora kinases have been validated as potential targets for therapy (Manfredi et al. 2007). For example, ENMD-2076, an inhibitor of Aurora kinase and multiple tyrosine kinases has been shown to have antimyeloma activity interfering with multiple pathways important in this disease (Wang et al. 2010). These examples of Aurora-kinase-inhibitor-mediated positive therapeutic response in human malignancies suggest that maximal efficacy of Aurora kinase inhibition in treating human cancers can be accomplished either through synergistic action of a drug targeting additional cancer-relevant pathways or through the "off target" action of the Aurora inhibitor itself. Significant amount of preclinical data supports the concept of combining Aurora inhibitors with agents targeting Aurora-A-interacting proteins and effectors or with less specific antimitotic agents such as microtubule poisons and ionizing radiation to achieve maximum sensitization of tumor cells (Venkataraman et al. 2012). Inhibition of Aurora-A sensitizes cancer cells to radiation therapy by attenuating negative regulatory phosphorylation on Aurora-A-interacting checkpoint activators, such as p53 (Dar et al. 2010). Synergistic inhibition of Aurora-A together with members of core proliferation and survival pathways, such as EGFR and SRC, enhancing therapeutic efficacy has also been reported (Astsaturov et al. 2010; Ratushny et al. 2012). These studies reinforce the idea of a systems biology approach towards developing targeted therapeutic strategies aimed at multiple aberrantly expressing components of the cancer-relevant cross-talking pathways. It is, therefore, logical to envision future targeted therapies combining Aurora kinase inhibition with EGFR inhibitor cetuximab and SRC inhibitor dasatinib (Nikonova et al. 2012). These combinatorial targeted therapeutic strategies will become progressively more effective as Aurora kinase functional interactions are elucidated in greater detail in different malignancies.

Conclusions

Aurora kinases are expressed at elevated levels in many cancers and shown to functionally interact with critical regulators of mitosis as well as in important signaling pathways deregulated in tumor cells. Besides their role in oncogenesis, Aurora kinases have been functionally implicated in some developmental disorders and chronic diseases. These findings have generated significant interest in evaluating Aurora kinases as therapeutic targets in cancer. As a result, many small-molecule Aurora kinase inhibitors have been investigated for their antitumor activities in preclinical studies and some have advanced to early phase clinical trials. It is proposed that therapeutic efficacy of Aurora kinase inhibiting small molecule drugs may be improved when combined with agents targeting additional cancer-relevant signaling pathways. This may be best achieved by designing combinatorial drug regimens of Aurora kinase inhibitors with agents targeting additional components of oncogenic pathways cross talking with Aurora kinases deregulated in cancer.

Acknowledgments Work in the author's laboratory is supported by grants of the National Institutes of Health and the University Cancer Foundation grants from UTMDACC to S.S.

References

Adams RR, Wheatley SP (2000) INCENP binds the Aurora-related kinase AIRK2 and is required to target it to chromosomes, the central spindle and cleavage furrow. Curr Biol 10:1075–1078

Astsaturov I, Ratushny V et al (2010) Synthetic lethal screen of an EGFR-centered network to improve targeted therapies. Sci Signal 3(140):ra67

Barford D (2011) Structural insights into anaphase-promoting complex function and mechanism. Philos Trans R Soc Lond B Biol Sci 366:3605–3624

Barros TP, Kinoshita K et al (2005) Aurora A activates D-TACC-Msps complexes exclusively at centrosomes to stabilize centrosomal microtubules. J. Cell Biol 170:1039–1046

Bayliss R, Sardon T et al (2003) Structural basis of Aurora-A activation by TPX2 at the mitotic spindle. Mol Cell 12:851–862

Bee C, Moshnikova A et al (2010) The growth and tumor suppressor novel ras effector 1A (NORE1A) is a regulatory node between ras signaling and microtubule nucleation. J Biol Chem 285:16258–16266

Birkenfeld J, Nalbant P et al (2007) GEF-H1 modulates localized RhoA activation during cytokinesis under the control of mitotic kinases. Dev Cell 12:699–712

Bischoff JR, Anderson L (1998) A homologue of Drosophila aurora kinase is oncogenic and amplified in human colorectal cancers. EMBO J 17:3052–3065

Borel F, Lohez OD et al (2002) Multiple centrosomes arise from tetraploidy checkpoint failure and mitotic centrosome clusters in p53 and RB pocket protein-compromised cells. Proc Natl Acad Sci U S A 99:9819–9824

Briassouli P, Chan F et al (2007) Aurora-A regulation of nuclear factor-kappaB signaling by phosphorylation of IkappaBalpha. Cancer Res 67:1689–1695

Brodie KM, Henderson BR (2012) Characterization of BRCA1 protein targeting, dynamics, and function at the centrosome: a role for the nuclear export signal, CRM1, and Aurora A kinase. J Biol Chem 287:7701–7716

Carlton JG, Caballe A et al (2012) ESCRT-III governs the Aurora B-mediated abscission checkpoint through CHMP4C. Science 336:220–225

Carmena M, Ruchaud S et al (2009) Making the Auroras glow: regulation of Aurora A and B kinase function by interacting proteins. Curr Opin Cell Biol 21:796–805

Chan YW, Jeyaprakash AA et al (2012) Aurora B controls kinetochore–microtubule attachments by inhibiting Ska complex–KMN network interaction. J Cell Biol 196:563–571

Cheeseman IM, Chappie JS et al (2006) The conserved KMN network constitutes the core microtubule-binding site of the kinetochore. Cell 127:983–997

Daniels MJ, Wang Y et al (2004) Abnormal cytokinesis in cells deficient in the breast cancer susceptibility protein BRCA2. Science 306:876–9

Dar AA, Belkhiri A et al (2009) The Aurora kinase A regulates GSK-3beta in gastric cancer cells. Oncogene 28:866–875

Dar AA, Goff LW et al (2010) Aurora kinase inhibitors–rising stars in cancer therapeutics? Mol Cancer Ther 9:268–278

Daum JR, Wren JD et al (2009) Ska3 is required for spindle checkpoint silencing and the maintenance of chromosome cohesion in mitosis. Curr Biol 19:1467–1472

DeLuca JG, Gall WE et al (2006) Kinetochore microtubule dynamics and attachment stability are regulated by Hec1. Cell 127:969–982

Douglas ME, Davies T et al (2010) Aurora B and 14–3-3 coordinately regulate clustering of centralspindlin during cytokinesis. Curr Biol 20:927–933

Eyers PA, Erikson E et al (2003) A novel mechanism for activation of the protein kinase Aurora A. Curr Biol 13:691–697

Farag SS (2011) The potential role of Aurora kinase inhibitors in haematological malignancies. Br J Haematol 155:561–579

Fernández-Miranda G, Trakala M et al (2011) Genetic disruption of aurora B uncovers an essential role for aurora C during early mammalian development. Development 138:2661–2672

Fu J, Bian M et al (2009) A single amino acid change converts Aurora-A into Aurora-B-like kinase in terms of partner specificity and cellular function. Proc Natl Acad Sci U S A 106:939–944

Garber K (2005) Divide and conquer: new generation of drugs targets mitosis. J Natl Cancer Inst 97:874–876

Gautschi O, Heighway J et al (2008) Aurora kinases as anticancer drug targets. Clin Cancer Res 14:1639–1648

Glotzer M (2009) The 3Ms of central spindle assembly: microtubules, motors and MAPs. Nat Rev Mol Cell Biol 10 9–20

Gruss OJ, Vernos I (2004) The mechanism of spindle assembly: functions of Ran and its target TPX2. J Cell Biol 166:949–955

Gully CP, Velazquez-Torres G et al (2012) Aurora B kinase phosphorylates and instigates degradation of p53. Proc Natl Acad Sci U S A 109:E1513–1522
Ha GH, Baek KH et al (2007) p53 activation in response to mitotic spindle damage requires signaling via BubR1-mediated phosphorylation. Cancer Res 67:7155–7164
Hans F, Skoufias DA et al (2009) Molecular distinctions between Aurora A and B: a single residue change transforms Aurora A into correctly localized and functional Aurora B. Mol Biol Cell 20:3491–3502
Hégarat N, Smith E et al (2011) Aurora A and Aurora B jointly coordinate chromosome segregation and anaphase microtubule dynamics. J Cell Biol 195:1103–1113
Huang YF, Chang MD et al (2009) TTK/hMps1 mediates the p53-dependent postmitotic checkpoint by phosphorylating p53 at Thr18. Mol Cell Biol 29:2935–2944
Irelan JT, Murphy TJ et al (2007) A role for IkappaB kinase 2 in bipolar spindle assembly. Proc Natl Acad Sci U S A 104:16940–5
Jelluma N, Brenkman AB et al (2008) Mps1 phosphorylates Borealin to control Aurora B activity and chromosome alignment. Cell 132:233–246
Jeyaprakash AA, Klein UR et al (2007) Structure of a Survivin-Borealin-INCENP core complex reveals how chromosomal passengers travel together. Cell 131:271–285
Joukov V, De Nicolo A et al (2010) Centrosomal protein of 192 kDa (Cep192) promotes centrosome-driven spindle assembly by engaging in organelle-specific Aurora A activation. Proc Natl Acad Sci U S A 107:21022–21027
Kanda A, Kawai H et al (2005) Aurora-B/AIM-1 kinase activity is involved in Ras-mediated cell transformation. Oncogene 24:7266–7272
Kantarjian HM, Sekeres MA et al (2010) Phase I study to assess the safety and tolerability of AZD1152 In Combination with Low Dose Cytosine Arabinoside In Patients with Acute Myeloid Leukemia (AML). Blood (ASH Annual Meeting Abstracts) 116:Abstract 656
Karthigeyan D, Prasad SB et al (2011) Biology of Aurora A kinase: Implications in cancer manifestation and therapy. Med Res Rev 31:757–793
Kashatus DF, Lim KH et al (2011) RALA and RALBP1 regulate mitochondrial fission at mitosis. Nat Cell Biol 13:1108–1115
Katayama H, Sasai K et al (2004) Phosphorylation by aurora kinase A induces Mdm2-mediated destabilization and inhibition of p53. Nat Genet 36: 55–62
Katayama H, Sasai K et al (2008) Aurora kinase-A regulates kinetochore/chromatin associated microtubule assembly in human cells. Cell Cycle 7:2691–2704
Katayama H, Wang J et al (2012).Aurora kinase-A inactivates DNA damage-induced apoptosis and spindle assembly checkpoint response functions of p73. Cancer Cell. 21:196–211
Kim Y, Holland AJ et al (2010) Aurora kinases and protein phosphatase 1 mediate chromosome congression through regulation of CENP-E. Cell 142:444–455
Kwon YW, Kim IJ et al (2012) Pten regulates Aurora-A and cooperates with Fbxw7 in modulating radiation-induced tumor development. Mol Cancer Res 10:834–844
Lim KH, Brady DC et al (2010) Aurora-A phosphorylates, activates, and relocalizes the small GTPase RalA. Mol Cell Biol 30:508–523
Littlepage LE, Ruderman JV (2002) Identification of a new APC/C recognition domain, the A box, which is required for the Cdh1-dependent destruction of the kinase Aurora-A during mitotic exit. Genes Dev 16:2274–2285
Liu Q, Kaneko S et al (2004) Aurora-A abrogation of p53 DNA binding and transactivation activity by phosphorylation of serine 215. J Biol Chem 279:52175–52182
Liu D, Vleugel M et al (2010) Regulated targeting of protein phosphatase 1 to the outer kinetochore by KNL1 opposes Aurora B kinase. J Cell Biol 188:809–820
Ma N, Matsunaga S et al (2011) The nuclear scaffold protein SAF-A is required for kinetochore-microtubule attachment and contributes to the targeting of Aurora-A to mitotic spindles. J Cell Sci 124:394–404
Macůrek L, Lindqvist A et al (2008) Polo-like kinase-1 is activated by aurora A to promote checkpoint recovery. Nature 455:119–123

Maldonado M, Kapoor TM (2011) Constitutive Mad1 targeting to kinetochores uncouples checkpoint signalling from chromosome biorientation. Nat Cell Biol 13:475–482
Manfredi MG, Ecsedy JA et al (2007) Antitumor activity of MLN8054, an orally active small-molecule inhibitor of Aurora A kinase. Proc Natl Acad Sci U S A 104:4106–4111
Maris JM, Morton CL et al (2010) Initial testing of the aurora kinase a inhibitor MLN8237 by the Pediatric Preclinical Testing Program (PPTP). Pediatr Blood Cancer 55:26–34
Mondal G, Rowley M (2012) BRCA2 localization to the midbody by Filamin A regulates CEP55 signaling and completion of cytokinesis. Dev Cell 23:137–152
Mori D, Yano Y et al (2007) NDEL1 phosphorylation by Aurora-A kinase is essential for centrosomal maturation, separation, and TACC3 recruitment. Mol Cell Biol 27:352–367
Mori D, Yamada M et al (2009) An essential role of the aPKC-Aurora A-NDEL1 pathway in neurite elongation by modulation of microtubule dynamics. Nat Cell Biol 11:1057–1068
Nair JS, Ho AL et al (2009) Aurora B kinase regulates the postmitotic endoreduplication checkpoint via phosphorylation of the retinoblastoma protein at serine 780. Mol Biol Cell 20: 2218–2228
Nikonova AS, Astsaturov I et al (2012) Aurora A kinase (AURKA) in normal and pathological cell division. Cell Mol Life Sci 70:661–687
Norden C, Mendoza M et al (2006) The NoCut pathway links completion of cytokinesis to spindle midzone function to prevent chromosome breakage. Cell 125:85–98
Ota T, Suto S et al (2002) Increased mitotic phosphorylation of histone H3 attributable to AIM-1/Aurora-B overexpression contributes to chromosome number instability. Cancer Res 62:5168–5177
Otto T, Horn S et al (2009) Stabilization of N-Myc is a critical function of Aurora A in human neuroblastoma. Cancer Cell 15:67–78
Ouchi M, Fujiuchi N et al (2004) BRCA1 phosphorylation by Aurora-A in the regulation of G2 to M transition. J Biol Chem 279:19643–19648
Plotnikova OV, Pugacheva EN et al (2010) Rapid calcium-dependent activation of Aurora-A kinase. Nat Commun 1:64. doi: 10.1038/ncomms1061
Plotnikova OV, Nikonova AS et al (2012) Calmodulin activation of Aurora-A kinase (AURKA) is required during ciliary disassembly and in mitosis. Mol Biol Cell 23:2658–2670
Pugacheva EN, Jablonski SA et al (2007) HEF1-dependent Aurora A activation induces disassembly of the primary cilium. Cell 129:1351–1363
Ratushny V, Pathak HB et al (2012) Dual inhibition of SRC and Aurora kinases induces postmitotic attachment defects and cell death. Oncogene 31:1217–1227
Rong R, Jiang LY et al (2007) Mitotic kinase Aurora-A phosphorylates RASSF1A and modulates RASSF1A-mediated microtubule interaction and M-phase cell cycle regulation. Oncogene 26:7700–7708
Ruchaud S, Carmena M et al (2007) Chromosomal passengers: conducting cell division. Nat Rev Mol Cell Biol 8:798–812
Ryser S, Dizin E et al (2009) Distinct roles of BARD1 isoforms in mitosis: full-length BARD1 mediates Aurora B degradation, cancer-associated BARD1beta scaffolds Aurora B and BRCA2. Cancer Res 69:1125–1134
Sasai K, Katayama H et al (2004) Aurora-C kinase is a novel chromosomal passenger protein that can complement Aurora-B kinase function in mitotic cells. Cell Motil Cytoskeleton 59:249–263
Saurin AT, van der Waal MS et al (2011) Aurora B potentiates Mps1 activation to ensure rapid checkpoint establishment at the onset of mitosis. Nat Commun 2:316
Scapin S, Carneiro F et al (2006) The crystal structure of the small GTPase Rab11b reveals critical differences relative to the Rab11a isoform. J Struct Biol 154:260–268
Schmidt JC, Kiyomitsu T et al (2010) Aurora B kinase controls the targeting of the Astrin–SKAP complex to bioriented kinetochores. J Cell Biol 191:269–280
Seki A, Coppinger JA (2008) Bora and the kinase Aurora a cooperatively activate the kinase Plk1 and control mitotic entry. Science 320:1655–1658
Shah NP, Skaggs BJ et al (2007) Sequential ABL kinase inhibitor therapy selects for compound drug-resistant BCR-ABL mutations with altered oncogenic potency. J Clin Invest 117:2562–2569

Steigemann P, Wurzenberger C et al (2009) Aurora B-mediated abscission checkpoint protects against tetraploidization. Cell 136:473–484

Thompson SL, Compton DA (2011) Chromosome missegregation in human cells arises through specific types of kinetochore–microtubule attachment errors. Proc Natl Acad Sci U S A 108:17974–8

Torchia EC, Caulin C et al (2012) Myc, Aurora Kinase A, and mutant p53(R172H) co-operate in a mouse model of metastatic skin carcinoma. Oncogene 31:2680–2690

Toya M, Terasawa M et al (2011) A kinase-independent role for Aurora A in the assembly of mitotic spindle microtubules in *Caenorhabditis elegans* embryos. Nat Cell Biol 13:708–714

Tsai MY, Zheng Y (2005) Aurora A kinase-coated beads function as microtubule-organizing centers and enhance RanGTP-induced spindle assembly. Curr Biol 15:2156–2163

van der Waal MS, Hengeveld RC et al (2012a) Cell division control by the chromosomal passenger complex. Exp Cell Res 318:1407–1420

van der Waal MS, Saurin AT et al (2012b) Mps1 promotes rapid centromere accumulation of Aurora B. EMBO Rep. doi:10.1038/embor.2012.93

Venkataraman S, Alimova I et al (2012) Targeting Aurora Kinase A enhances radiation sensitivity of atypical teratoid rhabdoid tumor cells. J Neurooncol 107:517–526

Wang X, Sinn AL et al (2010) Preclinical activity of a novel multiple tyrosine kinase and aurora kinase inhibitor, ENMD-2076, against multiple myeloma. Br J Haematol 150:313–325

Wang F, Ulyanova NP et al (2011) A positive feedback loop involving Haspin and Aurora B promotes CPC accumulation at centromeres in mitosis. Curr Biol 21:1061–1069

Welburn JP, Grishchuk EL et al (2009) The human kinetochore Ska1 complex facilitates microtubule depolymerization-coupled motility. Dev Cell 16:374–385

Welburn JP, Vleugel M et al (2010) Aurora B phosphorylates spatially distinct targets to differentially regulate the kinetochore–microtubule interface. Mol Cell 38:383–392

Williams JP, Stewart T et al (2006) The retinoblastoma protein is required for Ras-induced oncogenic transformation. Mol Cell Biol 26:1170–1182

Wu L, Ma CA (2011) Aurora B interacts with NIR-p53, leading to p53 phosphorylation in its DNA-binding domain and subsequent functional suppression. J Biol Chem 286:2236–2244

Wu JC, Chen TY et al (2005) Identification of V23RalA-Ser194 as a critical mediator for Aurora-A-induced cellular motility and transformation by small pool expression screening. J Biol Chem 280:9013–9022

Yang H, Ou CC et al (2004) Aurora-A kinase regulates telomerase activity through c-Myc in human ovarian and breast epithelial cells. Cancer Res 64:463–467

Yang C, Tang X et al (2011) Aurora-B mediated ATM serine 1403 phosphorylation is required for mitotic ATM activation and the spindle checkpoint. Mol Cell 44:597–608

Yu X, Minter-Dykhouse K et al (2005) Chfr is required for tumor suppression and Aurora A regulation. Nat Genet 37:401–406

Zhang X, Ems-McClung SC et al (2008) Aurora A phosphorylates MCAK to control ran-dependent spindle bipolarity. Mol Biol Cell 19:2752–2765

Zhou H, Kuang J et al (1998) Tumour amplified kinase STK15/BTAK induces centrosome amplification, aneuploidy and transformation. Nat Genet 20:189–193

Chapter 20
The Clinical Development of Aurora Kinase Inhibitors in Acute Myeloid Leukemia

Kevin R. Kelly, Ciara L. Freeman and Francis J. Giles

Abstract The Aurora family of serine/threonine kinases is essential for chromosome alignment, segregation, centrosomal maturation, mitotic spindle formation, and cytokinesis during mitosis. Their fundamental role in cell cycle regulation and aberrant expression in a broad range of malignancies prompted the development of small molecules that selectively inhibit their activity. Recent studies have revealed new insights into the cellular effects of Aurora kinase inhibition in the treatment of acute myeloid leukemia (AML). Moreover, early-phase clinical studies on AML have shown that these agents have therapeutic efficacy both alone and in combination with chemotherapy.

Keywords Aurora kinases · Acute myeloid leukemia

20.1 Introduction

The Aurora family of mitotic kinases was discovered while studying mutant alleles in drosophila melanogaster. The mutant allele caused failure of centrosome separation resulting in monopolar spindles reminiscent of the aurora borealis (Glover et al. 1995). In mammalian cells, there are three known members of the Aurora

F. J. Giles (✉)
HRB Clinical Research Facility, Galway & Trinity College Dublin,
National University of Ireland, Geata an Eolais University Road, Galway, Ireland
e-mail: frankgiles@aol.com

K. R. Kelly
Institute for Drug Development, Cancer Therapy, and Research Center,
University of Texas, Health Science Center, Hematology Clinic—Cancer
Therapy & Research Center, San Antonio, TX 78229, USA
e-mail: Kevrkelly77@hotmail.com

C. L. Freeman
Department of Haematology, Barts and The London NHS Trust, London, E1 2ES, UK
e-mail: Ciara.Freeman@bartshealth.nhs.co.uk

M. Andreeff (ed.), *Targeted Therapy of Acute Myeloid Leukemia*,
Current Cancer Research, DOI 10.1007/978-1-4939-1393-0_20

family: Aurora A, B, and C. Aurora A and B are expressed in all proliferating cells, whereas the expression of Aurora C is mainly restricted to the testes where it plays a role in spermatogenesis (Kimura Met al. 1999).

Aurora A and B regulate a diverse array of events throughout mitosis. Aurora A is critical for the maintenance of mitotic spindle bipolarity, separation of centrosomes, and accurate chromosome alignment (Crane et al 2004; Liu and Ruderman 2006). The regulation of the progression from G_2 to M by Aurora A has been proposed to occur, in part, through the relocalization of cyclin B1 to the nucleus and activation of cyclin-dependent kinase. Suppression of Aurora A expression leads to G_2/M arrest whereas ectopic expression abrogates the G_2/M checkpoint (Fu et al. 2007).

Aurora B forms the "enzymatic heart" of the chromosomal passenger complex (CPC) (Vader et al. 2006). The CPC includes Aurora-B, survivin, and inner centromere protein. This complex is crucial in the formation of correct microtubule–kinetochore attachments, the establishment of the spindle assembly checkpoint and cytokinesis (Hauf et al. 2003; Ditchfield et al. 2003; Giet and Glover 2001; Goto et al. 2003). In addition, Aurora B phosphorylates histone H3 at Ser^{10} and Ser^{28}, facilitating chromosome condensation and subsequent alignment during mitosis (Monier et al. 2007). Chemical inhibition of Aurora B impairs all CPC functions. Importantly, its depletion results in mitotic exit without cell separation causing a doubling of DNA (Fu et al. 2007).

Aurora C, with its specialized location in spermatocytes, functions similarly to Aurora B (Ducat and Zheng 2004; Sausville 2004). Although its role has not been clearly defined, there are some data to suggest that Aurora C is expressed in leukemic cells and interacts with survivin to prevent apoptosis and induce cell cycle progression (Kobayashi et al. 2006).

20.1.1 Aurora Kinases in Acute Myeloid Leukemia

Aurora kinases are aberrantly expressed in a variety of malignancies including acute myeloid leukemia (AML) and chronic myeloid leukemia (CML), lymphoma, prostate, colon, breast, lung, prostate, head and neck, and thyroid cancer as well as in most forms of leukemia (Bischoff et al. 1998; Sen et al. 1997; Smith et al. 2005; Chieffi et al. 2006; Reiter et al. 2006; Ulisse et al. 2006). A number of mechanisms lead to increased Aurora A expression in malignancies including amplification of the chromosomal region 20q13.2 which comprises the Aurora A kinase gene (Jeng 2004). Other mechanisms leading to Aurora A over expression in malignancy include posttranslational stabilization and transcriptional induction (Farruggio et al. 1999; Crane et al. 2004). In contrast to AURKA, the AURKB gene does not appear to be amplified in cancer. However, Aurora kinase B is highly expressed in many tumor types including prostate as well as head and neck squamous cell carcinoma (Chieffi et al. 2006; Reiter et al. 2006).

Aurora A may function as an oncogene through the induction of genetic instability and enhanced survival signaling. Ectopic over expression of AURKA leads to transformation of rodent fibroblasts (Bischoff et al. 1998; Huang et al. 2008; Zhou et al. 1998). Unsurprisingly, given its role during mitosis, high Aurora A expression is correlated with anueploidy and has also been associated with poor prognosis when present in malignant cells (Huang et al. 2008; Miyoshi et al. 2001; Crosio et al. 2002). Aurora A has been shown to interact with several crucial cell cycle regulators, including p53 (Katayama et al. 2004) and BRCA-1 (Ouchi et al. 2004). By impairing the regulatory function of these proteins, notably at the DNA checkpoint, genetic instability and carcinogenesis are more likely to occur (Huang et al. 2008). Aurora A aberrant expression may also prevent apoptosis in cancer cells by indirectly activating nuclear factor-kappa B (NF-κB) through phosphorylation of its inhibitor IκB, which targets IκB for degradation (Briassouli et al. 2007).

Aurora A and B are over expressed in AML. Aurora A expression by Western blot analysis and immunohistochemistry staining was found to be highly elevated in 65 of 98 (66 %) of AML cells from de novo AML patients with no significant difference between age, race, blood count, or French–American–British (FAB) classification. In contrast, stem cells from the normal "control" bone marrow specimens revealed negligible levels of Aurora A protein expression (12 of 12). Fewer patients had elevated Aurora B expression (40 of 98 patients, 40 %) (Huang et al. 2008). High Aurora kinase expression in a variety of leukemia types makes it a promising molecular target.

20.1.2 Aurora Kinase Inhibitors in Clinical Development

The awareness that Aurora kinases were aberrantly expressed in malignancies and were involved in tumorigenesis led to the development of a large number of Aurora kinase inhibitors (AKIs) for cancer therapy. Early studies revealed insights into the molecular consequences of Aurora A inhibition in cancer cells. Microinjection of Aurora A antibodies and/or RNAi-mediated knockdown of Aurora A kinase led G_2/M arrest, growth inhibition, and apoptosis (Hirota et al. 2003; Hata et al. 2005; Marumoto et al. 2002). Other consequences of targeted Aurora A inhibition include disruption of multiple mitotic events, culminating in failure of centrosome separation, monopolar spindle formation, and incomplete cytokinesis (Marumoto et al. 2003). On the other hand, RNAi-mediated inhibition of Aurora B expression disrupts chromosomal biorientation, cytokinesis, and the mitotic checkpoint (Goto et al. 2003; Severson et al. 2000). Abrogation of the mitotic checkpoint allows cells to go through multiple cycles of aberrant mitosis without cytokinesis resulting in massive polyploidy leading to cell death by mitotic catastrophe (Nair et al. 2009; Hauf et al. 2003). In the light of these multiple molecular sequelae, AKIs may be selectively more toxic to rapidly dividing cancer cells over nondividing cells.

20.2 Pan-AKIs

20.2.1 ZM447439

ZM44743 is a quinazoline derivative that is an ATP-competitive inhibitor of both Aurora A and B developed by AstraZeneca (Ditchfield et al. 2003). It was one of the first AKIs to be extensively characterized and has been a useful tool to study Aurora kinases in cancer. In preclinical models of AML and acute lymphocytic leukemia (ALL), ZM447439 induced growth inhibition, accumulation of polyploid cells, and apoptosis but had no effect on the clonogenic growth of myeloid stem cells harvested from healthy volunteers (Ikezoe et al. 2007). Consistent with its inhibition of Aurora B, ZM447439 treatment resulted in defects in chromosome condensation and alignment and impairment of the spindle checkpoint, phenotypes indicative of Aurora B inhibition (Ditchfield et al. 2003; Gadea and Ruderman 2005). While the use of ZM447439 has been important to study the preclinical consequences of Aurora inhibition in cancer, it has not been developed clinically.

20.2.2 VX-680/MK-0457

The first AKI to enter clinical evaluation was VX-680/MK-0457, initially developed by Vertex Pharmaceuticals. VX-680 is a pyrimidine derivative that has activity against all three Aurora kinases (inhibition constant values of 0.7, 18, and 4.6 nM for Aurora A, B, and C respectively) (Harrington et al. 2004). Treatment with VX-680 led to cell arrest and apoptosis in leukemia cell lines. As with other pan-AKIs, VX-680 induced a cellular phenotype consistent with Aurora B inhibition, i.e., inhibition of cytokinesis, polyploidy, and reduced phosphorylation of histone H3.

Of potentially important therapeutic implications, VX-680 has significant activity against wild-type and T315I-mutated BCR-ABL (Carter et al. 2005). In turn, it can inhibit the proliferation of cells expressing a wide variety of the clinically important BCR-ABL mutations with IC_{50} values of 100–200 nM. In CML, mutations in BCR-ABL can result in failure of tyrosine kinase inhibitor (TKI) therapy. Of particular importance is the T315I mutation at the base of the ATP-binding pocket of BCR-ABL that accounts for 10–15% of mutations and is highly resistant to currently available TKI therapy (Giles et al. 2007). Consistent with this preclinical data, three patients with the T315I mutation (two with CML and one with Philadelphia chromosome positive (Ph^+) ALL) had objective responses in a phase I/II study of VX-680/MK-0457 (Giles et al. 2007).

MK-0457 synergizes with the pan-histone deacetylase inhibitor (HDACI) vorinostat in preclinical models of leukemia (Dai et al. 2008; Fiskus et al. 2008). This synergy may be in part mediated through HDACIs disruption of the chaperone function of HSP90, a known client protein of BCR-ABL, and thus may make BCR-ABL more sensitive to MK-0457 inhibition. Both vorinostat and MK-0457 were shown to induce the expression of the BH3-only pro-apoptotic protein BIM and the lethal effects of the combination were shown to be particularly dependent on this pathway.

Many of the AKIs in clinical development have off-target inhibition of other clinically relevant tyrosine kinases. VX-680 also inhibits FMS like tyrosine kinase (FLT3) that is over expressed or mutated in many AML patients (Harrington et al. 2004; Carter et al. 2005; Schnittger et al. 2002). Several other AKIs including ZM447439 and AS703569 have demonstrated off-target inhibition of FLT3 *in vitro* (Huang et al. 2008; Ikezoe et al. 2007; Sarno et al. 2007; Shiotsu et al. 2007). In addition to off-target effect on FLT3, VX-680 also inhibits Janus kinase 2 (JAK2). The Val617Phe point mutation in the JAK2 gene leads to constitutive tyrosine phosphorylation activity and occurs in the majority of patients with myeloproliferative disorders (MPD) (Baxter et al. 2005). Consistent with this, VX-680 inhibits JAK2 *in vitro* and has normalized platelet counts and induced partial remission in patients with $JAK2^{V617F}$ positive MPD and AML transformed from MPD (Giles et al. 2006a, b, 2007). Despite these promising and intriguing clinical responses, the clinical development of VX-680/MK-0457 was halted due to concerns regarding cardiac toxicity.

20.2.3 PHA739358/Danusertib

PHA739358 is 3-aminopyrazole derivative developed by Nerviano/Pfizer (Fancelli et al. 2006). It is an ATP-competitive inhibitor of all three Aurora kinases (IC_{50} of 0.013, 0.079, and 0.061 for Aurora A, B, and C, respectively). As well as inhibiting the Aurora family of kinases, PHA-739358 inhibited other tumor-related kinases such as fibroblast growth factor receptor 1 (FGFR1), transforming tyrosine kinase protein (TRKA), ABL, and rearranged during transfection (RET) in the low nanomolar range. The parent compound of danusertib, PHA-680632, had demonstrated activity against a wide range of cancer cell lines *in vitro* and *in vivo* at well-tolerated doses (Fancelli et al. 2005).

Danusertib was rationally designed as a more potent Aurora A inhibitor by examining the x-ray co-crystal structure of PHA-680632 in complex with Aurora A. As is the case with other pan-AKIs, cells treated with PHA-739358 underwent endoreduplication and showed reduced phosphorylation of histone H3 (Fancelli et al. 2006). PHA739358 showed significant activity in preclinical models of cancer (Carpinelli and Moll 2008). Like VX-680/MK-0457, PHA-739358 inhibits BCR-ABL including T315I-mutated BCR-ABL and synergistically increased the efficacy of imatinib (Gontarewicz et al. 2008).

Danusertib has been evaluated in a number of early-phase clinical studies using various schedules (Cohen et al. 2009a; Steeghs et al. 2009a). The most common toxicity was neutropenia that was typically of short duration but dose limiting. Non-hematological toxicities reported included nausea, mucositis, and alopecia. Half-life is 30 h and the dosing regimen recommended for phase II study was 330 mg/m^2 as 6-h intravenous infusion was the recommended dose for phase II studies

(Steeghs et al. 2009b). Phase I single-agent data in patients with advanced-stage CML or Ph$^+$ ALL resistant or intolerant of imatinib or second-generation TKI therapy were encouraging. The cohort of 23 resistant/relapsed patients treated (11 Ph$^+$ ALL, eight CML blast crisis, four CML-accelerated phase) showed three cytogenetic (one complete, one partial, one minimal) and five hematological responses, and one clinical improvement (reduction in extramedullary disease mass) at the time of reporting (Cortes et al. 2009). Further phase I and II trials evaluating danusertib as single agent or in combination for both solid tumors and hematological malignancies are ongoing (Shiotsu et al. 2007).

20.2.4 R763/AS703569/MSC1992371A

R763 is a pan-AKI developed by Rigel Pharmaceuticals Inc that is orally available (Ahuja et al. 2007). In addition to inhibiting Aurora A and B, it also has activity against FLT3 kinase, vascular endothelial growth factor receptor (VEGFR) kinase, BCR-ABL, and T315I-mutated BCR-ABL. It also has activity against JAK2 kinase but at higher concentrations. Oral administration of R763/AS703569 markedly reduced tumor growth in xenograft models of breast, colon, pancreatic, lung, and ovarian tumors. MV4-11, an AML cell line, which harbors the FLT3 internal tandem duplication mutation, was particularly sensitive to R763 *in vitro* and *in vivo*. Consistent with Aurora kinase inhibition, R763 led to endoreduplication and an increase in the DNA content of the nuclei without subsequent cytokinesis. In addition, the inhibition of other non-Aurora kinases may enhance its antitumor effects. Interesting in a panel of cell lines representing FLT3-ITD AML, there was a trend toward the bypassing of endoreduplication and direct induction of apoptosis.

A phase I study of R763 has been completed in advanced hematologic malignancies evaluating different potential dosing schedules for the drug (Sonet et al. 2008). Dose-limiting toxicities were grade 3–4 mucositis/stomatitis severe neutropenia with infection, sepsis, and diarrhea. There were some responses to monotherapy including two complete responses (CRs) among 54 patients with AML and one among three patients with ALL. Partial responses (PRs) were seen in myelodysplastic syndromes (MDS), MPD, and CML. Four disease-specific expansion cohorts were initiated after the maximum tolerated dose (MTD) had been determined. A total of twenty-six patients were treated in the dose-expansion cohort. While three patients with MPD and two patients with accelerated phase CML had clinical benefit, none of the 20 patients with acute leukemia enrolled in the AML or ALL/CML blast-phase cohorts had clinical benefit. In addition, frequent toxicities were noted necessitating dose reductions and subsequent lack of efficacy at these reduced doses. As a result of these toxicities necessitating dose reductions to therapeutically ineffective doses, the clinical development of R763 was suspended.

20.2.5 ENMD2076

ENMD2076 is an orally available inhibitor of Aurora A and B. It is selectively more active against Aurora A compared to Aurora B (IC_{50} value of 14 nmol/L for Aurora A vs. IC_{50} of 350 nmol/L for Aurora B). It also inhibits kinases involved in angiogenesis (VEGFR and FGFRs) as well as FLT3, Kit, and CSF1R. In preclinical evaluation, ENMD2076 inhibits the growth of AML cell lines and suppresses tumor growth in AML xenograft models (MV4-11 and Hl-60) (Fletcher et al. 2011).

A phase I study of ENMD-2076 in patients with advanced myeloid malignancies (AML and chronic myelomoncytic leukemia) has been reported on (Yee et al. 2009). Several dose-limiting toxicities were reported including typhilits, fatigue, and syncope. However, some responses were seen in heavily pretreated patients. Of 20 evaluable patients, one patient achieved a CR with incomplete hematological recovery, three achieved a morphologic leukemia-free state, and four other patients had reductions in marrow blast count.

20.2.6 KW2249

KW-2449 is a small molecule TKI with known activity against Aurora A, FLT3, FGFR-1, and Abl kinase. Since it has potent FLT-3 as well as Aurora A activity (IC_{50} values of 0.007 and 0.048 µmol/L respectively), targeting AML has been an attractive approach for this agent (Shiotsu et al. 2007). In preclinical studies, KW-2449 was particularly effective at slowing the growth of FLT3-mutated xenograft model (Shiotsu et al. 2009). The sensitivity of FLT3 wild-type leukemia cell lines to KW-2449 was five- to tenfold lower than that in mutant FLT3-expressing leukemia cells. Treatment of FLT3 wild-type human leukemia cells induced a phenotype consistent with Aurora kinase inhibition with reduction of phosphorylated histone H3, G2/M arrest, and apoptosis.

KW2449 was evaluated in 37 patients with leukemia (31 patients had AML, five CML, and one ALL) (Shiotsu et al. 2009). Pharmacodynamic assays confirmed inhibition of FLT3 with reduction of P-FLT3 and P-STAT5. The most frequent severe toxicities were febrile neutropenia, pneumonia, and thrombocytopenia. Clinical responses were observed in eight of 31 patients with AML (26%; FLT3 mutation: five positive and three negative) and one patient with CML. Three of the eight AML patients who responded were FLT3 wild-type indicating that KW-2449 may be having activity independent of FLT3 inhibition.

However, correlative data from this phase I study showed that FLT3 inhibition to less than 20% of baseline was necessary for effective cytotoxic activity of the drug (Pratz et al. 2009). With the BID dosing schedule originally used, this level of FLT3 inhibition as evidenced by inhibition of P-STAT5 and P-FLT3 was not achieved at trough levels of the drug. Although the drug showed activity *in vitro* when cells were exposed at sufficient concentrations and significant FLT-3 inhibition at the 2-h time point *in vivo*, blast reduction was short lived. Given the above and the half-life of the drug, the trial was discontinued early before the maximal tolerated dose was determined and a modified dosing strategy was planned. Alternative TID and QID dosing schedules are now being evaluated to accommodate for the short half-life and to achieve sufficient target inhibition.

20.2.7 AT9283

AT9283 is a pyrazole–benzimidazole compound that inhibits both Aurora A and B equally (IC_{50} 0nzinM). It is being developed by ASTEX Pharmaceuticals. In addition to inhibiting Aurora A and B, AT9283 was also found to inhibit a number of other kinases including JAK2, FLT3, and ABL (T315I; IC_{50} = 1–30 nM) (Howard et al. 2009). Consistent with its profile as a pan-AKI, AT9283 treatment resulted in endoreduplication and reduction of histone H3 phosphorylation in tumor cells. AT9283 was evaluated as a continuous 72-h infusion in phase I study of 29 patients with refractory leukemia (Foran 2008; Arkenau et al. 2011). Very encouraging single-agent activity in leukemia was observed. One third of patients with refractory AML experienced a significant reduction in bone marrow blasts following treatment with AT9283. Two patients with refractory CML had a hematological response and one had a partial cytogenetic response after four cycles of treatment. Dose-limiting toxicities included elevated transaminases, non-cardiac creatine kinase and lactate dehydrogenase rises, tumor lysis syndrome, myelosuppression, and alopecia. Pharmacodynamic and pharmacokinetic analysis supported increasing the duration of infusion to increase the biological effect of AT9283.

Interestingly, leukemic cells with an Aurora A phenotype may prove even more sensitive to AT9283. In a study using a panel of AML cell lines with or without mutations of c-kit, FLT-3 and RAS two distinct phenotypes emerged. Those driven by the oncogenic mutations listed above showed accumulation of cells in G2/M (4N) arrest followed by apoptosis characteristic of Aurora A inhibition. Those without these mutations developed >4 N DNA content and polyploidy followed by apoptosis consistent with Aurora B inhibition. Authors went on to speculate that their findings indicated that patients with mutations in oncogenic signaling pathways, may prove more susceptible to inhibition by AT9283 due to the presence of an intact mitotic checkpoint and a dominant Aurora A phenotype (Goodall et al. 2008).

20.3 Aurora A-Selective Inhibitors

20.3.1 Alisertib (MLN8237)

Alisertib is a small molecule adenosine triphosphate (ATP)-competitive reversible inhibitor of Aurora A kinase that is being developed for the treatment of advanced malignancies. Alisertib inhibits Aurora A with an inhibition constant (Ki) of

0.43 nM. Unlike many other AKIs undergoing clinical evaluation it is approximately 200-fold more selective for Aurora A than Aurora B (IC_{50}=1534 nM) (Manfredi et al. 2011). Moreover, alisertib is selective for Aurora A kinase when compared to other kinases (at a minimum 250-fold more selective *in vitro*) and receptors. Like its parent molecule, MLN8054, alisertib has a benzazapine scaffold. The clinical development of MLN8054 was stopped due to excess somnolence that patients experienced. Somnolence is significantly less of a problem with alisertib as it has less affinity for benzodiazapine receptors. As would be expected from its kinase inhibitory profile, alisertib treatment results in the formation of abnormal mitotic spindles, an accumulation of mitotic cells, inductions of p53, p21, p27, and a decrease in the proliferation of a tumor cell lines grown in culture (Manfredi et al. 2011). Alisertib synergizes with docetaxol, another agent that disrupts mitosis in preclinical models of mantle cell lymphoma (Qi et al. 2011). Alisertib is currently being investigated in a large number of clinical studies in solid tumors and hematological malignancies. Several different formulations and dosing schedules have been evaluated in adult and pediatric patients.

The side effects of alisertib have been similar to those observed with other AKIs including myelosuppression, diarrhea, and alopecia. However, the use of a treatment-free period for bone marrow and gastrointestinal tract recovery between each cycle of drug administration allows repeated treatment cycles over periods extending beyond 12–24 months (Kelly et al. 2011a, b). Apart from alopecia, the predominant toxicities are largely reversible.

Objective responses to alisertib treatment have been observed across a broad range of malignancies including AML and lymphomas. In many cases these responses have been sustained more than six months (Dees et al. 2010). In a phase I study of alisertib in 56 patients with advanced hematological malignancies, four patients with lymphoma and one patient with multiple myeloma had PRs to therapy and a further 13 patients had prolonged stable disease (SD) (Kelly et al. 2011a). A phase II study of 48 patients with refractory aggressive non-Hodgkin's lymphoma (NHL) has been reported on. The overall response rate was 32 %, but 57 % of patients with T-cell lymphoma responded (Friedberg et al. 2010). These promising results have prompted a phase I/II study of alisertib in combination with rituximab and vincristine in B-cell lymphoma in the USA and a phase III registration study comparing alisertib with investigators choice is planned in peripheral T-cell lymphoma.

Alisertib has been investigated in preclinical models of leukemia. It is effective *in vitro* and *in vivo* in preclinical models of CML including T315I-mutated CML (Kelly et al. 2011). In an investigation into the activity and mechanism of action of alisertib in preclinical models of AML, alisertib in combination with cytarabine activated the transcription factor FOXO3a, which in turn led to induction of its targets BIM and P27 (Kelly et al. 2012). Furthermore, alisertib significantly increased the ability of cytarabine to impair tumor growth in xenograft models of AML suggesting that the combination of alisertib and cytarabine may represent an attractive therapeutic strategy in AML.

Based on the promising activity of alisertib in preclinical models of AML an open label phase II study of alisertib was performed in 46 patients with AML and 11 patients with MDS. Most of the patients were significantly pretreated. Consistent with prior clinical experience with alisertib, toxicities included febrile neutropenia, thrombocytopenia, anemia, fatigue, and neutropenia. Some treatment-related somnolence was observed. However, the response to single-agent alisertib in this population was modest with six AML patients (13 %) responding and no patients with MDS responding. Only one of these patients had a CR and five patients had a PR (Goldberg et al. 2010).

20.4 Aurora B-Selective Inhibitors

20.4.1 Barasertib (AZD1152)

Barasertib is an Aurora B specific inhibitor developed by AstraZeneca. It is a quinazoline prodrug that is converted to the active metabolite barasertib-HQPA in plasma. In recombinant enzyme assays, it is more than 1000-foldselective for inhibition of Aurora B over Aurora A kinase activity (Ki of 0.36 and 1.7 μM for Aurora B and A, respectively) (Lowenberg et al. 2011). Treatment of colorectal SW620 xenografts induces a phenotype consistent with Aurora B inhibition with reduced histone H3 phosphorylation, accumulation of polyploid cells, and subsequently apoptosis (Wilkinson et al. 2007). Barasertib has also been evaluated in preclinical models of leukemia (Yang et al. 2007). It can potently suppress the proliferation of leukemia cell lines (MOLM13, MV4-11, PALL2, NB4, HL60, and K562) with an IC_{50} ranging from 3 to 40 nM. Similar to the synergy observed following treatment with alisertib and docetaxol, barasertib can enhance the activity of the vincristine, a tubulin depolymerizing agent and the topoisomerase II inhibitor, daunorubicin.

Barasertib was initially investigated in a phase I study in 59 patients with various solid tumors (Boss et al. 2011). It was given as a 2-h intravenous infusion every week. The dose-limiting toxicity observed was neutropenia and while objective responses were not seen, a number of patients had prolonged SD. The drug was otherwise well tolerated.

Two early-phase clinical trials of barasertib in patients with AML have been reported on, one completed in Japan and the other in the USA and Europe (Lowenberg et al. 2011; Tsuboi et al. 2011). Barasertib was administered as a continuous 7-day infusion every 21 days in both studies and a MTD of 1200 mg was defined. Consistent with Aurora kinase inhibition in rapidly proliferating cells the most common toxicities observed were mucositis, stomatitis, and neutropenia. The drug appears to be more active in AML compared with solid tumors. In the AML studies, the overall response rate was 25 % in the US/EU study and 19 % in the Japanese study. Unlike other Aurora B inhibitors such as AT9283 and AS703569, barasertib and in particular its metabolite, barasertib-HQPA are highly specific for Aurora B over other kinases such as FLT3. Therefore, the efficacy observed in these studies would be expected to be a consequence of Aurora B inhibition therefore helping validate Aurora B as therapeutic target.

Barasertib has also been investigated in combination with low dose ara-C (LDA-C) in patients over the age of 60 who are considered unfit for intensive chemotherapy (Kantarjian et al. 2010). In this study, the MTD of AZD1152 was 1000 mg when given in combination with LDAC in older patients. Co-treatment with LDAC slightly increased the incidence of adverse events seen with either agent alone including febrile neutropenia, thrombocytopenia, and anemia. However, the overall investigator-reported clinical response rate (CR plus CRi) was 43 %.

20.5 Toxicities of Aurora Kinases Inhibitors

Because of the role that Aurora kinases plays in mitosis in normal cells, inhibition of this family of enzymes is expected to be toxic to proliferating tissues which would include rapidly dividing hematological cells and the cells lining the gastrointestinal tract (Dees et al. 2011). Clinical data from multiple phase I and II studies of Aurora kinases have demonstrated that the predominant toxicities of this class of drugs reflect the mechanism of action in proliferating tissues (bone marrow, gastrointestinal epithelium, and hair follicles). Consistent with this, the most common dose-limiting toxicity observed in phase one studies of AKIs has been grade 3 or 4 neutropenia.

Other reported toxicities of AKIs may reflect off-target effects. The ATP pocket is highly conserved across the more than 500 kinases contained in the human genome (Bantscheff et al. 2007), therefore it is little surprise that many of the agents designed to inhibit the Aurora kinases also influence other tyrosine kinase pathways leading to off-target toxicities.

Common non-hematological toxicities include somnolence, hypertension and liver toxicity. Cardiac toxicity was reported with VX-680/MK-0457 that subsequently halted development of the drug. MK-0457 causes QTc prolongation due to inhibition of the human ether-a-go-go related gene (hERG) channel (Moore et al. 2010), a potassium channel responsible for the "rapid" delayed rectifier current (I_{Kr}). A diverse range of drugs can potentially inhibit this channel and although great effort is taken in the development of a drug to avoid this potential interaction, the predictive value of preclinical testing is in the order of 80 %, thus not infallible (Pollard et al. 2008). QTc prolongation does not appear to be a class effect of Aurora kinases.

Some concern has been raised that Aurora kinases may have an essential role in cardiomyocyte cell cycle control and normal development in response to injury (Cheng and Force 2010; Ahuja et al. 2007), but how significant this is has not been highlighted in clinical trials with small molecule AKIs to date. It is likely that more long-term follow-up will be required to answer this question.

Importantly most of the toxicities observed with AKIs appear to be reversible. Indeed the marrow suppression associated with AZD1152 has been studied in animal models and has been shown to be reversible on withdrawal of the drug (Wilkinson et al. 2007) Consistent with these findings in animal models, most dosing schedules of AKIs involve administration of the drug for approximately 1 week followed by a 2-week break to allow bone marrow and gastrointestinal epithelial cells to recover. An alternative strategy is to use granulocyte colony stimulating factors (G-CSF) to aid marrow recovery. For example, in a phase I study of danusertib, the MTD without G-CSF was 500 mg/m^2 whereas with factor support the MTD was significantly higher at 750 mg/m^2 (Cohen et al. 2009b)

There are preclinical data to suggest that the chromosomal instability induced by deficiency of Aurora A can lead to enhanced tumorigenesis (Lu et al.2008). A recent study showed that Aurora-A-null mice die in utero whereas Aurora A heterozygosity leads to a significantly increased tumor incidence in mice. This suggests that Aurora A is important in maintaining chromosomal stability and acts as a tumor suppressor. Some concern has also been raised about the potential long-term effect of the AKIs on breast tissue, as polyploidy has been induced in treated normal mammary epithelial cell cultures (Ditchfield et al. 2003). Increased incidence of secondary malignancies have not yet been reported in the clinical studies completed to date but careful long-term follow-up of patients exposed to Aurora inhibitors will be necessary. The possible impact of Aurora C inhibition on spermatogenesis and fertility is also a question that needs to be addressed with longer follow-up data

Conclusion

Our understanding of the biology of Aurora kinases, their role in oncogenesis, and their suitability as anticancer targets has improved considerably over the past 10 years. Despite this, a number of important questions remain to be fully answered. Whether optimal therapeutic efficacy is achieved through the inhibition of Aurora A, Aurora B, or both kinases simultaneously is still unclear and is the subject of continuing research. Ongoing trials may provide new insights regarding whether there are any advantages to selectively targeting individual Aurora isoforms.

There are a large number of kinases in the human genome. Many of the kinase inhibitors in use in the clinic today were originally developed to inhibit a single kinase but their clinical utility has been attributed to the inhibition of other related kinases. It is likely therefore that some of the clinical efficacy observed with the AKIs currently under clinical investigation can be attributed to off-target kinase inhibition. MK-0457, for example, inhibits the activity of Aurora A and B along with BCR-ABL and FLT3. The clinical responses observed with MK-0457 treatment in highly refractory CML patients may be a consequence of off-target inhibition of BCR-ABL. Likewise, KW-2449 is a potent inhibitor of FLT-3 a factor that may account for some of the activity reported with this drug in AML.

It has yet to be determined whether the expression of Aurora A or Aurora B will predict response to treatment with AKIs. Indeed, this has not been clearly shown in clinical studies performed to date. It would appear that AML, a disease frequently characterized by the rapid proliferation of malignant cells appears to be a disease type associated with objective responses across various AKIs. While high Aurora A or B expression may not necessarily predict response, the expression of the oncogene MYC does appear to predict response to Aurora kinase B inhibitors at least in preclinical models (Hook et al. 2012).

As with many other targeted therapies that are in development the optimal use of AKIs may be in combination with currently available anticancer therapies. Indeed, several promising synergisms between AKIs and chemotherapy and radiotherapy have been demonstrated *in vitro* and *in vivo*. An attractive strategy is, combine Aurora inhibitors with tubulin-disrupting agents that arrest cells in mitosis, a stage in the cell cycle where Aurora kinases play a critical role. However, other potential combinations have been evaluated clinically and in preclinical models such as combinations with DNA-damaging agents and monoclonal antibodies.

Despite some of the limitations outlined above, a great deal of enthusiasm for further evaluation of AKIs in the clinic remains. Innovative preclinical science is providing rationale for the development of effective combination strategies that are now being evaluated in clinical trials in AML. It is hoped that these concerted efforts in Aurora kinase research will translate into novel anticancer strategies that will ultimately improve outcomes for patients with AML.

References

Ahuja P, Sdek P, MacLellan WR (2007) Cardiac myocyte cell cycle control in development, disease, and regeneration. Physiol Rev 87(2):521–544

Arkenau HT et al (2011) A phase I dose escalation study of AT9283, a small molecule inhibitor of aurora kinases, in patients with advanced solid malignancies. Ann Oncol 23(5):1307–1313

Bantscheff M et al (2007) Quantitative chemical proteomics reveals mechanisms of action of clinical ABL kinase inhibitors. Nat Biotechnol 25(9):1035–1044

Baxter EJ et al (2005) Acquired mutation of the tyrosine kinase JAK2 in human myeloproliferative disorders. Lancet 365(9464):1054–1061

Bischoff JR et al (1998) A homologue of Drosophila aurora kinase is oncogenic and amplified in human colorectal cancers. EMBO J17 (11):3052–3065

Boss DS et al (2011) Clinical evaluation of AZD1152, an i.v. inhibitor of Aurora B kinase, in patients with solid malignant tumors. Ann Oncol 22(2):431–437

Briassouli P et al (2007) Aurora-A regulation of nuclear factor-kappaB signaling by phosphorylation of Ikappa Balpha. Cancer Res 67(4):1689–1695

Carpinelli P, Moll J (2008) Aurora kinases and their inhibitors: more than one target and one drug. Adv Exp Med Biol 610:54–73

Carter TA et al (2005) Inhibition of drug-resistant mutants of ABL, KIT, and EGF receptor kinases. Proc Natl Acad Sci U S A 102:11011–11016

Cheng H, Force T (2010) Molecular mechanisms of cardiovascular toxicity of targeted cancer therapeutics. Circ Res 106(1):21–34

Chieffi P et al (2006) Aurora B expression directly correlates with prostate cancer malignancy and influence prostate cell proliferation. Prostate 66(3):326–333

Cohen RB et al (2009a) A phase I dose-escalation study of danusertib (PHA-739358) administered as a 24-hour infusion with and without granulocyte colony-stimulating factor in a 14-day cycle in patients with advanced solid tumors. Clin Cancer Res 15(21):6694–6701

Cohen RB et al (2009b) A phase I dose-escalation study of danusertib (PHA-739358) administered as a 24-hour infusion with and without granulocyte colony-stimulating factor in a 14-day cycle in patients with advanced solid tumors. Clin Cancer Res 15(21):6694–6701

Cortes-F, Dombret JH, Schafhausen P, Brummendorf TH, Boissel N, Latini F, Capolongo L, Laffranchi B, Comis S (2009) Danusertib Hydrochloride (PHA-739358), a multi-kinase aurora inhibitor, elicits clinical benefit in advanced chronic myeloid leukemia and philadelphia chromosome positive acute lymphoblastic leukemia. Blood (ASH Annual Meeting Abstracts) 114:864

Crane R et al (2004) Aurora A, meiosis and mitosis. Biol Cell 96(3):215–229

Crane R, Kloepfer A, Ruderman JV (2004) Requirements for the destruction of human Aurora-A. J Cell Sci 117(Pt 25):5975–5983

Crosio C et al (2002) Mitotic phosphorylation of histone H3: spatio-temporal regulation by mammalian Aurora kinases. Mol Cell Biol 22(3):874–885

Dai Y et al (2008) Vorinostat synergistically potentiates MK-0457 lethality in chronic myelogenous leukemia cells sensitive and resistant to imatinib mesylate. Blood 112(3):793–804

Dees EC, I.J., Burris H, Astsaturov IA, Stinchcombe T, Liu H, Galvin K, Venkatakrishnan K, Fingert HJ, Cohen RB (2010) Phase I study of the investigational drug MLN8237, an Aurora A kinase (AAK) inhibitor, in patients (pts) with solid tumors. J Clin Oncol 28:15s(15):abstr 3010

Dees EC et al (2011) Phase 1 study of MLN8054, a selective inhibitor of Aurora A kinase in patients with advanced solid tumors. Cancer Chemother Pharmacol 67(4):945–954

Ditchfield C et al (2003) Aurora B couples chromosome alignment with anaphase by targeting BubR1, Mad2, and Cenp-E to kinetochores. J Cell Biol 161(2):267–280

Ducat D, Zheng Y (2004) Aurora kinases in spindle assembly and chromosome segregation. Exp Cell Res 301(1):60–67

Fancelli D et al (2005) Potent and selective Aurora inhibitors identified by the expansion of a novel scaffold for protein kinase inhibition. J Med Chem 48(8):3080–3084

Fancelli D et al (2006) 1,4,5,6-tetrahydropyrrolo[3,4-c]pyrazoles: identification of a potent Aurora kinase inhibitor with a favorable antitumor kinase inhibition profile. J Med Chem 49(24):7247–7251

Farruggio DC, Townsley FM, Ruderman JV (1999) Cdc20 associates with the kinase aurora2/Aik. Proc Natl Acad Sci U S A 96(13):7306–7311

Fiskus W et al (2008) Cotreatment with vorinostat enhances activity of MK-0457 (VX-680) against acute and chronic myelogenous leukemia cells. Clin Cancer Res 14(19):6106–6615

Fletcher GC et al (2011) ENMD-2076 is an orally active kinase inhibitor with antiangiogenic and antiproliferative mechanisms of action. Mol Cancer Ther 10(1):126–137

Foran JM (2008) Phase I and pharmacodynamic trial of AT9283, an aurora kinase inhibitor, in patients with refractory leukemia. In: ASCO Annual Meeting

Friedberg J, M.D., Jung J, Persky DO, Lossos IS, Danaee H, Zhou X, Jane E, Leonard J, Bernstein SH (2011) Phase 2 Trial of Alisertib (MLN8237), An Investigational, Potent Inhibitor of Aurora A Kinase (AAK), in Patients (pts) with Aggressive B- and T-Cell Non-Hodgkin Lymphoma (NHL). Blood (Ash Annual Meeting Abstracts) 111(14):4211

Fu J et al (2007) Roles of Aurora kinases in mitosis and tumorigenesis. Mol Cancer Res 5(1):1–10

Gadea BB, Ruderman JV (2005) Aurora kinase inhibitor ZM447439 blocks chromosome-induced spindle assembly, the completion of chromosome condensation, and the establishment of the spindle integrity checkpoint in Xenopus egg extracts. Mol Biol Cell 16(3):1305–1318

Giet R, Glover DM (2001) Drosophila aurora B kinase is required for histone H3 phosphorylation and condensin recruitment during chromosome condensation and to organize the central spindle during cytokinesis. J Cell Biol 152(4):669–682

Giles F et al (2006) MK-0457 is a novel Aurora Kinase and Janus Kinase 2 (JAK2) inhibitor with activity in transformed JAK2-positive myeloproliferative disease (MPD). ASH Annual Meeting Abstracts 108(11):4893

Giles F et al (2006) MK-0457, a novel multikinase inhibitor, has activity in refractory aml, including transformed jak2 positive myeloproliferative disease (mpd), and in philadelphia-positive all. ASH Annual Meeting Abstracts 108(11):1967

Giles FJ et al (2007) MK-0457, a novel kinase inhibitor, is active in patients with chronic myeloid leukemia or acute lymphocytic leukemia with the T315I BCR-ABL mutation. Blood 109(2):500–502

Glover DM et al (1995) Mutations in aurora prevent centrosome separation leading to the formation of monopolar spindles. Cell 81(1):95–105

Goldberg S, Craig GE, Lister J, Kassis J, Pigneux A, Schiller GJ, Jung J, Leonard J, Fingert H, Westervelt P (2010) Phase 2 study of MLN8237, an investigational Aurora A Kinase (AAK) inhibitor in patients with acute myelogenous leukemia (AML) or myelodysplastic syndromes (MDS). In: Annual Meeting of American Society of Hematology

Gontarewicz A et al (2008) Simultaneous targeting of Aurora kinases and Bcr-Abl kinase by the small molecule inhibitor PHA-739358 is effective against imatinib-resistant BCR-ABL mutations including T315I. Blood 111(8):4355–4364

Goodall J, Squires MS, Lock V, Ravandi F, Kantarjian HM, Foran J, Thompson NT, Lyons JF (2008) Outcome of Aurora kinase inhibition of acute myeloid leukemia by AT9283 is dependent upon the presence or absence of mutations in type 1 oncogenic kinase signalling pathways. Blood (ASH Annual Meeting Abstracts) 112:1613

Goto H et al (2003) Aurora-B regulates the cleavage furrow-specific vimentin phosphorylation in the cytokinetic process. J Biol Chem 278(10):8526–8530

Harrington EA et al (2004) VX-680, a potent and selective small-molecule inhibitor of the Aurora kinases, suppresses tumor growth in vivo. Nat Med 10(3):262–267

Hata T et al (2005) RNA interference targeting aurora kinase a suppresses tumor growth and enhances the taxane chemosensitivity in human pancreatic cancer cells. Cancer Res 65(7):2899–2905

Hauf S et al (2003) The small molecule Hesperadin reveals a role for Aurora B in correcting kinetochore-microtubule attachment and in maintaining the spindle assembly checkpoint. J Cell Biol 161(2):281–294

Hirota T et al (2003) Aurora-A and an interacting activator, the LIM protein Ajuba, are required for mitotic commitment in human cells. Cell 114(5):585–598

Hook KE et al (2012) An integrated genomic approach to identify predictive biomarkers of response to the aurora kinase inhibitor PF-03814735. Mol Cancer Ther 11(3):710–719

Howard S et al (2009) Fragment-based discovery of the pyrazol-4-yl urea (AT9283), a multitargeted kinase inhibitor with potent aurora kinase activity. J Med Chem 52(2):379–388

Huang XF et al (2008) Aurora kinase inhibitory VX-680 increases BAX/BCL-2 ratio and induces apoptosis in Aurora-A-high acute myeloid leukemia. Blood 111(5):2854–2865

Ikezoe T et al (2007) A novel treatment strategy targeting Aurora kinases in acute myelogenous leukemia. Mol Cancer Ther 6(6):1851–1857

Jeng YM et al (2004) Overexpression and amplification of Aurora-A in hepatocellular carcinoma. Clin Cancer Res 10(6):2065–2071

Kantarjian HM, Sekeres MA, Vincent R et al (2010) Phase I study to assess the safety and tolerability of azd1152 in combination with low dose cytosine arabinoside in patients with acute myeloid leukemia (AML). ASH Annual Meeting Abstracts 116:656

Katayama H et al (2004) Phosphorylation by Aurora kinase A induces MDM2-mediated destabilization and inhibition of p53. Nat Genet 36(1):55–62

Kelly KR et al (2011) The novel Aurora A kinase inhibitor MLN8237 is active in resistant chronic myeloid leukaemia and significantly increases the efficacy of nilotinib. J Cell Mol Med 15(10):2057–2070

Kelly KR et al (2012) Targeting aurora a kinase activity with the investigational agent alisertib increases the efficacy of cytarabine through a FOXO-dependent mechanism. Int J Cancer 131(11):2693–2703

Kelly KR, Goy A, Berdeja JG, Reeder CB, McDonagh KT, Zhou X, Danaee H, Xiao H, Benaim Eand Shea TC (2011) Results from a phase 1 multicenter trial of alisertib (MLN8237)—an investigational Aurora A kinase inhibitor—in patients with advanced hematologic malignancies. ASH Annual Meeting Abstracts 118(118):4110

Kelly KR, Goy A, Berdeja JG, Reeder CB, McDonagh KT, Zhou X, Danaee H, Xiao H, Benaim Eand Shea TC (2011) Results from a phase 1 multicenter trial of alisertib (MLN8237)—an

investigational Aurora A kinase inhibitor—in patients with advanced hematologic malignancies. In: Proc Am Soc Hematol
Kimura M et al (1999) Cell cycle-dependent expression and centrosome localization of a third human aurora/Ipl1-related protein kinase, AIK3. J Biol Chem 274(11):7334–7340
Kobayashi M, Nakamura Satoki, Ono Takaaki, Sugimoto Yuya, Sahara Naohi, Shinjo Kaori, Shigeno Kazuyuki, Ohnishi Kazunori (2006) Analysis of aurora kinase expressions and cell cycle regulation by aurora-c in leukemia cells. Blood (ASH Annual Meeting Abstracts) 108:1366
Liu Q, Ruderman JV (2006) Aurora A, mitotic entry, and spindle bipolarity. Proc Natl Acad Sci U S A 103(15):5811–5816
Lowenberg B et al (2011) Phase 1/2 study to assess the safety, efficacy, and pharmacokinetics of barasertib (AZD1152) in patients with advanced acute myeloid leukemia. Blood 118(23):6030–6036
Lu LY et al (2008) Aurora A is essential for early embryonic development and tumor suppression. J Biol Chem 283(46):31785–3190
Manfredi MG et al (2011) Characterization of Alisertib (MLN8237), an investigational small-molecule inhibitor of Aurora A Kinase using novel in vivo pharmacodynamic assays. Clin Cancer Res 17(24):7614–7624
Marumoto T et al (2002) Roles of Aurora-A kinase in mitotic entry and G2 checkpoint in mammalian cells. Genes Cells 7(11):1173–1182
Marumoto T et al (2003) Aurora-A kinase maintains the fidelity of early and late mitotic events in HeLa cells. J Biol Chem 278(51):51786–51795
Miyoshi Y et al (2001) Association of centrosomal kinase STK15/BTAK mRNA expression with chromosomal instability in human breast cancers. Int J Cancer 92(3):370–3
Monier K, Mouradian S, Sullivan KF (2007) DNA methylation promotes Aurora-B-driven phosphorylation of histone H3 in chromosomal subdomains. J Cell Sci 120(Pt 1):101–114
Moore AS et al (2010) Aurora kinase inhibitors: novel small molecules with promising activity in acute myeloid and Philadelphia-positive leukemias. Leukemia 24(4):671–678
Nair JS et al (2009) Aurora B kinase regulates the postmitotic endoreduplication checkpoint via phosphorylation of the retinoblastoma protein at serine 780. Mol Biol Cell 20(8):2218–2228
Ouchi M et al (2004) BRCA1 phosphorylation by Aurora-A in the regulation of G2 to M transition. J Biol Chem 279(19):19643–19648
Pollard CE, Valentin JP, Hammond TG (2008) Strategies to reduce the risk of drug-induced QT interval prolongation: a pharmaceutical company perspective. Br J Pharmacol 154(7):1538–1543
Pratz KW et al (2009).A pharmacodynamic study of the FLT3 inhibitor KW-2449 yields insight into the basis for clinical response. Blood 113(17):3938–3946
Qi W et al (2011) Aurora inhibitor MLN8237 in combination with docetaxel enhances apoptosis and anti-tumor activity in mantle cell lymphoma. Biochem Pharmacol 81(7):881–890
Reiter R et al (2006) Aurora kinase A messenger RNA overexpression is correlated with tumor progression and shortened survival in head and neck squamous cell carcinoma. Clin Cancer Res 12(17):5136–5141
Sarno S et al (2007) The novel aurora kinase inhibitor as703569 shows potent anti-tumor activity in acute myeloid leukemia (AML). ASH Annual Meeting Abstracts 110(11):915
Sausville EA (2004) Aurora kinases dawn as cancer drug targets. Nat Med, 10(3):234–235
Schnittger S et al (2002) Analysis of FLT3 length mutations in 1003 patients with acute myeloid leukemia: correlation to cytogenetics, FAB subtype, and prognosis in the AMLCG study and usefulness as a marker for the detection of minimal residual disease. Blood 100(1):59–66
Sen S, Zhou H, White RA (1997) A putative serine/threonine kinase encoding gene BTAK on chromosome 20q13 is amplified and overexpressed in human breast cancer cell lines. Oncogene 14(18):2195–2200
Severson AF et al (2000) The aurora-related kinase AIR-2 recruits ZEN-4/CeMKLP1 to the mitotic spindle at metaphase and is required for cytokinesis. Curr Biol 10(19):1162–1171
Shiotsu Y et al (2007) KW-2449, a novel multi-kinase inhibitor against FLT3, Abl, FGFR1 and Aurora, suppresses the growth of aml both in vitro and in vivo. ASH Annual Meeting Abstracts 110(11):1832

Shiotsu Y et al (2009) KW-2449, a novel multikinase inhibitor, suppresses the growth of leukemia cells with FLT3 mutations or T315I-mutated BCR/ABL translocation. Blood 114(8):1607–1617

Shiotsu Y, Kiyoi Hitoshi, Ozeki Kazutaka, Umehara Hiroshi, Shimizu Makiko, Akinaga Shiro, Naoe Tomoki (2007) KW-2449, a novel multi-kinase inhibitor against FLT3, Abl, FGFR1 and Aurora, suppresses the growth of aml both in vitro and in vivo. Blood (ASH Annual Meeting Abstracts) 110:1832

Smith SL et al (2005) Overexpression of aurora B kinase (AURKB) in primary non-small cell lung carcinoma is frequent, generally driven from one allele, and correlates with the level of genetic instability. Br J Cancer 93(6):719–729

Sonet A, Graux C, Maertens J, Hartog C-M (2008) Phase I, dose-escalation study of 2 dosing regimens of AS703569, an inhibitor of aurora and other kinases, administered orally in patients with advanced hematological malignancies. Blood (ASH Annual Meeting Abstracts) 112:2963

Steeghs N et al (2009a) Phase I pharmacokinetic and pharmacodynamic study of the aurora kinase inhibitor danusertib in patients with advanced or metastatic solid tumors. J Clin Oncol 27(30):5094–5101

Steeghs N et al (2009b) Phase I pharmacokinetic and pharmacodynamic study of the aurora kinase inhibitor danusertib in patients with advanced or metastatic solid tumors. J Clin Oncol 27(30):5094–5101

Tsuboi K et al (2011) A Phase I study to assess the safety, pharmacokinetics and efficacy of barasertib (AZD1152), an Aurora B kinase inhibitor, in Japanese patients with advanced acute myeloid leukemia. Leuk Res 35(10):1384–1389

Ulisse S et al (2006) Expression of Aurora kinases in human thyroid carcinoma cell lines and tissues. Int J Cancer 119(2):275–282

Vader G, Medema RH, Lens SM (2006) The chromosomal passenger complex: guiding Aurora-B through mitosis. J Cell Biol 173(6):833–837

Wilkinson RW et al (2007) AZD1152, a selective inhibitor of Aurora B kinase, inhibits human tumor xenograft growth by inducing apoptosis. Clin Cancer Res 13(12):3682–3688

Yang J et al (2007) AZD1152, a novel and selective aurora B kinase inhibitor, induces growth arrest, apoptosis, and sensitization for tubulin depolymerizing agent or topoisomerase II inhibitor in human acute leukemia cells in vitro and in vivo. Blood 110(6):2034–2040

Yee KWL et al (2009) A phase i study of enmd-2076 in patients with relapsed or refractory leukemia. ASH Annual Meeting Abstracts 116:3307

Zhou H et al (1998) Tumour amplified kinase STK15/BTAK induces centrosome amplification, aneuploidy and transformation. Nat Genet 20(2):189–193

Part III
Epigenetics

Chapter 21
AML: Deacetylases

Margherita Ghisi and Ricky W. Johnstone

Abstract Human malignancies, including acute myeloid leukemia (AML), are driven by genetic lesions that result in loss of function of tumor suppressor genes and/or activation of oncogenes. In addition to changes in the genome, we now understand that aberrations to the epigenome, either through mutation of genes encoding epigenetic regulators, or inappropriate recruitment of epigenetic enzymes to genetic loci by oncogenic fusion proteins commonly found in AML, can lead to neoplastic transformation. Alterations in the function, expression, or localization of histone acetyltransferases (HATS) and histone deacetylases (HDACs) have been shown to be important for the development of AML. Compounds that target these epigenetic regulators have now been developed and therefore offer a new therapeutic option for patients with AML driven by alterations to the epigenome.

Keywords Deacetylases · Histone-modifying enzymes · Acute myeloid leukemia · Epigenetics · Epigenetic regulators · Epigenetic therapy · HDAC inhibitors

21.1 Introduction

Histone acetyltransferases (HATs) and histone deacetylases (HDACs) are two families of posttranslational modifying enzymes that mediate, respectively, the addition and the removal of acetyl groups to and from the amino-terminal lysine residues of proteins (Hildmann et al. 2007; Segre and Chiocca 2011; Yang and Seto 2007). The reversible acetylation of histones by HATs and HDACs, in concert with other posttranslational modifications, affects chromatin structure and controls the transcription of genes (Izzo and Schneider 2010; Kouzarides 2007; Li et al. 2007). Besides their important function as chromatin remodelers, both HATs and HDACs can target numerous nonhistone proteins, regulating their activity, localization, and stability (Yao and Yang 2011).

It is becoming increasingly apparent that genetic lesions that affect the expression, function or localization of epigenetic enzymes including HATs and HDACs

R. W. Johnstone (✉) · M. Ghisi
Cancer Therapeutics Program, Peter MacCallum Cancer Centre,
St. Andrews Place, East Melbourne, Victoria 3002, Australia
e-mail: ricky.johnstone@petermac.org

M. Andreeff (ed.), *Targeted Therapy of Acute Myeloid Leukemia*,
Current Cancer Research, DOI 10.1007/978-1-4939-1393-0_21

are important drivers on cancer onset and progression (Chi et al. 2010; Dawson and Kouzarides 2012). In the context of acute myeloid leukemia (AML), aberrant histone acetylation is frequent and is emerging to be an important mechanism in the pathogenesis of various classes of AML (Jankowska and Szpurka 2012; Linggi et al. 2005). Alterations of posttranslational histone modifications in leukemic cells thus represent an attractive novel therapeutic target. Moreover, since epigenetic modifications are, in principle, reversible, it is possible to develop small molecules that target the histone onco-modifications and reset the cancer cell gene expression, reestablishing a normal epigenetic pattern. In recent years, several compounds, that target histone-modifying enzymes, have been developed and tested preclinically giving promising results in different models of myeloid malignancies (Dawson et al. 2011). Some of these agents are already in clinical use as standard care or are undergoing clinical trials.

21.2 HDACs and HATs: Classification

21.2.1 HDACs

The HDAC family of enzymes is highly conserved from yeast to mammals. To date, 18 different HDACs have been identified in humans. According to a commonly used criterion of classification, HDACs can be grouped into four classes (I, II, III, IV), based on their homology to the corresponding yeast protein, their subcellular localization (primarily nuclear or cytoplasmatic), and their enzymatic activity (Zn^{2+}- or NAD^{+}-dependent; Hildmann et al. 2007; Mehnert and Kelly 2007; Ropero and Esteller 2007; Segre and Chiocca 2011; Thiagalingam et al. 2003; Yang and Seto 2007). The class I HDACs (HDAC1, 2, 3 and 8) are small (22–55 kDa) ubiquitously expressed proteins. They are primarily localized in the nucleus, where they are recruited in large multiprotein complexes associated to chromatin and they play important roles as histone modifiers and transcriptional regulators (Hodawadekar and Marmorstein 2007). Class II HDACs (HDAC5, 6, 7, 9, and 10) are larger enzymes (120–135 kDa) that show a more tissue-specific pattern of expression. They can shuttle between the nucleus and the cytoplasm in response to cellular signals and target both histone and nonhistone proteins such as Hsp90 (Kovacs et al. 2005), α-tubulin (Hubbert et al. 2002) and epidermal growth factor receptor (EGFR; Deribe et al. 2009). Class III HDACs, also known as sirtuins, comprises seven members (Sirt1-7). Unlike the other classes of HDACs that are all dependent on Zn^{2+} for their activity, sirtuins require the cofactor NAD^{+} in order to catalyze the deacetylation of their targets. As a consequence, this class of HDACs sensitively react to changes in the metabolic and redox status of the cell and link the intracellular energetics to a variety of cellular functions, such as cell cycle, response to DNA damage, metabolism, apoptosis, and autophagy (Gregoretti et al. 2004; Rajendran et al. 2011; Schwer and Verdin 2008). HDAC11 is the only member identified so far belonging to the class IV HDACs. It is a Zn^{2+}-dependent deacetylase sharing partial sequence similarity with the catalytic core of both class I and II HDACs, but not much is known yet about its cellular function (Gregoretti et al. 2004).

21.2.2 *HATs*

Based on their functional localization, HATs can be broadly grouped into type A HATs, which exert their function in the nucleus and generally play important roles in the regulation of gene expression by functioning as transcriptional co-activators, and type B HATs (HAT1, HAT2, Elp3), which are mainly cytoplasmic enzymes that modify the newly synthesized histones before their assembly into chromatin (Parthun 2007, 2012). Type B HATs can bind soluble histones, interact with chaperon proteins (Ai and Parthun 2004; Campos et al. 2010), and participate to the process of histone replacement and turnover from nucleosomes (Verzijlbergen et al. 2011). The type A HATs are involved in the acetylation of both histone and nonhistone proteins, such as p53, E2F1, and GATA1, and they can be further classified into 5 main families, including the GNAT family (GCN5, PCAF, ELP3), the MYST family (TIP60, MOZ, MOF, MORF, and HBO1), the p300/CBP family, the general transcription factor HAT family (ATF2, TAF1, and TFIIIC90), and the nuclear hormone-related HATs (SRC4 and ACTR; Hodawadekar and Marmorstein 2007; Nagy and Tora 2007; Yang and Seto 2007).

21.2.2.1 HDACs and HATs: Epigenetic Function

Chromatin is a dynamic structure that undergoes a continuous remodeling process that includes repositioning of nucleosomes along the DNA, incorporation of different histone variants at specific genomic locations and covalent modifications of histones and DNA (Campos and Reinberg 2009; Izzo and Schneider 2010; Kouzarides 2007; Kunert and Brehm 2009; Talbert and Henikoff 2010). In particular, histones undergo a series a posttranslational modifications, including acetylation of lysines, methylation of lysines and arginines, phosphorylation of serines and ubiquitination of lysines (Cosgrove et al. 2004). The acetylation/deacetylation of histones, together with the other posttranscriptional modifications of these proteins (methylation, phosphorylation, ubiquitination, sumoylation, propioylation and butyrylation), affects the structure and function of chromatin regulating active transcription of genes.

Histone modifications are regulated, monitored, and fine-tuned by distinct "writer," "eraser," and "reader" epigenetic enzymes. The "writers" (such as the HATs and the histone methyltransferases or HMT) and "erasers" (such as the HDACs and the histone demethylases or HDMT), respectively, add and remove specific posttranslational modifications to and from histones. The "readers" contain domains (such as bromodomains or chromodomains) that specifically recognize distinct histones modifications, and function as adaptor proteins, recruiting other enzymes that further contribute in the regulation of chromatin structure and gene transcription (Kouzarides 2007; Lee and Workman 2007; Seet et al. 2006; Yang and Seto 2007).

The histone acetylation status affects the affinity between the positively charged histones and the negatively charged DNA, thus creating a more condensed or open chromatin state and regulating the access of the transcriptional machinery to a particular locus (Choi and Howe 2009; Izzo and Schneider 2010). In general, HDAC activity is thought to condense chromatin and repress gene expression, while HAT activity is typically associated with transcriptional activation. Single histone modifications or a combination of modifications can act as "molecular targets" to facilitate recruitment of "reader" and effector proteins, such as chromatin-modifying enzymes and transcription factors that further modify chromatin to alter downstream transcriptional responses (Izzo and Schneider 2010; Strahl and Allis 2000). A complex interplay of different chromatin marks ultimately determines the chromatin structure and transcription status of a particular gene.

Tight control of histone modifications is critical for many physiological cellular processes, such as cell cycle progression, differentiation, development, apoptosis and response to diverse types of stress (Kouzarides 2007; Minucci and Pelicci 2006; Orkin and Hochedlinger 2011; Ropero and Esteller 2007). Loss of control can result in different pathological conditions, such as autoimmune disorders and cancer (Chi et al. 2010; Fullgrabe et al. 2011; Lawless et al. 2009; Martino and Prescott 2010; Meda et al. 2011).

The expression, localization, or function of many histone-modifying enzymes is altered in different cancers, and the alteration of histone marks has been linked to deregulated expression of many genes with important roles in neoplastic transformation, tumor progression, and resistance to therapy (Fullgrabe et al. 2011; Plass et al. 2008; Ropero and Esteller 2007). Genome-wide mapping of chromatin changes occurring during tumorigenesis showed that cancer cells undergo a global loss of acetylated H4-lysine 16 (H4K16ac) and H4-lysine 20 trimethylation (H4K20me3) in repetitive genomic DNA sequences, which are also known to be frequently hypomethylated in neoplastic cells. The de-repression of these repetitive regions of the genome by alteration of both DNA methylation and histone acetylation/methylation is thought to predispose the cells to genomic instability and further oncogenic changes (Feinberg and Vogelstein 1983; Fraga et al. 2005; Robertson 2005). Besides a global genomic hypomethylation, another common hallmark of cancer is CpG-island hypermethylation within the promoters of specific tumor-suppressor genes (Baylin and Ohm 2006). Interestingly, CpG-island hypermethylation has also been associated with a characteristic pattern of histone marks (deacetylation of histones H3 and H4, loss of histone H3 lysine trimethylation, and gain of H3K9 methylation and H3K27 trimethylation; Fullgrabe et al. 2011). At present, it is still not known which epigenetic modification comes first and is therefore the primary driver of the tumorigenic process; however, it is clear that the interplay between DNA methylation and histone modifications de-regulates gene transcription in tumors (Cedar and Bergman 2009; Vaissiere et al. 2008).

21.3 HDACs and HATs: Acetylation of Nonhistone Proteins

A global analysis of the human "acetylome" mapped more than 1750 proteins that can be acetylated at lysine sites (Choudhary et al. 2009). Thus, it is not surprising that regulation of HAT and HDAC function has a profound impact on many vital cellular processes. Indeed, HDACs and HATs have been shown to target and hence influence the activity and stability of transcription factors (p53, Rb, TFIIEb, TFIIF, NF-Kb, GATA1, and EKL), metabolic enzymes (Pyruvate kinase and Acetyl-CoA synthase), structural proteins (α-tubulin), chaperon proteins (Hsp90), enzymes involved in DNA dynamics (PCNA), as well as exogenous viral proteins (SV-40 T-antigen and HIV integrase; Yang and Seto 2007).

It is becoming increasingly clear that the tumor-promoting roles of deregulated HDACs and HATs can be mediated by their effects on nonhistonic proteins. For instance, Hsp90 is a molecular chaperone that stabilizes the folding and regulates the proteasomal degradation of multiple oncogenic kinases (FLT3, BCR-ABL, NPM-ALK, c-RAF, AKT and CDK4) involved in the development of AML. Deacetylation of Hsp90 by HDAC6 has been shown to influence its chaperone function. Indeed, depletion or inhibition of HDAC6 activity results in Hsp90 hyperacetylation, inhibition of ATP binding and subsequent destabilization and proteasomal degradation of its client proteins (Rao et al. 2008; Reikvam et al. 2009). Furthermore, many well-known oncogenes and tumor suppressors are regulated in their activity or stability by acetylation. Acetylation of the tumor suppressor p53 by p300 increases its DNA binding capability and transcriptional activity by affecting its interaction with specific co-factors. Moreover, lysine acetylation in the C-terminal domain of p53 prevents MDM2-mediated ubiquitination and subsequent degradation of the tumor-suppressor under DNA damage conditions (Gu and Roeder 1997; Iyer et al. 2004; Liu et al. 1999; Ropero and Esteller 2007; Taira and Yoshida 2012). In contrast, deacetylation of p53 by HDAC1 and SIRT1 has been shown to inhibit p53 transcriptional activity and its tumor suppressive functions (Luo et al. 2001). Recently, it was shown that acetylation by p300 of the oncogenic fusion protein AML1/ETO, found in t(8:21) AML, is essential for its self-renewal promoting effects in hematopoietic stem progenitor cells (HSPC) and is required for maintenance of leukemogenic potential in the mouse (Wang et al. 2011).

21.4 HDACs and HATs: Role in AML

A balanced and controlled activity of HDACs and HATs is essential for the normal development and homeostasis of hematopoietic cells. Many HATs (such as the MYST and the p300/CBP family) function as transcriptional co-activators with key hematopoietic transcription factors and are required for appropriate hematopoietic stem cells self-renewal and differentiation (Katsumoto et al. 2006; Kung et al. 2000; Perez-Campo et al. 2009; Rebel et al. 2002). Moreover, conditional

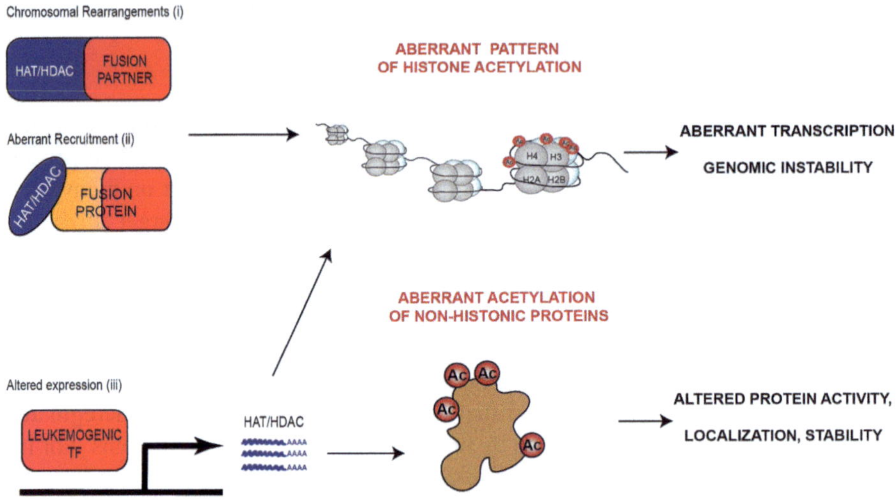

Fig. 21.1 Schematic representation of the three main mechanisms of histone acetyltransferases/histone deacetylases (HATs/HDACs) deregulation in the pathogenesis of AML. (*i*) Chromosomal rearrangement involving HDAC and HAT genes, leading to the generation of fusion or hyperactive oncogenic proteins, (*ii*) aberrant recruitment or ectopic association of HATs and HDACs by oncogenic fusion proteins or multiprotein complexes, and (*iii*) altered expression of HATs and HDACs genes mediated by oncogenic transcription factors

knock-out of HDAC1 or 2 in the hematopoietic compartment results in severe anemia and thrombocytopenia (Wilting et al. 2010). Aberrant acetylation of histone and nonhistone proteins associated to abnormal HAT and HDAC recruitment and activity is frequently observed in hematologic malignancies especially in AML. The aberrant HAT/HDAC activity in leukemia could be the consequence of (1) chromosomal rearrangement or genetic mutations involving HDAC and HAT genes, leading to the generation of fusion or hyperactive proteins, (2) aberrant recruitment or ectopic association of these enzymes to multiprotein complexes, and (3) altered gene expression (Fig. 21.1). In the following sections, we will provide examples of how these three mechanisms of de-regulated HAT/HDAC activity are involved in the pathogenesis of AML.

21.4.1 Genetic Mutations/Chromosomal Rearrangements Involving HAT/HDAC in AML

In AML, HAT genes can be targets of chromosomal rearrangements or mutations that generate chimeric or mutant proteins with altered activity. This mechanism of deregulation is exemplified by leukemias driven by translocations of the MOZ/MORF and p300/CBP genes. MOZ, CBP and p300 are HATs that, through chromosomal translocations, form fusion proteins associated with AML or therapy-related myelodysplastic syndromes. Overall, MOZ fusion proteins, such as MOZ-CBP,

MOZ-p300, MOZ-TIF2, and MOZ-NcoA3, derived from t(8;16) (p11;p13), t(8;22) (p11;q13), inv(8) (p11;q13) and t(8; 20) (p11;q13) translocations respectively, are present in 6.5% of AML. MOZ plays a critical role in the hematopoietic process and is required for hematopoietic stem cells (HSCs) self-renewal and maintenance (Thomas et al. 2006; Voss and Thomas 2009). This HAT can interact with several transcription factors such as PU.1, AML1, p53, Runx2, and NF-KB that play essential roles in myeloid differentiation and leukemogenesis, to cooperatively regulate the transcription of downstream target genes (Chan et al. 2007; Katsumoto et al. 2006; Katsumoto et al. 2008; Kitabayashi et al. 2001; Pelletier et al. 2002).

Most of MOZ fusion proteins retain the N-terminal nuclear localization domain, the PHD (plant homeobox-like domain) zinc finger domain involved in binding to methylated histone and the intact Myst-type HAT domain. Moreover, all MOZ fusion partners are either other HATs, like CBP and p300, or adaptor proteins able to recruit HATs or other epigenetic modifying enzymes (TIF2 and NcoA3; Katsumoto et al. 2008; Yang and Ullah 2007). Target genes of MOZ fusion proteins can be transcriptionally modulated both positively and negatively depending on the promoter context (Chan et al. 2007; Kitabayashi et al. 2001). The biological effect of MOZ fusion protein expression is often a block of blast differentiation and transformation of myeloid progenitors into leukemic cells endowed with self-renewal capacity (Avvakumov and Cote 2007; Huntly et al. 2004). The aberrant chromatin acetylation, which is a key molecular driver of leukemogenesis, is mostly mediated by the MOZ fusion partner or by aberrantly recruited HATs, as the HAT activity of the MYST domain, which is preserved in most MOZ translocations, is not always necessary for AML development (Fig. 21.2; Deguchi et al. 2003).

Overall, the importance of altered HAT activity for MOZ fusion protein-driven leukemogenesis suggests that restoration of an appropriate acetylation status by HAT inhibitor treatment might be therapeutically relevant in this context (Katsumoto et al. 2008; Shima and Kitabayashi 2011; Yang and Ullah 2007).

21.4.2 Aberrant Recruitment of HAT/HDAC in AML

Fusion proteins resulting from chromosomal translocations that are causative of hematologic malignancies can aberrantly interact with HDACs and HATs and recruit them to their target genes. Typical examples of this mechanism are recurrent fusion proteins that are frequently found in AML, such as PML-RARα, AML1/ETO and CBFβ-MYH11. Both the two latter fusion genes are generated by chromosomal rearrangements affecting the core binding factor (CBF) genes (Muller et al. 2008). The CBF is a heterodimeric transcription factor that is composed of a CBFα (also known as RUNX1 or AML1) and a CBFβ subunit. CBFα contains a DNA-binding domain, a heterodimerization domain and several transactivation domains. By contrast, CBFβ by itself does not bind directly to DNA, but, interacting with the α subunit, is able to enhance the DNA-binding and transcriptional activity of CBFα (Goyama and Mulloy 2011; Wang et al. 1996). The CBF complex is essential for normal hematopoiesis and disruption of any of the members of the complex leads to the development of hematological disorders, such as myelodysplastic syndrome (MDS), acute lymphoblastic

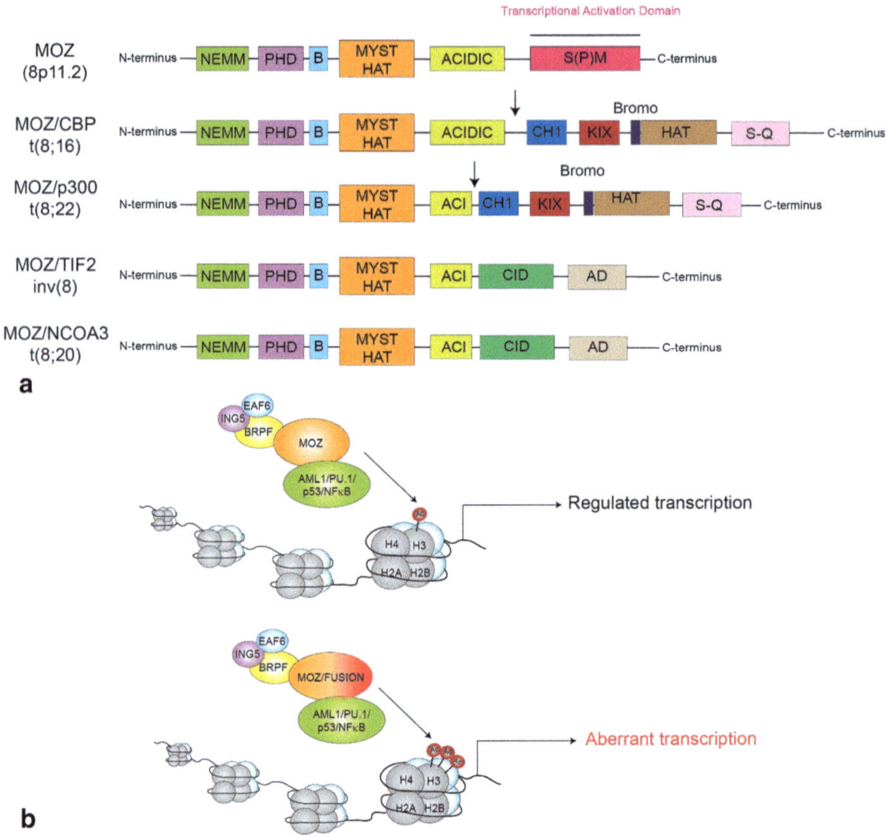

Fig. 21.2 MOZ and mechanisms of transformation by MOZ fusion proteins. **a** Structure of normal MOZ and of leukemia-associated MOZ fusion proteins. *MOZ* histone acetyltransferase contains, starting from its *N-terminus,* the N-terminal part of Enok, MOZ or MORF (*NEMM*) domain, including the H15-domain which promotes nuclear targeting, the plant homeobox-like domain (*PHD*) zinc finger involved in binding to methylated histones, a basic domain (*B*), the *MYST* acetyltransferase domain followed by an *ACIDIC* glutamate/aspartate-rich domain, and the serine-methionine-rich (*SM*) transactivation domain, which includes a proline/glutamine-stretch (*P*). In MOZ fusion proteins (*MOZ/CBP, MOZ/p300, MOZ/TIF2, MOZ/NCOA3*) the N-terminal part of MOZ gene (including the *NEMM, PHD* finger, *B* and *MYST* domain) is retained, and the breakpoints are located inside or after the *ACIDIC* domain (as indicated by the *black arrows*). Depending on the translocation partner, MOZ fusions acquire at their *C-terminus* a cysteine/histidine-rich domain 1 (*CH1*), kinase-inducible domain (*KIX*), a bromodomain (*Bromo*), an additional histone acetyltransferase (*HAT*) domain and a serine-/glutamine-rich domain, or a CBP interacting domain (*CID*) and an activation domain (*AD*). **b** Mechanism of transformation by MOZ fusion proteins. *MOZ* physiologically interacts with many critical hematopoietic transcription factors (such as *AML1, PU.1, p53, NFkB*) modulating the chromatin acetylation status in order to assure a balanced expression of target genes and normal cell differentiation. On the contrary, MOZ fusion proteins display enhanced acetyltransferase activity (due to the acquisition of either an additional HAT domain or of a HAT-binding domain), resulting in hyper-acetylation of histones, aberrant transcription of target genes and ultimately inducing the transformation of myeloid progenitors into leukemic cells

leukemia (ALL) and AML (Muller et al. 2008; Okuda et al. 1996; Sasaki et al. 1996; Speck and Gilliland 2002). The t(8;21) chromosomal translocation is one of the most frequent translocations found in *de novo* AML, with an occurrence of 4–12 % in adult and 12–30 % in pediatric patients. The fusion protein resulting from t(8;21) involves the N-terminal portion of the RUNX1 protein, containing the DNA-binding domain, with the almost full-length ETO protein (Erickson et al. 1992; Goyama and Mulloy 2011). The resulting fusion protein retains the capacity to bind to AML1-target genes, but lacks the RUNX trans-activation domain, which is replaced by transcriptional repressor ETO. ETO contains four nervy homology regions (NHR), that interact with several transcriptional co-repressors, including N-CoR, mSin3A, several members of the HDAC family (HDAC1, 2 and 3 by direct and HDAC6 and 8 by indirect interaction) and the DNA methyltransferase DNMT1 (Amann et al. 2001; Gelmetti et al. 1998; Hug and Lazar 2004). The recruitment of HDACs and DNMT1 by the ETO portion of the protein results in decreased histone acetylation, elevated levels of DNA methylation and in a less accessible chromatin state. Consequently, the AML1/ETO fusion protein functions as a dominant repressor for many AML1-responsive genes (Fig. 21.3). In particular, the aberrant recruitment of HDACs seems to have a crucial role in the leukemogenesis and in the inhibition of myeloid cell differentiation induced by the AML1/ETO fusion protein. Indeed, HDAC inhibitor (HDACi) treatment *in vitro* induces apoptosis and transcriptional reactivation of target genes in AML1-ETO-bearing leukemic cells (Barbetti et al. 2008; Liu et al. 2007). Moreover, recently published data from our laboratory Bots et al. (2014) indicate that HDACi treatment in a mouse model of AML1-ETO AML results in a very strong therapeutic response mediated by cellular differentiation.

The CBFβ-MYH11 fusion protein is generated by the inversion inv(16) and is present in up to 12 % of the AML cases (Look 1997; Muller et al. 2008). The fusion protein retains the CBFα-binding domain of CBFβ and acquires an additional CBFα binding-domain in the MYH11 portion. As a consequence, CBFβ-MYH11 heterodimerizes with higher affinity with CBFα, forming multimeric aggregates that sequester a part of the available CBFα preventing it from binding to its targets on the DNA. Following β-MYH11, CBFα can remain inactive in the nucleus in the form of large rod-like inclusion bodies (intranuclear sequestration), or can be exported in the cytoplasm as deposits on cytoskeletal filaments (cytoplasmatic sequestration; Shigesada et al. 2004). In addition to this "sequestration mechanism," CBFβ-MYH11 inhibits CBFα-mediated transcriptional activity by recruiting HDACs and other transcriptional repressors, thereby aberrantly suppressing the expression of the CBF-target genes (Durst et al. 2003).

Another example of aberrant recruitment of HDACs resulting in leukemia development is promyelocytic leukemia (PML) driven by the PML-RARα fusion which derives from the t(15;17) that juxtaposes the PML gene on chromosome 15 to the retinoic acid receptor α (RARα) gene on chromosome 17 (Insinga et al. 2005c; Lin et al. 1999). The resulting fusion protein retains the DNA and ligand binding activity of RARα, but functions as a constitutive transcriptional repressor of retinoic acid-responsive genes by recruiting epigenetic modifying enzymes, such as the nuclear co-repressor (N-CoR)-HDACs complex, and inducing chromatin modifications not

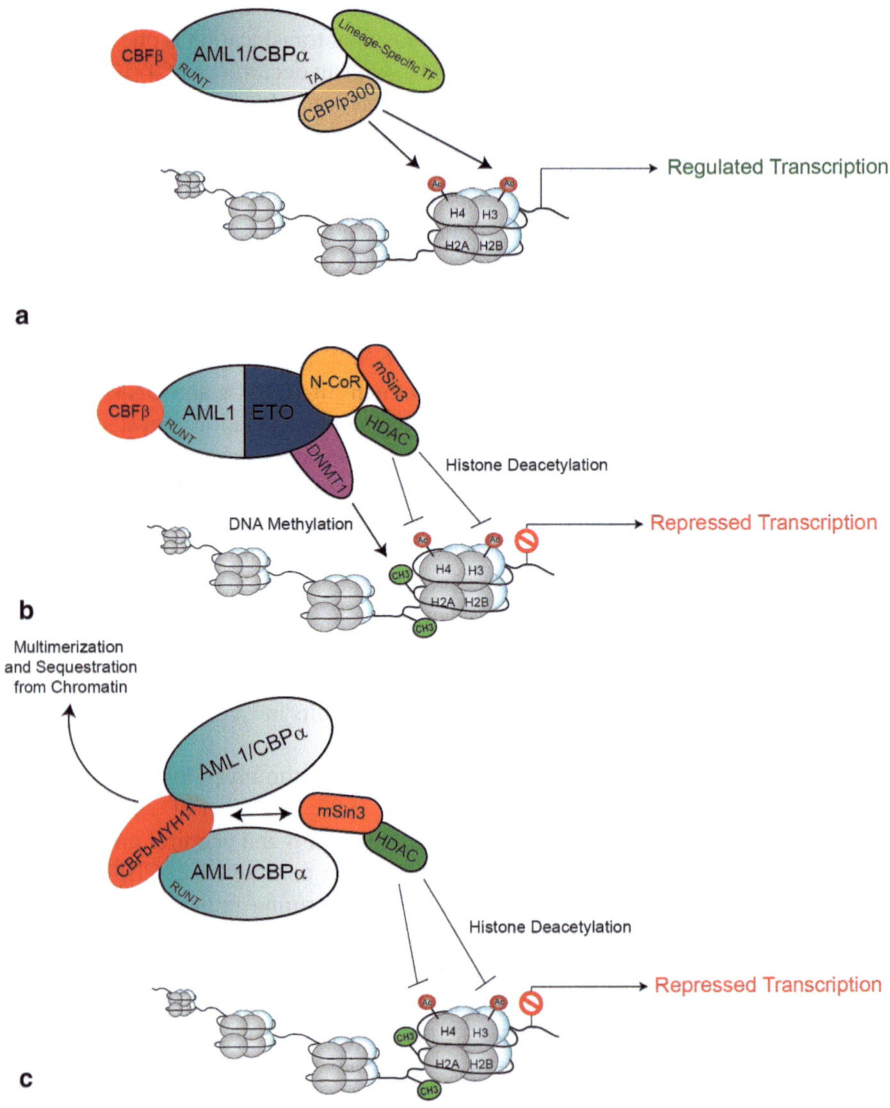

Fig. 21.3 Core binding factor (CBF) and mechanisms of transformation by CBF fusion proteins. The heterodimeric transcription factor CBF is composed by an α subunit (known as *AML1/CBFα* or RUNX1) and a β subunit (*CBFβ*). *CBFα* contains a heterodimerization domain, a DNA-binding domain (*RUNT*), and a transactivation domain (*TA*). The *TA* recruits to the complex transcriptional activators such as the HAT *CBP* and *p300* and promotes active transcription of AML1 target genes (**a**). In the *AML1/ETO* fusion protein [t(8;21)] the N-terminal portion of CBFα, including the heterodimerization domain and the *RUNT* domain, is fused to the almost full-length ETO repressor protein. The *ETO* portion of the resulting fusion protein recruits to the complex transcriptional co-repressors including *N-CoR, mSin3A,* several members of the *HDAC* family and the DNA methyltransferase *DNMT1.* The association of *AML1/ETO* with *HDAC*s and *DNMT1* causes a decrease in the level of histone acetylation and an increase in DNA methylation resulting in the repression of the transcriptional activity of AML1 target genes and in the triggering of the leukemogenic

permissive for transcription (Grignani et al. 1998; Lin et al. 1998). The recruitment of HDACs seems to be crucial for the transforming potential of PML-RARα, as inhibition of their activity or mutation of the N-CoR binding site converts the fusion protein from being a repressor to an activator of the RA pathway and rescues the characteristic block in hematopoietic differentiation (Insinga et al. 2005a; Saeed et al. 2011). Besides the effect on its transcriptional activity, the association with HDACs was shown to confer an additional oncogenic property to PML-RARα. The PML-RARα-HDACs complex can interact with the tumor suppressor p53, and causes its de-acetylation and subsequent proteasome-mediated degradation, resulting in an inhibition of the p53-dependent stress response (Insinga et al. 2005c). *In vivo* studies showed that treatment with the HDACi valproic acid (VPA) induced rapid tumor regression and prolonged survival associated with terminal granulocytic differentiation (Insinga et al. 2005b). However, unfortunately this effect was not associated with eradication of leukemia initiating cells (LIC; Leiva et al. 2012).

Overall, the case of PML-RARα is a good example of how the HDACs can be involved at different levels in the development of acute leukemia, and of how targeting the activity of histone-modifying enzymes could be a promising therapeutic approach for some types of AML.

21.4.3 Aberrant Expression of HATs/HDACs in AML

Altered expression of individual histone-modifying enzymes has been reported in a variety of cancers. For instance, increased expression of HDAC1 has been identified in gastric (Choi et al. 2001), prostate (Halkidou et al. 2004), colon (Wilson et al. 2006), and breast (Zhang et al. 2005) carcinomas. Increased expression of HDAC2 in colorectal cancer was found to be associated with loss of APC tumor suppressor gene and to confer protection of the tumor cells from apoptosis (Zhu et al. 2004). In 2006 Ozdag et al. performed an extensive analysis of gene expression patterns of histone-modifying enzymes in six different types of solid tumors (colorectal, renal, breast, ovarian, bladder cancer, and glioblastoma) showing a differential expression in the neoplastic samples compared to their normal counterparts (Ozdag et al. 2006). Recently, HDAC2, 3, 6, 7, and 8 were identified as highly expressed in a panel of 94 childhood ALL in respect to normal pediatric bone marrow samples. In the same study, a higher than median expression of HDAC3 was associated with a significantly higher 5-year event free survival (EFS), while high expression of HDAC7 and 9 correlated with poor prognosis and significantly lower 5-year EFS (Moreno et al. 2010). In addition, enhanced expression of HDAC1, 2, and 6 were

phenotype (**b**). The *CBFβ-MYH11* fusion protein [inv(16)] retains the original N-terminal *CBFα*-binding domain, while acquiring an additional *CBFα*-binding site in the *MYH11* portion. As a result, *CBFβ-MYH11* forms oligomers with *CBFα* and inhibits its functions by sequestering it from its DNA targets. Moreover, the *MYH11* portion of the fusion protein interacts with *HDACs* that deacetylate the histones at the promoter of CBFα target genes, suppressing their expression and promoting transformation (**c**) identified in peripheral T-cell lymphoma (PTCL) and diffuse large B-cell lymphoma (DLBCL) compared to normal lymphoid tissue (Marquard et al. 2009).

To date, very few studies have examined the expression levels of the different members of the HDAC and HAT families in AML. Bradbury et al. (2005) analyzed by quantitative polymerase chain reaction (qPCR) the expression of all the 18 members of the HDAC family, three HDAC-associated proteins (RBBP4, RBBP7, SIN3A) and two histone methyltransferases (SET7, MLL) in mononuclear cells from AML patients and a variety of control cell types (PBMC, CD34-positive cells from umbilical cord blood and $CD34^+$ PBMC from GCSF-mobilized adult donors). In this study, only HDAC6 and the NAD^+-dependent deacetylase SIRT1 were consistently overexpressed in the AML samples, while HDAC5 was downregulated. However, in the future, more studies will be required in order to better investigate the expression profiles of HATs and HDACs in the various AML subtypes.

21.5 Role of Other Histone-Modifying Enzymes in AML

21.5.1 Histone Lysine Methyltransferases and Demethylases and AML

Besides HATs and HDACs, many other histone-modifying enzymes (such as HMTs or HDMTs), as well as epigenetic "readers," have been associated with tumorigenesis in general and in particular with the pathogenesis of AML. Translocations that join the C-terminal PHD finger of the histone demethylase JARID1A or of PHF23 to the common leukemia fusion partner NUP98 have been reported in AML samples (Reader et al. 2007; van Zutven et al. 2006). A recent paper by Wang et al. demonstrated that these fusions generate potent oncoproteins that inhibit myeloid differentiation and induce AML *in vivo* by preventing the removal of H3K4me3 at many loci encoding critical hematopoietic transcription factors (Hox genes, Gata3, Meis1, Eya1, Pbx1; Chi et al. 2010; Wang et al. 2009).

Another histone-modifying enzyme that is the target of recurrent chromosomal translocations in acute leukemia is the mixed lineage leukemia (MLL) gene on chromosome 11q23 (Slany 2009). MLL is a H3K4 methyltransferase and a powerful epigenetic modifier that positively controls the expression of several important developmental genes, including the homeobox cluster (HOX) genes (Liu et al. 2009; Yu et al. 1995). MLL is important in body segment patterning, hematopoiesis, and cell cycle progression (Milne et al. 2005a; Ono et al. 2005; Takeda et al. 2006). The MLL N-terminal domain contains the binding site for the adaptor protein Menin, three AT-hooks and a CXXC transcription repressor domain, which are all essential elements that regulate the targeting of MLL to the DNA. The CXXC domain is also involved in the recruitment of repressive factors, such as HDACs and Polycomb group proteins (Xia et al. 2003). The rest of the MLL protein includes 4 PHD-fingers, which mediate protein–protein interaction, a transactivation domain interacting with CBP/p300, and a C terminal SET-domain, which mediates the methylation of histone 3 at K4 (Liu et al. 2009). Translocations of the MLL gene give rise to particularly aggressive acute leukemias characterized by dismal prognosis and the expression of surface markers of mixed lymphoid and myeloid lineages. MLL-rearrangements occur in 70 % of infant ALL, 10 % of de novo acute leukemias in both infant and adult and in more than 30 % of therapy-induced MDSs. All the MLL fusion proteins retain the MLL N-terminal portion (including the DNA binding domains and the domains responsible for HDACs and Polycomb group proteins recruitment), but lose the PHD finger domains, the transactivation domain, and the K4H3-metyltransferase domain (Fig. 21.4; Liu et al. 2009; Meyer et al. 2009; Slany 2009).

To date, 73 different chromosomal translocations involving the MLL gene and 54 different fusion partners have been identified. However, six main fusion partner proteins (AF4, AF9, ENL, AF10, ELL, AF6) account for more than 85 % of all clinical cases of MLL-rearranged leukemias (Meyer et al. 2009; Slany 2009). The fusion partners are indispensable for leukemogenic potential and individually determine the phenotype of the resulting leukemia. For example, MLL-AF4 gives rise preferentially to ALL, while MLL-AF9 generates AML (Chen et al. 2006; Dobson et al. 1999). Most of these fusion partners are part of the EAP (ENL Associated Protein) complex. This multiprotein complex interacts with the transcription elongation factor (pTEF), and with the histone methyltransferase DOT1L (Slany 2010). pTEF is an heterodimer composed by the cyclin-dependent kinase CDK9 and Cyclin T, which allows the activation of transcriptional elongation by phosphorylating the C-terminal domain of RNA polymerase II (Peterlin and Price 2006). To date, DOT1L is the only methyltransferase able to catalyze the methylation of H3 at lysine 79. This is a histone modification associated with open chromatin state and active transcription (Nguyen and Zhang 2011; Okada et al. 2005). It is believed that fusions involving MLL and components of the EAP complex induce leukemogenesis by aberrantly recruiting the multiprotein complex to the promoters of MLL-target genes. This causes a site-specific epigenetic remodeling of chromatin, involving aberrant methylation and acetylation of histones, as well as inappropriate transcription elongation that results in an over-activation of the transcription of MLL target genes. The deregulated expression of MLL-target genes, such as in particular HOXA9 and its co-factor Meis1, has been shown to mediate leukemic transformation by inhibiting cellular differentiation and conferring self-renewal capacity to lineage-committed progenitors (Ayton and Cleary 2003; Milne et al. 2005b; Wong et al. 2007; Zeisig et al. 2004). The fusion of MLL with CBP and p300 has been reported in few cases of therapy-induced secondary leukemias (Rowley et al. 1997; Sobulo et al. 1997). These rare leukemias represent another case in which the deregulated activity of these important HATs, resulting in hyperacetylation of chromatin and enhanced transcriptional activation, is involved in the pathogenesis of AML.

Various components of the polycomb repressive complex 2 (PRC2) have been identified as important in the pathogenesis of different cancers (Chi et al. 2010). PRC2 is composed of two essential non-catalytic subunits, Eed (Embryonic Ectoderm Development) and Suz12 (Suppressor of Zest 12), and by the SET-domain H3K27 methyltransferase Ezh1 or 2 (Enhancer of Zeste Homolog 1 or 2). PRC2

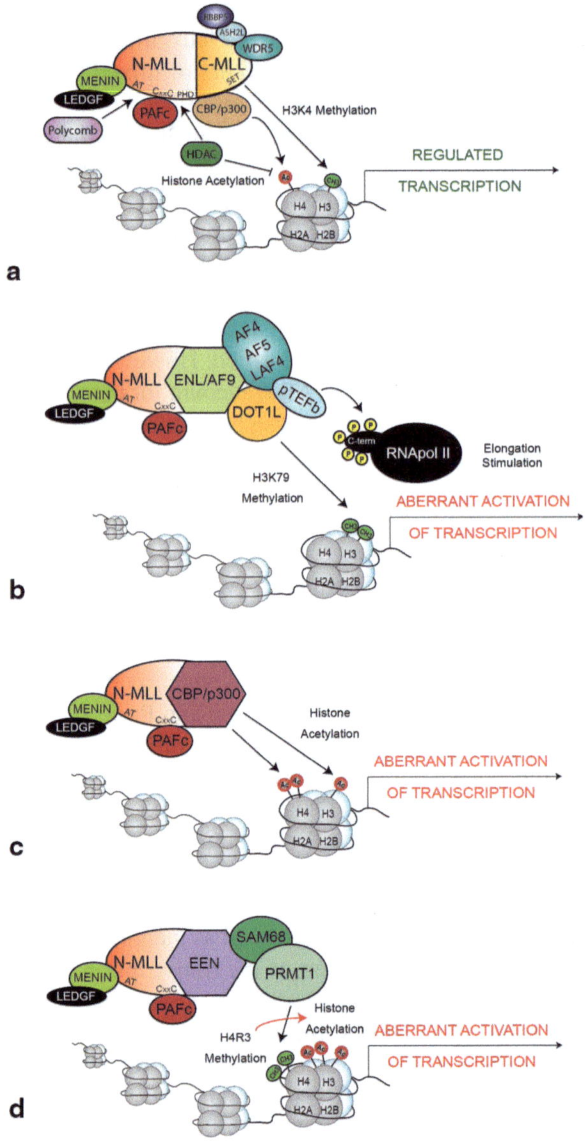

Fig. 21.4 Mixed lineage leukemia (MLL) and mechanisms of transformation by MLL fusion proteins. The wild-type MLL protein is proteolytically cleaved into an N-terminal (*N-MLL*) and a C-terminal (*C-MLL*) fragment that form the core components of the MLL complex. MLL has multiple functional domain including, starting from its N-terminal end, (1) the *Menin*-binding domain; (2) three *AT* hooks; (3) a *CxxC* transcription repressor domain, which is involved in the recruitment of transcriptional repressor, such as HDACs and *PolyComb* group proteins; (4) 4 *PHD*-fingers, (5) a transactivation domain, interacting with the HAT *CBP/p300,* (6) and a C terminal *SET*-domain, which mediates the methylation of histone 3 at K4. The recruitment of several histone-modifying enzymes to the DNA by MLL complex results in a balanced pattern of histone acetylation and methylation and in the regulated transcription of its target genes (**a**). All

promotes gene silencing by inducing histone H3K27 methylation (Cao et al. 2002; Chi et al. 2010; Shen et al. 2008). Although the H3K27-specific methyltransferase Ezh2 has been frequently found overexpressed in a variety of solid tumors (prostate, breast, colon, skin, and lung) and in some subtypes of lymphoma, loss-of-function mutations have been observed in MDSs, suggesting a tumor suppressor role for this PRC2 component in this context (Ernst et al. 2010; Nikoloski et al. 2010). However, by performing a shRNA-based functional screening, Shi et al. identified Eed, Suz112, Ezh1, and 2 as specifically required for the maintenance of MLL-rearranged AML. In particular, silencing of the above mentioned PCR2 components specifically induced proliferation arrest and differentiation of MLL-fusion leukemic cells, and inhibited the leukemia growth *in vivo* (Shi et al. 2012). The importance of PRC2 in MLL-driven leukemogenesis has been confirmed by another separate study that showed in an MLL/AF9 mouse model of AML that the loss of PRC2 function is incompatible with leukemia self-renewal and growth (Neff et al. 2012). Overall, these findings, while implicating PRC2 in myeloid leukemogenesis, suggest that the role of this complex might depend on the cellular context and on the specific genetic lesion.

21.5.2 Histone Arginine Methyltransferases and AML

Like other translocations of the MLL gene, MLL/EEN induces AML in mice, albeit with long latency (Kong et al. 2006). Protein Arginine Methyltransferase 1 (PRMT1) mediates the methylation of histone H4 at arginine 3, a modification that is generally associated with an open chromatin state and transcriptional activation (Pal and Sif 2007). PRMT1 was shown to be a critical component of the MLL/EEN oncogenic fusion protein complex (Liu et al. 2004; Shia et al. 2012). The EEN portion of the chimeric fusion protein indirectly recruits PRMT1 through interaction with the bridging molecule Sam68. The methylation of histone H4 at arginine 3 has been shown to promote histone acetylation within the promoter region of MLL-

---

the MLL fusion proteins retain the *Menin*-binding domain, the *AT* hooks and the *CxxC* domain of *N-MLL,* while losing the PHD finger domain, the transactivation domain and the SET domain. Most of the fusion partners of MLL (such as *AF9, ENL, AF4, AF5,* and *LAF4*) are part of the ENL associated protein (EAP) complex. This group of MLL fusion proteins mediates transformation by recruiting to the MLL complex one the positive transcription elongation factor (*pTEF*), and the *H3K79* methyltransferase *DOT1L*. The methylation of *H3K79* by *DOT1L* induces an open chromatin state and active transcription, while *pTEFb* promotes the activation of transcriptional elongation by phosphorylating the C-terminal domain of RNA polymerase II. These two cooperating mechanisms results in an aberrant activation of MLL target genes and in the block of cell differentiation (**b**). The fusion of *MLL* with the HAT *CBP* and *p300* generate a chimeric protein with hyper-acetylating activity. This results in an altered histone acetylation status and again in aberrant transcription of MLL-target genes and leukemogenesis (**c**). In MLL/EEN fusion protein, the *EEN* portion of the chimeric fusion protein recruits the *H4R3* methyltransferase *PRMT1* by binding with *Sam68*. *PRMT1* mediates the methylation of H4 at arginine 3 and this modification was shown to be linked to increased histone acetylation, enhanced transcriptional activity and ultimately MLL/EEN-associated transformation (**d**) target genes (Pal and Sif 2007; Slany 2009). The activity of PRMT1 was shown to be fundamental for MLL-mediated transformation of primary myeloid progenitors in this system (Cheung et al. 2007). In addition, PRMT1 was shown to interact with the oncogenic fusion protein AML1/ETO (Shia et al. 2012). As mentioned before, AML1/ETO is believed to induce leukemic transformation mainly by repressing the expression of AML1 target genes. However, AML1/ETO can also transactivate the expression of some specific genes (such as M-CSFR and BCL-2), and this was proposed to be also important for its leukemogenic potential (Frank et al. 1995; Rhoades et al. 1996). PRMT1 interaction with the truncated form of AML1/ETO (AML1/ETO9a (A/E9a)), was shown to be required for A/E9a-driven transcription activation function and for the maintenance of the self-renewal capacity of the leukemic cells, suggesting a potential role for PRMT1 in controlling the development of AML1/ETO-associated AML (Shia et al. 2012).

21.6 Targeted Therapy of AML

The overwhelming evidence that epigenetic de-regulation is a major pathogenic driver in AML provides a mechanistic rationale for treating AML with drugs that specifically target and reverse aberrant epigenetic activity. Thus, it is not surprising that much attention has been recently focused in the development of epigenetic drugs inhibiting the function of distinct histone-modifying enzymes. In particular, HDACi have been shown to exert a potent anticancer activity across a range of different hematological malignancies, including AML. So far, FDA has approved the use of the HDACi Vorinostat (Zolinza) and Romidepsin (Istodax) for the treatment of MDS and cutaneous T-cell lymphoma (CTCL), respectively (Verbrugge et al. 2011). However, many other epigenetic drugs, including HDACi, as well as compounds targeting other histone-modifying enzymes, have been developed and have demonstrated robust therapeutic potential in preclinical studies.

21.6.1 HDACi

Despite their proven anticancer activity in selected hematological malignancies, the molecular processes that mediate the therapeutic effects of HDACi are still not well understood. Many experimental studies have shown that the antitumor activity of HDACi can be mediated by their ability to induce differentiation, cell cycle arrest and/or cell death in the cancer cells (Bolden et al. 2006; Dickinson et al. 2010; Fathi et al. 2010), depending on the model and the HDACi used. HDACi-mediated cell death has been associated with the activation of the death receptor pathway (Nebbioso et al. 2005), the alteration in the expression of members of the intrinsic apoptotic cascade (such as Bcl-2, Bcl-XL, Bim, Bax, and Bak; Zhang et al. 2004), and with the accumulation of reactive oxygen species (ROS; Rosato et al. 2003;

Ungerstedt et al. 2005). Other postulated mechanisms of lethality are by impairment of the proteasome/aggresome pathway and misfolded protein response (MPR; Bali et al. 2005; Catley et al. 2006; Fotheringham et al. 2009), as well as by acetylation and disruption of the function of DNA repair proteins (e.g., Ku70; Chen et al. 2007). Interestingly, neoplastic cells have been shown to be more sensitive *in vitro* than normal cells to HDACi-induced cell death, making these compounds particularly appealing candidates for clinical applications (Dokmanovic and Marks 2005). In addition, HDACi has been proven to affect tumor microenvironment and indirectly inhibit tumor development and progression by exerting an antiangiogenic, anti-invasive and immunomodulatory effect (Dickinson et al. 2010).

HDACi can be divided into four main classes based on their chemical structure: short-chain fatty acids (such as VPA and butyrates), hydroximates (such as vorinostat, panobinostat and trichostatin-A), cyclic tetrapeptides (such as trapoxin and romidepsin), and benzamide derivatives (such as Mocetinostat and MS-275). HDACi can have varying degrees of selectivity towards different classes of HDACs. The majority of HDACi (such as vorinostat, panobinostat and trichostatin A) are considered pan-deacetylase inhibitors and target class I, II and IV HDACs. Other compounds instead inhibit preferentially class I HDACs (such as romidepsin, valproate and MS-275), or are HDAC6-specific (such as tubacin; Bertrand 2010; Dickinson et al. 2010).

In preclinical studies, He at al. showed that the treatment *in vitro* with vorinostat induces cell growth arrest and apoptosis, and enhances the differentiation effect induced by retinoid acid (RA) on t(15;17)-positive APL cells. Moreover, vorinostat and RA combination therapy was more effective in significantly prolonging the survival of PLZF-RARα/RARα-PLZF APL transgenic mice (He et al. 2001). Although only two HDACi (vorinostat and romidepsin) have been approved by the FDA for clinical practice, in the past 10 years, the number of HDACi developed and tested in preclinical studies and clinical trials has been constantly growing (Masetti et al. 2011; Wagner et al. 2010). In particular, vorinostat, MS-275 (entinostat), panobinostat, MG-0103, belinostat, sodium phenylbutyrate, VPA, and romidepsin have been evaluated or are in the process of being evaluated in clinical trials with patients with AML or MDS (http://www.cancer.gov/clinicaltrials).

In a phase I study in adult patients with leukemia or MDS, vorinostat monotherapy resulted in hematological improvement (defined as $>50\,\%$ decrease in blast counts) or complete response (linked to downregulation of proliferation-associated genes) in 17 % of the patients. All the subjects showed induction of H3 acetylation in the peripheral blood and bone marrow following vorinostat administration. Of note, all the patients responding to the treatment had AML (Garcia-Manero et al. 2008).

MS-275 treatment *in vitro* of human leukemic cell lines and primary leukemic cells from AML patients resulted in dose-dependent effect on the leukemic blasts, ranging from $p21^{WAF1/CIP1}$-dependent growth arrest and differentiation to the marked induction of ROS production, mitochondrial damage and apoptosis, respectively at low and high drug concentrations (Rosato et al. 2003). MS-275 potently induces growth arrest in AML cells carrying internal tandem duplication (ITD) in Flt3 gene and this effect is associated with the proteasome-mediated degradation of Flt3 protein (Nishioka et al. 2008b). MS-275 antileukemic activity is significantly enhanced by the concomitant inhibition of MEK/ERK or mTOR pathway (Nishioka et al. 2008a). These observations suggest that treatment with this compound, alone or in combination with inhibitors of the mTOR or MEK/ERK signaling, could be a promising and rational strategy to treat patients with AML carrying cooperating activating mutations of the Flt3 gene. So far, MS-275 demonstrated limited single-agent activity in the clinics in advanced acute leukemias (Gojo et al. 2007), but recently a large intergroup trial was initiated to test the combination with 5-azacytidine in patients with MDS, chronic myelomonocytic leukemia (CMMoL) and AML (http://www.clinicaltrials.gov under NCT00101179).

VPA in combination with RA can induce differentiation of primary leukemic blasts obtained from AML patients *in vitro* (Gottlicher et al. 2001). Subsequently, VPA was found not only to selectively inhibit the enzymatic activity of class I HDACs, but also to induce the ubiquitination and proteasomal degradation of HDAC2 (Kramer et al. 2003). In addition, VPA was reported to disrupt the physical interaction between AML1/ETO and HDAC1. This results in the dissociation of the AML1/ETO-HDAC1 complex from the promoter of AML1 target genes, eventually leading to their transcriptional activation and to a significant antileukemic effect, associated with partial cell differentiation and apoptosis (Liu et al. 2007). In another study, Insinga et al. investigated the effects of VPA treatment on AML expressing PML-RAR or AML1/ETO oncoproteins. Treatment with VPA significantly enhanced the survival of mice with transplanted APL associated with induction of leukemia-specific apoptosis in the bone marrow and the spleen of the animals. In this system, the pro-apoptotic effect exerted by VPA was partly mediated by the activation of the death receptor pathway (TRAIL and Fas signaling pathways), but independent from p53 response. Interestingly, the induction of apoptosis was observed specifically in leukemic blasts, but not in preleukemic or normal cells. Similarly, VPA treatment of mouse leukemias expressing AML1/ETO resulted in massive apoptosis of the leukemic cells associated with the induction of the death receptor pathway (Insinga et al. 2005b). In contrast to this study, results from our laboratory Bots et al. (2014) showed that treatment with the pan-deacetylase inhibitor panobinostat in a model of AML1/ETO9a-driven AML results in potent induction of myeloid differentiation, a drastic reduction of the leukemic burden and a significant survival benefit in the treated mice in the absence of apoptosis. Panobinostat induced proteasomal degradation of AML1/ETO9a and genetic studies using an inducible form of the fusion protein demonstrated that depletion of AML1/ETO9a alone was sufficient to phenocopy the molecular, biological, and therapeutic effects of panobinostat.

Given the promising results obtained *in vitro* and in the mouse models of leukemia, many phase I/II clinical trial with VPA administrated as monotherapy or in combination have been conducted in adult patients with AML and myelodysplastic syndrome (MDS; Corsetti et al. 2011; Garcia-Manero et al. 2006; Raffoux et al. 2010). Overall, VPA showed a good to modest therapeutic activity in the various clinical trials. In particular, the combination with all-*trans* retinoic acid (ATRA) in patients with MDS and AML resulted in hematological improvement (according to international working group criteria for MDS), and clinical response in 24 % of the patients (Bug et al. 2005; Kuendgen et al. 2005; Kuendgen et al. 2006; Soriano et al. 2007).

Romidepsin was tested in a phase I clinical trial on a cohort of adult patients with AML. Several patients had evidence of antileukemic activity following the treatment (even if no partial or complete response was observed), but the repeated administration of the drug was limited by progressive adverse symptoms (Byrd et al. 2005). Another clinical study indicated a differential antileukemic and molecular activity of romidepsin in CBF-driven AML compared to other classes of AML (Odenike et al. 2008).

Overall, despite the strong antileukemic effect obtained in preclinical models of leukemia, the success of HDACi used as single agents in patients with AML has so far been relatively minor. However, promising results have been obtained in the clinic from the combination of HDACi with hypomethylating agents or with standard chemotherapy (Fathi et al. 2010; Ornstein and Sekeres 2012; Quintas-Cardama et al. 2011; Shipley and Butera 2009; Soriano et al. 2007). From the results of these first clinical studies and also from many preclinical observations, it appears that there are particular classes of myeloid malignancies that strongly respond to HDACi treatment. This suggests that a more targeted use of these compounds for the treatment of AML with specific genetic lesions could have great therapeutic potential. The future challenge will be to identify the epigenetic modulators that are specifically deregulated in different groups of leukemias and, in parallel, to develop drugs able to target more selectively distinct HDACs in order to reduce the adverse side effects and potentiate the therapeutic efficacy.

21.6.2 Inhibitors of other Histone-modifying Enzymes

In addition to HDACi, inhibitors of other epigenetic readers, writers and erasers have been developed (Copeland et al. 2009; Neugebauer et al. 2008; Verbrugge et al. 2011). For instance, a specific small-molecule inhibitor of the H3K79 HMT Dot1L (EPZ004777) was recently developed. Dot1L, is recruited by several oncogenic fusion proteins involved in the pathogenesis of AML including MLL/AF9, MLL/ENL, MLL/AF10, MLL/AF4, CALM/AF10, and SET/NUP12 and is required for their leukemogenic potential (Barry et al. 2010). Dot1L recruitment to ectopic loci (such as Hox cluster genes) causes aberrant expression of target genes involved in the leukemogenic process. EPZ004777 efficiently inhibited H3K79 methylation *in vitro*, mediated decreased expression of MLL fusion-associated genes and selectively induced differentiation and apoptosis of MLL-rearranged cells (Daigle et al. 2011). *In vivo*, administration of this compound resulted in a limited, but significant survival advantage in a MLL-rearranged xenograft model. Unfortunately, the therapeutic efficacy of this compound is greatly affected by its poor pharmacokinetic properties. However, recently many efforts have been directed towards the structural design and development of new and more potent Dot1L inhibitors with better pharmacokinetics *in vivo* (Yao et al. 2011).

Inhibition of Brd4 by the small-molecule BET bromodomain inhibitor of JQ1 led to a robust antileukemic effect in a mouse model of AML driven by the fusion protein MLL/AF9 and an oncogenic form of Nras. Brd4 is a member of the BET family of bromodomain-containing proteins that recognizes acetylated histones and mediates the effect of this modification on transcription. In a shRNA screen performed by Zuber et al. Brd4 was identified as a critical cofactor for the onset and progression of MLL/AF9-positive AML. Treatment with JQ1 induced cell cycle arrest and apoptosis of MLL-rearranged leukemic cell lines and primary AML samples from patients *in vitro*. Moreover, treatment of established AML in mice with JQ1 resulted in a significant delay in the disease progression and in survival benefit, associated to features of terminal myeloid differentiation of the leukemic cells. These effects were identified to be at least in part dependent on the indirect blunting effect of JQ1 on c-Myc expression through Brd4 inhibition (Zuber et al. 2011).

Similarly, Dawson et al showed that treatment with another small molecule inhibitor of the BET family of proteins (GSK1210151A or I-BET151) induced G1/G0 cell cycle arrest and apoptosis in MLL-fusion cell lines *in vitro*, and mediated a significant survival benefit in mouse models of MLL/AF9 and MLL/AF4. In this study, these antileukemic effects were proven to be at least in part mediated by the displacement of Brd3/4, PAF and SEC complexes from chromatin, resulting in the suppression of transcription of critical anti-apoptotic and pro-proliferative genes, such as Bcl-2, c-Myc, and Cdk6 (Dawson et al. 2011). Taken together, these studies suggest the potential of targeting the BET family of proteins for the therapy of MLL fusion leukemias.

Conclusions

A growing amount of data indicates that alterations in the activity or expression of histone-modifying enzymes play a critical role in the development and progression of AML, and, thus, could constitute particularly promising therapeutic targets for these diseases. In the recent years, there has been a strong focus on the development and use of epigenetic drugs and in particular of HDACi for the treatment of hematologic malignancies, including AML and MDS. Many compounds have been tested in preclinical setting giving encouraging results, and some have proceeded to evaluation in phase I/II clinical trials on AML patients. Although some promising results have been obtained with HDACi (especially in combination studies) in patients with AML, so far, the therapeutic success of these drugs has been limited by a series of different factors. First of all, some histone-modifying enzymes seem to play very different and specific roles in distinct classes of AML, depending on the cellular origin and the genetic background of the leukemic cells. Thus, inhibitors of specific enzymes or classes of enzymes are likely to have very different effects on different subtypes of leukemias. Another important issue is that in many cases the exact mechanism by which these epigenetic drugs exert their therapeutic activity is still not clear, and this obviously complicates the optimization of their clinical use.

A better understanding of the cancer-specific dependency on different epigenetic regulators, the development of more selective inhibitors of epigenetic enzymes, and a better stratification of patients enrolled in the clinical studies will probably greatly improve in the future the performance and the impact of epigenetic drugs in the clinics.

Acknowledgments RWJ is a principal research fellow of the National Health and Medical Research Council of Australia (NHMRC) and supported by NHMRC Program and Project Grants, the Susan G Komen Breast Cancer Foundation, Cancer Council Victoria, The Victorian Cancer Agency, The Leukemia Foundation of Australia and the Victorian Breast Cancer Research Consortium. MG is a postdoctoral fellow supported by the Italian Cancer Research Foundation (FIRC).

References

Ai X., Parthun MR (2004) The nuclear Hat1p/Hat2p complex: a molecular link between type B histone acetyltransferases and chromatin assembly. Mol Cell 14:195–205

Amann JM, Nip J, Strom DK, Lutterbach B, Harada H, Lenny N, Downing JR, Meyers S, Hiebert SW (2001) ETO, a target of t(8;21) in acute leukemia, makes distinct contacts with multiple histone deacetylases and binds mSin3A through its oligomerization domain. Mol Cell Biol 21:6470–6483

Avvakumov N, Cote J (2007). The MYST family of histone acetyltransferases and their intimate links to cancer. Oncogene 26:5395–5407

Ayton PM, Cleary ML (2003) Transformation of myeloid progenitors by MLL oncoproteins is dependent on Hoxa7 and Hoxa9. Genes Dev 17:2298–2307

Bali P, Pranpat M, Bradner J, Balasis M, Fiskus W, Guo F, Rocha K, Kumaraswamy S, Boyapalle S, Atadja P et al (2005) Inhibition of histone deacetylase 6 acetylates and disrupts the chaperone function of heat shock protein 90: a novel basis for antileukemia activity of histone deacetylase inhibitors. J Biol Chem 280:26729–26734

Barbetti V, Gozzini A, Rovida E, Morandi A, Spinelli E, Fossati G, Mascagni P, Lubbert M, Dello Sbarba P, Santini V (2008). Selective anti-leukaemic activity of low-dose histone deacetylase inhibitor ITF2357 on AML1/ETO-positive cells. Oncogene 27:1767–1778

Barry ER, Corry GN, Rasmussen TP (2010) Targeting DOT1L action and interactions in leukemia: the role of DOT1L in transformation and development. Expert Opin Ther Targets 14:405–418

Baylin SB, Ohm JE (2006) Epigenetic gene silencing in cancer—a mechanism for early oncogenic pathway addiction? Nat Rev Cancer 6:107–116

Bertrand P (2010) Inside HDAC with HDAC inhibitors. Eur J Med Chem 45:2095–2116

Bolden J.E, Peart MJ, Johnstone RW (2006) Anticancer activities of histone deacetylase inhibitors. Nat Rev Drug Discov 5:769–784

Bots M et al (2014) Differentiation therapy for the treatment of t(8;21) acute myeloid leukemia using histone deacetylase inhibitors. Blood 123(9):1341–1352

Bradbury C.A, Khanim FL, Hayden R, Bunce CM, White DA, Drayson MT, Craddock C, Turner BM (2005) Histone deacetylases in acute myeloid leukaemia show a distinctive pattern of expression that changes selectively in response to deacetylase inhibitors. Leukemia 19:1751–1759

Bug G, Ritter M, Wassmann B, Schoch C, Heinzel T, Schwarz K, Romanski A, Kramer OH, Kampfmann M, Hoelzer D et al (2005) Clinical trial of valproic acid and all-*trans* retinoic acid in patients with poor-risk acute myeloid leukemia. Cancer 104:2717–2725

Byrd JC, Marcucci G, Parthun MR, Xiao JJ, Klisovic RB, Moran M, Lin TS, Liu S, Sklenar AR, Davis ME et al (2005) A phase 1 and pharmacodynamic study of depsipeptide (FK228) in chronic lymphocytic leukemia and acute myeloid leukemia. Blood 105:959–967
Campos EI, Reinberg D (2009) Histones: annotating chromatin. Annu Rev Genet 43:559–599
Campos EI, Fillingham J, Li G, Zheng H, Voigt P, Kuo WH, Seepany H, Gao Z, Day LA, Greenblatt JF et al (2010) The program for processing newly synthesized histones H3.1 and H4. Nat Struct Mol Biol 17:1343–1351
Cao R, Wang L, Wang H, Xia L, Erdjument-Bromage H, Tempst P, Jones RS, Zhang Y (2002) Role of histone H3 lysine 27 methylation in Polycomb-group silencing. Science 298:1039–1043
Catley L, Weisberg E, Kiziltepe T, Tai YT, Hideshima T, Neri P, Tassone P, Atadja P, Chauhan D, Munshi NC et al (2006) Aggresome induction by proteasome inhibitor bortezomib and alpha-tubulin hyperacetylation by tubulin deacetylase (TDAC) inhibitor LBH589 are synergistic in myeloma cells. Blood 108:3441–3449
Cedar H, Bergman Y (2009) Linking DNA methylation and histone modification: patterns and paradigms. Nat Rev Genet 10:295–304
Chan EM, Chan RJ, Comer EM, Goulet RJ, 3rd, Crean CD, Brown ZD, Fruehwald AM, Yang Z, Boswell HS, Nakshatri H et al (2007) MOZ and MOZ-CBP cooperate with NF-kappaB to activate transcription from NF-kappaB-dependent promoters. Exp Hematol 35:1782–1792
Chen W, Li Q, Hudson WA, Kumar A, Kirchhof N, Kersey JH (2006) A murine Mll-AF4 knock-in model results in lymphoid and myeloid deregulation and hematologic malignancy. Blood 108:669–677
Chen CS, Wang YC, Yang HC, Huang PH, Kulp SK, Yang CC, Lu YS, Matsuyama S, Chen CY (2007) Histone deacetylase inhibitors sensitize prostate cancer cells to agents that produce DNA double-strand breaks by targeting Ku70 acetylation. Cancer Res 67:5318–5327
Cheung N, Chan LC, Thompson A, Cleary ML, So CW (2007) Protein arginine-methyltransferase-dependent oncogenesis. Nat Cell Biol 9:1208–1215
Chi P, Allis CD, Wang GG (2010) Covalent histone modifications–miswritten, misinterpreted and mis-erased in human cancers. Nat Rev Cancer 10:457–469
Choi JK, Howe LJ (2009). Histone acetylation: truth of consequences? Biochem Cell Biol 87:139–150
Choi JH, Kwon HJ, Yoon BI, Kim JH, Han SU, Joo HJ, Kim DY (2001) Expression profile of histone deacetylase 1 in gastric cancer tissues. Jpn J Cancer Res 92:1300–1304
Choudhary C, Kumar C, Gnad F, Nielsen ML, Rehman M, Walther TC, Olsen JV, Mann M (2009) Lysine acetylation targets protein complexes and co-regulates major cellular functions. Science 325:834–840
Copeland RA, Solomon ME, Richon VM (2009) Protein methyltransferases as a target class for drug discovery. Nat Rev Drug Discov 8:724–732
Corsetti MT, Salvi F, Perticone S, Baraldi A, De Paoli L, Gatto S, Pietrasanta D, Pini M, Primon V, Zallio F et al (2011) Hematologic improvement and response in elderly AML/RAEB patients treated with valproic acid and low-dose Ara-C. Leuk Res 35:991–997
Cosgrove MS, Boeke JD, Wolberger C (2004) Regulated nucleosome mobility and the histone code. Nat Struct Mol Biol 11:1037–1043
Daigle SR, Olhava EJ, Therkelsen CA, Majer CR, Sneeringer CJ, Song J, Johnston LD, Scott MP, Smith JJ, Xiao Y et al (2011) Selective killing of mixed lineage leukemia cells by a potent small-molecule DOT1L inhibitor. Cancer Cell 20:53–65
Dawson MA, Kouzarides T (2012) Cancer epigenetics: from mechanism to therapy. Cell 150: 12–27
Dawson MA, Prinjha RK, Dittmann A, Giotopoulos G, Bantscheff M, Chan WI, Robson SC, Chung CW, Hopf C, Savitski MM et al (2011) Inhibition of BET recruitment to chromatin as an effective treatment for MLL-fusion leukaemia. Nature 478:529–533
Deguchi K, Ayton PM, Carapeti M, Kutok JL, Snyder CS, Williams IR, Cross NC, Glass CK, Cleary ML, Gilliland DG (2003) MOZ-TIF2-induced acute myeloid leukemia requires the MOZ nucleosome binding motif and TIF2-mediated recruitment of CBP. Cancer Cell 3:259–271

Deribe YL, Wild P, Chandrashaker A, Curak J, Schmidt MH, Kalaidzidis Y, Milutinovic N, Kratchmarova I, Buerkle L, Fetchko MJ et al (2009) Regulation of epidermal growth factor receptor trafficking by lysine deacetylase HDAC6. Sci Signal 2:ra84

Dickinson M, Johnstone RW, Prince HM (2010) Histone deacetylase inhibitors: potential targets responsible for their anti-cancer effect. Invest New Drugs 28(1)3–20

Dobson CL, Warren AJ, Pannell R, Forster A, Lavenir I, Corral J, Smith AJ, Rabbitts TH (1999) The mll-AF9 gene fusion in mice controls myeloproliferation and specifies acute myeloid leukaemogenesis. EMBO J 18:3564–3574

Dokmanovic M, Marks PA (2005) Prospects: histone deacetylase inhibitors. J Cell Biochem 96:293–304

Durst KL, Lutterbach B, Kummalue T, Friedman AD, Hiebert SW (2003) The inv(16) fusion protein associates with corepressors via a smooth muscle myosin heavy-chain domain. Mol Cell Biol 23:607–619

Erickson P, Gao J, Chang KS, Look T, Whisenant E, Raimondi S, Lasher R, Trujillo J, Rowley J, Drabkin H (1992) Identification of breakpoints in t(8;21) acute myelogenous leukemia and isolation of a fusion transcript, AML1/ETO, with similarity to Drosophila segmentation gene, runt. Blood 80:1825–1831

Ernst T, Chase AJ, Score J, Hidalgo-Curtis CE, Bryant C, Jones AV, Waghorn K, Zoi K, Ross FM, Reiter A et al (2010) Inactivating mutations of the histone methyltransferase gene EZH2 in myeloid disorders. Nat Genet 42:722–726

Fathi AT, Grant S, Karp JE (2010) Exploiting cellular pathways to develop new treatment strategies for AML. Cancer Treat Rev 36:142–150

Feinberg AP, Vogelstein B (1983) Hypomethylation distinguishes genes of some human cancers from their normal counterparts. Nature 301:89–92

Fotheringham S, Epping MT, Stimson L, Khan O, Wood V, Pezzella F, Bernards R, La Thangue NB (2009) Genome-wide loss-of-function screen reveals an important role for the proteasome in HDAC inhibitor-induced apoptosis. Cancer Cell 15:57–66

Fraga MF, Ballestar E, Villar-Garea A, Boix-Chornet M, Espada J, Schotta G, Bonaldi T, Haydon C, Ropero S, Petrie K et al (2005) Loss of acetylation at Lys16 and trimethylation at Lys20 of histone H4 is a common hallmark of human cancer. Nat Genet 37:391–400

Frank R, Zhang J, Uchida H, Meyers S, Hiebert SW, Nimer SD (1995) The AML1/ETO fusion protein blocks transactivation of the GM-CSF promoter by AML1B. Oncogene 11:2667–2674

Fullgrabe J, Kavanagh E, Joseph B (2011) Histone onco-modifications. Oncogene 30:3391–3403

Garcia-Manero G, Kantarjian HM, Sanchez-Gonzalez B, Yang H, Rosner G, Verstovsek S, Rytting M, Wierda WG, Ravandi F, Koller C et al (2006) Phase 1/2 study of the combination of 5-aza-2′-deoxycytidine with valproic acid in patients with leukemia. Blood 108:3271–3279

Garcia-Manero G, Yang H, Bueso-Ramos C, Ferrajoli A, Cortes J, Wierda WG, Faderl S, Koller C, Morris G, Rosner G et al (2008) Phase 1 study of the histone deacetylase inhibitor vorinostat (suberoylanilide hydroxamic acid [SAHA]) in patients with advanced leukemias and myelodysplastic syndromes. Blood 111:1060–1066

Gelmetti V, Zhang J, Fanelli M, Minucci S, Pelicci PG, Lazar MA (1998) Aberrant recruitment of the nuclear receptor corepressor-histone deacetylase complex by the acute myeloid leukemia fusion partner ETO. Mol Cell Biol 18:7185–7191

Gojo I, Jiemjit A, Trepel JB, Sparreboom A, Figg WD, Rollins S, Tidwell ML, Greer J, Chung EJ, Lee MJ et al (2007) Phase 1 and pharmacologic study of MS-275, a histone deacetylase inhibitor, in adults with refractory and relapsed acute leukemias. Blood 109:2781–2790

Gottlicher M, Minucci S, Zhu P, Kramer OH, Schimpf A, Giavara S, Sleeman JP, Lo Coco F, Nervi C, Pelicci PG et al (2001) Valproic acid defines a novel class of HDAC inhibitors inducing differentiation of transformed cells. EMBO J 20:6969–6978

Goyama S, Mulloy JC (2011) Molecular pathogenesis of core binding factor leukemia: current knowledge and future prospects. Int J Hematol 94:126–133

Gregoretti IV, Lee YM, Goodson HV (2004) Molecular evolution of the histone deacetylase family: functional implications of phylogenetic analysis. J Mol Biol 338:17–31

Grignani F, De Matteis S, Nervi C, Tomassoni L, Gelmetti V, Cioce M, Fanelli M, Ruthardt M, Ferrara FF, Zamir I et al (1998) Fusion proteins of the retinoic acid receptor-alpha recruit histone deacetylase in promyelocytic leukaemia. Nature 391:815–818

Gu W, Roeder RG (1997) Activation of p53 sequence-specific DNA binding by acetylation of the p53 C-terminal domain. Cell 90:595–606

Halkidou K, Gaughan L, Cook S, Leung HY, Neal DE, Robson CN (2004) Upregulation and nuclear recruitment of HDAC1 in hormone refractory prostate cancer. Prostate 59:177–189

He LZ, Tolentino T, Grayson P, Zhong S, Warrell RP, Jr, Rifkind RA, Marks PA, Richon VM, Pandolfi PP (2001) Histone deacetylase inhibitors induce remission in transgenic models of therapy-resistant acute promyelocytic leukemia. J Clin Invest 108:1321–1330

Hildmann C, Riester D, Schwienhorst A. (2007) Histone deacetylases—an important class of cellular regulators with a variety of functions. Appl Microbiol Biotechnol 75:487–497

Hodawadekar SC, Marmorstein R (2007) Chemistry of acetyl transfer by histone modifying enzymes: structure, mechanism and implications for effector design. Oncogene 26:5528–5540

Hubbert C, Guardiola A, Shao R, Kawaguchi Y, Ito A, Nixon A, Yoshida M, Wang XF, Yao TP (2002) HDAC6 is a microtubule-associated deacetylase. Nature 417:455–458

Hug BA, Lazar MA (2004) ETO interacting proteins. Oncogene 23:4270–4274

Huntly BJ, Shigematsu H, Deguchi K, Lee BH, Mizuno S, Duclos N, Rowan R, Amaral S, Curley D, Williams IR et al (2004) MOZ-TIF2, but not BCR-ABL, confers properties of leukemic stem cells to committed murine hematopoietic progenitors. Cancer Cell 6:587–596

Insinga A, Minucci S, Pelicci PG (2005a) Mechanisms of selective anticancer action of histone deacetylase inhibitors. Cell Cycle 4:741–743

Insinga A, Monestiroli S, Ronzoni S, Gelmetti V, Marchesi F, Viale A, Altucci L, Nervi C, Minucci S, Pelicci PG (2005b) Inhibitors of histone deacetylases induce tumor-selective apoptosis through activation of the death receptor pathway. Nat Med 11:71–76

Insinga A, Pelicci PG, Inucci S (2005c) Leukemia-associated fusion proteins. Multiple mechanisms of action to drive cell transformation. Cell Cycle 4:67–69

Iyer NG, Ozdag H, Caldas C (2004) p300/CBP and cancer. Oncogene 23:4225–4231

Izzo A, Schneider R (2010) Chatting histone modifications in mammals. Brief Funct Genom 9:429–443

Jankowska AM, Szpurka H (2012) Mutational determinants of epigenetic instablity in myeloid malignancies. Semin Oncol 39:80–96

Katsumoto T, Aikawa Y, Iwama A, Ueda S, Ichikawa H, Ochiya T, Kitabayashi I (2006) MOZ is essential for maintenance of hematopoietic stem cells. Genes Dev 20:1321–1330

Katsumoto T, Yoshida N, Kitabayashi I (2008) Roles of the histone acetyltransferase monocytic leukemia zinc finger protein in normal and malignant hematopoiesis. Cancer Sci 99:1523–1527

Kitabayashi I, Aikawa Y, Nguyen LA, Yokoyama A, Ohki M (2001) Activation of AML1-mediated transcription by MOZ and inhibition by the MOZ-CBP fusion protein. EMBO J 20:7184–7196

Kong CT, Sham MH, So CW, Cheah KS, Chen SJ, Chan LC (2006) The Mll-Een knockin fusion gene enhances proliferation of myeloid progenitors derived from mouse embryonic stem cells and causes myeloid leukaemia in chimeric mice. Leukemia 20:1829–1839

Kouzarides T (2007). Chromatin modifications and their function. Cell 128:693–705

Kovacs JJ, Murphy PJ, Gaillard S, Zhao X, Wu JT, Nicchitta CV, Yoshida M, Toft DO, Pratt WB, Yao TP (2005) HDAC6 regulates Hsp90 acetylation and chaperone-dependent activation of glucocorticoid receptor. Mol Cell 18:601–607

Kramer OH, Zhu P, Ostendorff HP, Golebiewski M, Tiefenbach J, Peters MA, Brill B, Groner B, Bach I, Heinzel T et al (2003) The histone deacetylase inhibitor valproic acid selectively induces proteasomal degradation of HDAC2. EMBO J 22:3411–3420

Kuendgen A, Knipp S, Fox F, Strupp C, Hildebrandt B, Steidl C, Germing U, Haas R, Gattermann N (2005) Results of a phase 2 study of valproic acid alone or in combination with all-*trans* retinoic acid in 75 patients with myelodysplastic syndrome and relapsed or refractory acute myeloid leukemia. Ann Hematol 84(1):61–66

Kuendgen A, Schmid M, Schlenk R, Knipp S, Hildebrandt B, Steidl C, Germing U, Haas R, Dohner H, Gattermann N (2006) The histone deacetylase (HDAC) inhibitor valproic acid as

monotherapy or in combination with all-*trans* retinoic acid in patients with acute myeloid leukemia. Cancer 106:112–119
Kunert N, Brehm A (2009) Novel Mi-2 related ATP-dependent chromatin remodelers. Epigenetics 4:209–211
Kung AL, Rebel VI, Bronson RT, Ch'ng LE, Sieff CA, Livingston DM, Yao TP (2000) Gene dose-dependent control of hematopoiesis and hematologic tumor suppression by CBP. Genes Dev 14:272–277
Lawless MW, Norris S, O'Byrne KJ, Gray SG (2009) Targeting histone deacetylases for the treatment of disease. J Cell Mol Med 13:826–852
Lee KK, Workman JL (2007) Histone acetyltransferase complexes: one size doesn't fit all. Nat Rev Mol Cell Biol 8:284–295
Leiva M, Moretti S, Soilihi H, Pallavicini I, Peres L, Mercurio C, Dal Zuffo R, Minucci S, de The H (2012) Valproic acid induces differentiation and transient tumor regression, but spares leukemia-initiating activity in mouse models of APL. Leukemia 26:1630–1637
Li B, Carey M, Workman JL (2007) The role of chromatin during transcription. Cell 128:707–719
Lin RJ, Nagy L, Inoue S, Shao W, Miller WH, Jr, Evans RM (1998) Role of the histone deacetylase complex in acute promyelocytic leukaemia. Nature 391:811–814
Lin RJ, Egan DA, Evans RM (1999) Molecular genetics of acute promyelocytic leukemia. Trends Genet 15:179–184
Linggi BE, Brandt SJ, Sun ZW, Hiebert SW (2005) Translating the histone code into leukemia. J Cell Biochem 96:938–950
Liu L, Scolnick DM, Trievel RC, Zhang HB, Marmorstein R, Halazonetis TD, Berger SL (1999) p53 sites acetylated in vitro by PCAF and p300 are acetylated in vivo in response to DNA damage. Mol Cell Biol 19:1202–1209
Liu H, Chen B, Xiong H, Huang QH, Zhang QH, Wang ZG, Li BL, Chen Z, Chen SJ (2004) Functional contribution of EEN to leukemogenic transformation by MLL-EEN fusion protein. Oncogene 23:3385–3394
Liu S, Klisovic RB, Vukosavljevic T, Yu J, Paschka P, Huynh L, Pang J, Neviani P, Liu Z, Blum W et al (2007) Targeting AML1/ETO-histone deacetylase repressor complex: a novel mechanism for valproic acid-mediated gene expression and cellular differentiation in AML1/ETO-positive acute myeloid leukemia cells. J Pharmacol Exp Ther 321:953–960
Liu H, Cheng EH, Hsieh JJ (2009) MLL fusions: pathways to leukemia. Cancer Biol Ther 8:1204–1211
Look AT (1997) Oncogenic transcription factors in the human acute leukemias. Science 278:1059–1064
Luo J, Nikolaev AY, Imai S, Chen D, Su F, Shiloh A, Guarente L, Gu W (2001) Negative control of p53 by Sir2alpha promotes cell survival under stress. Cell 107:137–148
Marquard L, Poulsen CB, Gjerdrum LM, de Nully Brown P, Christensen IJ, Jensen PB, Sehested M, Johansen P, Ralfkiaer E (2009) Histone deacetylase 1, 2, 6 and acetylated histone H4 in B- and T-cell lymphomas. Histopathology 54:688–698
Martino DJ, Prescott SL (2010) Silent mysteries: epigenetic paradigms could hold the key to conquering the epidemic of allergy and immune disease. Allergy 65:7–15
Masetti R, Serravalle S, Biagi C, Pession A (2011) The role of HDACs inhibitors in childhood and adolescence acute leukemias. J Biomed Biotechnol 2011:148046
Meda F, Folci M, Baccarelli A, Selmi C (2011) The epigenetics of autoimmunity. Cell Mol Immunol 8:226–236
Mehnert JM, Kelly WK (2007) Histone deacetylase inhibitors: biology and mechanism of action. Cancer J 13:23–29
Meyer C, Kowarz E, Hofmann J, Renneville A, Zuna J, Trka J, Ben Abdelali R, Macintyre E, De Braekeleer E, De Braekeleer M et al (2009) New insights to the MLL recombinome of acute leukemias. Leukemia 23:1490–1499
Milne TA, Hughes CM, Lloyd R, Yang Z, Rozenblatt-Rosen O, Dou Y, Schnepp RW, Krankel C, Livolsi VA, Gibbs D et al (2005a) Menin and MLL cooperatively regulate expression of cyclin-dependent kinase inhibitors. Proc Natl Acad Sci U S A 102:749–754

Milne TA, Martin ME, Brock HW, Slany RK, Hess JL (2005b) Leukemogenic MLL fusion proteins bind across a broad region of the Hox a9 locus, promoting transcription and multiple histone modifications. Cancer Res 65:11367–11374
Minucci S, Pelicci PG (2006) Histone deacetylase inhibitors and the promise of epigenetic (and more) treatments for cancer. Nat Rev Cancer 6:38–51
Moreno DA, Scrideli CA, Cortez MA, de Paula Queiroz R, Valera ET, da Silva Silveira V, Yunes JA, Brandalise SR, Tone LG (2010) Differential expression of HDAC3, HDAC7 and HDAC9 is associated with prognosis and survival in childhood acute lymphoblastic leukaemia. Br J Haematol 150:665–673
Muller AM, Duque J, Shizuru JA, Lubbert M (2008) Complementing mutations in core binding factor leukemias: from mouse models to clinical applications. Oncogene 27:5759–5773
Nagy, Z., and Tora, L. (2007). Distinct GCN5/PCAF-containing complexes function as co-activators and are involved in transcription factor and global histone acetylation. Oncogene 26, 5341–5357
Nebbioso A, Clarke N, Voltz E, Germain E, Ambrosino C, Bontempo P, Alvarez R, Schiavone EM, Ferrara F, Bresciani F et al (2005) Tumor-selective action of HDAC inhibitors involves TRAIL induction in acute myeloid leukemia cells. Nat Med 11:77–84
Neff T, Sinha AU, Kluk MJ, Zhu N, Khattab MH, Stein L, Xie H, Orkin SH, Armstrong SA (2012) Polycomb repressive complex 2 is required for MLL-AF9 leukemia. Proc Natl Acad Sci U S A 109:5028–5033
Neugebauer RC, Sippl W, Jung M (2008) Inhibitors of NAD+ dependent histone deacetylases (sirtuins). Curr Pharm Des 14:562–573
Nguyen AT, Zhang Y (2011) The diverse functions of Dot1 and H3K79 methylation. Genes Dev 25:1345–1358
Nikoloski G, Langemeijer SM, Kuiper RP, Knops R, Massop M, Tonnissen ER, van der Heijden A, Scheele TN, Vandenberghe P, de Witte T et al (2010) Somatic mutations of the histone methyltransferase gene EZH2 in myelodysplastic syndromes. Nat Genet 42:665–667
Nishioka C, Ikezoe T, Yang J, Koeffler HP, Yokoyama A (2008a) Inhibition of MEK/ERK signaling synergistically potentiates histone deacetylase inhibitor-induced growth arrest, apoptosis and acetylation of histone H3 on p21waf1 promoter in acute myelogenous leukemia cell. Leukemia 22:1449–1452
Nishioka C, Ikezoe T, Yang J, Takeuchi S, Koeffler HP, Yokoyama A (2008b) MS-275, a novel histone deacetylase inhibitor with selectivity against HDAC1, induces degradation of FLT3 via inhibition of chaperone function of heat shock protein 90 in AML cells. Leuk Res 32:1382–1392
Odenike OM, Alkan S, Sher D, Godwin JE, Huo D, Brandt SJ, Green M, Xie J, Zhang Y, Vesole DH et al (2008). Histone deacetylase inhibitor romidepsin has differential activity in core binding factor acute myeloid leukemia. Clin Cancer Res 14:7095–7101
Okuda T, van Deursen J, Hiebert SW, Grosveld G, Downing JR (1996) AML1, the target of multiple chromosomal translocations in human leukemia, is essential for normal fetal liver hematopoiesis. Cell 84:321–330
Okada Y, Feng Q, Lin Y, Jiang Q, Li Y, Coffield VM, Su L, Xu G, Zhang Y (2005) hDOT1L links histone methylation to leukemogenesis. Cell 121:167–178
Ono R, Nosaka T, Hayashi Y (2005) Roles of a trithorax group gene, MLL, in hematopoiesis. Int J Hematol 81:288–293
Orkin SH, Hochedlinger K (2011) Chromatin connections to pluripotency and cellular reprogramming. Cell 145:835–850
Ornstein MC, Sekeres MA (2012) Combination strategies in myelodysplastic syndromes. Int J Hematol 95:26–33
Ozdag H, Teschendorff AE, Ahmed AA, Hyland SJ, Blenkiron C, Bobrow L, Veerakumarasivam A, Burtt G, Subkhankulova T, Arends MJ et al (2006). Differential expression of selected histone modifier genes in human solid cancers. BMC Genom 7:90
Pal S, Sif S (2007) Interplay between chromatin remodelers and protein arginine methyltransferases. J Cell Physiol 213:306–315

Parthun MR (2007) Hat1: the emerging cellular roles of a type B histone acetyltransferase. Oncogene 26:5319–5328
Parthun MR (2012) Histone acetyltransferase 1: more than just an enzyme? Biochim Biophys Acta 1819:256–263
Pelletier N, Champagne N, Stifani S, Yang XJ (2002) MOZ and MORF histone acetyltransferases interact with the Runt-domain transcription factor Runx2. Oncogene 21:2729–2740
Perez-Campo FM, Borrow J, Kouskoff V, Lacaud G (2009) The histone acetyl transferase activity of monocytic leukemia zinc finger is critical for the proliferation of hematopoietic precursors. Blood 113:4866–4874
Peterlin BM, Price DH (2006) Controlling the elongation phase of transcription with P-TEFb. Mol Cell 23:297–305
Plass C, Oakes C, Blum W, Marcucci G (2008) Epigenetics in acute myeloid leukemia. Semin Oncol 35:378–387
Quintas-Cardama A, Santos FP, Garcia-Manero G (2011) Histone deacetylase inhibitors for the treatment of myelodysplastic syndrome and acute myeloid leukemia. Leukemia 25:226–235
Raffoux E, Cras A, Recher C, Boelle PY, de Labarthe A, Turlure P, Marolleau JP, Reman O, Gardin C, Victor M et al (2010) Phase 2 clinical trial of 5-azacitidine, valproic acid, and all-*trans* retinoic acid in patients with high-risk acute myeloid leukemia or myelodysplastic syndrome. Oncotarget 1:34–42
Rajendran R, Garva R, Krstic-Demonacos M, Demonacos C (2011) Sirtuins: molecular traffic lights in the crossroad of oxidative stress, chromatin remodeling, and transcription. J Biomed Biotechnol 2011:368276
Rao R, Fiskus W, Yang Y, Lee P, Joshi R, Fernandez P, Mandawat A, Atadja P, Bradner JE, Bhalla K (2008) HDAC6 inhibition enhances 17-AAG-mediated abrogation of hsp90 chaperone function in human leukemia cells. Blood 112:1886–1893
Reader JC, Meekins JS, Gojo I, Ning Y (2007) A novel NUP98-PHF23 fusion resulting from a cryptic translocation t(11;17)(p15;p13) in acute myeloid leukemia. Leukemia 21:842–844
Rebel VI, Kung AL, Tanner EA, Yang H, Bronson RT, Livingston DM (2002) Distinct roles for CREB-binding protein and p300 in hematopoietic stem cell self-renewal. Proc Natl Acad Sci U S A 99:14789–14794
Reikvam H, Ersvaer E, Bruserud O (2009). Heat shock protein 90 - a potential target in the treatment of human acute myelogenous leukemia. Curr Cancer Drug Targets 9:761–776
Rhoades KL, Hetherington CJ, Rowley JD, Hiebert SW, Nucifora G, Tenen DG, Zhang DE (1996) Synergistic up-regulation of the myeloid-specific promoter for the macrophage colony-stimulating factor receptor by AML1 and the t(8;21) fusion protein may contribute to leukemogenesis. Proc Natl Acad Sci U S A 93:11895–11900
Robertson KD (2005) DNA methylation and human disease. Nat Rev Genet 6:597–610
Ropero S, Esteller M (2007) The role of histone deacetylases (HDACs) in human cancer. Mol Oncol 1:19–25
Rosato RR, Almenara JA, Grant S (2003) The histone deacetylase inhibitor MS-275 promotes differentiation or apoptosis in human leukemia cells through a process regulated by generation of reactive oxygen species and induction of p21CIP1/WAF1 1. Cancer Res 63:3637–3645
Rowley JD, Reshmi S, Sobulo O, Musvee T, Anastasi J, Raimondi S, Schneider NR, Barredo JC, Cantu ES, Schlegelberger B et al (1997) All patients with the T(11;16)(q23;p13.3) that involves MLL and CBP have treatment-related hematologic disorders. Blood 90:535–541
Saeed S, Logie C, Stunnenberg HG, Martens JH (2011) Genome-wide functions of PML-RARalpha in acute promyelocytic leukaemia. Br J Cancer 104:554–558
Sasaki K, Yagi H, Bronson RT, Tominaga K, Matsunashi T, Deguchi K, Tani Y, Kishimoto T, Komori T (1996) Absence of fetal liver hematopoiesis in mice deficient in transcriptional coactivator core binding factor beta. Proc Natl Acad Sci U S A 93:12359–12363
Schwer B, Verdin E (2008) Conserved metabolic regulatory functions of sirtuins. Cell Metab 7:104–112
Seet BT, Dikic I, Zhou MM, Pawson T (2006) Reading protein modifications with interaction domains. Nat Rev Mol Cell Biol 7:473–483

Segre CV, Chiocca S (2011) Regulating the regulators: the post-translational code of class I HDAC1 and HDAC2. J Biomed Biotechnol 2011:690848
Shen X, Liu Y, Hsu YJ, Fujiwara Y, Kim J, Mao X, Yuan GC, Orkin SH (2008) EZH1 mediates methylation on histone H3 lysine 27 and complements EZH2 in maintaining stem cell identity and executing pluripotency. Mol Cell 32:491–502
Shi J, Wang E, Zuber J, Rappaport A, Taylor M, Johns C, Lowe SW, Vakoc CR (2012) The Polycomb complex PRC2 supports aberrant self-renewal in a mouse model of MLL-AF9;Nras(G12D) acute myeloid leukemia. Oncogene. doi:10.1038/onc.2012.110
Shia WJ, Okumura AJ, Yan M, Sarkeshik A, Lo MC, Matsuura S, Komeno Y, Zhao X, Nimer SD, Yates JR, 3rd et al (2012) PRMT1 interacts with AML1-ETO to promote its transcriptional activation and progenitor cell proliferative potential. Blood 119:4953–4962
Shigesada K, van de Sluis B, Liu PP (2004) Mechanism of leukemogenesis by the inv(16) chimeric gene CBFB/PEBP2B-MHY11. Oncogene 23:4297–4307
Shima Y, Kitabayashi I (2011) Deregulated transcription factors in leukemia. Int J Hematol 94:134–141
Shipley J, Butera JN (2009) Acute myelogenous leukemia. Exp Hematol 37:649–658
Slany RK (2009) The molecular biology of mixed lineage leukemia. Haematologica 94:984–993
Slany RK (2010) When speed matters: leukemogenic transformation by MLL fusion proteins. Cell Cycle 9:2475–2476
Sobulo OM, Borrow J, Tomek R, Reshmi S, Harden A, Schlegelberger B, Housman D, Doggett NA, Rowley JD, Zeleznik-Le NJ (1997) MLL is fused to CBP, a histone acetyltransferase, in therapy-related acute myeloid leukemia with a t(11;16)(q23;p13.3). Proc Natl Acad Sci U S A 94:8732–8737
Soriano AO, Yang H, Faderl S, Estrov Z, Giles F, Ravandi F, Cortes J, Wierda WG, Ouzounian S, Quezada A et al (2007) Safety and clinical activity of the combination of 5-azacytidine, valproic acid, and all-*trans* retinoic acid in acute myeloid leukemia and myelodysplastic syndrome. Blood 110:2302–2308
Speck NA, Gilliland DG (2002) Core-binding factors in haematopoiesis and leukaemia. Nat Rev Cancer 2:502–513
Strahl BD, Allis CD (2000) The language of covalent histone modifications. Nature 403:41–45
Taira N, Yoshida K (2012) Post-translational modifications of p53 tumor suppressor: determinants of its functional targets. Histol Histopathol 27:437–443
Takeda S, Chen DY, Westergard TD, Fisher JK, Rubens JA, Sasagawa S, Kan JT, Korsmeyer SJ, Cheng EH, Hsieh JJ (2006) Proteolysis of MLL family proteins is essential for taspase1-orchestrated cell cycle progression. Genes Dev 20:2397–2409
Talbert PB, Henikoff S (2010) Histone variants–ancient wrap artists of the epigenome. Nat Rev Mol Cell Biol 11:264–275
Thiagalingam S, Cheng KH, Lee HJ, Mineva N, Thiagalingam A, Ponte JF (2003) Histone deacetylases: unique players in shaping the epigenetic histone code. Ann N Y Acad Sci 983:84–100
Thomas T, Corcoran LM, Gugasyan R, Dixon MP, Brodnicki T, Nutt SL, Metcalf D, Voss AK (2006) Monocytic leukemia zinc finger protein is essential for the development of long-term reconstituting hematopoietic stem cells. Genes Dev 20:1175–1186
Ungerstedt JS, Sowa Y, Xu WS, Shao Y, Dokmanovic M, Perez G, Ngo L, Holmgren A, Jiang X, Marks PA (2005) Role of thioredoxin in the response of normal and transformed cells to histone deacetylase inhibitors. Proc Natl Acad Sci U S A 102:673–678
Vaissiere T, Sawan C, Herceg Z (2008) Epigenetic interplay between histone modifications and DNA methylation in gene silencing. Mutat Res 659:40–48
van Zutven LJ, Onen E, Velthuizen SC, van Drunen E, von Bergh AR, van den Heuvel-Eibrink MM, Veronese A, Mecucci C, Negrini M, de Greef GE et al (2006) Identification of NUP98 abnormalities in acute leukemia: JARID1A (12p13) as a new partner gene. Genes Chromosomes Cancer 45:437–446
Verbrugge I, Johnstone RW, Bots M (2011) Promises and challenges of anticancer drugs that target the epigenome. Epigenomics 3:547–565

Verzijlbergen KF, van Welsem T, Sie D, Lenstra TL, Turner DJ, Holstege FC, Kerkhoven RM, van Leeuwen F (2011) A barcode screen for epigenetic regulators reveals a role for the NuB4/HAT-B histone acetyltransferase complex in histone turnover. PLoS Genet 7:e1002284

Voss AK, Thomas T (2009) MYST family histone acetyltransferases take center stage in stem cells and development. Bioessays 31:1050–1061

Wagner JM, Hackanson B, Lubbert M, Jung M (2010) Histone deacetylase (HDAC) inhibitors in recent clinical trials for cancer therapy. Clin Epigenetics 1:117–136

Wang Q, Stacy T, Miller JD, Lewis AF, Gu TL, Huang X, Bushweller JH, Bories JC, Alt FW, Ryan G et al (1996) The CBFbeta subunit is essential for CBFalpha2 (AML1) function in vivo. Cell 87:697–708

Wang GG, Song J, Wang Z, Dormann HL, Casadio F, Li H, Luo JL, Patel DJ, Allis CD (2009) Haematopoietic malignancies caused by dysregulation of a chromatin-binding PHD finger. Nature 459:847–851

Wang L, Gural A, Sun XJ, Zhao X, Perna F, Huang G, Hatlen MA, Vu L, Liu F, Xu H et al (2011) The leukemogenicity of AML1-ETO is dependent on site-specific lysine acetylation. Science 333:765–769

Wilson AJ, Byun DS, Popova N, Murray LB, L'Italien K, Sowa Y, Arango D, Velcich A, Augenlicht LH, Mariadason JM (2006) Histone deacetylase 3 (HDAC3) and other class I HDACs regulate colon cell maturation and p21 expression and are deregulated in human colon cancer. J Biol Chem 281:13548–13558

Wilting RH, Yanover E, Heideman MR, Jacobs H, Horner J, van der Torre J, DePinho RA, Dannenberg JH (2010) Overlapping functions of Hdac1 and Hdac2 in cell cycle regulation and haematopoiesis. EMBO J 29:2586–2597

Wong P, Iwasaki M, Somervaille TC, So CW, Cleary ML (2007) Meis1 is an essential and rate-limiting regulator of MLL leukemia stem cell potential. Genes Dev 21:2762–2774

Xia ZB, Anderson M, Diaz MO, Zeleznik-Le NJ (2003) MLL repression domain interacts with histone deacetylases, the polycomb group proteins HPC2 and BMI-1, and the corepressor C-terminal-binding protein. Proc Natl Acad Sci U S A 100:8342–8347

Yang XJ, Seto E (2007) HATs and HDACs: from structure, function and regulation to novel strategies for therapy and prevention. Oncogene 26:5310–5318

Yang XJ, Ullah M (2007) MOZ and MORF, two large MYSTic HATs in normal and cancer stem cells. Oncogene 26:5408–5419

Yao YL, Yang WM (2011) Beyond histone and deacetylase: an overview of cytoplasmic histone deacetylases and their nonhistone substrates. J Biomed Biotechnol 2011:146493

Yao Y, Chen P, Diao J, Cheng G, Deng L, Anglin JL, Prasad BV, Song Y (2011) Selective inhibitors of histone methyltransferase DOT1L: design, synthesis, and crystallographic studies. J Am Chem Soc 133:16746–16749

Yu BD, Hess JL, Horning SE, Brown GA, Korsmeyer SJ (1995) Altered Hox expression and segmental identity in Mll-mutant mice. Nature 378:505–508

Zeisig BB, Milne T, Garcia-Cuellar MP, Schreiner S, Martin ME, Fuchs U, Borkhardt A, Chanda SK, Walker J, Soden R et al (2004) Hoxa9 and Meis1 are key targets for MLL-ENL-mediated cellular immortalization. Mol Cell Biol 24:617–628

Zhang XD, Gillespie SK, Borrow JM, Hersey P (2004) The histone deacetylase inhibitor suberic bishydroxamate regulates the expression of multiple apoptotic mediators and induces mitochondria-dependent apoptosis of melanoma cells. Mol Cancer Ther 3:425–435

Zhang Z, Yamashita H, Toyama T, Sugiura H, Ando Y, Mita K, Hamaguchi M, Hara Y, Kobayashi S, Iwase H (2005) Quantitation of HDAC1 mRNA expression in invasive carcinoma of the breast*. Breast Cancer Res Treat 94:11–16

Zhu P, Martin E, Mengwasser J, Schlag P, Janssen KP, Gottlicher M (2004) Induction of HDAC2 expression upon loss of APC in colorectal tumorigenesis. Cancer Cell 5:455–463

Zuber J, Shi J, Wang E, Rappaport AR, Herrmann H, Sison EA, Magoon D, Qi J, Blatt K, Wunderlich M, Taylor MJ, Johns C, Chicas A, Mulloy JC, Kogan SC, Brown P, Valent P, Bradner JE, Lowe SW, Vakoc CR (2011) RNAi screen identifies Brd4 as a therapeutic target in acute myeloid leukaemia. Nature 478(7370):524–528

Chapter 22
Methylation in AML: Clinical Applications

Naveen Pemmaraju, Guillermo Garcia-Manero and Farhad Ravandi

Abstract Epigenetic therapy is a recognized central therapeutic strategy widely used in treatment of myelodysplatic syndrome (MDS). The approved agents, decitabine and azacytidine, are DNA hypomethylating agents that have been shown in randomized phase III trials to have clinical benefit in MDS patients. New research is investigating the use of these drugs in the setting of acute myeloid leukemia (AML) treatment, especially in older patients. Early studies are demonstrating promising results for the use of decitabine and azacytidine in treatment of AML. Other novel epigenetic/hypomethylating agents as well as chemotherapy combinations including hypomethylating agents represent an active area of research in treatment of AML patients.

Keywords Hypomethylator · Decitabine · Azacytidine · AML

22.1 Introduction

One of the first reports of methylation, specifically, hypermethylation, identifying this process as an important component of hematologic malignancies in humans, including acute myeloid leukemia (AML), was by Baylin et al. in the 1980s (Baylin et al. 1987; Jones and Baylin 2007). Since these early reports, it has become apparent that DNA methylation is a common event in myelodysplatic syndrome (MDS) (Garcia-Manero 2011) and that this so-called epigenetic aberrancy has the potential to be targeted for pharmacologic intervention with the hypomethylator agents

N. Pemmaraju (✉) · G. Garcia-Manero
Department of Leukemia, Division of Cancer Medicine, University of Texas
MD Anderson Cancer Center, 1515 Holcombe Blvd. Unit 0428, Houston, TX 77030, USA
e-mail: npemmaraju@mdanderson.org

G. Garcia-Manero
e-mail: ggarciam@mdanderson.org

F. Ravandi
3Department of Leukemia, Division of Cancer Medicine, The University of Texas
MD Anderson Cancer Center, FC 4. 3052, Houston, TX 77030, USA
email: fravandi@mdanderson.org

M. Andreeff (ed.), *Targeted Therapy of Acute Myeloid Leukemia*,
Current Cancer Research, DOI 10.1007/978-1-4939-1393-0_22

(azacytidine, decitabine), now approved for MDS and being tested in AML settings (McDevitt 2012; Flotho et al. 2009). Indeed, it is the discovery of DNA methylation itself being one of the many fundamental methods by which the carcinogenesis of AML is possibly propagated, demonstrated by DNA methylation profiling and quantitative assays that is leading to many groups testing the hypothesis that hypomethylating agents in the clinic may be of clinical benefit to AML patients (Jain et al. 2009; Bullinger et al. 2010; Figueroa et al. 2010; Issa and Kantarjian 2009; Galm et al. 2005; Valk et al. 2004; Schoofs and Muller-Tidow 2011). This chapter will focus on the development of research efforts with regard to the clinical applications of hypomethylator therapy, with the two approved hypomethylating agents widely available today, decitabine and azacytidine, both approved for the treatment of MDS, and now being actively studied for use in patients with AML.

22.2 Decitabine in AML

22.2.1 Decitabine: Background

Decitabine (5-aza-2'-deoxycytidine, Dacogen, Eisai Inc, Astex Pharmaceauticals), is an analog of the DNA base cytidine. It is a DNA-hypomethylating agent, received its Food and Drug Administration (FDA) approval for use in MDS after studies including a large randomized phase III trial led by Kantarjian et al. demonstrated clinical activity in a wide range of MDS patients (Kantarjian et al. 2006). In this study, 170 patients with diagnosis of MDS experienced randomization to receive either decitabine 15 mg/m^2 intravenously every 8 h for 3 days for every 6 weeks versus best supportive care. An overall response rate of 17% in the decitabine arm, compared to 0% in the supportive care arm ($p<0.001$). Those treated in the decitabine arm had trend in direction of a longer median time to progression to AML or death when compared to those patients who received supportive care (12.1 months vs. 7.8 months, $p=0.16$). Importantly, 18% of the patients in this trial had bone marrow blasts 20–30% at diagnosis, actually qualifying them as AML as per revised WHO classification, indicating possibility for activity among AML patient population of single agent decitabine.

With regard to investigating the antileukemic potential for decitabine, one of the original studies to demonstrate in vitro rationale was performed by Momparler et al. (1984) in which decitabine treatment resulted in more than 70% inhibition of DNA methylation in leukemia cells suggesting that the process of DNA methylation inhibition to be the mechanism of action of decitiabine against leukemia cells.

Gattei et al. (1993) examined the in vitro sensitivity of blast progenitors to decitabine in nine AML patient samples clinically treated with decitabine, and demonstrated that the sensitivity of progenitor blasts in culture suspension correlated with the remission outcomes in AML patients.

Methylation profiling in AML patient samples has demonstrated a myriad of defects in methylation among AML patients. In one study, the frequency of hypermethylation of several genes, in particular, SDC4, GPR37, PITX2, and MYOD1 was high among AML patients (range 47–64%) in this study of 36 AML patient cases, suggesting that there might be a group of AML patients with potential to benefit from hypomethylating agents (Toyota et al. 2001).

In addition to these data, accumulating evidence is showing that the mechanism of action of decitabine in AML may also involve antileukemia activity that is independent of methylation effects via microarray analysis assays (Schmelz et al. 2005). One such mechanism is via decitabine-induced caspase activation elaborating antileukemia activity (Tamm et al. 2005).

Further investigations of dectiabine as a single agent therapy in AML have included pharmacokinetic evaluation of the drug in MDS and AML. Cashen et al. demonstrated that decitabine, when dosed at 15 mg/m^2 intravenous schedule every 8 h for 3 days, yielded a definable, tolerable side-effect profile in MDS/AML patients and that the pharmacokinetic (PK) value of decitabine measured over multiple cycles was unchanged and stable (Cashen et al. 2008). With demonstration of the low-dose schedule of decitabine and its overall safety and tolerable side-effect report in AML patients, it appears that decitabine might be especially promising for older or unfit patients with AML who are not candidates for intensive therapy (Bryan et al. 2011).

22.2.2 Clinical Trials of Decitabine in AML

22.2.2.1 Single-Agent Trials

Because of the relative dearth of therapeutically active agents available for the treatment of AML, especially among older patients, novel treatment strategies have been investigated, including promising early results of the hypomethylator agents such as decitabine in clinical trials (Kantarjian et al. 1996; Lubbert and Minden 2005).

In the initial treatment designs incorporating decitabine into AML therapies, early investigators explored high doses of the hypomethylator drug in a similar approach to cytarabine-based strategies. Dosing schedules featuring total doses as high as 3000 mg/m^2 were used in one of these initial studies (Richel et al. 1991).

One early study in 22 pediatric patients treated with decitabine with relapsed/refractory AML and acute lymphoblastic leukemia (ALL) demonstrated 14% overall response (9% complete remission, CR, and 5% partial remission, PR) (Rivard et al. 1981). Focusing on patients with AML, Momparler et al. treated a group of 27 patients with relapsed disease with decitabine 37–81 mg/m^2 per cycle (over 36–60 h) and found an overall clinical response rate of 85% (22% CR, 15% PR) (Momparler et al. 1985).

A key study involving patients with MDS, chronic myeloid leukemia (CML), and AML was an important early study to demonstrate the concept and feasibility of administering decitabine, over multiple cycles, in older patients (Pinto et al. 1989). In this study of 19 older-age patients (receiving total doses of 135–810 mg/m^2), the overall response rate was found to be 70% (15% CR, 30% PR) with median response duration of 12 months (range 6–>58) and overall survival of 19 (7–>64) months.

A pilot study by Petti et al. showed promise for decitabine in treatment of AML, especially for treatment possibilities in older or unfit patients (Petti et al. 1993). In this trial, the goal was to observe the efficacy and toxicity of decitabine in AML patients with particularly poor prognosis. The investigators evaluated 12 patients with AML; all received decitabine at dosing 90–120 mg/m^2 intravenously every 8 h for 3 days for every 4–6 weeks. With 10 out of 12 of these patients evaluable for response, it was demonstrated that CR occurred in 3 patients and PR in 1 patient. As demonstrated by laboratory correlative work (immunophenotype testing and kinetic of remission studies), achievement of CR and PR with decitabine may have resulted via induction of leukemia cell differentiation.

In a phase I trial, Issa et al. studied decitabine administered in low dose, repeated dosing schedule in a cohort of relapsed and refractory leukemia patients. Out of 50 patients, 44 had MDS/AML. The dosing schedule that resulted in most responses was 15 mg/m^2 for 10 days (11 of 17 patients, 65 %) (Issa et al. 2004).

A retrospective study by Ravandi et al. suggested that therapy with hypomethylating agents (decitabine or azacytidine) improved the outcomes of an especially poor cytogenetic group, high-risk MDS and AML patients with abnormalities of chromosome 5 and 7 (Ravandi et al. 2009). In this study, 81 patients with either high-risk MDS (54 %) or AML (46 %) received hypomethylator therapy and overall survival was favorable in those treated with hypomethylator as compared to a similar group who received intensive chemotherapy ($p=0.019$).

22.2.2.2 Phase II Trials: Decitabine in AML

Three separate phase II clinical trials of decitabine in AML patients have been recently reported, one study featuring a 5-day dosing schedule and the other a 10-day dosing schedule.

In the first phase II trial, a multicenter effort by Cashen et al. reported on a single-arm, open label, phase II trial of older AML patients (age greater than or equal to 60) receiving front-line therapy with decitabine (Cashen et al. 2010). A dosing schedule of 20 mg/m^2 intravenously over 5 days every 4 weeks was utilized. Median of three cycles with range of 1–25 cycles was administered. Out of 55 AML patients with a mean age of 74 years, and median bone marrow blasts at presentation of 50 % (range 0–99), the reported overall response rate was 25 % (CR 24 %), even in patients with particularly poor/high-risk disease status such as antecedent MDS or high-risk cytogenetic groups. In this study, the overall survival was found to be 7.7 months (range 5.7–11.6 months); median survival for those patients achieving CR or a CRi reached 14 months. The most commonly occurring side effects in this trial were myelosuppression (100 %), neutropenic fever (29 %) and fatigue (26 %). Based on these promising phase II results, the authors suggested that a phase III study would be warranted to continue to study decitabine in the treatment of older AML patients.

In the other phase II trial, a single center study by Blum et al. (2010), exploration of a 10-day dosing schema (given as 20 mg/m^2, intravenously, over 1 h, on 10 consecutive days, days 1–10 of each cycle) of decitabine in AML patients was conducted. With a median age of 74 years (range 60–85), investigators evaluated 53 AML patients for treatment with single-agent decitabine in the frontline setting. Median number of four cycles was administered. Overall response rate was 64 % (25 patients (47 %) achieved CR and 9 patients CRi). As in the Cashen study, responses were noted across subgroup types, including poor prognostic subgroups of antecedent MDS and high-risk cytogenetics. Besides myelosuppression, neutropenic fever and infection were most common side effects with 31 patients (58 %) experiencing a grade 3 or greater infection event in the first two cycles of the therapy. Laboratory correlatives were performed in this study, including the measurement of mir-29b expression. Among 23 patients who had samples prior to treatment available for analysis, investigators in this study found that 11 of these patients had achieved CR. The analysis demonstrated that those patients with response to therapy with decitabine had a higher pre-therapy level of mir-29b than those patients who did not respond ($p=0.02$).

In the third phase II trial, Lubbert et al. (2012) evaluated 227 frontline AML patients with high-risk features and median 72 years treated with decitabine alone (although, notably, this trial did include administration of all-*trans* retinoic acid (ATRA) given to 100 patients in course 2 of therapy). CR was found to be 26 %, and even patients with high-risk features had responses (e.g., monosomal karyotype) with this approach, and median overall survival was 5.5 months (range 0–>57.5). On the basis of the promising results of these phase II trials, investigators have sought to further evaluate in the phase III setting.

22.2.2.3 Phase III Trials: Decitabine in AML

Two large phase III trials with AML patients receiving decitabine in the upfront setting have been recently reported.

In the trial led by Kantarjian et al. (2012), a randomized phase III trial was conducted at multiple sites in which comparison was studied between decitabine versus best supportive care or low-dose cytarabine in untreated AML patients age 65 or older. The investigators evaluated 485 patients in a 1:1 randomization schema of one arm with decitabine 20 mg/m^2 daily for 5 days every 28 days versus an arm with either best supportive care (including blood product transfusions) or low-dose cytarabine given at 20 mg/m^2 subcutaneous injection for 10 days every 28 days. Primary endpoint of the clinical study was overall survival. With a median age of 73 years (range 64–91 years), and 36 % with poor risk cytogenetics, the authors demonstrated a significantly higher remission rate (CR/CRp) in the decitabine treated arm versus the non-dectiabine treated arm (17.8 vs. 7.8 %, respectively, with $p=0.001$) and a median overall survival of 7.7 months in the decitabine treated arm and 5 months in the non-decitabine treatment arm (the p value was not significant). Patients in this trial received a median of four cycles of therapy with over 26.5 % of the patients receiving nine or more cycles.

In another large phase III study, Lubbert et al. (2011) tested decitabine compared to best supportive care in older patients with MDS (high-risk disease). Decitabine was administered at low dosing schedule (15 mg/m^2 intravenously, three times a day for 3 days, during 6-week cycles). Importantly, in this clinical trial, 32% of the patients met criteria for AML by WHO classification (equal to or greater than 20% blasts). The trial was open for enrollment to patients aged 60 or greater. With 233 patients evaluated, median age was 70 years (range 60–90 years). Notably, 53% had poor risk cytogenetics. The investigators demonstrated that PFS was significantly increased for decitabine arm compared to best supportive care arm (median PFS=6.6 months compared to 3.0 months, $p=0.004$); in addition, the risk of AML transformation at 1 year timepoint was also significantly lowered in decitabine arm ($p=0.036$). In terms of overall survival, there was no significant difference between the two groups (10.1 months for decitabine arm, 8.5 months for best supportive care arm).

22.2.2.4 Combination Studies

Early studies showed promising results with possibilities for decitabine to be used in combination with other agents in treatment of AML patients. One approach of these earlier trials focused on combination of decitabine with either amsacrine or an anthracycline in relapsed/refractory acute leukemia patients (Richel et al. 1991; Willemze et al. 1993; Schwartsmann et al. 1997; Willemze et al. 1997) and these promising results suggested that decitabine had a more tolerable toxicity profile than Ara-C and that combination studies should be further studied.

More recently, there has been interest in combining hypomethylator agents with histone deacetylase inhibitors for AML patients. In a phase I/II trial conducted by Garcia-Manero et al. (Garcia-Manero et al. 2006), both untreated (older, unfit) and relapsed/refractory AML patients were treated with the combination of decitabine and valproic acid. Out of 54 patients evaluated, the investigators found 10 CRs (19%) and 2 CRp (3%) with remission duration of 7.2 months and overall survival of 15.3 months in responding patients, suggesting that this combined epigenetic chemotherapy approach was feasible. In another phase I study, Blum et al. (2007) also evaluated the combination of decitabine and valproic acid in an AML population. Out of 25 patients with median age of 70 years, 13 patients had relapsed AML. In this phase I study designed to find the optimal biologic dose of decitabine in this population and the maximum tolerated dose (MTD) of the valproic acid in combination with decitabine, the investigators reported a response rate of 44%, 4 with CR, 4 with CRi, 3 PR. Importantly, the authors concluded that the addition of valproic acid led to observed encephalopathy at even low doses and that the drug and dose to be studied further could be decitabine at 20 mg/m^2 per day for 10 days alone,with the possibility of another histone deacetylatalse agent..

Many other decitabine-based combinations are continuing to be tested in clinical trials. One study reviewed the combination of decitabine with gemtuzumab ozogamicin, and found among 12 AML patients treated with this combination, 5 (42%)

attained CR and went on to SCT with overall good side-effect profile (Chowdhury et al. 2009). Another combination approach featured clofarabine with addition of low dose Ara-C alternating with decitabine in older AML patients (Faderl et al. 2012). In this trial, Faderl et al. (2012) evaluated 60 newly diagnosed AML patients with median age of 70 years. In this strategy, patients were allowed to receive up to 17 cycles of consolidation therapy that included decitabine. CR was achieved in 58 %, median overall survival 12.7 months and mortality during induction was 7 % at 8 weeks with tolerable side-effect profile, suggesting a combined cytotoxic chemotherapy/hypomethylator approach could be utilized in an older AML patient population. Yet another combination approach was recently reported by Blum et al. with decitabine combined with bortezomib (Blum et al. 2012). The authors studied this combination in 19 AML patients determined to be higher-risk AML, with median age of 70 years; decitabine was given as 20 mg/m^2 with 10-day dosing days 1–10 and bortezomib was added at escalating doses to 1.3 mg/m^2 goal dose on days 5, 8, 12, and 15. The authors found five of ten patients (50 %) with untreated AML had CR/CRi. Of interest, researchers found that bortezomib contributed to downregulation of FLT3 at the end of cycle one as a pharmacodynamic endpoint.

22.2.3 Decitabine in AML and SCT

Several early studies have explored the possibility of utilizing decitabine for treatment of AML in the stem cell transplant (SCT) treatment setting. Giralt et al. (1997) performed two studies that sought to evaluate the safety and efficacy of decitabine in both the post-transplant leukemia relapse setting and with decitabine as a component of conditioning therapy regimen prior to SCT. In one small study, the authors examined three leukemia patients (two with AML) treated in the post-SCT relapse setting; treatment was with decitabine as a single agent at 1000 mg/m^2 total dose. This study demonstrated median time to relapse of 7 months. In another small study, the authors demonstrated that four leukemia patients (one with AML) were able to successfully tolerate decitabine as part of an SCT conditioning regimen.

In addition, a phase I study showed encouraging data that decitabine (doses of 100–150 mg/m^2 every 12 h for 5 days) plus a donor stem cell infusion from original donor in the post-SCT setting was a tolerable and feasible approach (Ravandi et al. 2001).

The combination of decitabine with cyclophosphamide and busulfan, to form a novel conditioning regimen for leukemia patients undergoing allogeneic SCT was investigated by de Lima et al. (2003). In this study, 12 AML patients were included out of 23 total leukemia patients evaluated. The authors described no instances of rate limiting toxicity from the decitabine and demonstrated 26 % of the patients still alive at 3.3 years after SCT.

To further explore the idea of utilizing decitabine in the SCT setting especially in older MDS and AML patients, Lubbert et al. (2009) evaluated 15 MDS/AML patients. With a median age of 69 years (range 60–75 years), 5 of 15 patients had AML.

The total group showed a very tolerable side-effect profile with use of decitabine in the frontline treatment setting, which was followed by conditioning chemotherapy in preparation for SCT. The authors suggested that decitabine-based approaches could be considered as a treatment option in those MDS/AML patients who are not candidates for intensive chemotherapy.

22.2.4 *Future Directions*

22.2.4.1 Epigenetic Priming

One promising approach to AML patients is the concept of epigenetic priming therapy. In this treatment strategy, investigators administer a hypomethylating agent prior to administration of cytotoxic chemotherapy. Recently, Scandura et al. (2011) reported early results of this approach in which they tested in phase I setting, decitabine (in two different dose schedules) followed by standard cytotoxic chemotherapy program of anthracyline (daunorubicin 60 mg/m^2 for three doses) and ARA-C (100 mg/m^2 for 7 days). Investigators found that 90% of patients had a response to therapy (27 patients) with 17 CR and 10 PR. These promising early results suggest that epigenetic priming therapy warrants further investigation in the AML setting.

22.2.4.2 Decitabine as Maintenance Therapy

The role for hypomethylator agents in the maintenance setting for AML patients is actively being tested. One such study featured a treatment approach with AML patients in CR or subsequent CR being randomized to maintenance with conventional care versus decitabine up to 12 cycles (Boumber et al. 2012). Among 50 patients enrolled with median age 57 years, this trial showed feasibility of the notion for decitabine as a potential maintenance strategy in AML patients in CR, with a trend to less patients relapsing in the decitabine treated group versus conventional care, although the difference was not reported to be statistically significant. Further studies will be warranted to investigate the use of hypomethylating drugs in the AML maintenance setting.

22.2.4.3 SGI-110

The development of novel DNA hypomethylating agents represents a new area of research in the field of MDS and AML. One such agent is SGI-110 (Astex Pharmaceuticals), which is a decitabine-guanosine dinucleotide that is linked by a phosphodiester linker. The pharmacokinetics are different that decitabine, in that its subcutaneous administration and slower release of its active metabolite (decitabine) has the potential to result in longer exposure to active drug. This agent is currently being tested in an ongoing multi-center phase I/II study testing different dose schedules of SGI-110 in relapsed/refractory AML patients.

22.3 Azacytidine in AML

22.3.1 Background: Azacytidine

5-azacytidine (azacytidine, Vidaza, Celgene) is a DNA-hypomethylating agent approved by the FDA for the treatment of MDS. This approval was based on the results of two large phase III studies, the CALGB 9221 trial (Silverman et al. 2002), and the AZA-001 trial, a large randomized phase III trial (Fenaux et al. 2009; Keating 2012). In the AZA-001 trial, Fenaux et al. (2009) evaluated 358 patients and demonstrated an increased overall survival in high-risk MDS patients receiving azacytidine versus other conventional care regimens (azacytidine cohort median overall survival 24.5 months vs. 15.0 months for conventional care cohort, $p=0.0001$). In the CALGB 9221 study, Silverman et al. (2002) randomized 191 MDS patients to evaluate azacytidine versus supportive care. In this trial, it was shown that overall response rates were higher in the azacytidine arm, along with landmark analysis at 6 months (performed to account for crossover confounding effects) demonstrating median overall survival of 18 months in azacytidine arm versus 11 months in supportive care arm ($p=0.03$).

22.3.2 Clinical Trials in AML

22.3.2.1 Phase I/II Clinical Trials, Single-Agent Azacytidine in AML

With regard to the development of azacytidine in the field of AML, the earliest clinical reports of this specific use are from Hrodek and Vesely (1972) in the 1970s, in which the first signal of azacytidine's potential antileukemic activity was demonstrated in pediatric leukemia patients (Santini et al. 2001). Building on these early results, Karon et al. and McCreadie et al. conducted clinical trials both published in 1973. In the trial by Karon et al., in which 37 pediatric patients (ages 2–17 years), with acute leukemias who have failed treatment with standard Ara-C based chemotherapy, were given 5-azacytidine (days 1–5 every 2 weeks). Among AML patients, the authors demonstrated that 6 out of 14 (43 %) patients achieved remission status in the bone marrow (Karon et al. 1973). In the trial by McCredie et al. (1973), a total of 28 patients with acute leukemias were treated with 5 days intravenous administration of azacytidine, demonstrating an overall response rate of 28 %. Both of these early studies suggested activity for azacytidine as an antileukemic drug even in Ara-C refractory patients. Further studies in the mid- and late-1970s contributed to mounting interest and evidence in azacytidine-based AML therapy with larger studies of larger number of patients continuing to show safety, feasibility, and tolerability in addition to high response rates and even CRs with this chemotherapeutic approach in AML patients (Von Hoff et al. 1976; Vogler et al. 1976; Saiki et al. 1978).

In the 1980s, one of the notable trials in this field was led by Saiki et al. (1981). In this trial, the authors analyzed 154 relapsed/refractory acute leukemia patients who were treated with azacytidine. The authors focused on evaluating different dosing schedules. Overall, CR was achieved in nine patients in this study. The authors concluded that there was clinical benefit and less toxicity profile with lower dose, continuous infusion dose schedules with duration of 5, 7, and 10 days.

A retrospective review of 20 AML patients (by WHO) with low blast count (21–38 %) and who were found to be poor candidates for intensive chemotherapy, were given therapy with azacytidine. This study reported four CR and an overall response rate of 60 % and the drug was administered safely overall in an outpatient setting with infection being the most common toxicity noted in eight patients (Sudan et al. 2006). A larger study of 82 AML patients treated with azacytidine (as first therapy (n=35) or salvage (n=47) was analyzed by Maurillo et al. (2012). In this study, investigators found a median age of 72 years with a third of the patients having secondary AML. The analysis demonstrated an overall response rate of 32 % with 12 CRs, 4 CRi, and 10 PR, with greater occurrence of these responses occurring in the frontline treatment group. Based on these promising results, the authors suggested that single-agent azacytidine could be a treatment approach for older AML patients..

A multicenter phase I/II study was undertaken to further study this concept. Al-Ali et al. (2012) analyzed 40 older AML patients treated with azacytidine. The median age of the cohort was 72 years. The overall response rate was reported as 50 % in the frontline treated patients and 10 % in the relapsed/refractory group (p=0.008), signaling the safety and feasibility of single-agent azacytidine in an older AML patient population, including newly diagnosed patients.

22.3.2.2 Phase III Clinical Trials, Single-Agent Azacytidine in AML

With the reclassification of blast percent in MDS and AML by the WHO from the FAB system, Fenaux et al. were interested to analyze the results from the AZA-001 trial with treatment for intermediate risk and high-risk MDS patients with azacytidine (Font 2011). In their analysis, Fenaux et al. found that one third of the patients treated were, in fact, now classified as AML (i.e., those patients had 21–30 % blasts) (Fenaux et al. 2010). The analysis compared azacytidine -treated patients to those in the other arms of the trial (best supportive care, low-dose ARA-C, or intensive chemotherapy). The authors demonstrated that treatment with azacytidine resulted in overall survival benefit for low blast count AML compared with conventional care regimens (24.5 months vs. 16.0 months p=0.005). Further evidence supporting this finding is from the previously reported Silverman et al. study which analyzed the results of a smaller number of AML patients (n=27) that were treated in three CALGB [A-Za-z_-'&;]{3,20} (8421, 8921, 9221) and this study had also seen benefit with azacytidine therapy with median overall survival of 19.3 months. Taken together, the subgroup analysis of these large phase III trials that originally treated MDS patients and included low blast count AML as classified by WHO, leads to the promise of azacytidine as a therapy consideration for AML patients, especially older AML patients and those unfit for intensive chemotherapy.

22.3.3 *Azacytidine in Stem Cell Transplant*

The promise of the application of azacytidine in treatment of patients with AML has included its potential for use in the SCT setting. In one early study, Zittoun et al. (1989) utilized azacytidine as one part of a multidrug post-remission AML treatment plan (with or without addition of SCT). Among a large cohort of patients, 60 of these patients ultimately underwent SCT (43 were allogeneic SCT, 17 autologous). The authors concluded that further studies were needed to help determine the effectiveness of multidrug consolidation, and various types of SCTs should be further studied.

One key study in this field was performed by de Lima et al. in the maintenance setting post-SCT (Lima and Thall 2010). In this study, the authors tested the hypothesis and potential for hypomethylator therapy with azacytidine to help reduce the rate of relapse in AML patients undergoing allogeneic SCT. Among 45 AML patients analyzed and with median age of 60 years, investigators began treatment with azacytidine at D+40 from SCT date for post-SCT maintenance schedule. They found that four cycles of azacytidine given days 1–5 on a 30-day cycle at dose 32 mg/m^2 per cycle was the best dose schedule/combination out of several tested in a relapsed/refractory MDS/AML population, with possibility for this approach yielding improvements in overall and event-free survival, with further exploration of dosing/scheduling/duration (Lima and Thall 2010).

Building on this concept, Platzbecker et al. also studied administration of azacytidine in the post-SCT setting but in this study, they focused their efforts on those patients exhibiting minimal residual disease positivity (MRD$^+$) (Platzbecker et al. 2012). A goal of four cycles of azacytidine was to be given to 20 MDS/AML patients with MRD$^+$ status while still in CR; 16 of these patients (80 %) had responses as defined by either a rising CD34$^+$ donor chimerism to greater than or equal to 80 % or stabilization, with no evidence of overt relapse of the disease. Thirteen patients ultimately experienced hematologic relapse (65 %); the authors found this occurrence was delayed to a median of 231 days after the initial decrease of CD34$^+$ donor chimerism to less than or equal to 80 % was originally noted, suggesting that the administration of azacytidine in the setting of complete remission but with MRD$^+$ status was able to effectively help to delay the occurrence of overt relapse of the patients' underlying MDS/AML and this approach could be safe and feasible in the MDS/AML population in the post-SCT setting. Future studies in SCT setting will aim to explore further the mechanism of action by which azacytidine elaborates its benefit, especially in the post-SCT setting; one such study in this regard hypothesized that azacytidine enhances the graft versus leukemia effect without actually increased graft versus host disease in AML patients post-SCT treated with azacytidine in the post-SCT setting (Goodyear et al. 2012).

22.3.4 Combination Clinical Trials and Future Directions

22.3.4.1 Combination Therapies

A number of combination chemotherapy clinical trials have been undertaken including azacytidine in the treatment of AML patients. One of the earliest of these efforts was undertaken by Winton and the SWOG in 1985 in which investigators demonstrated the clinical activity of sequentially administered azacytidine with amsacrine in a relapsed/refractory group of 128 patients, suggesting that drug combinations with azacytidine might be promising (Winton et al. 1985). Other early and more recent studies continued to test the drug combination hypothesis in both post-induction strategies of azacytidine combined with other drugs (Volger et al. 1995; Rees et al. 1996), and in relapsed settings with various other chemotherapeutic agents including mitoxantrone (Goldberg et al. 1993) amsacrine and etoposide (Steuber et al. 1996), hydroxyurea and gemtuzumab (Nand et al. 2008), thalidomide (Raza et al. 2008), Ara-C (Borthakur et al. 2010) all suggesting a possible role for azacytidine in various AML treatment settings.

One key azacytidine-based combination to highlight is with combination of valproic acid, a histone deacetylase inhibitor. The pairing of these two agents has led to observation of synergistic activity both in vitro and in vivo (Braiteh et al. 2008). Braiteh et al. demonstrated the feasibility of this combination in phase I setting in various advanced malignancies. A phase I/II study by Soriano et al. (2007) aimed to assess the safety of the combination of azacytidine and valproic acid, with the addition of ATRA specifically in MDS/AML patients. In this trial, 53 patients were analyzed with median age of 69 years. There was neurotoxicity established as the dose limiting toxicity (DLT) and overall response rate was reported as 42 %, suggesting safety of the three drug combination and suggestion for further study. A phase 2 trial led by Raffoux et al. (2010) was then undertaken, further studying the three drug combination (azacytidine, valproic acid, ATRA). Out of 65 MDS/AML patients with median age of 72 years, 13 patients were relapsed/refractory. The investigators found 14 CRs and 3 PRs and median overall survival of 12.4 months. Another study assessing the combination of DNA hypomethylation and histone deacetylase inhibition was conducted by Kuendgen et al. (2011). In this trial, the authors tested only the combination of azacytidine and valproic acid, leaving off the ATRA that had been included in the previous two trials. Among 24 patients that received the combination therapy, 17 patients had AML. A total of 7 patients were found to have CR with another two patients in PR. The authors concluded that there was indeed clinical benefit to the two-drug combination and that prospective clinical trials would be warranted to further assess this therapeutic approach.

22.3.4.2 Future Directions

There are many exciting applications, concepts, and drug combinations in the area of azacytidine research for patients in AML. This section will highlight a few of these promising ongoing approaches.

Oral Azacytidine In a pilot pharmacologic study, Garcia-Manero et al. evaluated a newly reformulated version of oral azacytidine (Garcia-Manero et al. 2008). In this feasibility study, investigators demonstrated that the possibility of an oral version of this SQ/iv drug was indeed feasible and has adequate bioavailability in humans. This led Garcia-Manero et al. (2011) to conduct and report on a phase I clinical trial of oral azacytidine in treatment of MDS/AML. In this study, the authors evaluated the safety of this approach in 41 patients (AML, $n=8$), who received one cycle of SQ administration followed by oral drug. Diarrhea was found to be most common grade 3 and 4 event. An overall response rate of 35 % was reported in relapsed/refractory patients and 73 % rate in frontline patients. The oral formulation was found to have adequate bioavailability and clinical activity in MDS/AML patients and further trials are being planned to study this novel approach.

Azacytidine in Combination with Lenalidomide A phase I study was conducted to evaluate the safety of a novel sequential therapy in treatment of AML patients with azacytidine and lenalidomide in older AML patients (Pollyea et al. 2012). The approach was executed as sequential treatment in a cohort of 18 older, untreated AML patients. Results showed no maximum tolerated dose being reached and rate of CR/CRi to be 7/16 patients (44 %) with median response duration of 6.2 months. These promising data have led multiple groups to further explore this therapy combination in both MDS and AML patients including Garcia-Manero et al. (2011) as presented at ASH 2011.

Azacytidine combination with FLT3 Inhibitor Yet another approach in treatment of AML is the combination of a hypomethylator agent with an inhibitor of the FLT3 inhibitor. FLT3, or FMS-like 3, is mutated in approximately 30% AML patients (Pemmaraju et al. 2011) and in the setting of potential therapeutic targeting, Sato et al. have shown that traditional cytotoxic chemotherapy might lead to FLT3 ligand levels actually being increased and thus leading to FLT3 resistance (Sato et al. 2011). It is precisely in this setting that novel approaches involving azacytidine combined with a FLT3 inhibitor have been undertaken. One such study, an ongoing phase II clinical trial reported by Alattar et al. (ASCO 2012 Abstract 6558) evaluated the combination of azacytidine with sorafenib (an FDA-approved drug for treatment of advanced renal cell cancer and hepatocellular cancer, which is a multi-kinase inhibitor with targets including FLT3) has enrolled 32 patients (median age 61 years) with relapsed/refractory AML. FLT3-ITD mutation was positive in 30/32 patients. The reported CR/CRi is 50 % with median duration of the CR/CRi lasting 3 months and three patients ultimately going onto SCT. In another phase I/II study, Nazha et al. (ASCO 2012 Abstract 6589) studied another FLT3 inhibitor, PKC-412, in combination with azacytidine. In this study, investigators enrolled 14 patients thus far, with overall response rate reported as 3/13 (21 %), including two CRi. One out of four patients with FLT3-ITD was able to demonstrate a CRi. The authors of both studies concluded that the safety and efficacy of the combination of azacytidine with a FLT3 inhibitor appeared to be promising and the further, larger prospective trials would be useful to continue to study this approach.

Conclusion

Outcomes in AML, especially older patients, remain overall poor. The backbone of traditional cytotoxic therapy, generally consisting of an anthracycline combined with Ara-C, is not able to be utilized in a large amount of AML patients due to age, co-morbidities, and toxicity profile, and novel therapies are warranted in treatment of these patients. Epigenetic therapy, including hypomethylating agents (decitabine, azacytidine) and investigational agents in this field represent an exciting new era in treatment of AML patients and offers patients opportunity for treatment with potentially less toxic alternatives to traditional cytotoxic therapy. Future work in this field will include better delineation of biomarkers for prediction of response of these agents and further studies utilizing these agents in combination with other chemotherapeutic therapies.

References

Al-Ali HK, Jaekel N, Junghanss C et al (2012) Azacitidine in patients with acute myeloid leukemia medically unfit for or resistant to chemotherapy: a multicenter phase I/II study. Leuk Lymphoma 53:110–117

Baylin SB, Fearon ER, Vogelstein B et al (1987) Hypermethylation of the 5′ region of the calcitonin gene is a property of human lymphoid and acute myeloid malignancies. Blood 70:412–417

Blum W, Klisovic RB, Hackanson B et al (2007) Phase I study of decitabine alone or in combination with valproic acid in acute myeloid leukemia. J Clin Oncol 25:3884–3891

Blum W, Garzon R, Klisovic RB et al (2010) Clinical response and miR-29b predictive significance in older AML patients treated with a 10-day schedule of decitabine. Proc Natl Acad Sci U S A 107:7473–7478

Blum W, Schwind S, Tarighat SS et al (2012) Clinical and pharmacodynamic activity of bortezomib and decitabine in acute myeloid leukemia. Blood 119:6025–6031

Borthakur G, Huang X, Kantarjian H et al (2010) Report of a phase 1/2 study of a combination of azacitidine and cytarabine in acute myelogenous leukemia and high-risk myelodysplastic syndromes. Leuk Lymphoma 51:73–78

Boumber Y, Kantarjian H, Jorgensen J et al (Jun 5 2012) A randomized study of decitabine versus conventional care for maintenance therapy in patients with acute myeloid leukemia in complete remission. Leukemia 26(11):2428–2431

Braiteh F, Soriano AO, Garcia-Manero G et al (2008) Phase I study of epigenetic modulation with 5-azacytidine and valproic acid in patients with advanced cancers. Clin Cancer Res 14:6296–6301

Bryan J, Kantarjian H, Garcia-Manero G, Jabbour E (2011) Pharmacokinetic evaluation of decitabine for the treatment of leukemia. Expert Opin Drug Metab Toxicol 7:661–672

Bullinger L, Ehrich M, Dohner K et al (2010) Quantitative DNA methylation predicts survival in adult acute myeloid leukemia. Blood 115:636–642

Cashen AF, Shah AK, Todt L, Fisher N, DiPersio J (2008) Pharmacokinetics of decitabine administered as a 3-h infusion to patients with acute myeloid leukemia (AML) or myelodysplastic syndrome (MDS). Cancer Chemother Pharmacol 61:759–766

Cashen AF, Schiller GJ, O'Donnell MR, DiPersio JF (2010) Multicenter, phase II study of decitabine for the first-line treatment of older patients with acute myeloid leukemia. J Clin Oncol 28:556–561

Chowdhury S, Seropian S, Marks PW (2009) Decitabine combined with fractionated gemtuzumab ozogamicin therapy in patients with relapsed or refractory acute myeloid leukemia. Am J Hematol 84:599–600

de Lima M, Ravandi, Shahjahan M et al (2003) Long-term follow-up of a phase I study of high-dose decitabine, busulfan, and cyclophosphamide plus allogeneic transplantation for the treatment of patients with leukemias. Cancer 97:1242–1247

de Lima M, Giralt, Thall PF et al (2010) Maintenance therapy with low-dose azacitidine after allogeneic hematopoietic stem cell transplantation for recurrent acute myelogenous leukemia or myelodysplastic syndrome: a dose and schedule finding study. Cancer 116:5420–5431

Faderl S, Ravandi F, Huang X et al (26 Jan 2012) Clofarabine plus low-dose cytarabine followed by clofarabine plus low-dose cytarabine alternating with decitabine in acute myeloid leukemia frontline therapy for older patients. Cancer 118(18):4471–4477

Fenaux P, Mufti GJ, Hellstrom-Lindberg E et al (2009) Efficacy of azacitidine compared with that of conventional care regimens in the treatment of higher-risk myelodysplastic syndromes: a randomised, open-label, phase III study. Lancet Oncol 10:223–232

Fenaux P, Mufti GJ, Hellstrom-Lindberg E et al (2010) Azacitidine prolongs overall survival compared with conventional care regimens in elderly patients with low bone marrow blast count acute myeloid leukemia. J Clin Oncol 28:562–569

Figueroa ME, Lugthart S, Li Y et al (2010) DNA methylation signatures identify biologically distinct subtypes in acute myeloid leukemia. Cancer Cell 17:13–27

Flotho C, Claus R, Batz C et al (2009) The DNA methyltransferase inhibitors azacitidine, decitabine and zebularine exert differential effects on cancer gene expression in acute myeloid leukemia cells. Leukemia 23:1019–1028

Font P (2011) Azacitidine for the treatment of patients with acute myeloid leukemia with 20%–30% blasts and multilineage dysplasia. Adv Ther 28(3):1–9

Galm O, Wilop S, Luders C et al (2005) Clinical implications of aberrant DNA methylation patterns in acute myelogenous leukemia. Ann Hematol 84(1):39–46

Garcia-Manero G (2011) Treatment of higher-risk myelodysplastic syndrome. Semin Oncol 38:673–681

Garcia-Manero G, Kantarjian HM, Sanchez-Gonzalez B et al (2006) Phase 1/2 study of the combination of 5-aza-2′-deoxycytidine with valproic acid in patients with leukemia. Blood 108:3271–3279

Garcia-Manero G, Stoltz ML, Ward MR, Kantarjian H, Sharma S (2008) A pilot pharmacokinetic study of oral azacitidine. Leukemia 22:1680–1684

Garcia-Manero G, Daver NG, Borthakur G et al (2011) Phase I Study of the Combination of 5-Azacitidine Sequentially with High-Dose Lenalidomide in Higher-Risk Myelodysplastic Syndrome (MDS) and Acute Myelogenous Leukemia (AML). ASH Annual Meeting Abstracts 118:2613 December 10–13, 2001

Garcia-Manero G, Gore SD, Cogle C et al. (2011) Phase I study of oral azacitidine in myelodysplastic syndromes, chronic myelomonocytic leukemia, and acute myeloid leukemia. J Clin Oncol 29:2521–2527

Gattei V, Aldinucci D, Petti MC, Da Ponte A, Zagonel V, Pinto A (1993) In vitro and in vivo effects of 5-aza-2′-deoxycytidine (Decitabine) on clonogenic cells from acute myeloid leukemia patients. Leukemia 7(1):42–48

Giralt S, Davis M, O'Brien S et al (1997) Studies of decitabine with allogeneic progenitor cell transplantation. Leukemia 11(1):S32–34

Goldberg J, Gryn J, Raza A et al (1993) Mitoxantrone and 5-azacytidine for refractory/relapsed ANLL or CML in blast crisis: a leukemia intergroup study. Am J Hematol 43:286–290

Goodyear OC, Dennis M, Jilani NY et al (2012) Azacitidine augments expansion of regulatory T cells after allogeneic stem cell transplantation in patients with acute myeloid leukemia (AML). Blood 119:3361–3369

Hrodek O, Vesely J (1971) 5-azacytidine in childhood leukemia. Neoplasma 18:493–503

Issa JP, Kantarjian HM (2009) Targeting DNA methylation. Clin Cancer Res 15:3938–3946

Issa JP, Garcia-Manero G, Giles FJ et al (2004) Phase 1 study of low-dose prolonged exposure schedules of the hypomethylating agent 5-aza-2′-deoxycytidine (decitabine) in hematopoietic malignancies. Blood 103:1635–1640

Jain N, Rossi A, Garcia-Manero G (2009) Epigenetic therapy of leukemia: An update. Int J Biochem Cell Biol 41:72–80

Jones PA, Baylin SB (2007) The epigenomics of cancer. Cell 128:683–692

Kantarjian HM, Estey EH, Keating MA (1996) New chemotherapeutic agents in acute myeloid leukemia. Leukemia 10(1):S4–6

Kantarjian H, Issa JP, Rosenfeld CS et al (2006) Decitabine improves patient outcomes in myelodysplastic syndromes: results of a phase III randomized study. Cancer 106:1794–1803

Kantarjian HM, Thomas XG, Dmoszynska A et al (2012) Multicenter, randomized, open-label, phase III trial of decitabine versus patient choice, with physician advice, of either supportive care or low-dose cytarabine for the treatment of older patients with newly diagnosed acute myeloid leukemia. J Clin Oncol 30:2670–2677

Karon M, Sieger L, Leimbrock S, Finklestein JZ, Nesbit ME, Swaney JJ (1973) 5-Azacytidine: a new active agent for the treatment of acute leukemia. Blood 42:359–365

Keating GM (2012) Azacitidine: a review of its use in the management of myelodysplastic syndromes/acute myeloid leukaemia. Drugs 72:1111–1136

Kuendgen A, Bug G, Ottmann OG et al (2011) Treatment of poor-risk myelodysplastic syndromes and acute myeloid leukemia with a combination of 5-azacytidine and valproic acid. Clin Epigenetics 2:389–399

Lubbert M, Minden M (2005) Decitabine in acute myeloid leukemia. Semin Hematol 42:S38–42

Lubbert M, Bertz H, Ruter B et al (2009) Non-intensive treatment with low-dose 5-aza-2′-deoxycytidine (DAC) prior to allogeneic blood SCT of older MDS/AML patients. Bone Marrow Transplant 44:585–588

Lubbert M, Suciu S, Baila L et al (2011) Low-dose decitabine versus best supportive care in elderly patients with intermediate- or high-risk myelodysplastic syndrome (MDS) ineligible for intensive chemotherapy: final results of the randomized phase III study of the European Organisation for Research and Treatment of Cancer Leukemia Group and the German MDS Study Group. J Clin Oncol 29:1987–1996

Lubbert M, Ruter BH, Claus R et al (2012) A multicenter phase II trial of decitabine as first-line treatment for older patients with acute myeloid leukemia judged unfit for induction chemotherapy. Haematologica 97:393–401

Maurillo L, Venditti A, Spagnoli A et al (2012) Azacitidine for the treatment of patients with acute myeloid leukemia: report of 82 patients enrolled in an Italian Compassionate Program. Cancer 118:1014–1022

McCredie KB, Bodey GP, Burgess MA et al (1973) Treatment of acute leukemia with 5-azacytidine (NSC-102816). Cancer Chemother Rep 57:319–323

McDevitt MA (2012) Clinical applications of epigenetic markers and epigenetic profiling in myeloid malignancies. Semin Oncol 39:109–122

Momparler RL, Bouchard J, Onetto N, Rivard GE (1984) 5-aza-2′-deoxycytidine therapy in patients with acute leukemia inhibits DNA methylation. Leuk Res 8:181–185

Momparler RL, Rivard GE, Gyger M (1985) Clinical trial on 5-aza-2′-deoxycytidine in patients with acute leukemia. Pharmacol Ther 30:277–286

Nand S, Godwin J, Smith S et al (2008) Hydroxyurea, azacitidine and gemtuzumab ozogamicin therapy in patients with previously untreated non-M3 acute myeloid leukemia and high-risk myelodysplastic syndromes in the elderly: results from a pilot trial. Leuk Lymphoma 49:2141–2147

Pemmaraju N, Kantarjian H, Ravandi F, Cortes J (2011) FLT3 inhibitors in the treatment of acute myeloid leukemia: the start of an era? Cancer 117:3293–3304

Petti MC, Mandelli F, Zagonel V et al (1993) Pilot study of 5-aza-2′-deoxycytidine (Decitabine) in the treatment of poor prognosis acute myelogenous leukemia patients: preliminary results. Leukemia 7(1):36–41

Pinto A, Zagonel V, Attadia V et al (1989) 5-Aza-2′-deoxycytidine as a differentiation inducer in acute myeloid leukaemias and myelodysplastic syndromes of the elderly. Bone Marrow Transplant 4(3):28–32

Platzbecker U, Wermke M, Radke J et al (2012) Azacitidine for treatment of imminent relapse in MDS or AML patients after allogeneic HSCT: results of the RELAZA trial. Leukemia 26:381–389

Pollyea DA, Kohrt HE, Gallegos L et al (2012) Safety, efficacy and biological predictors of response to sequential azacitidine and lenalidomide for elderly patients with acute myeloid leukemia. Leukemia 26:893–901

Raffoux E, Cras A, Recher C et al (2010) Phase 2 clinical trial of 5-azacitidine, valproic acid, and all-trans retinoic acid in patients with high-risk acute myeloid leukemia or myelodysplastic syndrome. Oncotarget 1:34–42

Ravandi F, Kantarjian H, Cohen A et al (2001) Decitabine with allogeneic peripheral blood stem cell transplantation in the therapy of leukemia relapse following a prior transplant: results of a phase I study. Bone Marrow Transplant 27:1221–1225

Ravandi F, Issa JP, Garcia-Manero G et al (2009) Superior outcome with hypomethylating therapy in patients with acute myeloid leukemia and high-risk myelodysplastic syndrome and chromosome 5 and 7 abnormalities. Cancer 115:5746–5751

Raza A, Mehdi M, Mumtaz M, Ali F, Lascher S, Galili N (2008) Combination of 5-azacytidine and thalidomide for the treatment of myelodysplastic syndromes and acute myeloid leukemia. Cancer 113:1596–1604

Rees JK, Gray RG, Wheatley K (1996) Dose intensification in acute myeloid leukaemia: greater effectiveness at lower cost. Principal report of the Medical Research Council's AML9 study. MRC Leukaemia in Adults Working Party. Br J Haematol 94:89–98

Richel DJ, Colly LP, Kluin-Nelemans JC, Willemze R (1991) The antileukaemic activity of 5-Aza-2 deoxycytidine (Aza-dC) in patients with relapsed and resistant leukaemia. Br J Cancer 64:144–148

Rivard GE, Momparler RL, Demers J et al (1981) Phase I study on 5-aza-2′-deoxycytidine in children with acute leukemia. Leuk Res 5:453–462

Saiki JH, McCredie KB, Vietti TJ et al (1978) 5-azacytidine in acute leukemia. Cancer 42:2111–2114

Saiki JH, Bodey GP, Hewlett JS et al (1981) Effect of schedule on activity and toxicity of 5-azacytidine in acute leukemia: a Southwest Oncology Group Study. Cancer 47:1739–1742

Santini V, Kantarjian HM, Issa JP (2001) Changes in DNA methylation in neoplasia: pathophysiology and therapeutic implications. Ann Intern Med 134:573–586

Sato T, Yang X, Knapper S et al (2011) FLT3 ligand impedes the efficacy of FLT3 inhibitors in vitro and in vivo. Blood 117:3286–3293

Scandura JM, Roboz GJ, Moh M et al (2011) Phase 1 study of epigenetic priming with decitabine prior to standard induction chemotherapy for patients with AML. Blood 118:1472–1480

Schmelz K, Sattler N, Wagner M, Lubbert M, Dorken B, Tamm I (2005) Induction of gene expression by 5-Aza-2′-deoxycytidine in acute myeloid leukemia (AML) and myelodysplastic syndrome (MDS) but not epithelial cells by DNA-methylation-dependent and -independent mechanisms. Leukemia 19:103–111

Schoofs T, Muller-Tidow C (2011) DNA methylation as a pathogenic event and as a therapeutic target in AML. Cancer Treat Rev 37(1):13–18

Schwartsmann G, Fernandes MS, Schaan MD et al (1997) Decitabine (5-Aza-2′-deoxycytidine; DAC) plus daunorubicin as a first line treatment in patients with acute myeloid leukemia: preliminary observations. Leukemia 11(1):S28–31

Silverman LR, Demakos EP, Peterson BL et al (2002) Randomized controlled trial of azacitidine in patients with the myelodysplastic syndrome: a study of the cancer and leukemia group B. J Clin Oncol 20:2429–2440

Soriano AO, Yang H, Faderl S et al (2007) Safety and clinical activity of the combination of 5-azacytidine, valproic acid, and all-trans retinoic acid in acute myeloid leukemia and myelodysplastic syndrome. Blood 110:2302–2308

Steuber CP, Krischer J, Holbrook T et al (1996) Therapy of refractory or recurrent childhood acute myeloid leukemia using amsacrine and etoposide with or without azacitidine: a Pediatric Oncology Group randomized phase II study. J Clin Oncol 14:1521–1525

Sudan N, Rossetti JM, Shadduck RK et al (2006) Treatment of acute myelogenous leukemia with outpatient azacitidine. Cancer 107:1839–1843

Tamm I, Wagner M, Schmelz K (2005) Decitabine activates specific caspases downstream of p73 in myeloid leukemia. Ann Hematol 84(1):47–53

Toyota M, Kopecky KJ, Toyota MO, Jair KW, Willman CL, Issa JP (2001) Methylation profiling in acute myeloid leukemia. Blood 97:2823–2829

Valk PJ, Verhaak RG, Beijen MA et al (2004) Prognostically useful gene-expression profiles in acute myeloid leukemia. N Engl J Med 350:1617–1628

Vogler WR, Miller DS, Keller JW (1976) 5-Azacytidine (NSC 102816): a new drug for the treatment of myeloblastic leukemia. Blood 48:331–7

Volger WR, Weiner RS, Moore JO, Omura GA, Bartolucci AA, Stagg M (1995) Long-term follow-up of a randomized post-induction therapy trial in acute myelogenous leukemia (a Southeastern Cancer Study Group trial). Leukemia 9:1456–1460

Von Hoff DD, Slavik M, Muggia FM (1976) 5-Azacytidine. A new anticancer drug with effectiveness in acute myelogenous leukemia. Ann Intern Med 85:237–245

Willemze R, Archimbaud E, Muus P (1993) Preliminary results with 5-aza-2′-deoxycytidine (DAC)-containing chemotherapy in patients with relapsed or refractory acute leukemia. The EORTC Leukemia Cooperative Group. Leukemia 7(1):49–50

Willemze R, Suciu S, Archimbaud E et al (1997) A randomized phase II study on the effects of 5-Aza-2′-deoxycytidine combined with either amsacrine or idarubicin in patients with relapsed acute leukemia: an EORTC Leukemia Cooperative Group phase II study (06893). Leukemia 11(1):S24–27

Winton EF, Hearn EB, Martelo O et al (1985) Sequentially administered 5-azacitidine and amsacrine in refractory adult acute leukemia: a phase I-II trial of the Southeastern Cancer Study Group. Cancer Treat Rep 69:807–811

Zittoun R, Jehn U, Fiere D et al (1989) Alternating v repeated postremission treatment in adult acute myelogenous leukemia: a randomized phase III study (AML6) of the EORTC Leukemia Cooperative Group. Blood 73:896–906

Part IV
Nuclear Receptors

Chapter 23
PML/RARA as the Master Driver of APL Pathogenesis and Therapy Response

Hugues de Thé, Jun Zhu, Rihab Nasr, Julien Ablain and Valérie Lallemand-Breittenbach

Abstract Acute promyelocytic leukemia (APL) is a model disease for targeted therapy. APL is caused by a variety of fusion proteins, all implicating the retinoic acid receptor alpha (RARA). The promyelocytic gene (PML)/RARA fusion is by far the most frequent, present in 99 % of patients. Two unconventional drugs, retinoic acid (RA) and arsenic trioxide were first shown to exhibit extraordinary clinical activity and later found to directly target PML/RARA. RA binds PML/RARA via its RARA moiety, activates transcription and degrades PML/RARA. Arsenic only degrades the fusion protein by targeting its PML part. Mouse modeling in APL has allowed an unprecedented level of understanding of the disease pathogenesis and basis for therapy response, highlighting the key role of PML/RARA degradation in the latter. The combination of RA and arsenic definitively eradicate the disease in mice and in most patients. APL thus represents a paradigm for oncoprotein-targeted cures.

Keywords Retinoic acid · As_2O_3 · RARA · PML · Leukemia · Promyelocytic · Mouse models

H. de Thé (✉)
Department Molecular Pathology and Virology, INSERM U 944 et CNRS UMR 7212, Institut Universitaire d'Hématologie, l'Université Paris VII, 75475 Paris cedex 10, France
e-mail: dethe@univ-paris-diderot.fr

H. de Thé · J. Zhu · J. Ablain · V. Lallemand-Breittenbach
Université Paris Diderot, Sorbonne Paris Cité, Hôpital St. Louis 1, Avenue Claude Vellefaux, 75475 Paris cedex 10, France

INSERM UMR 944, Equipe labellisée par la Ligue Nationale contre le Cancer, Institut Universitaire d'Hématologie, Hôpital St. Louis 1, Avenue Claude Vellefaux, 75475 Paris cedex 10, France

CNRS UMR 7212, Hôpital St. Louis 1, Avenue Claude Vellefaux, 75475 Paris cedex 10, France

H. de Thé · J. Zhu
Pole Sino-Francais des Sciences du Vivant et de Génomique de l'Hôpital Rui Jin, Rui-Jin Hospital affiliated with Jiao Tong University, 197 Rui Jin Road, Shanghai 200025, China

R. Nasr
Department of Anatomy, Cell Biology and Physiological Sciences, American University of Beirut, Beirut, Lebanon

M. Andreeff (ed.), *Targeted Therapy of Acute Myeloid Leukemia*,
Current Cancer Research, DOI 10.1007/978-1-4939-1393-0_23

23.1 Introduction

As detailed in the accompanying clinical chapter, few diseases have had such a dramatic change in treatment and prognosis as acute promyelocytic leukemia (APL). Indeed, it rose from being an hematologic emergency with less than 30% 5-year survival, to a disease with a 95% definitive cure rate, with some patient even no longer receiving chemotherapy.

Similarly, few diseases have been the focus of so many physio-pathological studies, as well as those addressing the basis for therapy response. A reasonably clear image of the disease is now emerging. Amusingly, in contrast to most kinase inhibitors, that were designed to block the activity of an oncogenic kinase activated by point mutation or translocation, the two miracle APL drugs were found empirically and only much later demonstrated to target the driving oncogene. They actually played a key role in unraveling the disease pathogenesis, starting with the identification of its underlying molecular defect. Finally, these drugs exert a curative activity allowing patients to go definitively off-treatment.

For all these reasons, despite its low frequency, APL is a key model for targeted therapies.

23.2 PML/RARA, the Driver of APL Pathogenesis

More than 98% APL are associated with the balanced reciprocal translocation t(15;17)(q22;q11-12), fusing the promyelocytic gene (*PML*) with the retinoic acid receptor alpha gene (*RARA*; Fig. 23.1a). Other APL patients harbor alternative translocations always involving *RARA*, the most common being t(11;17) that involves the promyelocytic leukemia zinc finger (*PLZF*) gene (Chen et al. 1993; Piazza et al. 2001).

Other rare lesions, often shared with other leukemias or malignancies, have been implicated in APL progression, such as *MYC* amplification, Fms-like tyrosine kinase 3 activation, or *RAS* mutations (Akagi et al. 2009). However, APL has an almost constant incidence with age, suggesting that it arises from a single rate-limiting genetic event (Vickers et al. 2000). The recurrent presence of X-RARA fusions in APL patients and the fact that their sole expression initiates typical APL in transgenic mice strongly argue for a hierarchy in these genomic abnormalities, X-RARA being the primary actor of leukemogenesis. Still, because PML/RARA transgenic mice develop leukemias with long latencies and incomplete penetrance, the potential requirement for cooperating mutations and/or additional epigenetic changes to yield the full APL phenotype has been the object of many discussions (Welch et al. 2011). But recent studies have clearly demonstrated the absence of mutations actually contributing to transformation (Welch et al. 2012), while studies in APL that develop following chemotherapy have all demonstrated a short (typically 1 year) time interval between DNA-damaging chemotherapy and disease onset (Mistry et al. 2005). Thus, the authors consider APL as a monogenic, X-RARA-driven, disease (de The and Chen 2010) .

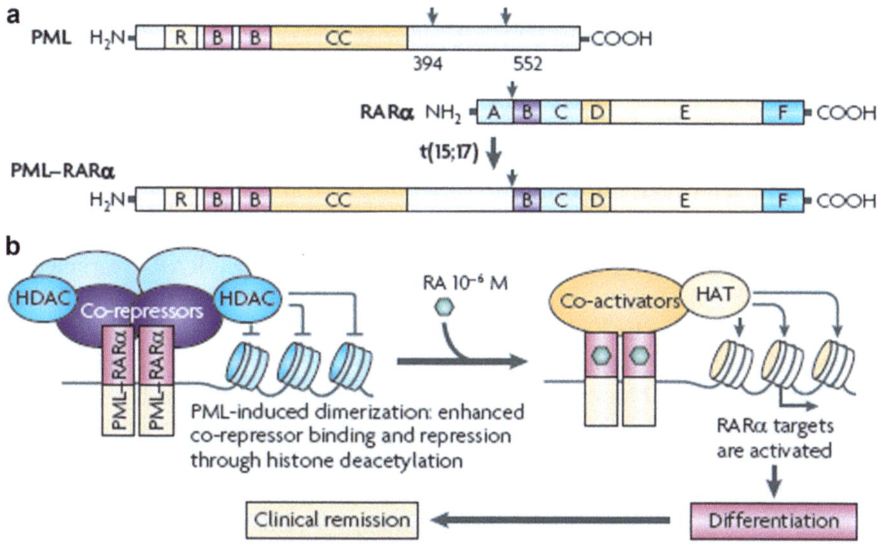

Fig. 23.1 **a** Schematic representation of promyelocytic gene/retinoic acid receptor alpha (*PML/RARA*). **b** The classic model of APL response to retinoic acid (*RA*). *HDAC* histone deacetylase. (Reprinted from de The and Chen 2010)

23.3 RARA, the Constant Partner of the Fusion

Retinoic acid (RA) has been involved in a variety of physiological regulatory mechanisms, in particular, morphogenesis, stem cell self-renewal, and myeloid differentiation (Kastner et al. 2001; Strickland and Mahdavi 1978).

RARA is a receptor for RA discovered in the early 1990. Like all other nuclear receptors, it has a highly conserved modular organization with a zinc-finger containing sequence-specific DNA-binding domain and a complex ligand-binding domain that also enables dimerization and transactivation (Kastner et al. 1995). RARA is normally bound to a member of the retinoid X receptor (RXR) family of nuclear receptors as an obligatory heterodimer. Within the retinoic acid receptor (RAR)/RXR complexes, both receptors contribute to DNA binding and transcriptional repression, but only RARs contribute to activation. The RAR and RXR DNA-binding domains each recognize a AGGTCA core motif, usually in a direct repeat orientation, separated by a spacing of 2 or 5 nucleotides. These retinoic acid response elements (RARE) confer RA sensitivity to the promoters where they are present (de The et al. 1990). Recent genomic experiments have outlined their distribution in the genome (Hua et al. 2009).

RARs are versatile transcriptional switches that can either repress or activate transcription. This is achieved by the binding of a family of proteins named coactivators or corepressors. At large, RAR/RXR complexes bind corepressors in their unliganded state and recruit coactivators in the presence of ligands. Interestingly, RARA appears to be a stronger binder for corepressors than other RARs. This may explain its constant implication in APL, as this will confer X-RARA fusions with stronger transcriptional repression (Farboud et al. 2003).

23.4 From the Classic to Refined Models for APL Pathogenesis

Different studies showed that PML/RARA or the rare PLZF/RARA variant bind co-repressors and histone deacetylase (HDAC) with higher affinity than RARA, due to their ability to homodimerize (Licht 2006). PML/RARA thus behaves as a super-repressor so that repression of basal retinoic acid signaling could contribute to the differentiation block (Fig. 23.1b). Pharmacological doses of RA could then release both the transcriptional and differentiation blocks (Melnick and Licht 1999).

This model, primarily based on cell-line studies, was progressively refined, notably with the findings that RXRA plays an important role in the transformation process, most likely by enhancing DNA binding of X-RARA fusions (Zhu et al. 2007; Zeisig et al. 2007). PML/RARA homodimers, together with the RXRA co-receptor, have 4 domains able to recognize the AGGTCA cores and accordingly display highly degenerated binding-site specificity (Kamashev et al. 2004). Consequently, the PML/RARA binding site repertoire is considerably enlarged when compared to the one of RARA, as demonstrated with natural PML/RARA target genes in human APL cells (Martens et al. 2010). Importantly, some of the recognized sequences are targets of other nuclear receptors (vitamin D receptor (VDR), thyroid receptor (TR), peroxisome proliferator-activated receptor (PPAR)...) controlling myeloid differentiation or stem cell self-renewal. Similar properties were described for other myeloid leukemia-associated oncogenic fusions, suggesting that dimerization-enforced relaxation in DNA-binding site specificity may be a general mechanism of leukemic transformation (So and Cleary 2004).

23.5 Is PML only a Dimerization Interface?

The first models for APL for APL pathogenesis viewed PML mainly as a provider of a strong dimerization interface. Indeed, all proteins fused to RARA in APL contain potent dimerization domains. While in cell lines RARA dimerization suffice to confer strong repressive ability on RARA signaling and some inhibition of differentiation, attempts to induce APL in vivo were largely unsuccessful (Sternsdorf et al. 2006). Importantly, these eventually succeeded only when using the PML dimerization domain (Occhionorelli et al. 2011), suggestive for an important contribution of PML beyond providing a dimerization interface.

PML protein initiates the formation of nuclear bodies (NBs), sub-nuclear spherical structures involved in the fine-tuning of several biological processes (Lallemand-Breitenbach and de The 2010) (Fig. 23.2, arrow in the left panel). A specific post-

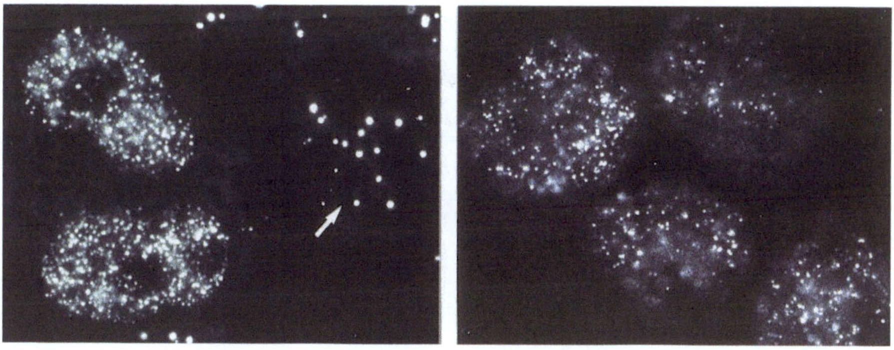

Fig. 23.2 *Left:* disruption of the normal PML pattern (cell on the right, *arrow*) by PML/RARA expression (cells on the left). *Right:* the micro-speckeled pattern typical of APL cells, a feature that may be used for diagnosis (Dyck et al. 1995). (Adapted from Koken et al. 1994)

translational modification, sumoylation, controls recruitment onto NBs of a wide variety of partner proteins. In turn, partner recruitment into PML NBs finely modulates their post-translational modification and may result in protein sequestration or activation. Importantly, PML loss and/or NB disruption seem to be associated with enhanced self-renewal (Ito et al. 2008; Regad et al. 2009). In APL, PML/RARA dimerizes with PML, leading to the replacement of the normal speckled nuclear distribution of PML by a micro-speckled one (Fig. 23.2) (Koken et al. 1994). Thus, in addition to transcriptional deregulation, this alteration in nuclear architecture could participate in APL pathogenesis, notably in fostering aberrant self-renewal.

Moreover, some studies have found that PML actually contributes to transcriptional repression by PML/RARA, through its modification by SUMO, a post-transcriptional modification that confers transcriptional repression ability to transcription factors (Zhu et al. 2005; Verger et al. 2003).

23.6 Two Drugs for one Disease

The introduction of RA for APL treatment in 1985 (Huang et al. 1988) constituted the first example of differentiation therapy (Degos et al. 1995). *Ex vivo* and in vivo, RA triggers rapid APL cell differentiation into granulocytes, which correlates with patient remissions (Fig. 23.3, top panel). With single-agent RA therapy, remissions are unfortunately usually transient (Warrell et al. 1993; Tallman and Altman 2009), suggesting that differentiation alone cannot abolish cancer cell self-renewal (Kogan 2009; de The and Chen 2010) (see the next chapter by M. Tallman).

The other potent anti-APL agent As_2O_3 (arsenic), is considerably more efficient than RA as single agent (Chen et al. 2011; Zhu et al. 2002). It induces both apoptosis and differentiation in vivo (Fig. 23.3, bottom panel) and, in combination with RA may yield 90 % definitive cures, even without DNA-damaging chemotherapies (Hu

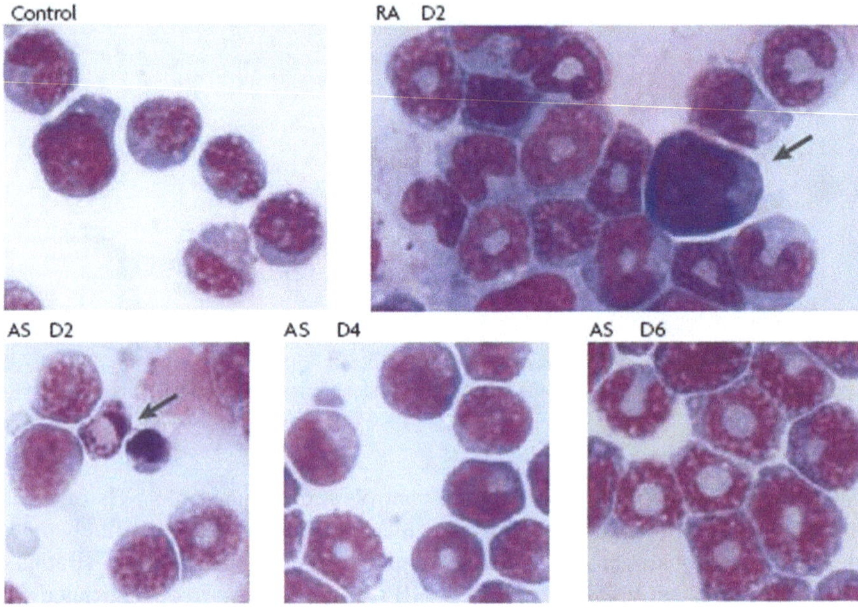

Fig. 23.3 Cellular effects of retinoic acid (*RA*) or arsenic (*AS*) in a mouse model of APL. Note that both drugs induce differentiation, but with different kinetics and that *AS* also induces apoptosis at 2 days (*arrow*). Note the reappearance of normal bone marrow cells after 2 days of *RA* (*arrow*). (Reprinted from de The and Chen 2010)

et al. 2009; Estey et al. 2006; de The and Chen 2010; Wang and Chen 2008). Clinical trials in non-APL cancer patients have been largely disappointing, unexpectedly demonstrating that this notoriously toxic compound has a great specificity for APL cells (Zhu et al. 2002).

23.7 Molecular Basis for PML/RARA-Targeted APL Therapies?

Molecular studies performed after demonstration of their clinical efficacy have revealed that both RA and arsenic directly target the PML/RARA oncoprotein for degradation (Quignon et al. 1997; de The and Chen 2010; Zhu et al. 2001) (Fig. 23.4a). In a remarkable and completely unexpected symmetry, RA targets the RARA part of PML/RARA, while arsenic directly targets its PML part (Quignon et al. 1997) (Fig. 23.4b). Thus, these two agents discovered by chance actually directly target PML/RARA through its two constitutive moieties, making them a posteriori targeted therapies. This strongly suggested an important, if not essential, contribution of PML/RARA degradation to therapy response.

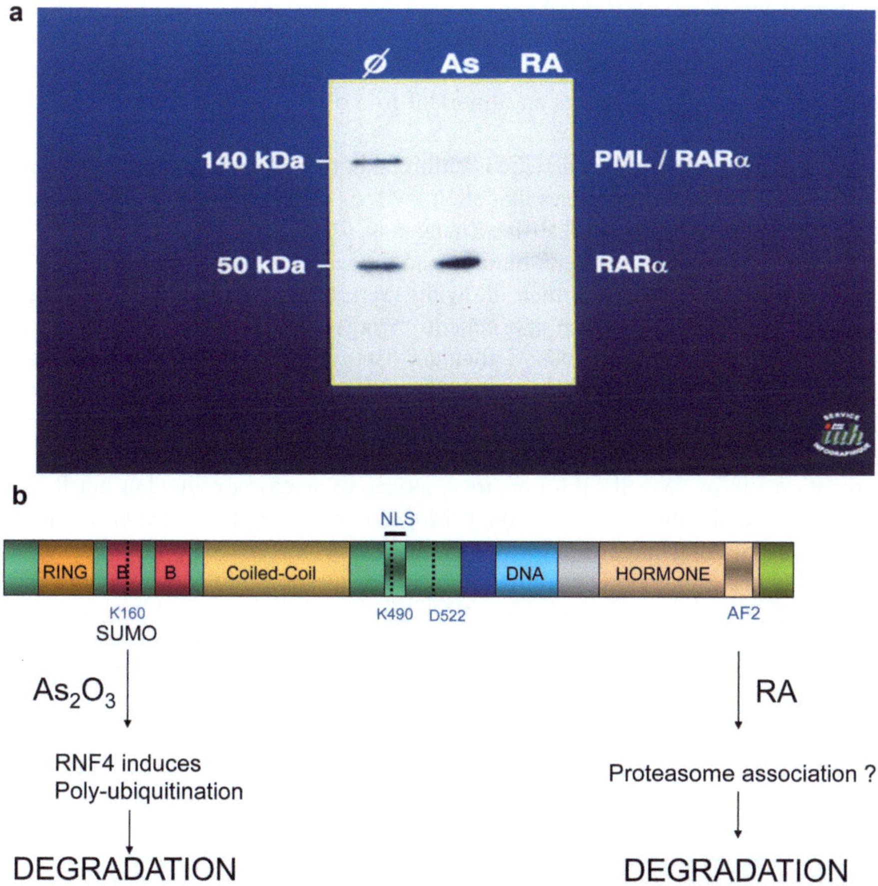

Fig. 23.4 **a** Retinoic acid and arsenic both degrade promyelocytic gene/retinoic acid receptor alpha (*PML/RARA*) in APL cells after an overnight treatment ex vivo at therapeutic concentrations. Note that retinoic acid (*RA*) also degrades *RARA,* while arsenic (*As*) does not, pointing to distinct mechanisms. **b** Pathways of PML/RARA degradation. Domains in PML/RARA are indicated. Note that *RA* targets the RARA part of the fusion, while arsenic targets in PML moiety through oxidation, direct binding, and sumoylation and RNF4-mediated poly-ubiquitination. See text and (Lallemand-Breitenbach et al. 2012) for details

Concerning RA targeting of PML/RARA, RA: 1) releases co-repressor binding from PML/RARA, 2) induces transactivation through the PML/RARA-mediated recruitment of co-activators, and 3) induces PML/RARA degradation. Thus, RA reverses all PML/RARA properties but the contribution of each to APL clearance remains debated (Ablain and de The 2011; Ablain et al. 2011). It must be noted that the therapeutic concentrations of RA used against APL are several orders of magnitudes higher than its physiologic concentrations or its binding affinity, and the decrease in plasma concentration is constantly associated to clinical RA resistance (Muindi et al. 1992). Other cases of RA-resistance are associated with mutations in the RARA moiety of PM/RARA that preclude RA-binding, transactivation and/or degradation (Gallagher et al. 2006; Gallagher 2002). The fact that only pharmacologic levels of RA are associated with therapy response and full PML/RARA degradation, strongly supports an important role of the latter in disease remission (Zhu et al. 2001).

PML/RARA targeting by arsenic is enforced both by direct binding and arsenic-induced reactive oxygen species that elicit PML oxidation through the formation of disulfide bridges (Jeanne et al. 2010; Zhang et al. 2010; de The and Chen 2010). As extensively reviewed elsewhere, binding and oxidation initiate formation of a PML mesh, its hypersumoylation, then allowing recruitment of the SUMO-dependent ubiquitin ligase RNF4, which subsequently triggers PML or PML/RARA degradation (Geoffrey and Hay 2009; Lallemand-Breitenbach et al. 2008; Lallemand-Breitenbach et al. 2001; Tatham et al. 2008).

The role of PML/RARA sumoylation and degradation in arsenic-based therapy is supported by significant genetic evidence. Mutation of the arsenic-sensitive sumoylation site in PML/RARA impairs response to arsenic *ex vivo* but not RA-induced differentiation (Zhu et al. 2005). Mutation of the cysteine residues required for arsenic binding impairs the response to As_2O_3 *ex vivo* (Jeanne et al. 2010) and neighboring mutations were observed in arsenic-resistant patients (Goto et al. 2011). Finally, in murine models of APL, vitamin E derivatives with mitochondrial toxicity generating oxidative stress induce prolonged remissions (Dos Santos et al. 2011), validating intracellular oxidation as a key anti-APL mechanism. Importantly, arsenic does not induce PLZF/RARA degradation and is accordingly inefficient in PLZF/RARA APL models (Rego et al. 2000; Jeanne et al. 2010).

23.8 Differentiation and/or Self-Renewal: Mice Come to the Stage

On the cellular side, differentiation-based therapy in APL primarily relies on the correlation between clinical remissions and morphological maturation of leukemia blasts (Warrell et al. 1993). However, this cannot explain why only few patients are cured by RA alone, neither why arsenic cures 70% of APL patients, although it does not induce differentiation *ex vivo*. Accordingly, there have been controversies as to the exact contribution of differentiation to APL cure (Kogan 2009; Ablain and de The 2011).

The most recently proposed models have uncoupled APL differentiation and loss of self-renewal (Kogan 2009). Indeed, blast differentiation does not necessarily trigger loss of leukemia-initiating activity and self-renewal (Nasr et al. 2008). Yet, these are the only features predicting disease eradication in vivo (Ablain and de The 2011).

Mouse models have played a key role in understanding the mechanisms of RA and arsenic therapies, pointing to the importance of PML/RARA degradation and challenging the sole role of differentiation (Ablain and de The 2011). In PML/RARA-driven APL, complete differentiation of the leukemia is achieved even at

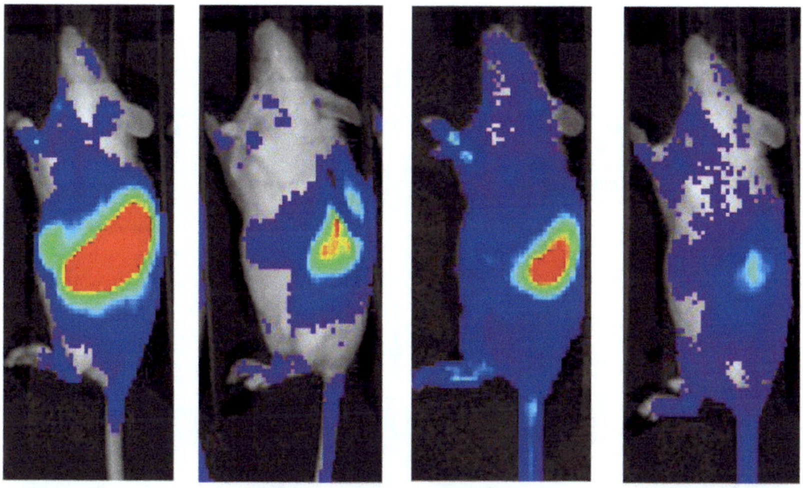

Fig. 23.5 Synergistic effects of RA and arsenic when combined. Luciferase imaging of APL in vivo after 3 days of treatment. (Reproduced from Nasr et al. 2008)

low RA doses, but complete APL clearance only appears with treatments at the highest (toxic) concentrations (Nasr et al. 2008). This might explain the efficiency of liposomal RA, which has led to cures as single agent in patients (Tsimberidou et al. 2006). Complete loss of clonogenic activity was observed in PML/RARA-driven APL mice treated with the RA/arsenic combination, although the combination actually delays morphologic differentiation (Shao et al. 1998; Lallemand-Breitenbach et al. 1999; Nasr et al. 2008). Indeed, this combination rapidly abolishes the ability of APL cells to be transplanted and induce new APL, which explains the dramatic synergy of RA/arsenic for tumor regression and survival in mice (Lallemand-Breitenbach et al. 1999; Nasr et al. 2008; Rego et al. 2000) (Fig. 23.5). Moreover, RA and arsenic induce PML/RARA degradation by different mechanisms, predicting absence of cross-resistance in vivo. As detailed in the companion chapter, this was later successfully transposed to patients, with over 90 % of them definitively cured by the RA/arsenic combination (Shen et al. 2004; Hu et al. 2009; Estey et al. 2006; de The and Chen 2010; Tallman and Altman 2009; Wang and Chen 2008).

23.9 What Actually Causes Loss of Clonogenic Activity?

Differentiation was proposed to reflect transcriptional activation while PML/RARA catabolism would entail loss of self-renewal (Kogan 2009; Ablain and de The 2011; Ablain et al. 2011). Yet, some recent studies have argued that even transcriptional derepression, for example through PML/RARA loss or reversal of histone deacetylation, may suffice for differentiation (Leiva et al. 2012). This may explain why

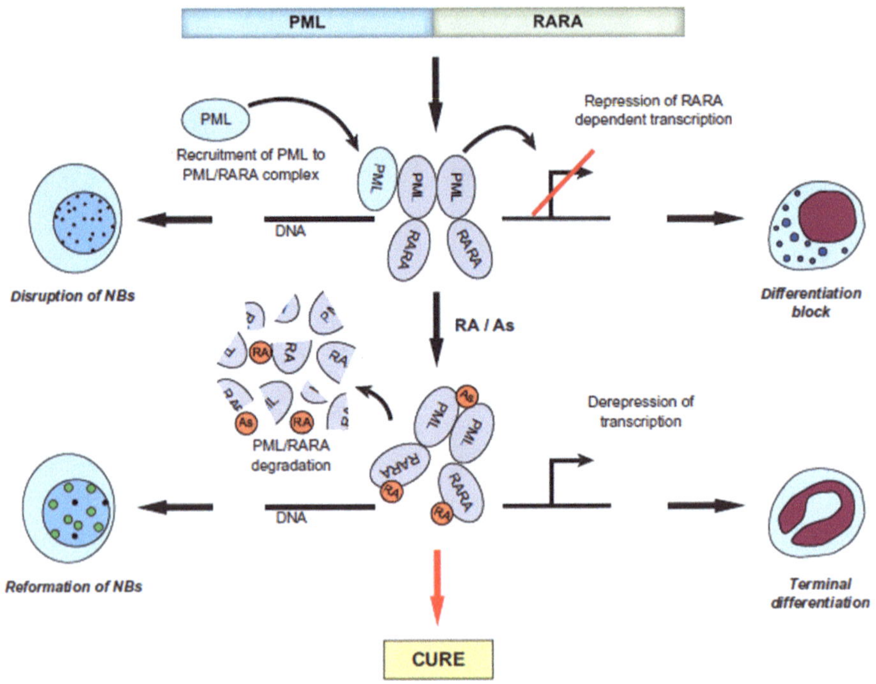

Fig. 23.6 A model for pathogenesis and treatment efficiency in APL. Promyelocytic gene/retinoic acid receptor alpha (*PML/RARA*) has a dual function to repress transcription and disrupt *PML* nuclear bodies (*NBs*). Therapies that degrade *PML/RARA* induce differentiation through derepression, while *PML/RARA* loss allows *NB* reformation. (Reproduced from Ablain et al. 2011)

arsenic (through PML/RARA degradation) may induce differentiation in vivo (Fig. 23.3) or *ex vivo* in combination with growth factors (Muto et al. 2001).

It remained to be determined how loss of clonogenic activity was entailed. Mechanistically, it is possible that degradation-induced derepression of some PML/RARA-specific target genes suffices for leukemia initiating cell (LIC) exhaustion. Alternatively, it is possible that efficacy of arsenic alone could rely on some other effects than PML/RARA degradation, either through PML itself or via other arsenic targets. One may thus envision that PML NB reassembly following PML-RARA degradation reassembly could activate specific pathways like apoptosis or modulate self-renewal.

In normal progenitors or in the context of other leukemic fusion proteins, PML controls self-renewal (Ito et al. 2008; Regad et al. 2009), consistent with the proposal that NBs tune several critical pathways involved in "stemness" and self-renewal, such as P53, AKT/PTEN, HIF1A (Ito et al. 2009; Song et al. 2008). Thus, PML/RARA loss and active enforcement of NB reformation (for example by arsenic) could directly contribute to loss of stemness. However, one should stress that studies on these biologically complex mechanisms will necessarily be performed in mice and that the regulation of self-renewal and the mode of interference by PML/RARA might be different between mice and humans (Fig. 23.6).

23.10 Could APL become a Model for other Targeted Therapies?

In APL, the extraordinary clinical potency of RA and arsenic reflects the fact that RARA and PML are both dispensable (in mice), while APL cells are addicted to the continuous expression of PML/RARA. These agents fully degrade RARA, PML, and PML/RARA, which exerts a maximal efficacy on APL cells without any toxicity on normal cells, hence the extremely high therapeutic index of these agents or their association (Lallemand-Breitenbach et al. 2005; de The and Chen 2010; Nardella et al. 2011).

APL is a paradigm for targeted therapies and probably the only leukemia where combination of non-cross reactive agents has led to definitive cures (Wang and Chen 2008; de The and Chen 2010; de The et al. 2012). It underscores the power of proteolysis rather than enzymatic inhibition. Indeed, complete proteolysis abolishes all of the functions of oncoproteins, including those linked to protein/protein interactions, which may be very important in LICs. Collectively, this suggests that agents targeting the stability of other dominant oncoproteins could be of high therapeutic value, in particular, in translocation-driven leukemias or sarcomas driven by a single dominant oncoprotein fusing two unessential genes (Ablain et al. 2011).

References

Ablain J, de The H (2011) Revisiting the differentiation paradigm in acute promyelocytic leukemia. Blood 117(22):5795–5802. doi:10.1182/blood-2011-02-329367

Ablain J, Nasr R, Bazarbachi A, de The H (2011) Oncoprotein proteolysis, an unexpected Achille's Heel of cancer cells? Cancer Discov 1:117–127

Akagi T, Shih LY, Kato M, Kawamata N, Yamamoto G, Sanada M, Okamoto R, Miller CW, Liang DC, Ogawa S, Koeffler HP (2009) Hidden abnormalities and novel classification of t(15;17) acute promyelocytic leukemia (APL) based on genomic alterations. Blood 113 (8):1741–1748. doi:blood-2007-12-130260 [pii] 10.1182/blood-2007-12-130260

Chen Z, Brand N, Chen A, Chen S, Tong J, Wang Z, Waxman S, Zelent A (1993) Fusion between a novel Kruppel-like zinc finger gene and the retinoic acid receptor a locus due to a variant t(11,17) translocation in acute promyelocytic leukemia. EMBO J 12:1161–1167

Chen SJ, Zhou GB, Zhang XW, Mao JH, de The H, Chen Z (2011) From an old remedy to a magic bullet: molecular mechanisms underlying the therapeutic effects of arsenic in fighting leukemia. Blood 117(24):6425–6437. doi:10.1182/blood-2010-11-283598

de The H, Chen Z (2010) Acute promyelocytic leukaemia: novel insights into the mechanisms of cure. Nat Rev Cancer 10(11):775–783. doi:nrc2943 [pii] 10.1038/nrc2943

de Thé H, Vivanco-Ruiz MdM, Tiollais P, Stunnenberg H, Dejean A (1990) Identification of a retinoic acid responsive element in the retinoic acid receptor beta gene. Nature 343:177–180

de The H, Le Bras M, Lallemand-Breitenbach V (2012) The cell biology of disease: acute promyelocytic leukemia, arsenic, and PML bodies. J Cell Biol 198(1):11–21. doi:10.1083/jcb.201112044

Degos L, Dombret H, Chomienne C, Daniel MT, Miclea JM, Chastang C, Castaigne S, Fenaux P (1995) All-trans retinoic acid as a differentiating agent in the treatment of acute promyelocytic leukemia. Blood 85(10):2643–2653.

Dos Santos GA, Abreu ELRS, Pestana CR, Lima AS, Scheucher PS, Thome CH, Gimenes-Teixeira HL, Santana-Lemos BA, Lucena-Araujo AR, Rodrigues FP, Nasr R, Uyemura SA, Falcao RP, de The H, Pandolfi PP, Curti C, Rego EM (2011) (+)alpha-Tocopheryl succinate inhibits the mitochondrial respiratory chain complex I and is as effective as arsenic trioxide or ATRA against acute promyelocytic leukemia in vivo. Leukemia: Off Leukemia Soc Am, Leukemia Res Fund, UK 26(3):451–460. doi:10.1038/leu.2011.216

Dyck JA, Warrell RP, Evans RM, Miller WH (1995) Rapid diagnosis of acute promyelocytic leukemia by immunohistochemical localization of PML/RAR alpha protein. Blood 86:862–867

Estey E, Garcia-Manero G, Ferrajoli A, Faderl S, Verstovsek S, Jones D, Kantarjian H (2006) Use of all-trans retinoic acid plus arsenic trioxide as an alternative to chemotherapy in untreated acute promyelocytic leukemia. Blood 107(9):3469–3473

Farboud B, Hauksdottir H, Wu Y, Privalsky ML (2003) Isotype-restricted corepressor recruitment: a constitutively closed helix 12 conformation in retinoic acid receptors beta and gamma interferes with corepressor recruitment and prevents transcriptional repression. Mol Cell Biol 23(8):2844–2858

Gallagher RE (2002) Retinoic acid resistance in acute promyelocytic leukemia. Leukemia: Off Leukemia Soc Am, Leukemia Res Fund, UK 16(10):1940–1958. doi:10.1038/sj.leu.2402719

Gallagher RE, Schachter-Tokarz EL, Zhou DC, Ding W, Kim SH, Sankoorikal BJ, Bi W, Livak KJ, Slack JL, Willman CL (2006) Relapse of acute promyelocytic leukemia with PML-RARalpha mutant subclones independent of proximate all-trans retinoic acid selection pressure. Leukemia: Off Leukemia Soc Am, Leukemia Res Fund, UK 20(4):556–562. doi:10.1038/sj.leu.2404118

Geoffroy MC, Hay RT (2009) An additional role for SUMO in ubiquitin-mediated proteolysis. Nat Rev Mol Cell Biol 10(8):564–568. doi:nrm2707 [pii] 10.1038/nrm2707

Goto E, Tomita A, Hayakawa F, Atsumi A, Kiyoi H, Naoe T (2011) Missense mutations in PML-RARA critical for the lack of responsiveness to arsenic trioxide treatment. Blood 118(6):1600–1609 doi:10.1182/blood-2011-01-329433

Hu J, Liu YF, Wu CF, Xu F, Shen ZX, Zhu YM, Li JM, Tang W, Zhao WL, Wu W, Sun HP, Chen QS, Chen B, Zhou GB, Zelent A, Waxman S, Wang ZY, Chen SJ, Chen Z (2009) Long-term efficacy and safety of all-trans retinoic acid/arsenic trioxide-based therapy in newly diagnosed acute promyelocytic leukemia. Proc Natl Acad Sci U S A 106(9):3342–3347. doi:10.1073/pnas.0813280106

Hua S, Kittler R, White KP (2009) Genomic antagonism between retinoic acid and estrogen signaling in breast cancer. Cell 137(7):1259–1271. doi:10.1016/j.cell.2009.04.043

Huang M, Ye Y, Chen R, Chai J, Lu J, Zhoa L, Gu L, Wang Z (1988) Use of all trans retinoic acid in the treatment of acute promyelocytic leukaemia. Blood 72:567–572

Ito K, Bernardi R, Morotti A, Matsuoka S, Saglio G, Ikeda Y, Rosenblatt J, Avigan DE, Teruya-Feldstein J, Pandolfi PP (2008) PML targeting eradicates quiescent leukaemia-initiating cells. Nature 453(7198):1072–1078. doi:nature07016 [pii] 10.1038/nature07016

Ito K, Bernardi R, Pandolfi PP (2009) A novel signaling network as a critical rheostat for the biology and maintenance of the normal stem cell and the cancer-initiating cell. Curr Opin Genet Dev 19(1):51–59. doi:S0959-437X(09)00017-3 [pii]10.1016/j.gde.2009.01.004

Jeanne M, Lallemand-Breitenbach V, Ferhi O, Koken M, Le Bras M, Duffort S, Peres L, Berthier C, Soilihi H, Raught B, de The H (2010) PML/RARA oxidation and arsenic binding initiate the antileukemia response of As2O3. Cancer Cell 18(1):88–98. doi:S1535-6108(10)00241-2 [pii] 10.1016/j.ccr.2010.06.003

Kamashev DE, Vitoux D, De Thé H (2004) PML/RARA-RXR oligomers mediate retinoid- and rexinoid-/cAMP in APL cell differentiation. J Exp Med 199:1–13

Kastner P, Mark M, Chambon P (1995) Nonsteroid nuclear receptors: what are genetic studies telling us about their role in real life? Cell 83(6):859–869

Kastner P, Lawrence HJ, Waltzinger C, Ghyselinck NB, Chambon P, Chan S (2001) Positive and negative regulation of granulopoiesis by endogenous RARalpha. Blood 97(5):1314–1320

Kogan SC (2009) Curing APL: differentiation or destruction? Cancer Cell 15 (1):7–8. doi:S1535-6108(08)00413-3 [pii] 10.1016/j.ccr.2008.12.012

Koken MHM, Puvion-Dutilleul F, Guillemin MC, Viron A, Linares-Cruz G, Stuurman N, de Jong L, Szostecki C, Calvo F, Chomienne C, Degos L, Puvion E, de Thé H (1994) The t(15;17) translocation alters a nuclear body in a RA-reversible fashion. EMBO J 13:1073–1083
Lallemand-Breitenbach V, de The H (2010) PML nuclear bodies. Cold Spring Harb Perspect Biol 2:a000661
Lallemand-Breitenbach V, Guillemin M-C, Janin A, Daniel M-T, Degos L, Kogan SC, Bishop JM, de The H (1999) Retinoic acid and arsenic synergize to eradicate leukemic cells in a mouse model of acute promyelocytic leukemia. J Exp Med 189:1043–1052
Lallemand-Breitenbach V, Zhu J, Puvion F, Koken M, Honore N, Doubeikovsky A, Duprez E, Pandolfi PP, Puvion E, Freemont P, de The H (2001) Role of Promyelocytic Leukemia (PML) sumolation in nuclear body formation, 11S proteasome recruitment, and As(2)O(3)-induced PML or PML/retinoic acid receptor alpha degradation. J Exp Med 193(12):1361–1372
Lallemand-Breitenbach V, Zhu J, Kogan S, Chen Z, de The H (2005) Opinion: how patients have benefited from mouse models of acute promyelocytic leukaemia. Nat Rev Cancer 5(10):821–827
Lallemand-Breitenbach V, Jeanne M, Benhenda S, Nasr R, Lei M, Peres L, Zhou J, Zhu J, Raught B, de The H (2008) Arsenic degrades PML or PML-RARalpha through a SUMO-triggered RNF4/ubiquitin-mediated pathway. Nat Cell Biol 10(5):547–555. doi:ncb1717 [pii] 10.1038/ncb1717
Lallemand-Breitenbach V, Zhu J, Chen Z, de The H (2012) Mechanisms of APL cure through PML/RARA degradation by As2O3. Trends Mol Med 18:36–42
Leiva M, Moretti S, Soilihi H, Pallavicini I, Peres L, Mercurio C, Dal Zuffo R, Minucci S, de The H (2012) Valproic acid induces differentiation and transient tumor regression, but spares leukemia-initiating activity in mouse models of APL. Leukemia: Off Leukemia Soc Am, Leukemia Res Fund, UK 26(7):1630–1637. doi:10.1038/leu.2012.39
Licht JD (2006) Reconstructing a disease: what essential features of the retinoic acid receptor fusion oncoproteins generate acute promyelocytic leukemia? Cancer Cell 9(2):73–74
Martens JH, Brinkman AB, Simmer F, Francoijs KJ, Nebbioso A, Ferrara F, Altucci L, Stunnenberg HG (2010) PML-RARalpha/RXR alters the epigenetic landscape in acute promyelocytic leukemia. Cancer Cell 17(2):173–185. doi:S1535-6108(10)00005-X [pii] 10.1016/j.ccr.2009.12.042
Melnick A, Licht JD (1999) Deconstructing a disease: RARalpha, its fusion partners, and their roles in the pathogenesis of acute promyelocytic leukemia. Blood 93:3167–3215
Mistry AR, Felix CA, Whitmarsh RJ, Mason A, Reiter A, Cassinat B, Parry A, Walz C, Wiemels JL, Segal MR, Ades L, Blair IA, Osheroff N, Peniket AJ, Lafage-Pochitaloff M, Cross NC, Chomienne C, Solomon E, Fenaux P, Grimwade D (2005) DNA topoisomerase II in therapy-related acute promyelocytic leukemia. N Engl J Med 352(15):1529–1538
Muindi J, Frankel SR, Miller WH Jr, Jakubowski A, Scheinberg DA, Young CW, Dmitrovsky E, Warrell RP, Jr. (1992) Continuous treatment with all-trans retinoic acid causes a progressive reduction in plasma drug concentrations: implications for relapse and retinoid "resistance" in patients with acute promyelocytic leukemia. Blood 79(2):299–303
Muto A, Kizaki M, Kawamura C, Matsushita H, Fukuchi Y, Umezawa A, Yamada T, Hata J, Hozumi N, Yamato K, Ito M, Ueyama Y, Ikeda Y (2001) A novel differentiation-inducing therapy for acute promyelocytic leukemia with a combination of arsenic trioxide and GM-CSF. Leukemia 15(8):1176–1184
Nardella C, Lunardi A, Patnaik A, Cantley LC (2011) The APL paradigm and the co-clinical trial project. Cancer Discov 1:108–116
Nasr R, Guillemin MC, Ferhi O, Soilihi H, Peres L, Berthier C, Rousselot P, Robledo-Sarmiento M, Lallemand-Breitenbach V, Gourmel B, Vitoux D, Pandolfi PP, Rochette-Egly C, Zhu J, de The H (2008) Eradication of acute promyelocytic leukemia-initiating cells through PML-RARA degradation. Nat Med 14(12):1333–1342. doi:nm.1891 [pii] 10.1038/nm.1891
Occhionorelli M, Santoro F, Pallavicini I, Gruszka A, Moretti S, Bossi D, Viale A, Shing D, Ronzoni S, Muradore I, Soncini M, Pruneri G, Rafaniello P, Viale G, Pelicci PG, Minucci S (2011) The self-association coiled-coil domain of PML is sufficient for the oncogenic conversion

of the retinoic acid receptor (RAR) alpha. Leukemia: Off Leukemia Soc Am, Leukemia Res Fund, UK 25(5):814–820. doi:10.1038/leu.2011.18

Piazza F, Gurrieri C, Pandolfi PP (2001) The theory of APL. Oncogene 20(49):7216–7222

Quignon F, Chen Z, de Thé H (1997) Retinoic acid and arsenic: towards oncogene targeted treatments of acute promyelocytic leukaemia. Biochim Biophys Acta 1333:M53–M61

Regad T, Bellodi C, Nicotera P, Salomoni P (2009) The tumor suppressor Pml regulates cell fate in the developing neocortex. Nat Neurosci 12(2):132–140. doi:nn.2251 [pii] 10.1038/nn.2251

Rego EM, He LZ, Warrell RP Jr, Wang ZG, Pandolfi PP (2000) Retinoic acid (RA) and As2O3 treatment in transgenic models of acute promyelocytic leukemia (APL) unravel the distinct nature of the leukemogenic process induced by the PML-RARalpha and PLZF-RARalpha oncoproteins. Proc Natl Acad Sci U S A 97:10173–10178

Shao W, Fanelli M, Ferrara FF, Riccioni R, Rosenauer A, Davison K, Lamph WW, Waxman S, Pelicci PG, Lo Coco F, Avvisati G, Testa U, Peschle C, Gambacorti-Passerini C, Nervi C, Miller WHJ (1998) Arsenic trioxide as an inducer of apoptosis and loss of PML/RARalpha protein in acute promyelocytic leukemia cells. J Natl Cancer Inst 90:124–133

Shen ZX, Shi ZZ, Fang J, Gu BW, Li JM, Zhu YM, Shi JY, Zheng PZ, Yan H, Liu YF et al (2004) All-trans retinoic acid/As2O3 combination yields a high quality remission and survival in newly diagnosed acute promyelocytic leukemia. Proc Natl Acad Sci U S A 101:5328–5335

So CW, Cleary ML (2004) Dimerization: a versatile switch for oncogenesis. Blood 104(4):919–922

Song MS, Salmena L, Carracedo A, Egia A, Lo-Coco F, Teruya-Feldstein J, Pandolfi PP (2008) The deubiquitinylation and localization of PTEN are regulated by a HAUSP-PML network. Nature 455(7214):813–817. doi:nature07290 [pii] 10.1038/nature07290

Sternsdorf T, Phan VT, Maunakea ML, Ocampo C, Sohal J, Siletto A, Galimi F, Le Beau MM, Evans R, Kogan S (2006) Forced retinoic acid receptor a homodimer prime mice for APL-like leukemia. Cancer Cell 9:81–94

Strickland S, Mahdavi V (1978) The induction of differentiation in teratocarcinoma stem cells by retinoic acid. Cell 15(2):393–403. doi:0092-8674(78)90008-9 [pii]

Tallman MS, Altman JK (2009) How I treat acute promyelocytic leukemia. Blood 114 (25):5126–5135. doi:blood-2009-07-216457 [pii] 10.1182/blood-2009-07-216457

Tatham MH, Geoffroy MC, Shen L, Plechanovova A, Hattersley N, Jaffray EG, Palvimo JJ, Hay RT (2008) RNF4 is a poly-SUMO-specific E3 ubiquitin ligase required for arsenic-induced PML degradation. Nat Cell Biol 10(5):538–546. doi:ncb1716 [pii] 10.1038/ncb1716

Tsimberidou AM, Tirado-Gomez M, Andreeff M, O'Brien S, Kantarjian H, Keating M, Lopez-Berestein G, Estey E (2006) Single-agent liposomal all-trans retinoic acid can cure some patients with untreated acute promyelocytic leukemia: an update of The University of Texas M. D. Anderson Cancer Center Series. Leuk Lymphoma 47(6):1062–1068. doi:K54M427781W6874X [pii] 10.1080/10428190500463932

Verger A, Perdomo J, Crossley M (2003) Modification with SUMO. A role in transcriptional regulation. EMBO Rep 4(2):137–142

Vickers M, Jackson G, Taylor P (2000) The incidence of acute promyelocytic leukemia appears constant over most of a human lifespan, implying only one rate limiting mutation. Leukemia 14 (4):722–726

Wang ZY, Chen Z (2008) Acute promyelocytic leukemia: from highly fatal to highly curable. Blood 111(5):2505–2515. doi:111/5/2505 [pii] 10.1182/blood-2007-07-102798

Warrell R, de Thé H, Wang Z, Degos L (1993) Acute promyelocytic leukemia. New Engl J Med 329:177–189

Welch JS, Yuan W, Ley TJ (2011) PML-RARA can increase hematopoietic self-renewal without causing a myeloproliferative disease in mice. J Clin Invest 121(4):1636–1645. doi:10.1172/JCI42953

Welch JS, Ley TJ, Link DC, Miller CA, Larson DE, Koboldt DC, Wartman LD, Lamprecht TL, Liu F, Xia J, Kandoth C, Fulton RS, McLellan MD, Dooling DJ, Wallis JW, Chen K, Harris CC, Schmidt HK, Kalicki-Veizer JM, Lu C, Zhang Q, Lin L, O'Laughlin MD, McMichael JF, Delehaunty KD, Fulton LA, Magrini VJ, McGrath SD, Demeter RT, Vickery TL, Hundal J,

Cook LL, Swift GW, Reed JP, Alldredge PA, Wylie TN, Walker JR, Watson MA, Heath SE, Shannon WD, Varghese N, Nagarajan R, Payton JE, Baty JD, Kulkarni S, Klco JM, Tomasson MH, Westervelt P, Walter MJ, Graubert TA, Dipersio JF, Ding L, Mardis ER, Wilson RK (2012) The origin and evolution of mutations in acute myeloid leukemia. Cell 150(2):264–278. doi:10.1016/j.cell.2012.06.023

Zeisig BB, Kwok C, Zelent A, Shankaranarayanan P, Gronemeyer H, Dong S, So CW (2007) Recruitment of RXR by homotetrameric RARalpha fusion proteins is essential for transformation. Cancer Cell 12(1):36–51. doi:S1535-6108(07)00175-4 [pii] 10.1016/j.ccr.2007.06.006

Zhang XW, Yan XJ, Zhou ZR, Yang FF, Wu ZY, Sun HB, Liang WX, Song AX, Lallemand-Breitenbach V, Jeanne M, Zhang QY, Yang HY, Huang QH, Zhou GB, Tong JH, Zhang Y, Wu JH, Hu HY, de The H, Chen SJ, Chen Z (2010) Arsenic trioxide controls the fate of the PML-RARalpha oncoprotein by directly binding PML. Science 328(5975):240–243. doi:328/5975/240 [pii] 10.1126/science.1183424

Zhu J, Lallemand-Breitenbach V, de The H (2001) Pathways of retinoic acid- or arsenic trioxide-induced PML/RARalpha catabolism, role of oncogene degradation in disease remission. Oncogene 20(49):7257–7265

Zhu J, Chen Z, Lallemand-Breitenbach V, de Thé H (2002) How acute promyelocytic leukemia revived arsenic. Nat Rev Cancer 2:705–713

Zhu J, Zhou J, Peres L, Riaucoux F, Honore N, Kogan S, de The H (2005) A sumoylation site in PML/RARA is essential for leukemic transformation. Cancer Cell 7(2):143–153

Zhu J, Nasr R, Peres L, Riaucoux-Lormiere F, Honore N, Berthier C, Kamashev D, Zhou J, Vitoux D, Lavau C, de The Hs (2007) RXR is an essential component of the oncogenic PML/RARA complex in vivo. Cancer Cell 12(1):23–35

Chapter 24
Arsenic Trioxide in Untreated APL

Elihu Estey

Abstract For many years, the combination of all-*trans* retinoic acid (ATRA) and an anthracycline (daunorubicin or idarubicin) constituted standard treatment for acute promyelocytic leukemia (APL). More recently, arsenic trioxide (ATO) has been found to be a more effective anti-APL drug than ATRA; used alone ATO produces cures in newly diagnosed APL much more frequently than ATRA. This observation has prompted use of the combination of ATO and ATRA in newly diagnosed APL. Although follow-up time with this treatment is much less than with the more standard treatment, results of a randomized trial suggest that ATO plus ATRA produces superior 2-year event-free survival than does ATRA plus anthracycline.

Keywords All-*trans* retinoic acid · Arsenic trioxide · Daunorubicin/idarubicin · Newly diagnosed acute promyelocytic leukemia

Abbreviations

ATO	Arsenic trioxide
ATRA	All-trans retinoic acid
CR	Complete remission
DFS	Disease-free survival
OS	Overall survival
AIDA	ATRA plus idarubicin
GIMEMA	Italian Group for Hematological Malignancies of the Adult
IDAC	Cytarabine at 1–2 g/m^2 per dose
CSF	Cerebrospinal fluid
Rx	Treatment
6MP	6-mercaptopurine,
Mtx	Methotrexate
PETHEMA/HOVON	Programa de Estudio y Tratamiento de las Hemopatías Malignas/Hemato-Oncologie voor Volwassenen Nederland
APL	Acute promyelocytic leukemia
EAPLG	European APL Group

E. Estey (✉)
University of Washington and Fred Hutchinson Cancer Research Center,
825 Eastlake Ave. E, Seattle, WA 98109, USA
e-mail: eestey@uw.edu

M. Andreeff (ed.), *Targeted Therapy of Acute Myeloid Leukemia*,
Current Cancer Research, DOI 10.1007/978-1-4939-1393-0_24

DNR	Daunorubicin
Ara-C	Cytarabine
CIR	Cumulative incidence of relapse
NALIG	North American Leukemia Intergroup

24.1 Introduction

This section will deal with use of arsenic trioxide (ATO) to treat newly diagnosed APL, for example the potential to use it, preferably with all-*trans* retinoic acid (ATRA) and in some patients gemtuzumab ozogamicin (GO), in place of chemotherapy, here meaning drugs such as daunorubicin, idarubicin, or mitoxantrone with or without cytarabine, used together with ATRA, as described in Dr. Tallman's section.

24.2 Why Replace Chemotherapy Plus ATRA

Given the 90% cure rates reported with chemotherapy plus ATRA, we will begin by discussing the rationale to replace chemotherapy in APL. A good part of the rationale is the desire to avoid toxicity. This can be short term or long term. The principal short-term toxicity is treatment-related mortality, particularly in patients aged 60 or above during the several courses of "consolidation therapy." Reports from France (Ades et al. 2005) and Spain (Sanz et al. 2004) indicate that this complication occurs in 8–19% of such patients. Development of a myelodysplastic syndrome or acute myeloid leukemia (AML) is a principal long-term complication. This has been reported to occur in 7% of 77 patients from Italy (Latagliata et al. 2002) although only in 1% of 677 patients followed for a median of 51 months in France (Lobe et al. 2003). Although it thus seems likely that myelodysplastic syndrome (MDS) can result from chemotherapy for APL, it is possible that secondary MDS represents part of the natural history of cured APL since the affected patients typically have the chromosome 5 and/or seven abnormalities usually seen after receipt of alkylating agents, although such agents are not used to treat APL.

While avoidance of toxicity is desirable, it is now widely accepted that early death remains a significant problem in APL. For example, Park et al. used the Surveillance Epidemiology and End Results (SEER) program of the National Cancer Institute and the New York Cancer Registry to collect data on 1400 patients treated between 1992 and 2007 (Park et al. 2011). They found that the early death rate was 17%, rising to 24% in patients aged over 55. Similarly, taking advantage of a population-based registry (Swedish Adult Acute Leukemia Registry), Juliusson et al. reported that 26% of 105 patients with APL treated from 1997 to 2006 died within 14 days of diagnosis, with 41% of the deaths due to hemorrhage (Lehmann et al. 2011). One very plausible explanation for the high early death rate is delay in administration of ATRA, as reported by Altman et al. (2011). However, as will be discussed below, it may be more important to use ATO than ATRA.

24.3 Data Suggesting Chemotherapy May Be Unnecessary in APL

Several pieces of information suggested the possibility of dispensing with chemotherapy in APL. First, liposomal ATRA (L-ATRA) used alone resulted in the likely cure of 42 % (12/28) patients with APL presenting with white blood cell (WBC) count < 10,000 (Tsimberidou et al. 2006), the criterion for low-risk (or largely equivalently intermediate-risk) APL (Sanz et al. 2009); these 12 patients have been followed > 5 years (Tsimberidou et al. 2006). L-ATRA, which is no longer available, was given intravenously in order to retard inducing hepatic enzymes as might occur with ATRA's oral administration. These enzymes might account for the decline in plasma ATRA levels seen with repeated administration of oral ATRA. Such decline has been postulated to underlie the observation that single-agent oral ATRA, while producing complete remission (CR) rates equivalent to those seen with chemotherapy, is highly unlikely to cure APL unless combined with chemotherapy (Muindi et al. 1992). Plasma ATRA levels were not evaluated in the L-ATRA studies; hence it is not known whether the superior results seen with L-ATRA than with oral ATRA result from more sustained ATRA levels with repeated administration.

Further evidence that chemotherapy is not needed to cure APL came from studies, conducted in India (Mathews et al. 2010) and Iran (Ghavamzadeh et al. 2011), in which patients received ATO without either chemotherapy (except in 11 % of the Indian patients who received daunorubicin during induction) or ATRA-baring relapse (Table 24.1). Both the Indian (in 197 patients) and the Iranian investigators (in 72 patients) reported CR rates of 86 % and with median follow-ups of 5 years (India) and 3 years (Iran) relapse-free survival (RFS) rates are 80 % (India) and 67 % (Iran). These RFS results may be somewhat inferior to those reported in Europe or the USA (Table 24.1), but such a conclusion is somewhat speculative in the absence of randomization and uncertainty as to the role of patient selection. In any event, the Iranian and Indian data indicate that ATO is a more effective anti-APL agent than oral ATRA.

The use of single-agent ATO in Iran and India reflected the high cost of ATRA and chemotherapy in these countries. However, Chinese data suggested that the combination of ATO plus ATRA is more effective than either alone (Shen et al. 2004). Chinese investigators randomized patients to either ATRA alone, ATO alone, or ATO plus ATRA. Although CR rates were similarly high with each option, reduction in transcripts for PML-RARα, the defining feature of APL, was greatest with the combination. Once in, CR patients received three cycles of chemotherapy, following which they received ATRA, ATO, or ATO plus ATRA as originally as-

Table 24.1 Comparison of single-agent ATO trials with cooperative group trials using ATRA, anthracyclines, and Ara-C

Study	Induction/consolidation	No. of patients	Age		WBC > 10,000/μL (high risk; %)	CR (%)	DFS	OS
			Median	Range				
AIDA (GIMEMA) (Lo-Coco et al. 2010)	AIDA/AIDA (WBC < 10,000/μL); AIDA + IDAC (WBC > 10,000/μL)	453	41	18–61	29	94	86 ± 3 % at 6 years low risk, 86 %; high risk, 85 %	87% ± 3 % at 6 years low risk, 89 %; high risk, 83 %
LPA (PETHEMA/ HOVON)(Sanz et al. 2010)	Similar to GIMEMA	402	42	3–83	29	93	92 ± 4 % at 3 years low risk, 94 %; high risk, 82 %	89 ± 4 % at 3 years low risk, 91 %; high risk, 79 %
APL (EAPLG) (Adès et al. 2006)	ATRA + DNR + ara-C/DNR + IDAC	169	All age < 60 (95 WBC < 10,000/ μL; 74 WBC > 10,000/μL)		44	99 low risk; 99 high risk	CIR at 5 years low risk, 13 %; high risk, 8 %	At 5 years low risk, 93 %; high risk, 90 %
C9710 (NALIG) (Powell et al. 2010)	ATRA + DNR + ara-C/ ATO × 2 → ATRA + DNR × 2	24	Not given		23	90	90 % at 3 years (similar in low/high risk)	86 % at 3 years
Ghavamzadeh et al. (2011)	ATO/ATO	197	29	11–71	19	86	67 ± 4 % at 5 years (similar in low/high risk)	64 ± 4 % at 5 years (similar in low/high risk)
Mathews et al. (2010)	ATO/ATO	72	28	3–75	24	86	80 ± 5 %	74 ± 5 %

signed. Disease-free survival (DFS) was best with ATO plus ATRA. Since the use of chemotherapy in each arm presumably reduced the differences in DFS, the observed superiority with ATO plus ATRA is likely an underestimate of the true difference.

24.4 MD Anderson Trial of ATO Plus ATRA with or without GO

With this background, MD Anderson investigators conducted a single-arm trial in which patients with low-risk disease (WBC < 10,000) received ATO (0.15 mg/kg daily) with ATRA (45 mg/m^2 daily) for induction. If WBC rose to > 30,000, patients received one dose of GO (9 mg/m^2). If bone marrow obtained 3–4 weeks after start of therapy showed an absence of abnormal promyelocytes, therapy was held. Once in, CR patients continued ATO at the same dose five times weekly during weeks 1–4, 9–12, 17–20, and 25–28 and received 45 mg/m^2 ATRA daily for 2 consecutive weeks out of every 4. Assuming the patient remained in CR, therapy was stopped 28 weeks after entering CR. Patients with high-risk APL (WBC > 10,000) received the same therapy except that they also received GO 9 mg/m^2 on day 1 of induction. GO, consisting of an antibody against the cell surface antigen CD33 combined with a toxin (Calechiamicin) is of interest in APL, given observations that the vast majority of APL cells are rich in CD33. Studies using GO as therapy of molecular relapsed APL (Lo-Coco et al. 2004) or combining GO with ATRA during induction therapy (Estey et al. 2002) confirmed its clinical activity.

Several features made the MD Anderson study possible. Among these were the availability of means to detect molecular relapse prior to hematologic relapse (Sanz et al. 2009) and the possibility that addition of GO at molecular relapse would avert hematologic relapse (Lo-Coco et al. 2004) and improve the outcome compared to what would transpire were the treatment only to be modified at the latter time. Thus, polymerase chain reaction (PCR) testing, at sensitivity 10 (−4), was to be done every month during the first year if patients were in CR and GO added if two consecutive tests were positive.

Another noteworthy feature of the study was its design, whose purpose was to stop the trial expeditiously, at the cost of a high probability of a false negative result, should initial data suggest patients were faring worse than historically seen at MD Anderson (Estey et al. 2006). Thus, the trial was examined after each cohort of five patients had been evaluated. Was there < 10 % probability that the true CR rate was < 80 % seen historically, the trial would close. With such a rule and, for example, a true CR rate of 60 %, the trial would stop after a median of 12 patients had received ATO plus ATRA (WBC < 10,000) or ATO plus ATRA plus GO (WBC > 10,000). Seven of these 12 would have entered CR contrasted with 10/12 assuming the 80 historical CR rate. Thus, the cost of the trial was $10-7=3$ patients. If the true CR rate was < 60 % the trial would stop earlier with the same

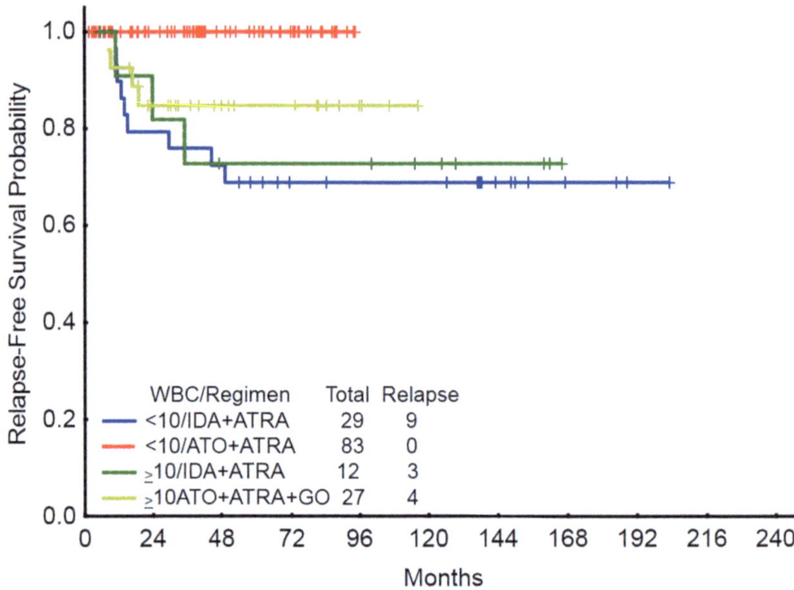

Fig. 24.1 Relapse-free survival in patients according to treatment as given at MD Anderson Cancer Center. *ATO* plus *ATRA*: as described in text. *IDA* plus *ATRA*: idarubicin plus *ATRA* for induction; once in CR 3 courses of idarubicin (without *ATRA*) and then 2 years of 6MP, Mtx, vincristine, and prednisone (*POMP*)

cost in patients. Similar considerations governed stopping for too low a rate of PCR negativity at 6 months.

Results of this MD Anderson trial have been published (Estey et al. 2006; Ravandi et al. 2010) and were updated in June 2012 (F. Ravandi, personal communication). Eighty-five patients (median age 48) with presenting WBC < 10,000 have been treated with ATO plus ATRA without chemotherapy; 83 have entered CR (98 %). Omission of patients given one dose of GO because of rise in WBC to > 30,000 alters this rate only slightly. Twenty-eight patients, median age 44, received ATO plus ATRA plus GO because of WBC > 10,000; 27 (96 %) have entered CR.

Of particular interest are the data for patients who entered CR. Median follow-up is about 3 years, but this is a bit misleading because patients are still being accrued. Perhaps more relevant is a follow-up time of > 3 years in 46 and of > 5 years in 24 of the patients with WBC < 10,000 who entered CR; corresponding figures are 17 and 9 for patients with WBC > 10,000. None of the 83 patients who presented with WBC < 10,000 and entered CR have relapsed (Fig. 24.1) while relapse has occurred in 9 of 29 who entered CR after presenting with WBC > 10,000. Figure 24.2 demonstrates that these RFS rates compare favorably with those seen at MD Anderson in patients given idarubicin plus ATRA for induction. However, none

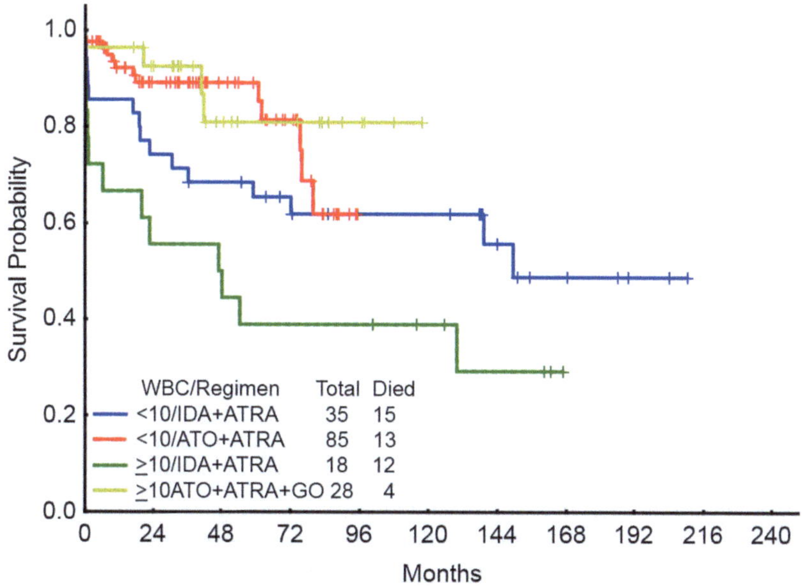

Fig. 24.2 Overall survival in patients according to treatment as given at MD Anderson Cancer Center. *ATO* plus *ATRA*: as described in text. *IDA* plus *ATRA*: idarubicin plus *ATRA* for induction; once in CR 3 courses of idarubicin (without *ATRA*) and then 2 years of 6MP, Mtx, vincristine, and prednisone (*POMP*)

of the latter patients received ATRA in remission; rather they received three cycles of idarubicin and 2 years of maintenance with 6-mercaptopurine (6MP), vincristine, methotrexate (Mtx), and prednisone (POMP). Although this clouds the comparison shown in Fig. 24.1, the RFS rates seen with ATO plus ATRA with or without GO appear very similar to those seen with anthracycline plus ATRA regimens in which ATRA was given both during induction and post-remission therapy.

Figure 24.2 illustrates survival data. Despite the 98 % CR rate and absence of known relapse in patients with WBC < 10,000, 13 of these patients have died. These deaths have typically occurred in patients who have been followed for >3 years; their cause is under investigation.

Several thoughts arise from this study. First, the absence of relapses in patients with WBC < 10,000 suggests that there may be no need for PCR monitoring in these patients if given ATO plus ATRA. However, the late deaths in CR need explanation. The results in general are not sufficiently good to dispense with a control group, which at MD Anderson does not exist since patients have not received anthracyclines plus ATRA during both induction and post-remission.

24.5 ATO Plus ATRA Versus ATRA Plus Chemotherapy: Randomized Studies

Two as yet unpublished randomized studies, one by the Medical Research Kingdom of the United Kingdom, the other under the auspices of the Italian Group for Hematological Malignancies of the Adult (GIMEMA) group in Italy, are addressing the need for a control group by randomizing patients with WBC < 10,000 between ATO plus ATRA, as done at MD Anderson, and the more standard approach, using anthracyclines plus ATRA and reserving ATO until relapse. More details are available for the GIMEMA study, whose primary objective is to compare event-free survival (EFS) at 2 years in the two arms (F. Lo Coco, personal communication May, 2012). As of May 2012, 102 patients had been entered in the ATO plus ATRA arm and 101 in the chemotherapy plus ATRA arm. Their median age, 47, was the same as in the MD Anderson study. The CR rate is 98 %, implying that it cannot be lower than 96 % in the ATO plus ATRA arm. At 24 months, median EFS is 88 %, which is very similar to that seen in patients with WBC < 10,000 at MD Anderson. All 103 patients tested after the third consolidation have been PCR negative, suggesting that rates of RFS are likely to be similar in the two arms. And indeed, the current EFS data suggest that ATO plus ATRA is not worse than ATRA plus chemotherapy for treatment of APL.

24.6 ATO in Consolidation

Further evidence of the usefulness of ATO in newly diagnosed AML comes from a North American Leukemia Intergroup (NALIG) study which randomized patients to receive two consolidation cycles with single-agent ATO (each administered 5 days weekly for 5 weeks) in addition to consolidation cycles with daunorubicin and ATRA (Powell et al. 2010). Patients in the ATO arm had superior RFS (85 % at 3 years) and survival (90 % at 3 years) than patients randomized to receive the same consolidation without ATO ($p < 0.001$ for RFS, $p = 0.07$ for survival).

Conclusions

The North American study has been criticized, given that results in the ATO arm are no better than those obtained in European studies without ATO. This brings up the fundamental point that several means, with or without ATO, undoubtedly exist to cure most patients with newly diagnosed APL. Although the desirability of curing APL without "chemotherapy" is rational enough, given some of the short- and long-term complications of chemotherapy, some of this desirability reflects the negative connotations surrounding the word "chemotherapy." It is certainly plausible that the long-term toxicity of treatments not considered chemotherapy are possibly greater than chemotherapy; in this connection the deaths beyond 3 years in the MD Anderson study (Fig. 24.2) are of interest.

What is striking, however, is that ATRA is used much more frequently in newly diagnosed patients than ATO despite the greater activity of the latter, as discussed above. This is perhaps most important in patients presenting with high WBC. The continued high death rate in such patients has been noted, leading to recommendations for immediate administration of ATRA should APL be suspected. The author wonders why this recommendation should not be extended to include ATO. Perhaps this will become more practical as oral ATO becomes widely available, as is almost certain to happen in the next several years (Au et al. 2003; Siu et al. 2006). A current NALIG study patterned after the MD Anderson study and giving ATO plus ATRA plus GO to newly diagnosed patients with WBC > 10,000 will likely be influential in this regard.

Last, the incorporation of ATO into therapy for newly diagnosed APL raises the question of the propriety of introducing new agents for disease where the success rate is already 90 % as it was (at least as reported in academic centers) with ATRA plus chemotherapy. To some, failure to investigate new treatments removes the possibility of a success rate > 90%. To others, such investigation implies that the ends justify the means, a principle often considered antithetical to the practice of medicine. It is likely that the tension between these philosophies will increase as therapy for various diseases becomes more successful.

References

Adès L, Chevret S, De Botton S, Thomas X, Dombret H, Beve B, Sanz M, Guerci A, Miguel JS, Dela Serna J, Garo C, Stoppa AM, Reman O, Stamatoulas A, Fey M, Cahn JY, Sotto JJ, Bourhis JH, Parry A, Chomienne C, Degos L, Fenaux P; European APL Group (2005) Outcome of acute promyelocytic leukemia treated with all trans retinoic acid and chemotherapy in elderly patients: the European group experience. Leukemia 19(2):230–233

Adès L, Chevret S, Raffoux E, de Botton S, Guerci A, Pigneux A, Stoppa AM, Lamy T, Rigal-Huguet F, Vekhoff A, Meyer-Monard S, Maloisel F, Deconinck E, Ferrant A, Thomas X, Fegueux N, Chomienne C, Dombret H, Degos L, Fenaux P; European Acute Promyelocytic Leukemia Group (2006) Is cytarabine useful in the treatment of acute promyelocytic leukemia? Results of a randomized trial from the European Acute Promyelocytic Leukemia Group. J Clin Oncol 24(36):5703–5710

Ades L, Raffoux E, Chevret S et al (2010) Is Ara-C required in the treatment of standard risk APL? Long term results of a randomized trial (APL 2000) from the French Belgian Swiss APL Group. Blood 116:3751–377

Altman JK, Rademaker A, Cull E, Weitner BB, Ofran Y, Rosenblat TL, Haidau A, Park JH, Ram S, Orsini J, Sandhu S, Catchatourian R, Trifilio SM, Adel NG, Frankfurt O, Mallios G, Deblasio T, Jurcic JG, Nimber S, Rowe JM, Douer D, Tallman MS (2011) Administration of all-trans retinoic acid (ATRA) to newly diagnosed patients (pts) with acute promyelocytic leukemia (APL) is delayed even at experienced centers and associated with an increased early death rate (EDR): a retrospective analysis of 205 pts. Blood 118:942

Au WY, Kumana CR, Kou M et al (2003) Oral arsenic trioxide in the treatment of relapsed acute promyelocytic leukemia. Blood 102:407–408

Estey EH, Giles FJ, Beran M, O'Brien S, Pierce SA, Faderl SH, Cortes JE, Kantarjian H (2002) Experience with gemtuzumab ozogamycin ("mylotarg") and all-trans retinoic acid in untreated acute promyelocytic leukemia. Blood 99(11):4222–4229

Estey E, Garcia-Manero G, Ferrajoli A, Faderl S, Verstovsek S, Jones D, Kantarjian H (2006) Use of all-trans retinoic acid plus arsenic trioxide as an alternative to chemotherapy in untreated acute promyelocytic leukemia. Blood 109(9):3469–3473

Ghavamzadeh A, Alimoghaddam K, Rostami S, Ghaffari SH, Jahani M, Iravani M, Mousavi SA, Bahar B, Jalili M (2011) Phase II study of single-agent arsenic trioxide for the front-line therapy of acute promyelocytic leukemia. J Clin Oncol 29(20):2753–2757

Latagliata R, Petti MC, Fenu S, Mancini M, Spiriti MA, Breccia M, Brunetti GA, Avvisati G, Lo Coco F, Mandelli F (2002) Therapy-related myelodysplastic syndrome-acute myelogenous leukemia in patients treated for acute promyelocytic leukemia: an emerging problem. Blood 99(3):822–824

Lehmann S, Ravn A, Carlsson L, Antunovic P, Deneberg S, Möllgård L, Derolf AR, Stockelberg D, Tidefelt U, Wahlin A, Wennström L, Höglund M, Juliusson G (2011) Continuing high early death rate in acute promyelocytic leukemia: a population-based report from the Swedish Adult Acute Leukemia Registry. Leukemia 25(7):1128–1134

Lobe I, Rigal-Huguet F, Vehhoff A, Desablens B, Bordessoule D, Mounier C, Ferrant A, Sanz M, Fey M, Chomienne C, Chevret S, Degos L, Fenaux P (2003) European APL group experience (Myelodysplastic syndrome after acute myeloid leukemia: the European APL group experience. Leukemia 17(8):1600–1604.

Lo-Coco F, Cimino G, Breccia M, Noguera NI, Diverio D, Finolezzi E, Pogliani EM, Di Bona E, Micalizzi C, Kropp M, Venditti A, Tafuri A, Mandelli F (2004) Gemtuzumab ozogamicin (Mylotarg) as a single agent for molecularly relapsed acute promyelocytic leukemia. Blood 104(7):1995–1999

Lo-Coco F, Avvisati G, Vignetti M, Breccia M, Gallo E, Rambaldi A, Paoloni F, Fioritoni G, Ferrara F, Specchia G, Cimino G, Diverio D, Borlenghi E, Martinelli G, Di Raimondo F, Di Bona E, Fazi P, Peta A, Bosi A, Carella AM, Fabbiano F, Pogliani EM, Petti MC, Amadori S, Mandelli F; Italian GIMEMA Cooperative Group (2010) Front-line treatment of acute promyelocytic leukemia with AIDA induction followed by risk-adapted consolidation for adults younger than 61 years: results of the AIDA-2000 trial of the GIMEMA Group. Blood 116(17):3171–3179

Mathews V, George B, Chendamarai E, Lakshmi KM, Desire S, Balasubramanian P, Viswabandya A, Thirugnanam R, Abraham A, Shaji RV, Srivastava A, Chandy M (2010) Single-agent arsenic trioxide in the treatment of newly diagnosed acute promyelocytic leukemia: long-term follow-up data. J Clin Oncol 28(24):3866–3871

Muindi J, Frankel SR, Miller WH Jr, Jakubowski A, Scheinberg DA, Young CW, Dmitrovsky E, Warrell RP Jr (1992) Continuous treatment with all-trans retinoic acid causes a progressive reduction in plasma drug concentrations: implications for relapse and retinoid "resistance" in patients with acute promyelocytic leukemia. Blood 79:299

Park JH, Qiao B, Panageas KS, Schymura MJ, Jurcic JG, Rosenblat TL, Altman JK, Douer D, Rowe JM, Tallman MS (2011) Early death rate in acute promyelocytic leukemia remains high despite all-trans retinoic acid. Blood 118(5):1248–1254

Powell BL, Moser B, Stock W et al (2010) Arsenic trioxide improves event-free and overall survival for adults with acute promyelocytic leukemia: North American Leukemia Intergroup Study C9710. Blood 116:3751–3757

Ravandi F, Estey EH, Cortes JE et al (2010) Phase II study of all-trans retinoic acid, arsenic trioxide, with or without gemtuzumab ozogamicin for the frontline therapy of patients with acute promyelocytic leukemia. Blood 116:472

Sanz MA, Vellenga E, Rayón C, Díaz-Mediavilla J, Rivas C, Amutio E, Arias J, Debén G, Novo A, Bergua J, de la Serna J, Bueno J, Negri S, Beltrán de Heredia JM, Martín G (2004) All-trans retinoic acid and anthracycline monochemotherapy for the treatment of elderly patients with acute promyelocytic leukemia. Blood 104(12):3490–3493

Sanz MA, Grimwade D, Tallman MS, Lowenberg B, Fenaux P, Estey EH, Naoe T, Lengfelder E, Büchner T, Döhner H, Burnett AK, Lo-Coco F (2009) Management of acute promyelocytic leukemia: recommendations from an expert panel on behalf of the European LeukemiaNet. Blood 113(9):1875–1891

Sanz MA, Montesinos P, Rayón C, Holowiecka A, de la Serna J, Milone G, de Lisa E, Brunet S, Rubio V, Ribera JM, Rivas C, Krsnik I, Bergua J, González J, Díaz-Mediavilla J, Rojas R, Manso F, Ossenkoppele G, González JD, Lowenberg B; PETHEMA and HOVON Groups (2010) Risk-adapted treatment of acute promyelocytic leukemia based on all-trans retinoic acid and anthracycline with addition of cytarabine in consolidation therapy for high-risk patients: further improvements in treatment outcome. Blood 115(25):5137–5146

Shen ZX, Shi ZZ, Fang J, Gu BW, Li JM, Zhu YM, Shi JY, Zheng PZ, Yan H, Liu YF, Chen Y, Shen Y, Wu W, Tang W, Waxman S, De Thé H, Wang ZY, Chen SJ, Chen Z (2004) All-trans retinoic acid/As2O3 combination yields a high quality remission and survival in newly diagnosed acute promyelocytic leukemia. Proc Natl Acad Sci U S A 101(15):5328–5335

Siu CW, Au WY, Yung C et al (2006) Effects of oral arsenic trioxide therapy on QT intervals in patients with acute promyelocytic leukemia: implications for long-term cardiac safety. Blood 108:103–106

Tsimberidou AM, Tirado-Gomez M, Andreeff M, O'Brien S, Kantarjian H, Keating M, Lopez-Berestein G, Estey E (2006) Single-agent liposomal all-trans retinoic acid can cure some patients with untreated acute promyelocytic leukemia: an update of The University of Texas M. D. Anderson Cancer Center Series. Leuk Lymphoma 47(6):1062–108

Chapter 25
Targeting PML-RARα with Retinoids

Eytan M. Stein and Martin S. Tallman

Abstract All-*trans* retinoic acid (ATRA), an oral retinoic acid derivative, revolutionized the treatment and prognosis of acute promyelocytic leukemia (APL) and anticipated the current age of targeted cancer therapy. The past 25 years have seen an international effort to design multicenter clinical trials that identify the optimal use of ATRA in patients with APL. Indeed, since ATRA is so effective, contemporary clinical trials aim to de-escalate conventional cytotoxic chemotherapy, as has been the case with other highly curable malignancies such as childhood acute lymphoblastic leukemia and Hodgkin's lymphoma. In this chapter, we review and reflect upon the seminal clinical trials that have led to the current adoption of ATRA as the standard of care for treating APL.

Keywords Acute promyelocytic leukemia · All-*trans* retinoic acid · Induction · Consolidation · Maintenance

25.1 Introduction

Acute promyelocytic leukemia (APL) is characterized by a reciprocal balanced translocation between chromosomes 15 and 17 that fuses the promyelocytic leukemia (PML) gene to the retinoic acid receptor alpha (RARα) gene and leads to a differentiation block at the promyelocyte stage of myeloid hematopoiesis. Until the introduction of the differentiating agent all-*trans* retinoic acid (ATRA), APL, which accounts for approximately 10% of acute myeloid leukemia (AML) in the USA, was one of the most fatal subtypes. Before the ATRA era, the treatment of APL

M. S. Tallman (✉)
Leukemia Service, Memorial Sloan-Kettering Cancer Center, Weill Cornell Medical Center, 1275 York Avenue, Box 380, New York, NY 10065, USA
e-mail: tallmanm@mskcc.org

E. M. Stein
Medical Oncology and Hematology, Memorial Sloan-Kettering Cancer Center, 1275 York Avenue, Box 380, New York, NY 10065, USA

M. Andreeff (ed.), *Targeted Therapy of Acute Myeloid Leukemia*, Current Cancer Research, DOI 10.1007/978-1-4939-1393-0_25

consisted of the combination of an anthracycline and cytarabine for induction and multiple cycles of intensive cytarabine-based consolidation, as used in other subtypes of AML. The recognition in the 1980s and 1990s that targeted therapy for APL was associated with remarkable changes in overall survival (OS) foreshadowed the explosion in targeted agents currently available in other hematologic malignancies and solid tumors.

25.2 Historical Overview: The Early Years

In 1980, the first report of ATRA as a differentiating agent for APL was described by Breitman and colleagues at the National Institutes of Health. Using HL-60 cells (initially thought to be an APL cell line), Breitman demonstrated that ATRA-induced terminal differentiation in more than 90 % of cultured cells (Breitman et al. 1980). This in vitro finding was repeated in multiple laboratory studies over the next 3 years and in 1983, Flynn and colleagues reported treatment of a patient with relapsed APL using an oral 13-*cis*-retinoic acid compound (Flynn et al. 1983). The patient, a 22-year-old man, initially presented with APL in 1980 and was treated with daunorubicin (DNR), cytarabine, and prednisone. He achieved a complete remission (CR), but relapsed 15 months after his initial therapy. His first salvage reinduction was attempted with DNR, cytarabine, and thioguanine without success and a subsequent reinduction regimen of high-dose cytarabine and m-AMSA was ineffective. A dose of 13-*cis*-retinoic acid at 100 mg/m^2/day was given as salvage therapy and, remarkably, the patient developed evidence of myeloid maturation with this vitamin A derivative. Although he eventually succumbed to disseminated candidal infection, the possibility that a relatively nontoxic therapy could treat relapsed acute leukemia ignited a worldwide interest among leukemia investigators.

In 1987, physicians in China published the first report describing the use of a *trans* (rather than *cis*) retinoic acid. Huang and colleagues described six patients with APL treated with ATRA and low-dose cytarabine (Huang et al. 1987) and updated this initial report in 1988 with a report of the treatment of 24 Chinese APL patients with ATRA (Huang et al. 1988). Sixteen of these patients were previously untreated and the remaining eight patients had relapsed or refractory disease. Remarkably, 23 of 24 of these patients given ATRA monotherapy at a dose of 45 mg/m^2/day, achieved either a partial remission (PR) or CR. While postinduction treatment varied among the responding patients, this initial report demonstrated the safety, tolerability, and efficacy of ATRA in a de novo as well as in a relapsed/refractory patient population. This study was followed by similar studies of small series of APL patients from France (Castaigne et al. 1990) and the USA (Warrell et al. 1991) that confirmed Huang's findings.

25.3 ATRA: Clinical Trials

25.3.1 What Is the Role of ATRA In Induction?

These initial reports of newly diagnosed and relapsed APL patients induced into remission with ATRA monotherapy led to the design of large, multicenter clinical trials that established ATRA as the standard of care for induction therapy in APL. For the first time, the treatment of AML included an agent directed at the apparent molecular abnormality responsible for the pathogenesis of the disease in all patients. The European APL 91 group conducted the first large randomized clinical trial of ATRA as remission induction therapy. Patients were randomized to induction chemotherapy—the standard of care at the time—or ATRA followed by induction chemotherapy (Fenaux et al. 1993). Patients were well balanced between the two treatment groups with 54 patients randomized to ATRA with chemotherapy and 47 randomized to chemotherapy alone. The ATRA group received ATRA at a dose of 45 mg/m^2/day until CR or day 90. Recognizing the apparent unusual sensitivity of leukemic promyelocytes to anthracyclines, consolidation consisted of DNR 60 mg/m^2/day for 3 days and cytarabine 200 mg/m^2/day for 7 days for two cycles followed by a final cycle of DNR 45 mg/m^2/day for 3 days and cytarabine 1 g/m^2 every 12 h for 4 days. ATRA was omitted from the induction regimen of the chemotherapy-only group.

Of the 54 patients assigned to receive ATRA, 91 % achieved CR while 81% of patients who received chemotherapy achieved CR—a difference which was statistically insignificant. Early death was relatively infrequent with five of the 54 patients in the ATRA group (9 %) and four of the 47 given chemotherapy (8.5 %) dying within 30 days of diagnosis. In the majority of patients (in both groups), early death was due to pulmonary or central nervous system hemorrhage. The 12-month event-free survival (EFS) was 79 % in the ATRA group and 50 % in the chemotherapy group. The estimated risk of relapse at 12 months was 19 % in the ATRA group and 40 % in the chemotherapy group. The 12-month OS was 91 % in the ATRA group and 80 % in the chemotherapy group, a difference that was statistically insignificant. The similar OS between groups was likely attributable to relapsed patients from the chemotherapy-only arm being salvaged with ATRA.

In 1995, the Japanese Adult Leukemia Study Group (JALSG) reported the results of the AML-92 study (Kanamaru et al. 1995). Patients with white blood cell counts (WBC) less than $3 \times 10^9/\mu l$ were induced with ATRA 45 mg/m^2/day until CR. Patients with WBC higher than $3 \times 10^9/\mu l$ were given ATRA in combination with DNR 40 mg/m^2/day for 3 days and behenoyl cytosine arabinoside (BHAC) 200 mg/m^2/day for 5 days. This was followed by three cycles of consolidation and six cycles of maintenance with multidrug chemotherapy. A historical cohort of patients from the JALSG AML-89 trial (where APL patients received chemotherapy alone) served as controls.

In the analysis, 89% of the 109 patients who received treatment with ATRA achieved CR compared with 70% of patients in the historical control group of the AML-89 trial who received intensive chemotherapy. This high rate of CR in the ATRA group was observed, whether patients received ATRA alone, ATRA with concomitant chemotherapy for an elevated WBC at diagnosis, or ATRA with later chemotherapy (in induction) for a rising WBC during initial treatment. The 23-month EFS in the AML-92 trial was 75% compared with 48% in the AML-89 trial. OS statistics were not provided.

In 1997, the Gruppo Italiano Malattie Ematologiche dell'Adulto (GIMEMA) reported the results of a multicenter, single-arm clinical trial (AIDA 0493) which used ATRA with concurrent Idarubicin for all patients during induction, followed by three cycles of multidrug consolidation chemotherapy (Mandelli et al. 1997). Patients who achieved a molecular CR after consolidation were randomized into one of four maintenance groups: chemotherapy with 6-MP and methotrexate, ATRA monotherapy, alternating 6-MP/methotrexate and ATRA, or observation. Those patients who did not achieve a molecular CR underwent allogeneic bone marrow transplantation.

Two hundred fifty-three patients were given ATRA at a dose of 45 mg/m^2/day until CR (for a maximum of 90 days) with idarubicin 12 mg/m^2 on days 2, 4, 6, and 8. In the per-protocol analysis, 95% of patients achieved a hematologic CR. The remaining 5% died during induction, with the majority of deaths from intracranial hemorrhage. The EFS at 2 years was 79% for all patients and OS was 87%. Of note, in this study, EFS was defined as an event after the achievement of CR, in distinction from the previously mentioned European APL-91 trial which considered failure to achieve CR as an "event."

The very long-term outcome and mature results of the GIMEMA AIDA-0493 protocol were reported in 2011 (Avvisati et al. 2011). From the time of the first report in 1997 to the second report in 2011, the protocol was amended in several ways. First, in the original protocol all patients were given three cycles of multi-agent consolidation chemotherapy. After 1997, patients older than age 60 were supposed to receive one cycle of consolidation and were removed from the study. However, some patients continued to receive the original AIDA protocol (three cycles of consolidation) because some centers chose not to follow the amendment to treatment. In addition, the maintenance arms containing observation and chemotherapy alone were dropped in 1997 and the treatment was amended so that patients were randomized to only the two ATRA-containing regimens: ATRA alone or ATRA alternating with chemotherapy.

While 995 patients were registered to the completed AIDA-0493 trial, only 807 patients were fully evaluable in the per-protocol analysis. Of these patients, 94.3% achieved a hematologic CR after induction and 5.45% died during induction. Primary resistance to ATRA was extremely uncommon with only two patients (0.25%) considered "ATRA resistant." Of those patients who died during induction, more than half succumbed to a hemorrhagic event. The 12-year EFS was 68.9% and the 12-year OS was 76.5%. Interestingly, disease-free survival (DFS) did not differ regardless of maintenance regimen suggesting that patients in molecular CR after consolidation do not derive benefit from maintenance treatment.

At the time GIMEMA was reporting the initial results of the AIDA protocol, the North American Intergroup published the results of a randomized trial using ATRA or chemotherapy as induction therapy. Patients were randomly assigned to receive ATRA 45 mg/m^2/day (in divided doses) until CR or 90 days (Tallman et al. 1997). Interestingly, patients were given hydroxyurea to lower the WBC to less than 10,000/µl before receiving ATRA. Those patients randomized to chemotherapy received DNR 45 mg/m^2/day for 3 days and cytarabine 100 mg/m^2/day for 7 days. Patients in the chemotherapy arm received a second cycle of DNR and cytarabine if they developed evidence of coagulopathy after cycle 1 or had a decreased fibrinogen with 50 % or greater abnormal promyelocytes in the bone marrow. Consolidation therapy was identical for both treatment arms and consisted of DNR 45 mg/m^2/day for 3 days and cytarabine 100 mg/m^2/day for 7 days for cycle 1 and 2 g/m^2 of cytarabine every 12 h for 4 days for cycle 2. Patients who were in CR after the two cycles of consolidation chemotherapy were rerandomized to receive maintenance therapy with 45 mg/m^2/day of ATRA for 1 year or observation.

Of the 346 evaluable patients, 70 % in the chemotherapy group and 73 % in the ATRA group achieved a CR (the actual CR rate was likely higher because a number of patients in a prolonged CR had no documented CR by marrow report and were therefore required to be counted as nonresponders in the intention to treat analysis). The number of induction deaths was similar in both groups with 14 and 11 % of patients, respectively, dying during induction. As in earlier trials, greater than 50 % of the induction deaths were attributable to bleeding. The 3-year DFS was 32 % for the chemotherapy arm and 67 % for the ATRA arm. The 3-year DFS of patients assigned to ATRA maintenance was 40 % in the observation arm and 65 % in the ATRA arm, regardless of induction regimen. When combining induction and maintenance therapies, patients who received ATRA during induction or consolidation (or both) had significantly fewer relapses than those who only received chemotherapy. Those patients who received ATRA during induction and consolidation had fewer relapses than the patients who received ATRA induction and observation or chemotherapy induction and ATRA maintenance. In the intention to treat analysis, the OS of patients assigned to induction chemotherapy was 50 % and was 67 % for those assigned to induction ATRA.

The long-term outcomes of these patients in the first North American Intergroup protocol were subsequently reported in 2002 (Tallman et al. 2002). The 5-year DFS was 29 % in the chemotherapy arm and 69 % in the ATRA arm. Five year OS was 45% in the chemotherapy arm and 69 % in the ATRA arm. When analyzed by both induction and maintenance arms, DFS was 16 % for patients given chemotherapy and observation, 47 % for patients given chemotherapy induction and ATRA maintenance, 55 % for ATRA induction followed by observation and 74 % for those patients randomized to ATRA for both induction and maintenance. Of the 23 patients with late relapses (after 2 years of continuous CR) only one patient had received ATRA in induction and ATRA maintenance. This study established the benefit of maintenance therapy in patients with APL in the ATRA era. This conclusion would be revisited a decade later as discussed below.

25.3.2 *Concurrent or Sequential ATRA in Induction?*

These early trials demonstrated the clear superiority of ATRA over chemotherapy when given as induction therapy. But aside from maintenance therapy, the studies differed over whether chemotherapy was given concurrent with ATRA or sequentially after ATRA. To address if concurrent ATRA/chemotherapy was better than sequential administration, the European APL Group designed the APL-93 trial, a randomized trial of ATRA with concurrent chemotherapy (ATRA 45 mg/m^2/day until CR or day 90 with DNR 60 mg/m^2 for 3 days and cytarabine 200 mg/m^2 for 7 days) or sequential chemotherapy with ATRA given before DNR and cytarabine. (Fenaux et al. 1999). However, only the lowest risk patients who presented with a white blood count of less than 5000/μl and were less than age 66 were randomized. Patients older than age 66 or who had a WBC greater than 5000/μl at the time of diagnosis received concurrent chemotherapy and ATRA. In addition, patients who developed an elevated WBC in the sequential group were immediately started on chemotherapy. Following induction, patients who achieved a CR received two cycles of consolidation with DNR 45 mg/m^2/days for 3 days and cytarabine 1 g/m^2 every 12 h for 4 days. After consolidation, patients who remained in CR were randomized to observation, maintenance with intermittent ATRA, continuous chemotherapy with 6-MP and methotrexate, or a combination of ATRA, 6-MP, and methotrexate.

The overall CR rate in all patients was 92 % and it did not differ significantly between the sequential versus concurrent chemotherapy groups. Early death occurred in 7 % of all patients. The EFS was 84 % in the concurrent therapy group and 77 % in the sequential therapy group, a difference that was not statistically significant. OS was 84 versus 81%, respectively, and was also statistically insignificant. When maintenance was analyzed, patients who received both chemotherapy and ATRA had fewer relapses than patients in the other maintenance groups.

With longer, 10-year follow-up, the outcomes of patients treated with sequential versus concurrent ATRA differed. The EFS was 64.4% in those treated sequentially while those who received concurrent ATRA/chemotherapy had an EFS of 76.3 %. The 10-year OS was statistically similar with 81.8 % of patients alive in the sequential group compared with 85 % in the concurrent therapy group.

In summary, the studies reported up until 1999 demonstrated unequivocally that the use of ATRA during induction resulted in improved EFS and OS compared to chemotherapy alone. In addition, EFS was improved with the use of concurrent rather than sequential ATRA and chemotherapy as in the European APL-93 trial. The use of concurrent ATRA/chemotherapy in the GIMEMA-0493 trial may explain the improved outcomes seen when compared to the North American Intergroup study. However, a number of issues remained unresolved. First, was maintenance a necessary part of therapy? While the APL-93 and North American Intergroup trials suggested that it was, the AIDA-0493 regimen did not show a benefit. Second, was there a role for ATRA in consolidation? Would the benefit of differentiating therapy with ATRA be sustained with long follow-up? Finally, could the amount of cytotoxic chemotherapy in consolidation be reduced without compromising patient outcomes?

25.3.3 During Consolidation, Is Monochemotherapy As Effective As Polychemotherapy?

The Spanish Programa para el Estudio de la Terapéutica en Hemopatía Maligna (PETHEMA) group sought to modify the intensity of consolidation chemotherapy used in the AIDA-0493 protocol in their LPA-96 trial (Sanz et al. 1999) by attempting to exploit the unusual sensitivity of leukemic promyelocytes to anthracyclines, an observation made by Bernard and colleagues three decades earlier (Bernard et al. 1973). 123 patients were given the AIDA-induction protocol of ATRA 45 mg/m^2/day until CR or day 90 in combination with idarubicin 12 mg/m^2 on days 2, 4, 6, and 8. However, consolidation treatment differed from AIDA and consisted of three cycles of single agent chemotherapy with idarubicin 5 mg/m^2 daily for 4 days (course one), mitoxantrone 10 mg/m^2 daily for 5 days (course two), and idarubicin 12 mg/m^2 for 1 day (course three) without any cytarabine. After consolidation, patients who were molecularly negative for the PML-RARα gene product were given 2 years of maintenance therapy with ATRA, 6-MP, and methotrexate.

Of the 123 patients eligible for analysis, 89 % achieved a hematologic CR. The majority of the induction failures, as seen in previous trials, were secondary to mortality related to the typical bleeding diathesis. The 2-year EFS and OS were 79 and 83 %, respectively, suggesting that less-intensive chemotherapy during consolidation while using an AIDA-0493 induction schema produced similar outcomes to intensive consolidation.

25.3.4 What Is the Role of ATRA in Consolidation? Can Therapy Be Risk Adapted?

While previous studies had explored chemotherapy in induction and maintenance, the possibility of using ATRA during consolidation had not been examined in a large clinical trial. Because of this the PETHEMA group designed a successor study to LPA 96, designated LPA 99, that introduced the use of ATRA during consolidation to those patients with intermediate- or high- risk disease (low-risk patients having a WBC less than 10,000/μl and platelet count greater than 40,000/μl—others were designated intermediate or high risk; Sanz et al. 2004). In addition to using ATRA in patients with intermediate- or high-risk disease, consolidation chemotherapy was intensified in the intermediate- and high-risk groups; cycle 1 of consolidation used idarubicin 7 mg/m^2 (as opposed to 5 mg/m^2) for 4 days and cycle 3 of chemotherapy used idarubicin 12 mg/m^2 for 2 (rather than 1) days. Results were compared to the historical group of patients who had been enrolled on the LPA-96 trial.

The CR rate for patients in LPA 99 was 90 %, which was the same as that seen in the LPA-96 trial. Intriguingly, the DFS rate for intermediate- and high-risk patients in LPA 99 was 90 % compared to 77 % in LPA 96. The 3-year OS was 78 % in the LPA-96 trial and 85 % in the LPA-99 trial, which was not statistically significant. However, when restricting the analysis to the intermediate- and high-risk groups,

OS was 86% in the LPA-99 trial compared to 73% in the LPA-96 trial. In an updated analysis published in 2008, the 5-year OS in the LPA-99 trial was 82% (Sanz et al. 2008). When only the intermediate- and high-risk groups were analyzed, the 5-year OS was 81%. This trial was among the first to demonstrate that risk-adapted therapy was feasible and effective in patients with high-risk disease.

25.3.5 Is Cytarabine Necessary for the Treatment of APL?

While the role of ATRA in consolidation was being evaluated, the European APL group was investigating cytarabine in consolidation in APL (Ades et al. 2006). As previously noted, both the European APL-91 and APL-93 trials incorporated the use of DNR and cytarabine in consolidation, while the PETHEMA LPA-99 trial sought to eliminate cytarabine from induction and consolidation regimens. In order to test whether cytarabine could be eliminated from the treatment of APL, the European APL group designed a trial to determine the need for cytarabine in patients with low-risk disease. Less than 60 patients, with low-risk disease were randomized to the "standard" arm of ATRA, DNR, and cytarabine for induction followed by two cycles of consolidation with DNR and cytarabine, versus the investigational arm that eliminated cytarabine from consolidation. Of note, unlike the LPA-99 trial, ATRA was not given during consolidation cycles. Patients with high-risk disease were not randomized.

Of the 356 patients enrolled on the trial, 196 patients were low-risk and less than age 60 and were randomized to cytarabine versus no cytarabine. While the hematologic CR rates after induction and molecular CR rates after consolidation were statistically equivalent in both treatment arms, the cumulative incidence of relapse (CIR) was significantly higher in the group not given cytarabine; the 2-year CIR was 15.9% in the no-cytarabine arm versus 4.7% in the group given cytarabine. In addition, the inclusion of cytarabine in this low-risk group of patients did not lead to a greater number of deaths as the OS of patients in the cytarabine arm at 2 years was 97.9 versus 89.6% in the no-cytarabine arm. While the results of this trial suggested that cytarabine is a necessary part of anti-APL therapy even for patients with low-risk disease, the absence of ATRA use in consolidation, as in the LPA-99 trial, made it difficult to interpret the results. Was cytarabine needed if ATRA was used in consolidation? Is it possible that cytarabine and ATRA in combination were only needed for high-risk disease?

To help determine the optimal role of cytarabine in APL, the PETHEMA group in conjunction with the Dutch HOVON group designed the LPA-05 trial (Sanz et al. 2010). The design of this trial was similar to LPA 99, but patients with high-risk disease were given cytarabine in consolidation with results compared to the historical control group of high-risk patients in the LPA-99 trial. While the CIR was higher in the historical control group with high-risk disease (14% in LPA 05 vs. 27% in LPA

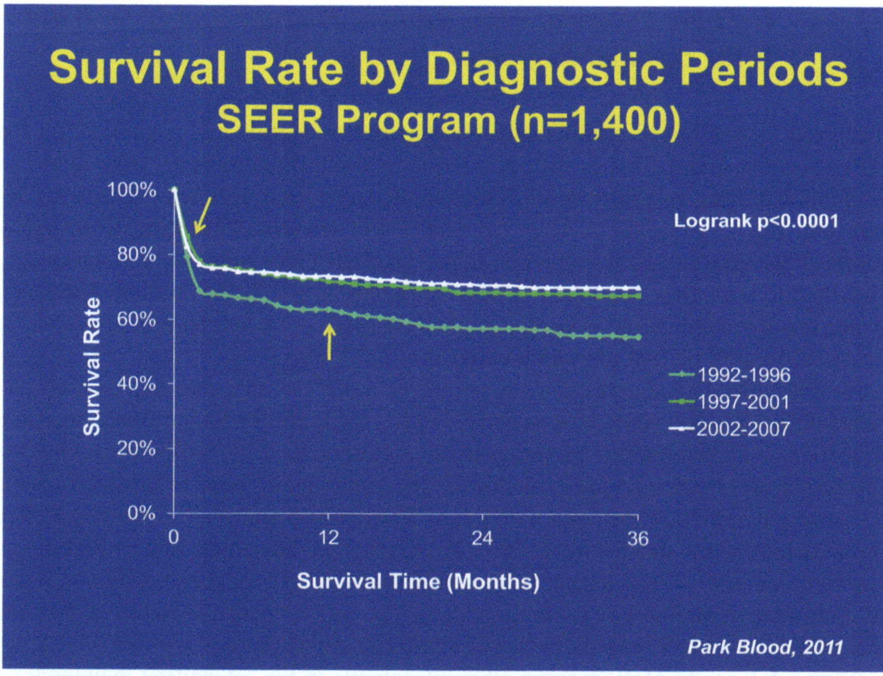

Fig. 25.1 Survival Rate by Diagnostic Periods SEER Program (n=1400). (Park 2011)

99 at 4 years), the DFS and OS were statistically equivalent, suggesting that patients who did relapse could be salvaged with subsequent therapy.

The GIMEMA group in the AIDA-2000 trial, published in 2010, sought to answer a similar question as the PETHEMA group, specifically whether cytarabine could be eliminated from consolidation in low- and intermediate-risk patients and whether introducing ATRA during consolidation was an effective treatment strategy.(Lo-Coco et al. 2010). In this study, patients received standard induction chemotherapy as given in the AIDA-0493 regimen (ATRA and idarubicin). Consolidation in the low- and intermediate-risk groups contained three cycles of monochemotherapy combined with ATRA while in the high-risk group patients received three cycles of polychemotherapy (with cytarabine in cycles 1 and 3) with ATRA. Results were compared to the historical controls of the AIDA-0493 trial.

As would be expected from results of prior trials, the outcomes in the AIDA-2000 groups from both low-risk and intermediate/high-risk patients were improved over the AIDA-0493 trials. The 6-year DFS in the overall study cohort was 85.6% in AIDA 2000 versus 69.5% in AIDA 0493. The OS was 87.4% in AIDA 2000 versus 78.1% in AIDA 0493. Much of this was driven by improved outcomes in the high-risk group with DFS of 84.5% in AIDA 2000 and 49.6% in AIDA 0493 and OS of 83.4% in AIDA 2000 and 61.3% in AIDA 0493 (Fig. 25.1).

25.3.6 Is There a Role of Maintenance Therapy and If So What Is the Optimal Regimen?

The need for prolonged maintenance therapy after induction and consolidation for APL is controversial. This controversy stems from conflicting results from randomized clinical trials. In addition, all of the maintenance trials enrolled patients in CR, but some trials, before the widespread adoption of sensitive PCR tests, used hematologic CR to define remission while others used the more sensitive molecular CR. Finally, the induction and consolidation regimens in trials that evaluated maintenance therapy were varied making it difficult to draw any between-trial conclusions about the overall benefit of maintenance.

In 2007, the JALSG reported the results of a clinical trial that randomized patients with APL who were PCR negative for the PML-RARα fusion transcript after induction and consolidation to "intensified" maintenance with combination chemotherapy or observation (Asou et al. 2007). The induction regimen in this trial used a risk-adapted approach: patients with a low WBC (less than 3×10^9) received ATRA monotherapy while those with a higher WBC received idarubicin and cytarabine in combination with ATRA during induction. Consolidation consisted of three cycles of multidrug chemotherapy (without ATRA). Patients randomized to the intensified maintenance arm received six cycles of multidrug chemotherapy without ATRA. At a median follow-up of 49 months, 28 % of patients in the intensified maintenance group had relapsed and 15 % died as compared to only 20 % of patients relapsing in the observation group and only 3 % dying. The results of this study suggested that intensified maintenance after induction with ATRA and consolidation with combination chemotherapy may be harmful.

As discussed earlier, one question addressed by the North American Intergroup was the role of maintenance ATRA after achieving CR. In the long-term outcome data published in 2002, there was a distinct DFS advantage for patients assigned to ATRA maintenance; DFS in the ATRA maintenance group (regardless of induction regimen) was 61 % and was only 36 % in the observation arm (Tallman et al. 2002). Those patients who received both ATRA in induction and ATRA maintenance had a DFS of 74 % compared to a DFS of only 16% for those who received chemotherapy and observation.

In the European APL trial, patients who were in hematologic CR after consolidation were randomized to one of four maintenance arms: observation, ATRA (45 mg/m^2/day for 15 days once every 3 months), chemotherapy (methotrexate 15 mg/m^2/week and 6-MP 90 mg/m^2/day), or ATRA with chemotherapy for a total of 2 years with chemotherapy dose modifications based on blood counts. In the final analysis, the 10-year EFS and OS in the concurrent chemotherapy/ATRA and chemotherapy-alone maintenance groups were significantly better than the results in the ATRA and observation arms. The CIR was lowest in the ATRA/chemotherapy group (13.4 %) followed by the chemotherapy group (23.4 %), ATRA group (33.0 %), and observation group (43.2 %).

The AIDA-0493 trial used a similar maintenance scheme as the European APL trial, but only enrolled patients who were in a molecular CR at the end of consolidation (Avvisati et al. 2011). Interestingly, this trial saw no benefit to the use of ATRA maintenance for those patients in molecular CR at the end of consolidation, contradicting the results of the previous European and North American Intergroup trials. Despite the negative results, subsequent, modern trials have adopted the widespread use of maintenance with ATRA, 6-MP, and methotrexate after consolidation. In addition, the utility of maintenance therapy in the era when ATRA is given during consolidation has never been prospectively evaluated.

25.3.7 What Is the Appropriate Dose of ATRA?

For adults, the currently recommended dose of ATRA is 45 mg/m^2/day. However, there are data to suggest that lowering the dose of ATRA does not decrease efficacy. In 1993, Castaigne and colleagues published the results of a multicenter study using ATRA at a dose of 25 mg/m^2/day in 30 patients with APL, 12 who were previously untreated, 14 in first relapse, and four who had not achieved CR with induction chemotherapy alone (Castaigne et al. 1993). Twenty-four of 30 patients achieved a CR. In a similar study published by Shen and colleagues in 1996, 27 patients with de novo APL were treated with ATRA at a dose of 15–20 mg/m^2/day (Chen et al. 1996). 92 % of these patients achieved a CR. These results suggest that in patients who are having intolerable side effects from ATRA it may be possible to lower the dose without a decrease in efficacy. Indeed, pediatric patients are now routinely treated with a dose of 25 mg/m^2 day with survival rates that rival those seen in adults.

25.4 How Have Contemporary Treatment Regimens Affected the Early-Death Rate in APL?

The introduction of ATRA into induction regimens clearly improves the outcomes of patients with APL by preventing relapse but, in randomized clinical trials, not by decreasing the incidence of early death which is attributable to hemorrhage in more than 50% of patients in most series. In most trials performed after the introduction of ATRA, the early-death rate (defined as death within 30 days) ranges between 5 and 10 % with approximately half of those deaths attributable to hemorrhage.

For unclear reasons, the "real-world" early-death rate outside of clinical trials continues to be markedly elevated. This has been demonstrated in a number of publications analyzing population-based databases of patients with APL. Using the United States Surveillance Epidemiology and End Results (SEER) database, Park and colleagues observed an overall early-death rate of 17.3 % between 1992

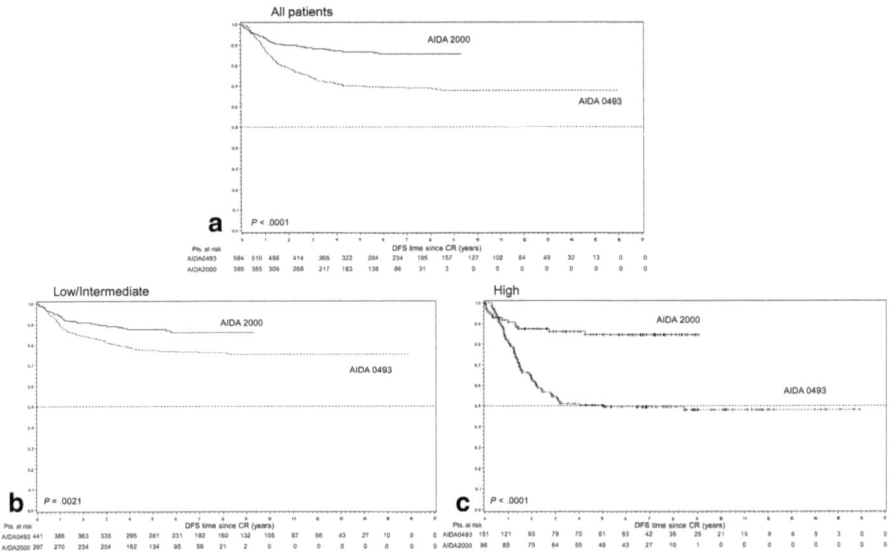

Fig. 25.2 AIDA 2000—DFS according to protocol. (Lo-Coco et al. 2010)

and 2007 (Fig. 25.2; Park et al. 2011). Interestingly, the early-death rate between 2001 and 2007, when ATRA had been introduced into clinical practice, was 17.5 %, and between 1992 and 1995, when ATRA was only available on clinical trials, was 22.1 %. Patients older than 54 years fared particularly poorly with an early-death rate of 24.2 %. An elevated early-death rate was also seen in a review of a Swedish National Database that found an early-death rate of 29 % for patients diagnosed with APL between 1997 and 2006 (Lehmann et al. 2011). Forty-one percent of these patients with early death died from a hemorrhagic event. Similarly, a retrospective review of patients with APL treated at the Stanford Hospital between 1997 and 2009 demonstrated an early-death rate of 26 % (McClellan et al. 2012).

Therefore, although the rate of death from relapsed disease has dramatically decreased in the ATRA era, the early-death rate has remained stubbornly elevated when looking at population-based registries of all patients diagnosed with APL. The reasons for this remain unclear and are an area of active clinical investigation.

25.5 What Are Toxicities Associated with ATRA?

Although the toxicities of ATRA are relatively minimal, the use of ATRA can lead to the APL differentiation syndrome, characterized by noncardiogenic pulmonary edema, weight gain, and peripheral edema. Systemic steroids appear to treat this process and should be initiated at the first suspicion of the APL differentiation syndrome.

Other toxicities seen with ATRA include elevated transaminases, hypercholesterolemia and hypertriglyceridemia and, in children, pseudotumor cerebri. These effects are generally transient and resolve with ATRA discontinuation.

25.6 Are the Benefits of Differentiating Therapy with ATRA Sustained?

The remarkable short-term outcomes seen with ATRA therapy during induction appear to be sustained over the long term. As mentioned earlier, both the European APL-93 trial and the AIDA-0493 trial have reported the very long-term outcomes of ATRA induction. In the APL-93 trial, 81.8 % of the patients in the sequential arm, and 85 % of patients in the concurrent arm were alive at 10 years (Ades et al. 2010). At the 12-year follow-up of the AIDA-0493 trial, 76.5 % of the patients remained alive (Avvisati et al. 2011). The very long-term outcomes of the first North American Intergroup protocol were presented in abstract form in 2011 (Douer et al. 2011). The 12-year OS rate for patients induced with chemotherapy was 42% while those induced with ATRA had an OS of 60 %. Of the 245 patients who received ATRA at any time (induction, maintenance, or both), 195 (80 %) achieved CR and 69 are known to have relapsed: Nine (4.6 %) patients had a late relapse (after 3 years), all of whom had received ATRA at some time during treatment, and 60 (31 %) patients relapsed early. Taken together, these trials strongly suggest that ATRA treatment results in a prolonged and durable response, with few late relapses and no evidence of late ATRA toxicity.

25.7 The Modern Use of ATRA

Based on the well-designed clinical trials discussed in the preceding sections, ATRA has been universally adopted and incorporated into practice guidelines for the treatment of APL (Table 25.1). In the USA, the National Comprehensive Cancer Network has included the most recent GIMEMA, PETHEMA, and European APL protocols, with adapted, risk-stratified therapy, as treatment options for newly diagnosed APL. In addition, the general practice among leukemia experts is to prescribe maintenance treatment with ATRA, 6-MP, and methotrexate to patients with high-risk disease for a total of 2 years. In fact, many clinicians prescribe maintenance therapy to all patients regardless of relapse risk. However, the necessity of maintenance in the modern era remains to be validated in a well-conducted randomized clinical trial. Indeed, the treatment of APL has become so effective with ATRA that modern clinical trials are attempting to remove chemotherapy from the treatment of APL and substitute arsenic trioxide (ATO—discussed in the next chapter) with minimal chemotherapy. While some trials have attempted to eliminate ATRA from treatment regimens and use ATO alone, the wisdom of this approach is questionable, given the remarkable efficacy of ATRA and its relatively safe side-effect profile.

Table 25.1 Contemporary clinical trials using ATRA for the treatment of APL

Group	Publication date	Number of patients	Treatment regimen	CR	Early death	EFS/DFS	Overall Survival
European APL-91 Group	December 1993	101	*Induction:* one cycle of DNR 60 mg/m^2 ′ 3 days and Ara-C 200 mg/m^2 ′ 7 days vs. ATRA 45 mg/m^2 until CR (or 90 days) → DNR 60 mg/m^2 ′ 3 days and ARA-C 200 mg/m^2 ′ 7 days *Consolidation:* DNR and ARA-C for two cycles *Maintenance:* none	81 % 91 %	9 % 8 %	50 % (1 year) 79 % (1 year)	80 % (1 year) 91 % (1 year)
JALSG	March 1995	110	*Induction:* ATRA 45 mg/m^2 and DNR 40 mg/m^2 ′ 3 days with BHOC 200 mg/m^2 ′ 5 days for white count greater than 3000/µl. *Consolidation:* three cycles of multidrug chemotherapy *Maintenance:* six cycles of multidrug chemotherapy	89 %	–	75 %	–
GIMEMA AIDA 0493	August 1997	253	*Induction:* ATRA 45 mg/m^2/day until CR (or 90 days) and idarubicin 12 mg/m^2 days 2,4,6,8 *Consolidation:* three cycles of multidrug chemotherapy *Maintenance:* Arm A: 6MP and MTX Arm B: ATRA Arm C: chemo alternating with ATRA Arm D: observation	94.3 %	5.45 %	68.9 % (12 years)	76.5 % (12 years)
North American Intergroup	October 1997	401	*Induction:* DNR 45 mg/m^2 days 1–3 and Ara-C 100 mg/m^2 days 1–7 vs. ATRA 45 mg/m^2 until CR (or 90 days) *Consolidation:* two cycles of chemotherapy *Maintenance:* Arm A: ATRA Arm B: observation	70% 73%	14% 11%	29 %(5 years) 69 %(5 years)	45 %(5 years) 69 %(5 years)

Table 25.1 (continued)

Group	Publication date	Number of patients	Treatment regimen	CR	Early death	EFS/DFS	Overall Survival
European APL Group APL-93	August 1999		*Induction:* ATRA 45 mg/m^2 à DNR 60 $mg/m^2/d$ and AraC 200 $mg/m^2/d$ for 7 days vs. ATRA 45 mg/m^2 with DNR 60 $mg/m^2/d$ and AraC 200 $mg/m^2/d$ for 7 days *Consolidation:* DNR 45 $mg/m^2/d$ for 3 days and AraC 1 g/m^2 AraC every 12 h for 4 days. *Maintenance:* Arm A: 6MP and MTX Arm B: ATRA Arm C: Chemo alternating with ATRA Arm D: Observation	92.6 % 96.2 %		64.4 % (10 years) 73 %(10 year)	81.8 % (10 years) 85 % (10 years)
PETHEMA LPA 96	November 1999		*Induction:* ATRA 45 mg/m^2 and IDA 12 mg/m^2 days 2,4,6,8 *Consolidation*: three cycles of single agent chemotherapy *Maintenance:* ATRA/6-MP/methotrexate x 2 years	89 %	9.7 %	79 % (2 years)	82 % (2 years)
PETHEMA LPA 99	Feb 2004		*Induction:* ATRA 45 mg/m^2 and IDA 12 mg/m^2 days 2,4,6,8 *Consolidation:* three cycles of single agent chemotherapy with ATRA for intermediate/high-risk patients *Maintenance:* ATRA/6-MP/methotrexate x 2 years	91 %	10.6 %	84 % (5 years)	82 % (5 years)
European APL Group	December 2006		*Induction:* ATRA 45 $mg/m^2/d$ with DNR 60 $mg/m^2/d$ x 3 days and Ara-C 200 $mg/m^2/d$ x 7 days. vs. ATRA 45 $mg/m^2/d$ with DNR 60 $mg/m^2/d$ x 3 days *Consolidation:* DNR/AraC x two cycles vs. DNR x two cycles *Maintenance:* ATRA/6-MP/methotrexate x 2 years	99 % 94 %	1 % 3.9 %	93.3 % (2 years) 77.2 % (2 years)	97.9 % (2 years) 89.6 % (2 years)

Table 25.1 (continued)

Group	Publication date	Number of patients	Treatment regimen	CR	Early death	EFS/DFS	Overall Survival
PETHEMA and HOVON LPA05	April 2010		*Induction:* ATRA 45 mg/m^2 and IDA 12 mg/m^2 days 2,4,6,8 *Consolidation:* *Low risk:* ATRA/single agent chemotherapy *Intermed risk:* ATRA/single agent chemotherapy *High risk:* ATRA/single agent chemo/Ara-C *Maintenance:* ATRA/6-MP/methotrexate x 2 years	92.5 %	7.4 %	90 % (4 years) 93 % (4 years) 92 % (4 years) 82 % (4 years)	88 % (4 years) 96 % (4 years) 91 % (4 years) 79 % (4 years)
GIMEMA AIDA 2000	July 2010		*Induction:* ATRA 45 mg/m^2 and IDA 12 mg/m^2 days 2,4,6,8 *Consolidation:* *Low/intermed risk:* ATRA with single-agent chemotherapy *High risk:* ATRA + multi-agent chemo and Ara-C *Maintenance:* ATRA/6-MP/methotrexate x 2 years	94.4 %	5.6 %	85.6 % (6 years) 85.9 % (6 years) 84.5 % (6 years)	87.4% (6 years) 89.1%(6 years) 83.4% (6 years)

Conclusion

The discovery of ATRA has led to the remarkable reversal of APL from one of the most fatal or most curable subtype of acute leukemia in adults. It is gratifying that besides being remarkably effective, ATRA is well tolerated by patients with few toxic side effects, and is given orally, which allows for ease of administration and patient satisfaction. The ability to effectively incorporate ATRA was only achieved because of an international effort to conduct large well-designed clinical trials.

References

Ades L, Chevret S, Raffoux E, de Botton S, Guerci A, Pigneux A, Stoppa AM, Lamy T, Rigal-Huguet F, Vekhoff A, Meyer-Monard S, Maloisel F, Deconinck E, Ferrant A, Thomas X, Fegueux N, Chomienne C, Dombret H, Degos L, Fenaux P (2006) Is cytarabine useful in the treatment of acute promyelocytic leukemia? Results of a randomized trial from the European Acute Promyelocytic Leukemia Group. J Clin Oncol Res 24(36):5703–5710. doi:10.1200/JCO.2006.08.1596

Ades L, Guerci A, Raffoux E, Sanz M, Chevallier P, Lapusan S, Recher C, Thomas X, Rayon C, Castaigne S, Tournilhac O, de Botton S, Ifrah N, Cahn JY, Solary E, Gardin C, Fegeux N, Bordessoule D, Ferrant A, Meyer-Monard S, Vey N, Dombret H, Degos L, Chevret S, Fenaux P (2010) Very long-term outcome of acute promyelocytic leukemia after treatment with all-trans retinoic acid and chemotherapy: the European APL Group experience. Blood 115(9):1690–1696. doi:10.1182/blood-2009-07-233387

Asou N, Kishimoto Y, Kiyoi H, Okada M, Kawai Y, Tsuzuki M, Horikawa K, Matsuda M, Shinagawa K, Kobayashi T, Ohtake S, Nishimura M, Takahashi M, Yagasaki F, Takeshita A, Kimura Y, Iwanaga M, Naoe T, Ohno R (2007) A randomized study with or without intensified maintenance chemotherapy in patients with acute promyelocytic leukemia who have become negative for PML-RARalpha transcript after consolidation therapy: the Japan Adult Leukemia Study Group (JALSG) APL97 study. Blood 110(1):59–66. doi:10.1182/blood-2006-08-043992

Avvisati G, Lo-Coco F, Paoloni FP, Petti MC, Diverio D, Vignetti M, Latagliata R, Specchia G, Baccarani M, Di Bona E, Fioritoni G, Marmont F, Rambaldi A, Di Raimondo F, Kropp MG, Pizzolo G, Pogliani EM, Rossi G, Cantore N, Nobile F, Gabbas A, Ferrara F, Fazi P, Amadori S, Mandelli F (2011) AIDA 0493 protocol for newly diagnosed acute promyelocytic leukemia: very long-term results and role of maintenance. Blood 117(18):4716–4725. doi:10.1182/blood-2010-08-302950

Bernard J, Weil M, Boiron M, Jacquillat C, Flandrin G, Gemon MF (1973) Acute promyelocytic leukemia: results of treatment by daunorubicin. Blood 41(4):489–496

Breitman TR, Selonick SE, Collins SJ (1980) Induction of differentiation of the human promyelocytic leukemia cell line (HL-60) by retinoic acid. PNAS 77(5):2936–2940

Castaigne S, Chomienne C, Daniel MT, Ballerini P, Berger R, Fenaux P, Degos L (1990) All-trans retinoic acid as a differentiation therapy for acute promyelocytic leukemia. I. Clinical results. Blood 76(9):1704–1709

Castaigne S, Lefebvre P, Chomienne C, Suc E, Rigal-Huguet F, Gardin C, Delmer A, Archimbaud E, Tilly H, Janvier M et al (1993) Effectiveness and pharmacokinetics of low-dose all-trans retinoic acid (25 mg/m2) in acute promyelocytic leukemia. Blood 82(12):3560–3563

Chen GQ, Shen ZX, Wu F, Han JY, Miao JM, Zhong HJ, Li XS, Zhao JQ, Zhu J, Fang ZW, Chen SJ, Chen Z, Wang ZY (1996) Pharmacokinetics and efficacy of low-dose all-trans retinoic acid in the treatment of acute promyelocytic leukemia. Leukemia : official journal of the Leukemia Society of America, Leukemia Research Fund, UK 10(5):825–828

Douer D, Zickl L, Schiffer CA, Appelbaum FR, Feusner JH, Shepherd LE, Willman CL, Bloomfield CD, Rowe JM, Wiernik PH, Tallman MS (2011) Late Relapses Following All-Trans Retinoic Acid for Acute Promyelocytic Leukemia Are Uncommon, Respond Well to Salvage Therapy and Occur Independently of Prognostic Factors At Diagnosis: Long-Term Follow-up of North American Intergroup Study I0129. ASH Annual Meeting Abstracts 118(21):83

Fenaux P, Chastang C, Chevret S, Sanz M, Dombret H, Archimbaud E, Fey M, Rayon C, Huguet F, Sotto JJ, Gardin C, Makhoul PC, Travade P, Solary E, Fegueux N, Bordessoule D, Miguel JS, Link H, Desablens B, Stamatoullas A, Deconinck E, Maloisel F, Castaigne S, Preudhomme C, Degos L (1999) A randomized comparison of all transretinoic acid (ATRA) followed by chemotherapy and ATRA plus chemotherapy and the role of maintenance therapy in newly diagnosed acute promyelocytic leukemia. The European APL Group. Blood 94(4):1192–1200

Fenaux P, Le Deley MC, Castaigne S, Archimbaud E, Chomienne C, Link H, Guerci A, Duarte M, Daniel MT, Bowen D et al (1993) Effect of all transretinoic acid in newly diagnosed acute promyelocytic leukemia. Results of a multicenter randomized trial. European APL 91 Group. Blood 82(11):3241–3249

Flynn PJ, Miller WJ, Weisdorf DJ, Arthur DC, Brunning R, Branda RF (1983) Retinoic acid treatment of acute promyelocytic leukemia: in vitro and in vivo observations. Blood 62(6):1211–1217

Huang ME, Ye YC, Chen SR, Zhao JC, Gu LJ, Cai JR, Zhao L, Xie JX, Shen ZX, Wang ZY (1987) All-trans retinoic acid with or without low dose cytosine arabinoside in acute promyelocytic leukemia. Report of 6 cases. JCMA 100(12):949–953

Huang ME, Ye YC, Chen SR, Chai JR, Lu JX, Zhoa L, Gu LJ, Wang ZY (1988) Use of all-trans retinoic acid in the treatment of acute promyelocytic leukemia. Blood 72(2):567–572

Kanamaru A, Takemoto Y, Tanimoto M, Murakami H, Asou N, Kobayashi T, Kuriyama K, Ohmoto E, Sakamaki H, Tsubaki K et al (1995) All-trans retinoic acid for the treatment of newly diagnosed acute promyelocytic leukemia. Japan Adult Leukemia Study Group. Blood 85(5):1202–1206

Lehmann S, Ravn A, Carlsson L, Antunovic P, Deneberg S, Mollgard L, Derolf AR, Stockelberg D, Tidefelt U, Wahlin A, Wennstrom L, Hoglund M, Juliusson G (2011) Continuing high early death rate in acute promyelocytic leukemia: a population-based report from the Swedish Adult Acute Leukemia Registry. Leukemia : official journal of the Leukemia Society of America, Leukemia Research Fund. UK 25(7):1128–1134. doi:10.1038/leu.2011.78

Lo-Coco F, Avvisati G, Vignetti M, Breccia M, Gallo E, Rambaldi A, Paoloni F, Fioritoni G, Ferrara F, Specchia G, Cimino G, Diverio D, Borlenghi E, Martinelli G, Di Raimondo F, Di Bona E, Fazi P, Peta A, Bosi A, Carella AM, Fabbiano F, Pogliani EM, Petti MC, Amadori S, Mandelli F (2010) Front-line treatment of acute promyelocytic leukemia with AIDA induction followed by risk-adapted consolidation for adults younger than 61 years: results of the AIDA-2000 trial of the GIMEMA Group. Blood 116(17):3171–3179. doi:10.1182/blood-2010-03-276196

Mandelli F, Diverio D, Avvisati G, Luciano A, Barbui T, Bernasconi C, Broccia G, Cerri R, Falda M, Fioritoni G, Leoni F, Liso V, Petti MC, Rodeghiero F, Saglio G, Vegna ML, Visani G, Jehn U, Willemze R, Muus P, Pelicci PG, Biondi A, Lo Coco F (1997) Molecular remission in PML/RAR alpha-positive acute promyelocytic leukemia by combined all-trans retinoic acid and idarubicin (AIDA) therapy. Gruppo Italiano-Malattie Ematologiche Maligne dell'Adulto and Associazione Italiana di Ematologia ed Oncologia Pediatrica Cooperative Groups. Blood 90(3):1014–1021

McClellan JS, Kohrt HE, Coutre S, Gotlib JR, Majeti R, Alizadeh AA, Medeiros BC (2012) Treatment advances have not improved the early death rate in acute promyelocytic leukemia. Haematologica 97(1):133–136. doi:10.3324/haematol.2011.046490

Park JH, Qiao B, Panageas KS, Schymura MJ, Jurcic JG, Rosenblat TL, Altman JK, Douer D, Rowe JM, Tallman MS (2011) Early death rate in acute promyelocytic leukemia remains high despite all-trans retinoic acid. Blood 118(5):1248–1254. doi:10.1182/blood-2011-04-346437

Sanz MA, Martin G, Rayon C, Esteve J, Gonzalez M, Diaz-Mediavilla J, Bolufer P, Barragan E, Terol MJ, Gonzalez JD, Colomer D, Chillon C, Rivas C, Gomez T, Ribera JM, Bornstein R, Roman J, Calasanz MJ, Arias J, Alvarez C, Ramos F, Deben G (1999) A modified AIDA protocol with anthracycline-based consolidation results in high antileukemic efficacy and reduced toxicity in newly diagnosed PML/RARalpha-positive acute promyelocytic leukemia. PETHEMA group. Blood 94(9):3015–3021

Sanz MA, Martin G, Gonzalez M, Leon A, Rayon C, Rivas C, Colomer D, Amutio E, Capote FJ, Milone GA, De La Serna J, Roman J, Barragan E, Bergua J, Escoda L, Parody R, Negri S, Calasanz MJ, Bolufer P (2004) Risk-adapted treatment of acute promyelocytic leukemia with all-trans-retinoic acid and anthracycline monochemotherapy: a multicenter study by the PETHEMA group. Blood 103(4):1237–1243. doi:10.1182/blood-2003-07-2462

Sanz MA, Montesinos P, Vellenga E, Rayon C, de la Serna J, Parody R, Bergua JM, Leon A, Negri S, Gonzalez M, Rivas C, Esteve J, Milone G, Gonzalez JD, Amutio E, Brunet S, Garcia-Larana J, Colomer D, Calasanz MJ, Chillon C, Barragan E, Bolufer P, Lowenberg B (2008) Risk-adapted treatment of acute promyelocytic leukemia with all-trans retinoic acid and anthracycline monochemotherapy: long-term outcome of the LPA 99 multicenter study by the PETHEMA Group. Blood 112(8):3130–3134. doi:10.1182/blood-2008-05-159632

Sanz MA, Montesinos P, Rayon C, Holowiecka A, de la Serna J, Milone G, de Lisa E, Brunet S, Rubio V, Ribera JM, Rivas C, Krsnik I, Bergua J, Gonzalez J, Diaz-Mediavilla J, Rojas R, Manso F, Ossenkoppele G, Gonzalez JD, Lowenberg B (2010) Risk-adapted treatment of acute promyelocytic leukemia based on all-trans retinoic acid and anthracycline with addition of cytarabine in consolidation therapy for high-risk patients: further improvements in treatment outcome. Blood 115(25):5137–5146. doi:10.1182/blood-2010-01-266007

Tallman MS, Andersen JW, Schiffer CA, Appelbaum FR, Feusner JH, Ogden A, Shepherd L, Willman C, Bloomfield CD, Rowe JM, Wiernik PH (1997) All-trans-retinoic acid in acute promyelocytic leukemia. NEJM 337(15):1021–1028. doi:10.1056/NEJM199710093371501

Tallman MS, Andersen JW, Schiffer CA, Appelbaum FR, Feusner JH, Woods WG, Ogden A, Weinstein H, Shepherd L, Willman C, Bloomfield CD, Rowe JM, Wiernik PH (2002) All-trans retinoic acid in acute promyelocytic leukemia: long-term outcome and prognostic factor analysis from the North American Intergroup protocol. Blood 100(13):4298–4302. doi:10.1182/blood-2002-02-0632

Warrell RP, Jr., Frankel SR, Miller WH, Jr., Scheinberg DA, Itri LM, Hittelman WN, Vyas R, Andreeff M, Tafuri A, Jakubowski A et al. (1991) Differentiation therapy of acute promyelocytic leukemia with tretinoin (all-trans-retinoic acid). NEJM 324(20):1385–1393. doi:10.1056/NEJM199105163242002

Chapter 26
NR4A Orphan Receptors as Drug Targets

Stephen Safe, Syng-Ook Lee, Cong Meng and Beiyan Zhou

Abstract The orphan receptor NR4Al (Nur77, TR3, NGFI-B) is overexpressed in many different cancer cell lines and tumors, and, in leukemia cells, NR4Al is also more highly expressed than in B- or T-cell. In contrast, we observed the opposite pattern of NR4A2 (Nurrl) expression in these same cells. NR4Al is primarily oncogenic in cancer cell lines and is a target for inhibitors or inactivators and treatment of Jurkat cells with an NR4Al inactivator-induced apoptosis and inhibited cell growth. The in vitro characterization of NR4Al as pro-oncogenic in leukemia cells is in contrast to the tumor suppressor-like activity of NR4Al in transgenic mouse models, and these differences are currently being investigated.

Keywords NR4A receptors · Leukemia cell · Overexpression · Oncogenic · NR4A1 NR4A2 · NR4A3

26.1 Introduction

The nuclear receptor (NR) superfamily of ligand-activated transcription factors is composed of 48 different genes that are characterized by their common structural features or domains with different but sometimes overlapping functions (Bookout et al. 2006; Gronemeyer et al. 2004; McKenna et al. 2009). The N-terminal A/B domain among receptors can be highly variable and usually contains one or more activation functions (AFs, e.g., AF1) that are involved in transactivation. NRs function primarily as transcription factors that bind DNA, and the DNA-binding domains

S. Safe (✉)
Department of Veterinary Physiology and Pharmacology, Texas A&M University,
4466 TAMU, College Station, TX 77843, USA
e-mail: ssafe@cvm.tamu.edu

S. Safe · S.-O. Lee
Institute of Biosciences and Technology, Texas A&M Health Science Center,
2121 W. Holcombe Blvd, Houston, TX 77030, USA

C. Meng · B. Zhou
Department of Veterinary Physiology and Pharmacology, Texas A&M University,
4466 TAMU, College Station, TX 77843, USA

M. Andreeff (ed.), *Targeted Therapy of Acute Myeloid Leukemia*,
Current Cancer Research, DOI 10.1007/978-1-4939-1393-0_26

(DBDs) of NRs are the most conserved regions among the different families. X-ray crystallographic analyses have shown that NR DBDs typically contain two zinc-finger motifs which involve chelation of zinc ions with specific cysteine residues. Bourguet and coworkers (Bourguet et al. 2000) indicated that some of the DBD structures exhibit differences in terms of their diverse DNA response elements and their binding as homodimers (e.g., steroid hormone receptors), heterodimers (e.g., retinoic X receptor (RXR)—receptor partner), and monomers (NR4A1 orphan receptor). The modular structure of NRs also contains a highly variable hinge region (D) and a C-terminal E/F domain which contains the ligand-binding domain and AF2. Two NRs, namely dosage-sensitive sex reversal, adrenal hyperplasia critical region on chromosome X, gene 1 (DAX1, NR0B1) and small heterodimerization partner (SHP, NR0B2), do not have classical DBDs and can only bind DNA in combination with a partner (Burris et al. 1996; Seol et al. 1996).

The human NRs have been arbitrarily subdivided into three categories which include: (1) the endocrine receptors that bind steroid and thyroid hormones, retinoids and vitamin D, (2) several adopted orphan receptors for which low-to-medium molecular weight agonists have been identified, and (3) orphan receptors which have no known endogenous ligands. Bookout and coworkers (Bookout et al. 2006) have profiled expression patterns of mouse NRs in 39 tissues and their analysis, coupled with known functions of individual NRs, demonstrates that clusters of mouse NRs regulate reproduction and development; lipid metabolism and energy homeostasis; CNS, circadian, and basal metabolic functions; steroidgenesis; bile acid; and xenobiotic metabolism. For some individual functions, NRs may act coordinately, and there are examples of individual receptors such as RXR and RAR that play a role in multiple functions. Many NRs are expressed as full-length and truncated forms which have significantly different functions, and their differential response to levels of endogenous or exogenous ligands contributes to the complex roles of NRs in "normal" physiology.

26.2 NR4A Orphan Receptors

Among the NR superfamily, the NR4A subfamily is a small subgroup of orphan receptors, namely NR4A1 (Nur77, TR3, NGFI-B), NR4A2 (Nurr1), and NR4A3 (Nor1) (Maxwell and Muscat 2006; Pearen and Muscat 2010; Lee et al. 2012; To et al. 2012; Zhang 2007; Shi 2007). These receptors were initially identified as immediate-early response genes induced in PC12 cells treated with nerve growth factor (Milbrandt 1988), and subsequent studies have reported induction of one or more of these genes in other cells and tissues exposed to various types of stressors (Maxwell and Muscat 2006; Pearen and Muscat 2010). Extensive evidence indicates an important role for NR4A in multiple physiological pathways and disease, suggesting that modulation of NR4A receptor expression may have important therapeutic implications (Maxwell and Muscat 2006; Pearen and Muscat 2010; Lee et al. 2011; To et al. 2012).

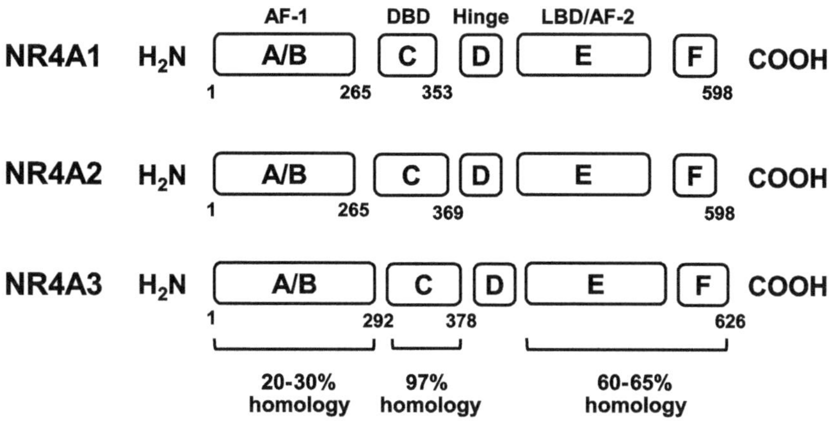

Fig. 26.1 NR4A receptors, their sequence homologies, and domains. The sequence homologies for NR4A1, NR4A2, and NR4A3 vary from 97 % (DBD), 60–65 % (E/F) to 20–30 % (A/B)

26.2.1 NR4A Receptor Structure

Figure 26.1 illustrates the typical domain structures of NR4A1, NR4A2, and NR4A3 (Maruyama et al. 1998). Their DBDs exhibit 97 % homology and, not surprisingly, this correlates with their binding to a common set of DNA response elements. The NGFI-B response element (NBRE) is an octanucleotide that binds NR4A members as monomers and homodimers (Maira et al. 1999; Paulsen et al. 1995; Philips et al. 1997; Wilson et al. 1992; Wilson et al. 1991). NR4A homodimers bind Nur-responsive elements (NuRE) which are everted repeats containing NBREs (Philips et al. 1997), and there is also evidence for heterodimer binding to NuREs. The pro-opiomelanocortin (POMC) gene promoter contains a well-characterized NuRE (Philips et al. 1997). Recent studies in this laboratory demonstrate that NR4A2 activator 1,1-bis(3'-indolyl)-1-(p-bromophenyl)methane (DIM-C-pPhBr) induces vasoactive intestinal peptide (VIP) in pancreatic cancer cells, and this response is significantly blocked after knockdown of NR4A2 (Li et al. 2012). A third mechanism of NR4A-mediated transactivation is due to the formation of NR4A1 and NR4A2 (but not NR4A3) heterodimers with RXR, which in turn induce transactivation from a DR5 element (Perlmann and Jansson 1995; Zetterstrom et al. 1996).

The similarities of the DBDs of NR4A receptors contrast to the decreased homologies of their C-terminal regions (60–65 %) and their highly divergent N-terminal domains (20–30 % homology; Pearen and Muscat 2010). X-ray crystallographic analysis of NR4A2 shows that this receptor is in a conformationally "active" form with a minimal hydrophobic pocket or cleft (Wang et al. 2003), and similar results have been observed in modeling studies (Wansa et al. 2002; Wansa et al. 2003). Nevertheless, the NR4A receptors interact with coactivators and as indicated below, there is extensive evidence for ligand-dependent activation of these orphan receptors.

26.2.2 NR4A Receptor Function: Noncancer Endpoints

Like other orphan NR families, there is increasing evidence that NR4A receptors play an essential role in maintaining cellular homeostasis (Maxwell and Muscat 2006; Pearen and Muscat 2010). NR4A1-/- mice do not exhibit any obvious phenotypical changes (Lee et al. 1995; Cheng et al. 1997), and this may be due to compensation by NR4A2 or NR4A3 since these receptors exhibit some overlapping tissue-specific activities. However, there is also evidence that NR4A-/- mice are compromised in conditions of stress. For example, a comparison of the effects of a high-fat diet on wild-type versus NR4A1-/- mice shows that the knockout animals have enhanced susceptibilities to increased insulin resistance and diet-induced obesity; glucose metabolism was also altered and tissue-specific damage included hepatic steatosis (Chao et al. 2009). The loss of NR4A2 in mice results in death within 24 h after birth, and NR4A2-/+ heterozygotes exhibit neuronal problems due, in part, to malfunctions in dopaminergic signaling (Nsegbe et al. 2004; Eells et al. 2006; Kadkhodaei et al. 2009). NR4A2 regulates differentiation of CD4+ T-cells through induction of Foxp3 and deletion of NR4A2 decreases Foxp3 and impairs the functions of Treg cells in mice and in vitro cell culture models (Sekiya et al. 2011), thus demonstrating an important role for NR4A2 in maintaining T-cell homeostasis. NR4A3-/- mice generated by two different groups were remarkably different (embryolethal vs. viable) (DeYoung et al. 2003; Ponnio et al. 2002; Ponnio and Conneely 2004; Zhao et al. 2010; Woronicz et al. 1994; Woronicz et al. 1995) and several deficits were observed in surviving mice. These included vascular injury, decreased response to stressors, and damage to the semicircular canals of the inner ear. Some of the early in vivo studies using NR4A knockdown or overexpression showed the importance of NR4A1 and NR4A3 in T-cell receptor-mediated apoptosis, whereas recent in vivo studies on Treg cells and T-cell homeostasis (Sekiya et al. 2011) showed that NR4A2 expression induced Foxp3 and minimal effects were observed for NR4A1 or NR4A3. However, a recent report using NR4A-/- mice showed a complex role for this receptor in thymocytes where this receptor “appears to act as a balancing molecule in fate determination at a critical juncture of T-cell differentiation” (Fassett et al. 2012).

NR4A receptors are immediate-early response genes, and their diversity of function is evident from the various stressors that induce NR4A expression, and these include neurotransmitters, growth factors, phorbol esters, peptide hormones, membrane depolarization, sheer and mechanical stress, fatty acids, prostaglandins, magnetic fields, calcium, and inflammatory cytokines (reviewed in Maxwell and Muscat 2006; Pearen and Muscat 2010). Not surprisingly, NR4A receptors play an important role in maintaining homeostasis due to their regulation of metabolism in skeletal muscle, liver, and pancreas which is due, in part, to β-adrenoreceptor-mediated regulation of NR4A expression in some of these tissues (Pearen and Muscat 2010). In addition to their function in metabolism, NR4A receptors are involved in neuronal function (e.g., dopamine synthesis), T-cell and thymocytes, the hypothalamus–pituitary axis, and macrophage and cardiovascular functions

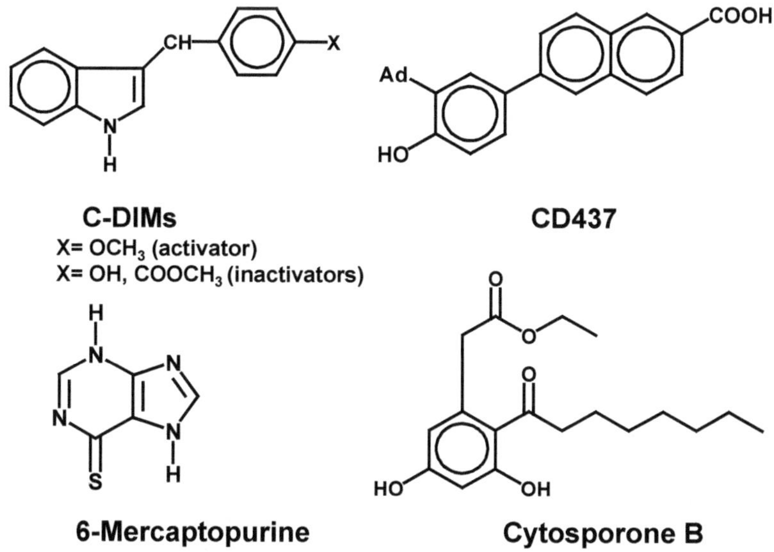

Fig. 26.2 Ligands that activate or inactivate NR4A-mediated transactivation

(Maxwell and Muscat 2006; Pearen and Muscat 2010). For example, NR4A1 is expressed in neonatal smooth muscle cells (SMCs) (humans) and NR4A1 reduces vascular lesions by inhibiting SMC proliferation (Arkenbout et al. 2002; Pires et al. 2007). It is possible that cardiovascular and other diseases may be targets for NR4A agonists or antagonists.

26.3 Drugs that Activate NR4A Receptors

26.3.1 Drug-induced Nuclear Export of NR4A1

Although the crystal structure and modeling studies suggest that the ligand-binding pocket in the C-terminal D domain is constitutively occupied, there is evidence that NR4A receptors can be "activated" or "deactivated" by ligands (Fig. 26.2), and some compounds directly bind NR4A1. One of the most unique "activation" mechanisms discovered among all NRs has been extensively characterized for NR4A1 in cancer cells treated with 12-tetradecanoylphorbol-13-acetate (TPA) or the caged retinoid 6-[3-1-adamantyl)-4-hydroxyphenyl]-2-naphthalene carboxylic acid (CD437 or APHN; Cao et al. 2004; Dawson et al. 2001; Holmes et al. 2003a, b, 2004; Kolluri et al. 2008; Li et al. 2000; Li et al. 1998; Lin et al. 2004). Both compounds induce apoptosis in several cancer cell lines, and this activity is associated with nuclear export of the receptor and formation of a pro-apoptotic NR4A1-bcl-2 complex through release of cytochrome c and activation of the intrinsic apoptosis pathway (Li et al. 2000; Lin et al. 2004). Similar results have been observed with a peptidomimetic and the anticancer drug paclitaxel that binds and/or activates bcl-2 resulting in the induction of apoptosis (Kolluri et al. 2008; Ferlini et al. 2009). This unusual NR4A1 export pathway has been linked to the pro-apoptotic activities of multiple agents including sulindac, HDAC inhibitors, calcium ionophores, cadmium, tin derivatives, cytosporone B (Csn-B) and related analogs, n-butylenephthalide, acetylshikonin derivatives, butyrate, viruses, VP16, insulin-like growth-factor-binding protein 3 (IGFBP3), synthetic chenodeoxycholic acid derivatives, and 5-fluorouracil (Chen et al. 2008; Chinnaiyan et al. 2006; Gennari et al. 2002; Jeong et al. 2003; Lee et al. 2004; Lee et al. 2002; Lee et al. 2005; Liu et al. 2008; Liu et al. 2010; Liu et al. 2002; Shin et al. 2004; Wilson et al. 2003; Wu et al. 2002a; Zhan et al. 2008).

Csn-B and synthetic structurally related analogs induce nuclear export of NR4A1 and induce apoptosis via mitochondrial–protein interactions as described above for other apoptosis-inducing agents (Liu et al. 2010; Zhan et al. 2008). However, in contrast to these compounds, Csn-B induces NR4A1 interactions with genomic NBRE sites in the NR4A1 promoter, and results of fluorescence-quenching experiments show that Csn-B directly binds to the C-terminal ligand-binding domain (aa 354–598) of NR4A1. Csn-B and related compounds represent a unique structural class of NR4A1 agonists with clinical potential for treating cancer- and noncancer-related diseases.

26.3.2 C-DIM Compounds Activate/Inactivate NR4A1 and NR4A2

Studies in this laboratory have utilized the anticancer agent 1,1-bis(3'-indolyl) methane (DIM) as a scaffold for development of ligands that activate or deactivate NRs. Initial studies showed that several 1,1-bis(3'-indolyl)-1-(p-substituted phenyl) methane compounds (C-DIMs) inhibited cancer cell and tumor growth using multiple tumor types, and structure-activity studies demonstrated that some C-DIMs activated peroxisome proliferator-activated receptor γ (PPARγ) (Qin et al. 2004; Chintharlapalli et al. 2005b; Chintharlapalli et al. 2006; Hong et al. 2004; Chintharlapalli et al. 2007; Chintharlapalli et al. 2004; Kassouf et al. 2006; Lei et al. 2008a; Lei et al. 2008b; Lei et al. 2006; Hong et al. 2008; Ichite et al. 2009; Guo et al. 2010). Compounds containing p-trifluoromethyl (DIM-C-pPhCF3), p-t-butyl (DIM-C-pPhtBu), and p-phenyl (DIM-C-pPhC6H5) groups were the most potent activators of PPARγ-dependent transactivation and exhibited potencies similar to that observed for prostaglandin J2 (PGJ2) (Qin et al. 2004; Hong et al. 2004; Chintharlapalli et al. 2004). A comparison of C-DIM analogs with other PPARγ agonists showed that there were differences in their recruitment of various coactivators, suggesting that C-DIMs were selective PPARγ modulators (Chintharlapalli et al. 2004). These compounds also selectively induced several PPARγ-dependent genes such as caveolin-1, p21, caveolin-2 and Krüppel-like factor 4 in various cancer cell lines; however, it was apparent from RNA interference (RNAi) studies and PPARγ antagonists that the anticancer activity of C-DIMs and other PPARγ agonists was primarily PPARγ-independent.

Subsequent screening studies identified C-DIMs that modulate NR4A-mediated transactivation. Initial screening used NR4A1 (mouse)-GAL4 (NR4A1m-GAL4) constructs and showed that at least two p-phenyl-substituted C-DIMs, namely the p-methoxy (DIM-C-pPhOCH3) and unsubstituted (DIM-C-Ph) analogs, induced transactivation in pancreatic cancer cells transfected with the NR4A1-GAL4 chimera and a GAL4 response element reporter gene (UASx5-luc) (Chintharlapalli et al. 2005a). Subsequent studies confirmed that DIM-C-pPhOCH3 activated the NR4A1 (human)-GAL4 (NR4A1-GAL4) construct, approximately twofold, which was significantly lower than observed in the initial studies (Chintharlapalli et al. 2005a; Safe et al. 2011). Despite these differences, DIM-C-pPhOCH3 induced several NR4A1-dependent responses and genes in different cancer cell lines, and the role of NR4A1 was confirmed by RNAi studies (Cho et al. 2010; Cho et al. 2008; Cho et al. 2007; Yoon et al. 2011; Lee et al. 2009). For example, in pancreatic cancer cells, DIM-C-pPhOCH3 induced cleaved poly-ADP-ribose polymerase (PARP), a marker of apoptosis, and tumor necrosis factor-related apoptosis-inducing ligand (TRAIL), a death receptor ligand, and these responses were decreased after NR4A1 knockdown (Chintharlapalli et al. 2005a). Treatment with DIM-C-pPhOCH3 did not affect expression of NR4A1 or induce nuclear export of the receptor, and the pro-apoptotic effects of DIM-C-pPhOCH3 were observed in the presence or absence of the nuclear export inhibitor leptomycin B (Chintharlapalli et al. 2005a). These observations clearly distinguished between the NR4A1-active C-DIMs and other apoptosis-inducing agents that induce apoptosis through nuclear export of NR4A1 and subsequent formation of extranuclear pro-apoptotic complexes, such as the mitochondrial bcl-2/NR4A1 heterodimer. Subsequent studies in several different cancer cell lines confirmed that DIM-C-pPhOCH3 and DIM-C-Ph induced apoptosis through activation of nuclear NR4A1 and several NR4A1-regulated genes, and the in vitro and in vivo anticancer activities of these compounds were also enhanced by receptor-independent pathways (Chintharlapalli et al. 2005a; Safe et al. 2011; Cho et al. 2010; Cho et al. 2008; Cho et al. 2007; Yoon et al. 2011; Lee et al. 2009).

26.3.3 NR4A2 (Nurr1) and NR4A3 (Nor1) Ligands

6-Mercaptopurine (6-MP) is an anticancer agent that activates the N-terminal domains of both NR4A2 and NR4A3, even though there is minimal structural similarity between these domains (Lee et al. 1995; Ordentlich et al. 2003). 6-MP and structurally related analogs were highly specific among purine and pyrimidine analogs in their activation of NR4A2 and NR4A3 (but not NR4A1), and the mechanisms of 6-MP-mediated transactivation also involved TRAP220 coactivator recruitment to the AF1 domain (Wansa and Muscat 2005). A recent report (Li et al. 2012) showed that 6-MP also induced NR4A2-dependent transactivation in Panc28, but not in Panc1 pancreatic cancer cells transfected with NBRE-luc and NuRE-luc constructs. However, 6-MP induced transactivation in both cell lines after transfection with GAL4-NR4A2/UASx5-luc; 6-MP also activated GAL-NR4A2(A–B) and GAL4-NR4A2(C–F) which contain N-terminal (A/B) and C-terminal (C/F) domains of NR4A2. Thus, the activity and mechanisms of action of 6-MP is highly dependent on cell context, the transactivation system, and specific domains (of NR4A2).

The structure-dependent activity of C-DIMs containing both substituted phenyl or heteroaromatic groups as activators of NR4A2 was investigated in Panc1 and Panc28 pancreatic cells using multiple reporter gene assays (NBRE-luc, NuRE-luc, and UASx5-luc) (Li et al. 2012). The results were consistent among the assays in both cell lines, and p-phenyl-substituted compounds were more active than several different C-DIMs containing heteroaromatic groups. C-DIMs containing para-CF3, t-butyl, and CN substituents that activate PPARγ also activated NR4A1, and the para- Br, iodo, and trifluoromethoxy were also NR4A2 activators. Using the p-bromophenyl analog (DIM-C-pPhBr) as a model, it was shown that the order of potency was p-bromo > m-bromo > o-bromo for isomers with variable bromo substitution, and methylation of the indole nitrogen completely abrogated activity. Like 6-MP, DIM-C-pPhBr induced transactivation in cells transfected with GAL4-NR4A2(A-B) and GAL4-NR4A2(C-F), and inhibition of transactivation was observed in pancreatic cancer cells cotreated with the mitogen-activated protein kinase (MAPK) inhibitor PD98059. This study identified a subclass of C-DIMs that also activated NR4A2-dependent transactivation (this includes several genes), and the mechanism was due to nuclear NR4A2 which does not undergo nuclear translocation after treatment with DIM-C-pPhBr (Li et al. 2012). There are also reports that substituted benzimidazoles and isoxazolo-pyridinones are NR4A2 agonists (Dubois et al. 2006; Hintermann et al. 2007), and prostaglandin-A2 activates NR4A3 (Kagaya et al. 2005); however, their mechanism of action and effects in cancer cell lines have not been determined. The use of these and NR4A2-active C-DIM compounds as anticancer agents clearly requires further study; however, there is evidence that NR4A2-active C-DIMs inhibit bladder cancer cell and tumor growth.

26.4 Drugs that Inactivate NR4A Receptors

26.4.1 Pro-oncogenic Activity of NR4A1

Although apoptosis-inducing agents that induce nuclear export of NR4A1 and C-DIMs that activate nuclear NR4A1 are effective anticancer agents, there is strong evidence that NR4A1 is pro-oncogenic in many cancer cell lines (Kolluri et al. 2003; Ke et al. 2004; Li et al. 2006; Bras et al. 2000; Lee et al. 2010; Lee et al. 2011; Wu et al. 2011; Zhao et al. 2006). For example, knockdown of NR4A1 in HeLa (cervical), M14 (melanoma), A20B (lymphoma), HCT116 (colon), Panc1,

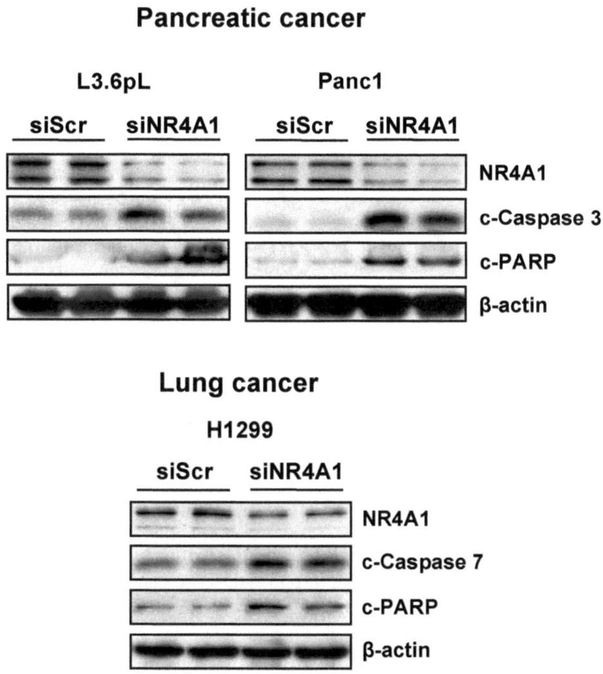

Fig. 26.3 Knockdown of NR4A1 by RNAi activates caspases and induces apoptosis in pancreatic and lung cancer cells

L3.6pL and MiaPaca2 (pancreatic), and H460 and A549 (lung) cancer induces one or more markers of apoptosis (Kolluri et al. 2003; Ke et al. 2004; Li et al. 2006; Bras et al. 2000). Studies in this laboratory showed that knockdown of NR4A1 by RNAi in pancreatic and lung cancer cells increased cleavage (activation) of caspases and PARP (Lee et al. 2010; Lee et al. 2011; Fig. 26.3), and similar results were observed in breast and colon cancer cell lines. Moreover, NR4A1 knockdown also decreased cell proliferation and angiogenesis in cancer lines, confirming the "oncogene-like" activity of NR4A1, and suggesting that deactivation or decreased expression of this receptor could represent a strategy for developing novel mechanism-based anticancer agents.

26.4.2 C-DIMs as NR4A1 Deactivators

The p-hydroxyphenyl DIM analog (DIM-C-pPhOH) was initially identified as an inhibitor of NR4A1-dependent transactivation induced by DIM-C-pPhOCH3 in pancreatic cancer cells, and this compound was subsequently investigated as a potential NR4A1 deactivator (Chintharlapalli et al. 2005a). DIM-C-pPhOH decreased NR4A1-dependent transactivation, and this was AF1 (N-terminal domain)-dependent in pancreatic cancer cells (Lee et al. 2010). In subsequent studies, the comparative effects of NR4A1 knockdown and DIM-C-pPhOH showed that both treatments decreased pancreatic cell growth, induced apoptosis, and decreased the expression

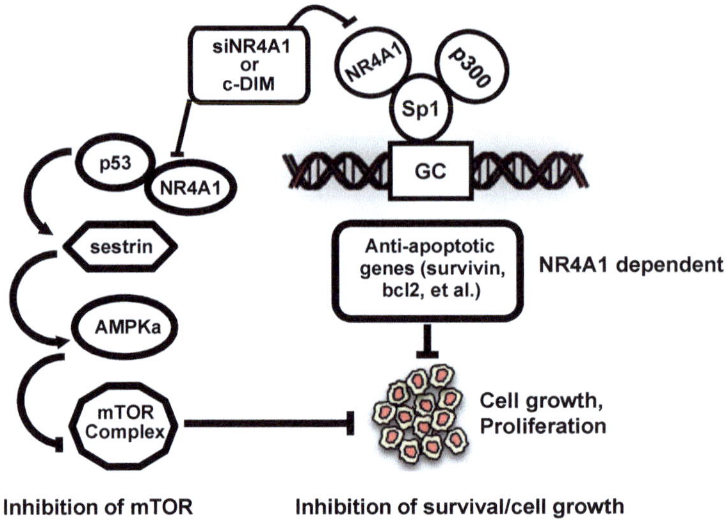

Fig. 26.4 Inactivation of NR4A1 by RNA interference of C-DIMs decreases survival pathways and inhibits mTOR signaling in p53 wild-type cancer cells

of several genes including both survivin and bcl-2 (Lee et al. 2010). Using survivin as a model, it was shown that survivin expression in pancreatic cancer cells was dependent on formation of an NR4A1–p300 complex with Sp1 bound to the proximal GC-rich region of the survivin promoter, and knockdown of NR4A1, p300, or Sp1 decreased survivin levels. Treatment of cells with DIM-C-pPhOH decreased survivin expression, and chromatin immunoprecipitation (ChIP) assays showed that the loss of gene expression was associated with dissociation of p300 from the "active" complex, and this presumably was due to DIM-C-pPhOH-induced conformational change in NR4A1 which decreased interactions with p300 (Fig. 26.4).

NR4A1 interacts with p53 and blocks function of the tumor suppressor gene (Zhao et al. 2006), and recent studies in lung cancer cells demonstrated that knockdown of NR4A1 by RNAi or treatment with DIM-C-pPhOH-activated p53 and p53-dependent inactivation of the mammalian target of rapamycin (mTOR) (Fig. 26.4; Lee et al. 2011). p53 activates sestrin which in turn activates AMPKα, and this results in inhibition of mTOR and downstream signaling pathways. mTOR inhibition by NR4A1 knockdown or DIM-C-pPhOH was observed in p53 wild-type A549 and H460 cells, but not in p53 mutant H1299 lung cancer cells. DIM-C-pPhOH also inhibited lung tumor growth in an orthotopic lung model (A549 cells), mTOR signaling, and expression of survivin/bcl-2 in tumors from mice treated with DIM-C-pPhOH. These results show that the pro-oncogenic activity of NR4A1 is due to the role of this transcription factor in maintaining high expression of genes associated with survival and cell proliferation and inhibition of p53 (Lee et al. 2011). Moreover, in ongoing studies in pancreatic and other cancer cell lines, there is evidence that NR4A1 suppresses both endoplasmic reticulum (ER) and oxidative stress, and inactivation (by DIM-C-pPhOH) or knockdown of NR4A1 induces genes and other stress markers which activate caspases and cell death.

26.5 NR4A Receptors and Leukemia

The expression and function of NR4A1 and other NR4A receptors have not been extensively investigated in leukemia cell lines or patients, and the potential for these receptors as drug targets for treatment of different types of leukemia is not well defined. Nevertheless, there is some evidence that this subfamily of nuclear receptors plays a role in leukemia and drugs that activate or deactivate NR4A receptors or induced NR4A nuclear export may have potential for clinical applications.

26.5.1 Knockout Mouse Models

NR4A1 and NR4A3 exhibit overlapping properties, and knockout of each receptor in mice did not result in significant changes in these animals. However, double-knockout NR4A1-/-/NR4A3-/- mice resulted in a dramatic phenotype in which mice exhibited splenomegaly, hepatomegaly, decreased body weight, and lethality by 2–4 weeks of age (Mullican et al. 2007). Multiple genomic and cellular changes were observed in these mice including increased hematopoietic stem cells and myeloid progenitors, and the cause of death was attributed to lethal acute myeloid leukemia (AML). A recent study reported the effects of decreased expression of NR4A1 (NR4A1+/-/NR4A3-/-) and NR4A3 (NR4A1-/-/NR4A3-/+), and these mice exhibited “a chronic myeloid malignancy that closely recapitulates the pathologic features of mixed myeloplastic/myoproliferative neoplasms (MDS/MPNs)” (Ramirez-Herrick et al. 2011). It was also reported that NR4A1 and NR4A3 expression was significantly lower in leukemia-enriched blasts from patients diagnosed with AML compared to bone marrow from healthy donors (Mullican et al. 2007). Based on results from mouse knockout studies and the human expression data, both NR4A1 and NR4A3 appear to be tumor suppressors; however, with few exceptions, most previous RNAi studies in cancer cells suggest that NR4A1 is oncogenic (Kolluri et al. 2003; Ke et al. 2004; Li et al. 2006; Bras et al. 2000; Lee et al. 2010; Lee et al. 2011; Wu et al. 2011; Zhao et al. 2006), and further research is required to address the oncogenic versus tumor suppressor-like activity of NR4A receptors in leukemia.

26.5.2 PML-NR4A1 Interactions

The promyelocytic leukemia gene (PML) exhibits tumor suppressor activity and regulates expression and apoptosis. PML is lost or inactivated in many hematopoietic malignancies and solid tumors (Salomoni and Pandolfi 2002; Gurrieri et al. 2004);

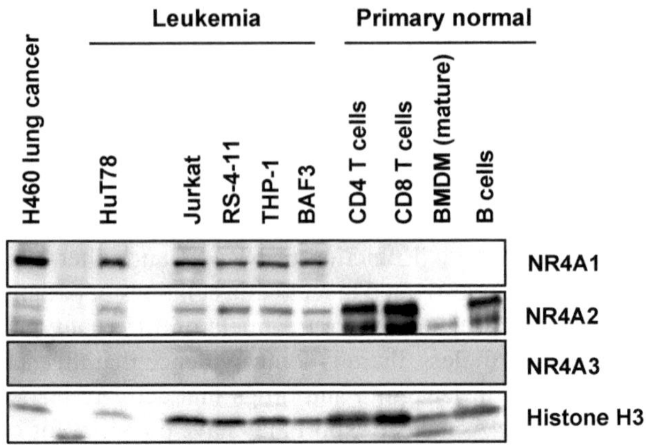

Fig. 26.5 NR4A receptor expression in leukemia and non-leukemia cell lines

however, in acute promyelocytic leukemia (APL) patients, an APL fusion protein with the retinoic acid receptor α (RARα) is formed (PML/RARα) due to a t(15;17) chromosomal translocation. In contrast to PML, the PML/RARα fusion protein is oncogenic and significantly contributes to the APL phenotype. PML is a potent repressor of NR4A1 through interactions with the DBD of the receptor; however, despite PML-mediated suppression of NR4A1, PML overexpression enhances phorbol ester/NR4A1-dependent apoptosis (Wu et al. 2002b). The interaction of NR4A1–PML/RARα in leukemia cells has not been determined and could potentially be a possible target for drugs that activate or inactivate NR4A1.

26.5.3 NR4A Receptor Expression in Leukemia Cells

A comparison between DNA damage-induced apoptosis-resistant and -nonresistant B cell chronic lymphoid leukemia (B-CLL) cells' lines showed that NR4A1 was among thirteen genes that were overexpressed, suggesting a possible role for NR4A1 in drug resistance (Vallat et al. 2003). NR4A1 mRNA was also expressed in T-cell acute lymphoblastic leukemia (T-ALL) PEER cells (Nagel et al. 2008), and it was shown that NR4A1 was downregulated by knockdown of NKX2-5, and this was due, in part, to the inverse regulation of MEF2C and NR4A1. Since there is minimal information on levels of NR4A receptor proteins in leukemia cells, we compared protein expression in noncancer B cells, CD4+ and CD8+ T-cells, and bone marrow-derived macrophages (BMDM) to human lung H460, Jurkat, HuT-78, THP1 and RS-411, and mouse BAF3 leukemia cell lines (Fig. 26.5). The results showed low-to-non-detectable NR4A1 protein levels in the T-cells and B cells, whereas relatively high levels were observed in the leukemia cell lines. In contrast, NR4A2 was highly expressed in B cells and CD4+ and CD8+ T-cells (but not in BMDM) and significantly decreased in leukemia cells and H460 lung cancer cells, whereas NR4A3 was not detectable in any of the cancer or noncancer cells.

There are only a limited number of studies showing the effects of activated or deactivated NR4A1 on leukemia cell growth and survival. Histone deacetylase

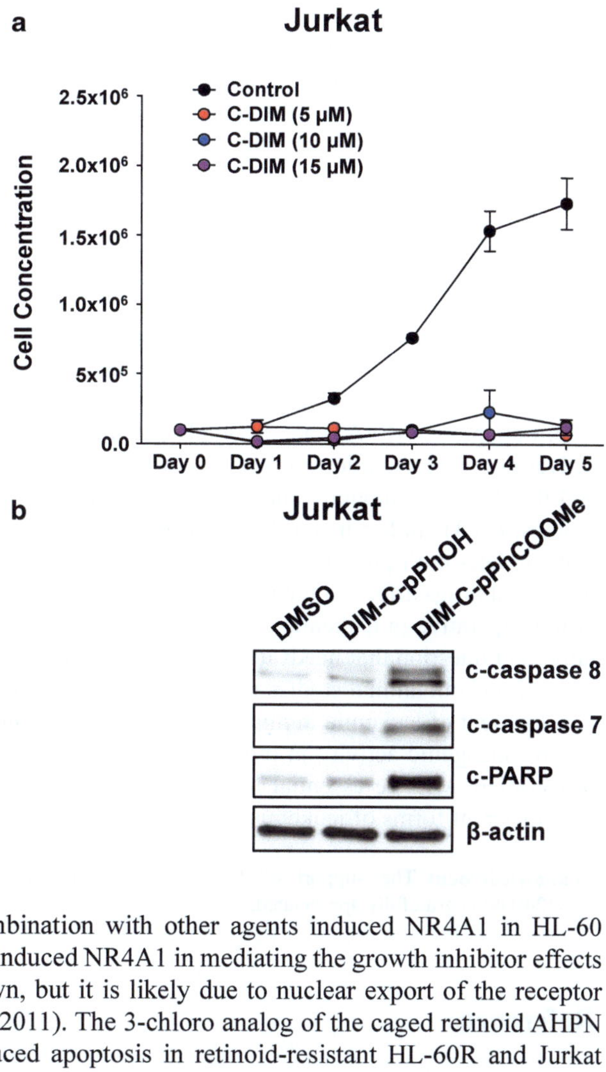

Fig. 26.6 The NR4A1 inactivator DIM-C-pPhCOOMe inhibits Jurkat cell growth (a) and induces caspase cleavage (activation) and PARP cleavage (b)

inhibitors alone or in combination with other agents induced NR4A1 in HL-60 cells; however, the role of induced NR4A1 in mediating the growth inhibitor effects of these drugs is not known, but it is likely due to nuclear export of the receptor (Liu et al. 2009; Liu et al. 2011). The 3-chloro analog of the caged retinoid AHPN inhibited growth and induced apoptosis in retinoid-resistant HL-60R and Jurkat leukemic cells, and these effects were associated with nuclear export of NR4A1 since some of the effects in Jurkat cells were attenuated by overexpression of bcl-2 (Dawson et al. 2001). These results were also consistent with the drug-induced nuclear export of NR4A1 which forms a mitochondrial pro-apoptotic NR4A1-bcl-2 complex (Li et al. 2000).

Recent studies in this laboratory have focused on development of C-DIMs that inactivate NR4A1 (Lee et al. 2010; Lee et al. 2011). In addition to DIM-C-pPhOH, we have identified several other compounds that inactivate NR4A1 including the p-carboxymethylphenyl analog (DIM-C-pPhCOOMe). Treatment of Jurkat cells with DIM-C-pPhCOOMe resulted in a dose-dependent inhibition of cell proliferation, and 10 μM DIM-C-pPhCOOMe also induced cleavage of caspase 8, caspase 7, and PARP, demonstrating activation of apoptosis (Fig. 26.6). DIM-C-pPhOH had minimal activity due to the low concentration used. These results suggest that NR4A1 may also be a drug target in leukemia cells and current studies are focused confirming these initial observations. Previous studies also showed that C-DIMs substituted on both the phenyl and indole rings inhibited AML cell growth and induced apoptosis, and the role of NR4A receptors in mediating these effects is also being investigated (Contractor et al. 2005).

26.6 Summary

The NR4A family of NRs and particularly NR4A1 are potential therapeutic drug targets, and studies with solid tumors show that some apoptosis-inducing anticancer agents that induce nuclear export of NR4A1 decrease tumor survival by formation of pro-apoptotic NR4A1–bcl-2 complexes. In contrast, there is also evidence that activation or deactivation of NR4A1 by specific C-DIMs also induces apoptosis and that these compounds inhibit tumor growth in both xenograft and orthotopic models. RNAi studies in solid tumors show that NR4A1 is a pro-oncogenic factor that regulates multiple survival and growth-promoting pathways. There is some evidence that drugs affecting the nuclear export or nuclear inactivation of NR4A1 may also be operative in leukemia cells, and this requires further investigation. Another important question that needs to be resolved is the function of NR4A1 in leukemia cells. Knockout animal studies indicate that NR4A1 acts as a tumor suppressor for development of leukemia in mice, whereas RNAi studies show that NR4A1 exhibits pro-oncogenic activity in most other cancers. Resolution of these paradoxical observations will be essential for future development of NR4A1-active drugs to treat different forms of leukemia.

Acknowledgment The support of Texas AgriLife and the National Institutes of Health (CA124998) is gratefully appreciated.

References

Arkenbout EK, de Waard V, van Bragt M, van Achterberg TA, Grimbergen JM, Pichon B, Pannekoek H, de Vries CJ (2002) Protective function of transcription factor TR3 orphan receptor in atherogenesis: decreased lesion formation in carotid artery ligation model in TR3 transgenic mice. Circulation 106(12):1530–1535

Bookout AL, Jeong Y, Downes M, Yu RT, Evans RM, Mangelsdorf DJ (2006) Anatomical profiling of nuclear receptor expression reveals a hierarchical transcriptional network. Cell 126(4):789–799

Bourguet W, Germain P, Gronemeyer H (2000) Nuclear receptor ligand-binding domains: three-dimensional structures, molecular interactions and pharmacological implications. Trends Pharmacol Sci 21(10):381–388

Bras A, Albar JP, Leonardo E, de Buitrago GG, Martinez AC (2000) Ceramide-induced cell death is independent of the Fas/Fas ligand pathway and is prevented by Nur77 overexpression in A20 B cells. Cell Death Differ 7(3):262–271

Burris TP, Guo W, McCabe ER (1996) The gene responsible for adrenal hypoplasia congenita, DAX-1, encodes a nuclear hormone receptor that defines a new class within the superfamily. Recent Prog Horm Res 51:241-259; discussion 259–260

Cao X, Liu W, Lin F, Li H, Kolluri SK, Lin B, Han YH, Dawson MI, Zhang XK (2004) Retinoid X receptor regulates Nur77/TR3-dependent apoptosis [corrected] by modulating its nuclear export and mitochondrial targeting. Mol Cell Biol 24(22):9705–9725

Chao LC, Wroblewski K, Zhang Z, Pei L, Vergnes L, Ilkayeva OR, Ding SY, Reue K, Watt MJ, Newgard CB, Pilch PF, Hevener AL, Tontonoz P (2009) Insulin resistance and altered systemic glucose metabolism in mice lacking Nur77. Diabetes 58(12):2788–2796

Chen YL, Jian MH, Lin CC, Kang JC, Chen SP, Lin PC, Hung PJ, Chen JR, Chang WL, Lin SZ, Harn HJ (2008) The induction of orphan nuclear receptor Nur77 expression by n-butyl-enephthalide as pharmaceuticals on hepatocellular carcinoma cell therapy. Mol Pharmacol 74(4):1046–1058

Cheng LE, Chan FK, Cado D, Winoto A (1997) Functional redundancy of the Nur77 and Nor-1 orphan steroid receptors in T-cell apoptosis. EMBO J 16(8):1865–1875

Chinnaiyan P, Varambally S, Tomlins SA, Ray S, Huang S, Chinnaiyan AM, Harari PM (2006) Enhancing the antitumor activity of ErbB blockade with histone deacetylase (HDAC) inhibition. Int J Cancer 118(4):1041–1050

Chintharlapalli S, Burghardt R, Papineni S, Ramaiah S, Yoon K, Safe S (2005a) Activation of Nur77 by selected 1,1-Bis(3'-indolyl)-1-(p-substituted phenyl)methanes induces apoptosis through nuclear pathways. J Biol Chem 280(26):24903–24914

Chintharlapalli S, Papineni S, Baek SJ, Liu S, Safe S (2005b) 1,1-Bis(3'-indolyl)-1-(p-substitutedphenyl)methanes are peroxisome proliferator-activated receptor gamma agonists but decrease HCT-116 colon cancer cell survival through receptor-independent activation of early growth response-1 and nonsteroidal anti-inflammatory drug-activated gene-1. Mol Pharmacol 68(6):1782–1792

Chintharlapalli S, Smith R, 3rd, Samudio I, Zhang W, Safe S (2004) 1,1-Bis(3'-indolyl)-1-(p-substitutedphenyl)methanes induce peroxisome proliferator-activated receptor gamma-mediated growth inhibition, transactivation, and differentiation markers in colon cancer cells. Cancer Res 64(17):5994–6001

Chintharlapalli S, Papineni S, Safe S (2006) 1,1-Bis(3'-indolyl)-1-(p-substituted phenyl)methanes inhibit colon cancer cell and tumor growth through PPARgamma-dependent and PPARgamma-independent pathways. Mol Cancer Ther 5(5):1362–1370

Chintharlapalli S, Papineni S, Safe S (2007) 1,1-bis(3'-indolyl)-1-(p-substitutedphenyl)methanes inhibit growth, induce apoptosis, and decrease the androgen receptor in LNCaP prostate cancer cells through peroxisome proliferator-activated receptor gamma-independent pathways. Mol Pharmacol 71(2):558–569

Cho SD, Yoon K, Chintharlapalli S, Abdelrahim M, Lei P, Hamilton S, Khan S, Ramaiah SK, Safe S (2007) Nur77 agonists induce proapoptotic genes and responses in colon cancer cells through nuclear receptor-dependent and nuclear receptor-independent pathways. Cancer Res 67(2):674–683

Cho SD, Lei P, Abdelrahim M, Yoon K, Liu S, Guo J, Papineni S, Chintharlapalli S, Safe S (2008) 1,1-bis(3'-indolyl)-1-(p-methoxyphenyl)methane activates Nur77-independent proapoptotic responses in colon cancer cells. Mol Carcinog 47(4):252–263

Cho SD, Lee SO, Chintharlapalli S, Abdelrahim M, Khan S, Yoon K, Kamat AM, Safe S (2010) Activation of nerve growth factor-induced B alpha by methylene-substituted diindolylmethanes in bladder cancer cells induces apoptosis and inhibits tumor growth. Mol Pharmacol 77(3):396–404

Contractor R, Samudio IJ, Estrov Z, Harris D, McCubrey JA, Safe SH, Andreeff M, Konopleva M (2005) A novel ring-substituted diindolylmethane,1,1-bis[3'-(5-methoxyindolyl)]-1-(p-t-butylphenyl) methane, inhibits extracellular signal-regulated kinase activation and induces apoptosis in acute myelogenous leukemia. Cancer Res 65(7):2890–2898

Dawson MI, Hobbs PD, Peterson VJ, Leid M, Lange CW, Feng KC, Chen G, Gu J, Li H, Kolluri SK, Zhang X, Zhang Y, Fontana JA (2001) Apoptosis induction in cancer cells by a novel analogue of 6-[3-(1-adamantyl)-4-hydroxyphenyl]-2-naphthalenecarboxylic acid lacking retinoid receptor transcriptional activation activity. Cancer Res 61(12):4723–4730

DeYoung RA, Baker JC, Cado D, Winoto A (2003) The orphan steroid receptor Nur77 family member Nor-1 is essential for early mouse embryogenesis. J Biol Chem 278(47):47104–47109

Dubois C, Hengerer B, Mattes H (2006) Identification of a potent agonist of the orphan nuclear receptor Nurr1. ChemMedChem 1(9):955–958

Eells JB, Misler JA, Nikodem VM (2006) Early postnatal isolation reduces dopamine levels, elevates dopamine turnover and specifically disrupts prepulse inhibition in Nurr1-null heterozygous mice. Neuroscience 140(4):1117–1126

Fassett MS, Jiang W, D'Alise AM, Mathis D, Benoist C (2012) Nuclear receptor Nr4a1 modulates both regulatory T-cell (Treg) differentiation and clonal deletion. Proc Natl Acad Sci U S A 109(10):3891–3896

Ferlini C, Cicchillitti L, Raspaglio G, Bartollino S, Cimitan S, Bertucci C, Mozzetti S, Gallo D, Persico M, Fattorusso C, Campiani G, Scambia G (2009) Paclitaxel directly binds to Bcl-2 and functionally mimics activity of Nur77. Cancer Res 69(17):6906–6914

Gennari A, Bleumink R, Viviani B, Galli CL, Marinovich M, Pieters R, Corsini E (2002) Identification by DNA macroarray of nur77 as a gene induced by di-n-butyltin dichloride: its role in organotin-induced apoptosis. Toxicol Appl Pharmacol 181(1):27–31

Gronemeyer H, Gustafsson JA, Laudet V (2004) Principles for modulation of the nuclear receptor superfamily. Nat Rev Drug Discov 3(11):950–964

Guo J, Chintharlapalli S, Lee SO, Cho SD, Lei P, Papineni S, Safe S (2010) Peroxisome proliferator-activated receptor gamma-dependent activity of indole ring-substituted 1,1-bis(3'-indolyl)-1-(p-biphenyl)methanes in cancer cells. Cancer Chemother Pharmacol 66(1):141–150

Gurrieri C, Capodieci P, Bernardi R, Scaglioni PP, Nafa K, Rush LJ, Verbel DA, Cordon-Cardo C, Pandolfi PP (2004) Loss of the tumor suppressor PML in human cancers of multiple histologic origins. J Natl Cancer Inst 96(4):269–279

Hintermann S, Chiesi M, von Krosigk U, Mathe D, Felber R, Hengerer B (2007) Identification of a series of highly potent activators of the Nurr1 signaling pathway. Bioorg Med Chem Lett 17(1):193–196

Holmes WF, Soprano DR, Soprano KJ (2003a) Comparison of the mechanism of induction of apoptosis in ovarian carcinoma cells by the conformationally restricted synthetic retinoids CD437 and 4-HPR. J Cell Biochem 89(2):262–278

Holmes WF, Soprano DR, Soprano KJ (2003b) Early events in the induction of apoptosis in ovarian carcinoma cells by CD437: activation of the p38 MAP kinase signal pathway. Oncogene 22(41):6377–6386

Holmes WF, Soprano DR, Soprano KJ (2004) Synthetic retinoids as inducers of apoptosis in ovarian carcinoma cell lines. J Cell Physiol 199(3):317–329

Hong J, Samudio I, Liu S, Abdelrahim M, Safe S (2004) Peroxisome proliferator-activated receptor gamma-dependent activation of p21 in Panc-28 pancreatic cancer cells involves Sp1 and Sp4 proteins. Endocrinology 145(12):5774–5785

Hong J, Samudio I, Chintharlapalli S, Safe S (2008) 1,1-bis(3'-indolyl)-1-(p-substituted phenyl) methanes decrease mitochondrial membrane potential and induce apoptosis in endometrial and other cancer cell lines. Mol Carcinog 47(7):492–507

Ichite N, Chougule MB, Jackson T, Fulzele SV, Safe S, Singh M (2009) Enhancement of docetaxel anticancer activity by a novel diindolylmethane compound in human non-small cell lung cancer. Clin Cancer Res 15(2):543–552

Jeong JH, Park JS, Moon B, Kim MC, Kim JK, Lee S, Suh H, Kim ND, Kim JM, Park YC, Yoo YH (2003) Orphan nuclear receptor Nur77 translocates to mitochondria in the early phase of apoptosis induced by synthetic chenodeoxycholic acid derivatives in human stomach cancer cell line SNU-1. Ann N Y Acad Sci 1010:171–177

Kadkhodaei B, Ito T, Joodmardi E, Mattsson B, Rouillard C, Carta M, Muramatsu S, Sumi-Ichinose C, Nomura T, Metzger D, Chambon P, Lindqvist E, Larsson NG, Olson L, Bjorklund A, Ichinose H, Perlmann T (2009) Nurr1 is required for maintenance of maturing and adult midbrain dopamine neurons. J Neurosci 29(50):15923–15932

Kagaya S, Ohkura N, Tsukada T, Miyagawa M, Sugita Y, Tsujimoto G, Matsumoto K, Saito H, Hashida R (2005) Prostaglandin A2 acts as a transactivator for NOR1 (NR4A3) within the nuclear receptor superfamily. Biol Pharm Bull 28(9):1603–1607

Kassouf W, Chintharlapalli S, Abdelrahim M, Nelkin G, Safe S, Kamat AM (2006) Inhibition of bladder tumor growth by 1,1-bis(3'-indolyl)-1-(p-substitutedphenyl)methanes: a new class of peroxisome proliferator-activated receptor gamma agonists. Cancer Res 66(1):412–418

Ke N, Claassen G, Yu DH, Albers A, Fan W, Tan P, Grifman M, Hu X, Defife K, Nguy V, Meyhack B, Brachat A, Wong-Staal F, Li QX (2004) Nuclear hormone receptor NR4A2 is involved in cell transformation and apoptosis. Cancer Res 64(22):8208–8212

Kolluri SK, Bruey-Sedano N, Cao X, Lin B, Lin F, Han YH, Dawson MI, Zhang XK (2003) Mitogenic effect of orphan receptor TR3 and its regulation by MEKK1 in lung cancer cells. Mol Cell Biol 23(23):8651–8667

Kolluri SK, Zhu X, Zhou X, Lin B, Chen Y, Sun K, Tian X, Town J, Cao X, Lin F, Zhai D, Kitada S, Luciano F, O'Donnell E, Cao Y, He F, Lin J, Reed JC, Satterthwait AC, Zhang XK (2008) A short Nur77-derived peptide converts Bcl-2 from a protector to a killer. Cancer Cell 14(4):285–298

Lee SL, Wesselschmidt RL, Linette GP, Kanagawa O, Russell JH, Milbrandt J (1995) Unimpaired thymic and peripheral T cell death in mice lacking the nuclear receptor NGFI-B (Nur77). Science 269(5223):532–535

Lee JM, Lee KH, Weidner M, Osborne BA, Hayward SD (2002) Epstein-Barr virus EBNA2 blocks Nur77- mediated apoptosis. Proc Natl Acad Sci U S A 99(18):11878–11883

Lee JM, Lee KH, Farrell CJ, Ling PD, Kempkes B, Park JH, Hayward SD (2004) EBNA2 is required for protection of latently Epstein-Barr virus-infected B cells against specific apoptotic stimuli. J Virol 78(22):12694–12697

Lee KW, Ma L, Yan X, Liu B, Zhang XK, Cohen P (2005) Rapid apoptosis induction by IGFBP-3 involves an insulin-like growth factor-independent nucleomitochondrial translocation of RXRalpha/Nur77. J Biol Chem 280(17):16942–16948

Lee SO, Chintharlapalli S, Liu S, Papineni S, Cho SD, Yoon K, Safe S (2009) p21 expression is induced by activation of nuclear nerve growth factor-induced Balpha (Nur77) in pancreatic cancer cells. Mol Cancer Res 7(7):1169–1178

Lee SO, Abdelrahim M, Yoon K, Chintharlapalli S, Papineni S, Kim K, Wang H, Safe S (2010) Inactivation of the orphan nuclear receptor TR3/Nur77 inhibits pancreatic cancer cell and tumor growth. Cancer Res 70(17):6824–6836

Lee SO, Li X, Khan S, Safe S (2011) Targeting NR4A1 (TR3) in cancer cells and tumors. Expert Opin Ther Targets 15(2):195–206

Lee SO, Andey T, Jin UH, Kim K, Sachdeva M, Safe S (2012) The nuclear receptor TR3 regulates mTORC1 signaling in lung cancer cells expressing wild-type p53. Oncogene 31(27):3265–3276

Lei P, Abdelrahim M, Safe S (2006) 1,1-Bis(3'-indolyl)-1-(p-substituted phenyl)methanes inhibit ovarian cancer cell growth through peroxisome proliferator-activated receptor-dependent and independent pathways. Mol Cancer Ther 5(9):2324–2336

Lei P, Abdelrahim M, Cho SD, Liu S, Chintharlapalli S, Safe S (2008a) 1,1-Bis(3'-indolyl)-1-(p-substituted phenyl)methanes inhibit colon cancer cell and tumor growth through activation of c-jun N-terminal kinase. Carcinogenesis 29(6):1139–1147

Lei P, Abdelrahim M, Cho SD, Liu X, Safe S (2008b) Structure-dependent activation of endoplasmic reticulum stress-mediated apoptosis in pancreatic cancer by 1,1-bis(3'-indoly)-1-(p-substituted phenyl)methanes. Mol Cancer Ther 7(10):3363–3372

Li Y, Lin B, Agadir A, Liu R, Dawson MI, Reed JC, Fontana JA, Bost F, Hobbs PD, Zheng Y, Chen GQ, Shroot B, Mercola D, Zhang XK (1998) Molecular determinants of AHPN (CD437)-induced growth arrest and apoptosis in human lung cancer cell lines. Mol Cell Biol 18(8):4719–4731

Li H, Kolluri SK, Gu J, Dawson MI, Cao X, Hobbs PD, Lin B, Chen G, Lu J, Lin F, Xie Z, Fontana JA, Reed JC, Zhang X (2000) Cytochrome c release and apoptosis induced by mitochondrial targeting of nuclear orphan receptor TR3. Science 289(5482):1159–1164

Li QX, Ke N, Sundaram R, Wong-Staal F (2006) NR4A1, 2, 3–an orphan nuclear hormone receptor family involved in cell apoptosis and carcinogenesis. Histol Histopathol 21(5):533–540

Li X, Lee SO, Safe S (2012) Structure-dependent activation of NR4A2 (Nurr1) by 1,1-bis(3'-indolyl)-1-(aromatic)methane analogs in pancreatic cancer cells. Biochem Pharmacol 83(10):1445–1455

Lin B, Kolluri SK, Lin F, Liu W, Han YH, Cao X, Dawson MI, Reed JC, Zhang XK (2004) Conversion of Bcl-2 from protector to killer by interaction with nuclear orphan receptor Nur77/TR3. Cell 116(4):527–540

Liu S, Wu Q, Ye XF, Cai JH, Huang ZW, Su WJ (2002) Induction of apoptosis by TPA and VP-16 is through translocation of TR3. World J Gastroenterol 8(3):446–450

Liu J, Zhou W, Li SS, Sun Z, Lin B, Lang YY, He JY, Cao X, Yan T, Wang L, Lu J, Han YH, Cao Y, Zhang XK, Zeng JZ (2008) Modulation of orphan nuclear receptor Nur77-mediated apoptotic pathway by acetylshikonin and analogues. Cancer Res 68(21):8871–8880

Liu HB, Voso MT, Gumiero D, Duong J, McKendrick JJ, Dear AE (2009) The anti-leukemic effect of a novel histone deacetylase inhibitor MCT-1 and 5-aza-cytidine involves augmentation of Nur77 and inhibition of MMP-9 expression. Int J Oncol 34(2):573–579

Liu JJ, Zeng HN, Zhang LR, Zhan YY, Chen Y, Wang Y, Wang J, Xiang SH, Liu WJ, Wang WJ, Chen HZ, Shen YM, Su WJ, Huang PQ, Zhang HK, Wu Q (2010) A unique pharmacophore for activation of the nuclear orphan receptor Nur77 in vivo and in vitro. Cancer Res 70(9):3628–3637

Liu HB, Mayes PA, Perlmutter P, McKendrick JJ, Dear AE (2011) The anti-leukemic effect and molecular mechanisms of novel hydroxamate and benzamide histone deacetylase inhibitors with 5-aza-cytidine. Int J Oncol 38(5):1421–1425

Maira M, Martens C, Philips A, Drouin J (1999) Heterodimerization between members of the Nur subfamily of orphan nuclear receptors as a novel mechanism for gene activation. Mol Cell Biol 19(11):7549–7557

Maruyama K, Tsukada T, Ohkura N, Bandoh S, Hosono T, Yamaguchi K (1998) The NGFI-B subfamily of the nuclear receptor superfamily (review). Int J Oncol 12(6):1237–1243

Maxwell MA, Muscat GE (2006) The NR4A subgroup: immediate early response genes with pleiotropic physiological roles. Nucl Recept Signal 4:e002

McKenna NJ, Cooney AJ, DeMayo FJ, Downes M, Glass CK, Lanz RB, Lazar MA, Mangelsdorf DJ, Moore DD, Qin J, Steffen DL, Tsai MJ, Tsai SY, Yu R, Margolis RN, Evans RM, O'Malley BW (2009) Minireview: Evolution of NURSA, the Nuclear Receptor Signaling Atlas. Mol Endocrinol 23(6):740–746

Milbrandt J (1988) Nerve growth factor induces a gene homologous to the glucocorticoid receptor gene. Neuron 1(3):183–188

Mullican SE, Zhang S, Konopleva M, Ruvolo V, Andreeff M, Milbrandt J, Conneely OM (2007) Abrogation of nuclear receptors Nr4a3 and Nr4a1 leads to development of acute myeloid leukemia. Nat Med 13(6):730–735

Nagel S, Meyer C, Quentmeier H, Kaufmann M, Drexler HG, MacLeod RA (2008) MEF2C is activated by multiple mechanisms in a subset of T-acute lymphoblastic leukemia cell lines. Leukemia 22(3):600–607

Nsegbe E, Wallen-Mackenzie A, Dauger S, Roux JC, Shvarev Y, Lagercrantz H, Perlmann T, Herlenius E (2004) Congenital hypoventilation and impaired hypoxic response in Nurr1 mutant mice. J Physiol 556(Pt 1):43–59

Ordentlich P, Yan Y, Zhou S, Heyman RA (2003) Identification of the antineoplastic agent 6-mercaptopurine as an activator of the orphan nuclear hormone receptor Nurr1. J Biol Chem 278(27):24791–24799

Paulsen RF, Granas K, Johnsen H, Rolseth V, Sterri S (1995) Three related brain nuclear receptors, NGFI-B, Nurr1, and NOR-1, as transcriptional activators. J Mol Neurosci 6(4):249–255

Pearen MA, Muscat GE (2010) Minireview: Nuclear hormone receptor 4A signaling: implications for metabolic disease. Mol Endocrinol 24(10):1891–1903

Perlmann T, Jansson L (1995) A novel pathway for vitamin A signaling mediated by RXR heterodimerization with NGFI-B and NURR1. Genes Dev 9(7):769–782

Philips A, Lesage S, Gingras R, Maira MH, Gauthier Y, Hugo P, Drouin J (1997) Novel dimeric Nur77 signaling mechanism in endocrine and lymphoid cells. Mol Cell Biol 17(10):5946–5951

Pires NM, Pols TW, de Vries MR, van Tiel CM, Bonta PI, Vos M, Arkenbout EK, Pannekoek H, Jukema JW, Quax PH, de Vries CJ (2007) Activation of nuclear receptor Nur77 by 6-mercaptopurine protects against neointima formation. Circulation 115(4):493–500

Ponnio T, Conneely OM (2004) nor-1 regulates hippocampal axon guidance, pyramidal cell survival, and seizure susceptibility. Mol Cell Biol 24(20):9070–9078

Ponnio T, Burton Q, Pereira FA, Wu DK, Conneely OM (2002) The nuclear receptor Nor-1 is essential for proliferation of the semicircular canals of the mouse inner ear. Mol Cell Biol 22(3):935–945

Qin C, Morrow D, Stewart J, Spencer K, Porter W, Smith R, 3rd, Phillips T, Abdelrahim M, Samudio I, Safe S (2004) A new class of peroxisome proliferator-activated receptor gamma (PPAR-gamma) agonists that inhibit growth of breast cancer cells: 1,1-Bis(3'-indolyl)-1-(p-substituted phenyl)methanes. Mol Cancer Ther 3(3):247–260

Ramirez-Herrick AM, Mullican SE, Sheehan AM, Conneely OM (2011) Reduced NR4A gene dosage leads to mixed myelodysplastic/myeloproliferative neoplasms in mice. Blood 117(9):2681–2690

Safe S, Kim K, Li X, Lee SO (2011) NR4A orphan receptors and cancer. Nucl Recept Signal 9:e002

Salomoni P, Pandolfi PP (2002) The role of PML in tumor suppression. Cell 108(2):165–170

Sekiya T, Kashiwagi I, Inoue N, Morita R, Hori S, Waldmann H, Rudensky AY, Ichinose H, Metzger D, Chambon P, Yoshimura A (2011) The nuclear orphan receptor Nr4a2 induces Foxp3 and regulates differentiation of CD4+ T cells. Nat Commun 2:269

Seol W, Choi HS, Moore DD (1996) An orphan nuclear hormone receptor that lacks a DNA binding domain and heterodimerizes with other receptors. Science 272(5266):1336–1339

Shi Y (2007) Orphan nuclear receptors in drug discovery. Drug Discov Today 12(11-12):440–445

Shin HJ, Lee BH, Yeo MG, Oh SH, Park JD, Park KK, Chung JH, Moon CK, Lee MO (2004) Induction of orphan nuclear receptor Nur77 gene expression and its role in cadmium-induced apoptosis in lung. Carcinogenesis 25(8):1467–1475

To SK, Zeng JZ, Wong AS (2012) Nur77: a potential therapeutic target in cancer. Expert Opin Ther Targets 16(6):573–585

Vallat L, Magdelenat H, Merle-Beral H, Masdehors P, Potocki de Montalk G, Davi F, Kruhoffer M, Sabatier L, Orntoft TF, Delic J (2003) The resistance of B-CLL cells to DNA damage-induced apoptosis defined by DNA microarrays. Blood 101(11):4598–4606

Wang Z, Benoit G, Liu J, Prasad S, Aarnisalo P, Liu X, Xu H, Walker NP, Perlmann T (2003) Structure and function of Nurr1 identifies a class of ligand-independent nuclear receptors. Nature 423(6939):555–560

Wansa KD, Muscat GE (2005) TRAP220 is modulated by the antineoplastic agent 6-Mercaptopurine, and mediates the activation of the NR4A subgroup of nuclear receptors. J Mol Endocrinol 34(3):835–848

Wansa KD, Harris JM, Muscat GE (2002) The activation function-1 domain of Nur77/NR4A1 mediates trans-activation, cell specificity, and coactivator recruitment. J Biol Chem 277(36):33001–33011

Wansa KD, Harris JM, Yan G, Ordentlich P, Muscat GE (2003) The AF-1 domain of the orphan nuclear receptor NOR-1 mediates trans-activation, coactivator recruitment, and activation by the purine anti-metabolite 6-mercaptopurine. J Biol Chem 278(27):24776–24790

Wilson TE, Fahrner TJ, Johnston M, Milbrandt J (1991) Identification of the DNA binding site for NGFI-B by genetic selection in yeast. Science 252(5010):1296–1300

Wilson TE, Day ML, Pexton T, Padgett KA, Johnston M, Milbrandt J (1992) In vivo mutational analysis of the NGFI-A zinc fingers. J Biol Chem 267(6):3718–3724

Wilson AJ, Arango D, Mariadason JM, Heerdt BG, Augenlicht LH (2003) TR3/Nur77 in colon cancer cell apoptosis. Cancer Res 63(17):5401–5407

Woronicz JD, Calnan B, Ngo V, Winoto A (1994) Requirement for the orphan steroid receptor Nur77 in apoptosis of T-cell hybridomas. Nature 367(6460):277–281

Woronicz JD, Lina A, Calnan BJ, Szychowski S, Cheng L, Winoto A (1995) Regulation of the Nur77 orphan steroid receptor in activation-induced apoptosis. Mol Cell Biol 15(11):6364–6376

Wu Q, Liu S, Ye XF, Huang ZW, Su WJ (2002a) Dual roles of Nur77 in selective regulation of apoptosis and cell cycle by TPA and ATRA in gastric cancer cells. Carcinogenesis 23(10):1583–1592

Wu WS, Xu ZX, Ran R, Meng F, Chang KS (2002b) Promyelocytic leukemia protein PML inhibits Nur77-mediated transcription through specific functional interactions. Oncogene 21(24):3925–3933

Wu H, Lin Y, Li W, Sun Z, Gao W, Zhang H, Xie L, Jiang F, Qin B, Yan T, Chen L, Zhao Y, Cao X, Wu Y, Lin B, Zhou H, Wong AS, Zhang XK, Zeng JZ (2011) Regulation of Nur77 expression by beta-catenin and its mitogenic effect in colon cancer cells. FASEB J 25(1):192–205

Yoon K, Lee SO, Cho SD, Kim K, Khan S, Safe S (2011) Activation of nuclear TR3 (NR4A1) by a diindolylmethane analog induces apoptosis and proapoptotic genes in pancreatic cancer cells and tumors. Carcinogenesis 32(6):836–842

Zetterstrom RH, Solomin L, Mitsiadis T, Olson L, Perlmann T (1996) Retinoid X receptor heterodimerization and developmental expression distinguish the orphan nuclear receptors NGFI-B, Nurr1, and Nor1. Mol Endocrinol 10(12):1656–1666

Zhan Y, Du X, Chen H, Liu J, Zhao B, Huang D, Li G, Xu Q, Zhang M, Weimer BC, Chen D, Cheng Z, Zhang L, Li Q, Li S, Zheng Z, Song S, Huang Y, Ye Z, Su W, Lin SC, Shen Y, Wu Q (2008) Cytosporone B is an agonist for nuclear orphan receptor Nur77. Nat Chem Biol 4 (9):548–556

Zhang XK (2007) Targeting Nur77 translocation. Expert Opin Ther Targets 11(1):69–79

Zhao BX, Chen HZ, Lei NZ, Li GD, Zhao WX, Zhan YY, Liu B, Lin SC, Wu Q (2006) p53 mediates the negative regulation of MDM2 by orphan receptor TR3. EMBO J 25(24):5703–5715

Zhao Y, Howatt DA, Gizard F, Nomiyama T, Findeisen HM, Heywood EB, Jones KL, Conneely OM, Daugherty A, Bruemmer D (2010) Deficiency of the NR4A orphan nuclear receptor NOR1 decreases monocyte adhesion and atherosclerosis. Circ Res 107(4):501–511

Part V
Cell Surface Receptors

Chapter 27
Antibody-Based Therapeutics Targeting CD33, CD45, and CD66

Roland B. Walter, Oliver W. Press and Irwin D. Bernstein

Abstract Targeted therapy of acute myeloid leukemia (AML) with monoclonal antibodies (MoAbs) has thus far been mostly directed at CD33, CD45, and CD66. The notion that some AMLs may predominantly or entirely involve committed $CD33^+$ myeloid precursors provided the rationale for eradicating underlying stem cells with anti-CD33 antibodies. Emerging data demonstrating efficacy of the anti-CD33 immunoconjugate, gemtuzumab ozogamicin, in acute promyelocytic leukemia and other favorable- and intermediate-risk AMLs validate CD33 as drug target. Unlike CD33, CD45 and CD66 are noninternalizing antigens and have primarily been targeted to deliver radionuclides to sites of hematopoietic tissue, particularly to augment pre-transplant conditioning regimens. Early studies document the feasibility of this approach in selected patients, although future controlled studies will need to assess whether this strategy indeed leads to improved outcomes. Given this encouraging clinical experience, the use of MoAbs is likely expanding significantly in the future with identification of additional AML (stem) cell-associated antigens.

Keywords Antibody · CD33 · CD45 · CD66 · Gemtuzumab ozogamicin · Immunotherapy · Leukemic stem cells · Lintuzumab · Radioimmunotherapy

R. B. Walter (✉)
Clinical Research Division, Fred Hutchinson Cancer Research Center; Department of Medicine/Division of Hematology, University of Washington, 1100 Fairview Ave N, D2-190, 98109-1024 Seattle, WA, USA
e-mail: rwalter@fhcrc.org

O. W. Press
Clinical Research Division, Fred Hutchinson Cancer Research Center, Department of Medicine/Division of Medical Oncology, University of Washington, 1100 Fairview Ave N, D3-190, 98109-1024 Seattle, WA, USA
e-mail: press@uw.edu

I. D. Bernstein
Clinical Research Division, Fred Hutchinson Cancer Research Center, Department of Pediatrics, University of Washington, 1100 Fairview Ave N, D2-373, 98109-1024 Seattle, WA, USA
e-mail: ibernste@fhcrc.org

M. Andreeff (ed.), *Targeted Therapy of Acute Myeloid Leukemia,*
Current Cancer Research, DOI 10.1007/978-1-4939-1393-0_27

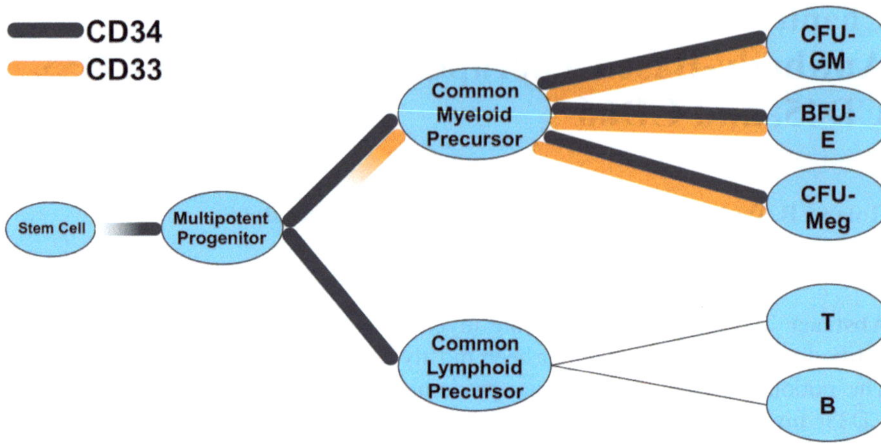

Fig. 27.1 CD33 and CD34 as differentiation antigens. Simplified hypothetical model of stem and progenitor cells in the human hematopoietic system, showing expression patterns of CD33 and CD34. (This figure was initially published in *Blood* (Walter et al. 2012). © The American Society of Hematology)

27.1 Introduction

As most current patients with acute myeloid leukemia (AML) will die of their disease, the need for effective yet relatively well-tolerated new therapies is unquestioned. Possibly unlike any other modality, monoclonal antibodies (MoAbs) have raised high expectations to accomplish this goal. Indeed, AML has been a paradigm for the therapeutic use of MoAbs because of well-defined cell-surface antigens and easy tumor cell accessibility. However, while AMLs often express aberrant types or abundances of cell-surface antigens relative to normal blood cell counterparts, true leukemia-specific antigens have yet to be identified. Nonetheless, several antigens have been exploited in AML, most notably CD33, CD45, and CD66. Herein, we discuss emerging clinical data on the effectiveness of unconjugated, toxin-loaded, and radiolabeled antibodies directed against these antigens, and highlight some of the limitations of current approaches and how these might be overcome.

27.2 Native and Toxin-Conjugated Anti-CD33 Antibodies

In healthy individuals, CD33 (SIGLEC-3) is primarily found on early multi-lineage hematopoietic progenitors and committed myeloid precursors, but not on hematopoietic stem cells (HSCs; Fig. 27.1) or outside the hematopoietic system (Andrews et al. 1983, 1986; Griffin et al. 1984). Leukemic blasts of the vast majority (85–90 %) of adult and pediatric AML patients express CD33 (Dinndorf et al. 1986; Griffin et al. 1984), but the impetus to pursue CD33 as therapeutic target emanated from the notion that some AMLs may predominantly or entirely involve committed

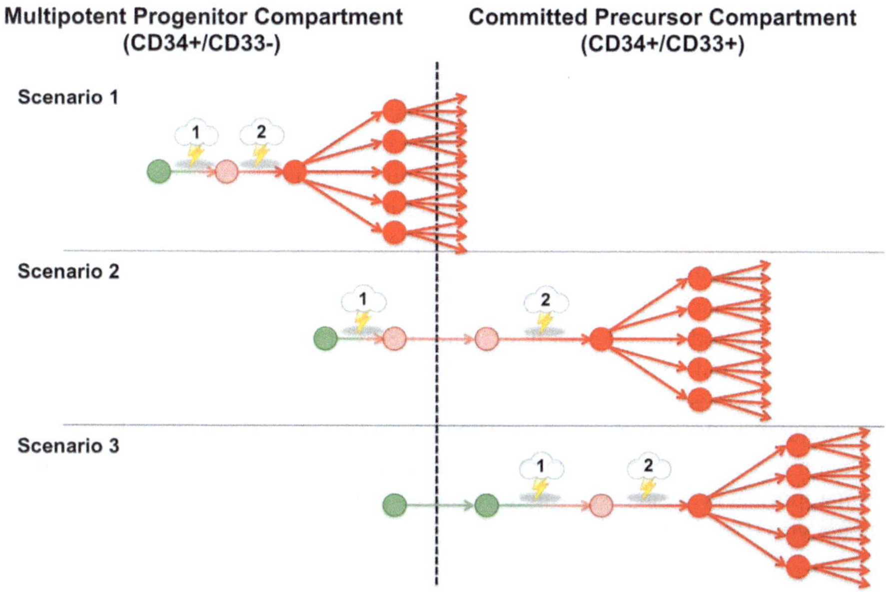

Fig. 27.2 Proposed models of AML transformation. Depicted are three proposed simplified scenarios of step-wise transformation in human AML leading from a normal cell (*green*) to a premalignant cell (*light red*) and, eventually, to a malignant cell (*dark red*) with clonal expansion. Similar models have been proposed by others (Jordan and Guzman 2004). (This figure was initially published in *Blood* (Walter et al. 2012). © The American Society of Hematology)

$CD33^+$ myeloid precursors, suggesting that this antigen could serve to eradicate underlying leukemic stem cells (LSCs) in such leukemias (Walter et al. 2012).

27.2.1 AML Stem/Progenitor Cells and CD33

The cellular origin of AML remains controversial (Dick 2008; Lane and Gilliland 2010; Majeti 2011; Passegué et al. 2003; Stubbs and Armstrong 2007; Walter et al. 2012). However, increasing evidence suggests heterogeneity among the LSCs across the spectrum of the human disease. The first hint of this diversity came from early studies on X-chromosome inactivation patterns that showed multi-lineage clonal dominance in some leukemias, reflecting origination and expansion at the level of $CD33^-$ HSCs. In others, dominance was limited to granulocytes and monocytes, suggesting that expansion of the malignant clone could occur at the committed $CD33^+$ myeloid precursor cell stage (Fialkow et al. 1981, 1987). While initial xenotransplantation models implied that HSCs were the general target of AML transformation, more recent studies indicate that transformation may occur in precursor cells beyond the HSC stage, consistent with the notion of LSC heterogeneity (Walter et al. 2012). Together with data supporting a multistep pathogenesis (Kelly and Gilliland 2002), these findings suggest three scenarios of AML transformation (Fig. 27.2; Walter et al. 2012). In the first scenario, initial and subsequent mutations occur in pluripotent $CD33^-$ HSCs (for simplicity referred to as "immature" leukemias). In contrast, in "mature" leukemias, the collaborating mutation leading to frank transformation occurs in a committed $CD33^+$ myeloid precursor, while the initial mutation is acquired in an HSC (scenario 2) or a committed precursor (scenario 3). An example for a "mature" leukemia may be APL, as small studies indicate that this disease is mainly expressed in granulocytes/monocytes and predominant involves $CD33^+$ precursors (Grimwade and Enver 2004).

As the LSC origin likely defines the display of surface antigens, the concept of LSC heterogeneity is essential for the design and success of LSC-directed therapies targeting such antigens. "Mature" leukemias may be particularly suited for stem-cell-directed therapies as underlying LSCs are expected to be distinct from normal HSCs and possibly intrinsically less resistant to therapy, as illustrated by the favorable prognosis of APL (Walter et al. 2012). A first attempt to eliminate LSCs in "mature" leukemias focused on CD33 based on in vitro studies showing that ablation of $CD33^+$ cells contained in human AML specimens could restore normal hematopoiesis in some cases where clonal dominance was limited to granulocytes/monocytes (Bernstein et al. 1987, 1992).

27.2.2 Treatment with Unconjugated Anti-CD33 MoAbs

Initial studies were conducted with a murine antibody (M195) in ten patients with AML. Although saturation of CD33 binding sites was observed with doses of ≥ 5 mg/m^2, only some patients had transient decreases in peripheral blast counts, even at saturating MoAb doses (Scheinberg et al. 1991). Subsequent studies employed a humanized IgG_1 construct of M195, lintuzumab (HuM195; SGN-33), which had greater than eightfold higher binding avidity than the parent antibody and, unlike M195, demonstrated antibody-dependent cell-mediated cytotoxicity and was not immunogenic (Caron et al. 1992, 1994). Still, as single agent, lintuzumab is only modestly active in overt non-APL AML, as suggested by a small trial on nine patients with relapsed/refractory disease, in which supersaturating doses of 12–36 mg/m^2 per day for 4 days for two courses resulted in sustained complete remission (CR) in one patient with low tumor burden ($<10\%$ marrow blasts) and reductions in marrow leukemia burden in three others despite saturation of CD33 binding sites throughout the entire 4-week trial period (Caron et al. 1998). Similar observations were made in a later study on 50 adults with relapsed/refractory AML randomized to receive lintuzumab at either 12 or 36 mg/m^2 per day on days 1–4 and 15–18: Only two CRs and one partial remission (PR) were achieved among 49 evaluable patients, all at the lower dose level and in patients with baseline marrow disease involvement of $<30\%$ (Feldman et al. 2003). More recently, higher doses of lintuzumab (1.5–8 mg/kg/week) for 5 weeks, followed by every other week treatment for those who experienced clinical benefit, were investigated in patients with $CD33^+$ myeloid malignancies. Among the 17 patients with AML, seven had an objective response (four morphologic CRs, two PRs, and one morphologic leukemia-free state) with a median duration of therapy of 25.1 (range, 4.1–57.1) weeks (Raza et al. 2009).

Unconjugated anti-CD33 MoAbs have also been investigated in APL. For example, when lintuzumab (3 mg/m^2 twice weekly for six doses) was added to all-*trans* retinoic acid (ATRA) in patients who entered morphological CR with ATRA alone, eight of the 18 evaluable subjects in first morphologic remission were RT-PCR negative before further consolidative chemotherapy when assessed 30 days after achievement of CR. This appeared to compare favorably to results from a previous trial where seven or 34 patients who continued ATRA alone were RT-PCR negative at the same time point ($p=0.07$) (Jurcic et al. 2000), but better-controlled data are required for conclusive assessment of lintuzumab's efficacy in APL.

Two randomized trials have tested lintuzumab together with conventional chemotherapy. In the first, 191 patients with relapsed/refractory AML were randomly assigned to receive mitoxantrone, etoposide, and cytarabine with or without lintuzumab (12 mg/m^2). Addition of lintuzumab was associated with an insignificantly better overall response rate (ORR; CR+CR with incomplete platelet recovery [CRp]: 36 vs. 28 %, $p=0.28$) but unchanged overall survival (OS) (Feldman et al. 2005). In the second, 211 patients older than age 60 with untreated AML were randomized to receive low-dose cytarabine with either lintuzumab (600 mg/week for four doses in the first cycle and every other week for two doses in subsequent cycles) or placebo in a double-blinded phase 2b study. Again, addition of lintuzumab did not improve OS (median, 4.7 vs. 5.1 months, $p=0.76$) (Sekeres et al. 2013). Because of these negative results, the clinical development of lintuzumab was terminated in 2010.

27.2.3 *Gemtuzumab Ozogamicin (GO)*

27.2.3.1 Development of GO

The cumulative experience with lintuzumab suggests that unconjugated anti-CD33 MoAbs are largely ineffective in patients with overt non-APL AML. As CD33 is an internalizing antigen, it was suspected early on that addition of a toxic payload might improve the therapeutic efficacy of anti-CD33 MoAbs (Walter et al. 2012). A murine antibody (P67.6) was thus conjugated to enediyne antitumor antibiotics ("calicheamicins"), which were initially detected in a screen for potent DNA-damaging drugs. A more stable disulfide derivative of calicheamicin-γ_1 (*N*-acetyl gamma calicheamicin dimethyl hydrazide), which remained 100–1000-fold more potent than clinically used anticancer agents, was ultimately selected for further testing. A key aspect for good potency and selectivity was the inclusion of a hydrolysable linker for rapid release of the toxic moiety under acidic conditions within the target cell. Subsequently, P67.6 was humanized into a human IgG_4 kappa framework (hP67.6) to minimize immunogenicity and conjugated with the calicheamicin derivative via an acid-labile linker to yield gemtuzumab ozogamicin (GO). GO showed selective cytotoxicity against CD33$^+$ AML cells and effectively inhibited colony-forming cells in human AML specimens in vitro, and caused regression of CD33$^+$ AML cell line xenografts in athymic mice (Walter et al. 2012).

Table 27.1 Cellular parameters implicated in gemtuzumab ozogamicin (GO) efficacy. (Modified from a table that was initially published in *Blood* (Walter et al. 2012). © The American Society of Hematology)

Factor	Comment
Uptake of CD33–GO complexes	
Receptor-mediated uptake	
CD33 expression levels	Quantitative relationship between CD33 expression and GO efficacy in engineered cell lines; expression levels associated with cytogenetic risk of AML and CD33 SNPs
CD33 saturation	In vitro evidence linking reduced CD33 saturation to reduced GO cytotoxicity
CD33 internalization	Relatively slow process, controlled by intracellular tyrosine motifs and likely tyrosine phosphorylation status of CD33
Re-expression of CD33 binding sites	Surface CD33 levels return to pretreatment levels within 72 h after CD33 antibody administration; could contribute to amount of internalized GO, in particular if given in fractionated doses
Nonreceptor mediated uptake	Suggested by experimental studies (clinical role unknown)
Intracellular trafficking of GO	Hypothetical
Activation of GO	Low pH in lysosomes required
Extrusion of GO	
ABC family of drug transporters	Established role of Pgp and MRP1; role of other transporters unknown
Induction of cytotoxicity	
Generation of SS- and DS-DNA breaks	Hypersensitivity of cell lines with defects in DNA repair to calicheamicins
Mitochondrial pathways of apoptosis	Good experimental evidence for role of pro- and anti-apoptotic Bcl-2 protein family members
Other downstream pro- or anti-apoptotic	Not examined in detail signaling pathways
Cell cycle status	Limited in vitro data suggesting that resting cells are relatively less susceptible to GO

DS double-stranded, *SNP* single-nucleotide polymorphism, *SS* single-stranded

27.2.3.2 Mechanism of Action of GO

hP67.6 is largely nontoxic and primarily facilitates uptake of the calicheamicin derivative into $CD33^{+}$ cells (Table 27.1). Internalized CD33–GO complexes are then routed to lysosomes, where the toxic moiety is presumably released. Calicheamicins have been shown to bind in the minor grove of DNA and, after activation by cellular thiols, initiate single- and double-stranded breaks. This DNA damage elicits a strong cellular response with cell cycle arrest followed by either DNA repair or, if damage is overwhelming, apoptosis and cell death, predominantly through mitochondrial pathways with cytochrome c release, involvement of Bcl-2 family proteins, and activation of caspases (Linenberger 2005; Walter et al. 2012).

This mechanism points to intracellular concentrations of the calicheamicin-γ_1 derivative as well as the cellular response to the toxin's DNA-damaging effect as critical determinants of GO efficacy (Table 27.1). Conceptually, the intracellular load of activated calicheamicin is affected by the amount of GO uptake, the efficacy of toxin release from the antibody and subsequent activation via cellular thiols as well as toxin inactivation/metabolism or extrusion. Consistent with this concept, efflux mediated by ATP-binding cassette transporters, predominantly P-glycoprotein (Pgp; ABCB1) and to a lesser degree multidrug resistance protein 1 (MRP 1), mediate resistance to GO (Linenberger 2005; Walter et al. 2012). Furthermore, experimental studies revealed a striking, quantitative relationship between CD33 expression/uptake and GO efficacy in engineered human AML cell lines (Jager et al. 2011; Walter et al. 2005), but demonstration of this relationship in clinical AML specimens has been more challenging. Explanations for this difficulty could include the recent observation that CD33 expression levels are inversely associated with favorable-risk disease features (Pollard et al. 2012) as well as the possibility that CD33 abundance on disease-relevant stem/progenitor subsets of AML cells may not be adequately reflected by the average CD33 expression of bulk blasts. On the other hand, the toxicity of the calicheamicin moiety will be modulated by the ability of the cell to repair DNA damage and the activity of downstream pro- and anti-apoptotic pathways. Unlike drug efflux and target antigen expression levels, these factors have not been studied in detail; however, their clinical relevance is suggested by the observation that the sensitivity to the calicheamicin derivative varies over 100,000-fold between individual primary AML cell samples (Goemans et al. 2008).

27.2.3.3 Early Clinical Studies and Accelerated Approval of GO

A phase 1 study in 40 adults with relapsed or refractory CD33$^+$ AML who received up to three doses of GO in 2-week intervals found morphologic elimination of leukemia in eight (20%) patients, with three (7.5%) achieving CR and two (5%) achieving CRp. As dose-limiting, nonhematologic toxicity was not reached, the highest dose level of 9 mg/m^2, which provided almost complete saturation of CD33 binding sites, was chosen for further study (Sievers et al. 1999). Three open-label, multicenter single-arm phase 2 trials then evaluated GO in a larger cohort of adults with CD33$^+$ de novo AML in first relapse. An interim analysis included 142 patients with a median age of 61 (range: 22 to 84) years who typically received two doses of GO 14 days apart. Based on an ORR of 29.6% (CR: 16.2%; CRp: 13.4%) (Sievers et al. 2001), GO was given accelerated US marketing approval in 2000 for treatment of adults > 60 years of age with CD33$^+$ AML in first relapse that were not candidates for standard cytotoxic chemotherapy (Bross et al. 2001). The final report on 277 patients confirmed the early results in relapsed AML, with 35 (12.6%) and 36 (13.0%) patients achieving CR and CRp, respectively, for an ORR of 25.6%, although remission durations were relatively short (Larson et al. 2005).

27.2.3.4 Subsequent Clinical Trials and Discontinuation of Commercial Availability of GO

Numerous studies have subsequently investigated GO in various clinical situations. As results from most of these trials have been reviewed previously in detail (Abutalib and Tallman 2006; Breccia and Lo-Coco 2011; Fenton and Perry 2005; Giles et al. 2003; Pagano et al. 2007; Stasi et al. 2008; Tsimberidou et al. 2006), we will only highlight those trials that are most informative regarding the potential clinical utility of GO.

Treatment of APL

GO is likely most effective in APL. For example, single agent GO resulted in molecular CR in nine of 11 patients with molecularly relapsed APL tested after two doses and in 13 of 13 patients tested after the third dose (Lo-Coco et al. 2004). Additional case reports corroborate the notion of very high efficacy of GO in the setting of minimal residual disease (MRD) in APL, including patients with very advanced disease (Breccia et al. 2007; Petti et al. 2001). Furthermore, phase 2 trials suggest that GO can substitute for anthracycline therapy, even in high-risk disease, and add benefit to therapy with ATRA (Estey et al. 2002; Ravandi et al. 2009).

Treatment of Pediatric AML

So far, only few clinical studies have explored GO in childhood AML, primarily in patients with relapsed/refractory disease, and GO has no established role in this patient population. However, a large randomized trial testing the addition of GO to induction chemotherapy was led by the Children's Oncology Group (AAML0531) and has recently completed accrual of >1000 participants, with preliminary outcome results expected for 2012 or 2013.

Treatment of Adult Non-APL AML

Several phase 2 studies investigated GO monotherapy in unselected patients with newly diagnosed and/or relapsed/refractory non-APL AML. While these studies confirmed single-agent activity of GO, the ORRs have typically not exceeded 25–35%, and were occasionally quite disappointing, particularly in heavily pretreated patients. Of note, most of these studies followed a twice-weekly schedule of GO administration. Less well explored is the use of GO given at shorter intervals. However, as new CD33 binding sites are continuously re-expressed and surface CD33 levels return to pretreatment levels within 72 h after antibody administration (Caron et al. 1994; Van Der Velden et al. 2001), repeated administration of lower, (near-) saturating doses of GO every 3 days may enhance intracellular drug accumulation beyond what can be accomplished with twice-weekly administration. The Acute Leukemia French Association (ALFA) group has relatively recently reported the use of GO in such fractionated doses with promising efficacy and acceptable toxicity (Farhat et al. 2012; Taksin et al. 2007), but no direct comparisons with the traditional administration schedule are available.

A wealth of studies has explored GO with other therapeutics or chemosensitizers with mixed results overall. As most were small, single-arm trials, no firm conclusions can be drawn from these studies as to the efficacy of GO relative to other drugs, or whether the addition of GO provided any benefit over what would have been achieved with the other therapeutics alone. Recently, however, several well-controlled large studies have been conducted that congruently show that GO improves survival in a significant and relatively well-defined subset of patients with newly diagnosed non-APL AML when added to conventional chemotherapy (Table 27.2; Walter et al. 2012). In the Medical Research Council (MRC) AML-15 trial, 1113 predominantly adult patients were randomized to receive a single dose of GO (3 mg/m^2) on day 1 of the first of two induction courses with one of three induction regimens (cytarabine/daunorubicin/etoposide, daunorubicin/cytarabine, fludarabine/cytarabine/G-CSF/idarubicin). While adding GO did not affect ORRs or survival across the entire study cohort, subgroup analyses showed that addition of GO improved OS at 5 years in patients with favorable cytogenetics (79 vs. 51%; $p=0.0003$) but not in those with unfavorable cytogenetics (8 vs. 11%; $p=0.4$). There was also a survival benefit for some intermediate-risk patients, as indicated by an internally validated index using cytogenetics, age, and performance status, which predicted that ~70% of these patients had an improved 5-year OS if given GO (Burnett et al. 2011).

In the ALFA-0701 trial, 278 patients with primary AML aged 50–70 years received daunorubicin/cytarabine with or without GO (3 mg/m^2) on days 1, 4, and 7; a second course of daunorubicin/cytarabine was given for residual disease on day 15. Patients in remission then received two courses of daunorubicin/cytarabine with or without GO (3 mg/m^2) on day 1 of each cycle. While there was no statistically significant difference in ORRs and treatment-related mortality (TRM) between the two arms, the event-free survival (EFS) at 2 years was significantly superior in the GO arm (40.8 vs. 17.1%, $p=0.0003$), as was disease-free survival (DFS; 50.3 vs. 22.7%, $p=0.0003$) and OS (53.2 vs. 41.9%, $p=0.037$). Subgroup analysis showed that the EFS benefit occurred in patients with favorable/intermediate cytogenetics but not in those with an adverse karyotype (Castaigne et al. 2012). In the GOELAMS AML 2006 IR study, adults aged 18–60 years with de novo AML and intermediate karyotype received daunorubicin/cytarabine with or without GO (6 mg/m^2) on day 4; GO was also added to consolidation therapy according to the initial randomization (Delaunay et al. 2011). Among 238 analyzed patients, there was no difference in ORRs or TRM between the two treatment arms. Overall, there was also no statistically significant difference in EFS (GO vs. control: 51 vs. 33%) or OS (53 vs. 46%) at 3 years. However, subgroup analyses showed that in patients who did not undergo allogeneic hematopoietic cell transplantation (HCT), EFS was significantly higher in the GO group (53.7 vs. 27%, $p=0.0308$). Finally,

Table 27.2 Phase 3 studies of gemtuzumab ozogamicin (GO) in newly diagnosed non-APL AML. (This table was initially published in *Blood* (Walter et al. 2012). © The American Society of Hematology)

Study	Disease	*N*	Age (median)	Treatment	Results
MRC AML15 (Burnett et al. 2011)	AML	1113	0–71 (49)	± GO (3 mg/m^2) on day 1 of the first of two induction courses with either ADE or DA or FLAG-IDA	No difference in ORR, TRM, relapse, or survival. Improved 5-year OS for favorable-risk subgroup (79 vs. 51 %; $p=0.0003$) with GO; predicted 10 % OS benefit for ~70 % of patients with intermediate-risk disease
ALFA 0701 (Castaigne et al. 2012)	AML	278	59–66 (62)	± GO (3 mg/m^2) on days 1, 4, and 7 of DA induction and day 1 of each of two courses of DA consolidation	No difference in ORR or mortality; improved 2-year EFS (40.8 vs. 17.1 %, $p=0.0003$), DFS (50.3 vs. 22.7 %, $p=0.0003$), and OS (53.2 vs. 41.9 %, $p=0.037$) with GO; survival benefit seen in favorable/intermediate, but not adverse-risk disease
GOELAMS AML 2006 IR(Delaunay et al. 2011)	AML (intermediate risk)	238	18–60 (50)	± GO (6 mg/m^2) on day 1 with DA induction and MA consolidation	No difference in ORR, TRM, or 3-year EFS; improved EFS with GO in patients who did not undergo allogeneic HCT (53.7 vs. 27 %, $p=0.0308$)
NCRI AML16 (Burnett et al. 2012)	AML, high-risk MDS	1115	51–84 (67)	± GO (3 mg/m^2) on day 1 of the first of two induction courses with either DA or DCLo	No difference in TRM; trend towards reduced risk of persistent disease with GO (17 vs. 21 %, $p=0.06$); reduced 3-year relapse risk (68 vs. 76 %; $p=0.007$) and superior DFS (21 vs. 16 %; $p=0.04$) and OS (25 vs. 20 %; $p=0.05$) with GO
SWOG S0106 (Petersdorf et al. 2009; SWOG 2010)	AML	596	18–60 (47)	± GO (6 mg/m^2) on day 4 of the first of up to two induction courses with DA[a]	Increased TRM in GO arm (5.7 vs. 1.4 %; $p=0.01$); no difference in ORR, DFS, or OS; possible trend towards improved OS in favorable-risk subgroup with GO (hazard ratio: 0.49 [0.12–2.04])

ADE cytarabine/daunorubicin/etoposide, *DA* daunorubicin/cytarabine, *DClo* daunorubicin/clofarabine, *DFS* disease-free survival, *EFS* event-free survival, *FLAG-Ida* fludarabine/cytarabine/G-CSF/idarubicin, *HCT* hematopoietic cell transplantation, *MA* mitoxantrone/cytarabine, *ORR* overall response rate (CR+CRi), *OS* overall survival, *TRM* treatment-related mortality

[a] Daunorubicin was used at 60 mg/m^2 /day in the control (-GO) arm and 45 mg/m^2 /day in the GO arm

in the NCRI AML-16 trial, 1115 adults aged 51–84 years with AML or high-risk myelodysplastic syndromes (MDS; >10% marrow blasts) received either daunorubicin/cytarabine or daunorubicin/clofarabine for two cycles with or without GO (3 mg/m^2) on day 1 of the first induction course (Burnett et al. 2012). Similar to the other trials, there was no significant difference in ORRs and TRM between the treatment arms, although the patients receiving GO had a slightly lower likelihood of persistent disease (17 vs. 21%, $p=0.06$). However, use of GO was associated with reduced relapse risk (at 3 years: 68 vs. 76%; $p=0.007$) and superior DFS (21 vs. 16%; $p=0.04$) as well as OS (25 vs. 20%; $p=0.05$). A meta-analysis of AML15 and AML16 on 2224 patients showed a significant benefit of GO for risk of relapse (odds ratio [OR]: 0.82 [95% CI: 0.72–0.93], $p=0.002$) and survival (OR: 0.88 [0.79–0.98], $p=0.02$); the survival benefit was seen in patients with favorable-risk (OR: 0.47 [0.28–0.77]) and intermediate-risk (OR: 0.84 [0.73–0.97]) disease but not adverse-risk (OR: 1.02 [0.81–1.27]) disease (Burnett et al. 2012).

The results from these European studies differ from the Southwest Oncology Group (SWOG) trial S0106, which was developed with the drug sponsor to fulfill the post-approval commitment to the US Food and Drug Administration (FDA). Six hundred thirty-seven patients aged 18–60 years with de novo AML were accrued to receive up to two cycles of induction chemotherapy with daunorubicin/cytarabine with or without a single dose of GO (6 mg/m^2) on day 4 of the first induction (Petersdorf et al. 2009). Unlike the European studies in which identical doses of conventional chemotherapeutics were used in both arms, S0106 used a lower daunorubicin dose with GO (45 vs. 60 mg/m^2). Overall, among the 596 evaluable patients, S0106 showed no difference in response or survival in either induction of post-consolidation with the addition of GO, but a trend towards improved OS was seen in patients with favorable-risk leukemias (hazard ratio: 0.49 [0.12–2.04], although S0106 was not powered to detect important outcome differences in this patient subset. As there was an attempt to achieve equitoxicity with the lower dose of daunorubicin in the GO arm, the S0106 trial is confounded by the lower dose of daunorubicin administered to patients receiving GO, which might mask a greater benefit of the immunoconjugate. Nonetheless, based on the lack of pre-specified overall improvement in outcome, S0106 was prematurely terminated (SWOG 2010). Following discussions with the FDA, Pfizer voluntarily withdrew the new drug application for GO in the US in 2010 (Pfizer Inc 2010). GO was withdrawn from the US and European markets, but continues to be commercially available in Japan, where it has received full regulatory approval (Ricart 2011). Undoubtedly, however, with the emerging data on the benefit of GO for defined subsets of AML, reconsideration of the drug's value by the manufacturer and regulatory authorities is warranted (Walter et al. 2012).

27.2.4 Treatment with Other Conjugated Anti-CD33 MoAbs

More recently, a humanized IgG_1 anti-CD33 MoAb (huMy9-6) conjugated to a thiol-containing maytansinoid derivative (DM4), a potent tubulin inhibitor, has entered clinical testing. Three phase 1 studies of AVE9633 (huMy9-6-DM4) as a single agent have been performed in 54 patients with refractory/relapsed AML to determine the best schedule of administration of this immunoconjugate. However, efficacy was only modest, with one CRp, one PR, and biological activity (>50% decrease of blasts either in the peripheral blood with low blast counts or in the bone marrow) in five other patients. Two of the three schedules were discontinued early because of drug inactivity at doses significantly higher than CD33 saturation, and clinical development of Ave9633 was terminated (Lapusan et al. 2012).

27.2.5 Synthesis of Clinical Experience with Native and Toxin-Conjugated Anti-CD33 MoAbs

Despite the disappointing activity of unconjugated anti-CD33 MoAbs or AVE9633, the clinical experience with GO supports the conclusion that CD33 is a valid target for some AMLs. The long-term benefit achieved with GO in APL and other favorable-risk leukemias, and possibly some patients with intermediate-risk disease, would be consistent with an ablative effect on $CD33^+$ AML stem or progenitor cells. The most compelling data to support this notion come from the treatment of APL, although the high and homogeneous expression of CD33 and low activity of Pgp may partially explain the exquisite sensitivity of APL towards GO. However, studies to date have not determined whether GO, besides "debulking" of mature $CD33^+$ progeny, can directly kill $CD33^+$ LSCs, and whether long-term benefit from GO is related to successful targeting of LSCs (Walter et al. 2012).

It has become clear that CD33 is challenging to target with immunotoxins because of antigen expression at relatively low abundance (about 10^4 CD33 molecules/cell), slow conjugate internalization, and ABC drug transporter activity in AML cells (Walter et al. 2012). Given these limiting factors, rational strategies to improve CD33-targeted immunotherapy would therefore include maneuvers that increase CD33 expression/turnover or inhibit drug efflux (Jager et al. 2011). An alternative entails the development of novel immunoconjugates employing more potent toxins that are not (or less) susceptible to ABC transporter activity. Additionally, approaches that interfere with pro-/anti-apoptotic signaling downstream of the toxin-inflicted cellular damage (e.g., DNA damage in the case of calicheamicins) may lead to a lowering of the apoptotic threshold and increased cellular toxicity of immunotoxins.

27.3 Treatment with Radiolabeled Antibodies Targeting CD33, CD45, and CD66

Early HCT experience established the anti-leukemia efficacy of radiation (Thomas et al. 1975). However, while relapse rates in AML decrease upon escalation of total body irradiation (TBI) in a dose-dependent manner, this benefit is offset by increased toxicities and higher early nonrelapse mortality (NRM) (Clift et al. 1990, 1998). Nevertheless, this observation suggests an improved outcome if additional radiation could be directed specifically to hematopoietic tissue and directly provides the rationale for the exploration of radioimmunotherapy (RIT).

So far, CD33, CD45, and CD66 have been exploited most for this purpose. Since normal and malignant cells express CD33 and CD45, antibodies targeting these antigens can deliver radiation to marrow, spleen, and lymph nodes in patients with high or low tumor burden. The types of emissions employed have primarily focused on β-particles (^{131}I, ^{90}Y, ^{188}Re), which deposit energy over a relatively long distance and allow irradiation of many cells adjacent to the decay position if sufficient radiation is deposited. This offers the advantage of targeting cells that express limited amounts of the target antigen, or lack it completely, by relying on "crossfire" for cell kill of neighboring antigen-negative cells; this mechanism is particularly important for CD66-targeted therapies as the malignant clone commonly lacks CD66. A disadvantage of this approach, however, is that stem cell rescue may be required, since not only targeted but also surrounding normal hematopoietic cells are exposed to radiotherapy (Appelbaum 1999). More recent studies have used α-emitters (e.g., ^{213}Bi or ^{225}Ac): the short-range energy deposition may destroy targeted cells due to high linear transfer energy but spare neighboring cells, a potential advantage if avoidance of ablative marrow toxicity is a goal (Matthews and Appelbaum 2004).

27.3.1 CD33-Targeted RIT

Radionuclides are an attractive alternative to toxins to increase the efficacy of anti-CD33 MoAbs, and several studies have explored such antibodies as carriers for β- as well as α-emitters in AML (Walter and Pagel 2011).

27.3.1.1 β-Emitting Anti-CD33 MoAbs

Initial studies, conducted at Fred Hutchinson Cancer Research Center (FHCRC) and Memorial Sloan-Kettering Cancer Center (MSKCC), focused on murine and humanized anti-CD33 MoAbs labeled with ^{131}I, a radionuclide chosen for its ready availability, low cost, and simple radiochemistry. With ^{131}I, the MoAbs delivered greater amounts of radiation to marrow and spleen than the liver (usually the normal organ receiving the highest radiation dose), lungs, or kidneys. When used alone, cytoreductive effects on AML cells were seen in many patients, and profound cytopenias occurred with higher doses of the radionuclide. In conjunction with conditioning regimens, CRs were achieved in the majority of patients transplanted without engraftment complications and with little toxicity related to the radioisotope except perhaps some increase in liver toxicity (Caron et al. 1992, 1994; Pagel et al. 2006b, 2008, 2009a, b).

While these early investigations overall demonstrated the feasibility of using these radiolabeled antibodies alone or as part of a standard HCT preparative regimen for advanced AML, they identified important limitations, in particular the short marrow residence time (Appelbaum et al. 1992; Scheinberg et al. 1991), a phenomenon likely explained by antibody internalization with subsequent deiodination and release of ^{131}I (Van Der Jagt et al. 1992). As ^{90}Y provided superior retention of radioactivity relative to ^{125}I when conjugated to anti-CD33 antibodies in pre-clinical studies (Press et al. 1996), limited clinical efforts were undertaken to study the potential use of ^{90}Y-HuM195 (Jurcic 2001). Furthermore, the low abundance of CD33 resulted in saturation at doses of 3–5 mg/m^2, rendering dose-escalations via higher individual doses impossible, although repeated dosing in 48–72-h intervals could partly overcome this limitation by taking advantage of re-expressed CD33 binding sites during antibody infusions (Caron et al. 1994; Scheinberg et al. 1991).

27.3.1.2 α-Emitting Anti-CD33 MoAbs

To reduce bystander toxicity and allow selective leukemia cell kill, e.g., in the setting of MRD, the MKSCC group explored the use of high energy, short-range α-emitters, initially utilizing ^{212}Bi-HuM195 and ^{213}Bi-HuM195 (Mcdevitt et al. 1999; Nikula et al. 1999). Despite similar internalization, these bismuth radioimmunoconjugates were retained about two- to threefold better than ^{131}I-HuM195, likely due to binding to transferrin and other intracellular proteins, and showed efficient killing of a CD33$^+$ AML cell line with as little as two bismuth atoms bound per cell. A phase 1 trial in 18 patients with advanced myeloid leukemias treated with ^{213}Bi-HuM195 demonstrated rapid localization to, and retention in, areas of disease involvement; there was no significant extramedullary toxicity, and absorbed dose ratios between sites of leukemic involvement and the whole body were 1000-fold greater than those seen with β-emitting anti-CD33 MoAbs. Although ^{213}Bi-HuM195 had clear antileukemic activity—14 of 15 evaluable patients had reductions in circulating blasts, and 14 of 18 evaluable patients had reductions in the percentage of bone marrow blasts—no remissions were achieved. It was thought that this limited clinical efficacy was due to the difficulty of delivering the one to two ^{213}Bi atoms necessary for cell kill to every leukemia cell at the specific activities (Jurcic et al. 2002). A phase 1/2 study then evaluated the effects of ^{213}Bi-HuM195 after partial cytoreduction with cytarabine (200 mg/m^2/day for 5 days) in 31 patients with newly diagnosed or relapsed/refractory AML. Clinical responses were seen in six of the 25 patients, all of which with poor-risk disease, who received ^{213}Bi-HuM195 at doses greater or equal the maximum tolerated dose (MTD; $\geq$37 MBq/kg), including two CR, two CRp, and two PR, whereas none of the six patients receiving $<$37 MBq/kg had a response. Besides severe cytopenias and myelosuppression-associated febrile episodes, drug-related toxicities were relatively mild and transient.

A major hurdle for the use of ^{213}Bi is the radionuclide's short half-life, which results in short marrow residence time and requires an on-site ^{225}Ac/^{213}Bi generator. As one strategy to overcome this limitation, investigators at MSKCC took advantage of a single ^{225}Ac atom as "nanogenerator" attached to a therapeutic MoAb. ^{225}Ac decays with a half-life of 10 days and generates therapeutic, short-lived daughter α-particles (^{221}Fr [$t_{1/2}$=4.8 min], ^{217}At [$t_{1/2}$=32.3 ms, and ^{213}Bi [$t_{1/2}$=45.6 min]) at or inside the targeted cell. In proof-of-principle studies, such ^{225}Ac-containing immunoconjugates could kill tumor cell lines at radioactivity doses 1000 times lower than ^{213}Bi analogs in vitro and prolong survival of animals in several xenograft models (Mcdevitt et al. 2001). As the daughter α-particles may no longer be bound to the antibody, however, the likelihood of toxicities increases, and studies in cynomolgus monkeys indeed indicated that high doses of ^{225}Ac-labeled MoAbs could cause renal toxicity and anemia (Miederer et al. 2004). Preliminary results from a phase 1 trial on seven patients with relapsed/refractory AML indicate that a single dose of ^{225}Ac-HuM195 (0.5–2 μCi/kg) is safe and has antileukemic activity, with elimination of circulating blasts in three of six evaluable patients and dose-related reductions of >33 % of bone marrow blasts in four patients at 4 weeks following treatment (Rosenblat et al. 2007). This trial is ongoing, and a follow-up trial using ^{225}Ac-HuM195 in combination with conventional chemotherapy is currently being initiated.

27.3.2 CD45-Targeted RIT in AML

CD45 (leukocyte common antigen) is found on almost all hematopoietic cell lineages except mature thrombocytes, mature erythrocytes, and some of their progenitors. Most hematologic malignancies, including the vast majority of acute leukemias, express CD45 (Andres and Kadin 1983; Dahlke et al. 2004; Nakano et al. 1990). It is expressed in a relatively high copy number (200,000/cell) and is not appreciably internalized or shed after ligand binding (Press et al. 1994; Van Der Jagt et al. 1992). CD45 exists in several isoforms but the CD45-targeting antibodies used so far for RIT approaches are pan specific (Dahlke et al. 2004).

27.3.2.1 Conventional RIT Targeting CD45

The lack of internalization and much higher abundance of CD45 on normal and malignant blood cells provide two important advantages over CD33 as target for RIT. Indeed, early preclinical studies by investigators at FHCRC demonstrated that ^{131}I-anti-CD45 MoAbs bound AML cells with longer cell surface retention and provided superior tumor targeting of AML xenografts in athymic mice compared to ^{131}I-anti-CD33 antibodies (Van Der Jagt et al. 1992). With these antibodies, radiation could be delivered relatively specifically to hematopoietic and lymphoid tissues (two- to threefold more to marrow, up to 12-fold more to spleen, and two- to eightfold more to lymph nodes than to lung, liver, or kidney) in mice and macaques (Matthews et al. 1991, 1992). Because of significant myelosuppression, clinical studies with radiolabeled anti-CD45 MoAbs have thus far relied on HCT to reconstitute hematopoiesis.

Initial phase 1 investigations in 44 patients with advanced acute leukemia or MDS were conducted with ^{131}I-BC8 (murine IgG_1 anti-CD45 MoAb) followed by standard conditioning with cyclophosphamide (120 mg/kg) and TBI (12 Gy) and HLA-matched related allogeneic or autologous HCT (Matthews et al. 1999). On average, estimated radiation doses to marrow and spleen were 2.3-fold and 4.8-fold greater than to the liver, respectively, and 37 of the subjects (84 %) had favorable MoAb biodistribution, findings that compared favorably with the institution's previous experience using ^{131}I-labeled anti-CD33 MoAbs. The MTD was found to be 10.5 Gy delivered to the normal organ receiving the highest dose (liver in all cases), with occurrence of grade 3/4 mucositis as the dose-limiting toxicity; at the MTD level, this approach yielded an average of 24 Gy and 50 Gy more radiation delivered to the marrow and spleen, respectively, without excessive toxicity. Of 25 treated patients with AML/MDS, seven survived disease-free 15–89 (median, 65) months post transplant. Later analyses of 18 patients with high-risk AML/MDS undergoing HLA-matched related or unrelated allogeneic HCT treated with this regimen suggested that a greater radiation-absorbed dose delivered to marrow may lead to lower post-transplant relapse rates: specifically, only one of nine patients relapsed in the high (>7 cGy/mCi) marrow-absorbed dose, while six of nine patients relapsed in the lower dose (<7 cGy/mCi) cohort, with hazard of relapse in this group more than six times higher than that seen in the group that received the higher marrow dose (Pagel et al. 2006b).

Given this early clinical experience, a phase 1/2 trial then incorporated ^{131}I-BC8 into a conditioning regimen with targeted busulfan (area-under-curve: 600–900 ng/mL) and cyclophosphamide (120 mg/kg) for patients with AML in first CR (Pagel et al. 2006a). Of 59 patients evaluated, 52 (88 %) had favorable antibody biodistribution, and 46 received therapeutic doses of ^{131}I-BC8, which delivered 5.3–19.0 Gy radiation to the marrow, 17–72 Gy to the spleen, and 3.5–5.25 Gy to the liver. The estimated 3-year NRM and DFS were 21 and 61 %, respectively. These results were compared with those from 509 similar patients from the International Bone Marrow Transplant Registry, and it was estimated that the hazard of mortality was 0.65 (95 % CI: 0.39–1.08, $p=0.09$) relative to that of the registry patients.

More recently, ^{131}I-BC8 was explored as addition to reduced-intensity conditioning with fludarabine (30 mg/m^2/day for 3 days) and low-dose TBI (2 Gy) in 58 patients >50 years of age with advanced AML or MDS (Pagel et al. 2009a). This study demonstrated that a maximum dose of 24 Gy delivered by ^{131}I-BC8 to the liver could be tolerated; this 24 Gy dose was estimated to provide 36 Gy radiation to marrow and 102 Gy to spleen. The 100-day NRM in this study was 12 %, and estimated 1-year probabilities of relapse and survival were 40 and 41 %, respectively, for the entire cohort, with 1-year survival estimated at 48 % for the 21 patients who received ^{131}I-BC8 at the MTD (Pagel et al. 2009a).

27.3.2.2 Anti-CD45 MoAb-Based Pretargeted RIT (PRIT)

While these clinical studies with conventional anti-CD45 RIT have yielded encouraging results, many patients still relapse despite myeloablative doses of radionuclides. Conventional RIT has often been limited by a suboptimal therapeutic index (target-to-nontarget ratio) and an inability to selectively deliver sufficient radiation doses to tumors. Conceptually, PRIT circumvents these limitations by separating the targeting vehicle from the subsequently administered therapeutic radioisotope, which binds to the tumor-localized antibody or is quickly excreted if unbound. A growing number of preclinical studies demonstrate that PRIT is feasible, safe, and provides improved radionuclide delivery to malignant cells compared to conventional RIT while sparing normal cells from nonspecific radiotoxicity (Walter et al. 2010). In AML, PRIT efforts have so far focused on CD45. Preclinical murine studies directly compared conventional anti-CD45 RIT with an anti-CD45 streptavidin conjugate in combination with a dendrimeric synthetic biotinylated clearing agent and radiolabeled biotin. These studies demonstrated the superiority of PRIT over conventional RIT with regard to tumor localization and tumor-to-blood and tumor-to-normal organ radioactivity concentration ratios and markedly enhanced therapeutic efficacy; radiation dose calculations suggested that PRIT could deliver twice as much radiation-absorbed dose to the marrow and five times more to the spleen relative to conventional RIT (Pagel et al. 2008, 2009b). These data indicated that anti-CD45 PRIT might allow intensification of targeted radiotherapy to sites of leukemic involvement with diminished toxicity. These findings were corroborated by studies in nonhuman primates showing that administration of an anti-CD45 streptavidin fusion protein [BC8 $(scFv)_4SA$] with radio-DOTA-biotin significantly improved the target-to-normal organ ratio of absorbed radiation compared to directly radiolabeled bivalent anti-CD45 antibody. The kidney was the only nontarget organ that experienced higher levels of measured radiation in the pretargeted animal, suggesting that nonspecific radiation exposure to the kidney may define the dose-limiting toxicity with this 2-step PRIT approach (Green et al. 2009). Overall, the results from these experimental animal studies would support translation of the anti-CD45 PRIT strategy to the clinic for patients with AML.

27.3.3 CD66-Targeted RIT in AML

Similar to CD45, CD66 antigens are neither internalized nor shed (Becker et al. 1994). They are expressed on members of the carcinoembryonic-antigen-related cell-adhesion molecule (CEACAM) family of proteins (Gray-Owen and Blumberg 2006). Some of these glycoproteins are found on hematopoietic cells as well as on epithelial or endothelial cells. However, while they are present on myeloid cells from the late myeloblast or early promyelocyte stage, early myeloid progenitors or multipotent progenitors do not express CD66 antigens, and they are only occasionally found on AML cells. Thus, successful RIT of AML will primarily depend on bystander effects. Although the expression patterns may differ between the individual CD66 antigens, this is currently of limited significance for RIT as many anti-CD66 antibodies recognize several CD66 antigens (Walter and Pagel 2011).

Most clinical experience has so far been gained with an anti-CD66 antibody (BW 250/183) labeled with ^{188}Re, which was chosen because of its high energy β-emissions and, relative to ^{131}I, shorter half-life. An initial phase 1/2 study conducted by investigators at the University of Ulm explored the feasibility of adding ^{188}Re-labeled BW 250/183 to various conditioning regimens in patients with high-risk AML or MDS undergoing allogeneic or autologous HCT, primarily with T-cell-depleted stem cell grafts (Bunjes et al. 2001). A favorable dosimetry was observed in all patients. RIT had minimal infusion-related toxicities and delivered a mean ± SD of 15.3 ± 4.9 Gy of additional radiation to the marrow, 19.5 ± 17.5 Gy to the spleen, 6.0 ± 2.3 Gy to the liver, and 7.4 ± 2.3 Gy to the kidney. Thus, in contrast to radiolabeled anti-CD45 MoAbs, the normal organ that received the highest radiation was the kidney, and late renal toxicity was observed in 17 % of the patients. In fact, as demonstrated later, the application of ^{188}Re-BW 250/183 at the doses used leads to excess renal toxicity despite a concomitant 50 % reduction of the external-beam contribution to the kidney dose (Rottinger et al. 2003). An update of this study focused on 57 patients with acute leukemia, 53 of which had AML. In this cohort, the TRM was 30 % after a median follow-up of 26 months, and 14 % of the patients had late renal toxicity. This study demonstrated a significant difference between the 64 % disease-free survival for 44 patients either that were in CR or that had < 15 % blasts in the marrow at the time of transplant as compared to 8 % for 13 patients that had > 15 % blasts at the time of transplant (Bunjes 2002)

Less encouraging were the results from a comparable study on 19 patients treated with similar doses of ^{188}Re-BW 250/183 in combination with various preparative regimens followed by HCT (Klein et al. 2002). In this study, 13 patients received unmanipulated grafts in combination with immunosuppressive therapy, whereas six patients received allografts with CD34$^+$ selected cells but were given T-cells back every 4–6 weeks starting at 28 days post transplant to facilitate the "graft-versus-leukemia" effect. Fifteen of 19 patients developed acute graft versus host disease (GVHD), with eight having grade three or four disease. Eleven of the 19 patients had intestinal GVHD, which was fatal in at least four cases. Overall, TRM was high (9 of 19 patients). Given the apparent differences in outcome between this study and the study using T-cell-depleted grafts, it was initially thought that the presence of T-cells in donor grafts might account for the high TRM and incidence of severe GVHD (Klein et al. 2002). However, data from a more recent study, in which 21 patients with AML or MDS were treated with ^{188}Re-BW 250/183 followed by myeloablative HCT with busulfan/cyclophosphamide or reduced-intensity HCT with fludarabine/busulfan or fludarabine/melphalan with unmanipulated allogeneic graft may contradict this hypothesis (Koenecke et al. 2008). On average, the radioimmunoconjugate delivered 10.9 (5.0–21.3) Gy to the bone marrow, 17.1 (7.3–41.0) Gy to the spleen, 3.6 (1.1–11.5) Gy to the liver, and 7.8 (3.9–12.2) Gy to the kidney. Interestingly, three patients developed veno-occlusive disease of the liver. Six patients died from treatment-related causes. Acute and chronic GVHD developed in

8 of 21 and 9 of 16 patients, respectively. Six patients relapsed 41 to 367 days post transplant. At a median follow-up of 42 months, the disease-free survival was 43 % (Koenecke et al. 2008).

For older high-risk AML or MDS patients, BW 250/183 labeled with either ^{188}Re or ^{90}Y was combined with a reduced-intensity conditioning regimen. A phase 1/2 study comprised 20 patients, in which fludarabine or fludarabine/melphalan-based conditioning was used for matched related and matched unrelated donor grafts that were T-cell depleted with alemtuzumab-based strategies (Ringhoffer et al. 2005). The radioimmunoconjugates provided a mean of 21.9 ± 8.4 Gy radiation to the marrow space. While toxicity was relatively low with a 100-day mortality of 15 %, the cumulative incidence of relapse remained high (55 % at 30 months post transplant). The risk of relapse was significantly higher for patients not in remission compared to those in either first or second remission (60 vs. 42 %) (Ringhoffer et al. 2005), possibly highlighting the difficulty of targeting low-energy β-emissions to AML blasts by relying on crossfire from surrounding CD66$^+$ normal myeloid cells. Of note, relative to ^{188}Re, ^{90}Y delivered a significantly higher radiation dose to bone marrow and liver, whereas the kidney dose was nonsignificantly lower, possibly due to a higher in vivo stability of the ^{90}Y conjugate, which would reduce exposure of the renal cortex to dissociated radionuclides and indirectly increase the bone marrow dose (Ringhoffer et al. 2005). A later study confirmed that the absorbed renal dose was lower in patients receiving BW 250/183 labeled with ^{90}Y relative to those receiving ^{188}Re-labeled antibody, which may have clinical significance as only patients receiving ^{188}Re-BW 250/183 developed post-transplant nephropathy (incidence about 20 %), particularly when used with TBI (Zenz et al. 2006).

27.3.4 Synthesis of Clinical Experience with RIT in AML

Thus far, the clinical investigations have mainly pursued the strategy of integrating RIT into myeloablative or reduced-intensity conditioning regimens to augment the delivery of radiation to hematopoietic sites to improve disease control without increasing TRM. A number of proof-of-principle studies demonstrate that RIT for AML is safe, feasible, and has acceptable toxicities, even in older adults. However, controlled trials will ultimately be required to determine whether RIT indeed leads to improved patient outcome. Still, this approach is complex, and many questions remain, including which antigen should be targeted and which radionuclide should be used. Undoubtedly, future studies will explore alternative targets and novel radionuclides, while innovative strategies such as pretargeting may optimize the administration of radionuclide therapy and allow more widespread use of very short half-life materials such as the α-emitters. With continued research, RIT holds the potential to establish itself as first-line treatment in AML for small volume disease or as a radiation boost in combination with other treatments, such as stem cell rescue or as adjuvant therapy (Walter and Pagel 2011).

Conclusion

The clinical studies conducted so far with unconjugated, toxin-loaded, and radio-labeled MoAbs have demonstrated that CD33 and perhaps CD45 and CD66 are valid targets for the treatment of AML. Given this encouraging experience, the use of antibodies is likely expanding significantly in the future with identification of additional AML (stem) cell-associated antigens. However, AML comprises a heterogeneous group of diseases, and antibody-based therapeutics likely only benefit defined patient subsets, as exemplified by the clinical experience with GO. This will pose formidable challenges to the identification of adequate numbers of suitable patients in whom specific therapies can be tested in a controlled fashion.

Acknowledgements This work was supported by grants P30-CA015704-35S6, P01-CA044991, R01-CA109663, and R01-CA136639 from the National Cancer Institute/National Institutes of Health (NCI/NIH) and a Specialized Center of Research grant (#7008-08) from the Leukemia & Lymphoma Society. R.B.W. is a Leukemia & Lymphoma Scholar in Clinical Research.

References

Abutalib SA, Tallman MS (2006) Monoclonal antibodies for the treatment of acute myeloid leukemia. Curr Pharm Biotechnol 7:343–369

Andres TL, Kadin ME (1983) Immunologic markers in the differential diagnosis of small round cell tumors from lymphocytic lymphoma and leukemia. Am J Clin Pathol 79:546–552

Andrews RG, Torok-Storb B, Bernstein ID (1983) Myeloid-associated differentiation antigens on stem cells and their progeny identified by monoclonal antibodies. Blood 62:124–132

Andrews RG, Takahashi M, Segal GM et al (1986) The L4F3 antigen is expressed by unipotent and multipotent colony-forming cells but not by their precursors. Blood 68:1030–1035

Appelbaum FR (1999) Antibody-targeted therapy for myeloid leukemia. Semin Hematol 36:2–8

Appelbaum FR, Matthews DC, Eary JF et al (1992) The use of radiolabeled anti-CD33 antibody to augment marrow irradiation prior to marrow transplantation for acute myelogenous leukemia. Transplantation 54:829–833

Becker W, Goldenberg DM, Wolf F (1994) The use of monoclonal antibodies and antibody fragments in the imaging of infectious lesions. Semin Nucl Med 24:142–153

Bernstein ID, Singer JW, Andrews RG et al (1987) Treatment of acute myeloid leukemia cells in vitro with a monoclonal antibody recognizing a myeloid differentiation antigen allows normal progenitor cells to be expressed. J Clin Invest 79:1153–1159

Bernstein ID, Singer JW, Smith FO et al (1992) Differences in the frequency of normal and clonal precursors of colony-forming cells in chronic myelogenous leukemia and acute myelogenous leukemia. Blood 79:1811–1816

Breccia M, Lo-Coco F (2011) Gemtuzumab ozogamicin for the treatment of acute promyelocytic leukemia: mechanisms of action and resistance, safety and efficacy. Expert Opin Biol Ther 11:225–234

Breccia M, Cimino G, Diverio D et al (2007) Sustained molecular remission after low dose gemtuzumab-ozogamicin in elderly patients with advanced acute promyelocytic leukemia. Haematologica 92:1273–1274

Bross PF, Beitz J, Chen G et al (2001) Approval summary: gemtuzumab ozogamicin in relapsed acute myeloid leukemia. Clin Cancer Res 7:1490–1496

Bunjes D (2002) 188Re-labeled anti-CD66 monoclonal antibody in stem cell transplantation for patients with high-risk acute myeloid leukemia. Leuk Lymphoma 43:2125–2131

Bunjes D, Buchmann I, Duncker C et al (2001) Rhenium 188-labeled anti-CD66 (a, b, c, e) monoclonal antibody to intensify the conditioning regimen prior to stem cell transplantation for patients with high-risk acute myeloid leukemia or myelodysplastic syndrome: results of a phase I–II study. Blood 98:565–572

Burnett AK, Hills RK, Milligan D et al (2011) Identification of patients with acute myeloblastic leukaemia who benefit from the addition of gemtuzumab ozogamicin: results of the MRC AML15 trial J Clin Oncol 29:369–377

Burnett AK, Russell N, Hill RK et al (2012) The addition of gemtuzumab ozogamicin to induction chemotherapy improves survival in older patients with acute myeloid leukemia. J Clin Oncol 30(32):3924–3931

Caron PC, Co MS, Bull MK et al (1992) Biological and immunological features of humanized M195 (anti-CD33) monoclonal antibodies. Cancer Res 52:6761–6767

Caron PC, Jurcic JG, Scott AM et al (1994) A phase 1B trial of humanized monoclonal antibody M195 (anti-CD33) in myeloid leukemia: specific targeting without immunogenicity. Blood 83:1760–1768

Caron PC, Dumont L, Scheinberg DA (1998) Supersaturating infusional humanized anti-CD33 monoclonal antibody HuM195 in myelogenous leukemia. Clin Cancer Res 4:1421–1428

Castaigne S, Pautas C, Terré C et al (2012) Effect of gemtuzumab ozogamicin on survival of adult patients with de-novo acute myeloid leukaemia (ALFA-0701): a randomised, open-label, phase 3 study. Lancet 379:1508–1516

Clift RA, Buckner CD, Appelbaum FR et al (1990) Allogeneic marrow transplantation in patients with acute myeloid leukemia in first remission: a randomized trial of two irradiation regimens. Blood 76:1867–1871

Clift RA, Buckner CD, Appelbaum FR et al (1998) Long-term follow-up of a randomized trial of two irradiation regimens for patients receiving allogeneic marrow transplants during first remission of acute myeloid leukemia. Blood 92:1455–1456

Dahlke MH, Larsen SR, Rasko JE et al (2004) The biology of CD45 and its use as a therapeutic target. Leuk Lymphoma 45:229–236

Delaunay J, Recher C, Pigneux A et al (2011) Addition of gemtuzumab ozogamycin to chemotherapy improves event-free survival but not overall survival of AML patients with intermediate cytogenetics not eligible for allogeneic transplantation. Results of the GOELAMS AML 2006 IR study [abstract #79]. Blood 118:37–38

Dick JE (2008) Stem cell concepts renew cancer research. Blood 112:4793–4807

Dinndorf PA, Andrews RG, Benjamin D et al (1986) Expression of normal myeloid-associated antigens by acute leukemia cells. Blood 67:1048–1053

Estey EH, Giles FJ, Beran M et al (2002) Experience with gemtuzumab ozogamycin ("mylotarg") and all-trans retinoic acid in untreated acute promyelocytic leukemia. Blood 99:4222–4224

Farhat H, Reman O, Raffoux E et al (2012) Fractionated doses of gemtuzumab ozogamicin with escalated doses of daunorubicin and cytarabine as first acute myeloid leukemia salvage in patients aged 50-70-year old: a phase 1/2 study of the acute leukemia French association. Am J Hematol 87:62–65

Feldman E, Kalaycio M, Weiner G et al (2003) Treatment of relapsed or refractory acute myeloid leukemia with humanized anti-CD33 monoclonal antibody HuM195. Leukemia 17:314–318

Feldman EJ, Brandwein J, Stone R et al (2005) Phase III randomized multicenter study of a humanized anti-CD33 monoclonal antibody, lintuzumab, in combination with chemotherapy, versus chemotherapy alone in patients with refractory or first-relapsed acute myeloid leukemia. J Clin Oncol 23:4110–4116

Fenton C, Perry CM (2005) Gemtuzumab ozogamicin: a review of its use in acute myeloid leukaemia. Drugs 65:2405–2427

Fialkow PJ, Singer JW, Adamson JW et al (1981) Acute nonlymphocytic leukemia: heterogeneity of stem cell origin. Blood 57:1068–1073

Fialkow PJ, Singer JW, Raskind WH et al (1987) Clonal development, stem-cell differentiation, and clinical remissions in acute nonlymphocytic leukemia. N Engl J Med 317:468–473
Giles F, Estey E, O'brien S (2003) Gemtuzumab ozogamicin in the treatment of acute myeloid leukemia. Cancer 98:2095–2104
Goemans BF, Zwaan CM, Vijverberg SJ et al (2008) Large interindividual differences in cellular sensitivity to calicheamicin may influence gemtuzumab ozogamicin response in acute myeloid leukemia. Leukemia 22:2284–2285
Gray-Owen SD, Blumberg RS (2006) CEACAM1: contact-dependent control of immunity. Nature Rev 6:433–446
Green DJ, Pagel JM, Nemecek ER et al (2009) Pretargeting CD45 enhances the selective delivery of radiation to hematolymphoid tissues in nonhuman primates. Blood 114:1226–1235
Griffin JD, Linch D, Sabbath K et al (1984) A monoclonal antibody reactive with normal and leukemic human myeloid progenitor cells. Leuk Res 8:521–534
Grimwade D, Enver T (2004) Acute promyelocytic leukemia: where does it stem from? Leukemia 18:375–384
Jager E, Van Der Velden VHJ, Te Marvelde JG et al (2011) Targeted drug delivery by gemtuzumab ozogamicin: mechanism-based mathematical model for treatment strategy improvement and therapy individualization. PLoS One 6:e24265
Jordan CT, Guzman ML (2004) Mechanisms controlling pathogenesis and survival of leukemic stem cells. Oncogene 23:7178–7187
Jurcic JG (2001) Antibody therapy for residual disease in acute myelogenous leukemia. Crit Rev Oncol Hematol 38:37–45
Jurcic JG, Deblasio T, Dumont L et al (2000) Molecular remission induction with retinoic acid and anti-CD33 monoclonal antibody HuM195 in acute promyelocytic leukemia. Clin Cancer Res 6:372–380
Jurcic JG, Larson SM, Sgouros G et al (2002) Targeted alpha particle immunotherapy for myeloid leukemia. Blood 100:1233–1239
Kelly LM, Gilliland DG (2002) Genetics of myeloid leukemias. Annu Rev Genomics Hum Genet 3:179–198
Klein SA, Hermann S, Dietrich JW et al (2002) Transplantation-related toxicity and acute intestinal graft-versus-host disease after conditioning regimens intensified with Rhenium 188-labeled anti-CD66 monoclonal antibodies. Blood 99:2270–2271
Koenecke C, Hofmann M, Bolte O et al (2008) Radioimmunotherapy with [(188)Re]-labelled anti-CD66 antibody in the conditioning for allogeneic stem cell transplantation for high-risk acute myeloid leukemia. Int J Hematol 87:414–421
Lane SW, Gilliland DG (2010) Leukemia stem cells. Semin Cancer Biol 20:71–76
Lapusan S, Vidriales MB, Thomas X et al (2012) Phase I studies of AVE9633, an anti-CD33 antibody-maytansinoid conjugate, in adult patients with relapsed/refractory acute myeloid leukemia. Invest New Drugs 30:1121–1131
Larson RA, Sievers EL, Stadtmauer EA et al (2005) Final report of the efficacy and safety of gemtuzumab ozogamicin (Mylotarg) in patients with CD33-positive acute myeloid leukemia in first recurrence. Cancer 104:1442–1452
Linenberger ML (2005) CD33-directed therapy with gemtuzumab ozogamicin in acute myeloid leukemia: progress in understanding cytotoxicity and potential mechanisms of drug resistance. Leukemia 19:176–182
Lo-Coco F, Cimino G, Breccia M et al (2004) Gemtuzumab ozogamicin (Mylotarg) as a single agent for molecularly relapsed acute promyelocytic leukemia. Blood 104:1995–1999
Majeti R (2011) Monoclonal antibody therapy directed against human acute myeloid leukemia stem cells. Oncogene 30:1009–1019
Matthews DC, Appelbaum FR (2004) Radioimmunotherapy and hematopoietic cell transplantation. In: Blume KG, Forman SJ, Appelbaum FR (eds) Thomas' hematopoietic cell transplantation, 3rd edn. Blackwell Publishing Ltd, Massachusetts, pp 198–208
Matthews DC, Appelbaum FR, Eary JF et al (1991) Radiolabeled anti-CD45 monoclonal antibodies target lymphohematopoietic tissue in the macaque. Blood 78:1864–1874

Matthews DC, Badger CC, Fisher DR et al (1992) Selective radiation of hematolymphoid tissue delivered by anti-CD45 antibody. Cancer Res 52:1228–1234

Matthews DC, Appelbaum FR, Eary JF et al (1999) Phase I study of (131)I-anti-CD45 antibody plus cyclophosphamide and total body irradiation for advanced acute leukemia and myelodysplastic syndrome. Blood 94:1237–1247

Mcdevitt MR, Finn RD, Ma D et al (1999) Preparation of alpha-emitting 213Bi-labeled antibody constructs for clinical use. J Nucl Med 40:1722–1727

Mcdevitt MR, Ma D, Lai LT et al (2001) Tumor therapy with targeted atomic nanogenerators. Science 294:1537–1540 (New York)

Miederer M, Mcdevitt MR, Sgouros G et al (2004) Pharmacokinetics, dosimetry, and toxicity of the targetable atomic generator, 225Ac-HuM195, in nonhuman primates. J Nucl Med 45:129–137

Nakano A, Harada T, Morikawa S et al (1990) Expression of leukocyte common antigen (CD45) on various human leukemia/lymphoma cell lines. Acta Pathol Jpn 40:107–115

Nikula TK, Mcdevitt MR, Finn RD et al (1999) Alpha-emitting bismuth cyclohexylbenzyl DTPA constructs of recombinant humanized anti-CD33 antibodies: pharmacokinetics, bioactivity, toxicity and chemistry. J Nucl Med 40:166–176

Pagano L, Fianchi L, Caira M et al (2007) The role of gemtuzumab ozogamicin in the treatment of acute myeloid leukemia patients. Oncogene 26:3679–3690

Pagel JM, Appelbaum FR, Eary JF et al (2006a) 131I-anti-CD45 antibody plus busulfan and cyclophosphamide before allogeneic hematopoietic cell transplantation for treatment of acute myeloid leukemia in first remission. Blood 107:2184–2191

Pagel JM, Gooley T, Rajendran JG et al (2006b) Targeted radiotherapy using 131I-anti-CD45 antibody followed by allogeneic hematopoietic cell transplantation (HCT): the relationships among dosimetry, bone marrow uptake, and relapse [abstract]. Eur J Nucl Med Mol Imaging 33:S193

Pagel JM, Hedin N, Drouet L et al (2008) Eradication of disseminated leukemia in a syngeneic murine leukemia model using pretargeted anti-CD45 radioimmunotherapy. Blood 111:2261–2268

Pagel JM, Gooley TA, Rajendran J et al (2009a) Allogeneic hematopoietic cell transplantation after conditioning with 131I-anti-CD45 antibody plus fludarabine and low-dose total body irradiation for elderly patients with advanced acute myeloid leukemia or high-risk myelodysplastic syndrome. Blood 114:5444–5453

Pagel JM, Matthews DC, Kenoyer A et al (2009b) Pretargeted radioimmunotherapy using anti-CD45 monoclonal antibodies to deliver radiation to murine hematolymphoid tissues and human myeloid leukemia. Cancer Res 69:185–192

Passegué E, Jamieson CH, Ailles LE et al (2003) Normal and leukemic hematopoiesis: are leukemias a stem cell disorder or a reacquisition of stem cell characteristics? Proc Natl Acad Sci U S A 100(Suppl 1):11842–11849

Petersdorf S, Kopecky K, Stuart RK et al (2009) Preliminary results of Southwest Oncology Group Study S0106: an international intergroup phase 3 randomized trial comparing the addition of gemtuzumab ozogamicin to standard induction therapy versus standard induction therapy followed by a second randomization to post-consolidation gemtuzumab ozogamicin versus no additional therapy for previously untreated acute myeloid leukemia [abstract #790]. Blood 114:326–327

Petti MC, Pinazzi MB, Diverio D et al (2001) Prolonged molecular remission in advanced acute promyelocytic leukaemia after treatment with gemtuzumab ozogamicin (Mylotarg CMA-676). Br J Haematol 115:63–65

Pfizer Inc (2010) http://media.pfizer.com/files/products/mylotarg_hcp_letter.pdf. Accessed 23 Jan 2012

Pollard JA, Alonzo TA, Loken M et al (2012) Correlation of CD33 expression level with disease characteristics and response to gemtuzumab ozogamicin containing chemotherapy in childhood AML. Blood 119:3705–3711

Press OW, Howell-Clark J, Anderson S et al (1994) Retention of B-cell-specific monoclonal antibodies by human lymphoma cells. Blood 83:1390–1397

Press OW, Shan D, Howell-Clark J et al (1996) Comparative metabolism and retention of iodine-125, yttrium-90, and indium-111 radioimmunoconjugates by cancer cells. Cancer Res 56:2123–2129

Ravandi F, Estey E, Jones D et al (2009) Effective treatment of acute promyelocytic leukemia with all-trans-retinoic acid, arsenic trioxide, and gemtuzumab ozogamicin. J Clin Oncol 27:504–510

Raza A, Jurcic JG, Roboz GJ et al (2009) Complete remissions observed in acute myeloid leukemia following prolonged exposure to lintuzumab: a phase 1 trial. Leuk Lymphoma 50:1336–1344

Ricart AD (2011) Antibody-drug conjugates of calicheamicin derivative: gemtuzumab ozogamicin and inotuzumab ozogamicin. Clin Cancer Res 17:6417–6427

Ringhoffer M, Blumstein N, Neumaier B et al (2005) 188Re or 90Y-labelled anti-CD66 antibody as part of a dose-reduced conditioning regimen for patients with acute leukaemia or myelodysplastic syndrome over the age of 55: results of a phase I-II study. Br J Haematol 130:604–613

Rosenblat TL, Mcdevitt MR, Pandit-Taskar N et al (2007) Phase I trial of the targeted alpha-particle nano-generator actinium-225 (225Ac)-HuM195 (anti-CD33) in acute myeloid leukemia (AML) [abstract #910]. Blood 110:277a

Rottinger EM, Bartkowiak D, Bunjes D et al (2003) Enhanced renal toxicity of total body irradiation combined with radioimmunotherapy. Strahlenther Onkol 179:702–707

Scheinberg DA, Lovett D, Divgi CR et al (1991) A phase I trial of monoclonal antibody M195 in acute myelogenous leukemia: specific bone marrow targeting and internalization of radionuclide. J Clin Oncol 9:478–490

Sekeres MA, Lancet JE, Wood BL et al (2013) Randomized phase IIb study of low-dose cytarabine and lintuzumab versus low-dose cytarabine and placebo in older adults with untreated acute myeloid leukemia. Haematologica 98:119–129

Sievers EL, Appelbaum FR, Spielberger RT et al (1999) Selective ablation of acute myeloid leukemia using antibody-targeted chemotherapy: a phase I study of an anti-CD33 calicheamicin immunoconjugate. Blood 93:3678–3684

Sievers EL, Larson RA, Stadtmauer EA et al (2001) Efficacy and safety of gemtuzumab ozogamicin in patients with CD33-positive acute myeloid leukemia in first relapse. J Clin Oncol 19:3244–3254

Stasi R, Evangelista ML, Buccisano F et al (2008) Gemtuzumab ozogamicin in the treatment of acute myeloid leukemia. Cancer Treat Rev 34:49–60

Stubbs MC, Armstrong SA (2007) Therapeutic implications of leukemia stem cell development. Clin Cancer Res 13:3439–3442

SWOG (2010) http://www.swogstat.org/ROS/ROSBooks/Spring2010/Leukemia.pdf. Accessed 23 Jan 2012

Taksin AL, Legrand O, Raffoux E et al (2007) High efficacy and safety profile of fractionated doses of Mylotarg as induction therapy in patients with relapsed acute myeloblastic leukemia: a prospective study of the alfa group. Leukemia 21:66–71

Thomas ED, Storb R, Clift RA et al (1975) Bone-marrow transplantation. N Engl J Med 292:832–843

Tsimberidou AM, Giles FJ, Estey E et al (2006) The role of gemtuzumab ozogamicin in acute leukaemia therapy. Br J Haematol 132:398–409

Van Der Jagt RH, Badger CC, Appelbaum FR et al (1992) Localization of radiolabeled antimyeloid antibodies in a human acute leukemia xenograft tumor model. Cancer Res 52:89–94

Van Der Velden VH, Te Marvelde JG, Hoogeveen PG et al (2001) Targeting of the CD33-calicheamicin immunoconjugate Mylotarg (CMA-676) in acute myeloid leukemia: in vivo and in vitro saturation and internalization by leukemic and normal myeloid cells. Blood 97:3197–3204

Walter RB, Pagel JM (2011) Targeted radionuclide therapy for leukemia. In: Speer TW (ed) Targeted radionuclide therapy. Walters Kluwer/Lippincott Williams & Wilkins, Philadelphia, pp 441–457

Walter RB, Raden BW, Kamikura DM et al (2005) Influence of CD33 expression levels and ITIM-dependent internalization on gemtuzumab ozogamicin-induced cytotoxicity. Blood 105:1295–1302

Walter RB, Press OW, Pagel JM (2010) Pretargeted radioimmunotherapy for hematologic and other malignancies. Cancer Biother Radiopharm 25:125–142
Walter RB, Appelbaum FR, Estey EH et al (2012) Acute myeloid leukemia stem cells and CD33-targeted immunotherapy. Blood 119:6198–6208
Zenz T, Schlenk RF, Glatting G et al (2006) Bone marrow transplantation nephropathy after an intensified conditioning regimen with radioimmunotherapy and allogeneic stem cell transplantation. J Nucl Med 47:278–286

Part VI
Stem Cells

Chapter 28
New Heterogeneity of the Leukemic Stem Cells

Dominique Bonnet

Abstract The cancer stem cell (CSC) hypothesis proposes that cancer is propagated by a subpopulation of cells with stem cell properties, namely the ability to proliferate while balancing self-renewal with differentiation, to maintain the cancer. The presence of CSCs provides an explanation for the biological heterogeneity of cells within a cancer. In this chapter, we will focus mostly on acute myeloid leukemia, which was the first described model of CSCs. We will summarize recent data demonstrating a novel complexity of the leukemic stem cells compartment and discuss the potential implications this might have on targeting these cells.

Keywords Hematopoietic stem cell (HSC) · Xenotransplantation model · Leukemic stem cell (LSC) · Cancer stem cell (CSC) · Self-renewal · Acute myeloid leukemia · Quiescency · Pre-leukemic stem cell (pre-LSC)

28.1 Introduction

The cancer stem cell (CSC) model was first described in the blood cancer, acute myeloid leukemia (AML; Lapidot et al. 1994; Bonnet and Dick 1997), and since then has evolved in complexity with descriptions of CSCs in other tumors (Al-Hajj et al. 2003; Ricci-Vitiani et al. 2007; Singh et al. 2004; Schatton et al. 2008).

As these rare CSCs are both required and sufficient to reconstitute a new tumor, they have immediate and important clinical implications. A better identification and characterization of CSCs should provide a better understanding of tumor developmental biology and the genetic events involved in the transformation process. Studies have shown that CSCs, as normal somatic stem cells, express high levels of multidrug resistance (MDR) pumps that efficiently efflux cytotoxic drugs (Zhou et al. 2001), making them particularly difficult to target with cytotoxic therapies. Similarly, they appear to be less immunogenic than more differentiated cancer cells in the tumor (Costello et al. 2000), and therefore also more difficult to eradicate with immunotherapy. Thus, it has become evident that CSCs might prove not only

D. Bonnet (✉)
Haematopoietic Stem Cell Laboratory Cancer Research UK, London Research Institute, 44 Lincoln's Inn Fields, WC2A 3PX London, UK
e-mail: dominique.bonnet@cancer.org.uk

M. Andreeff (ed.), *Targeted Therapy of Acute Myeloid Leukemia*,
Current Cancer Research, DOI 10.1007/978-1-4939-1393-0_28

to be the most important but also the most difficult cancer cells to eliminate with conventional therapies, and that a specific monitoring and targeting of these elusive CSCs could become an important tool towards identification and characterization of improved cellular and molecular targets for development of cancer therapies.

Based on these studies, the paradigm of cancer as a hierarchical disease, sustained by a rare population of CSCs has emerged. Implicit in this model of cancer development is the notion that CSCs are biologically distinct from other cells in the tumor and are able to initiate and sustain tumor growth in vivo whereas the bulk cells are not.

28.2 Models of Tumor Heterogeneity

At least two models have been proposed to account for the heterogeneity of tumors—the hierarchical model and the stochastic model. In the hierarchical model, tumors are thought to be aberrant representations of the hierarchy seen in normal tissues, whereby there is a developmental progression between specific subpopulations, which have distinct intrinsic functions and potentials. By contrast, in the stochastic model, all cells are biologically equivalent, but respond stochastically to intrinsic cues such as transcription factor levels, or to extrinsic cues such as the microenvironment, or immune response, to give rise to alterations in the characteristics of individual cells. More recently, a branching evolutionary model has been described in which different sets of genetic and epigenetic alterations confer stem-like properties in multiple clones (Fig. 28.1a). As the disease progresses, genetic heterogeneity develops but eventually a dominant subclone emerges which has acquired a survival advantage over other clones (Greaves 2009).

However, when considering a CSC, the distinction between the models is blurred, as they all accommodate the concept of a functional CSC, but what differs is the organization of the CSCs in relation to the other cancer cells. The stochastic model proposes that all cells are able to behave functionally as a CSC, given the right cues, and this occurs at random. Significantly, all progeny of the CSC are themselves able to act as cancer-propagating cells. However, in the hierarchical model, only a distinct subpopulation at the apex of the hierarchy has stem cell properties. In this model, it should be possible to prospectively isolate the CSC population, whereas in the stochastic model, all cells have CSC potential and therefore cannot be distinguished from the bulk population. The evolutionary model allows for the coexistence of multiple CSC clones. At any one time point, certain CSC clones may be dominant and others minor, but these may vary depending on disease stage; for example, relapse may occur from a clone which is minor at diagnosis but becomes dominant later (Yang et al. 2008; Mullighan et al. 2008). Each of these CSCs may, in theory, also sit at the apex of a hierarchy. A further dimension to the hierarchical model is potentially a bidirectional flow, whereby cells lower down in the hierarchy are not only able to act as CSCs but are also able to give rise to more primitive cells, thus uncoupling lineage differentiation from CSC potential (Gibbs et al. 2012; Fig. 28.1b).

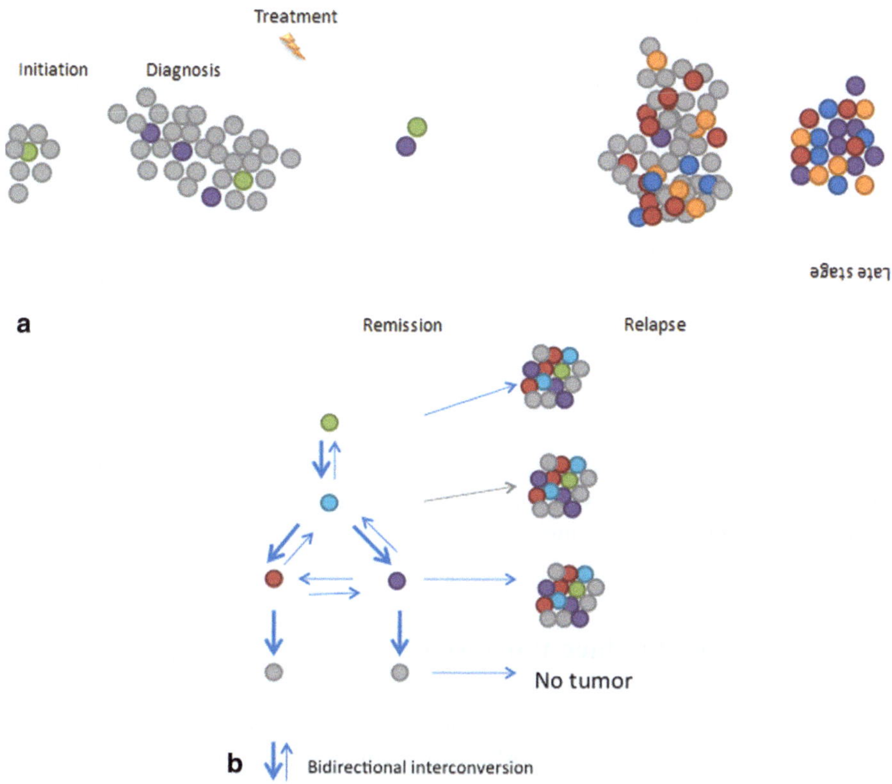

Fig. 28.1 **a** Branching evolutionary model. Multiple CSC subclones arise as a result of genetic or epigenetic alterations. In this picture, *colored circles* represent non-CSCs. A hierarchical structure can exist in this model. Different subclones may be dominant at different time points (e.g., diagnosis and relapse), while some may disappear depending on selective genetic and environmental pressures. As the disease progresses, genetically distinct CSC subclones accumulate. This may ultimately result in an increase in CSC frequency and breakdown of hierarchical structure, as all cells now have CSC capacity. **b** Reversible phenotypic hierarchy. A phenotypically defined hierarchical structure exists in a cancer, but cells at multiple points in the hierarchy are functional CSCs. The CSC capacity (and frequency) may also be ordered hierarchically, such that the population at the top is more enriched for CSCs than those further down the hierarchy. However, the CSCs lower down the hierarchy are able to recapitulate the whole tumor, including the more primitive CSCs that precede them in the hierarchy, suggesting a certain "plasticity" in the hierarchy. Thus, CSCs need to be defined functionally rather than by surface phenotype

28.3 Identification of the Leukemic Stem Cell in AML

The adaptation of xenotransplantation assays to examine the propagation of AML in vivo has allowed the phenotypic identification of the AML-initiating cells (AML-IC). Transplantation of primary AML cells into nonobese/severe combined immunodeficiency (NOD/SCID) mice led to the finding that only rare cells, termed AML-IC, are capable of initiating and sustaining growth of the leukemic clone in vivo, and serial transplantation experiments showed that AML-IC possess high self-renewal capacity, and thus can be considered to be the leukemic stem cells (LSCs). By using the term LSC, we refer to a cell that has self-renewal and differentiation potential and is able to reinitiate the leukemia when transplanted into NOD/SCID. This definition does not preclude the nature of the cells that is being transformed (i.e., normal HSC, progenitors or mature cells).

In the original papers (Lapidot et al. 1994; Bonnet and Dick 1997), in the samples examined, only cells, which were $CD34^{+}CD38^{-}$, were able to initiate leukemia in NOD/SCID recipients, even when this represented a minor population within the total cells. Furthermore, this population of cells was able to proliferate extensively in vivo, to recapitulate all the phenotypic populations seen in the original sample, while also maintaining a $CD34^{+}CD38^{-}$ population. This data supported the hypothesis that leukemia originates in a primitive cell, which resembles normal HSC, and that leukemia, like normal hematopoiesis is organized as a hierarchy. These cells were defined as SCID-leukemia-initiating cells (SL-ICs), which accurately describes their observed in vivo capacities. Over the years, the term leukemic stem cell has come into more common usage.

28.4 The LSC Surface Phenotype

As described above, the LSC was originally described as arising from the $CD34^{+}CD38^{-}$ population within AML, which also defines the HSC compartment in normal hematopoiesis. However, more recent work has made it clear that the picture is not as simple as this.

One of the confounding factors in earlier studies on the AML LSC is the use of fluorescence-labeled antibodies in flow cytometric isolation of subpopulations. Work from our laboratory demonstrated that some antibodies reduced the engraftment capability of some AML cells (Taussig et al. 2008). We specifically assessed the effect of anti-CD38 antibodies on the engraftment of AML mononuclear cells and indeed saw a reduction in engraftment. This was shown to be due to Fc-receptor-mediated clearance of the coated cells by residual innate immunity in NOD/SCID mice. This inhibition of AML engraftment by anti-CD38 antibody suggested that the CD38 antigen is expressed on some SL-ICs.

Using methods to abrogate the effect of innate immunity, we went on to look specifically at engraftment of subpopulations, sorted by CD34 and CD38 expression, from seven AML samples. Unexpectedly, seven out of seven samples gave rise to leukemic engraftment from the $CD34^{+}/CD38^{+}$ population at 8–12 weeks after transplantation. It seems therefore, in these samples, SL-ICs reside in the $CD34^{+}/CD38^{+}$ population.

This work also highlighted the heterogeneity of AML LSCs as the $CD34^{+}/CD38^{-}$ population varied in size between samples, and indeed was not always able to give rise to leukemic engraftment. Subsequent studies (Taussig et al. 2010; Martelli et al. 2010) focusing specifically on AML with nucleophosmin (NPM) mutations, which accounts for approximately one third of samples (Schnittger et al. 2005), and in which CD34 expression is characteristically low (Chou et al. 2006; Haferlach et al. 2009), showed that SL-ICs exist in both $CD34^+$ and $CD34^-$ fractions. Further confirmation of the existence of SL-ICs in other compartments came from the Carroll's laboratory. They showed that when using the more immunodeficient NSG mouse, SL-ICs, although enriched in the Lin-$CD38^-$ in most samples, can also be found with all combinations of CD34 and CD38 expression (Sarry et al. 2011; Vargaftig et al. 2011).

Therefore, it is likely that the SL-IC population may vary between AML samples, and SL-IC activity may reside in more than one population within a sample (Sarry et al. 2011; Taussig et al. 2010). Another possibility raised by these findings is that the SL-IC phenotype is not stable over the course of the disease. Indeed it was shown that engrafted leukemic cells sometimes differed from the original transplanted population, for example, $CD34^+$ AML cells gave rise to $CD34^-$ leukemia in the mouse (Taussig et al. 2010) or $CD34^+/CD38^+$ can revert to give rise to $CD34^+/CD38^-$ in vivo in the transplanted mice [D. Bonnet, unpublished information]. Further support for this concept comes from work by Gibbs et al. (2012), who demonstrated that even in a genetically uniform HoxA9-Meis1 murine model of AML, leukemia-initiating cells (LICs) exist in three immunophenotypically distinct compartments. These populations correspond phenotypically to an HSC, myeloid progenitor, and lymphoid progenitor compartment, not only could each population initiate leukemia but they each recapitulated the spectrum of LICs as well, indicating a plasticity in the LIC phenotype (Fig. 28.1b).

It is clear that using phenotypes that correspond to normal populations is inadequate for reliably defining the LSC in AML, as the normal differentiation pattern is likely to be disrupted in leukemia, resulting in aberrant expression of markers. Effort is ongoing to try to identify a marker, which will enable isolation, and eventually therapeutic targeting, of the LSC, across a wide range of AMLs, while also distinguishing them from normal healthy HSCs or progenitors. Several markers have been identified as being preferentially expressed on the LSC including CD90 (Blair et al. 1997), CLL-1 (van Rhenen et al. 2007), CD44 (Jin et al. 2009), CD25 (Saito et al. 2010), CD32 (Saito et al. 2010), and CD47 (Jaiswal et al. 2009; Majeti et al. 2009a). Two recently described LSC markers are T-cell Ig mucin-3 (TIM3; Jan et al. 2011) and CD96 (Hosen et al. 2007), which are both expressed on LSCs (defined in these studies as $CD34^+/CD38^-/CD90^-$), but only at low levels on normal bone marrow (BM) HSCs ($CD34^+/CD38^-/CD90^+$). However, it is likely that due to the heterogeneity of AML LSCs, no single marker will identify the LSC in all AML cases.

28.5 Cell of Origin of the LSC

From the above, the CSC should be a functionally defined entity and does not necessarily imply a stem cell, which has become cancerous. However, the normal cell in which the LSC originates is still a subject of debate. The original description of

LSCs as being $CD34^+/CD38^-$ were suggestive of an HSC cell of origin; however, as the phenotype of LSCs has been shown to be more diverse than this, the possibility of a progenitor cell of origin has been postulated, at least in some AML cases. In a study looking at a subset of $CD34^+$ AML, the LSC was shown to lie in two progenitor populations, the newly defined $CD34^+/CD38^-/CD90^-/CD45RA^+$ lymphoid-primed multipotent progenitor (LMPP) and the $CD34^+/CD38^+/CD110^+/CD45RA^+$ granulocyte-macrophage progenitor (GMP), which seems to be arranged hierarchically (Goardon et al. 2011). By comparing microarray gene expression profiles, these populations were found to resemble normal progenitors with the corresponding phenotype, rather than normal HSCs. However, the AML LSC populations expressed a self-renewal program which is not normally active in progenitors. This raises the possibility that the LSC may derive from a progenitor population, which has acquired stem cell properties, or from a cell, which has failed to coordinate the switching off of "stemness" genes while differentiating into a progenitor. The other option described before is that the phenotype does not necessary preclude of the origin of the cell that is being transformed. Indeed, some plasticity in the phenotype of the LSC might occur via intrinsic or extrinsic environment clues, which do not preclude the origin of the cells.

The only direct evidence of the origin of the cell that is being transformed is when purified stem or progenitor cells are being transduced by oncogenes or fusion proteins. Several studies have demonstrated that mouse myeloid progenitors can be transformed into AML by transduction with the oncogenes MLL-ENL (Cozzio et al. 2003) or MOZ-TIF2 (Huntly et al. 2004). These oncogenes confer self-renewal ability on the myeloid progenitors. On the other hand, when MLL-AF9 fusion gene was expressed under the endogenous murine promoter of MLL gene, only the HSC-enriched population could be transformed, while the GMP population was resistant (Chen et al. 2008). However, at high gene dosage as achieved by retroviral transduction, GMPs could also be transduced to give rise to AML with a GMP phenotype (Chen et al. 2008; Krivtsov et al. 2006; Somervaille and Cleary 2006). Leukemic cells with a GMP phenotype (L-GMP) could be isolated from the primary leukemia and these were able to initiate AML in secondary recipients at a high frequency, indicating self-renewal. Gene expression analysis showed that the L-GMP had a program more similar to normal GMP, but with reactivation of a subset of genes that are normally restricted to HSCs (Krivtsov et al. 2006). These studies suggest the LSC that maintains the leukemia arises downstream of the normal HSC, but are nonetheless hierarchical in organization.

In another MLL-fusion protein, MLL-GAS9, only transduced HSCs, but not CMP or GMP could give rise to leukemia. These results show that the cell type is important for the leukemogenic effect of certain fusion proteins. However, retroviral-mediated transduction of fusion proteins into various stages of hematopoiesis is a very artificial model, which may not represent the leukemogenic event in vivo and hence the leukemia obtained may not behave in a similar fashion.

It should be noted that studies using the MLL-AF9 and MLL-ENL fusion proteins in human cells have resulted in different types of leukemia, with a more mixed lineage or lymphoid bias than seen in murine studies, suggesting that differences exist between murine and human models (Barabe et al. 2007; Wei et al. 2008).

Therefore, even these murine models of AML which probably represent a far more simple process than primary human AML demonstrate that it is still far from clear as to in which cell the LSC arises. In reality, it is likely that either HSCs or progenitors can be transformed into the LSC, and this varies from case to case. As discussed earlier, phenotype switching or the existence of more than one LSC within a sample, also adds confusion to the identification of the cell of origin.

To add further complexity, it may be that as the leukemia progresses the nature of the LSC changes. For instance, in chronic myeloid leukemia (CML), which arises as a result of the fusion protein, BCR-ABL (from the t (9; 22) (q34; 11) chromosomal translocation), it is known that the chronic phase is maintained by a $CD34^+$/$CD38^-$ BCR-ABL expressing stem cell. However, when the disease transforms into a more aggressive acute phase, which is essentially AML, GMPs acquire self-renewal and drive the disease, independent of the BCL-ABL fusion protein (Jamieson et al. 2004). Therefore, there appears to be two distinct LSCs at different phases of the disease, which differ phenotypically and genetically. The concept of genomic variegation has been more recently described in ALL (Andersson et al. 2011; Notta et al. 2011). It appears that different leukemic clones are present in a given patient and that novel clones can appear during the evolution of the disease when other might become minor clones.

28.6 Notion of Pre-LSCs

The first evidence of the presence of a pre-leukemic clone comes from the description by Prof M Greaves's group of the existence of leukemia-associated fusion oncogene sequences in neonatal blood spots taken years before the onset of leukemia in children (Gale et al. 1997). More recent work suggests that the source of this DNA may be pre-LSCs—cells that contain the leukemia-associated sequences and have self-renewal potential but without the ability to induce leukemia (Hong et al. 2008). Twins were identified, one who had ALL with the TEL-AML1 fusion oncogene and one who was healthy but had a long lived population of progenitor B-cells with the same TEL-AML1 oncogene. TEL-AML1 conferred the ability to self-renew on progenitor B-cells (as well as altering the phenotype). Presumably pre-LSCs lack the additional DNA hits to allow full leukemic transformation. Although TEL-AML1 confers phenotypic and functional changes there is no reason to think that all pre-LSCs would be phenotypically and functionally different to the normal HSCs. Thus, many pre-LSCs may not be readily detectable in other patients except using next generation sequencing technique now available (Corces-Zimmerman and Majeti 2014).

The existence of a common (JAK2 wild type) pre-LSC would explain the JAK2 wild-type secondary AMLs that arise in patients with JAK2-mutated myeloproliferative disorders (the JAK2 mutation being a later event; Jelinek et al. 2005). A common (Philadelphia chromosome negative) pre-LSC may also be the cause of the Philadelphia negative chromosomal abnormalities seen in patients with CML treated successfully with imatinib (Bumm et al. 2003). Clonal-tracking analysis in patients over time using the next generation sequencing provided important insight into the genetic variegation of the leukemic clones overtime and allowed the analysis the origin of recurrent clones at relapse. Clonal-tracking analysis provided an explanation of the origin of the recurrence of malignant disease many years after remission has been attained. In some cases of TEL-AML1 indeed, the clones responsible for recurrence were detectable as minority populations at diagnosis of leukemia (Konrad et al. 2003). Between diagnosis and recurrence, further mutations occur in the pre-LSCs that transform them into fully malignant cells. Eliminating these pre-cancerous stem cells may be important for true cure.

28.7 Frequency of LSCs

One of the debates surrounding the existence of CSCs is whether, by definition they should be rare, and if they are not, then does this invalidate the model? Several studies have assessed this question. However, it is important to state that the CSC model does not necessarily predict frequency of CSCs within the tumor bulk; it merely describes cancer cells with properties such as self-renewal, which enable it to propagate the cancer. The question of frequency is of relevance; however, if functional CSCs are common (as predicted by the stochastic model), efforts to identify specific markers and to develop CSC-targeted therapies become redundant, as clearance of the bulk cells will suffice. The evidence so far suggests that the frequency of CSCs probably varies hugely depending on the cancer type and the assay conditions used.

This latter point is demonstrated by a series of studies in the skin cancer, melanoma. Melanoma-initiating cells were reported to be a rare population, comprising only 0.0001% of human melanoma cells as assessed by tumorigenesis in NOD/SCID mice. However, a subsequent study (Quintana et al. 2008) challenged this finding with the advent of more immunodeficient mouse strains. When equal numbers of human melanoma cells were injected subcutaneously into NOD/SCID and NSG mice, tumors formed with a shorter latency and with a higher frequency of melanoma-initiating cells in NSG compared to NOD/SCID. The frequency of tumorigenic cells was further enhanced by coinjection of Matrigel, and using this improved assay, the authors demonstrated that by limiting dilution assays, one in four melanoma cells have tumor-initiating ability. However, no marker could be identified to distinguish tumorigenic from nontumorigenic cells. These data suggest that in some cancers, such as melanoma, CSCs are less rare than others such as AML.

A potential confounding factor in the assessment of LSC frequency is the use of the xenotransplantation model. This may select out human populations, which are true LSCs, but lack properties required to engraft in the recipient mouse environment. In this situation, the frequency of LSCs may be underestimated. One way to avoid the issues of the species and immune barriers is to perform syngeneic transplants using mouse BM cells. Eμ-myc transgenic mice develop spontaneous lymphoma, a B-lymphocyte neoplasm. Injection of as few as one to ten cells from these lymphomas into histocompatible recipient mice are able to initiate new tumors, which recapitulate the original lymphoma (Kelly et al. 2007). This study suggests that the transplant model may affect the perceived frequency of CSCs, which may not in fact be rare. However, other work has demonstrated that similar frequencies are obtained for a given cancer independent of the model. For example, Barabe et al. (2007) showed that cells from a B-ALL generated by expression of the fusion oncogene MLL-ENL in human umbilical cord blood cells demonstrated an LSC frequency of 1% in NOD/SCID mice, which is comparable to that found in a murine model of MLL-AF9 leukemia (Krivtsov et al. 2006).

It seems therefore impossible to generalize when talking about the frequency of CSCs, which is likely to vary widely between cancers, and indeed within a cancer type. The frequency may also change depending on the stage or aggressiveness of the disease. In the original studies, the frequency of LSCs in human AML was calculated to be in the order of 1 in 10^4–10^6 (Bonnet and Dick 1997). However, the debate over the frequency of CSCs in other cancers has encouraged investigators to reevaluate the AML LSC frequency in the light of the newer, more immunodeficient mice. In our earlier work, approximately half of AML samples could not engraft NOD/SCID mice (Pearce et al. 2006); however, transplantation of some of these samples into NSG mice also failed to generate engraftment (Vargaftig et al. 2011). The frequency of SL-ICs in engrafting samples was also assessed in NSG mice in a limiting dilution analysis, and compared to that found in NOD/SCID mice. Interestingly, for samples with a high frequency in NSG mice (above 1 in 50,000 unfractionated cells), a 12- to 111-fold increase was seen when compared to the frequency in the NOD/SCID model. This was shown to be partly due to residual innate immunity in the NOD/SCID mouse. However, below this threshold, there was little difference between the two models. The main impact of the increased immunodeficiency of NSG mice appears to be a more robust AML engraftment, such that a higher level of chimerism is achieved with a given number of cells. Therefore, despite the more permissive environment, there remains a wide heterogeneity in the LSC frequency between patient samples, but the range remains of a similar magnitude independent of the model used. This frequency in AML is still rare, of the order of 1 in 400 to 1 in 10 million (Vargaftig et al. 2011; Sanchez et al. 2009).

28.8 Clinical Relevance

One of the major tenets of the CSC hypothesis is that in order to cure a patient of the disease, the CSCs would have to be eliminated. Furthermore relapse is thought to arise from therapy-resistant CSCs, which may remain quiescent for prolonged periods after the bulk tumor has been cleared. Although an effective CSC-directed therapy as yet remains elusive, the model has provided not only an understanding of the biology of cancer but also insights into the clinical progression of the disease.

28.8.1 Prognostic Implications

At present, the key prognostic indicator in AML is the presence of certain recurrent, nonrandom cytogenetic abnormalities, which enables the classification of AML into three risk groups (Grimwade et al. 2010). Patient age and the presence of known molecular mutations add further risk stratification (Smith et al. 2011). However, despite these parameters, the relapse risk and clinical outcome in many patients cannot be reliably predicted. It would be logical that the LSC compartment has some impact on prognosis that may be able to be quantified and contribute to prognostication. As techniques in stem cell biology develop, data are accumulating which support this hypothesis.

The xenotransplantation model, which represents the gold standard assay in the assessment of LSC activity, is in itself able to provide prognostic information. The biological determinants of AML engraftment remain unclear; however, it has been demonstrated that engraftment does correlate with the cytogenetically defined risk group (Pearce et al. 2006; Vargaftig et al. 2011). All the poor prognosis samples, but none of the good prognosis samples engrafted. Of the normal karyotype intermediate risk group, half the samples engrafted. When the clinical outcomes of 25 normal karyotype cases were analyzed, overall survival was found to be significantly shorter in patients whose samples engrafted compared to those which did not engraft (31 vs. 76 %, 2-year survival, respectively). The NOD/SCID assay therefore is able to detect biological differences that reflect clinical outcome, although it is not clear whether this is a question of frequency or intrinsic LSC differences.

Clinically, it is this intermediate risk group that poses the most difficult dilemmas when designing a treatment strategy, due to the heterogeneity of outcomes. NOD/SCID engraftment is one means of distinguishing high- and low-risk patients within this group; however, it is not feasible as a test in the clinical setting, due to its long-time course, required for large cell numbers and the cost of animal experimentation. Therefore, work is focused on trying to develop tools, which may be applicable in the clinic.

Recent years have seen the development of techniques, which allow whole genome analysis of transcriptional profiles using microarray platforms. However, one of the issues when applying this technique to cancer is the heterogeneity of cell populations, each of which may have a distinct gene expression signature. Further difficulties arise when looking at LSCs as their phenotype is variable, and so isolation of this rare population remains controversial. Two studies have generated an LSC signature from a purified population using either $CD34^+/CD38^-/CD123^+$ (Majeti et al. 2009b) or a functionally defined LSC (Eppert et al. 2011) populations from AML samples. In the latter study, the engraftment capacity of all populations defined by CD34 and CD38 from 16 AML samples was tested, and found that although LSCs could be found in all four subfractions, in 50 % of samples, LSCs contained $CD34^+/CD38^-$ cells and 50 % had $CD34^+/CD38^+$ LSCs. The LSC signature was compared to a signature generated from normal HSCs, and found to have considerable overlap, although there were some differentially regulated genes.

In both studies, the clinical significance of the level of expression of the LSC signature was examined, and a higher LSC signature score was associated with a worse prognosis, independent of other known factors such as cytogenetic group. It is perhaps surprising that the gene signature of a rare LSC population could be detectable in the bulk cells used for the analysis. This may be because the LSC frequency is in fact higher than is currently detectable, or that the aberrant gene expression pattern is transmitted to a certain degree to the progeny of the LSC, and so the overall signature reflects that of the LSCs to some degree (Eppert et al. 2011).

Therefore, inferior patient outcome appears to be related in some way to a higher frequency of LSCs. As disease relapse is a major factor in the survival rates of AML, it is tempting to hypothesize that this is due to the resistance of LSCs to conventional chemotherapy, and the greater the pool of LSCs, the harder it is to cure the disease.

28.8.2 Therapeutic Targeting

CSCs are thought to be exquisitely resistant to conventional chemo- and radiotherapy. A number of mechanisms have been suggested to account for this (Nurrul Abdullah and Kai-Hua Chow 2013). These include expression of ABC transporters, which pump small molecules including drugs out of the cell, efficient DNA repair mechanisms, and the ability to reside in hypoxic niches, which provide radioprotection and avoidance of high drug concentrations. Furthermore, like normal stem cells, CSCs are highly quiescent and cycle infrequently, and thus are unaffected by antiproliferative drugs. Therefore, effective treatments may need to target not only bulk tumor cells but also the rare CSC population, as there is evidence that relapsed or refractory disease arises from the surviving CSCs (Aguirre-Ghiso 2007). Indeed, the CSC paradigm has important implications for future drug design and testing, as the killing of rare, quiescent CSCs as well as of the bulk tumor needs to be taken into consideration. Therefore, systems will need to be developed which robustly identify CSCs and also have readout to assess the specific impact of treatments on CSC activities. Novel therapies are being developed. Thus include immunotherapy against cell surface marker (CD33, CD123, CD47, CD44, etc.), specific pathway inhibitors such as NF-κB, AKT/PI3K. For more details on the potential therapeutic targeting of LSC available, we will refer the readers to Dr. Guzman's chapter in this book.

Conclusions

The CSC field has evolved rapidly over the past 2 decades, however it is becoming clear that the CSC or LSC is not a single entity, and more work is required to identify the LSC (or indeed LSCs) in a given patient. It is likely that a functional definition of the LSC will be required, in addition to a combination of surface markers and a characteristic gene expression program, in order to encompass the diversity of LSCs both within and between patients. This heterogeneity poses challenges in developing an effective LSC-directed therapeutic approach.

References

Aguirre-Ghiso JA (2007) Models, mechanisms and clinical evidence for cancer dormancy. Nat Rev Cancer 7:834–846

Al-Hajj M, Wicha MS, Benito-Hernandez A et al (2003) Prospective identification of tumorigenic breast cancer cells. Proc Natl Acad Sci U S A 100:3983–3988

Anderson K, Lutz C, van Delft FW et al (2011) Genetic variegation of clonal architecture and propagating cells in leukemia. Nature 469:356–361

Barabe F, Kennedy JA, Hope KJ, Dick JE (2007) Modeling the initiation and progression of human acute leukemia in mice. Science 316:600–604

Blair A, Hogge DE, Ailles LE et al (1997) Lack of expression of thy-1 (cd90) on acute myeloid leukemia cells with long-term proliferative ability in vitro and in vivo. Blood 89:3104–3112

Bonnet D, Dick JE (1997) Human acute myeloid leukemia is organized as a hierarchy that originates from a primitive hematopoietic cell. Nat Med 3:730–737

Bumm T, Muller C, Al-Ali HK et al (2003) Emergence of clonal cytogenetic abnormalities in Ph- cells in some CML patients in cytogenetic remission to imatinib but restoration of polyclonal hematopoiesis in the majority. Blood 101:1941–1949

Chen W, Kumar AR, Hudson WA et al (2008) Malignant transformation initiated by mll-af9: gene dosage and critical target cells. Cancer Cell 13:432–440

Chou CW, Tang JL, Lin LL et al (2006) Nucleophosmin mutations in de novo acute myeloid leukemia: the age-dependent incidences and the stability during disease evolution. Cancer Res 66:3310–3316

Corces-Zimmerman MR, Majeti R (2014) Pre-leukemic evolution of hematopoietic stem cells: the importance of early mutations in leukemogenesis. Leukemia Jul 9. doi: 10.1038/leu.2014.211

Costello RT, Mallet F, Gaugler B et al (2000) Human acute myeloid leukemia CD34+/CD38- progenitor cells have decreased sensitivity to chemotherapy and Fas-induced apoptosis, reduced immunogenicity, and impaired dendritic cell trans- formation capacities. Cancer Res 60:4403–4411

Cozzio A, Passegue E, Ayton PM et al (2003) Similar MLL-associated leukaemias arising from self-renewing stem cells and short-lived myeloid progenitors. Genes Dev 17:3029–3035

Eppert K, Takenaka K, Lechman ER et al (2011) Stem cell gene expression programs influence clinical outcome in human leukemia. Nat Med 17:1086–1093

Gale KB, Ford AM, Repp AM et al (1997) Backtracking leukemia to birth: identification of clonotypic gene fusion sequences in neonatal blood spots. Proc Natl Acad Sci U S A 94(25):13950–13954

Gibbs KD, Jager A, Crespo O et al (2012) Decoupling of tumor-initiating activity from stable immunophenotype in hoxa9-meis1-driven AML. Cell Stem Cell 10:210–217

Goardon N, Marchi E, Atzberger A et al (2011) Coexistence of LMPP-like and GMP-like leukemia stem cells in acute myeloid leukemia. Cancer Cell 19:138–152

Greaves M (2009) Darwin and evolutionary tales in leukemia. Hematology Am Soc Hematol Educ Program 2009:3–12

Grimwade D, Hills RK, Moorman AV et al (2010) Refinement of cytogenetic classification in acute myeloid leukemia: Determination of prognostic significance of rare recurring chromosomal abnormalities among 5876 younger adult patients treated in the United Kingdom Medical Research Council trials. Blood 116:354–365

Haferlach C, Mecucci C, Schnittger S et al (2009) AML with mutated NPM1 carrying a normal or aberrant karyotype show overlapping biologic, pathologic, immunophenotypic, and prognostic features. Blood 114:3024–3032

Hong D, Gupta R, Ancliff P et al (2008) Initiating and cancer-propagating cells in TEL-AML1-associated childhood leukemia. Science 319(5861):336–339

Hosen N, Park CY, Tatsumi N et al (2007) CD96 is a leukemic stem cell-specific marker in human acute myeloid leukemia. Proc Natl Acad Sci U S A 104:11008–11013

Huntly B, Shigematsu H, Deguchi K et al (2004) MOZ-TIF2, but not BCR-ABL, confers properties of leukemic stem cells to committed murine hematopoietic progenitors. Cancer Cell 6:587–596

Jaiswal S, Jamieson CH, Pang WW et al (2009) CD47 is upregulated on circulating hematopoietic stem cells and leukemia cells to avoid phagocytosis. Cell 138:271–285

Jamieson CH. Ailles LE, Dylla SJ et al (2004) Granulocyte-macrophage progenitors as candidate leukemic stem cells in blast-crisis CML. N Engl J Med 351:657–667

Jan M, Chao MP, Cha AC et al (2011) Prospective separation of normal and leukemic stem cells based on differential expression of TIM3, a human acute myeloid leukemia stem cell marker. Proc Natl Acad Sci U S A 108:5009–5014

Jelinek J, Oki Y, Gharibyan V et al (2005) JAK2 mutation 1849G>T is rare in acute leukemias but can be found in CMML, Philadelphia chromosome-negative CML, and megakaryocytic leukemia. Blood 106:3370–3373

Jin L, Lee EM, Ramshaw HS et al (2009) Monoclonal antibody-mediated targeting of CD123, IL-3 receptor alpha chain, eliminates human acute myeloid leukemic stem cells. Cell Stem Cell 5:31–42

Kelly PN, Dakic A, Adams JM et al (2007) Tumor growth need not be driven by rare cancer stem cells. Science 317:337

Konrad M, Metzler M, Panzer S et al (2003) Late relapses evolve from slow-responding subclones in t(12, 21)-positive acute lymphoblastic leukemia: evidence for the persistence of a preleukemic clone. Blood 101(9):3635–3640

Krivtsov AV, Twomey D, Feng Z et al (2006) Transformation from committed progenitor to leukaemia stem cell initiated by MLL,ÄìAF9. Nature 442:818–822

Lapidot T, Sirard C, Vormoor J et al (1994) A cell initiating human acute myeloid leukaemia after transplantation into SCID mice. Nature 367:645–648

Majeti R, Chao MP, Alizadeh AA et al (2009a) CD47 is an adverse prognostic factor and therapeutic antibody target on human acute myeloid leukemia stem cells. Cell 138:286–299

Majeti R, Becker MW, Tian Q et al (2009b) Dysregulated gene expression networks in human acute myelogenous leukemia stem cells. Proc Natl Acad Sci U S A 106:3396–3401

Martelli MP, Pettirossi V, Thiede C et al (2010) CD34+ cells from AML with mutated NPM1 harbor cytoplasmic mutated nucleophosmin and generate leukemia in immunocompromised mice. Blood 116:3907–3922

Mullighan RC, Phillips LA et al (2008) Genomic analysis of the clonal origins of relapsed acute lymphoblastic leukemia. Science 322(5906):1377–1380

Notta F, Mullighan CG, Wang JC et al (2011) Evolution of human BCR-ABL1 lymphoblastic leukaemia-initiating cells. Nature 46(7330):362–367

Nurrul Abdullah L, Kai-Hua Chow E (2013) Mechanisms of chemoresistance in cancer stem cells. Clin Transl Med 2:3

Pearce DJ, Taussig DC, Zibara K et al (2006) AML engraftment in the NOD/SCID assay reflects the outcome of AML: implications for our understanding of the heterogeneity of AML. Blood 107:1166–1173

Quintana E, Shackleton M, Sabel MS et al (2008) Efficient tumour formation by single human melanoma cells. Nature 456:593–598

Ricci-Vitiani L, Lombardi DG, Pilozzi E et al (2007) Identification and expansion of human colon-cancer-initiating cells. Nature 445:111–115

Saito Y, Kitamura H, Hijikata A et al (2010) Identification of therapeutic targets for quiescent, chemotherapy-resistant human leukemia stem cells. Sci Transl Med 2(17):17ra9

Sanchez PV, Perry RL, Sarry JE et al (2009) A robust xenotransplantation model for acute myeloid leukemia. Leukemia 23:2109–2117

Sarry JE, Murphy K, Perry R et al (2011) Human acute myelogenous leukemia stem cells are rare and heterogeneous when assayed in NOD/SCID/IL2Rgammac-deficient mice. J Clin Invest 121:384–395

Schatton T. Murphy GF, Frank NY et al (2008) Identification of cells initiating human melanomas. Nature 451:345–349

Schnittger S, Schoch C, Kern W et al (2005) Nucleophosmin gene mutations are predictors of favorable prognosis in acute myelogenous leukemia with a normal karyotype. Blood 106:3733–3739

Singh SK, Hawkins C, Clarke ID et al (2004) Identification of human brain tumour initiating cells. Nature 432:396–401

Smith ML, Hills RK, Grimwade D (2011) Independent prognostic variables in acute myeloid leukaemia. Blood Rev 25:39–51

Somervaille TCP, Cleary ML (2006) Identification and characterization of leukemia stem cells in murine MLL-AF9 acute myeloid leukemia. Cancer Cell 10:257–268

Taussig DC, Miraki-Moud F, Anjos-Afonso F et al (2008) Anti-CD38 antibody-mediated clearance of human repopulating cells masks the heterogeneity of leukemia-initiating cells. Blood 112:568–575

Taussig DC, Vargaftig J, Miraki-Moud F et al (2010) Leukemia-initiating cells from some acute myeloid leukemia patients with mutated nucleophosmin reside in the CD34(-) fraction. Blood 115:1976–1984

Van Rhenen A, van Dongen GA, Kelder A et al (2007) The novel AML stem cell associated antigen CLL-1 aids in discrimination between normal and leukemic stem cells. Blood 110:2659–2666

Vargaftig J, Taussig DC, Griessinger E et al (2011) Frequency of leukemic initiating cells does not depend on the xenotransplantation model used. Leukemia 4:858–860

Wei J, Wunderlich M, Fox C et al (2008) Microenvironment determines lineage fate in a human model of MLL-AF9 leukemia. Cancer Cell 13:483–495

Yang JJ, Bhojwani D, Yang W et al (2008) Genome-wide copy number profiling reveals molecular evolution from diagnosis to relapse in childhood acute lymphoblastic leukemia. Blood 112:4178–4183

Zhou S, Schuetz D, Bunting KD et al (2001) The ABC transporter Bcrp1/ABCG2 is expressed in a wide variety of stem cells and is a molecular determinant of the side- population phenotype. Nat Med 7:1028–1034

Chapter 29
Targeting Leukemia Stem Cells

Duane C. Hassane and Monica L. Guzman

Abstract Leukemia stem cells (LSCs) are chemoresistant and able to initiate and perpetuate disease. There is increasing evidence that LSCs contribute to poor prognosis and disease relapse (van Rhenen et al., Clin Cancer Res 11(18):6520–6527, 2005; Eppert et al., Nat Med 17(9):1086–1093, 2011; Jan et al., Proc Natl Acad Sci U S A 108(12):5009–5014, 2011; Gerber et al., Blood 119(15):3571–3577, 2012). Given their clinical relevance, LSCs are a highly pursued target for improving therapeutic outcomes. Efforts in the past decade have been rapidly expanded to discover novel therapeutic strategies for eliminating LSCs that simultaneously spare their normal hematopoietic counterparts.

Keywords Leukemia stem cells · Cancer stem cells · Survival mechanisms · Microenvironment · Chemoresistance · Chemical genomics · Cancer genomics

29.1 Introduction

Leukemia stem cells (LSCs) are critical to the pathology of acute myeloid leukemias (AMLs) and thus represent an essential therapeutic target for improving disease outcomes. Indeed, there are ever-increasing data demonstrating that residual LSCs are not effectively eliminated by current standard of care antileukemia therapies and thus are likely the source of disease relapse (Fig. 29.1). The ability to isolate and characterize LSC populations has greatly improved in the past decade and, with these data the number of therapeutic approaches proposed to eliminate this malignant population has grown. These approaches encompass the blocking of

M. L. Guzman (✉)
Department of Medicine, Hematology, and Medical Oncology,
Pharmacology in Medicine, Weill Medical College of Cornell University,
1300 York Avenue, A-605B, Box 113, New York, NY 10065, USA
e-mail: mlg2007@med.cornell.edu

D. C. Hassane
Department of Medicine, Hematology, and Medical Oncology,
Institute for Computational Biomedicine, Weill Medical College of Cornell University,
1305 York Avenue, Box 140, New York, NY 10021, USA
e-mail: dhassane@med.cornell.edu

M. Andreeff (ed.), *Targeted Therapy of Acute Myeloid Leukemia*,
Current Cancer Research, DOI 10.1007/978-1-4939-1393-0_29

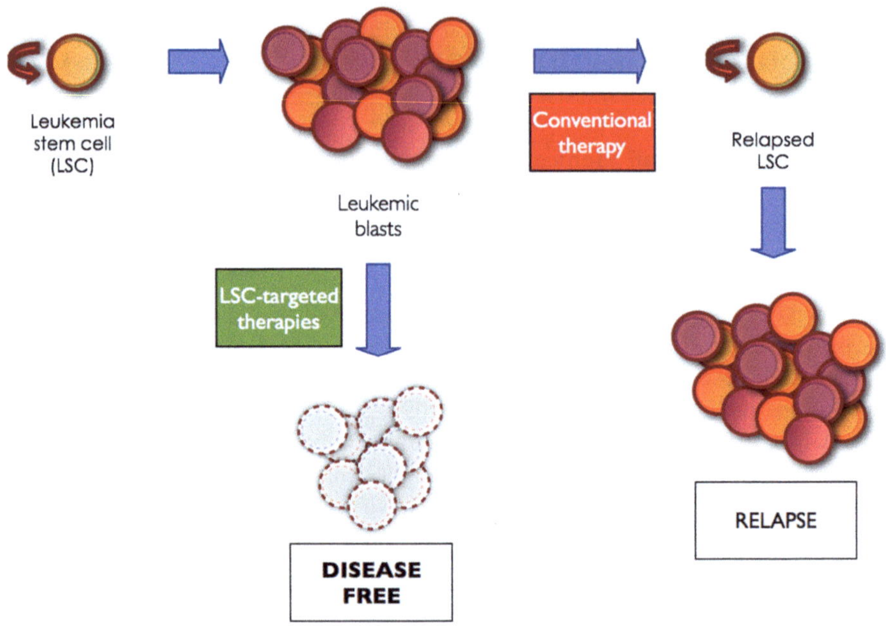

Fig. 29.1 Leukemia stem cell model for therapeutic responseLeukemia stem cells (*LSC*) give rise and maintain leukemic blasts. Conventional chemotherapy drugs fail to eliminate LSCs, which can result in disease relapse. Current therapeutic focus for AML has been to identify targets in LSCs that can result in improved therapies and better disease outcomes

survival pathways, disruption of niche-LSC interaction, antibody-based targeting, perturbation of the epigenetic machinery, disruption of stem cell renewal, iron metabolism or mitochondrial function, and targeting mutant tyrosine kinases or altered cell signaling cascades (Fig. 29.2). Thus, to date, there is an extensive therapeutic arsenal to reach the elusive LSC target. As these agents move to clinical trials, we will be able to determine the feasibility of using any of these therapeutic approaches to eliminate LSCs in patients.

29.2 Targeting LSC Unique Features

29.2.1 Immunophenotype

LSCs can share immunophenotypic markers with their normal counterparts, such as $CD34^+$ and CD38− (Bonnet and Dick 1997; Guzman and Jordan 2004). Recent data have suggested that the LSC immunophenotype can be different from patient to patient (Sarry et al. 2011; Taussig et al. 2008) with LSCs also being detected within $CD34^+/CD38^+$ AML cell populations in some patients. Despite this

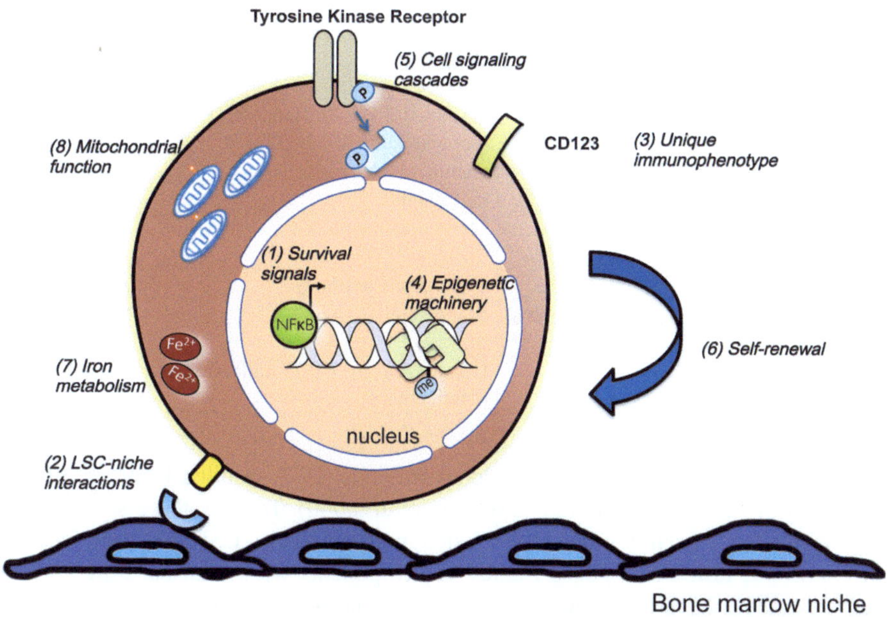

Fig. 29.2 Identified therapeutic targets in LSCs. LSCs depend on different key regulatory mechanisms for their survival. These mechanisms have thus become potential therapeutic targets to ablate LSCs. These mechanisms include: (*1*) survival pathways, (*2*) niche-LSC interaction, (*3*) unique cell surface antigens, (*4*) epigenetic regulators, (*5*) cell signaling cascades, (*6*) stem-cell renewal, (*7*) iron metabolism, and (*8*) mitochondrial function

diversity, antibodies targeting leukemia populations that include LSCs have been reported and are the basis of targeted antibody-based anti-LSC therapies. The list of LSC-immunophenotypic markers keeps growing, and a summary of these cell surface antigens is shown in Table 29.1. To date, several of these antigens have become the target for emerging therapeutic strategies that are been explored in clinical trials or preclinical laboratory studies. These include: CD123 (e.g., DT388IL-3; CSL360 Jordan et al. 2000; Roberts 2008), CD33 (gemtuzumab ozogamicin (GO) (Hauswirth et al. 2007)), CD44 (see Chap. 31; Jin et al. 2006; Krause et al. 2006), and CLL-1 (van Rhenen et al. 2007).

Strategies for targeting cell surface antigens include monoclonal antibodies (Ab), antibodies conjugated to toxins or radioisotypes, chimeric antigen receptor modified Tcells (CAR Tcells), and nanoparticles containing antibodies directed at the targeted surface antigen.

29.2.1.1 CD123

CD123 (IL-3Rα) was found preferentially expressed in the $CD34^{+}/CD38-$ compartment of AML samples when compared to their normal counterparts (Leyton et al. 2011; Jin et al. 2009; Jordan et al. 2000). More recently, it was shown that targeting

Table 29.1 Immunophenotypic markers and approaches for leukemia stem cell therapies

Immunophenotypic markers			
Target	Reference	Therapeutic approach	Stage
CD123	Jordan et al. 2000; Jin et al. 2009	DT388-IL3 (SL-401); anti-CD123 monoclonal antibody (CSL360)	Phase I/II
CD33	Hauswirth et al. 2007	Gemtuzumab ozogamicin	Phase III
CD44	Jin et al. 2006	huARH460	N/A
CLL-1	van Rhenen et al. 2007; Bakker et al. 2004	Monoclonal antibody	N/A
CD47	Majeti et al. 2009b	Monoclonal antibody	N/A
TIM3	Jan et al. 2011, Kikushige et al. 2010	Monoclonal antibody	N/A
CD25	Saito et al. 2010	Humanized anti-CD25 antibody (daclizumab)	Phase I/II trial

CD123 using the anti-CD123 monoclonal antibody, 7G3, resulted in decreased engraftment of LSCs (Leyton et al. 2011; Jin et al. 2009). To date, there are two biologic agents have been developed and tested in phase I trials. CSL360, a monoclonal antibody, was tested in a phase I study for 26 patients with relapsed, refractory or high-risk AML. Minimal toxicity was observed and one patient achieved CR (Roberts et al. 2008). DT388IL3 (SL-401), a recombinant protein consisting of diphtheria toxin-interleukin-3 fusion, was also tested in phase I clinical trials for 74 patients with relapsed or refractory adult AML, de novo poor-risk elderly AML, or high-risk MDS (Cohen et al. 2005; Hall et al. 1998; Konopleva et al. 2010).

29.2.2 *CONCLUSION*

29.2.2.1 CD33

Leukemic blasts express CD33 in 85–90% of AML patients (Griffin et al. 1984; Dinndorf et al. 1986). Its expression in $CD34^{+}$/CD38-/$CD123^{+}$ subpopulations was also reported for AML patients with $CD33^{+}$ AML (Hauswirth et al. 2007). Furthermore, BM autografts depleted of $CD33^{+}$ cells in vitro resulted in sustained durable engraftment. These data suggest that CD33 is not expressed in normal HSCs (Robertson et al. 1992). Therefore, targeting CD33 has been of interest to eliminate LSCs in $CD33^{+}$leukemias. The anti-CD33 antibody–calicheamicin conjugate GO demonstrated selective toxicity against $CD33^{+}$ cells, inhibiting leukemia colony formation in a significant subset of AML specimens (Hamann et al. 2002a, 2002b).

29.2.2.2 Other Novel Phenotypic Markers Under Preclinical Investigation

Other antigens of interest are CD47, CLL-1, TIM-3, and CD25 (Majeti et al. 2009b; Majeti 2011; van Rhenen et al. 2007; Jan et al. 2011; Kikushige et al. 2010; Saito et al. 2010). The anti-CD47 monoclonal antibody, B6H12.2, reduced tumor burden and leukemia was not observed in secondary xenotransplants (Majeti 2011). Anti-CLL-1 monoclonal antibody was reported to induce toxicity in AML blasts (Bakker et al. 2004). In addition, CLL-1-targeted nanoparticles loaded with daunorubin have been also explored as a selective therapeutic strategy against LSCs (Zhang et al. 2012). TIM-3 expression was used to prospectively separate HSC from leukemic cells in the AML specimens that possess residual normal HSC function suggesting selectivity for TIM-3 in LSCs (Jan et al. 2011).

An antihuman TIM-3 antibody was developed (ATIK2a) and demonstrated complement-dependent and antibody-dependent cellular cytotoxic activities. This antibody did not harm reconstitution of normal human HSCs, but blocked engraftment of AML after xenotransplantation (Kikushige et al. 2010).

Taken together, these studies offer great promise in eradicating LSCs using antibody-targeted therapies while minimizing toxicity to normal functioning cells.

29.2.3 Survival Pathways

Pathways controlling survival and apoptosis are aberrantly regulated in LSCs and are critical for LSC survival. Thus, molecules involved in such pathways have been explored as potential therapeutic targets. These pathways include NF-κB, PI3 kinase/Akt/mTOR (mammalian target of rapamycin), and the intrinsic apoptosis pathway.

29.2.3.1 NF-κB

NF-κB is a transcription factor complex that plays an important role in the regulation of cell growth and survival. Under normal conditions, NF-κB is excluded from the nucleus by interaction with its cellular inhibitor, IκBα. Upon stimulation, a signaling cascade results in the phosphorylation, ubiquitination, and subsequent degradation of IκBα, and results in the liberation of NF-κB with subsequent nuclear translocation where it activates transcription of a variety of genes (Hayden and Ghosh 2008; Mercurio and Manning 1999). NF-κB has been found constitutively active in most hematologic malignancies (Karin et al. 2002). LSCs from human AML specimens have shown constitutive NF-κB activity, when compared to normal hematopoietic progenitor/stem cells, suggesting that NF-κB represents a selective anti-LSC target in that it can discriminate HSCs from LSCs (Guzman et al. 2001, 2002). In addition, AML chemotherapeutic drugs, such as cytarabine and anthracyclines, further enhance NF-κB activity resulting in a survival advantage to the leukemia cells (Mayo and Baldwin 2000). Different strategies have been used to inhibit NF-kB in LSCs, including proteasome inhibitors and other small molecules such as the sesquiterpene lactone parthenolide (PTL), derived from a medicinal plant know as feverfew (Guzman et al. 2005). An orally bioavailable derivative of PTL, dimethylaminoparthenolide (DMAPT, LC-1), is currently being tested in a phase I clinical trial in AML (Guzman et al. 2007).

29.2.3.2 PI3 Kinase/Akt/mTOR

The PI3 kinase/Akt/mTOR signaling pathway has also been shown to be important target for LSCs (Xu et al. 2003, 2005). Constitutive PI3K activity has been reported in primary AML samples and is associated with proliferation, survival, antiapoptotic responses, and drug resistance (Xu et al. 2003, 2005). The PI3K/mTOR network includes many transcription factors and survival processes such as the FOXO family of proteins, important in ROS regulation; the BCL-2 family of proteins (apoptosis regulators); and the wnt/β-catenin signaling pathway (implicated in LSC self-renewal) (Martelli et al. 2011). Treatment with PI3K/mTOR inhibitors (e.g., wortmannin and rapamycin) results in decreased LSC activity evaluated in xenotransplants (Xu et al. 2003). Importantly, the activation of PI3K/mTOR pathway has been associated with Nrf-2 cyto-protective responses. PI3K/mTOR pathway inhibitors can potentiate the effects of parthenolide on LSCs with concomitant inhibition of Nrf2 (Hassane et al. 2010).

29.2.3.3 BH3 Mimetics

Bcl-2 family members play an important role in the control of the intrinsic apoptosis pathway. These proteins regulate the mitochondrial outer membrane permeabilization, which is a key event for caspase activation and commitment to cell death (Lomonosova and Chinnadurai 2008; Strasser 2005). Aberrant expression of some of the Bcl-2 family members has been described in AML. An elevated ratio of BCL2 to BAX ratio (resulting in antiapoptotic signals) has been reported as an unfavorable prognostic factor in AML patients (Del Poeta et al. 2003; Glaser et al. 2012). Thus, inhibitory molecules for Bcl-2-related antiapoptotic proteins have been developed as an antileukemia strategy. Importantly, it was shown that the BH3-mimetic that interferes with BCL-2 family members, ABT-737, also has activity against LSCs (Cragg et al. 2009; Konopleva et al. 2006).

29.3 Genomic Approaches

Efforts to target LSCs are impeded by the absence of a clearly delineated LSC phenotype. Though surface markers have proven instrumental in defining leukemia subpopulations in which the probability of malignant cells bearing stem-like properties is higher, the ability to point to a specific leukemia cell and unambiguously declare it an LSC is currently not feasible in the absence of corroborating functional assays. Therefore, high throughput small molecule screens for LSC-targeted drugs from cultured cells remain suboptimal. Given that the requisite functional assays are expensive and time consuming, capturing a complete blueprint of the molecular changes in leukemia resulting from exposure to anti-LSC drugs can greatly assist drug discovery. Indeed, continuing efforts at the Broad Institute have demonstrated the capability of using drug-induced transcriptional perturbations to define functional relationships between compounds in the absence of structural similarity using a compendium of drug-induced gene expression changes, publicly available in the Connectivity Map database (Hieronymus et al. 2006; Wei et al. 2006; Lamb et al. 2006).

Toward identifying drugs that circumvent the suboptimal pharmacology of parthenolide, previous studies applied and extended the Connectivity Map approach by recording the transcriptional perturbations caused by parthenolide in $CD34^+$ AML cells and using this signature to probe the gene expression space of the Connectivity Map and Gene Expression Omnibus (Hassane et al 2008). This effort produced two compounds with previously unreported toxicity against LSCs (i.e., celastrol and 4-hydroxy-2-nonenal). Celastrol acts as an Hsp90 inhibitor (Hieronymus et al. 2006), thereby indirectly mediating parthenolide's NF-κB-inhibitory function since NF-κB activity is dependent on Hsp90 (Broemer et al. 2004; Chen et al. 2002). The lipid peroxidation product, 4-hydroxy-2-nonenal, is a reported inhibitor of the proteasome (Ferrington and Kapphahn 2004) and NF-κB (Ji et al. 2001; Page et al. 1999).

In addition to finding new drugs, it was subsequently demonstrated that transcriptional signatures of drugs can be use to define unexpected synergies between drugs. Indeed, evident within the transcriptional signature of parthenolide was a cytoprotective response involving a therapeutically undesirable activation of the PI3K/mTOR pathway and antioxidant genes controlled by Nrf2 (Hassane et al. 2010). It was predicted that inhibitors of the PI3K/mTOR pathways would thereby enhance parthenolide function, allowing it to be functional at lower doses. Consistent with the data generated by the transcriptional signature, the combination of parthenolide and PI3K/mTOR inhibitors was synergistic and continued to preserve the function of normal hematopoietic cells. Therefore, as more drugs are profiled, the power of using drug signatures to predict combinations is promising. Compounds can be selected that improve standard of care therapies so that they target LSCs. Moreover, compounds with more effective pharmacology are entering clinical trials and, eventually, may represent anti-LSC therapeutic options to patients.

29.4 Microenvironment

HSCs reside within highly specialized compartments of the bone marrow that are defined into endosteal and vascular regions. These regions provide critical support to promote HSC survival, self-renewal and differentiation. LSCs, like HSCs, have been shown to rely on their interactions with these compartments for their self-renewal and survival functions (Lane et al. 2009). In addition, LSCs have been shown to suppress normal hematopoiesis by taking over the HSC niches (Colmone et al. 2008). The endosteum region of the BM has been shown protect LSCs from AML patients, defined as $CD34^+/CD38^-$, against chemotherapy in xenotransplant models (Ishikawa et al. 2007). Therefore, another anti-LSC strategy that has been explored is to interfere with the interactions of LSCs with their niche.

29.4.1 Disruption of CXCR4/SDF-1 Interactions

Granulocyte-colony-stimulating factor (G-CSF) was one of the first cytokines reported to have an effect on cell cycle and mobilization of HSCs (Metcalf 1985). G-CSF induces mobilization by the release of proteolytic enzymes that result in the disruption of stem cell anchorages between HSCs and their niche, such as CXCR4/SDF-1, N-cadherin, and VCAM-1/VLA-4 interactions. For AML, it was also shown that treatment of AML-xenografts with G-CSF-induced mobilization of LSCs as well as cell cycle entry resulting in increased susceptibility to chemotherapy. These findings suggested that priming with G-CSF may improve the ability of chemotherapy drugs to eliminate the quiescent LSC population (Metcalf 1985).

Additional strategies that abrogate the CXCR4/SDF-1 interactions include the use of CXCR4 neutralizing antibodies or CXCR4-antagonists (e.g., AMD3100). Both approaches have shown decreased AML survival in vitro and decreased engraftment of primary AML cells. In addition, these strategies result in loss of quiescence and mobilization (Tavor et al. 2004; Liesveld et al. 2007). Currently, there are several phase I clinical trials testing the use of anti-CXCR4 agents with or without G-CSF and proteasome inhibition in AML (http://clinicaltrials.gov/). Therapies that are designed to disrupt the interactions between the niche and LSCs may be more beneficial in the postremission setting to mobilize protected residual LSCs.

29.5 Self-Renewal

Interfering with the mechanisms that regulate self-renewal of LSCs has also been considered as a feasible strategy to ablate LSCs. Pathways such as Notch, Hedgehog (Hh), and WNT have been reported as potential targets as they are important regulators in stem cell self-renewal and differentiation.

29.5.1 Notch Pathway

The Notch signaling pathway is involved in cell-fate decisions. Delta and Jagged are the two evolutionary conserved families of Notch ligands. Notch activation initiates when receptors engage ligand, ultimately leading to proteolytic cleavage events mediated by the ADAM-type metalloproteases and γ-secretase (Kopan and Ilagan 2009). In AML, Jagged-2 was found to be overexpressed using gene expression profiles of LSCs. It was also shown that a treatment with DAPT, a γ-secretase inhibitor, resulted in decreased colony formation (Gal et al. 2006). To date gamma-secretase inhibitors (GSIs) are being tested Please take out in preclinical models and clinical trials for solid tumors and Tcell leukemias (http://clinicaltrials.gov/). However, GSIs have been shown to not only inhibit the cleavage of Notch but also other proteins, such as CD44 (Lammich et al. 2002).

29.5.2 Hedgehog (Hh) Pathway

In mammals, Hh signaling is initiated by the binding of a Hh ligand [Sonic Hh (Shh), Indian Hh (Ihh), or Desert Hh (Dhh)] to their surface receptor Patched (Ptch). Upon binding, Smoothened (Smo) inhibition by Ptch is alleviated, resulting in the activation of the Gli transcription factors (Taipale and Beachy 2001). Loss of Smo was shown to result in decreased transformation of HSCs to CML when transduced with a BCR-ABL retrovirus mediated by decreased self-renewal. Moreover, pharmacologic inhibition of Hh signaling with cyclopamine (Smo inhibitor) also demonstrated that Hh pathway is required for maintenance of CML stem cells including imatinib-resistant cells (Zhao et al. 2009). Importantly, it has been shown that Hh inhibition has no significant effect on normal hematopoiesis. In contrast to CML, Hh signaling was not required for the development of AML mediated by MLL-AF9 (Hoffmann et al. 2009). In B-ALL, Hh inhibition resulted in decreased clonogenic activity (Lin et al. 2010). Thus, these data suggest that Hh signaling may be dispensable for normal hematopoietic stem cell function and that targeting of Hh may not be an ideal LSC target for all leukemias.

29.5.3 Wnt/β-Catenin Pathway

Dysregulation of the canonical Wnt/β-catenin signaling pathway has been found to contribute to the development of numerous types of human cancers including AML (Clevers 2006). In AML, expression of β-catenin has been correlated with poor prognosis (Clevers 2006; Majeti et al. 2009a). Moreover, Wnt/β-catenin signaling is required for the oncogenic transformation of HSCs into LSCs (Wang et al. 2010). Thus, Wnt/β-catenin pathway is an attractive therapeutic target to ablate LSCs. The targeting of β-catenin in LSCs has demonstrated promising results in preclinical studies of mixed lineage leukemia-translocated AML (Yeung et al. 2010). However, it has also been shown that while the Wnt/β-catenin pathway is aberrantly regulated, not all LSCs are dependent on this pathway for survival (Gandillet et al. 2011). Thus, identification of patients with LSCs that are "addicted" to the Wnt/β-catenin pathway will be of importance for successful clinical trials.

29.6 Epigenetic Modifiers

Recent reports have suggested that, in AML, genetic mutations can lead to epigenetic changes that produce loss of function from aberrant activation of epigenetic mechanisms (Toyota et al. 2001; Lund and Lohuizen 2004; Figueroa et al. 2010). Such epigenetic regulatory events include DNA methylation and posttranslational histone modifications. Pharmacologic inhibitors of DNA methyltransferases or histone deacetylases, thus, represent approaches to interfere with these mechanisms and restore normal gene expression. In addition to inhibitors of DNA methyltransferases and histone deacetylases, new emerging therapeutic target have been identified for LSCs such as DOT1L and LSD1 (KDM1A).

29.6.1 DOT1L

DOT1L is a histone H3K79 methyltransferase. In normal cells, methylation of H3K79 by DOT1L plays a role in cell cycle regulation and transcriptional elongation and has been found to be highly conserved among species (Barry et al. 2010). In leukemia, DOT1L is involved in the regulation of genes critical to LSCs self-renewal and survival, such as HOXA9, MEIS1, FLT3, and BCL2. DOT1L was also found to be required for initiation and maintenance of MLL-AF9-induced leukemia. Downregulation of DOT1L leads to the decreased expression of HOXA genes (Chang et al. 2010; Jo et al. 2011; Bernt et al. 2011). Inhibitors for DOT1L are currently under development, such as EPZ004777 (Daigle et al. 2011). Treatment with this inhibitor demonstrated selective targeting of DOT1L leading to decreased transcription of key genes associated with leukemogenesis in MLL, such as HOXA9 and MEIS1. Thus, DOT1L represents a new target of great interest to ablate LSCs in MLL driven leukemias.

29.6.2 LSD1 (KDM1A)

LSD1 (KDM1/AOF2), a lysine-specific demethylase shown to demethylate H3K4 and H3K9, is highly expressed in patients with AML. In a recent report, an epigenetic study demonstrated that loss of RARα2 expression in AML was associated with a reduction in H3K4me2 on the RARα2 promoter. It was hypothesized that targeting LSD1 would restore the expression of RARα2 and thus cells would be sensitive to ATRA differentiation. Indeed, it was observed that treatment with LSD1 inhibitors in combination with ATRA resulted in differentiation and cell death of AML cells (Schenk et al. 2012).

29.7 Other Approaches

29.7.1 HIF-1a

LSCs ($CD34^+/CD38^-$) obtained from AML patients demonstrated increased HIF1α expression. Importantly, HIF1α shRNA resulted in a significant reduction in colony formation. Treatment with echinomycin, a HIF-1α inhibitor, eliminated LSCs demonstrated in a serial xenograft transplant model (Wang et al. 2011). The specificity of echinomycin was further corroborated with the HIF1α knockdown, as the transduction of AML cells with Hif1α shRNA resulted in decreased sensitivity to echinomycin. While echinomycin had dose-limiting gastrointestinal toxicities in prior clinical trials, other HIF-1α inhibitors such as EZN-2968 and PX-478 are currently being tested in other malignancies (Greenberger et al. 2008; Lee and Kim 2011; http://clinicaltrials.gov/).

29.7.2 BCL-6

BCL-6 is a transcriptional repressor found to be an important mediator of CML and ALL stem cell chemoresistance (Duy et al. 2010; Hurtz et al. 2011; Pellicano and Holyoake 2011). BCL-6 is required for the formation of germinal centers and directly inhibits genes involved in apoptosis and DNA damage sensing, including ATR, CHEK1, TP53, CDKN1, p21, and BCL-2 (Basso and Dalla-Favera 2012). BCL-6 is a member of the BTB-POZ family of proteins and mediates transcriptional silencing through its BTB repression domain. Constitutive expression of BCL-6 drives the pathogenesis of diffuse large Bcell lymphomas (DLBCLs) (Ahmad et al. 2003; Ye 2000). In BCR-ABL$^+$ B-ALL and in CML, BCL-6 was shown to be essential for imatinib-resistance through p53- and ARF-mediated mechanisms. BCL-6 inhibition with a novel peptidomimetic inhibitor, RI-BPI, resulted in ablation of LSCs in B-ALL and CML that are not eliminated by imatinib (Polo et al. 2004; Duy et al. 2010; Hurtz et al. 2011). These studies suggest that simultaneous targeting of BCR-ABL and BCL-6 represent a therapeutic strategy to ablate LSCs and prevent TKI resistance and relapse in BCR-ABL positive B-ALL and CML patients.

29.7.3 Chemical Screen

Chemical screens have been used to identify and repurpose FDA-approved drugs to eliminate LSCs with the goal of accelerating clinical translation. These approaches have identified drugs such as salinomycin, tigecycline, and kinetin riboside (KR) (Skrtic et al. 2011; McDermott et al. 2012). Tigecycline is a broad-spectrum antibiotic that was found to interfere with the mitochondrial translation machinery in a leukemia-specific manner. The increased sensitivity of leukemic cells to tigecycline led to the observation that leukemia cells possess a larger mitochondrial mass, a larger mitochondrial DNA copy number, and an increased rate of oxygen consumption (Skrtic et al. 2011). KR, in vitro, was found effective in a large number of primary AML samples. However, LSCs were not inhibited in all samples tested in xenotransplants suggesting heterogeneity in the response (McDermott et al. 2012). These results emphasize the need to evaluate a broad spectrum of patient samples with both in vitro and in vivo assays to identify the characteristics of AML patients whose LSCs are more likely to respond to KR.

29.7.4 Iron Chelators

Iron is essential for the cellular functions of cancer cells and therefore cancer cells demonstrate sensitivity to iron depletion (Richardson et al. 2009). Iron chelators display antitumor activity through both their pro-oxidant activities and iron depletion itself. In AML, iron deprivation therapies (i.e., iron chelators or transferrin-receptor-blocking antibodies) result in mature differentiation of leukemic cells into monocytes (Callens et al. 2010). Cell death has also been observed after exposure to iron chelators such as the antifungal agent ciclopirox in leukemia and myeloma cells mediated by the inhibition of the iron-dependent enzyme ribonucleotide reductase (Eberhard et al. 2009). These effects are also observed in stem/progenitor cells.

References

Ahmad KF, Melnick A, Lax S, Bouchard D, Liu J, Kiang CL, Mayer S, Takahashi S, Licht JD, Prive GG (2003) Mechanism of SMRT corepressor recruitment by the BCL6 BTB domain. Mol Cell 12(6):1551–1564. doi:S1097276503004544 [pii]

Bakker AB, van den Oudenrijn S, Bakker AQ, Feller N, van Meijer M, Bia JA, Jongeneelen MA, Visser TJ, Bijl N, Geuijen CA, Marissen WE, Radosevic K, Throsby M, Schuurhuis GJ, Ossenkoppele GJ, de Kruif J, Goudsmit J, Kruisbeek AM (2004) C-type lectin-like molecule-1: a novel myeloid cell surface marker associated with acute myeloid leukemia. Cancer Res 64(22):8443–8450. doi:64/22/8443 [pii] 10.1158/0008-5472.CAN-04-1659

Barry ER, Corry GN, Rasmussen TP (2010) Targeting DOT1L action and interactions in leukemia: the role of DOT1L in transformation and development. Expert Opin Ther Targets 14(4):405–418. doi:10.1517/14728221003623241

Basso K, Dalla-Favera R (2012) Roles of BCL6 in normal and transformed germinal center B cells. Immunol Rev 247(1):172–183. doi:10.1111/j.1600-065X.2012.01112.x

Bernt KM, Zhu N, Sinha AU, Vempati S, Faber J, Krivtsov AV, Feng Z, Punt N, Daigle A, Bullinger L, Pollock RM, Richon VM, Kung AL, Armstrong SA (2011) MLL-rearranged leukemia is dependent on aberrant H3K79 methylation by DOT1L. Cancer Cell 20(1):66–78. doi:S1535-6108(11)00228-5 [pii] 10.1016/j.ccr.2011.06.010

Bonnet D, Dick JE (1997) Human acute myeloid leukemia is organized as a hierarchy that originates from a primitive hematopoietic cell. Nat Med 3:730–737

Broemer M, Krappmann D, Scheidereit C (2004) Requirement of Hsp90 activity for IkappaB kinase (IKK) biosynthesis and for constitutive and inducible IKK and NF-kappaB activation. Oncogene 23(31):5378–5386. doi:10.1038/sj.onc.1207705 1207705 [pii]

Callens C, Coulon S, Naudin J, Radford-Weiss I, Boissel N, Raffoux E, Wang PH, Agarwal S, Tamouza H, Paubelle E, Asnafi V, Ribeil JA, Dessen P, Canioni D, Chandesris O, Rubio MT, Beaumont C, Benhamou M, Dombret H, Macintyre E, Monteiro RC, Moura IC, Hermine O (2010) Targeting iron homeostasis induces cellular differentiation and synergizes with differentiating agents in acute myeloid leukemia. J Exp Med 207(4):731–750. doi:jem.20091488 [pii] 10.1084/jem.20091488

Chang MJ, Wu H, Achille NJ, Reisenauer MR, Chou CW, Zeleznik-Le NJ, Hemenway CS, Zhang W (2010) Histone H3 lysine 79 methyltransferase Dot1 is required for immortalization by MLL oncogenes. Cancer Res 70(24):10234–10242. doi:70/24/10234 [pii] 10.1158/0008-5472.CAN-10-3294

Chen G, Cao P, Goeddel DV (2002) TNF-induced recruitment and activation of the IKK complex require Cdc37 and Hsp90. Mol Cell 9(2):401–410. doi:S1097276502004501 [pii]

Clevers H (2006) Wnt/beta-catenin signaling in development and disease. Cell 127(3):469–480. doi:S0092-8674(06)01344-4 [pii] 10.1016/j.cell.2006.10.018

Cohen KA, Liu TF, Cline JM, Wagner JD, Hall PD, Frankel AE (2005) Safety evaluation of DT388IL3, a diphtheria toxin/interleukin 3 fusion protein, in the cynomolgus monkey. Cancer Immunol Immunother 54(8):799–806. doi:10.1007/s00262-004-0643-4

Colmone A, Amorim M, Pontier AL, Wang S, Jablonski E, Sipkins DA (2008) Leukemic cells create bone marrow niches that disrupt the behavior of normal hematopoietic progenitor cells. Science 322(5909):1861–1865. doi:322/5909/1861 [pii] 10.1126/science.1164390

Cragg MS, Harris C, Strasser A, Scott CL (2009) Unleashing the power of inhibitors of oncogenic kinases through BH3 mimetics. Nat Rev Cancer 9(5):321–326. doi:nrc2615 [pii] 10.1038/nrc2615

Daigle SR, Olhava EJ, Therkelsen CA, Majer CR, Sneeringer CJ, Song J, Johnston LD, Scott MP, Smith JJ, Xiao Y, Jin L, Kuntz KW, Chesworth R, Moyer MP, Bernt KM, Tseng JC, Kung AL, Armstrong SA, Copeland RA, Richon VM, Pollock RM (2011) Selective killing of mixed lineage leukemia cells by a potent small-molecule DOT1L inhibitor. Cancer Cell 20(1):53–65. doi:S1535-6108(11)00227-3 [pii] 10.1016/j.ccr.2011.06.009

Del Poeta G, Venditti A, Del Principe MI, Maurillo L, Buccisano F, Tamburini A, Cox MC, Franchi A, Bruno A, Mazzone C, Panetta P, Suppo G, Masi M, Amadori S (2003) Amount of spontaneous apoptosis detected by Bax/Bcl-2 ratio predicts outcome in acute myeloid leukemia (AML). Blood 101(6):2125–2131. doi:10.1182/blood-2002-06-1714 2002-06-1714 [pii]

Dinndorf PA, Andrews RG, Benjamin D, Ridgway D, Wolff L, Bernstein ID (1986) Expression of normal myeloid-associated antigens by acute leukemia cells. Blood 67(4):1048–1053

Duy C, Yu JJ, Nahar R, Swaminathan S, Kweon SM, Polo JM, Valls E, Klemm L, Shojaee S, Cerchietti L, Schuh W, Jack HM, Hurtz C, Ramezani-Rad P, Herzog S, Jumaa H, Koeffler HP, de Alboran IM, Melnick AM, Ye BH, Muschen M (2010) BCL6 is critical for the development of a diverse primary B cell repertoire. J Exp Med 207(6):1209–1221. doi:jem.20091299 [pii] 10.1084/jem.20091299

Eberhard Y, McDermott SP, Wang X, Gronda M, Venugopal A, Wood TE, Hurren R, Datti A, Batey RA, Wrana J, Antholine WE, Dick JE, Schimmer AD (2009) Chelation of intracellular iron with the antifungal agent ciclopiroxolamine induces cell death in leukemia and myeloma cells. Blood 114(14):3064–3073. doi:blood-2009-03-209965 [pii] 10.1182/blood-2009-03-209965

Eppert K, Takenaka K, Lechman ER, Waldron L, Nilsson B, van Galen P, Metzeler KH, Poeppl A, Ling V, Beyene J, Canty AJ, Danska JS, Bohlander SK, Buske C, Minden MD, Golub TR, Jurisica I, Ebert BL, Dick JE (2011) Stem cell gene expression programs influence clinical outcome in human leukemia. Nat Med 17(9):1086–1093. doi:nm.2415 [pii] 10.1038/nm.2415

Ferrington DA, Kapphahn RJ (2004) Catalytic site-specific inhibition of the 20S proteasome by 4-hydroxynonenal. FEBS Lett 578(3):217–223. doi:S0014-5793(04)01351-1 [pii] 10.1016/j.febslet.2004.11.003

Figueroa ME, Lugthart S, Li Y, Erpelinck-Verschueren C, Deng X, Christos PJ, Schifano E, Booth J, van Putten W, Skrabanek L, Campagne F, Mazumdar M, Greally JM, Valk PJ, Lowenberg B, Delwel R, Melnick A (2010) DNA methylation signatures identify biologically distinct subtypes in acute myeloid leukemia. Cancer Cell 17(1):13–27. doi:S1535-6108(09)00420-6 [pii] 10.1016/j.ccr.2009.11.020

Gal H, Amariglio N, Trakhtenbrot L, Jacob-Hirsh J, Margalit O, Avigdor A, Nagler A, Tavor S, Ein-Dor L, Lapidot T, Domany E, Rechavi G, Givol D (2006) Gene expression profiles of AML derived stem cells; similarity to hematopoietic stem cells. Leukemia 20(12):2147–2154. doi:2404401 [pii] 10.1038/sj.leu.2404401

Gandillet A, Park S, Lassailly F, Griessinger E, Vargaftig J, Filby A, Lister TA, Bonnet D (2011) Heterogeneous sensitivity of human acute myeloid leukemia to beta-catenin down-modulation. Leukemia 25(5):770–780. doi:leu201117 [pii] 10.1038/leu.2011.17

Gerber JM, Smith BD, Ngwang B, Zhang H, Vala MS, Morsberger L, Galkin S, Collector MI, Perkins B, Levis MJ, Griffin CA, Sharkis SJ, Borowitz MJ, Karp JE, Jones RJ (2012) A clinically relevant population of leukemic CD34(+)CD38(−) cells in acute myeloid leukemia. Blood 119(15):3571–3577. doi:blood-2011-06-364182 [pii] 10.1182/blood-2011-06-364182

Glaser SP, Lee EF, Trounson E, Bouillet P, Wei A, Fairlie WD, Izon DJ, Zuber J, Rappaport AR, Herold MJ, Alexander WS, Lowe SW, Robb L, Strasser A (2012) Anti-apoptotic Mcl-1 is essential for the development and sustained growth of acute myeloid leukemia. Genes Dev 26(2):120–125. doi:26/2/120 [pii] 10.1101/gad.182980.111

Greenberger LM, Horak ID, Filpula D, Sapra P, Westergaard M, Frydenlund HF, Albaek C, Schroder H, Orum H (2008) A RNA antagonist of hypoxia-inducible factor-1alpha, EZN-2968, inhibits tumor cell growth. Mol Cancer Ther 7(11):3598–3608. doi:1535-7163.MCT-08-0510 [pii] 10.1158/1535-7163.MCT-08-0510

Griffin JD, Linch D, Sabbath K, Larcom P, Schlossman SF (1984) A monoclonal antibody reactive with normal and leukemic human myeloid progenitor cells. Leuk Res 8(4):521–534

Guzman ML, Jordan CT (2004) Considerations for targeting malignant stem cells in leukemia. Cancer Control 11(2):97–104

Guzman ML, Neering SJ, Upchurch D, Grimes B, Howard DS, Rizzieri DA, Luger SM, Jordan CT (2001) Nuclear factor-kappaB is constitutively activated in primitive human acute myelogenous leukemia cells. Blood 98(8):2301–2307

Guzman ML, Swiderski CF, Howard DS, Grimes BA, Rossi RM, Szilvassy SJ, Jordan CT (2002) Preferential induction of apoptosis for primary human leukemic stem cells. Proc Natl Acad Sci U S A 99(25):16220–16225

Guzman ML, Rossi RM, Karnischky L, Li X, Peterson DR, Howard DS, Jordan CT (2005) The sesquiterpene lactone parthenolide induces apoptosis of human acute myelogenous leukemia stem and progenitor cells. Blood 105(11):4163–4169

Guzman ML, Rossi RM, Neelakantan S, Li X, Corbett CA, Hassane DC, Becker MW, Bennett JM, Sullivan E, Lachowicz JL, Vaughan A, Sweeney CJ, Matthews W, Carroll M, Liesveld JL, Crooks PA, Jordan CT (2007) An orally bioavailable parthenolide analog selectively eradicates acute myelogenous leukemia stem and progenitor cells. Blood 110(13):4427–4435

Hall PD, Kreitman RJ, Willingham MC, Frankel AE (1998) Toxicology and pharmacokinetics of DT388-GM-CSF, a fusion toxin consisting of a truncated diphtheria toxin (DT388) linked to human granulocyte-macrophage colony-stimulating factor (GM-CSF) in C57BL/6 mice. Toxicol Appl Pharmacol 150(1):91–97. doi:S0041-008X(98)98389-2 [pii] 10.1006/taap.1998.8389

Hamann PR, Hinman LM, Beyer CF, Lindh D, Upeslacis J, Flowers DA, Bernstein I (2002a) An anti-CD33 antibody-calicheamicin conjugate for treatment of acute myeloid leukemia. Choice of linker. Bioconjug Chem 13(1):40–46. doi:bc0100206 [pii]

Hamann PR, Hinman LM, Hollander I, Beyer CF, Lindh D, Holcomb R, Hallett W, Tsou HR, Upeslacis J, Shochat D, Mountain A, Flowers DA, Bernstein I (2002b) Gemtuzumabozogamicin, a potent and selective anti-CD33 antibody-calicheamicin conjugate for treatment of acute myeloid leukemia. Bioconjug Chem 13(1):47–58. doi:bc010021y [pii]

Hassane DC, Guzman ML, Corbett C, Li X, Abboud R, Young F, Liesveld JL, Carroll M, Jordan CT (2008) Discovery of agents that eradicate leukemia stem cells using an in silico screen of public gene expression data. Blood 111(12):5654–5662

Hassane DC, Sen S, Minhajuddin M, Rossi RM, Corbett CA, Balys M, Wei L, Crooks PA, Guzman ML, Jordan CT (2010) Chemical genomic screening reveals synergism between parthenolide and inhibitors of the PI-3 kinase and mTOR pathways. Blood 116(26):5983–5990. doi:blood-2010-04-278044 [pii] 10.1182/blood-2010-04-278044

Hauswirth AW, Florian S, Printz D, Sotlar K, Krauth MT, Fritsch G, Schernthaner GH, Wacheck V, Selzer E, Sperr WR, Valent P (2007) Expression of the target receptor CD33 in CD34+/CD38-/CD123+ AML stem cells. Eur J Clin Invest 37(1):73–82

Hayden MS, Ghosh S (2008) Shared principles in NF-kappaB signaling. Cell 132(3):344–362. doi:S0092-8674(08)00120-7 [pii] 10.1016/j.cell.2008.01.020

Hieronymus H, Lamb J, Ross KN, Peng XP, Clement C, Rodina A, Nieto M, Du J, Stegmaier K, Raj SM, Maloney KN, Clardy J, Hahn WC, Chiosis G, Golub TR (2006) Gene expression signature-based chemical genomic prediction identifies a novel class of HSP90 pathway modulators. Cancer Cell 10(4):321–330. doi:S1535-6108(06)00282-0 [pii] 10.1016/j.ccr.2006.09.005

Hofmann I, Stover EH, Cullen DE, Mao J, Morgan KJ, Lee BH, Kharas MG, Miller PG, Cornejo MG, Okabe R, Armstrong SA, Ghilardi N, Gould S, de Sauvage FJ, McMahon AP, Gilliland DG (2009) Hedgehog signaling is dispensable for adult murine hematopoietic stem cell function and hematopoiesis. Cell Stem Cell 4(6):559–567. doi:S1934-5909(09)00152-0 [pii] 10.1016/j.stem.2009.03.016

http://clinicaltrials.gov/http://clinicaltrials.gov/.052313

Hurtz C, Hatzi K, Cerchietti L, Braig M, Park E, Kim YM, Herzog S, Ramezani-Rad P, Jumaa H, Muller MC, Hofmann WK, Hochhaus A, Ye BH, Agarwal A, Druker BJ, Shah NP, Melnick AM, Muschen M (2011) BCL6-mediated repression of p53 is critical for leukemia stem cell survival in chronic myeloid leukemia. J Exp Med 208(11):2163–2174. doi:jem.20110304 [pii] 10.1084/jem.20110304

Ishikawa F, Yoshida S, Saito Y, Hijikata A, Kitamura H, Tanaka S, Nakamura R, Tanaka T, Tomiyama H, Saito N, Fukata M, Miyamoto T, Lyons B, Ohshima K, Uchida N, Taniguchi S, Ohara O, Akashi K, Harada M, Shultz LD (2007) Chemotherapy-resistant human AML stem cells home to and engraft within the bone-marrow endosteal region. Nat Biotechnol 25(11):1315–1321

Jan M, Chao MP, Cha AC, Alizadeh AA, Gentles AJ, Weissman IL, Majeti R (2011) Prospective separation of normal and leukemic stem cells based on differential expression of TIM3, a human acute myeloid leukemia stem cell marker. Proc Natl Acad Sci U S A 108(12):5009–5014. doi:1100551108 [pii] 10.1073/pnas.1100551108

Ji C, Kozak KR, Marnett LJ (2001) IkappaB kinase, a molecular target for inhibition by 4-hydroxy-2-nonenal. J BiolChem 276(21):18223–18228. doi:10.1074/jbc.M101266200 M101266200 [pii]

Jin L, Hope KJ, Zhai Q, Smadja-Joffe F, Dick JE (2006) Targeting of CD44 eradicates human acute myeloid leukemic stem cells. Nat Med 12(10):1167–1174

Jin L, Lee EM, Ramshaw HS, Busfield SJ, Peoppl AG, Wilkinson L, Guthridge MA, Thomas D, Barry EF, Boyd A, Gearing DP, Vairo G, Lopez AF, Dick JE, Lock RB (2009) Monoclonal antibody-mediated targeting of CD123, IL-3 receptor alpha chain, eliminates human acute myeloid leukemic stem cells. Cell Stem Cell 5(1):31–42. doi:S1934-5909(09)00207-0 [pii] 10.1016/j.stem. 2009.04.018

Jo SY, Granowicz EM, Maillard I, Thomas D, Hess JL (2011) Requirement for Dot1l in murine postnatal hematopoiesis and leukemogenesis by MLL translocation. Blood 117(18):4759–4768. doi:blood-2010-12-327668 [pii] 10.1182/blood-2010-12-327668

Jordan CT, Upchurch D, Szilvassy SJ, Guzman ML, Howard DS, Pettigrew AL, Meyerrose T, Rossi R, Grimes B, Rizzieri DA, Luger SM, Phillips GL (2000) The interleukin-3 receptor alpha chain is a unique marker for human acute myelogenous leukemia stem cells. Leukemia 14(10):1777–1784

Karin M, Cao Y, Greten FR, Li ZW (2002) NF-kappaB in cancer: from innocent bystander to major culprit. Nat Rev Cancer 2(4):301–310. doi:10.1038/nrc780

Kikushige Y, Shima T, Takayanagi S, Urata S, Miyamoto T, Iwasaki H, Takenaka K, Teshima T, Tanaka T, Inagaki Y, Akashi K (2010) TIM-3 is a promising target to selectively kill acute myeloid leukemia stem cells. Cell Stem Cell 7(6):708–717. doi:S1934-5909(10)00636-3 [pii] 10.1016/j.stem.2010.11.014

Konopleva M, Contractor R, Tsao T, Samudio I, Ruvolo PP, Kitada S, Deng X, Zhai D, Shi YX, Sneed T, Verhaegen M, Soengas M, Ruvolo VR, McQueen T, Schober WD, Watt JC, Jiffar T, Ling X, Marini FC, Harris D, Dietrich M, Estrov Z, McCubrey J, May WS, Reed JC, Andreeff M (2006) Mechanisms of apoptosis sensitivity and resistance to the BH3 mimetic ABT-737 in acute myeloid leukemia. Cancer Cell 10(5):375–388. doi:S1535-6108(06)00313-8 [pii] 10.1016/j.ccr.2006.10.006

Konopleva M, Rizzieri DA et al (2010) Phase I trial results for SL-401, a novel cancer stem cell (CSC) Targeting agent, demonstrate clinical efficacy at tolerable doses in patients with heavily pre-treated AML, poor risk elderly AML, and high risk MDS. Paper presented at the 52nd American Society of Hematology, Orlando, FL, 4–7 Dec.

Kopan R, Ilagan MX (2009) The canonical Notch signaling pathway: unfolding the activation mechanism. Cell 137(2):216–233. doi:S0092-8674(09)00382-1 [pii] 10.1016/j.cell.2009.03.045

Krause DS, Lazarides K, von Andrian UH, Van Etten RA (2006) Requirement for CD44 in homing and engraftment of BCR-ABL-expressing leukemic stem cells. Nat Med 12(10):1175–1180

Lamb J, Crawford ED, Peck D, Modell JW, Blat IC, Wrobel MJ, Lerner J, Brunet JP, Subramanian A, Ross KN, Reich M, Hieronymus H, Wei G, Armstrong SA, Haggarty SJ, Clemons PA, Wei R, Carr SA, Lander ES, Golub TR (2006) The connectivity map: using gene-expression signatures to connect small molecules, genes, and disease. Science 313(5795):1929–1935. doi:313/5795/1929 [pii] 10.1126/science.1132939

Lammich S, Okochi M, Takeda M, Kaether C, Capell A, Zimmer AK, Edbauer D, Walter J, Steiner H, Haass C (2002) Presenilin-dependent intramembrane proteolysis of CD44 leads to the liberation of its intracellular domain and the secretion of an Abeta-like peptide. J Biol Chem 277(47):44754–44759. doi:10.1074/jbc.M206872200 M206872200 [pii]

Lane SW, Scadden DT, Gilliland DG (2009) The leukemic stem cell niche: current concepts and therapeutic opportunities. Blood 114(6):1150–1157. doi:blood-2009-01-202606 [pii] 10.1182/blood-2009-01-202606

Lee K, Kim HM (2011) A novel approach to cancer therapy using PX-478 as a HIF-1alpha inhibitor. Arch Pharm Res 34(10):1583–1585. doi:10.1007/s12272-011-1021-3

Leyton JV, Hu M, Gao C, Turner PV, Dick JE, Minden M, Reilly RM (2011) Auger electron radioimmunotherapeutic agent specific for the CD123+/CD131- phenotype of the leukemia stem cell population. J Nucl Med 52(9):1465–1473. doi:jnumed.111.087668 [pii] 10.2967/jnumed.111.087668

Liesveld JL, Bechelli J, Rosell K, Lu C, Bridger G, Phillips G 2nd, Abboud CN (2007) Effects of AMD3100 on transmigration and survival of acute myelogenous leukemia cells. Leuk Res 31(11):1553–1563. doi:S0145-2126(07)00088-4 [pii] 10.1016/j.leukres.2007.02.017

Lin TL, Wang QH, Brown P, Peacock C, Merchant AA, Brennan S, Jones E, McGovern K, Watkins DN, Sakamoto KM, Matsui W (2010) Self-renewal of acute lymphocytic leukemia cells is limited by the Hedgehog pathway inhibitors cyclopamine and IPI-926. PLoS One 5(12):e15262. doi:10.1371/journal.pone. 0015262

Lomonosova E, Chinnadurai G (2008) BH3-only proteins in apoptosis and beyond: an overview. Oncogene 27(Suppl 1):S2–S19. doi:onc200939 [pii] 10.1038/onc.2009.39

Lund AH, van Lohuizen M (2004) Epigenetics and cancer. Genes Dev 18(19):2315–2335. doi:18/19/2315 [pii] 10.1101/gad.1232504

Majeti R (2011) Monoclonal antibody therapy directed against human acute myeloid leukemia stem cells. Oncogene 30(9):1009–1019. doi:onc2010511 [pii] 10.1038/onc.2010.511

Majeti R, Becker MW, Tian Q, Lee TL, Yan X, Liu R, Chiang JH, Hood L, Clarke MF, Weissman IL (2009a) Dysregulated gene expression networks in human acute myelogenous leukemia stem cells. Proc Natl Acad Sci U S A 106(9):3396–3401. doi:0900089106 [pii] 10.1073/pnas.0900089106

Majeti R, Chao MP, Alizadeh AA, Pang WW, Jaiswal S, Gibbs KD Jr, van Rooijen N, Weissman IL (2009b) CD47 is an adverse prognostic factor and therapeutic antibody target on human acute myeloid leukemia stem cells. Cell 138(2):286–299. doi:S0092-8674(09)00650-3 [pii] 10.1016/j.cell.2009.05.045

Martelli AM, Evangelisti C, Chappell W, Abrams SL, Basecke J, Stivala F, Donia M, Fagone P, Nicoletti F, Libra M, Ruvolo V, Ruvolo P, Kempf CR, Steelman LS, McCubrey JA (2011) Targeting the translational apparatus to improve leukemia therapy: roles of the PI3K/PTEN/Akt/mTOR pathway. Leukemia 25 (7):1064–1079. doi:leu201146 [pii] 10.1038/leu.2011.46

Mayo MW, Baldwin AS (2000) The transcription factor NF-kappaB: control of oncogenesis and cancer therapy resistance. Biochim Biophys Acta 1470(2):M55–M62

McDermott SP, Eppert K, Notta F, Isaac M, Datti A, Al-Awar R, Wrana J, Minden MD, Dick JE (2012) A small molecule screening strategy with validation on human leukemia stem cells uncovers the therapeutic efficacy of kinetin riboside. Blood 119(5):1200–1207. doi:blood-2011-01-330019 [pii] 10.1182/blood-2011-01-330019

Mercurio F, Manning AM (1999) Multiple signals converging on NF-kappaB. Curr Opin Cell Biol 11(2):226–232

Metcalf D (1985) The granulocyte-macrophage colony-stimulating factors. Science 229(4708):16–22

Page S, Fischer C, Baumgartner B, Haas M, Kreusel U, Loidl G, Hayn M, Ziegler-Heitbrock HW, Neumeier D, Brand K (1999) 4-Hydroxynonenal prevents NF-kappaB activation and tumor

necrosis factor expression by inhibiting IkappaB phosphorylation and subsequent proteolysis. J Biol Chem 274(17):11611–11618
Pellicano F, Holyoake TL (2011) Assembling defenses against therapy-resistant leukemic stem cells: Bcl6 joins the ranks. J Exp Med 208(11):2155–2158. doi:jem.20112087 [pii] 10.1084/jem.20112087
Polo JM, Dell'Oso T, Ranuncolo SM, Cerchietti L, Beck D, Da Silva GF, Prive GG, Licht JD, Melnick A (2004) Specific peptide interference reveals BCL6 transcriptional and oncogenic mechanisms in B-cell lymphoma cells. Nat Med 10(12):1329–1335. doi:nm1134 [pii] 10.1038/nm1134
Richardson DR, Kalinowski DS, Lau S, Jansson PJ, Lovejoy DB (2009) Cancer cell iron metabolism and the development of potent iron chelators as anti-tumour agents. Biochim Biophys Acta 1790(7):702–717. doi:S0304-4165(08)00098-6 [pii] 10.1016/j.bbagen.2008.04.003
Roberts AW (2008) A Phase 1 and correlative biological study of CSL360 (anti-CD123 mAb) in AML. Paper presented at the American Society of Hematology 50th Annivesary San Francisco, CA, vol 114, Issue 22 Abstract 883, 5–8 Dec. 2009
Roberts A, He S, Bradstock K, Hertzberg M, Durrant S, Ritchie D, Lewis I, Marlton P, McLachlan A, Yeadon T, Busfield S, Barriden M, Davis R, Hosback S, Mirosa D, Biondo M, Bamford S, DeWitte M, Basser R (2008) A Phase 1 and correlative biological study of CSL360 (anti-CD123 mAb) in AML. Paper presented at the 50th Annual Meeting of the American- Society-of-Hematology, San Francisco, CA, 6–9 Dec. 2008
Robertson MJ, Soiffer RJ, Freedman AS, Rabinowe SL, Anderson KC, Ervin TJ, Murray C, Dear K, Griffin JD, Nadler LM et al. (1992) Human bone marrow depleted of CD33-positive cells mediates delayed but durable reconstitution of hematopoiesis: clinical trial of MY9 monoclonal antibody-purged autografts for the treatment of acute myeloid leukemia. Blood 79(9):2229–2236
Saito Y, Kitamura H, Hijikata A, Tomizawa-Murasawa M, Tanaka S, Takagi S, Uchida N, Suzuki N, Sone A, Najima Y, Ozawa H, Wake A, Taniguchi S, Shultz LD, Ohara O, Ishikawa F (2010) Identification of therapeutic targets for quiescent, chemotherapy-resistant human leukemia stem cells. Sci Transl Med 2(17):17ra19. doi:2/17/17ra9 [pii] 10.1126/scitranslmed.3000349
Sarry JE, Murphy K, Perry R, Sanchez PV, Secreto A, Keefer C, Swider CR, Strzelecki AC, Cavelier C, Recher C, Mansat-De Mas V, Delabesse E, Danet-Desnoyers G, Carroll M (2011) Human acute myelogenous leukemia stem cells are rare and heterogeneous when assayed in NOD/SCID/IL2Rgammac-deficient mice. J Clin Invest 121(1):384–395. doi:41495 [pii] 10.1172/JCI41495
Schenk T, Chen WC, Gollner S, Howell L, Jin L, Hebestreit K, Klein HU, Popescu AC, Burnett A, Mills K, Casero RA Jr, Marton L, Woster P, Minden MD, Dugas M, Wang JC, Dick JE, Muller-Tidow C, Petrie K, Zelent A (2012) Inhibition of the LSD1 (KDM1A) demethylase reactivates the all-trans-retinoic acid differentiation pathway in acute myeloid leukemia. Nat Med 18(4):605–611. doi:nm.2661 [pii] 10.1038/nm.2661
Skrtic M, Sriskanthadevan S, Jhas B, Gebbia M, Wang X, Wang Z, Hurren R, Jitkova Y, Gronda M, Maclean N, Lai CK, Eberhard Y, Bartoszko J, Spagnuolo P, Rutledge AC, Datti A, Ketela T, Moffat J, Robinson BH, Cameron JH, Wrana J, Eaves CJ, Minden MD, Wang JC, Dick JE, Humphries K, Nislow C, Giaever G, Schimmer AD (2011) Inhibition of mitochondrial translation as a therapeutic strategy for human acute myeloid leukemia. Cancer Cell 20(5):674–688. doi:S1535-6108(11)00398-9 [pii] 10.1016/j.ccr.2011.10.015
Strasser A (2005) The role of BH3-only proteins in the immune system. Nat Rev Immunol 5(3):189–200. doi:nri1568 [pii] 10.1038/nri1568
Taipale J, Beachy PA (2001) The Hedgehog and Wnt signalling pathways in cancer. Nature 411(6835):349–354. doi:10.1038/35077219 35077219 [pii]
Taussig DC, Miraki-Moud F, Anjos-Afonso F, Pearce DJ, Allen K, Ridler C, Lillington D, Oakervee H, Cavenagh J, Agrawal SG, Lister TA, Gribben JG, Bonnet D (2008) Anti-CD38 antibody-mediated clearance of human repopulating cells masks the heterogeneity of leukemia-initiating cells. Blood 112(3):568–575

Tavor S, Petit I, Porozov S, Avigdor A, Dar A, Leider-Trejo L, Shemtov N, Deutsch V, Naparstek E, Nagler A, Lapidot T (2004) CXCR4 regulates migration and development of human acute myelogenous leukemia stem cells in transplanted NOD/SCID mice. Cancer Res 64(8):2817–2824

Toyota M, Kopecky KJ, Toyota MO, Jair KW, Willman CL, Issa JP (2001) Methylation profiling in acute myeloid leukemia. Blood 97(9):2823–2829

van Rhenen A, Feller N, Kelder A, Westra AH, Rombouts E, Zweegman S, van der Pol MA, Waisfisz Q, Ossenkoppele GJ, Schuurhuis GJ (2005) High stem cell frequency in acute myeloid leukemia at diagnosis predicts high minimal residual disease and poor survival. Clin Cancer Res 11(18):6520–6527. doi:11/18/6520 [pii] 10.1158/1078-0432.CCR-05-0468

van Rhenen A, van Dongen GA, Kelder A, Rombouts EJ, Feller N, Moshaver B, Stigter-van Walsum M, Zweegman S, Ossenkoppele GJ, Jan Schuurhuis G (2007) The novel AML stem cell associated antigen CLL-1 aids in discrimination between normal and leukemic stem cells. Blood 110(7):2659–2666

Wang Y, Krivtsov AV, Sinha AU, North TE, Goessling W, Feng Z, Zon LI, Armstrong SA (2010) The Wnt/beta-catenin pathway is required for the development of leukemia stem cells in AML. Science 327(5973):1650–1653. doi:327/5973/1650 [pii] 10.1126/science.1186624

Wang Y, Liu Y, Malek SN, Zheng P (2011) Targeting HIF1alpha eliminates cancer stem cells in hematological malignancies. Cell Stem Cell 8(4):399–411. doi:S1934-5909(11)00057-9 [pii] 10.1016/j.stem.2011.02.006

Wei G, Twomey D, Lamb J, Schlis K, Agarwal J, Stam RW, Opferman JT, Sallan SE, den Boer ML, Pieters R, Golub TR, Armstrong SA (2006) Gene expression-based chemical genomics identifies rapamycin as a modulator of MCL1 and glucocorticoid resistance. Cancer Cell 10(4):331–342. doi:S1535-6108(06)00283-2 [pii] 10.1016/j.ccr.2006.09.006

Xu Q, Simpson SE, Scialla TJ, Bagg A, Carroll M (2003) Survival of acute myeloid leukemia cells requires PI3 kinase activation. Blood 102(3):972–980

Xu Q, Thompson JE, Carroll M (2005) mTOR regulates cell survival after etoposide treatment in primary AML cells. Blood 106(13):4261–4268

Ye BH (2000) BCL-6 in the pathogenesis of non-Hodgkin's lymphoma. Cancer Invest 18(4):356–365

Yeung J, Esposito MT, Gandillet A, Zeisig BB, Griessinger E, Bonnet D, So CW (2010) beta-Catenin mediates the establishment and drug resistance of MLL leukemic stem cells. Cancer Cell 18(6):606–618. doi:S1535-6108(10)00434-4 [pii] 10.1016/j.ccr.2010.10.032

Zhang H, Luo J, Li Y, Henderson PT, Wang Y, Wachsmann-Hogiu S, Zhao W, Lam KS, Pan CX (2012) Nanomedicine 8(7):1116–1124.

Zhao C, Chen A, Jamieson CH, Fereshteh M, Abrahamsson A, Blum J, Kwon HY, Kim J, Chute JP, Rizzieri D, Munchhof M, VanArsdale T, Beachy PA, Reya T (2009) Hedgehog signalling is essential for maintenance of cancer stem cells in myeloid leukaemia. Nature 458(7239):776–779. doi:nature07737 [pii] 10.1038/nature07737

Part VII
Microenvironment

Chapter 30
Regulation of Hematopoiesis by CXCL12/CXCR4 Signaling

Daniel C. Link

Abstract CXCL12 (stromal-derived factor 1, SDF-1) is a chemokine that is constitutively expressed at high levels in the bone marrow and is a key component of the stem cell niche. Its major receptor CXCR4 is broadly expressed on hematopoietic cells, including hematopoietic stem cells (HSCs). The CXCL12/CXCR4 axis plays a key role in regulating many aspects of hematopoiesis. It is essential for normal B-cell lymphopoiesis and is a key determinant of neutrophil release from the bone marrow. Activating mutations of *CXCR4* are causative of the warts, hypogammaglobulinemia, infections, and myelokathexis (WHIM) syndrome, which is characterized by abnormal neutrophil retention in the bone marrow and impaired B-cell development. The CXCL12/CXCR4 axis also plays an essential role in HSC maintenance and trafficking, and pharmaceuticals that inhibit CXCR4 signaling result in rapid mobilization.

Keywords CXCL12 · CXCR4 · Stromal-derived factor 1 · Hematopoietic stem cells · Stem cell niche · Mobilization · Quiescence · B-cell lymphopoiesis · WHIM syndrome

30.1 Introduction

The chemokine CXCL12 (stromal-derived factor 1, SDF-1) is highly expressed in the bone marrow and is a key component of the stem cell niche. Through interaction with its major receptor CXCR4, CXCL12 regulates many aspects of hematopoiesis, including B-cell development, neutrophil release from the bone marrow, and hematopoietic stem cell (HSC) maintenance and trafficking. Pharmaceutical agents that target CXCR4 have been developed and are already used clinically to induce HSC mobilization. However, to fully realize the therapeutic potential of these agents, a detailed understanding of the biology of CXCL12 and CXCR4 are needed. In this chapter, preclinical studies are reviewed that define the biological activities of CXCL12 and CXCR4 in hematopoiesis.

D. C. Link (✉)
Division of Oncology, Department of Medicine, Washington University School of St. Louis, 660 S. Euclid Avenue, Campus Box 8007, Saint Louis, MO 63110, USA
e-mail: dlink@dom.wustl.edu

M. Andreeff (ed.), *Targeted Therapy of Acute Myeloid Leukemia*, Current Cancer Research, DOI 10.1007/978-1-4939-1393-0_30

30.2 CXCL12

CXCL12 is a C-X-C chemokine that was first identified in 1993 as a protein produced by bone marrow stromal cells that stimulated proliferation of B cells (Nagasawa et al. 1994; Tashiro et al. 1993). It was originally termed SDF-1 or pre-B-cell growth-stimulating factor. There are two major alternatively spliced forms of CXCL12. CXCL12α is the predominantly expressed isoform consisting of 89 amino acids. CXCL12β contains an additional four amino acids at its carboxy terminus. Both CXCL12 isoforms bind CXCR4 with similar affinity and elicit similar biological responses (Hesselgesser et al. 1998; Shirozu et al. 1995).

CXCL12 is constitutively expressed at high levels by several stromal cell populations in the bone marrow. Studies using transgenic mice in which the green fluorescent protein is inserted into the *Cxcl12* locus identified a perivascular cell population with very high CXCL12 expression, termed CXCL12-abundant reticular (CAR) cells (Tokoyoda et al. 2004). Subsequent studies showed that CAR cells are mesenchymal progenitors that have the capacity for both osteoblast and adipocyte differentiation and are in direct contact with a subset of HSCs (Omatsu et al. 2010; Sugiyama et al. 2006). High level CXCL12 expression also has been reported in mature osteoblasts (Christopher et al. 2009; Petit et al. 2002; Semerad et al. 2005), mesenchymal stem cells (Mendez-Ferrer et al. 2010; Sacchetti et al. 2007), and endothelial cells (Ponomaryov et al. 2000; Semerad et al. 2005) in mice. Likewise, in human bone marrow, CXCL12 expression in osteoblasts, endothelial cells, and scattered reticular stromal cells (possibly the human equivalent of CAR cells) has been reported (Petit et al. 2002).

30.3 CXCR4 and CXCR7

The major receptor for CXCL12 is CXCR4, a G-protein-coupled receptor that also serves as a co-receptor for human immunodeficiency virus. CXCR4 is broadly expressed on hematopoietic cells, including HSCs, neutrophils, lymphocytes, monocytes, and megakaryocytes. It is also expressed on stromal cell populations including osteoblast progenitors and endothelial cells (Dar et al. 2005). Interestingly, a recent study showed that conditional deletion of CXCR4 in osteoblast precursors results in impaired osteoblast development, suggesting that CXCL12/CXCR4 signaling may regulate the stem cell niche (Zhu et al. 2011). CXCL12 also has an additional receptor, CXCR7 (Balabanian et al. 2005a), which is believed to act as a decoy receptor to sequester the ligand (Boldajipour et al. 2008; Hartmann et al. 2008); however, CXCR7 does not appear to play a major role in hematopoiesis (Sierro et al. 2007). There are several low-affinity ligands of CXCR4, such as trefoil factor 2 (Dubeykovskaya et al. 2009), macrophage migration inhibitory factor (MIF; Bernhagen et al. 2007), and defensin (Feng et al. 2006), but their significance in hematopoiesis is unclear.

Constitutive deletion of *Cxcr4* or *Cxcl12* is perinatal lethal and produces an identical phenotype that includes neurological, vascular, and hematopoietic defects (Ma et al. 1998; Nagasawa et al. 1996; Tachibana et al. 1998; Zou et al. 1998). The similar phenotypes of these mice confirm that CXCL12 is the primary ligand for CXCR4 and that CXCR4 is the major receptor for CXCL12.

30.4 CXCL12/CXCR4 Signaling is a Key Regulator of Hematopoiesis

As predicted by the broad expression of CXCR4, CXCL12/CXCR4 signaling has been implicated in the proliferation, survival, and/or trafficking of most hematopoietic lineages. In this section, we review the contribution of CXCR4 signaling to the regulation of B lymphocytes and other mature hematopoietic lineages; hematopoietic stem/progenitors are covered in the next section.

30.4.1 B Lymphocytes

CXCR4 is expressed at all stages of B-cell development from HSCs to plasma cells (Nie et al. 2004). In *Cxcr4-/-* or *Cxcl12-/-* mice or in chimeric mice reconstituted with *Cxcr4-/-* fetal liver cells, there is marked reduction in B cells that extends to the pre-pro-B-cell stage, indicating that CXCR4/CXCL12 signaling is required for B-cell development at the earliest stages (Egawa et al. 2001). To study the contribution of CXCR4 later in B-cell development, *Cxrcr4* was conditionally deleted using transgenic mice expressing Cre-recombinase under control of the CD19 promoter. Studies of these mice indicate that CXCR4 signaling also is required for the retention of B-cell precursors (Nie et al. 2008) and the homing of plasma cells to the bone marrow (Tokoyoda et al. 2004).

30.4.2 Neutrophils

CXCR4 is expressed on neutrophils and neutrophil precursors (Eash et al. 2009). In Cxcr4-/- fetal liver chimeras, there is a reduction of myeloid cells, though not as severe as observed with B cells (Kawabata et al. 1999; Ma et al. 1999). In mice with a myeloid-specific deletion of *Cxcr4*, constitutive mobilization of mature neutrophils is observed, indicating that CXCR4 signaling negatively regulates neutrophil release from the bone marrow (Eash et al. 2009). Consistent with this observation, treatment with granulocytic colony-stimulating factor (G-CSF), a potent neutrophil mobilizing agent, is associated with downregulation of CXCR4 expression of neutrophils (Kim et al. 2006; Levesque et al. 2003; Semerad et al. 2002). In addition to regulating neutrophil release, CXCR4 may also contribute to the homing and clearance of senescent neutrophils in the bone marrow (Martin et al. 2003; Suratt et al. 2004).

30.4.3 Other Hematopoietic Cell Populations

There is evidence suggesting that CXCR4 signaling also plays an important role in the production and/or function of megakaryocytes, T-cells, and natural killer cells. CXCL12, by localizing megakaryocytes to the perivascular niche, may promote their maturation and platelet release into the circulation (Avecilla et al. 2004). Studies of *Cxcr4-/-* fetal liver chimeras or adult *Cxcr4* conditionally deleted mice suggest that CXCR4 signaling is required for the normal proliferation, survival, and function of T-cells (Ara et al. 2003) and natural killer cells (Noda et al. 2011).

30.5 WHIM Syndrome

The importance of CXCR4 signaling in neutrophil and B-cell homeostasis is highlighted by the identification of mutations of *CXCR4* in warts, hypogammaglobulinemia, infections, and myelokathexis (WHIM) syndrome. A defining feature of WHIM syndrome is myelokathexis, which refers to the abnormal retention of neutrophils in the bone marrow, resulting in peripheral blood neutropenia (Gorlin et al. 2000). Hypogammaglobulinemia is variable and when present is usually associated with low or absent B cells in the blood. Almost all cases of WHIM syndrome are inherited in an autosomal dominant fashion. In the majority of cases, WHIM syndrome is caused by truncation mutations of *CXCR4* (Fig. 30.1; Hernandez et al. 2003). These mutations confer enhanced responsiveness to CXCL12, suggesting a model in which increased CXCR4 signaling leads to altered B-cell trafficking and the abnormal retention of neutrophils in the bone marrow (Balabanian et al. 2005b; Gorlin et al. 2000; Gulino et al. 2004; Hernandez et al. 2003). Consistent with this model, two recent studies demonstrated the therapeutic efficacy of CXCR4 antagonists in WHIM syndrome (Dale et al. 2011; McDermott et al. 2011).

30.6 CXCL12/CXCR4 Signaling Is Required for HSC Maintenance and Trafficking

CXCL12, which is constitutively expressed at high level in the bone marrow, is the most potent known chemoattractant for HSCs (Aiuti et al. 1997; Wright et al. 2002). The CXCL12/CXCR4 axis regulates multiple aspects of HSC biology, which are reviewed in this section.

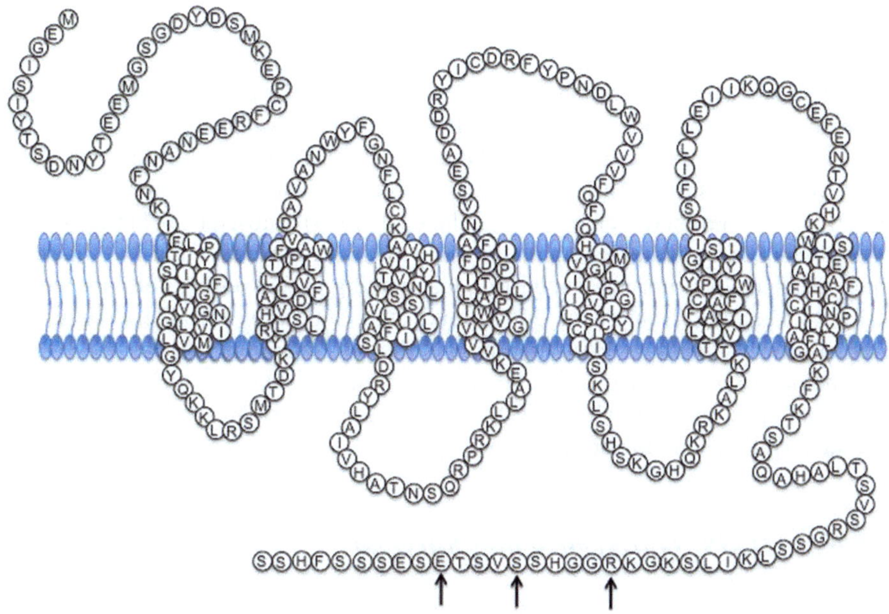

Fig. 30.1 *CXCR4 mutations in WHIM syndrome*. CXCR4 is a seven transmembrane spanning G protein-coupled chemokine receptor. Mutations of CXCR4 are found in the majority of patients with WHIM syndrome and typically truncate the carboxy-terminal tail (*arrows*). The most common mutation is R334X, which removes the carboxy-terminal 19 amino acids

30.6.1 HSC Retention in the Bone Marrow

There is compelling evidence that CXCL12/CXCR4 signaling provides a key retention signal for HSCs in the bone marrow. During embryonic development, CXCR4 signaling is required for the efficient translocation of hematopoiesis from the fetal liver to the bone marrow. In mice lacking CXCR4 or CXCL12, while fetal hematopoiesis is relatively normal, there is a defect in the migration of HSCs to the bone marrow (Ma et al. 1998; Tachibana et al. 1998; Zou et al. 1998). In *Cxcr4-/-* fetal liver chimeras or after conditional deletion of *Cxcr4* in adult mice, there is constitutive mobilization of hematopoietic stem and progenitor cells (HSPCs) into the blood (Kawabata et al. 1999; Ma et al. 1999; Sugiyama et al. 2006). Finally, treatment of mice or humans with a CXCR4 antagonist results in the rapid mobilization of HSPCs into the blood (Broxmeyer et al. 2005; Liles et al. 2003). Together, these data suggest that the CXCL12/CXCR4 axis provides a key retention signal for HSPCs in the bone marrow. Of note, Tokoyoda and colleagues showed that the majority of HSPCs are in direct contact with CAR cells (Sugiyama et al. 2006; Tokoyoda et al. 2004). Thus, CXCL12 may serve to localize and retain HSCs within the stem cell niche in the bone marrow.

30.6.2 Homing to the Bone Marrow

Initial studies showed that that CXCR4 was required for efficient engraftment of hematopoietic progenitor stem cells (HPSCs) after transplantation into irradiated recipients (Kawabata et al. 1999; Peled et al. 1999). Conversely, enforced expression of CXCR4 in human HSPCs is associated with increased hematopoietic engraftment in immunodeficient mice (Kahn et al. 2004). Subsequent studies showed that homing of hematopoietic progenitors to the bone marrow within 24 h in irradiated recipients was normal, suggesting that homing in this setting does not require CXCR4 signaling (Bonig et al. 2004; Foudi et al. 2006). Of note, these studies did not specifically measure HSC homing and did not determine whether the HSPCs that homed to bone marrow were localized within the stem cell niche. Thus, it remains an open question whether CXCR4 signaling contributes to HSC homing to the stem cell niche in the bone marrow.

30.6.3 HSPC Quiescence and Survival

There are contrasting data on the role of CXCR4 signaling in regulating HSPC proliferation. Treatment of humans $CD34^{+}$ cells with CXCL12 induces proliferation and enhanced responsiveness to cytokines in vitro (Lataillade et al. 2002; Lataillade et al. 2000). However, conditional deletion of *Cxcr4* in adult mice results in sustained proliferation of primitive HPSCs, suggesting that CXCR4 signaling may be required to maintain HSC quiescence (Nie et al. 2008). CXCR4 signaling also has been implicated in regulating cell survival. Treatment of human HSPCs with CXCL12 enhances their survival in vitro (Lataillade et al. 2002; Lee et al. 2002), and HSPCs from transgenic mice with enforced expression of CXCL12 display enhanced survival in vitro (Broxmeyer et al. 2003).

30.7 Disruption of CXCL12/CXCR4 Signaling Is a Key Step in G-CSF-Induced HPSC Mobilization

Numerous lines of evidence suggest that the dominant mechanism by which G-CSF induces HSPC mobilization is through suppression of the CXCL12/CXCR4 axis. While G-CSF treatment leads to an early, transient increase in CXCL12 production (Petit et al. 2002), prolonged treatment causes a progressive decrease in CXCL12 mRNA in the bone marrow (Semerad et al. 2005), with a highly correlated concurrent drop in CXCL12 protein (Levesque et al. 2003; Petit et al. 2002; Semerad et al. 2005). Indeed, the level of CXCL12 is inversely correlated with the degree of mobilization, with the lowest levels of CXCL12 mRNA or protein correlating with maximal mobilization (Katayama et al. 2006; Semerad et al. 2005). Likewise, HPSC mobilization in response to Flt3 ligand or stem cell factor is associated with decreased CXCL12 expression in the bone marrow (Christopher et al. 2009). G-CSF administration also is associated with decreased CXCR4 expression on mobilized HSPC, further disrupting CXCL12/CXCR4 signaling (Dlubek et al. 2006; Levesque et al. 2003). Direct evidence for the requirement for CXCR4 signaling in HSPC mobilization comes from the study of *Cxcr4-/-* bone marrow chimeras (Christopher et al. 2009). These mice, while displaying constitutive mobilization, do not further mobilize HSPC in response to G-CSF but maintain the ability to further mobilize with a VLA-4 antagonist.

30.7.1 G-CSF Suppresses Expression of CXCL12 in Osteolineage Cells

As noted previously, osteoblast lineage cells, including mature osteoblasts and CAR cells are the major source of CXCL12 production in the bone marrow (Petit et al. 2002; Semerad et al. 2002; Tokoyoda et al. 2004). Strikingly, treatment with G-CSF results in a marked suppression of mature osteoblast number and function (Froberg et al. 1999; Katayama et al. 2006; Semerad et al. 2005; Winkler et al. 2010), due, in part, to increased apoptosis (Christopher and Link 2008). Moreover, the surviving osteoblasts reduce their expression of CXCL12 mRNA by approximately one half (Semerad et al. 2005). Similarly, Mayack and colleagues reported that during HSPC mobilization with cyclophosphamide plus G-CSF, there is an initial expansion of osteoblast lineage cells that is followed by their suppression (Mayack and Wagers 2008). Interestingly, osteoblast suppression is a common feature of HSPC mobilization by other cytokines, including Flt3 ligand and stem cell factor (Christopher et al. 2009). We also have observed that G-CSF treatment results in decreased CXCL12 expression in CAR cells (Link unpublished observations). Though alterations in osteoblast lineage cells are likely to be a key step in G-CSF-induced mobilization, the mechanisms by which this leads to HSPC egress from the bone marrow remain to be elucidated.

30.7.2 G-CSF Signaling in Bone Marrow Monocytes/ Macrophages is Sufficient to Suppress CXCL12 Expression and Induce HSPC Mobilization

Two recent studies showed that conditional ablation of bone marrow monocytes/ macrophages induces strong HSPC mobilization and suppression of CXCL12 expression (Chow et al. 2011; Winkler et al. 2010). Moreover, our group recently showed that expression of the G-CSF receptor on monocyte-lineage cells in the bone marrow is sufficient to induce G-CSF HSPC mobilization, osteoblasts suppression, and decreased CXCL12 expression (Christopher et al. 2011). Together, these studies suggest a model in which bone marrow monocytes/macrophages produce trans-acting factor(s) that maintain CXCL12 expression in osteolineage cells

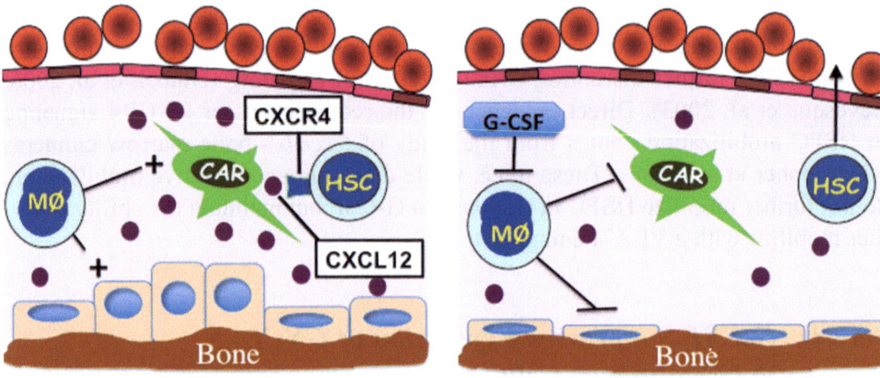

Fig. 30.2 *Model of G-CSF-induced HSPC mobilization*. Under steady-state conditions (*left panel*), bone marrow monocytes/macrophages (*Mø*) produce transacting factor(s) that support osteoblasts and *CAR* cells. These cells produce large amounts of *CXCL12* that help maintain *HSC* quiescence and retention in the bone marrow. Treatment with *G-CSF* (*right panel*) results in direct monocyte/macrophage suppression that, in turn, results in a decreased *CXCL12* expression from osteoblasts and *CAR* cells. Ultimately, this disruption of *CXCL12/CXCR4* signaling contributes to *HSC* mobilization

in the bone marrow. Genetic or pharmacologic ablation of monocytes/macrophage or G-CSF suppression of these cells leads to decreased CXCL12 expression in the bone marrow and ultimately HSPC mobilization (Fig. 30.2).

30.8 Modulators of CXCR4/CXCL12 Signaling

Numerous pathways have been shown to modulate mobilization either by altering baseline CXCL12 production or modulating CXCR4 responsiveness. Pharmacologically inhibiting or genetically reducing bone marrow synthesis of catecholamines blunts mobilization by G-CSF (Katayama et al. 2006), most likely by regulating bone marrow stromal cell production of CXCL12 (Mendez-Ferrer et al. 2010). Cell surface molecules can also modulate CXCL12 responsiveness. For example, CD26 (dipeptidylpeptidase 4, DPP4) and carboxypeptidase M are proteases present on the surface HSPCs that can cleave CXCL12, thereby inhibiting CXCR4 signaling (Christopherson et al. 2004). Pharmacologic or genetic inhibition of CD26 is associated with increased CXCR4 signaling and impaired G-CSF-induced HSPC mobilization (Christopherson et al. 2003). Similarly, platelet endothelial cell adhesion molecule-1 (PECAM-1/CD31) is an adhesion molecule expressed on both stromal and hematopoietic cells. Mice lacking PECAM-1 exhibit constitutive mobilization, perhaps due to reduced HSPC responsiveness to CXCL12 (Ross et al. 2008). Finally, a recent report from Tesio et al. showed that G-CSF induces hepatocyte growth factor/c-Met signaling in HSPC, potentially contributing to mobilization by inhibiting the ability of HSPC to chemotax in response to CXCL12 (Tesio et al. 2011).

Conclusions

Preclinical studies have established that the CXCL12/CXCR4 axis plays an essential role in regulating many aspects of hematopoiesis. It is required for normal B-, T-, and NK cell development and trafficking. It regulates neutrophil release from the bone marrow. Finally, it is a key component of the stem cell niche, where it helps maintain HSCs quiescence and retention in the niche. Pharmacologic inhibitors of CXCR4 already are used to mobilize HSCs for stem cell transplantation (Devine et al. 2004). However, the broad hematopoietic activity of the CXCL12/CXCR4 axis suggests other potential therapeutic uses of CXCR4 inhibitors. In particular, by disrupting interactions of leukemic stem cells with the stem cell niche, CXCR4 inhibitors may induce apoptosis and proliferation, thereby rending these cells more sensitive to chemotherapy. Indeed, preclinical and early clinical data (discussed in detail in Chap. 34) suggest this may be true for acute myelogenous leukemia. The data reviewed in this chapter, suggests that chemosensitization with CXCR4 inhibitors also may be relevant for other hematopoietic malignancies, such as B-cell ALL, T-cell ALL, and multiple myeloma.

References

Aiuti A, Webb IJ, Bleul C, Springer T, Gutierrez-Ramos JC (1997) The chemokine SDF-1 is a chemoattractant for human CD34+ hematopoietic progenitor cells and provides a new mechanism to explain the mobilization of CD34+ progenitors to peripheral blood. J Exp Med 185:111–120

Ara T, Itoi M, Kawabata K, Egawa T, Tokoyoda K, Sugiyama T, Fujii N, Amagai T, Nagasawa T (2003) A role of CXC chemokine ligand 12/stromal cell-derived factor-1/pre-B cell growth stimulating factor and its receptor CXCR4 in fetal and adult T cell development in vivo. J Immunol 170:4649–4655

Avecilla, ST, Hattori K, Heissig B, Tejada R, Liao F, Shido K, Jin DK, Dias S, Zhang F, Hartman TE et al (2004) Chemokine-mediated interaction of hematopoietic progenitors with the bone marrow vascular niche is required for thrombopoiesis. Nat Med 10:64–71

Balabanian K, Lagane B, Infantino S, Chow KY, Harriague J, Moepps B, Arenzana-Seisdedos F, Thelen M, Bachelerie F (2005a) The chemokine SDF-1/CXCL12 binds to and signals through the orphan receptor RDC1 in T lymphocytes. J Biol Chem 280:35760–35766

Balabanian K, Lagane B, Pablos JL, Laurent L, Planchenault T, Verola O, Lebbe C, Kerob D, Dupuy A, Hermine O et al (2005b) WHIM syndromes with different genetic anomalies are accounted for by impaired CXCR4 desensitization to CXCL12. Blood 105:2449–2457

Bernhagen J, Krohn R, Lue H, Gregory JL, Zernecke A, Koenen RR, Dewor M, Georgiev I, Schober A, Leng L et al (2007) MIF is a noncognate ligand of CXC chemokine receptors in inflammatory and atherogenic cell recruitment. Nat Med 13:587–596

Boldajipour B, Mahabaleshwar H, Kardash E, Reichman-Fried M, Blaser H, Minina S, Wilson D, Xu Q, Raz E (2008) Control of chemokine-guided cell migration by ligand sequestration. Cell 132:463–473

Bonig H, Priestley GV, Nilsson LM, Jiang Y, Papayannopoulou T (2004) PTX-sensitive signals in bone marrow homing of fetal and adult hematopoietic progenitor cells. Blood 104:2299–2306

Broxmeyer HE, Cooper S, Kohli L, Hangoc G, Lee Y, Mantel C, Clapp DW, Kim CH (2003) Transgenic expression of stromal cell-derived factor-1/CXC chemokine ligand 12 enhances myeloid progenitor cell survival/antiapoptosis in vitro in response to growth factor withdrawal and enhances myelopoiesis in vivo. J Immunol 170:421–429

Broxmeyer HE, Orschell CM, Clapp DW, Hangoc G, Cooper S, Plett PA, Liles WC, Li X, Graham-Evans B, Campbell TB et al (2005) Rapid mobilization of murine and human hematopoietic stem and progenitor cells with AMD3100, a CXCR4 antagonist. J Exp Med 201:1307–1318

Chow A, Lucas D, Hidalgo A, Mendez-Ferrer S, Hashimoto D, Scheiermann C, Battista M, Leboeuf M, Prophete C, van Rooijen N et al (2011) Bone marrow CD169+ macrophages promote the retention of hematopoietic stem and progenitor cells in the mesenchymal stem cell niche. J Exp Med 208:261–271

Christopher MJ, Link DC (2008) Granulocyte colony-stimulating factor induces osteoblast apoptosis and inhibits osteoblast differentiation. J Bone Miner Res 23:1765–1774

Christopher MJ, Liu F, Hilton MJ, Long F, Link DC (2009) Suppression of CXCL12 production by bone marrow osteoblasts is a common and critical pathway for cytokine-induced mobilization. Blood 114:1331–1339

Christopher MJ, Rao M, Liu F, Woloszynek JR, Link DC (2011) Expression of the G-CSF receptor in monocytic cells is sufficient to mediate hematopoietic progenitor mobilization by G-CSF in mice. J Exp Med 208:251–260

Christopherson KW, 2nd, Cooper S, Broxmeyer HE (2003) Cell surface peptidase CD26/DPPIV mediates G-CSF mobilization of mouse progenitor cells. Blood 101:4680–4686

Christopherson KW, 2nd, Hangoc G, Mantel CR, Broxmeyer HE (2004) Modulation of hematopoietic stem cell homing and engraftment by CD26. Science 305:1000–1003

Dale DC, Bolyard AA, Kelley ML, Westrup EC, Makaryan V, Aprikyan A, Wood B, Hsu FJ (2011) The CXCR4 antagonist plerixafor is a potential therapy for myelokathexis, WHIM syndrome. Blood 118:4963–4966

Dar A, Goichberg P, Shinder V, Kalinkovich A, Kollet O, Netzer N, Margalit R, Zsak M, Nagler A, Hardan I et al (2005) Chemokine receptor CXCR4-dependent internalization and resecretion of functional chemokine SDF-1 by bone marrow endothelial and stromal cells. Nat Immunol 6:1038–1046

Devine SM, Flomenberg N, Vesole DH, Liesveld J, Weisdorf D, Badel K, Calandra G, DiPersio JF (2004) Rapid mobilization of CD34+ cells following administration of the CXCR4 antagonist AMD3100 to patients with multiple myeloma and non-Hodgkin's lymphoma. J Clin Oncol 22:1095–1102

Dlubek D, Drabczak-Skrzypek D, Lange A (2006) Low CXCR4 membrane expression on CD34(+) cells characterizes cells mobilized to blood. Bone Marrow Transplant 37:19–23

Dubeykovskaya Z, Dubeykovskiy A, Solal-Cohen J, Wang TC (2009) Secreted trefoil factor 2 activates the CXCR4 receptor in epithelial and lymphocytic cancer cell lines. J Biol Chem 284:3650–3662

Eash KJ, Means JM, White DW, Link DC (2009) CXCR4 is a key regulator of neutrophil release from the bone marrow under basal and stress granulopoiesis conditions. Blood 113:4711–4119

Egawa T, Kawabata K, Kawamoto H, Amada K, Okamoto R, Fujii N, Kishimoto T, Katsura Y, Nagasawa T (2001) The earliest stages of B cell development require a chemokine stromal cell-derived factor/pre-B cell growth-stimulating factor. Immunity 15:323–334

Feng Z, Dubyak GR, Lederman MM, Weinberg A (2006) Cutting edge: human beta defensin 3–a novel antagonist of the HIV-1 coreceptor CXCR4. J Immunol 177:782–786

Foudi A, Jarrier P, Zhang Y, Wittner M, Geay JF, Lecluse Y, Nagasawa T, Vainchenker W, Louache F (2006) Reduced retention of radioprotective hematopoietic cells within the bone marrow microenvironment in CXCR4-/- chimeric mice. Blood 107:2243–2251

Froberg MK, Garg UC, Stroncek DF, Geis M, McCullough J, Brown DM (1999) Changes in serum osteocalcin and bone-specific alkaline phosphatase are associated with bone pain in donors receiving granulocyte-colony-stimulating factor for peripheral blood stem and progenitor cell collection. Transfusion 39:410–414

Gorlin RJ, Gelb B, Diaz GA, Lofsness KG, Pittelkow MR, Fenyk JR, Jr (2000) WHIM syndrome, an autosomal dominant disorder: clinical, hematological, and molecular studies. Am J Med Genet 91:368–376

Gulino AV, Moratto D, Sozzani S, Cavadini P, Otero K, Tassone L, Imberti L, Pirovano S, Notarangelo LD, Soresina R et al (2004) Altered leukocyte response to CXCL12 in patients with warts hypogammaglobulinemia, infections, myelokathexis (WHIM) syndrome. Blood 104:444–452

Hartmann TN, Grabovsky V, Pasvolsky R, Shulman Z, Buss EC, Spiegel A, Nagler A, Lapidot T, Thelen M, Alon R (2008) A crosstalk between intracellular CXCR7 and CXCR4 involved in rapid CXCL12-triggered integrin activation but not in chemokine-triggered motility of human T lymphocytes and CD34+ cells. J Leukoc Biol 84:1130–1140

Hernandez PA, Gorlin RJ, Lukens JN, Taniuchi S, Bohinjec J, Francois F, Klotman ME, Diaz GA (2003) Mutations in the chemokine receptor gene CXCR4 are associated with WHIM syndrome, a combined immunodeficiency disease. Nat Genet 34:70–74

Hesselgesser J, Liang M, Hoxie J, Greenberg M, Brass LF, Orsini MJ, Taub D, Horuk R (1998) Identification and characterization of the CXCR4 chemokine receptor in human T cell lines: ligand binding, biological activity, and HIV-1 infectivity. J Immunol 160:877–883

Kahn J, Byk T, Jansson-Sjostrand L, Petit I, Shivtiel S, Nagler A, Hardan I, Deutsch V, Gazit Z, Gazit D et al (2004) Overexpression of CXCR4 on human CD34+ progenitors increases their proliferation, migration, and NOD/SCID repopulation. Blood 103:2942–2949

Katayama Y, Battista M, Kao WM, Hidalgo A, Peired AJ, Thomas SA, Frenette, PS (2006) Signals from the sympathetic nervous system regulate hematopoietic stem cell egress from bone marrow. Cell 124:407–421

Kawabata K, Ujikawa M, Egawa T, Kawamoto H, Tachibana K, Iizasa H, Katsura Y, Kishimoto T, Nagasawa T (1999) A cell-autonomous requirement for CXCR4 in long-term lymphoid and myeloid reconstitution. Proc Natl Acad Sci U S A 96:5663–5667

Kim HK, De La Luz Sierra M, Williams CK, Gulino AV, Tosato G (2006) G-CSF down-regulation of CXCR4 expression identified as a mechanism for mobilization of myeloid cells. Blood 108:812–820

Lataillade JJ, Clay D, Dupuy C, Rigal S, Jasmin C, Bourin P, Le Bousse-Kerdiles MC (2000) Chemokine SDF-1 enhances circulating CD34(+) cell proliferation in synergy with cytokines: possible role in progenitor survival. Blood 95:756–768

Lataillade JJ, Clay D, Bourin P, Herodin F, Dupuy C, Jasmin C, Le Bousse-Kerdiles MC (2002) Stromal cell-derived factor 1 regulates primitive hematopoiesis by suppressing apoptosis and by promoting G(0)/G(1) transition in CD34(+) cells: evidence for an autocrine/paracrine mechanism. Blood 99:1117–1129

Lee Y, Gotoh A, Kwon HJ, You M, Kohli L, Mantel C, Cooper S, Hangoc G, Miyazawa K, Ohyashiki K, Broxmeyer HE (2002) Enhancement of intracellular signaling associated with hematopoietic progenitor cell survival in response to SDF-1/CXCL12 in synergy with other cytokines. Blood 99:4307–4317

Levesque JP, Hendy J, Takamatsu Y, Simmons PJ, Bendall LJ (2003) Disruption of the CXCR4/CXCL12 chemotactic interaction during hematopoietic stem cell mobilization induced by GCSF or cyclophosphamide. J Clin Invest 111:187–196

Liles WC, Broxmeyer HE, Rodger E, Wood B, Hubel K, Cooper S, Hangoc G, Bridger GJ, Henson GW, Calandra G, Dale DC (2003) Mobilization of hematopoietic progenitor cells in healthy volunteers by AMD3100, a CXCR4 antagonist. Blood 102:2728–2730

Ma Q, Jones D, Borghesani PR, Segal RA, Nagasawa T, Kishimoto T, Bronson RT, Springer TA (1998) Impaired B-lymphopoiesis, myelopoiesis, and derailed cerebellar neuron migration in CXCR4- and SDF-1-deficient mice. Proc Natl Acad Sci U S A 95:9448–9453

Ma Q, Jones D, Springer TA (1999) The chemokine receptor CXCR4 is required for the retention of B lineage and granulocytic precursors within the bone marrow microenvironment. Immunity 10:463–471

Martin C, Burdon PC, Bridger G, Gutierrez-Ramos JC, Williams TJ, Rankin SM (2003) Chemokines acting via CXCR2 and CXCR4 control the release of neutrophils from the bone marrow and their return following senescence. Immunity 19:583–593

Mayack SR, Wagers AJ (2008) Osteolineage niche cells initiate hematopoietic stem cell mobilization. Blood 112:519–531

McDermott DH, Liu Q, Ulrick J, Kwatemaa N, Anaya-O'Brien S, Penzak SR, Filho JO, Priel DA, Kelly C, Garofalo M et al (2011) The CXCR4 antagonist plerixafor corrects panleukopenia in patients with WHIM syndrome. Blood 118:4957–4962

Mendez-Ferrer S, Michurina TV, Ferraro F, Mazloom AR, Macarthur BD, Lira SA, Scadden DT, Ma'ayan A, Enikolopov GN, Frenette PS (2010) Mesenchymal and haematopoietic stem cells form a unique bone marrow niche. Nature 466:829–834

Nagasawa T, Kikutani H, Kishimoto T (1994) Molecular cloning and structure of a pre-B-cell growth-stimulating factor. Proc Natl Acad Sci U S A 91:2305–2309

Nagasawa T, Hirota S, Tachibana K, Takakura N, Nishikawa S, Kitamura Y, Yoshida N, Kikutani H, Kishimoto T (1996) Defects of B-cell lymphopoiesis and bone-marrow myelopoiesis in mice lacking the CXC chemokine PBSF/SDF-1. Nature 382:635–638

Nie Y, Waite J, Brewer F, Sunshine MJ, Littman DR, Zou YR (2004) The role of CXCR4 in maintaining peripheral B cell compartments and humoral immunity. J Exp Med 200:1145–1156

Nie Y, Han YC, Zou YR (2008) CXCR4 is required for the quiescence of primitive hematopoietic cells. J Exp Med 205:777–783

Noda M, Omatsu Y, Sugiyama T, Oishi S, Fujii N, Nagasawa T (2011) CXCL12-CXCR4 chemokine signaling is essential for NK-cell development in adult mice. Blood 117:451–458

Omatsu Y, Sugiyama T, Kohara H, Kondoh G, Fujii N, Kohno K, Nagasawa T (2010) The essential functions of adipo-osteogenic progenitors as the hematopoietic stem and progenitor cell niche. Immunity 33:387–399

Peled A, Petit I, Kollet O, Magid M, Ponomaryov T, Byk T, Nagler A, Ben-Hur H, Many A, Shultz L et al (1999) Dependence of human stem cell engraftment and repopulation of NOD/SCID mice on CXCR4. Science 283:845–848

Petit I, Szyper-Kravitz M, Nagler A, Lahav M, Peled A, Habler L, Ponomaryov T, Taichman RS, Arenzana-Seisdedos F, Fujii N et al (2002) G-CSF induces stem cell mobilization by decreasing bone marrow SDF-1 and up-regulating CXCR4.[erratum appears in Nat Immunol 2002 Aug;3(8):787]. Nat Immunol 3:687–694

Ponomaryov T, Peled A, Petit I, Taichman RS, Habler L, Sandbank J, Arenzana-Seisdedos F, Magerus A, Caruz A, Fujii N et al (2000) Induction of the chemokine stromal-derived factor-1 following DNA damage improves human stem cell function. J Clin Invest 106:1331–1339

Ross EA, Freeman S, Zhao Y, Dhanjal TS, Ross EJ, Lax S, Ahmed Z, Hou TZ, Kalia N, Egginton S et al (2008) A novel role for PECAM-1 (CD31) in regulating haematopoietic progenitor cell compartmentalization between the peripheral blood and bone marrow. PLoS One 3:e2338

Sacchetti B, Funari A, Michienzi S, Di Cesare S, Piersanti S, Saggio I, Tagliafico E, Ferrari S, Robey PG, Riminucci M, Bianco P (2007) Self-renewing osteoprogenitors in bone marrow sinusoids can organize a hematopoietic microenvironment. Cell 131:324–336

Semerad CL, Liu F, Gregory AD, Stumpf K, Link DC (2002) G-CSF is an essential regulator of neutrophil trafficking from the bone marrow to the blood. Immunity 17:413–423

Semerad CL, Christopher MJ, Liu F, Short B, Simmons PJ, Winkler I, Levesque JP, Chappel J, Ross FP, Link DC (2005) G-CSF potently inhibits osteoblast activity and CXCL12 mRNA expression in the bone marrow. Blood 106:3020–3027

Shirozu M, Nakano T, Inazawa J, Tashiro K, Tada H, Shinohara T, Honjo T (1995) Structure and chromosomal localization of the human stromal cell-derived factor 1 (SDF1) gene. Genomics 28:495–500

Sierro F, Biben C, Martinez-Munoz L, Mellado M, Ransohoff RM, Li M, Woehl B, Leung H, Groom J, Batten M et al (2007) Disrupted cardiac development but normal hematopoiesis in mice deficient in the second CXCL12/SDF-1 receptor, CXCR7. Proc Natl Acad Sci U S A 104:14759–14764

Sugiyama T, Kohara H, Noda M, Nagasawa T (2006) Maintenance of the hematopoietic stem cell pool by CXCL12-CXCR4 chemokine signaling in bone marrow stromal cell niches. Immunity 25:977–988

Suratt BT, Petty JM, Young SK, Malcolm KC, Lieber JG, Nick JA, Gonzalo JA, Henson PM, Worthen GS (2004) Role of the CXCR4/SDF-1 chemokine axis in circulating neutrophil homeostasis. Blood 104:565–571

Tachibana K, Hirota S, Iizasa H, Yoshida H, Kawabata K, Kataoka Y, Kitamura Y, Matsushima K, Yoshida N, Nishikawa S et al (1998) The chemokine receptor CXCR4 is essential for vascularization of the gastrointestinal tract. Nature 393:591–594

Tashiro K, Tada H, Heilker R, Shirozu M, Nakano T, Honjo T (1993) Signal sequence trap: a cloning strategy for secreted proteins and type I membrane proteins. Science 261:600–603

Tesio M, Golan K, Corso S, Giordano S, Schajnovitz A, Vagima Y, Shivtiel S, Kalinkovich A, Caione L, Gammaitoni L et al (2011) Enhanced c-Met activity promotes G-CSF-induced mobilization of hematopoietic progenitor cells via ROS signaling. Blood 117:419–428

Tokoyoda K, Egawa T, Sugiyama T, Choi BI, Nagasawa T (2004) Cellular niches controlling B lymphocyte behavior within bone marrow during development. Immunity 20:707–718

Winkler IG, Sims NA, Pettit AR, Barbier V, Nowlan B, Helwani F, Poulton IJ, van Rooijen N, Alexander KA, Raggatt LJ, Levesque JP (2010) Bone marrow macrophages maintain hematopoietic stem cell (HSC) niches and their depletion mobilizes HSCs. Blood 116:4815–4828

Wright DE, Bowman EP, Wagers AJ, Butcher EC, Weissman IL (2002) Hematopoietic stem cells are uniquely selective in their migratory response to chemokines. J Exp Med 195:1145–1154

Zhu W, Liang G, Huang Z, Doty SB, Boskey AL (2011) Conditional inactivation of the CXCR4 receptor in osteoprecursors reduces postnatal bone formation due to impaired osteoblast development. J Biol Chem 286:26794–26805

Zou YR, Kottmann AH, Kuroda M, Taniuchi I, Littman DR (1998) Function of the chemokine receptor CXCR4 in haematopoiesis and in cerebellar development. Nature 393:595–599

Chapter 31
CXCR4/CXCL12 as a Therapeutic Target

Geoffrey L. Uy and John F. DiPersio

Abstract The interaction of acute myeloid leukemia (AML) blasts with the bone marrow microenvironment regulates critical processes important in sustaining the malignant clone including self-renewal, cell growth, and proliferation as well as resistance to chemotherapy. The chemokine receptor, CXC receptor 4 (CXCR4), with its ligand chemokine ligand 12 (CXCL12), is the principal regulator of trafficking of normal and malignant stem cells to the bone marrow. High expression of CXCR4 on AML blasts has been identified as an adverse prognostic feature in AML. As a target for therapeutic intervention, preclinical studies have demonstrated in animal models the potential for CXCR4 antagonists to potentiate the effect of chemotherapy in AML and tyrosine kinase inhibitors in AML. Based on these studies, clinical trials have been initiated in AML with CXCR4 inhibitors in combination with either chemotherapy or FMS-like tyrosine kinase 3 (FLT3) inhibitors.

Keywords Chemokine · Chemokine receptor · Plerixafor · Bone marrow microenvironment · Stem cell mobilization · G-protein-coupled receptor

31.1 Introduction

The interaction of acute myeloid leukemia (AML) blasts with the bone marrow microenvironment is an important mediator of chemotherapy resistance and disease relapse. CXC receptor 4 (CXCR4) is a seven transmembrane G-protein-coupled receptor present on hematopoietic cells including AML blasts. In conjunction with its ligand, chemokine ligand 12 (CXCL12), CXCR4 regulates homing and retention of both normal and malignant hematopoietic stem cells to the bone marrow. Binding

J. F. DiPersio (✉)
Division of Oncology, Washington University School of Medicine, 660 S. Euclid Ave, Campus Box 8007, Saint Louis, MO 63110, USA
e-mail: jdipersi@dom.wustl.edu

G. L. Uy
Washington University School of Medicine, 660 S. Euclid Ave, Campus Box 8007, Saint Louis, MO 63110, USA
e-mail: guy@dom.wustl.edu

M. Andreeff (ed.), *Targeted Therapy of Acute Myeloid Leukemia*,
Current Cancer Research, DOI 10.1007/978-1-4939-1393-0_31

of CXCL12 increases intracellular calcium, induces cell migration, and activates downstream signaling pathways including phosphoinositide 3-kinase (PI3K), mitogen-activated protein kinase (MEK)/extracellular-signal-regulated kinase (ERK), and protein kinase C (PKC).

Targeting the bone marrow microenvironment in AML via disruption of the CXCR4/CXCL12 interaction is attractive as a therapeutic approach in AML. Physical detachment of leukemic cells can disrupt drug resistance mediated by engagement of specific adhesion molecules present on the surface of the leukemic blast and marrow stromal elements. Furthermore, disruption of CXCL12-induced signaling may promote apoptosis and enhance the sensitivity of AML blasts to chemotherapy.

31.2 Prognostic Impact of CXCR4 Expression in AML

Similar to normal hematopoietic stem cells, AML blasts migrate in response to CXCL12 produced by osteoblasts and marrow stromal cells. While CXCR4 is universally expressed on normal $CD34^+$ hematopoietic progenitors, its expression on primary AML blasts is more heterogeneous (Mohle et al. 1998). Consequently, several groups have explored whether CXCR4 expression in AML is associated with patient outcomes. A study of 90 adult patients with AML examined surface expression of CXCR4 in cryopreserved mononuclear cells by flow cytometry. While the expression of CXCR4 itself had a borderline prognostic value, subjects with greater than 1 % of cells co-expressing CD34 and CXCR4 had significantly reduced relapse free and overall survival. Furthermore, CXCR4 expression was found to be significantly higher in subjects harboring an FMS-like tyrosine kinase 3-internal tandem duplication (FLT3-ITD) mutation and proposed a potential link between the two (Rombouts et al. 2004). In a separate study, CXCR4 expression was evaluated in fresh samples by flow cytometry using mean fluorescence intensity to stratify AML patients by low, intermediate, and high CXCR4 intensity. Low CXCR4 expression was associated with longer relapse-free and overall survival compared to higher CXCR4 levels and was independent of cytogenetics, lactate dehydrogenase (LDH), leukocyte count, or age (Spoo et al. 2007). Because FLT3 signaling is known to be an adverse prognostic indicator and could account for the prognostic impact of CXCR4 expression, CXCR4 expression was evaluated in FLT3-ITD negative normal karyotype patients by immunohistochemistry. In this study, CXCR4 expression was associated with both event free and overall survival (Konoplev et al. 2007). Finally, a recent report also showed that lower CXCR4 expression as measured by flow cytometry was associated with achievement of CR and OS (Tavernier-Tardy et al. 2009).

Although these studies indicate that high expression of CXCR4 is an adverse prognostic feature, they need to be interpreted with caution due to their retrospective nature as well as variability in patient samples and assay techniques. Furthermore, validation of CXCR4 expression as a biomarker may be problematic as it has been reported that exposure of AML blasts to room air oxygen tension (21 %)

results in a rapid decrease in the expression of CXCR4 which may lead to an underestimation of the overall surface expression of CXCR4 (Fiegl et al. 2009).

31.3 Preclinical Models of CXCR4 Inhibitors in AML

Early work with AML cell lines suggested that treatment with plerixafor inhibited the proliferation and triggered changes mimicking differentiation (Tavor et al. 2008). Treatment of AML cell lines also resulted in decreased migration of AML blasts in response to CXCL12, and decreased the outgrowth of leukemia colony forming units in methylcellulose assays (Liesveld et al. 2007).

Additional studies examined the ability of CXCR4 inhibitors to augment the effect of chemotherapy in AML. A peptide inhibitor of CXCR4, RCP168, downregulated both phosphor-AKT and phosphor-ERK signaling in response to CXCL12 in vitro using Jurkat cells. The addition of either RCP168 or a small molecule inhibitor of CXCR4, AMD3465, reversed stroma-mediated protection of apoptosis in primary AML samples exposed to cytarabine. This protection was only observed in a subset of samples tested and interestingly, of these samples, 7/7 harbored an FLT3-ITD mutation (Zeng et al. 2006). Using a mouse model of AML, Nervi et al. examined the use of the CXCR4 antagonist plerixafor (formerly AMD3100) as a chemosensitizing agent in AML. Plerixafor mobilized leukemic blasts into the peripheral circulation with peak mobilization occurring approximately 4–6 h after administration. Furthermore, the addition of plerixafor improved the overall survival of genetically AML mice harboring a PML-RARA transgene when administered concurrently with chemotherapy (Nervi et al. 2009).

The combination of CXCR4 inhibitors with tyrosine kinase inhibitors and histone deacetylase inhibitors has also been explored. To investigate a possible link between FLT3 and CXCR4, a small molecule CXCR4 inhibitor, AMD3465, was combined with an FLT3 inhibitor, sorafenib. In a Ba/F3 FLT3-ITD viral transduction model of leukemia, AMD3465 facilitated sorafenib-induced apoptosis of these cells (Zeng et al. 2009). In a separate study, FLT3-ITD-transduced hematopoietic progenitors showed reduced surface and CXCR4 expression and decreased migration and homing in immunodeficient nonobese diabetic (NOD)/severe combined immunodeficiency (SCID) xenografts. Plerixafor also reduced proliferation and colony formation of FLT-ITD positive AML blasts compared to ITD negative cells (Jacobi et al. 2010).

Finally, the combination of a CXCR4 inhibitor with a pan histone deacetylase inhibitor, panobinostat, has been studied in AML. Acetylation of hsp90 increased proteasome-mediated degradation of CXCR4. Furthermore, panobinostat induced apoptosis of primary AML cells when combined with CXCR4 inhibitors (Mandawat et al. 2010).

Table 31.1 CXCR4 inhibitors in clinical trials

Drug	Manufacturer	Type	Clinical indication
Plerixafor	Sanofi	Small molecule	Stem cell mobilization, AML, multiple myeloma, solid tumors
ALX-0651	Ablynx	Single domain antibody fragment	Stem cell mobilization (development discontinued)
BKT140	Biokine Therapeutics	Small molecule	Multiple myeloma
BMS-936594	Bristol-Myers Squibb	Monoclonal antibody	Multiple myeloma, AML
LY-2510924	Eli Lilly	Peptidomimetic	Renal cell, small cell lung CA
MSX-122	Altiris Therapeutics	Small molecule	Metastatic or locally advanced solid tumors
NOX-A12	Noxxon Pharma	RNA oligomer	Chronic lymphocytic leukemia; multiple myeloma; glioblastoma multiforme
POL6326	Polyphor Ltd.	Peptidomimetic	Stem cell mobilization
TG-0054	Taigen Biotechnology	Small molecule	Stem cell mobilization

31.4 CXCR4 Antagonists in Clinical Development

The development of CXCR4 inhibitors for clinical use was initially based on its identification as the co-receptor for HIV in T-cell tropic strains. Elucidation of its role in hematopoietic stem cell retention and trafficking for CXCR4 led to the development of these agents for use in mobilization of cells for hematopoietic stem cell transplantation. The first of these agents to be developed, plerixafor, is a bicyclam small molecule inhibitor of CXCR4. In clinical trials, plerixafor was found to be efficacious when combined with granulocyte-colony-stimulating factor (G-CSF) for the mobilization of hematopoietic stem cells (HSCs) prior to autologous transplantation in patients with either multiple myeloma or non-Hodgkin lymphoma and is approved in both the USA and Europe for this indication (DiPersio et al. 2009a; DiPersio et al. 2009b; Devine et al. 2004)

A number of other agents which target the CXCR4/CXCL12 axis have been developed for clinical purposes. These include small molecule inhibitors, monoclonal antibodies, and peptide and RNA mimetics. While the majority are being evaluated for hematopoietic stem cell mobilization, some of these agents are also being tested in tumor sensitization models including AML (Table 31.1). A monoclonal antibody to CXCR4, BMS-936564 has been demonstrated to exert anti-tumor activity in AML xenograft models (Kuhne et al. 2011). POL6326, a protein epitope mimetic, has been demonstrated to mobilize leukemic cells in a Ba/F3 FLT3-ITD NOD/SCID xenograph model mouse (Chen et al. 2010).

31.5 Clinical Trials with CXCR4 Antagonists for AML

Theoretical concerns exist in combining CXCR4 antagonists with chemotherapy. Normal hematopoietic stem and progenitor cells universally express CXCR4 and are mobilized into the peripheral circulation in response to administration of CXCR4 antagonists. It is conceivable that these agents may also sensitize normal HSCs to chemotherapy leading to prolonged marrow aplasia and delayed hematopoietic recovery. In addition, significant mobilization of leukemic blasts into the peripheral circulation may precipitate symptomatic or even life-threatening hyperleukocytosis. An early report of the combination of G-CSF and plerixafor resulted in the massive mobilization of leukemic cells from the marrow of a patient with AML in complete remission undergoing autologous transplantation (Andreeff et al. 2006).

The first clinical trial of plerixafor in AML was phase I/II study of 52 subjects with relapsed or refractory AML. In this study, plerixafor at doses of up to 0.24 mg/kg/d were administered subcutaneously 4 h prior to mitoxantrone, etoposide, and cytarabine. In the phase II, complete remissions (CR plus CRi) were observed in 46% of patients. The regimen appeared safe. Although this study limited eligible patients to circulating blast counts of less than 20,000/mm^3, no symptomatic hyperleukocytosis occurred. In addition, there was no evidence of delayed hematopoietic recovery observed. Correlative studies demonstrated that plerixafor could result in a modest (approximately 2.5-fold) mobilization of blasts into the peripheral circulation. No correlation was identified between surface expression of CXCR4 and response (Uy et al. 2012).

In a subsequent trial, newly diagnosed adult patients with AML were treated with the combination of cytarabine 100 mg/m$_2$ CIVI on days 1–7 and daunorubicin 90 mg/m^2 on days 1–3. In contrast to the initial study, plerixafor was given as a 30-min IV infusion, 4–5 h before chemotherapy at dose levels of 0.32, 0.40, and 0.48 mg/kg. Of 21 subjects with available data, 14 (67%) had complete response (CR), 2 had CR with incomplete count recovery (CRi), 2 had residual leukemia (RL), 2 had treatment failure (TF) due to resistant disease and 1 was not evaluable (NE) due to early death (Uy et al. 2011).

Although clinical experience with CXCR4 inhibitors in AML is limited, it should be noted that other agents which affect the CXCR4/CXCL12 axis have been used in studies of patients with AML. G-CSF and GM-CSF are routinely administered to patients in order to hasten neutrophil recovery after the administration of chemotherapy. A number of studies have also explored G-CSF or GM-CSF as "priming" agents (Rowe et al. 2004; Lowenberg et al. 2003; Pabst et al. 2012; Thomas et al. 2010). Although the rationale for growth factor priming is to induce cycling of quiescent leukemic blasts to render them more sensitive to cell cycle specific agents such as cytarabine, G-CSF and to a lesser extent GM-CSF are potent stem cell mobilizing agents whose effects occur via down regulation of CXCL12. G-CSF reduces CXCL12 mRNA and protein expression in the bone marrow (Semerad et al. 2005; Christopher et al. 2009; Petit et al. 2002). G-CSF has also been reported to downregulate CXCR4 expression on hematopoietic cells (Kim et al. 2006).

Based on these initial results, clinical trials are currently underway examining the role of CXCR4 inhibitors in AML in a number of clinical settings (Table 31.2).

Table 31.2 Clinical trials of CXCR4 inhibitors in AML. (From ClinicalTrials.gov. Bethesda (MD): National Library of Medicine (USA). 2000–2012 Available at: http://www.clinicaltrials.gov/ Accessed 1, Jun 2012. GOELAMS Groupe Ouest Est d'etude des Leucemies et Autres Maladies du Sang; PETHEMA Programa Español de Tratamientos en Hematología)

Study title	Sponsor	Planned enrolment
A Phase 1, Open-label, multicenter study of BMS-936564 (MDX-1338) in subjects with relapsed/refractory acute myelogenous leukemia and selected B-cell malignancies	Bristol-Myers Squibb	82
An NCRI acute myeloid leukaemia working group pilot trial under the auspices of the cardiff experimental cancer medicine centre to establish the feasibility of combining either the tyrosine kinase inhibitor AC220 or the CXCR4 inhibitor plerixafor with chemotherapy in older patients with acute myeloid leukaemia and high risk myelodysplastic syndrome	Cardiff University	90
Chemosensitization with plerixafor plus G-CSF in relapsed or refractory acute myeloid leukemia	Washington University	67
A phase I pediatric study of a plerixafor-containing regimen in second allogeneic stem cell transplantation	St. Jude Children's Research Hospital	18
A phase I study using plerixafor as a chemosensitizing agent for relapsed acute leukemia and MDS in pediatric patients	Emory University	21
Phase I/II study of plerixafor and clofarabine in previously untreated older (>/=60 years) adult patients with Acute Myelogenous Leukemia (AML) With two or more unfavorable prognostic factors for whom standard induction chemotherapy is unlikely to be of benefit	M.D. Anderson cancer center	66
Induction therapy with decitabine and plerixafor priming for patients ≥ 60 years with acute myeloid leukemia	Weill medical college of cornell university	48
A phase 1, dose escalation study of plerixafor in combination with induction and consolidation chemotherapy in patients with relapsed acute myeloid leukemia	GOELAMS	24
G-CSF and plerixafor with busulfan and fludarabine for allogeneic stem cell transplantation for myeloid leukemias	M.D. Anderson cancer center	72
Multicenter, prospective, open-label, single-arm, phase I–II clinical trial to analyze induction therapy with a combination of fludarabine, idarubicin, cytarabine, G-CSF, and plerixafor for the treatment of young patients with relapsed or refractory AML	PETHEMA	55
G-CSF and plerixafor with sorafenib for acute myelogenous leukemia with FLT3 mutations	M.D. Anderson cancer center	28

CXCR4 inhibitors are being incorporated with traditional cytotoxic regimens in both the salvage setting and in the upfront treatment of AML. Additional studies combine plerixafor with the hypomethylating agent decitabine and with the FLT3 inhibitor, sorafenib. Plerixafor is also being incorporated into a preparative regimen prior to allogeneic stem cell transplantation. While the majority of studies of CXCR4 inhibitors in AML use plerixafor, additional inhibitors are being developed and tested in early phase I/II studies primarily in the relapsed refractory setting.

Conclusions

While initial clinical trials have demonstrated the safety and feasibility of combining CXCR4 inhibitors with chemotherapy in AML, several critical questions remain. These early studies were performed with relatively low numbers of patients and are insufficient to determine whether expression of CXCR4 may serve as a biomarker of response to treatment with these agents. Although mobilization of AML blasts has been observed in vivo, the effect is relatively modest compared to the mobilization of normal HSCs. Since only a small percentage of the total leukemic burden is mobilized in response to treatment, mobilization of blasts per se is unlikely to be required for the potential therapeutic benefit. However, the magnitude of mobilization may still serve a surrogate marker for CXCR4/CXCL12 disruption. Finally, these initial studies were conducted as nonrandomized studies in biologically and clinically heterogeneous patient populations. To determine the true contribution of CXCR4 inhibition to clinical outcomes will require larger randomized phase III studies.

References

Andreeff M, Konoplev S, Wang R-Y, Zeng Z, McQueen T, Shi Y-X, Medeiros LJ, Estey E, McCarty JM, Elkins S, Champlin R, Calandra G, Bridger G, Konopleva M (2006) Massive mobilization of AML cells into circulation by disruption of leukemia/stroma cell interactions using CXCR4 antagonist AMD3100: first evidence in patients and potential for abolishing bone marrow microenvironment-mediated resistance. ASH Annu Meet Abstr 108(11):568

Chen Y, Zeng Z, Shi Y, Jacamo R, Ludin C, Dembowsky K, Frenette PS, Konopleva M, Andreeff M (2010) Targeting CXCR4, SDF1 and beta-adrenergic receptors in the AML microenvironment by novel antagonist POL6326, G-CSF and isoproterenol. ASH Annu Meet Abstr 116(21):2179

Christopher MJ, Liu F, Hilton MJ, Long F, Link DC (2009) Suppression of CXCL12 production by bone marrow osteoblasts is a common and critical pathway for cytokine-induced mobilization. Blood 114(7):1331–1339. doi:10.1182/blood-2008-10-184754

Devine SM, Flomenberg N, Vesole DH, Liesveld J, Weisdorf D, Badel K, Calandra G, DiPersio JF (2004) Rapid mobilization of CD34+ cells following administration of the CXCR4 antagonist AMD3100 to patients with multiple myeloma and non-Hodgkin's lymphoma. J Clin Oncol 22(6):1095–1102. doi:10.1200/JCO.2004.07.131

DiPersio JF, Micallef IN, Stiff PJ, Bolwell BJ, Maziarz RT, Jacobsen E, Nademanee A, McCarty J, Bridger G, Calandra G (2009a) Phase III prospective randomized double-blind placebo-controlled trial of plerixafor plus granulocyte colony-stimulating factor compared with placebo plus granulocyte colony-stimulating factor for autologous stem-cell mobilization and transplantation for patients with non-Hodgkin's lymphoma. J Clin Oncol 27(28):4767–4773. doi:10.1200/JCO.2008.20.7209

DiPersio JF, Stadtmauer EA, Nademanee A, Micallef IN, Stiff PJ, Kaufman JL, Maziarz RT, Hosing C, Fruehauf S, Horwitz M, Cooper D, Bridger G, Calandra G (2009b) Plerixafor and G-CSF versus placebo and G-CSF to mobilize hematopoietic stem cells for autologous stem cell transplantation in patients with multiple myeloma. Blood 113(23):5720–5726. doi:10.1182/blood-2008-08-174946

Fiegl M, Samudio I, Clise-Dwyer K, Burks JK, Mnjoyan Z, Andreeff M (2009) CXCR4 expression and biologic activity in acute myeloid leukemia are dependent on oxygen partial pressure. Blood 113(7):1504–1512. doi:10.1182/blood-2008-06-161539

Jacobi A, Thieme S, Lehmann R, Ugarte F, Malech HL, Koch S, Thiede C, Muller K, Bornhauser M, Ryser M, Brenner S (2010) Impact of CXCR4 inhibition on FLT3-ITD-positive human AML blasts. Exp Hematol 38(3):180–190. doi:10.1016/j.exphem.2009.12.003

Kim HK, De La Luz Sierra M, Williams CK, Gulino AV, Tosato G (2006) G-CSF down-regulation of CXCR4 expression identified as a mechanism for mobilization of myeloid cells. Blood 108(3):812–820. doi:10.1182/blood-2005-10-4162

Konoplev S, Rassidakis GZ, Estey E, Kantarjian H, Liakou CI, Huang X, Xiao L, Andreeff M, Konopleva M, Medeiros LJ (2007) Overexpression of CXCR4 predicts adverse overall and event-free survival in patients with unmutated FLT3 acute myeloid leukemia with normal karyotype. Cancer 109(6):1152–1156. doi:10.1002/cncr.22510

Kuhne MR, Mulvey T, Chen S, Pan C, Chong C, Niekro W, Kempe T, Henning KA, Cohen L, Korman A, Cardarelli PM (2011) BMS-936564/MDX-1338: a fully human anti-cxcr4 antibody induces apoptosis in vitro and shows anti tumor activity in vivo. ASH Annu Meet Abstr 118(21):1543

Liesveld JL, Bechelli J, Rosell K, Lu C, Bridger G, Phillips G, 2nd, Abboud CN (2007) Effects of AMD3100 on transmigration and survival of acute myelogenous leukemia cells. Leuk Res 31(11):1553–1563. doi:10.1016/j.leukres.2007.02.017

Lowenberg B, van Putten W, Theobald M, Gmur J, Verdonck L, Sonneveld P, Fey M, Schouten H, de Greef G, Ferrant A, Kovacsovics T, Gratwohl A, Daenen S, Huijgens P, Boogaerts M (2003) Effect of priming with granulocyte colony-stimulating factor on the outcome of chemotherapy for acute myeloid leukemia. N Engl J Med 349(8):743–752. doi:10.1056/NEJMoa025406

Mandawat A, Fiskus W, Buckley KM, Robbins K, Rao R, Balusu R, Navenot JM, Wang ZX, Ustun C, Chong DG, Atadja P, Fujii N, Peiper SC, Bhalla K (2010) Pan-histone deacetylase inhibitor panobinostat depletes CXCR4 levels and signaling and exerts synergistic antimyeloid activity in combination with CXCR4 antagonists. Blood 116(24):5306–5315. doi:10.1182/blood-2010-05-284414

Mohle R, Bautz F, Rafii S, Moore MA, Brugger W, Kanz L (1998) The chemokine receptor CXCR-4 is expressed on CD34+ hematopoietic progenitors and leukemic cells and mediates transendothelial migration induced by stromal cell-derived factor-1. Blood 91(12):4523–4530

Nervi B, Ramirez P, Rettig MP, Uy GL, Holt MS, Ritchey JK, Prior JL, Piwnica-Worms D, Bridger G, Ley TJ, DiPersio JF (2009) Chemosensitization of acute myeloid leukemia (AML) following mobilization by the CXCR4 antagonist AMD3100. Blood 113(24):6206–6214. doi:10.1182/blood-2008-06-162123

Pabst T, Vellenga E, van Putten W, Schouten HC, Graux C, Vekemans MC, Biemond B, Sonneveld P, Passweg J, Verdonck L, Legdeur MC, Theobald M, Jacky E, Bargetzi M, Maertens J, Ossenkoppele GJ, Lowenberg B (2012) Favorable effect of priming with granulocyte colony-stimulating factor in remission induction of acute myeloid leukemia restricted to dose-escalation of cytarabine. Blood 119(23):5367–5373. doi:10.1182/blood-2011-11-389841

Petit I, Szyper-Kravitz M, Nagler A, Lahav M, Peled A, Habler L, Ponomaryov T, Taichman RS, Arenzana-Seisdedos F, Fujii N, Sandbank J, Zipori D, Lapidot T (2002) G-CSF induces stem cell mobilization by decreasing bone marrow SDF-1 and up-regulating CXCR4. Nat Immunol 3(7):687–694. doi:10.1038/ni813

Rombouts EJ, Pavic B, Lowenberg B, Ploemacher RE (2004) Relation between CXCR-4 expression, Flt3 mutations, and unfavorable prognosis of adult acute myeloid leukemia. Blood 104(2):550–557. doi:10.1182/blood-2004-02-0566

Rowe JM, Neuberg D, Friedenberg W, Bennett JM, Paietta E, Makary AZ, Liesveld JL, Abboud CN, Dewald G, Hayes FA, Tallman MS, Wiernik PH (2004) A phase 3 study of three induction regimens and of priming with GM-CSF in older adults with acute myeloid leukemia: a trial by the Eastern Cooperative Oncology Group. Blood 103(2):479–485. doi:10.1182/blood-2003-05-1686

Semerad CL, Christopher MJ, Liu F, Short B, Simmons PJ, Winkler I, Levesque JP, Chappel J, Ross FP, Link DC (2005) G-CSF potently inhibits osteoblast activity and CXCL12 mRNA expression in the bone marrow. Blood 106(9):3020–3027. doi:10.1182/blood-2004-01-0272

Spoo AC, Lubbert M, Wierda WG, Burger JA (2007) CXCR4 is a prognostic marker in acute myelogenous leukemia. Blood 109(2):786–791. doi:10.1182/blood-2006-05-024844

Tavernier-Tardy E, Cornillon J, Campos L, Flandrin P, Duval A, Nadal N, Guyotat D (2009) Prognostic value of CXCR4 and FAK expression in acute myelogenous leukemia. Leuk Res 33(6):764–768. doi:10.1016/j.leukres.2008.10.014

Tavor S, Eisenbach M, Jacob-Hirsch J, Golan T, Petit I, Benzion K, Kay S, Baron S, Amariglio N, Deutsch V, Naparstek E, Rechavi G (2008) The CXCR4 antagonist AMD3100 impairs survival of human AML cells and induces their differentiation. Leukemia 22(12):2151–5158. doi:10.1038/leu.2008.238

Thomas X, Raffoux E, Renneville A, Pautas C, de Botton S, Terre C, Gardin C, Hayette S, Preudhomme C, Dombret H (2010) Which AML subsets benefit from leukemic cell priming during chemotherapy? Long-term analysis of the ALFA-9802 GM-CSF study. Cancer 116(7):1725–1732. doi:10.1002/cncr.24943

Uy GL, Avigan D, Cortes JE, Becker PS, Chen RW, Liesveld JL, Hewes B, Johns D, Erba HP (2011) Safety and tolerability of plerixafor in combination with cytarabine and daunorubicin in patients with newly diagnosed acute myeloid leukemia- preliminary results from a phase I study. ASH Annu Meet Abstr 118(21):82

Uy GL, Rettig MP, Motabi IH, McFarland K, Trinkaus KM, Hladnik LM, Kulkarni S, Abboud CN, Cashen AF, Stockerl-Goldstein KE, Vij R, Westervelt P, Dipersio JF (2012) A phase 1/2 study of chemosensitization with the CXCR4 antagonist plerixafor in relapsed or refractory acute myeloid leukemia. Blood 119(17):3917–3924. doi:10.1182/blood-2011-10-383406

Zeng Z, Samudio IJ, Munsell M, An J, Huang Z, Estey E, Andreeff M, Konopleva M (2006) Inhibition of CXCR4 with the novel RCP168 peptide overcomes stroma-mediated chemoresistance in chronic and acute leukemias. Mol Cancer Ther 5(12):3113–3121. doi:10.1158/1535-7163.MCT-06-0228

Zeng Z, Shi YX, Samudio IJ, Wang RY, Ling X, Frolova O, Levis M, Rubin JB, Negrin RR, Estey EH, Konoplev S, Andreeff M, Konopleva M (2009) Targeting the leukemia microenvironment by CXCR4 inhibition overcomes resistance to kinase inhibitors and chemotherapy in AML. Blood 113(24):6215–6224. doi:10.1182/blood-2008-05-158311

Chapter 32
VLA-4: A Cell's Consequential Encounter

Thalia Papayannopoulou

Abstract The role of integrins as modulators of hematopoietic cell function is continuously evolving. After the initial establishment of integrins as the primary forces necessary for adhesion and migration of mature leucocytes, new roles in hematopoietic stem/progenitor cell retention and/or function in bone marrow (BM) were uncovered; more recently, attention has shifted to neoplastic hematopoiesis and cancer with instrumental integrin-dependent influences on reciprocal interactions of neoplastic cells with their microenvironment. The ever-expanding knowledge of integrin-mediated signaling and their multitude of signaling mediators and other cooperating molecules provide ample opportunities for harnessing this multifaceted influence of integrins for therapeutic benefit. In this context, a brief account of very late antigen 4 (VLA4) integrin in normal and neoplastic hematopoiesis is presented.

Keywords Integrins · VLA4 · Mobilization · Microenvironment · Leukemia

32.1 Introduction

Integrins are heterodimeric transmembrane proteins composed of multiple types of alpha subunits combined with a limited number of beta subunits. There are a total of 8 β subunits heterodimerized with 18 alpha partners to yield at least 24 different heterodimers, each with distinct ligand-binding and signaling properties (Hynes 1992; Takagi and Springer 2002). Very late antigen 4 (VLA-4) or α4β1 integrin (integrin alpha 4, CD49d/integrin beta 1, CD29) belongs to the β1 family, the largest among integrins with 12 alpha partners and with wide expression among hematopoietic and non-hematopoietic cells alike. β1 integrins are the most utilized β integrin subunits in mammals, and they are highly evolutionarily conserved and among the most extensively studied (Arroyo et al. 1999, 2000; Brakebusch et al. 2002a, b; Brakebusch and Fassler 2005).

T. Papayannopoulou (✉)
Department of Medicine, Division of Hematology, Fred Hutchinson Cancer Research Center, 1705 NE Pacific St. Seattle, Box 357710, WA 98195-7710, USA
e-mail: thalp@u.washington.edu

M. Andreeff (ed.), *Targeted Therapy of Acute Myeloid Leukemia*,
Current Cancer Research, DOI 10.1007/978-1-4939-1393-0_32

The main function of all integrins is to integrate the inside of the cell to its outside environment. Each cell expresses its own repertoire of integrins that act as cell sensors, integrating cues from the microenvironment, and tells the cell how to behave. Consequently, integrins have pleiotropic functions affecting many cellular processes both in development and in adult life. To carry out their dynamic and pleiotropic functions, integrin's structure is specialized to be highly responsive to sudden signals and is regulated; integrins are able to switch from an inactive (low affinity, highly bent configuration) form present in many cell types to an active (high affinity, extended) conformational state and vice versa in response to binding events, both at a ligand-binding pocket and the cytoplasmic domains, thereby eliciting conformational changes between and within the integrin subunits (Luo and Springer 2006; Shattil et al. 2010). It is important to emphasize that, in addition to providing forces needed for adhesion and migration, integrins are bidirectional signaling machines through inside-out and outside-in signaling. Since integrins lack intrinsic kinase activity, in order to propagate their signals, they recruit downstream tyrosine kinases, which enable a crosstalk of integrins with many other intracellular signaling partners. There is also a crosstalk of integrins with other integrins so that signaling by one integrin can influence the affinity or avidity of another. Furthermore, the crosstalk of integrins with growth factor receptors and their ability to assemble different signaling complexes downstream provides many layers for signaling specificity and cooperativity (Legate et al. 2009). Nevertheless, several integrins have unique, non-overlapping functions depending on cell context.

All functional data concerning integrins were originally derived through the use of anti-functional or activating antibodies and later on through genetic studies (Papayannopoulou and Nakamoto 1993; Papayannopoulou et al. 1995; Craddock et al. 1997; Arroyo et al. 1999). Currently, all integrin subunits have been targeted in the mouse and functional consequences have been described. Through these studies, several in vivo functions of integrins have been uncovered, both during development and in adult life, the latter studied in conditionally ablated animals.

The function of VLA-4 integrin is among the most widely studied. A detailed knowledge of its adhesion and multifunctional signaling properties, along with its diverse ligand partners, has been accumulated by studying multiple preclinical models. Successful outcomes in the latter models has subsequently led to several efficacious clinical trials in inflammatory (Crohn's disease) or autoimmune diseases (Multiple sclerosis, Ulcerative colitis; Yonekawa and Harlan 2005; Staunton et al. 2006; Woodside and Vanderslice 2008). It is important to emphasize that the antibody data regarding VLA-4, in contrast to data with other integrins, have been highly concordant to genetic data.

In addition to the importance of anti-functional VLA-4 antibodies in the migratory properties of mature lymphocytes in vivo, the same Abs have shown an important function in the retention of stem/progenitor cells within the bone marrow (BM) environment (Papayannopoulou and Nakamoto 1993; Papayannopoulou et al. 1995; Priestley et al. 2006). In this context, VLA4 displays cooperatively with other integrins (Papayannopoulou et al. 2001; Katayama et al. 2003) and its effects are also additive or enhanced by another important parallel pathway, that of the chemokine CXC receptor 4 (CXCR4)/ chemokine ligand 12 (CXCL12; Burger et al. 2003; Papayannopoulou 2003, 2004; Papayannopoulou and Scadden 2008; Bonig et al. 2009; Rettig et al. 2012). Although the antibody properties have led to enhanced mobilization in preclinical models, their use in humans can only be implemented clinically, when an effective small molecule blocking VLA4 function becomes available, to avoid the impact of stable antibody attachment on the subsequent homing of donor cells (at present several small molecule antagonists of VLA-4 (BIO1211, IVL745, HM1031, GSK-683699, and R411/Valategrast; Woodside and Vanderslice 2008) tested for other indications with a single study assessing mobilization (Rettig et al,2012).

The integrin's impact on hematopoietic cell biology continues to evolve. Expanded knowledge on the role of integrins in neoplastic hematopoiesis and in cancer has currently created new frontiers in integrin biology and what follows is an attempt to briefly highlight the relevance of VLA4 functional features in leukemic hematopoiesis along with its potential as a therapeutic target.

32.2 VLA4 in Neoplastic Hematopoiesis

As integrins have been recognized as playing critical roles in the proliferation and survival of normal cells by relaying and integrating environmental cues into downstream signaling pathways, a similar sensory role of integrins has been found to be operable, or to go awry in neoplastic cells creating survival signals and protection of neoplastic cells from chemotherapy. Thus, several in vitro studies using chronic myeloid leukemia (CML) cells have demonstrated an abnormal β1 and α4 function in CML cells with a normalization following treatment with Interferon α or STI571 (Gleevec) or arsenic trioxide (Bhatia et al. 1994, 1995; Jiang et al. 2000; Guo-Bao et al. 2010). Other in vitro studies using myeloma cells have shown chemo-protection by VLA-4, which was overturned by Bortezomid (Hatano et al. 2008; Nαoborio-Hatano et al. 2009). Propelled by these in vitro studies, several in vivo studies were initiated to uncover, based on VLA-4 expression, a prognostic value for treatment and therapeutic efficacy and survival advantage. Several such studies in chronic lymphocytic leukemia (CLL) have found that high VLA-4 (CD49d) expression predicts poor survival as an independent risk factor either alone or in a protein complex with CD44/chemokines/MMPs (Redondo-Munoz et al. 2008; Rossi et al. 2008; Shanafelt et al. 2008a, b; Nuckel et al. 2009; Zucchetto et al. 2009; Rossi et al. 2010; Majid et al. 2011). Similar studies in acute myeloid leukemia (AML) have not shown consistent results. Although heterogeneity in VLA-4 expression and a chemo-protective effect with anti-VLA4 treatments in vitro was found in all studies, the conclusions regarding the prognostic value of VLA-4 expression were discordant (Matsunaga et al. 2003; Becker et al. 2009; Walter et al. 2010).

One can only speculate about the behavior that integrins may display in neoplastic hematopoiesis. Several mechanistic scenarios may be entertained. One possibility is that integrins in these neoplastic cells respond to the outside microenvironmental cues like the normal cells do. According to this scenario, the spread and retention patterns of leukemic cells to tissues will depend on the functional expression levels of integrins; cells with high integrin expression may have a higher migratory ability to metastasize and be retained in extramedullary tissues leading to a spread of leukemia in many tissues. Examples for this scenario are CLL cells with a heavy tissue load of CLL cells, or the migratory properties and metastatic potential of melanoma cells with high integrin expression (Etoh et al. 1992; Barkan and Chambers 2011; Sil et al. 2011). If the integrins are hyperactive, then the high adhesion forces may keep them confined to one tissue rather than metastasize. Integrins in this scenario will also be responsible for survival signals and chemo-protection under chemotherapy, as was shown for CLL (see References above), MM (Hata et al. 1993; Podar et al. 2011), Macroglobulinemia (Ngo et al. 2008). At the other end of the spectrum is the situation where the normal responses of integrins to outside cues are lost, either because of neoplastic transformation, or other coexisting abnormalities that exert a trans-dominant control over the integrin function. In this case, retention mechanisms are no longer integrin-dependent, and cells circulate freely in blood irrespective of their maturation stage or their integrin expression levels and distribute freely to tissues. A paradigm in this case is CML where integrin function is lost and can be restored after treatment. (See CML references above.)

In AML, all in vitro studies have shown a chemo-protective effect of $\alpha 4\beta 1$, through the use of anti-functional Abs, or by using CS1 peptides (Matsunaga et al. 2003, 2008; Becker et al. 2009). Yet, in vivo studies differed in their conclusions regarding the prognostic significance of VLA-4 expression. In this context, the findings with pediatric AML are particularly insightful. This well designed prospective study makes several important points of relevance to the core function of VLA-4: (a) None of the patients with high VLA-4 expression was positive with FMS-like tyrosine kinase 3 (FLT3)/ internal tandem duplication (ITD); (b) all the standard risk patients had longer disease-free survival (DFS) and event-free survival (EFS) than the low VLA-4 expressors; (c) high VLA-4 expressors had more extramedullary disease; (d) the outcome of all high-risk patients was independent of VLA-4 expression/function; and finally (e) induction rates were irrelevant to VLA-4 expression. In other words, this study uncovered that the expression of VLA-4 in pediatric AML (Walter et al. 2010) cells could be dictated by a particular neoplastic phenotype (FLT3/ITD) and that other factors in high-risk patients may override VLA-4 function. These features are particularly instructive for the design and expectations of clinical trials aiming to harness the adhesive/homing potential, or counteract the chemo-protection of VLA-4 expressing AML cells. It is also of interest that the same conclusions regarding prognostic significance of VLA-4 were reached in adult AML patients with an additional twist (Becker et al. 2009). Here the functional status of VLA-4, not the overall expression level, was relevant for prognostic purposes. It was previously known that the capacity of VLA-4 to bind to soluble vascular cell adhesion molecule-1 (VCAM-1, sVCAM-1) is cell type specific, is actively regulated and controls VLA4-dependent functions (Rose et al. 2000). Thus, sVCAM-1 may act as a competitive ligand/inhibitor-blocking adhesion to endothelium, or to BM stroma. Similarly, soluble fibronectin (at physiologic concentrations) occupies the high-affinity VLA-4 receptor and becomes an effective inhibitor of adhesion to immobilized fibronectin (Faull et al. 1993). The presence of high levels of sVCAM-1 within the BM microenvironment in AML (Sudhoff et al. 1996) could dictate VLA-4-dependent function. Coexistence of both high and low affinity integrins in each cell further complicates interpretations of expression levels.

Beyond these surface-related considerations, integrin function can also be down-regulated by integrin-suppression pathways (Hras or Raf1), or by caspase-dependent proteolysis of integrin activators or of integrin cytoplasmic domains important in signaling. Further detailed molecular studies are warranted to understand altered integrin function in leukemic cells both at presentation and at relapse to understand the altered biology.

32.3 VLA-4 as a Clinical Surrogate Target in Leukemia

All the aforementioned studies highlight the fact that integrins are active players in leukemic cell behavior. This realization provides a rationale for carefully designed clinical studies to harness any detrimental aspects of integrin behavior in leukemia. In these studies, the influence of integrins is envisioned as only adjuvant to the rest of the panoply used to attack leukemia. It is clear that there is no compelling need to include integrin inhibitors in initial remission-induction studies. Their relevance may become important in order to address leukemia stem cells (LSC), or rather leukemia initiating cells (LIC) of any flavor. Therefore, such studies should be initiated at the consolidation phases or at repeat chemotherapeutic regimens. Integrin functional behavior should be studied by appropriate tests before and after relapse, since such studies are insightful and may differ in each case. To what extent the normal cells are influenced should also be studied.

An important consideration influencing efficacy as well as selectivity of action is the choice of integrin inhibitory molecules. These can be in the form of antibodies, peptides, or small molecules and only their in vivo performance should be the decisive factor. Thus far, all the molecules designed for clinical use have targeted the extracellular region of integrins that interfere with ligand binding. Inhibitors designed to target the intracellular portion of integrins, potentially even distinguishing selective signaling pathways that are integrin-dependent, have thus far not been explored. This should be an area of active future research with potentially fruitful outcomes.

It has been previously amply demonstrated that, in addition to the VLA-4-dependent pathway in cell trafficking, another parallel and important pathway with independent function is that of CXCR4/SDF-1-dependent signaling. These two pathways and their respective ligands are functional across species barriers and have validated their recognition as important key players in cell trafficking. A combination of these two molecules, at least for mobilization purposes, has shown additive effects (Bonig et al. 2009; Rettig et al. 2012). One could also envision that combination of these two pathways in leukemia—now only separately being tried— is warranted and should yield higher efficiency. In addition, combinations of cooperating adhesion molecules, like VLA-4 with E-selectins already active in normal cells, are also effective in the case of AML cells (Chien et al. 2011). We have previously shown that VLA-4, in combination with E-Selectin, inhibits homing of Stem/Progenitor cells by > 90% (Papayannopoulou et al. 2001). Thus, it may be expected that such a combination may have greater potency in inhibiting the retention of cells in BM and making them vulnerable to chemotherapeutic attack.

Although all data so far have been dealt with retention in BM, an untested parameter in humans is whether retention in other tissues is similar to BM using an anti-functional VLA-4 strategy. For example, in mice, splenic homing and retention of hematopoietic cells is VLA-4-independent (Papayannopoulou et al. 1995). Whether this is also true in humans is unclear. This issue is relevant to the broader concept of hematopoietic cell interactions with their microenvironment, or “niche.” Perturbations of the interactions of neoplastic hematopoietic cells with their supportive “niches” may have consequences for both normal and leukemic cells, according to recently presented evidence (Lane et al. 2009; Zeng et al. 2009, 2012). Not only were leukemic cells able to disrupt the normal “niche,” selectively supporting normal stem cells, but also were able to create “alternative niches” by manipulating the cells around them (Zhang et al. 2012).

Conclusion

In summary, newer tools for interfering with integrin function in leukemia or other cancers, and further studies on the functional behavior of integrins in neoplastic hematopoiesis are needed in order to translate our knowledge into well-designed clinical studies with therapeutic dividends.

References

Arroyo AG, Yang JT, Rayburn H, Hynes RO (1999) Alpha4 integrins regulate the proliferation/differentiation balance of multilineage hematopoietic progenitors in vivo. Immunity 11(5):555–566

Arroyo AG, Taverna D, Whittaker CA, Strauch UG, Bader BL, Rayburn H, Crowley D, Parker CM, Hynes RO (2000) In vivo roles of integrins during leukocyte development and traffic: insights from the analysis of mice chimeric for alpha 5, alpha v, and alpha 4 integrins. J Immunol 165(8):4667–4675

Barkan D, Chambers AF (2011) beta1-integrin: a potential therapeutic target in the battle against cancer recurrence. Clin Cancer Res 17(23):7219–7223. doi:10.1158/1078-0432.CCR-11-0642

Becker PS, Kopecky KJ, Wilks AN, Chien S, Harlan JM, Willman CL, Petersdorf SH, Stirewalt DL, Papayannopoulou T, Appelbaum FR (2009) Very late antigen-4 function of myeloblasts correlates with improved overall survival for patients with acute myeloid leukemia. Blood 113(4):866–874. doi:10.1182/blood-2007-12-124818

Bhatia R, Wayner EA, McGlave PB, Verfaillie CM (1994) Interferon-alpha restores normal adhesion of chronic myelogenous leukemia hematopoietic progenitors to bone marrow stroma by correcting impaired beta 1 integrin receptor function. J Clin Invest 94(1):384–391. doi:10.1172/JCI117333

Bhatia R, McGlave PB, Verfaillie CM (1995) Treatment of marrow stroma with interferon-alpha restores normal beta 1 integrin-dependent adhesion of chronic myelogenous leukemia hematopoietic progenitors. Role of MIP-1 alpha. J Clin Invest 96(2):931–939. doi:10.1172/JCI118141

Bonig H, Watts KL, Chang KH, Kiem HP, Papayannopoulou T (2009) Concurrent blockade of alpha4-integrin and CXCR4 in hematopoietic stem/progenitor cell mobilization. Stem Cells 27(4):836–837. doi:10.1002/stem.9

Brakebusch C, Fassler R (2005) beta 1 integrin function in vivo: adhesion, migration and more. Cancer Metastasis Rev 24(3):403–411. doi:10.1007/s10555-005-5132-5

Brakebusch C, Bouvard D, Stanchi F, Sakai T, Fassler R (2002a) Integrins in invasive growth. J Clin Invest 109(8):999–1006. doi:10.1172/JCI15468

Brakebusch C, Fillatreau S, Potocnik AJ, Bungartz G, Wilhelm P, Svensson M, Kearney P, Korner H, Gray D, Fassler R (2002b) Beta1 integrin is not essential for hematopoiesis but is necessary for the T cell-dependent IgM antibody response. Immunity 16(3):465–477

Burger JA, Spoo A, Dwenger A, Burger M, Behringer D (2003) CXCR4 chemokine receptors (CD184) and $\alpha4\beta1$ integrins mediate spontaneous migration of human CD34+ progenitors and acute myeloid leukaemia cells beneath marrow stromal cells (pseudoemperipolesis). Br J Haematol 122(4):579–589. doi:10.1046/j.1365-2141.2003.04466.x

Chien S, Zhao X, Patton JT, Smith T, Sarkar A, Magnani JL, Appelbaum FR, Becker PS (2011) Novel dual E-selectin-CXCR4 Inhibitors Mobilize Human Acute Myeloid Leukemia (AML) cells in the NODscid IL2R{gamma}c-/- Xenograft and confer susceptibility to cytarabine. ASH Annual Meeting Abstracts 118(21):579

Craddock CF, Nakamoto B, Andrews RG, Priestley GV, Papayannopoulou T (1997) Antibodies to VLA4 integrin mobilize long-term repopulating cells and augment cytokine-induced mobilization in primates and mice. Blood 90(12):4779–4788

Etoh T, Byers HR, Mihm MC Jr (1992) Integrin expression in malignant melanoma and their role in cell attachment and migration on extracellular matrix proteins. J Dermatol 19(11):841–846

Faull RJ, Kovach NL, Harlan JM, Ginsberg MH (1993) Affinity modulation of integrin alpha 5 beta 1: regulation of the functional response by soluble fibronectin. J Cell Biol 121(1):155–162. doi:10.1083/jcb.121.1.155

Guo-Bao W, Xiao-Qin C, Qi-Rong G, Jie L, Gui-Nan L, Yue L (2010) Arsenic Trioxide overcomes cell adhesion-mediated drug resistance through down-regulating the expression of beta(1)-integrin in K562 chronic myelogenous leukemia cell line. Leuk Lymphoma 51(6):1090–1097. doi:10.3109/10428191003746315

Hata H, Matsuzaki H, Matsuno F, Yoshida M, Sonoki T, Takemoto S, Takatsuki K (1993) Expression of fibronectin and adhesion to fibronectin in myeloma cell lines. Acta Haematol 89(1):26–31

Hatano K, Kikuchi J, Takatoku M, Shimizu R, Wada T, Ueda M, Nobuyoshi M, Oh I, Sato K, Suzuki T, Ozaki K, Mori M, Nagai T, Muroi K, Kano Y, Furukawa Y, Ozawa K (2008) Bortezomib overcomes cell adhesion-mediated drug resistance through downregulation of VLA-4 expression in multiple myeloma. Oncogene 28(2):231–242. doi:http://www.nature.com/onc/journal/v28/n2/suppinfo/onc2008385s1.html

Hynes RO (1992) Integrins: versatility, modulation, and signaling in cell adhesion. Cell 69(1):11–25

Jiang Y, Zhao RC, Verfaillie CM (2000) Abnormal integrin-mediated regulation of chronic myelogenous leukemia CD34+ cell proliferation: BCR/ABL up-regulates the cyclin-dependent kinase inhibitor, p27Kip, which is relocated to the cell cytoplasm and incapable of regulating cdk2 activity. Proc Natl Acad Sci U S A 97(19):10538–10543. doi:10.1073/pnas.190104497

Katayama Y, Hidalgo A, Furie BC, Vestweber D, Furie B, Frenette PS (2003) PSGL-1 participates in E-selectin-mediated progenitor homing to bone marrow: evidence for cooperation between E-selectin ligands and alpha4 integrin. Blood 102(6):2060–2067. doi:10.1182/blood-2003-04-1212

Lane SW, Scadden DT, Gilliland DG (2009) The leukemic stem cell niche: current concepts and therapeutic opportunities. Blood 114(6):1150–1157. doi:10.1182/blood-2009-01-202606

Legate KR, Wickstrom SA, Fassler R (2009) Genetic and cell biological analysis of integrin outside-in signaling. Genes Dev 23(4):397–418. doi:10.1101/gad.1758709

Luo B-H, Springer TA (2006) Integrin structures and conformational signaling. Curr Opin Cell Biol 18(5):579–586. doi:10.1016/j.ceb.2006.08.005

Majid A, Lin TT, Best G, Fishlock K, Hewamana S, Pratt G, Yallop D, Buggins AG, Wagner S, Kennedy BJ, Miall F, Hills R, Devereux S, Oscier DG, Dyer MJ, Fegan C, Pepper C (2011) CD49d is an independent prognostic marker that is associated with CXCR4 expression in CLL. Leukemia Res 35(6):750–756. doi:10.1016/j.leukres.2010.10.022

Matsunaga T, Takemoto N, Sato T, Takimoto R, Tanaka I, Fujimi A, Akiyama T, Kuroda H, Kawano Y, Kobune M, Kato J, Hirayama Y, Sakamaki S, Kohda K, Miyake K, Niitsu Y (2003) Interaction between leukemic-cell VLA-4 and stromal fibronectin is a decisive factor for minimal residual disease of acute myelogenous leukemia. Nat Med 9(9):1158–1165. doi:http://www.nature.com/nm/journal/v9/n9/suppinfo/nm909_S1.html

Matsunaga T, Fukai F, Miura S, Nakane Y, Owaki T, Kodama H, Tanaka M, Nagaya T, Takimoto R, Takayama T, Niitsu Y (2008) Combination therapy of an anticancer drug with the FNIII14 peptide of fibronectin effectively overcomes cell adhesion-mediated drug resistance of acute myelogenous leukemia. Leukemia 22(2):353–360. doi:10.1038/sj.leu.2405017

Ngo HT, Leleu X, Lee J, Jia X, Melhem M, Runnels J, Moreau AS, Burwick N, Azab AK, Roccaro A, Azab F, Sacco A, Farag M, Sackstein R, Ghobrial IM (2008) SDF-1/CXCR4 and VLA-4 interaction regulates homing in Waldenstrom macroglobulinemia. Blood 112(1):150–158. doi:10.1182/blood-2007-12-129395

Noborio-Hatano K, Kikuchi J, Takatoku M, Shimizu R, Wada T, Ueda M, Nobuyoshi M, Oh I, Sato K, Suzuki T, Ozaki K, Mori M, Nagai T, Muroi K, Kano Y, Furukawa Y, Ozawa K (2009) Bortezomib overcomes cell-adhesion-mediated drug resistance through downregulation of VLA-4 expression in multiple myeloma. Oncogene 28(2):231–242. doi:10.1038/onc.2008.385

Nuckel H, Switala M, Collins CH, Sellmann L, Grosse-Wilde H, Duhrsen U, Rebmann V (2009) High CD49d protein and mRNA expression predicts poor outcome in chronic lymphocytic leukemia. Clin Immunol 131(3):472–480. doi:10.1016/j.clim.2009.02.004

Papayannopoulou T (2003) Bone marrow homing: the players, the playfield, and their evolving roles. Curr Opin Hematol 10(3):214–219

Papayannopoulou T (2004) Current mechanistic scenarios in hematopoietic stem/progenitor cell mobilization. Blood 103(5):1580–1585. doi:10.1182/blood-2003-05-1595

Papayannopoulou T, Nakamoto B (1993) Peripheralization of hemopoietic progenitors in primates treated with anti-VLA4 integrin. Proc Natl Acad Sci U S A 90(20):9374–9378

Papayannopoulou T, Scadden DT (2008) Stem-cell ecology and stem cells in motion. Blood 111(8):3923–3930. doi:10.1182/blood-2007-08-078147

Papayannopoulou T, Craddock C, Nakamoto B, Priestley GV, Wolf NS (1995) The VLA4/VCAM-1 adhesion pathway defines contrasting mechanisms of lodgement of transplanted murine hemopoietic progenitors between bone marrow and spleen. Proc Natl Acad Sci U S A 92(21):9647–9651

Papayannopoulou T, Priestley GV, Nakamoto B, Zafiropoulos V, Scott LM (2001) Molecular pathways in bone marrow homing: dominant role of α4β1 over β2-integrins and selectins. Blood 98(8):2403–2411. doi:10.1182/blood.V98.8.2403

Podar K, Zimmerhackl A, Fulciniti M, Tonon G, Hainz U, Tai YT, Vallet S, Halama N, Jager D, Olson DL, Sattler M, Chauhan D, Anderson KC (2011) The selective adhesion molecule inhibitor Natalizumab decreases multiple myeloma cell growth in the bone marrow microenvironment: therapeutic implications. Br J Haematol 155(4):438–448. doi:10.1111/j.1365-2141.2011.08864.x

Priestley GV, Scott LM, Ulyanova T, Papayannopoulou T (2006) Lack of alpha4 integrin expression in stem cells restricts competitive function and self-renewal activity. Blood 107(7):2959–2967. doi:10.1182/blood-2005-07-2670

Redondo-Munoz J, Ugarte-Berzal E, Garcia-Marco JA, del Cerro MH, Van den Steen PE, Opdenakker G, Terol MJ, Garcia-Pardo A (2008) Alpha4beta1 integrin and 190-kDa CD44v constitute a cell surface docking complex for gelatinase B/MMP-9 in chronic leukemic but not in normal B cells. Blood 112(1):169–178. doi:10.1182/blood-2007-08-109249

Rettig MP, Ansstas G, DiPersio JF (2012) Mobilization of hematopoietic stem and progenitor cells using inhibitors of CXCR4 and VLA-4. Leukemia 26(1):34–53. doi:10.1038/leu.2011.197

Rose DM, Cardarelli PM, Cobb RR, Ginsberg MH (2000) Soluble VCAM-1 binding to α4 integrins is cell-type specific and activation dependent and is disrupted during apoptosis in T cells. Blood 95(2):602–609

Rossi D, Zucchetto A, Rossi FM, Capello D, Cerri M, Deambrogi C, Cresta S, Rasi S, De Paoli L, Bodoni CL, Bulian P, Del Poeta G, Ladetto M, Gattei V, Gaidano G (2008) CD49d expression is an independent risk factor of progressive disease in early stage chronic lymphocytic leukemia. Haematologica 93(10):1575–1579. doi:10.3324/haematol.13103

Rossi D, Bodoni CL, Zucchetto A, Rasi S, De Paoli L, Fangazio M, Rossi FM, Ladetto M, Gattei V, Gaidano G (2010) Low CD49d expression and long telomere identify a chronic lymphocytic leukemia subset with highly favourable outcome. Am J Hematol 85(8):619–622. doi:10.1002/ajh.21756

Shanafelt TD, Geyer SM, Bone ND, Tschumper RC, Witzig TE, Nowakowski GS, Zent CS, Call TG, Laplant B, Dewald GW, Jelinek DF, Kay NE (2008a) CD49d expression is an independent predictor of overall survival in patients with chronic lymphocytic leukaemia: a prognostic parameter with therapeutic potential. Br J Haematol 140(5):537–546. doi:10.1111/j.1365-2141.2007.06965.x

Shanafelt TD, Hanson C, Dewald GW, Witzig TE, LaPlant B, Abrahamzon J, Jelinek DF, Kay NE (2008b) Karyotype evolution on fluorescent in situ hybridization analysis is associated with short survival in patients with chronic lymphocytic leukemia and is related to CD49d expression. J Clin Oncol 26(14):e5–e6. doi:10.1200/JCO.2008.16.7874

Shattil SJ, Kim C, Ginsberg MH (2010) The final steps of integrin activation: the end game. Nat Rev Mol Cell Biol 11(4):288-300. doi:10.1038/nrm2871

Sil H, Sen T, Chatterjee A (2011) Fibronectin-integrin (alpha5beta1) modulates migration and invasion of murine melanoma cell line B16F10 by involving MMP-9. Oncol Res 19(7):335–348

Staunton DE, Lupher ML, Liddington R, Gallatin WM (2006) Targeting integrin structure and function in disease. Adv Immunol 91:111–157. doi:10.1016/S0065-2776(06)91003-7

Sudhoff T, Wehmeier A, Kliche KO, Aul C, Schlomer P, Bauser U, Schneider W (1996) Levels of circulating endothelial adhesion molecules (sE-selectin and sVCAM-1) in adult patients with acute leukemia. Leukemia 10(4):682–686

Takagi J, Springer TA (2002) Integrin activation and structural rearrangement. Immunol Rev 186:141–163

Walter RB, Alonzo TA, Gerbing RB, Ho PA, Smith FO, Raimondi SC, Hirsch BA, Gamis AS, Franklin JL, Hurwitz CA, Loken MR, Meshinchi S (2010) High expression of the very late antigen-4 integrin independently predicts reduced risk of relapse and improved outcome in pediatric acute myeloid leukemia: a report from the children's oncology group. J Clin Oncol 28(17):2831–2838. doi:10.1200/JCO.2009.27.5693

Woodside DG, Vanderslice P (2008) Cell adhesion antagonists: therapeutic potential in asthma and chronic obstructive pulmonary disease. BioDrugs 22(2):85–100

Yonekawa K, Harlan JM (2005) Targeting leukocyte integrins in human diseases. J Leukoc Biol 77(2):129–140. doi:10.1189/jlb.0804460

Zeng Z, Shi YX, Samudio IJ, Wang RY, Ling X, Frolova O, Levis M, Rubin JB, Negrin RR, Estey EH, Konoplev S, Andreeff M, Konopleva M (2009) Targeting the leukemia microenvironment by CXCR4 inhibition overcomes resistance to kinase inhibitors and chemotherapy in AML. Blood 113(24):6215–6224. doi:10.1182/blood-2008-05-158311

Zhang B, Ho Yin W, Huang Q, Maeda T, Lin A, Lee S-u, Hair A, Holyoake Tessa L, Huettner C, Bhatia R (2012) Altered microenvironmental regulation of leukemic and normal stem cells in chronic myelogenous leukemia. Cancer Cell 21(4):577–592

Zucchetto A, Benedetti D, Tripodo C, Bomben R, Dal Bo M, Marconi D, Bossi F, Lorenzon D, Degan M, Rossi FM, Rossi D, Bulian P, Franco V, Del Poeta G, Deaglio S, Gaidano G, Tedesco F, Malavasi F, Gattei V (2009) CD38/CD31, the CCL3 and CCL4 chemokines, and CD49d/vascular cell adhesion molecule-1 are interchained by sequential events sustaining chronic lymphocytic leukemia cell survival. Cancer Res 69(9):4001–4009. doi:10.1158/0008-5472.CAN-08-4173

Chapter 33
VLA-4 Function and Prognosis in Acute Myeloid Leukemia

Pamela S. Becker and Frederick R. Appelbaum

Abstract Acute myeloid leukemia (AML) cells are dependent on integrins and chemokine receptors for migration and marrow homing. Moreover, adhesion within the marrow microenvironment results in chemotherapy resistance for AML as well as other hematologic malignancies, by activating survival pathways or interfering with apoptosis. Retention within the marrow is dependent on chemokine receptor type 4 (CXCR4), which mediates migration toward stromal derived factor (SDF)-1, also known as CXCL12, whereas attachment is largely attributed to very late antigen (VLA)-4, an integrin that binds to both alternatively spliced fibronectin, a component of the extracellular matrix, and vascular cell adhesion molecule-1 (VCAM-1) on marrow stromal cells. There are correlations of the level of expression of both these receptors with outcomes in AML after treatment. Disruption of these two types of interactions can enhance chemotherapy-mediated cytotoxicity in laboratory studies. The anti-VLA-4 antibody, natalizumab, was approved for treatment of multiple sclerosis and Crohn's disease several years ago, and this antibody or similar inhibitors of these types of interactions have now been included in clinical trials for different types of cancer. In the future, small molecule inhibitors or antibodies that interfere with adhesion proteins may become critical additions to standard chemotherapy approaches in AML.

Keywords Very late antigen-4 · Vascular cell adhesion molecule · Homing · Migration · Fibronectin · Cell adhesion mediated chemotherapy resistance

---

F. R. Appelbaum (✉)
Seattle Cancer Care Alliance and Clinical Research Division, Fred Hutchinson Cancer Research Center 1100 Fairview Avenue North, P.O. Box 19024 D5-310, Seattle, WA 98109-1024, USA
e-mail: fappelba@fhcrc.org

P. S. Becker
Division of Hematology and Institute for Stem Cell and Regenerative Medicine, University of Washington School of Medicine, University of Washington, 850 Republican Street, Rm N415, Campus Box 358056 Seattle, WA 98109 USA
e-mail: pbecker@u.washington.edu

M. Andreeff (ed.), *Targeted Therapy of Acute Myeloid Leukemia*,
Current Cancer Research, DOI 10.1007/978-1-4939-1393-0_33

List of Abbreviations

AML	acute myeloid leukemia
CXCR4	CXC receptor 4, CD184
Flt3ITD	fms-like tyrosine kinase 3 internal tandem duplication
MFI	mean fluorescence intensity
SDF-1	stromal derived factor-1, CXCL12
sVCAM-1	soluble vascular cell adhesion molecule-1
VCAM-1	vascular cell adhesion molecule-1
VLA-4	very late antigen-4, CD49d
VLA-5	very late antigen-5, CD49e

33.1 Introduction

Adhesion of acute myeloid leukemia (AML) blasts within the bone marrow microenvironment confers chemotherapy drug resistance. Migration toward and retention within the marrow is mediated by CXCR4, but attachment involves integrins, the most significant of which is likely VLA-4. Both CXCR4 and VLA-4 play critical roles in homing and migration of hematopoietic cells, and inhibition of adhesion with small molecules or antibodies may overcome leukemia drug resistance. These inhibitors are now being incorporated into clinical trials for AML in both the relapsed and upfront setting.

33.2 VLA-4 and Homing/Migration of Hematopoietic Stem Cells

VLA-4 is a heterodimer of the α_4 and β_1 integrins that mediates attachment of normal hematopoietic stem cells to the alternatively spliced CS1 peptide of fibronectin in the bone marrow extracellular matrix (For further information, see Chap. 34 by Dr. Papayannopoulou). VLA-5 is another integrin that mediates binding to a different peptide of fibronectin. Inhibition of VLA-4 with anti-VLA-4 antibody causes migration of normal hematopoietic stem cells from the bone marrow into the peripheral blood in mice (Papayannopoulou et al. 1995) and nonhuman primates (Craddock et al. 1997). Similarly, a small molecule inhibitor of VLA-4 also mobilized hematopoietic stem cells in mice (Ramirez et al. 2009). Patients undergoing treatment with natalizumab, an FDA approved anti-VLA-4 antibody for the treatment of multiple sclerosis and Crohn's disease, also exhibit mobilization of hematopoietic progenitor cells (Bonig et al. 2008; Zohren et al. 2008). Homing of human hematopoietic progenitors to the marrow of immunodeficient mice is dependent on both CXCR4 and VLA-4 (Pelled et al. 2000). Migration of normal stem cells toward stromal derived factor-1 (SDF-1 or CXCL12) via CXCR4 is also dependent on VLA-4 (Bryant et al. 2012).

While cells are retained in the marrow by the chemokine receptor CXCR4 that responds to gradients of SDF-1 (For further information, see Chap. 32, by Dr. Di Persio), attachment to extracellular matrix and stromal cells occurs via integrins such as VLA-4 and other receptors.

33.3 VLA-4 and Chemotherapy Drug Resistance in AML

VLA-4 is highly expressed by normal hematopoietic stem cells and AML blasts. For example, for 175 cryopreserved AML patient bone marrow samples examined by flow cytometry gating on blasts, the mean expression was 60.3% for α_4 and 39.9% for β_1 (Becker et al. 2009). For 53 AML patients blood and 26 bone marrow samples analyzed prior to cryopreservation, the percent expression by gated blasts of α_4 was 92.4% (blood) and 83.9% (bone marrow), and expression of β_1, 93.1% (blood) and 83.8% (bone marrow; Chien et al. 2011). These high levels of α_4 and β_1 expression were also maintained by the $CD34^+/CD38^-/CD123^+$ putative leukemia stem cell population (Chien et al 2011). For 19 patient peripheral blood samples analyzed for VLA-4 expression by the blasts versus putative leukemia stem cells, the mean α_4 expression was 88.5% (blasts) versus 87% (stem cells) and for β_1, 81.1% (blood) versus 78.6% (stem cells; Chien et al. 2011). Studies of AML cell lines and patient AML cells have demonstrated the critical role of VLA-4 in migration in vitro (Burger et al. 2003). The migration of both patient AML cells and normal $CD34^+$ cells beneath adherent stromal cells known as pseudoemperipolesis can be blocked by antibody to VLA-4 or synthetic CS-1 peptide, as well as antibody to CXCR-4 (Burger et al. 2003). Adhesion of U937 to fibronectin via β1 integrins inhibited mitoxantrone and etoposide-induced apoptosis (Hazelhurst et al. 2001). Adhesion of U937 or HL60 to fibronectin inhibits daunorubicin or Ara-C induced apoptosis, and the adhesion-mediated chemotherapy resistance can be mediated by VLA-4 or VLA-5 (Matsunaga et al. 2003). VLA-4 levels expressed by HL60 fluctuate with cell cycle status, decreasing during S phase (Becker et al. 1999), an interesting finding given that cycling cells are most susceptible to cytarabine. Antibody to VLA-4, small molecule inhibitors of VLA-4, or fibronectin peptide that inhibits β_1 integrin, all overcome adhesion-mediated drug resistance for leukemia cell lines or AML patient cells (Matsunaga et al. 2003; Matsunaga et al. 2008; Becker et al. 2008; Becker et al. 2009). For patient bone marrow leukemia cells adherent to alternatively spliced fibronectin or immobilized vascular cell adhesion molecule-1 (VCAM-1), either antibody to VLA-4 or sVCAM-1 soluble vascular cell adhesion molecule-1 (sVCAM-1) overcome adhesion-mediated drug resistance and restore cytotoxicity to either cytarabine alone, or for some cytarabine-resistant samples, both daunorubicin and cytarabine (Becker et al. 2009). Similarly, a small molecule inhibitor of VLA-4 effectively restores chemotherapy sensitivity of AML patient blasts adherent to recombinant fibronectin peptide CH296 (Becker et al. 2008).

The mechanism by which adhesion confers drug resistance has been attributed to activation of survival pathways or inhibition of apoptotic pathways. Signal transduction pathways that may be utilized include phosphatidylinositol-3-kinase (PI3K)/Akt/bcl2, PI3K/integrin linked kinase (ILK)/Akt, or Wnt/glycogen synthase kinase 3β (GSK3β)/nuclear factor-kappa B (NFκB). The first pathway was implicated by experiments that showed that antibody to VLA-4, inhibitors to PI3K, or antisense oligonucleotides to bcl2 were all able to restore chemotherapy-induced apoptosis (Matsunaga et al. 2003). Another group demonstrated activation of the PI3K/ILK/Akt pathway that promotes survival by adhesion of NB4, Mo7e, and REH leukemia cell lines and primary AML samples to stroma (Tabe et al. 2007). De Toni and colleagues performed an exhaustive analysis of cell adhesion-mediated drug resistance of AML blasts and cell lines (De Toni et al. 2006). They found that adhesion to fibronectin confers resistance to daunorubicin in the absence of serum for 50% of 35 patient samples, with no relationship to patient clinical characteristics such as age, cytogenetic risk group, or response to treatment. They examined whether Wnt antagonists might induce resistance to apoptosis, as had been demonstrated for other cancer cell types, and found that incubation with the Wnt antagonist sFRP increases survival, and with the Wnt agonist Wnt3a increases chemo sensitivity. Studies with pharmacological inhibitors or siRNA demonstrate that adhesion or Wnt-induced chemoresistance is dependent on GSK3β, and adhesion of the leukemia cells results in dephosphorylation of Ser (9), a critical step in the activation of GSK3β. They proposed that the mechanism by which GSK3β promotes survival is via NFκB. In addition, adhesion to β1 integrins may also confer drug resistance by interference with apoptosis. For example, there is reduction of the amount of the isoforms of the pro-apoptotic bcl2 family member, Bim, upon adhesion of K562, U937 or KG1 leukemia cells to fibronectin (Hazlehurst et al. 2006). The reduction of Bim alternatively spliced variants occurs via proteosomal degradation. K562 cells with reduced Bim levels induced by shRNA exhibit resistance to imatinib or mitoxantrone.

33.4 Prognosis in AML and VLA-4 Expression

One study shows that expression of VLA-4 correlates with poor survival in AML (Matsunaga et al. 2003), but this is not supported by two other larger studies, one in adults (Becker et al 2009), and one in pediatric patients (Walter et al. 2010). In fact, these latter studies implied that higher expression as analyzed by mean fluorescence intensity or functional binding capacity correlates with improved survival in AML. The pediatric study (Walter et al. 2010) is a correlative study of bone marrow specimens obtained from 216 of 340 patients treated on a Children's Oncology Group protocol for de novo AML. The authors describe a 35-fold range of mean fluorescence intensity (MFI) of VLA-4 expression by flow cytometry. Low VLA-4 MFI correlates with the presence of the fms-like tyrosine kinase 3 internal tandem duplication (Flt3ITD), and high VLA-4 MFI with lower age. Higher VLA-4 MFI is associated with longer event-free survival and lower relapse rate. The adult study is comprised of 175 bone marrow samples obtained from patients undergoing induction chemotherapy on four Southwest Oncology Group trials (Becker et al. 2009). Functionally activated VLA-4 is assessed by binding of soluble VCAM-1 (sVCAM-1). There is no statistically significant difference between higher and lower percent expression of VLA-4 and survival, as had been found in the study by Matsunaga et al., and no relationship between mean fluorescence intensity and survival. However, there is a strong correlation between the percent binding of sVCAM-1 and survival; for patients with the highest quartile of sVCAM-1 binding, the 5-year estimated survival is 34%, compared with 12, 9, and 11% in the lower three quartiles in order from next highest to lowest. In addition, the maximum percent activated β_1 integrin negatively correlates with age (higher age, less potential expression of β_1 integrin). This latter finding is intriguing, since older patients have worse prognosis. It may seem paradoxical and counter-intuitive that expression of an integrin receptor that mediates chemotherapy resistance might be associated with improved survival. However, VLA-4 mediates both adhesion and migration, can be functionally activated, or not, and these attributes may be continuously modulated as cells move from the bone marrow to the blood and tissues. Moreover, durability of remission may be related to adhesion properties of the leukemia stem cell population, and not the blast population per se. As CXCR4 also plays a role in retention in the marrow and prognosis in AML, as well as other receptors, there may be an intricate and complex interplay between adhesion and release in a well-choreographed ballet that ultimately mediates drug sensitivity or resistance in the bone marrow microenvironment, and ultimately, prognosis and survival.

33.5 Therapeutic Potential of VLA-4 Inhibitors for AML

The VLA-4 receptor has already been targeted therapeutically with an antibody to α_4 integrin that functionally inhibits both $\alpha_4\beta_1$ and $\alpha_4\beta_7$ called natalizumab (Tysabri). It has been FDA approved for multiple sclerosis (Miller et al 2003) and Crohn's disease (Gordon et al 2001). Patients with multiple sclerosis have now been treated with Tysabri for several years. Another blocking antibody to $\alpha_4\beta_7$ has been studied in a phase I/II trial in ulcerative colitis and Crohn's disease (Feagan et al. 2001). There are also several small molecule inhibitors that have undergone laboratory evaluation that functionally inhibit VLA-4 (Becker et al. 2008; Ramirez et al. 2009; Setogochi et al. 2012), as well as inhibitors that have been utilized in clinical trials. These small molecule inhibitors demonstrate the ability to reduce inflammation in preclinical models such as asthma and experimental autoimmune encephalomyelitis. Because of the availability of these anti-VLA-4 antibody and small molecule inhibitors, this inhibitor approach to overcome adhesion mediated drug resistance is testable in AML. In fact, natalizumab decreases multiple myeloma cell growth alone or in combination with chemotherapy in preclinical studies (Podar et al. 2011), and has already been utilized as a single agent in a clinical trial in relapsed/refractory myeloma (clinicaltrials.gov identifier NCT00675428). However, caution

Table 33.1 Integrin antagonists with clinical application or undergoing preclinical evaluation

Agent	Status	Application	Target or mechanism	FDA-approved indication	References
Natalizumab	Clinical use	Multiple sclerosis, Crohn's disease	$\alpha_4\beta_1$ or $\alpha_4\beta_7$ (α_4 subunit)	Yes	Gordon et al. 2001
Natalizumab	Clinical trial and preclinical	Multiple myeloma	$\alpha_4\beta_1$ or $\alpha_4\beta_7$ (α_4 subunit)	No	Podar et al. 2011
Volociximab	Clinical trial	Cancer (e.g., renal cell or prostate ca, melanoma,) Macular degeneration	$\alpha_5\beta_1$	No	Almokadem & Belani 2012
PF-04605412	Clinical trial and preclinical	Cancer	$\alpha_5\beta_1$	No	Li et al. 2010
Etaratuzumab	Clinical trial	Melanoma, colorectal or prostate ca	$\alpha_V\beta_3$	No	Hersey et al. 2010
Intetumumab	Clinical trial	Prostate ca, melanoma	α_V	No	O'Day et al. 2011
Firategrast	Clinical trial	Multiple sclerosis	$\alpha_4\beta_1$	No	Miller et al. 2012
BIO1211	Clinical trial	Asthma	$\alpha_4\beta_1$	No	Abraham et al. 2000
BIO5192	Preclinical	Stem cell mobilization, experimental autoimmune encephalomyelitis	$\alpha_4\beta_1$	No	Ramirez et al. 2009; Theien et al. 2003
DW-908e	Preclinical	AML	$\alpha_4\beta_1$	No	Becker et al. 2008
BIO8809	Preclinical	Dry eye disease	$\alpha_4\beta_1$	No	Ecoiffier et al. 2008

needs to be exercised when natalizumab is utilized in the clinical setting, as progressive multifocal leukoencephalopathy has occurred in recipients of natalizumab, both in multiple sclerosis and Crohn's (Kleinschmidt –DeMasters and Tyler 2005; Langer-Gould et al. 2005; van Assche et al. 2005), and this potential risk has limited its study to date in patients who are otherwise immunocompromised.

Another target for cancer therapy is the tumor neovasculature in solid tumors. Antibodies to integrins such as the fibronectin receptor $\alpha_5\beta_1$ or the vitronectin receptor $\alpha_V\beta_3$ have been used for this purpose (Table. 33.1). An IgG4 chimeric antibody to $\alpha_5\beta_1$, volociximab, has been tested in clinical trials in solid tumors and antibody to $\alpha_V\beta_3$, monoclonal antibody specific for alphavbeta3 integrin (MEDI 522), has been used in clinical trials in melanoma. The rationale is that these integrins appear to be upregulated by the process of tumor neoangiogenesis, and specifically target the tumor vasculature. In the phase I dose escalation study of voloxicimab, there was one case of renal carcinoma that exhibited reduction of pulmonary nodules, and one case of metastatic melanoma that remained stable for the treatment course of 14 months (Ricart et al. 2008). Such studies suggest potential efficacy of the approach of targeting integrins for cancer therapy, and although hematological malignancies do appear to be dependent on neoangiogenesis, these studies to date have been confined to solid tumors.

In addition to the use of specific inhibitors of VLA-4, it is possible that other therapeutic agents might downregulate VLA-4 expression or function. For example, treatment of leukemia cell line KG1 or primary AML cells, as well as normal hematopoietic progenitors, with histone deacetylase inhibitors SAHA and VPA caused downregulation of VLA-4 and decreased adhesion to mesenchymal stromal cells (Mahlknecht und Schönbein 2008).

Lastly, a combined approach of mobilization of leukemia with both CXCR4 and VLA-4 inhibitors may be more effective than a single agent, as the combination blockade of these two receptors in the macaque results in greater than an additive effect on hematopoietic stem cell mobilization (Bonig et al. 2009).

Conclusion

VLA-4 integrin plays a critical role in adhesion-mediated chemotherapy resistance. Disruption of its activity, either directly by specific inhibitors, or by other mechanisms that secondarily reduce its expression or function, may play an adjunctive role in improving chemotherapy sensitivity and outcomes in AML.

References

Abraham WM, Gill A, Ahmed A, Sielczak MW, Lauredo IT, Botinnikova Y, Lin KC, Pepinsky B, Leone DR, Lobb RR, Adams SP (2000) A small-molecule, tight-binding inhibitor of the integrin alpha(4)beta(1) blocks antigen-induced airway responses and inflammation in experimental asthma in sheep. Am J Respir Crit Care Med 162(2 Pt 1):603–611

Almokadem S, Belani CP (2012) Volociximab in cancer. Expert Opin Biol Ther 12: 251–257

Becker PS, Nilsson SK, Li Z, Berrios VM, Dooner MS, Cooper C, Hsieh, CC, Quesenberry PJ (1999) Adhesion receptor expression by hematopoietic cell lines and murine progenitors: modulation by cytokines and cell cycle status. Exp Hematol 27:533–541

Becker PS, Appelbaum FR, Chien S, Zhao X, Bonig H, Mukasa R, Papayannopoulou T (2008) Oral small molecule inhibitor of VLA-4 overcomes adhesion mediated chemotherapy resistance of acute myeloid leukemia (AML) blasts in vitro, without impairment of normal blood cell recovery when combined with chemotherapy in vivo. Blood (Abstr) 112:858

Becker PS, Kopecky KJ, Wilks AN, Chien S, Harlan JM, Willman CL, Petersdorf SH, Stirewalt DL, Papayannopoulou T, Appelbaum FR (2009) Very late antigen-4 function of myeloblasts correlates with improved overall survival for patients with acute myeloid leukemia. Blood 113:866–874

Bonig H, Wundes A, Chang KH, Lucas S, Papayannopoulou T (2008) Increased numbers of circulating hematopoietic stem/progenitor cells are chronically maintained in patients treated with the CD49d blocking antibody natalizumab. Blood 111:3439–3441

Bonig H, Watts KL, Chang K-H, Kiem H-P, Papayannopoulou T (2009) Concurrent blockade of α4-integrin and CXCR4 in hematopoietic stem/progenitor cell mobilization. Stem Cells 27:836–837

Bryant J, Ahern DJ, Brennan FW (2012) CXCR4/SDF-1 and VLA-4/VCAM-1 are key chemokine/adhesion molecules in the migration of cytokine-activated T cells. Arth Rheum 64(7):2137-2146 doi:10.1002/art.34394

Burger JA, Spoo A, Dwenger A, Burger M, Behringer D. CXCR4 (2003) Chemokine receptors (CD184) and alpha4beta1 integrins mediate spontaneous migration of human CD34_ progenitors and acute myeloid leukaemia cells beneath marrow stromal cells (pseudoemperipolesis) Br J Haematol 122:579–589

Chien S, Basu S, Zhao X, Becker PS (2011) Adhesion receptor expression by leukemia stem cells in acute myeloid leukemia: relevance to interaction with the bone marrow stromal microenvironment. Exp Hematol 39 (Suppl 1):S47.

Craddock CF, Nakamoto B, Andrews RG, Priestley GV, Papayannopoulou T (1997) Antibodies to VLA-4 integrin mobilize long-term repopulating cells and augment cytokine-induced mobilization in primates and mice. Blood 90:4779–4788

De Toni F, Racaud-Sultan C, Chicanne G, Mas VM, Cariven C, Mesange F, Salles JP, Demur C, Allouche M, Payrastre B, Manenti S, Ysebaert L (2006) A crosstalk between the Wnt and the adhesion-dependent signaling pathways governs the chemosensitivity of acute myeloid leukemia. Oncogene 25:3113–3122

Ecoiffier T, El Annan J, Rashid S, Schaumberg D, Dana R (2008) Modulation of integrin α4β1 (VLA-4) in dry eye disease. Arch Ophthalmol 126:1695–1699

Feagan B, McDonald J, Greenberg G. (2001) An ascending dose trial of a humanized alpha4beta7 antibody in ulcerative colitis (UC) Gastroenterol 118:A874

Gordon FH, Lai CW, Hamilton MI, Allison MC, Srivastava ED, Fouweather MG et al (2001) A randomized placebo controlled trial of a humanized monoclonal antibody to alpha4 integrin in active Crohn's disease. Gastroenterol 121:268–274

Hazlehurst LA, Argilagos RF, Dalton WS (2006) β1 integrin mediated adhesion increases Bim protein degradation and contributes to drug resistance in leukaemia cells. Br J Haematol 136:269–275

Hazlehurst LA, Valkov N, Wisner L, Storey JA, Boulware D, Sullivan DM, Dalton WS (2001) Reduction in drug-induced DNA double-strand breaks associated with beta1 integrin-mediated adhesion correlates with drug resistance in U937 cells. Blood 98:1897–1903

Hersey P, Sosman J, O'Day S, Richards J, Bedikian A, Gonzalez R, Sharfman W, Weber R, Logan T, Buzoianu M, Hammershaimb L, Kirkwood JM, Etaracizumab Melanoma Study Group (2010) A randomized phase 2 study of etaracizumab, a monoclonal antibody against integrin alpha(v)beta(3), + or - dacarbazine in patients with stage IV metastatic melanoma. Cancer 116:1526–1534

Kleinschmidt-DeMasters BK, Tyler KL (2005) Progressive multifocal leukoencephalopathy complicating treatment with natalizumab and interferon beta-1a for multiple sclerosis. N Engl J Med 353:369–374

Langer-Gould A, Atlas SW, Green AJ, Bollen AW, Pelletier D (2005) Progressive multifocal leukoencephalopathy in a patient treated with natalizumab. N Engl J Med 353:375–381

Li G, Zhang L, Chen E, Wang J, Jiang X, Chen JH, Wickman G, Amundson K, Bergqvist S, Zobel J, Buckman D, Baxi SM, Bender SL, Casperson GF, Hu-Lowe DD (2010) Dual functional monoclonal antibody PF-04605412 targets integrin alpha5beta1 and elicits potent antibody-dependent cellular cytotoxicity. Cancer Res 70:10243–10254

Mahlknecht U, Schönbein C (2008) Histone deacetylase inhibitor treatment downregulates VLA-4 adhesion in hematopoietic stem cells and acute myeloid leukemia blast cells. Haematologica 93:443–446

Matsunaga T, Takemoto N, Sato T, Takimoto R, Tanaka I, Fujimi A, Akiyama T, Kuroda H, Kawano Y, Kobune M, Kato J, Hirayama Y, Sakamaki S, Kohda K, Miyake K, Niitsu Y (2003) Interaction between leukemic-cell VLA-4 and stromal fibronectin is a decisive factor for minimal residual disease of acute myelogenous leukemia. Nat Med 9:1158–1165

Matsunaga T, Fukai F, Miura S, Nakane Y, Owaki T, Kodama H, Tanaka M, Nagaya T, Takimoto R, Takayama T, Niitsu Y (2008) Combination therapy of an anticancer drug with the FNIII14 peptide of fibronectin effectively overcomes cell adhesion-mediated drug resistance of acute myelogenous leukemia. Leukemia 22:353–360

Miller DH, Khan OA, Sheremata WA, Blumhardt LD, Rice GP, Libonati MA et al (2003) A controlled trial of natalizumab for relapsing multiple sclerosis. N Engl J Med 348:15–23

Miller DH, Weber T, Grove R, Wardell C, Horrigan J, Graff O, Atkinson G, Dua P, Yousry T, Macmanus D, Montalban X (2012) Firategrast for relapsing remitting multiple sclerosis: a phase 2, randomised, double-blind, placebo-controlled trial. Lancet Neurol 11:131–139

Papayannopoulou T, Craddock C, Nakamoto B, Priestley GV, Wolf NS (1995) The VLA4/VCAM-1 adhesion pathway defines contrasting mechanisms of lodgement of transplanted murine hemopoietic progenitors between bone marrow and spleen. Proc Natl Acad Sci USA 92: 9647–9651

Peled A, Kollet O, Ponomaryov T, Petit I, Franitza S, Grabovsky V, Slav MM, Nagler A, Lider O, Alon R, Zipori D, Lapidot T (2000) The chemokine SDF-1 activates the integrins LFA-1, VLA-4, and VLA-5 on immature human CD34(+) cells: role in transendothelial/stromal migration and engraftment of NOD/SCID mice. Blood 95:3289–3296

Podar K, Zimmerhackl A, Fulciniti M, Tonon G, Hainz U, Tai YT, Vallet S, Halama N, Jäger D, Olson DL, Sattler M, Chauhan D, Anderson KC (2011) The selective adhesion molecule inhibitor Natalizumab decreases multiple myeloma cell growth in the bone marrow microenvironment: therapeutic implications. Br J Haematol 155:438–448

Ramirez P, Rettig MP, Uy GL, Deych E, Holt MS, Ritchey JK, DiPersio JF (2009) BIO5192, a small molecule inhibitor of VLA-4, mobilizes hematopoietic stem and progenitor cells. Blood 114:1340–1343

Ricart AD, Tolcher AW, Liu G, Holen K, Schwartz G, Albertini M, Weiss G, Yazji S, Ng C, Wilding G (2008) Volociximab, a chimeric monoclonal antibody that specifically binds alpha5beta1 integrin: a phase I, pharmacokinetic, and biological correlative study. Clin Cancer Res 14(23):7924–7929

Setoguchi M, Iimura S, Sugimoto Y, Yoneda Y, Chiba J, Watanabe T, Muro F, Iigo Y, Takayama G, Yokoyama M, Taira T, Aonuma M, takashi T, Nakayama A, Machinaga N (2012) Identification of *trans*-4-[1-[[7-fluoro-2-(1-methyl-3-indolyl)-6-benzoxazolyl]acetyl]-(4*S*)-fluoro-(2*S*)-pyrrolidinylmethoxy]cyclohexanecarboxylic acid as a potent, orally active VLA-4 antagonist. Bioorganic Med Chem 20:1201–1212

Tabe Y, Jin L, Tsutsumi-Ishii Y, Xu Y, McQueen T, Priebe W, Mills GB, Ohsaka A, Nagaoka I, Andreeff M, Konopleva M (2007) Activation of integrin-linked kinase is a critical prosurvival pathway induced in leukemic cells by bone marrow-derived stromal cells. Cancer Res 67:684–694

Theien BE, Vanderlugt CL, Nickerson-Nutter C, ornebise M, Scott DM, Perper SJ, Whalley, ET, Miller SD (2003) Differential effects of treatment with a small-molecule VLA-4 antagonist before and after onset of relapsing EAE. Blood 102:4464–4471

Van Assche G, Van Ranst M, Sciot R, Dubois B, Vermeire S, Noman M et al (2005) Progressive multifocal leukoencephalopathy after natalizumab therapy for Crohn's disease. N Engl J Med 353:362–368

Walter RB, Alonzo TA, Gerbing RB, Ho PA, Smith FO, Raimondi SC, Hirsch BA, Gamis AS, Franklin JL, Hurwitz CA, Loken MR, Meshinchi S (2010) High expression of the very late antigen-4 integrin independently predicts reduced risk of relapse and improved outcome in pediatric acute myeloid leukemia: a report from the children's oncology group. J Clin Oncol 28:2831–2838

Zohren F, Toutzaris D, Klarner V, Hartung HP, Kieseier B, Haas R (2008) The monoclonal anti-VLA-4 antibody natalizumab mobilizes CD34+ hematopoietic progenitor cells in humans. Blood 111:3893–3895

Chapter 34
VLA4 in Acute Lymphoblastic Leukemia

Halvard Bönig and Yong-Mi Kim

Abstract Chemotherapy resistance and off-target long-term toxicity of the cytoreductive drugs remain significant problems of current therapies for pre-B cell acute lymphoblastic leukemia (ALL). It was shown that chemotherapy resistance is promoted in part by leukemia cell-extrinsic host-derived signals, i.e., by signals emanating from the bone marrow environment. Detailed studies using function-blocking antibodies and genetic models have identified the $\alpha_4\beta_1$ integrin, very-late-acting integrin-4 (VLA4), as a critical molecule regulating bone marrow homing and active retention of normal and leukemia cells. Preclinical evidence has been provided that interference with VLA4-mediated adhesion of ALL cells can thus sensitize the cells to chemotherapy. If these data can be corroborated in clinical trials, targeting VLA4 could improve cure rates for chemotherapy-resistant ALL, as well as it might allow for reduction of chemotherapy doses for standard-risk leukemias, thus attenuating acute and long-term sequelae of off-target toxicity.

Keywords VLA4 · Adhesion · Drug resistance · Acute lymphoblastic leukemia

34.1 Introduction

In the treatment of pediatric acute lymphoblastic leukemia (ALL), much progress has been made over the last decades. With current chemotherapy regimen, hematological remissions are achieved in >95 % of cases. This translates to an overall

Y.-M. Kim (✉)
Department of Pediatrics, Division of Hematology and Oncology,
Children's Hospital Los Angeles, University of Southern California,
4650 Sunset Boulevard, MS#57, Los Angeles, CA 90027, USA
e-mail: YmKim@chla.usc.edu

H. Bönig
University of Washington, Seattle, WA, USA
e-mail: hbonig@uw.edu; H.Boenig@blutspende.de

Institute for Transfusion Medicine and Immunohematology, Department of Cellular Therapeutics/Cell Processing, Goethe University School of Medicine
and German Red Cross Blood Service Baden-Württemberg-Hesse,
Sandhofstrasse 1, 60528 Frankfurt am Main, Germany

M. Andreeff (ed.), *Targeted Therapy of Acute Myeloid Leukemia*,
Current Cancer Research, DOI 10.1007/978-1-4939-1393-0_34

survival of children with ALL of 75–85 % (Pui et al. 2009, 2011;Salzer et al. 2010), albeit with sometimes severe long-term side effects in many survivors (Pui und Evans 2006). For children with ALL who relapse, therapeutic options remain limited and 50–95 % will succumb to their illness (Gaynon et al. 1998; 2000). Overall survival rates in adults are much poorer and have reached approximately 40 % (Larson und Stock 2008; Pui und Jeha 2007; Faderl et al. 2010). Residual leukemia cells, which manage to evade the toxicity of cytoreductive chemotherapy, are routinely monitored throughout the therapy and during follow-up. Minimal residual disease (MRD), a term reserved to describe a status where residual leukemia cell burden is so low that it can only be detected with sensitive multi-parametric flow cytometry or molecular methods (Campana 2009; Cave et al. 1998), portends an unfavorable prognosis and, in fact, MRD positivity at the end of induction is one of the most discriminating parameters predicting long-term outcome of ALL patients (Campana 2012). Recently, new markers for MRD have been proposed, which could increase the sensitivity of MRD detection by flow cytometry (Coustan-Smith et al. 2011) or molecular methods (Kang et al. 2010) so that identification of those patients at high risk of relapse may improve. Unfortunately, innovative therapeutic interventions that can be offered to these patients are lacking. The elimination of relapse-initiating ALL cells thus remains a significant challenge and novel approaches for targeting of residual leukemia cells are warranted (Konopleva et al. 2009).

Anoikis, Greek for “homelessness,” describes a principle by which cells which after de-adhesion are deprived of direct cell contact and undergo spontaneous apoptosis (Frisch und Ruoslahti 1997; Frisch und Screaton 2001; Meredith et al. 1993; Ruoslahti und Reed 1994). For a number of hematologic malignancies, it was shown that *anoikis* can be harnessed to sensitize resistant malignant cells to chemotherapy and the first clinical data are now coming forth (Uy et al. 2012).

The promise of de-adhesion mediated chemosensitization is significant. If targets that are ideally specific for the malignant cells or at least expressed at higher expression levels than on the nonmalignant counterparts can be identified, one might manage to limit chemotoxicity only or predominantly to the malignant cells. Any excess off-target toxicity achieved by de-adhesion drugs should be limited to cells expressing the targeted adhesion molecule. In that fashion, chemosensitization by de-adhesion might not only benefit patients with resistant disease, but also facilitate treatment of standard risk disease in patients who cannot tolerate standard dose chemotherapy, and ultimately may achieve dose reductions for all standard risk patients with the added benefit of reduced off-target toxicity and long-term sequelae of chemotherapy.

Molecular targets for leukemia cell de-adhesion from stroma cells include chemokines/chemokine receptors, which can activate integrin avidity and tethering through inside-out signaling, integrins themselves or their ligands and selectins or their ligands. To this end, a considerable body of data has accumulated for the CXC chemokine receptor 4 (CXCR4) (Sipkins et al. 2005; Zeng et al. 2006; Zeng et al. 2009; Bonig et al. 2009). Further targets including CD44 (Krause et al. 2006; Jin et al. 2006) and very-late-acting integrin-4 (VLA4) (Manabe et al.1994; Matsunaga et al. 2003) have been described but explored less vigorously. Clinical trials with

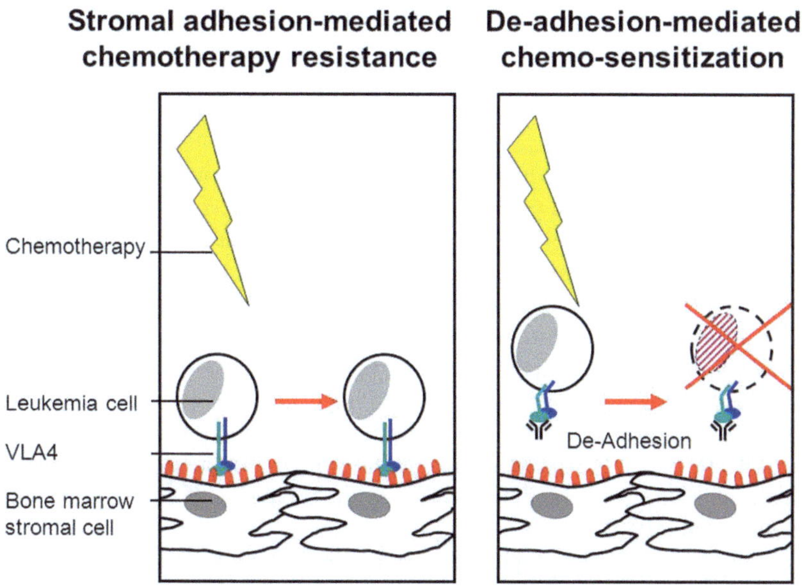

Fig. 34.1 Schematic model of VLA4-targeted therapy in leukemia. *Left panel* Adhesion of leukemia cells to stromal cells mediates drug resistance. A critical adhesion molecule in cell adhesion-mediated chemoprotection is VLA4. *Right panel* Interrupting VLA4-mediated adhesion of leukemia cells to chemoprotective stromal cells mediates chemosensitization of leukemia cells

therapies targeting VLA4 integrin could go forth immediately, since clinical-grade antifunctional humanized antibodies against VLA4 licensed for use in auto-immune disease exist (Bonig et al. 2008; Polman et al. 2006). Therefore, we review here VLA4 integrin as an important molecule in adhesion and drug resistance of pre-B-ALL cells and propose interference with VLA4 as a putative target for the treatment of ALL (Fig. 34.1).

34.2 VLA4 in normal HSC and in ALL

34.2.1 VLA4 in Normal HSC

Integrins are a family of 16 heterodimers sharing significant structural and functional commonalities. They consist of an α- and a β-chain, which are noncovalently bound. The α-chain predominantly defines the ligand specificity of the integrin (Hynes 2002). Conformational changes induced by approximation of the transmembrane regions of the integrin open the ligand-binding pocket. Ligand binding further affects integrin conformation, stretching the heterodimer from a bent to an extended conformation (Chigaev et al. 2004). In this conformation, integrins elicit intracellular signals through a variety of cell-type dependent pathways, which include survival/apoptosis, cell cycle, metabolism, among others (Boudreau und Jones 1999; Parsons 1996). Integrins thus mediate physical adhesion to suitable matrices/ligands while at the same time providing outside-in signals regulating cell fate.

Evidence abounds supporting prominent roles of integrins in the interaction of hematopoietic cells with the bone marrow microenvironment. β1 integrins, a member of which is VLA4, are expressed on a variety of cells besides hematopoietic cells. On hematopoietic cells, the integrin expression pattern changes during differentiation. We and others provided evidence that VLA4 blockade or ablation resulted in significant mobilization of stem and progenitor cells (Bonig et al. 2008; Scott et al. 2003). Mobilization of immature cells was relatively more enhanced than mobilization of more mature myeloid precursors, despite similar integrin surface expression, likely because of an altered integrin hierarchy and altered integrin activation status. With respect to the relevant stromal ligand, vascular cell adhesion molecule 1 (VCAM-1) was identified as physiologically relevant, whereas interference with fibronectin binding did not result in stem cell egress from marrow (Craddock et al. 1997).

A small number of hematopoietic stem/progenitor cells circulate in blood at all times. Their number is augmented during hematopoietic stress (i.e., during infection or inflammation) after chemotherapy or irradiation, and further increased during enforced mobilization with pharmacological mobilizing agents. A certain need-adapted leakiness of stem cell release is probably adaptive, but how the bone marrow manages to retain the bulk of stem cells remains only rudimentarily understood. Our improved understanding of the molecular targets of pharmacological stem cell mobilization has, by extrapolation, served to define factors involved in stem cell retention. Among these is the alpha4 integrin, VLA4 (Bonig et al. 2008, Bonig und Papayannopoulou 2004). Excellent reviews looking at stem cell retention and release provide a more comprehensive picture (Ratajczak und Kim 2012; Ehninger und Trumpp 2011; Hoggatt und Pelus 2011; Korbling und Freireich 2011). The focus of this review is to provide a brief overview of how VLA4 figures into stem cell retention and release, and to set the scene for a review of some current ideas about how VLA4 can be harnessed to treat ALL.

34.2.2 VLA4 Expression on ALL Cells

ALL cells express a considerable panel of integrins, importantly including several members of the β1 class of "very-late-acting" integrins. Homing and engraftment of normal hematopoietic stem cell (HSC) have been shown to be dependent on VLA4 (α4β1 or CD49d/CD29) integrin (Scott et al. 2003). Similarly, VLA4 mediates homing of lymphocytes to sites of inflammation (Lobb et al. 1996). These data were deemed meaningful for leukemia, since ALL cells share properties of both HSCs and lymphocytes. A role of VLA4 has been shown for MRD in acute myeloid leukemia (AML) (Matsunaga et al. 2003). However, whether VLA4 expression levels portend a better or worse prognosis in AML remains unclear since diametrically opposing data have been published (Matsunaga et al. 2003, Becker et al. 2009).

VLA4 is expressed on pre-B-ALL cells (Hara et al. 2000; Liesveld et al. 1993). Moreover, in a recent study, α_4 integrin expression was quantified by real-time polymerase chain reaction in leukemia cells from 56 patients with relapsed ALL enrolled in the ALL-REZ BFM 2002 trial of the Berlin–Frankfurt–Münster study group (Shalapour et al. 2011), and high alpha4 integrin mRNA expression was identified as an adverse risk factor in childhood ALL at first relapse. As gene expression may not necessarily predict protein expression, and particularly cannot determine whether a gene is expressed in all or only a fraction of the cell population studied, VLA4 expression was also studied in a panel of pre-B-ALL blasts. VLA4 expression was highly homogeneous, at least as high as on normal pre-B cells, and VLA4 negative ALL blasts were never observed (unpublished). Using in vivo adhesion assays, both static and under flow, on isolated matrix proteins and on bone marrow stroma cells, we confirmed that not only is VLA4 expressed but is also functional. Using antifunctional antibodies against VLA4, we could in fact demonstrate that VLA4 is a dominant adhesion molecule, since antifunctional VLA4 antibody blocked adhesion not only to defined, VLA4-specific matrix proteins but also to stroma cells which express ligands for a variety of integrins, including VLA5 and VLA6.

34.3 VLA4 Ligands for ALL Cells

VLA4 binds VCAM-1, a cell-surface protein expressed on activated vascular endothelium and a host of other cells, as well as several extracellular matrix (ECM) proteins, including osteopontin (OPN) and fibronectin (Hynes 2002; Juneja et al. 1992). These ligands are briefly introduced here:

VCAM-1 is a type I membrane protein encoded by the VCAM1 gene and expressed in two different splice isoforms (Kumar et al. 1994; Ryan et al. 1991). As the name suggests, it is expressed mainly on the surface of vascular endothelial cells throughout the vascular tree (Elices et al. 1990). VCAM-1 expression on blood vessels is induced after cytokine stimulation, which is regulated either by increasing VCAM1 mRNA expression or mRNA stabilization reported (Iademarco et al. 1995). However, even at steady state, VCAM-1-deficient mice have a distinct hematopoietic phenotype with increased numbers of circulating stem/progenitor cells as well as mature blood cells, and diffuse expression of VCAM-1 in the bone marrow, notably on bone marrow fibroblasts, has been reported (Papayannopoulou et al. 1995; Ulyanova et al. 2007). These observations may indicate a distinct regulation of VCAM1 in bone-marrow and on hematopoietic cells compared to other organs, but more data are clearly needed. An alternative receptor for VCAM-1 to VLA4 is MadCAM, the alpha4 beta7 integrin heterodimer (Katayama et al. 2004).

Fibronectin, a large glycoprotein dimer, is an integral component of ECM in all connective tissues including bone marrow (Lobb und Hemler 1994). It binds primarily other matrix proteins, such as collagens and heparan sulfates, but contains moieties for VLA4 binding (within the V-domain) as well as for VLA5 and the platelet integrin alphaV-beta-3 (the RGD domain) (Garcia-Pardo et al. 1990). Due to its critical role during all steps of embryogenesis, the specific role of fibronectin for immature and mature hematopoiesis could not be separately analyzed. Experiments with molecules antagonizing fibronectin versus VCAM-1 binding, however, did not detect apparent effects of fibronectin-blockade on the trafficking of immature hematopoietic cells, whereas VCAM-1 inhibition interfered quantitatively with bone between hematopoietic stem/progenitor cells and bone marrow (Craddock et al. 1997).

OPN is a negatively charged ECM glycoprotein, expressed in many tissues besides bone, even though bone is where it is most abundantly reported (Anborgh et al. 2011). It has been described as a ligand for VLA4 (Bayless et al. 1998; Bogetto et al. 2000), but alternative receptors include the alpha9 integrins (Yokosaki et al. 1994). Like VCAM-1, OPN expression is also enhanced by inflammatory cytokines. Besides its role as an anchorage protein specifically for osteoblasts and osteoclasts, as well as for hematopoietic cells, including lymphocytes, OPN is also a signaling molecule, mediating migration, antiapoptosis and survival, among others. A role for OPN as a negative regulator of HSC proliferation and a mediator of HSC localization within bone marrow has been proposed (Nilsson et al. 2005; Stier et al. 2005). Besides functioning as an ECM molecule, OPN apparently also exists as an intracellular protein with roles in signaling. Its precise role for hematopoiesis has therefore been difficult to pinpoint.

All three VLA4 ligands, fibronectin, VCAM-1, and OPN, are highly abundant in bone marrow as well as in many other tissues. Which of them is most critical for leukemia cell attachment remains elusive.

34.4 Role of VLA4 as a Chemoprotectant for ALL Cells

After cytoreductive chemotherapy of pre-B ALL patients, residual leukemia cells are predominantly found in the bone marrow (Gaynon et al. 1998; Eckert et al. 2001; Goulden et al. 1994; Miniero et al. 1995; Tallen et al. 2010). Therefore, the bone marrow environment has been regarded as a shelter for drug-resistant leukemia cells; from which leukemia relapse originate (Bradstock und Gottlieb 1995). Critically to our hypothesis for VLA4 targeting in ALL treatment, VLA4 mediates adhesion of ALL cells (Filshie et al. 1998) and β1 integrin-mediated cell adhesion to stromal cells has been shown to lead to *cell adhesion-mediated drug resistance (CAM-DR)* of AML (Hazlehurst et al. 2007).

We propose that, based on accumulated knowledge about integrins, stroma, and ECM, at least four hypothetical mechanisms for stroma-cell-derived pro-survival signals can be envisioned. One option for the underlying mechanism could be attributed to inhibition of apoptosis through outside-in signals emanating from stromal integrin ligands. Several groups have shown that adhesion to bone marrow (BM) stromal cells protects pre-B ALL cell survival in vitro. This effect on leukemia cell survival and apoptosis prevention would require direct cell contact between leukemia and stroma cells (Bradstock und Gottlieb 1995; Bradstock et al. 1993; Hall et al. 2004; Hall und Gibson 2004; Miyake et al. 1992; Mudry et al. 2000; Shah et al. 1998). Alternatively, soluble factors produced by stroma cells could support leukemia cell survival, and integrins would anchor leukemia cells in environments rich in these soluble factors (Ashley et al. 1994). A third possibility would be that stroma cell-surface- or ECM-bound molecules which are not, however, adhesion ligands, generate survival signals, so that cell–cell contact would still be required for the pro-survival effect to occur, but that integrin-mediated adhesion solely serves to approximate leukemia cells to stroma. Finally, simple physical protection from chemotherapy by stromal matrix and stroma cells could also contribute, for it was shown that leukemia cells can, in an integrin-mediated fashion, migrate underneath or into stromal/matrix layers, where they would likely be exposed to lower local concentrations of chemotherapeutic agents. These four options are, of course, not mutually exclusive and any or all could contribute additively or synergistically.

While several studies have considered these possibilities in AML and other hematological malignancies (Gibson 2002; Hehlgans et al. 2007), a small number of papers have been published which can be used to begin to explore these different possibilities also in ALL. ALL cells have been described to migrate and adhere to bone marrow stromal cells (Makrynikola und Bradstock 1993; Makrynikola et al. 1994). Mudry et al. investigated co-culture systems of stroma cells or isolated stromal matrix proteins and three pre-B-ALL cell lines and one T-ALL cell line in vitro (Mudry et al. 2000). Stromal cell co-culture sustained leukemia cell proliferation even in the presence of ALL chemotherapies, cytarabine or etoposide, but only when ALL cells were in direct contact with the stroma cells. This effect was mediated through VCAM-1, whereas suspension of leukemia cells above stroma cells or culture on fibronectin was not sufficient to mediate chemotherapy protection.

In addition to bone marrow as a site of leukemia relapse, central nervous system (CNS) infiltration by leukemia cells is an important reservoir for therapy refractoriness. In similar studies as above, but with a selection of CNS stroma cells instead of bone marrow stroma cells, in vitro chemoprotection by CNS stroma cells was also demonstrated (Akers et al. 2011). The exact molecular mechanism of these observations remains to be clarified.

These observations are in agreement with a later publication by Astier and colleagues who showed that ß1 integrin stimulation inhibits caspase-3 and -7 induction in pre-B ALL cells (Astier et al. 2004). B cell lymphoma-2 (Bcl-2) is the antiapoptotic protein which was most exhaustively evaluated in studies aimed at understanding programmed cell death. Modulation of the biological activity of Bcl-2 by proteolytic cleavage, with a 23 kDa cleaved Bcl-2 product having pro-apoptotic activity, has been reported. Co-culture of three pre-B ALL cell lines with bone marrow stromal cells during chemotherapy exposure resulted in reduced levels of 23 kDa Bcl-2 protein indicating that the bone marrow microenvironment may contribute to the maintenance of residual leukemic disease during treatment by reducing generation of pro-apoptotic 23 kDa Bcl-2 (Fortney et al. 2002). Also to the topic of modulation of apoptosis mediators in chemotherapy-exposed, stroma-co-cultured ALL cells, Wang et al. assessed which molecular mediators of apoptosis were involved (Wang et al. 2004). They described reduced levels of poly ADP-ribose polymerase (PARP) and Bcl-2 cleavage products in ALL cells after chemotherapy when these were cultured on stroma, and provided evidence that this effect was Akt-mediated, since in leukemia cells with blunted Akt activity such stroma-induced attenuation of death pathways was not observed.

Whether different integrin ligands can elicit different outside-in signals is not clear but would seem possible. Fortney et al. reported that in ALL cells, caspase 3 activity was induced by cytarabine or etoposide (Fortney et al. 2001). This effect was blunted in ALL cells cultured with human bone marrow stroma cell layers. While this observation does not explain the origin of the survival signal (stroma-derived or leukemia cell-autonomous), it suggests that at the level of the leukemia cell survival may be due, at least in part, to modulation of the caspase 3 pathway.

The prolongation of AML cell survival has been shown to be dependent on interactions between stromal and AML cells, with a direct involvement of β1 integrins (Manabe et al. 1994). Adhesion of VLA-4 to fibronectin was shown to induce activation of the PI-3K/AKT pathway and upregulation of BCL2 in leukemia cells (de la Fuente et al. 2003), resulting in drug resistance. Similarly, for pre-B ALL, β1 integrin stimulation by fibronectin (Fn), a ligand for β1 integrin, reversed apoptosis is induced by doxorubicin in human pre-B leukemia cells (Sarkar et al. 2002). Stroma cells genetically engineered to overexpress VCAM-1 were more efficient at promoting leukemia cell survival than the parental stroma cell line, which expressed much lower levels of VCAM-1 (Hall et al. 2004; Hall und Gibson 2004). In these experiments, leukemia survival appeared to be mediated through the PI3 kinase pathway.

Several pathways in mediation of bone marrow stromal cell-mediated cell survival and chemoprotection of ALL cells have been proposed; however, the exact mechanism remains elusive:

Data supporting a role for PI3K/Akt in stroma-mediated protection during chemotherapy have been reported (Wang et al. 2004). Thus, ALL cells made deficient for Akt signaling by virtue of a dominant-negative Akt were not efficiently rescued from chemotherapy (cytarabine, etoposide) under stromal cell co-culture conditions. Similarly, blockade of the PI3 kinase pathway with specific inhibitors of PI3 kinase activity resulted in attenuated stroma-mediated survival effects on leukemia cells. The authors also showed that stromal co-culture maintained Akt-phosphorylation during chemotherapy and reduced PP2A activity relative to ALL cells treated with chemotherapy under stroma-free conditions. Comparing gene expression profiles of leukemia cells with high versus low VLA-4 expression, we identified 27 differentially expressed genes primarily involved in the PI3K/Akt, ephrin, and Rho GTPase pathways (Shalapour et al. 2011). Recently, Notch-3 and 4 signaling has been implicated in the bone-marrow-derived mesenchymal stromal cells (MSCs)–B-ALL cell interaction (Nwabo Kamdje und Krampera 2011).

Bertrand et al. showed in a stroma cell co-culture dependent B-ALL line BLIN-2 that concurrent inhibition of MEK and either PI3 K or mTOR results in rapid leukemia cell apoptosis (Bertrand et al. 2005). The same effect was achieved by blocking epidermal growth factor receptor (EGFR) (Spengeman et al. 2005), which is expressed by many immature hematopoietic cells (Johnston et al. 2006), on BLIN-2 cells co-cultured with stroma cells, and inversely, EGF treatment or engineering of cells to express a constitutively active EGFR variant was sufficient to partly maintain BLIN-2 cells without stroma. In aggregate, these data implicate the EGF/EGFR pathway as a survival pathway by which stroma cells support leukemia cells, at least in vitro.

Similar data were presented by Pilozzi and colleagues, who in a co-culture model containing bone marrow MSCs and ALL cells identified a role for human ether-à-go-go-related gene 1 (hERG1) in leukemia cell survival, in that hERG1 channels contribute to a signaling complex containing, in addition, β1 integrin and CXCR4 (Pillozzi et al. 2011). The assembly of (and likely, engagement of cognate ligands for) this protein complex triggers ERK and PI3 K signaling, with apparent consequences for cell survival. However, the effect seemed to be at least partially mediated through hERG1 channel function, since specific channel blockers interfered with pro-survival signals emanating from MSCs. Some benefit of these channel blockers in a xenograft model of ALL was also demonstrated. These data connect cell survival and ß1 integrin and may therefore partly explain how integrin blockade can interfere with leukemia cell survival.

34.5 Translational: Targeting VLA4 to Sensitize ALL Cells to Chemotherapy

Considering the accumulated evidence of an important role of VLA4 as an effector of stromal adhesion-mediated survival, it seemed reasonable to query its potential for targeting ALL cells for "*anoikis*" in vivo. Conveniently, VLA4 has previously been identified as a critical mediator of lymphocyte homing to inflammatory sites so that VLA4-targeting drugs have been and remain a subject of enormous pharmaceutical interest with a view to the treatment of autoimmune disease. For several years, an almost fully humanized antifunctional VLA4 antibody has been available in the clinic for treatment of relapsing–remitting multiple sclerosis as well as of complicated refractory inflammatory bowel disease (Polman et al. 2006; Kappos et al. 2011; Sandborn et al. 2005; MacDonald und McDonald 2007). Evidence has been provided that the antibody exerts its effect not by depleting T-cells, but by interfering with their adhesion to relevant matrices in vivo (Planas et al. 2012). The acute tolerability of the antibody is, for the most part, excellent (Pucci et al. 2011; Elices 1995). Long-term treatment has sometimes been associated with induction of neutralizing antibodies against the few remaining murine amino acid residues of the antibody, and with mild infusion-associated serum sickness (Fox et al. 2007; Oliver et al. 2011; Sorensen et al. 2011; Weber et al. 2012). JC-virus associated promyelocytic leukemia (PML), although grave, was infrequently observed and only after prolonged VLA4 blockade (Pucci et al. 2011; Clifford et al. 2010), and should thus not be expected in leukemia patients who would likely not receive more than one or two doses of antifunctional antibody. Long-term effects on hematopoiesis have been studied; elevated numbers of circulating stem/progenitor cells are sustained, without any evidence for hematopoietic exhaustion (Bonig et al. 2008). Leukocytosis is regularly observed and is stably maintained throughout treatment. General immunosuppression and proneness to infectious complications is not a hallmark of anti-VLA4 antibody treatment.

Alternative molecules targeting VLA4 have also been reported (Becker et al. 2008; Ramirez et al. 2009; Vanderslice et al. 2010; Hsieh et al. 2014; Shishido et al. 2014), although none are universally available for preclinical and clinical evaluation. The long half-life of VLA4 antibodies is considered an advantage by clinicians caring for patients with autoimmune disease, since once-monthly treatments suffice to maintain a clinical effect. By contrast, leukemia patients would probably benefit from drugs with very short half-lives—in the range of hours or days. The reported small molecule antagonists likely fulfill this expectation.

Evidence has been provided that leukemia blast survival is stroma cell dependent (see above). While it remains unclear whether the critical survival signals are mediated directly through outside-in signaling through adhesion receptors, or by approximating leukemia cells to secreted and ECM-bound survival-promoting factors like cytokines and chemokines, or even simply by leukemia cells migrating into ECM and thus hiding out in places less accessible to chemotherapy, the therapeutic principle of chemo-sensitization by de-adhesion has been well documented for a variety of malignant cells in vitro and in in vivo models, generally xenograft models (Troeger et al. 2012). In vitro, we had been able to sensitize chemotherapy-resistant ALL cells to chemotherapy by treatment of stroma cell adhering ALL cells with antifunctional VLA4 antibody. We, therefore, hypothesized that by perturbing VLA4-mediated adhesion in vivo, the interaction of resistant ALL cells with niches conducive to leukemia cell survival might be disrupted and ALL cells might become sensitive to chemotherapy.

To test whether ALL cell niche engagement in vivo was VLA4 dependent, we used two models of leukemia, one genetic (conditional VLA4 ablation of p210-induced murine leukemia) and one pharmacological (antifunctional VLA4 antibody treatment of xenografted human pre-B ALL) (Hsieh et al. 2013). Conditional ablation of VLA4 decreased adhesion and re-plating activity of murine BCR-ABL1 ($p210^+$) leukemia cells in vitro, and was associated with leukemia eradication in vivo when combined with tyrosine kinase inhibitors. Primary human ALL blasts were pretreated with antifunctional VLA4 antibodies and ALL cell seeding to bone marrow of low-dose irradiated non-obese diabetic (NOD)/severe combined immunodeficiency (SCID) recipients was tested. Much more so than normal immature hematopoietic cells, seeding (or homing) of ALL cells to bone marrow was blocked almost completely by VLA4-blockade, and ALL cells accumulated in spleen instead. As a result of this treatment, the survival of recipients of ALL cells was significantly prolonged, but eventual engraftment and leukemia progress was not prevented in any of the eight primary leukemias tested. These studies established VLA4 as a relevant receptor for niche engagement of ALL cells. To mimic a minimal residual disease situation, we next transplanted primary pre-B-ALL cells, allowed them to home and engraft for several days, and subsequently treated recipient mice with antifunctional VLA4 antibody, conventional ALL chemotherapy (VDL = vincristine, dexamethasone, L-asparaginase), or a combination thereof. Recipients of primary leukemia cells without any treatment served as controls. These mice developed a rapidly lethal leukemia, with an average time until death of slightly more than 1 month. Anti-VLA4 treatment alone (i.e., disruption of normal niche interaction alone) sufficed to modestly but already significantly prolong survival, although the mechanism remains elusive. A 4-week cycle of VDL chemotherapy induced a clinical remission in all mice. All mice eventually relapsed, and although survival time was almost doubled, compared to untreated mice, not a single mouse was cured. It was therefore most remarkable to observe that mice treated with the combination therapy of VLA4 antibody plus VDL chemotherapy achieved a sustained molecular remission for at least 150 days. At the time of sacrifice, we could detect no trace of residual leukemia cells, including by sensitive methods as are used for MRD diagnostics. We propose, although we can currently not prove this, that the mechanism by which the combination therapy managed to eliminate leukemia cells was indeed by disruption of interactions of leukemia cells with bone marrow stroma, resulting in chemo-sensitization. It has to be admitted that the therapy was only this successful when it was used early, 3 days after leukemia cell transfer. The fact that VLA4 antibody treated and chemotherapy-only treated mice rapidly succumbed to their illness clearly indicates that treatment at that point does not critically interfere with leukemia cell engraftment. If treatment was delayed for several days, then the combination treatment still managed to significantly delay relapse and death from leukemia, but no cures were observed. We therefore believe that an excessive number of leukemic blasts exceeds the therapeutic potential of *anoikis,* or in other words, that a minimal residual disease situation must be achieved in order for *anoikis* to be optimally successful.

Given the critical role of VLA4 for normal immature hematopoiesis, whether VLA4 blockade in combination with chemotherapy might not result in excessive toxicity also to normal hematopoietic cells must be addressed. As mentioned above, VLA4 blockade is tolerated over many years with apparent effects on mature hematopoiesis, although four- to six-fold increased numbers of circulating stem/progenitor cells are continuously maintained during treatment (Bonig et al. 2008). In VLA4-deficient mice, which have the same hematopoietic phenotype as chronically anti-VLA4-treated multiple sclerosis patients, we also had the opportunity to test the effect of chemotherapy. The VDL chemotherapy used for leukemia eradication in our xenotransplants was only very mildly myelotoxic, with no difference in regeneration between WT and VLA4-deficient mice. Even the much more toxic treatment with 5-fluorouracil was well tolerated by both strains, and only platelet regeneration was delayed by 1 day in VLA4 ablated versus WT mice. This observation should be reassuring with respect to the combination treatment in ALL patients.

Which, then, would be the logical treatment phases where anti-VLA4 antibody could be introduced into clinical protocols, and which patients would benefit the most? The vast majority of patients achieve a clinical remission after induction. Among them, however, those with less than a molecular remission are at significant risk for relapse and ultimately, treatment failure. As a first, we would propose to add VLA4 antibody to the consolidation chemotherapy in those patients.

Conclusion

Significant evidence has been provided using relevant in vitro as well as preclinical in vivo models that targeting of VLA4 can modify chemotherapy responsiveness of hematological malignancies, including ALL. The overall good tolerability of clinically available VLA4 targeting drugs by themselves has been demonstrated. Murine data did not demonstrate excessive hematopoietic toxicity from chemotherapy in the absence of VLA4, suggesting that the combination of anti-VLA4 AB and chemotherapy should also be safe and tolerable. Clinical trials can be initiated immediately. Whether standard-risk ALL can be successfully treated with anti-VLA4 AB and reduced-dose chemotherapy should be formally tested in suitable preclinical models. The expectation is, moreover, that other VLA4-positive high-risk leukemias can similarly benefit from addition of VLA4 targeted therapies. VLA4-targeting drugs with more favorable pharmacokinetics than a humanized antibody with a week-long half-life are likely even more suitable for leukemia treatment. To this end, significant progress has been made in the development of small-molecule VLA4 antagonists (Becker et al. 2008; Ramirez et al. 2009).

Acknowledgments HB acknowledges funding by the German Federal Ministry for Education and Research (BMBF, Ci3), the Federal State of Hesse (LOEWE CGT and LOEWE OSF), Deutsche Krebshilfe and DFG. YK acknowledges funding by the National Institute of Health (R01 CA172896) and the St. Baldrick's Foundation.

References

Akers SM, Rellick SL, Fortney JE, Gibson LF (2011) Cellular elements of the subarachnoid space promote ALL survival during chemotherapy. Leuk Res 35:705–711

Anborgh PH, Mutrie JC, Tuck AB, Chambers AF (2011) Pre- and post-translational regulation of osteopontin in cancer. J Cell Commun Signal 5:111–122

Ashley DM, Bol SJ, Kannourakis G (1994) Human bone marrow stromal cell contact and soluble factors have different effects on the survival and proliferation of paediatric B-lineage acute lymphoblastic leukaemic blasts. Leuk Res 18:337–346

Astier AL, Svoboda M, Hinds E, De BR, Munoz O, Freedman AS (2004) Integrins regulate survival of pre-B-ALL cells through differential IAP and caspase-7 ubiquitination and degradation. Leukemia 18:873–875

Bayless KJ, Meininger GA, Scholtz JM, Davis GE (1998) Osteopontin is a ligand for the alpha-4beta1 integrin. J Cell Sci 111:1165–1174

Becker PS, Appelbaum FR, Chien S, Zhao X, Bonig H., Mukasa R, Papayannopoulou T (2008) Oral Small Molecule Inhibitor of VLA-4 Overcomes Adhesion Mediated Chemotherapy

Resistance of Acute Myeloid Leukemia (AML) Blasts in Vitro, without Impairment of Normal Blood Cell Recovery When Combined with Chemotherapy in Vivo. Blood 112a:858

Becker PS, Kopecky KJ, Wilks AN, Chien S, Harlan JM, Willman CL, Petersdorf SH, Stirewalt DL, Papayannopoulou T, Appelbaum FR (2009) Very late antigen-4 function of myeloblasts correlates with improved overall survival for patients with acute myeloid leukemia. Blood 113:866–874

Bertrand FE, Spengemen JD, Shelton JG, McCubrey JA (2005) Inhibition of PI3 K, mTOR and MEK signaling pathways promotes rapid apoptosis in B-lineage ALL in the presence of stromal cell support. Leukemia 19:98–102

Bogetto L, Gabriele E, Cariati R, Dolcetti R, Spessotto P, Doglioni C, Boiocchi M, Perris R, Colombatti A (2000) Bidirectional induction of the cognate receptor-ligand alpha4/VCAM-1 pair defines a novel mechanism of tumor intravasation. Blood 95:2397–2406

Bonig H, Papayannopoulou T (2004) VLA-4: An updated account of its role in hemopoietic stem/progenitor cell trafficking. In: Keller J (ed) Pluripotent hematopoietic stem cells. Landes Bioscience Publishing, Austin, 1–10

Bonig H, Wundes A, Chang KH, Lucas S, Papayannopoulou T (2008) Increased numbers of circulating hematopoietic stem/progenitor cells are chronically maintained in patients treated with the CD49d blocking antibody natalizumab. Blood 111:3439–3441

Bonig H, Chudziak D, Priestley G, Papayannopoulou T (2009) Insights into the biology of mobilized hematopoietic stem/progenitor cells through innovative treatment schedules of the CXCR4 antagonist AMD3100. Exp Hematol 37:402–415

Boudreau NJ, Jones PL (1999) Extracellular matrix and integrin signalling: the shape of things to come. Biochem J 339:481–488

Bradstock KF, Gottlieb DJ (1995) Interaction of acute leukemia cells with the bone marrow microenvironment: implications for control of minimal residual disease. Leuk Lymphoma 18:1–16

Bradstock K, Makrynikola V, Bianchi A, Byth K (1993) Analysis of the mechanism of adhesion of precursor-B acute lymphoblastic leukemia cells to bone marrow fibroblasts. Blood 82:3437–3444

Campana D (2009) Role of minimal residual disease monitoring in adult and pediatric acute lymphoblastic leukemia. Hematol Oncol Clin North Am 23:1083–1098, vii

Campana D (2012) Should minimal residual disease monitoring in acute lymphoblastic leukemia be standard of care? Curr Hematol Malig Rep 7:170–177

Cave H, van der Werff ten Bosch, Suciu S, Guidal C, Waterkeyn C, Otten J, Bakkus M, Thielemans K, Grandchamp B, Vilmer E (1998) Clinical significance of minimal residual disease in childhood acute lymphoblastic leukemia. European Organization for Research and Treatment of Cancer—Childhood Leukemia Cooperative Group. N Engl J Med 339:591–598

Chigaev A, Zwartz GJ, Buranda T, Edwards BS, Prossnitz ER, Sklar LA (2004) Conformational regulation of alpha 4 beta 1-integrin affinity by reducing agents. "Inside-out" signaling is independent of and additive to reduction-regulated integrin activation. J Biol Chem 279:32435–32443

Clifford DB, De LA, Simpson DM, Arendt G, Giovannoni G, Nath A (2010) Natalizumab-associated progressive multifocal leukoencephalopathy in patients with multiple sclerosis: lessons from 28 cases. Lancet Neurol 9:438–446

Coustan-Smith E, Song G, Clark C, Key L, Liu P, Mehrpooya M, Stow P, Su X, Shurtleff S, Pui CH, Downing JR, Campana D (2011) New markers for minimal residual disease detection in acute lymphoblastic leukemia. Blood 117:6267–6276

Craddock CF, Nakamoto B, Elices M, Papayannopoulou T (1997) The role of CS1 moiety of fibronectin in VLA mediated haemopoietic progenitor trafficking. Br J Haematol 97:15–21

de la Fuente MT, Casanova B, Cantero E, Hernandez del CM, Garcia-Marco J, Silva A, Garcia-Pardo A (2003) Involvement of p53 in alpha4beta1 integrin-mediated resistance of B-CLL cells to fludarabine. Biochem Biophys Res Commun 311:708–712

Eckert C, Biondi A, Seeger K, Cazzaniga G, Hartmann R, Beyermann B, Pogodda M, Proba J, Henze G (2001) Prognostic value of minimal residual disease in relapsed childhood acute lymphoblastic leukaemia. Lancet 358:1239–1241

Ehninger A, Trumpp A (2011) The bone marrow stem cell niche grows up: mesenchymal stem cells and macrophages move in. J Exp Med 208:421–428
Elices MJ (1995) The integrin alpha 4 beta 1 (VLA-4) as a therapeutic target. Ciba Found Symp 189:79–85
Elices MJ, Osborn L, Takada Y, Crouse C, Luhowskyj S, Hemler ME, Lobb RR (1990) VCAM-1 on activated endothelium interacts with the leukocyte integrin VLA-4 at a site distinct from the VLA-4/fibronectin binding site. Cell 60:577–584
Faderl S, O'Brien S, Pui CH, Stock W, Wetzler M, Hoelzer D, Kantarjian HM (2010) Adult acute lymphoblastic leukemia: concepts and strategies. Cancer 116:1165–1176
Filshie R, Gottlieb D, Bradstock K (1998) VLA-4 is involved in the engraftment of the human pre-B acute lymphoblastic leukaemia cell line NALM-6 in SCID mice. Br J Haematol 102:1292–1300
Fortney JE, Zhao W, Wenger SL, Gibson LF (2001) Bone marrow stromal cells regulate caspase 3 activity in leukemic cells during chemotherapy. Leuk Res 25:901–907
Fortney JE, Hall BM, Bartrug L, Gibson LF (2002) Chemotherapy induces bcl-2 cleavage in lymphoid leukemic cell lines. Leuk Lymphoma 43:2171–2178
Fox EJ, Vartanian TK, Zamvil SS (2007) The immunogenicity of disease-modifying therapies for multiple sclerosis: clinical implications for neurologists. Neurologist 13:355–362
Frisch SM, Ruoslahti E (1997) Integrins and anoikis. Curr Opin Cell Biol 9:701–706
Frisch SM, Screaton RA (2001) Anoikis mechanisms. Curr Opin Cell Biol 13:555–562
Garcia-Pardo A, Wayner EA, Carter WG, Ferreira OC, Jr. (1990) Human B lymphocytes define an alternative mechanism of adhesion to fibronectin. The interaction of the alpha 4 beta 1 integrin with the LHGPEILDVPST sequence of the type III connecting segment is sufficient to promote cell attachment. J Immunol 144:3361–3366
Gaynon PS, Qu RP, Chappell RJ, Willoughby ML, Tubergen DG, Steinherz PG, Trigg ME (1998) Survival after relapse in childhood acute lymphoblastic leukemia: impact of site and time to first relapse—the Children's Cancer Group Experience. Cancer 82:1387–1395
Gaynon PS, Trigg ME, Heerema NA, Sensel MG, Sather HN, Hammond GD, Bleyer WA (2000) Children's Cancer Group trials in childhood acute lymphoblastic leukemia: 1983–1995. Leuk 14:2223–2233
Gibson LF (2002) Survival of B lineage leukemic cells: signals from the bone marrow microenvironment. Leuk Lymphoma 43:19–27
Goulden N, Langlands K, Steward C, Katz F, Potter M, Chessells J, Oakhill A (1994) PCR assessment of bone marrow status in 'isolated' extramedullary relapse of childhood B-precursor acute lymphoblastic leukaemia. Br J Haematol 87:282–285
Hall BM, Gibson LF (2004) Regulation of lymphoid and myeloid leukemic cell survival: role of stromal cell adhesion molecules. Leuk Lymphoma 45:35–48
Hall BM, Fortney JE, Taylor L, Wood H, Wang L, Adams S, Davis S, Gibson LF (2004) Stromal cells expressing elevated VCAM-1 enhance survival of B lineage tumor cells. Cancer Lett 207:229–239
Hara J, Matsuda Y, Fujisaki H, Tokimasa S, Ohta H, Osagi Y, Takai K (2000) Expression of adhesion molecules in childhood B-lineage-cell neoplasms. Int J Hematol 72:69–73
Hazlehurst LA, Argilagos RF, Dalton WS (2007) Beta1 integrin mediated adhesion increases Bim protein degradation and contributes to drug resistance in leukaemia cells. Br J Haematol 136:269–275
Hehlgans S, Haase M, Cordes N (2007) Signalling via integrins: implications for cell survival and anticancer strategies. Biochim Biophys Acta 1775:163–180
Hoggatt J, Pelus LM (2011) Many mechanisms mediating mobilization: an alliterative review. Curr Opin Hematol 18:231–238
Hsieh YT, Gang EJ, Geng HM, Park E, Huantes S, Chudziak D, Dauber K, Schaefer P, Scharman C, Shimada H, Shojaee S, Parameswaran R, Klemm L, Loh M, Kang ES, Koo HH, Hofmann WK, Andrade J, Crooks GM, Willman CL, Müschen M, Papayannopoulou T, Heisterkamp N, Bönig H, Kim YM (2013) Integrin alpha4 blockade eradicates drug resistant pre-B acute lymphoblastic leukemia. Blood 121(10):1814–1818

Hsieh YT, Gang EJ, Shishido S, Kim HN, Pham J, Khazal S, Osborne A, Esguerra ZA, Kwon E, Jang J, Bonig H, Biediger RJ, Vanderslice P, Kim YM (2014) Effects of the small-molecule inhibitor of integrin α4, TBC3486, on pre-B-ALL cells. Leukemia doi: 10.1038/leu.2014.182
Hynes RO (2002) Integrins: bidirectional, allosteric signaling machines. Cell 110:673–687
Iademarco MF, Barks JL, Dean DC (1995) Regulation of vascular cell adhesion molecule-1 expression by IL-4 and TNF-alpha in cultured endothelial cells. J Clin Invest 95:264–271
Jin L, Hope KJ, Zhai Q, Smadja-Joffe F, Dick JE (2006) Targeting of CD44 eradicates human acute myeloid leukemic stem cells. Nat Med 12:1167–1174
Johnston JB, Navaratnam S, Pitz MW, Maniate JM, Wiechec E, Baust H, Gingerich J, Skliris GP, Murphy LC, Los M (2006) Targeting the EGFR pathway for cancer therapy. Curr Med Chem 13:3483–3492
Juneja HS, Rajaraman S, Gay RE, Gay S, Schmalsteig FC (1992) Characterization of heterotypic adherence between transformed human lymphoblastic cells and marrow stromal cells: VCAM-1 is a ligand for one of the leukemia cell adhesion proteins. Exp Hematol 20:1263–1270
Kang H, Chen IM, Wilson CS, Bedrick EJ, Harvey RC, Atlas SR, Devidas M, Mullighan CG, Wang X, Murphy M, Ar K, Wharton W, Borowitz MJ, Bowman WP, Bhojwani D, Carroll WL, Camitta BM, Reaman GH, Smith MA, Downing JR et al (2010) Gene expression classifiers for relapse-free survival and minimal residual disease improve risk classification and outcome prediction in pediatric B-precursor acute lymphoblastic leukemia. Blood 115:1394–1405
Kappos L, Bates D, Edan G, Eraksoy M, Garcia-Merino A, Grigoriadis N, Hartung HP, Havrdova E, Hillert J, Hohlfeld R, Kremenchutzky M, Lyon-Caen O, Miller A, Pozzilli C, Ravnborg M, Saida T, Sindic C, Vass K, Clifford DB, Hauser S et al (2011) Natalizumab treatment for multiple sclerosis: updated recommendations for patient selection and monitoring. Lancet Neurol 10:745–758
Katayama Y, Hidalgo A, Peired A, Frenette PS (2004) Integrin alpha4beta7 and its counterreceptor MAdCAM-1 contribute to hematopoietic progenitor recruitment into bone marrow following transplantation. Blood 104:2020–2026
Konopleva M, Tabe Y, Zeng Z, Andreeff M (2009) Therapeutic targeting of microenvironmental interactions in leukemia: mechanisms and approaches. Drug Resist Updat 12:103–113
Korbling M, Freireich EJ (2011) Twenty-five years of peripheral blood stem cell transplantation. Blood 117:6411–6416
Krause DS, Lazarides K, von Andrian UH, Van Etten RA (2006) Requirement for CD44 in homing and engraftment of BCR-ABL-expressing leukemic stem cells. Nat Med 12:1175–1180
Kumar AG, Dai XY, Kozak CA, Mims MP, Gotto AM, Ballantyne CM (1994) Murine VCAM-1. Molecular cloning, mapping, and analysis of a truncated form. J Immunol 153:4088–4098
Larson S, Stock W (2008) Progress in the treatment of adults with acute lymphoblastic leukemia. Curr Opin Hematol 15:400–407
Liesveld JL, Winslow JM, Frediani KE, Ryan DH, Abboud CN (1993) Expression of integrins and examination of their adhesive function in normal and leukemic hematopoietic cells. Blood 81:112–121
Lobb RR, Hemler ME (1994) The pathophysiologic role of alpha 4 integrins in vivo. J Clin Invest 94:1722–1728
Lobb RR, Abraham WM, Burkly LC, Gill A, Ma W, Knight JA, Leone DR, Antognetti G, Pepinsky RB (1996) Pathophysiologic role of alpha 4 integrins in the lung. Ann N Y Acad Sci 796:113–123
MacDonald JK, McDonald JW (2007) Natalizumab for induction of remission in Crohn's disease. Cochrane Database Syst Rev CD006097
Makrynikola V, Bradstock KF (1993) Adhesion of precursor-B acute lymphoblastic leukaemia cells to bone marrow stromal proteins. Leukemia 7:86–92
Makrynikola V, Bianchi A, Bradstock K, Gottlieb D, Hewson J (1994) Migration of acute lymphoblastic leukemia cells into human bone marrow stroma. Leukemia 8:1734–1743
Manabe A, Murti KG, Coustan-Smith E, Kumagai M, Behm FG, Raimondi SC, Campana D (1994) Adhesion-dependent survival of normal and leukemic human B lymphoblasts on bone marrow stromal cells. Blood 83:758–766

Matsunaga T, Takemoto N, Sato T, Takimoto R, Tanaka I, Fujimi A, Akiyama T, Kuroda H, Kawano Y, Kobune M, Kato J, Hirayama Y, Sakamaki S, Kohda K, Miyake K, Niitsu Y (2003) Interaction between leukemic-cell VLA-4 and stromal fibronectin is a decisive factor for minimal residual disease of acute myelogenous leukemia. Nat Med 9:1158–1165

Meredith JE, Jr., Fazeli B, Schwartz MA (1993) The extracellular matrix as a cell survival factor. Mol Biol Cell 4:953–961

Miniero R, Saracco P, Pastore G, Zurlo MG, Terracini B, Rosso P, Masera G (1995) Relapse after first cessation of therapy in childhood acute lymphoblastic leukemia: a 10-year follow-up study. Italian Association of Pediatric Hematology-Oncology (AIEOP). Med Pediatr Oncol 24:71–76

Miyake K, Hasunuma Y, Yagita H, Kimoto M (1992) Requirement for VLA-4 and VLA-5 integrins in lymphoma cells binding to and migration beneath stromal cells in culture. J Cell Biol 119:653–662

Mudry RE, Fortney JE, York T, Hall BM, Gibson LF (2000) Stromal cells regulate survival of B-lineage leukemic cells during chemotherapy. Blood 96:1926–1932

Nilsson SK, Johnston HM, Whitty GA, Williams B, Webb RJ, Denhardt DT, Bertoncello I, Bendall LJ, Simmons PJ, Haylock DN (2005) Osteopontin, a key component of the hematopoietic stem cell niche and regulator of primitive hematopoietic progenitor cells. Blood 106:1232–1239

Nwabo Kamdje AH, Krampera M (2011) Notch signaling in acute lymphoblastic leukemia: any role for stromal microenvironment? Blood 118:6506–6514

Oliver B, Fernandez O, Orpez T, Alvarenga MP, Pinto-Medel MJ, Guerrero M, Leon A, Lopez-Madrona JC, Maldonado-Sanchez R, Garcia-Leon JA, Luque G, Fernandez V, Leyva L (2011) Kinetics and incidence of anti-natalizumab antibodies in multiple sclerosis patients on treatment for 18 months. Mult Scler 17:368–371

Papayannopoulou T, Craddock C, Nakamoto B, Priestley GV, Wolf NS (1995) The VLA4/VCAM-1 adhesion pathway defines contrasting mechanisms of lodgement of transplanted murine hemopoietic progenitors between bone marrow and spleen. Proc Natl Acad Sci USA 92:9647–9651

Parsons JT (1996) Integrin-mediated signalling: regulation by protein tyrosine kinases and small GTP-binding proteins. Curr Opin Cell Biol 8:146–152

Pillozzi S, Masselli M, De LE, Accordi B, Cilia E, Crociani O, Amedei A, Veltroni M, D'Amico M, Basso G, Becchetti A, Campana D, Arcangeli A (2011) Chemotherapy resistance in acute lymphoblastic leukemia requires hERG1 channels and is overcome by hERG1 blockers. Blood 117:902–914

Planas R, Jelcic I, Schippling S, Martin R, Sospedra M (2012) Natalizumab treatment perturbs memory- and marginal zone-like B-cell homing in secondary lymphoid organs in multiple sclerosis. Eur J Immunol 42:790–798

Polman CH, O'Connor PW, Havrdova E, Hutchinson M, Kappos L, Miller DH, Phillips JT, Lublin FD, Giovannoni G, Wajgt A, Toal M, Lynn F, Panzara MA, Sandrock AW (2006) A randomized, placebo-controlled trial of natalizumab for relapsing multiple sclerosis. N Engl J Med 354:899–910

Pucci E, Giuliani G, Solari A, Simi S, Minozzi S, Di PC, Galea I (2011) Natalizumab for relapsing remitting multiple sclerosis. Cochrane Database Syst Rev CD007621

Pui CH, Evans WE. (2006) Treatment of acute lymphoblastic leukemia N Engl J Med 354:166–178

Pui CH, Jeha S (2007) New therapeutic strategies for the treatment of acute lymphoblastic leukaemia. Nat Rev Drug Discov 6:149–165

Pui CH, Campana D, Pei D, Bowman WP, Sandlund JT, Kaste SC, Ribeiro RC, Rubnitz JE, Raimondi SC, Onciu M, Coustan-Smith E, Kun LE, Jeha S, Cheng C, Howard SC, Simmons V, Bayles A, Metzger ML, Boyett JM, Leung W et al (2009) Treating childhood acute lymphoblastic leukemia without cranial irradiation. N Engl J Med 360:2730–2741

Pui CH, Carroll WL, Meshinchi S, Arceci RJ (2011) Biology, risk stratification, and therapy of pediatric acute leukemias: an update. J Clin Oncol 29:551–565

Ramirez P, Rettig MP, Uy GL, Deych E, Holt MS, Ritchey JK, DiPersio JF (2009) BIO5192, a small molecule inhibitor of VLA-4, mobilizes hematopoietic stem and progenitor cells. Blood 114:1340–1343

Ratajczak MZ, Kim C (2012) The use of chemokine receptor agonists in stem cell mobilization. Expert Opin Biol Ther 12:287–297

Ruoslahti E, Reed JC (1994) Anchorage dependence, integrins, and apoptosis. Cell 77:477–478

Ryan DH, Nuccie BL, Abboud CN, Winslow JM (1991) Vascular cell adhesion molecule-1 and the integrin VLA-4 mediate adhesion of human B cell precursors to cultured bone marrow adherent cells. J Clin Invest 88:995–1004

Salzer WL, Devidas M, Carroll WL, Winick N, Pullen J, Hunger SP, Camitta BA (2010) Long-term results of the pediatric oncology group studies for childhood acute lymphoblastic leukemia 1984-2001: a report from the children's oncology group. Leuk 24:355–370

Sandborn WJ, Colombel JF, Enns R, Feagan BG, Hanauer SB, Lawrance IC, Panaccione R, Sanders M, Schreiber S, Targan S, van DS, Goldblum R, Despain D, Hogge GS, Rutgeerts P (2005) Natalizumab induction and maintenance therapy for Crohn's disease. N Engl J Med 353:1912–1925

Sarkar S, Svoboda M, De BR, Freedman AS (2002) The role of Aktand RAFTK in beta1 integrin mediated survival of precursor B-acute lymphoblastic leukemia cells. Leuk Lymphoma 43:1663–1671

Scott LM, Priestley GV, Papayannopoulou T (2003) Deletion of alpha4 integrins from adult hematopoietic cells reveals roles in homeostasis, regeneration, and homing. Mol Cell Biol 23:9349–93460

Shah N, Oseth L, LeBien TW (1998) Development of a model for evaluating the interaction between human pre-B acute lymphoblastic leukemic cells and the bone marrow stromal cell microenvironment. Blood 92:3817–3828

Shalapour S, Hof J, Kirschner-Schwabe R, Bastian L, Eckert C, Prada J, Henze G, von Stackelberg A, Seeger K (2011) High VLA-4 expression is associated with adverse outcome and distinct gene expression changes in childhood B-cell precursor acute lymphoblastic leukemia at first relapse. Haematol 96:1627–1635

Shishido S, Bönig H, Kim YM (2014) Role of Integrin Alpha4 in Drug Resistance of Leukemia. Frontiers of Oncology 4:99

Sipkins DA, Wei X, Wu JW, Runnels JM, Cote D, Means TK, Luster AD, Scadden DT, Lin CP (2005) In vivo imaging of specialized bone marrow endothelial microdomains for tumour engraftment. Nature 435:969–973

Sorensen PS, Jensen PE, Haghikia A, Lundkvist M, Vedeler C, Sellebjerg F, Koch-Henriksen N, Fogdell-Hahn A, Myhr KM, Hillert J, Gold R (2011) Occurrence of antibodies against natalizumab in relapsing multiple sclerosis patients treated with natalizumab. Mult Scler 17:1074–1078

Spengeman JD, Green TD, McCubrey JA, Bertrand FE (2005) Activated EGFR promotes the survival of B-lineage acute leukemia in the absence of stromal cells. Cell Cycle 4:483–487

Stier S, Ko Y, Forkert R, Lutz C, Neuhaus T, Grunewald E, Cheng T, Dombkowski D, Calvi LM, Rittling SR, Scadden DT (2005) Osteopontin is a hematopoietic stem cell niche component that negatively regulates stem cell pool size. J Exp Med 201:1781–1791

Tallen G, Ratei R, Mann G, Kaspers G, Niggli F, Karachunsky A, Ebell W, Escherich G, Schrappe M, Klingebiel T, Fengler R, Henze G, von Stackelberg, A (2010) Long-term outcome in children with relapsed acute lymphoblastic leukemia after time-point and site-of-relapse stratification and intensified short-course multidrug chemotherapy: results of trial ALL-REZ BFM 90. J Clin Oncol 28:2339–2347

Troeger A, Johnson AJ, Wood J, Blum WG, Andritsos LA, Byrd JC, Williams DA (2012) RhoH is critical for cell-microenvironment interactions in chronic lymphocytic leukemia in mice and humans. Blood 119:4708–4718

Ulyanova T, Priestley GV, Nakamoto B, Jiang Y, Papayannopoulou T (2007) VCAM-1 ablation in nonhematopoietic cells in MxCre+ VCAM-1f/f mice is variable and dictates their phenotype. Exp Hematol 35:565–571

Uy GL, Rettig MP, Motabi IH, McFarland K, Trinkaus KM, Hladnik LM, Kulkarni S, Abboud CN, Cashen AF, Stockerl-Goldstein KE, Vij R, Westervelt P, DiPersio JF (2012) A phase 1/2 study of chemosensitization with the CXCR4 antagonist plerixafor in relapsed or refractory acute myeloid leukemia. Blood 119:3917–3924

Vanderslice P, Woodside DG, Caivano AR, Decker ER, Munsch CL, Sherwood SJ, Lejeune WS, Miyamoto YJ, McIntyre BW, Tilton RG, Dixon RA (2010) Potent in vivo suppression of inflammation by selectively targeting the high affinity conformation of integrin alpha4beta1. Biochem Biophys Res Commun 400:619–624

Wang L, Fortney JE, Gibson LF (2004) Stromal cell protection of B-lineage acute lymphoblastic leukemic cells during chemotherapy requires active Akt. Leuk Res 28:733–742

Weber MS, Menge T, Lehmann-Horn K, Kronsbein HC, Zettl U, Sellner J, Hemmer B, Stuve O (2012) Current treatment strategies for multiple sclerosis—efficacy versus neurological adverse effects. Curr Pharm Des 18:209–219

Yokosaki Y, Palmer EL, Prieto AL, Crossin KL, Bourdon MA, Pytela R, Sheppard D (1994) The integrin alpha 9 beta 1 mediates cell attachment to a non-RGD site in the third fibronectin type III repeat of tenascin. J Biol Chem 269:26691–26696

Zeng Z, Samudio IJ, Munsell M, An J, Huang Z, Estey E, Andreeff M, Konopleva M (2006) Inhibition of CXCR4 with the novel RCP168 peptide overcomes stroma-mediated chemoresistance in chronic and acute leukemias. Mol Cancer Ther 5:3113–3121

Zeng Z, Shi YX, Samudio IJ, Wang RY, Ling X, Frolova O, Levis M, Rubin JB, Negrin RR, Estey EH, Konoplev S, Andreeff M, Konopleva M (2009) Targeting the leukemia microenvironment by CXCR4 inhibition overcomes resistance to kinase inhibitors and chemotherapy in AML. Blood 113:6215–6224

Part VIII
Hypoxia

Chapter 35
The Leukemic Bone Marrow Microenvironment: Targeting Hypoxia with Hypoxia-Activated Prodrugs

Juliana Benito, Marina Konopleva and William R. Wilson

Abstract Survival and maintenance of normal hematopoietic stem cells (HSCs) are regulated by bone marrow (BM) microenvironment. The experimental evidence indicates the vital role of hypoxia in maintenance of the primitive quiescent HSCs within the discrete subendosteal BM niches. Recent findings demonstrate that in the setting of hematologic malignancies BM is highly hypoxic, and that progression of the disease is associated with expansion of hypoxic niches and stabilization of the oncogenic hypoxia-inducible factor-1α (HIF-1α). This chapter will provide an overview of normal and leukemic BM microenvironment with an especial emphasis on pathological hypoxia. We will further describe the development of the hypoxia-activated prodrugs (HAPs) and their applicability in hematological malignancies.

Keywords Bone marrow · Microenvironment · HSCs · Hematological malignancies · Hypoxia · HAP

35.1 Introduction

Bone marrow (BM) residents, the hematopoietic stem cells (HSCs), are the root cells for hematopoiesis, producing both lymphoid and myeloid cells, which are necessary for the body's immune integrity, oxygen (O_2) delivery, blood clotting, waste removal,

M. Konopleva (✉)
Division of Cancer Medicine—Unit 448, The University of Texas MD Anderson Cancer Center, 1515 Holcombe Blvd, Houston TX 77030, USA
e-mail: mkonople@mdanderson.org

J. Benito
Department of Leukemia, Division of Cancer Medicine, The University of Texas MD Anderson Cancer Center, 1515 Holcombe Blvd, Houston TX 77030, USA
e-mail: jbenito@mdanderson.org

W. R. Wilson
Auckland Cancer Society Research Centre, University of Auckland, Private Bag 92019, Auckland NZ USA
e-mail: wr.wilson@auckland.ac.nz

M. Andreeff (ed.), *Targeted Therapy of Acute Myeloid Leukemia*, Current Cancer Research, DOI 10.1007/978-1-4939-1393-0_35

and a wide variety of physiologic processes that are necessary for survival. Despite the prevailing concept by Schofield regarding the indispensable role of the BM microenvironment in supporting hematopoiesis (Schofield 1983), the difficulty in accessing the BM led to the fact that for quite some time, the intracellular regulatory mechanisms of HSCs have been studied in isolation of its confines in the bone marrow, with little emphasis on the effects this environment might have on this essential process (Spradling et al. 2001). Introduction of new in vivo imaging tools, such as intravital multiphoton microscopy, provided real-time visualization of cellular interactions (Schofield 1983; Lo Celso et al. 2009) and resulted in a radical revolution in the way stem cells are studied in the bone marrow. Such studies have shown that hematopoiesis depends not only on HSCs but also on the microenvironment where they reside, strengthening the "stem cell niche" hypothesis (Schofield 1983).

Stemming from its anatomy, the BM is a complex system where HSCs are believed to reside in two distinct niches, although not clearly separated, namely the endosteal and vascular (sinusoidal) niches (Adams et al. 2006; Calvi et al. 2003; Ding et al. 2012; Kiel et al. 2005; Nilsson et al. 2001; Visnjic et al. 2004; Zhang et al. 2003). Cellular components of these niches include bone lining cells (osteoblasts and osteoclasts), mesenchymal stem cells, sinusoidal endothelium and perivascular stromal cells, immune cells, and other cell types that play different roles in HSC homeostasis (Fujisaki et al. 2011).

By means of examining the expression of the signaling lymphocyte activation molecule (SLAM) family receptors, including $CD150^{+}/CD244^{-}/CD48^{-}$ (Kiel et al. 2005), most HSCs were shown to reside adjacent to sinusoidal endothelium in the spleen and bone marrow, while few showed preference for the BM endosteum. Therefore, two specialized niches in the BM support HSCs. Elucidating the functions of these two niches is crucial to understanding the behavior of HSCs and to exploit them for clinical applications.

Formed by both cortical and trabecular bone, the endosteum is the inner surface of the BM cavity. This surface is lined by bone cells such as osteoblasts and osteoclasts; however, the later are less well characterized in the context of their role in HSC regulation. Intravital microscopy has revealed that fluorescently labeled HSCs home to the BM endosteum, suggesting a preference for this anatomical site for their survival and maintenance (Fujisaki et al. 2011). Important for HSCs regulation, osteoblasts have been suggested to express several cytokines, chemokines, and adhesion molecules such as CXCL12, angiopoietin 1, stem cell factor (SCF), and thrombopoietin. Several lines of evidence suggest that the HSCs that associate with the osteoblasts are quiescent in nature, giving them the ability to survive and contribute to hematopoiesis over a long period of time (Kataoka et al. 2011; Mullally et al. 2012; Notta et al. 2011; Morrison and Spradling 2008).

The BM vascular niche consists mainly of sinusoids that are thin-walled vessels connecting the marrow cavity with the systemic circulation. They are lined by a single layer of endothelium and arise directly from arterioles. Functionally similar to the endosteal niche, cytokines produced by cells supporting the vascular compartment, including SCF and CXCL12, are also important for HSC regulation. In an effort to elucidate the specific contribution of each BM niche to HSC maintenance, Ding et al. conducted a study in which cre-lox conditional knockout mice were used to delete SCF from osteoblasts, sinusoidal endothelium, perivascular stromal cells, and nestin-positive mesenchymal stem cells (Ding et al. 2012). They found that sinusoidal endothelium and leptin receptor (lepr)-expressing perivascular stomal cells but not osteoblasts or nestin-cre or nestin-creER-expressing cells are directly responsible for the expression of SCF and functionally regulate HSCs in the sinusoidal niche. This pivotal study paves the way for defining the functional specificity of cells forming the HSC niches.

35.2 BM Hypoxia and HSC

In normal tissues, O_2 levels are tightly regulated due to the detrimental consequences that having high or low O_2 levels can cause. While in the atmosphere, the partial pressure of O_2 (pO_2) is 21 % (or 160 mmHg), the levels in the human tissues exhibit great variations and are far below the atmospheric values. Tissue or cellular hypoxia is defined therefore as a drop in O_2 levels below the physiological values (Ivanovic 2009). In arterial blood, PO_2 varies between 75 and 100 mmHg that is equivalent to 10 and 13.3 % partial atmospheric O_2 pressure (Shapiro 1995; Malatesha et al. 2007; Chu et al. 2003) while in venous blood is between 30 and 40 mmHg or 4–5.3 % partial atmospheric O_2 pressure (Chu et al. 2003; Walkey et al. 2010). In tissues, O_2 concentration depends on vascularization, ranging from 4 to 14 % in well-perfused tissues and between 0.7 and 7 % in low-perfused organs (Ivanovic 2009; Wolfle and Jungermann 1985; Jungermann and Kietzmann 1997; Roy et al. 2003; Welch et al. 2001; Mik et al. 2004). Direct oxygen measurements in the BM have been obscured by the difficulty in accessing the tissue. Studies performed in patients with chronic pulmonary disease (COPD) or healthy volunteers showed mean values of BM O_2 between 48 and 54.9 mm Hg, respectively (Harrison et al. 2002; Skouby 1976). Subsequent studies by Parmar et al. utilizing Hoechst 33342 (Ho) perfusion dye suggested the existence of an O_2 gradient from the vasculature to the endosteal niche (Parmar et al. 2007), whereby majority of HSCs resided in the group with the lowest Ho intensity. These HSC bound the hypoxia marker pimonidazole and were sensitive to the hypoxia-activated drug, tirapazamine. It was proposed that hypoxia promotes maintenance of HSCs in a quiescent and pluripotent state (Semenza 2007). In contrast, HSCs that reside in close proximity to the vascular niche are short-term (ST), actively cycling, and are responsible for replenishing circulating cells by rapidly differentiating into leukocytes, erythrocytes, and platelets (Winkler et al. 2010). The metabolic state of ST and long-term (LT) HSCs is thought to be due to an O_2-rich sinusoidal HSC niche relative to the hypoxic endosteal niche, respectively (Parmar et al. 2007; Chow et al. 2001a, b). LT-HSCs residing in the endosteal niche are dependent on the master transcriptional regulator of hypoxia response, hypoxia-inducible factor-1 (HIF-1) which maintains them in a quiescent state by promoting expression of different target genes including VEGFa, Cripto/GRP78, and CXCR4 (Miharada et al. 2011; Rehn et al. 2011). HIF-1 is the master transcriptional regulator of hypoxic cellular response. In the presence of O_2, HIF-1α subunits are normally degraded by the protein encoded by the von Hippel–Lindau tumor suppressor gene VHL (pVHL) but are stabilized under hypoxic conditions. A vast array of gene products controlling energy metabolism, glycolysis, angiogenesis, apoptosis, and the cell cycle (Semenza 2010; Weidemann and Johnson 2008) are induced as a result of HIF-1 activation. Recent genetic studies have shown that HSCs stabilize HIF-1α protein and that this is critical for maintaining cell cycle quiescence (Takubo et al. 2010) and utilizing glycolysis instead of mitochondrial oxidative phosphorylation (Simsek et al. 2010).

35.2.1 Pathophysiologic Role of Hypoxia in Solid Tumors

Hypoxia is a common phenomenon shared by multiple solid tumors and it is often associated with poor prognosis. The significance of hypoxia in the context of solid tumors is attested by an extensive literature. The O_2 levels determined with O_2 electrodes show that tumors can be very heterogeneous with regions exhibiting less than 5 mmHg pO_2 (corresponding to approximately 0.7 % O_2 in the gas phase) and with median values that are far below those found in the normal tissues. As examples, median O_2 levels in glioblastoma were reported in the range of 4.9–5.6 mmHg (Kayama et al. 1991; Rampling et al. 1994; Evans et al. 2004a, b; Collingridge et al. 1999; Vaupel et al. 2007), in head and neck carcinoma between 12.2 and 14.7 mmHg (Nordsmark et al. 1994, 2007; Becker et al. 1998; Le et al. 2003), while pancreatic and prostate tumors exhibit the lowest levels with values around 2.5 mmHg (Koong et al. 2000; Thomlinson and Gray 1955; Ahn and Brown 2007). Hypoxia is associated with increased aggressiveness and invasiveness, metastatic behavior, and resistance to chemotherapy (Semenza 2010; Brown and William 2004; Anderson et al. 1996; Borch et al. 2001). Several mechanisms have been described to explain hypoxia effects and in many cases HIF transcription factors (HIF-1, HIF-2) and their downstream targets are the culprits. Indeed, immunohistochemical analysis of biopsies showed that HIF-1α or HIF-2α protein (or both) are overexpressed relative to the normal tissue counterparts, in a vast number of human cancers (Zhong et al. 1999). Downstream of HIF-1 is a long list of genes that are involved in key processes leading to tumor progression including angiogenesis (Chen and Hu 2009) and metabolism regulation (Ganjoo et al. 2011; Guise et al. 2007; Semenza 2009; Chen and Russo 2012), and additionally cell survival, chemo- and radiotherapy resistance, genetic instability, immortalization, immune evasion, invasion, and metastasis and stem cell maintenance (Moeller et al. 2007; Lukashev et al. 2007; Chan and Giaccia 2007; Sullivan and Graham 2007; Barnhart and Simon 2007; Das et al. 2008; Lu and Kang 2010; Zhu et al. 2012; Jiao and Nan 2012; Rohwer and Cramer 2011; Schults et al. 2010).

35.2.2 Role of Hypoxia in Hematologic Malignancies

Even though the importance of hypoxia in cancer has been proposed over 50 years ago (Thomlinson and Gray 1955), studies of its role in hematological malignancies have been impeded mostly because of the difficulty of accessing the BM. In a rat leukemia model, Jensen et al. used in vivo pulse labeling with 2-nitroimidazole linked to theophylline (NITP) and showed by flow cytometry that leukemic cells infiltrating the BM were markedly hypoxic (Jensen et al. 2000). Using conventional gas assay methodology pO_2 values in the BM aspirates from 19 AML patients were found to be on average 6.06% (Fiegl et al. 2009), a value that likely underestimates the hypoxia levels within the endosteal niches. Through utilization of the hypoxia tracer, pimonidazole (PIM), hypoxia was recently reported to constitute a widespread component of the BM environment. As such, vast expansion of hypoxic areas was detected by PIM staining in the BM of mice bearing xenografts or syngenic acute leukemia models (Fig. 35.1) which increased with disease progression (Benito et al. 2011). Further support for widespread hypoxia in leukemic marrows came from the report of frequent HIF-1 expression in patients with acute lymphoblastic leukemia (ALL) which in turn was associated with inferior outcomes (Frolova et al. 2012a). Importantly, while HIF-1α was expressed in BM from ALL patients at the time of diagnosis, it routinely was undetectable in subsequent BM biopsies collected at the time when remission was obtained following induction chemotherapy (Benito et al. 2011). Hypoxia may contribute to leukemia progression by inducing chemoresistance of leukemic cells (Benito et al. 2011; Frolova et al. 2012b) and by promoting expression of chemokine receptor CXCR4 which in turn induces homing of leukemic blasts to the protective BM environment (Fiegl et al. 2009; Benito et al. 2011). Importantly, Matsunaga et al. recently suggested that hypoxia, through HIF-1α, promotes quiescence of leukemic stem cells (LSC) residing in the endosteal niches and that this could contribute to persistence of residual leukemic cells (commonly defined as minimal residual disease, MRD) in AML (Matsunaga et al. 2012).

The importance of hypoxia and of HIFs in disease progression has been lately uncovered in multiple myeloma (MM; reviewed in (Martin et al. 2011)). In the first report pinpointing the importance of a hypoxic microenvironment, Asosinghet et al. used a 5T2MM murine in vivo MM model and, by utilization of PIM, showed that a heavily infiltrated MM BM is less hypoxic than the normal BM due to the presence of active angiogenesis which is in turn stimulated by initial hypoxia (Asosingh et al. 2005). They also indicated that hypoxia may have a role in establishment of the disease since they demonstrated that the hypoxic BM environment selects for tumor-initiating and apoptosis-resistant $CD45^+$ MM cells. In contrast, a more recent study in mice harboring 5T33MM cells demonstrated significantly higher positive PIM areas in the BM of diseased mice compared to the healthy control animals (Hu et al. 2010), indicating more pronounced hypoxia associated with MM progression. In a provocative recent study, Azab et al. (2012) reported that BM hypoxia induces epithelial-to-mesenchymal transition (EMT)-like features in MM cells and thereby promoting their dissemination and re-homing to new sites in the BM. Altogether, these early findings indicate the previously unrecognized role of hypoxia in the hematologic malignancies (Fig. 35.1).

The irrefutable role that hypoxia plays not only in solid tumors but also in liquid tumors has made it a "universal" target for the treatment of cancer. In the next section, we will discuss the development and advances of new therapies that seek to

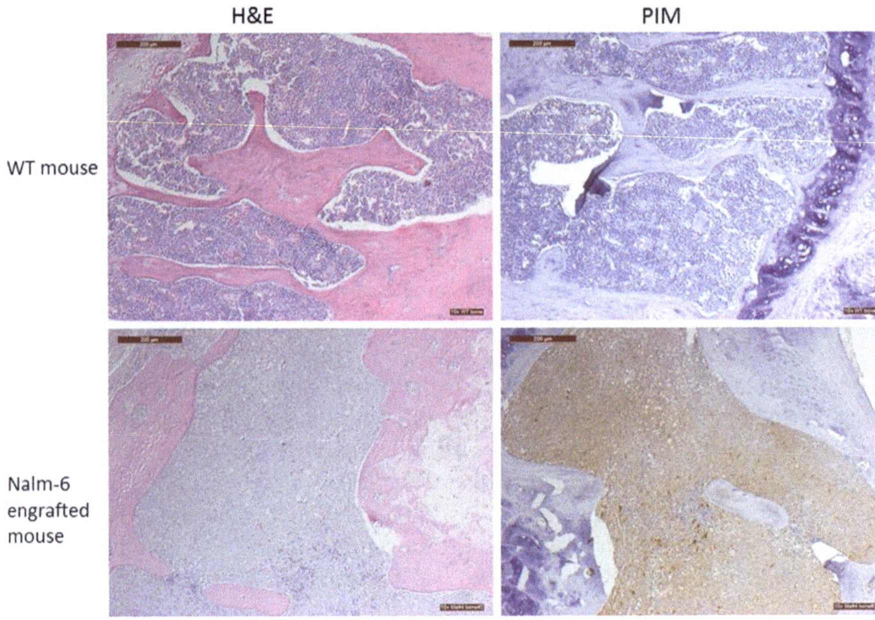

Fig. 35.1 Hypoxia in the leukemic BM microenvironment. Wild-type (WT) or Nalm-6 pre-B-ALL engrafted NSG mice received pimonidazole and were sacrificed 2 h later. Representative images are shown of BM stained with hematoxylin and eosin (H&E) or PIM antibody. Original magnification 100×

eliminate cancer cells residing in hypoxic environments with a specific emphasis in hematological malignancies.

35.3 Hypoxia-Activated Prodrugs

Two general approaches are being explored for targeting hypoxic cells in solid tumors, both of which are potentially applicable to hypoxia in AML and other leukemias. One approach focuses on inhibiting molecular targets required for survival of hypoxic cells, particularly HIF-1 and its downstream targets or upstream signaling partners. HIF-1 itself continues to be a challenging target for drug development (Nordgren and Tavassoli 2011; Xia et al. 2012). In contrast, development of inhibitors of the HIF-1 target carbonic anhydrase-IX (CAIX), a critically important regulator of intracellular pH and cell survival in hypoxic microenvironments (Dubois et al. 2011; Swietach et al. 2007), is well advanced with several antibodies and small molecule inhibitors currently in preclinical and clinical development for solid tumors (McDonald et al. 2012). However, understanding of the molecular targets that underpin survival of hypoxic cells in leukemias, and their chemoresistance, is still limited.

A second broad approach uses hypoxia-activated prodrugs (HAPs) to target hypoxic cells; these prodrugs act as direct O_2 sensors, with O_2 suppressing their metabolic activation. HAP development is a relatively mature field, although not without its own significant challenges as discussed in several reviews (Ahn and Brown 2007; Wardman 2001; Chen and Hu 2009; Denny 2010; Wilson and Hay 2011). These prodrugs are activated by enzymatic reduction in hypoxic cells, and typically (most nitro compounds, quinones, and aromatic *N*-oxide HAPs) selectivity for hypoxia is a consequence of one-electron reduction of the prodrug by NADPH:cytochrome P450 oxidoreductase (POR; Belcourt et al. 1996; Guise et al. 2007; Wang et al. 2012) and other flavoprotein reductases (Guise et al. 2012), and redox cycling of the resulting prodrug radical (Mason and Holtzman 1975) in the presence of O_2 (illustrated for PR-104A in Fig. 35.2). In the absence of O_2, the prodrug radical has a much longer lifetime and can fragment, or be reduced further, to generate the active drug in hypoxic cells. Two-electron reductases such as NAD(P)H dehydrogenase [quinone]-1 (NQO-1, DT-diaphorase) can in some cases bypass the O_2-sensitive free radical intermediate resulting in prodrug activation independently of hypoxia, especially with quinone bioreductive prodrugs such as mitomycin C and apaziquone (EO9) which lack hypoxic selectivity in cell lines with high NQO-1 activity (Plumb et al. 1994). In some cases, hypoxic selectivity is a consequence of competition between the HAP and O_2 for reductants rather than redox cycling, as illustrated by the tertiary amine N-oxide banoxantrone (AQ4N; Patterson 2002) and by cobalt(III) complexes (Anderson et al. 1996).

The following outline does not describe the full range of HAP chemical classes under investigation, which are discussed in the above reviews, but provides an overview of the compounds of greatest current interest in the context of AML.

35.3.1 Nitrobenzamide Mustards: PR-104

The original concept for this class of prodrugs was to exploit the large change in electron density in a benzene ring when a strongly electron-withdrawing nitro group is reduced to an electron-donating hydroxylamine or amine, using this as a switch to activate a nitrogen mustard moiety that is in electronic conjugation with the nitro group (Denny and Wilson 1986). PR-104 (Fig. 35.2), the most recent embodiment of this concept, was developed at the University of Auckland and was licensed to Proacta.

PR-104 is a water-soluble phosphate ester pre-prodrug that is rapidly hydrolyzed systemically to the corresponding alcohol, PR-104A, both in nonclinical species and humans (Patel et al. 2011). PR-104A acts as an HAP through one-electron reduction of the nitro group *para* to its asymmetrical mustard, a reaction that is catalyzed by POR and other diflavin reductases (Guise et al. 2012) in hypoxic cells. The resulting hydroxylamine (PR-104H) and amine (PR-104M) are activated nitrogen mustards that kill hypoxic cells through formation of long-lived DNA interstrand cross-links (Gu et al. 2009; Singleton et al. 2009). In addition to killing hypoxic

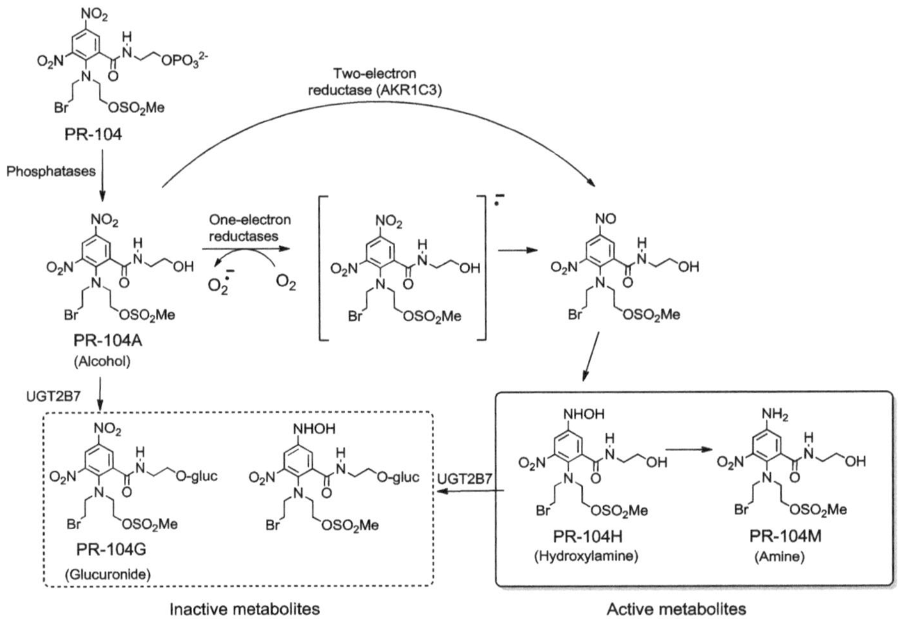

Fig. 35.2 Structure and mechanism of activation of the nitrobenzamide mustard PR-104. Reduction of the electron-withdrawing nitro group of PR-104A to electron-donating hydroxylamine or amine activates the nitrogen mustard moiety. One-electron reductases, such as NADPH:cytochrome P450 oxidoreductase and other flavoenzymes, form a nitro radical that is reoxidized by O_2. Two-electron reduction by aldo-keto reductase 1C3 can bypass the nitro radical to effect O_2-insensitive activation

cells efficiently in human tumor xenografts (as demonstrated by marked activity in combination with ionizing radiation, which spares hypoxic cells), PR-104 showed surprising monotherapy activity in some preclinical models (Patterson et al. 2007). It was subsequently found that PR-104A is a substrate for an O_2-insensitive two-electron reductase, aldo-keto reductase 1C3 (AKR1C3), which had not previously been known to act as a nitroreductase and which does not activate other HAP classes (Guise et al. 2010). Activation by AKR1C3 appears to be a major contributor to the monotherapy activity of PR-104 in preclinical models (Guise et al. 2010), suggesting the potential for PR-104 to exploit the upregulation of AKR1C3 via the activated Nrf2/Keap1/ARE transcriptional response in some tumors (Sporn and Liby 2012). In addition, local diffusion of PR-104H and PR-104 M from hypoxic zones also appears to provide a "bystander effect" that contributes to monotherapy activity of PR-104 in some tumors (Hicks et al. 2007). Given that activation of PR-104A by one-electron reductases requires severe hypoxia (50 % inhibition at ~0.1 μM O_2; Hicks et al. 2007), as is typical of other nitro compounds (Taylor and Rauth 1982; Siim et al. 1994; Koch 1993a), this bystander effect likely also contributes to killing of more moderately hypoxic cells that are important contributors to radioresistance and chemoresistance.

Initial phase I clinical trials of PR-104 in patients with advanced solid tumors established a maximum tolerated dose (MTD) of 1100 mg/m^2 for a once-every-3-week schedule (Jameson et al. 2010) and 675 mg/m^2/week for dosing weekly for 3 weeks of a 4-week cycle, with little evidence of therapeutic activity in these unselected tumors. However, emerging evidence for hypoxia in human leukemias (see above) suggested that evaluation against leukemias might be warranted. In addition, hematological toxicity was dose limiting in the above phase I studies, with no other significant toxicities, suggesting the potential for dose escalation in leukemias. Although use of alkylators in the treatment of AML has declined with the advent of anthracycline/cytarabine combinations, their activity in this setting is well attested and they are still often used in conditioning regimes for allogeneic or autologous stem cell transplantation; thus imposing additional selectivity through targeting hypoxia in BM (and potentially disease in peripheral tissue) is attractive. In turn, alkylators are routinely included in chemotherapy regimens for lymphoid malignancies including ALL. Preclinical studies demonstrated impressive single-agent activity of PR-104 in xenograft models of human pre-B-ALL cell line NALM6, primary human adult AML and pediatric ALL xenografts in immunodeficient mice supporting feasibility of exploiting HAPs in leukemia setting (Benito et al. 2011). In these models, BM and spleen of animals harboring late-stage leukemia infiltrates harbored vast hypoxic areas with leukemic cells expressing HIF-1α and its target gene CXCR4, which was eliminated following reduction of disease burden by PR-104.

Additional pharmacologic advantages may argue for potential utility of PR-104 in a leukemia setting. The active metabolites of PR-104A are well able to exploit defects in DNA cross-link repair (Gu et al. 2009; Hunter et al. 2012) which may sensitize some AML subgroups to cross-linking agents (D'Andrea 2010). In addition, PR-104A is glucuronidated by a specific UDP-glucuronyltransferase (UGT2B7), the resulting glucuronide being the predominant PR-104 metabolite in human plasma (Gu et al. 2011; McKeage et al. 2011). PR-104H is also a substrate for UGT2B7 identifying this as a potential mechanism of resistance and making the very low expression of UGT2B7 in AML an advantage (Gu et al. 2011). Finally, high expression of AKR1C3 mRNA in AML blasts in some patients (Birtwistle et al. 2009) suggests the possibility of dual activation of PR-104A by AKR1C3 and hypoxic (one-electron) reductases.

Supporting this concept, high-dose PR-104 (550 mg/kg, q7dx6) showed remarkable monotherapy activity against a panel of T- and B-ALL xenografts (Houghton et al. 2011). Reducing the dose into a range providing plasma pharmacokinetics achievable in solid tumor oncology patients retained strong activity in an AKR1C3-expressing T-ALL xenograft but not a low-AKR1C3 B-ALL xenograft (Benito et al. 2011). Recent studies demonstrate consistently higher expression and enzyme activity of AKR1C3 in cells from a panel of T-ALL than B-ALL xenografts (Manesh et al. 2012) suggesting possibly utility in this subtype. In addition, the recent development of AKR1C3-specific inhibitors (Jamieson et al. 2012) has made it possible to modify the coumberone fluorogenic probe assay for the AKR1C enzyme family (Guise et al. 2010; Halim et al. 2008) to provide a functional assay for AKR1C3 activity in leukemia cells. This may ultimately make it feasible to evaluate AKR1C3 activity in leukemic blasts as part of an individualized patient biomarker profile for PR-104.

The preclinical data summarized above paved the way for moving PR-104 into early phase I/II clinical trial in relapsed and/or refractory acute leukemias (AML/ALL; Benito et al. 2010; Konopleva et al. 2011; PR104 2009-0772: clinicaltrials.gov ID:NCT01037556). PR-104 was administered as a 1-hour intravenous infusion q 2 weeks, at escalating doses of 1.1, 1.6, 2.2, 3, and 4 gm/m^2, with cohort expansion at 3 and 4 gm/m^2. Interim analysis demonstrated that PR-104 was tolerated at doses three to four times higher than the solid tumor MTD; however, myelosuppression, primarily neutropenia and thrombocytopenia, can be prolonged at doses $\geq$3 gm/m^2. Analysis of BM biopsies following pimonidazole administration in consenting patients confirmed that hypoxia is a prevalent feature of the leukemic microenvironment. Evidence of activity at doses of 3 or 4 gm/m^2 support continued evaluation of this regimen in acute leukemias.

35.3.2 Phosphoramidate Mustards: TH-302

TH-302 (Fig. 35.3) is currently in clinical development as an HAP, in a broad series of indications (including AML/ALL), by Threshold Pharmaceuticals and its licensee Merck Serono. Like PR-104, TH-302 generates a DNA-cross-linking mustard-alkylating agent following reduction of a nitro group in the prodrug, but differs in several other respects.

TH-302 originated from earlier investigations of phosphoramidate mustards by Borch and co-workers who identified nitroheterocyclic prodrugs that fragment upon reduction to release cytotoxic phosphoramide mustard with modest selectivity for hypoxic cells (Borch et al. 2000, 2001). Development of symmetrical (achiral) analogs at Threshold led to prodrugs with improved hypoxic selectivity in culture, particularly the 2-nitroimidazole TH-302 which showed a 400-fold anoxia/oxic potency differential for H460 cells and promising activity with gemcitabine in a xenograft model (Duan et al. 2008). High hypoxic selectivity of TH-302 has been confirmed in a wide range of tumor cell lines, and its fragmentation on reduction to release the potent cytotoxin bromo-isophosphoramide mustard (Br-iPM) has been shown to proceed largely from the initial 2-nitroimidazole free radical (Meng et al. 2012; Fig. 35.3). TH-302 is a substrate for POR (Meng et al. 2012), but not for AKR1C3 (Guise et al. 2012), and generates DNA cross-links in hypoxic cells (Meng et al. 2012). The latter study also demonstrated a bystander effect in multicellular layer cultures when TH-302 is reduced using a bacterial nitroreductase. Significant monotherapy activity was seen in 8 of 11 xenograft models, with evidence that this is hypoxia-dependent and greater than can be accounted for by direct hypoxic cell killing (Azab et al. 2012). These results suggest the contribution of a bystander effect emanating from the severely hypoxic regions needed to activate the prodrug efficiently. The ability of Br-iPM to diffuse from the cells in which it is generated is supported by its detection in plasma of nonclinical species (Jung et al.

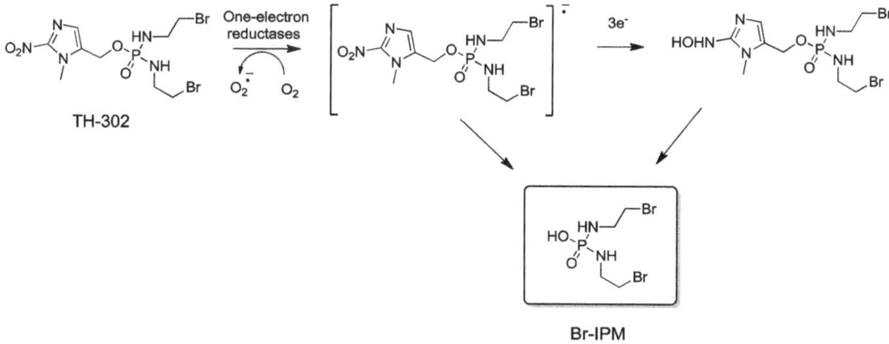

Fig. 35.3 Structure and mechanism of reductive activation of TH-302. Fragmentation to release the DNA cross-linking agent bromo-isophosphoramide mustard (Br-IPM) occurs primarily from the initial nitro radical

2012) and humans (Saunders et al. 2011). Impressive activity of TH-302 has been reported in combination with conventional chemotherapy in multiple preclinical solid tumor models (Liu et al. 2012). TH-302 induced apoptosis of multiple myeloma cells in the 5T33MM murine model, both under anoxia in vitro and within the BM microenvironment in the xenograft in vivo MM model (Hu et al. 2010), and exhibited hypoxia-selective cytotoxicity against acute leukemia cell lines, primary leukemia samples, and in the in vivo AML xenograft model (Benito et al. 2012).

In humans, the dose-limiting toxicity of TH-302 as monotherapy on a once every 3 weeks or weekly for 3 weeks schedule was skin and mucosal toxicity with minimal hematological toxicity (Saunders et al. 2011), although hematological toxicity has been more prominent in combination with gemcitabine in advanced pancreatic cancer (Borad et al. 2011) and with doxorubicin in soft tissue sarcomas (Ganjoo et al. 2011). In both these settings, response rates were higher than for historical comparisons, and a randomized phase II trial in combination with gemcitabine has recently demonstrated significant activity in combination with gemcitabine in advanced pancreatic cancer (median PFS 5.6 months for the combination vs. 3.6 months for gemcitamine alone, hazard ratio 0.61, $p=0.005$; Borad et al. 2012). In September 2011, a phase III trial of TH-302 in combination with doxorubicin was initiated in soft tissue sarcoma, and in March 2012, the US FDA and EMA both granted Orphan Drug designation to TH-302 for this indication. An interim analysis of this pivotal trial is expected early in 2013. Preliminary results of ongoing phase I clinical trial of single-agent TH-302 in patients with advanced leukemias ($n=34$) were reported at ASCO 2012 (Konopleva et al. 2012). TH-302 was administered IV over 30–60 min daily on days 1–5 of a 21-day cycle, at escalating doses of 120, 170, 240, 330, 460, or 550 mg/m^2. MTD was determined to be 460 mg/m^2, with dose-limiting toxicity of esophagitis at higher dose (TH-302 2010-0268: clinicaltrials.gov ID:NCT01149915). Clinical activity has been noted with a few objective responses, but majority of the cyto-reductions were transient, arguing for future strategies combining HAPs with chemotherapeutics targeting AML and ALL cells.

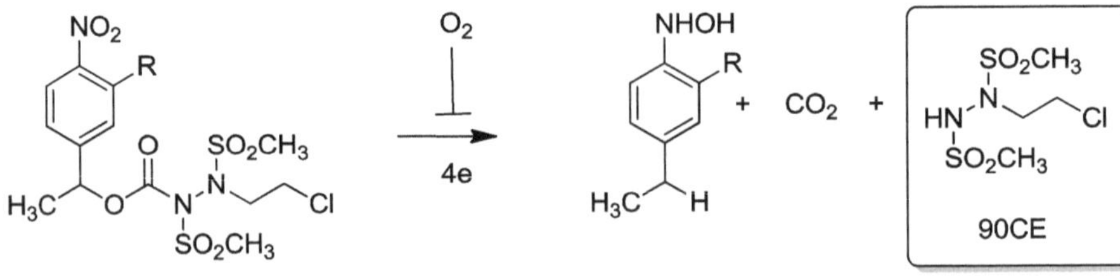

KS119: R=H
KS119W: R=OPO_3H_2

Fig. 35.4 Structure and mechanism of reductive activation of KS119 and its water-soluble analogue KS119W. Reduction of the nitro group to the hydroxylamine or amine triggers fragmentation to release the DNA-cross-linking chloroethylating agent 90CE

35.3.3 Chloroethyl Hydrazine Prodrugs: KS119 and KS119W

These HAPs originate from Yale University and were licensed initially to Vion Pharmaceuticals. KS119 (Fig. 35.4) is a nitrobenzyl prodrug with a sulfonyl hydrazine linker that fragments on reduction of the nitro group to release the DNA alkylating agent 1,2-bis(methylsulfonyl)-1-(2-chloroethyl)hydrazine (90CE; Seow et al. 2005). 90CE chloroethylates the O^6-position of guanine, leading to subsequent formation of an interstrand G-C cross-link. Reduction of KS119 is catalyzed by POR, NADH:cytochrome b_5 reductase, and xanthine oxidase, but not by NQO-1. It has been shown to kill EMT6 and CHO cells selectively under hypoxic conditions in culture, and is active against hypoxic cells in EMT6 tumors in mice (Seow et al. 2005). A notable feature is that KS119 cytotoxicity under hypoxic conditions appears to be relatively insensitive to the repair protein O^6 alkylguanine-DNA alkyltransferase (AGT, also known as O^6methylguanine methyltransferase), despite the ability of AGT to remove the initial O^6guanine chloroethyl adduct from 90CE (Baumann et al. 2010).

Although not itself in clinical development, KS119 is of significant interest in relation to targeting hypoxia in AML because 90CE is also the active cytotoxic moiety from the untargeted chloroethyl hydrazine prodrug laromustine (cloretazine, Onrigin™). The latter compound showed efficacy in previously untreated poor risk elderly AML patients as monotherapy (Giles et al. 2007; Giles 2009). A phase III trial in combination with high dose cytarabine (NCT00112554) showed a higher CR rate on the experimental arm (35 vs. 19 %, $p=0.005$), but at the expense of marked myelotoxicity and reduced PFS (Giles et al. 2009). However, laromustine releases 90CE systemically by a spontaneous (base catalyzed) reaction, whereas KS119 provides hypoxia-selective formation of 90CE and thus has potential for improved therapeutic ratio through selective release in hypoxic niches. KS119 has limited aqueous solubility (40 μM), which has led to development of a more soluble phosphate ester pre-prodrug, KS119W (Fig. 35.4), which retains the hypoxia-selective cytotoxicity of KS119 and also shows activity against hypoxic cells in solid tumors as determined by studies in combination with radiation (Kim et al. 2012).

35.3.4 Benzotriazine di-N-oxides: Tirapazamine and SN30000

The benzotriazine di-*N*-oxide tirapazamine (TPZ, Fig. 35.5) differs from the above classes of HAP in that cytotoxicity is predominantly due to oxidative DNA damage by free radicals that arise from the initial one-electron reduction product, rather than DNA alkylation. Although TPZ has not been investigated in the context of AML, it is the most intensively studied HAP and has been instructive in several respects.

The hypoxia-selective cytotoxicity of TPZ was discovered by Martin Brown at Stanford University (Brown 1993), and it has undergone extensive clinical studies against solid tumors (Reddy and Williamson 2009; McKeown et al. 2007). Most recently, it failed to meet the primary endpoint of overall survival in a pivotal phase III trial in combination with standard-of-care chemoradiation of advanced head and neck squamous cell carcinoma (HNSCC; Jameson et al. 2010), despite encouraging phase II data in the same indication (Rischin et al. 2005). The negative phase III results have been attributed in part to poor radiotherapy quality control in some contributing centers (Peters et al. 2010), and to the lack of stratification for the presence of the target (i.e., hypoxia). Retrospective analysis of potential biomarkers for hypoxia, in patients in the phase III without radiotherapy deviations, suggested an advantage for TPZ in patients with higher plasma levels of hepatocyte growth factor, which is a hypoxia-induced cMet ligand (Le et al. 2012). The failure to evaluate hypoxia prospectively may have been a lost opportunity, especially given earlier studies suggesting that TPZ activity is restricted to patients with hypoxia as detected by PET imaging with the 2-nitroimidazole bioreductive probe [^{18}F]-FMISO (Rischin et al. 2006). In the context of biomarkers for TPZ (and its analogue SN30000), 2-nitroimidazole probes have the advantage that they are reporters of the one-electron reductases that activate the HAPs, as well as hypoxia (Wang et al. 2012), and thus may be more informative predictive biomarkers than hypoxia-regulated molecular markers. However, important progress has been made recently in defining hypoxic gene signatures with prognostic value (Buffa et al. 2010; Toustrup et al. 2012), and one such signature has been shown to predict response to the hypoxia-selective radiosensitizer nimorazole in a retrospective study (Toustrup et al. 2011).

One of the limitations of TPZ as an HAP is that its rapid metabolic reduction limits its ability to penetrate efficiently into hypoxic tissue (Rischin et al. 2006). An analog from the University of Auckland with improved extravascular penetration characteristics, SN30000 (Fig. 35.5), provides superior hypoxic cell killing in xenograft models (Hicks et al. 2010) and is currently in clinical development through Cancer Research UK. SN30000 provides an opportunity to develop an improved TPZ analogue with stronger biomarker support, utilizing recent developments in both bioreductive probes and molecular markers of hypoxia.

Although the benzotriazine di-N-oxides have not been explored for treatment of leukemias, it is noteworthy that these agents have significant myelotoxicity in mice and humans (Rischin et al. 2005; Hicks et al. 2010) and that TPZ is selectively toxic to presumptively hypoxic HSC in the BM in mice (Parmar et al. 2007). Thus, expanded hypoxia in the BM and other involved tissues in advanced leukemia may

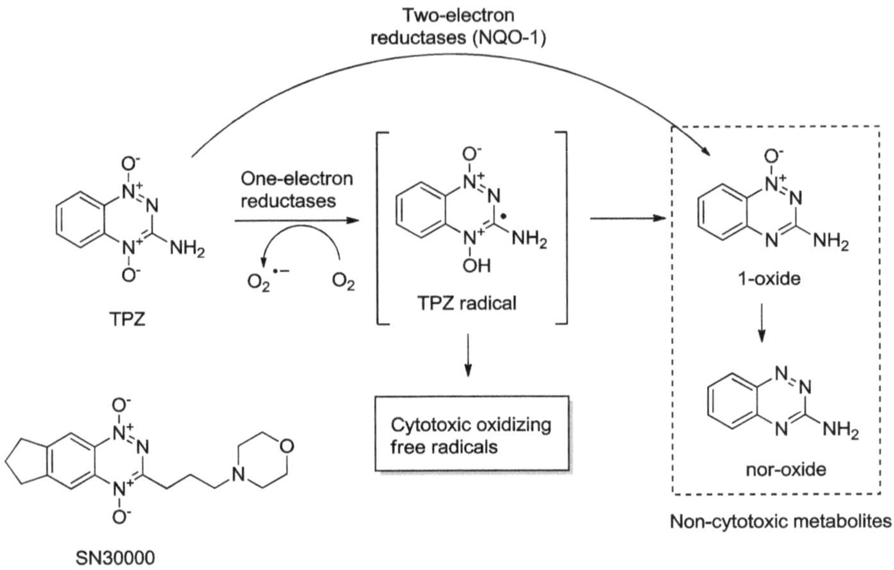

Fig. 35.5 Structure and mechanism of reductive activation of the benzotriazine di-*N*-oxide tirapazamine (*TPZ*). The initial free radical spontaneously gives rise to oxidizing radicals capable of inducing DNA strand breaks. The two-electron (1-oxide) and four-electron (nor-oxide) reduction products are not cytotoxic. SN30000 is activated by a similar bioreductive mechanism

be exploitable by these agents. Unlike the above nitro compounds, which provide relatively stable cytotoxic agents on reduction, the free radical cytotoxic agents formed on reduction of the benzotriazine di-*N*-oxides do not appear to provide a bystander effect (Wilson et al. 2007). However, this possible disadvantage is offset by activation at more moderate levels of hypoxia than for nitroaromatic HAPs (Hicks et al. 2010; Koch 1993b; Hicks et al. 2004), potentially involving much of the BM in advanced leukemia. In addition, the large pool of circulating blasts in advanced leukemia will be beyond the reach of local bystander effects, and will need to be addressed using different strategies such as repeat dosing to eliminate cells as they reseed into hypoxic niches and/or combination with agents effective against oxic leukemic cells.

Conclusion

Emerging preclinical data indicate the previously unrecognized existence of expanded hypoxic areas in the tumor-infiltrated BM in hematologic malignancies. Early reports indicate that hypoxia, similar to solid tumors, functions as a mediator of chemoresistance through regulation of angiogenesis, metabolic switching, modulation of microenvironment, and other mechanisms which represent active areas of ongoing investigation. Identification of the key downstream targets and pathways mediating resistance of hematopoietic tissue-derived tumors will facilitate development of the molecularly targeted therapies addressing this major component of diseased microenvironment. While targeting of transcription factors such as HIF-1/2 remains a challenge, exploiting tumor hypoxia with HAPs appears to be feasible and has already moved from preclinical experiments into early human clinical trials. In the longer term, development of HAP that is triggered by hypoxia to release molecularly targeted inhibitors (rather than DNA-reactive cytotoxins) may provide better-tolerated strategies for exploiting hypoxia. One example of this approach is provided by bioreductive prodrugs of AGT inhibitors; these HAPs are designed to overcome the lack of tumor specificity of current AGT inhibitors such as O^6-benzylguanine (Penketh et al. 2012; Patterson et al. 2009). A further example is HAPs that release irreversible panErbB kinase inhibitors when reduced in hypoxic tissue (Patterson et al. 2009). Similar approaches may have potential for addressing key pathways in hypoxic AML cells to provide highly selective targeted therapy in the future.

References

Adams GB, Chabner KT, Alley IR, Olson DP, Szczepiorkowski ZM, Poznansky MC et al (2006) Stem cell engraftment at the endosteal niche is specified by the calcium-sensing receptor. Nature 439(7076):599–603

Ahn GO, Brown M (2007) Targeting tumors with hypoxia-activated cytotoxins. Front Biosci 12:3483–3501

Anderson RF, Denny WA, Ware DC, Wilson WR (1996) Pulse radiolysis studies on the hypoxia-selective toxicity of a colbalt- mustard complex. Br J Cancer 27:48–51

Asosingh K, De Raeve H, de Ridder M, Storme GA, Willems A, Van Riet I et al (2005) Role of the hypoxic bone marrow microenvironment in 5T2MM murine myeloma tumor progression. Haematologica 90(6):810–817. Epub 2005/06/14

Azab AK, Hu J, Quang P, Azab F, Pitsillides C, Awwad R et al (2012) Hypoxia promotes dissemination of multiple myeloma through acquisition of epithelial to mesenchymal transition-like features. Blood 119(24):5782–5794. Epub 2012/03/08

Barnhart BC, Simon MC (2007) Metastasis and stem cell pathways. Cancer Metastasis Rev 26(2):261–271. Epub 2007/07/25

Baumann RP, Penketh PG, Ishiguro K, Shyam K, Zhu YL, Sartorelli AC (2010) Reductive activation of the prodrug 1,2-bis(methylsulfonyl)-1-(2-chloroethyl)-2-[[1-(4-nitrophenyl)ethoxy]car bonyl]hydrazine (KS119) selectively occurs in oxygen-deficient cells and overcomes O(6)-alkylguanine-DNA alkyltransferase mediated KS119 tumor cell resistance. Biochem Pharmacol 79(11):1553–1561

Becker A, Hansgen G, Bloching M, Weigel C, Lautenschlager C, Dunst J (1998) Oxygenation of squamous cell carcinoma of the head and neck: comparison of primary tumors, neck node metastases, and normal tissue. Int J Radiat Oncol Biol Phys 42(1):35–41. Epub 1998/09/25

Belcourt MF, Hodnick WF, Rockwell S, Sartorelli AC (1996) Differential toxicity of mitomycin C and porfiromycin to aerobic and hypoxic Chinese hamster ovary cells overexpressing human NADPH:cytochrome c (P-450) reductase. Proc Natl Acad Sci U S A 93(1):456–460

Benito J, Shi Y, Szymanska BCHBI, Lu H, Konoplev S, Fang W et al (2010) Targeting the leukemia-associated hypoxic microenvironment with hypoxia-activated prodrug PR-104. Blood (ASH Annual Meeting Abstracts) 116:868

Benito J, Shi Y, Szymanska B, Carol H, Boehm I, Lu H et al (2011) Pronounced hypoxia in models of murine and human leukemia: high efficacy of hypoxia-activated prodrug PR-104. Plos One 6(8):e23108

Benito J, Lu H, Shi Y, Jacamo R, McQueen T, Konoplev S, Bueso-Ramos C, Zweidler-McKay P, Handisides D, Hart CP, Andreeff M, Konopleva M (2012). Hypoxia activated pro-drug TH-302 induces hypoxia-dependent anti-leukemia activity in vitro and in vivo Proc AACR 103rd Annual Meeting, abstract #4641

Birtwistle J, Hayden RE, Khanim FL, Green RM, Pearce C, Davies NJ et al (2009) The aldo-keto reductase AKR1C3 contributes to 7,12-dimethylbenz(a)anthracene-3,4-dihydrodiol mediated oxidative DNA damage in myeloid cells: implications for leukemogenesis. Mutat Res 662(1–2):67–74

Borad MJ, Chiorean EG, Molina JR, Mita AC, Infante JR, Schelman WR et al (2011) Clinical benefits of TH-302, a tumor selective hypoxia-activated prodrug, and gemcitabline in first-line pancreatic cancer (PanC). J Clin Oncol 29:(4) (abstract 265)

Borad MJ, Reddy S, Uronis H, Sigal DS, Cohn AL, Schelman WR et al (2012) Randomized phase II study of the efficacy and safety of gemcitabine + TH-302 (G + T) vs gemcitabine (G) alone in previously untreated patients with advanced pancreatic cancer. Proc AACR 103rd Annual Meeting:Abstract LB-121

Borch RF, Liu J, Schmidt JP, Marakovits JT, Joswig C, Gipp JJ et al (2000) Synthesis and evaluation of nitroheterocyclic phosphoramidates as hypoxia-selective alkylating agents. J Med Chem 43(11):2258–2265

Borch RF, Liu J, Joswig C, Baggs RB, Dexter DL, Mangold GL (2001) Antitumor activity and toxicity of novel nitroheterocyclic phosphoramidates. J Med Chem 44(1):74–77

Brown JM (1993) SR 4233 (tirapazamine): a new anticancer drug exploiting hypoxia in solid tumours. Br J Cancer 67(6):1163–1170

Brown JM, William WR (2004) Exploiting tumour hypoxia in cancer treatment. Nat Rev Cancer 4(6):437–447

Buffa FM, Harris AL, West CM, Miller CR (2010) Large meta-analysis of multiple cancers reveals a common, compact and highly prognostic hypoxia metagene. Br J Cancer 102:428–435

Calvi LM, Adams GB, Weibrecht KW, Weber JM, Olson DP, Knight MC et al (2003) Osteoblastic cells regulate the haematopoietic stem cell niche. Nature 425(6960):841–846

Chan DA, Giaccia AJ (2007) Hypoxia, gene expression, and metastasis. Cancer Metastasis Rev 26(2):333–339. Epub 2007/04/27

Chen Y, Hu L (2009) Design of anticancer prodrugs for reductive activation. Med Res Rev 29(1):29–64

Chen JQ, Russo J (2012) Dysregulation of glucose transport, glycolysis, TCA cycle and glutaminolysis by oncogenes and tumor suppressors in cancer cells. Biochim Biophys Acta 1826(2):370–384. Epub 2012/07/04

Chow DC, Wenning LA, Miller WM, Papoutsakis ET (2001a) Modeling pO(2) distributions in the bone marrow hematopoietic compartment. II. Modified Kroghian models. Biophys J 81(2):685–696. Epub 2001/07/21

Chow DC, Wenning LA, Miller WM, Papoutsakis ET (2001b) Modeling pO(2) distributions in the bone marrow hematopoietic compartment. I. Krogh's model. Biophys J 81(2):675–684. Epub 2001/07/21

Chu YC, Chen CZ, Lee CH, Chen CW, Chang HY, Hsiue TR (2003) Prediction of arterial blood gas values from venous blood gas values in patients with acute respiratory failure receiving mechanical ventilation. JFormos Med Assoc 102(8):539–543. Epub 2003/10/22

Collingridge DR, Piepmeier JM, Rockwell S, Knisely JP (1999) Polarographic measurements of oxygen tension in human glioma and surrounding peritumoural brain tissue. Radiotherapy and oncology : ESTRO 53(2):127–131. Epub 2000/02/09

D'Andrea AD (2010) Targeting DNA repair pathways in AML. Best Pract Res Clin Haematol. 23:469–473

Das B, Tsuchida R, Malkin D, Koren G, Baruchel S, Yeger H (2008) Hypoxia enhances tumor stemness by increasing the invasive and tumorigenic side population fraction. Stem Cells 26(7):1818–1830

Denny WA (2010) Hypoxia-activated prodrugs in cancer therapy: progress to the clinic. Future Oncol 6(3):419–428

Denny WA, Wilson WR (1986) Considerations for the design of nitrophenyl mustards as agents with selective toxicity for hypoxic tumor cells. J Med Chem 29(6):879–887

Ding L, Saunders TL, Enikolopov G, Morrison SJ (2012) Endothelial and perivascular cells maintain haematopoietic stem cells. Nature 481(7382):457–462. Epub 2012/01/28

Duan JX, Jiao H, Kaizerman J, Stanton T, Evans JW, Lan L et al (2008) Potent and highly selective hypoxia-activated achiral phosphoramidate mustards as anticancer drugs. J Med Chem 51(8):2412–2240

Dubois L, Peeters S, Lieuwes NG, Geusens N, Thiry A, Wigfield S et al (2011) Specific inhibition of carbonic anhydrase IX activity enhances the in vivo therapeutic effect of tumor irradiation 9. Radiother Oncol 99(3):424–431

Evans SM, Judy KD, Dunphy I, Jenkins WT, Nelson PT, Collins R et al (2004a) Comparative measurements of hypoxia in human brain tumors using needle electrodes and EF5 binding. Cancer Res 64(5):1886–1892. Epub 2004/03/05

Evans SM, Judy KD, Dunphy I, Jenkins WT, Hwang WT, Nelson PT et al (2004b) Hypoxia is important in the biology and aggression of human glial brain tumors. Clin Cancer Res 10(24):8177–8184. Epub 2004/12/30

Fiegl M, Samudio I, Clise-Dwyer K, Burks JK, Mnjoyan Z, Andreeff M (2009) CXCR4 expression and biologic activity in acute myeloid leukemia are dependent on oxygen partial pressure. Blood 113(7):1504–1512

Frolova O, Samudio I, Benito J, Jacamo R, Kornblau SM, Markovic A et al (2012a) Regulation of HIF-1alpha signaling and chemoresistance in acute lymphocytic leukemia under hypoxic conditions of the bone marrow microenvironment. Cancer Biol Ther 13(10):1–13

Frolova O, Samudio I, Benito JM, Jacamo R, Kornblau SM, Markovic A et al (2012b) Regulation of HIF-1α signaling and chemoresistance in acute lymphocytic leukemia under hypoxic conditions of the bone marrow microenvironment. Cancer Biol Ther 13(10):858–870

Fujisaki J, Wu J, Carlson AL, Silberstein L, Putheti P, Larocca R et al (2011) In vivo imaging of Treg cells providing immune privilege to the haematopoietic stem-cell niche. Nature 474(7350):216–219. Epub 2011/06/10

Ganjoo KN, Cranmer LD, Butrynski JE, Rushing D, Adkins D, Okuno SH et al (2011) A phase I study of the safety and pharmacokinetics of the hypoxia-activated prodrug TH-302 in combination with doxorubicin in patients with advanced soft tissue sarcoma. Oncology 80(1–2):50–56

Giles F (2009) The return of alkylators to non-myeloablative therapy of AML. Leuk Res 33:1022–3

Giles F, Rizzieri D, Karp J, Vey N, Ravandi F, Faderl S et al (2007) Cloretazine (VNP40101M), a novel sulfonylhydrazine alkylating agent, in patients age 60 years or older with previously untreated acute myeloid leukemia. J Clin Oncol 25(1):25–31

Giles F, Vey N, DeAngelo D, Seiter K, Stock W, Stuart R et al (2009) Phase 3 randomized, placebo-controlled, double-blind study of high-dose continuous infusion cytarabine alone or with laromustine (VNP40101M) in patients with acute myeloid leukemia in first relapse. Blood 114(19):4027–4033

Gu Y, Patterson AV, Atwell GJ, Chernikova SB, Brown JM, Thompson LH et al (2009) Roles of DNA repair and reductase activity in the cytotoxicity of the hypoxia-activated dinitrobenzamide mustard PR-104A. Mol Cancer Ther 8(6):1714–1723

Gu Y, Tingle MD, Wilson WR (2011) Glucuronidation of anticancer prodrug PR-104A: Species differences, identification of human UDP-glucuronosyltransferases and implications for therapy. J Pharmacol Exp Ther 337(3):692–702

Guise CP, Wang A, Thiel A, Bridewell D, Wilson WR, Patterson AV (2007) Identification of human reductases that activate the dinitrobenzamide mustard prodrug PR-104A: a role for NADPH:cytochrome P450 oxidoreductase under hypoxia. Biochem Pharmacol 74(6):810–820

Guise CP, Abbattista MR, Singleton RS, Holford SD, Connolly J, Dachs GU et al (2010) The bioreductive prodrug PR-104A is activated under aerobic conditions by human aldo-keto reductase 1C3. Cancer Res 70(4):1573–1584

Guise CP, Abbattista MR, Tipparaju SR, Lambie NK, Su J, Li D et al (2012) Diflavin oxidoreductases activate the Bioreductive Prodrug PR-104A under Hypoxia. Mol Pharmacol 81(1):31–40

Halim M, Yee DJ, Sames D (2008) Imaging induction of cytoprotective enzymes in intact human cells: coumberone, a metabolic reporter for human AKR1C enzymes reveals activation by panaxytriol, an active component of red ginseng. J Am Chem Soc 130(43):14123–14128

Harrison JS, Rameshwar P, Chang V, Bandari P (2002) Oxygen saturation in the bone marrow of healthy volunteers. Blood 99(1):394

Hicks KO, Siim BG, Pruijn FB, Wilson WR (2004) Oxygen dependence of the metabolic activation and cytotoxicity of tirapazamine: implications for extravascular transport and activity in tumors. Radiat Res 161(6):656–666

Hicks KO, Myint H, Patterson AV, Pruijn FB, Siim BG, Patel K et al (2007) Oxygen dependence and extravascular transport of hypoxia-activated prodrugs: Comparison of the dinitrobenzamide mustard PR-104A and tirapazamine. Int J Radiat Oncol Biol Phys 69(2):560–571

Hicks KO, Siim BG, Jaiswal JK, Pruijn FB, Fraser AM, Patel R et al (2010) Pharmacokinetic/pharmacodynamic modeling identifies SN30000 and SN29751 as tirapazamine analogues with improved tissue penetration and hypoxic cell killing in tumors Clin Cancer Res 16(20):4946–4957

Houghton PJ, Lock R, Carol H, Morton CL, Phelps D, Gorlick R et al (2011) Initial testing of the hypoxia activated prodrug PR-104 by the pediatric preclinical testing program. Pediatr Blood Cancer 57(3):443–453

Hu J, Handisides DR, Van VE, De RH, Menu E, Vande BI et al (2010) Targeting the multiple myeloma hypoxic niche with TH-302, a hypoxia-activated prodrug. Blood 116(9):1524–7

Hunter FW, Wang J, Patel R, Hsu HL, Hickey AJ, Hay MP et al (2012) Homologous recombination repair-dependent cytotoxicity of the benzotriazine di-N-oxide CEN-209: comparison with other hypoxia-activated prodrugs. Biochem Pharmacol 83(5):574–585

Ivanovic Z (2009) Hypoxia or in situ normoxia: the stem cell paradigm. J Cell Physiol 219(2):271–275. Epub 2009/01/23

Jameson MB, Rischin D, Pegram M, Gutheil J, Patterson AV, Denny WA et al (2010) A phase I trial of PR-104, a nitrogen mustard prodrug activated by both hypoxia and aldo-keto reductase 1C3, in patients with solid tumors. Cancer Chemother Pharmacol 65(4):791–801

Jamieson SMF, Brooke D, Heinrich D, Atwell GJ, Silva S, Hamilton EJ et al (2012) 3-(3,4-Dihydroisoquinolin-2(1H)-ylsulfonyl)benzoic acids; highly potent and selective inhibitors of thetype 5 17-#-hydroxysteroid dehydrogenase AKR1C3. J Med Chem. Epub 2012/08/09

Jensen PO, Mortensen BT, Hodgkiss RJ, Iversen PO, Christensen IJ, Helledie N et al (2000) Increased cellular hypoxia and reduced proliferation of both normal and leukaemic cells during progression of acute myeloid leukaemia in rats. Cell Prolif 33(6):381–395

Jiao M, Nan KJ (2012) Activation of PI3 kinase/Akt/HIF-1alpha pathway contributes to hypoxia-induced epithelial-mesenchymal transition and chemoresistance in hepatocellular carcinoma. Int J Oncol 40(2):461–468. Epub 2011/09/17

Jung D, Lin L, Jiao H, Cai X, Duan JX, Matteucci M (2012) Pharmacokinetics of TH-302: a hypoxically activated prodrug of bromo-isophosphoramide mustard in mice, rats, dogs and monkeys. Cancer Chemother Pharmacol 69(3):643–654

Jungermann K, Kietzmann T (1997) Role of oxygen in the zonation of carbohydrate metabolism and gene expression in liver. Kidney Int 51(2):402–412. Epub 1997/02/01

Kataoka K, Sato T, Yoshimi A, Goyama S, Tsuruta T, Kobayashi H et al (2011) Evi1 is essential for hematopoietic stem cell self-renewal, and its expression marks hematopoietic cells with long-term multilineage repopulating activity. J Exp Med 208(12):2403–2416. Epub 2011/11/16

Kayama T, Yoshimoto T, Fujimoto S, Sakurai Y (1991) Intratumoral oxygen pressure in malignant brain tumor. J Neurosurg 74(1):55–59. Epub 1991/01/01

Kiel MJ, Yilmaz OH, Iwashita T, Terhorst C, Morrison SJ (2005) SLAM family receptors distinguish hematopoietic stem and progenitor cells and reveal endothelial niches for stem cells. Cell 121(7):1109–1121

Kim EY, Liu Y, Akintujoye OM, Shyam K, Grove TA, Sartorelli AC et al (2012) Preliminary studies with a new hypoxia-selective cytotoxin, KS119W, in vitro and in vivo. Radiat Res 178:126–137. Epub ahead of print

Koch CJ (1993a) Unusual oxygen concentration dependence of toxicity of SR-4233, a hypoxic cell toxin. Cancer Res 53(17):3992–3997. Epub 1993/09/01

Koch CJ (1993b) Unusual oxygen concentration dependence of toxicity of SR-4233, a hypoxic cell toxin. Cancer Res 53(17):3992–3997

Konopleva M, Borthakur G, Thall PF, Coveler A, Ravandi F, Jabbour E et al (2011) Phase I/II study of PR104, a bioreductive prodrug, in patients with relapsed/refractory Acute Myeloid Leukemia (AML) using patient-specific adaptive dose selection. Blood 118(21):661

Konopleva M, Handisides D, Lorente GA, Benito JM, Borthakur G, Jabbour E, Faderl S, Cortes JE, Kroll S, Andreeff M, Kantarjian H, Thomas DA (2012) A Phase 1 study of TH-302, a hypoxia-activated cytotoxic prodrug, in subjects with advanced leukemias. ASCO 2012 annual meeting. J Clin Oncol 30:2314–2326 (suppl; abstr 6585)

Koong AC, Mehta VK, Le QT, Fisher GA, Terris DJ, Brown JM et al (2000) Pancreatic tumors show high levels of hypoxia. Intern J Radiat Oncol Biol Phys 48(4):919–922

Le QT, Kovacs MS, Dorie MJ, Koong A, Terris DJ, Pinto HA et al (2003) Comparison of the comet assay and the oxygen microelectrode for measuring tumor oxygenation in head-and-neck cancer patients. Int J Radiat Oncol Biol Phys 56(2):375–383. Epub 2003/05/10

Le QT, Fisher R, Oliner KS, Young RJ, Cao H, Kong C et al (2012) Prognostic and predictive significance of plasma HGF and IL-8 in a phase III trial of chemoradiation with or without tirapazamine in locoregionally advanced head and neck cancer. Clin Cancer Res 18(6):1798–1807

Liu Q, Sun JD, Wang J, Ahluwalia D, Baker AF, Cranmer LD et al (2012) TH-302, a hypoxia-activated prodrug with broad in vivo preclinical combination therapy efficacy: optimization of dosing regimens and schedules. Cancer Chemother Pharmacol 69(6):1487–1498

Lo Celso C, Fleming HE, Wu JW, Zhao CX, Miake-Lye S, Fujisaki J et al (2009) Live-animal tracking of individual haematopoietic stem/progenitor cells in their niche. Nature 457(7225):92–96. Epub 2008/12/05

Lu X, Kang Y (2010) Hypoxia and hypoxia-inducible factors: master regulators of metastasis. Clin Cancer Res 16(24):5928–5935. Epub 2010/10/22

Lukashev D, Ohta A, Sitkovsky M (2007) Hypoxia-dependent anti-inflammatory pathways in protection of cancerous tissues. Cancer Metastasis Rev 26(2):273–279. Epub 2007/04/04

Malatesha G, Singh NK, Bharija A, Rehani B, Goel A (2007) Comparison of arterial and venous pH, bicarbonate, PCO2 and PO2 in initial emergency department assessment. Emerg Med J 24(8):569–571. Epub 2007/07/27

Manesh DM, Szymanska B, Carol H, Boehm I, Evans K, Jamieson S, Wilson WR, Houghton PJ, Smith MA, Lock RB (2012) Molecular determinants of T-cell acute lymphoblastic leukaemia sensitivity to the pre-prodrug PR-104. 24th Lorne Cancer Conference 2012. 8–10 Feb 2012, 259

Martin SK, Diamond P, Gronthos S, Peet DJ, Zannettino AC (2011) The emerging role of hypoxia, HIF-1 and HIF-2 in multiple myeloma. Leukemia 25(10):1533–1542. Epub 2011/06/04

Mason RP, Holtzman JL (1975) The role of catalytic superoxide formation in the O_2 inhibition of nitroreductase. Biochem Biophys Res Commun 67(4):1267–1274

Matsunaga T, Imataki O, Torii E, Kameda T, Shide K, Shimoda H et al (2012) Elevated HIF-1alpha expression of acute myelogenous leukemia stem cells in the endosteal hypoxic zone may be a cause of minimal residual disease in bone marrow after chemotherapy. Leuk Res 36(6):e122–e124. Epub 2012/03/27

McDonald PC, Winum JY, Supuran CT, Dedhar S (2012) Recent developments in targeting carbonic anhydrase IX for cancer therapeutics. Oncotarget 3(1):84–97

McKeage MJ, Gu Y, Wilson WR, Hill A, Amies K, Melink TJ et al (2011) A phase I trial of PR-104, a pre-prodrug of the bioreductive prodrug PR-104A, given weekly to solid tumour patients. BMC Cancer 11:432. Epub 2011/10/11

McKeown SR, Cowen RL, Williams KJ (2007) Bioreductive drugs: from concept to clinic. Clin Oncol 19:427–442

Meng F, Evans JW, Bhupathi D, Banica M, Lan L, Lorente G et al (2012) Molecular and cellular pharmacology of the hypoxia-activated prodrug TH-302. Mol Cancer Ther 11(3):740–751

Miharada K, Karlsson G, Rehn M, Rorby E, Siva K, Cammenga J et al (2011) Cripto regulates hematopoietic stem cells as a hypoxic-niche-related factor through cell surface receptor GRP78. Cell Stem Cell 9(4):330–344. Epub 2011/10/11

Mik EG, van Leeuwen TG, Raat NJ, Ince C (2004) Quantitative determination of localized tissue oxygen concentration in vivo by two-photon excitation phosphorescence lifetime measurements. J Appl Physiol 97(5):1962–1969. Epub 2004/07/13

Moeller BJ, Richardson RA, Dewhirst MW (2007) Hypoxia and radiotherapy: opportunities for improved outcomes in cancer treatment. Cancer Metastasis Rev 26(2):241–248. Epub 2007/04/19

Morrison SJ, Spradling AC (2008) Stem cells and niches: mechanisms that promote stem cell maintenance throughout life. Cell 132(4):598–611

Mullally A, Poveromo L, Schneider RK, Al-Shahrour F, Lane SW, Ebert BL (2012) Distinct roles for long-term hematopoietic stem cells and erythroid precursor cells in a murine model of Jak2V617F polycythemia vera. Blood 121(18):3692–3702. Epub 2012/05/26

Nilsson SK, Johnston HM, Coverdale JA (2001) Spatial localization of transplanted hemopoietic stem cells: inferences for the localization of stem cell niches. Blood 97(8):2293–2299

Nordgren IK, Tavassoli A (2011) Targeting tumour angiogenesis with small molecule inhibitors of hypoxia inducible factor. Chem Soc Rev 40(8):4307–4317

Nordsmark M, Bentzen SM, Overgaard J (1994) Measurement of human tumour oxygenation status by a polarographic needle electrode. An analysis of inter- and intratumour heterogeneity. Acta Oncol 33(4):383–389. Epub 1994/01/01

Nordsmark M, Eriksen JG, Gebski V, Alsner J, Horsman MR, Overgaard J (2007) Differential risk assessments from five hypoxia specific assays: the basis for biologically adapted individualized radiotherapy in advanced head and neck cancer patients. Radiother Oncol 83(3):389–397

Notta F, Doulatov S, Laurenti E, Poeppl A, Jurisica I, Dick JE (2011) Isolation of single human hematopoietic stem cells capable of long-term multilineage engraftment. Science 333(6039):218–221. Epub 2011/07/09

Parmar K, Mauch P, Vergilio JA, Sackstein R, Down JD (2007) Distribution of hematopoietic stem cells in the bone marrow according to regional hypoxia. Proc Natl Acad Sci U S A 104(13):5431–5436. Epub 2007/03/22

Patel K, Choy SF, Hicks KO, Melink TJ, Holford NHG, Wilson WR (2011) A combined pharmacokinetic model for the hypoxia-targeted prodrug PR-104A in humans, dogs, rats and mice predicts species differences in clearance and toxicity. Cancer Chemother Pharmacol 67:1145–1155

Patterson LH (2002) Bioreductively activated antitumor N-oxides: the case of AQ4N, a unique approach to hypoxia-activated cancer chemotherapy. Drug Metab Rev 34(3):581–592

Patterson AV, Ferry DM, Edmunds SJ, Gu YC, Singleton RS, Patel K et al (2007) Mechanism of action and preclinical antitumor activity of the novel hypoxia-activated DNA cross-linking agent PR-104. Clinical Cancer Res 13(13):3922–3932

Patterson AV, Jaiswal J, Syddall SP; Abbattista M, van Leeuwen W, Puryer M, Thompson A, Hsu A, Mehta S, Pruijn A, Lu GL, Donate F, Denny WA, Wilson WR, Smaill JB (2009) Cellular metabolism, murine pharmacokinetics and preclinical antitumor activity of SN29966, a novel hypoxia-activated irreversible pan-HER inhibitor. Mol Cancer Ther 8(12 Suppl 1):B79 (abstract)

Penketh PG, Shyam K, Baumann RP, Ishiguro K, Patridge EV, Zhu R et al (2012) A Strategy for Selective O(6) -Alkylguanine-DNA Alkyltransferase Depletion Under Hypoxic Conditions. Chem Biol Drug Design 80(2):279–290. Epub 2012/05/05

Peters LJ, O'Sullivan B, Giralt J, Fitzgerald TJ, Trotti A, Bernier J et al (2010) Critical impact of radiotherapy protocol compliance and quality in the treatment of advanced head and neck cancer: Results from TROG 02.02. J Clin Oncol 28:2996–3001

Plumb JA, Gerritsen M, Workman P (1994) DT-diaphorase protects cells from the hypoxic cytotoxicity of indoloquinone EO9. Br J Cancer 70(6):1136–1143

Rampling R, Cruickshank G, Lewis AD, Fitzsimmons SA, Workman P (1994) Direct measurement of pO2 distribution and bioreductive enzymes in human malignant brain tumors. Int J Radiat Oncol Biol Phys 29(3):427–431. Epub 1994/06/15

Reddy SB, Williamson SK (2009) Tirapazamine: a novel agent targeting hypoxic tumor cells. Expert Opin Investig Drugs 18(1):77–87

Rehn M, Olsson A, Reckzeh K, Diffner E, Carmeliet P, Landberg G et al (2011) Hypoxic induction of vascular endothelial growth factor regulates murine hematopoietic stem cell function in the low-oxygenic niche. Blood 118(6):1534–1543. Epub 2011/06/15

Rischin D, Peters L, Fisher R, Macann A, Denham J, Poulsen M et al (2005) Tirapazamine, cisplatin, and radiation versus fluorouracil, cisplatin, and radiation in patients with locally advanced head and neck cancer: a randomized phase II trial of the Trans-Tasman Radiation Oncology Group (TROG 98.02). J Clin Oncol 23(1):79–87

Rischin D, Hicks RJ, Fisher R, Binns D, Corry J, Porceddu S et al (2006) Prognostic significance of [18F]-misonidazole positron emission tomography-detected tumor hypoxia in patients with advanced head and neck cancer randomly assigned to chemoradiation with or without tirapazamine: a substudy of Trans-Tasman Radiation Oncology Group Study 98.02. J Clin Oncol 24(13):2098–2104

Rohwer N, Cramer T (2011) Hypoxia-mediated drug resistance: novel insights on the functional interaction of HIFs and cell death pathways. Drug Resist Updat 14(3):191–201. Epub 2011/04/07

Roy S, Khanna S, Wallace WA, Lappalainen J, Rink C, Cardounel AJ et al (2003) Characterization of perceived hyperoxia in isolated primary cardiac fibroblasts and in the reoxygenated heart. J Biol Chem 278(47):47129–47135. Epub 2003/09/04

Saunders P, Cisterne A, Weiss J, Bradstock KF, Bendall LJ (2011) The mammalian target of rapamycin inhibitor RAD001 (everolimus) synergizes with chemotherapeutic agents, ionizing radiation and proteasome inhibitors in pre-B acute lymphocytic leukemia. Haematologica 96(1):69–77

Schofield R (1983) The stem cell system. Biomed Pharmacother 37(8):375–380. Epub 1983/01/01

Schults MA, Timmermans L, Godschalk RW, Theys J, Wouters BG, van Schooten FJ et al (2010) Diminished carcinogen detoxification is a novel mechanism for hypoxia-inducible factor 1-mediated genetic instability. J Biol Chem 285(19):14558–14564. Epub 2010/03/17

Semenza GL (2007) Oxygen-dependent regulation of mitochondrial respiration by hypoxia-inducible factor 1. Biochem J 405(1):1–9. Epub 2007/06/09

Semenza GL (2009) Regulation of cancer cell metabolism by hypoxia-inducible factor 1. Semin Cancer Biol 19(1):12–16. Epub 2008/12/31

Semenza G (2010) Defining the role of hypoxia-inducible factor 1 in cancer biology and therapeutics. Oncogene 29(5):625–634. Epub 2009/12/01

Seow HA, Penketh PG, Shyam K, Rockwell S, Sartorelli AC (2005) 1,2-Bis(methylsulfonyl)-1-(2-chloroethyl)-2-[[1-(4-nitrophenyl)ethoxy]carbonyl]hydrazine: an anticancer agent targeting hypoxic cells. Proc Natl Acad Sci U S A 102(26):9282–9287

Shapiro BA (1995) Temperature correction of blood gas values. Respir Care Clin N Am 1(1):69–76. Epub 1995/09/01

Siim BG, Atwell GJ, Wilson WR (1994) Oxygen dependence of the cytotoxicity and metabolic activation of 4-alkylamino-5-nitroquinoline bioreductive drugs. Br J Cancer 70(4):596–603. Epub 1994/10/01

Simsek T, Kocabas F, Zheng J, DeBerardinis RJ, Mahmoud AI, Olson EN et al (2010) The distinct metabolic profile of hematopoietic stem cells reflects their location in a hypoxic niche. Cell Stem Cell 7(3):380–390

Singleton RS, Guise CP, Ferry DM, Pullen SM, Dorie MJ, Brown JM et al (2009) DNA crosslinks in human tumor cells exposed to the prodrug PR-104A: relationships to hypoxia, bioreductive metabolism and cytotoxicity. Cancer Res 69(9):3884–3891

Skouby AR (1976) Haematologic adaptation in patients with chronic bronchitis and pulmonary insufficiency. Acta Med Scand 199(3):185–190. Epub 1976/01/01

Sporn MB, Liby KT (2012) NRF2 and cancer: the good, the bad and the importance of context. Nat Rev Cancer 12(8):564–571. Epub 2012/07/20

Spradling A, Drummond-Barbosa D, Kai T (2001) Stem cells find their niche. Nature 414(6859):98–104. Epub 2001/11/02

Sullivan R, Graham CH (2007) Hypoxia-driven selection of the metastatic phenotype. Cancer Metastasis Rev 26(2):319–331. Epub 2007/04/27

Swietach P, Vaughan-Jones RD, Harris AL (2007) Regulation of tumour pH and the role of carbonic anhydrase 9. Cancer Metastasis Rev 26:299–310

Takubo K, Goda N, Yamada W, Iriuchishima H, Ikeda E, Kubota Y et al (2010) Regulation of the HIF-1α level is essential for hematopoietic stem cells. Cell Stem Cell 7(3):391–402

Taylor YC, Rauth AM (1982) Oxygen tension, cellular respiration, and redox state as variables influencing the cytotoxicity of the radiosensitizer misonidazole. Radiat Res 91(1):104–123. Epub 1982/07/01

Thomlinson RH, Gray LH (1955) The histological structure of some human lung cancers and the possible implications for radiotherapy. Br J Cancer 9(4):539–549. Epub 1955/12/01

Toustrup K, Sorensen BS, Nordsmark M, Busk M, Wiuf C, Alsner J et al (2011) Development of a hypoxia gene expression classifier with predictive impact for hypoxic modification of radiotherapy in head and neck cancer. Cancer Res 71(17):5923–5931

Toustrup K, Sorensen BS, Alsner J, Overgaard J (2012) Hypoxia gene expression signatures as prognostic and predictive markers in head and neck radiotherapy. Semin Radiat Oncol 22(2):119–127

Vaupel P, Hockel M, Mayer A (2007) Detection and characterization of tumor hypoxia using pO2 histography. Antioxid Redox Signal 9(8):1221–1235. Epub 2007/06/01

Visnjic D, Kalajzic Z, Rowe DW, Katavic V, Lorenzo J, Aguila HL (2004). Hematopoiesis is severely altered in mice with an induced osteoblast deficiency. Blood 103(9):3258–3264. Epub 2004/01/17

Walkey AJ, Farber HW, O'Donnell C, Cabral H, Eagan JS, Philippides GJ (2010) The accuracy of the central venous blood gas for acid-base monitoring. J Intensive Care Med 25(2):104–110. Epub 2009/12/19

Wang J, Foehrenbacher A, Su J, Patel R, Hay MP, Hicks KO et al (2012) The 2-nitroimidazole EF5 is a biomarker for oxidoreductases that activate bioreductive prodrug CEN-209 under hypoxia. Clin Cancer Res 18(6):1684–1695

Wardman P (2001) Electron transfer and oxidative stress as key factors in the design of drugs selectively active in hypoxia. Curr Med Chem 8(7):739–761

Weidemann A, Johnson RS (2008) Biology of HIF-1alpha. Cell Death Differ 15(4):621–627. Epub 2008/02/09

Welch WJ, Baumgartl H, Lubbers D, Wilcox CS (2001) Nephron pO2 and renal oxygen usage in the hypertensive rat kidney. Kidney Int 59(1):230–237. Epub 2001/01/03

Wilson WR, Hay MP (2011) Targeting hypoxia in cancer therapy. Nat Rev Cancer 11(6):393–410

Wilson WR, Hicks KO, Pullen SM, Ferry DM, Helsby NA, Patterson AV (2007) Bystander effects of bioreductive drugs: potential for exploiting pathological tumor hypoxia with dinitrobenzamide mustards. Radiat Res 167(6):625–636

Winkler IG, Barbier V, Wadley R, Zannettino AC, Williams S, Levesque JP (2010) Positioning of bone marrow hematopoietic and stromal cells relative to blood flow in vivo: serially reconstituting hematopoietic stem cells reside in distinct nonperfused niches. Blood 116(3):375–385. Epub 2010/04/16

Wolfle D, Jungermann K (1985) Long-term effects of physiological oxygen concentrations on glycolysis and gluconeogenesis in hepatocyte cultures. Eur J Biochem 151(2):299–303. Epub 1985/09/02

Xia Y, Choi HK, Lee K (2012) Recent advances in hypoxia-inducible factor (HIF)-1 inhibitors. Eur J Med Chem 49:24–40

Zhang J, Niu C, Ye L, Huang H, He X, Tong WG et al (2003) Identification of the haematopoietic stem cell niche and control of the niche size. Nature 425(6960):836–841

Zhong H, De Marzo AM, Laughner E, Lim M, Hilton DA, Zagzag D et al (1999) Overexpression of hypoxia-inducible factor 1alpha in common human cancers and their metastases. Cancer Res 59(22):5830–5835

Zhu Y, Zhao T, Itasaka S, Zeng L, Yeom CJ, Hirota K et al (2012) Involvement of decreased hypoxia-inducible factor 1 activity and resultant G(1)-S cell cycle transition in radioresistance of perinecrotic tumor cells. Oncogene 2012:1–11. Epub 2012/06/20

Part IX
Micro-RNAs

Chapter 36
MicroRNAs as Therapeutic Targets

Maitri Y. Shah and George A. Calin

Abstract MicroRNAs (miRNAs) are evolutionarily conserved class of small, regulatory RNAs that negatively regulate gene expression, and control multiple cellular processes involved in cancer initiation, progression, and metastasis. Extensive research in the past decade has identified vital role of miRNAs in hematopoiesis and leukemogenesis, with miRNAs functioning as oncogenes or tumor suppressors. Several miRNAs are extensively dysregulated in acute myeloid leukemia (AML), and their expression patterns are useful diagnostic and prognostic markers in AML. Furthermore, miRNA expression patterns have been associated with specific cytogenetic and molecular aberrations in AML subsets. Based on these observations, miRNAs have emerged as promising therapeutic tools for AML. In this chapter, we focus on the roles of miRNAs in AML tumorigenesis, their widespread deregulation, the rationale and strategies for their use as AML therapeutic targets, and the advantages and current challenges to their use in the clinic.

Keywords MicroRNA · AML · Therapeutics · Anti-miRs · MiRNA mimics

36.1 Introduction

Acute myeloid leukemia (AML) is a heterogeneous disease of hematopoietic progenitor cells, characterized by proliferation of maturation-arrested myeloid blasts in bone marrow and blood. These blasts lose the ability to differentiate normally

G. A. Calin (✉) · M. Y. Shah
Department of Experimental Therapeutics, Houston, USA
e-mail: gcalin@mdanderson.org

Graduate School of Biomedical Sciences, Houston, USA

G. A. Calin
Center for RNA Interference and Non-Coding RNAs, The University of Texas
MD Anderson Cancer Center, Houston, TX 77030, USA

Department of Experimental Therapeutics,
Division of Cancer Medicine
The University of Texas MD Anderson Cancer Center 1515 Holcombe
Blvd. Houston, Texas 77030, USA

M. Andreeff (ed.), *Targeted Therapy of Acute Myeloid Leukemia*,
Current Cancer Research, DOI 10.1007/978-1-4939-1393-0_36

and thus do not respond to normal regulators of proliferation, resulting in cancer. It is the most common malignant myeloid disorder in adults, and its incidence increases with age. The pathogenesis of AML is complex and constitutes distinct genotypic entities that can be recognized and categorized based on their clinical, cytogenetic, and molecular features (Marcucci et al. 2011a). Recurrent chromosomal aberrations and genetic mutations contribute to the increased myeloid cell proliferation and have been identified as one of the most important prognostic factors in AML. The most common genetic alterations observed clinically include translocations (i.e., t(15;17); t(8;21); t(9;11)), inversions (i.e., inv(16)), deletions (i.e., del(7q)), trisomies (i.e., +8), and monosomies (i.e., −5/−7) (Havelange et al. 2009). Nevertheless, about 45 % of AML patients present with a normal karyotype, with no evident chromosomal instability, are thus classified as cytogenetically normal (CN). For the past several decades, the overarching focus of AML research and therapy has been on the chromosomal and proteomic perturbations that are rampant in AML. However, in the recent years, the development of novel targeted therapeutics has fueled the assimilation of genetic, transcriptomic, and proteomic platforms to better understand the heterogeneity of the disease and guide better therapeutic decisions. With the seminal discovery of microRNAs (miRNAs) and their imperative role in human diseases, a large body of evidence has accumulated that confirms them as active players in tumorigenesis.

The discovery of miRNAs in the past decade has been one of the most fascinating breakthroughs of recent times. The past decade has shown remarkable advances in our knowledge of miRNAs and their functional importance in a wide range of developmental and physiological processes. These are small (19–24 nt), endogenous highly conserved noncoding RNAs that negatively regulate gene expression by binding to the 3′-untranslated region (3′-UTR) of the target mRNAs, and cause mRNA degradation or translational repression (Ambros 2004). Thus, they are involved in posttranscriptional gene silencing and are predicted to target about 30 % of the protein-coding genes (Lewis et al. 2003). miRNAs are usually transcribed from miRNA genes by RNA polymerase II (RNAPII) to give a long, capped, and polyadenylated primary miRNA transcript (pri-miRNA; Fig. 36.1). This stem–loop structure is then cleaved by a microprocessor complex (composed of Rnase III Drosha and double-stranded DNA binding domain protein DGCR8) to give hairpin-shaped miRNA precursors (pre-miRNAs; Bartel 2004; Kim 2005). These pre-miRNAs are exported to the cytoplasm by the nuclear export factor Exportin-5/Ran GTP, followed by further processing to approximately 22 nucleotide miRNA:miRNA* duplexes in the cytoplasm by Rnase III enzyme Dicer, and its double-stranded RNA binding domain protein TRBP. The mature miRNA then gets incorporated into RNA-induced silencing complex (RISC), usually resulting in the repression of target gene expression. In humans, miRNAs usually bind to specific sequences with partialcomplementarity in the 3′-UTR of target mRNAs transcripts, called "seed sequences" or miRNA response elements (MREs), which then result in translational repression (Ambros 2004). These miRNAs play an important role in multiple biological processes, including apoptosis, cell differentiation, developmental timings, embryogenesis, metabolism, organogenesis, and various diseases,

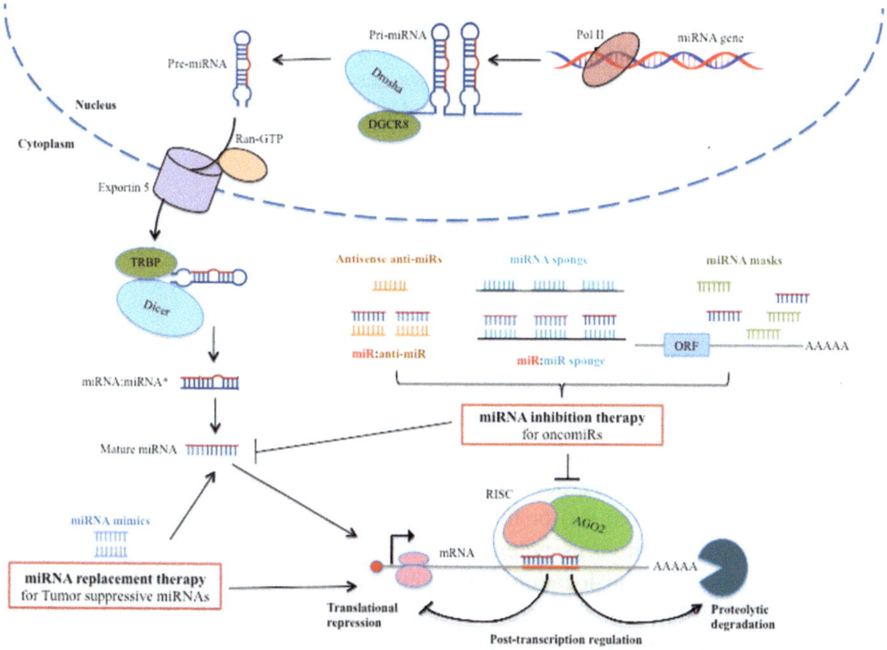

Fig. 36.1 miRNA biogenesis and miRNA-based therapy for management of cancer

including cancers (He and Hannon 2004). miRNA genes are frequently located at fragile sites, minimal regions of loss of heterozygosity, regions of amplification, and common chromosomal breakpoints (Calin et al. 2004), thus acting as tumor suppressor genes or oncogenes (Esquela-Kerscher and Slack 2006). The expression and role of miRNA in AML are just starting to be revealed, and further understanding of the molecular nature and functions of these small regulatory RNAs would help identify novel therapeutic tools for AML.

36.2 miRNAs as Therapeutic Targets for AML

Despite extensive research, AML biology remains poorly understood. The integration of the whole-genome-profiling (WGP) approach including noncoding RNAs in the large-scale gene expression studies has identified an important role of miRNAs in AML pathogenesis. AML represents a complex multistep genetic disorder, with the confederation of a few pervasive driver mutations and several less penetrant passenger mutations. The genomic deletion of miR-15a/16-1 in chronic lymphocytic leukemia (CLL) patients provided the first link between altered miRNA expression and hematological malignancies (Calin et al. 2002). Since then, several groups have reported unique miRNA expression signatures associated with recurrent cytogenetic and molecular aberrations and clinical outcomes in AML; thus recognizing miRNAs as novel therapeutic interventions for AML patients (Havelange et al. 2009; Marcucci et al. 2011b). Below, we describe a few pioneering studies that identified the most significantly deregulated miRNAs and established the translational potential of miRNAs in AML.

36.2.1 miRNA Signatures Associated with Cytogenetic, Morphological, and Molecular Characteristics

AML is a cytogenetically heterogeneous disease with abundant recurrent chromosomal abnormalities. Using integrated gene and miRNA expression profiling, several independent large-scale studies reported distinct miRNA expression signatures associated with cytogenetic and molecular abnormalities in AML. Interestingly, different research groups have documented the deregulation of the same unique oncogenic miRNAs (oncomiRs) or tumor suppressive miRNAs in different subtypes of AML, identifying them as potential therapeutic targets in AML (Schotte et al. 2012). Patients with translocations involving the mixed-lineage leukemia (*MLL*) gene in 11q23 [t(11q23)/*MLL*] are associated with a distinctive miRNA signature that includes upregulation of the oncogenic miR-17-92 cluster and miR-196b and downregulation of miR-29 family members (Dixon-McIver et al. 2008; Garzon et al. 2008b; Jongen-Lavrencic et al. 2008; Li et al. 2008; Popovic et al. 2009). In patients with acute promyelocytic leukemia (APL), a t(15;17)(q22;q12) results in a fusion protein called PML-RARα (Mrozek and Bloomfield 2008). Upregulation of miR-382 (Jongen-Lavrencic et al. 2008) and miR-181 family members (Li et al. 2008) was reported in this subset of leukemia by multiple reports. Abnormalities of chromosome 16, which affect the core binding factor beta subunit (CBFB), include t(8;21)(q22;q22), inv(16)(p13;q22), or t(16;16)(p13;q22) (Marcucci et al. 2011b). In these core binding factor (CBF)-related AMLs, let-7b and let-7c were downregulated while miR-126/126* were upregulated (Jongen-Lavrencic et al. 2008; Li et al. 2008).

Concerning CN-AMLs, miRNA expression levels correlate with the molecular subtypes, mutational status, and associated gene expression patterns in CN-AML (Havelange et al. 2009; Marcucci et al. 2011b). Upregulation of miR-155, miR-10a, and miR-10b was described in CN-AML with internal tandem duplication (ITD) of the *FLT3* gene (*FLT3-ITD*; Garzon et al. 2008b; Jongen-Lavrencic et al. 2008). Overexpression of miR-335 and miR-181 family members is related to CN-AML with *C/EBPα* mutations and is associated with good clinical outcome in high-risk CN-AML (Jongen-Lavrencic et al. 2008; Marcucci et al. 2008a). Likewise, upregulation of several miRNAs including miR-10a, miR-10b, miR-196a, miR-196b, and let-7a; and downregulation of miR-126, miR-130a, and miR-451 are associated with favorable nucleophosmin (*NPM1*)-mutations in CN-AML (Becker et al., 2010; Jongen-Lavrencic et al. 2008). On the other hand, CN-AML patients with unfavorable isocitrate dehydrogenase 2 (*IDH2*)-mutations display high levels of miR-1, miR-125b, and miR-133a (Marcucci et al. 2010). Finally, in patients with CN-AML overexpressing meningioma 1(*MN1)*, miR-126 is overexpressed and has a negative impact on the outcome (Langer et al. 2009).

Interestingly, several of the oncomiRs associated with AML are embedded in the homeotic (*HOX*) gene clusters and play an important role in regulation of the *HOX* genes. These include miR-196b in *HOXA* (at chromosome 7p15), miR-10a and 196a-1 in *HOXB* (at 17q21), miR-196a-2 in *HOXC* (at 12q13), and miR-10b in *HOXD* (at 2q31) clusters (Havelange et al. 2009). Notably, *HOX* gene family upregulation is a prominent characteristic of gene expression signatures identified in *NMP1*-mutated CN-AMLs, thus explaining the overexpression of these miRNAs in this subtype of AML (Garzon et al. 2008a). Similarly, miR-196b is co-expressed with its neighboring genes *HOXA9* and *HOX10A* in *MLL*-rearranged AML (Popovic et al. 2009). Alternatively, other oncomiRs, such as miR-126 (Shen et al. 2008) and miR-181 family (Naguibneva et al. 2006) are also important in regulation of *HOX* gene expression in AML. HOX genes are important transcription factors that regulate the early steps of hematopoiesis, and deregulation of these genes contributes to leukemogenesis in AML (Sauvageau et al. 1995). Thus, targeting of these HOX-associated miRNAs represent a novel therapeutic strategy for AML management.

36.2.2 miRNAs as Diagnostic and Prognostic Biomarkers in AML

After the pioneering study (Bullinger et al. 2004) demonstrating that gene expression profiling can be used to distinguish AML from acute lymphoblastic leukemia (ALL), several studies used miRNA profiling to identify miRNA signatures that can classify lineage-specific leukemia. Mi et al. (Mi et al. 2007) first showed that miRNA expression profiling can separate patients with AML from those with ALL on the basis of a unique miRNA signature including 21 upregulated and six downregulated miRNAs. Likewise, research groups have reported that miRNA expression pattern in AML blasts differs from the normal $CD34^+$ cells (Dixon-McIver et al. 2008; Garzon et al. 2008b; Jongen-Lavrencic et al. 2008; Li et al. 2008). Interestingly, several of these miRNAs play an important role in the regulation of hematopoiesis. However, there is a lack of uniformity in the miRNA expression profiles derived from these miRNAs.

Several studies have shown that miRNA expression changes can influence the clinical outcome of AML patients. One of the earlier studies by Garzon et al. (2008b) reported that high expression levels of miR-191 and miR-199a correlated with poor overall survival and progression-free survival. These miRNAs were identified to independently predict clinical outcome in AML patients. Another study by Marcucci et al. (2008b) reported a miRNA signature that correlated with high-risk CN-AML patients with *FLT3-ITD* and/or wild-type *NPM1*. The prognostic signature included miR-124, miR-128-1, miR-194, miR-219-5p, miR-220a, and miR-320 which associated with increased risk; while miR-181a and miR-181b were inversely correlated with risk of event. Interestingly, higher expression of miR-181a is an independent prognostic marker in CN-AML and was found to be associated with longer overall and disease-free survival (Schwind et al. 2010). These data suggest that miR-181a treatment might be a useful therapeutic option for CN-AML with low miR-181a expression.

Interestingly, miRNAs are also emerging as regulators of chemoresistance in the current front line antileukemia chemotherapies (such as the anthracycline daunorubicin). Several studies have demonstrated that drug-resistant subtypes of leukemia display characteristic miRNA signatures (Schotte et al. 2011; Zhao et al. 2010; Zimmerman et al. 2010). For example, administration of anti-miR-21 oligonucleotide along with cytarabine has shown increased potency in AML cell lines (Li et al. 2010). Thus, it will be vital to understand the role of miRNAs in AML chemoresistance in order to develop novel ways to tackle this major clinical problem.

36.3 miRNA-Based Therapeutic Approaches for AML

Despite considerable advances in our understanding of the molecular carcinogenesis of AML and extensive research on combined and targeted therapies, the overall benefit for the AML patients has been dismal. There is still an invariable need for the development of novel therapeutic tools with better specificity, higher efficacy, and fewer side effects. By the virtue of their ability to regulate important cellular processes, target multiple protein-coding genes and their aberrant perturbations in AML, miRNA modulation has emerged to be a viable therapeutic strategy and a powerful intervention tool. Two strategies have been described for the treatment of cancer using miRNA therapy. First, by targeting multiple molecular defects accumulated in the multistep pathway of a specific cancer—"multiplex RNA inhibition"; or second, by using multiple agents to target one specific molecular defect linked with cancer pathogenesis—"sandwich RNA inhibition" (Spizzo et al. 2009). Table 36.1 shows several candidate miRNAs that can be therapeutically targeted for AML management.

Current approaches for AML therapy involve either (1) the restoration of tumor suppressive miRNA using "miRNA mimics," or (2) the inhibition of oncogenic miRNA using "anti-miRNAs." Additionally, epigenetic drugs can also be used to successfully alter the expression levels of miRNAs that are epigenetically regulated (Schotte et al. 2012).

36.3.1 miRNA Mimics: miRNA Replacement Therapy

Mutations or downregulation of tumor suppressive miRNAs contribute to oncogenic transformation in tumor cells. An effective option to inhibit this is to restore the normal function of these miRNAs by substituting the lost miRNA using synthetic miRNA-like molecules called "miRNA mimics." These are small, 2′-O′-methoxy modified RNA duplexes that can be loaded into RISC just like mature miRNAs and can achieve the downstream inhibition of the target mRNAs (Garzon et al. 2010). The efficacy of miRNA replacement therapy has been reported in AML using in

Table 36.1 Targetable miRNA in AML

miRNA	miR expression	AML chromosomal/ genetic aberration	Known gene target	Reference
Let-7a	Upregulation	CN-AML with *NPM1*-mutation	KRAS; HMGA2	(Becker et al. 2010; Garzon et al. 2008a; Johnson et al. 2005; Lee and Dutta 2007)
Let-7b	Downregulation	AML with inv(16); AML with t(8;21)	KRAS; HMGA2	(Johnson et al. 2005; Jongen-Lavrencic et al. 2008; Lee and Dutta 2007)
Let-7c	Downregulation	AML with inv(16); AML with t(8;21)	KRAS; HMGA2	(Johnson et al. 2005; Jongen-Lavrencic et al. 2008; Lee and Dutta 2007)
miR-1	Upregulation	CN-AML with *IDH2*-mutation	MEIS1	(Marcucci et al. 2010; Starczynowski et al. 2011)
miR-10a	Upregulation	CN-AML with *NPM1*-mutation	HOXA1	(Becker et al. 2010; Garzon et al. 2008a; Garzon et al. 2006)
miR-10b	Upregulation	CN-AML with *NPM1*-mutation	TIAM1; HOXD10	(Becker et al. 2010; Garzon et al. 2008a; Ma et al. 2007; Moriarty et al. 2010)
miR-17~92 cluster	Upregulation	AML with t(11q23)/MLL	CDKN1A (p21); PTEN	(Garzon et al. 2008b; Jongen-Lavrencic et al. 2008; Mi et al. 2010; Wong et al. 2010; Xiao et al. 2008)
miR-29 family	Downregulation	AML with t(11q23)/*MLL*; CN-AML with *NPM1*-wild-type	MCL1; CDK6; SP1; DNMT3A/B; TCL1	(Garzon et al. 2008a; Garzon et al. 2009a; Garzon et al. 2009b; Garzon et al. 2008b; Mott et al. 2007)
miR-125b	Upregulation	CN-AML with *IDH2*-mutation	p53; DICER1; ST18	(Klusmann et al. 2010; Le et al. 2009; Marcucci et al. 2010)

Table 36.1 (continued)

miRNA	miR expression	AML chromosomal/genetic aberration	Known gene target	Reference
miR-126	Upregulation Downregulation	AML with inv(16) or AML with t(8;21); CN-AML with *NPM1*-mutation	PLK2	(Dixon-McIver et al. 2008; Garzon et al. 2008a; Garzon et al. 2008b; Li et al. 2008)
miR-126*	Upregulation	AML with inv(16) or AML with t(8;21);	PLK2	(Dixon-McIver et al. 2008; Garzon et al. 2008b; Li et al. 2008)
miR-130a	Downregulation	CN-AML with *NPM1*-mutation	SMAD4	(Becker et al. 2010; Hager et al. 2011)
miR-133a	Upregulation	CN-AML with *IDH2*-mutation	N/A	(Marcucci et al. 2010)
miR-155	Upregulation	CN-AML with *FLT3-ITD*	SHIP1; C/EPBα	(Costinean et al. 2009; Garzon et al. 2008b; Jongen-Lavrencic et al. 2008; O'Connell et al. 2009)
miR-181 family	Upregulation	CN-AML with *C/EBPα*-mutation	TCL1; p27Kip1; HOXA11	(Cuesta et al. 2009; Jongen-Lavrencic et al. 2008; Marcucci et al. 2008a; Naguibneva et al. 2006; Wang et al. 2009)
miR-196b	Upregulation	AML with t(11q23)/MLL; CN-AML with *NPM1*-mutation	HOXB8	(Jongen-Lavrencic et al. 2008; Li et al. 2008; Popovic et al. 2009)
miR-223	Downregulation	AML with t(8;21)	MEF2C	(Fazi et al. 2007; Johnnidis et al. 2008)
miR-382	Upregulation	AML with t(15;17)	Kallikrein 5	(Jongen-Lavrencic et al. 2008; Kriegel et al. 2012)
miR-451	Downregulation	CN-AML with *NPM1*-mutation	MYC	(Becker et al. 2010; Li et al. 2011)

vitro models. An example of such an approach was introduction of synthetic mimics for miR-193a, which induced marked apoptosis and blocked proliferation in AML cells by reducing the levels of c-KIT oncogene (Gao et al. 2011). Another study reported that introduction of miR-29b mimics downregulated the oncogenic MCL-1 protein, reducing cell proliferation and inducing apoptosis in AML in vitro as well as in vivo models (Garzon et al. 2009a; Garzon et al. 2009b). Additionally, treatment of AML blasts with miR-223 mimetics restored cell differentiation (Marcucci et al. 2011b). However, the in vivo efficiency of these mimics still needs to be evaluated. Certain modifications that have been applied to improve the stability and increase the half-life of these mimics are i) introduction of double-stranded (instead of standard single-stranded) miRNA mimetics; and ii) expression of longer pre-miRNA-like shRNAs from plasmid or viral vectors driven by strong promoters (Garzon et al. 2010).

36.3.2 Anti-miRs:Antisense Inhibition of Mature miRNA

OncomiRs that are frequently overexpressed in AML need to be inhibited to help restore the normal function of its target tumor suppressive genes. This can be achieved using complementary single stranded oligonucleotides that sequester the endogenous miRNA in an unrecognized conformation that is not processed by the RISC, resulting in exclusion of the mature miRNA from the RISC. These include antisense anti-miR oligonucleotides (AMOs), locked nucleic acid (LNA) anti-miRNAs, antagomirs, miRNA sponges, miRNA masks, and small molecule inhibitors of miRNAs (Garzon et al. 2010).

AMOs are single-stranded RNA molecules complementary to the target miRNA (Hutvagner et al. 2004). They work as competitive inhibitors of miRNAs by annealing to the mature miRNA and inhibiting the interaction of that miRNA with its target mRNAs. These are usually 2′-O-methyl or 2′-O-methoxyethyl modified AMOs, which confer nuclease resistance and improved stability and specificity. Another optimal modification to AMOs is the addition of cholesterol functionality at the 3′-position to generate “antagomirs” (Krutzfeldt et al. 2005). In leukemia cells, use of antagomirs against miR-19 inhibited leukemogenesis by reactivating the tumor suppressor PTEN (Mavrakis et al. 2010). Similarly, treatment with miR-196b antagomirs inhibited growth in *MLL-AF9* transformed bone marrow cells (Popovic et al. 2009). Another example of modified AMOs are the LNA anti-miRs, in which a methylene bridge connecting the 2′-O atom and the 4′-C atom “locks” the ribose ring in a C3′-endo or C2′-endo conformation (Vester and Wengel 2004). LNA-modified oligonucleotides exhibit improved mismatch discrimination, higher thermal stability, and high-affinity Watson–Crick hybridization with their RNA target molecules. A recent study reported that miR-21 could be effectively silenced in vitro using LNA-modified antisense oligonucleotides (ASOs), leading to a significant reduction in cell viability (Chan et al. 2005). Likewise, intravenous injection of LNA anti-miR against miR-122 induced marked decrease of mature miRNA levels in liver and other tissues except brain (Vester and Wengel 2004). Phase I clinical trials for these LNA-anti-miRs are currently underway to study their pharmacokinetic and pharmacodynamic profiles in humans.

Another mode to inhibit oncomirs can be achieved by the use of "miRNA sponges" or "miRNA decoys" (Ebert et al. 2007). These are RNA transcripts with multiple tandem repeat of miRNA binding sites for an endogenous miRNA, which can stably interact with or "soak up" the corresponding miRNA and prevent its interaction with its target mRNAs. An important attribute of this approach is its ability to target closely related miRNA family members that share overlapping targets. In vivo administration of miR-326 sponge resulted in depletion of miR-326 in the CD^+ T-cells, and inhibited the development of autoimmune disease (Du et al. 2009). On the other hand, "miRNA masks" can be used to achieve gene-specific anti-miRNA therapy that could "mask" the specific target mRNA from the endogenous miRNA, and thus prevent the repressive action of miRNAs in an mRNA-selective manner (Xiao et al. 2007). These "miRNA masks" are single-stranded 2′-O′methyl-modified ASOs with locked 5′ and 3′ ends that are entirely complementary to the miRNA binding sites in the 3′-UTR of the target mRNA.

Some small molecules, which have been widely used against proteins involved in cancers, may also target certain miRNAs. Examples include clinically successful imatinib mesylate (Gleevec) for the treatment of chronic myeloid leukemia (CML) (Kantarjian et al. 2006). Additionally, nanoparticle formulations have been used for in vitro delivery of siRNAs and miRNAs (Chen et al. 2010). Identification of other novel inhibitors of miRNAs might be a useful supplement to the conventional chemotherapeutic agents.

36.3.3 Use of Epigenetic Drugs

The expression levels of miRNAs that are epigenetically regulated in AML can be modulated using epigenetic drugs (Schotte et al. 2012). Introduction of demethylating drugs can help reactivate downregulated miRNAs, and thus restore their expression levels. An example includes miR-124a, which is hypermethylated in ALL patients, and this has been linked to higher relapse and mortality rates (Agirre et al. 2009). Treatment of ALL cells with trichostatin A or 5-aza-2′-deoxycytidine results in demethylation of the miR-124a promoter, and thus reexpression of the miRNA (Schotte et al. 2012).

36.4 Advantages and Disadvantages of miRNA-Based Therapy

One of the most attractive advantages of miRNA as therapeutic targets is their ability to regulate multiple genes, often in the perspective of a common network. This is important in AML since multiple biological processes often get deregulated and hence cannot be treated by a single-agent therapy. Instead, miRNAs can target multiple genes at the same time enabling a more cumulative and stronger effect on a set of related target proteins at multiple levels. The incessant discovery and identification of new miRNAs important in different AML processes adds to the pool of novel druggable targets amenable to pharmacological intervention. Finally, miRNA therapy affords fewer side effects and reduced nonself-immune response as compared to chemotherapy or antibody therapy. Thus miRNAs make a promising therapeutic modality for AML treatment.

One of the major challenges to miRNA-based therapeutics is the biological instability of unmodified naked RNAs and AMOs due to rapid degradation by cellular and serum nucleases makes their direct delivery an unfeasible option (Aagaard and Rossi 2007; Zhao et al. 2009). This can be prevented through the use of targeted delivery vehicle to achieve sufficient tissue specificity and cellular uptake. Both viral options (adeno-associated virus (AAV) vectors or lentiviral vectors, LVs) and nonviral options (chemical modifications of the oligonucleotides, lipid complexes with small molecules, liposomes, polymers, antibodies, and nanoparticles) have been considered, each with certain advantages and disadvantages (Garzon et al. 2010). Targeted delivery can also be achieved by tagging the modified oligonucleotides with antibodies. This is specifically useful in AML, since immunoconjugates directed against CD19 or CD33 can be exploited for leukemia malignancies (Mian and Zeleznik-Le 2010). Finally, the route of administration is another important clinical consideration for AML therapy. Systemic delivery is the only route of administration for AML since local administration cannot be used in hematological malignancies. Although systemic delivery provides a more effective bio distribution to the target tissues, barriers to in vivo delivery include renal clearance, failure to cross the capillary endothelium (for RNAs conjugated with large molecules), limited passage through extracellular matrix, and destruction by scavenging macrophages (Bader et al. 2011). Furthermore, long-term safety of these agents also needs to be evaluated before their transition into clinical use. Thus, the major issues to the translational potential of miRNA-based therapeutics from bench to bedside in cancer care include its stability, mode of delivery, potential off-targets effects, and safety concerns (Castanotto and Rossi 2009).

Conclusion

The discovery of miRNAs has opened new perspectives in the treatment of AML and other hematological malignancies. miRNA-based therapeutics hold great promise, both as novel biomarkers and as highly specific, targeted therapies. The plethora of new data available on miRNAs involved in AML pathogenesis provides us with great opportunities for therapeutic intervention, but with concomitant challenges. Nevertheless, a better understanding of its functions in hematopoiesis, leukemogenesis, epigenetic regulation, and cytogenetic aberrations in AML will help fine-tune our understanding of the miRNA networks in AML, and in the development of novel targeted therapies.

References

Aagaard L, Rossi JJ (2007) RNAi therapeutics: principles, prospects and challenges. Adv Drug Deliv Rev 59:75–86

Agirre X, Vilas-Zornoza A, Jimenez-Velasco A, Martin-Subero JI, Cordeu L, Garate L, San Jose-Eneriz E, Abizanda G, Rodriguez-Otero P, Fortes P et al (2009) Epigenetic silencing of the tumor suppressor microRNA Hsa-miR-124a regulates CDK6 expression and confers a poor prognosis in acute lymphoblastic leukemia. Cancer Res 69:4443–4453

Ambros V (2004) The functions of animal microRNAs. Nature 431:350–355

Bader AG, Brown D, Stoudemire J, Lammers P (2011) Developing therapeutic microRNAs for cancer. Gene Ther 18:1121–1126

Bartel DP (2004) MicroRNAs: genomics, biogenesis, mechanism, and function. Cell 116:281–297

Becker H, Marcucci G, Maharry K, Radmacher MD, Mrozek K, Margeson D, Whitman SP, Wu YZ, Schwind S, Paschka P et al (2010) Favorable prognostic impact of NPM1 mutations in older patients with cytogenetically normal de novo acute myeloid leukemia and associated gene- and microRNA-expression signatures: a Cancer and Leukemia Group B study. J Clin Oncol 28:596–604

Bullinger L, Dohner K, Bair E, Frohling S, Schlenk RF, Tibshirani R, Dohner H, Pollack JR (2004) Use of gene-expression profiling to identify prognostic subclasses in adult acute myeloid leukemia. N Engl J Med 350:1605–1616

Calin GA, Dumitru CD, Shimizu M, Bichi R, Zupo S, Noch E, Aldler H, Rattan S, Keating M, Rai K et al (2002). Frequent deletions and down-regulation of micro- RNA genes miR15 and miR16 at 13q14 in chronic lymphocytic leukemia. Proc Natl Acad Sci U S A 99:15524–15529

Calin GA, Sevignani C, Dumitru CD, Hyslop T, Noch E, Yendamuri S, Shimizu M, Rattan S, Bullrich F, Negrini M et al (2004) Human microRNA genes are frequently located at fragile sites and genomic regions involved in cancers. Proc Natl Acad Sci U S A 101: 2999–3004

Castanotto D, Rossi JJ (2009) The promises and pitfalls of RNA-interference-based therapeutics. Nature 457:426–433

Chan JA, Krichevsky AM, Kosik KS (2005) MicroRNA-21 is an antiapoptotic factor in human glioblastoma cells. Cancer Res 65:6029–6033

Chen Y, Zhu X, Zhang X, Liu B, Huang L (2010). Nanoparticles modified with tumor-targeting scFv deliver siRNA and miRNA for cancer therapy. Mol Ther 18:1650–1656

Costinean S, Sandhu SK, Pedersen IM, Tili E, Trotta R, Perrotti D, Ciarlariello D, Neviani P, Harb J, Kauffman LR et al (2009) Src homology 2 domain-containing inositol-5-phosphatase and CCAAT enhancer-binding protein beta are targeted by miR-155 in B cells of Emicro-MiR-155 transgenic mice. Blood 114:1374–1382

Cuesta R, Martinez-Sanchez A, Gebauer F (2009) miR-181a regulates cap-dependent translation of p27(kip1) mRNA in myeloid cells. Mol Cell Biol 29:2841–2851

Dixon-McIver A, East P, Mein CA, Cazier JB, Molloy G, Chaplin T, Lister A, Young BD, Debernardi S (2008) Distinctive patterns of microRNA expression associated with karyotype in acute myeloid leukaemia. PLoS One 3:e2141

Du C, Liu C, Kang J, Zhao G, Ye Z, Huang S, Li Z, Wu Z, Pei G (2009) MicroRNA miR-326 regulates TH-17 differentiation and is associated with the pathogenesis of multiple sclerosis. Nat Immunol 10:1252–1259

Ebert MS, Neilson JR, Sharp PA (2007) MicroRNA sponges: competitive inhibitors of small RNAs in mammalian cells. Nat Methods 4:721–726

Esquela-Kerscher A, Slack FJ (2006) Oncomirs—microRNAs with a role in cancer. Nat Rev Cancer 6:259–269

Fazi F, Racanicchi S, Zardo G, Starnes LM, Mancini M, Travaglini L, Diverio D, Ammatuna E, Cimino G, Lo-Coco F et al (2007) Epigenetic silencing of the myelopoiesis regulator microRNA-223 by the AML1/ETO oncoprotein. Cancer Cell 12:457–466

Gao XN, Lin J, Li YH, Gao L, Wang XR, Wang W, Kang HY, Yan GT, Wang LL, Yu L (2011) MicroRNA-193a represses c-kit expression and functions as a methylation-silenced tumor suppressor in acute myeloid leukemia. Oncogene 30:3416–3428

Garzon R, Pichiorri F, Palumbo T, Iuliano R, Cimmino A, Aqeilan R, Volinia S, Bhatt D, Alder H, Marcucci G et al (2006) MicroRNA fingerprints during human megakaryocytopoiesis. Proc Natl Acad Sci USA 103:5078–5083

Garzon R, Garofalo M, Martelli MP, Briesewitz R, Wang L, Fernandez-Cymering C, Volinia S, Liu CG, Schnittger S, Haferlach T et al (2008a) Distinctive microRNA signature of acute myeloid leukemia bearing cytoplasmic mutated nucleophosmin. Proc Natl Acad Sci U S A 105:3945–3950

Garzon R, Volinia S, Liu CG, Fernandez-Cymering C, Palumbo T, Pichiorri F, Fabbri M, Coombes K, Alder H, Nakamura T et al (2008b) MicroRNA signatures associated with cytogenetics and prognosis in acute myeloid leukemia. Blood 111:3183–3189

Garzon R, Heaphy CE, Havelange V, Fabbri M, Volinia S, Tsao T, Zanesi N, Kornblau SM, Marcucci G, Calin GA et al (2009a) MicroRNA 29b functions in acute myeloid leukemia. Blood 114:5331–5341

Garzon R, Liu S, Fabbri M, Liu Z, Heaphy CE, Callegari E, Schwind S, Pang J, Yu J, Muthusamy N et al (2009b) MicroRNA-29b induces global DNA hypomethylation and tumor suppressor gene reexpression in acute myeloid leukemia by targeting directly DNMT3A and 3B and indirectly DNMT1. Blood 113:6411–6418

Garzon R, Marcucci G, Croce CM (2010) Targeting microRNAs in cancer: rationale, strategies and challenges. Nat Rev Drug Discov 9:775–789

Hager M, Pedersen CC, Larsen MT, Andersen MK, Hother C, Gronbaek K, Jarmer H, Borregaard N, Cowland JB (2011) MicroRNA-130a-mediated down-regulation of Smad4 contributes to reduced sensitivity to TGF-beta1 stimulation in granulocytic precursors. Blood 118:6649–6659

Havelange V, Garzon R, Croce CM (2009) MicroRNAs: new players in acute myeloid leukaemia. Br J Cancer 101:743–748

He L, Hannon GJ (2004) MicroRNAs: small RNAs with a big role in gene regulation. Nat Rev Genet 5:522–531

Hutvagner G, Simard MJ, Mello CC, Zamore PD (2004) Sequence-specific inhibition of small RNA function. PLoS Biol 2:E98

Johnnidis JB, Harris MH, Wheeler RT, Stehling-Sun S, Lam MH, Kirak O, Brummelkamp TR, Fleming MD, Camargo FD (2008) Regulation of progenitor cell proliferation and granulocyte function by microRNA-223. Nature 451:1125–1129

Johnson SM, Grosshans H, Shingara J, Byrom M, Jarvis R, Cheng A, Labourier E, Reinert KL, Brown D, Slack FJ (2005) RAS is regulated by the let-7 microRNA family. Cell 120:635–647

Jongen-Lavrencic M, Sun SM, Dijkstra MK, Valk PJ, Loewenberg B (2008) MicroRNA expression profiling in relation to the genetic heterogeneity of acute myeloid leukemia. Blood 111:5078–5085

Kantarjian HM, Talpaz M, Giles F, O'Brien S, Cortes J (2006) New insights into the pathophysiology of chronic myeloid leukemia and imatinib resistance. Ann Intern Med 145:913–923

Kim VN (2005) MicroRNA biogenesis: coordinated cropping and dicing. Nat Rev Mol Cell Biol 6:376–385.

Klusmann JH, Li Z, Bohmer K, Maroz A, Koch ML, Emmrich S, Godinho FJ, Orkin SH, Reinhardt D (2010) miR-125b-2 is a potential oncomiR on human chromosome 21 in megakaryoblastic leukemia. Genes Dev 24:478–490

Kriegel AJ, Liu Y, Cohen B, Usa K, Liu Y, Liang M (2012) MiR-382 targeting of kallikrein 5 contributes to renal inner medullary interstitial fibrosis. Physiological genomics 44:259–267

Krutzfeldt J, Rajewsky N, Braich R, Rajeev KG, Tuschl T, Manoharan M, Stoffel M (2005) Silencing of microRNAs in vivo with 'antagomirs'. Nature 438:685–689

Langer C, Marcucci G, Holland KB, Radmacher MD, Maharry K, Paschka P, Whitman SP, Mrozek K, Baldus CD, Vij R et al (2009) Prognostic importance of MN1 transcript levels, and biologic insights from MN1-associated gene and microRNA expression signatures in cytogenetically normal acute myeloid leukemia: a cancer and leukemia group B study. J Clin Oncol 27:3198–3204.

Le MT, Teh C, Shyh-Chang N, Xie H, Zhou B, Korzh V, Lodish HF, Lim B (2009) MicroRNA-125b is a novel negative regulator of p53. Genes Dev 23:862–876

Lee YS, Dutta A (2007) The tumor suppressor microRNA let-7 represses the HMGA2 oncogene. Genes Dev 21:1025–1030

Lewis BP, Shih IH, Jones-Rhoades MW, Bartel DP, Burge CB (2003) Prediction of mammalian microRNA targets. Cell 115:787–798

Li X, Sanda T, Look AT, Novina CD, von Boehmer H (2011) Repression of tumor suppressor mi R-451 is essential for NOTCH1-induced oncogenesis in T-ALL. J Exp Med 208:663–675

Li Y, Zhu X, Gu J, Hu H, Dong D, Yao J, Lin C, Fei J (2010) Anti-miR-21 oligonucleotide enhances chemosensitivity of leukemic HL60 cells to arabinosylcytosine by inducing apoptosis. Hematology 15:215–221

Li Z, Lu J, Sun M, Mi S, Zhang H, Luo RT, Chen P, Wang Y, Yan M, Qian Z et al (2008) Distinct microRNA expression profiles in acute myeloid leukemia with common translocations. Proc Natl Acad Sci USA 105:15535–15540

Ma L, Teruya-Feldstein J, Weinberg RA (2007) Tumour invasion and metastasis initiated by microRNA-10b in breast cancer. Nature 449:682–688

Marcucci G, Maharry K, Radmacher MD, Mrozek K, Vukosavljevic T, Paschka P, Whitman SP, Langer C, Baldus CD, Liu CG et al. (2008a) Prognostic significance of, and gene and microRNA expression signatures associated with, CEBPA mutations in cytogenetically normal acute myeloid leukemia with high-risk molecular features: a Cancer and Leukemia Group B Study. J Clin Oncol 26:5078–5087

Marcucci G, Radmacher MD, Maharry K, Mrozek K, Ruppert AS, Paschka P, Vukosavljevic T, Whitman SP, Baldus CD, Langer C et al (2008b) MicroRNA expression in cytogenetically normal acute myeloid leukemia. N Engl J Med 358,:1919–1928

Marcucci G, Maharry K, Wu YZ, Radmacher MD, Mrozek K, Margeson D, Holland KB, Whitman SP, Becker H, Schwind S et al (2010) IDH1 and IDH2 gene mutations identify novel molecular subsets within de novo cytogenetically normal acute myeloid leukemia: a Cancer and Leukemia Group B study. J Clin Oncol 28:2348–2355

Marcucci G, Haferlach T, Doehner H (2011a) Molecular genetics of adult acute myeloid leukemia: prognostic and therapeutic implications. J Clin Oncol 29:475–486

Marcucci G, Mrozek K, Radmacher MD, Garzon R, Bloomfield CD (2011b) The prognostic and functional role of microRNAs in acute myeloid leukemia. Blood 117:1121–1129

Mavrakis KJ, Wolfe AL, Oricchio E, Palomero T, de Keersmaecker K, McJunkin K, Zuber J, James T, Khan AA, Leslie CS et al (2010) Genome-wide RNA-mediated interference screen identifies miR-19 targets in Notch-induced T-cell acute lymphoblastic leukaemia. Nature Cell Biology 12:372–379

Mi S, Lu J, Sun M, Li Z, Zhang H, Neilly MB, Wang Y, Qian Z, Jin J, Zhang Y et al (2007) MicroRNA expression signatures accurately discriminate acute lymphoblastic leukemia from acute myeloid leukemia. Proc Natl Acad Sci U S A 104:19971–19976

Mi S, Li Z, Chen P, He C, Cao D, Elkahloun A, Lu J, Pelloso LA, Wunderlich M, Huang H et al (2010) Aberrant overexpression and function of the miR-17-92 cluster in MLL-rearranged acute leukemia. Proc Natl Acad Sci U S A 107:3710–3715

Mian YA, Zeleznik-Le NJ (2010) MicroRNAs in leukemias: emerging diagnostic tools and therapeutic targets. Current Drug Targets 11:801–811

Moriarty CH, Pursell B, Mercurio AM (2010) miR-10b targets Tiam1: implications for Rac activation and carcinoma migration. J Biol Chem 285:20541–20546

Mott JL, Kobayashi S, Bronk SF, Gores GJ (2007) mir-29 regulates Mcl-1 protein expression and apoptosis. Oncogene 26:6133–6140

Mrozek K, Bloomfield CD (2008) Clinical significance of the most common chromosome translocations in adult acute myeloid leukemia. JNCI Monographs 39:52–57

Naguibneva I, Ameyar-Zazoua M, Polesskaya A, Ait-Si-Ali S, Groisman R, Souidi M, Cuvellier S, Harel-Bellan A (2006) The microRNA miR-181 targets the homeobox protein Hox-A11 during mammalian myoblast differentiation. Nature Cell Biol 8:278–284

O'Connell RM, Chaudhuri AA, Rao DS, Baltimore D (2009) Inositol phosphatase SHIP1 is a primary target of miR-155. Proc Natl Acad Sci U S A 106:7113–7118

Popovic R, Riesbeck LE, Velu CS, Chaubey A, Zhang J, Achille NJ, Erfurth FE, Eaton K, Lu J, Grimes HL et al (2009). Regulation of mir-196b by MLL and its overexpression by MLL fusions contributes to immortalization. Blood 113:3314–3322

Sauvageau G, Thorsteinsdottir U, Eaves CJ, Lawrence HJ, Largman C, Lansdorp PM, Humphries RK (1995) Overexpression of HOXB4 in hematopoietic cells causes the selective expansion of more primitive populations in vitro and in vivo. Genes Dev 9:1753–1765

Schotte D, De Menezes RX, Moqadam FA, Khankahdani LM, Lange-Turenhout E, Chen C, Pieters R, Den Boer ML (2011) MicroRNA characterize genetic diversity and drug resistance in pediatric acute lymphoblastic leukemia. Haematologica 96:703–711

Schotte D, Pieters R, Den Boer ML (2012) MicroRNAs in acute leukemia: from biological players to clinical contributors. Leukemia 26:1–12

Schwind S, Maharry K, Radmacher MD, Mrozek K, Holland KB, Margeson D, Whitman SP, Hickey C, Becker H, Metzeler KH et al (2010) Prognostic significance of expression of a single microRNA, miR-181a, in cytogenetically normal acute myeloid leukemia: a Cancer and Leukemia Group B study. J Clin Oncol 28:5257–5264

Shen WF, Hu YL, Uttarwar L, Passegue E, Largman C (2008) MicroRNA-126 regulates HOXA9 by binding to the homeobox. Mol Cell Biol 28:4609–4619

Spizzo R, Rushworth D, Guerrero M, Calin GA (2009) RNA inhibition, microRNAs, and new therapeutic agents for cancer treatment. Clin Lymphoma Myeloma 9(3):313–318

Starczynowski DT, Morin R, McPherson A, Lam J, Chari R, Wegrzyn J, Kuchenbauer F, Hirst M, Tohyama K, Humphries RK et al (2011) Genome-wide identification of human microRNAs located in leukemia-associated genomic alterations. Blood 117:595–607

Vester B, Wengel J (2004) LNA (locked nucleic acid): high-affinity targeting of complementary RNA and DNA. Biochemistry 43:13233–13241

Wang X, Gocek E, Liu CG, Studzinski GP (2009) MicroRNAs181 regulate the expression of p27Kip1 in human myeloid leukemia cells induced to differentiate by 1,25-dihydroxyvitamin D3. Cell Cycle 8:736–741

Wong P, Iwasaki M, Somervaille TC, Ficara F, Carico C, Arnold C, Chen CZ, Cleary M (2010) The miR-17-92 microRNA polycistron regulates MLL leukemia stem cell potential by modulating p21 expression. Cancer Res 70:3833–3842

Xiao C, Srinivasan L, Calado DP, Patterson HC, Zhang B, Wang J, Henderson JM, Kutok JL, Rajewsky K (2008) Lymphoproliferative disease and autoimmunity in mice with increased miR-17-92 expression in lymphocytes. Nat Immunol 9:405–414

Xiao J, Yang B, Lin H, Lu Y, Luo X, Wang Z (2007) Novel approaches for gene-specific interference via manipulating actions of microRNAs: examination on the pacemaker channel genes HCN2 and HCN4. J Cell Physiol 212:285–292

Zhao X, Pan F, Holt CM, Lewis AL, Lu JR (2009) Controlled delivery of antisense oligonucleotides: a brief review of current strategies. Expert Opin Drug Deliv 6:673–686

Zhao X, Yang L, Hu J, Ruan J (2010) miR-138 might reverse multidrug resistance of leukemia cells. Leukemia Res 34:1078–1082

Zimmerman EI, Dollins CM, Crawford M, Grant S, Nana-Sinkam SP, Richards KL, Hammond SM, Graves LM (2010) Lyn kinase-dependent regulation of miR181 and myeloid cell leukemia-1 expression: implications for drug resistance in myelogenous leukemia. Mol Pharmacol 78:811–817

Chapter 37
Clinical Implications of MicroRNAs in AML

Parvathi Ranganathan and Ramiro Garzon

Abstract It is now established that deregulated microRNA (miRNA) expression is a prominent feature in acute myeloid leukemia (AML). Subsequently, functional studies have shown that miRNAs play an important role in the pathogenesis of AML by acting as tumor suppressors or oncogenes. Recent data indicate that distinctive miRNA expression signatures are associated with chemotherapy response and clinical outcome and that targeting miRNAs in AML is a novel emerging therapeutic strategy. In this chapter, we will discuss the clinical application of miRNAs as biomarkers for diagnosis and prognosis in AML and review the current strategies to target miRNAs in this disease.

Keywords MicroRNAs · Biomarkers · Antisense

37.1 Introduction

MicroRNAs (miRNAs) are naturally occurring, small, noncoding 18–25-nucleotide RNAs that are cleaved from larger hairpin precursors by a complex protein system that includes the RNase III Drosha and Dicer, members of the Argonaute family (Bartel 2004; Lee et al. 2004). Mature miRNAs regulate gene expression by hybridizing to the 3′ untranslated regions of protein-coding messenger RNAs (mRNAs) and causing their posttranscriptional repression and/or degradation (Bartel 2004). miRNAs are involved in the regulation of cellular homeostasis and in many other physiologic processes, including hematopoiesis (Bartel 2004; Croce 2009; Garzon and Croce 2008a). Several studies have established that miRNA expression is widely deregulated in cancer, including acute myeloid leukemia (AML) (Garzon et al. 2010; Marcucci et al. 2011). Distinctive miRNA signatures have been identified in association with cytogenetic and molecular subsets of AML and mechanistic studies have proven that miRNAs regulate critical protein-coding gene networks in myeloid leukemogenesis (reviewed by Marcucci et al. 2011). Recent data indicate that few miRNAs could be used as biomarkers for outcome prediction and treatment

R. Garzon (✉) · P. Ranganathan
College of Medicine, Department of Internal Medicine, Comprehensive Cancer Center, The Ohio State University, 460 W 12th Avenue, Columbus, OH 43210, USA
e-mail: ramiro.garzon@osumc.edu

M. Andreeff (ed.), *Targeted Therapy of Acute Myeloid Leukemia*,
Current Cancer Research, DOI 10.1007/978-1-4939-1393-0_37

response (Marcucci et al. 2011). There is also a great enthusiasm in targeting miRNAs in AML. In this chapter, we will discuss the current and potential clinical applications of miRNAs in AML.

37.2 MiRNAs as Biomarkers for Outcome Prediction

Using a custom microarray miRNA platform, our group has reported a miRNA signature comprised by five miRNAs; miR-199a, miR-199b, miR-191, miR-20, and miR-25, that when over expressed adversely affected overall survival in AML patients with intermediate and poor cytogenetic risk. This signature was subsequently validated in an independent cohort of patients with similar baseline and cytogenetic/molecular characteristics but using real-time polymerase chain reaction (RT-PCR). Only miR-199 and miR-191 remained significantly associated with overall survival, independent from other variables, including cytogenetics (Garzon et al. 2008b). Marcucci et al., reported a miRNA signature, characterized by upregulation of miR-181a and miR-181b family members and downregulation of miR-124, miR-204, miR-194, miR-320, and 219-5p associated with overall and disease-free survival in a cohort of high-risk cytogenetically normal AML patients (CN-AML; Marcucci et al. 2008; Schwind et al. 2010). More importantly, a follow-up study focusing only on miR-181a expression showed that higher miR-181a expression is associated with a higher complete remission (CR) rate, longer overall survival, and a trend for longer disease-free survival. The impact of miR-181a was most striking in poor molecular risk patients with *FLT3* internal tandem duplication (FLT3-ITD) and/or *NPM1* wild type. In multivariable analyses, higher miR-181a expression was significantly associated with better outcome, both in the whole patient cohort and in patients with FLT3-ITD and/or *NPM1* wild type (Schwind et al. 2010). Thus, in AML a miRNA signature and a single miRNA (miR-181a), are associated with clinical outcome of patients with CN-AML and may refine their molecular risk classification.

Another group reported low expression of miR-let7b and miR-9 in patients with favorable cytogenetic risk, (t(8;21), inv(16), and t(15;17)), whereas patients with poor or intermediate cytogenetic risk groups had higher expression of these miRNAs (Dixon-McIver et al. 2008). These studies show that miRNA profiling can be used to predict outcome, however, further independent validation using large cohorts of patients is needed. In addition, it would be important to show that miRNAs can improve the current available biomarkers.

37.2.1 miRNAs as Predictors of Treatment Response

A provocative study by Blum et al., recently reported that high pretreatment expression levels of miR-29b in bone marrow mononuclear cells was associated with high chances of complete remission to the hypomethylating agent decitabine as a single agent, in elderly (> 60 years) AML patients (Blum et al. 2010). If these results are replicated, it is possible to use miR-29b expression levels to stratify patients to this agent. Interestingly, miR-29b is a miRNA intimately involved in the regulation of DNA methylation. Our group has reported that miR-29b directly targets DNA methyltransferases 3A and 3B and indirectly DNMT1 through SP1. Overexpressing miR-29b in AML cell lines and primary AML samples decreases DNMT3A, 3B, and DNMT1 levels and induces global DNA hypomethylation and previously silenced tumor suppressor re-expression (Garzon et al. 2009).

37.3 MiRNAs as Therapeutic Targets

The consistent deregulation of a small group of miRNAs including miR-155, let-7a, miR-21, miR29b, and the miR17-92 cluster in multiple cancers (Calin and Croce 2006; Garzon and Croce 2008a; Garzon et al. 2010; Sandhu and Garzon 2011), coupled with the fact that targeting a single miRNA can have effects in multiple oncogenic/tumor suppressor pathways, have sparked interest in developing miRNA-based therapies to fight cancer. In AML, few miRNAs such as miR-155 and miR-196b, are upregulated in leukemic blasts from specific cytogenetic and molecular subsets and promote leukemogenesis by acting as oncogenes (Marcucci et al. 2011). In contrast, other group of miRNAs such as miR-29b or let-7a is absent or expressed at very low levels in leukemic blasts and act as tumor suppressor miRNAs (Garzon et al. 2010; Marcucci et al. 2011). Depending on the function of the miRNAs in myeloid leukomogenesis, the two primary ways of targeting miRNAs would be to suppress an oncogenic miRNA or restore a miRNA with tumor suppressor function (Calin and Croce 2006; Garzon et al. 2010; Sandhu and Garzon 2011).

37.3.1 Targeting an OncomiRNA

MiRNA expression can be silenced using synthetic antisense oligonucleotides (oligos) known as antimiRs or antagomiRs. Antisense oligonucleotides work as competitive inhibitors of miRNAs, presumably by annealing to the mature miRNA guide strand and inducing degradation or stoichiometric duplex formation (Garzon et al. 2010). Researchers have introduced modifications to the chemical structure of the oligos to increase stability, binding affinity, and specificity. These chemical modifications include the introduction of 2-O-methyl groups, 2′-O-methoxyethyl groups (2′-MOE), 2-*O*Me phosphoramidites, and locked nucleic acid (LNA; Stenvang et al. 2012; Garzon et al. 2010). In the LNA modification, the ribose ring is "locked" by a methylene bridge connecting the 2-O atom and the 4-C atom. By locking the molecule with the methylene bridge, LNA oligonucleotides display unparallel hybridization affinity. The LNA-antimiRs have been successful in silencing endogenous miRNA in vitro and in vivo (Elmen et al. 2008a, b; Krutzfeldt et al. 2005). Remarkably, the simple systemic delivery of a naked unconjugated LNA antimiR-122 has also been shown to effectively antagonize liver-expressed miR-122 in nonhuman primates (Elmen et al. 2008a). Based on these studies, Santaris Pharma has performed a safety phase 1 and phase 2a clinical trials in humans with hepatitis C, where miR-122 is critical to sustain viral replication. These trials have shown that the use of LNA antimiR-122 in hepatitis C patients is safe and long lasting. In addition, the authors found that it is effective in decreasing hepatitis C viral load (Stenvang et al. 2012). This certainly has paved the way for the use of LNA antimiR in cancer, including AML. Up to date, LNA antimiR development in AML is at the preclinical stage. It is likely that in a year or two, several phase 1 clinical trials will be started.

Another strategy to target and downmodulate oncogenic miRNAs is to use sponges (Ebert et al. 2007), which are transcripts derived from expression vectors that contain multiple tandem binding sites to the miRNA of interest. This approach has been shown to be feasible for in vitro and in vivo studies, though it will be unlikely that this strategy could be applicable to humans because of safety concerns of using expression vectors in vivo. Another approach is to use drugs, including small molecules that could modulate miRNA expression. Gumireddy et al., using a luciferase reporter assay, screen small molecule libraries searching for a compound that could inhibit miR-21 expression, an oncomiRNA upregulated in several solid tumors and hematological malignancies. A compound was identified and was subsequently tested in vitro, confirming that it could inhibit miR-21 expression and elicit antitumoral effects (Gumireddy et al. 2008). Using the same approach, other groups recently identified small molecule inhibitors that blocked miR-122 expression (Young et al. 2010). Therefore, this strategy is promising and hopefully many small molecule inhibitors will be identified in the near future that could be tested in the preclinical and clinical arena.

37.3.2 Restoring a Tumor Suppressor miRNA

The restoration of a lost tumor suppressor miRNA expression can be achieved by using synthetic oligonucleotides that mimic mature miRNAs, adenovirus vectors, and pharmacological agents, including small molecule compounds (Kota et al. 2009; Garzon et al. 2010; Sandhu and Garzon 2011). Similar to antisense oligos, miRNA mimics usually have different chemistry modifications that allow for better stability and enhanced target specificity. One of the problems of this modality is that in contrast to the antisense approach, where the oligos are usually small (even only 8 nucleotides), the mimic oligos length are much longer (at least 18–24 nucleotides) since the full miRNA mature length is required to mimic the endogenous miRNAs. Frequently, even full pre-miRNA sequences length (80–100 nucleotides) is used, in particularly for in vitro work. In addition, given the larger length of these oligos, they are less stable, more prone for degradation and have poor tissue penetration (Garzon et al. 2010). For all these reasons, mimic oligos technologies are more difficult to develop.

One way to overcome these problems will be to use adenovirus-associated vectors, which are a popular choice due to the fact that they do not integrate into the human genome, and have minimal toxicity. Recently, it was reported that systemic administration of miR-26a in a mouse model of hepatocellular carcinoma using adeno-associated virus (AAV) resulted in protection from disease by inhibition of tumor cell proliferation and induction of apoptosis of the cancer cells (Kota et al. 2009). Last, small molecule compounds, chemotherapy agents, or targeted therapies could be used to restore the expression of a tumor suppressor miRNA. For example, our group recently identified that the proteosome inhibitor bortezomib and the histone deacetylase inhibitor HDAC42, are able to restore the expression of the tumor suppressor miR-29b by disrupting the Sp1-NFkB-HDAC inhibitory regulation of miR-29b transcription (Liu et al. 2010). The advantage of this approach is some of these drugs are currently FDA approved for cancer treatment and there are extensive phase 1 and 2 clinical trial data, paving the way for developing clinical trials targeting miRNA expression. Blum et al. recently reported a phase 1 trial of bortezomib and decitabine. Nineteen poor-risk AML patients (median age 70 years; range, 32–84 years) were enrolled. Induction with decitabine (20 mg/m^2 intravenously on days 1–10) plus bortezomib (escalated up to the target 1.3 mg/m^2 on days 5, 8, 12, and 15) was tolerable with a complete response rate of 50 % in previously untreated patients. Pharmacodynamic analysis showed miR-29b upregulation and FLT3 downregulation posttreatment (Blum et al. 2012). One of the disadvantages of this approach is the lack of specificity of these drugs that could result in some unwanted effects.

37.4 Future Directions

To establish miRNAs as biomarkers for prognosis in AML, further independent biomarker validation using large cohort of patients, and a different and easy to use technology to measure them are needed. These studies also need to show that there is an added benefit of using miRNA to the current biomarkers. The finding of a specific miRNA profile that could predict treatment response to certain drugs is extremely interesting, since this may allow a more individualized cancer treatment. To bring miRNA therapeutics from the bench to the clinic, researchers must overcome some of technical hurdles that face the effective use of these molecules as therapeutic agents; such as oligos stability and efficient delivery to the target tissues. Promising candidates will be selected for preclinical studies including extensive pharmacokinetics and pharmacodynamics studies in mice using different schedules and formulations and toxicity studies. It is expected also that new small molecule compounds targeting miRNAs will be identified, tested in vitro and in vivo for effectivity and moved to early phase 1 safety clinical trials.

Conclusions

MiRNAs are widely deregulated in AML and promote myeloid leukemogenesis. Distinctive signatures and single miRNAs have been shown to associate with outcome, including treatment response to specific drugs. As a proof of principle, targeting miRNAs in AML elicit strong antileukemic effects, and therapeutic strategies to silence an oncomiRNA or restore a tumor suppressor miRNA are being developed. We foresee in the near future, several phase 1 safety clinical trials in man using anti-miRNA oligonucleotides, small molecule compounds, and chemotherapies/targeted therapies agents but focused on miRNA targeting.

References

Bartel DP (2004) MicroRNAs: genomics, biogenesis, mechanism, and function. Cell 116:281–297

Blum W, Garzon R et al (2010) Clinical response and miR-29b predictive significance in older AML patients treated with a 10-day schedule of decitabine. Proc Natl Acad Sci U S A 107:7473–7478

Blum W, Schwind S et al (2012) Clinical and pharmacodynamic activity of bortezomib and decitabine in acute myeloid leukemia. Blood 119:6025–6031

Calin GA, Croce CM (2006) MicroRNA signatures in human cancers. Nat Rev Cancer 6:857–866

Croce CM (2009) Causes and consequences of microRNA dysregulation in cancer. Nat Rev Genet 10:704–714

Dixon-McIver A, East P et al (2008) Distinctive patterns of microRNA expression associated with karyotype in acute myeloid leukaemia. PLoS One 3:e2141

Ebert MS, Neilson JR et al (2007) MicroRNA sponges: competitive inhibitors of small RNAs in mammalian cells. Nat Methods 4:721–726

Elmen J, Lindow M et al (2008a) LNA-mediated microRNA silencing in non-human primates. Nature 452:896–899

Elmen J, Lindow M et al (2008b) Antagonism of microRNA-122 in mice by systemically administered LNA-antimiR leads to up-regulation of a large set of predicted target mRNAs in the liver. Nucleic Acids Res 36:1153–1162

Garzon R, Croce CM (2008a) MicroRNAs in normal and malignant hematopoiesis. Curr Opin Hematol 15:352–358

Garzon R, Volinia S et al (2008b) MicroRNA signatures associated with cytogenetics and prognosis in acute myeloid leukemia. Blood 111:3183–3189

Garzon R, Liu S et al (2009) MicroRNA-29b induces global DNA hypomethylation and tumor suppressor gene reexpression in acute myeloid leukemia by targeting directly DNMT3A and 3B and indirectly DNMT1. Blood 113:6411–6418

Garzon R, Marcucci G et al (2010) Targeting microRNAs in cancer: rationale, strategies and challenges. Nat Rev Drug Discov 9:775–789

Gumireddy K, Young DD et al (2008) Small-molecule inhibitors of microrna miR-21 function. Angew Chem Int Ed Engl 47:7482–7484

Kota J, Chivukula RR et al (2009) Therapeutic microRNA delivery suppresses tumorigenesis in a murine liver cancer model. Cell 137:1005–1017

Krutzfeldt J, Rajewsky N et al (2005) Silencing of microRNAs *in vivo* with 'antagomirs'. Nature 438:685–689

Lee Y, Kim M et al (2004) MicroRNA genes are transcribed by RNA polymerase II. EMBO J 23: 4051–4060

Liu S, Wu LC et al (2010) Sp1/NFkappaB/HDAC/miR-29b regulatory network in KIT-driven myeloid leukemia. Cancer Cell 17:333–347

Marcucci G, Mrozek K et al (2011) The prognostic and functional role of microRNAs in AML. Blood 117:1121–1219

Marcucci G, Radmacher MD et al (2008) MicroRNA expression in cytogenetically normal acute myeloid leukemia. N Engl J Med 358:1919–1928

Sandhu S, Garzon R (2011) Potential applications of microRNAs in cancer diagnosis, prognosis, and treatment. Semin Oncol 38:781–787

Schwind S, Maharry K et al (2010) Prognostic significance of expression of a single microRNA, miR-181a, in cytogenetically normal acute myeloid leukemia: a Cancer and Leukemia Group B study. J Clin Oncol 28:5257–5264

Stenvang J, Petri A et al (2012) Inhibition of microRNA function by antimiR oligonucleotides. Silence 3:2–17

Young DD, Connely CM et al (2010) Small molecule modifiers of microRNA miR-122 function for the treatment of hepatitis C virus infection and hepatocellular carcinoma. J Am Chem Soc 132:7976–7981

Part X
Metabolism

Chapter 38
Inhibition of Glycolysis as a Therapeutic Strategy in Acute Myeloid Leukemias

Rob C. Laister, Mark D. Minden and Tak W. Mak

Abstract The metabolic pathways in cancer cells are reprogrammed such that they utilize nutrients quite differently than their normal, nonmalignant counterparts. It has been known for some time that the cancer phenotype results in alterations to glucose metabolism and, more recently, modifications to both glutamine and fatty acid metabolism have been noted. So prevalent is this altered metabolism in malignancy that many now consider it a hallmark of the cancer phenotype. As such, the metabolic discrepancies between cancer cells and normal cells provide a therapeutic window for the potential development of targeted anticancer agents. A number of pharmacological agents that either directly target the enzymes driving tumor glycolysis or the upstream mediators of the glycolytic pathway are currently under investigation with the hope of combining them with existing clinical protocols. Akin to the cytokines and chemokines produced by cancer cells, the intermediates and byproducts of altered tumor glycolysis, upon secretion from cancer cells, are also capable of modulating the phenotypes of normal cells located in the tumor microenvironment. Thus, glycolytic inhibitors may also rescue the effects that tumor derived metabolites have on surrounding cells.

Keywords Glucose metabolism · Leukemia · Targeted therapy · Kinase inhibitors · Bone marrow microenvironment · Lactic acid

R. C. Laister (✉)
Ontario Cancer Institute, Princess Margaret Hospital, Toronto Medical Discovery Tower, 101 College St, Toronto, ON M5G 1L7, Canada
e-mail: rlaister@uhnres.utoronto.ca

M. D. Minden
Ontario Cancer Institute, Princess Margaret Hospital, 610 University Ave, Toronto ON M5G 2M9, Canada

T. W. Mak
Campbell Family Institute for Breast Cancer Research, Ontario Cancer Institute, 620 University Ave, Toronto, ON M5G 2C1 Canada
e-mail: tmak@oci.utoronto.ca

M. Andreeff (ed.), *Targeted Therapy of Acute Myeloid Leukemia*,
Current Cancer Research, DOI 10.1007/978-1-4939-1393-0_38

38.1 Introduction: Glycolysis and Tumor Cell Metabolism

The observation of a fundamental alteration in tumor cell metabolism initially identified by Otto Warburg has experienced renewed interest in recent years (Warburg et al. 1927; Warburg 1956). While the ultimate fate of glucose carbon atoms in normal cells is to drive energy production via the tricarboxylic acid (TCA)/Krebs cycle, the central tenet of the Warburg effect lies in the fact that even under conditions of ample oxygen, tumor cells tend to convert glucose into lactate (glycolysis) rather than diverting glucose carbons into the TCA cycle for use in oxidative phosphorylation (OXPHOS). Thus, in most documented cases of tumor cell metabolism, the vast majority of the glucose taken up is used to produce lactate and nucleotides, while other carbon sources, such as glutamine, are used for the anaplerotic reactions many of which lead into and out of the TCA cycle (DeBerardinis et al. 2007). As to why cancer cells take the path of a seemingly paradoxical, energy-inefficient use of glucose remains poorly understood and is an active area of research. Furthermore, it is likely that the prevalence of glycolysis over OXPHOS in tumor cells is not absolute and that tumor cells are capable of using a portion of their intracellular glucose for OXPHOS. The extent to which they do this has not been studied extensively and is most likely, highly variable, depending on the tumor's specific metabolic demands.

In the context of glucose utilization, this potentiation of a cancer cell's metabolic rate forms the foundation of positron emission tomography (PET) imaging wherein ^{18}F18 labeled glucose uptake in cancer cells far outweighs that of most normal tissues, providing a diagnostic marker that contrasts healthy cells from tumor cells. Taking into consideration the fact that tumor cells are under the control of a metabolic program that differs from their normal, nonmalignant counterparts, it stands to reason that some of the essential elements that drive the tumor metabolic phenotype would serve as potential candidates for targeted therapy.

The genetic basis of the Warburg effect, like many other features of the cancer phenotype, can be traced to the dysregulation of oncogenes and loss of tumor suppressor function with the usual suspects myelocytomatosis (MYC), p53, and hypoxia inducible factors (HIF) playing key roles (Tennant et al. 2009; Bayley and Devilee 2012; Jones and Thompson 2009; Dang 2012). A rigorous characterization of the Warburg effect and the genetic events that drive it has identified key targets for the development of anticancer therapeutics that inhibit glucose metabolism. A complete and thorough characterization of glucose metabolism in acute myeloid leukemia (AML) is, however, still lacking. A number of groups, including our own, are striving towards a better understanding of AML metabolism with the intent of moving these findings towards clinical applications.

Three potential strategies to leverage control over glucose metabolism in AML are (1) use of inhibitors that directly target enzymes in the glycolytic pathway, (2) inhibition of the pathways that activate aerobic glycolysis, and (3) use of drugs that drive AML cells away from aerobic glycolysis and towards OXPHOS.

38.2 Direct Targeting of Glycolytic Enzymes

If metabolic reprogramming towards aerobic glycolysis is indeed a hallmark of cancer, it stands to reason that this phenomenon could serve as a foundation for developing drugs that target the enzymes of the glycolytic pathway. A number of pharmacological agents that directly target different nodes of the glycolytic pathway have been reported; many of these agents are currently in the preclinical stage of development.

38.2.1 Hexokinase

Once glucose is transported into the cell, it is phosphorylated at the 6 position by hexokinase, generating glucose-6-phosphate (Fig. 38.1). This is a crucial transformation as it is the first and rate-limiting step in the glycolytic pathway. 3-Bromo-2oxopropionate-1-propyl ester (BrPO) is a pro-drug that is enzymatically hydrolyzed inside cells to generate 3-bromopyruvate (3-BrPA), which is a potent inhibitor of hexokinase. This compound was also found to have synergistic effects in leukemia and lymphoma cell lines when used in combination with rapamycin (Xu et al. 2005). A potential drawback or added benefit of this class of compounds is their ability to react nonspecifically with protein thiol groups (Scatena et al. 2008). Such lack of selectivity will make it difficult to further develop BrPO, BrPA, or related molecules as clinically useful anticancer agents; however, they at least show proof of principle for glycolytic inhibitors targeted against hexokinase. 2-Deoxyglucose (2-DG) is a glucose mimic lacking the 2 hydroxyl group. It is taken up by the cell and phosphorylated in the same manner as glucose, generating 2-deoxyglucose-6-phosphate which cannot be metabolized by the downstream enzyme phosphohexose isomerase. As it accumulates inside the cell, hexokinase activity for glucose is depressed via competitive inhibition, and adenosine triphosphate (ATP) levels drop (Maher et al. 2004). One caveat to this molecule is that the presence of glucose counters its action as a hexokinase inhibitor. As such, constantly high intracellular concentrations of 2-DG are required for efficacy. Lonidamine is yet another small molecule known to inhibit hexokinase activity. In vitro experiments revealed that lonidamine was able to decrease glycolysis, oxygen consumption, and lactate production in leukemia cell lines (Floridi et al. 1981). Conversely, the molecule had the opposite effects on normal germ cells and submaxillary gland cells. As hexokinase is upstream of the branch point into the pentose phosphate pathway for nucleotide generation, inhibition of hexokinase should starve both the catabolic glycolytic pathway and the anabolic nucleotide biosynthesis pathway of substrate.

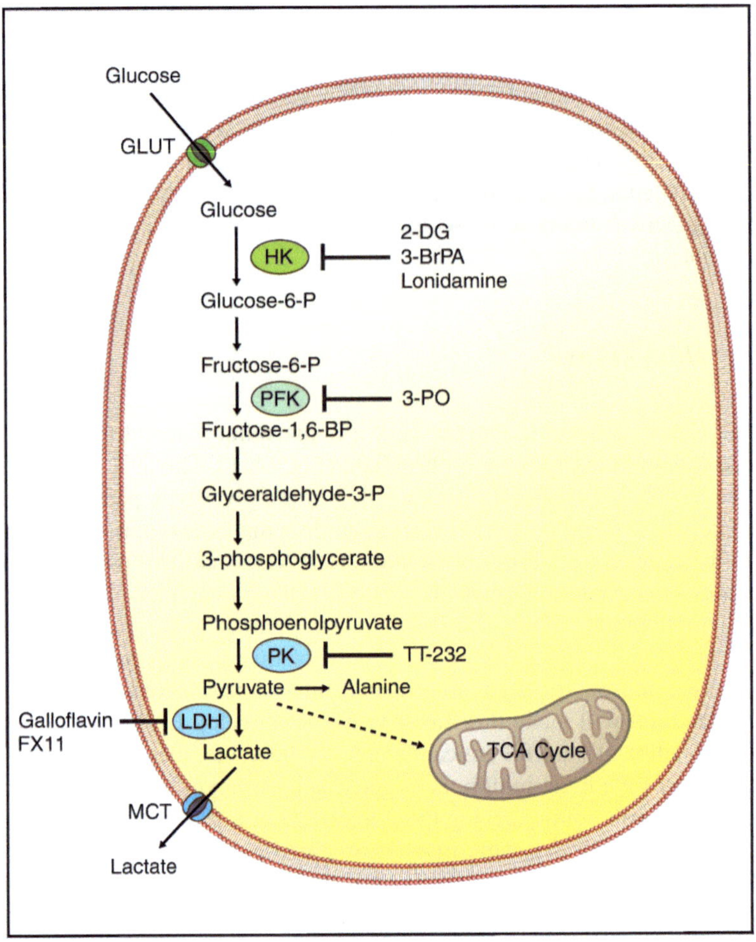

Fig. 38.1 Direct inhibition of glycolytic enzymes. An abridged diagram of the glycolytic pathway showing some of the key players involved in glucose metabolism and molecules that can be used to inhibit them. The majority of the glucose is shuttled through to lactate and excreted (*solid arrows*) whereas the branch point from pyruvate into the TCA cycle is used sparingly (*dashed arrow*). *GLUT* glucose transporter, *HK* hexokinase, *PFK* phosphofructokinase, *PK* pyruvate kinase, *LDH* lactate dehydrogenase, *MCT* monocarboxylic acid transporter. (Professional illustration by Takami Iijima)

38.2.2 Phosphofructokinase

Suppression of glycolysis has also been achieved via the downregulation of phosphofructokinase (PFK) activity. This kinase converts fructose-6-phosphate into fructose-1,6-bisphosphate. PFK is strongly activated by the metabolite fructose-2,6-bisphophate (F2,6BP). F2,6BP is the product of the enzyme PFKFB3, which is highly expressed in cancer cells. The small molecule 3PO inhibits PFKFB3 and reduces intracellular levels of F2,6BP (Clem et al. 2008). The reduction of phosphofructokinase activity by 3PO results in a reduction of glucose uptake, lactate secretion, and tumor growth in a xenograft model. In cell culture, 3PO has been shown to have IC_{50} values in the range of 0.5–1 μM against AML (HL60) and chronic myelogenous leukemia (CML) (K562) cell lines.

38.2.3 Pyruvate Kinase

In the penultimate step of aerobic glycolysis, pyruvate kinase (PK) transfers a phosphate group from phosphenolpyruvate to adenosine diphosphate (ADP), thus generating ATP and pyruvate. PK exists as four splice forms (L, R, M1, and M2) with the PKM2 isoform being most prevalent in cancer. The PKM2 splice form has reduced catalytic activity which is believed to allow for the accumulation of glycolytic intermediates upstream of PK, so they can enter anabolic processes such as the pentose phosphate pathway for nucleotide biosynthesis (Cairns et al. 2011). TT-232 (TLN-232, CAP-232) is a cyclic heptapeptide derivative of somastatin that targets PKM2 thereby compromising tumor cell glycolysis. In an AML xenograft mouse model, administering TT-232 was shown to reduce tumor volume; a phase II clinical trial using this drug in renal cell carcinoma has been completed (Tejada et al. 2006; NCT00422786).

38.2.4 Lactate Dehydrogenase

The last step in aerobic glycolysis is the conversion of pyruvate into lactate, a process catalyzed by the hetero-tetrameric lactate dehydrogenase (LDH) that exists as five isoforms (Fig. 38.1). The link between the LDH-A isoform and malignancies is highlighted by the fact that LDHA levels are controlled by the *MYC* oncogene, which is highly expressed in many cancers. Indeed, high levels of serum LDHA are characteristic of lymphomas and leukemias in which *MYC* is dysregulated by a chromosomal translocation (Shim et al. 1997). In addition to being a marker of transformation, LDH-A has been found to play an important role in tumor maintenance (Fantin et al. 2006). Using a virtual screening protocol to search for LDH inhibitors, Manerba and colleagues identified the molecule galloflavin as having activity against both LDH-A and B isoforms (Manerba et al. 2012). This compound was capable of inhibiting cellular respiration, lactate production, and triggered apoptosis in cell lines, albeit with a rather high IC_{50} of ~150 μM. Another LDH-A inhibitor, FX11, showed activity against P493 lymphoma cell lines both in culture and mouse xenograft models (Le et al. 2010). This study reconfirmed the need of LDH-A for tumor maintenance and showed that inhibition of LDH-A results in significantly higher levels of intracellular reactive oxygen species along with decreased ATP levels. Thus, FX11-mediated inhibition of LDH-A results in both bioenergetic and oxidative stresses. FX11 was also found to be weakly synergistic with the nicotinamide adenine dinucleotide (NAD) synthesis inhibitor FK866 when used in combination studies in a xenograft model with a Burkitt's lymphoma cell line.

38.3 Targeting the Regulators of the Glycolytic Pathway

An obvious caveat in the clinical application of agents that directly inhibit glycolysis lies in the fact that healthy cells also need glucose for energy production, albeit via OXPHOS and not aerobic glycolysis. As such, direct inhibition of glycolytic enzymes could portend unwanted toxicity in healthy cells. Indeed, some of the initial trials using 2-DG as an anticancer therapeutic were halted due to toxicity. An alternative to directly targeting glycolytic enzymes, while maintaining an impact on glucose flux, lies in the use of agents that inhibit the proteins that activate the effectors of the glycolytic pathway. Perhaps not surprisingly, a number of these activators have also been characterized as oncoproteins that drive cell proliferation and survival. Historically, the bulk of the literature on oncogene and tumor suppressor function has focused on characterizing the signaling pathways these proteins control with respect to their roles in apoptotic pathways, pro-survival pathways, transcription, etc. More recent studies have emerged that link many of these proteins with various aspects of cellular metabolism including glycolysis, which is a critical component of cellular proliferation.

38.3.1 Oncogenic Receptor Tyrosine Kinases

Upon binding ligand, receptor tyrosine kinases (RTKs) propagate growth signals from the extracellular milieu through to nuclear transcriptional programs. In AML, the kinases c-KIT, CSF1R (c-FMS) and FMS-like tyrosine kinase-3 (FLT-3) have key roles in driving oncogenesis. These kinases are crucial regulators of normal hematopoiesis; however, their aberrant activation can contribute to leukemogenesis. While these receptors are traditionally activated by ligand-mediated dimerization, ligand independent activating mutations have been reported for both c-Kit and FLT-3. The FLT3 internal tandem duplication (ITD) alters the juxtamembrane region of the protein resulting in a constitutively activated kinase; the presence of this mutation is associated with a high relapse rate and poor prognosis in AML patients (Kioyi 1999). Mutations to c-KIT active site residues can also lead to constitutive, ligand-independent activation. The association between the RTKs and glycolysis is mediated through their downstream activation of PI3K and subsequently AKT, which directly activates a number of glycolytic enzymes by phosphorylation (Fig. 38.2). There are a multitude of small molecules that have activity against AML-related RTKs. For the sake of brevity, we mention only two that have recently been used in AML clinical trials. Imatinib is an effective inhibitor of the ABL kinase, but also has off-target activity against c-KIT. In a phase I/II trial using imatinib in combination with standard AML chemotherapy, the loss of phopho-AKT, a downstream target of the c-KIT cascade, was shown to be an important correlative of efficacy. Nilotinib/AMN107 is another BCR-ABL inhibitor approved for use in CML that also has c-KIT off-target activity (Weisber et al. 2006). This molecule in combination with the mTOR inhibitor everolimus/RAD001 has recently been tested in a phase I/II trial in AML (NCT00762632).

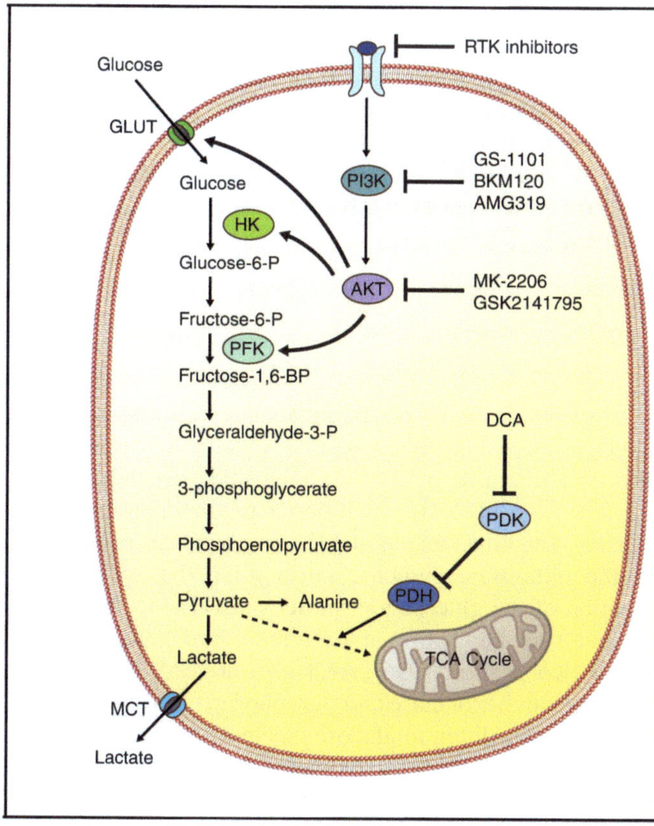

Fig. 38.2 Targeted inhibition of the upstream regulators of glycolysis as in Fig. 38.1 and highlighting some of the upstream regulators of the enzymes in the glycolytic pathway, their downstream targets, and some of the small molecule inhibitors that can be used to target them. AKT directly regulates several enzymes that modulate glucose flux and forms a hub downstream of PI3K and RTKs. PDH activity can shunt glucose-derived pyruvate into the TCA cycle by converting it into acetyl-CoA (*dashed arrow*). DCA mediated inhibition of the negative regulator of PDH, PDK shifts glucose metabolism towards OXPHOS (*dashed arrow*) and away from lactate production. *DCA* dichloroacetate, *GLUT* glucose transporter, *HK* hexokinase, *MCT* monocarboxylic acid transporter, *PDH* pyruvate dehydrogenase, *PDK* pyruvate dehydrogenase kinase, *PFK* phosphofructokinase; *PI3K* phosphatidyl inositol-3 kinase, *RTK* receptor tyrosine kinase. (Professional illustration by Takami Iijima)

38.3.2 PI3K

The PI3K pathway is frequently activated in cancer cells due to activating mutations to PI3K or due to functional loss of its negative regulator PTEN. PI3K is a lipid kinase that adds a phosphate group to the inositol C3 position of phosphatidylinositol(4,5)diphosphate. The product of this reaction serves to activates the downstream kinase AKT (discussed below), which has direct consequences on glucose metabolism (Fig. 38.2; Vivanco and Sawyers 2002). Drug-mediated inactivation of PI3K using GS-1101 was toxic to chronic lymphocytic leukemia (CLL) patient samples and lowered phospho-AKT levels. GS-1101 triggered caspase-mediated apoptosis and showed no toxicity towards normal B, T, and NK cells (Herman et al. 2010). The use of PI3K inhibitors in clinical trials is ongoing. A phase III clinical trial using GS-1101 as a single agent in CLL is currently in progress (NCT01539291). Phase I studies employing the PI3K inhibitors BKM120 in AML/ALL (NCT01396499) and AMG319 in B cell leukemias and lymphomas (NCT01300026) are also at the stage of recruitment.

38.3.3 AKT

Activation of the PI3K pathway results in AKT recruitment to the plasma membrane via its phosphoinositide binding, PH domain. At the membrane, AKT is phosphorylated and activated by 3-phosphoinositide-dependent protein kinase 1 (PDK1) where it drives glycolysis in addition to regulating transcriptional and anti-apoptotic cascades. The direct action of AKT on glycolysis occurs at several levels (Fig. 38.2). At the initial step of glucose transport into the cell, AKT activity results in the translocation of the glucose transporter (GLUT4) to the plasma membrane and increased levels of the GLUT1 transporter, culminating in increased glucose flux into the cell (Kohn et al. 1996). AKT also drives glucose flux at two other steps by activating both hexokinase and phosphofructokinase (Gottlob et al. 2001; Deprez et al. 1997). In addition to glucose metabolism, AKT also positively regulates fatty acid metabolism by phosphorylating ATP-citrate lyase, an enzyme that converts citrate to acetyl-CoA for lipid biosynthesis (Bauer et al. 2005).

In preclinical studies using T-cell acute lymphoblastic leukemia (ALL) cell lines, the allosteric AKT inhibitor MK-2206 induced apoptosis and caused rapid loss of AKT phosphorylation and its targets GSK3 and mTOR (Simioni et al. 2012). This study also showed synergy between the DNA damaging anthracycline doxorubicin and MK-2206. A phase II clinical trial using single agent MK-2206 in AML is currently in the process of recruiting patients; whether or not correlative studies are planned to assess the glycolytic state of MK-2206 treated AML patient samples is unknown (NCT01253447). A first in human phase I clinical trial of another oral AKT inhibitor, GSK2141795, in solid tumors and lymphomas is also ongoing (NCT00920257). Unfortunately, preclinical data for this compound are sparse.

38.3.4 PDK

If being wed to the Warburg effect is a driving factor in leukemic cells, a strategy for inhibiting aerobic glycolysis in cancer cells could lie in using pharmacologics to drive glucose away from lactate and into the TCA cycle, hence promoting OXPHOS. This would, in effect, redirect the path of glucose metabolism to that found in nonleukemic cells. Phosphorylation of pyruvate dehydrogenase (PDH) by the pyruvate dehydrogenase kinase (PDK) leads to its inactivation, thus halting the conversion of pyruvate into Acetyl-CoA which would, otherwise, enter the TCA for energy production and lipid biosynthesis (Fig. 38.2). Consequently, this phosphorylation event, in part, drives the Warburg effect by preventing glucose-derived carbon from entering the TCA. Preclinical investigations have shown that the small molecule dichloroacetate (DCA) is an inhibitor of mitochondrial PDK and capable of rewiring metabolism from aerobic glycolysis to OXPHOS, resulting in cancer cell apoptosis (Bonnett et al. 2007). Notably, DCA effects were specific to cancer cells with no toxicity to normal cells.

38.4 Glycolysis and Toxic Tumor Effluents

Protein crosstalk between leukemic cells and their surrounding environment is known to influence many facets of leukemic cell biology and has been an active area of interest for quite some time. AML cells produce an array of secreted cytokines and chemokines that, upon release into the extracellular matrix, can trigger signaling cascades via paracrine mechanisms in neighboring cells. Similarly, other cell types in the bone marrow niche produce proteins that modulate leukemic cell growth, survival, migration, and drug resistance. For example, the interaction between soluble, stromal derived factor 1 (SDF1) and its receptor CXCR4 can activate MAP kinase and PI3K signaling pathways on leukemic cells (Burger et al. 2000; Nishio et al. 2005). In addition to these soluble signaling proteins, molecules involved in cell-adhesion can play major roles in transmitting signals between AML cells and cells in the bone marrow, via direct cell–cell interactions. Of emerging importance is the observation that the potentiation of glycolysis in AML not only exerts a direct metabolic consequence on the tumor cell population itself, but also influences other cell types in the tumor microenvironment. These cell types can include immune response cells, stromal cells, and even other cells of the tumor population that exist in different metabolic states. Furthermore, tumor cells can be metabolically coupled to these cells in the microenvironment such that a transfer of metabolites can maintain tumor progression and survival. Metabolites derived either directly from glucose or as consequences of glycolysis that are excreted from AML cells have adverse effects on the phenotypes of surrounding cells. By far, the most widely studied of these molecules is lactate.

Catabolism of glucose through the nonoxidative portion of the glycolytic pathway culminates in the generation of pyruvate which can have four fates. In cells engaged in OXPHOS, pyruvate is converted into acetyl-CoA following which it can enter the TCA cycle for ATP production. Pyruvate can also enter the TCA cycle via pyruvate carboxylase, which generates 4 carbon oxaloacetate. The amino acid alanine is also generated from pyruvate via transamination; this requires glutamate as a nitrogen donor. Lastly, lactate dehydrogenase A (LDHA) can reduce pyruvate into lactate in a process that also consumes a molecule of NADH and produces a molecule of NAD. Lactate is then transported out of the cell into the extracellular

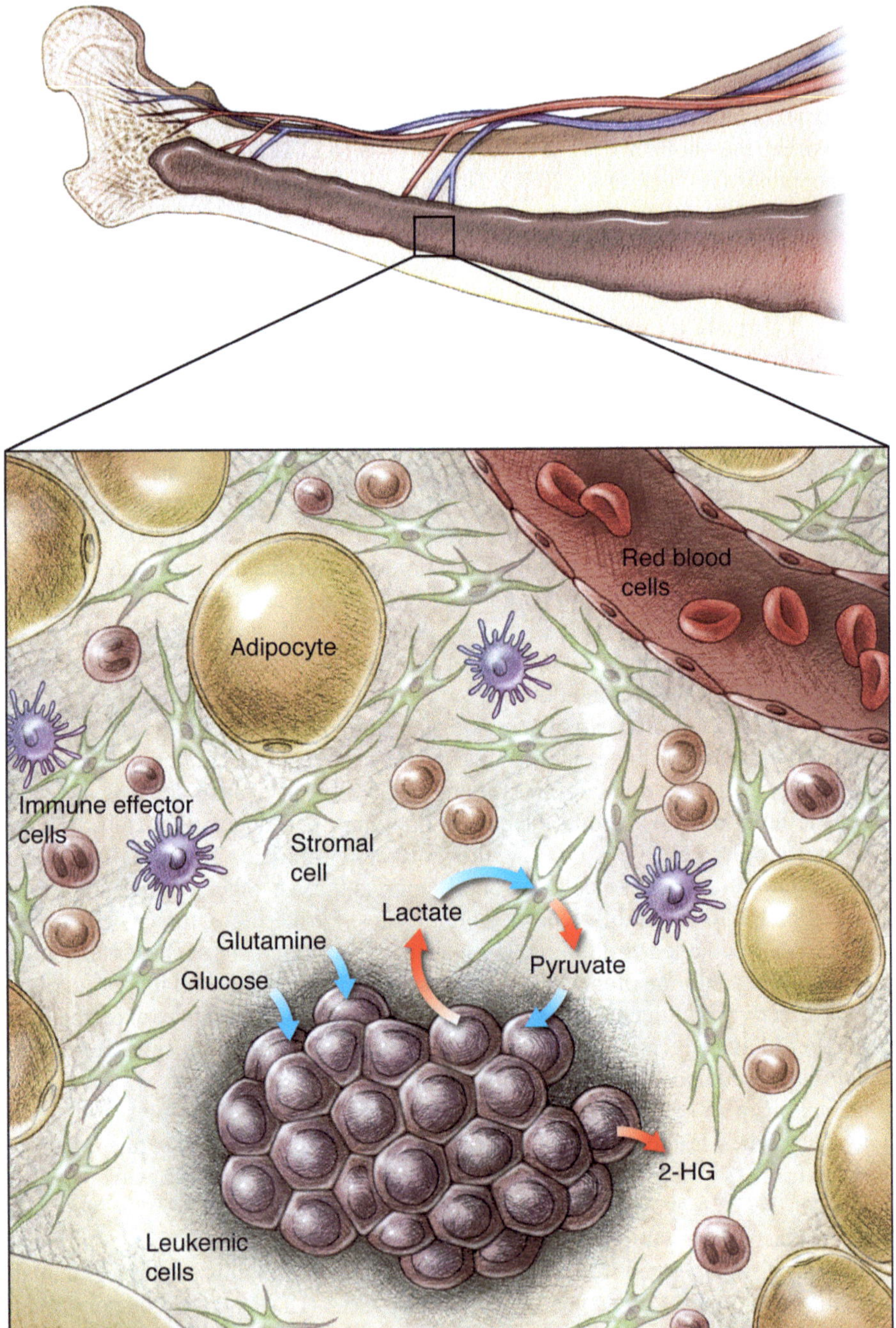

Fig. 38.3 AML, metabolites, and the bone marrow. Schematic of leukemic cells as they might reside in the bone marrow environment. Leukemic cells (*deep purple*) take up glucose and

compartment via the mono-carboxylic transporter (MCT). AML cells exhibiting high rates of glycolysis exude lactic acid into the extracellular space which can have deleterious effects on other members of the bone marrow microenvironment (Fig. 38.3). From a clinical perspective, tumor biopsies containing high levels of lactate were found to correlate with a higher incidence of metastasis and reduced overall survival (Walenta et al. 2000; Walenta et al. 2004).

Dendritic cell differentiation has been shown to be modulated by tumor cell derived lactate. Lactate from multicellular tumor spheroids was found to impart specific tumor-associated phenotypes to dendritic cells under high lactate conditions (Gottfried et al. 2006). In these experiments, lactate reduced dendritic cell IL-12 production, strongly inhibited antigen presentation and altered antigen expression. The authors suggest that lactate in combination with cytokines results in a dendritic cell phenotype that may contribute to the tumor's ability to escape detection by the immune system. Lactate can also affect several important aspects of cytotoxic T-lymphocyte (CTL) function. When CTLs are exposed to concentrations of lactate mimicking those found in the tumor environment, their proliferation, cytokine release, and cytotoxic activity are all inhibited (Fischer et al. 2007). In contrast to dendritic cells, the surface antigen profile of CTLs exposed to lactic acid is not perturbed. Furthermore, in these studies, it was shown that high levels of lactic acid blocks CTL lactate metabolism. Extracellular lactate is taken up by infiltrating T-cells and inhibits their lactate export. Finally, this study also revealed that, in 140 cancer patients, higher serum lactate levels correlate with a higher tumor burden.

Monocytes are components of the innate immune system; a crucial aspect of their normal function involves the secretion of inflammatory cytokines such as TNF alpha, IL-1, IL-6, and IL-12. The loss of activity of tumor infiltrating monocytes has been shown to correlate with lactate concentrations, in that tumor-derived lactate and acidification reduced secretion of TNF (Dietl et al. 2010). Inhibition of tumor lactate excretion subsequently can rescue TNF secretion by monocytes. Furthermore, the addition of lactate to the culture medium strongly increases the flux of lactate into monocytes, consequently suppressing glycolysis. Because glycolysis is crucial for monocyte function, the authors conclude that extracellular lactate confers an immunosuppressive phenotype by altering glucose metabolism in the monocytes. Similarly, dendritic cells, when activated, become increasingly reliant on glucose metabolism as part of their normal, biological response (Krawczyk et al. 2009). When dendritic cells are stimulated with a toll-like receptor agonist, they shift their glucose metabolism from OXPHOS to aerobic glycolysis. Efficient glucose metabolism in this new state requires efficient secretion of lactate to prevent

glutamine and secrete lactic acid. Stromal cells (*green*) can take up tumor derived lactic, convert it back into pyruvate which can be taken up by leukemic cells and used again. The excess lactic acid excreted by the AML cells also has deleterious effects on other cells in the bone marrow. In AML with IDH1 or IDH2 mutations, 2HG is released into the bone marrow environment, the effect of which is, as yet, unknown. For simplicity, we depict only the bone marrow; however, we assume these events to hold for leukemic cells that have infiltrated other tissues. (Professional illustration by Takami Iijima) the build-up of intracellular concentrations. Increase in intracellular lactate derived from and secreted by tumor cells appears to hinder glucose metabolism in immune cells. Still lacking, however, is evidence into the mechanism by which lactate affects the altered phenotypes of the infiltrating immune effector cells. One possibility is that the lactate, through its ability to disrupt the quaternary structure of PFK-1, inactivates the enzyme (Costa 2007). Such an event would be predicted to impede glycolysis and hence immune cell function, although exactly how or why this would inhibit secretion of certain cytokines and not others requires further investigation.

Thus, in addition to the altered cytokine patterns in and around tumor cells, tumor derived metabolites can also have an effect on immune effector cells. Taken together, these studies conclude that high levels of lactate in the tumor microenvironment contributes to immune evasion. In the case of AML, where immune function is already compromised due to impaired hematopoiesis, further inhibition of any remaining immune cells by tumor effluents would be especially egregious. Thus, agents that are capable of reducing the amount of lactate immune cells are exposed to, for instance by inhibiting glycolysis, may also play an immunomodulatory role. Whether or not targeting glycolytic pathways would serve to boost the effects of conventional immunotherapy such as regimes employing monoclonal antibodies, remains untested. The inhibitory role of lactate in immune cell function also raises an important issue in considering the use of monocarboxylate transporter inhibitors as chemotherapeutics. It should be noted that directly inhibiting MCT lactate exporters may result in intracellular lactate accumulation in immune cells. Presumably this could mimic the adverse effects tumor derived lactate has on the immune response. Thus, judicious targeting of glycolysis as an anticancer therapeutic should also take into account potential effects on immune effector cells, their metabolic requirements, and the potential for off-target metabolic effects on these cells.

A more recent case of altered tumor metabolism revolves around the production of the so-called "oncometabolite" 2-hydroxyglutarate (2-HG) by mutant isocitrate dehydrogenase (IDH) enzymes. The IDH enzymes are responsible for the oxidation of isocitrate to alpha-ketoglutarate (aKG), a key component of the TCA cycle and a co-factor for a large number of enzymes. It was recently reported that recurrent point mutations in the IDH1 active site, rather than conferring a loss of function phenotype, induced a change of function (Dang et al. 2009). This neomorphic activity of the IDH mutants confers the ability to stereospecifically reduce alpha-ketoglutarate into 2-HG. Mutations to both IDH1 and IDH2 resulting in 2-HG production have been identified in AML (Mardis et al. 2010; Chou et al. 2010; Gross et al. 2010). 2-HG is structurally very similar to alpha-ketoglutarate and has been shown to compete for the alpha-ketoglutarate binding sites on alpha-ketoglutarate-dependent enzymes. Through this mechanism, 2-HG can inhibit the function of proteins such as ten-eleven translocation-2 (TET2), which is involved in DNA demethylation and the histone demethylase jumonji C (JMJC) (Figueroa et al. 2010; Chang et al. 2007). In keeping with the loss of TET2 function, high levels of 2-HG results in an increase in DNA and histone hypermethylation. In addition to the inhibition of aKG-dependent enzymes, recent work has shown that 2-HG can activate the alpha-ketoglutarate-dependent prolylhydroxylase egg-laying defective nine homolog-2 (EGLN) (Koivunen et al. 2012). As the EGLN family of hydroxylases target HIF-1 for degradation, 2-HG-mediated activation of EGLN1 and 2 results in lower levels of HIF-1 protein. Thus, this tumor metabolite is capable of affecting the epigenetic status and the kinetics of the response to hypoxia in the cells in which it accumulates. It is of note that 2-HG is released from AML cells harboring IDH1 mutations and, as such, AML patients with 2-HG producing mutations have substantially higher levels of 2-HG in their sera, compared to patients with wild type IDH (Gross et al. 2010). Is it possible that 2-HG is taken up by surrounding normal cells where it can subsequently affect biochemical processes? While this has not been tested per se, the addition of 2-HG to the media of wild-type IDH glioma cells, albeit at relatively high concentrations, was capable of increasing intracellular 2-HG concentrations to the levels observed in IDH mutant cells (Reitman et al. 2011). Therefore, like lactate, it appears that 2-HG can indeed enter cells from the extracellular milieu. Therapeutics capable of reducing 2-HG production and secretion, such as inhibitors specific for mutant IDH, may not only rescue the effects of 2-HG on the AML cells in which it is produced, but also other cell types in which it can accumulate.

Metabolites that are secreted by one cell type are free to be taken up and utilized by other opportunistic cells. The large amounts of lactate produced by cancer cells can be taken up by surrounding stromal cells by MCTs. Once inside these stromal cells, this lactate can be oxidized back to pyruvate which is then secreted back into the extracellular matrix. The MCT transporters on cancer cells can then take up this pyruvate where it can further be metabolized (Koukourakis et al. 2006). This intriguing cycle metabolically couples cancer cells to proximal stromal cells where they can recycle glucose carbon by shunting lactate and pyruvate back and forth between the two cell types via the bidirectional action of MCTs (Fig. 38.3). Similarly, metabolic symbiosis has been observed between tumor cells in different oxidative niches. Here, hypoxic tumor cells can secrete lactate that is taken up by oxidative tumor cells and used to produce energy (Sonveaux et al. 2008). As the aerobic tumor cells are driven by lactate oxidation, this spares glucose stores for use by tumor cells in the hypoxic niche.

The ketone body 3-hydroxybutyrate (3-HB) is another product of aerobic glycolysis which, like lactate, can be secreted by MCT transporters. Derived from pyruvate in stromal cells, secreted 3-HB, once taken up by cancer cells, can enter the TCA cycle where it can be utilized to fuel cancer cell growth (Bonucelli et al. 2010). When administered systemically through intraperitoneal injection, 3-HB and to a lesser extent lactate can promote tumor growth in a mouse xenograft model. This phenomenon is accomplished without additional blood vessel formation in the tumor. In this sense, ketone bodies may circumvent the need for angiogenesis. In addition, in cell migration assays, both 3-HB and lactate can act as chemo-attractants for cancer cells.

Another example of the interdependence of tumor cells and the stroma relates to the sensitivity of ALL stem cells to the effects of l-asparaginase. While the ALL cells have reduced levels of the enzyme asparagine synthetase, the abutting stromal cells have normal levels of the enzyme and can excrete asparagine into the extracellular space. The drug L-asparaginase, which catalyzes the degradation of asparagine into aspartic acid, is commonly used in the treatment of ALL. In co-culture experiments, ALL cells were able to take stromal-derived asparagine to supplement their own supplies, thus shielding them from L-asparaginase treatment (Iwamoto et al. 2007). Targeted knockdown of asparagine synthetase in these stromal cells restored ALL sensitivity to L-asparaginase. Conversely, upregulation of stromal cell asparagine synthetase offered further protection. In this study, the transfer of asparagine from stromal cells to ALL cells represents a basis for metabolic protection and a mechanism of drug resistance.

Conclusion

Leukemia is a disease of the bone marrow, a dynamic environment consisting of a variety of cell types that rely on each other for their growth, survival, and function. Their high glycolytic activity and the presence of mutations such as IDH1/2 results in an altered microenvironment that can change the support capacity of stromal cells for normal blood forming cells, as well as alter the state of immune responsive cells, changing them from effector cells to negative regulatory cells. The development of pharmacologic agents that reduce the glycolytic rate of leukemic cells is predicted to not only alter the leukemic cells, but other cells within the bone marrow, allowing the return of normal hematopoiesis. As the effects of the leukemic cells on marrow stroma may be long-lived, dual therapy aimed at the leukemic cells and stroma may be needed. These may be targeted at metabolic or epigenetic changes within the stroma. The development of these strategies requires the development of a deeper understanding of the leukemic cells and importantly the cells that they interact with.

References

Bauer DE, Hatzivassiliou G, Zhao F et al (2005) Oncogene 24:6314–6322
Bayley JP, Devilee P (2012) The Warburg effect in 2012. Curr Opin Onco 24:62–67
Bonnett S, Archer SL, Allalunis-Turner J et al (2007) Cancer Cell 11:37–51
Bonucelli G, Tsirigos A, Whitaker-Menezes D et al (2010) Cell Cycle 9:3506–3513
Burger JA, Tsukada N, Burger M et al (2000) Blood 96:2655–2663
Cairns RA, Harris IS, Mak TW (2011) Nat Rev Cancer 11:85–95
Chang B, Chen Y, Zhao Y et al (2007) Science 318:444–447
Chou WC, Hou HA, Chen CY et al (2010) Blood 115:2749–2754
Clem B, Teland S, Clem A et al (2008) Mol Cancer Ther 7:110–120
Costa LT, Da Silva D, Guimaraes CR, Zancan P et al (2007) Biochem J 408:123–130
Dang CV (2012) Cell 149:22–35
Dang L, White DW, Gross S et al (2009) Nature 462:739–744
DeBerardinis RJ, Mancuso A, Daikhin E (2007) PNAS 104:19345–19350
Deprez J, Vertommen D, Alessi DR et al (1997) J Biol Chem 272:17269–17275
Dietl K, Renner K, Dettmer K (2010) J Immunol 184:1200–1209

Fantin VR, St-Pierre J, Leder P (2006) Cancer Cell 9:425–434
Figueroa ME, Abdel-Wahab O, Lu C et al (2010) Cancer Cell 18:553–567
Fischer K, Hoffman P, Voelkl S et al (2007) Blood 109:3812–3819
Floridi A, Paggi MG, Marcante ML et al (1981) J Natl Cancer Inst 66:497–499
Gottfried E, Kunz-Schughart LA, Ebner S et al (2006) Blood 107:2013–2021
Gottlob K, Majewski N, Kennedy S et al (2001) Genes Dev 15:1406–1418
Gross S, Cairns RA, Minden MD et al (2010) J Exp Med 207:339–344
Herman SE, Gordon AL, Wagner AJ et al (2010) Blood 116:2078–2088
Iwamoto S, Mihara K, Downing JR et al (2007) J Clin Invest 117:1049–1057
Jones RG, Thompson CB (2009) Genes Dev 23:537–548
Kioyi H, Naoe T, Nakano, Y et al (1999) Blood 93:3074–3080
Kohn AD, Summers SA, Birnbaum MJ et al (1996) J Biol Chem 271:31372–31378
Koivunen P, Lee S, Duncan CG (2012) Nature 483:484–488
Koukourakis MI, Giatromanolaki A, Harris AL et al (2006) Cancer Res 66:632–637
Krawczyk CM, Holowka T, Sun J et al (2009) Blood 115:4742–4749
Le A, Cooper CR, Gouw AM et al (2010) PNAS 107:2037–2042
Maher JC, Krishan A, Lampidis TJ (2004) Cancer Chemother Pharmacol 53:1745–1751
Manerba M, Vettraino, M, Fiume L et al (2012) Chem Med Chem 7:311–317
Mardis ER, Ding L, Dooling DJ et al (2010) N Engl J Med 361:1058–1066
Murray CM, Bundick ID, Cook, RI et al (2005) Nat Chem Biol 1:371–376
Nishio M, Endo T, Tsukada et al (2005) Blood 106:1012–1020
Rathmell JC, Farkash EA, Gao W, Thompson CB (2001) J Immunol 167:6869–6876
Reitman ZJ, Jón G, Karoly Ed et al (2011) PNAS 108:3270–3275
Scatena R, Bottoni P, Pontoglio A et al (2008) Expert Opin Invertis Drugs 17:1533–1545
Shim, H, Dolde, C, Lweis BC et al (1997) PNAS 94:6658–6663
Simioni C, Neri LM, Tabellini G et al (2012) Leukemia Epub ahead of publication 1–7
Sonveaux P, Vegran F, Schroeder T et al (2008) J Clin Invest 118:3930–3942
Tejada M, Gaal D, Hullan L et al (2006) Anticancer Red 26:3477–3483
Tennant DA, Duran R, Boulahbel H et al (2009) Carcinogenesis 30:1269–1280
Vivanco I, Sawyers CL (2002) Nat Rev Cancer 2:489–501
Walenta S, Wetterling M, Lehrke M et al (2000) Can Res 60:916–921
Walenta S, Schroeder T, Mueller-Klieser W (2004) Curr Med Chem 11:2195–2204
Warburg O (1956) Science 123:309–314
Warburg O, Wind F, Negelin E (1927) J Gen Physiol 8:519–530
Weisber E, Manley P, Mestan J et al (2006) Br J Cancer 94:1765–1769
Wise DR, DeBerardinis RJ, Mancuso A et al (2008) PNAS 105:18782–18787
Xu RH, Pelicano H, Zhang H, Giles FJ et al (2005) Leukemia 19:2153–2158

Chapter 39
Molecular and Biochemical Basis for the Reprogramming of Intermediary Metabolism in Leukemia Cells

Ismael Samudio and Juliana Vélez

Abstract Recent evidence suggests that leukemia cell metabolism is profoundly altered to promote cell survival and resistance to apoptotic stimuli. It is intriguing that albeit leukemia cells display de Warburg effect (increased glycolysis in the presence of adequate amounts of oxygen), this metabolic phenotype is not associated with impairment to the oxidative capacity of mitochondria, but instead represents a shift to oxidize substrates other than pyruvate. This metabolic shift is also associated with increased resistance to apoptotic stimuli, and appears to be promoted, at least in part, by interactions with the bone marrow microenvironment. This chapter reviews some of the most compelling evidence for profound metabolic alterations in leukemia cells, and presents a model of leukemia cell intermediary metabolism that reconciles classical and novel metabolic paradigms with microenvironmental cues that promote leukemia cell survival.

Keywords Metabolism · Mitochondrial uncoupling · Anaplerosis · Krebs cycle · Warburg effect · Fatty acid metabolism · Chemoresistance · Bone marrow microenvironment

39.1 Introduction

Much has changed in the past 80 years since Otto Warburg proposed that the origin of cancer relied on an irreparable damage to the oxidative capacity of normal cells that supported the switch to glycolysis for energy generation and a malignant phenotype (Warburg 1956a). Curiously, this phenotype was also observed by Warburg in embryonic tissues, and recent research has demonstrated that normal stem cells may indeed rely on aerobic glycolysis—fermenting pyruvate to lactate in the

I. Samudio (✉) · J. Vélez
Biochemistry Laboratory # 301, Pontificia Universidad Javeriana,
Cra 7 # 40–62, Bogota, Colombia
e-mail: Ismael.samudio@gmail.com

J. Vélez
email: jvelez0411@gmail.com

M. Andreeff (ed.), *Targeted Therapy of Acute Myeloid Leukemia*,
Current Cancer Research, DOI 10.1007/978-1-4939-1393-0_39

presence of ample oxygen (O_2)—rather than on the complete oxidation of pyruvate in the Krebs cycle to generate cellular energy (Folmes et al. 2012). However, it remains to be determined whether this phenotype is causative in neoplastic development, or whether it is just a required characteristic for malignant transformation. In addition, Warburg's proposed theory of irreparable damage to the mitochondria has been substantially modified. In light of recent evidence it may be tempting to speculate that cancer progenitors arise from normal progenitors that already had a Warburg phenotype; however, that would be an oversimplification of cancer stem cell biology and metabolism in general. More important, it does not take into account that the tumor microenvironment may contribute as much to the Warburg phenotype as the tumor cell itself—perhaps even more. In this chapter, we will address some peculiarities of cancer and leukemia cell metabolism that promote cell survival and chemoresistance.

39.2 Leukemia Cell Metabolism

A preponderance of evidence supports the notion that leukemia cells display a glycolytic phenotype even in the presence of adequate amounts of O_2 (Samudio et al. 2008, 2010). However, in contrast to what Warburg hypothesized, it is now evident that leukemia cells do not have an impaired O_2 consumption capacity (Samudio et al. 2008, 2010). Rather, it appears as if they have acquired the capacity to oxidize substrates other than pyruvate in the Krebs cycle (Samudio et al. 2009, 2010). This finding is intriguing in light that in other tumor types, like glioblastoma for instance, it has been reported that anaplerotic entry of glutamine-derived glutamate fuels in large part the activity of the Krebs cycle and oxidative phosphorylation (OXPHOS) (DeBerardinis et al. 2007), suggesting that shunting pyruvate away from mitochondrial oxidation is an acquired trait of different types of cancer. In leukemia cells recent evidence suggests that fatty-acid-derived acetyl-CoA fuels Krebs cycle activity and the molecular reduction of O_2 (Samudio et al. 2010), and unpublished observations suggest that adequate amounts of glutamine-derived oxaloacetate are essential to maintain O_2 consumption in these cells. These findings support the notion that leukemia cell mitochondria are less prone to oxidize pyruvate or convert it to oxaloacetate for anaplerotic entry to the Krebs cycle (Fig. 39.1). But what would be the selective advantage of this metabolic shift?

39.2.1 Bioenergetic Considerations

First, although OXPHOS is metabolically more efficient than glycolysis in terms of ATP generation, OXPHOS only occurs in the mitochondria, which represents a small volume of the cytoplasm. In contrast, glycolysis occurs in the cytosol which typically represents >70% of the cell volume and thus has the potential of matching net ATP contributions from OXPHOS, depending on the cell type. In addition,

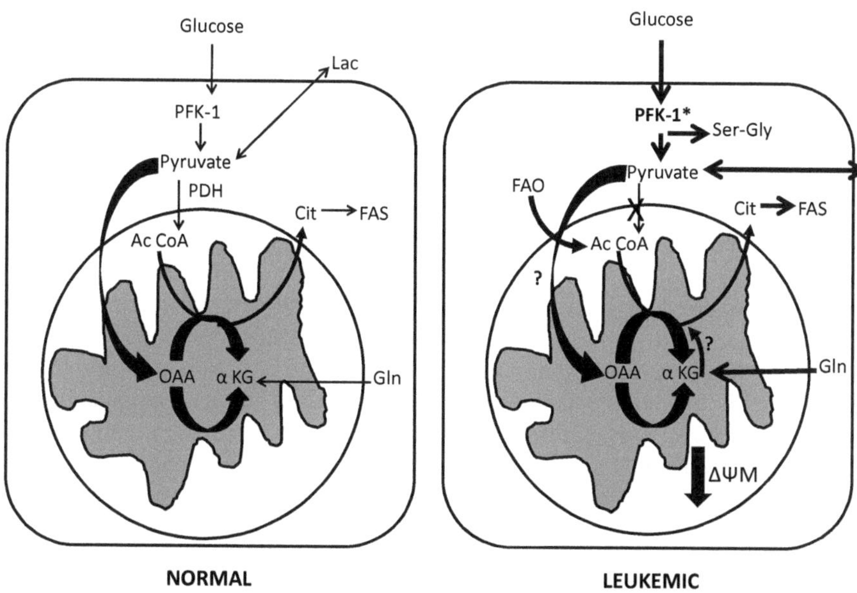

Fig. 39.1 Reported and potential metabolic alterations of leukemia versus normal cells. Leukemia cells appear to depend on increased oxidation of fatty acids (*FAO*) to replace pyruvate-derived acetylCoA to maintain Krebs cycle function. It is unknown if pyruvate carboxylation to oxaloacetate does occur in leukemia cells, but unpublished observations support the notion that glutamine may enter anaplerotically into the Krebs cycle as α-ketoglutarate. Intriguingly, this metabolic phenotype is associated with a dramatic loss in ΔΨM which via incompletely identified mechanisms promotes resistance to apoptotic stimuli. *PFK1* phosphofructokinase 1, *PFK1** phosphofructokinase 1 shorter (more active) polypeptide, *Lac* lactate, *AcCoA* acetyl-CoA, *Cit* citrate, *FAS* fatty acid synthase, *OAA* oxaloacetate, *αKG* α-ketoglutarate, *Gln* glutamine, *Ser* serine, *Gly* glycine

OXPHOS depends on adequate amounts of O_2 which may not be readily achievable in most tumor types, including leukemia (see Hypoxia section of this book). Lastly, glycolysis depends on the availability of NAD^+ for the reaction catalyzed by glyceraldehyde-3-phosphate dehydrogenase, which can be readily formed by the fermentation of pyruvate to lactate, without the need to rely on the glycerol-3-phosphate or malate–aspartate shunts that depend on OXPHOS to oxidize $FADH_2$ or NADH, respectively. Altogether, it is improbable that glycolysis is at an energetic disadvantage to OXPHOS in terms of net ATP generated.

Here, it is important to point out that leukemia cells must have also evolved mechanisms to bypass the net inhibitory effect of ATP accumulation on glycolysis, particularly at the point of phosphofructokinase 1 (PFK-1), the canonical regulator of glycolysis that carries out the conversion of fructose-6-phosphate into fructose-1,6-bisphosphate. Recent evidence suggests that in one lymphoma cell line (Nb2-11) and in one leukemia cell line (TF-1), PFK-1 has been posttranslationally modified via a proteinase K-type of mechanism, into a shorter polypeptide which maintains catalytic activity, but is not allosterically inhibited by citrate or ATP, and superactivated by fructose-2,6-bisphosphate (Smerc 2011). Although similar modifications have not been reported in primary leukemia samples, it is intriguing to speculate that bypassing inhibitory effects of intracellular ATP accumulation would be a critical growth advantage of leukemia versus normal progenitors.

In addition to resisting the inhibitory effects of ATP accumulation, leukemia cells also maximize the use of what little mitochondrial ATP may be produced. More than 2 decades ago, Arany et al. reported that when associated to the mitochondria, the glycolytic enzyme hexokinase promoted O_2 consumption in leukemia cells (Arany et al. 1988) suggesting perhaps that mitochondrially derived ATP enhanced glycolysis in these cells. More recent evidence (reviewed in Mathupala et al. 2010) indeed supports the notion that a particular isoform of hexokinase, hexokinase II (HKII)—one of the four isoforms of hexokinase, found overexpressed in cancer cells—associates with the mitochondria to attain direct access to OXPHOS generated ATP for the phosphorylation of glucose. Additionally, it has been reported that mitochondrial bound HKII inhibits apoptosis in leukemia and lymphoma cell lines (Chen et al. 2009). HKII is targeted to mitochondria by its association with voltage-dependent anion channel (VDAC), the outer mitochondrial membrane anion channel that allows entry of most metabolites, but only permits egress of proapoptotic proteins when the intrinsic pathway of programmed cell death is activated (Shoshan-Barmatz and Golan 2012). AKT-dependent phosphorylation of HKII potentiates its binding to mitochondria (Majewski et al. 2004a, b), providing another mechanism by which AKT regulates cell survival. Mechanistically, it is possible that HKII prevents cytochrome *c* release by blocking the channel space in VDAC, necessary for release of larger molecules during apoptosis (Abu-Hamad 2008), although conflicting reports suggest that the antiapoptotic effect of HKII may be independent of VDAC, but depends on the presence and/or activity of cyclophilin D and adenine nucleotide translocator (Chiara 2008). Regardless of the precise mechanisms, it is clear that HKII plays a central role in regulating mitochondrial metabolism and glycolysis in leukemia cells.

39.2.2 Anabolic and Anaplerotic Considerations

While complete oxidation of glucose in the Krebs cycle yields more ATP per mole of glucose, it fails to safeguard carbon skeletons for anabolic reactions. On the one hand, increased glycolytic rates with resulting accumulation of pyruvate can actually inhibit glycolysis at key points to redirect glucose carbons for biomass generation. For instance, pyruvate transamination produces alanine which inhibits allosterically pyruvate kinase, and results in accumulation of phosphoenolpyruvate which via equilibrium reactions is converted to 3-phosphoglycerate which serves as a precursor to serine and glycine (DeBerardinis 2011). Phosphoenolpyruvate is also an inhibitor of PFK, which would shunt fructose-6-phosphate to glucose-6-phosphate that can now enter the pentose phosphate shunt to produce nicotinamide adenine dinucleotide phosphate (NADPH) for biosynthesis and ribose-5-phosphate for nucleotide biosynthesis. On the other hand, to maintain an adequate supply of carbon skeletons to replenish oxaloacetate in the Krebs cycle, pyruvate can be carboxylated or alternatively, several Krebs cycle intermediates can be generated anaplerotically through the metabolism of odd chain fatty acids (producing succinyl-CoA) or the metabolism of amino acids (which can generate α-ketoglutarate, succinyl-CoA, fumarate, or oxaloacetate). Here it is important to consider that replenishing oxaloacetate in the Krebs cycle—directly by carboxylation of pyruvate or transamination of aspartate, or indirectly by replenishing α-ketoglutarate, succinyl-CoA, or fumarate—is essential to maintain the capacity of oxidizing substrates other than glucose, in particular fatty acids. The above considerations support the notion that increased glycolytic flux would paradoxically result in a mitochondrial adaptation to utilize other carbon sources to support Krebs cycle activity, both in terms of the regeneration of oxaloacetate via anaplerotic reactions and the formation of citrate by condensing for instance fatty-acid-derived acetyl CoA with oxaloacetate. Figure 39.1 illustrates some of the major reported and potential metabolic alterations of leukemia versus normal cells.

39.2.3 The Crabtree Effect: Another Name for the Warburg Effect?

Intriguingly, high concentrations of glucose have been observed to transiently inhibit O_2 consumption in yeast and some mammalian cells, a phenomenon known as the Crabtree effect—named after the English biochemist Herbert Grace Crabtree (Diaz-Ruiz 2011). Recent work demonstrates that the Crabtree effect results from saturating the respiratory capacity of cells—the increased amounts of glucose-derived NADH saturates the electron transport chain (Agrimi et al. 2011)—as well as a direct inhibitory effect of fructose 1,6-bisphosphate on mitochondrial respiration (Diaz-Ruiz et al. 2008). As such, the major difference between the Crabtree effect and the Warburg effect may simply lie on how transient is the suppression of O_2 consumption in the Crabtree effect; a rather long-lasting suppression would mimic the Warburg effect, in the absence of permanent and transmissible defects in OXPHOS. Although no recent evidence highlights the role of the Crabtree effect in leukemias, it is intriguing to consider the possibility that the inhibitory effect of fructose 1,6-bisphosphate may prompt a metabolic switch to the oxidation of fatty acids which generate greater amounts of NADH and $FADH_2$ as an adaptation to maintain O_2 consumption capacity.

39.3 Metabolism and the Microenvironment

39.3.1 Mesenchymal Stromal Cells and Mitochondrial Uncoupling

Recent evidence has demonstrated that mitochondrial uncoupling—a short circuit in the electrochemical gradient of the mitochondrial membrane—promotes resistance to intrinsic apoptosis in leukemia cells, in part, via antagonism of bax/bak oligomerization (Samudio et al. 2008, 2010). Moreover, this metabolic phenotype, originally reported in leukemia cells cultured on mesenchymal stromal feeder layers (Samudio et al. 2008), results in decreased entry of pyruvate into the Krebs cycle and a shift to fatty acid oxidation to support O_2 consumption, presumably permitting utilization of glucose carbon skeletons for the generation of biomass (Samudio et al. 2009). Herein it is important to point out that this metabolic shift, albeit a phenocopy of the Warburg effect does not support Otto Warburg's hypothesis of a permanent respiratory damage to cancer cell mitochondria (Warburg 1956a, b). Instead, these observations are in agreement with Feodor Lynen's suggestion that the continuing reduction of O_2 without the synthesis of ATP could mimic the Warburg effect in the absence of permanent, transmissible alterations to the oxidative capacity of cells (Lynen 1951). While it has been suggested that mitochondrial uncoupling in leukemia cells co-cultured with 2-mesitylenesulfonyl chloride Mesenchymal Stromal Cells (MSC) could be potentiated by uncoupling protein 2 (UCP2; Fig. 39.2), it is clear—based on siRNA interference experiments targeting UCP2—that expression of this protein is not an absolute requirement for short circuiting the mitochondrial electrochemical gradient (Samudio et al. 2008). Since uncoupling may in fact promote the increased expression of UCP2 (Li et al. 2002; Turner et al. 2007), it is still unclear what precise molecular events initiate MSC-dependent mitochondrial uncoupling in leukemia cells.

Additionally, Zhang W et al. have recently demonstrated that MSC take up cysteine, the spontaneously formed disulfide (oxidized) of the amino acid cysteine, from the extracellular milieu, and reduce it back to cysteine (Zhang et al. 2012). This cysteine can then be released from the MSC to be taken up by leukemia cells and used in the synthesis of the potent antioxidant glutathione (GSH) (Zhang et al. 2012). The results of Zhang and colleagues suggest that the increased chemoprotection afforded by MSC on leukemia cells is mediated, at least in part, by the increased levels of GSH. As illustrated in Fig. 39.2, this mechanism may be complementary to the antiapoptotic effects of mitochondrial uncoupling by increasing resistance to oxidative damage. More importantly perhaps, this mechanism provides proof of the concept that the microenvironment can support leukemia cell viability by providing nutrients in their biologically relevant form. Whether other metabolites are exchanged in between the leukemia and stromal compartment remains to be determined, but it is likely that continuous nutrient exchange between these compartments will play a key role in metabolic adaptations of the leukemic bone marrow.

39.3.2 Hypoxia in the Bone Marrow and the Reverse Krebs Cycle

It is intriguing that although the bone marrow is a highly vascularized sinusoidal cavity, O_2 concentration appears to be similar to that found in venous blood (6%) (Fiegl et al. 2009), suggesting the notion that the bone marrow may be ipso facto hypoxic—at least physiologically. However, if indeed leukemia cell mitochondria are uncoupled in vivo, the mechanism by which O_2 tension is reduced in the bone marrow may be directly related to the increased O_2-reducing capacity of leukemia cells.

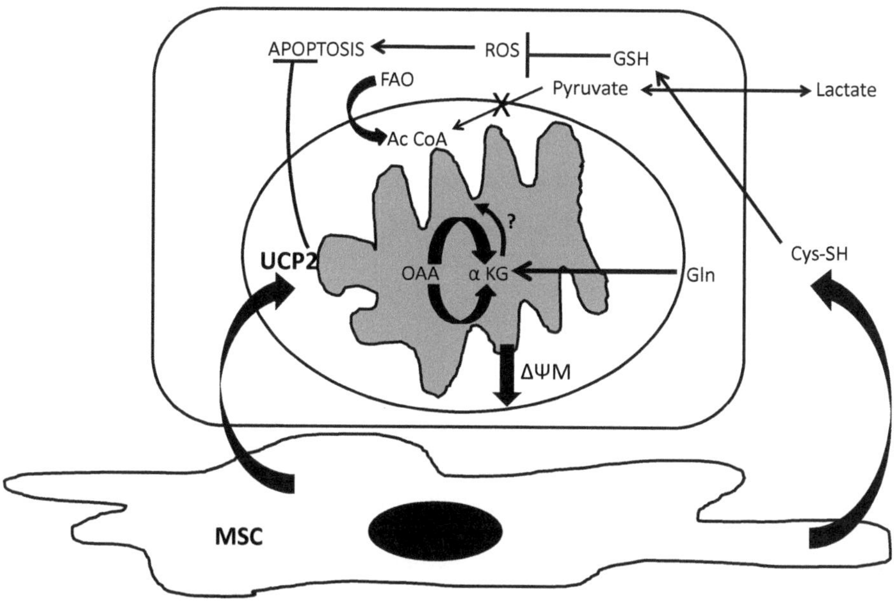

Fig. 39.2 Metabolism and microenvironment in leukemia cells. Stromal mesenchymal cells promote leukemia cell survival via complimentary pathways. On the one hand, they promote mitochondrial uncoupling via increased expression of UCP2, on the other hand, they supply the reduced form of cystine (cysteine) for synthesis of GSH in leukemia cells. Both of these events antagonize the onset of apoptosis. *GSH* reduced glutathione, *ROS* reactive oxygen species, *UCP2* uncoupling protein 2, *Cys-SH* cysteine, *MSC* mesenchymal stromal cells

Still, this presents an interesting paradox: The reduced O_2 tension would eventually lead to an inhibition of Krebs cycle activity—via accumulation of NADH—and the reduced generation of ATP and biomass. Intriguing recent research has suggested that under hypoxic conditions the Krebs cycle can actually go in reverse, starting from glutamine-derived α-ketoglutarate to form citrate via reductive carboxylation, to support the de novo synthesis of fatty acids (Mullen et al. 2012). Albeit no evidence of this effect has been found in leukemia cells under hypoxia (Fig. 39.2 speculates on the possibility of this event), it is tempting to speculate that reductive carboxylation of α-ketoglutarate may rescue biomass generation from the Krebs cycle in leukemia cells under reduced O_2 tension, whether this resulted from cell autonomous consumption of available O_2, or an as of yet unidentified anatomical and/or microenvironment effect. Nevertheless, the reverse Krebs cycle faces some theoretical hurdles that would have to be addressed. First, it is unclear if sufficient amounts of α-ketoglutarate, CO_2, and NADPH (reductive carboxylation of α-ketoglutarate requires NADPH via mitochondrial IDH2) can be present in the mitochondrial matrix to promote the thermodynamically spontaneous formation of isocitrate. Second, it is kinetically improbable that isocitrate dehydrogenase, an enzyme that does not contain biotin, can catalyze carboxylation of α-ketoglutarate to form isocitrate. Perhaps a human ortholog of 2-oxoglutarate carboxylase (found in a variety of microorganisms; reaction: ATP + 2-oxoglutarate + HCO_3^- = ADP + phosphate + oxalosuccinate; EC 6.4.1.7) that can more readily catalyze carboxylation of α-ketoglutarate (also known as 2-oxoglutarate) into oxalosuccinate which is then converted to isocitrate by IDH1 could account for the observed reversal of the Krebs cycle. However, whether this enzyme activity is expressed in leukemia cells, and whether leukemia cells can in fact generate isocitrate from α-ketoglutarate, remains to be determined.

Conclusions

Metabolism is critically linked to leukemia cell survival in multiple interacting ways. First, metabolism promotes adequate energy balance and biomass generation to support cell proliferation; second, metabolism can orchestrate functional changes in mitochondria that accompany resistance to intrinsic apoptotic stimuli; third, metabolism is regulated by microenvironmental cues—interactions with stroma, hypoxia, etc.—that are commonly associated with leukemia progression and chemoresistance. To fully grasp the impact of metabolism in all these interacting ways, and to understand how metabolism is regulated in leukemia cells, it will be important to recognize that the classical models of regulation need to be thoroughly modified and updated to fit recent data illustrating the effects of the microenvironment, anaplerosis, posttranslational protein modification, and reductive carboxylation, among others. Whether or not this regulation can be targeted for therapeutic benefit in patients with leukemia will depend largely on a thorough understanding of the limiting reactions that support the altered metabolic behavior of leukemia cells.

References

Abu-Hamad S, Zaid H, Israelson A, Nahon E, Shoshan-Barmatz V (2008) Hexokinase-I protection against apoptotic cell death is mediated via interaction with the voltage-dependent anion channel-1: mapping the site of binding. J Biol Chem 283:13482–13490

Agrimi G, Brambilla L, Frascotti G, Pisano I, Porro D, Vai M et al (2011) Deletion or overexpression of mitochondrial NAD+ carriers in Saccharomyces cerevisiae alters cellular NAD and ATP contents and affects mitochondrial metabolism and the rate of glycolysis. Appl Environ Microbiol 77:2239–2246

Arany I, Ember I, Rady P (1988) Subcellular distribution of hexokinase in leukemic and stimulated lymphoid cells of mice. Haematologia (Budap) 21:109–114

Chen Z, Zhang H, Lu W, Huang P (2009) Role of mitochondria-associated hexokinase II in cancer cell death induced by 3-bromopyruvate. Biochim Biophys Acta 1787:553–560

Chiara F, Castellaro D, Marin O, Petronilli V, Brusilow WS, Juhaszova M et al (2008) Hexokinase II detachment from mitochondria triggers apoptosis through the permeability transition pore independent of voltage-dependent anion channels. PLoS One 3:e1852

DeBerardinis RJ (2011) Serine metabolism: some tumors take the road less traveled. Cell Metab 14:285–6

DeBerardinis RJ, Mancuso A, Daikhin E, Nissim I, Yudkoff M, Wehrli S et al (2007) Beyond aerobic glycolysis: transformed cells can engage in glutamine metabolism that exceeds the requirement for protein and nucleotide synthesis. Proc Natl Acad Sci U S A 104:19345–19350

Diaz-Ruiz R, Averet N, Araiza D, Pinson B, Uribe-Carvajal S, Devin A et al (2008) Mitochondrial oxidative phosphorylation is regulated by fructose 1,6-bisphosphate. A possible role in Crabtree effect induction? J Biol Chem 283:26948–26955

Diaz-Ruiz R, Rigoulet M, Devin A (2011) The Warburg and Crabtree effects: on the origin of cancer cell energy metabolism and of yeast glucose repression. Biochim Biophys Acta 1807:568–576

Fiegl M, Samudio I, Clise-Dwyer K, Burks JK, Mnjoyan Z, Andreeff M (2009) CXCR4 expression and biologic activity in acute myeloid leukemia are dependent on oxygen partial pressure. Blood 113:1504–1512

Folmes CD, Nelson TJ, Dzeja PP, Terzic A (2012) Energy metabolism plasticity enables stemness programs. Ann N Y Acad Sci 1254:82–9

Li LX, Skorpen F, Egeberg K, Jorgensen IH, Grill V (2002) Induction of uncoupling protein 2 mRNA in beta-cells is stimulated by oxidation of fatty acids but not by nutrient oversupply. Endocrinology 143:1371–7

Lynen F (1951) Die Rolle der Phosphorsaeure bei Dehydrierungsvorgaengen und ihre biologische Bedeutung. Naturwissenschaften 30:398–406

Majewski N, Nogueira V, Bhaskar P, Coy PE, Skeen JE, Gottlob K et al (2004a) Hexokinase-mitochondria interaction mediated by Akt is required to inhibit apoptosis in the presence or absence of Bax and Bak. Mol Cell 16:819–830

Majewski N, Nogueira V, Robey RB, Hay N (2004b) Akt inhibits apoptosis downstream of BID cleavage via a glucose-dependent mechanism involving mitochondrial hexokinases. Mol Cell Biol 24:730–740

Mathupala SP, Ko YH, Pedersen PL (2010) The pivotal roles of mitochondria in cancer: Warburg and beyond and encouraging prospects for effective therapies. Biochim Biophys Acta 1797:1225–1230

Mullen AR, Wheaton WW, Jin ES, Chen PH, Sullivan LB, Cheng T et al (2012) Reductive carboxylation supports growth in tumour cells with defective mitochondria. Nature 481:385–8

Samudio I, Fiegl M, McQueen T, Clise-Dwyer K, Andreeff M (2008) The Warburg effect in leukemia-stroma cocultures is mediated by mitochondrial uncoupling associated with uncoupling protein 2 activation. Cancer Res 68:5198–5205

Samudio I, Fiegl M, Andreeff M (2009) Mitochondrial uncoupling and the Warburg effect: molecular basis for the reprogramming of cancer cell metabolism. Cancer Res 69:2163–2166

Samudio I, Harmancey R, Fiegl M, Kantarjian H, Konopleva M, Korchin B et al (2010) Pharmacologic inhibition of fatty acid oxidation sensitizes human leukemia cells to apoptosis induction. J Clin Invest 120:142–156

Shoshan-Barmatz V, Golan M (2012) Mitochondrial VDAC1: function in cell life and death and a target for cancer therapy. Curr Med Chem 19:714–735

Smerc A, Sodja E, Legisa M (2011) Posttranslational modification of 6-phosphofructo-1-kinase as an important feature of cancer metabolism. PLoS One 6:e19645

Turner N, Bruce CR, Beale SM, Hoehn KL, So T, Rolph MS et al (2007) Excess lipid availability increases mitochondrial fatty acid oxidative capacity in muscle: evidence against a role for reduced fatty acid oxidation in lipid-induced insulin resistance in rodents. Diabetes 56:2085–2092

Warburg O (1956a) On the origin of cancer cells. Science 123:309–314

Warburg O (1956b) On respiratory impairment in cancer cells. Science 124:269–270

Zhang W, Trachootham D, Liu J, Chen G, Pelicano H, Garcia-Prieto C et al (2012) Stromal control of cystine metabolism promotes cancer cell survival in chronic lymphocytic leukaemia. Nat Cell Biol 14:276–286

Part XI
Cell Therapy

Chapter 40
NK-Cell Immunotherapy for AML

Dean Anthony Lee, Laurence James Neil Cooper and Elizabeth J. Shpall

Abstract Natural killer (NK) cells are large granular lymphocytes of the innate immune system first described four decades ago for their ability to recognize and kill target cells a priori—without prior experience of the target—and in a manner that is antigen unrestricted. Because of recent advances in understanding the mechanisms of NK-cell target recognition and the diversity within the NK-cell repertoire, we now recognize that NK cells play an important role in cancer immunosurveillance. Acute myeloid leukemias are among the most well characterized of NK-cell-sensitive malignancies, and differences in NK-cell genotype, phenotype, and function correlate with disease progression and predict therapeutic responses. The ability to generate clinical-grade therapeutic products of sufficient purity, number, and function has only recently allowed NK-cell immunotherapy to be pursued in clinical trials as monotherapy, in combination with monoclonal antibodies or immunomodulating drugs, and within the setting of hematopoietic stem cell transplantation.

Keywords NK cells · Cellular therapy · Haploidentical stem cell transplantation · Adoptive immunotherapy · Innate immunity

40.1 Introduction

Natural killer (NK) cells were initially defined, as their name suggests, by their ability to recognize and kill cancer cells without prior sensitization. Acute myeloid leukemia (AML) is one the most well-characterized NK-cell-sensitive

D. A. Lee (✉) · L. J. N. Cooper
Division of Pediatrics Cell Therapy Section, The University of Texas MD Anderson Cancer Center, 1515 Holcombe Boulevard, Unit 0853, Houston, TX 77030, USA
e-mail: dalee@mdanderson.org

L. J. N. Cooper
e-mail: ljncooper@mdanderson.org

E. J. Shpall
Department of Stem Cell Transplantation and Cellular Therapy, Division of Cancer Medicine, The University of Texas MD Anderson Cancer Center, 1515 Holcombe Boulevard, Unit 0423, Houston, TX 77030, USA
e-mail: eshpall@mdanderson.org

M. Andreeff (ed.), *Targeted Therapy of Acute Myeloid Leukemia*,
Current Cancer Research, DOI 10.1007/978-1-4939-1393-0_40

malignancies. Though NK cells were identified and defined functionally nearly 40 years ago, their application to cancer immunotherapy was delayed by a lack of identifying phenotypic markers, understanding of their mechanisms of target recognition, and an inability to acquire or propagate large numbers of activated cells. With many of these obstacles now resolved or being addressed, NK-cell immunotherapy is receiving renewed attention for its potential as a targeted therapy for myeloid malignancy. In this chapter, we will discuss the basics of NK-cell biology, present laboratory and clinical data supporting the antileukemic role of NK cells, discuss results of early-phase clinical trials of adoptive immunotherapy with NK cells for AML, and present the next steps needed for translating NK-cell immunotherapy to the clinic.

40.2 Mechanisms of NK-Cell Recognition and Killing of Myeloid Malignancies

NK cells are lymphocytes of the innate immune system with cytotoxic and regulatory functions, which mediate critical direct and indirect immune responses against infections and malignancies (Raulet and Guerra 2009). NK cells are stimulated by several classes of activating receptors and are negatively regulated by human leukocyte antigen (HLA)-binding inhibitory receptors, primarily in the killer immunoglobulin-like receptor (KIR) family. KIR–ligand interactions, or the lack thereof, result in enhanced cytotoxicity against HLA class I mismatched targets. Alloreactive NK cells from donors that are haploidentical at the HLA locus have been reported to enhance engraftment, reduce graft-versus-host disease (GVHD), and prevent relapse of myeloid leukemias after allogeneic hematopoietic stem-cell transplantation (HSCT).

Human NK cells comprise approximately 10–15 % of all peripheral blood lymphocytes and are identified by the lack of cluster of differentiation 3 (CD3) and the presence of CD56 and/or CD16 expression on the cell surface. NK cells are characterized functionally on the basis of their ability to recognize targets without prior sensitization to those targets. They do so on the basis of a balance between inhibitory receptors, which primarily act through recognition of self, and activating receptors, which recognize proteins that evidence cell stress or nonself origin. NK-cell cytotoxicity is triggered when the balance of activating and inhibiting receptor signals is tipped toward activation. The main group of NK-cell inhibitory receptors is KIR, which primarily recognize HLA class I molecules, preventing NK-mediated cell lysis of normal HLA-expressing cells. Several families of activating receptors have been characterized, including natural cytotoxicity receptors (NCR), and NKG2-family lectin-like receptors. There are at least two subsets of NK cells that differ phenotypically and functionally. The majority of circulating NK cells (90 %) express CD16 with low CD56 surface expression, termed CD56^{dim}, and are important in direct and antibody-dependent cell cytotoxicity (ADCC) of targets (Cooper et al. 2001). ADCC is mediated by CD16, the

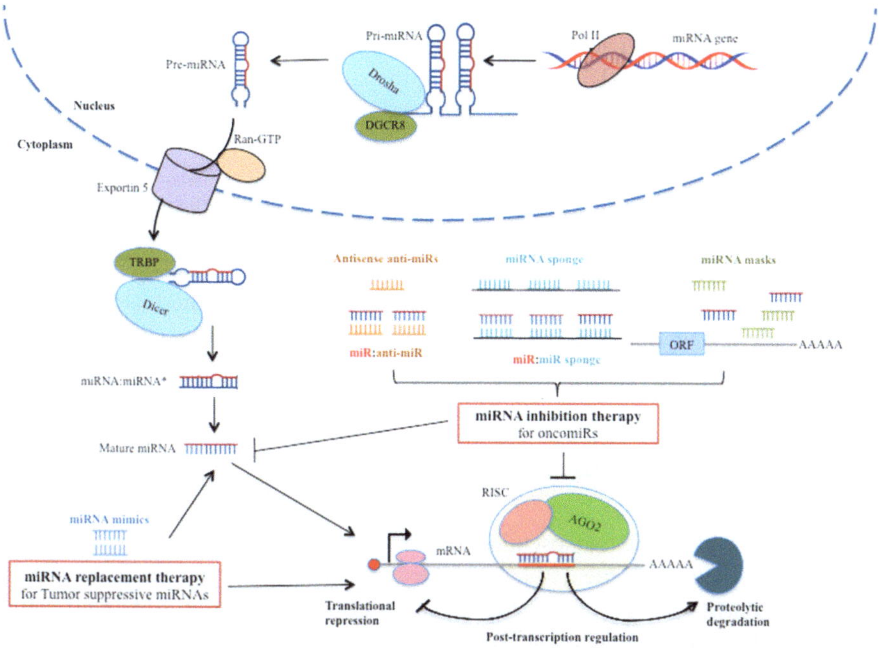

Fig. 40.1 The antileukemic benefit of KIR mismatch. NK cells in a donor who is heterozygous for HLA-C with respect to KIR ligands (having both a Group C1 allele that interacts with KIR2DL2 and a Group C2 allele that interacts with KIR2DL1) will be skewed toward no activation when receiving both inhibitory signals. However, when the NK cells are transferred to a recipient who is homozygous at HLA-C (e.g., HLA-C alleles are Group C1) the NK cells will lack the inhibitory signal through KIR2DL1, further tipping the balance toward activation when encountering tumor cells that also express activating ligands

low-affinity Fc receptor, which binds the Fc portion of immunoglobulin G (IgG) antibodies (Trinchieri 1989). Upon recognizing target cells through a predominance of activating signals, NK cells release granules containing perforins, which form a pore in the cell membrane, and granzymes that enter target cells through the pore and mediate apoptosis by activation of the caspase cascade (Browne et al. 1999). They can also kill target cells via death receptor ligands such as Fas ligand (FasL) and tumor necrosis factor (TNF)-related apoptosis-inducing ligand (TRAIL; Zamai et al. 1998).

40.2.1 *Inhibitory KIR Mismatch*

In humans, a subset of HLA class I molecules expressed on the target cells have been identified as ligands of the NKG2- and KIR-family receptors. The ligands for these NK receptors are specific for different HLA isotypes. For example, the KIR2DL4 interacts with HLA-G, KIR2DL1 interacts with alleles of the HLA-C locus that have the amino acid lysine (K) at residue 80, and KIR2DL2 and KIR2DL3 interact with alleles of HLA-C that have asparagine (N) at residue 80. Similarly, KIR3DL1 interacts only with HLA-B alleles carrying the supertypic serologic epitope HLA-Bw4 (defined primarily by threonine (T) at residue 80).

NK-cell reactivity is governed through a balance of activating and inhibitory receptors. Based on the "missing ligand model" (Karre et al. 1986), killing of tumor cells by NK cells is greatly enhanced when there is a missing inhibitory signal (i.e., missing self) on the target. Therefore, in the setting of allogeneic HSCT in which the donor and recipient are mismatched in HLA-B or HLA-C alleles, the NK cells from the donor may not be inhibited by the leukemic cells of the recipient if they lack the appropriate HLA alleles to bind these inhibitory receptors (Fig. 40.1). This results in an improved graft-versus-leukemia effect, as described further below.

40.2.2 Activating KIR A/B Haplotypes

Fifteen KIR genes and two pseudogenes have been identified so far, which are all located on chromosome 19q13.4, are extremely polymorphic, and are present or absent in many allelic combinations. These genes are present in defined haplotypes that differ in gene content between individuals, such that most individuals lack some of the KIR genes. The NK-cell repertoire therefore may vary between individuals, and NK-cell reactivity depends on interactions between the HLA and KIR haplotypes.

KIRs have cytoplasmic domains comprised of a short immunoreceptor tyrosine-based activating motif (ITAM) or a long immunoreceptor tyrosine-based inhibition motif (ITIM; Pegram et al. 2011). The allelic variations in KIR haplotypes have been clustered into two groups, referred to as haplotypes A and B (Uhrberg et al. 1997). The "A" haplotype has only one activating gene (KIR2DS4), whereas the "B" haplotype has more than one activating KIR gene, including KIR2DS1 and KIR2DS2. Because the KIR "B" haplotype carries more activating genes than the "A" haplotype, allogeneic transplants from donors with the "B" haplotype are predicted to have superior NK-cell-mediated antitumor effects (McQueen et al. 2007). In most studies, the presence or absence of each KIR gene within the genome is reported, rather than full delineation of KIR genes in each allele. Thus, the complete absence of relevant activating genes is labeled "AA," while presence of the genes is termed "Bx," as the second allele cannot be determined with confidence.

The "missing-self hypothesis," which postulates that NK cells recognize and destroy autologous cells with lost or altered self-HLA class I molecules (Ljunggren and Karre 1990), is now considered an oversimplification of NK-cell regulation, since classical HLA class I is not always required to protect from NK-cell-mediated cytotoxicity and nor is it always sufficient to prevent NK-cell cytotoxicity (Farag and Caligiuri 2006). There have been several models used for predicting NK-cell alloreactivity. Ruggeri et al. initially described the ligand–ligand mismatch model when there is incompatibility between the donor KIR and recipient KIR ligands (i.e., ligand–ligand mismatch occurs if the donor has a ligand that is absent in the recipient; Ruggeri et al. 2002). It is now recognized that KIR genotype has a profound effect on the KIR repertoire and that the expression of KIR genes is not entirely dependent on expression of their HLA ligands, but is controlled by promoter methylation (Santourlidis et al. 2002). In addition, it is important to note that possession of a KIR gene does not imply its expression on NK cells (Torkar et al. 1998). Thus, Leung et al. defined mismatch in the receptor–ligand model as incompatibility between the donor KIR and recipient KIR ligands on the basis of expression (i.e., receptor–ligand mismatch occurs if the donor expresses an inhibitory receptor for which the cognate ligand expression is absent in the recipient; Leung et al. 2004). A third model, described by Gagne et al., defines incompatibility as a mismatch between the donor KIR and recipient KIR (i.e., receptor–receptor mismatch if the donor has a receptor that is absent in the recipient; Gagne et al. 2002). The last model can be considered the prototype for the subsequent "KIR haplotype model" proposed by the Stanford group, which is founded on the notion that the more activating KIR genes the donor carries, the higher the potential for alloreactivity.

40.2.3 NCRs/NKG2D

Other important NK-cell receptors include those in the NKG2 and NCR families. The NCR (NKp30, NKp44, NKp46, and NKp80) are activating receptors with expression mostly restricted to NK cells (Moretta et al. 2001). The C-type lectin receptors confined to the NKG2 family are expressed as heterodimers with CD94, except for NKG2D which is expressed as homodimer (Pegram et al. 2011). These receptors of this family include both activating and inhibitory members (Table 40.1). NKG2D, the major activating receptor in this family, recognizes two families of proteins that increased under cellular stress—the major histocompatibility complex (MHC)-class-I-related chains A or B (MICA/B) or members of the UL-16-binding protein (ULBP) family. Pende et al. (Pende et al. 2005) showed that myeloid leukemia cells express the poliovirus receptor (PVR) and nectin, suggesting a critical role for NK-cell-mediated recognition through DNAM-1 rather than NKG2D. Importantly, AML may repress expression of ligands for activating receptors on NK cells as a mechanism of immune escape (Nowbakht et al. 2005).

40.2.4 NK-Cell Subsets

NK cells have both cytotoxic and immunoregulatory functions. Human NK cells are characterized by expression of CD56 with or without expression of CD16, and the absence of T-cell receptor and CD3 expression. There are at least two subtypes of NK cells: $CD56^{dim}$ $CD16^{bright}$ cells, which compose 90 % of the NK cells from circulation which have cytotoxic function, and express HLA class I allele-specific KIR, and the $CD56^{bright}CD16^{neg}$ cells present in small numbers in the peripheral blood,

Table 40.1 Activating and inhibitory receptors on NK cells and their known corresponding ligands on target cells

NK cell receptor	Corresponding ligand
Inhibitory	
KIR2DL1 (CD158A)	HLA-C group C2 (K80)
KIR2DL2/3 (CD158B1, B2)	HLA-C group C1 (N80)
KIR2DL4 (CD158D)	HLA-G
KIR2DL5A/B (CD158F)	
KIR3DL1 (CD158E1)	HLA-Bw4 (N77 and/or T80)
KIR3DL2 (CD158K)	HLA-A3/A11
KIR3DL3 (CD158Z)	
NKG2A (CD159A)	HLA-E
Activating	
KIR2DS1 (CD158 H)	HLA-C group 2
KIR2DS2 (CD158 J)	HLA-C group 1? (very low affinity)
KIR2DS3	
KIR2DS4 (CD158I)	
KIR2DS5 (CD158G)	
KIR3DS1 (CD158E2)	
NKG2C (CD159C)	HLA-E
NKG2D (CD314)	MICA, MICB, ULBP1-5
NKG2E	HLA-E
NKp30 (CD337, NCR3)	B7-H6, BAT3
NKp44 (CD336, NCR2)	
NKp46 (CD335, NCR1)	HSPG
NKp80 (KLRF1, CLEC5C)	CLEC2B (AICL)
DNAM1 (CD226)	CD112 (Nectin-2), CD155 (PVR)
FcγRIIIa (CD16a)	IgG1, IgG3
CD160	HLA-C

but 100 % in lymphoid tissues, and secrete immunoregulatory cytokines, such as interferon-γ (IFN-γ), and promote adaptive immune responses. NK-cell cytotoxicity can be upregulated by increased secretion of cytokines including interleukin (IL)-2 and IL-15, whereas IL-12 and IL-18 are reported to increase INF-γ secretion by NK cells (Terme et al. 2008).

40.2.5 FcR Polymorphisms Effect Cytotoxicity

Genetic polymorphisms encoding phenylalanine or valine at FcγRIIIa (CD16a) amino acid position 158 are common in humans (Koene et al. 1997; Sondermann et al. 2000), and the presence of phenylalanine results in a receptor with a low affinity for IgG1 and IgG3, while valine increases the receptor affinity dramatically (Wu et al. 1997). The high-affinity polymorphism of FcγRIIIa correlates with improved clinical and molecular responses to CD20-specific rituximab monoclonal antibody therapy (Cartron et al. 2002; Persky et al. 2012; Veeramani et al. 2011). Indeed, the polymorphism status may be predictive of response to and survival in follicular lymphoma patients receiving rituximab (Persky et al. 2012; Weng and Levy 2003). Thus, the design of targeting antibodies against AML will need to consider the role of endogenous or adoptively transferred NK cells in ADCC with regard to the polymorphisms of FcγRIIIa.

40.2.6 *Mechanisms of Cytotoxicity*

Activated NK cells exhibit an augmented ability to attack tumor cells using two cytotoxic mechanisms. In the first pathway, target-directed exocytosis of lytic granules releases perforin at the immunological synapse to form pores in the target cell membrane, allowing co-secreted granzymes to enter target cells and activate the caspase cascade leading to cell death. In the second pathway, ligation of TRAIL, TNF, or Fas-ligand to their cognate receptors on target cells induces cell death in sensitive or sensitized target cells (Kayagaki et al. 1999; Smyth et al. 2001; Takeda et al. 2001). In addition to these pathways, activated NK cells are also known to produce IFN-γ, which is required for NK-cell function against multiple cancer types (Smyth et al. 2002).

40.3 Clinical Data supporting the Antileukemic Role of NK Cells

40.3.1 *Cytokine Therapy*

Positive antitumor responses to administering pharmacologic doses of IL-2 in patients with renal cell carcinoma and melanoma (Rosenberg et al. 1993) led to studies of infusing in vitro IL-2-activated "LAK" cells (Rosenberg et al. 1994). Since IL-2-activated NK cells have strong lytic potential against AML, subsequent studies assessed the effect of IL-2 administration alone as an immune/NK stimulant. However, a beneficial effect on overall survival was not demonstrated in individual trials alone or in meta-analysis (Buyse et al. 2011).

40.3.2 *The Beneficial Role of NK Cells in the Setting of Allogeneic HSCT*

40.3.2.1 NK-Cell Immune Reconstitution

NK cells are the first lymphocytes to recover after allogeneic stem cell transplantation, with both NK-cell number and function recovering within the first few weeks, achieving supranormal numbers for the first few months (Storek et al. 2008), perhaps in compensation for the much more delayed T-cell recovery. Donor-derived

NK cells initially emerging from the allograft have an immature phenotype with low KIR expression and overexpression of NKG2A. By 6–9 months post engraftment, NK cells acquire phenotypic characteristics of mature cells (Shilling et al. 2003). The ratio of $CD56^{bright}/CD56^{dim}$ cells, used to discriminate mature cytotoxic cells from less mature cytokine-secreting cells, may be reversed for the first 4–6 months post transplant resulting in increased production of IFN-γ with less cytotoxicity. However, NK cells in the first few months post-transplant have increased activity against targets lacking KIR ligands, particularly AML (Leung et al. 2004; Ruggeri et al. 2007). However, the killing potential of NK cells derived from transplanted patients is fourfold lower than that of NK cells isolated from normal donors (Chklovskaia et al. 2004). This is in contrast to NK cells that are activated and propagated ex vivo which exhibit high cytotoxicity (Fujisaki et al. 2009; North et al. 2007; Denman et al. 2012; Somanchi et al. 2011). Thus, administration of numerically expanded NK cells early in the engraftment period may improve the control of both pathogens and malignant disease post transplant.

40.3.2.2 Overall Effects on Survival and Engraftment

While the role of activating and inhibitory receptors on NK cells, expression of activating and inhibitory ligands on tumor cells, and various subsets of NK cells have received significant attention, the clinical benefit of NK-cell alloreactivity has not been uniformly demonstrated. This likely reflects differences between reports in the predictive model used for determining NK-cell alloreactivity (as described above), conditioning regimens, graft sources and processing, and differing approaches to prophylaxis of GVHD (reviewed in Benjamin et al. 2010; Passweg et al. 2005; Gill et al. 2009). With this variability in mind, some important observations nonetheless suggest that NK-cell antileukemic effect is beneficial in certain settings. Recipients of haplotransplantation, in which the donor KIR were licensed against an HLA group missing in the recipient ("KIR mismatched"), had higher engraftment rates and produced leukemia-reactive NK cells that had higher activity against cells lacking the relevant HLA ligand (Ruggeri et al. 1999).

Despite the variability of retrospective data with regard to patient population, underlying diseases, conditioning regimens, graft composition and cell dose, degree of T-cell depletion, post transplant immunosuppressive regimens, and differing methods of determining alloreactivity, several groups demonstrated that increased numbers of alloreactive NK cells have benefit in the transplant setting. Donor-derived NK-cell chimerism 1 month after transplant was shown to correlate with late graft rejection in children receiving reduced intensity conditioning (Matthes-Martin et al. 2003), with increased disease-free survival in patients receiving reduced-intensity conditioning (Dunbar et al. 2008), and with relapse-free and overall survival in patients receiving T-depleted grafts (Savani et al. 2007), resulting in an adjusted hazard ratio of 20 for individuals with low numbers of NK cells. In addition to NK-cell recovery after transplant, the NK-cell composition of the allogeneic hematopoietic stem-cell infusion product itself may also predict overall survival (Clausen et al. 2007).

40.3.2.3 Specific Benefit Attributed to KIR Mismatch

The most consistent beneficial effect of NK cells as predicted by donor–recipient mismatch or KIR gene content has been observed in patients with myeloid leukemias receiving haploidentical transplants. Ruggeri initially showed a benefit in engraftment in patients receiving KIR-mismatched haplotransplants (Ruggeri et al. 1999). Based on the hypothesis that a mismatch between inhibitory KIR on NK cells and HLA molecules on tumor would lead to higher cytotoxicity, this group then analyzed 92 patients with high-risk leukemia (57 with AML) who received a haplotype-mismatched HSCT. They reported that NK-cell alloreactivity (defined as the presence of KIR ligands in the donor which are absent in recipients) improved engraftment, protected from GVHD, and reduced the rate of relapse (0 vs. 75 %), a stratification that also correlated with finding anti-recipient NK-cell clones in these patients (Ruggeri et al. 2002). Based on these data, they proposed the infusion of NK cells after transplant to target disease relapse in a minimal residual disease setting. Large retrospective trials have confirmed the benefit of KIR mismatch in a variety of adoptive immunotherapy settings (Miller et al. 2007; Giebel et al. 2003). In mouse xenograft models, it was demonstrated that the role of alloreactive NK cells in eradicating human leukemia improved engraftment by targeting host T lymphocytes and reduced GVHD by eliminating recipient-type dendritic cells (Gill et al. 2009). In addition, the NK-cell dose of the infusion product has been associated with better outcomes following matched sibling transplants (Kim et al. 2005).

In the setting of HLA-matched transplants, analysis of 202 donor–recipient pairs revealed that event-free survival was significantly lower for BX donors to AA recipients than for AA donors to BX recipients (hazard ratio 3.8, $p = 0.046$; McQueen et al. 2007). Similarly, evaluation of 86 patients with advanced hematologic malignancies (25 with AML) showed improved survival with mismatch in inhibitory KIRs. Importantly, this study also showed a dose effect for KIR genes, demonstrating a correlation of increased survival with increased mismatches (Symons et al. 2010). However, in the setting of cord blood transplantation KIR mismatch has been associated with increased GVHD and lower survival (Brunstein et al. 2009). This difference in stem cell sources may be attributed to subsequent differences in NK-cell reconstitution (Verneris and Miller 2009; Lu et al. 2008).

40.3.2.4 Benefit to Antitumor Response Related to Specific KIR Genes

In addition to the findings that KIR mismatch, KIR haplotype, and NK-cell number are associated with improved outcomes, the presence or absence of specific KIR genes in the donor may also be important. The presence of three specific B-haplotype KIR genes—KIR2DL5A, KIR2DS1, and KIR3DS1—were associated with reduced relapse rates after HLA-matched sibling donor transplants. In 68 patients with AML, those who received transplants from donors with these three KIR genes present had a relapse rate of only 13 %, compared to 57 % for those with donors who lacked the three-gene set. Importantly, this finding was shown only for patients AML, and not other hematologic malignancies (Stringaris et al. 2010).

Among 49 AML patients receiving T-depleted haplotransplants, KIR mismatch for 2DL1 and 3DL1 were associated with significant improvements in overall survival and relapse-free survival (Chen et al. 2012). However, this stratification did not predict for survival when applied to matched-sibling transplants.

40.3.3 Adoptive Transfer

40.3.3.1 NK-Cell Products Derived by Apheresis

A number of studies have identified mechanisms of NK-cell dysfunction in AML that are attributed to initial development, progression, or relapse. Decreased production of cytokines and decreased cytotoxicity activity (Tajima et al. 1996), decreased expression of activating receptors (Costello et al. 2002), and defective expression of intracellular signaling pathways (Buggins et al. 1998) have all been described in NK cells derived from patients with AML. For this reason, in addition to the potential advantage of KIR mismatch described earlier, adoptive immunotherapy with NK cells has been studied primarily using NK cells obtained from normal healthy donors, usually haploidentical related individuals. Importantly, the dysfunction of patient-derived NK cells can be restored by activation and/or propagation ex vivo (Siegler et al. 2005), setting the stage for adoptive transfer of revitalized autologous NK cells.

In a nontransplant cell therapy setting in which poor-prognosis AML patients received adoptive immunotherapy with haploidentical NK cells (Miller et al. 2005) 3 of 4 (75 %) KIR ligand mismatched patients achieved a complete remission compared to only 2 of 15 (13 %) KIR ligand matched patients, further strengthening the association between KIR ligand mismatch and NK-cell efficacy against AML. The NK-cell products were obtained by steady-state leukapheresis followed by immunomagnetic depletion of T-cells and overnight IL-2 activation. The cells were delivered after high-dose cyclophosphamide and fludarabine (Hi-Cy/Flu) lymphodepletion, and all patients received subcutaneous IL-2 after the infusions. Compared with the low-intensity cyclophosphamide/fludarabine (Cy/Flu) regimen in non-AML patients, infusions after the more intense Hi-Cy/Flu resulted in detectable expansion of NK cells in vivo as shown in PCR-based chimerism assays, and in vivo expansion correlated with response.

In a similar study, Rubnitz et al. reported the safety of infusing haploidentical KIR-mismatched NK cells as consolidation therapy for children with AML in remission (Rubnitz et al. 2010). However, these patients, unlike those in the Miller study, had no evidence of disease at the time of adoptive immunotherapy. In this trial, patients also received a Hi-Cy/Flu regimen, NK cells, and IL-2. With a median follow up of 964 days, all ten patients remained in remission. No GVHD was observed.

Table 40.2 Clinical trials of adoptive immunotherapy with NK cells in patients with AML

Reference	Method of NK cell production	NK cells delivered (mean/median as reported)	Patients Treated, Responses
Passweg et al. 2004	Apheresis, CD3-depletion, CD56-selection	1.6×10^7/kg	Five patients with high-risk myeloid malignancy. NK cells infused post transplant for low chimerism. Two with improved chimerism, one stable, one decreased chimerism, one relapse
Miller et al. 2005	Apheresis, CD3-depletion, ± CD56-selection	1×10^7/kg	Nineteen patients with relapsed/refractory poor-prognosis AML. NK cells infused after cyclophosphamide/fludarabine. 5 achieved CR
Rubnitz et al. 2010	Apheresis, CD3-depletion, CD56-selection	2.9×10^7/kg	Ten pediatric AML patients. NK cells infused post-transplant. 10 in CR at median f/u 964 days
Rizzieri et al. 2010	Apheresis, CD56-selection	1.1×10^7/kg	Nineteen patients with myeloid malignancy. 1–3 NK cell infusions delivered after transplant. 50 % 1-year survival
Nguyen et al. 2011	Apheresis, CD3-depletion	3.2×10^7/kg	One AML patient. NK cells infused for relapse after haplotransplant. Achieved CR
Brehm et al. 2011	Apheresis, CD3-depletion, CD56-selection	1.2×10^7/kg	Six AML patients. 1–3 NK cell infusions delivered after haplotransplant. 50 % survival at median 160 days

Several other studies, summarized in Table (40.2), have demonstrated safety and feasibility of NK-cell infusions delivered in a variety of contexts. Although the results of survival and/or remission are encouraging, none of the studies were powered or designed to demonstrate efficacy. Koehl et al. treated one pediatric patient with AML after haploidentical HSCT with two infusions of haploidentical KIR-mismatched NK cells (Koehl et al. 2004). The cell infusions were well tolerated without significant adverse events or GVHD, but the patient died of early relapse on day +80. Passweg et al. infused NK cells following T-cell-depleted haploidentical HSCT to four patients with AML (Passweg et al. 2004). Rizzieri et al. recently reported on a total of 51 NK-cell-enriched donor lymphocyte infusions (DLIs) to 30 patients following a 3-6/6 HLA-matched T-cell-depleted nonmyeloablative allogeneic transplant with alemtuzumab (Rizzieri et al. 2010). Eight weeks following transplantation, donor NK-cell-enriched DLIs were processed using a $CD56^+$ selecting column with up to three fresh infusions allowed. Evaluating outcomes by disease, 1-year survival was 50 % for the 19 patients with myeloid diseases. All of these trials are small pilot studies in diverse populations that included myeloid malignancies.

MDACC recently reported preliminary results from a phase I study demonstrating the safety of haploidentical alloreactive NK-cell infusion after allogeneic matched related and unrelated donor transplants added to the preparative regimen (Champlin RE, et al Blood. 2010;116:489a). Patients received escalating doses of NK cells (10^6/kg to 3×10^7/kg) followed by infusion of IL-2 for 5 days. All patients had rapid engraftment and tolerated the treatment well. Only three patients developed grade 2 acute GVHD while none had grade 3–4 acute GVHD. Only one patient died of treatment-related mortality. This trial confirmed that infusion of haploidentical NK cells post transplant is feasible and was not associated with increased incidence of GVHD, delayed engraftment, or higher treatment-related mortality (at NK-cell doses of 10 to 3×10^7/kg).

40.3.3.2 NK-Cell Products Numerically Expanded ex vivo

In order to increase the cell dose for improved antileukemic effect, NK cells have been expanded ex vivo and delivered to patients for adoptive cancer immunotherapy. Barkholt et al. treated five patients after allogeneic HSCT with recipient-derived allogeneic NK/NKT cells that had been expanded for 19 days using OKT3 and IL-2 (Barkholt et al. 2009). Three to six infusions were given to each patient in escalating doses for a total of 19 product infusions, with a median $CD56^+$ cell dose of 13.2×10^6/kg (8.1–40.3) at the highest dose level. Infusion-related toxicities were limited to fever and rash, and one case of *Staphylococcus epidermidis* septicemia.

The highest doses of NK cells yet reported to be infused into patients were derived by expanding autologous CD3-depleted apheresis products for 3 weeks with IL-2, OKT3, and irradiated PBMC feeder cells (Parkhurst et al. 2011). Cell doses per recipient weight (kg) were not reported in this study, but with eight patients receiving $1.88–7.60 \times 10^{10}$ total nucleated cells containing 93–96% NK cells, the only toxicity reported was one patient who had transient shortness of breath requiring supplemental oxygen.

40.3.3.3 Novel Methods for Improved Expansion of NK-Cell Products

Ex vivo propagation of NK cells for the purpose of adoptive immunotherapy has been achieved with various combinations of cytokines, antibodies, and feeder cells. Cytokines have been used alone (Koehl et al. 2004; Koehl et al. 2005; Clausen et al. 2001; Clausen et al. 2003; Clausen et al. 2004; Klingemann and Martinson 2004; Decot et al. 2010; de Rham et al. 2007; Spanholtz et al. 2010), or in combination with other activating ligands or antibodies (Tomala et al. 2009; Wu and Xu 2010; Alici et al. 2008; Barkholt et al. 2009; Carlens et al. 2001; Satwani et al. 2011; Siegler et al. 2010 2005; Sutlu et al. 2010; Boissel et al. 2008; Reim et al. 2009). Feeder cells have typically been derived from EBV-lymphoblastoid cell lines (Torelli et al. 2002; Torelli et al. 2005; Berg et al. 2009; Clemenceau et al. 2006; Perussia et al. 1987) or genetically modified artificial antigen-presenting cells (aAPC) based on the K562 cell line (Cho and Campana 2009; Fujisaki et al. 2009; Gong et al. 2010; Zhang et al. 2011; Denman et al. 2012). The highest numeric expansion ex vivo of NK cells yet reported has been achieved using K562 expressing 4-1BBL and membrane-bound IL-21, achieving a mean expansion at 21 days of more than 30,000-fold. Despite a more defined role for IL-15 in NK-cell proliferation in vivo, IL-21 in this setting appears to maintain proliferative potential through telomere extension in the NK cells (Denman et al. 2012). Finally, the differentiation and propagation of NK cells from alternative sources such as cord blood or pluripotent stem cells may expand the availability of NK cells to recipients who otherwise lack suitable related donors.

Clinical trials with the expanded NK cells described above are expected to exceed the previous limitations in cell number using apheresis-derived NK-cell products. Several clinical trials are reported in ClinicalTrials.gov as accepting new patients.

Conclusion

Adoptive transfer of NK cells is experiencing a renaissance in basic science and applied clinical trials as a potential immunotherapeutic for AML. With new methods now available to generate large numbers of NK for higher cell doses and repeated infusions, there are myriad opportunities for adoptive immunotherapy to be combined with chemotherapy, antibodies, or allogeneic HSCT.

References

Alici E, Sutlu T, Bjorkstrand B, Gilljam M, Stellan B, Nahi H, Quezada HC, Gahrton G, Ljunggren HG, Dilber MS (2008) Autologous antitumor activity by NK cells expanded from myeloma patients using GMP-compliant components. Blood 111(6):3155–3162. doi:blood-2007-09-110312 [pii]10.1182/blood-2007-09-110312

Barkholt L, Alici E, Conrad R, Sutlu T, Gilljam M, Stellan B, Christensson B, Guven H, Bjorkstrom NK, Soderdahl G, Cederlund K, Kimby E, Aschan J, Ringden O, Ljunggren HG, Dilber MS (2009) Safety analysis of ex vivo-expanded NK and NK-like T cells administered to cancer patients: a phase I clinical study. Immunotherapy 1(5):753–764. doi:10.2217/imt.09.47

Benjamin JE, Gill S, Negrin RS (2010) Biology and clinical effects of natural killer cells in allogeneic transplantation. Curr Opin Oncol 22(2):130–137. doi:10.1097/CCO.0b013e328335a559

Berg M, Lundqvist A, McCoy P, Jr., Samsel L, Fan Y, Tawab A, Childs R (2009) Clinical-grade ex vivo-expanded human natural killer cells up-regulate activating receptors and death receptor ligands and have enhanced cytolytic activity against tumor cells. Cytotherapy 11 (3):341–355. doi:909785497 [pii]10.1080/14653240902807034

Boissel L, Tuncer HH, Betancur M, Wolfberg A, Klingemann H (2008) Umbilical cord mesenchymal stem cells increase expansion of cord blood natural killer cells. Biol Blood Marrow Transplant 14(9):1031–1038

Brehm C, Huenecke S, Quaiser A, Esser R, Bremm M, Kloess S, Soerensen J, Kreyenberg H, Seidl C, Becker PS, Muhl H, Klingebiel T, Bader P, Passweg JR, Schwabe D, and Koehl U (2011) IL-2 stimulated but not unstimulated nk cells induce selective disappearance of peripheral blood cells: Concomitant results to a phase i/ii study. PLoS One 6: e27351

Browne KA, Blink E, Sutton VR, Froelich CJ, Jans DA, Trapani JA (1999) Cytosolic delivery of granzyme B by bacterial toxins: evidence that endosomal disruption, in addition to transmembrane pore formation, is an important function of perforin. Mol Cell Biol 19(12):8604–8615

Brunstein CG, Wagner JE, Weisdorf DJ, Cooley S, Noreen H, Barker JN, DeFor T, Verneris MR, Blazar BR, Miller JS (2009) Negative effect of KIR alloreactivity in recipients of umbilical cord blood transplant depends on transplantation conditioning intensity. Blood 113(22):5628–5634. doi:10.1182/blood-2008-12-197467

Buggins AG, Hirst WJ, Pagliuca A, Mufti GJ (1998) Variable expression of CD3-zeta and associated protein tyrosine kinases in lymphocytes from patients with myeloid malignancies. Br J Haematol 100(4):784–792

Buyse M, Squifflet P, Lange BJ, Alonzo TA, Larson RA, Kolitz JE, George SL, Bloomfield CD, Castaigne S, Chevret S, Blaise D, Maraninchi D, Lucchesi KJ, Burzykowski T (2011) Individual patient data meta-analysis of randomized trials evaluating IL–2 monotherapy as remission maintenance therapy in acute myeloid leukemia. Blood 117(26):7007–7013. doi:10.1182/blood-2011-02-337725

Carlens S, Gilljam M, Chambers BJ, Aschan J, Guven H, Ljunggren HG, Christensson B, Dilber MS (2001) A new method for in vitro expansion of cytotoxic human CD3-CD56 + natural killer cells. Hum Immunol 62(10):1092–1098

Cartron G, Dacheux L, Salles G, Solal-Celigny P, Bardos P, Colombat P, Watier H (2002) Therapeutic activity of humanized anti-CD20 monoclonal antibody and polymorphism in IgG Fc receptor FcgammaRIIIa gene. Blood 99(3):754–758

Chen DF, Prasad VK, Broadwater G, Reinsmoen NL, DeOliveira A, Clark A, Sullivan KM, Chute JP, Horwitz ME, Gasparetto C, Long GD, Yang Y, Chao NJ, Rizzieri DA (2012) Differential impact of inhibitory and activating Killer Ig-Like Receptors (KIR) on high-risk patients with myeloid and lymphoid malignancies undergoing reduced intensity transplantation from haploidentical related donors. Bone Marrow Transplant 47(6):817–823. doi:10.1038/bmt.2011.181

Chklovskaia E, Nowbakht P, Nissen C, Gratwohl A, Bargetzi M, Wodnar-Filipowicz A (2004) Reconstitution of dendritic and natural killer-cell subsets after allogeneic stem cell transplantation: effects of endogenous flt3 ligand. Blood 103(10):3860–3868. doi:10.1182/blood-2003-04-1200

Cho D, Campana D (2009) Expansion and activation of natural killer cells for cancer immunotherapy. Korean J Lab Med 29(2):89–96. doi:200904089 [pii]10.3343/kjlm.2009.29.2.89

Clausen J, Petzer AL, Vergeiner B, Enk M, Stauder R, Gastl G, Gunsilius E (2001) Optimal timing for the collection and in vitro expansion of cytotoxic CD56(+) lymphocytes from patients undergoing autologous peripheral blood stem cell transplantation. J Hematother Stem Cell Res 10(4):513–521. doi:10.1089/15258160152509127

Clausen J, Vergeiner B, Enk M, Petzer AL, Gastl G, Gunsilius E (2003) Functional significance of the activation-associated receptors CD25 and CD69 on human NK-cells and NK-like T-cells. Immunobiology 207(2):85–93

Clausen J, Enk M, Vergeiner B, Eisendle K, Petzer AL, Gastl G, Gunsilius E (2004) Suppression of natural killer cells in the presence of CD34 + blood progenitor cells and peripheral blood lymphocytes. Biol Blood Marrow Transplant 10(10):691–697. doi:10.1016/j.bbmt.2004.06.009S1083879104003064 [pii]

Clausen J, Wolf D, Petzer AL, Gunsilius E, Schumacher P, Kircher B, Gastl G, Nachbaur D (2007) Impact of natural killer cell dose and donor killer-cell immunoglobulin-like receptor (KIR) genotype on outcome following human leucocyte antigen-identical haematopoietic stem cell transplantation. Clin Exp Immunol 148(3):520–528. doi:10.1111/j.1365-2249.2007.03360.x

Clemenceau B, Gallot G, Vivien R, Gaschet J, Campone M, Vie H (2006) Long-term preservation of antibody-dependent cellular cytotoxicity (ADCC) of natural killer cells amplified in vitro from the peripheral blood of breast cancer patients after chemotherapy. J Immunother 29(1):53–60. doi:00002371-200601000-00006 [pii]

Cooper MA, Fehniger TA, Caligiuri MA (2001) The biology of human natural killer-cell subsets. Trends Immunol 22 (11):633–640

Costello RT, Sivori S, Marcenaro E, Lafage-Pochitaloff M, Mozziconacci MJ, Reviron D, Gastaut JA, Pende D, Olive D, Moretta A (2002) Defective expression and function of natural killer cell-triggering receptors in patients with acute myeloid leukemia. Blood 99 (10):3661–3667 de Rham C, Ferrari-Lacraz S, Jendly S, Schneiter G, Dayer JM, Villard J (2007) The proinflammatory cytokines IL-2, IL-15 and IL-21 modulate the repertoire of mature human natural killer cell receptors. Arthritis Res Ther 9(6):R125. doi:ar2336 [pii]10.1186/ar2336

Decot V, Voillard L, Latger-Cannard V, Aissi-Rothe L, Perrier P, Stoltz JF, Bensoussan D (2010) Natural-killer cell amplification for adoptive leukemia relapse immunotherapy: comparison of three cytokines, IL-2, IL-15, or IL-7 and impact on NKG2D, KIR2DL1, and KIR2DL2 expression. Exp Hematol 38(5):351–362. doi:S0301-472X(10)00051-2 [pii]10.1016/j.exphem.2010.02.006

Denman CJ, Senyukov VV, Somanchi SS, Phatarpekar PV, Kopp LM, Johnson JL, Singh H, Hurton L, Maiti SN, Huls MH, Champlin RE, Cooper LJ, Lee DA (2012) Membrane-Bound IL-21 Promotes Sustained Ex Vivo Proliferation of Human Natural Killer Cells. PLoS One 7(1):e30264. doi:10.1371/journal.pone.0030264

Dunbar EM, Buzzeo MP, Levine JB, Schold JD, Meier-Kriesche HU, Reddy V (2008) The relationship between circulating natural killer cells after reduced intensity conditioning hematopoietic stem cell transplantation and relapse-free survival and graft-versus-host disease. Haematologica 93(12):1852–1858. doi:10.3324/haematol.13033

Farag SS, Caligiuri MA (2006) Human natural killer cell development and biology. Blood Rev 20(3):123–137. doi:S0268-960X(05)00055-X [pii]10.1016/j.blre.2005.10.001

Fujisaki H, Kakuda H, Shimasaki N, Imai C, Ma J, Lockey T, Eldridge P, Leung WH, Campana D (2009) Expansion of highly cytotoxic human natural killer cells for cancer cell therapy. Cancer Res 69(9):4010–4017. doi:0008-5472.CAN-08-3712 [pii]10.1158/0008-5472.CAN-08-3712

Gagne K, Brizard G, Gueglio B, Milpied N, Herry P, Bonneville F, Cheneau ML, Schleinitz N, Cesbron A, Follea G, Harrousseau JL, Bignon JD (2002) Relevance of KIR gene polymorphisms in bone marrow transplantation outcome. Hum Immunol 63(4):271–280

Giebel S, Locatelli F, Lamparelli T, Velardi A, Davies S, Frumento G, Maccario R, Bonetti F, Wojnar J, Martinetti M, Frassoni F, Giorgiani G, Bacigalupo A, Holowiecki J (2003) Survival advantage with KIR ligand incompatibility in hematopoietic stem cell transplantation from unrelated donors. Blood 102(3):814–819

Gill S, Olson JA, Negrin RS (2009) Natural killer cells in allogeneic transplantation: effect on engraftment, graft- versus-tumor, and graft-versus-host responses. Biol Blood Marrow Transplant 15(7):765–776. doi:10.1016/j.bbmt.2009.01.019

Gong W, Xiao W, Hu M, Weng X, Qian L, Pan X, Ji M (2010) Ex vivo expansion of natural killer cells with high cytotoxicity by K562 cells modified to co-express major histocompatibility complex class I chain-related protein A, 4–1BB ligand, and interleukin–15. Tissue Antigens. doi:TAN1535 [pii]10.1111/j.1399-0039.2010.01535.x

Karre K, Ljunggren HG, Piontek G, Kiessling R (1986) Selective rejection of H –2-deficient lymphoma variants suggests alternative immune defence strategy. Nature 319 (6055):675–678. doi:10.1038/319675a0

Kayagaki N, Yamaguchi N, Nakayama M, Takeda K, Akiba H, Tsutsui H, Okamura H, Nakanishi K, Okumura K, Yagita H (1999) Expression and function of TNF-related apoptosis-inducing ligand on murine activated NK cells. J Immunol 163(4):1906–1913

Kim DH, Sohn SK, Lee NY, Baek JH, Kim JG, Won DI, Suh JS, Lee KB, Shin IH (2005) Transplantation with higher dose of natural killer cells associated with better outcomes in terms of non-relapse mortality and infectious events after allogeneic peripheral blood stem cell transplantation from HLA-matched sibling donors. Eur J Haematol 75(4):299–308. doi:EJH514 [pii]10.1111/j.1600-0609.2005.00514.x

Klingemann HG, Martinson J (2004) Ex vivo expansion of natural killer cells for clinical applications. Cytotherapy 6(1):15–22

Koehl U, Sorensen J, Esser R, Zimmermann S, Gruttner HP, Tonn T, Seidl C, Seifried E, Klingebiel T, Schwabe D (2004) IL-2 activated NK cell immunotherapy of three children after haploidentical stem cell transplantation. Blood Cells Mol Dis 33(3):261–266. doi:S1079-9796(04)00171-8 [pii]10.1016/j.bcmd.2004.08.013

Koehl U, Esser R, Zimmermann S, Tonn T, Kotchetkov R, Bartling T, Sorensen J, Gruttner HP, Bader P, Seifried E, Martin H, Lang P, Passweg JR, Klingebiel T, Schwabe D (2005) Ex vivo expansion of highly purified NK cells for immunotherapy after haploidentical stem cell transplantation in children. Klinische Padiatrie 217(6):345–350

Koene HR, Kleijer M, Algra J, Roos D, von dem Borne AE, de Haas M (1997) Fc gammaRIIIa –158 V/F polymorphism influences the binding of IgG by natural killer cell Fc gammaRIIIa, independently of the Fc gammaRIIIa –48 L/R/H phenotype. Blood 90 (3):1109–1114

Leung W, Iyengar R, Turner V, Lang P, Bader P, Conn P, Niethammer D, Handgretinger R (2004) Determinants of antileukemia effects of allogeneic NK cells. J Immunol 172 (1):644–650

Ljunggren HG, Karre K (1990) In search of the 'missing self': MHC molecules and NK cell recognition. Immunol Today 11(7):237–244

Lu X, Kondo Y, Takamatsu H, Ohata K, Yamazaki H, Takami A, Akatsuka Y, Nakao S (2008) CD16 + CD56- NK cells in the peripheral blood of cord blood transplant recipients: a unique subset of NK cells possibly associated with graft-versus-leukemia effect. Eur J Haematol 81(1):18–25. doi:EJH1073 [pii]10.1111/j.1600-0609.2008.01073.x

Matthes-Martin S, Lion T, Haas OA, Frommlet F, Daxberger H, Konig M, Printz D, Scharner D, Eichstill C, Peters C, Lawitschka A, Gadner H, Fritsch G (2003) Lineage-specific chimaerism after stem cell transplantation in children following reduced intensity conditioning: potential predictive value of NK cell chimaerism for late graft rejection. Leukemia 17(10):1934–1942. doi:10.1038/sj.leu.2403087

McQueen KL, Dorighi KM, Guethlein LA, Wong R, Sanjanwala B, Parham P (2007) Donor-recipient combinations of group A and B KIR haplotypes and HLA class I ligand affect the outcome of HLA-matched, sibling donor hematopoietic cell transplantation. Hum Immunol 68(5):309–323. doi:10.1016/j.humimm.2007.01.019

Miller JS, Soignier Y, Panoskaltsis-Mortari A, McNearney SA, Yun GH, Fautsch SK, McKenna D, Le C, Defor TE, Burns LJ, Orchard PJ, Blazar BR, Wagner JE, Slungaard A, Weisdorf DJ, Okazaki IJ, McGlave PB (2005) Successful adoptive transfer and in vivo expansion of human haploidentical NK cells in patients with cancer. Blood 105(8):3051–3057

Miller JS, Cooley S, Parham P, Farag SS, Verneris MR, McQueen KL, Guethlein LA, Trachtenberg EA, Haagenson M, Horowitz MM, Klein JP, Weisdorf DJ (2007) Missing KIR ligands are associated with less relapse and increased graft-versus-host disease (GVHD) following unrelated donor allogeneic HCT. Blood 109(11):5058–5061

Moretta A, Bottino C, Vitale M, Pende D, Cantoni C, Mingari MC, Biassoni R, Moretta L (2001) Activating receptors and coreceptors involved in human natural killer cell-mediated cytolysis. Annu Rev Immunol 19:197–223. doi:19/1/197[pii]10.1146/annurev.immunol.19.1.197

Nguyen S, Béziat V, Norol F, Uzunov M, Trebeden-Negre H, Azar N, Boudifa A, Bories D, Debré P, Vernant J-P, Vieillard V, Dhédin N (2011) Infusion of allogeneic natural killer cells in a patient with acute myeloid leukemia in relapse after haploidentical hematopoietic stem cell transplantation. Transfusion 51:1769–1778

North J, Bakhsh I, Marden C, Pittman H, Addison E, Navarrete C, Anderson R, Lowdell MW (2007) Tumor-primed human natural killer cells lyse NK-resistant tumor targets: evidence of a two-stage process in resting NK cell activation. J Immunol 178(1):85–94. doi:178/1/85 [pii]

Nowbakht P, Ionescu MC, Rohner A, Kalberer CP, Rossy E, Mori L, Cosman D, De Libero G, Wodnar-Filipowicz A (2005) Ligands for natural killer cell-activating receptors are expressed upon the maturation of normal myelomonocytic cells but at low levels in acute myeloid leukemias. Blood 105(9):3615–3622

Parkhurst MR, Riley JP, Dudley ME, Rosenberg SA (2011) Adoptive transfer of autologous natural killer cells leads to high levels of circulating natural killer cells but does not mediate tumor regression. Clin Cancer Res 17(19):6287–6297. doi:10.1158/1078-0432.CCR-11-1347

Passweg JR, Tichelli A, Meyer-Monard S, Heim D, Stern M, Kuhne T, Favre G, Gratwohl A (2004) Purified donor NK-lymphocyte infusion to consolidate engraftment after haploidentical stem cell transplantation. Leukemia 18(11):1835–1838. doi:10.1038/sj.leu.2403524

Passweg JR, Stern M, Koehl U, Uharek L, Tichelli A (2005) Use of natural killer cells in hematopoetic stem cell transplantation. Bone Marrow Transplant 35(7):637–643. doi:10.1038/sj.bmt.1704810

Pegram HJ, Andrews DM, Smyth MJ, Darcy PK, Kershaw MH (2011) Activating and inhibitory receptors of natural killer cells. Immunol Cell Biol 89(2):216–224. doi:10.1038/icb.2010.78

Pende D, Spaggiari GM, Marcenaro S, Martini S, Rivera P, Capobianco A, Falco M, Lanino E, Pierri I, Zambello R, Bacigalupo A, Mingari MC, Moretta A, Moretta L (2005) Analysis of the receptor-ligand interactions in the natural killer-mediated lysis of freshly isolated myeloid or lymphoblastic leukemias: evidence for the involvement of the Poliovirus receptor (CD155) and Nectin–2 (CD112). Blood 105(5):2066–2073

Persky DO, Dornan D, Goldman BH, Braziel RM, Fisher RI, Leblanc M, Maloney DG, Press OW, Miller TP, Rimsza LM (2012) Fc gamma receptor 3a genotype predicts overall survival in follicular lymphoma patients treated on SWOG trials with combined monoclonal antibody plus chemotherapy but not chemotherapy alone. Haematologica 97 (6):937–942. doi:10.3324/haematol.2011.050419

Perussia B, Ramoni C, Anegon I, Cuturi MC, Faust J, Trinchieri G (1987) Preferential proliferation of natural killer cells among peripheral blood mononuclear cells cocultured with B lymphoblastoid cell lines. Nat Immun Cell Growth Regul 6(4):171–188

Raulet DH, Guerra N (2009) Oncogenic stress sensed by the immune system: role of natural killer cell receptors. Nat Rev Immunol 9(8):568–580. doi:10.1038/nri2604

Reim F, Dombrowski Y, Ritter C, Buttmann M, Hausler S, Ossadnik M, Krockenberger M, Beier D, Beier CP, Dietl J, Becker JC, Honig A, Wischhusen J (2009) Immunoselection of breast and ovarian cancer cells with trastuzumab and natural killer cells: selective escape of CD44high/CD24low/HER2low breast cancer stem cells. Cancer Res 69 (20):8058–8066. doi:0008-5472.CAN-09-0834 [pii]10.1158/0008-5472.CAN-09-0834

Rizzieri DA, Storms R, Chen DF, Long G, Yang Y, Nikcevich DA, Gasparetto C, Horwitz M, Chute J, Sullivan K, Hennig T, Misra D, Apple C, Baker M, Morris A, Green PG, Hasselblad V, Chao NJ (2010) Natural killer cell-enriched donor lymphocyte infusions from A 3–6/6 HLA matched family member following nonmyeloablative allogeneic stem cell transplantation. Biol Blood Marrow Transplant 16(8):1107–1114. doi:10.1016/j.bbmt.2010.02.018

Rosenberg SA, Lotze MT, Yang JC, Topalian SL, Chang AE, Schwartzentruber DJ, Aebersold P, Leitman S, Linehan WM, Seipp CA, et al. (1993) Prospective randomized trial of high-dose interleukin –2 alone or in conjunction with lymphokine-activated killer cells for the treatment of patients with advanced cancer. J Natl Cancer Inst 85(8):622–632

Rosenberg SA, Yannelli JR, Yang JC, Topalian SL, Schwartzentruber DJ, Weber JS, Parkinson DR, Seipp CA, Einhorn JH, White DE (1994) Treatment of patients with metastatic melanoma with autologous tumor-infiltrating lymphocytes and interleukin 2. J Natl Cancer Inst 86(15):1159–1166

Rubnitz JE, Inaba H, Ribeiro RC, Pounds S, Rooney B, Bell T, Pui CH, Leung W (2010) NKAML: a pilot study to determine the safety and feasibility of haploidentical natural killer cell transplantation in childhood acute myeloid leukemia. J Clin Oncol 28(6):955–999. doi:JCO.2009.24.4590 [pii]10.1200/JCO.2009.24.4590

Ruggeri L, Capanni M, Casucci M, Volpi I, Tosti A, Perruccio K, Urbani E, Negrin RS, Martelli MF, Velardi A (1999) Role of natural killer cell alloreactivity in HLA-mismatched hematopoietic stem cell transplantation. Blood 94(1):333–339

Ruggeri L, Capanni M, Urbani E, Perruccio K, Shlomchik WD, Tosti A, Posati S, Rogaia D, Frassoni F, Aversa F, Martelli MF, Velardi A (2002) Effectiveness of donor natural killer cell alloreactivity in mismatched hematopoietic transplants. Science 295(5562):2097–2100

Ruggeri L, Mancusi A, Capanni M, Urbani E, Carotti A, Aloisi T, Stern M, Pende D, Perruccio K, Burchielli E, Topini F, Bianchi E, Aversa F, Martelli MF, Velardi A (2007) Donor natural killer cell allorecognition of missing self in haploidentical hematopoietic transplantation for acute myeloid leukemia: challenging its predictive value. Blood 110 (1):433–440. doi:10.1182/blood-2006-07-038687

Santourlidis S, Trompeter HI, Weinhold S, Eisermann B, Meyer KL, Wernet P, Uhrberg M (2002) Crucial role of DNA methylation in determination of clonally distributed killer cell Ig-like receptor expression patterns in NK cells. J Immunol 169(8):4253–4261

Satwani P, van de Ven C, Ayello J, Cairo D, Simpson LL, Baxi L, Cairo MS (2011) Interleukin (IL)-15 in combination with IL-2, fms-like tyrosine kinase-3 ligand and anti-CD3 significantly enhances umbilical cord blood natural killer (NK) cell and NK-cell subset expansion and NK function. Cytotherapy. doi:10.3109/14653249.2011.563292

Savani BN, Mielke S, Adams S, Uribe M, Rezvani K, Yong AS, Zeilah J, Kurlander R, Srinivasan R, Childs R, Hensel N, Barrett AJ (2007) Rapid natural killer cell recovery determines outcome after T-cell-depleted HLA-identical stem cell transplantation in patients with myeloid leukemias but not with acute lymphoblastic leukemia. Leukemia 21 (10):2145–2152. doi:2404892 [pii]10.1038/sj.leu.2404892

Shilling HG, McQueen KL, Cheng NW, Shizuru JA, Negrin RS, Parham P (2003) Reconstitution of NK cell receptor repertoire following HLA-matched hematopoietic cell transplantation. Blood 101(9):3730–3740. doi:10.1182/blood -2002-08-2568

Siegler U, Kalberer CP, Nowbakht P, Sendelov S, Meyer-Monard S, Wodnar-Filipowicz A (2005) Activated natural killer cells from patients with acute myeloid leukemia are cytotoxic against autologous leukemic blasts in NOD/SCID mice. Leukemia 19 (12):2215–2222

Siegler U, Meyer-Monard S, Jorger S, Stern M, Tichelli A, Gratwohl A, Wodnar-Filipowicz A, Kalberer CP (2010) Good manufacturing practice-compliant cell sorting and large-scale expansion of single KIR-positive alloreactive human natural killer cells for multiple infusions to leukemia patients. Cytotherapy 12(6):750–763. doi:10.3109/14653241003786155

Smyth MJ, Cretney E, Takeda K, Wiltrout RH, Sedger LM, Kayagaki N, Yagita H, Okumura K (2001) Tumor necrosis factor-related apoptosis-inducing ligand (TRAIL) contributes to interferon gamma-dependent natural killer cell protection from tumor metastasis. J Exp Med 193(6):661–670

Smyth MJ, Hayakawa Y, Takeda K, Yagita H (2002) New aspects of natural-killer-cell surveillance and therapy of cancer. Nat Rev Cancer 2 (11):850–861. doi:10.1038/nrc928

Somanchi SS, Senyukov VV, Denman CJ, Lee DA (2011) Expansion, purification, and functional assessment of human peripheral blood NK cells. J Vis Exp (48). doi:2540 [pii]10.3791/2540

Sondermann P, Huber R, Oosthuizen V, Jacob U (2000) The 3.2-A crystal structure of the human IgG1 Fc fragment-Fc gammaRIII complex. Nature 406(6793):267–273. doi:10.1038/35018508

Spanholtz J, Tordoir M, Eissens D, Preijers F, van der Meer A, Joosten I, Schaap N, de Witte TM, Dolstra H (2010) High log-scale expansion of functional human natural killer cells from umbilical cord blood CD34-positive cells for adoptive cancer immunotherapy. PLoS One 5(2):e9221. doi:10.1371/journal.pone.0009221

Storek J, Geddes M, Khan F, Huard B, Helg C, Chalandon Y, Passweg J, Roosnek E (2008) Reconstitution of the immune system after hematopoietic stem cell transplantation in humans. Semin Immunopathol 30(4):425–437. doi:10.1007/s00281-008-0132-5

Stringaris K, Adams S, Uribe M, Eniafe R, Wu CO, Savani BN, Barrett AJ (2010) Donor KIR Genes 2DL5A, 2DS1 and 3DS1 are associated with a reduced rate of leukemia relapse after HLA-identical sibling stem cell transplantation for acute myeloid leukemia but not other hematologic malignancies. Biol Blood Marrow Transplant 16(9):1257–1264. doi:10.1016/j.bbmt.2010.03.004

Sutlu T, Stellan B, Gilljam M, Quezada HC, Nahi H, Gahrton G, Alici E (2010) Clinical-grade, large-scale, feeder-free expansion of highly active human natural killer cells for adoptive immunotherapy using an automated bioreactor. Cytotherapy 12(8):1044–1055. doi:10.3109/14653249.2010.504770

Symons HJ, Leffell MS, Rossiter ND, Zahurak M, Jones RJ, Fuchs EJ (2010) Improved survival with inhibitory killer immunoglobulin receptor (KIR) gene mismatches and KIR haplotype B donors after nonmyeloablative, HLA-haploidentical bone marrow transplantation. Biol Blood Marrow Transplant 16(4):533–542. doi:10.1016/j.bbmt.2009.11.022

Tajima F, Kawatani T, Endo A, Kawasaki H (1996) Natural killer cell activity and cytokine production as prognostic factors in adult acute leukemia. Leukemia 10(3):478–482

Takeda K, Smyth MJ, Cretney E, Hayakawa Y, Yamaguchi N, Yagita H, Okumura K (2001) Involvement of tumor necrosis factor-related apoptosis-inducing ligand in NK cell-mediated and IFN-gamma-dependent suppression of subcutaneous tumor growth. Cell Immunol 214 (2):194–200. doi:10.1006/cimm.2001.1896

Terme M, Ullrich E, Delahaye NF, Chaput N, Zitvogel L (2008) Natural killer cell–directed therapies: moving from unexpected results to successful strategies. Nature Immunol 9 (5):486–494. doi:10.1038/ni1580

Tomala J, Chmelova H, Mrkvan T, Rihova B, Kovar M (2009) In vivo expansion of activated naive CD8 + T cells and NK cells driven by complexes of IL-2 and anti-IL-2 monoclonal antibody as novel approach of cancer immunotherapy. J Immunol 183(8):4904–4912. doi:183/8/4904 [pii]10.4049/jimmunol.0900284

Torelli GF, Guarini A, Palmieri G, Breccia M, Vitale A, Santoni A, Foa R (2002) Expansion of cytotoxic effectors with lytic activity against autologous blasts from acute myeloid leukaemia patients in complete haematological remission. Br J Haematol 116(2):299–307

Torelli GF, Guarini A, Maggio R, Alfieri C, Vitale A, Foa R (2005) Expansion of natural killer cells with lytic activity against autologous blasts from adult and pediatric acute lymphoid leukemia patients in complete hematologic remission. Haematologica 90(6):785–792

Torkar M, Norgate Z, Colonna M, Trowsdale J, Wilson MJ (1998) Isotypic variation of novel immunoglobulin-like transcript/killer cell inhibitory receptor loci in the leukocyte receptor complex. Eur J Immunol 28(12):3959–3967. doi:10.1002/(SICI)1521-4141(199812)28:12 & #60;3959::AID-IMMU3959 & #62;3.0.CO;2-2

Trinchieri G (1989) Biology of natural killer cells. Adv Immunol 47:187–376

Uhrberg M, Valiante NM, Shum BP, Shilling HG, Lienert-Weidenbach K, Corliss B, Tyan D, Lanier LL, Parham P (1997) Human diversity in killer cell inhibitory receptor genes. Immunity 7(6):753–763

Veeramani S, Wang SY, Dahle C, Blackwell S, Jacobus L, Knutson T, Button A, Link BK, Weiner GJ (2011) Rituximab infusion induces NK activation in lymphoma patients with the high-affinity CD16 polymorphism. Blood 118(12):3347–3349. doi:10.1182/blood-2011-05-351411

Verneris MR, Miller JS (2009) The phenotypic and functional characteristics of umbilical cord blood and peripheral blood natural killer cells. Br J Haematol 147(2):185–191. doi:10.1111/j.1365-2141.2009.07768.x

Weng WK, Levy R (2003) Two immunoglobulin G fragment C receptor polymorphisms independently predict response to rituximab in patients with follicular lymphoma. J Clin Oncol 21(21):3940–3947. doi:10.1200/JCO.2003.05.013

Wu Z, Xu Y (2010) IL-15R alpha-IgG1-Fc enhances IL-2 and IL-15 anti-tumor action through NK and CD8 + T cells proliferation and activation. J Mol Cell Biol 2(4):217–222. doi:mjq012 [pii]10.1093/jmcb/mjq012

Wu J, Edberg JC, Redecha PB, Bansal V, Guyre PM, Coleman K, Salmon JE, Kimberly RP (1997) A novel polymorphism of FcgammaRIIIa (CD16) alters receptor function and predisposes to autoimmune disease. J Clin Invest 100(5):1059–1070. doi:10.1172/JCI119616

Zamai L, Ahmad M, Bennett IM, Azzoni L, Alnemri ES, Perussia B (1998) Natural killer (NK) cell-mediated cytotoxicity: differential use of TRAIL and Fas ligand by immature and mature primary human NK cells. J Exp Med 188(12):2375–2380

Zhang H, Cui Y, Voong N, Sabatino M, Stroncek DF, Morisot S, Civin CI, Wayne AS, Levine BL, Mackall CL (2011) Activating signals dominate inhibitory signals in CD137 L/IL-15 activated natural killer cells. J Immunother 34(2):187–195. doi:10.1097/CJI.0b013e31820d2a21

Chapter 41
Allogeneic and Autologous T-cell Strategies to Enhance Targeting of Acute Myeloid Leukemias

Gerrit Weber, Catherine M. Bollard and A. John Barrett

Abstract Acute myelogenous leukemia (AML) is a realistic target for T-cell immunotherapy because clinically powerful T-cell-mediated graft-versus leukemia (GVL) effects are well characterized following allogeneic stem cell transplantation (SCT). T-cell therapies in combination with SCT include donor lymphocyte infusions with or without added cytokines, infusions of selected T-cells deprived of graft-versus-host disease (GVHD) activity, and T-cells expressing a suicide gene permitting GVL but preventing GVHD by T-cell elimination. The identification of a series of minor histocompatibility antigens (mHAG) restricted to the myeloid lineage has made it possible to generate leukemia-reactive T-cells by expanding donor lymphocytes specific for a mHAG present in the recipient but absent in the donor. Key to the development of treatments that extend targeted T-cell strategies beyond SCT has been the identification of a range of leukemia-associated antigens (LAA). LAA can be used, either as inserts of the parent gene of a tumor antigen or as peptide mixes, to arm antigen-presenting cells (APCs). Dendritic cell APCs are used to induce LAA-specific T-cells either in vivo using the APCs as vaccines or in vitro generating LAA-specific T-cells in culture. Initial clinical results with mHAG-specific T-cells in transplanted patients show promising disease control but have identified off-target effects of some T-cell infusions, and reveal a tendency for tumor escape. Future developments for T-cell therapy for AML include the association of T-cell infusions with agents that can upregulate LAA, and the genetic modification of CTL to overcome tumor resistance.

C. M. Bollard (✉)
Blood+Marrow Transplantation, Children's National Health System,
111 Michigan Ave, NW, Suite 7225 Washington, DC 20010, USA
e-mail: cbollard@cnmc.org

G. Webert
University Children's Hospital, Pediatric Hematology/Oncology, University of Wuerzburg,
1102 Bates Street, Suite 1770.01, Houston, TX 77030, USA
e-mail: Weber_g1@ukw.de

A. J. Barrett
Stem Cell Allogenic Transplantation Section, Hematology Branch, National Heart,
Lung and Blood Institute, National Institutes of Health P.O. Box 30105,
Bethesda, MD 20892-1202, USA
e-mail: barrettj@nih.gov

M. Andreeff (ed.), *Targeted Therapy of Acute Myeloid Leukemia*,
Current Cancer Research, DOI 10.1007/978-1-4939-1393-0_41

Keywords Adoptive immunotherapy · Cytotoxic T-lymphocytes · Leukemia-associated antigens · Tissue-restricted minor histocompatibility antigens · Donor lymphocyte infusions · Suicide gene transfer · Dendritic cell vaccines · Tumor escape

41.1 Introduction

AML is a susceptible target for T-cell attack because the leukemia expresses self-antigens through both major histocompatibility complex (MHC) class I and II molecules and typically expresses the costimulatory molecules CD81 and CD86. Evidence for an antileukemic effect of T-cell on acute myelogenous leukemia (AML) comes from several observations: (1) there is an increased risk of acquiring AML during prolonged immunosuppression (e.g., renal transplant recipients have a two-fold increase in the risk of developing AML and more immunosuppressed heart transplant recipients have a fivefold increased risk of developing AML)(Gale and Opelz 2012). (2) after allogeneic stem cell transplantation (SCT) for AML a strong graft-versus-leukemia (GVL) reaction, accompanied by graft-versus-host disease (GVHD) is mediated by donor T-cells and contributes to the cure of AML by SCT; and (3) leukemia antigen-specific T-cells are present in patients with AML in remission, and a more favorable outcome is observed for patients with leukemia that expresses multiple leukemia-associated antigens (Greiner et al. 2005; Schmitt et al. 2008). These observations form the rational basis for using antileukemia T-cell therapies to control AML in the context of preventing or treating relapse after chemotherapy or SCT, using either autologous or allogeneic T-cells. Here, we describe the allo- and self-antigens targeted by T-cells and the strategies used to select or expand leukemia-specific T-cells for treatment of AML.

41.2 Leukemia Antigens

The suitability of antigenic targets on AML cells is determined by how closely the candidate antigen conforms to the criteria of an ideal tumor antigen: For effective elimination of the leukemia, the antigens must be well expressed by the true leukemic stem cell progenitors that fall within the CD34+ CD38–- lineage negative subset. Leukemia-associated antigens (LAA) should be uniquely expressed by the leukemia to reduce the risk of bystander damage to normal stem cells. Detection of low frequencies of LAA-specific T-cells during remission or after SCT is important evidence that a particular antigen can elicit a T-cell response. T-cell target antigens in AML fall into two categories: (i) The so-called minor histocompatibility antigens (mHAG) and other allogeneic antigens recognized by donor T-cells and (ii) LAA'S which can elicit both autologous and allogeneic T-cell cytotoxic T-cell (CTL) responses.

Table 41.1 Characteristics of common human mHAG

mHAG	HLA restriction	% Disparity	% Disparity
		MSD	URD
HA-1	A*0201; A*0206	6.4	11.6
LRH-1	B*0702	4.9	7.5
ACC-2	B*4403	3.6	6.7
HEATR 1	B*0801	3.1	5.6
ACC-1	A*2402	2.8	5.2
ACC-6	B*4402; B*4403	2.3	5.9
HA-2	A*0201	1.8	2.5
SP110	A*0301	1.6	2.5
LBLY75-1K	DRB1*1301	1.2	2.4
HA-1	B*60; B*40012	<1	<1
ACC1	A*2402	<1	<1
CD19	DQ, A1*05; B1*02	10.4	15.9
HB-1	B*4402; B*4403	3.9	6.8
HB1	B*4402; B*4403	1.2	1.3

41.3 Minor Antigens

In the setting of allogeneic SCT, antigens that differ between donor and recipient cause an immune reaction between the graft and host tissue, even in HLA-identical donor pairs. These mHAGs are peptides derived from intracellular proteins. Polymorphisms in the amino acid sequence of intracellular proteins lead to the differential expression of peptides presented by HLA molecules, which can induce an alloimmune T-cell response. mHAGs are presented by both MHC class I and II with variable tissue specificity and immune responses which are similarly tissue-restricted or ubiquitous. The majority of mHAGs remain unknown in man, and the repertoire of characterized ubiquitous or tissue-restricted expression patterns remains less than 50. Y-chromosomal antigens usually show a broad expression pattern whereas mHAGs on autosomal genes usually have a more localized tissue expression. Alloreactions against ubiquitous mHAGs induce GVHD and GVL, but if the mHAGs are restricted to hematopoietic tissue (as is the case with HA-1, the most well-studied mHAG) the alloresponse is restricted to a GVL effect. Minor antigens with an expression pattern restricted to leukemic cells or certain cell lineages thus represent ideal targets for leukemia-specific T-cell immunotherapy.

A number of MHC class I- and class II-restricted mHAGs are now described (Table 41.1; Griffioen et al. 2012) Because class II molecules are predominantly restricted in their expression to hematopoietic cells, these class II mHAGs are of particular interest because of the potential leukemic specificity of $CD4^+$ T-cell alloresponses (Stumpf et al. 2009; Falkenburg et al. 2003).

41.4 Leukemia-Associated Antigens

Aside from alloantigens, there exists a broad spectrum of LAA, which can be grouped as follows.

41.4.1 Antigens Common for Many Malignancies

In this category are the cancer–testis antigens (e.g., MAGE, BAGE, GAGE, and NY-ESO-1). They are physiologically expressed in testis, ovary, and placenta cells (trophoblast), but also in a variety of tumor cells and myeloid malignancies. Members of the MAGE family have been detected in AML and ALL (Martinez et al. 2007), BAGE expression in AML (Adams et al. 2002; Greiner et al. 2004; Schmitt et al. 2006) and HAGE in AML and CML (Adams et al. 2002; Schmitt et al. 2006).

41.4.2 Antigens Overexpressed by Malignant Cells

These antigens are widely expressed at low level in normal tissues but overexpressed by the malignant cell. Examples of these antigens found also in hematological malignancies are Her2/neu, AFP, telomerase, WT1, RHAMM, and PRAME. Hematologic myeloid malignancies most frequently overexpress WT1, PRAME, and RHAMM (Greiner et al. 2005; Oka et al. 2004).

41.4.3 Antigens Specific for Hematopoietic Lineages

These include primary granule proteins proteinase 3, neutrophil elastase, and cathepsin G, aberrantly expressed by early progenitors in myeloid malignancies, and the BMI-1 polycomb proteins in CML. Some antigens from the novel protein sequences encoded by chromosome translocations BCR-ABL and PML-RARA have also been characterized, although while such antigens are truly leukemia-specific, they have not proved in general to have great immunotherapeutic potential. Table 41.2 summarizes LAA found in AML. To be immunogenic, LAA must be processed and presented in the context of MHC molecules on the cell surface of the malignant cell, in high enough concentration to function as targets for T-cell-based therapies. Unlike mHAGs which represent single epitopes, LAA have the potential advantage of being able to generate CD4 helper and cytotoxic CD8 T-cells against multiple epitopes presented by a diversity of HLA molecules. Such broad T-cell responses reduce the risk of immune escape by leukemia cells. Furthermore, since T-cell responses are not restricted to a particular HLA group, LAA represent an attractive universal antigen source for immunotherapy (Barrett and Rezvani 2007).

Naturally occurring immune responses against a variety of tumor antigens have been detected in leukemia patients as well as healthy individuals. However, an

Table 41.2 Tumor antigens occurring in AML

Year	LAA	HLA	Leukemia	Finding
Molldrem et al. 1996	PR-1	A*0201	CML AML	PR-1-specific CTL is associated with better response to interferon alpha response
Rezvani et al. 2003	PR1 WT1	A*0201	CML AML	WT-1- and PR-1-specific memory CTL detected in 60–70 % patients
Schmitt et al. 2006	RHAMM	A2	CML	High expression of RHAMM, PRAME, and PR3 in blast phase CML. RHAMM-specific CD8 T-cells detected in 4/5 CML-CP
Graf et al. 2007	FLT3	A*0101	AML	Ex vivo generation of FLT3-specific CTL from FLT3 ITD AML Pt
Yong et al. 2007	PR3, ELA2	A*0201	CML	High PR3/ELA expression in pre-TX advanced CML. Donor PR1-CTL associated better LFS, OS
Rezvani et al. 2009	PRAME	A*0201	ALL, AML, CML	PRAME-specific memory $CD8^+$ T-cell in AML/CML/ALL patients
Melenhorst et al. 2009	WT1 PR1	A*0201	AML,CML	LAA-specific CD8 high avidity in bone marrow in CR patients
Ochi et al. 2009	Aurora-A kinase	A*0201	CML, AML	Aurora-A kinase-specific T-cell lines specifically killed $CD34^+$/$CD38^-$ CML leukemia stem cell population
Yong et al. 2011	BMI-1	A*0201	CML	BMI expression better posttransplant LFS/higher BMI-1 expression in $CD34^+$ cells associated with CTL response
Sloand et al. 2011	WT1	A*0201	MDS	High WT1 expression in trisomy 8, WT1-CTL in IST-responsive MDS

increased frequency of cytotoxic T-lymphocytes (CTLs) occurs in leukemia patients, but importantly also in patients after SCT, indicating a spontaneous reactivity with LAA in vivo (Schmitt et al. 2006; Oka et al. 2004; Gannage et al. 2005; Ochi et al. 2009; Ohminami et al. 2000; Rezvani et al. 2003; Rezvani et al. 2009). The occurrence of CTL responses against the proteinase 3 peptide PR1 in CML patients has been associated with a better long-term survival of these patients (Molldrem et al. 2000). While frequencies of LAA-specific T-cells are low in the circulation, the bone marrow appears to be a reservoir of increased frequencies of high avidity antigen-specific CTLs (Melenhorst et al. 2009)

The detection of PR1-specific CTLs in the blood of SC donors was associated with a better LFS and OS in CML patients after SCT indicating that there is a positive correlation between graft-derived LAA-specific cytotoxic T-cells and outcome (Yong et al. 2007). To evaluate this approach for further clinical application, several groups have shown that antigen-specific CTLs can be generated from healthy donor and patient peripheral blood in vitro with high efficacy. These CTLs have cytolytic activity against leukemia cells as well as solid tumors in vitro and in vivo, and clinical trials of adoptive T-cell therapy are under way (Yong et al. 2007; Bornhauser et al. 2011; Gerdemann et al. 2011; Weber et al. 2009; Weber et al. 2013a, b).

41.5 Targeted Immunotherapy for AML in the Context of Allogeneic SCT

41.5.1 Donor Lymphocyte Infusion

The infusion of unmanipulated allogeneic lymphocytes (donor lymphocyte infusion, DLI) can eliminate residual leukemia cells in CML and induce complete remission after SCT, without further administration of chemotherapy. Best responses have been observed in patients with relapsed chronic phase CML after SCT (Kolb et al. 1995; Porter et al. 1994). DLI has also been applied to patients with acute leukemia but with less frequent responses and only rare complete remissions. Studies in AML implementing DLI as relapse treatment did not observe a response in the majority of patients. Only a small group of patients achieved partial or complete remission (Kolb et al. 1995; Levine et al.1995, 2008). Even after an initial response, the majority of AML patients later relapse (Porter et al. 1999; Porter et al. 2000). Similar results were obtained when DLI was applied for the prevention of relapse: Only a small proportion of the patients treated showed a response or clinical benefit (Schmid et al. 2007; Bethge et al. 2004). Approaches combining DLI with chemotherapeutic agents or cytokine administration offer the best chance of a response but are associated with a high treatment-related mortality and nondurable remission (Levine et al. 2002; Takami et al. 2005).

To augment the potency of the DLI, co-administration of cytokines, such as interleukin (IL) -2, alone or in combination with cellular therapies, has been explored. IL-2 may confer an antileukemic effect in the nontransplant setting, and after hematopoietic SCT with reduced relapse rates in patients with AML (Meloni et al 1994; Soiffer et al. 1994; Schlegel et al. 2011). The combination of DLI and IL2 administration showed a beneficial effect in some patients not responding to cellular therapy alone (Slavin et al. 1996). More recently, Chinese investigators have reported remarkably strong antileukemic effects from a combination of interferon-alpha and DLI in patients with AML relapsing after SCT, with prolonged leukemia-free survival in the order of 50 % (Tang et al. 2011).

Another technique to increase the efficacy of DLI has been to activate T-cell ex vivo with anti-CD3 and anti-CD28 antibody has shown some benefit in patients with CML (Porter et al. 2006). A promising improvement to DLI is the generation of antigen-specific CTLs to target leukemia selectively as discussed below.

41.5.2 Suicide-Gene-Transduced T-Cells

A major risk of unmanipulated DLI infusion is the development of acute GVHD requiring immunosuppressive treatment. GVHD is caused by alloreactive T-cells in the graft attacking the skin, gut, and liver. One possibility to control GVHD after DLI administration is to transduce donor lymphocytes with a suicide gene that can be activated to eliminate alloreacting T-cells at the onset of severe GVHD. The most commonly used suicide gene is the herpes simplex virus thymidine kinase (HSV-tk), activated by ganciclovir administration. Transduced T-cells are rapidly eliminated following ganciclovir-induced cytotoxicity. HSV-tk-transduced T-cell have been tested in clinical trials (Bonini et al. 1997; Ciceri et al. 2007; Traversari et al. 2007) but have limited persistence in vivo due to CD4 and CD8 T-cell immune responses against viral antigens (Berger et al. 2006; Traversari et al. 2007). However, the graft versus leukemia effect was not impaired by the underlying immunologic reaction (Ciceri et al. 2007).

More recently, other suicide genes based on apoptosis have been explored in vitro and in clinical trials (Berger et al. 2004; Thomis et al. 2001). An intracellular Fas domain coupled to an extracellular FK506-binding domain is inserted into the T-cell. AP1903, a bivalent drug, causes dimerization of the binding domain followed by Fas-mediated apoptosis and effective elimination of the transduced cells upon administration of the drug. Another approach is the use of inducible Caspase 9 (iCasp9) as a suicide switch (Di Stasi and Dotti 2011; Straathof et al. 2005; Tey et al. 2007). Caspase activation leads to rapid elimination of transduced alloreactive cells. Even though iCasp9 is downregulated in quiescent cells, expression is rapidly upregulated in alloreactivated cells. Clinical application in five pediatric patients, four of whom developed GVHD, showed an effective elimination of more than 90 % of cells after administration of AP1903 (Di Stasi and Dotti 2011). Lastly, human CD20 has also been investigated as a nontoxic selection marker and suicide gene (Bonini et al. 2003; Griffioen et al. 2009). Transduced cells can be targeted by anti-CD20 antibody (rituximab) in the case of GVHD. Despite the different strategies being evaluated, all studies show that the T-cell products transduced with suicide genes can maintain in vitro and in vivo function against viral and tumor antigens.

41.5.3 Selective Depletion of Alloreactive Cells

Risks from DLI aside, outcome after SCT is constrained by morbidity and mortality from severe GVHD. Donor T-cells in the graft targeting mHAG are the main source of GVHD. Destructive T-cell responses against donor tissue can be limited by immunosuppression or T-cell depletion in the graft, but both methods reduce antiviral reactivity and antitumor activity causing severe infectious complications and favoring relapse of residual leukemia. Several approaches to target and selectively eliminate alloreactive T-cells from the graft or the DLI product have been explored in vitro and in clinical trials. Depletion strategies based on selective depletion of upregulated extracellular activation markers induced by alloantigens presented by recipient PBMC, T-cells, or EBV-LCL have been used targeting the activation markers, CD25, CD69, CD134, CD137, and CD95.

Clinical selective allodepletion trials targeting CD25 with anti-CD25 immunotoxin showed accelerated T-cell reconstitution with a low rate of a GVHD in patients receiving allodepleted T-cells (Amrolia et al. 2005; Amrolia et al. 2006; Solomon et al. 2005). After allodepletion targeting CD95-mediated activation-induced cell death antiviral and antitumor activity was conserved (Hartwig et al. 2008). Depletion of CD134-expressing alloreactive cells was effective in vitro without impairing the function of CMV and tumor antigen (WT1)-specific T-cells as shown by pentamer analysis (Ge et al. 2008).

An alternative approach to selective depletion is to target activated alloreactive T-cells with pharmacological inhibitors, such as the chaperone heat shock protein of 90 kDa, which has a crucial role in signaling pathways of activated T-cells (Stuehler et al. 2009). Alternatively, a photodepletion process, targeting photosensitized activated T-cells has been tested in clinical trials. Activation-based changes in p-glycoprotein alter the efflux of the photosensitizer TH9402, a phototoxic rhodamine-like dye. After incubating alloactivated T-cells with the dye, the cell product is exposed to fluorescent light, inducing apoptosis of activated cells (Mielke et al. 2008; Perruccio et al. 2008). A clinical trial using a photodepleted T-cell product showed protection against the occurrence of GVHD, but was associated with delayed immune reconstitution and a high rate of infectious complications (Mielke et al. 2011). The photodepleted graft T-cell product seemed to be affected, showing a reduced frequency of CD4 central memory cells and persisting CD4 deficiency (McIver et al. 2011). Functional analyses of the photodepleted T-cell products showed a reduced antiviral and antileukemic potential (Perruccio et al. 2011).

41.5.4 Selection and Expansion of mHA-Specific T-Cells

While mHA-specific T-cells can be manufactured with increasing facility, the limitation of all mHA-based methodologies is the need for a “boutique approach” for each patient/donor pair. HA-1, the best-characterized mHA, is for example restricted to HLA-A2+ individuals and only HA1 positive recipients of an SCT from an HA1 negative donor can benefit from the alloresponse. Given that the frequency of HLA-A2 in the Caucasian population rarely exceeds 30% and that HA1 has a relatively

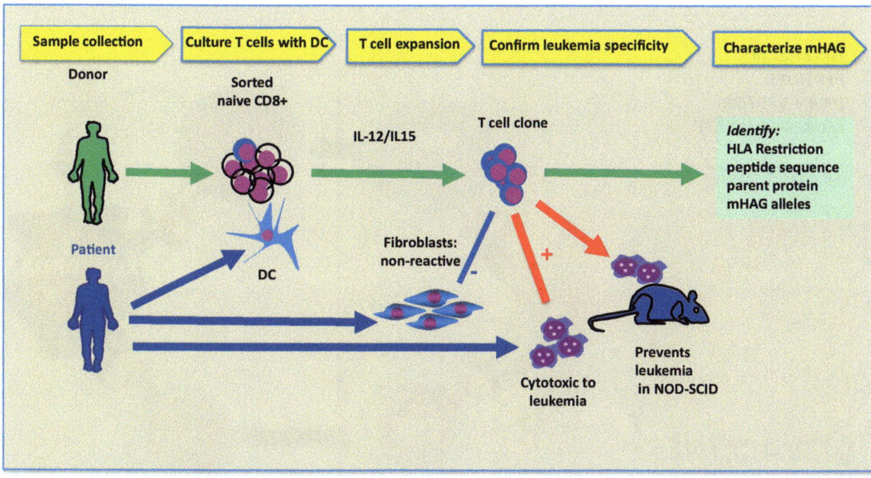

Fig. 41.1 Generation of mHAG-specific CTLs from healthy donors targeting patient leukemia cells. mHAG-specific CTLs can be generated and expanded from sorted naïve donor CD8^{+} T-cells by stimulation with patient DC and IL-12 substitution and further expansion with IL-15. T-cell clones are then used to test leukemia specificity. Clones not recognizing patient fibroblasts are further evaluated for their antileukemia efficacy in vivo and can be used to define the recognized protein as well as HLA-restriction elements of the recognized peptide. For several antigens that have been identified so far, expression of the encoding gene, as well as naturally occurring CTLs targeting the mHAG, have been detected in patients after SCT or DLI

balanced polymorphism, there is only a one in four chance that any donor–recipient pairing will be favorable (about 8 % of the entire transplant population). Currently, there are not enough mHA characterized to provide a spectrum of mHA-specific T-cells broad enough to cover the majority of the transplant recipients.

Bleakeley et al. (Bleakley et al. 2010) demonstrated a feasible approach to generate mHAG CTLs in vitro from unsensitized donors by stimulating naïve CD8+ T-cells with recipient dendritic cell (DC; Fig. 41.1). Leukemia-reactive T-cell clones are selected, tested to exclude recognition of patient fibroblasts, and tested for the ability to eliminate leukemia in a mouse model. Leukemia-recognizing T cell clones are further analyzed for their recognition element as well as the HLA restriction of these CTLs. This approach successfully identified mHAG-reactive T-cells in every donor–recipient pair tested leading to the description of new mHAGs defined at the genetic and protein sequence level (Bleakley et al. 2010). However, these sophisticated, time consuming, and expensive approaches are not widely available at present. Furthermore, the narrow repertoire of mHAG-specific CTLs recognizing only hematologic cells that can effectively eliminate leukemia without off-target side effects on nonhematopoietic tissue limits clinical applicability (Warren et al. 2010).

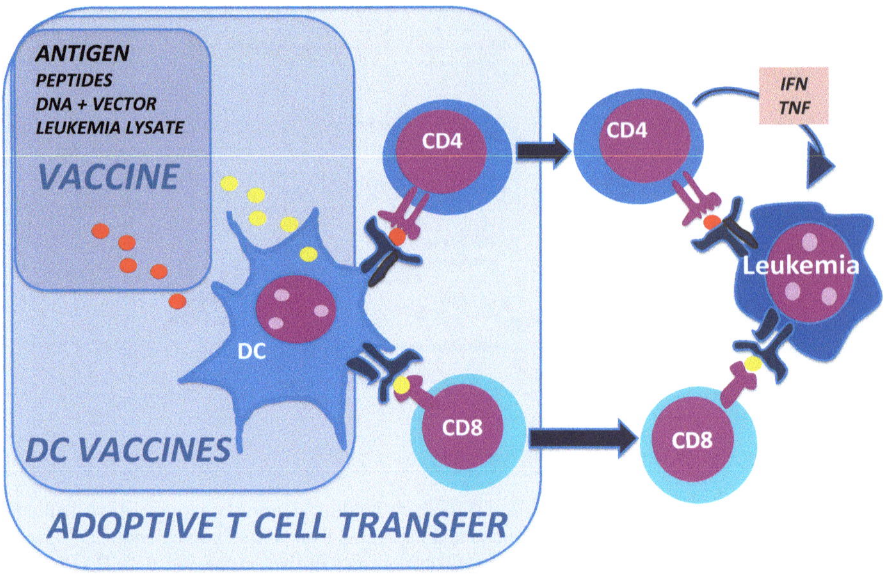

Fig. 41.2 Immunotherapeutic treatment approaches using vaccine strategies or adoptive T-cell transfer. Schematic overview of immunotherapeutic treatment approaches using vaccine strategies, cellular vaccines, or the adoptive transfer of in vitro generated or primed T-cells targeting leukemia cells. A combinatory approach activating both $CD4^+$ and $CD8^+$ T-cells has been shown to be most effective and able to achieve a sustained immune response in vivo

41.6 Targeted Therapies Applicable in the Autologous and Allogeneic Setting

41.6.1 DC Strategies and Vaccination

Other immune-based therapies to boost the antileukemia effect in both the allogeneic and autologous setting are vaccination and adoptive cellular immunotherapy (Fig. 41.2). Vaccines eliciting broad antileukemic activity using irradiated leukemic cells, cell lysates (Zeng et al. 2003), or leukemia-derived DCs (Houtenbos et al. 2006; Rosenblatt and Avigan 2006) have been shown to be feasible and do not require prior knowledge of antigen expression (Anguille et al. 2011). On the other hand, immunotherapies targeting defined antigens, such as mHAGs or LAA, are under exploration (Rezvani and Barrett 2008; Rezvani 2011). Clinical trials using peptides or whole protein are under investigation, and RNA and DNA vaccines are also being explored (Rice et al. 2008; Chaise et al. 2008).

DC-based approaches with autologous DCs loaded exogenously with peptides or electroporated with mRNA have been used in clinical studies and shown to be safe and able to elicit an immune response in patients with AML. First observations showed that the immune activation and detection of antigen-specific T-cells elicited by the vaccine might predict improved outcome (Van et al. 2009; Van T et al. 2010). DCs electroporated with full-length WT1 mRNA have been subcutaneously administered to ten AML patients in partial or complete remission after chemotherapy. This approach resulted in molecular remissions in five of ten patients, with tumor markers returning to normal levels. Three of these patients became long-term responders with complete remission lasting for at least 3 years. Moreover, these clinical responses were accompanied by an increased frequency of peptide-specific T-cells as shown by tetramer staining and IFNγ secretion as well as a general increased immune activation (Van T et al. 2010).

A similar approach utilizing a DC-based vaccination therapy is the administration of autologous, leukemia-derived DCs, generated in vitro (Kremser et al. 2010; Royer et al. 2008). DCs express a variety of autologous antigens, in combination with activation markers for the presentation of those antigens. DCs have been generated from both leukemia cells and autologous monocytes with the latter approach resulting in a better DC yield and efficacy (Kremser et al 2010; Royer et al 2008).

41.6.2 Peptide Vaccination

Peptide vaccines have the attraction of being a readily applicable "off the shelf" approach to generating LAA-specific T-cells. Peptide vaccination with peptides derived from BCR-ABL, WT1, PR-1, and RHAMM can increase the frequency of antigen-specific CTLs in the recipient. Furthermore, immune activation, measured as an increased frequency of T-cells detected by tetramer analysis or an increased frequency of cytokine-producing T-cells. Functional analyses demonstrated that there is an antileukemic effect in vitro and clinical trials were able to show a reduction of tumor markers correlating with a clinical benefit. (Oka et al. 2004; Royer et al. 2008; Pinilla-Ibarz et al. 2000; Rezvani 2008; Rezvani et al. 2008; Rezvani et al. 2011; Schmitt et al. 2008; Royer et al. 2008; Qazilbash et al. 2007; Keilholz et al. 2009; Scheibenbogen et al. 2002; Maslak et al. 2010; Greiner et al. 2010) Combined vaccines of two or more peptides (PR1 and WT1 or four WT1-derived peptides) have also been tested in clinical trials and been shown to be safe, elicit T-cell responses and result in prolonged disease-free survival of AML patients receiving the vaccine (Maslak et al. 2010; Rezvani et al. 2008). Further, to achieve a persisting immune response of high avidity T-cells as well as a combined immune response of CD4 and CD8 T-cells, a combination of MHC class I- and II-restricted peptides have been used (Maslak et al. 2010; Fujiki et al. 2010). This study also included WT1 peptides with a modified amino acid sequence (heteroclitic peptides) with a higher binding affinity able to confer greater cytotoxicity than T-cells induced with the native peptide sequence (Maslak et al. 2010).

41.6.3 LAA-Specific T-Cells

Another strategy to enhance the antileukemia effect in vivo without relying on an immune response to a vaccine is the ex vivo generation and expansion of antigen-specific CTLs for adoptive T-cell transfer. This approach may overcome the critical obstacle of vaccine therapy when there are limited immune responses as a result of increased immunosuppression after chemotherapy or SCT. Indeed, adoptively transferring T-cells into the lymphopenic milieu induced by chemotherapy or SCT can favor the selective and clonal expansion of transferred LAA-specific T-cells. Furthermore, immune responses against weak antigens seem to be enhanced favoring this outgrowth of tumor antigen-specific T-cells (Rezvani et al. 2011; Maslak et al. 2010; Fujiki et al. 2010; Krupica et al. 2006; Cho et al. 2000; Clarke and Rudensky 2000; Wrzesinski et al. 2010; Williams et al. 2007). During the recovery phase from lymphopenia, an increased potential for activation from co-stimulation leads to highly effective antigen presentation and increased cytokine production (mainly IL7 and IL15)—factors leading to a beneficial T-cell expansion in vivo (Schluns and Lefrancois 2003).

Tumor antigen-specific T-cell have been detected in patients with various hematological malignancies as well as solid tumors, and importantly occur at increased frequencies in patients after SCT (Rezwani et al. 2008; Weber et al. 2013a, b).

(Rezvani and Barrett 2008) Autologous immune responses against leukemia antigens may not be efficient enough to prevent relapse or eradicate disease. Therefore, the ex vivo generation of virus and tumor antigen-specific T-cells has been evaluated and adoptive transfer of CTLs targeting virus or tumor have been tested in clinical trials showing responses in CML and melanoma patients, and eradication of virus-associated lymphomas with reconstitution of antiviral immunity (Dudley et al. 2002; Falkenburg et al. 1999; Riddell and Greenberg 2000; Rooney et al. 1998; Walter et al. 1995). Importantly, most approaches of adoptive T-cell therapy show that a combined CD4 and CD8 T-cell response is critical for a robust and durable immune response and improved outcome (Walter et al. 1995; Rooney et al. 1998).

Several studies have shown that T-cells targeting leukemia- and lymphoma-associated antigens can be generated from donor and patient PBMC and show cytolytic activity against lymphoma cell lines and primary tumor cells in vitro (Weber et al. 2009; Gerdemann et al. 2011; Quintarelli et al. 2008, 2011; Weber et al. 2013a, b). WT1-, PR1-, and/or bcr-abl-specific CTLs have been prophylactically administered to CML patients after SCT without the induction of GVHD. Of the 14 patients who received CTL, 13 were alive at the time of publication, and 7 of them were in molecular remission. Peptide-reactive T-cells were detected in the peripheral blood of patients (Bornhauser et al. 2011). In contrast, treatment of leukemia relapse after SCT with mHAG CTLs (HA-1 and HA-2) in a pilot trial did not show a clinical response (Meij et al. 2012).

41.7 Immune Escape and Strategies to Overcome it

Tumor cells have developed various mechanisms to escape from the host's immune system and are also able to overcome chemo- or immunotherapeutic therapies making them ineffective in the treatment and eradication of leukemia cells (Fig. 41.3). A major obstacle of vaccine as well as adoptive T cell therapies is the ability of malig-

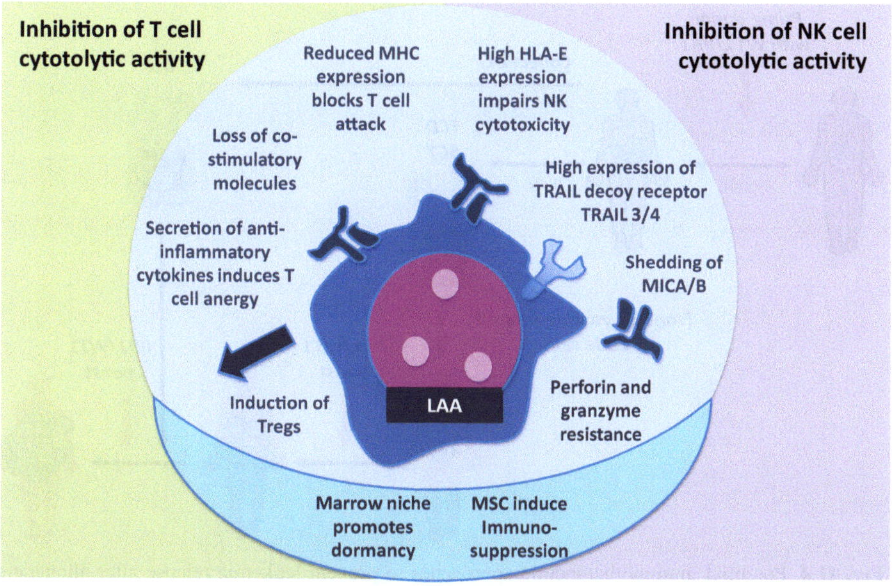

Fig. 41.3 Leukemia immune escape strategies. Immune evasion of malignant cells is one of the main obstacles which reduces treatment efficacy of anticancer therapies. Changes in antigen and MHC expression, secretion of inhibitory cytokines as well as an increased Treg population can cause an inhibition of naturally occurring T-cells as well as adoptively transferred cells. Approaches to overcome the immune escape of tumor cells are necessary to achieve a prolonged antitumor effect in vivo. This may be achieved by overcoming multiple leukemia-specific immune evasion mechanisms simultaneously, e.g., using epigenetically modifying drugs in combination with adoptive T-cell therapies targeting multiple antigens

nant cells to change expression of antigens by downregulation mechanisms, due to selective pressure or selective elimination of cells expressing the targeted antigen. AML cells exhibit a number of features of immune escape from cytotoxic T-cells. They downregulate MHC expression notably after haploidentical SCT when the relapsed leukemia loses expression of the entire mismatched MHC haplotype. AML can also evade T-cell attack by becoming resistant to perforin and granzyme, preventing immune synapse formation and downregulating tumor antigen expression. AML cells can secrete immunosuppressive cytokines, such as VEGF, TGF-β, galectin, or IDO (2,3-dioxygenase) or attraction of Tregs and MDSCs (myeloid-derived suppressor cells) (Schreiber et al. 2011). Further, through some poorly understood mechanisms, regulatory T-cells (Tregs) are increased. To overcome AML resistance to cytotoxic T-cells, gene modifications have been made to improve efficacy and specificity of in vitro generated autologous or allogeneic CTLs (Nagai et al. 2012; Ghorashian et al. 2011; Nicholson et al. 2012; Stauss et al. 2007; Xue et al. 2010). Finally, tumor cell lines and patients treated with demethylating agents such as azacytidine/sodium valproate can upregulate TAA expression with an increased frequency of TAA-specific T-cells detected in the peripheral blood of treated patients (Cruz et al. 2011; Goodyear et al. 2010).

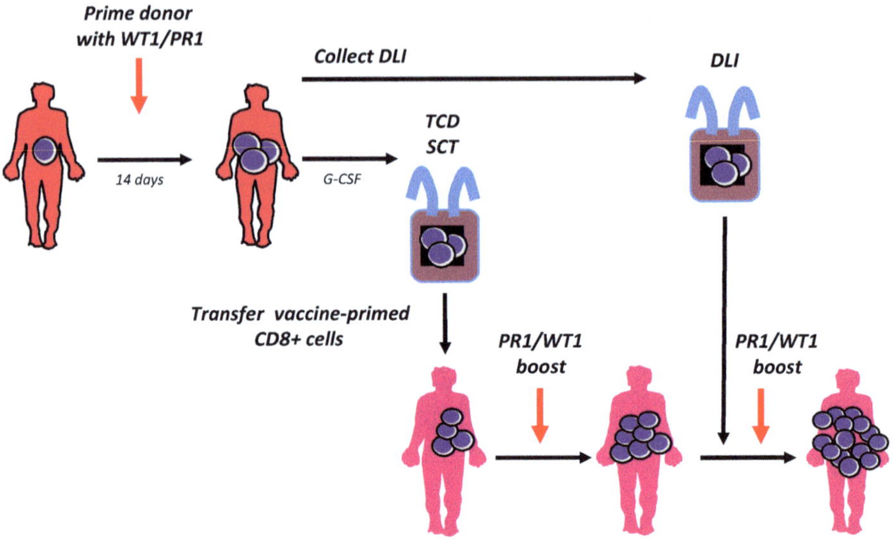

Fig. 41.4 Potential immunotherapeutic approaches to prevent leukemia relapse after allogeneic SCT. As shown in this schema, leukemia vaccines can be administered to the donor prior to stem cell apheresis and leukapheresis for DLI acquisition to achieve an enriched cell product containing T-cells already primed to leukemia antigens. After administration of the primed DLI product to the patient, these T-cells could be further boosted in vivo by administering the leukemia vaccine to the patient. Such a combination strategy might result in an enhanced proliferation and activation of leukemia-specific CTLs to eradicate residual leukemic cells

Conclusions

While immunotherapeutic approaches for AML show promise, these strategies are rarely successful when treating patients with a large disease burden. Recent clinical approaches have therefore focused on applying immunotherapy to prevent relapse or treat minimal residual disease after either chemotherapy or SCT. Future developments combining immunotherapy with conventional chemotherapy or induction treatment may prove more successful than immunotherapy alone. Targeting multiple tumor antigens might also allow the production of a more generalized immunotherapeutic cell product and a wider applicability as well as facilitating off-the-shelf administration of cell products. Another approach would be to prime the allogeneic donor with leukemia peptide vaccines prior to stem cell apheresis (Fig. 41.4). Finally, future applications will combine multiple immunomodulatory treatment approaches, such as: (1) T-cell therapies with epigenetically modifying drugs to upregulate tumor antigen expression prior to T-cell transfer, (2) adoptive T cell transfer with vaccine strategies to boost the immune response, and (3) T-cell therapies with antibodies. Hence, the ultimate goal will be to increase the targeted response against the leukemia cells using combination immune-based therapies, small molecules, and immune modulating agents to decrease bystander toxicity and improve outcomes for all patients with leukemia.

References

Adams SP, Sahota SS, Mijovic A et al (2002) Frequent expression of HAGE in presentation chronic myeloid leukaemias. Leukemia 16:2238–2242

Amrolia PJ, Mucioli-Casadei G, Huls H et al (2005) Add-back of allodepleted donor T cells to improve immune reconstitution after haplo-identical stem cell transplantation. Cytotherapy 7:116–125

Amrolia PJ, Muccioli-Casadei G, Huls H et al (2006) Adoptive immunotherapy with allodepleted donor T-cells improves immune reconstitution after haploidentical stem cell transplantation. Blood 108:1797–1808

Anguille S, Lion E, Smits E, Berneman ZN, Van Tendeloo VFI (2011) Dendritic cell vaccine therapy for acute myeloid leukemia: questions and answers. Hum Vaccin 7:579–584

Barrett AJ, Rezvani K (2007) Translational mini-review series on vaccines: peptide vaccines for myeloid leukaemias. Clin Exp Immunol 148:189–198

Berger C, Blau CA, Huang ML et al (2004) Pharmacologically regulated Fas-mediated death of adoptively transferred T cells in a nonhuman primate model. Blood 103:1261–1269

Berger C, Flowers ME, Warren EH, Riddell SR (2006) Analysis of transgene-specific immune responses that limit the in vivo persistence of adoptively transferred HSV-TK-modified donor T cells after allogeneic hematopoietic cell transplantation. Blood 107:2294–2302

Bethge WA, Hegenbart U, Stuart MJ et al (2004) Adoptive immunotherapy with donor lymphocyte infusions after allogeneic hematopoietic cell transplantation following nonmyeloablative conditioning. Blood 103:790–795

Bleakley M, Otterud BE, Richardt JL et al (2010) Leukemia-associated minor histocompatibility antigen discovery using T-cell clones isolated by in vitro stimulation of naive CD8+ T cells. Blood 115:4923–4933

Bonini C, Ferrari G, Verzeletti S et al (1997) HSV-TK gene transfer into donor lymphocytes for control of allogeneic graft versus leukemia. Science 276:1719–1724

Bonini C, Grez M, Traversari C et al (2003) Safety of retroviral gene marking with a truncated NGF receptor. Nat Med 9:367–369

Bornhauser M, Thiede C, Platzbecker U et al (2011) Prophylactic transfer of BCR-ABL-, PR1-, and WT1-reactive donor T cells after T cell-depleted allogeneic hematopoietic cell transplantation in patients with chronic myeloid leukemia. Blood 117:7174–7184

Chaise C, Buchan SL, Rice J et al (2008) DNA vaccination induces WT1-specific T-cell responses with potential clinical relevance. Blood 112:2956–2964

Cho BK, Rao VP, Ge Q, Eisen HN, Chen J (2000) Homeostasis-stimulated proliferation drives naive T cells to differentiate directly into memory T cells. J Exp Med 192:549–556

Ciceri F, Bonini C, Marktel S et al (2007) Antitumor effects of HSV-TK-engineered donor lymphocytes after allogeneic stem-cell transplantation. Blood 109:4698–4707

Clarke SR, Rudensky AY (2000) Survival and homeostatic proliferation of naive peripheral CD4+ T cells in the absence of self peptide:MHC complexes. J Immunol 165:2458–2464

Cruz CR, Gerdemann U, Leen AM et al (2011) Improving T-cell therapy for relapsed EBV-negative Hodgkin lymphoma by targeting upregulated MAGE-A. Clin Cancer Res 17:7058–7066

Di Stasi A., Tey, Dotti G et al (2011) Inducible apoptosis as a safety switch for adoptive cell therapy. N Engl J Med 365:1673–1683

Dudley ME, Wunderlich JR, Robbins PF et al (2002) Cancer regression and autoimmunity in patients after clonal repopulation with antitumor lymphocytes. Science 298:850–854

Falkenburg JH, Wafelman AR, Joosten P et al (1999) Complete remission of accelerated phase chronic myeloid leukemia by treatment with leukemia-reactive cytotoxic T lymphocytes. Blood 94:1201–1208

Falkenburg JH, van de Corp, Marijt EW, Willemze R (2003) Minor histocompatibility antigens in human stem cell transplantation. Exp Hematol 31:743–751

Fujiki F, Oka Y, Kawakatsu M et al (2010) A clear correlation between WT1-specific Th response and clinical response in WT1 CTL epitope vaccination. Anticancer Res 30:2247–2254

Gale RP, Opelz G (2012) Commentary: does immune suppression increase risk of developing acute myeloid leukemia? Leukemia 26:422–423

Gannage M, Abel M, Michallet AS et al (2005) Ex vivo characterization of multiepitopic tumor-specific CD8 T cells in patients with chronic myeloid leukemia: implications for vaccine development and adoptive cellular immunotherapy. J Immunol 174:8210–8

Ge X, Brown J, Sykes M, Boussiotis VA (2008) CD134-allodepletion allows selective elimination of alloreactive human T cells without loss of virus-specific and leukemia-specific effectors. Biol Blood Marrow Transplant 14:518–530

Gerdemann U, Katari U, Christin AS et al (2011) Cytotoxic T lymphocytes simultaneously targeting multiple tumor-associated antigens to treat EBV negative lymphoma. Mol Ther 19:2258–2268

Ghorashian S, Nicholson E, Stauss HJ (2011) T cell gene-engineering to enhance GVT and suppress GVHD. Best Pract Res Clin Haematol 24:421–433

Goodyear O, Agathanggelou A, Novitzky-Basso I et al (2010) Induction of a CD8+ T-cell response to the MAGE cancer testis antigen by combined treatment with azacitidine and sodium valproate in patients with acute myeloid leukemia and myelodysplasia. Blood 116:1908–1918

Graf C, Heidel F, Tenzer S et al (2007) A neoepitope generated by an FLT3 internal tandem duplication (FLT3-ITD) is recognized by leukemia-reactive autologous CD8+ T cells. Blood 109:2985–2988

Greiner J, Ringhoffer M, Taniguchi M et al (2004) mRNA expression of leukemia-associated antigens in patients with acute myeloid leukemia for the development of specific immunotherapies. Int J Cancer 108:704–711

Greiner J, Li L, Ringhoffer M et al (2005) Identification and characterization of epitopes of the receptor for hyaluronic acid-mediated motility (RHAMM/CD168) recognized by CD8+ T cells of HLA-A2-positive patients with acute myeloid leukemia. Blood 106:938–945

Greiner J, Schmitt A, Giannopoulos K et al (2010) High-dose RHAMM-R3 peptide vaccination for patients with acute myeloid leukemia, myelodysplastic syndrome and multiple myeloma. Haematologica 95:1191–1197

Griffioen M, van Egmond EH, Kester MG et al (2009) Retroviral transfer of human CD20 as a suicide gene for adoptive T-cell therapy. Haematologica 94:1316–1320

Griffioen M, Honders MW, van der Meijden ED et al (2012) Identification of 4 novel HLA-B*40:01 restricted minor histocompatibility antigens and their potential as targets for graft-versus-leukemia reactivity. Haematologica 97: 1196–1204

Hartwig UF, Nonn M, Khan S et al (2008) Depletion of alloreactive donor T lymphocytes by CD95-mediated activation-induced cell death retains antileukemic, antiviral, and immunoregulatory T cell immunity. Biol Blood Marrow Transplant 14:99–109

Houtenbos I, Westers TM, Ossenkoppele GJ, van de Loosdrecht AA (2006) Leukemia-derived dendritic cells: towards clinical vaccination protocols in acute myeloid leukemia. Haematologica 91:348–355

Keilholz U, Letsch A, Busse A et al (2009) A clinical and immunologic phase 2 trial of Wilms tumor gene product 1 (WT1) peptide vaccination in patients with AML and MDS. Blood 113:6541–6548

Kolb H-J, Schattenberg A, Goldman JM et al (1995) Graft-versus-leukemia effect of donor lymphocyte infusions in marrow grafted patients. Blood 86:2041–2050

Kremser A, Dressig J, Grabrucker C et al (2010) Dendritic cells (DCs) can be successfully generated from leukemic blasts in individual patients with AML or MDS: an evaluation of different methods. J Immunother 33:185–199

Krupica T, Jr., Fry TJ, Mackall CL (2006) Autoimmunity during lymphopenia: a two-hit model. Clin Immunol 120:121–128

Levine BL, Ueda Y, Craighead N, Huang ML, June CH (1995) CD28 ligands CD80 (B7-1) and CD86 (B7-2) induce long-term autocrine growth of CD4+ T cells and induce similar patterns of cytokine secretion in vitro. Int Immunol 6:891–904

Levine JE, Braun T, Penza SL et al (2002) Prospective trial of chemotherapy and donor leukocyte infusions for relapse of advanced myeloid malignancies after allogeneic stem-cell transplantation. J Clin Oncol 20:405–412

Levine JE, Barrett AJ, Zhang MJ et al (2008) Donor leukocyte infusions to treat hematologic malignancy relapse following allo-SCT in a pediatric population. Bone Marrow Transplant 42:201–205

Martinez A, Olarte I, Mergold MA et al (2007) mRNA expression of MAGE-A3 gene in leukemia cells. Leuk Res 31:33–37

Maslak PG, Dao T, Krug LM et al (2010) Vaccination with synthetic analog peptides derived from WT1 oncoprotein induces T-cell responses in patients with complete remission from acute myeloid leukemia. Blood 116:171–179

McIver ZA, Melenhorst JJ, Grim A et al (2011) Immune reconstitution in recipients of photodepleted HLA-identical sibling donor stem cell transplantations: T cell subset frequencies predict outcome. Biol Blood Marrow Transplant 17:1846–1854

Meij P, Jedema I, van der Hoorn MA et al (2012) Generation and administration of HA-1-specific T-cell lines for the treatment of patients with relapsed leukemia after allogeneic stem cell transplantation: a pilot study. Haematologica 97:1205–1208

Melenhorst JJ, Scheinberg P, Chattopadhyay PK et al (2009) High avidity myeloid leukemia-associated antigen-specific CD8+ T cells preferentially reside in the bone marrow. Blood 113:2238–2244

Meloni G, Foa R, Vignetti M et al (1994) Interleukin-2 may induce prolonged remissions in advanced acute myelogenous leukemia. Blood 84:2158–2163

Mielke S, Nunes R, Rezvani K et al (2008) A clinical-scale selective allodepletion approach for the treatment of HLA-mismatched and matched donor-recipient pairs using expanded T lymphocytes as antigen-presenting cells and a TH9402-based photodepletion technique. Blood 111:4392–4402

Mielke S, McIver ZA, Shenoy A et al (2011) Selectively T cell-depleted allografts from HLA-matched sibling donors followed by low-dose posttransplantation immunosuppression to improve transplantation outcome in patients with hematologic malignancies. Biol Blood Marrow Transplant 17:1855–1861

Molldrem J, Dermime S, Parker K et al (1996) Targeted T-cell therapy for human leukemia: cytotoxic T lymphocytes specific for a peptide derived from Proteinase 3 preferentially lyse human myeloid leukemia cells. Blood 88:2450–2457

Molldrem JJ, Lee PP, Wang C et al (2000) Evidence that specific T lymphocytes may participate in the elimination of chronic myelogenous leukemia. Nat Med 6:1018–1023

Nagai K, Ochi T, Fujiwara H et al (2012) Aurora kinase A-specific T-cell receptor gene transfer redirects T lymphocytes to display effective antileukemia reactivity. Blood 119:368–376

Nicholson E, Ghorashian S, Stauss H (2012) Improving TCR Gene Therapy for Treatment of Haematological Malignancies. Adv Hematol. Epub ahead of print

Ochi T, Fujiwara H, Suemori K et al (2009) Aurora-A kinase: a novel target of cellular immunotherapy for leukemia. Blood 113:66–74

Ohminami H, Yasukawa M, Fujita S (2000) HLA class I-restricted lysis of leukemia cells by a CD8(+) cytotoxic T-lymphocyte clone specific for WT1 peptide. Blood 95:286–293

Oka Y, Tsuboi A, Taguchi T et al (2004) Induction of WT1 (Wilms' tumor gene)-specific cytotoxic T lymphocytes by WT1 peptide vaccine and the resultant cancer regression. Proc Natl Acad Sci U S A 101:13885–13890

Perruccio K, Topini F, Tosti A et al (2008) Photodynamic purging of alloreactive T cells for adoptive immunotherapy after haploidentical stem cell transplantation. Blood Cells Mol Dis 40:76–83

Perruccio K, Topini F, Tosti A et al (2011) Optimizing a photoallodepletion protocol for adoptive immunotherapy after haploidentical SCT. Bone Marrow Transplant. Epub ahead of print

Pinilla-Ibarz J, Cathcart K, Korontsvit T et al (2000) Vaccination of patients with chronic myelogenous leukemia with bcr-abl oncogene breakpoint fusion peptides generates specific immune responses. Blood 95:1781–1787

Porter DL, Orloff GJ, Antin JH (1994) Donor mononuclear cell infusions as therapy for B-cell lymphoproliferative disorder following allogeneic bone marrow transplant. Transplant Sci 4:12–14

Porter DL, Collins RH, Jr., Shpilberg O et al (1999) Long-term follow-up of patients who achieved complete remission after donor leukocyte infusions. Biol Blood Marrow Transplant 5:253–261
Porter DL, Collins RH, Hardy C et al (2000) Treatment of relapsed leukemia after unrelated donor marrow transplantation with unrelated donor leukocyte infusions [In Process Citation]. Blood 95:1214–1221
Porter DL, Levine BL, Bunin N et al (2006) A phase 1 trial of donor lymphocyte infusions expanded and activated ex vivo via CD3/CD28 costimulation. Blood 107:1325–1331
Qazilbash MH, Saliba RM, Hosing C et al (2007) Autologous stem cell transplantation is safe and feasible in elderly patients with multiple myeloma. Bone Marrow Transplant 39:279–283
Quintarelli C, Dotti G, De AB et al (2008) Cytotoxic T lymphocytes directed to the preferentially expressed antigen of melanoma (PRAME) target chronic myeloid leukemia. Blood 112:1876–1885
Quintarelli C, Dotti G, Hasan ST et al (2011) High-avidity cytotoxic T lymphocytes specific for a new PRAME-derived peptide can target leukemic and leukemic-precursor cells. Blood 117:3353–3362
Rezvani K (2008) PR1 vaccination in myeloid malignancies. Expert Rev Vaccines 7:867–875
Rezvani K (2011) Peptide vaccine therapy for leukemia. Int J Hematol 93:274–280
Rezvani K, Barrett AJ (2008) Characterizing and optimizing immune responses to leukaemia antigens after allogeneic stem cell transplantation. Best Pract Res Clin Haematol 21:437–453
Rezvani K, Grube M, Brenchley JM et al (2003) Functional leukemia-associated antigen-specific memory CD8+ T cells exist in healthy individuals and in patients with chronic myelogenous leukemia before and after stem cell transplantation. Blood 102:2892–2900
Rezvani K, Yong AS, Mielke S et al (2008) Leukemia-associated antigen-specific T-cell responses following combined PR1 and WT1 peptide vaccination in patients with myeloid malignancies. Blood 111:236–242
Rezvani K, Yong AS, Tawab A et al (2009) Ex vivo characterization of polyclonal memory CD8+ T-cell responses to PRAME-specific peptides in patients with acute lymphoblastic leukemia and acute and chronic myeloid leukemia. Blood 113:2245–2255
Rezvani K, Yong AS, Mielke S et al (2011) Repeated PR1 and WT1 peptide vaccination in Montanide-adjuvant fails to induce sustained high-avidity, epitope-specific CD8+ T cells in myeloid malignancies. Haematologica 96:432–440
Rice J, Ottensmeier CH, Stevenson FK (2008) DNA vaccines: precision tools for activating effective immunity against cancer. Nat Rev Cancer 8:108–120
Riddell SR, Greenberg PD (2000) T-cell therapy of cytomegalovirus and human immunodeficiency virus infection. J Antimicrob Chemother 45(T3):35–43
Rooney CM, Smith CA, Ng CY et al (1998) Infusion of cytotoxic T cells for the prevention and treatment of Epstein-Barr virus-induced lymphoma in allogeneic transplant recipients. Blood 92:1549–1555
Rosenblatt J, Avigan D (2006) Can leukemia-derived dendritic cells generate antileukemia immunity? Expert Rev Vaccines 5:467–472
Royer PJ, Bougras G, Ebstein F et al (2008) Efficient monocyte-derived dendritic cell generation in patients with acute myeloid leukemia after chemotherapy treatment: application to active immunotherapy. Exp Hematol 36:329–339
Scheibenbogen C, Letsch A, Thiel E et al (2002) CD8 T-cell responses to Wilms tumor gene product WT1 and proteinase 3 in patients with acute myeloid leukemia. Blood 100:2132–2137
Schlegel P, Teltschik HM, Pfeiffer M et al (2011) Long-term IL-2 therapy after transplantation of T cell depleted stem cells from alternative donors in children. Best Pract Res Clin Haematol 24:443–452
Schluns KS, Lefrancois L (2003) Cytokine control of memory T-cell development and survival. Nat Rev Immunol 3:269–279
Schmid C, Labopin M, Nagler A et al (2007) Donor lymphocyte infusion in the treatment of first hematological relapse after allogeneic stem-cell transplantation in adults with acute myeloid leukemia: a retrospective risk factors analysis and comparison with other strategies by the EBMT Acute Leukemia Working Party. J Clin Oncol 25:4938–4945

Schmitt M, Li L, Giannopoulos K et al (2006) Chronic myeloid leukemia cells express tumor-associated antigens eliciting specific CD8+ T-cell responses and are lacking costimulatory molecules. Exp Hematol 34:1709–1719

Schmitt M, Schmitt A, Rojewski MT et al (2008) RHAMM-R3 peptide vaccination in patients with acute myeloid leukemia, myelodysplastic syndrome, and multiple myeloma elicits immunologic and clinical responses. Blood 111:1357–1365

Schreiber RD, Old LJ, Smyth MJ (2011) Cancer immunoediting: integrating immunity's roles in cancer suppression and promotion. Science 331:1565–1570

Slavin S, Naparstek E, Nagler A et al (1996) Allogeneic cell therapy with donor peripheral blood cells and recombinant human Interleukin-2 to treat leukemia relapse after allogeneic bone marrow transplantation. Blood 87:2195–2204

Sloand EM, Melenhorst JJ, Tucker ZC et al (2011) T-cell immune responses to Wilms tumor 1 protein in myelodysplasia responsive to immunosuppressive therapy. Blood 117:2691–2699

Soiffer RJ, Murray C, Gonin R, Ritz J (1994) Effect of low-dose interleukin-2 on disease relapse after T-cell-depleted allogeneic bone marrow transplantation. Blood 84:964–971

Solomon SR, Mielke S, Savani BN et al (2005) Selective depletion of alloreactive donor lymphocytes: a novel method to reduce the severity of graft-versus-host disease in older patients undergoing matched sibling donor stem cell transplantation. Blood 106:1123–1129

Stauss HJ, Cesco-Gaspere M, Thomas S et al (2007) Monoclonal T-cell receptors: new reagents for cancer therapy. Mol Ther 15:1744–1750

Straathof KC, Pule MA, Yotnda P et al (2005) An inducible caspase 9 safety switch for T-cell therapy. Blood 105:4247–4254

Stuehler C, Mielke S, Chatterjee M et al (2009) Selective depletion of alloreactive T cells by targeted therapy of heat shock protein 90: a novel strategy for control of graft-versus-host disease. Blood 114:2829–2836

Stumpf AN, van der Meijden ED, van Bergen CA et al (2009) Identification of 4 new HLA-DR-restricted minor histocompatibility antigens as hematopoietic targets in antitumor immunity. Blood 114:3684–3692

Takami A, Okumura H, Yamazaki H et al (2005) Prospective trial of high-dose chemotherapy followed by infusions of peripheral blood stem cells and dose-escalated donor lymphocytes for relapsed leukemia after allogeneic stem cell transplantation. Int J Hematol 82:449–455

Tang W, Zhou Q, Jin Z et al (2011) Novel Therapy with Interferon-a in Combination with Donor Lymphocyte Infusion for High Risk Acute Leukemia Patients Who Relapsed After Allogeneic Hematopoietic Stem Cell Transplantation [abstract]. Blood 118:658

Tey SK, Dotti G, Rooney CM, Heslop HE, Brenner MK (2007) Inducible caspase 9 suicide gene to improve the safety of allodepleted T cells after haploidentical stem cell transplantation. Biol Blood Marrow Transplant 13:913–924

Thomis DC, Marktel S, Bonini C et al (2001) A Fas-based suicide switch in human T cells for the treatment of graft-versus-host disease. Blood 97:1249–1257

Traversari C, Marktel S, Magnani Z et al (2007) The potential immunogenicity of the TK suicide gene does not prevent full clinical benefit associated with the use of TK-transduced donor lymphocytes in HSCT for hematologic malignancies Blood 109:4708–4715

Van T, V, Van D, V, Van DA et al (2010) Induction of complete and molecular remissions in acute myeloid leukemia by Wilms' tumor 1 antigen-targeted dendritic cell vaccination. Proc Natl Acad Sci U S A 107:13824–13829

Van DA, Van D, V, Nijs G et al (2009) Clinical-grade manufacturing of autologous mature mRNA-electroporated dendritic cells and safety testing in acute myeloid leukemia patients in a phase I dose-escalation clinical trial. Cytotherapy 11:653–668

Walter EA, Greenberg PD, Gilbert MJ et al (1995) Reconstitution of cellular immunity against cytomegalovirus in recipients of allogeneic bone marrow by transfer of T-cell clones from the donor. N Engl J Med 333:1038–1044

Warren EH, Fujii N, Akatsuka Y et al (2010) Therapy of relapsed leukemia after allogeneic hematopoietic cell transplantation with T cells specific for minor histocompatibility antigens. Blood 115:3869–3878

Weber G, Karbach J, Kuci S et al (2009) WT1 peptide-specific T cells generated from peripheral blood of healthy donors: possible implications for adoptive immunotherapy after allogeneic stem cell transplantation. Leukemia 23:1634–1642

Weber G, Gerdemann U, Caruana I, Savoldo B, Hensel NF, Rabin KR, Shpall EJ, Melenhorst JJ, Leen AM, Barrett AJ, Bollard CM (2013a) Generation of multi-leukemia antigen-specific T cells to enhance the graft-versus-leukemia effect after allogeneic stem cell transplant. Leukemia 27(7):1538–1547

Weber G, Caruana I, Reuce RH, Barrett AJ, Gerdemann U, Leen AM, Rabin KR, Bollard CM (2013b) Generation of tumor antigen-specific T cell lines from pediatric patients with acute lymphoblastic leukemia-implications for immunotherapy. Clinical Cancer Research 19(18):5079–5091

Williams KM, Hakim FT, Gress RE (2007) T cell immune reconstitution following lymphodepletion. Semin Immunol 19:318–330

Wrzesinski C, Paulos CM, Kaiser A et al (2010) Increased intensity lymphodepletion enhances tumor treatment efficacy of adoptively transferred tumor-specific T cells. J Immunother 33:1–7

Xue SA, Gao L, Thomas S et al (2010) Development of a Wilms' tumor antigen-specific T-cell receptor for clinical trials: engineered patient's T cells can eliminate autologous leukemia blasts in NOD/SCID mice. Haematologica 95:126–134

Yong AS, Rezvani K, Savani BN et al (2007) High PR3 or ELA2 expression by CD34+ cells in advanced-phase chronic myeloid leukemia is associated with improved outcome following allogeneic stem cell transplantation and may improve PR1 peptide-driven graft-versus-leukemia effects. Blood 110:770–775

Yong AS, Stephens N, Weber G et al (2011) Improved outcome following allogeneic stem cell transplantation in chronic myeloid leukemia is associated with higher expression of BMI-1 and immune responses to BMI-1 protein. Leukemia 25:629–637

Zeng Y, Feng H, Graner MW, Katsanis E (2003) Tumor-derived, chaperone-rich cell lysate activates dendritic cells and elicits potent antitumor immunity. Blood 101:4485–4491

Chapter 42
Indications for Hematopoietic Transplantation for AML

Sairah Ahmed and Richard Champlin

Abstract Allogeneic hematopoietic cell transplantation (AHCT) is an effective treatment for AML. This approach has been extensively evaluated in patients with AML and is indicated for patients in first remission with intermediate- to high-risk features, those with primary refractory disease or early relapse. Patients have traditionally received transplants from a human leukocyte antigen (HLA)-matched sibling donor. Improvements in HLA typing allowing identification of complete allele level matches have improved the results of unrelated donor transplants and results are now comparable with matched siblings. Recently, there have been major advances using HLA-nonidentical donors including haploidentical, one antigen mismatched unrelated donors or cord blood units, and an acceptable donor can now be identified for nearly all patients in need of a hematopoietic transplant.

In this chapter, we will review the indications for transplantation in acute myelogenous leukemia:

- First complete remission; consideration for risk categories for relapse:
  - Favorable risk acute myeloid leukemia (AML)
  - Intermediate-risk AML
  - Unfavorable-risk AML
  - Therapy-related AML
- Transplant in relapsed or refractory settings
- Autologous versus allogeneic transplantation
- Allogeneic donor categories
- Relapse post transplant, consideration for second allogeneic transplant
- Reduced intensity conditioning

R. Champlin (✉) · S. Ahmed
Department of Stem Cell Transplantation and Cellular Therapy, The University of Texas MD Anderson Cancer Center, 1515 Holcombe Blvd., Box 423, Houston, TX 77030, USA
e-mail: rchampli@mdanderson.org

S. Ahmed
e-mail: sahmed3@mdanderson.org

M. Andreeff (ed.), *Targeted Therapy of Acute Myeloid Leukemia*,
Current Cancer Research, DOI 10.1007/978-1-4939-1393-0_42

Keywords Allogeneic hematopoietic cell transplantation · Myeloablative · Nonmyeloablative · Core binding factor acute leukemia · Haploidentical transplant · Umbilical cord blood transplant · Secondary leukemia · Alternative donor groups

42.1 Introduction

Acute myeloid leukemia (AML) is a clonal disorder, with the leukemia cells in an individual patient descending from a common transformed cell. According to the SEER database from 2004 to 2008, the median age at diagnosis for AML is 67 years (Howlader et al. 2011). Approximately 30 % were under the age of 54, while 69.3 % of people diagnosed were above the age of 55 years (Howlader et al. 2011). It is estimated that 12,950 men and women (6830 men and 6120 women) will be diagnosed with and 9050 men and women will die of AML annually.

Allogeneic hematopoietic cell transplantation (AHCT) is an effective treatment for AML. This approach has been extensively evaluated in patients with AML and is indicated for patients in first remission with intermediate- to high-risk features, those with primary refractory disease or early relapse (Hubel et al. 2011). Patients have traditionally received transplants from an HLA-matched sibling donor. Improvements in HLA typing allowing identification of complete allele level matches have improved the results of unrelated donor transplants and results are now comparable with matched siblings. Recently, there have been major advances using HLA-nonidentical donors including haploidentical, one antigen mismatched unrelated donors or cord blood units, and an acceptable donor can now be identified for nearly all patients in need of a hematopoietic transplant (Barker et al. 2010; Schlenk et al. 2010). There has been progressive improvement in supportive care and this has resulted in significantly lower treatment-related mortality (TRM) (Gooley et al. 2010; Horan et al. 2011).

Initial studies of hematopoietic transplantation utilized high-dose myeloablative therapy. This can be followed by autologous or allogeneic hematopoietic stem cell transplantation (HSCT) to restore hematopoiesis. Myeloablative therapy is associated with considerable toxicity and the procedure needed to be restricted to young, medically fit patients. Much of the benefit of allogeneic hematopoietic transplantation is due to the immune-related graft-versus-leukemia effect (Horowitz et al. 1990; Truitt and Johnson et al. 1995). Recently, reduced intensity or nonmyeloablative preparative regimens have been utilized to achieve engraftment with little toxicity and allow graft-versus-malignancy effects to occur (Champlin et al. 2000a; Slavin et al. 1998; McSweeney et al. 1999). This has allowed treatment of older patients as well as those with comorbidities (Alatrash et al. 2011); survival is similar as with myeloablative transplants (Gyurkocza et al. 2010).

We review the indications for transplantation in acute myelogenous leukemia:

- First complete remission (CR); consideration for risk categories for relapse:
  - Favorable-risk AML
  - Intermediate-risk AML

Table 42.1 National Comprehensive Cancer Network (NCCN) consensus cytogenetic and molecular risk categories for acute myelogenous leukemia

Risk status	Cytogenetics	Molecular abnormalities
Favorable risk	Inv(16) or t(16;16) t(8;21) t(15;17)	Normal cytogenetics: With NPM1 mutation or isolated CEBPA mutation in the absence of FLT3-ITD
Intermediate risk	Normal cytogenetics+8 t(9;11)	t(8;21), inv (16), t(16;16): with c-KIT mutation
Unfavorable risk	Complex (≥3 clonal chromosomal abnormalities) −5, 5q−, −7, 7q− 11q23—non t(9;11) inv(3), t(3;3) t(6;9) t(9;22)	Normal cytogenetics with FLT3-ITD mutation

  - Unfavorable-risk AML
  - Therapy-related AML
- Transplant in relapsed or refractory settings
- Autologous versus allogeneic transplantation
- Allogeneic donor categories
- Relapse post transplant, consideration for second allogeneic transplant
- Reduced intensity conditioning

42.2 Classification of AML

Historically, AML has been organized by the French–American–British (FAB) system which was the first consistent morphologic and cytochemical outline for leukemia and has been widely used since it was published in 1976 (Bennett et al. 1976). In 2002 and again in 2008, the World Health Organization (WHO) modified the FAB classification to incorporate cytogenetic and molecular findings in addition to morphologic, cytochemical, and immunophenotypic features (Vardiman et al. 2009). According to the WHO, AML is classified into four groups, including secondary AML with preceding myelodysplastic syndrome (MDS) (Vardiman et al. 2009)..

It is now well established that cytogenetic and molecular abnormalities identified at diagnosis are the most important predictive factor for outcome in AML patients (Byrd et al. 2002; Grimwade et al. 2010). On the basis of specific cytogenetic and molecular abnormalities, patients can be divided into favorable-risk, intermediate-risk, and adverse-risk groups which is important for clinical decision making to determine if the patient is best served with a hematopoietic transplant or standard-dose consolidation chemotherapy (Slovak et al. 2000; Dohner et al. 2010) (Table 42.1).

42.3 Favorable-Risk AML

Favorable-risk AML, which accounts for approximately 9% (Dohner et al. 2010) of all AML patients, is defined as those patients with core binding factor (CBF) leukemia, namely, t[8;21], inv[16] or t[16;16], or those patients with acute promyelocytic leukemia (APL) with t(15;17). The t(8;21) abnormality seen in 8% of adult AML cases results in fusion of the gene encoding CBF-α on chromosome 8 with a gene (MTG8) on chromosome 21, while inv(16) is seen in approximately 7% of AMLs, and results in the fusion of the gene encoding CBF-β to the smooth muscle myosin heavy chain gene (Frederick et al. 2011). Patients with CBF-related AML have a favorable prognosis and the majority can achieve long-term remission with standard-dose chemotherapy; allogeneic hematopoietic transplants have not improved survival in these patients (Slovak et al. 2000; Koreth et al. 2009). Post-remission therapy with multiple cycles of HiDAC have shown a beneficial effect on CBF AML, although the exact number of cycles is still an open question (Byrd et al. 2002; Dohner et al. 2010). Currently, we do not proceed to allogeneic transplant in first CR for patients with CBF AML, or APL, the exception being KIT mutations (mKIT) in CBF AML which will be discussed in detail further. Autologous hematopoietic transplants have been successful in achieving durable remissions in a large fraction of patients, but controlled studies are needed to definitively determine its role.

APL accounts for 10–15% of patients diagnosed with AML (Tallman and Altman 2009). The remission rate and disease-free survival (DFS) for APL is very high with more than 70% of patients cured with ATRA-based treatment. There appears to be no role for HSCT in patients who are in a first molecular remission if achieved by the end of consolidation therapy. However, hematopoietic stem cell transplant could be considered in those patients who remain PCR positive as an approach to achieve molecular second CR by using salvage therapy followed by an autologous stem cell transplant, this can be expected to result in a 5-year DFS or relapse-free survival of 70–80% (Tallman and Altman 2009; Linker et al 2009; Dvorak et al 2008). An allogeneic approach is largely recommended if a second CR is unattainable due to the success of using arsenic as a salvage therapy (Tallman and Altman 2009; Botton and Chevret 2005).

42.4 Intermediate- and Unfavorable-Risk AML by Cytogenetics

The unfavorable-risk group, which accounts for 30% (Dohner et al. 2010) of all AML patients, can be categorized into complex karyotype, monosomal karyotype, and uncommon genetic rearrangements with poor prognoses (Vardiman et al. 2009). Complex karyotype, which occurs in 10–12% of patients, has consistently been associated with a very poor outcome (Dohner et al. 2010; Mrozek et al. 2008).

Complex karyotype has been defined as the presence of three or more (in some studies ≥ 5) chromosome abnormalities in the absence of t(8;21), inv(16) or t(16;16), and t(15;17). Monosomal karyotype is defined as two or more distinct autosomal monosomies or a single autosomal monosomy in the presence of other structural abnormalities and has an extremely poor prognosis (Medeiros et al. 2010) and had a 4-year overall survival (OS) of 4%, compared with a 21% 4-year OS in patients with other unfavorable karyotypes but without monosomal karyotype. In data from Medeiros et al., the only patients who were alive and disease free more than 6 years from diagnosis were those treated with allogeneic hematopoietic cell transplantation, although the study was not designed to test the utility of any specific therapy (Medeiros et al. 2010). With conventional consolidation chemotherapy, patients with unfavorable cytogenetics have dismal outcomes, with OS at 2 years of approximately 20% (Schlenk et al. 2010; Slovak et al. 2000; Bloomfield et al. 1998).

AML patients without favorable or unfavorable cytogenetic aberrations are classified as the intermediate-risk cytogenetic subclass. Typically, this group includes cytogenetically normal disease and accounts for approximately 50% of all AML patients. It is within this group that great progress has been made with molecular testing to further stratify risk.

An allogeneic HSCT has been found to improve leukemia-free survival for patients with intermediate and unfavorable-risk cytogenetics in CR1 based on extensive data from both multiple studies (Cornelissen et al. 2007; Suciu et al. 2003; Jourdan et al. 2005) as well as from meta-analyses (Yanada et al. 2005). Cornelissen et al. (2007) performed a meta-analysis based on published data including more than 4000 patients with AML in first remission enrolled in the HOVON/SAKK studies, the Medical Research Council (MRC) studies, the trials of the European Organization for Research and Treatment of Cancer (EORTC), and the BGMT studies and revealed that although there is substantial TRM in the HSCT group compared with chemotherapy (21 vs. 4%, $P<0.001$), the DFS was significantly better in the HSCT group 48% versus 37. These studies compared patients with an HLA-matched related donor versus those with no matched donor available (Cornelissen et al. 2007). Overall, transplant from a family donor statistically significantly prolonged DFS in patients with intermediate- and high-risk profiles, but not in patients with favorable-risk disease. In the EORTC AML 10 trial, these results were confirmed with patients in the high-risk group receiving matched related donor (MRD) allogeneic transplants and the OS at 4 years (50.2%) was significantly better than after chemotherapy or autologous transplant (29.4%) (Suciu et al. 2003). Earlier data without molecular information from Slovak et al. confirmed an OS benefit of allogeneic transplant over autologous transplant for intermediate-risk disease; however, there was no benefit over consolidation chemotherapy with both arms having a 50% OS at 5 years (Slovak et al. 2000). Although there continues to be some controversy regarding management of patients with intermediate-risk AML, we recommend allogeneic transplant in first CR from either a matched related or matched unrelated donor (Hubel et al. 2011). There is a general consensus that the optimal treatment for unfavorable-risk AML in first CR is an allogeneic stem cell transplant. Promising results have been seen with cord blood transplants and haploidentical transplants in patients with intermediate- or high-risk AML in CR1.

Koreth et al. reported in a meta-analysis across 14 trials a statistically significant survival benefit with allogeneic transplant in intermediate-risk AML with a combined HR of 0.83 (95% CI, 0.74–0.93). These trials primarily used matched sibling donors with genetic randomization (Koreth et al. 2009).

42.5 Unfavorable-/Favorable-Risk AML by Molecular Testing

Within the normal cytogenetic AML group, there have been significant advances in defining molecular abnormalities involved with pathogenesis and prognosis which have impacted treatment guidelines. In this group, molecular techniques are useful to detect somatic mutations of several genes such as internal tandem duplication (ITD) of fms-related tyrosine kinase 3 (FLT3) gene (Gilliland and Griffin 2002) and D835 (Yamamoto et al. 2001) mutation of FLT3, the nucleophosmin (NPM1) gene (Thiede et al. 2006), and the CCAAT/enhancer binding protein alpha (CEBPA) (Green et al. 2010) and have refined risk stratification. Among these mutations, FLT3-ITD is a powerful adverse prognostic factor which increases the risk of relapse, has a negative impact on survival, and is detected in approximately one third of patients with normal cytogenetics (Moreno et al. 2003). While having no significant impact on response to induction treatment, patients with FLT3 mutations have 2-year OS of approximately 20% (Moreno et al. 2003). The Johns Hopkins group has reported on allogeneic HSCT in CR1 at their institution showing that the survival for patients with FLT3-ITD AML is equivalent to non- FLT3-mutated AML (DeZern et al. 2011). More recently, Brunet et al. describe outcomes of 120 FLT3/ITD-positive patients after matched unrelated donor (MUD) or MRD allogeneic transplant. While the mutation continues to adversely affect the outcome of HSCT similarly as it does after chemotherapy, more than 50% of the patients harboring this mutation who received transplants were alive and leukemia-free at 2 years (Brunet et al. 2012).

Both NPM1 and CEBPA are thought to confer an improved OS in normal cytogenetic AML; however, this is only if there is no mutation in FLT3. Mutations in the CEBPA gene have been described in approximately 10% of patients with AML and CEBPA mutations predict statistically significant better event-free, disease-free, and OS independent of other mutations (Marcucci et al. 2008). However, Green and colleagues reported findings from an analysis of 1427 AML patients confirming only the presence of a CEBPA-double mutation should be used for therapy risk stratification and only in the absence of a FLT3 mutation is there a positive impact on disease-free and OS (Green et al. 2010).

NPM1 gene mutations are among the most frequent molecular alterations in AML with normal karyotype, and generally denote a good response to induction chemotherapy and favorable prognosis with higher CR rates and better event-free survival (EFS; in the absence of a concomitant FLT3-ITD mutation) (Thiede et al. 2006; Falini et al. 2011). As Schlenk reported in 2008 in patients with mutant NPM1 without a FLT3 mutation, there was no OS benefit for an allogeneic transplant due to the increased TRM and this group appeared to have an OS of 60% at 5 years without transplant (Schlenk et al. 2008).

Over the past several years, other novel recurrent somatic mutations have been found that may be important in risk stratification, including RAS (Mardis et al. 2009), TET2 (Delhommeau et al. 2009; Metzeler et al. 2011b), ASXL (Metzeler et al. 2011a), IDH1 or IDH2 (Marcucci et al. 2010; Ward et al. 2010), DNMT3A (Yan et al. 2011), MLL-PTD (Whitman et al. 2012; Caligiuri et al. 1998), and PHF6 (Vlierberghe et al. 2011) although further work needs to be done in order to verify their influence on disease progression. Interestingly, Patel and colleagues recently reported on 398 patients with AML and found DNMT3A and NPM1 mutations and MLL translocations predicted an improved outcome with high-dose induction chemotherapy, however the favorable effect of NPM1 mutations was restricted to patients with co-occurring NPM1 and IDH1 or IDH2 mutations (Patel et al. 2012).

These data are rapidly evolving and there continues to be some controversy regarding the use of these identified mutations for treatment planning. Current guidelines suggest that isolated NPM1 or CEBPA patients should not receive an allogeneic transplant in first CR (Hubel et al. 2011), while patients with FLT3 ITD mutations have a poor prognosis and should receive a transplant in first CR regardless of the donor type (Paun and Lazarus 2012). Further validation of the Patel et al. data may change these recommendations in terms of patients with NPM1 mutations.

42.6 Molecular Abnormalities in CBF AML

The role of mKIT has been investigated to identify a high-risk subset in the otherwise favorable CBF group. Two commonly identified KIT mutations (exon 17(mKIT 17) and exon 8 (mKIT 8)) in CBF AML appear to have prognostic relevance as shown in the data from a Cancer and Leukemia Group B Study (CALGB) by Paschka and colleagues which revealed that CBF AML with mKIT have a higher risk of relapse as well as worse OS despite equal CR rates (Paschka et al. 2006). Currently, these patients are considered intermediate risk and thus candidates for HSCT in CR1 due to the increased risk of relapse although there are no data yet to show benefit of transplant.

42.7 AML in the Elderly and Reduced Intensity Conditioning

AML is predominantly a disease of older adults. Age is an independent adverse prognostic factor in AML and the prognosis of patients greater than 60 years of age remains poor. High-dose myeloablative therapy cannot be considered for the vast majority of these patients due to preparative regimen-related toxicities resulting in a high risk of treatment-related mortality. Reduced intensity conditioning (RIC) has decreased the associated toxicity for older patients as well as patients with organ impairment. The reduced intensity preparative regimen produces sufficient immune suppression to prevent graft rejection and allows the graft-versus-leukemia effect to occur. Reduced intensity regimens such as busulfan–fludarabine or melphalan fludarabine (Alatrash et al. 2011) produce more direct cytoreduction than truly non-myeloablative regimens, such as low-dose total body radiation, but is unclear which regimen is optimal. Graft versus host disease (GVHD) remains a major cause of TRM (Schmid et al. 2012). Addition of alemtuzumab reduces the risk of GVHD, but the risk of relapse is increased and OS is unchanged (Soiffer et al. 2011). Relapse rates between 20 and 55 % have been reported after RIC HSCT in CR (Martino et al. 2006; Shimoni et al. 2005), however most of the information comparing RIC to MAC is retrospective in nature. Ringden et al. retrospectively assessed 401 patients between 1999 and 2005 with AML who were treated with RIC and 1154 patients received MAC with a MUD donor source and their analysis demonstrated that RIC transplants were associated with higher relapse in AML patients younger than 50 years of age and decreased NRM in those ≥50 years compared with MAC, nevertheless LFS was similar after both conditioning regimens, regardless of age (Ringden et al. 2009). The European Society for Blood and Marrow Transplantation (EBMT) looked at registry data for 2815 RIC transplants performed for AML in CR between 1999 and 2008 with the majority of patients being in first CR and found the cumulative incidence of relapse was 32 %. On multivariate analysis the important factors linked with improved OS were: remission after HSCT > 5 months, bone marrow (BM) blasts less than 27 %, and absence of acute GVHD after HSCT (Schmid et al. 2012). A recent study of the CALGB found significantly improved progression-free survival for RIC allogeneic transplantation compared to chemotherapy in patients 60–70 years of age with AML (32 vs. 15 % at 3 years) (Farag et al. 2011) and this approach can be considered a standard of care (Champlin 2011).

42.8 Minimal Residual Disease

Minimal residual disease can be detected by multiparameter flow cytometry or quantitative detection of residual cells expressing aberrant genes or leukemia-related mutations (Kern et al. 2004; Buccisano et al. 2010; Candoni et al. 2011; Kronke et al. 2011; Buccisano et al. 2012). Detection of minimal residual disease identifies patients at highest risk to relapse and allogeneic transplants are clearly indicated in these patients. It is uncertain if this information can be used to identify patients with intermediate- or high-risk prognostic features who could be managed conservatively with conventional chemotherapy. Minimal residual disease is also being studied post transplant as a potential indication for interventions like donor lymphocyte infusion or hypomethylating agent treatment.

42.9 Therapy-Related Leukemia

As more patients survive therapy for cancer after being treated with chemotherapy and ionizing radiation, the rate of therapy-related MDS (t-MDS) and therapy-related AML (t-AML) is increasing (Hubel et al. 2011). t-AML accounts for 10–20% of AML and has a particularly poor prognosis with conventional chemotherapy, with a median survival of 1 year or less (Borthakur and Estey 2007). The poor outcomes associated with t-AML are primarily due to resistance to chemotherapy and BM failure; these patients are often older and have comorbidities which also adversely impact prognosis. Several cytogenetic abnormalities that are associated with unfavorable-risk AML are also common in t-AML including del (7/7q) or a complex karyotype (Armand et al. 2008). Treatment-related AML may be preceded by a myelodysplastic phase; however, t-AML arising from antecedent topoisomerase-II inhibitors typically does not present with MDS and is characterized by balanced translocations involving 11q23 and 21q22 (Borthakur and Estey 2007).

Litzow et al. analyzed outcomes in 545 patients with t-AML receiving allogeneic transplants from 1990 to 2004. Approximately 77% were treated using a myeloablative regimen for conditioning and TRM was 41% while relapse was 27% at 1 year and at 5 years was at 48 and 31%, respectively. Disease-free survival (DFS) was 32% and OS was 37% at 1 year and 21 and 22% at 5 years, respectively. In that retrospective review, they further analyzed risk factors for adverse outcomes and upon multivariate analysis, four risk factors were found to influence DFS and OS: (1) age older than 35 years, (2) poor-risk cytogenetics, (3) t-AML not in remission, and (4) donor other than an HLA-identical sibling or a partially or well-matched unrelated donor (Litzow et al. 2010). Patients without any of these risk factors had an outcome comparable with transplantation for de novo AML/MDS while 5-year survival for subjects with three or four of these risk factors was 10% (5–15) and 4% (0–16), respectively ($P<0.001$) (Litzow et al. 2010).

A recent report from the EBMT described the outcome of allogeneic HCT for t-MDS/t-AML in 461 patients: 308 (67%) had AML and 57% were in CR at the time of transplant. The 3-year relapse-free survival and OS were 33 and 35%, respectively. Significant variables for EFS or OS were older age, abnormal cytogenetics, and not being in CR (Kroger et al. 2009).

Allogeneic HCT remains the only curative therapy for t-AML and considering the poor prognosis of this patient group, HSCT in CR1 should be considered for patients for whom suitable donors can be identified. For those patients with multiple risk factors including disease not in remission, the OS even with a transplant is unpromising.

42.10 Donor Groups

A patient has one in four chance of a given sibling being an HLA-matched donor and a 50% chance of a sibling being a haplo-identical match. Only 30% of patients with an indication of an allogeneic HSCT will have a HLA-identical sibling available to be a donor for HSCT (Ruggeri et al. 2010). Other sources of hematopoietic cells must be identified for the remaining patients.

42.10.1 Matched Related Versus Unrelated Donor Grafts

A HLA A,B, C, DR matched unrelated donor can be identified for approximately half of patients. Initially, treatment-related complications and TRM were worse with unrelated donor transplants (Szydlo et al. 1997), but with the advent of high-resolution allele-level HLA typing, results have improved and are now comparable to those with matched sibling donors (Lee et al. 2007; Saber et al. 2012).

In 2010, Gupta et al. for the US Center for International Blood and Marrow Transplant Research (CIBMTR) reported on 584 patients with unfavorable cytogenetics AML in first remission receiving either matched related or matched unrelated donor grafts and found similar outcomes for both types of transplants (Gupta et al. 2010). In smaller numbers, Walter et al. described comparable outcomes between matched related and unrelated donor graft source in intermediate-risk AML CR1 patients (Paun and Lazarus et al. 2012). Recently, a retrospective analysis of 2223 adult AML patients who underwent allogeneic HCT between 2002 and 2006 was published by Saber and colleagues through the CIBMTR (MRD, $n=624$; 8 of 8 HLA locus matched MUD, $n=1193$; 7 of 8 MUD, $n=406$). Although the incidence of acute graft versus host disease was significantly lower in MRD HSCT recipients than 8 of 8 MUD and 7 of 8 MUD HCT recipients (33, 51, and 53 %, respectively; $p<0.001$), the multivariate analysis revealed that the fully matched (8 of 8) MUD recipients had a similar survival rate compared to MRD HSCT recipients (relative risk (RR), 1.03; $p=0.62$). The 1-antigen mismatched (7 of 8) MUD HCT recipients had higher early mortality than the matched related donor transplants, but beyond 6 months post HSCT, their survival rates were similar (RR, 0.88; $p=0.30$). These results suggest that transplantation from MUD and MRD donors produces similar survival for patients with AML (Saber et al. 2012).

42.10.2 Peripheral Blood Stem Cells Versus BM Stem Cells as Donor Product

Hematopoietic stem cells are resident in the BM and only small numbers circulate in the blood. Historically, BM was the standard source of hematopoietic stem cells for transplantation. G-CSF can mobilize stem cells into the blood and clinical trials of allogeneic mobilized peripheral blood stem cell (PBSC) transplants demonstrated faster engraftment and hematologic recovery than BM grafts (Korbling et al. 1995; Bensinger et al. 1995; Schmitz et al. 1995). There is a similar incidence rate of acute GVHD and transplantation-related mortality in most studies (Schmitz et al. 1998) but there is generally a higher rate of chronic GVHD with PBSC transplants (Champlin et al. 2000b) and similar survival (Schmitz et al. 2006; Pidala et al. 2009).

Some studies report improved disease-free survival and OS in the PBSC group, primarily in patients with advanced disease (Bensinger et al. 2001; Mielcarek et al. 2012). These results differed from a prospective trial done by the EBMT with 350 patients which revealed the incidence of not only grades II–IV acute GVHD was significantly greater in patients who underwent PBSCT but also a greater incidence of chronic GVHD after PBSCT. However, they found no significant difference in OS or leukemia-free survival between the BMT and PBSCT groups at the 3-year follow-up (Schmitz et al. 2002). Subsequently, large retrospective EBMT registry analyses of both AML and acute lymphoblastic leukemia again revealed a greater incidence of chronic GVHD after PBSCT with rates of acute GVHD, leukemia-free survival and OS that were similar after BMT and PBSCT (Ringden et al. 2002). This question was recently revisited in a prospective trial by the Blood and Marrow Transplant Clinical Trials Network; patients were randomized between unrelated donor PBSC versus BM transplants. Survival and DFS were similar between the two groups, but there was a significantly higher incidence of chronic GVHD with PBSC (46 vs. 31 %) (Anasetti et al. 2011).

42.10.3 Alternative Donors

Recent data from the NMDP donor registry showed that the probability of finding an HLA matched adult donor was 51 % for Caucasians, 30 % for Hispanics, 20 % for Asians, and 17 % for African Americans, related primarily to the numbers of donors from each racial/ethnic group in registries. Therefore, many patients need an alternative donor source for HSCT (Ruggeri et al. 2010). One antigen mismatched unrelated donors, haploidentical related donors, and unrelated umbilical cord blood (UCB) grafts have been evaluated as donor sources. One antigen mismatched unrelated donor grafts are associated with increased treatment-related mortality, but similar survival in patients with advanced AML (Saber et al. 2012).

UCB is a rich source of hematopoietic stem cells (Broxmeyer 1992). Cord blood lymphocytes are less likely to produce GVHD than hematopoietic cells from adult donors (Kurtzberg et al. 2008). This allows partially matched cord blood cells to be effectively transplanted using donors matched for four or five of the six HLA A, B, and DR antigens. Cord blood transplants effectively restore hematopoiesis and immunity similar to other stem cell sources. However given the relatively low stem cell dose, hematopoietic recovery tends to be slower than BM or PBPC transplants. Initial studies reported higher TRM than with matched unrelated donor transplants. Eapen and colleagues performed a retrospective analysis from the CIBMTR on 1525 patients transplanted between 2002 and 2006 showed that despite using UCB grafts ($n=165$) that were not as well HLA matched as those obtained from unrelated adult donors ($n=888$ peripheral blood stem cells, 472=BM stem cells), leukemia-free survival rates were comparable (Eapen et al. 2010). It was also noted that TRM was higher in the UCB graft group which was likely due to slower engraftment rates, commonly seen in cord blood transplants especially in adults where smaller numbers of hematopoietic progenitor cells are infused in relation to body mass. Many investigators (Brunstein et al. 2010) have begun to infuse double UCB grafts to address this limitation since cell dose has been found to be the single most important determinant of successful UCB transplantation in adult patients. (Barker et al. 2005; Gluckman et al. 1997; Rubinstein et al. 1998) Brunsetin et al. looked retrospectively at data from the University of Washington and the University of Minnesota of 536 patients who were transplanted from an HLA-matched related donor (n=204), HLA allele-matched unrelated donor (n=152) or 1-antigen-mismatched unrelated adult donor (MMUD, n=52) or 4-6/6 HLA matched double UCB (n=128) graft after myeloablative conditioning for a variety of hematologic malignancies including AML, ALL, CML, and MDS, where AML made up 53 % of patients transplanted with a double UCB graft. Their data reveal leukemia-free survival at 5 years was similar for each donor type (51 %), with risk of relapse being lower (15 %) in the UCB group but risk of nonrelapse mortality being increased (34 %) secondary to GVHD and infectious complications (Brunstein et al. 2010). Of particular advantage is the ability to perform RIC treatments with umbilical cord grafts which could allow the graft-versus-leukemia effect of the immunocompetent cells in the graft to act, while the lower intensity of chemotherapy would allow patients who are older or with comorbidities to gain the opportunity to obtain HSCT (Brunstein and Laughlin 2010). Today, more than 450,000 cord blood grafts are available in more than 50 cord blood banks and more than 20,000 UCB transplants have been performed worldwide (Ruggeri et al. 2010). Given these data, cord blood transplants are a reasonable option for patients with AML and poor prognostic features in CR1 or with more advanced disease.

Advantages of haploidentical-related donors include: (1) Most patients will have a parent or sibling as a potential donor, (2) related donors are typically readily available and highly motivated, and (3) additional progenitor and immune cells are generally available from the donor for cellular therapies if needed. These transplants have historically had a very high risk of graft rejection and GVHD. GVHD can be overcome by either ex vivo or in vivo T-cell depletion or aggressive immune-suppression regimens, but these approaches delay immune reconstitution and potentially increase the risk of opportunistic infections and relapse (Anasetti et al. 2012). There has been a high risk of treatment-related mortality, particularly in adult recipients. An EBMT retrospective study, analyzing the outcome of HLA-haploidentical transplantation in several European centers, reported 48 % EFS in high-risk AML patients in CR1 (Ciceri et al. 2008). Ruggeri and colleagues (Ruggeri et al. 2007) reported on 112 high-risk AML patients receiving a haploidentical HSCT from natural killer (NK)-alloreactive versus non-NK-alloreactive donors, NK cell alloreactivity having been found to possess a powerful and effective antileukemic activity. Transplantation from NK-alloreactive donors was associated with a significantly lower relapse rate in patients transplanted in CR (3vs. 47 %, $p<0.003$), better EFS in patients transplanted in remission (67 vs. 18 %, $p<0.02$) but also in relapse (34 vs. 6 %, $p<0.04$), and overall reduced risk of relapse or death ($p<0.001$). The 67 % probability of surviving event-free for AML patients transplanted in remission from NK-alloreactive donors is in the range of best survival rates after transplantation from unrelated donors and cord blood units (Ruggeri et al. 2010).

More recently, T-cell replete haploidentical transplants have been done in the myeloablative setting with the addition of posttransplantation cyclophosphamide (Cy) to eliminate alloreactive T-cells responsible for both graft rejection and GVHD. The Johns Hopkins group has described its experience with 17 patients after haploidentical SCT using myeloablative conditioning using busulfan, Cy, and total body irradiation, followed by posttransplantation Cy (Symons et al. 2010). The cumulative incidence of nonrelapse mortality at 100 days was higher than average at 18 %, whereas GVHD rates were acceptable and none of the evaluable patients had graft rejection with although the follow up is short (Symons et al. 2010). The MD Anderson Cancer Center has an ongoing phase II study with the same conditioning schedule for haploidentical HSCT and found after a median follow-up of 6 months that OS was 71 % for first transplants and progression-free survival was 80 % for patients in remission at the time of transplantation with no death related to nonrelapse mortality in the first 6 months (Bayraktar et al. 2012). Longer follow-up is required to confirm if this approach will help abrogate the risk of GVHD while maintaining improved immune reconstitution.

In summary, there has been substantial improvement in the use of alternative donor transplants with OS approaching that seen with HLA matched related and unrelated HSCT. It is uncertain between cord blood or haploidentical donor transplants which will have superior results. The choice of the best alternative donor will depend on urgency of the transplantation, status and risk of the disease, and availability of donor source (Kroger et al. 2009).

42.11 Transplants for Relapsed or Refractory Disease

Patients with acute leukemia and primary induction failure, who do not achieve remission with induction chemotherapy have dismal prognosis, but up to 30 % can achieve prolonged remission with allogeneic hematopoietic stem-cell transplantation (Duval et al. 2010). Patients with relapsed AML also have a very poor prognosis with salvage chemotherapy. Results are improved with allogeneic hematopoietic transplantation (Armistead et al. 2009), but only a minority of patients have long-term remissions. Results are similar for patients transplanted in early relapse or in a second remission (Clift et al. 1992). Duval et al. (2010) analyzed 2255 patients who underwent transplantation for acute leukemia (both myeloid and lymphoid) in relapse or with primary induction failure after a myeloablative conditioning regimen between 1995 and 2004 through the CIBMTR. At time of transplant, 38 % of patients with AML were in primary induction failure, 45 % of AML patients were in first relapse, and 19 % of these patients with AML in first relapse were in first untreated relapse. For patients with AML, survival was worse for patients with any circulating blasts, a mismatched unrelated donor, a related donor other than an HLA-matched sibling, a Karnofsky score less than 90, and poor-risk cytogenetics

(Duval et al. 2010). The 3-year OS was 19% for AML and they reported a 100-day mortality of 39%. However, by using the above five risk factors they discovered a valuable prognostic scoring system which established that AML patients who had a predictive score of 0 had 42% OS at 3 years, whereas OS was 6% for a score ≥ 3. Transplant in relapse or primary induction failure can induce cure of leukemia, although its use has been controversial. These data show that by effectively stratifying patients, one can maximize the efficacy of transplant in this difficult disease group.

42.11.1 Relapse Post Transplant, Role of Second Allogeneic Transplants

Prognosis is poor for patients who relapse after allogeneic transplant. Treatment options consist of chemotherapy, clinical trials with novel agents, donor lymphocyte infusion, or a second allogeneic transplant. Intensive chemotherapy can result in transient second CR in some patients, particularly those who were in a remission at the time of their initial transplant. Few patients achieve long survival without either donor lymphocyte infusion or second transplant. Donor lymphocyte infusions without additional chemotherapy have been disappointing, but 10–20% of patients who receive intensive chemotherapy with a DLI achieve a durable CR (Carlens et al. 2001).

Eapen et al. presented the outcomes for 125 adult AML patients who received a second allo-SCT for recurrent or persistent leukemia after initial HLA-matched sibling SCT. Overall, 85% of the patients received allografts from the same donor for both transplants. Median follow-up of survivors was 93 months and the 3-year LFS and OS were both 27% (95% CI, 19–35%). In multivariate analyses, risks of treatment failure and mortality were lower in younger patients and patients who relapsed after 6 months from first transplantation. Risks of relapse were lower in patients who relapsed after 6 months from their first transplant, those in CR prior to second transplant, and higher risk of relapse was seen with reduced intensity conditioning (Eapen et al. 2004). According to their data, there was no benefit to using a different donor with the second transplant. It is our policy to consider a second allogeneic transplant only for patients who achieve a second CR to salvage chemotherapy.

42.12 Autologous Hematopoietic Transplantation for AML

The use of autologous transplantation in acute leukemia has always been controversial. Patients receive high-dose myeloablative chemotherapy followed by infusion of cryopreserved autologous hematopoietic cells. The major concern is potential contamination of the autograft with malignant cells. In addition, one lacks the immune graft-versus-leukemia effect against allogeneic target antigens. Nonetheless, autologous stem cell transplants can result in long-term CRs for patients transplanted in a first or second remission. It has been considered a treatment option available for patients in first CR who lacked an HLA-matched donor as an alternative to traditional consolidation with chemotherapy. Results have been variable; many studies have a major limitation that an autologous transplant could only be performed in a fraction of patients. The landmark large-scale study reported by Zittoun et al (1995). showed a 4-year DFS of 55, 48, and 30 % in recipients of allogeneic transplant, autologous transplant, and intensive chemotherapy in first CR, respectively. In this study, patients were first assigned on a donor versus no-donor basis and those patients without a sibling donor were randomized between autologous transplant and intensive chemotherapy. Those patients who received an auto transplant for consolidation had an improved survival compared to standard chemotherapy, but at the expense of more toxicity. A large meta-analysis was conducted including 13 studies (12 randomized controlled trials), 9 were in adults. When addressing using autologous transplant as consolidation for adults with AML in CR1, they reported a lower relapse and higher TRM associated with autologous SCT. There was a significant DFS benefit with autologous SCT, but this did not translate into an OS benefit (Wang et al. 2010). Currently, there is no indication for an autologous transplant for AML beyond those done on clinical trial.

Conclusion

In conclusion, allogeneic hematopoietic transplants from matched related and unrelated donors are considered standard of care for patients with AML who have failed to achieve remission or have more advanced disease. Patients in first CR who are at high risk to relapse are also appropriate candidates and ongoing research is directed at determining factors to identify patients who should receive standard consolidation chemotherapy or allogeneic transplantation. Most patients with favorable-risk cytogenetics and favorable molecular abnormalities such as CEBPA or NPM1 mutations should receive conventional chemotherapy rather than allogeneic transplants. At present, we recommend transplants for patients with intermediate- or high-risk cytogenetics, FLT3-ITD mutations, and other molecular abnormalities. Reduced intensity allogeneic transplants are indicated in medically fit elderly patients. Results with alternative donor transplants (cord blood and haploidentical transplants) are improving and should be considered for patients eligible for transplantation but lacking a matched donor.

References

Alatrash G, de Lima M, Hamerschlak N et al (2011) Myeloablative reduced-toxicity i.v. busulfan-fludarabine and allogeneic hematopoietic stem cell transplant for patients with acute myeloid leukemia or myelodysplastic syndrome in the sixth through eighth decades of life. Biol Blood Marrow Transplant 17:1490–1496

Anasetti C, Aversa F, Brunstein CG (2012) Back to the future: mismatched unrelated donor, haploidentical related donor, or unrelated umbilical cord blood transplantation? Biol Blood Marrow Transplant 18:161–165

Anasetti C, Logan B, Lee SJ et al (2011) increased incidence of chronic graft-versus-host disease (GVHD) and no survival advantage with filgrastim-mobilized peripheral blood stem cells (PBSC) compared to bone marrow (BM) transplants from unrelated donors: results of blood and marrow transplant clinical trials network (BMT CTN) protocol 0201, a phase iii, prospective, randomized trial. Blood 118:3

Armand P, Kim HT, Cutler CS et al (2008) A prognostic score for patients with acute leukemia or myelodysplastic syndromes undergoing allogeneic stem cell transplantation. Biol Blood Marrow Transplant 14:28–35

Armistead PM, de Lima M, Pierce S et al (2009) Quantifying the survival benefit for allogeneic hematopoietic stem cell transplantation in relapsed acute myelogenous leukemia. Biol Blood Marrow Transplant 15:1431–1438

Barker JN, Weisdorf DJ, DeFor TE et al (2005) Transplantation of 2 partially HLA-matched umbilical cord blood units to enhance engraftment in adults with hematologic malignancy. Blood 105:1343–1347

Barker JN, Byam CE, Kernan NA et al (2010) Availability of cord blood extends allogeneic hematopoietic stem cell transplant access to racial and ethnic minorities. Biol Blood Marrow Transplant 16:1541–1548

Bayraktar UD, Champlin RE, Ciurea SO (2012) Progress in haploidentical stem cell transplantation. Biol Blood Marrow Transplant 18:372–380

Bennett JM, Catovsky D, Daniel MT et al (1976) Proposals for the classification of the acute leukaemias. French-American-British (FAB) co-operative group. Br J Haematol 33:451–458

Bensinger WI, Weaver CH, Appelbaum FR et al (1995) Transplantation of allogeneic peripheral blood stem cells mobilized by recombinant human granulocyte colony-stimulating factor. Blood 85:1655–1658

Bensinger WI, Martin PJ, Storer B et al (2001) Transplantation of bone marrow as compared with peripheral-blood cells from HLA-identical relatives in patients with hematologic cancers. N Engl J Med 344:175–181

Bloomfield CD, Lawrence D, Byrd JC et al (1998) Frequency of prolonged remission duration after high-dose cytarabine intensification in acute myeloid leukemia varies by cytogenetic subtype. Cancer Res 58:4173–4179

Borthakur G, Estey AE (2007) Therapy-related acute myelogenous leukemia and myelodysplastic syndrome. Curr Oncol Rep 9:373–377

Broxmeyer HE, Hangoc G, Cooper S et al (1992) Growth characteristics and expansion of human umbilical cord blood and estimation of its potential for transplantation in adults. Proc Natl Acad Sci U S A 89:4109–4113

Brunet S, Labopin M, Esteve J et al (2012) Impact of FLT3 internal tandem duplication on the outcome of related and unrelated hematopoietic transplantation for adult acute myeloid leukemia in first remission: a retrospective analysis. J Clin Oncol 30:735–741

Brunstein CG, Laughlin MJ (2010) Extending cord blood transplant to adults: dealing with problems and results overall. Semin Hematol 47:86–96.

Brunstein CG, Gutman JA, Weisdorf DJ et al (2010) Allogeneic hematopoietic cell transplantation for hematologic malignancy: relative risks and benefits of double umbilical cord blood. Blood 116:4693–4699

Buccisano F, Maurillo L, Spagnoli A et al (2010) Cytogenetic and molecular diagnostic characterization combined to postconsolidation minimal residual disease assessment by flow cytometry improves risk stratification in adult acute myeloid leukemia. Blood 116:2295–2303

Buccisano F, Maurillo L, Del Principe MI et al (2012) Prognostic and therapeutic implications of minimal residual disease detection in acute myeloid leukemia. Blood 119:332–341

Byrd JC, Mrozek K, Dodge RK et al (2002) Pretreatment cytogenetic abnormalities are predictive of induction success, cumulative incidence of relapse, and overall survival in adult patients with de novo acute myeloid leukemia: results from Cancer and Leukemia Group B (CALGB 8461). Blood 100:4325–4336

Caligiuri MA, Strout MP, Lawrence D et al (1998) Rearrangement of ALL1 (MLL) in acute myeloid leukemia with normal cytogenetics. Cancer Res 58:55–59

Candoni A, Toffoletti E, Gallina R et al (2011) Monitoring of minimal residual disease by quantitative WT1 gene expression following reduced intensity conditioning allogeneic stem cell transplantation in acute myeloid leukemia. Clin Transplant 25:308–316

Carlens S, Remberger M, Aschan J, Ringden O (2001) The role of disease stage in the response to donor lymphocyte infusions as treatment for leukemic relapse. Biol Blood Marrow Transplant 7:31–38

Champlin R (2011) Reduced-intensity allogeneic hematopoietic transplantation should be considered a standard of care for older patients with acute myeloid leukemia. Biol Blood Marrow Transplant 17:1723–1724

Champlin R, Khouri I, Shimoni A et al (2000a) Harnessing graft-versus-malignancy: non-myeloablative preparative regimens for allogeneic haematopoietic transplantation, an evolving strategy for adoptive immunotherapy. Br J Haematol 111:18–29

Champlin RE, Schmitz N, Horowitz MM et al (2000b) Blood stem cells compared with bone marrow as a source of hematopoietic cells for allogeneic transplantation. IBMTR Histocompatibility and Stem Cell Sources Working Committee and the European Group for Blood and Marrow Transplantation (EBMT). Blood 95:3702–3709

Ciceri F, Labopin M, Aversa F et al (2008) A survey of fully haploidentical hematopoietic stem cell transplantation in adults with high-risk acute leukemia: a risk factor analysis of outcomes for patients in remission at transplantation. Blood 112:3574–3581

Clift RA, Buckner CD, Appelbaum FR et al (1992) Allogeneic marrow transplantation during untreated first relapse of acute myeloid leukemia. J Clin Oncol 10:1723–1729

Cornelissen JJ, van Putten WL, Verdonck LF et al (2007) Results of a HOVON/SAKK donor versus no-donor analysis of myeloablative HLA-identical sibling stem cell transplantation in first remission acute myeloid leukemia in young and middle-aged adults: benefits for whom? Blood 109:3658–3666

de Botton S, Fawaz, Chevret S et al (2005) Autologous and allogeneic stem-cell transplantation as salvage treatment of acute promyelocytic leukemia initially treated with all-trans-retinoic acid: a retrospective analysis of the European acute promyelocytic leukemia group. J Clin Oncol 23:120–126

Delhommeau F, Dupont S, Della Valle V et al (2009) Mutation in TET2 in myeloid cancers. N Engl J Med 360:2289–2301

De Zern AE, Sung A, Kim S et al (2011) Role of allogeneic transplantation for FLT3/ITD acute myeloid leukemia: outcomes from 133 consecutive newly diagnosed patients from a single institution. Biol Blood Marrow Transplant 17:1404–1409

Dohner H, Estey EH, Amadori S et al (2010) Diagnosis and management of acute myeloid leukemia in adults: recommendations from an international expert panel, on behalf of the European LeukemiaNet. Blood 115:453–474

Duval M, Klein JP, He W et al (2010) Hematopoietic stem-cell transplantation for acute leukemia in relapse or primary induction failure. J Clin Oncol 28:3730–3738

Dvorak CC, Agarwal R, Dahl GV, Gregory JJ, Feusner JH (2008) Hematopoietic stem cell transplant for pediatric acute promyelocytic leukemia. Biol Blood Marrow Transplant 14:824–830

Eapen M, Giralt SA, Horowitz MM et al (2004) Second transplant for acute and chronic leukemia relapsing after first HLA-identical sibling transplant. Bone Marrow Transpl 34:721–727

Eapen M, Rocha V, Sanz G et al (2010) Effect of graft source on unrelated donor haemopoietic stem-cell transplantation in adults with acute leukaemia: a retrospective analysis. Lancet Oncol 11:653–660

Falini B, Martelli MP, Bolli N et al (2011) Acute myeloid leukemia with mutated nucleophosmin (NPM1): is it a distinct entity? Blood 117:1109–1120

Farag SS, Maharry K, Zhang MJ et al (2011) Comparison of reduced-intensity hematopoietic cell transplantation with chemotherapy in patients age 60–70 years with acute myelogenous leukemia in first remission. Biol Blood Marrow Transplant 17:1796–1803

Frederick RA, Stephen JF, Robert SN, Karl GB (eds) (2011) Thomas' hematopoietic cell transplantation, 4th edn. Wiley-Blackwell, Malden, pp 761–774

Gilliland DG, Griffin JD (2002) The roles of FLT3 in hematopoiesis and leukemia. Blood 100:1532–1542
Gluckman E, Rocha V, Boyer-Chammard A et al (1997) Outcome of cord-blood transplantation from related and unrelated donors. Eurocord Transplant Group and the European Blood and Marrow Transplantation Group. N Engl J Med 337:373–381
Gooley TA, Chien JW, Pergam SA et al (2010) Reduced mortality after allogeneic hematopoietic-cell transplantation. N Engl J Med 363:2091–2101
Green CL, Koo KK, Hills RK, Burnett AK, Linch DC, Gale RE (2010) Prognostic significance of CEBPA mutations in a large cohort of younger adult patients with acute myeloid leukemia: impact of double CEBPA mutations and the interaction with FLT3 and NPM1 mutations. J Clin Oncol 28:2739–2747
Grimwade D, Hills RK, Moorman AV et al (2010) Refinement of cytogenetic classification in acute myeloid leukemia: determination of prognostic significance of rare recurring chromosomal abnormalities among 5876 younger adult patients treated in the United Kingdom Medical Research Council trials. Blood 116:354–365
Gupta V, Tallman MS, He W et al (2010) Comparable survival after HLA-well-matched unrelated or matched sibling donor transplantation for acute myeloid leukemia in first remission with unfavorable cytogenetics at diagnosis. Blood 116:1839–1848
Gyurkocza B, Storb R, Storer BE et al (2010) Nonmyeloablative allogeneic hematopoietic cell transplantation in patients with acute myeloid leukemia. J Clin Oncol 28:2859–2867
Horan JT, Logan BR, Agovi-Johnson MA et al (2011) Reducing the risk for transplantation-related mortality after allogeneic hematopoietic cell transplantation: how much progress has been made? J Clin Oncol 29:805–813
Horowitz MM, Gale RP, Sondel PM et al (1990) Graft-versus-leukemia reactions after bone marrow transplantation. Blood 75:555–562
Howlader N NA, Krapcho M, Neyman N, Aminou R, Waldron W, Altekruse SF, Kosary CL, Ruhl J, Tatalovich Z, Cho H, Mariotto A, Eisner MP, Lewis DR, Chen HS, Feuer EJ, Cronin KA, Edwards BK (2011) SEER Cancer Statistics Review, 1975-2008, National Cancer Institute. Bethesda, MD. http://seer.cancer.gov/csr/1975_2008/, based on November 2010 SEER data submission. posted to the SEER web site, 2011
Hubel K, Weingart O, Naumann F et al (2011) Allogeneic stem cell transplant in adult patients with acute myelogenous leukemia: a systematic analysis of international guidelines and recommendations. Leuk Lymphoma 52:444–457
Jourdan E, Boiron JM, Dastugue N et al (2005) Early allogeneic stem-cell transplantation for young adults with acute myeloblastic leukemia in first complete remission: an intent-to-treat long-term analysis of the BGMT experience. J Clin Oncol 23:7676–7684
Kern W, Voskova D, Schoch C, Hiddemann W, Schnittger S, Haferlach T (2004) Determination of relapse risk based on assessment of minimal residual disease during complete remission by multiparameter flow cytometry in unselected patients with acute myeloid leukemia. Blood 104:3078–3085
Korbling M, Przepiorka D, Huh YO et al (1995) Allogeneic blood stem cell transplantation for refractory leukemia and lymphoma: potential advantage of blood over marrow allografts. Blood 85:1659–1665
Koreth J, Schlenk R, Kopecky KJ et al (2009) Allogeneic stem cell transplantation for acute myeloid leukemia in first complete remission: systematic review and meta-analysis of prospective clinical trials. JAMA 301:2349–2361
Kroger N, Brand R, van Biezen AF et al (2009) Risk factors for therapy-related myelodysplastic syndrome and acute myeloid leukemia treated with allogeneic stem cell transplantation. Haematologica 94:542–549
Kronke J, Schlenk RF, Jensen KO et al (2011) Monitoring of minimal residual disease in NPM1-mutated acute myeloid leukemia: a study from the German-Austrian acute myeloid leukemia study group. J Clin Oncol 29:2709–2716
Kurtzberg J, Prasad VK, Carter SL et al (2008) Results of the cord blood transplantation study (COBLT): clinical outcomes of unrelated donor umbilical cord blood transplantation in pediatric patients with hematologic malignancies. Blood 112:4318–4327

Lee SJ, Klein J, Haagenson M et al (2007) High-resolution donor-recipient HLA matching contributes to the success of unrelated donor marrow transplantation. Blood 110:4576–4583

Linker CA, Owzar K, Powell B et al (2009) Auto-SCT for AML in second remission: CALGB study 9620. Bone Marrow Transplant 44:353–359

Litzow MR, Tarima S, Perez WS et al (2010) Allogeneic transplantation for therapy-related myelodysplastic syndrome and acute myeloid leukemia. Blood 115:1850–1857

Marcucci G, Maharry K, Radmacher MD et al (2008) Prognostic significance of, and gene and microRNA expression signatures associated with, CEBPA mutations in cytogenetically normal acute myeloid leukemia with high-risk molecular features: a Cancer and Leukemia Group B Study. J Clin Oncol 26:5078–5087

Marcucci G, Maharry K, Wu YZ et al (2010) IDH1 and IDH2 gene mutations identify novel molecular subsets within de novo cytogenetically normal acute myeloid leukemia: a Cancer and Leukemia Group B study. J Clin Oncol 28:2348–2355

Mardis ER, Ding L, Dooling DJ et al (2009) Recurring mutations found by sequencing an acute myeloid leukemia genome. N Engl J Med 361:1058–1066

Martino R, Iacobelli S., Brand R et al (2006) Retrospective comparison of reduced-intensity conditioning and conventional high-dose conditioning for allogeneic hematopoietic stem cell transplantation using HLA-identical sibling donors in myelodysplastic syndromes. Blood 108:836–846

McSweeney PA, Niederwieser D, Shizuru JA et al (1999) Outpatient allografting with minimally myeosuppressive immunosuppressive conditioning of low-dose TBI and postgrafting cyclosporine and mycophenolate mofetil. Blood 94(1):393a

Medeiros BC, Othus M, Fang M, Roulston D, Appelbaum FR (2010) Prognostic impact of monosomal karyotype in young adult and elderly acute myeloid leukemia: the Southwest Oncology Group (SWOG) experience. Blood 116:2224–2228

Metzeler KH, Becker H, Maharry K et al (2011a) ASXL1 mutations identify a high-risk subgroup of older patients with primary cytogenetically normal AML within the ELN Favorable genetic category. Blood 118:6920–6929

Metzeler KH, Maharry K, Radmacher MD et al (2011b) TET2 mutations improve the new European LeukemiaNet risk classification of acute myeloid leukemia: a Cancer and Leukemia Group B study. J Clin Oncol 29:1373–1381

Mielcarek M, Storer B, Martin PJ et al (2012) Long-term outcomes after transplantation of HLA-identical related G-CSF-mobilized peripheral blood mononuclear cells versus bone marrow. Blood 119:2675–2678

Moreno I, Martin G, Bolufer P et al (2003) Incidence and prognostic value of FLT3 internal tandem duplication and D835 mutations in acute myeloid leukemia. Haematologica 88:19–24

Mrozek K (2008) Cytogenetic, molecular genetic, and clinical characteristics of acute myeloid leukemia with a complex karyotype. Semin Oncol 35:365–377

Paschka P, Marcucci G, Ruppert AS et al (2006) Adverse prognostic significance of KIT mutations in adult acute myeloid leukemia with inv(16) and t(8;21): a Cancer and Leukemia Group B Study. J Clin Oncol 24:3904–3911

Patel JP, Gonen M, Figueroa ME et al (2012) Prognostic relevance of integrated genetic profiling in acute myeloid leukemia. N Engl J Med 366:1079–1089

Paun O, Lazarus HM (2012) Allogeneic hematopoietic cell transplantation for acute myeloid leukemia in first complete remission: have the indications changed? Curr Opin Hematol 19:95–101

Pidala J, Anasetti C, Kharfan-Dabaja MA, Cutler C, Sheldon A, Djulbegovic B (2009) Decision analysis of peripheral blood versus bone marrow hematopoietic stem cells for allogeneic hematopoietic cell transplantation. Biol Blood Marrow Transplant 15:1415–1421

Ringden O, Labopin M, Bacigalupo A et al (2002) Transplantation of peripheral blood stem cells as compared with bone marrow from HLA-identical siblings in adult patients with acute myeloid leukemia and acute lymphoblastic leukemia. J Clin Oncol 20:4655–4664

Ringden O, Labopin M, Ehninger G et al (2009) Reduced intensity conditioning compared with myeloablative conditioning using unrelated donor transplants in patients with acute myeloid leukemia. J Clin Oncol 27:4570–4577

Rubinstein P, Carrier C, Scaradavou A et al (1998) Outcomes among 562 recipients of placental-blood transplants from unrelated donors. N Engl J Med 339:1565–1577
Ruggeri L, Mancusi A, Capanni M et al (2007) Donor natural killer cell allorecognition of missing self in haploidentical hematopoietic transplantation for acute myeloid leukemia: challenging its predictive value. Blood 110:433–440
Ruggeri A, Ciceri F, Gluckman E, Labopin M, Rocha V (2010) Alternative donors hematopoietic stem cells transplantation for adults with acute myeloid leukemia: umbilical cord blood or haploidentical donors? Best Pract Res Clin Haematol 23:207–216
Saber W, Opie S, Rizzo JD, Zhang MJ, Horowitz MM, Schriber J (2012) Outcomes after matched unrelated donor versus identical sibling hematopoietic cell transplantation in adults with acute myelogenous leukemia. Blood 119:3908–3916
Schlenk RF, Dohner K, Krauter J et al (2008) Mutations and treatment outcome in cytogenetically normal acute myeloid leukemia. N Engl J Med 358:1909–1918
Schlenk RF, Dohner K, Mack S et al (2010) Prospective evaluation of allogeneic hematopoietic stem-cell transplantation from matched related and matched unrelated donors in younger adults with high-risk acute myeloid leukemia: German-Austrian trial AMLHD98A. J Clin Oncol 28:4642–4648
Schmid C, Labopin M, Nagler A et al (2012) Treatment, risk factors, and outcome of adults with relapsed AML after reduced intensity conditioning for allogeneic stem cell transplantation. Blood 119:1599–1606
Schmitz N, Bacigalupo A, Hasenclever D et al (1998) Allogeneic bone marrow transplantation vs filgrastim-mobilised peripheral blood progenitor cell transplantation in patients with early leukaemia: first results of a randomised multicentre trial of the European Group for Blood and Marrow Transplantation. Bone Marrow Transplant 21:995–1003
Schmitz N, Dreger P, Suttorp M et al (1995) Primary transplantation of allogeneic peripheral blood progenitor cells mobilized by filgrastim (granulocyte colony-stimulating factor). Blood 85:1666–1672
Schmitz N, Beksac M, Hasenclever D et al (2002) Transplantation of mobilized peripheral blood cells to HLA-identical siblings with standard-risk leukemia. Blood 100:761–767
Schmitz N, Eapen M, Horowitz MM et al (2006) Long-term outcome of patients given transplants of mobilized blood or bone marrow: a report from the International Bone Marrow Transplant Registry and the European Group for Blood and Marrow Transplantation. Blood 108:4288–4290
Shimoni A, Kroger N, Zabelina T et al (2005) Hematopoietic stem-cell transplantation from unrelated donors in elderly patients (age > 55 years) with hematologic malignancies: older age is no longer a contraindication when using reduced intensity conditioning. Leukemia 19:7–12
Slavin S, Nagler A, Naparstek E et al (1998) Nonmyeloablative stem cell transplantation and cell therapy as an alternative to conventional bone marrow transplantation with lethal cytoreduction for the treatment of malignant and nonmalignant hematologic diseases. Blood 91:756–763
Slovak ML, Kopecky KJ, Cassileth PA et al (2000) Karyotypic analysis predicts outcome of preremission and postremission therapy in adult acute myeloid leukemia: a Southwest Oncology Group/Eastern Cooperative Oncology Group Study. Blood 96:4075–4083
Soiffer RJ, Lerademacher J, Ho V et al (2011) Impact of immune modulation with anti-T-cell antibodies on the outcome of reduced-intensity allogeneic hematopoietic stem cell transplantation for hematologic malignancies. Blood 117:6963–6970
Suciu S, Mandelli F, de Witte T et al (2003) Allogeneic compared with autologous stem cell transplantation in the treatment of patients younger than 46 years with acute myeloid leukemia (AML) in first complete remission (CR1): an intention-to-treat analysis of the EORTC/GIMEMAAML-10 trial. Blood 102:1232–1240
Symons H, Chen AR, Leffell MS et al (2010) HLA-Haploidentical Bone Marrow Transplantation (BMT) for High Risk Hematologic Malignancies Using Myeloablative. Conditioning and High-Dose, Posttransplantation Cyclophosphamide. Blood 116:977
Szydlo R, Goldman JM, Klein JP et al (1997) Results of allogeneic bone marrow transplants for leukemia using donors other than HLA-identical siblings. J Clin Oncol 15:1767–1777
Tallman MS, Altman JK (2009) How I treat acute promyelocytic leukemia. Blood 114:5126–5135

Thiede C, Koch S, Creutzig E et al (2006) Prevalence and prognostic impact of NPM1 mutations in 1485 adult patients with acute myeloid leukemia (AML). Blood 107:4011–4020

Truitt RL, Johnson BD (1995) Principles of graft-vs.-leukemia reactivity. Biol Blood Marrow Transplant 1:61–68

Van Vlierberghe P, Patel J, Abdel-Wahab O et al (2011) PHF6 mutations in adult acute myeloid leukemia. Leukemia 25:130–134

Vardiman JW, Thiele J, Arber DA et al (2009) The 2008 revision of the World Health Organization (WHO) classification of myeloid neoplasms and acute leukemia: rationale and important changes. Blood 114:937–951

Wang J, Ouyang J, Zhou R, Chen B, Yang Y (2010) Autologous hematopoietic stem cell transplantation for acute myeloid leukemia in first complete remission: a meta-analysis of randomized trials. Acta Haematol 124:61–71

Ward PS, Patel J, Wise DR et al (2010) The common feature of leukemia-associated IDH1 and IDH2 mutations is a neomorphic enzyme activity converting alpha-ketoglutarate to 2-hydroxyglutarate. Cancer Cell 17:225–234

Whitman SP, Caligiuri MA, Maharry K et al (2012) The MLL partial tandem duplication in adults aged 60 years and older with de novo cytogenetically normal acute myeloid leukemia. Leukemia 26:1713–1717

Yamamoto Y, Kiyoi H, Nakano Y et al (2001) Activating mutation of D835 within the activation loop of FLT3 in human hematologic malignancies. Blood 97:2434–2439

Yan XJ, Xu J, Gu ZH et al (2011) Exome sequencing identifies somatic mutations of DNA methyltransferase gene DNMT3A in acute monocytic leukemia. Nat Genet 43:309–315

Yanada M, Matsuo K, Emi N, Naoe T (2005) Efficacy of allogeneic hematopoietic stem cell transplantation depends on cytogenetic risk for acute myeloid leukemia in first disease remission: a metaanalysis. Cancer 103:1652–1658

Zittoun RA, Mandelli F, Willemze R et al (1995) Autologous or allogeneic bone marrow transplantation compared with intensive chemotherapy in acute myelogenous leukemia. European Organization for Research and Treatment of Cancer (EORTC) and the Gruppo Italiano Malattie Ematologiche Maligne dell'Adulto (GIMEMA) Leukemia Cooperative Groups. N Engl J Med 332:217–223

Chapter 43
Mesenchymal Stem/Stromal Cell-Targeted Therapies for Solid Tumors and Hematological Malignancies

Christopher H. Booth, Lysette Mutkus, Karen Bussard, Erika Spaeth, Michael Andreeff and Frank C. Marini

Abstract Mesenchymal stem/stroma cells (MSCs) exhibit an innate tropism and capacity to migrate to sites of inflammation including wounding damage and the tumor microenvironment, where they have been shown to contribute to the development of tumor-associated stroma. Recent data suggest the tumor microenvironment plays a critical role in tumor progression in that both local tissue- and bone marrow-derived mesenchyme interact with tumor cells, which then provokes signaling responses that is thought to support tumor growth. Numerous reports demonstrate that sites of tissue damage secrete elevated levels of inflammatory mediators much like those found in the tumor microenvironment, which then stimulates recruitment and engraftment of MSC. This chapter will describe the emerging role of MSCs in tumor stroma development, describe their ability to home to and engraft in the tumor microenvironment, as well as how we can best employ stromal cells by using them as gene delivery vehicles directing them to sites of tumor formation to control tumor growth.

Keywords Tumor microenvironment · Tumor homing · Stroma · Anticancer · Cell vehicle · Inflammation

---

Christopher H. Booth and Lysette Mutkus contributed as the first author.

F. C. Marini (✉) · C. H. Booth · L. Mutkus · K. Bussard
Wake Forest Institute of Regenerative Medicine, Wake Forest Comprehensive Cancer Center, 391 Technology Way, Winston-Salem, NC 27157-1083, USA
e-mail: fmarini@wakehealth.edu

E. Spaeth · M. Andreeff · F. C. Marini
Department of Leukemia, University of Texas M.D. Anderson Cancer Center, Houston, TX 77030, USA

M. Andreeff (ed.), *Targeted Therapy of Acute Myeloid Leukemia*,
Current Cancer Research, DOI 10.1007/978-1-4939-1393-0_43

43.1 Introduction

Bone marrow (BM)-derived mesenchymal stem/stromal cells are an adult progenitor population with differentiation potential and self-renewal capacity. These cells appear to be the precursor population for tumor-associated stromal elements and as such are recruited into the tumor microenvironment where they form tumor-associated fibroblasts, pericytes, and fibrovascular stroma. Because of this innate physiological ability of MSCs to be recruited into tumors, and to participate in tumor stroma, we have demonstrated that injected MSCs home to tumors throughout the body, and preferentially incorporate into the tumor tissue. This chapter describes the potential of using of MSC as a tumor-targeted delivery vehicle in cancer gene therapy and can serve as an effective platform for delivering biological agents into tumors.

43.2 Tumor Development and Progression

The progression and development of cancer is believed to be a complicated cascade of events in which the cells accumulate repeated genetic and epigenetic modifications to their genome, ultimately leading to phenotypic changes in the cell with an increased uncontrolled proliferation and growth (Cho 2009; Palmero et al. 2010; Feinberg and Vogelstein 1983; Fraga et al. 2005; Seligson et al. 2005). A multi-staged hypothesis has been suggested specifically for how hematological malignancies arise and progress (Winkler et al. 2010). The tumor cell-centric view has led to many new important discoveries that have been paramount to further understanding the origins and progression of cancer (Wood et al. 2007; Campbell et al. 2010). This approach has also been structured within the human cancer genome project that has led to extremely rapid identification of new mutations. Even though these developments have led to many great and important discoveries, tumor-centric analysis is an incomplete model to explain the phenotypic changes in their entirety. The tumor-centric approach is oversimplified and relies on the simplicity and convenience of in vitro cell culture systems. This is mostly due to the fact that this approach does not take into consideration or provide the complexity of the tissue microenvironment itself and everything involved within in vivo tumor growth and metastasis.

To further this tumor-centric philosophy, an appealing idea of tumor "stem cells" have been developed over the years and the stem cell concept appears to be sufficient enough to produce tumors in animal models (Reya et al. 2001; Weaver et al. 1995). This tumor "stem cell" is a stem-like parental cell containing an increased tumor-initiation potential that is also capable of generating the bulk of the tumor cells required for tumor development. The reductionist view of cancer stem cells unfortunately does not fully address the intricacy of the interplay between all of the different types of cells that are necessary for generating tumors, for example, the relationship between the tumorigenic epithelial cells and nonneoplastic microenvironment cells, including the cells that are required to generate matrix, vasculature, and the characteristic immune suppression that is a prerequisite for the development of a tumor (Bissell and Labarge 2005; Kenny et al. 2007). In similar regard, it is also important to understand that it is still unclear whether or not the connective tissue that surrounds the epithelial tumor malignancies remains static, despite the different numbers of genetic alterations that cascade the development of malignant tumors. This microenvironment was previously believed to be an indirect subsidiary neighbor that tumors were able to thrive and develop in; (Park et al. 2000) however, more recent evidence has supported that the role of the microenvironment is vital for the development of tumors. These neighboring cells have been shown to actively shape the frequency and biology of tumors, which are unable to exist in isolation from these cells or sufficiently generate a malignant tumor by themselves. With the support of Paget's "seed and soil" hypothesis, Fidler et al. (Kopfstein and Christofori 2006) and Folkman et al.'s insights into the role of angiogenesis in tumor development (Lorusso and Ruegg 2008), this new concept that tumor progression is dependent on a controlled and specialized interaction between tumor cells and their actively supportive stroma is now recognized (Mahadevan and Von Hoff 2007; Zipori et al. 1985; Harrison et al. 2002; Jensen et al. 2000).

43.3 The Emerging Role of the Stroma in Cancer

Tumor cells survive in the tumor stroma: an environment that is surrounded by the extracellular matrix (ECM) and nonneoplastic microenvironmental cells, such as fibroblasts, endothelial cells, and various cells of the immune system (Petrulio 2006). The development and progression of tumors rely on the multitude of dynamic interactions that exist between the tumor parenchyma (the neoplastic cells) and the tumor stroma (Anton and Glod 2009; Guturu et al. 2009; Liao et al. 2009a; Zumsteg and Christofori 2009). Specifically, the tumor stroma responds to the signals and factors that are produced by the tumor cells, in turn providing the necessary components that are necessary for tumor survival, including the extracellular matrices, structural and organized support, vasculature, the removal of waste, and paracrine and juxtacrine growth factor production (Kidd et al. 2008; Pollard 2008).

In the specific case of the BM microenvironment, the stroma consists of fibroblasts, BM-derived MSCs, endothelial cells, macrophages, adipocytes, osteoblasts, and osteoclasts. In relation, it has been experimentally shown that most hematological malignant cells rapidly die when they are cultured in vivo, which supports the concept that environmental factors are critical for tumor cell survival. It has also been experimentally shown that hematological malignant cells survive much longer in vivo when they are cultured on stomal cells. It is clear that the interactions facilitated by direct contact between the tumor and BM stomal cells, as well as the tumor–ECM interactions, provide the necessary means for the survival or tumor cells, while the BM microenvironment also provides many soluble factors that stimulate tumor cell growth and survival.

It has been shown more recently that the stroma also acts as a reciprocal signaling partner in tumor progression. Specifically, there are processes that resemble normal inflammatory and wound-healing responses that typically cause changes in the stromal compartment (Dvorak 1986). Based on this, chronic inflammation and other pathological conditions can induce structural and functional changes within the tumor stroma. These changes of the stroma, which are capable of promoting the genetic alterations that are necessary for starting tumorigenesis, directly influence dormant tumors that are regulated by this "normal" stroma, causing the tumors to become active again and proliferate (Radisky and Radisky 2007). There are several transgenic and knockout animal models that block or alter key paracrine regulators that support this statement (Bierie and Moses 2006; Biswas et al. 2004). One murine model showed that the suppression of the transforming growth factor-β (TGF-β) response in stromal fibroblasts, through the ablation of the TGF-β type II receptor, modifies stromal–epithelial interactions, which disrupts the normal mammary development while also enhancing tumor growth and cell motility (Cheng et al. 2005). Tumors are also able to alter their microenvironments and induce them to recruit supplementary stromal cells that remain activated after being embedded, resulting in additional support of the tumor. The functions and steps in tumor angiogenesis clearly demonstrate these interactions. Angiogenesis, the growth and development of new blood vessels, is required to overcome the limitations of a solid tumor that has grown larger than a few millimeters in size, making the ability to receive oxygen and nutrients required for survival through the existing blood vessels more difficult. Tumor cells and the microenvironment assist each other in recruiting several types of stromal cells that are required for blood vessel formation, including fibroblasts, endothelial cells, pericytes, macrophages, and inflammatory cells (Anton and Glod 2009). When the local environment is under specific inflammatory conditions, for example, during tumor development or wound healing, the formation of fibrovascular structures starts when the local fibroblasts are recruited from the neighboring tissues and begin to divide (Nyberg et al. 2008). In doing so, local endothelial cells are also recruited and activated to divide and form new vasculature (Dirk et al. 2006). Tumor-associated macrophages (TAMs) and tumor-associated fibroblasts (TAFs) are also two important contributors to tumor angiogenesis. In the areas of hypoxia and neovascularization, it has been shown the there is TAM buildup and high levels of TAM-secreted pro-angiogenic factors, such as vascular endothelial growth factor (VEGF) and platelet-derived growth factor (PDGF; (Balkwill and Mantovani 2001). TAMs also produce an endothelial chemotactic factor, thymidine phosphorylase. In human solid tumors, higher levels of this factor have been connected with the formation of new vasculature, invasiveness, metastasis, and decreased patient survival (Hotchkiss et al. 2003). TAFs, on the other hand, may come from multiple sources, including resident tissue fibroblasts, pericytes, vascular smooth muscle cells, and endothelial cells (Micke and Ostman 2004). It is believed that the tumor microenvironment promotes the local fibroblasts to undertake a TAF phenotype (Kalluri and Zeisberg 2006). Particularly, there is recent evidence showing that BM-derived MSCs are a source of TAFs (Mishra et al. 2008). In regard to MSCs, TAFs are characterized as a morphologically homogeneous but functionally heterogeneous group of mesenchymal cells, which have been shown to produce large number of tumor-promoting growth factors, cytokines, and chemokines, while also modulating the local immune response of the body against the tumor (De Wever et al. 2008). When the cancer cells invade the surrounding tissue, there is substantial devastation and regeneration of intercellular elements. This newly synthesized stroma, also known as "cancer induced stroma," is comprised of pericytes, fibroblasts, endothelial cells of blood and lymph vessels, immature myeloid and inflammatory cells, including lymphocytes, granulocytes, and macrophages (Micke and Ostman 2004). Through multiple mechanisms, including matrix synthesis/degradation (Sugimoto et al. 2006) and the release of multiple cytokine and proteinases, specifically the activation of matrix metalloproteinases (MMPs; (Singer et al. 2002), TAFs are able create an invasive-supported environment. In invasive human breast carcinomas, it has been shown that TAFs release higher levels of stromal-derived factor-1 (SDF-1) than in normal fibroblasts (Orimo 2005). This has also been shown in the BM microenvironment, in which chemokines, specifically SDF-1 and its equivalent receptor CXCR4, have an important role in the homing of hematopoietic cells, which is critical for the development of the hematopoietic compartment and the regulation of normal and malignant hematopoiesis.

43.4 Epithelial to Mesenchyme Transition Generates MSC-Like Cells

Recent publications are providing evidence that tumors have the ability to produce their own inherent mesenchyme through a developmental reprogramming called "epithelial-to-mesenchymal transition (EMT)" (Colmone et al. 2008; Gaunder et al. 2009; Zeng et al. 2006): A latent embryonic process that triggers epithelial cells to lose their epithelial traits and develop the characteristic of mesenchymal cells (Matsunaga et al. 2003; Bendall et al. 1993). During this process, epithelial cells lose their cellular polarity through the downregulation of cytokeratin expression and the expression of cell–cell adhesion molecules, specifically E-cadherin (Chen et al. 2008b). Along with the decrease in epithelial gene expression is an increased expression of mesenchymal genes, namely vimentin and fibronectin (Ma et al. 2010). The induction of the EMT event can be initiated by a multitude of factors that maintain a role during tumor progression; specifically, through several cytokines and chemokines, such as TGF-β, through the expression of several developmentally critical transcription factors, such as Twist and Snail, or through exposure to extremely low oxygen levels (Eisen et al. 1998; Subramanian et al. 2005; Jakt et al. 2001; Knappskog et al. 2011). Throughout the EMT process, it has also been experimentally shown that epithelial cells not only display mesenchymal phenotypes but also express stem cell properties (Schofield 1978). Experimentally, researchers induced EMT in immortalized human mammary epithelial cells through the overexpression of specific EMT inducers, such as the Twist and Snail genes, with the result of an enhanced expression of stem cell markers ($CD44^{high}/CD24^{low}$) in these cells. These EMT-derived cells also have the ability to quickly form mammospheres, soft agar colonies, and tumors (Schofield 1978). These are specific properties that are typically correlated with mammary epithelial stem cells. These EMT-derived cells were also compared to BM-derived MSCs with the EMT-derived cells exhibiting phenotypic and functional properties of MSCs (Perry and Li 2007). When compared to their epithelial equivalents, the EMT-derived cells expressed increased levels of MSC markers on their cell surface, such as CD44 and PDGF-Rβ (CD140b). Resembling the abilities of MSCs, the EMT-derived cells were also able to differentiate into osteoblasts, adipocytes, and chondrocytes when cultured under the correct environmental conditions. These cells are also shown to invade and migrate towards MDA-MB-231 breast cancer cells, unlike the control cells used. In in vivo wound homing assays in nude mice, the EMT-derived cells are also shown to migrate to wound sites, which represent the functional similarities with MSCs. Through the co-culturing of breast cancer epithelial cells and MSCs, researchers have also reported increased expression levels of oncogenes (NCOA4, FOS), proto-oncogenes (FYN, JUN), as well as genes connected to invasion (MMP11), angiogenesis (VEGF), and anti-apoptosis (insulin-like growth factor, IGF1R, BCL2; (Li and Li 2006). Furthermore, it was also observed that an upregulation of EMT-specific markers (e.g., N-cadherin, vimentin, Twist, and Snail) occurs after coculturing of the epithelial cells with MSCs. This suggests that MSCs may possibly stimulate breast cancer metastasis with the facilitation of EMT. The summation of the current experimental reports proposes that MSC-like cells are possibly produced through EMT in epithelial cells, including that these EMT-derived cells express stem cell properties and enhance tumor progression.

43.5 Fibroblasts, Myofibroblasts, and Stromal Precursors

The growth of a solid tumor cannot be supported without the tumor cells attracting and stimulating fibroblasts, which provides a rich source of growth factors, including TGF-β, IGF-1, and bFGF (Denys et al. 2008; Wang et al. 2008). These stimulated fibroblasts are also required to provide an organizational structure of the tumor stroma through the production of ECM components. Extrapolating from the current understanding of stroma development (Castello-Cros and Cukierman 2009), it is believed that the stroma-genesis process occurs as a multistep cascade. Through combined regulation of proliferation, pro-invasive, and differentiation signals, this process drives the local recruitment of BM-derived stem cells or MSCs into the stroma (Amatangelo et al. 2005). Phenotypically similar to TAFs, these mesenchymal cells have the ability to differentiate into fibroblast-like cells, which generate ECM components and/or contribute to perivascular or vascular structures (Sund and Kalluri 2009). This model is further supported by the similarities that have been experimentally seen between reactive tumor stroma and stroma that is involved with wound healing (Dvorak1986). There are also many biological processes that are involved in wound repair that have also been observed in the reactive stroma during tumor progression, such as stromal cell development of a myofibroblast phenotype, deposition of type I collagen, and the induction of angiogenesis (Brown et al. 1989). Based on the crucial role that these fibroblasts and stromal precursor cells have in the construction of tumor stroma, we can interpret the inherent tropism of these cells in regard to tumor and wound healing environments.

43.6 Inherent Tropism of MSCs for Tumors and Sites of Injury

Numerous studies involve an assortment of animal models and a multitude of experimental scenarios that experimentally demonstrate the homing of MSCs after systemic or local infusion (Herrera et al. 2004; Shake et al. 2002). Pereira et al. reported in a study that systemic infusion of marker gene-expressing MSCs in irradiated syngeneic mice produced marker gene expression that was detected a month later in 5 % of lung cells and 8 % of BM cells (Pereira et al. 1995). This homing process of the MSCs to the lung and the bone when injected could possibly be explained by an inherent ability of MSCs to attach to matrix components. In this scenario, multiple reports have demonstrated that under different pathological conditions, MSCs selectively target the sites of injury regardless of the type of tissue or organ (Kumamoto et al. 2009; Semedo et al. 2009). This has further been supported by other studies that utilize fluorescent protein-labeled MSCs (Devine et al. 2001) or luciferase-labeled MSCs (Kidd et al. 2009) and show that these exogenously delivered MSCs are discovered at the sites of injury. There are experimental data in wound healing models (Satoh et al. 2004), tissue repair and regeneration models (Kahler et al. 2007; Le Blanc 2006; Mansilla et al. 2006), and brain injury models (Liao et al. 2009b; van Velthoven 2009) that confirm the innate tropism of MSCs towards wound healing areas. Jacamo et al. was the first research group to demonstrate that BM-derived MSCs proficiently home to different tumor types in xenograft models, including melanoma, glioma, colon, and breast cancers (Nakamizo et al. 2005; Studeny et al. 2002, 2004).

Accordingly, the collection of evidence now insinuates that specific conditioning regimens prior to cell transplantation, such as chemotherapy and focal/total body irradiation treatments, are able to improve the efficiency of MSC homing to areas of engraftment (Allers et al. 2004; Erices et al. 2003; Kim et al. 2010; Klopp et al. 2007). There are also many factors, including interleukin (IL)-1, IL-8, TNF-α, TGF-β, EGF, PDGF, and SDF-1, which are typically secreted from tumors and sites of injury, that have the ability to enhance MSC migration. With this current understanding, it can be considered that increased concentrations of chemoattractant mediators secreted at the injury site are required for MSCs to replace the damaged niche through migration, engraftment, and proliferation. In this regard, tumors are considered to be sites of tissue damage or specifically, "wounds that never heal" according to Dvorak (Dvorak 1986). The demonstrated inherent tropism that MSCs have for inflammatory sites, including tumors, and their ability to actively home and engraft into these pathological locations, while also delivering therapeutics, provides insight into the greater potential use of these cells as anticancer drug delivery vehicles. This idea has been demonstrated by many groups utilizing multiple xenograft mouse tumor models, all showing that MSCs that are modified by a viral transduction method, which stably expresses tumor necrosis factor-related apoptosis-inducing ligand (TRAIL), have the ability to home to tumors and reduce the burden of the tumor (Grisendi et al. 2010; Kim et al. 2008; Loebinger et al. 2009). Regardless of the debated concerns regarding the use of viral transduction for MSC-mediated gene delivery and an elevated risk of oncogenic transformation, there is a plethora of significant in vivo findings that support the utilization of these methods for therapies in the future.

43.7 How to Choose which Stroma Precursor Cells to Target Tumors

The invasion progression of cancer cells is connected with the development of tumor-specific stroma, which includes the recruitment of fibroblasts into the tumor stroma that have the ability to convert into smooth muscle actin-positive fibroblasts, specifically myofibroblasts, or activated fibroblasts. These "activated" fibroblasts play important roles in tumor formation and progression through production of collagens and other ECM proteins in response to numerous extracellular stimuli (Petrulio et al. 2006; Sangai et al. 2005; Reddy et al. 2008). The Strunk group has shown how the tumor microenvironment preferentially supports the engraftment of intravenously injected BM-derived mesenchymal stromal cells (Hofmann et al. 2012). Roni et al. have also supported these statements through the interperitoneal injection of β-galactosidase-transduced human fibroblasts into severe combined immunodeficiency (SCID) mice alongside ovarian cancer cells. The β-galactosidase-positive fibroblasts were preferentially localized within the cancer stroma but were not found within the normal tissue stroma (Anton and Glod 2009). Furthermore, the Ischii group utilized a BM transplant model to demonstrate that the stroma generated from invasive cancer cells consists of both BM-derived and non-BM-derived activated fibroblasts, while the BM-derived activated fibroblasts are recruited into cancer-induced stroma at a later stage (Guturu et al. 2009). Based on all of these findings, we can propose that BM-derived myofibroblasts provide for the cancer-induced desmoplastic reaction, as well as influencing changes in the microenvironment that influences tumor growth. These findings also directly establish the detailed involvement of BM-derived cells in regard to the formation of tumor stroma, suggesting that BM-derived cells are capable of targeting tumors due to physiological signs that are inducted by the tumor (Sangai et al. 2005; Reddy et al. 2008). All of these previous findings ultimately provide insight into the specific advantages for using BM-derived cells as cellular delivery vehicles. Most invasive cancers are most likely the cause of a desmoplastic reaction to a certain extent, which provides a mutual target for the treatment of multiple types of cancers. Also, the low intrinsic mutation rates of MSCs make them less predisposed to acquiring a drug-resistant phenotype when compared to unpredictable cancer cells. Even though there are publications that demonstrate the development of sarcomas in mice that were implanted with MSCs with high passage numbers, to date, there is still no report of any such event occurring in the many thousands of clinically treated patients who have received MSCs, typically in the context of tissue repair studies and for the treatment of graft-versus-host disease (Le Blanc et al. 2008). Based on all of the current knowledge of MSCs and the previous related findings, it is a strong possibility to see that the use of BM-derived myofibroblastic progenitor cells as transporters of unique therapies to prevent or inhibit local and metastatic growth of numerous cancer cells is an obtainable treatment in the future.

43.8 Using Cells as Vehicles Targeted at Tumors

In order for cells to be considered for use as an ideal cell-based delivery system for targeting tumors, there are specific characteristics that must be in place for this to even be a possibility (Willmon et al. 2009). First of all, the carrier cell used as a vehicle will need to express tumor-selective migratory capacity (i.e., these cells have to be extremely site specific exclusivity towards tumor cells with no tropism towards any other cell types or locations within the body). Secondly, it is required to be able to upregulate the expression of the therapeutic agents involved through genetic manipulation of the cells ex vivo. Lastly, it is vital that the cell vehicle be able to carry and deliver the therapeutic gene or agent to the tumor site without initiating a host immune response.

There have been many research groups over the past few years that have been able to demonstrate the unique ability of neural stem cells (NSC) to readily migrate throughout the brain (Aboody et al. 2000; Benedetti et al. 2000). The Aboody group, specifically, has demonstrated that when NSCs are transplanted into brain neoplasia animal models, they have the innate ability to migrate. These cells were detected near metastatic tumor beads, which were not localized or near the area of the original transplant site. Due to the innate tumor-tropic properties of the NSCs, the use of these cells is now being utilized as a tumor-targeted strategy for anti-glioma gene therapy. NSCs have the ability to be extremely effective in limiting tumor progression if they are engineered to express genes that have potent antitumor effects, such as IL-4 (Benedetti et al. 2000), IL-12, IFN-β (Ehtesham et al. 2002), IL-23 (Yuan et al. 2006), and TRAIL (Kim et al. 2006). In regard to central nervous system (CNS) malignancies, NSCs are more suitable for use as therapeutic delivery vehicles due to their intrinsic characteristics instead of MSCs, which are currently the preferred cell carrier for other organs (Mader et al. 2009). Unfortunately, despite the clear advantage of using these neural stem cells, the isolation and expansion of the cells for clinical application is currently technically challenging; therefore, other sources for therapeutic cells also needs to be studied.

MSCs not only have the ability to be involved in regenerative and immunomodulatory applications, but can also function as efficient gene delivery vehicles. There are many reports that have already proven the effective use of MSCs as the cell delivery vehicles for the in vivo delivery of several clinically significant anticancer agents post-engraftment within the tumor site (Studeny et al. 2004; Loebinger et al. 2009; Nakamura et al. 2004; Ren et al. 2008b; Xin et al. 2007; Zischek et al. 2009). Unfortunately, due to the absence of information for the homeostatic maintenance of MSCs in vivo and the possibility that they could transform and promote the growth of an existing tumor, or initiate a new one, the use of MSCs for gene delivery vehicles also poses a multitude of potential issues that could arise. This concern is also based on the fact that the MSCs have an innate ability for self-renewal with many similarities existing between stem cells and cancer cells. Researchers have demonstrated that mouse MSCs are capable of spontaneous transformation when cultured under specific conditions; however, human MSCs cultured under similar conditions do not exhibit this behavior (Lepperdinger et al. 2008). Furthermore, there are some studies in which researchers have expanded on this idea and now show that post-senescent human MSCs are often transformed (Rubio et al. 2005); however, some researchers have further characterized this behavior, showing how lower-passage human MSCs actually do not form tumors in vivo (Studeny et al. 2004). Further supporting this statement, there have recently been three publications reporting malignant transformations of MSCs that have had to be withdrawn, due to the contamination of malignant cells in the cell cultures (Chen et al. 2012; Vogel 2010).

43.9 MSC-Mediated Delivery of Therapeutic Agents

Our group published the first reports demonstrating that MSCs that are engineered to express IFN-β inhibit the growth of malignant cells in vivo, while also demonstrating that MSCs can produce biological agents locally at tumor sites (Studeny et al. 2004). Since then, numerous groups have also provided reports of tumor-targeted delivery of therapeutic agents using MSCs, further supporting our hypothesis. These studies have further been able to support the idea of using MSCs for targeted therapy due to their ease of being harvested, isolated, and expanded from multiple sources; for instance, BM, adipose tissue (Zuk et al. 2001), and placenta (In't anker et al. 2004). The MSCs are also easy to transduce with viral vectors while maintaining the ability to selectively home to the tumor microenvironment. Unfortunately, since most of the mouse models utilize immunodeficient mice, it is extremely difficult to assess the immune responses towards gene-modified MSCs. Recently, in order to avoid this problem of potential immunogenicity against adenoviral vectors, insertional mutagenesis using lenti/retroviral vectors our group has now focused on using electroporation to transfer the genes into MSCs.

43.9.1 Interferons

In multiple tumor models, it has been demonstrated that MSCs expressing interferon beta (IFN-β) can decrease tumor burden while also increasing animal survival. Interferons have the ability to mediate direct cytotoxic and immune stimulatory effects; however, it is ineffective when systemically delivered remotely from the tumor due to the short half-life of interferons, which concludes that the interferons need to be delivered regionally in order to therapeutically work (Nakamizo et al. 2005; Studeny et al. 2002; Kidd et al. 2010). Our group was able to demonstrate through xenograft models that breast cancer and melanoma tumors could be controlled in vivo through weekly injections of IFN-β-expressing MSCs (Studeny et al. 2004); an effect not seen with the injection of recombinant IFN-β protein. This study is the first proof of principle for using MSCs as an innovative and acceptable platform for the selective in situ production of biological agents within tumors. Ren et al. have also showed these results with the delivery of IFN-α, instead of IFN-β, by means of MSCs. Through systemic administration of MSCs producing IFN- α, they were able to show growth reduction of B16F10 melanoma cells, while also significantly prolonging survival in a lung metastasis melanoma model (Ren et al. 2008a). Similarly, using an aggressive model of blast crisis CLL (KBM-5), we were able to demonstrate the homing of systemically delivered MSCs to all of the experimental sites of disease, such as the BM, spleen, and lymph nodes. The migrated MSCs produced IFN-α that was secreted by a mifepristone (RU486)-activated adeno-associated virus (AAV), ultimately resulting in leukemia regression and increased survival: unlike the remote injection of recombinant IFN-α.

43.9.2 Interleukins

An alternative therapeutic strategy has also been proposed, suggesting that the use to MSCs engineered to express ILs would improve the antitumor surveillance through the activation of the host cells. Significant therapeutic effects have been observed when matrix-embedded MSCs, which are engineered to secrete IL-12, are injected adjacent to tumors (Eliopoulos et al. 2008). When the MSC-matrix plugs were implanted in the opposite flank of the tumor, there was no reduction in the tumor growth, as seen in the remote injection experiments with IFN-β, which suggests that regional administration of the therapeutic agent is required. Many researchers have also reported that when MSCs engineered to secrete IL-2 (Nakamura et al. 2004; Stagg et al. 2004) or IL-12 (Eliopoulos et al. 2008; Chen et al. 2008a; Parmar et al. 2007) are injected into a tumor-bearing mouse, they produced an immunological reaction and stimulated cells to infiltrate the tumor tissue. This immunological reaction proved to influence no systemic toxicity or increased levels of IL-12 in the serum and tumor area. In contrast, when the free Ad-IL-12 adenovector was administered alone, there was in elevated serum IL-12 level, which induced systemic toxicity (Cheng et al. 2005).

43.9.3 Conditionally Replicating Adenoviruses, Retroviruses, and Lentiviruses

An alternative approach for the treatment of tumors is through the use of naturally or genetically modified oncolytic viruses. Although systematically tested in clinical trials, the viruses are unsuccessful at delivering to the tumor sites. Since the viruses also have a nonspecific uptake in the liver and spleen, they produced an increased systemic toxicity. Due to these unfortunate characteristics, Komarova et al. have demonstrated that MSCs infected with a genetically modified adenovirus were able to replicate within the carrier MSCs and successfully delivered into the tumor site (Komarova et al. 2006). In the ovarian cancer model that they used, the researchers showed that the successful delivery of adenoviruses through MSC-based delivery increased the survival of tumor-bearing mice, unlike direct viral injection. The Curiel group used a SCID mouse xenograft model of breast cancer and demonstrated that MSCs transduced with conditionally replicating adenoviruses (CRAds) targeted the tumor site, increased the oncolytic effects, and ultimately increased the survival of the tumor-bearing animals, unlike the animals treated with CRAds alone (Stoff-Khalili et al. 2007). It has also been recently reported that intraperitoneal injection of MSC-CRAds created a reduced nonspecific organ infection compared to the systemic injection of CRAds (Dembinski et al. 2010). All of the results from these reports verify that MSCs can mediate CRAds specific oncolysis through their innate ability to act as cellular vehicles and transport the CRAds to the tumor sites.

MSCs have also been modified to express therapeutic molecules through retroviral transduction. In a mouse melanoma model, the antitumor properties of IL-12 have been observed by injecting human MSCs that are stably transduced with a retroviral vector expressing IL-12 (Elzaouk et al. 2006). The MSC–IL-12 complex drastically reduced the formation of lung metastases of B16F10 melanoma cells through the facilitation, primarily by $CD8^+$ cells and partly with $CD4^+$ T and natural killer (NK) cells.

Another viral approach for tumor therapy is through the transduction of MSCs with lentiviral vectors since human MSCs can be transduced with lentiviral vectors without losing any of their stem cell properties. A recent report using a mouse model of human glioblastoma showed that transduced MSCs effectively migrated to the tumor sites (Sasportas et al. 2009). They also reported that when they transduced MSCs to express recombinant TRAIL, they were resistant to the cytotoxic effects of TRAIL. When the MSCs were implanted into the mice bearing malignant GBM8 gliomas in vivo, the secreted TRAIL from the MSCs also provided a positive antitumor effect. Groups that have utilized the MSC–TRAIL complex in order to deliver higher concentrations of TRAIL in tumor-bearing mice have also reported similar results (Mohr et al. 2008).

43.9.4 Growth Factor Antagonists, Chemokines, and Suicide Gene Therapy

The use of MSCs transduced to express growth factor antagonists and chemokines have been proposed due to their inherent role observed in the treatments of tumors. Xin et al. reported a reduced metastasis and prolonged survival of mice bearing C26 and B16F10 lung metastases when they were injected with MSCs transduced with an adenoviral vector that expresses the immunostimulatory chemokine, CX3CL1 (fractalkine; (Xin et al. 2007). In a different report by Kanehira et al., they observed the antitumor effects of MSCs expressing NK4, an antagonist of hepatocyte growth factor (HGF). In tumor-bearing mice, they demonstrated that systemically injected NK4-MSCs migrated to the sites of the lung tumors with no presence in any areas absent of the tumors (Xin et al. 2007; Kanehira et al. 2007). NK4-MSCs also hindered the progression of lung metastases in a C26 lung metastasis model, while also extending survival. These results were driven by the inhibition of tumor-associated angiogenesis, lymphangiogenesis, and the induction of apoptosis.

There are some groups starting to support the incorporation of suicide genes into MSCs to target tumors and infect the tumor cells. This therapy provides the possibility of targeted chemotherapy through the modification of MSCs to produce a tumor-specific prodrug-converting cellular vehicle. This has successfully been performed by the Kucerova group with the successful retroviral transduction of the fusion gene cytosine deaminase/uracil phosphoribosyltransferase (CD) into adipose-derived MSCs (Kucerova et al. 2007). This gene converts cytosine to uracil while ammonia but also converts the antifungal agent 5-fluorocutosine (5FC) into 5-fluorouracil (5FU), a potent antimetabolite drug. When adipose tissue-derived MSCs expressing the CD gene product (CD–AT–MSC) were systemically administered with 5-FC, they observed intensified cytotoxic effects of 5-FC on HT-29 tumor cells in vitro. They supported this result in vivo with a tumor xenograft model by using the CD–AT–MSC complex to deliver the CD transgene to tumors with noticeable antitumor effects. Similarly, Uchibori et al. transduced MSCs with retroviral vectors that expressed the thymidine kinase of herpes simplex virus (HSK-tk). With the systemic injection of the modified MSCs in mice bearing 9L malignant glioma tumors, they observed an increase in HSK-tv transgene expression, which is associated with the reduction of tumor growth (Uchibori et al. 2009).

43.10 An Alternative to BM-Derived MSCs

There are additional mesenchymal tissues in the body in which MSC-like cells have been isolated. For example, tissue mesenchyme isolated from fetal lung, adipose, umbilical cord, placenta, and cardiac muscle appear to have differentiation and homing properties comparable to that of BM-derived MSCs. These other cell types should be able to serve as alternative cellular delivery vehicles for antitumor therapies to target sites of tumor formation. Some of these tissues may even possess better tumor homing or less tumor-supportive abilities than BM-derived MSC, as described in (Subramanian et al. 2012).

43.11 Future Directions in the Design of MSC-Based Delivery Strategies

Potential concerns over the use of MSCs as delivery vehicles originates from the confounding data described in the literature that both confirms and refutes the ability of MSCs to support tumor growth and progression. Our limited understanding of the homeostatic maintenance of MSC around and within the tumor microenvironment suggests a greater need to define their roles. Therefore, an important focus of future research should be on elucidating the roles of mesenchyme in both wound healing, in which MSCs are recruited, support rapid wound healing and then disappear, and tumor formation in which MSCs are recruited but then transition into a pro-inflammatory type cells in the process.

Although stem cells for use in cancer gene therapy are usually capable of some degree of tumor selectivity, one future aim could be to improve tumor homing or tumor selectivity, which would greatly increase the applicability and success of tumor-targeted cell delivery vehicle. Alternatively, the use of tumor-specific or tumor-activated gene promoters such as the CCL5 promoter, which could restrict the activation of genes delivered to only the tumor microenvironment, could potentially increase specificity and decrease any unwanted activity.

Conclusions

In this chapter, we have described the current state of the art for using MSCs as cellular delivery vehicles with specific selectivity for solid tumors and their respective tumor microenvironments. This selectivity is based on the innate physiological tropism of MSCs for sites of inflammation, be it wound repair or tumor development. The most obvious advantage offered by MSC is the highly selective local delivery and intratumoral release of therapeutic agents, especially proteins with poor pharmacokinetic profiles, such as the interferons. In addition to intratumorally delivering cancer-related proteins, cellular vehicles could also serve as protective "coatings" for the delivery of payloads such oncoselective viruses (e.g., Onyx-15 or Delta 24). Not only would such cells protect these viruses from the immune system but they would also allow the amplification of the initial viral load, with the possibility of increasing viral spread through cell-to-cell contact specifically at tumor sites.

Much work remains before the potential therapeutic benefits of MSC-based approaches can be fully realized. Even though we described the concept back in 1998 in an abstract to gene therapy, only one clinical trial utilizing tumor-targeted MSC has ever been conducted. This trial done in Spain used MSC to deliver the oncoselective ICOVIR adenovirus to three patients with advanced neuroblastoma. While the outcome of the trial suggested that MSC infused into three cancer patients were deemed safe (i.e., no adverse side effects were observed), there were limited clinical endpoints and unfortunately no biopsies were performed to determine if the MSC had integrated into tumor tissue. In the meantime, more trials have been proposed, but have never materialized; it is currently unclear why this field has not advanced. The development of optimally targeted cellular vectors will require advances in cell science and virology and will certainly benefit from the continued innovative application of the knowledge gained from basic biology, tissue stem cell engineering, and gene therapy.

References

Aboody KS, Brown A, Rainov NG, Bower KA, Liu S, Yang W et al (2000) Neural stem cells display extensive tropism for pathology in adult brain: evidence from intracranial gliomas. Proc Natl Acad Sci USA 97(23):12846–12851

Allers C, Sierralta WD, Neubauer S, Rivera F, Minguell JJ, Conget PA (2004) Dynamic of distribution of human bone marrow-derived mesenchymal stem cells after transplantation into adult unconditioned mice. Transplantation 78(4):503–508

Amatangelo MD, Bassi DE, Klein-Szanto AJ, Cukierman E (2005) Stroma-derived three-dimensional matrices are necessary and sufficient to promote desmoplastic differentiation of normal fibroblasts. Am J Pathol 167(2):475–488

Anton K, Glod J (2009) Targeting the tumor stroma in cancer therapy. Curr Pharm Biotechnol 10(2):185–191

Balkwill F, Mantovani A (2001) Inflammation and cancer: back to Virchow? Lancet 357(9255):539–545

Bendall LJ, Kortlepel K, Gottlieb DJ (1993) Human acute myeloid leukemia cells bind to bone marrow stroma via a combination of beta-1 and beta-2 integrin mechanisms. Blood 82(10):3125–3132

Benedetti S, Pirola B, Pollo B, Magrassi L, Bruzzone MG, Rigamonti D et al (2000) Gene therapy of experimental brain tumors using neural progenitor cells. Nat Med 6(4):447–450

Bierie B, Moses HL (2006) TGF-beta and cancer. Cytokine Growth Factor Rev 17(1–2):29–30

Bissell MJ, Labarge MA (2005) Context, tissue plasticity, and cancer: are tumor stem cells also regulated by the microenvironment? Cancer Cell 7(1):17–23

Biswas S, Chytil A, Washington K, Romero-Gallo J, Gorska AE, Wirth PS et al (2004) Transforming growth factor beta receptor type II inactivation promotes the establishment and progression of colon cancer. Cancer Res 64(14):4687–4692

Brown LF, Dvorak AM, Dvorak HF (1989) Leaky vessels, fibrin deposition, and fibrosis: a sequence of events common to solid tumors and to many other types of disease. Am Rev Respir Dis 140(4):1104–1107

Campbell PJ, Yachida S, Mudie LI, Stephens PJ, Pleasance ED, Stebbings LA et al (2010) The patterns and dynamics of genomic instability in metastatic pancreatic cancer. Nature 467(7319):1109–1113

Castello-Cros R, Cukierman E (2009) Stromagenesis during tumorigenesis: characterization of tumor-associated fibroblasts and stroma-derived 3D matrices. Methods Mol Biol 522:275–305

Chen X, Lin X, Zhao J, Shi W, Zhang H, Wang Y et al (2008a) A tumor-selective biotherapy with prolonged impact on established metastases based on cytokine gene-engineered MSCs. Mol Ther 16(4):749–756

Chen Y, Shao JZ, Xiang LX, Dong XJ, Zhang GR (2008b) Mesenchymal stem cells: a promising candidate in regenerative medicine. Int J Biochem Cell Biol 40(5):815–820

Chen Y, Jacamo R, Shi YX, Wang RY, Battula VL, Konoplev S et al (2012) Human extramedullary bone marrow in mice: a novel in vivo model of genetically controlled hematopoietic microenvironment. Blood 119(21):4971–4980

Cheng N, Bhowmick NA, Chytil A, Gorksa AE, Brown KA, Muraoka R et al (2005) Loss of TGF-beta type II receptor in fibroblasts promotes mammary carcinoma growth and invasion through upregulation of TGF-alpha-, MSP- and HGF-mediated signaling networks. Oncogene 24(32):5053–5068

Cho KR (2009) Ovarian cancer update: lessons from morphology, molecules, and mice. Arch Pathol Lab Med 133(11):1775–1781

Colmone A, Amorim M, Pontier AL, Wang S, Jablonski E, Sipkins DA (2008) Leukemic cells create bone marrow niches that disrupt the behavior of normal hematopoietic progenitor cells. Science 322(5909):1861–5

De Wever O, Demetter P, Mareel M, Bracke M (2008) Stromal myofibroblasts are drivers of invasive cancer growth. Int J Cancer 123(10):2229–2238

Dembinski JL, Spaeth EL, Fueyo J, Gomez-Manzano C, Studeny M, Andreeff M et al (2010) Reduction of nontarget infection and systemic toxicity by targeted delivery of conditionally replicating viruses transported in mesenchymal stem cells. Cancer Gene Ther 17(4):289–297

Denys H, Derycke L, Hendrix A, Westbroek W, Gheldof A, Narine K et al (2008) Differential impact of TGF-beta and EGF on fibroblast differentiation and invasion reciprocally promotes colon cancer cell invasion. Cancer Lett 266(2):263–274

Devine SM, Bartholomew AM, Mahmud N, Nelson M, Patil S, Hardy W et al (2001) Mesenchymal stem cells are capable of homing to the bone marrow of non-human primates following systemic infusion. Exp Hematol 29(2):244–255

Dirk AE, Oude Egbrink MG, Wagstaff J, Griffioen AW (2006) Monocyte/macrophage infiltration in tumors: modulators of angiogenesis. J Leukoc Biol 80(6):1183–1196

Dvorak HF (1986) Tumors: wounds that do not heal. Similarities between tumor stroma generation and wound healing. N Engl J Med 315(26):1650–9

Ehtesham M, Kabos P, Kabosova A, Neuman T, Black KL, Yu JS (2002) The use of interleukin 12-secreting neural stem cells for the treatment of intracranial glioma. Cancer Res 62(20):5657–5663

Eisen MB, Spellman PT, Brown PO, Botstein D (1998) Cluster analysis and display of genome-wide expression patterns. Proc Natl Acad Sci USA 95(25):14863–8

Eliopoulos N, Francois M, Boivin MN, Martineau D, Galipeau J (2008) Neo-organoid of marrow mesenchymal stromal cells secreting interleukin-12 for breast cancer therapy. Cancer Res 68(12):4810–8

Elzaouk L, Moelling K, Pavlovic J (2006) Anti-tumor activity of mesenchymal stem cells producing IL-12 in a mouse melanoma model. Exp Dermatol 15(11):865–874

Erices AA, Allers CI, Conget PA, Rojas CV, Minguell JJ (2003) Human cord blood-derived mesenchymal stem cells home and survive in the marrow of immunodeficient mice after systemic infusion. Cell Transplant 12(6):555–561

Feinberg AP, Vogelstein B (1983) Hypomethylation distinguishes genes of some human cancers from their normal counterparts. Nature 301(5895):89–92

Fraga MF, Ballestar E, Villar-Garea A, Boix-Chornet M, Espada J, Schotta G et al (2005) Loss of acetylation at Lys16 and trimethylation at Lys20 of histone H4 is a common hallmark of human cancer. Nat Genet 37(4):391–400

Gaundar SS, Bradstock KF, Bendall LJ (2009) p38MAPK inhibitors attenuate cytokine production by bone marrow stromal cells and reduce stroma-mediated proliferation of acute lymphoblastic leukemia cells. Cell Cycle 8(18):2975–2983

Grisendi G, Bussolari R, Cafarelli L, Petak I, Rasini V, Veronesi E et al (2010) Adipose-derived mesenchymal stem cells as stable source of tumor necrosis factor-related apoptosis-inducing ligand delivery for cancer therapy. Cancer Res 70(9):3718–3729

Guturu P, Shah V, Urrutia R (2009) Interplay of tumor microenvironment cell types with parenchymal cells in pancreatic cancer development and therapeutic implications. J Gastrointest Cancer 40(1–2):1–9

Harrison JS, Rameshwar P, Chang V, Bandari P (2002) Oxygen saturation in the bone marrow of healthy volunteers. Blood 99(1):394

Herrera MB, Bussolati B, Bruno S, Fonsato V, Romanazzi GM, Camussi G (2004) Mesenchymal stem cells contribute to the renal repair of acute tubular epithelial injury. Int J Mol Med 14(6):1035–1041

Hofmann NA, Ortner A, Jacamo RO, Reinisch A, Schallmoser K, Rohban R, Etchart N, Fruehwirth M, Beham-Schmid C, Andreeff M, Strunk D (2012) Oxygen sensing mesenchymal progenitors promote neo-vasculogenesis in a humanized mouse model in vivo. PLoS One 7(9):e44468

Hotchkiss KA, Ashton AW, Klein RS, Lenzi ML, Zhu GH, Schwartz EL (2003) Mechanisms by which tumor cells and monocytes expressing the angiogenic factor thymidine phosphorylase mediate human endothelial cell migration. Cancer Res 63(2):527–533

In't Anker PS, Scherjon SA, Kleijburg-van der Keur C, de Groot-Swings GM, Claas FH Fibbe WE et al (2004) Isolation of mesenchymal stem cells of fetal or maternal origin from human placenta. Stem Cells 22(7):1338–1345

Jakt LM, Cao L, Cheah KS, Smith DK (2001) Assessing clusters and motifs from gene expression data. Genome Res 11(1):112–123

Jensen PO, Mortensen BT, Hodgkiss RJ, Iversen PO, Christensen IJ, Helledie N et al (2000) Increased cellular hypoxia and reduced proliferation of both normal and leukaemic cells during progression of acute myeloid leukaemia in rats. Cell Prolif 33(6):381–395

Kahler CM, Wechselberger J, Hilbe W, Gschwendtner A, Colleselli D, Niederegger H et al (2007) Peripheral infusion of rat bone marrow derived endothelial progenitor cells leads to homing in acute lung injury. Respir Res 8:50

Kalluri R, Zeisberg M (2006) Fibroblasts in cancer. Nat Rev Cancer 6(5):392–401

Kanehira M, Xin H, Hoshino K, Maemondo M, Mizuguchi H, Hayakawa T et al (2007) Targeted delivery of NK4 to multiple lung tumors by bone marrow-derived mesenchymal stem cells. Cancer Gene Ther 14(11):894–903

Kenny PA, Lee GY, Bissell MJ (2007) Targeting the tumor microenvironment. Front Biosci 12:3468–3474

Kidd S, Spaeth E, Klopp A, Andreeff M, Hall B, Marini FC (2008) The (in) auspicious role of mesenchymal stromal cells in cancer: be it friend or foe. Cytotherapy 10(7):657–667

Kidd S, Spaeth E, Dembinski JL, Dietrich M, Watson K, Klopp A et al (2009) Direct evidence of mesenchymal stem cell tropism for tumor and wounding microenvironments using in vivo bioluminescent imaging. Stem Cells 27(10):2614–2623

Kidd S, Caldwell L, Dietrich M, Samudio I, Spaeth EL, Watson K et al (2010) Mesenchymal stromal cells alone or expressing interferon-beta suppress pancreatic tumors in vivo, an effect countered by anti-inflammatory treatment. Cytotherapy 12(5):615–625

Kim SK, Kim SU, Park IH, Bang JH, Aboody KS, Wang KC et al (2006) Human neural stem cells target experimental intracranial medulloblastoma and deliver a therapeutic gene leading to tumor regression. Clin Cancer Res 12(18):5550–6

Kim SM, Lim JY, Park SI, Jeong CH, Oh JH, Jeong M et al (2008) Gene therapy using TRAIL-secreting human umbilical cord blood-derived mesenchymal stem cells against intracranial glioma. Cancer Res 68(23):9614–9623

Kim SM, Oh JH, Park SA, Ryu CH, Lim JY, Kim DS et al (2010) Irradiation enhances the tumor tropism and therapeutic potential of tumor necrosis factor-related apoptosis-inducing ligand-secreting human umbilical cord blood-derived mesenchymal stem cells in glioma therapy. Stem Cells 28(12):2217–2228

Klopp AH, Spaeth EL, Dembinski JL, Woodward WA, Munshi A, Meyn RE et al (2007) Tumor irradiation increases the recruitment of circulating mesenchymal stem cells into the tumor microenvironment. Cancer Res 67(24):11687–11695

Knappskog S, Myklebust LM, Busch C, Aloysius T, Varhaug JE, Lonning PE et al (2011) RINF (CXXC5) is overexpressed in solid tumors and is an unfavorable prognostic factor in breast cancer. Ann Oncol 22(10):2208–2215

Komarova S, Kawakami Y, Stoff-Khalili MA, Curiel DT, Pereboeva L (2006) Mesenchymal progenitor cells as cellular vehicles for delivery of oncolytic adenoviruses. Mol Cancer Ther 5(3):755–766

Kopfstein L, Christofori G (2006) Metastasis: cell-autonomous mechanisms versus contributions by the tumor microenvironment. Cell Mol Life Sci 63(4):449–468

Kucerova L, Altanerova V, Matuskova M, Tyciakova S, Altaner C (2007) Adipose tissue-derived human mesenchymal stem cells mediated prodrug cancer gene therapy. Cancer Res 67(13):6304–6313

Kumamoto M, Nishiwaki T, Matsuo N, Kimura H, Matsushima K (2009) Minimally cultured bone marrow mesenchymal stem cells ameliorate fibrotic lung injury. Eur Respir J 34(3):740–8

Le Blanc K (2006) Mesenchymal stromal cells: Tissue repair and immune modulation. Cytotherapy 8(6):559–561

Le Blanc K, Frassoni F, Ball L, Locatelli F, Roelofs H, Lewis I et al (2008) Mesenchymal stem cells for treatment of steroid-resistant, severe, acute graft-versus-host disease: a phase II study. Lancet 371(9624):1579–1586

Lepperdinger G, Brunauer R, Jamnig A, Laschober G, Kassem M (2008) Controversial issue: is it safe to employ mesenchymal stem cells in cell-based therapies? Exp Gerontol 43(11):1018–1023

Li Z, Li L (2006) Understanding hematopoietic stem-cell microenvironments. Trends Biochem Sci 31(10):589–595

Liao D, Luo Y, Markowitz D, Xiang R, Reisfeld RA (2009a) Cancer associated fibroblasts promote tumor growth and metastasis by modulating the tumor immune microenvironment in a 4T1 murine breast cancer model. PLoS One 4(11):e7965

Liao W, Zhong J, Yu J, Xie J, Liu Y, Du L et al (2009b) Therapeutic benefit of human umbilical cord derived mesenchymal stromal cells in intracerebral hemorrhage rat: implications of anti-inflammation and angiogenesis. Cell Physiol Biochem 24(3–4):307–316

Loebinger MR, Eddaoudi A, Davies D, Janes SM (2009) Mesenchymal stem cell delivery of TRAIL can eliminate metastatic cancer. Cancer Res 69(10):4134–4142

Lorusso G, Ruegg C (2008) The tumor microenvironment and its contribution to tumor evolution toward metastasis. Histochem Cell Biol 130(6):1091–1103

Ma W, Wang M, Wang ZQ, Sun L, Graber D, Matthews J et al (2010) Effect of long-term storage in TRIzol on microarray-based gene expression profiling. Cancer Epidemiol Biomarkers Prev 19(10):2445–2452

Mader EK, Maeyama Y, Lin Y, Butler GW, Russell HM, Galanis E et al (2009) Mesenchymal stem cell carriers protect oncolytic measles viruses from antibody neutralization in an orthotopic ovarian cancer therapy model. Clin Cancer Res 15(23):7246–7255

Mahadevan D, Von Hoff DD (2007) Tumor-stroma interactions in pancreatic ductal adenocarcinoma. Mol Cancer Ther 6(4):1186–1197

Mansilla E, Marin GH, Drago H, Sturla F, Salas E, Gardiner C et al (2006) Bloodstream cells phenotypically identical to human mesenchymal bone marrow stem cells circulate in large amounts under the influence of acute large skin damage: new evidence for their use in regenerative medicine. Transplant Proc 38(3):967–9

Matsunaga T, Takemoto N, Sato T, Takimoto R, Tanaka I, Fujimi A et al (2003) Interaction between leukemic-cell VLA-4 and stromal fibronectin is a decisive factor for minimal residual disease of acute myelogenous leukemia. Nat Med 9(9):1158–1165

Micke P, Ostman A (2004) Tumour-stroma interaction: cancer-associated fibroblasts as novel targets in anti-cancer therapy? Lung Cancer 45(2):S163–S175

Mishra PJ, Humeniuk R, Medina DJ, Alexe G, Mesirov JP, Ganesan S et al (2008) Carcinoma-associated fibroblast-like differentiation of human mesenchymal stem cells. Cancer Res 68(11):4331–4339
Mohr A, Lyons M, Deedigan L, Harte T, Shaw G, Howard L et al (2008) Mesenchymal stem cells expressing TRAIL lead to tumour growth inhibition in an experimental lung cancer model. J Cell Mol Med 12(6B):2628–4263
Mueller LP, Luetzkendorf J, Widder M, Nerger K, Caysa H, Mueller T (2010) TRAIL-transduced multipotent mesenchymal stromal cells (TRAIL-MSC) overcome TRAIL resistance in selected CRC cell lines in vitro and in vivo. Cancer Gene Ther 18(4):229–239
Nakamizo A, Marini F, Amano T, Khan A, Studeny M, Gumin J et al (2005) Human bone marrow-derived mesenchymal stem cells in the treatment of gliomas. Cancer Res 65(8):3307–3318
Nakamura K, Ito Y, Kawano Y, Kurozumi K, Kobune M, Tsuda H et al (2004) Antitumor effect of genetically engineered mesenchymal stem cells in a rat glioma model. Gene Ther 11(14):1155–1164
Nyberg P, Salo T, Kalluri R (2008) Tumor microenvironment and angiogenesis. Front Biosci 13:6537–6553
Orimo A (2005) Stromal fibroblasts present in invasive human breast carcinomas promote tumor growth and angiogenesis through elevated SDF-1/CXCL12 secretion. Cell 121:335–348
Palmero EI, Achatz Mi, Ashton-Prolla P, Olivier M, Hainut P (2010) Tumor protein 53 mutations and inherited cancer; beyond Li-Fraumeni syndrome. Curr Opin Oncol 22(1):64–69
Park CC, Bissell MJ, Barcellos-Hoff MH (2000) The influence of the microenvironment on the malignant phenotype. Mol Med Today 6(8):324–329
Parmar K, Mauch P, Vergilio JA, Sackstein R, Down JD (2007) Distribution of hematopoietic stem cells in the bone marrow according to regional hypoxia. Proc Natl Acad Sci USA 104(13):5431–5436
Pereira RF, Halford KW, O'Hara MD, Leeper DB, Sokolov BP, Pollard MD et al (1995) Cultured adherent cells from marrow can serve as long-lasting precursor cells for bone, cartilage, and lung in irradiated mice. Proc Natl Acad Sci USA 92(11):4857–4861
Perry JM, Li L (2007) Disrupting the stem cell niche: good seeds in bad soil. Cell 129(6):1045–1047
Petrulio CA, Kim-Schulze S, Kaufman HL (2006) The tumour microenvironment and implications for cancer immunotherapy. Expert Opin Biol Ther 6(7):671–684
Pollard JW (2008) Macrophages define the invasive microenvironment in breast cancer. J Leukoc Biol 84(3):623–630
Radisky ES, Radisky DC (2007) Stromal induction of breast cancer: inflammation and invasion. Rev Endocr Metab Disord 8(3):279–287
Reddy K, Zhou Z, Schadler K, Jia SF, Kleinerman ES (2008) Bone marrow subsets differentiate into endothelial cells and pericytes contributing to Ewing's tumor vessels. Mol Cancer Res 6(6):929–936
Ren C, Kumar S, Chanda D, Chen J, Mountz JD, Ponnazhagan S (2008a) Therapeutic potential of mesenchymal stem cells producing interferon-alpha in a mouse melanoma lung metastasis model. Stem Cells 26(9):2332–2338
Ren C, Kumar S, Chanda D, Kallman L, Chen J, Mountz JD et al (2008b) Cancer gene therapy using mesenchymal stem cells expressing interferon-beta in a mouse prostate cancer lung metastasis model. Gene Ther 15(21):1446–1453
Reya T, Morrison SJ, Clarke MF, Weissman IL (2001) Stem cells, cancer, and cancer stem cells. Nature 414(6859):105–111
Rubio D, Garcia-Castro J, Martin MC, de la Fuente R, Cigudosa JC, Lloyd AC et al (2005) Spontaneous human adult stem cell transformation. Cancer Res 65(8):3035–3039
Sangai T, Ishii G, Kodama K, Miyamoto S, Aoyagi Y, Ito T et al (2005) Effect of differences in cancer cells and tumor growth sites on recruiting bone marrow-derived endothelial cells and myofibroblasts in cancer-induced stroma. Int J Cancer 115(6):885–892

Sasportas LS, Kasmieh R, Wakimoto H, Hingtgen S, van de Water JA, Mohapatra G et al (2009) Assessment of therapeutic efficacy and fate of engineered human mesenchymal stem cells for cancer therapy. Proc Natl Acad Sci USA 106(12):4822–4827

Satoh H, Kishi K, Tanaka T, Kubota Y, Nakajima T, Akasaka Y et al (2004) Transplanted mesenchymal stem cells are effective for skin regeneration in acute cutaneous wounds. Cell Transplant 13(4):405–412

Schofield R (1978) The relationship between the spleen colony-forming cell and the haemopoietic stem cell. Blood Cells 4(1-2):7–25

Seligson DB, Horvath S, Shit T, Yu H, Tze S, Grunstein M et al (2005) Global histone modification patterns predict risk of prostate cancer recurrence. Nature 435(7046):1262–6

Semedo P, Correa-Costa M, Antonio Cenedeze M, Maria Avancini Costa Malheiros D, Antonia dos Reis M, Shimizu MH et al (2009) Mesenchymal stem cells attenuate renal fibrosis through immune modulation and remodeling properties in a rat remnant kidney model. Stem Cells 27(12):3063–3073

Shake JG, Gruber PJ, Baumgartner WA, Senechal G, Meyers J, Redmond JM et al (2002) Mesenchymal stem cell implantation in a swine myocardial infarct model: engraftment and functional effects. Ann Thorac Surg 73(6):1919–1925 (discussion 26)

Singer CF, Kronsteiner N, Marton E, Kubista M, Cullen KJ, Hirtenlehner K et al (2002) MMP-2 and MMP-9 expression in breast cancer-derived human fibroblasts is differentially regulated by stromal-epithelial interactions. Breast Cancer Res Treat 72(1):69–77

Stagg J, Lejeune L, Paquin A, Galipeau J (2004) Marrow stromal cells for interleukin-2 delivery in cancer immunotherapy. Hum Gene Ther 15(6):597–608

Stoff-Khalili MA, Rivera AA, Mathis JM, Banerjee NS, Moon AS, Hess A et al (2007) Mesenchymal stem cells as a vehicle for targeted delivery of CRAds to lung metastases of breast carcinoma. Breast Cancer Res Treat 105(2):157–167

Studeny M, Marini FC, Champlin RE, Zompetta C, Fidler IJ, Andreeff M (2002) Bone marrow-derived mesenchymal stem cells as vehicles for interferon-beta delivery into tumors. Cancer Res 62(13):3603–3608

Studeny M, Marini FC, Dembinski JL, Zompetta C, Cabreira-Hansen M, Bekele BN et al (2004) Mesenchymal stem cells: potential precursors for tumor stroma and targeted-delivery vehicles for anticancer agents. J Natl Cancer Inst 96(21):1593–1603

Subramanian A, Tamayo P, Mootha VK, Mukherjee S, Ebert BL, Gillette MA et al (2005) Gene set enrichment analysis: a knowledge-based approach for interpreting genome-wide expression profiles. Proc Natl Acad Sci U S A 102(43):15545–15550

Subramanian A, Shu-Uin G, Kae-Siang N, Gauthaman K, Biswas A, Choolani M, Bongso A, Chui-Yess F (2012) Human umbilical cord Wharton Jelly mesenchyme stem cells do not transform to tumor associated fibroblasts in the presence of breast and ovarian cancer cells unlike bone marrow mesenchymal stem cells. J.Cell Biochem 113(6):1886–1895

Sugimoto H, Mundel TM, Kieran MW, Kalluri R (2006) Identification of fibroblast heterogeneity in the tumor microenvironment. Cancer Biol Ther 5(12):1640–6

Sund M, Kalluri R (2009) Tumor stroma derived biomarkers in cancer. Cancer Metastasis Rev 28(1–2):177–183

Uchibori R, Okada T, Ito T, Urabe M, Mizukami H, Kume A et al (2009) Retroviral vector-producing mesenchymal stem cells for targeted suicide cancer gene therapy. J Gene Med 11(5):373–381

van Velthoven CT, Kavelaars A, van Bel F, Heijnen CJ (2009) Mesenchymal stem cell treatment after neonatal hypoxic-ischemic brain injury improves behavioral outcome and induces neuronal and oligodendrocyte regeneration. Brain Behav Immun 24(3):387–393

Vogel G (2010) Cell biology. To scientists' dismay, mixed-up cell lines strike again. Science 329(5995):1004

Wang P, Bowl MR, Bender S, Peng J, Farber L, Chen J et al (2008) Parafibromin, a component of the human PAF complex, regulates growth factors and is required for embryonic development and survival in adult mice. Mol Cell Biol 28(9):2930–2940

Weaver VM, Howlett AR, Langton-Webster B, Petersen OW, Bissell MJ (1995) The development of a functionally relevant cell culture model of progressive human breast cancer. Semin Cancer Biol 6(3):175–184

Willmon C, Harrington K, Kottke T, Prestwich R, Melcher A, Vile R (2009) Cell carriers for oncolytic viruses: Fed Ex for cancer therapy. Mol Ther 17(10):1667–1676

Winkler IG, Barbier V, Wadley R, Zannettino AC, Williams S, Levesque JP (2010) Positioning of bone marrow hematopoietic and stromal cells relative to blood flow in vivo: serially reconstituting hematopoietic stem cells reside in distinct nonperfused niches. Blood 116(3):375–385

Wood LD, Parsons DW, Jones S, Lin J, Sjoblom T, Leary RJ et al (2007) The genomic landscapes of human breast and colorectal cancers. Science 318(5853):1108–1113

Xin H, Kanehira M, Mizuguchi H, Hayakawa T, Kikuchi T, Nukiwa T et al (2007) Targeted delivery of CX3CL1 to multiple lung tumors by mesenchymal stem cells. Stem Cells 25(7):1618–1626

Yuan X, Hu J, Belladonna ML, Black KL, Yu JS (2006) Interleukin-23-expressing bone marrow-derived neural stem-like cells exhibit antitumor activity against intracranial glioma. Cancer Res 66(5):2630–2638

Zeng Z, Samudio IJ, Munsell M, An J, Huang Z, Estey E et al (2006) Inhibition of CXCR4 with the novel RCP168 peptide overcomes stroma-mediated chemoresistance in chronic and acute leukemias. Mol Cancer Ther 5(12):3113–3121

Zipori D, Reichman N, Arcavi L, Shtalrid M, Berrebi A, Resnitzky P (1985) In vitro functions of stromal cells from human and mouse bone marrow. Exp Hematol 13(7):603–609

Zischek C, Niess H, Ischenko I, Conrad C, Huss R, Jauch KW et al (2009) Targeting tumor stroma using engineered mesenchymal stem cells reduces the growth of pancreatic carcinoma. Ann Surg 250(5):747–753

Zuk PA, Zhu M, Mizuno H, Huang J, Futrell JW, Katz AJ et al (2001) Multilineage cells from human adipose tissue: implications for cell-based therapies. Tissue Eng 7(2):211–228

Zumsteg A, Christofori G (2009) Corrupt policemen: inflammatory cells promote tumor angiogenesis. Curr Opin Oncol 21(1):60–70

Erratum to: NK-Cell Immunotherapy for AML

Dean Anthony Lee, Laurence James Neil Cooper and Elizabeth J. Shpall

Erratum to:
Chapter 40 in: M. Andreeff (ed.), *Targeted Therapy of Acute Myeloid Leukemia*, Current Cancer Research, DOI 10.1007/978-1-4939-1393-0_40

The Publisher regrets that in chapter 40, figure 40.1 is incorrect. The correct figure is given below.

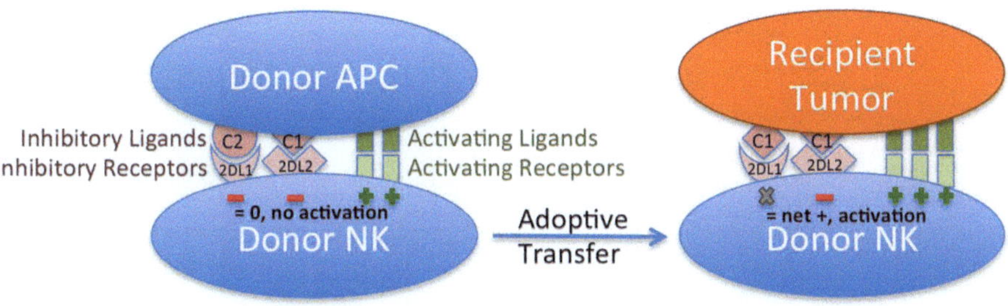

Fig. 40.1 The antileukemic benefit of KIR mismatch. NK cells in a donor who is heterozygous for HLA-C with respect to KIR ligands (having both a Group C1 allele that interacts with KIR2DL2 and a Group C2 allele that interacts with KIR2DL1) will be skewed toward no activation when receiving both inhibitory signals. However, when the NK cells are transferred to a recipient who is homozygous at HLA-C (e.g., HLA-C alleles are Group C1) the NK cells will lack the inhibitory signal through KIR2DL1, further tipping the balance toward activation when encountering tumor cells that also express activating ligands

The online version of the original chapter can be found under
DOI 10.1007/978-1-4939-1393-0_40

D. A. Lee (✉) · L. J. N. Cooper
Division of Pediatrics Cell Therapy Section, The University of Texas MD Anderson Cancer Center, 1515 Holcombe Boulevard, Unit 0853, Houston, TX 77030, USA
e-mail: dalee@mdanderson.org

L. J. N. Cooper
e-mail: ljncooper@mdanderson.org

E. J. Shpall
Department of Stem Cell Transplantation and Cellular Therapy, Division of Cancer Medicine, The University of Texas MD Anderson Cancer Center, 1515 Holcombe Boulevard, Unit 0423, Houston, TX 77030, USA
e-mail: eshpall@mdanderson.org

M. Andreeff (ed.), *Targeted Therapy of Acute Myeloid Leukemia*,
Current Cancer Research, DOI 10.1007/978-1-4939-1393-0_44

Index

M. Andreeff (ed.), *Targeted Therapy of Acute Myeloid Leukemia*,
Current Cancer Research, DOI 10.1007/978-1-4939-1393-0

MIX
Papier aus verantwortungsvollen Quellen
Paper from responsible sources
FSC® C105338

If you have any concerns about our products,
you can contact us on
ProductSafety@springernature.com

In case Publisher is established outside the EU,
the EU authorized representative is:
Springer Nature Customer Service Center GmbH
Europaplatz 3, 69115 Heidelberg, Germany

Printed by Libri Plureos GmbH
in Hamburg, Germany